5th International Conference on Mechatronics 2013

(ICOM'13)

IOP Conference Series: Materials Science and Engineering Volume 53

Kuala Lumpur, Malaysia
2 – 4 July 2013

Editors:

A.A. Shafie
R. Khan
M. Rashid

ISBN: 978-1-63439-864-0
ISSN: 1757-8981

Printed from e-media with permission by:

Curran Associates, Inc.
57 Morehouse Lane
Red Hook, NY 12571

proceedings.com

Some format issues inherent in the e-media version may also appear in this print version.

Printed by Curran Associates, Inc. (2015)

For permission requests, please contact the Institute of Physics
at the address below.

Institute of Physics
Dirac House, Temple Back
Bristol BS1 6BE UK

Phone: 44 1 17 929 7481
Fax: 44 1 17 920 0979

techtracking@iop.org

Additional copies of this publication are available from:

Curran Associates, Inc.
57 Morehouse Lane
Red Hook, NY 12571 USA
Phone: 845-758-0400
Fax: 845-758-2634
Email: curran@proceedings.com
Web: www.proceedings.com

TABLE OF CONTENTS

PREFACE

The Fifth International Conference on Mechatronics (ICOM2013), took place in Kuala Lumpur Malaysia from 2-4th July 2013. The biannual conference which started in 2001 is regularly organized by Faculty of Engineering, International Islamic University Malaysia (IIUM) with the aims to serve as a platform for exchange of ideas on advances of in mechatronics and their applications as well as to foster research and worldwide collaboration. The theme for the 2013 conference was "Mechatronics: Sustainable Development through Innovative Solutions". The ICOM 2013 Conference consisted of Keynote Speech (5) and oral contributions (150). The topics of the conference were:

- Mechatronic systems and Applications
- Intelligent Systems
- Control and Instrumentation
- Signal and Image Processing
- Machine Vision
- Robotics and Automation
- Manufacturing Mechatronics
- Green Mechatronics
- Mechatronic Education
- Smart Materials and Structures
- Active Vibration Control
- Computer and Information Technology
- MEMS and NEMS
- Biomechatronics and Rehabilitation Engineering
- Autonomous Systems
- Energy and Sustainability
- Transportation System

It is our great pleasure to present this volume of *IOP Conference Series: Materials Science and Engineering* (MSE) to the scientific community to promote further research in these areas. We believe that this volume will be both an excellent source of scientific material in the fast evolving fields that were covered by ICOM 2013.

We thank the authors for their enthusiastic and high-grade contributions. We would also like to express our gratitude to the Organizing Committee, the Institutions and Sponsors and everyone who contributed to this conference through their supports and invaluable efforts.

Editors

Amir A. Shafie
Department of Mechatronics, International Islamic University Malaysia (IIUM), Kuala Lumpur Malaysia
aashafie@iium.edu.my

Raisuddin Khan
Department of Mechatronics, International Islamic University Malaysia (IIUM), Kuala Lumpur Malaysia
raisuddin@iium.edu.my

Mahbubur Rashid
Department of Mechatronics, International Islamic University Malaysia (IIUM), Kuala Lumpur Malaysia
mahbub@iium.edu.my

Organizing Committee

- Md. Raisuddin Khan
- Md Mozasser Rahman
- Shahrul Naim Sidek
- Muhammad Mahbubur Rashid
- Amir Akramin Shafie
- Nahrul Khair Alang Md Rasyid
- Rini Akmeliati
- Roslizar Mat Ali
- Hazlina Md Yusof
- Asan Gani Abdul Muthalif
- Wahju Sediono
- Salmiah Ahmad
- Iskandar Al-Thani Mahmood
- Yasir Mohd Mustafah
- Moinul Bhuiyan
- Zulkifli Zainal Abidin
- Nadzril Sulaiman
- Momoh J. Salami

IOP Conf. Series: Materials Science and Engineering **53** (2013) 012001 doi:10.1088/1757-899X/53/1/012001

The Concept Sustainable Agriculture: Challenges and Prospects

M S Abubakar and M L Attanda
Department of Agricultural Engineering, Faculty of Engineering,
Bayero University, Kano-Nigeria

E-mail: abubakarms@gmail.com

Abstract. Agriculture has changed dramatically, especially since the end of World War II. Food and fibre productivity raised due to new technologies, mechanization, increased chemical use, specialization and government policies that favoured maximizing production. Sustainable agriculture is a subject of great interest and lively debate in many parts of the world. Most agriculturalists agree that the concept of sustainable agriculture is of paramount importance to the sustainability of our biosphere and its ever increasing human population. This paper is an effort to identify the ideas, practices and policies that constitute concept of sustainable agriculture.

1. Introduction

Sustainable agriculture is a subject of great interest and lively debate in many segments of the world. The debates stem largely from differing viewpoints as to what is sustainable agriculture [1-5]. Sustainable agriculture is defined as a system that, "over the long term, enhances environmental quality and the resource base on which agriculture depends; provides for basic human food and fibre needs; is economically viable; and enhances the quality of life for farmers and society as a whole"[1, 6, 7]. From this statement numerous definitions emerged but the concept surrounding agricultural sustainability remains the same. Also sustainable agriculture is defined as a commitment to satisfy human food and fibre needs and to enhance the quality of life for farmers and society as a whole, now and into the future. Consequently, there is that no brief, universally acceptable definition of sustainable agriculture has yet emerged. This is because sustainable agriculture is viewed more often as a management philosophy rather than a method of operation [8], and as such acceptance or rejection of any definition is linked to one's value system [9]. But regardless of its precise definition, most agriculturalists agree that the concept of sustainable agriculture is of paramount importance to the sustainability of our biosphere and its ever increasing human population. Sustainability hinges on a combined focus on environmental changes and their impact on society (people), the environment (planet) and economic value (profit) known as the '*Triple Bottom Line (TBL)*'. Increasingly it is being recognised that people, profit and the planet dimensions are interlinked and an important challenge for public and private policy is to take them jointly into account (Figure 1) [7].

Dr. M.S. Abubakar: Email-abubakarms@gmail.com

IOP Conf. Series: Materials Science and Engineering **53** (2013) 012001 doi:10.1088/1757-899X/53/1/012001

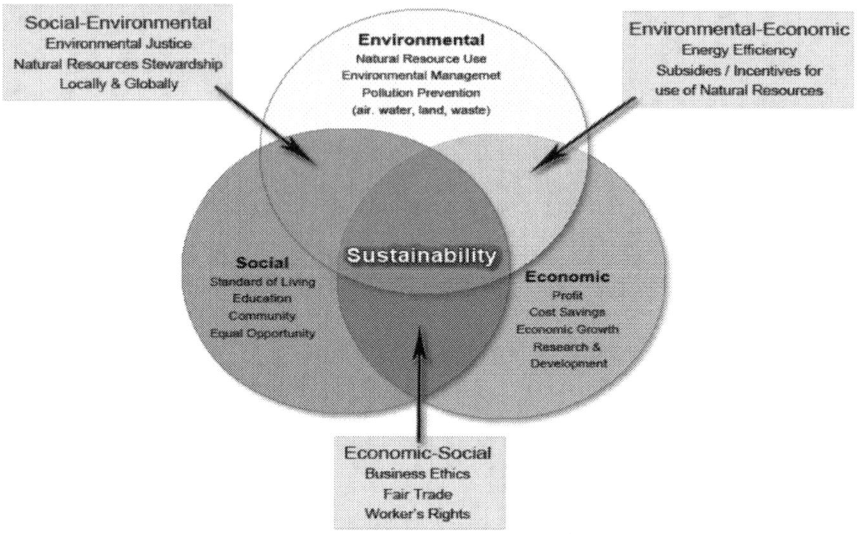

Figure 1- Triangle of sustainability

Agriculture has changed dramatically, especially since the end of World War II. Food and fibre productivity raised due to new technologies, mechanization, increased chemical use, specialization and government policies that favoured maximizing production. Agriculture is highly susceptible to climate variability and its related effects. Food security and maintenance of sustainable ecological balance are major challenges for thinkers, researchers, conservationists and policy makers. Sustainable agriculture should be taken as an eco-system approach, where soil-water-plants- environment-living beings live in harmony with a well balanced equilibrium of food chains and their related energy balances. The goal is to address environmental issues of natural resource management to sustain significant increases in farm productivity through the efficient use of land and other resources and provide better economic returns to individuals and contribute to the quality of life and economic development. It is essential that innovative technologies are used to ensure sustainable agriculture and productivity using, modern irrigation systems, improved varieties, improved soil quality and conserving the environment using resource conservation technologies [2]. Although these changes have had many positive effects and reduced many risks in farming, there have also been significant costs. Prominent among these are topsoil depletion, groundwater contamination, the decline of family farms, continued neglect of the living and working conditions for farm labourers, increasing costs of production, and the disintegration of economic and social conditions in rural communities. Sustainable agriculture emerged as part of a growing critique of the negative environmental consequences of unquestioned modern farming methods [4]. Even thought, the concept of sustainable agriculture is still new, however, the issue for sustainable agriculture is gathering increasing support and acceptance within mainstream agriculture. Not only does sustainable agriculture address many environmental and social concerns, but it offers innovative and economically viable opportunities for farmers, consumers, policymakers. This paper is an effort to identify the ideas, practices and policies that constitute our concept of sustainable agriculture. The challenges to agriculturalists circulate around the abilities of agricultural engineers to develop and implement new technologies that will allow the farmers to maintain and increase yields of agriculture products while increasing ecological efficiencies.

2. Environmental Perspective

Increasingly, the idea of sustainability agricultural has come to be associated with the maintenance of environmental quality both on and off the farm. One of the environment related activities is modern commercial agricultural systems, characterized by intensive tillage and cropping practices, and high rates of mechanization and chemical input use have resulted in excessive amounts of soil erosion and nutrient loss, impairing the productive capacity of soil resources and placing greater emphasis on the use of purchased inputs [2]. Such systems of agriculture are increasingly regarded as unsustainable. In these cases, sustainability is defined as the preservation of the productive capacity of the land resource. Another environment-related interpretation of sustainability concerns the effect of agriculture on the natural environment beyond the farm gate and apart from the productivity of cropland. Pollution of surface and ground water resources due to transport of chemical fertilizers from farm fields are two commonly expressed concerns related to sustainable agriculture. Loss of species' habitat and reduction in biological diversity are also concerns of those with an ecological orientation toward sustainability. In contrast to the concerns about reduced productive capacity, this ecological perspective on sustainability suggests that natural resources should be protected in their own right, and should not be placed at risk by agricultural activities which are not environmentally friendly [10].

3. Economic and Social Perspective

In addition to ensuring long term food production and environmental quality, the concept of sustainability is also applied to producers and rural communities. For many analysts, the sustainability of agriculture can be described by assessing the economic returns of farming [11]. In commercial economies, farms which are unable to generate sufficient profits, because of low farm product prices, reduced yields, higher costs of production, or other reasons, that are not self staining. Consequently, a requirement of agricultural sustainability is the sustainable agriculture existence of economic returns which are sufficient to sustain farm businesses, and to adequately reward producers [12]. Also a broader interpretation of sustainable agriculture extends the farm viability concept to the maintenance of rural community systems. This view holds that prospects for environmentally friendly agriculture which is efficient in its production of food and fair in its distribution of benefits are greatest when agriculture functions within a healthy rural community system which supports local decision-making and stewardship values [13]. The idea of equity transcends food, environment and producer objectives, and is frequently adopted as the central characteristic to define sustainable agriculture. For many, the preservation over time of productive capacity, or environmental integrity, or family farming represents a fundamental element of sustainability. Hence intergenerational equity in agriculture refers to the protection of the rights and opportunities of future generations to derive benefits from resources which are in use today [4]. Agricultural practices which diminish long term prospects for food production or impair water quality or other natural resources, regardless of their short term benefits, are not considered sustainable. Equity concerns are not limited to safeguarding the future of agriculture. The principle has been applied to describe the rights of less advantaged groups in society to basic food supplies, and to the opportunities and resources required to farm in ways which enhance prospects for sustainability [14]. Thus, intragenerational equity refers to the fair and equitable distribution of benefits from agriculture among and between countries, regions or social groups.

4. Challenges and Prospective

This study has explored several techniques of sustainable agriculture, including interpretations of the sustainability concept, alternative methods of pursuing sustainability in the practice of agriculture, and analytical options for assessing the performance of food production systems in the face of changing conditions [15]. Agricultural practices should be consistent with certain principles of sustainability, including the preservation of the productive capacity of soil and the avoidance of off-farm pollution and environmental degradation. Yet, the literature suggests that the adoption of more environmentally friendly farming practices should be adopted. However, the benefits of sustaining agriculture accrue not only to farmers, but also to societies generally, now and in the future. Also, sustainable agriculture is not just an analytical question, defined in terms of environmental quality, productivity, sufficiency or whatever, but it is also a philosophy toward farming. By its nature, the concept is goal oriented and value loaded. However, it is unlikely that all goals and values associated with sustainable agriculture will be achievable simultaneously, or even sequentially. For example, should farmers be required, or even expected, to adopt practices which would yield environmental benefits at the possible expense of economic viability? Notwithstanding the value of analytical procedures in providing information, much future action in pursuing agricultural sustainability will be determined by decisions and trade-offs made among and between goals, both individual and societal [11]. Again, much of the thinking regarding sustainable agriculture has been done in the context of the 'supply side' of the food production equation. Analysts have sought to assess the productive capacity of land and the impacts of various phenomena in order to measure the potential for food production. While much is now known about the implications of such supply side factors as soil erosion and land conversion, the more important and intractable questions may be on the demand side of agriculture. For example, world population growth, or rapidly increasing demand for food because of improvements in income may have a far greater impact on the ability of agriculture to meet demands for food than do the biophysical limits to production. Consequently, understanding of the prospects for sustainable agriculture globally requires further progress in forecasting, and perhaps dampening, future levels of demand [11, 16]. This is complicated by uncertainties regarding population growth, and even more so by several factors such as government policies which affect access to foreign markets, and the role of regional agricultural systems in supplying food aid in the developing world. Finally, there is a need to better understand the nature of agricultural system response to disruptions of all types in order to anticipate how agriculture might be affected by various changes in the future [17]. In addition, the impacts of some forces, such as climatic warming, on the prospects for sustaining food production have typically been predicted with some bold assumptions on how farmers, or the agricultural system generally, will respond to changed conditions. For example, it is often assumed that technological innovation will allow agriculture to continue to achieve high outputs of food, or that farmers operating in an altered climate will know precisely how best to adapt to new opportunities for food production, and that they will do so. In reality, little is known about the processes of response and adjustment in agriculture, particularly at the aggregate scale [11, 14, 16]. Improved understanding of these adaptive processes in agriculture would enhance assessments of agricultural sustainability and improve prospects for achieving it.

5. Conclusion

Agriculture has changed dramatically, especially since the end of World War II. Food and fibre productivity raised due to new technologies, mechanization, increased chemical use, specialization and government policies that favoured maximizing production. New policies for ensuring the sustainability of agriculture will be crucial if we are to meet the demands of improving yields without comprising environmental integrity or public health. Given emerging pressures and resource constraints, agricultural policies need to help meet the objectives of poverty reduction, agricultural production and environmental sustainability. A major challenge is to create policies, institutions and technologies that make the goals more compatible.

References

[1] T. Crews, C. Mohler, and A. Power, "Energetics and Ecosystem Integrity: The Defining Principles of Sustainable Agriculture " *American Journal Alternative Agriculture,* vol. 6, p. 146, 1991.

[2] M. Dover and L. Talbot, "To Feed the Earth: Agro-Ecology for Sustainable Development," 1987.

[3] USDA, "Report and Recommendations on Organic Farming," U.S. Gov. Printing Office, Washington, DC1980.

[4] D. R. Keeney, "Toward a Sustainable Agriculture: Need for Clarification of Concepts and Terminology," *American Journal Alternative Agriculture* vol. 4, p. 101, 1989.

[5] R. Lowrance, P. F. Hendrix, and E. P. Odum, "A Hierarchical Approach to Sustainable Agriculture," *American Journal Alternative Agriculture,* vol. 1, p. 169, 1986.

[6] C. B. Flora, "Building Sustainable Agriculture: A New Application of Farming Systems Research and Extension," *Journal of Sustainable Agriculture,* vol. 2, pp. 37-50, 1992.

[7] E. V. Kambewa, *Contracting For Sustainability: An Analysis of the Lake Victoria-EU.* Nile Perch Chain: Wageningen Pers, 2007.

[8] R. J. MacRae, J. Henning, and S. B. Hill, "Strategies to Overcome Barriers to the Development of Sustainable Agriculture in Canada: The Role of Agribusiness," *Journal Agricultural Environmental Ethics* vol. 6, p. 21, 1993.

[9] E. A. Clark and S. F. Weise, *A Forage-Based Vision of Sustainable Agriculture*: Madison, WI, 1993.

[10] D. Neher, "Ecological Sustainability in Agricultural Systems," *Journal of Sustainable Agriculture,* vol. 2, pp. 51-62, 1992.

[11] M. Brklacich, C. Bryant, and B. Smit, "Review and Appraisal of Concept of Sustainable Food Production Systems," *Environmental Management,* vol. 15, pp. 1-14, 1991.

[12] J. E. Ikerd, "Agriculture's Search for Sustainability and Profitability," *Journal of Soil and Water Conservation,* vol. 45, pp. 18-23, 1990.

[13] C. A. Edwards, "The Importance of Integration in Sustainable Agricultural Systems," *Agriculture, Ecosystems and Environment,* vol. 27, pp. 25-35, 1989.

[14] C. A. Francis, "Practical Applications of Low-Input Agriculture in the Midwest," *Journal of Soil and Water Conservation,* vol. 45, pp. 65-67, 1990.

15] P. Crosson, "Sustainable Agriculture," *Resources,* vol. 106, pp. 14-17, 1992.

[16] J. Burkhardt, "The Morality Behind Sustainability," *Journal of Agricultural Ethics,* vol. 2, pp. 113-128, 1989.

[17] K. P. George, "Do We Have a Moral Responsibility to Practice a Sustainable Agriculture?," *Journal of Sustainable Agriculture,* vol. 1, pp. 81-95, 1990.

Modelling the surface roughness behaviour of an EDMed workpiece with different tool electrodes using DoE

M B Ndaliman[1,2,3], A A Khan[1], M Y Ali[1] and M A Mohammad Hambiyah[1]

[1]Department of Manufacturing and Materials Engineering, International Islamic University Malaysia, 53100 Kuala Lumpur, Malaysia
[2]Department of Mechanical Engineering, Federal University of Technology, Minna, Nigeria

E-mail: mbndaliman@yahoo.com

Abstract. Surface roughness of a workpiece is one of the measures used in evaluating the performance of any machining technique. In this paper, the behaviour of a post-EDM roughness of mild steel material subjected to machining with two different electrodes is evaluated through modelling using three machining variables. The variables are peak current, pulse on-time and pulse off-time, while the electrodes used are Cu-TaC compact green electrode and metallic Cu. The investigation was planned and analysed with Design of Experiment (DoE) in which the output response obtained with the two electrodes were compared. Results indicate that the roughness obtained with Cu-TaC electrode is a function of peak current, pulse on-time, pulse off-time with interaction between on-time and off time, whereas that of metallic Cu electrode is related with current and pulse on-time only. Further more, the metallic Cu electrode produced comparatively lower roughness under all the machining conditions. The lowest roughness obtained with Cu-TaC electrode is over 100% higher than that of Cu-electrode.

Keywords. EDM; Surface roughness, Cu-TaC electrode; Metallic Cu electrode; DoE

1. Introduction

The roughness of machined surface expresses the degree of irregularities of it texture. In most production processes finer irregularities of the surface texture are desired. In electro-discharge machining (EDM), many factors contribute to the level of surface roughness attained by the workpiece during machining. These include the machining input parameters, the dielectric fluid and the tool electrode. It is possible to optimize the EDM process input parameters to obtain better surface roughness [1].

Electrode is one of the vital components of EDM, because it is the cutting tool whose form is replicated on the workpiece. Its performance and cost depends on the material, design and fabrication technique used in producing it. The conventional electrodes used in the early days of EDM were mainly metallic or graphite. The most commonly metallic electrode materials were copper, brass, silver, tungsten, copper-tungsten, silver-tungsten and tungsten carbide [2]. The major shortcomings of electrodes are the production difficulty and low cutting speed. The two electrode materials with competitive usage among EDM users are copper and graphite.

[3] To whom any correspondence should be addressed.

Practically the choice of EDM electrode material is based on the tool size, type, the work material requirement and the electrode fabrication technique. Other important factors of consideration are the work removal rate, tool's resistance to wear, the desired workpiece surface roughness, the tool's machinability and its cost [3]. Research interests have been generated on production tool electrode from such materials like Al, Cr, Cr/Ni, Cu/Co, Cu/Mn, Cu/Sn, Cu/W, Ni, Ni/Co, Ni/Fe, Ni/Mn, Ni/Si, Ti, Ti/Al, TiC/Ni and W/CrC/Cu [4-5] because of their various individual and combine properties. But their fabrication through conventional means can be difficult. This makes the need for alternative design and manufacturing methods of the electrodes to become necessary. The manufacturing process of EDM electrodes is such as an important aspect as it accounts for over 50% of the machining cost depending on their level of complexity in geometry [3]. The more complex geometries are expensive to produce. Though, conventional methods like machining, casting and forming have been used to manufacture electrodes over the years, modern techniques have sprang up for their production. These include the use of powder metallurgy (*PM*) and rapid tooling.

In their continuous search for better EDM tooling alternatives, researchers have found *PM* method of electrodes' fabrication to be faster and more economical. Through this method, large number of EDM tools with even complex geometries can be manufactured in short period of time [6]. The properties of the electrodes can be controlled by regulating the *PM* process variables like composition, compacting pressures and sintering conditions. These possibilities were established in the mid 1990's [7-8]. Zaw et al. [9] investigated the performance of sintered ZrB_2/Cu and TiSi/Cu. After comparing them with conventional tools (Cu, Gr and CuW), they submitted that both the composite electrodes could not measure up to the standard electrodes under those sintering conditions. However, working with different compositions of the same ZrB_2/Cu, Khanra et al. [10] observed that ZrB2-40 wt% Cu gave higher material removal rate (MRR) with reduced tool wear rate (TWR) than pure solid Cu. *PM* compacted electrodes can be produced from single powder [8, 11-12] or mixture of different powders [13-14]. They are used as green or sintered electrodes in machining. Higher compacting pressures and sintering condition favour the basic machining outputs. Thus, better MRR, lower tool wear and good surface finish for the work material.

In this work, the surface roughness trends of mild steel machined with *PM* compacted Cu-TaC electrode is modelled with EDM process variables. In addition, its behaviour was compared with the EDM carried out with metallic copper (Cu) electrode under the same conditions. This research is conducted in order to consolidate the performance of the electrode on the steel material since its properties (the electrode) has indicated that it can be used for EDM work [15-16]. Also, previous research results on the electrode's performance showed that it can reasonably be use in material removal process [17-18]. The input variables of investigation are peak current (I), pulse-on time (t_{on}) and pulse-off time (t_{off}).

2. Design of Experiment (DoE)

The investigation was planned and implemented with design of experiment (DoE). Using two-level full factorial design involving three factors, a total of eight experimental runs were generated. The factors and their levels are presented in table 1. The design matrix alongside the response is presented in table 2. A first-order model (linear relation) was proposed for the output response (Ra) of the study as given in equation (1).

$$\hat{y} = y - \varepsilon = a_0 + a_1A + a_2B + a_3C + a_4AB + a_5AC + a_6BC \qquad (1)$$

Where $\hat{y}$ = predicted value of the response; y = measured value of the response; ε = experimental error; a_0, a_1, $a_{2...}$ a_6 are the model parameters to be estimated using the experimental data; A, B, and C are peak current, pulse duration and pulse interval respectively; and AB, AC and BC are the interaction effects of the variables. The significant 'a's to be evaluated and the non-significant ones would be

IOP Conf. Series: Materials Science and Engineering **53** (2013) 012002 doi:10.1088/1757-899X/53/1/012002

neglected in the final equations. Design Expert version 6.0.8 [19] software was used to conduct the analysis. Analysis of variance (ANOVA) was utilized to identify the significant model terms. Finally, the models were subjected to diagnostic test to ascertain their adequacy.

Table 1. The input factors and the levels used.

Factor	Level	
	-1 (Low)	+1 (High)
Current, I (A)	3.50	7.50
On-time, t_{on} (μs)	6.00	7.50
Off-time, t_{off} (μs)	6.50	8.50

3. Materials and Methods

The workpiece material used in the investigation is mild steel The EDM input parameters used in the investigation were peak current (*I*), pulse-on time (t_{on}) and pulse-off time (t_{off}), while surface roughness is the output variable studied. The Cu-TaC Powder metallurgy (*PM*) electrode was produced by compaction of the mixed Cu and TaC powders with composition of 25% and 75% respectively in the matrix. Similar investigation was conducted with metallic copper electrode for use in comparison. The EDM was conducted according to the run order presented in Table 2, with kerosene dielectric fluid. The surface roughness of the work material was determined with Mitutoyo SURFTEST Machine (the surface measuring instrument). During this measurement, the average of ten readings was record for each surface in order to obtain the most accurate result. 3-D surface plots showing the behaviour of the Ra for the different machining are presented. The graphical comparison of the performance of the two types of electrodes is also presented.

4. Results and Discussions

The roughness of the surface obtained after the EDM are designated as Ra_1 and Ra_2 for the *PM* and metallic electrodes respectively in table 2. These results were used to model the behavior of response with respect to the input variables.

Table 2. Design matrix with the responses.

Run No	A: Peak Current (A)	B: On time (μs)	C: Off time (μs)	Ra_1 (μm)	Ra_2 (μm)
1	7.50	6.00	6.50	21.82	11.68
2	7.50	7.50	6.50	23.10	18.27
3	3.50	7.50	6.50	19.62	12.63
4	3.50	6.00	6.50	18.48	8.77
5	7.50	6.00	8.50	21.20	9.59
6	3.50	6.00	8.50	17.61	8.07
7	7.50	7.50	8.50	22.99	15.90
8	3.50	7.50	8.50	19.88	12.94

4.1. Analysis and the Models

Tables 3 and 4 present the improved ANOVAs of Ra_1 and Ra_2 with their statistical significant terms. In table 3, all the input factors are found to be significant as their Prob > F are less than 0.05. Additionally, interaction is found to exist between *B* and *C*. This implied therefore that the model with the factors *A*, *B*, *C* and *BC* as its terms is statistically significant at 5% level of significance. It is also observed from table 4 that the Ra_2 model has factors *A* and *B* as the only significant terms in it. These models obtained on the basis of improved ANOVAs are given in equations (2) and (3).

Table 3. The improved ANOVA of Ra_1 with Cu-TaC electrode.

Source	Sum of Squares	DF	Mean Square	F Value	Prob > F
Model	28.66	4	7.17	341.62	0.0003
A	22.84	1	22.84	1088.93	0.0001
B	5.26	1	5.26	250.69	0.0005
C	0.23	1	0.23	10.73	0.0466
BC	0.34	1	0.34	16.14	0.0277
Residual	0.63	3	0.021		
Cor Total	28.73	7			
	$R^2 = 0.9978$	Adj R^2 = 0.9949	Pred R^2 = 0.9844	Adeq Precision = 47.271	

Table 4. The improved ANOVA of Ra_2 with metallic Cu electrode.

Source	Sum of Squares	DF	Mean Square	F Value	Prob > F
Model	82.65	3	27.55	24.39	0.0049
A	21.22	1	21.22	15.79	0.0123
B	58.48	1	58.48	51.78	0.0020
C	2.94	1	2.94	2.60	0.1819
Residual	4.52	4	1.13		
Cor Total	87.16	7			
	$R^2 = 0.9482$	Adj R^2 = 0.9093	Pred R^2 = 0.7927	Adeq Precision = 13.144	

$$Ra_1 = +23.78994 + 0.84487I - 0.97650t_{on} - 2.01950t_{off} + 0.27433It_{off} \quad (2)$$

$$Ra_2 = -12.03519 + 0.81437I + 3.60500t_{on} - 0.60625t_{off} \quad (3)$$

The R^2 obtained in machining with the two electrode types are over 0.9 (tables 3 and 4). Also, the Predicted and Adjusted R^2 for each of the two electrodes are reasonably close with the difference between them less than 0.20. Finally, their signal-to-noise are greater than 4.0 which is the minimum acceptable value [20]. Based on these statistical features, the models seem adequate to predict the *Ra* behaviors of the two different electrodes. Further confirmation of the models' adequacy would depend on the diagnostic tests.

For the model equations to be adequate, the residuals of response should be normally distributed, unstructured (randomly distributed) and with constant variance [20-21]. The normal plots of residuals in figure 1 indicate that they (residuals) approximately follow straight lines. Therefore normality condition is satisfied for the two electrode types. Another condition to be satisfied is that the residual plots against the predicted values should not show any obvious pattern. The fitted values in figure 2 are scattered randomly on display, suggesting that the variances of the original observations are constant for all values. Hence, based on this (constant variance), the models of the data which produced the two plots can be considered adequate.

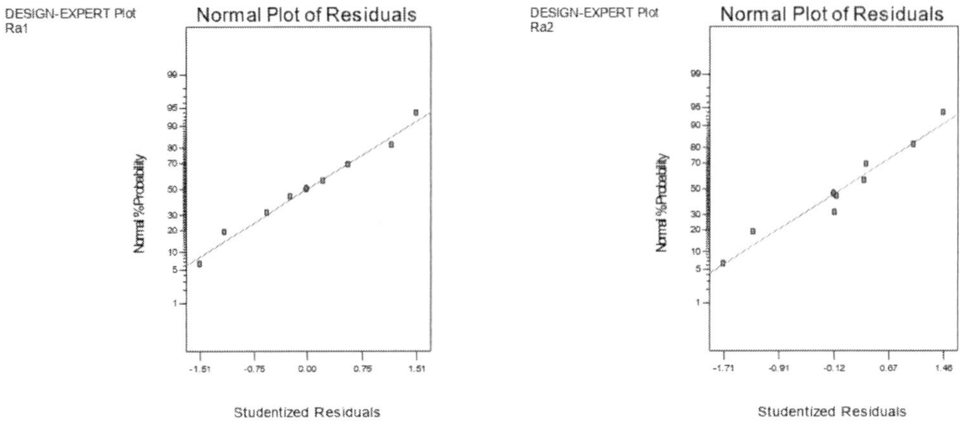

(a) (b)

Figure 1. Normal probability of residuals: (a) Ra_1; (b) Ra_2.

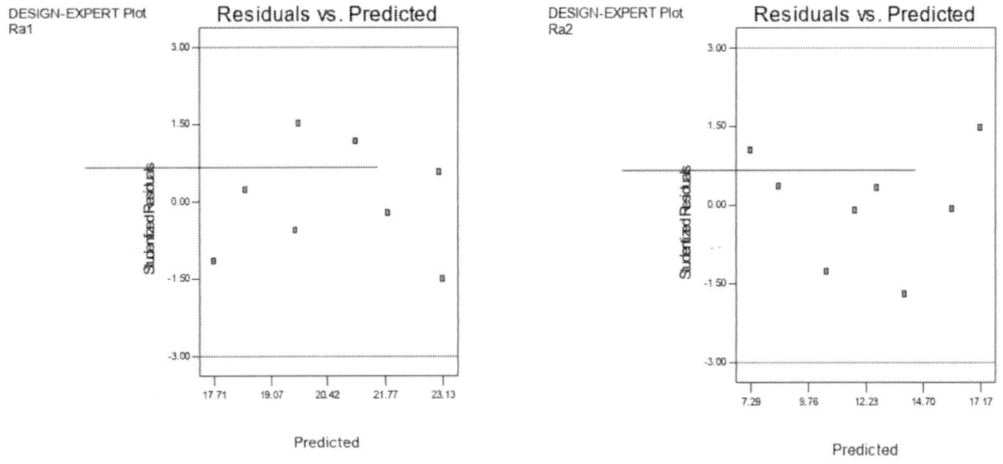

(a) (b)

Figure 2. Plot of residuals against predicted values; (a) Ra_1; (b) Ra_2.

4.2. Discussions

An observation of table 2 shows that the Ra_1 are generally higher than Ra_2. Thus machining with Cu-TaC electrodes produce rougher surfaces. The highest Ra_1 of 23.10 μm was obtained under extreme machining conditions. Similarly, the highest Ra_2 of 18.27 μm resulted from the same machining conditions. The reason for high Ra_1 could be due to the tendency of the *PM* electrode to wear faster and deposit lump particles on to the workpiece under these machining conditions. Researches have supported the fact that high I and t_{on} normally result in increase tool wear [22]. In the case of the Cu-TaC electrode, its high tool wear coupled with material deposition of the work surface would result in irregular profiles on it (workpiece) with a direct consequent on the roughness [13]. Therefore, the *PM* compact electrodes are associated with rougher surfaces compare to metallic ones [11].

The 3-D surface plots generated by the models for the two electrode type are presented in figures 3 to 5. From the analysis, the Ra of the mild steel was found to be a function of I, t_{on} and t_{off} with the interaction effect between t_{on} and t_{off} for the case of Cu-TaC electrode machining. On the other hand, the Ra_2 depends only on I and t_{on} with metallic Cu electrode. In figure 3, the contour plot show the values on the lines for the Ra_1 behaviour (figure 3a), whereas the trend in the experimental region is illustrated by 3-D response surface (figure 3b). The Ra_1 increases with increasing current and pulse on-time. However the influence of on-time on the Ra_1 seems to be mild compare to that of the current. Also, the effect of interaction between the pulse on-time and off-time is revealed by figure 4, which show lower Ra_1 are obtainable only with combination of low discharge duration at high pulse interval. Under this condition, the contour plot show a low Ra_1 of 19.74μm (figure 4a). In practical terms, it implies that machining efficiency would have to be compromised in order to obtain lower roughness with Cu-TaC electrode, since more time would be spent with material removal in the cutting process. The surface roughness obtained with metallic Cu machining increases with increasing I_p and t_{on} and vice versa (figure 5). This is agreement with the normal Ra behaviour of copper electrodes [1, 23]. Increasing these variables imply that the discharge intensity would increase at longer striking period on the workpiece. Therefore, the craters on the EDMed surface would increase in number, size and depth, and thus lead to higher roughness.

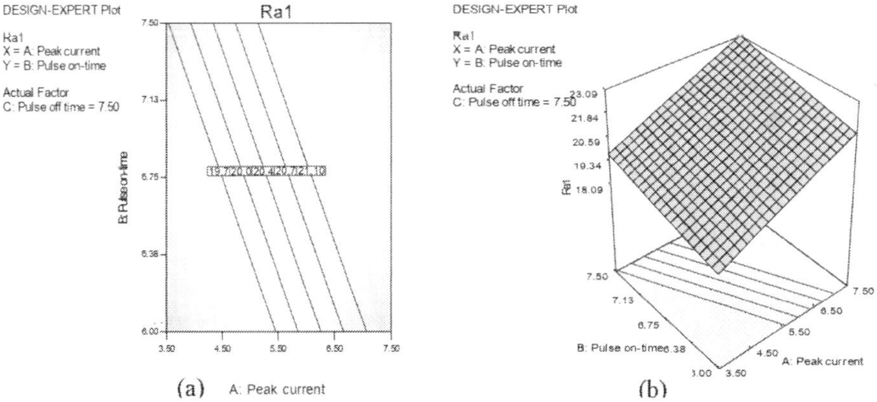

(a) (b)

Figure 3. 3-dimensional plots of Ra_1 with I and t_{on}: (a) contour plot; (b) surface plot.

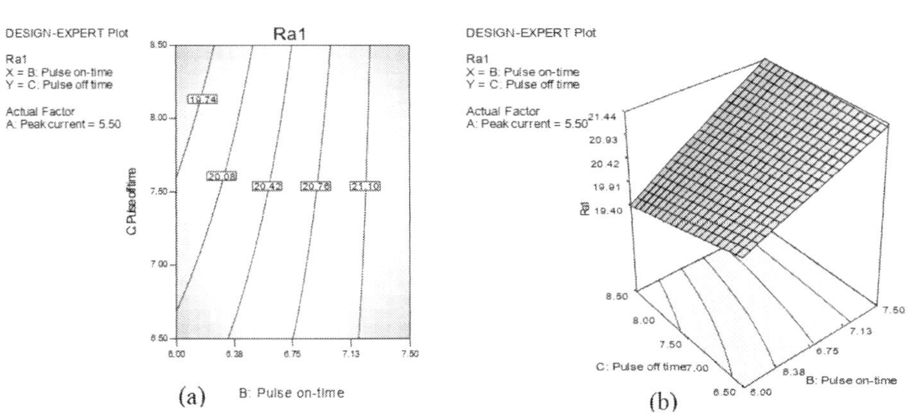

(a) (b)

Figure 4. 3-dimensional plots of Ra_1 with t_{on} and t_{off}: (a) contour plot; (b) surface plot.

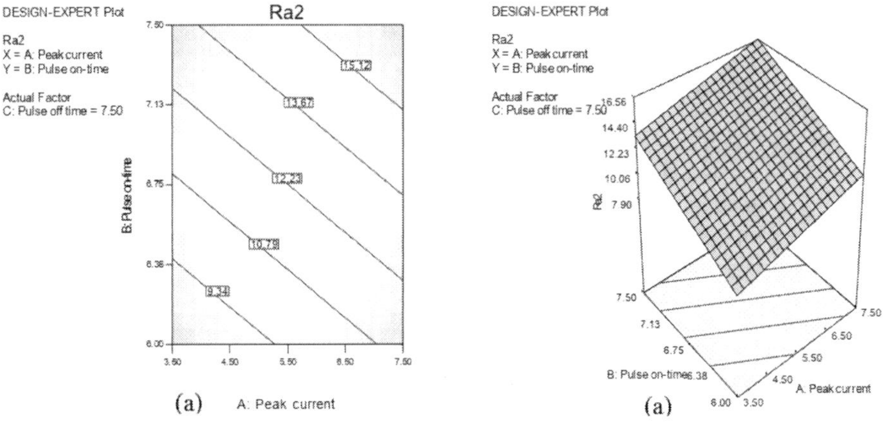

Figure 5. 3-dimensional plots of Ra_2 with I and t_{on}: (a) contour plot; (b) surface plot.

The graphical comparison of the experimental runs with the two electrodes is shown in figure 6. The surface roughness curve obtained with Cu-TaC *PM* electrode can be observed to be clearly higher than that of Cu electrode. The lowest Ra obtained with the Cu-TaC electrode is over 110% higher than the lowest Ra with metallic electrode.

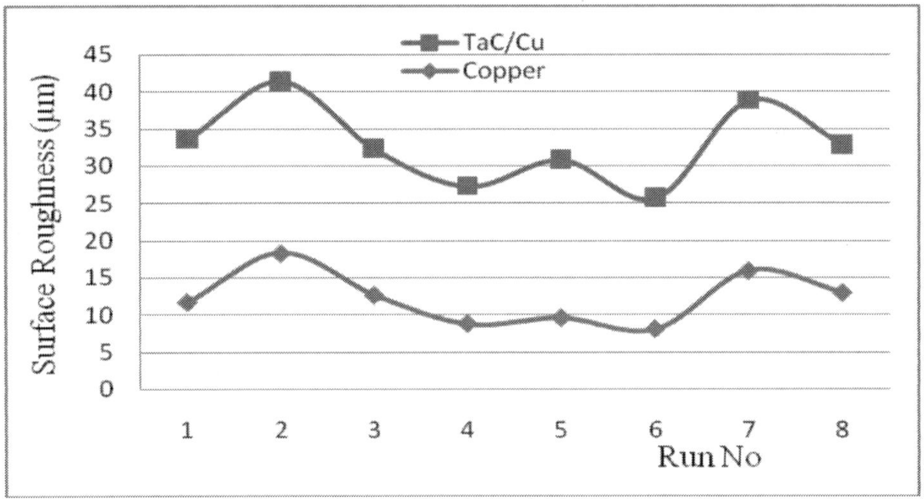

Figure 6. Comparison of the Ra generated by Cu-TaC compacted electrode with that of metallic Cu electrode

5. Conclusions

The following conclusions can be drawn from the investigation.

1. From the results and analysis, the surface roughness of the mild steel machined with Cu-TaC electrode was found to be a function of I, t_{on} and t_{off} with the interaction effect between t_{on} and t_{off}, while metallic Cu electrode depends only on I and t_{on}. Therefore, their models are made up of the terms involving the variables.

2. Direct relationship exists between the surface roughness with current and pulse on-time as variables in both electrode types, though the influence of on-time is mild as compare to that of the current in Cu-TaC machining.
3. As a result of the interaction effect between the pulse on-time and off-time lower roughness are obtainable with combination of low discharge duration and high pulse interval.
4. Machining with Cu electrode produced lower roughness under all conditions compared that of Cu-TaC electrode, with the lowest roughness obtained with the Cu-TaC electrode is over 100% higher than the lowest roughness with the metallic electrode.

References

[1] Kiyak M and Cakir O 2007 *Journal of Materials Processing Technology* **191** 141
[2] Kern R 2008 TechTips: Sinker electrode material selection. In *EDM Today*
[3] Jha B, Ram K and Rao M 2011 *Journal of Engineering Science and Technology Review* **4** 118
[4] Beri N, Maheshwari S, Sharma C and Kumar A 2010 *Materials and Manufacturing Processes* **25** 1186
[5] Lee H, Simao J, Aspinwall D, Dewes R and Voice W 2004 *Journal of Materials Processing Technology* **149** 334
[6] Patowari P K, Saha P and Mishra P K 2011 *The International Journal of Advanced Manufacturing Technology* **54** 593
[7] Samuel M P and Philip P K 1997 *International Journal of Machine Tools and Manufacture* **37** 1625
[8] Samuel M P and Philip P K 1996 *Indian Journal of Engineering Materials Science* **3** 229
[9] Zaw H, Fuh J, Nee A and Lu L 1999 *Journal of materials processing technology* **89** 182
[10] Khanra A K, Sarkar B R, Bhattacharya B, Pathak L C and Godkhindi M M 2007 *Journal of Materials Processing Technology* **183** 122
[11] Ho S, Aspinwall D and Voice W 2007 *Journal of Materials Processing Technology* **191** 123
[12] Wang Z L, Fang Y, Wu P N, Zhao W S and Cheng K 2002 *Journal of Materials Processing Technology* **129** 139
[13] Patowari P, Mishra U, Saha P and Mishra P 2011 *Materials and Manufacturing Processes* **26** 668
[14] Chiang K-T 2008 *The International Journal of Advanced Manufacturing Technology* **37** 523
[15] Ndaliman M B and Khan A A 2011 *Journal of Mechanics Engineering and Automation* **1** 385
[16] Ndaliman M B, Hazza M, Khan A A and Ali M Y 2012 *Australian Journal of Basic and Applied Sciences* **6** 192
[17] Khan A A, Mohammad Hambiyah M A and Nazi Nadin M F 2011 In *Advanced Machining Process* ed. M Y Ali, A K M NurulAmin, E Y T Adesta pp. 23-7 Kuala Lumpur: IIUM Press
[18] Khan A A, Nadin M F N and Hambiyah M A M 2011 In *Advanced Machining Process* ed. M Y Ali, A K M N Amin, E Y T Adesta pp. 49-53 Kuala Lumpur: IIUM Press
[19] Design Expert version (6.0.8) 2002
[20] Myers R H, Montgomery D C and Anderson-Cook C M 2009 *Response Surface Methodology: Process and Product Optimization Using Designed Experiments* New Jersey: John Wiley and Sons Inc. 680 pp.
[21] Montgomery D C 2009 *Design and Analysis of Experiments* New York: John Wiley & Sons Inc
[22] Khan A 2008 *The International Journal of Advanced Manufacturing Technology* **39** 482
[23] Beri N, Maheshwari S, Sharma C and Kumar A 2008 *International Journal of Mechanical, Industrial and Aerospace Engineering* **2** 167
[24] Kiyak M and Cakir O 2007 *Journal of Materials Processing Technology* **191** 141

IOP Conf. Series: Materials Science and Engineering **53** (2013) 012003 doi:10.1088/1757-899X/53/1/012003

Cauchy Integral Formula

M Azram[1] and F A M Elfaki
Department of Science, Faculty of Engineering
IIUM, Kuala Lumpur 50728, Malaysia
azram50@hotmail.com

ABSTRACT: Cauchy-Goursat integral theorem is a pivotal, fundamentally important, and well celebrated result in complex integral calculus. It requires analyticity of the function inside and on the boundary of the simple closed curve. In this study we will investigate the condition(s) under which $\oint_c f(z)dz = 0$ even though the function is not analytic at a point inside C. Consequently, we will extend the above notion to a finite numbers of points and will present an easy and simple proof of unquestionably the most important, significant and pivotal result known as Cauchy integral formula.

1. INTRODUCTION

Complex variable is an extension of real calculus which in fact discloses everything hidden in the real calculus. Complex integration is pivotal in the study of complex variables. As in calculus, the fundamental theorem of calculus is significant because it relates integration with differentiation and at the same time provides method of evaluating integral so is the complex analog to develop integration along arcs and contours is complex integration. Complex integration is elegant, powerful and a useful tool for mathematicians, physicists and engineers.

Cauchy-Goursat theorem is pivotal, fundamentally important and well celebrated theorem of the complex integral calculus. This theorem is not only a pivotal result in complex integral calculus but is frequently applied in quantum mechanics, electrical engineering, conformal mappings, method of stationary phase, mathematical physics and many other areas of mathematical sciences and engineering. It provides a convenient tool for evaluation of a wide variety of complex integration. It also forms the cornerstone of the development of results $f'(z)$ of an analytic function $f(z)$ is analytic, Cauchy's integral formula and many advance topics in complex integration. Due to its pivotal role and importance, mathematicians have discussed it in all respects [3-5,7,8].

M Azram, Department of Science, Faculty of Engineering, IIUM, Kuala Lumpur 50728, Malaysia azram50@hotmail.com

Historically, it was firstly established by A.L. Cauchy (1789-1857) [1] and later on extended by E. Goursat (1858-1936) [1] without assuming the continuity of $f'(z)$. Azram [6] has given a variational but simple proof of Cauchy-Goursat integral theorem. Cauchy-Goursat integral theorem has laid down the deeper foundations for Cauchy- Riemann theory of complex variables. The line integral of a complex function is mostly dependent on the endpoints of the path of integration as well as on the choice of the path. This dependence often complicates situations. Hence, condition(s) under which this does not occur are of considerable importance and is due to Cauchy-Goursat integral theorem.

Cauchy-Goursat integral theorem leads to a remarkable Cauchy integral formula which shows that the values of an analytic function at the interior points can be determined by the values of the function on the boundary. Another important and useful consequence of Cauchy's integral formula is that every analytic function is infinitely differentiable. Guy Jumarie [4] has studied the Cauchy integral formula via the modified Riemann–Liouville derivative. The following are also important results resulting as a consequence of Cauchy Integral formula; Cauchy's estimate Cayley-Hamilton theorem, Cauchy's inequality , Liouville's theorem, Fundamental theorem of algebra, Gauss' mean value theorem, Maximum modulus theorem and Minimum modulus theorem etc.

Cauchy-Goursat integral theorem requires analyticity of the function inside and on the boundary of the simple closed curve. In this study we will investigate the condition(s) under which $\oint_C f(z)dz = 0$ even though the function is not analytic at a point inside C. Consequently, we will extend the above notion to a finite numbers of points and will present an easy and simple proof of unquestionably the most important, significant and pivotal result known as Cauchy integral formula.

2. RESULTS AND DISCUSSION

Theorem 1. If a function $f(z)$ is analytic inside and on a simple closed curve C except at an interior point z_0 such that $\lim_{z\to z_0} f(z) = 0$ then $\oint_C f(z)dz = 0$

Proof. Let $f(z)$ be analytic inside and on a simple closed clockwise oriented curve C except at an interior point z_0. Since $\lim_{z\to z_0} f(z) = 0 \Rightarrow$ given $\varepsilon > 0\ \exists\ \delta > 0 \ni$ $|f(z)| < \varepsilon$ *whenever* $0 < |z - z_0| < \delta$. Now, choose a clockwise oriented circle C_1 of radius r centre at z_0 that is C_1 : $|z - z_0| = r$. The circle C_1 is completely inside C and $0 < r < \delta$ (Figure 1).

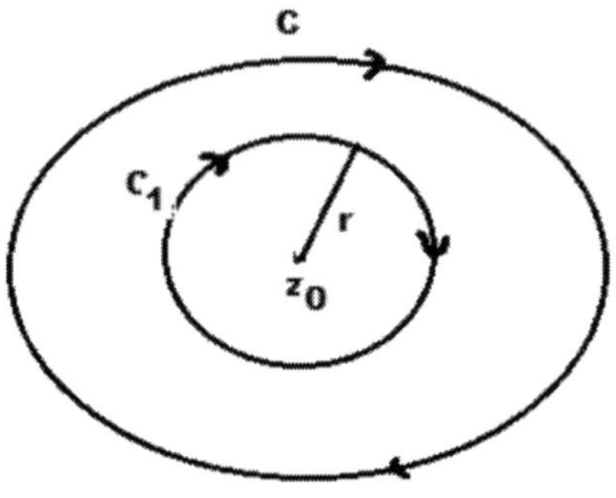

Figure 1

Since $\lim_{z \to z_0} f(z) = 0 \Rightarrow \lim_{z \to z_0} (z - z_0) f(z) = 0 \Rightarrow$

$$|(z - z_0) f(z)| < \varepsilon \text{ whenever } 0 < |z - z_0| < \delta \Rightarrow |f(z)| < \frac{\varepsilon}{|z - z_0|} = \frac{\varepsilon}{r}$$

By well-known results in complex integration $\oint_c f(z)dz = \oint_{c_1} f(z)dz \Rightarrow$

$\left|\oint_c f(z)dz\right| = \left|\oint_{c_1} f(z)dz\right| \le \oint_{c_1} |f(z)dz| < \frac{\varepsilon}{r} 2\pi\, r = 2\pi\varepsilon \Rightarrow \left|\oint_c f(z)dz\right| = 0$. Hance, $\oint_c f(z)dz = 0$. This completes the proof.

Analogue to the above result, we can extend the result to a finite numbers of points that is;

Theorem 2. If a function $f(z)$ is analytic inside and on a simple closed curve C except at a finite numbers of interior points $z_i : i = 0, 1, 2\, 3, --- n$ such that $\lim_{z \to z_i} f(z) = 0$ then

$$\oint_c f(z)dz = \sum_{i=0}^{n} \oint_{c_i} f(z)\, dz = 0$$

CAUCH INTEGRAL FORMULA

Theorem 3. Let $f(z)$ be analytic inside and on a simple closed curve C and z_0 is a point inside C then $\oint_c \frac{f(z_0)}{z - z_0} dz = 2\pi\, i\, f(z_0)$.

Proof. Since $f(z)$ is analytic inside and on C so is at $z_0 \Rightarrow f'(z_0) = \lim_{z \to z_0} \frac{f(z) - f(z_0)}{z - z_0}$ exist.

$$\lim_{z \to z_0} (z - z_0) f'(z_0) = \lim_{z \to z_0} (z - z_0) \frac{f(z) - f(z_0)}{z - z_0} = 0$$

Applying the result of theorem 1, we can get $\oint_C \frac{f(z) - f(z_0)}{z - z_0} dz = 0$. Now

$$\oint_C \frac{f(z) - f(z_0)}{z - z_0} dz = 0 \Rightarrow \oint_C \frac{f(z)}{z - z_0} dz = \oint_C \frac{f(z_0)}{z - z_0} dz = f(z_0) \oint_C \frac{1}{z - z_0} dz \quad (1)$$

Now, choose a circle C_1 inside C of radius r centre at z_0 that is $C_1 : |z - z_0| = r$.

$$|z - z_0| = r \Rightarrow z - z_0 = re^{i\theta} \quad where \quad 0 \le \theta < 2\pi \Rightarrow dz = i\, re^{i\theta}\, d\theta$$

Consequently result in eq (1) changes as

$$\oint_C \frac{f(z)}{z - z_0} dz = f(z_0) \oint_C \frac{1}{z - z_0} dz = f(z_0) \oint_{C_1} \frac{1}{z - z_0} dz = f(z_0) \int_0^{2\pi} \frac{i\, r\, e^{i\theta}}{re^{i\theta}} d\theta = 2\pi\, i\, f(z_0).$$

This completes the proof.

3. CONCLUSION

Cauchy-Goursat theorem is the basic pivotal theorem of the complex integral calculus. Theorem 1 provides us a kind of extension to Cauchy-Goursat theorem. Significance of theorem 1 is that it provides us a simple proof of Cauchy integral formula avoiding strict and rigor mathematical requirements.

REFERENCES

[1] Churchill R V and James W B 2003 Complex Variables and Applications *7th Ed.* McGraw Hill Inc.

[2] Jumarie G 2010 Cauchy's integral formula via the modified Riemann–Liouville derivative for analytic functions of fractional order *Applied Mathematical Letters*, **23**(12) 1444-1450

[3] Gario P 1981 Cauchy's Theorem on the Rigidity of Convex Polyhrdra *Archimede* **33**(1-2) 53-69

[4] Martins and Luiz C 1976 On Cauchy's Theorem in Classical Physics, *Arch. Rational Mech. Anal.* **60**(4) 305-324

IOP Conf. Series: Materials Science and Engineering **53** (2013) 012003 doi:10.1088/1757-899X/53/1/012003

[5] Mibu Y 1959 Cauchy's Theorem in Banach Spaces *J. Math. Soc.* Japan **11** 76-82

[6] Azram M, Daoud J I and Elfaki F A M **2010** On the Cauchy-Goursat Theorem *J. Applied Sci.* **10**(13) 1349-135

[7] Blaya R A, Reyes J B, Brackx F, Schepper H D and Sommen F 2012 Cauchy Integral Formulae in Quaternionic Hermitean Clifford Analysis *Complex Analysis and Operator Theory* **6**(5) 971-985

[8] Segev R and Rodnay G 2000 Cauchy Theorem's on Manifolds *J. Elasticity* **56**(2) 129-144

Analysis of Shot Boundary Based on Color and Texture Features of Frame

Chuen-Horng Lin[1], Muh-Don Hsiao[2] and Li-Jung Fu[3]

[1]Department of Computer Science and Information Engineering, [2,3]Department of Information Management, National Taichung University of Science and Technology, No. 129, Sec. 3, Sanmin Rd., Taichung, Taiwan, R.O.C.

E-mail:[1]linch@nutc.edu.tw,[2]paul@nutc.edu.tw and [3]j0916637996@gmail.com

Abstract. This paper proposed a shot boundary detection method based on color and texture features of frames. The shot boundary detection is employed in order to build a highly efficient method for video description. The research methods include feature extraction using color and texture feature, and shot boundary detection offering an adaptive measuring method. In this paper, the color and texture features are Traditional Color Histogram (TCH) and Histogram of Gradient Directions (HGD), respectively. The adaptive measuring method was proposed for the threshold of distance as the basis of identifying shot changes between frames. In order to validate the shot boundary detection method proposed herein, the video database provided by Carleton University website was used for the experiment. The experimental and comparison results of shot boundary detection were validated, and the experimental results were analyzed. This study carried out a series of comparisons and analyses for the above video database, and proved that the shot boundary detection method proposed could effectively detect the shot boundary.
Keywords: shot boundary detection, texture feature, color feature, adaptive measuring.

1. Introduction

Under the rapid development of information technology and multimedia technology, many electronic imaging equipments have become popular, resulting in the growing amount of multimedia information. Facing such massive multimedia information, how to create a mechanism to search for multimedia information effectively is a topic worth exploring. As most multimedia information can be obtained from the internet, it is very convenient for users; however, in the case of fuzzy keyword definition and different classification rules when users are selecting the type of information, the search result may not be satisfactory, and the search efficiency is low.

At present, there are three common methods to search for multimedia information (including word, excel, power point, pdf). The first method uses keywords to search for e-articles, using keywords to search for the related information; the second method uses keywords to search for images, comparing the description of video with the keywords; the third method uses keywords to search for videos. These three search methods can result in thousands of data, images or videos as search results in a very short time. However, the keywords cannot explicitly describe the content of the data to be searched, so the users have to screen the data.

[1] To whom any correspondence should be addressed

If the keyword is replaced by video description, the problems in the keyword-based search can be reduced. The video description mode detects the features from the video content to represent the video; however, as the technology advances, the video can generate diversified shot change automatically by software, so that the video cut becomes more convenient. The video production modes have no specific standard, they vary with the video characteristics; thus, how to define the video content is not defined, as long as the user requirements are met.

A video is composed of a series of consecutive frames, and a frame is a static image. As the relation between consecutive frames is very complex, and the video is composed of different consecutive frames, general studies on video retrieval are divided into feature extraction, video cut/shot boundary detection, key feature frame selection and video retrieval/indexing.

Previous studies on image retrieval and video retrieval are mostly based on the image processing and computer vision theory. In terms of image retrieval, many studies use the content based image retrieval (CBIR) method to replace the keyword retrieval mode. At present, the CBIR method is mostly used, and the technology has been mature. In terms of video retrieval, Deng et al.[2,3] focused on the content-based image and video retrieval (CBIVR) field. The features in the frame data are extracted and compared with the frame features in the image database for the similarity to find out the most similar frame to be retrieved. This method no longer requires manual annotation.

The CBIVR system extracts the features from the frames directly, and then identifies the boundary of frame conversion by comparing the frames. The mostly used frame features are color[1,4], texture[1,5-6], spatial relation[7], shape[8], space-time feature[9], key point feature[10] and motion mode feature[11].

This paper proposes an efficient video description mode, and creates a new content-based video retrieval algorithm. The main processing procedure of video content description in this paper is to extract the frame features, detect different shots, filter the shot noise with fuzzy content definition, cluster the shots, select the representative key feature set of shot, and classify the inconsecutive shots with similar contents. The shot description of video is then established based on the key feature set for redescribing the entire video.

This paper proposed a shot boundary detection method based on color and texture features of frames. Section 2.1 presents the Video Assumption. Feature Extraction will be introduced in section 2.2. Section 2.3 presents the Distance Measurement of Feature. Experiments and comparisons with other approaches are presented in Section 3. Conclusions will be offered in Section 4.

2. The Proposed Method

The query processing in a video retrieval is as shown in Figure 1, including videos video assumption, feature extraction, distance measurement of feature, shot boundary detection, shot noise filter, shot cluster, key feature set of cluster and shot, shot classification and video retrieval.

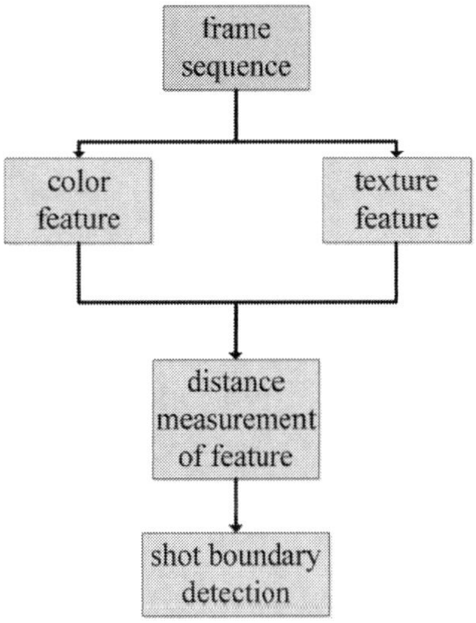

Figure 1. The workflow of proposed system

2.1. Video Assumption

A video is composed of multiple different scenes, a scene is composed of multiple shots, a shot is composed of a series of consecutive frames, and a frame is a static image, as shown in Figure 2.

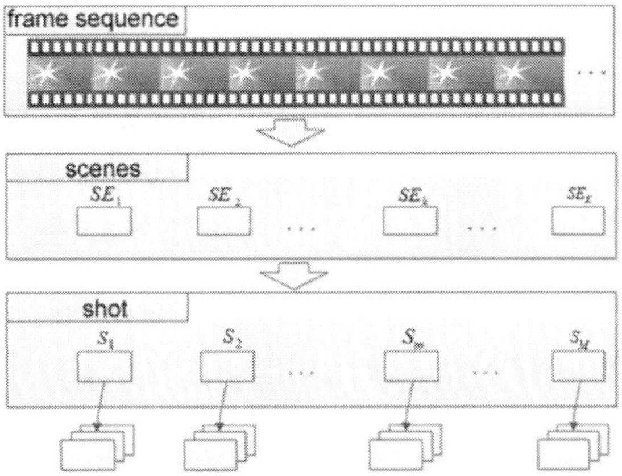

Figure 2. The video basic architecture

First, each shot is detected from the frame sequence and expressed as $S_i, i=1,2,...,M$, there are M shots, the shot with fuzzy content definition is expressed as $\bar{S}_i$, $i=1,2,...\bar{M}$, the total number is $\bar{M}$. After shot noise filter, there are N $(=M-\bar{M})$ effective shots. *video* is expressed as follows:

$$video = V \cup \bar{V},\ V = \bigcup_{i=1}^{M} S_i,\ \bar{V} = \bigcup_{i=1}^{\bar{M}} \bar{S}_i, \tag{1}$$

Then, the shots are clustered according to their characteristics, if No. n shot is classified as L_n cluster, expressed as $C_{n,}L_n$. The key feature set of cluster is taken out of each cluster. If the key feature set of No. L_n cluster of No. n shot is expressed as $k(C_n, L_n)$, and the key feature set of shot of No. n shot is expressed as $k(S_n)$, the shots of the video are classified finally, No. R class is expressed as CS_R.

2.2. Feature Extraction

A video is composed of a series of consecutive frames, and the processing time for extracting complex features may be too long. Therefore, this paper uses universally calculated and simple features, which are Traditional Color Histogram (TCH) and Histogram of Gradient Directions (HGD).

2.2.1. Traditional Color Histogram (TCH). This paper uses the color histogram as one of frame features, as the frame has three components of RGB. The histogram of three components of j-th frame can be expressed as follows:

$$H_R^j = (H_{R,0}^j, H_{R,1}^j, \dots, H_{R,255}^j),\ H_G^j = (H_{G,0}^j, H_{G,1}^j, \dots, H_{G,255}^j),$$
$$H_B^j = (H_{B,0}^j, H_{B,1}^j, \dots, H_{B,255}^j), \tag{2}$$

where $H_{R,i}^j$ $H_{G,i}^j$, $H_{B,i}^j$ denote the total number of pixels of R, G and B components of j-th frame on i-th grey respectively, and there are 768 features.

2.2.2. Histogram of Gradient Directions (HGD). The color histogram difference method is very sensitive to the variance in pixel tone scale. There may be excessive detection, so the texture feature is included. Considering the texture of frame, the color frame is converted into gray scale frame, RGB to YCbCr conversion is the most commonly used color coordinate system for image processing. Y is the luminance component and C_b and C_r are the chrominance components. The luminance edge is determined using Y information to perform the edge detection method described above. RGB of the pixel (x, y) was transformed into the YCbCr domain using the following formula:

$$\begin{bmatrix} Y \\ C_b \\ C_r \end{bmatrix} = \begin{bmatrix} 0.209 & 0.587 & 0.114 \\ -0.168 & -0.331 & 0.5 \\ 0.5 & -0.418 & -0.081 \end{bmatrix} \times \begin{bmatrix} R \\ G \\ B \end{bmatrix} + \begin{bmatrix} 0 \\ 128 \\ 128 \end{bmatrix} \tag{3}$$

Sobel edge detection was employed to detect the gradient feature. The gradient magnitude ∇Y and direction $\theta(x, y)$ for $Y(x, y)$ of pixel at coordinates (x, y) are given by:

$$\nabla Y = g(x,y) = \sqrt{g_x^2(x,y) + g_y^2(x,y)} \text{ and } \theta(x,y) = \tan^{-1}\left(\frac{g_y(x,y)}{g_x(x,y)}\right) \tag{4}$$

In most cases, the edge variation can be expressed as horizontal direction g_x and vertical direction g_y. A 3×3 block was used to compute the variations in the horizontal and vertical directions g_x and g_y.

The gradient direction histogram[12] calculates the cumulative distribution of gradient size in gradient direction as one of frame features. In order to avoid too sensitive gradient direction, the gradient direction is selected in a region, the gradient direction is between 0 and 360°, δ ° is one unit. The gradient sizes of pixels in this direction range are calculated, a frame can establish the histogram of values in $\frac{360°}{\delta}$ gradient direction regions ($\frac{360°}{\delta}$ regions), expressed as follows:

$$H_k^j(\delta) = \sum_{i=1}^{N^j} HGD(H_k^j, g_i^j, \theta_i^j), \tag{5a}$$

and

$$HGD(H_k^j, g, \theta) = \begin{cases} H_k^j + g_i^j & ,\text{if } (k-1)\times\delta \le \theta_i^j \le k\times\delta \\ 0 & ,\text{otherwise} \end{cases}, \tag{5b}$$

H_k^j denotes the total gradient size in j-th frame and gradient direction in k-th region, N^j is the total number of pixels in j-th frame, g_i^j and θ_i^j are the gradient size and direction on i-th pixel of j-th frame respectively, the range value of k is $0 < k \le \frac{360°}{\delta}$. H_d^j is regarded as one feature of the frame in this study, so there are 12 features.

2.3. Distance Measurement of Feature

TCH and HGD are useful to describe the relationship between color and texture of an image. Due to significantly complementary, these two features are integrated to establish a color- texture and color-histogram based frame matching between frames. TCH of the $j-1$ frame and the j frame as (H_R^{j-1}, H_G^{j-1}, H_B^{j-1}) and (H_R^j, H_G^j, H_B^j) are obtained from equation (6). The frame matching distance Δ_j^{TCH} between frames based on TCH is formulated as the following:

$$\Delta_j^{TCH} = \sum_{i=1}^{255}\left(\left|\frac{H_{R,i}^{j-1} - H_{R,i}^j}{\max(H_{R,i}^{j-1} + H_{R,i}^j, v)}\right| + \left|\frac{H_{G,i}^{j-1} - H_{G,i}^j}{\max(H_{G,i}^{j-1} + H_{G,i}^j, v)}\right| + \left|\frac{H_{B,i}^{j-1} - H_{B,i}^j}{\max(H_{B,i}^{j-1} + H_{B,i}^j, v)}\right|\right), \tag{6}$$

where v is any small number that avoids denominator is 0.

Considering HGD $H_k^{j-1}(\delta)$ and $H_k^j(\delta)$ of the $j-1$ and j frame are obtained from equation (7). The definition of the frame matching distance Δ_j^{HGD} between frames based on the HGD is shown as the following equation:

$$\Delta_j^{HGD} = \sum_{k=1}^{\frac{360°}{\delta}}\left|\frac{H_k^{j-1} - H_k^j}{\max(H_k^{j-1} + H_k^j, v)}\right|. \tag{7}$$

The frame matching combines the TCH and HGD to quantize the similarity of the $j-1$ and j frame. Using such frame matching, one can define the frame matching distance Δ_j^{TH} between the $j-1$ and j frame as:

$$\Delta_j^{TH} = \Delta_j^{TCH} + \Delta_j^{HGD} \tag{8}$$

where Δ_j^{TH} decreases with increasing similarity of Q and D.

2.4. Shot Boundary Detection

This paper takes the distance measurement Δ_j^{TH} between two frames as the basis of identifying shot change. The distance measurement between two similar frames is small, and the distance measurement of shot change is large.

The measured distance value between two frames determines the shot change. If the measured distance is greater than the threshold T_H, it is regarded as shot change, as shown in Figures 3 and 4; otherwise, the two frames are similar.

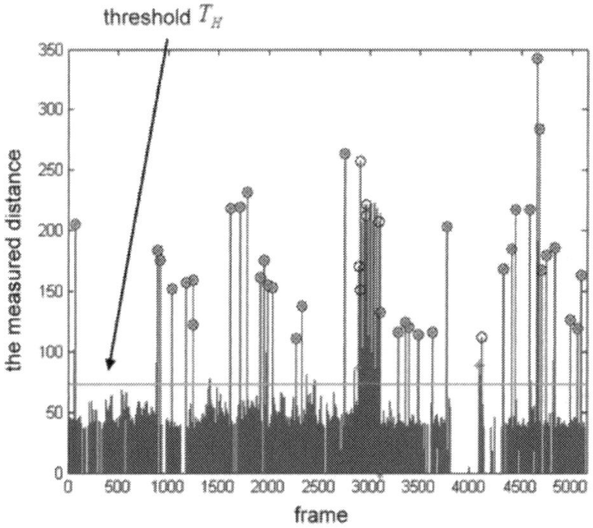

Figure 3. The measured distance value between two frames.

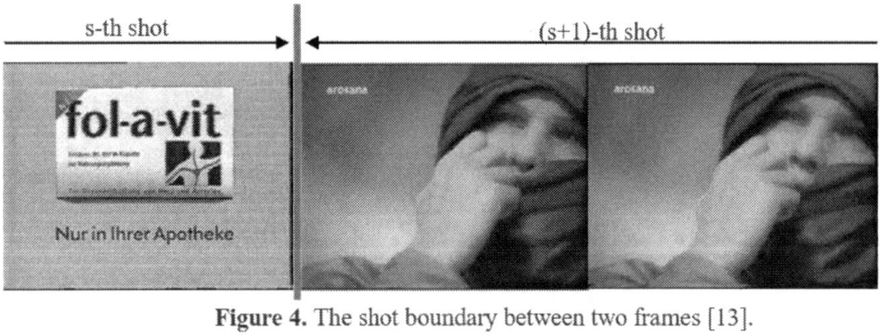

Figure 4. The shot boundary between two frames [13].

As the conversion of frame is not always sudden conversion, the conversion of some frames is slow. If only the difference between two consecutive frames is used for detection, the detection of shot will be influenced. Therefore, the shot boundary detection proposed in this paper can be divided into three steps to detect the boundary of shot change.

Step 1: if the measured distance value of j-th frame $\Delta_j^{TH} > w_1 \mu$, the inter frame is called candidate shot change boundary; otherwise, it is called non-shot change boundary, μ is the measured mean value of distance of all the frames in the video, expressed as follows:

$$\mu = \frac{1}{N-1}\sum_{j=2}^{N}\Delta_j^{TH}, \tag{9}$$

where N the total number of frames of the video, Δ_j^{TH} is the difference between j-th and $(j-1)$-th, w_1 is the weight. For the candidate shot change boundary, considering the problems resulted from slow change, a frame difference value correction method is proposed. The difference value of frame is not merely the difference between two consecutive frames. Thus, this paper proposes three levels of thresholds to make four intervals, as shown in Figure 5.

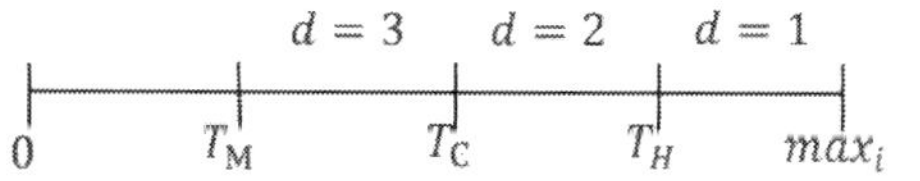

Figure 5. Three levels of thresholds.

If the difference value is between T_M to T_C, the difference values are in bell-shaped distribution and expressed as low width, as shown in Figure 6 (a), the frame range value of this interval is set as three frames $(d = 3)$ in front and behind respectively. If the difference value is between T_C to T_H, the difference values are in normal distribution and the bell shape is expressed in medium size, as shown in Figure 6 (b). The frame range value of this interval is set as two frames$(d = 2)$ in front and behind respectively. If the difference value is greater than T_H, the difference values are in steep shape and the bell shape is expressed in high narrow, as shown in Figure 6 (c). The frame range value of this interval is set as one frame$(d = 1)$ in front and behind respectively.

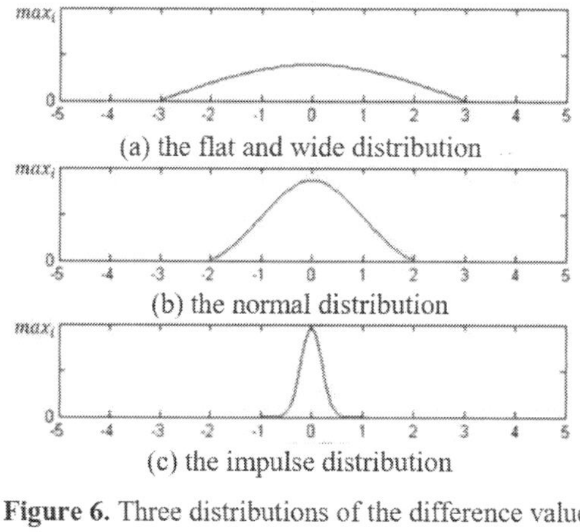

Figure 6. Three distributions of the difference value.

Three difference thresholds T_M, T_C and T_H are defined as high difference threshold T_H, medium high difference threshold T_C and mean difference threshold T_H respectively, expressed as follows:

$$T_M = w_2\mu, T_C = w_3\mu \text{ and } T_H = w_4\mu, \tag{10}$$

where w_2, w_3 and w_4 are the weights. The maximum difference value of the region is recalculated according to the designed frame range, expressed as follows:

$$\Delta^{\max_L} = \max[\Delta_{j-d}^{TH}, \Delta_{j-d+1}^{TH}, \ldots, \Delta_{j+d-1}^{TH}, \Delta_{j+d}^{TH}], \tag{11}$$

Finally, the difference value Δ_j^{TH} on candidate shot change boundary must meet the following conditions:

$$\Delta_j^{TH} = \Delta^{max_L}, \Delta_j^{TH} \geq w_5, \Delta_j^{TH} \geq \Delta^{T_L}, \quad (12)$$

where w_5 is the weight of basic difference value and $\Delta^{T_L} = \frac{w_6 \Delta^{max}}{10}$ and $\Delta^{max} = \max[\Delta_j^{TH}], j = 2, \dots, N$. If the difference value on the candidate shot change boundary does not meet the aforesaid conditions, the candidate condition is cancelled.

Step 2: The range of frame will be expanded in this stage, and three levels of thresholds are proposed to distinguish four intervals, as shown in Figure 7.

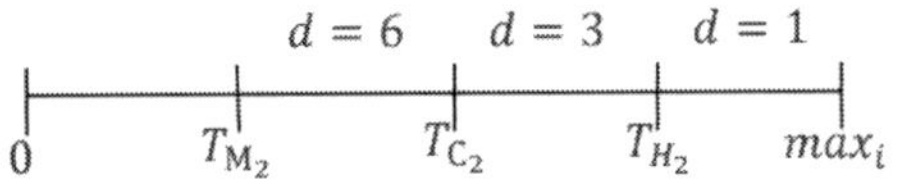

Figure 7. Three levels of thresholds.

If the difference value is between T_{M_2} to T_{C_2}, the difference values are in bell-shaped distribution and expressed in low width, as shown in Figure 6 (a), the frame range value of this interval is increased to six frames $(d = 6)$ in front and behind respectively. If the difference value is in T_{C_2} to T_{H_2}, the difference values are in normal distribution and the bell shape is expressed in medium size, as shown in Figure 6 (b), and the frame range value of this interval is increased to three frames $(d = 3)$ in front and behind respectively. If the difference value is greater than T_{H_2}, the difference values are in steep shape and the bell shape is expressed in high narrow, as shown in Figure 6 (c). The frame range value of this interval is also one frame$(d = 1)$ in front and behind respectively.

Three difference thresholds T_{M_2}, T_{C_2} and T_{H_2} are defined as Stage II high difference threshold T_{H_2}, Stage II medium high difference threshold T_{C_2} and Stage II mean difference threshold T_{M_2} respectively, expressed as follows:

$$T_{M_2} = w_7 \mu, T_{C_2} = w_8 \mu \text{ and } T_{H_2} = w_9 \mu \quad (13)$$

where w_7, w_8 and w_9 are the weights of parameters.

The final step is to check whether the difference value on the candidate shot change boundary is the maximum value in the frame range. If yes, it is defined as candidate shot change boundary; otherwise, the candidacy is cancelled.

Step 3: This stage is the last stage eliminating the candidate shot change boundary, considering the removal of non-shot change boundary. The j-th candidate shot change boundary must meet the following conditions:

$$|\Delta_j^{TH} - \Delta_{j-1}^{TH}| \leq w_{10}\mu \text{ or } |\Delta_j^{TH} - \Delta_{j+1}^{TH}| \leq w_{10}\mu \quad (14)$$

where Δ_j^{TH} and Δ_{j-1}^{TH} are the absolute difference of difference of j-th and $(j-1)$-th frame, Δ_j^{TH} and Δ_{j+1}^{TH} are the absolute difference of difference of j-th and $(j+1)$-th frame, and w_{10} is the weight. If the candidate shot change boundary meets the above conditions, the candidacy is cancelled, and the rest are the shot change boundaries in this paper.

3. Experiments

Two types of video database are used for this experiment. The first type is the video database provided by Carleton University[13] website, for validating the shot boundary detection proposed in this paper. This type of video database contains ten videos, as shown in Table 1, and the objects in all the videos move very fast. The colors of No. 2, 3 and 5 videos are plain and relatively dark. Although the content of No. 10 video is relatively dark, the colors are multiple. The quantity of consecutive frames of most shots is less than 10, and some dissolution effects are usually added in shot change, so that the detection of shot boundary becomes more difficult.

Table 1. The first types of video database.

No.1

No.2

No.3

No.4

No.5

No.6

No.7

No.8

No.9

No.10

As the first type of video database provides diversified videos, including cartoons, dramas and black and white videos, the threshold parameter of each video is adjusted slightly to conform to different types of videos. Table 2 shows the setting of parameter weights w_1 and w_6. The content of No. 10 video is relatively dark and the shot change is implemented by dissolve. Thus, all the feature differences are relatively small, and w_1 is reduced to 1. As for the setting of parameter weights w_7 to w_9, No. 1, 2, 4, 6, 7 and 9 videos have the same setting. As No. 3 video is of gray image, the content of No. 5 and 8 videos is relatively dark, and the background is black. The objects in No. 8 video move fast, the set values must be adjusted. As for the setting of parameter weight w_{10}, No. 3, 7 and 10 are adjusted down properly for slow frame conversion, relatively dark video content and black background.

Table 2. The weight values in this paper.

No.	w_1	w_2	w_3	w_4	w_5	w_6	w_7	w_8	w_9	w_{10}
1	2	1	2	3	0.1	3	2	4	6	1
2	2	1	2	3	0.1	3	2	4	6	1
3	2	1	2	3	0.1	3	1	2	3	0.5
4	2	1	2	3	0.1	3	2	4	6	1
5	2	1	2	3	0.1	3	1	2	3	1
6	2	1	2	3	0.1	3	2	4	6	1
7	2	1	2	3	0.1	3	2	4	6	0.5
8	2	1	2	3	0.1	3	1.5	3	5	1
9	2	1	2	3	0.1	3	2	4	6	1
10	1	1	2	3	0.1	3	0.5	1	2	0.1

The shot boundary detection results in this study are divided into Fuzzy Logic Approach[14], Edge Tracking[15], Histogram[16] and Pixel Localization[17] for comparison and analysis. The Fuzzy Logic Approach applies color, texture, edge difference and motion compensation to video shot boundary. The Edge Tracking uses texture feature and edge tracking to generate different frames to detect the boundary of slow conversion. The Histogram calculates the difference between two consecutive frames. Pixel Localization uses texture feature and the inter frame in image compression to calculate the difference between regional pixels to detect the shot boundary.

Three efficiency evaluation standards are used in this paper for the efficiency evaluation of experimental results of shot boundary detection, which are precision, recall[17] rate and F1 index[14], expressed as follows:

$$\text{precision} = \frac{\text{correct}}{\text{correct} + \text{falsepositive}}, \ \text{recall} = \frac{\text{correct}}{\text{correct} + \text{missed}}, \tag{15}$$

$$\text{F1} = \frac{2 \times \text{precision} \times \text{recall}}{\text{precision} + \text{recall}}, \tag{16}$$

where correct is the ground truth of shot boundary, falsepositive is not a shot boundary, missed refers to the shot boundary is not detected. The result of shot boundary and the efficiency evaluation are shown in Table 3. The percentage sign of numerical values is omitted, and the red bold numerical values represent better results. The precision, recall rate and comprehensive F1 index of No. 1, 2, 4, 6, 7 and 9 videos detected using this method have reached 100%. The precision, recall rate and comprehensive F1 index of No. 3, 5, 8 and 10 videos are lower than that by other methods. The factors influencing the precision and recall rate and the correction of the ground truth are described below.

Table 3. The precision of these methods.

	Present method			**Fuzzy logic**			**Edge tracking**			**Histogram**			**Pixel localization**		
No.	P	R	F1	P	R	F1	P	R	F1	P	R	F1	P	R	F1
1	100	100	100	100	100	100	100	100	100	100	100	100	100	100	100
2	100	100	100	100	100	100	100	100	100	100	37.5	54.5	82.5	82.5	82.5
3	90.7	86.0	88.3	88.1	100	93.7	59.5	87.0	70.7	93.6	53.6	68.2	76.4	77.8	77.1
4	100	100	100	100	100	100	100	100	100	100	94.1	97.0	100	100	100
5	90.0	96.4	93.1	93.8	100	96.8	93.8	100	96.8	95.5	70.0	80.8	86.7	86.7	86.7
6	100	100	100	100	100	100	100	100	100	100	100	100	0.0	0.0	0.0
7	100	100	100	95.0	100	97.4	81.0	94.4	87.2	100	66.7	80.0	70.8	99.4	82.7
8	87.8	90.0	88.9	90.9	100	95.2	89.5	89.5	89.5	97.1	89.5	93.0	92.7	100	96.2
9	100	100	100	100	100	100	100	100	100	100	50.0	75.0	100	100	100
10	72.2	72.2	72.2	92.9	89.7	91.3	49.7	89.7	64.0	85.0	39.5	54.0	62.3	54.0	57.9

The result of false shot boundary detection of No. 3 video is shown in Figure 8. As the content of consecutive frames is relatively dark or relatively bright, the detection of shot boundary is influenced, as shown in Figure 8 (a) and Figure 8 (b). The red frame is the false boundary detected by this method. The traveling speed of objects in the video results in falsely detected shot boundary, as shown in Figure 8 (c) and Figure 8 (d).

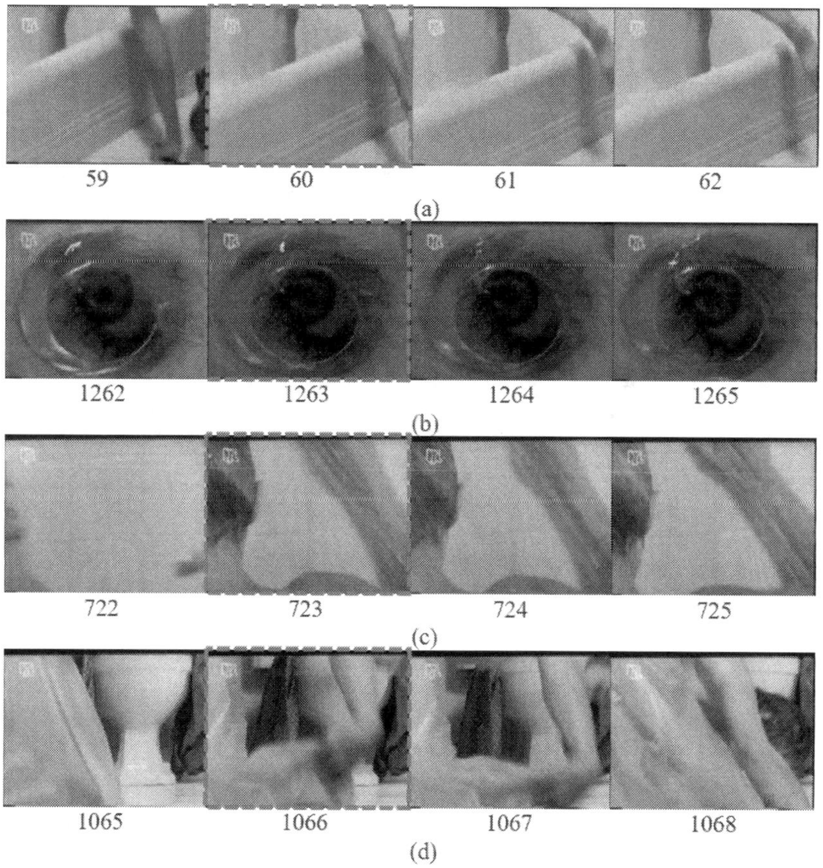

Figure 8. The results of false shot boundary detection.

Different users have different opinions on the shot boundary, so that the recall rate in the experimental results is reduced. As a result, the videos with problems are proposed and described. The real shot data of No. 3 video are compared with the data after practical viewing for analysis. It is found that the difference between the first frame and the second frame is larger than that between the third frame and the fourth frame. Therefore, it is appropriate to change the fourth frame of the ground truth to the second frame. The yellow frame is changed the ground truth as shown in Figure 9.

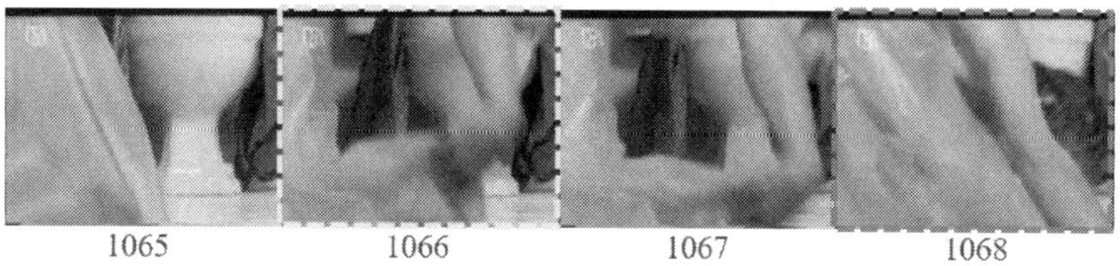

Figure 9. The frames between two shot.

The ground truth of No. 5 video are compared with the data after practical viewing for analysis. As shown in Figure 10 (a), there is a short dissolution, and the ground truth of the shot boundary define the frame after dissolution (No. 249). This paper argues that the frame (No. 248) in the change between shots should be defined. Figure 10 (b) has the same situation, namely the first and the second frame can be identified as the same shot; thus, this paper suggests that the third frame is appropriate. The yellow frame in Figure 10. is the changed the ground truth of the shot boundary.

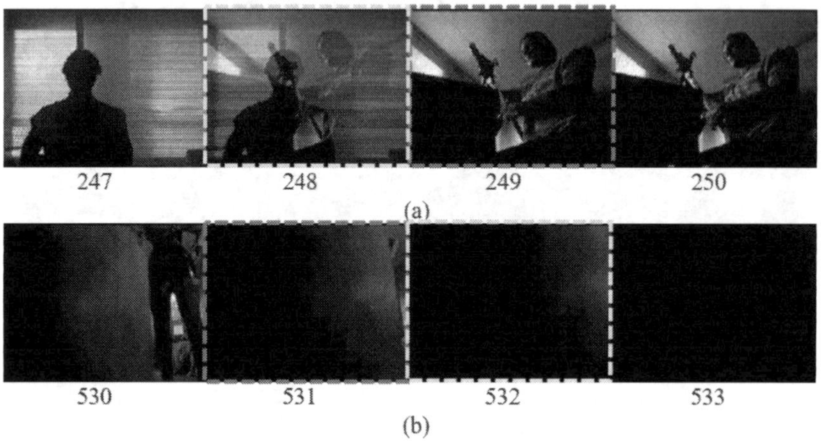

Figure 10. The frames between two shot.

The ground truth of the shot boundary in No. 10 video are compared with the data after practical viewing for analysis. It is found that the consecutive frames are the course of dissolution, as shown in Figure 11 (a). It is appropriate to change the ground truth to yellow frame. Many boundaries in No. 10 video are converted by dissolution, as the ground truth have not indicated these dissolutions. This paper selects a set of consecutive frames as indication. The yellow frame in Figure 11 (b). is the increased ground truth.

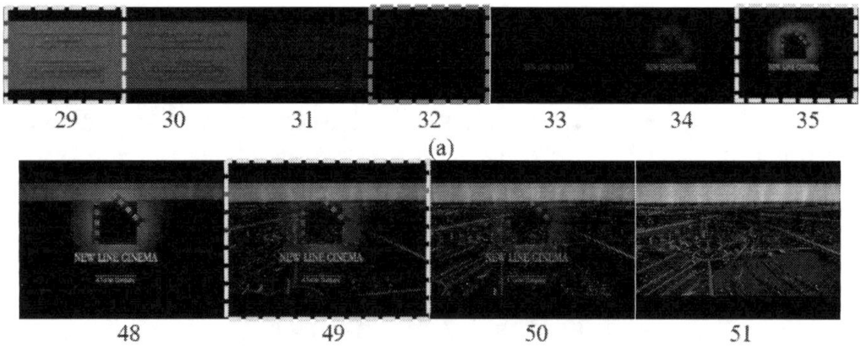

Figure 11. The frames between the different shot.

According to the aforesaid modification of real shot data, the precision, recall rate and comprehensive F1 index of the 10 videos are recalculated, as shown in Table 4. The numerical values of No. 3, 5 and 10 videos are still increased after the real shot data are modified, thus proving that this method can detect the short dissolution of sudden change and slow change.

Table 4. The precision of these methods for redefined the ground truth.

	Present method			Fuzzy logic			Edge tracking			Histogram			Pixel localization		
No.	P	R	F1	P	R	F1	P	R	F1	P	R	F1	P	R	F1
1	100	100	100	100	100	100	100	100	100	100	100	100	100	100	100
2	100	100	100	100	100	100	100	100	100	100	37.5	54.5	82.5	82.5	82.5
3	92.6	87.7	90.1	88.1	100	93.7	59.5	87.0	70.7	93.6	53.6	68.2	76.4	77.8	77.1
4	100	100	100	100	100	100	100	100	100	100	94.1	97.0	100	100	100
5	93.3	100	96.5	93.8	100	96.8	93.8	100	96.8	95.5	70.0	80.8	86.7	86.7	86.7
6	100	100	100	100	100	100	100	100	100	100	100	100	0.0	0.0	0.0
7	100	100	100	95.0	100	97.4	81.0	94.4	87.2	100	66.7	80.0	70.8	99.4	82.7
8	87.8	90.0	88.9	90.9	100	95.2	89.5	89.5	89.5	97.1	89.5	93.0	92.7	100	96.2
9	100	100	100	100	100	100	100	100	100	100	50.0	75.0	100	100	100
10.	88.9	96.7	92.6	92.9	89.7	91.3	49.7	89.7	64.0	85.0	39.5	54.0	62.3	54.0	57.9

4. Conclusions

This paper proposed a shot boundary detection method based on color and texture features of frames, and performed a series of analyses. The experimental results of comparison and analysis of this shot boundary detection method were compared with Fuzzy Logic Approach, Edge Tracking, Histogram and Pixel Localization. The analysis of experimental results showed that the ground truth provided by website has obvious inaccuracy in recognition, so that the analysis result is influenced. The boundary of shot was detected effectively, proving that the proposed method can be used for shot filter, cluster, classification and retrieval in the future. This method can also be developed into a rapid and convenient video retrieval system.

5. Acknowledgements

This work was supported in part by National Science Council, Taiwan, under GRANT NO. NSC 101-2221-E-025 -009.

References

[1] C. H. Lin, R. T. Chen, and Y. K. Chan 2009 A smart content-based image retrieval system based on color and texture feature *Image and Vision Computing* vol 27, pp 658-665
[2] A. F. Smeaton, P. Over, and A. R. Doherty 2010 Video shot boundary detection: Seven years of TRECVid activity *Computer Vision and Image Understanding* Vol 114, pp 411-418
[3] H. Lee, J. Yu, Y. Im, J. M. Gil, and D. Park 2011 A unified scheme of shot boundary detection and anchor shot detection in news video story parsing *Journal Multimedia Tools and Applications* Vol 51, pp 1127-1145
[4] M. H. Hung, C. H. Hsieh, C. M. Kuo, and J. S. Pan 2011 Generalized playfield segmentation of sport videos using color features *Pattern Recognition Letters* Vol 32, pp 987-1000
[5] A. Hafiane and B. Zavidovique 2008 Local relational string and mutual matching for image retrieval *Information Processing and Management* Vol 44, pp. 1201-1213
[6] C. H. Lin and Y. J. Syu 2010 Fast segmentation of porcelain images based on texture features *Journal of Visual Communication and Image Representation* Vol 21, pp 707-721
[7] C. H. Lin, Y. K. Chan, K. H. Chen, D. C. Huang, and Y. J. Chang 2011 Fast color-spatial feature based image retrieval methods *Expert Systems with Applications* Vol 38, pp 11412-11420

[8] C. H. Wei, Y. Li, W. Y. Chau, and C. T. Li 2009 Trademark image retrieval using synthetic features for describing global shape and interior structure *Pattern Recognition* Vol 42, pp 386-394

[9] W. Chen and Y. J. Zhang 2008 Parametric model for video content analysis *Pattern Recognition Letters* Vol 29, pp 181-191

[10] Y. G. Jiang, J. Yang, C. W. Ngo, and A. G. Hauptmann 2010 Representations of keypoint-based semantic concept detection: A comprehensive study *IEEE Transactions on Multimedia* Vol 12, pp 42-53

[11] A. M. Amel, B. A. Abdessalem, and M. Abdellatif 2010 Video shot boundary detection using motion activity descriptor *Journal of Telecommunications* Vol 2, pp 54-59

[12] S. R. Fountain and T. N. Tan 1998 Efficient rotation invariant texture features for content-based image retrieval *Pattern Recognition* Vol 31, pp 1725-1732

[13] P. Bose, R. Laganiere,and A. Whitehead, VIVA Research Lab, University of Ottawa, School of Computer Science, Carleton University,Some results in video segmentation, Retrieved on January, 2012, from http://www.site.uottawa.ca/~laganier/videoseg/.

[14] H. Fang, J. Jiang, and Y. Feng 2006 A fuzzy logic approach for detection of video shot boundaries *Pattern Recognition* Vol 39, pp 2092-2100

[15] A. Whitehead, P. Bose, and R. Laganiere 2004 Feature based cut detection with automatic threshold selection *Image and Video Retrieval* pp 2064-2065

[16] S. Pfeiffer, R. Lienhart, G. Kühne, and W. Effelsberg 1998 The MoCA project-movie content analysis research at the university of mannheim *GI Jahrestagung* pp 329-338

[17] J. S. Boreczky and L. A. Rowe 1996 Comparison of video shot boundary detection techniques *Journal of Electronic Imaging* Vol 5, pp 122-128

IOP Conf. Series: Materials Science and Engineering 53 (2013) 012005 doi:10.1088/1757-899X/53/1/012005

Development of Solar Powered Irrigation System

AI Abdelkerim[1], MMR Sami Eusuf[1,4], MJE Salami[1], A. Aibinu[2] and M A Eusuf[3]

[1]International Islamic University Malaysia, Department of Mechatronics Engineering, 50728 Kuala Lumpur, Malaysia
[2]Federal University of Technology, Mina, Nigeria
[3] International Islamic University Malaysia, Department of Architecture, 50728 Kuala Lumpur, Malaysia

E-mail: samieusuf@gmail.com

Abstract. Development of a solar powered irrigation system has been discussed in this paper. This system would be SCADA-based and quite useful in areas where there is plenty of sunshine but insufficient water to carry out farming activities, such as rubber plantation, strawberry plantation, or any plantation, that requires frequent watering. The system is powered by solar system as a renewable energy which uses solar panel module to convert Sunlight into electricity. The development and implementation of an automated SCADA controlled system that uses PLC as a controller is significant to agricultural, oil and gas monitoring and control purpose purposes. In addition, the system is powered by an intelligent solar system in which solar panel targets the radiation from the Sun. Other than that, the solar system has reduced energy cost as well as pollution. The system is equipped with four input sensors; two soil moisture sensors, two level detection sensors. Soil moisture sensor measures the humidity of the soil, whereas the level detection sensors detect the level of water in the tank. The output sides consist of two solenoid valves, which are controlled respectively by two moistures sensors.

1. Introduction

The irrigation system is defined as a system that distributes water to targeted area. Basically, it is meant for agriculture purposes. The efficiency of the irrigation is based on the system used. Since antiquity, the human life is based on agriculture and the irrigation system is one of the tools that boost agriculture. There are many other types of irrigation system all over the world but these irrigations are encountering many problems. In fact, there are few modern systems but they mostly fail in one way to another. The automation plays an important role in the world economy; therefore, engineers struggle to come out with combined automatic devices in order to create complex systems that help human in its activities so that the system automatically processes itself without any human intervention. So we would like to develop an automatic irrigation system.

Basically, the paper consists of electrical part and mechanical part. The electrical part consists of photovoltaic, which is meant to generate power and the power is stored in the rechargeable battery. The mechanical part consists of pump to pump out the water from the water source. There are two solenoid valves that are used to control the water flow. Additionally, Programmable Logic Controller (PLC) is interfaced with SCADA (Supervisory Control Data Acquisition) system developed in a personal computer (PC) using LABVIEW. SCADA system is used for controlling the automation system via computer so there is a need to develop HMI human-machine interface employed to communicate with automation

[4] To whom any correspondence should be addressed.

device like PLCs and other computers, like water computers. Water is recognized as a source of human life, as well as plants and animals. Therefore, the water needs to be managed properly without any waste. The proper management of any liquid is very significant, especially in our case of water management of the irrigation system. The best liquid management is provided by SCADA.

The parameters in the project are soil humidity condition, water level condition, the position of the Sun and SCADA system. Humidity of the environment plays a major role in agriculture development. Due to the high cost and the effectiveness of the system, that has yet to be proven, watering process based on these parameters is not widely used in the World, as it is still a very new technology that tends to post modernize the agriculture. Most of the existing systems are manual system. The manual system needs labor for monitoring the productivity and health crop. Considering labor's salary, the system will cost much more than the automatic system, in which there is no assistance to the system. The solar system is used to generate the power to the entire system and the solar system is much cheaper than the electrical system. It is suitable to the rural area that is why the solar system is used as a power supplier to replace DC motor electricity source. In fact the initial cost of solar installation is higher than use of DC electrical motor but the solar system has no bill compared to electrical which has bill to pay every month. It is a versatile source of renewable energy that can be used in any application. Solar power is also clean and pollution free.

As it is stated in the paper, the discussion is about the development of an intelligent irrigation system, which can be used in plantation. The utilization of greenhouse concept is used in the paper for reducing atmosphere pollution. The system consists of hardware and software and, finally, the integration of the two parts to provide the results. The hardware system consists of the sensors, actuators, drivers and PLC. The software is all about designing SCADA system, in which, it has the ability to read from the automatic system and write into the system. In software design, there is a need for full development of NI instruments, such as DSC module, OPC server, MODBUS server. In hardware design, we need all the components that are necessary to accomplish the project, and these components are PLC, DC water pump motor, actuators, sensors and some minor components like tank and reservoir.

2. Related Work

In this section, we are going to review papers and books related to SCADA based irrigation system. The system is divided in three different branches, as SCADA, PLC and Solar Panel, and the literature review consists of the aforementioned three independents parts.

SCADA, which is known as the supervisory control and data acquisition, happens to be a computer system for assembling and analysing real time data. Furthermore, SCADA is frequently regarded as the central system that controls and monitors an entire system that is expansively distributed over an extended distance. Remote Terminal Unit (RTU) and Programmable Logic Controller (PLC) are responsible for performing automatically in regards to the size of the site. However, there is always a restriction by the host control functions towards the central site over-ride or supervisory level capability. Essentially, the SCADA system consists of hardware and the software parts.

Remote Terminal Unit (RTU) gathers information from the inputs devices such as sensors, valves, motors, alarms and many more. RTU can be a programmable logic controller (PLC) as well as computer. RTU collect all the system information and keep it the memory until the master terminal unit (MTU) send the request to transmit data [9].

MTU collects all the data stored and sends information to the device for interference operation. MTU can be a computer consisting of Human Machine Interface (HMI) as medium of communication for processing data and information. HMI human machine interface provides a centralized monitoring and controlling system for the various inputs and outputs [10].

The communication is very important part of SCADA system, as it is a way of establishing connection between MTU and RTU. Through communication server the device send the signal from the site to the server device so that the data can be processed and the system can make decision based on the processed data. The basic communication types are LAN local area network, wireless and radio signal [12].

NI LABVIEW software can communicate with a Programmable Logic Controller (PLC) in a variety of ways. In order to make a successful communication with PLC, Modbus serial communication protocol was made available by Modicon in 1979, and it was then expanded to the TCP protocol. Modbus turned into one of the regular communications medium in the industry, as it included real-time and historical trending, tools for logging data to a networked historical database, organization of events and alarms. Modbus also assists in networking Lab-VIEW Real-Time targets and Modbus devices into a whole system, and, lastly, Modbus adjoins security into the user interfaces [13]. The second part is for the Solar Irrigation system.

a) Rain harvesting b) Solar powered system

Figure 1. Solar powered irrigation system.

Figure 1 describes the component of irrigation system, which consists of solar power and rain harvesting. A solar powered agricultural system is used for sustaining the availability of food production; it is powered by a solar system, which is recognized as the most suitable source in case of irrigation [11]. The system is based on time that is to say from 7.30 to 3pm is peak watering demand time. And from Sunrise to Sunset is called peak energy production time or the time in which the battery is charged. The energy charged in the battery will power up the engine when the specified time comes [1].

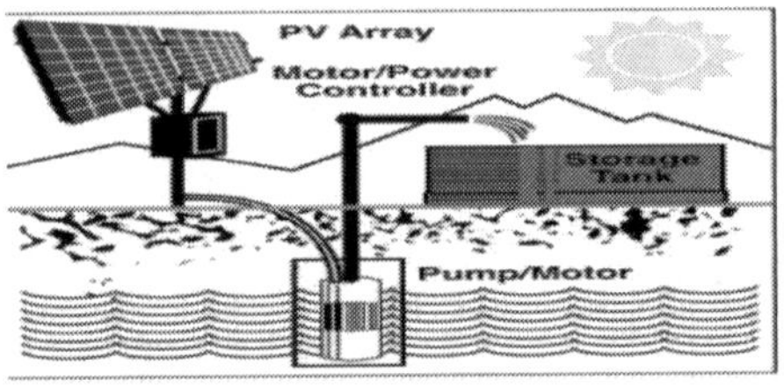

Figure 2. Photovoltaic system [2]

The system function is based on time and it has a system that predicts the weather condition but it fail to predict sometime due to climate change. It is quite difficult to forecast the weather condition due to climate change. The area happens to be in a condition of deficiency of water, which rationally contains negative irregularity in the watering pattern that may be over watered. The release of water to the seedlings in trays, by means of watering hoses, has resulted in dislodged seedlings from trays, thus lowering the production and augmented the cost linked with weekend employment of labour [1]. The developed system in [1] is not intelligent system. Additionally, the weakness of the system is outnumbered, and one of the biggest weaknesses is about failure of predicting the weather. Furthermore, there is no sensor used in the system. It is advisable to use rain sensor as well, as soil moisture sensor is in work by [1]. Figure 2 below shows a solar powered irrigation system which consists of photo voltaic, controller and storage tank.

The paper provides a preface to solar-powered livestock watering systems, as well as the discussions of price, mechanism, and terms, with some suggestions for the installment and design of these systems. The potency and flaw of the solar pumping system are evaluated in contrast to the key options for pumping in far-flung locations: mechanical windmills [2].

The system utilized solar panels, mechanical windmills, and portable generators. The objective is to pump water from the water source for livestock areas, where there is an unavailability of electricity from the power lines. These systems bestow livestock with immense access potential of investigation by encouraging the animals to leave from watercourse and lakes. Livestock pressure is reduced on stream banks, avoiding nutrient loading, damaging streamside erosion, vegetation, and pollution [2].

Solar powered systems have high initial cost and low maintenance cost compared with other remote systems. There are two mounting structures, such as fixed and tracking mounting structures. For fixed mounts, there is a tolerance for high wind and it is less expensive but the tilt angle needs to be adjusted to the south. Tracking arrays follow Sun across the sky. Batteries are not used because tanks are so big enough to keep water for few days, hence reducing the maintenance and installation cost [2].

Partially, it is a good initiative to omit the batteries which reduce the cost. The tracking mount system for photovoltaic is better idea since it follows Sun across the sky. There is an important controller which helps the pumps from low and high harmful voltage and maximizing the water pumped [2]. Some great ideas are present in the system, such as using tracking mount system and reducing the cost by omitting the batteries, even though the system is not intelligent enough [2].

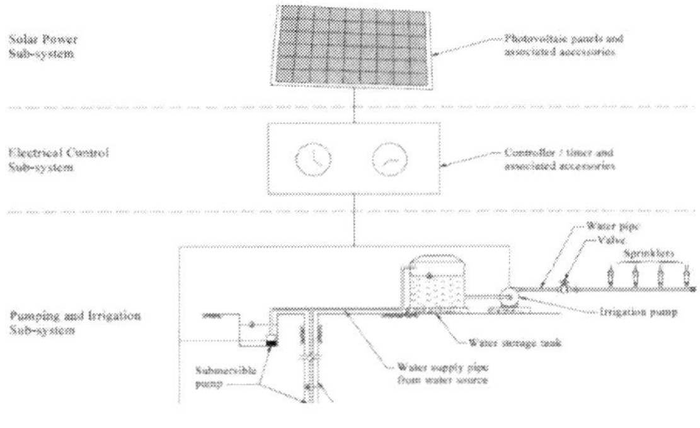

Figure 3. Solar Installation System [2]

Figure 3 displays the simulation of solar powered irrigation system. This system is to help the organization of freshly planted vegetation on man-made slopes, where man made vegetation slopes green during the dry seasons; an automatic irrigation system, utilizing solar power for watering vegetation, is applied. Photovoltaic panels are utilized to generate electricity to produce electricity, that is later stored in rechargeable batteries. Power is supplied by the batteries for operating the system. Water is pumped from a submersible pump from a natural watercourse to a storage tank and the water is then drawn by an irrigation pump to the sprinklers installed at the slope toe for watering the vegetation on the slope [11]. Initially, there were some quandaries involved with the solar powered automatic irrigation system. Improvements to the system were instantly executed to develop the efficiency of the system.

Therefore, it is learnt that the utilization of an automatic irrigation system, which is functioned by solar power, is the best option in designing the irrigation system. In fact the greatest problem that is facing the system is due to the defection of batteries. It is necessary to come up with great idea about batteries.

The system happens to be in a very adequate position as it is nicely structured with all the equipments of the solar powered automatic irrigation system, where a control unit controls the control relays, electrical timers, an inverter and a charge controller [3].

The period for the operations is maintained by the control of relays and timers, which influences the pumping and the irrigation system as well. The charging of the batteries is controlled by the charge controller, with the assistance of the PV panels. For the pump operations, rechargeable batteries help the inverter to convert direct current (DC) into alternating current (AC).

For the rainfall detection, rain detector sensor is connected to the control unit of the system for the detection, and if there a detection of rain within a range of three hours before a scheduled irrigation, the system will skip the irrigation function for that day [3].

The topic is about generating electricity using solar system. It is known as photovoltaic system. This form of system is the most important especially in rural [4]. Areas where there is luck of foil. It is found better system compare to other system such as motor and so on. The Sun's rays are able to generate energy by making the rays strike onto the surface of the panel, thus releasing energy in the form of heat. Special materials also influence the release of energy from the surface of the panels.

This material is used by the Solar Panels to transform the light energy from the Sun directly into electricity, with the assistance of the photovoltaic process. The power produced by the solar panels is directly proportional to the intensity of Sunlight striking the Solar Panel. Furthermore, bigger Solar Panels are able to seize more rays, thus producing more power. Four different components are able to be found in a Solar Electric System and they are: Solar panel, battery, regulator (controller), and one or more lights or other [4].

However, solar electric systems in [4] comparisons to other lighting systems are:
1. Vivid and Stable Light
2. Radio and TV is able to run with it
3. Monthly Electricity Bill is not required
4. Secured; No Fire Hazards required
5. No Pollution and Noise
6. Lifelong

Using photovoltaic to generate power or electricity is the best idea because it is found that it has many advantage such less expensive. It is good for rural areas. One of the good points is about how to increase the power of photovoltaic; just it is simply to put in parallel in parallel to each other [4].

The function of solar powered water pumping system is the crucial aspiration of this paper, as well explaining the differences with the other energy sources [5]. Two fundamental types of solar powered water pumping systems are found and they are: Battery-Coupled and Direct-Coupled. These two fundamental types help us to determine the optimum system for a particular application, and they are explained briefly below [5]:

2.1 Battery-Coupled Solar Water Pumping System:

Battery-coupled solar water pumping system is composed of charge-control regulator, photovoltaic panels, and batteries, pump controller, pressure switch and tank, and DC water pump. During the daytime, the batteries are charges with the electric current produced by the PV panels, and the batteries in return supply power to the pump whenever water is required.

2.2 Direct-Coupled Solar Pumping System:

Electricity from the PV modules is sent directly to the pump in this pumping system, and the electricity helps the water to pump water through a pipe to where it is required. This water pumping mechanism is designed to be executed only in the daytime, as the quantity of water pumped is entirely dependent on the type of pump used and the amount of Sunlight striking the PV panels [5].

- Battery-Coupled Solar Water Pumping System:
  Using battery is the most trusted since there might be a day where the radiation coming from the Sun will not be available. For better precaution it is advisable to use batteries in solar system in order the store energy for future use.

- Direct Coupled Solar Pumping System:
  Direct connection is not advisable because there will be a day without storage and radiation coming from Sun.

The configuration, which is battery-coupled solar water pumping system, is accepted in our case. The coupled battery is to be used again if it becomes necessary [5].

The system is really pertinent, because we are going to use photo voltaic system to generate electricity to our irrigation system. This article is consecrated only for the Photovoltaic system, therefore we will implement the points that we have learned from this article. Photovoltaic cells are capable of transforming the energy from the solar radiation into electricity due to the energy transfer transpiring at the sub-atomic level [6].

Small packages, called photons, make the very nature of solar energy. Open circuit voltage and short circuit current capabilities make the relevant analysis required for a photovoltaic cell.

The advantage of utilizing a direct solar radiation is also found in [7]. There is an importance in the article due to the cost competitiveness of photovoltaic power for irrigation with conventional energy sources for petite, remote applications. Lastly, if the entire system designs and utilization timing is circumspectly considered and organized, there is always a presence of opportunity to utilize the solar energy as efficiently [7].

3. Methodology

The system consist of hardware and software. The harware part involves PLC (Programmable Logic Controller), smart solar tracker, motors, sensors and valves. The software part is about building SCADA (Supervisory Control and Dada Aquisation) system which is the interface medium between hardware and computer.

Hardware

The solar tracker system is equipped with LDR (light detecting resistor), gear box and charger. System tracks the direction of the radiation reflected from the Sun. LDR varies the voltage value based on the intensity of the light. There are two LDR circuit positioned on end- right and end- left of the solar panel. The microcontroller receives the reading of the voltage from the LDRs, then it will give instruction to the drivers of the motor to rotate in direction of the LDR that has the highest voltage. The system charges the baterries as the other parts of the system utilizes it.

Figure 4. Solar Tracker

Fig.4 shows the solar tracker system. The solar tracker system is monitored by using the DC motor in which it rotates the panel according to the desired location. The PLC is used as the brain of the system. Fig.5 shows the PLC that has been used in the experiment for acquiring the data. It takes the input from the sensors and gives instructions to the output side devices to operate.

Basically, for this system there are four input sensors and three outputs actuators. The four input sensors are two level sensors and two moisture sensors. The outputs are motor (water pump DC motor) and valves (12 VDC valves).

Figure 5. PLC

Software

Supervisory Control and Data Acquisition, also known as SCADA, is a computer system for the congregation and analysis of real time data. SCADA is also referred to as the central system that scrutinizes and controls an entire set of a system, which is extended over a long distance. The bulk of the site is usually executed involuntarily by a remote terminal unit, RTU, or by PLC, which is known as the Programmable Logic Controller. Furthermore, there is a constraint present in the host control functions towards the supervisory level capability [8].

For instance, the control of cooling water flow is done by the PLC during the industrial process but the SCADA system might permit a machinist to alter the control set point for the flow and will permit any alarm conditions, such as flow loss. However, the feedback control loop is congested through the PLC or RTU; the monitoring of the total performance of the loop is executed by the SCADA system. The system overview has been shown in Figure 6. The system consists of hardware (data acquisition card) and the software parts. SCADA consists of many components such as RTU, MTU.RTU, remote terminal unit gather information from the inputs devices such as sensors, valves, motors, alarms and many more [8].

RTU collect all the system information and keep it the memory until the MTU (Master Terminal Unit) send the request to transmit data. MTU collects all the data stored and sends information to the device for interference operation. MTU can be a computer consisting of Human Machine Interface (HMI) as medium of communication for processing data and information. HMI interface provides a centralized monitoring and controlling system for the various inputs and outputs. NI LABVIEW software can communicate with PLC in a variety of ways. Modbus is a serial communication protocol published by Modicon in 1979 to communicate with PLC, and was then extended to the TCP protocol. Modbus became one of the standard communications in the industries. Thus Modbus communications server is used as medium of communication between RTU (PLC) and MTU.

The SCADA system is protected and requires password to access, according to Fig.7. There are two types of account, the administrator account and the operator account.

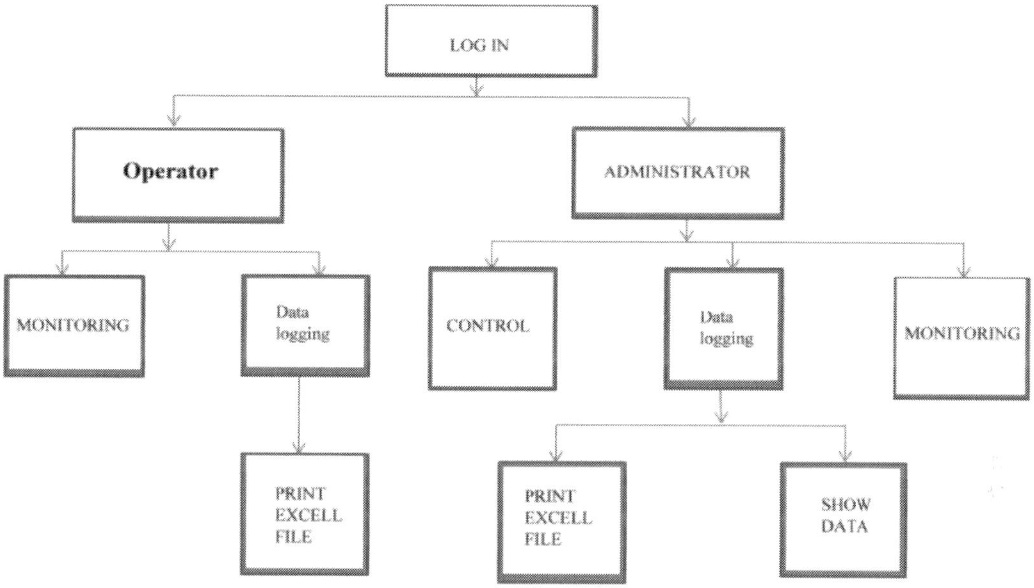

Figure 6. SCADA System Overview

The operation principle as shown below in Figure 4, there is a need to enter the password to the provided NI security login system.

Figure 7. SCADA Operating Principle

Upon entering a valid password, the interface will take us to project explorer (Fig.8) where there are many components of the project under the named folders. Furthermore, the user has anoption of selecting any kind of project folder according to the objective of the user.

Figure 8. Project explorer

If the user selects the HMI folder, the following Fig.9 is displayed. The HMI is configured and connected with the RTU (PLC). This HMI shows the state of the variables in PLC and from the HMI there is a possibility of changing the value read from the PLC, since PLC acts like slave and HMI acts like the master. Firstly, the system displays the values in the PLC (RTU) and secondly the value can be changed.

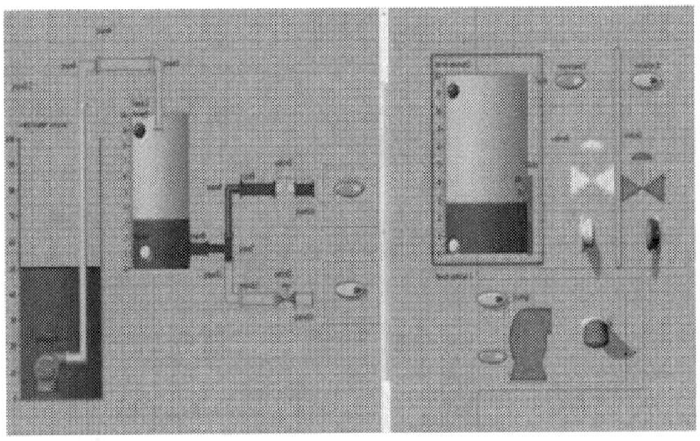

Figure 9. HMI

4. Results and Discussions

The process of recording proceedings with an automated computer is known as data logging. Furthermore, data logging is also used to provide an audit trail that can be utlized to comprehend the actions of the system and to analyze the qaundaries. Data logging tends to be constructive when it comes to the combination of log file entries from mutiple different sources. Correlations between seemingly dissimilar occurrences on different servers happen to capitulate from the earlier mentioned approach, in amalgamation with the statistical analysis. Other solution employ network-wide quering and reporting. The function of the datalog is to record all the parameters value measure durring each operation. The datalog can save file in the project explorer under the database file in form of notepad, according to Fig.10.

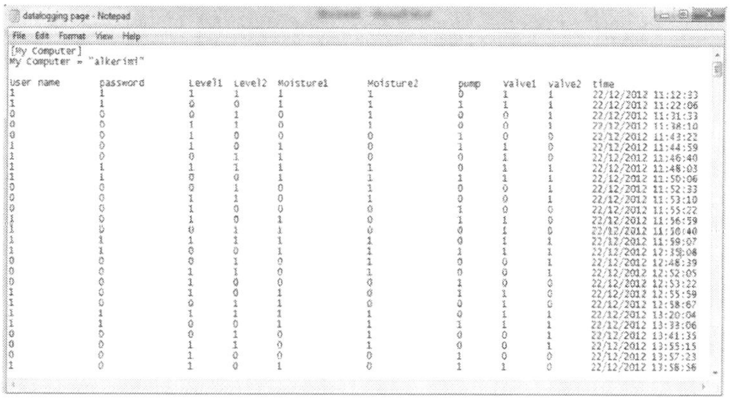
datalogging page - Notepad
File Edit Format View Help
[My Computer]
My Computer = "alkerimi"

user name	password	Level1	Level2	Moisture1	Moisture2	pump	valve1	valve2	time
1	1	1	1	1	1	0	1	1	22/12/2012 11:12:33
1	1	0	0	1	1	1	1	1	22/12/2012 11:22:06
0	0	0	1	0	1	0	0	1	22/12/2012 11:31:33
0	0	1	1	0	1	0	0	1	22/12/2012 11:38:10
0	0	1	0	0	0	1	0	0	22/12/2012 11:43:22
1	0	1	0	1	0	1	1	0	22/12/2012 11:44:59
1	0	0	1	1	0	0	1	0	22/12/2012 11:46:40
1	1	1	1	1	1	0	1	1	22/12/2012 11:48:03
1	1	0	0	1	1	1	1	1	22/12/2012 11:50:06
0	0	0	1	0	1	0	0	1	22/12/2012 11:52:33
0	0	1	1	0	1	0	0	1	22/12/2012 11:53:10
0	0	1	0	0	0	1	0	0	22/12/2012 11:55:22
1	0	1	0	1	0	1	1	0	22/12/2012 11:56:59
1	0	0	1	1	0	0	1	0	22/12/2012 11:58:40
1	1	1	1	1	1	0	1	1	22/12/2012 11:59:07
1	1	0	0	1	1	1	1	1	22/12/2012 12:35:08
0	0	0	1	0	1	0	0	1	22/12/2012 12:48:39
0	0	1	1	0	1	0	0	1	22/12/2012 12:52:05
0	0	1	0	0	0	1	0	0	22/12/2012 12:53:22
1	0	1	0	1	0	1	1	0	22/12/2012 12:55:59
1	0	0	1	1	0	0	1	0	22/12/2012 12:58:67
1	1	1	1	1	1	0	1	1	22/12/2012 13:20:04
1	1	0	0	1	1	1	1	1	22/12/2012 13:33:06
0	0	0	1	0	1	0	0	1	22/12/2012 13:41:35
0	0	1	1	0	1	0	0	1	22/12/2012 13:55:15
0	0	1	0	0	0	1	0	0	22/12/2012 13:57:23
1	0	1	0	1	0	1	1	0	22/12/2012 13:58:56

Figure 10. Data log

Upon saving the data in database for every specified cycle, the system uses the recorded data in database in order to plot the behavior of the variables. It helps to monitor the system status by displaying the real status of the control variable. In addition to this, the system is able to display graphically the status of controlled variable for previous days or hours as long as in the scope of database. In order to provide an everlasting log of facility performance, graphing is used as the data storage capability in the SCADA system.

However, most of the significant system parameters, such as moisture sensors, level sensors, pump and valves, happens to be stored in every fixed cycle of time. The system have the capability to record critical all the state of operation of all defined variable. Furthermore, data storage uses a separated server from that which is utilized for the primary system control software and is backed up sporadically.

Fig.11 represents the behavior of inputs and outputs variable for a specified cycle of time. Here the variables are two level sensors, two moisture sensors, two solenoid valves and water pump motor. The graph is based on the data on database. The inputs and outputs are digital signals.

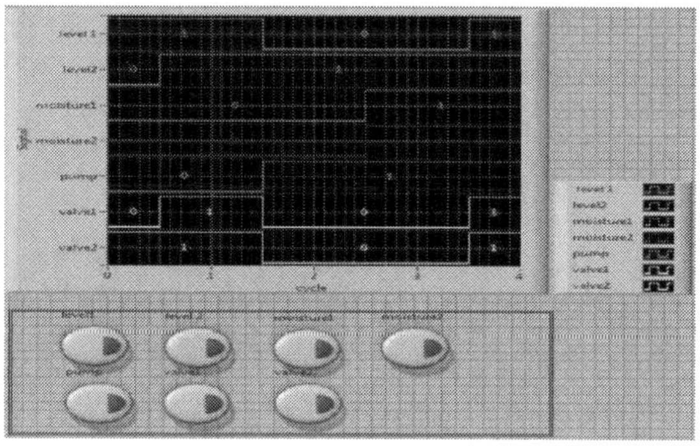

Figure 11. Behaviour of inputs &outputs

The above graph can be converted into excel data sheet on order to observe clearly for monitoring purpose (Table 1). In excel format the values are seen clearly for every cycle, in reference to the table shown below-

Table 1. Excel Data Show, representing the Graph of Input & Output behavior.

	B	C	D	E	F	G	H	I	J	K	L
1	Signal - level1	cycle - level2	Signal - level2	cycle - moisture1	Signal - moisture1	cycle - moisture2	Signal - moisture2	cycle - pump	Signal - pump	cycle - valve1	Signal - valve1
2	1	0	0	0	0	0	0	0	0	0	0
3	1	1	1	1	0	1	0	1	0	1	1
4	0	2	1	2	0	2	0	2	1	2	0
5	0	3	1	3	1	3	1	3	1	3	0
6	1	4	1	4	1	4	1	4	1	4	1

5. Conclusion

This paper managed to stumble upon the fact that the largest advantage of solar energy is its attribute as being free and unlimited source of energy. We have also found out that the research of the development of solar irrigation system requires vast knowledge and familiarity about renewable energy, as well as other parameters of control. The parameters of the system that have been used in the project are soil moisture sensor, light detecting sensor and level sensors. The design of SCADA system is meant for adding an operator on automatic irrigation system. Through SCADA system, the operator can read and write data from the controller (PLC) in addition to this there is a report of the system in excels form and that can be printed. After continuous work and effort, it was manageable to test the system by making it function properly. This project is very vital for all systems that deal with liquid monitoring and controlling specially in the irrigation field. The SCADA system helps administrators to control and monitor irrigation system. As stated, the project can be expanded to oil and gas monitoring system and greenhouse implementation system. Based on the result obtained from measurement, the system performance is quite reliable and accurate. The system has successfully overcome quite a few failings of the existing system by reducing the power consumption, maintenance and automatic system interface with SCADA system.

Acknowledgments
We would like to thank Dr. Momoh and Dr. Musa for their immense support in helping us with this paper. Appreciation goes to the Faculty of Engineering, International Islamic University Malaysia, for giving us the opportunity to implement the necessary experiment for acquiring the result. Our sincere appreciation goes to the Mechatronics System Design Laboratory for giving us a space to accomplish the work. Appreciation goes to Dr. Muhammad Abu Eusuf for sharing his knowledge from the Civil and Environmental Engineering perspective.

References

[1] Harris 2009 *Solar Powered Automated Greenhouse Irrigation System* (LumiTech International Limited)
[2] Morris And Lynne 2008 *Solar-Powered Livestock Watering Systems* (ATTRA)
[3] B.L.S. Lui 2006 *Trial Use of Solar Power Automatic Irrigation System* 7-11
[4] Noko And Road 2005 *Solar International Botswana (Pty) Ltd.* (InWEnt)
[5] B. Eker 2005 *Solar Powered Water Pumping Systems* **3** 7-11
[6] Vick And Almas 2008 *Developing Wind And/ Or Solar Powered Crop Irrigation Systems For The Great Plains* **27** 235-245
[7] Helikson ,Haman And Baird 2009 *Pumping Water For Irrigation Using Solar Energy* (Florida Energy Extension Service)
[8] Boyer, Stuart A 2010 *SCADA Supervisory Control and Data Acquisition* (International Society of Automation)
[9] Gordon R. Clarke, Deon Reyders, Edwin Wright 2004 *Practical Modern SCADA Protocols: DNP3, 60870.5 and Related Systems Newnes* (Elsevier)
[10] Richard Hantula 2009 *How Do Solar Panels Work* (Science in the Real World)
[11] Irrigation *https://en.wikipedia.org/wiki/irrigation* (Wikipedia)
[12] James M. Ebeling 2010 *Computer Monitoring & Control Technology* (Freshwater Institute)
[13] Halvorsen 2012 *Opc And Real-Time Systems in Labview* (Telemark University College)

IOP Conf. Series: Materials Science and Engineering 53 (2013) 012006 doi:10.1088/1757-899X/53/1/012006

Time Domain Feature Extraction Technique for earth's electric field signal prior to the Earthquake

W Astuti[1], **W Sediono, R Akmeliawati, M J E Salami**

Intelligent Mechatronics System Research Units (IMSRU)
Department of Mechatronics Engineering,
International Islamic University Malaysia,
Gombak, Selangor Darul Ehsan, Malaysia.

E-mail: winda1977@gmail.com, wsediono@iium.edu.my

Abstract. Earthquake is one of the most destructive of natural disasters that killed many people and destroyed a lot of properties. By considering these catastrophic effects, it is highly important of knowing ahead of earthquakes in order to reduce the number of victims and material losses. Earth's electric field is one of the features that can be used to predict earthquakes (EQs), since it has significant changes in the amplitude of the signal prior to the earthquake. This paper presents a detailed analysis of the earth's electric field due to earthquakes which occurred in Greece, between January 1, 2008 and June 30, 2008. In that period of time, 13 earthquakes had occurred. 6 of them were recorded with magnitudes greater than Ms=5R (5R), while 7 of them were recorded with magnitudes greater than Ms=6R (6R). Time domain feature extraction technique is applied to analyze the 1st significant changes in the earth's electric field prior to the earthquake. Two different time domain feature extraction techniques are applied in this work, namely Simple Square Integral (SSI) and Root Mean Square (RMS). The 1st significant change of the earth's electric field signal in each of monitoring sites is extracted using those two techniques. The feature extraction result can be used as input parameter for an earthquake prediction system.

1. Introduction

One of the major earthquakes in the world after an Alaskan earthquake with a magnitude of 9.2R in 1964 is the Northern Sumatra earthquake with a magnitude of 9.1R on December 26, 2004. The Northern Sumatra earthquake killed over 230.000 people [1]. The most recent giant earthquake with a magnitude of 9R on March 11, 2011 occurred in Tohoku, Japan [2]. This earthquake was followed by a tsunami that killed 18,000 people [3]. Considering these disastrous effects of an earthquake, it is highly important of knowing ahead of earthquakes in order to reduce the number of victims and material losses.

Geophysical phenomena that can be observed prior to the earthquake, such as seismological, geodetic, geochemical, hydrological or electro field [5], can be used as input for short term earthquake prediction system. One of the methods is the VAN method, which is named after the initials of the researchers' initials, P. Varotsos, K. Alexopoulos, and K. Nomicos [6]. This method was carried out by continuously monitoring the earth's electric field potential changes and their East-West (E-W) and North-South (N-S) polarity gradients.

---

[1] To whom any correspondence should be addressed.

In this work, two difference time domain feature extraction techniques are applied for extracting the 1st significant change of the earth's electric field signal prior to the earthquake (EQ). This feature can be used as the input parameter to the earthquake prediction system.

This paper is organized as follows; Section II describes the earth's electric field. Section III discussed the proposed feature extraction method which is based on the SSI and RMS techniques of the earth's electric field signal. The experimental results are presented in Section IV. Finally, the conclusion is presented in Section V.

2. Earth's Electric Field

When a strong earthquake is in the preparation, the activity of tectonic stress load change of the lithosphere specifically in the seismogenic area is increasing [6, 7]. There are a great number of geophysical and geochemical phenomena changes prior to the earthquake. The implementation of the earth's electric field to predict the earthquake has been widely used in Greece. There are three monitoring sites in different areas, i.e. Athens (ATH), Pyrgos (PYR) and Hios (HIO).

Based on the measurements of ground surface earth's electric field, it is assumed that the x- and y-components of the electric field, i.e. E_x and E_y, are registered by horizontal dipoles. The z component, contains the same quality of information as the x- and y- components. However, due to technical difficulties, the measurement of the z-component required a vertical dipole with 150-200 m depth in the ground. Because of this, the z-component has been ignored [11]. The total magnitude of the measurement electric field is given by

$$|\vec{E}| = \sqrt{E_x^{\ 2} + E_y^{\ 2}} \qquad (1)$$

Earth's electric field data used in this work was collected from the database three different monitoring sites in Greece available at www.earthquakeprediction.gr [9]. These monitoring sites are located in areas shown in Figure 1:

- Athens (ATH) monitoring site, installed on May 23, 2003, operating up to now.
- Pyrgos (PYR) monitoring site, installed on April 15, 2003, operating up to now.
- Hios (HIO) monitoring site, installed on March 18, 2006, operating up to 2010.

As shown in Figure 1, the distances between each of the monitoring are sited as follow:

- PYR monitoring site to ATH monitoring site is 216.6 km.
- ATH monitoring site to HIO monitoring site is 209.5 km.
- PYR monitoring site to HIO monitoring site is 425.6 km.

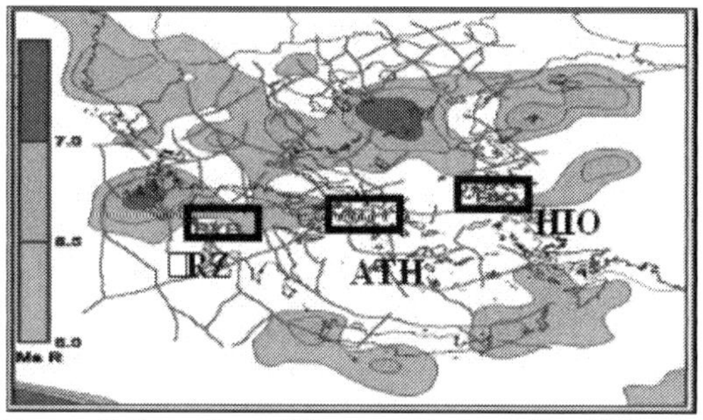

Figure 1. Location of the monitoring sites PYR (A), ATH (B), and HIO (C) [9]

3. Proposed Feature Extraction Technique

The proposed technique in this work is based on time domain feature extraction technique; SSI and RMS. Figure 2 shows the flowchart of the proposed technique. The raw data of the earth's electric field are read properly. The differencing technique is then applied to the earth's electric field signal. The peak detection technique is used to find the 1st significant change in the Earth's electric field. SSI and RMS feature extraction are applied to extract the earth's electric field signal.

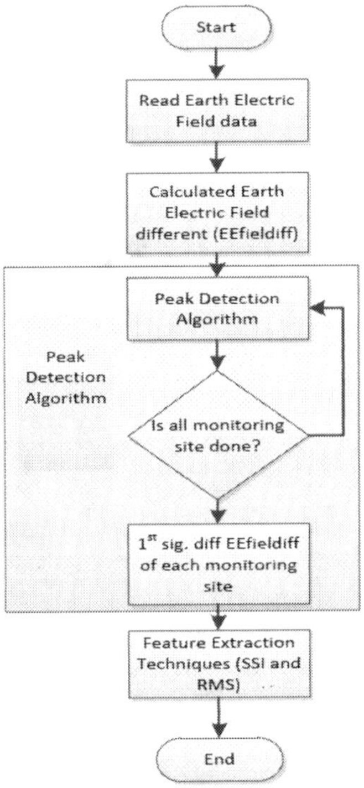

Figure 2. Flowchart of the proposed feature extraction technique

3.1. Reading of earth's electric field data

This stage involves the process of reading and interpreting data extracted from the earth electric field database. The raw data of the earth's electric field consists of two pairs of polarities, i.e. East-West (E-W) and North-South (N-S) polarities. The total amplitude of the earth's electric field can be found by using the following expression;

$$|\vec{E}| = \sqrt{E_{EW}^{\ 2} + E_{NS}^{\ 2}} \quad (2)$$

Examples of earth's electric field data from two polarities E-W and N-S are plotted in Figure 3. The first row (a) contains data from E-W polarity. The second row (b) contains data from N-S polarity. This sample of data was taken on February 6, 2008 prior to the 6.7R EQ on February 14, 2008. The differencing technique is then applied to the earth's electric field as shown in Figure 3.

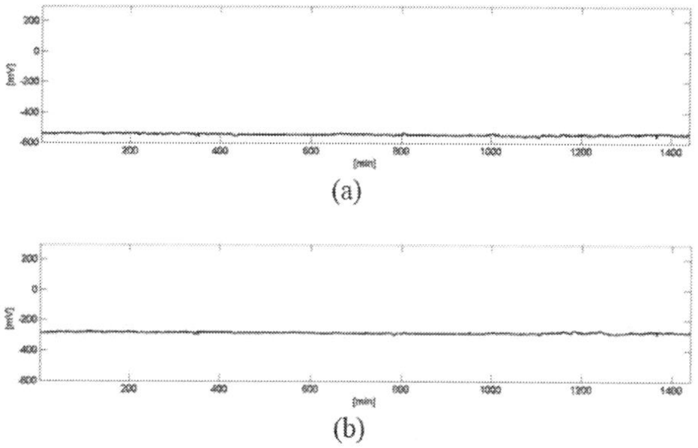

Figure 3. Raw earth's electric field data: (a) E-W and (b) N-S polarity prior to 6.7R EQ on Febuary 14, 2008.

3.2. The differencing technique of the earth electric field (EEFieldiff) signal

The differencing method is implemented to the earth's electric field. This method is used to observe the change in the earth's electric field over a certain period of time. *E [n]* denotes the value of the amplitude of the earth's electric field at discrete time *[n]*, and *E [n-1]* at discrete time *[n-1]*. The differencing of the earth's electric field, also called earth's electric field difference (EEfieldiff), is given by

$$\Delta E = E[n] - E[n-1] \tag{3}$$

where ΔE is the resulting EEfieldiff. An example of the EEfieldiff is shown in Figure 4.

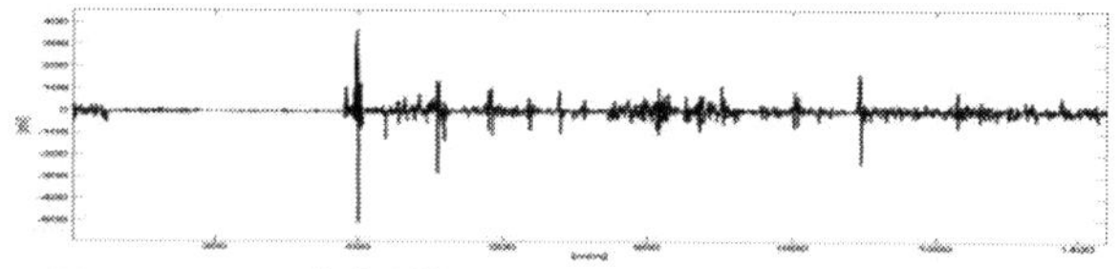

Figure 4. EEfieldiff prior to 6.7R EQ on Febuary 14, 2008.

3.3. Peak Detection Algorithm

The idea of peak detection algorithm is finding the 1st maximum amplitude of EEfieldiff prior to the earthquake. This first maximum amplitude of EEfieldiff is called the signal's peak. The time when the peak of EEfieldiff occurred is referred to as the peak time. As illustrated in Figure 4, the 1st significant change of the earth's electric field with the amplitude of 368 [mV/min] is found at the peak time of 397 minutes. The peak detection of EEfieldiff is shown in Figure 5. The maximum amplitude is detected by segmenting EEfieldiff per day. The calculated slope of the maximum amplitude of the segmented signal and the next segmented signal is assigned as the 1st significant change of EEfieldiff as shown in Figure 6. Once the 1stsignificant change of the SES is detected in one of the monitoring sites, the same procedure would be applied to the rest data from other monitoring sites. This procedure begins at the point where the 1stsignificant change from the previous monitoring sites is detected.

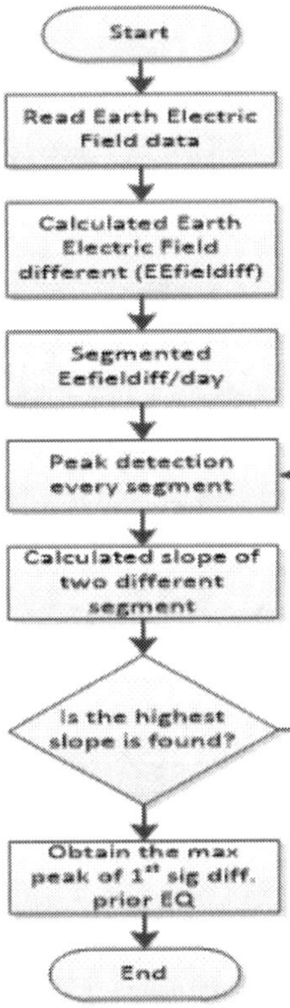

Figure 5. Algorithm of the determination of the 1st significant change of earth electric field prior to the earthquake.

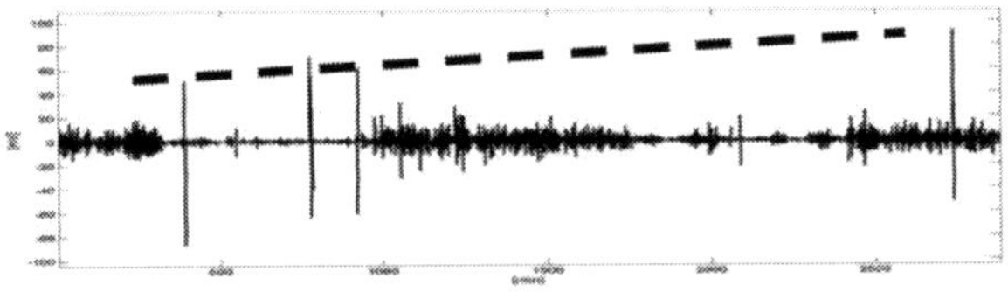

Figure 6. Slope between February 5, 2008 and February 6, 2008.

3.4. Time domain features extraction technique

Feature extraction is a process of extracting the 1st significant change of the earth's electric field to the feature vector that can be used as input to the identification/classification system. Simple Square Integral (SSI) and Root Mean Square (RMS) techniques are implemented in this work [10].

Simple Square Integral (SSI). This technique applied the absolute square of the signal that will result as the energy of the signal as shown in equation below:

$$SSI_k = \sum_{i=1}^{N} \left(\left| x_i^2 \right| \right) \tag{4}$$

where x is the amplitude of the signal. N is number of sampled signals.

IOP Conf. Series: Materials Science and Engineering **53** (2013) 012006 doi:10.1088/1757-899X/53/1/012006

Root Mean Square (RMS). This technique is modeled as amplitude modulated Gaussian Random process which RMS is related to the constant force and non-fatiguing contractual, as shown in equation (5):

$$RMS_k = \sqrt{\frac{1}{N}\sum_{i=1}^{N} x_i^2} \tag{5}$$

3.5 Classification of earthquake prone area.

Thirteen earthquake data registered in Greece between from January 1, 2008 to June 30, 2008 are used in this work. Six of them were recorded with magnitudes greater than Ms=5R (5R), while 7 of them were recorded with magnitudes greater than Ms=6R (6R). The earthquakes are classified based on the location where the earthquake has happened. Figure 7 shows that 13 earthquake can be grouped with five different locations. Two different time domain feature extraction techniques are applied to extract 1st significant difference of EEfieldiff.

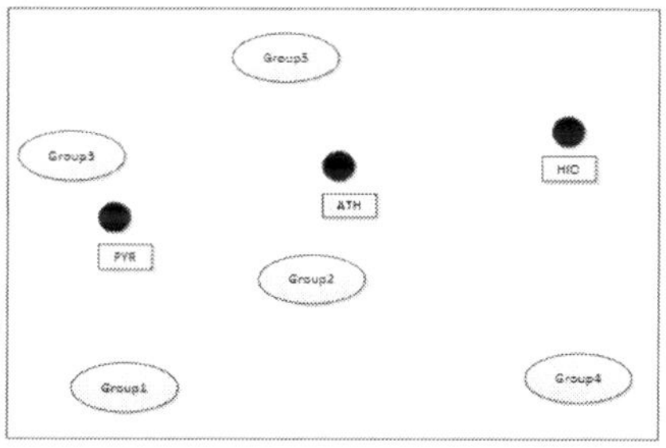

Figure 7. Grouping based on the area of 13 earthquakes in the period of January 1, to June 30, 2008.

4. Experimental Result and Discussion

The detailed analysis is performed on EQ data between January 1, 2008 and June 30, 2008, as shown in Table 1. The 1st significant change signal prior to the earthquake registered at three monitoring sites are investigated, part of the results are shown in Table 2. For the analysis, the EEfieldiff data are extracted using SSI and RMS methods, as shown in Table 3. The result shows the average energy and RMS among three monitoring sites. It shows that

- The smaller the magnitude of the earthquake, the short time duration of the 1st significant change signal EEfieldiff detected prior to EQ, with greater energy and smaller RMS for the same epicenter area.
- The greater the magnitude of the earthquake, the longer time duration of the 1st significant change signal EEfieldiff detected prior to the EQ, with smaller energy and greater RMS for the same epicenter area.

Table 1. List of EQs registered in Greece between January 1, 2008 and June 30, 2008

No.	Group	Mag. (R)	Date
1	1	5.6	2/19/08
2	1	5.7	2/26/08
3	1	5.5	3/19/08
4	1	5.0	1/14/08
6	1	6.7	2/14/08
7	1	6	6/21/08
8	1	6.5	2/20/08
9	2	5.1	1/29/08
10	2	6.6	1/6/08
11	3	5.5	2/1/08
12	3	7	6/8/08
13	4	5.6	3/28/08

Table 2. List of 1st significant EEfieldiff prior to the earthquake registered at three monitoring sites (part of the result).

Class	Mag.	Date of EQ	1st sig SESD	Days prior EQ (days)
1	5.6	2/19/08	2/14/08 2/15/08	4
			(PYR)	
			2/16/08 2/17/08	2
			(ATH)	
			2/16/08 2/17/08	2
			(HIO)	

Table 3. The energy and RMS result of the 1st sig. diff EEfieldiff prior to EQ.

Cls	Mag.(R)	Date of EQ	Days Prior to EQ	Energy (E)	RMS
1	5.6	2/19/08	2	$6.55x10^5$	12.65
1	5.7	2/26/08	4	$5.57x10^5$	11.94
1	5.5	3/19/08	1	$11.19x10^5$	14.6
1	6.7	2/14/08	17	$33.17x10^5$	23.56
1	6	6/21/08	9	$1.85x10^3$	4.62
1	6.5	2/20/08	3	$6.24x10^5$	11.85
2	5.1	1/29/08	4	$7.85x10^5$	13.85
2	6.6	1/6/08	20	$13.07x10^5$	16.97
3	5.5	2/1/08	1	$7.46x10^5$	13.25
3	7	6/8/08	41	$43.55x10^5$	25.92
4	5.6	3/28/08	5	$2.896x10^5$	9.15
4	6.7	7/15/08	17	$0.81x10^5$	5.208
5	6.1	10/14/08	21	$0.51x10^5$	4.16

Table 3 shows that the 1stsig. Diff of EEFieldiff has certain characteristics in its energy and RMS feature. Therefore, it is highly probable that the 1st significant difference of EEfilediff could be used in predicting and determining the location and magnitude of EQs. This characteristic can be applied to the identification system to predict any incoming earthquakes.

5. Conclusion

In this work, the extraction of the earth's electric signal prior to the earthquake has been investigated. We first generate the data of EEFieldiff (earth's electric field difference) which represents the change of the earth's electric field over a certain period of time. In this case, the 1stsignificant EEfieldiff has been detected.

The SSI and RMS techniques are applied to the 1st significant EEfieldiff signal. The result shows that the duration of the 1st significant EEfieldiff is detected prior to EQ and its resulting energy will determine the magnitude of the incoming earthquake. The RMS feature extraction technique has the same characteristic as SSI feature extraction. It is very likely that these characteristics can be used as input parameters to the prediction system of the incoming earthquake.

Acknowledgment
We thank Dr Constantine Thanassoulas for his valuable advice, information, encouragement and discussion in support of this work.

References

[1] Haryadi M 2001 *Earthquake and Tsunami kill more than 400 people in Aceh and North Sumatra*
[2] USGS 2011 Magnitude 9.0-Near the East Coast of Honshu (Japan)
[3] Gavinfarreli Earthquake and Tsunami in Japan: The number of deaths and missing victims more than 18000
[4] Bhargava N, Katiyar V K, Sharma M L, and Pradhan P 2009 Earthquakes Prediction through Animal Behavior: A Review *Indian J. of Biomechanics, Special Issue*
[5] Ikeya M 2004 Earthquake and Animal: from Folk Legend to Science *World Scientific*
[6] Lazarus D 1995 Physical Mechanisms for Generation and Propagation of SESs. A critical review of VAN: Earthquake Prediction From Seismic Electric Signals (Lighthill, S., J., (editor). University College London. World Scientific Publishing)
[7] Thanassoulas C and Tselentis G 1993 Periodic variation in the Earth's electric field as earthquake precursors: result from recent experiment in Greece Tectonophysics **224** 103-111p
[8] Uyeda S 1996 Introduction to the VAN method of Earthquake Prediction. A critical review of VAN: Earthquake Prediction From Seismic Electric Signals Lighthill, S., J., (editor). University College London World Scientific Publishing
[9] Thanassoulas C 2007 Short-term Earthquake PredictionH.Dounias& Co
[10] Ericka R-J R 2011 Stage for Developing Control System using EMG and EEg Signals: A survey (Technical report: CES-513. ISSN 1744-8050)

Prototyping of a Situation Awareness System in the Maritime Surveillance

D O D Handayani [1], **W Sediono**[2], **A Shah**[1]
[1]Department of Computer Science, Information and Communication Technology
[2]Department of Mechatronics, Engineering
International Islamic University of Malaysia
Gombak, Selangor Darul Ehsan, Malaysia

E-mail: dini.handayani@gmail.com, {wsediono,asadullah}@iium.edu.my

Abstract. This paper discusses about the design of a Situation Awareness (SA) system to support vessel crews and control room operators in improving the decision making process. The architecture of the system is ontology based. The vessel crews and control room operators may face a loss of SA. They may have limited cognitive abilities which make it difficult to make a decision in a high stress level, short time availability and continuously evolving situation with incomplete information. In this work, we describe the application of Semantic Web Rule Language to represent corresponding knowledge in the maritime surveillance domain. The result of this research will demonstrate that an ontology based system can be used to remodel the information into a meaningful and valuable form to predict the future states of SA and improve the decision making process.

1. Introduction

Decision making is a crucial part in performing maritime surveillance. Decision is made in various conditions, from threatening to normal situations. Vessel crews and control room operators need to fuse information from diverse sources to observe and analyze the situation before making a decision.

Experience, training, intelligence and a healthy physical condition will influence the process of interpreting information. Although the incoming information is the same, different people may have different interpretations (1).

This research proposes an SA system to improve the vessel crew and control room operator in analyzing the incoming information to provide an understanding of the situation to assist in making a good decision in a threatening situation. For the purpose of this research, the Semantic Web Rule Language (SWRL) will be employed to represent the knowledge and model the threatening scenario to support the vessel crew and control room operator in the decision making process.

This paper is organized as follows. Section two discusses the SA in the decision making process. In section three, we describe the proposed architecture of system for the analysis of threatening vessel behavior. Meanwhile, section four discusses about the R-Scene prototype and case study whereas Section five presents the ontology based SA. Finally, the conclusion will be presented in section six.

[1] To whom any correspondence should be addressed.

2. SA in the decision making process

As cited in (2) Endsley defined SA as "*The perception of the elements in the environment within a volume of time and space, the comprehension of their meaning and the projection of their status in the near future*".

This research proposes a method to extend the concept of SA using ontology and inference to make projections of the future. Figure 1 illustrates four layers describing the different levels of abstraction.

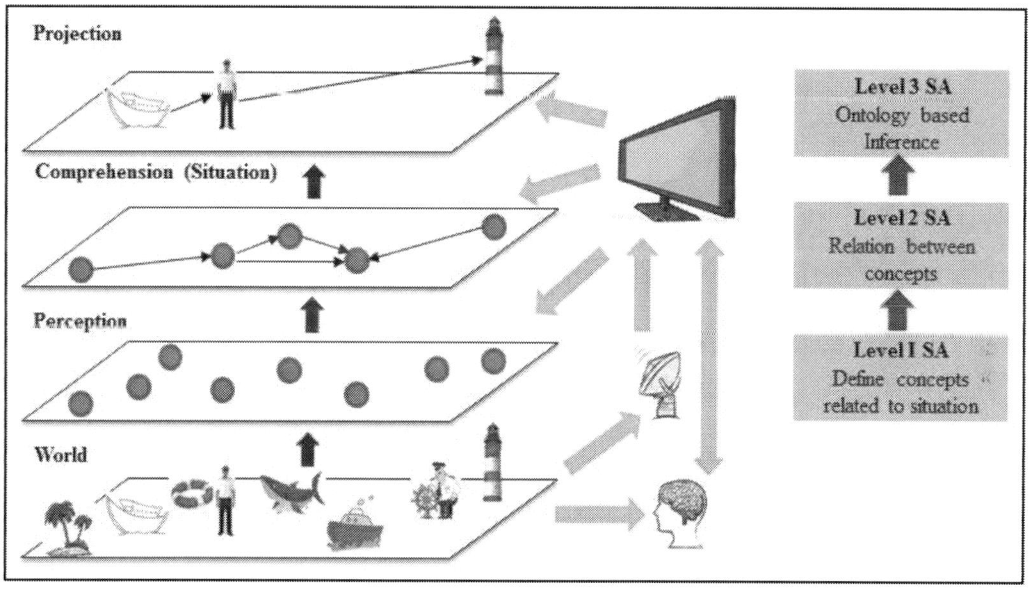

Figure 1. Ontology based SA, adopted and extended from (1,3)

The bottom layer is the "World". It is a symbol for physical or conceptual things, or both that may be the object of situation. On the right side of the World layer is a human head that shows that the SA actually takes place in the human's brain. Hence, the human observes the information of the world and human gets input from the computer, as displayed in Figure 1.

The second layer is the "Perception" layer. The dots in this layer represent the objects from the World layer that are obtained from sensors and represented in the computer memory. The arrow from the World layer to the radar image represents the sensory process, providing data to the computer. This layer is compatible with the output of the Perception process in the Endsley's model.

The next layer is "Comprehension". The layer illustrates the lines that represent the relation of each point. The process of filtering irrelevant information in order to obtain useful data by integrating some relevant information occurs in this layer. This layer represents also the Comprehension stage in the Endsley's model of SA.

The top layer is "Projection". This layer has a direct relationship with the Endsley's model in which projection is defined as the capability to predict future state situation based on the events of comprehension.

3. The Proposed Architecture System of R-Scene

In the analysis of threatening vessel behavior, we proposed for a system that consists of four main processes as demonstrated in figure 3.

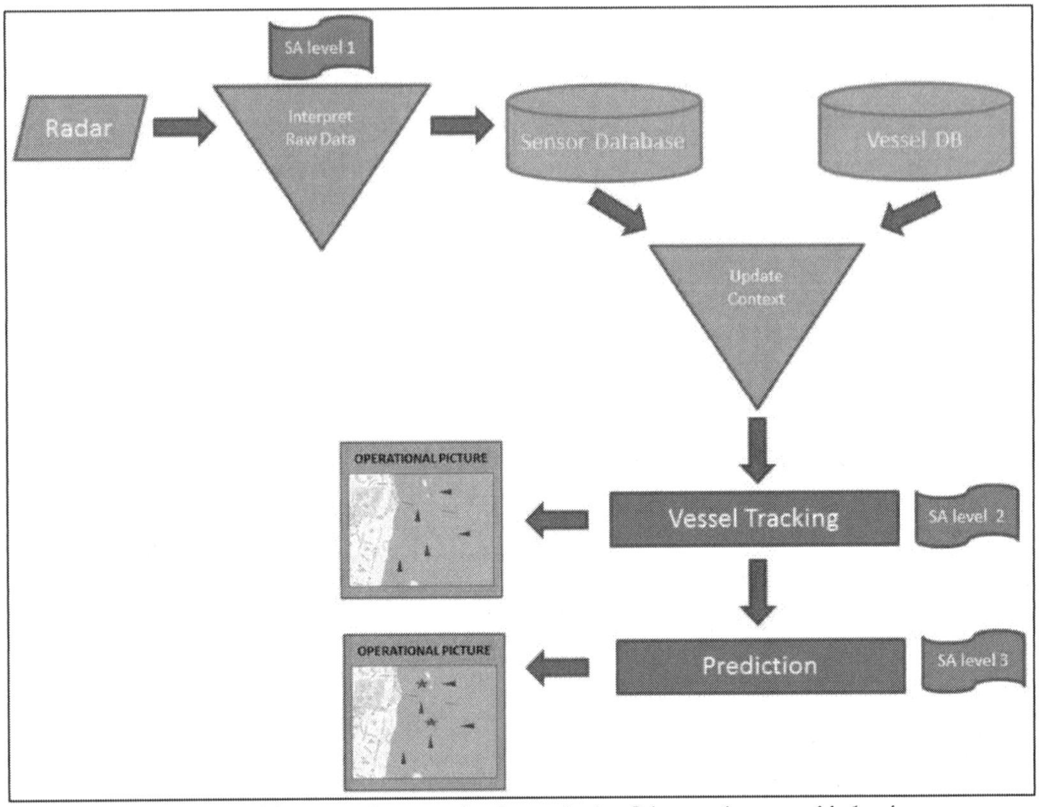

Figure 3. Architecture of system for the analysis of threatening vessel behaviour

Generally, the four processes will include:

1. The process of interpreting raw data into readable form to obtain the pattern from previous scenarios. The obtained pattern will be used to predict the future state.

2. The process of merging sensors and vessel database. Data from sensors and vessel database will be fed back to the ontology processing engine to update the context ontology. This step is compatible with the SA level 1

3. The tracking process. In this process, the system will filter all related data to model the situation. This step is compatible with the SA level 2.

4. The process of predicting possible future states of the current situation. In this stage, the inference engine will process all new information using the predefined SWRL rules and characterize the behavior of the vessel. This step is compatible with the SA level 3.

4. R-Scene Prototype and Case Study

Our prototype system, the R-Scene Prototype, is illustrated in figure 4. As depicted, it shows the digital map of Penang island with the latitude of 5.354632° and longitude of 100.348349°.

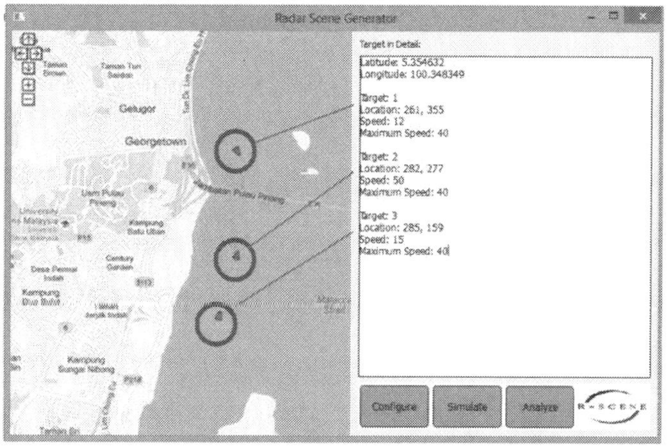

Figure 4. Screenshot of the prototype R-Scene.

QT 4.7.2 was chosen to create the Graphical User Interface front end. Figure 4 demonstrates our preliminary results. The figure consists of two main parts. The left side indicates the digital image. As shown, three targets are displayed in the current scenario. The right side consists of a window for displaying details of targets and three buttons to activate the tasks 'Configure', 'Simulate' and 'Analyze'. The display window indicates the coordinates of the digital map and detailed information of each target. In our preliminary results, we have detected 3 targets in three different positions with various speeds. Speed and maximum speed are needed for the scenario of high speed vessel. The speeds of targets are 12 knots for the first target, 40 knots for the second target and 15 knots for the third target, respectively. Knowing that the maximum speed allowed in that area is 40 knots, the alert class will give the warning message and point out the second target as a suspicious target.

5. Ontology-Based SA

Vandecasteele & Napoli (4) mentioned that to understand the phenomenon, it is necessary to reflect the information in a simple way. In order to be able to discriminate between normal and threatening situation, the use of context is inevitable. For the purpose of our research, we will utilize the ontology to reach our objectives. This section will briefly describe the terms ontology, knowledge base and rule base.

5.1 Ontology

This paper will discuss the ontology approach from the artificial intelligence (AI) perspective. The ontology developed in this research is used as a tool to automatically integrate the knowledge and information part. Previous study by (5) emphasized on the importance for one to bear in mind that ontology merely serves as a specification of a conceptualization. For the AI systems, ontology is defined as representational vocabulary for a researcher who needs to share information in a domain, what exists is that which can be represented (5,6).

5.2 The knowledge base

To implement the ontology we first need to translate the expert knowledge in such a way that it can be applied in our system. We plan to conduct interviews with expert to obtain knowledge about the threatening situations, how to identify it and how to solve the corresponding problems. The Protégé 4.2.0 software is employed to process the information gathered. Figure 5 presents an example of ontology for maritime surveillance.

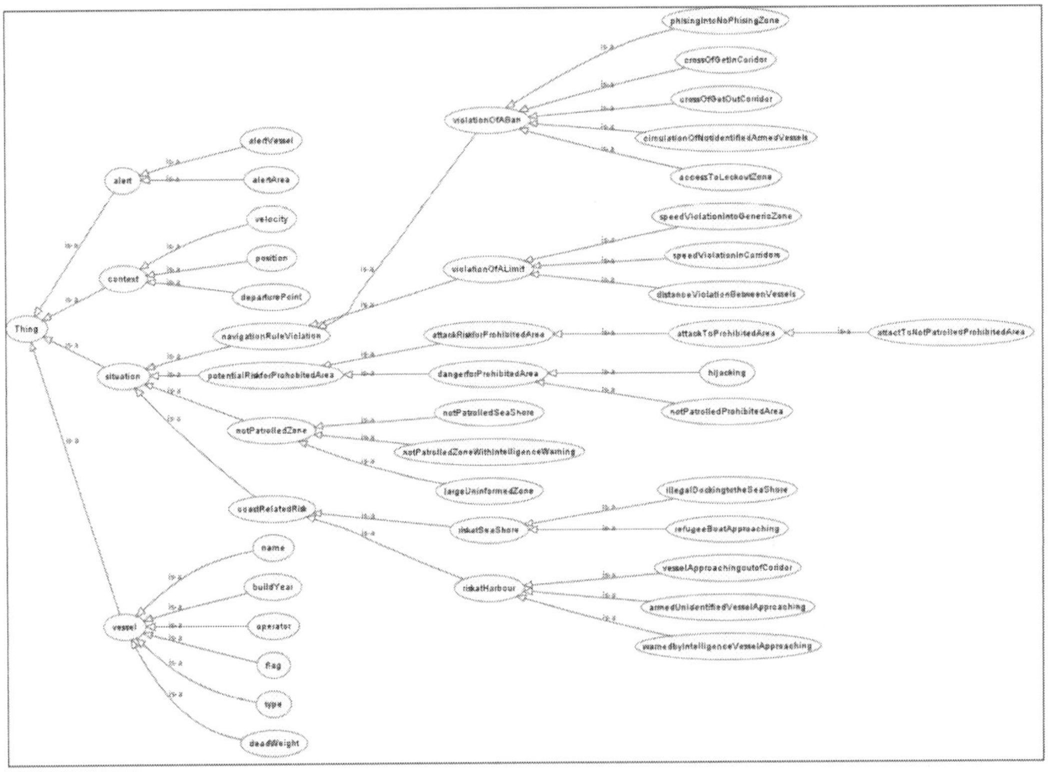

Figure 2. Maritime surveillance ontology, adopted and extended from (4,7)

The maritime ontology consists of 4 main classes:

- Vessel: this class consists of all basic information about a vessel. These include the vessel id, vessel name, type of vessel (fishing boat, cargo, etc.), maximum speed, deadweight, operator, flag and build year. Vessel class data will be derived from the vessel database.

- Situation: this class represents the possible interpretation of a situation in the maritime surveillance. The results from interviews will be used to generate the situation class.

- Alert: this class defines various alerts related to the maritime surveillance. The interview results will be utilized to generate the alert class.

- Context: the context class consists of elements that enrich alerts. It provides information related to vessel characteristics, its departure point, velocity and position. The context class data will be derived from radar data.

However, to make it more useful in the process of analysis, ontology requires for the addition of a rule base for the identification and characterization of threatening situations.

5.3. Rule based system

To represent and integrate rules into our ontology, we will utilize the Semantic Web Rule Language (SWRL), a combination of OWL-DL and RuleML (8), that will enrich the semantics of an ontology.

The rule was constructed by combining the experts knowledge and useful information from literature researches (9). Suppose the aim of the rule is to detect a ship moving at a speed that is excessive for its type. The request, translated into the SWRL is shown below and read as follows, "If a vessel (?vId) of an identified type (?vType) has a speed (?vSpeed) greater than (greaterThan) the maximum speed allowed for that type of vessel (?vMaxSpeed), then trigger an alert (Alert_Speed_HighSpeed)".

```
Vessel(?vId) ^ hasVesselType(?vId, ?vType) ^
hasSpeed(?vId, ?vSpeed) ^ hasSpeed(?vId, ?vTypeSpeed) ^
greaterThan(?vMaxSpeed, ?vSpeed)
hasAlert(?vId, Alert_Speed_HighSpeed)
```

In this example, we attempt to make comparison between the speed of the vessel and the maximum speed allowed for the vessel. If the vessel has a speed greater than the maximum speed, then the system will issue the alert of high speed.

6. Conclusion

This paper describes the development of a prototype for the detection of threatening vessel behaviour. The proposed prototype system R-Scene will integrate maritime surveillance and maritime SA. The prototype will provide maritime SA with a simple and easy user interface for maritime security and surveillance applications. The feature of the prototype will integrate vessel information database and radar data. The tracking process enables track analysis and target query by location, date and/or vessel specific information. In the prediction process, the analysis will denote the historical and prediction of the vessel movement. The example of scenario shows that the warning message will appear when the speed of target exceeds the specified maximum speed. For future works, some other scenarios will be tested to enrich the maritime ontology. Scenario of the maritime navigation in a prohibited area and prediction of collisions are planned to be completed in the next stages of our work.

References

[1]. Javed Y, Norris T 2011 Prototyping and Evaluation of a Computerized Emergency Management System Based on an Ontological Inference Design *IEEE Global Humanitarian Tech. Conf.* **1** 462–6

[2]. Wickens CD 2008 Situation Awareness : Review of Mica Endsley ' s 1995 Articles on Situation Awareness Theory and Measurement **3** 397–403.

[3]. Kokar MM, Matheus CJ, Baclawski K 2009 Ontology-based situation awareness *Information Fusion* **10** 83–98

[4]. Vandecasteele A, Napoli A 2012 Spatial Ontologies for Detecting Abnormal Maritime Behaviour *OCEANS* (Yeosu) 1 – 7

[5]. Gruber TR 1993 A translation approach to portable ontology specifications *Knowl. Acquis.* (London, UK: Academic Press Ltd.) **5** 199–220

[6]. Noy NF, McGuinness 2001 DL. Ontology Development 101: A Guide to Creating Your First Ontology

[7]. Farinelli A, Nardi D, Pigliacampo R, Rossi M, Settembre GP 2012 Cooperative Situation Assessment in a Maritime Scenario *Int. J. of Intelligent Systems* **27** 477–501

[8]. Horrocks I, Patel-Schneider PF, Boley H, Tabet S, Grosof B, Dean M 2004 SWRL: A Semantic Web Rule Language Combining OWL and RuleML

[9]. Laere J Van, Nilsson M 2009 Evaluation of a workshop to capture knowledge from subject matter experts in maritime surveillance *12th Int. Conf. on Information Fusion Seattle (*Seattle, WA, USA) 171–8

IOP Conf. Series: Materials Science and Engineering **53** (2013) 012008 doi:10.1088/1757-899X/53/1/012008

Design and Development of Mopping Robot-'HotBot'

M. R. Khan[1], N. M. L. Huq, M. M. Billah, S. M. Ahmmad
Department of Mechatronics Engineering, Kullyyah of Engineering, International Islamic University Malaysia, Jalan Gombak, Kuala Lumpur – 53100, Malaysia.

E-mail: raisuddin@iium.edu.my

Abstract. To have a healthy, comfortable, and fresh civilized life we need to do some unhealthy households. Cleaning the dirty floor with a mop is one of the most disgusting and scary house hold jobs. Mopping robots are a solution of such problem. However, these robots are still on the way to be smart enough. Many points limit their efficiency, i.e. cleaning sticky dirt, having dry floor after cleaning, monitoring, cost etc. 'HotBot' is a mopping robot that can clean dirty floor with nice efficiency leaving no sticky dirt. Hot water can be used for heavy stains or normal water for usual situation and economy. It needs neither to be monitored during mopping nor to wipe the floor after it. 'HotBot' has sensors to detect obstacles and a control mechanism to avoid those. Moreover, it cleans sequentially and equipped with several accident-protection-systems. Moreover, it is also cost effective compared to the robots available so far.

1. Introduction

Throughout the years, many engineering applications and devices has made our lives easier and more comfortable. Although devices such as washing machines and dishwashers have served this purpose, it still requires some degree of human input. In developing a cleaning robot, there are several challenges that one has to confront, i.e. the method of cleaning, path planning, covering the whole floor, cleaning the sticky dirt, cleaning it completely, leaving a dry floor behind, maintenance so on and so forth. It becomes even more complicated when safety, economy, energy consumption are considered to be in optimum level. Moreover, making it acceptable socially becomes the highest challenge especially when it is a new product.

The development of cleaning robots started almost three decades before [1]. Since then several industries came out with different ideas of cleaning robots and got popularities in developed countries and a try to make these robots acceptable at home continued. The early age was focused on the development and optimization of these robots. Hofner, C., and Schmidt, G. (1995) [2] used 2D map and kinematic and geometric robot model for a cleaning path planning. At the same time in another research Horn, J., and Schmidt, G. (1995) [3] described a localization system for mobile robots. The position and orientation was determined through a 3D-laser-range-image by comparing it to a 3D-environment model. For the autonomy of a mobile robot, it should know about its working region. Thus, a model of the environment is very much useful for it. So, Magin, G., Ruβ, A., Burschka, D., and Färber, G. (1995) [4] developed a dynamic 3D model of the environment using computer graphics

[1] To whom any correspondence should be addressed.

and data base technology. Besides the modelling and mapping of the environment the control over movement is equally important for a mobile robot. Palacin, J., Salse, J. A., Valganon, I., and Clua, X. (2004) [5] built a mobile robot for floor cleaning considering motor control, robot control, safety etc. They concluded that the robot cleaning process must be isolated from the human beings, which are due to interruption in its environment. In case of having no map of the workspace for the robot, it is very difficult for it to work and decide according to Magin, G. *et al.* [4]. Moreover, crowded area is even more difficult to work in, as stated by [5]. For such conditions Jansri, A., Klomkarn, K., and Sooraksa, P. (2004) [6] developed and simulated several chaotic patterns for robots path and compared the performance of the robots. Martins-Filho, L. S., and Macau, E. E. (2007) [7] studied the robots and their dynamic features of chaotic system and proposed a planning for robots that those robots should be introduced to the chaotic motion behavior by a planner goal position sequence.

The above developments took place in the technical sides of the cleaning robots. Meanwhile some researchers were engaged in analysing the acceptance level of these robots in society. Jodi Forlizzi, J., & DiSalvo, C. (2006) [8] worked on the effect of the design on human-robot interaction at home and found that the acceptability of these robots to the society varies according to their introduction to the family but it is mostly welcomed. Several other works had been carried out on social effects and acceptability of these robots in home. Sung, J. Y., Grinter, R. E., Christensen, H. I., & Guo, L. (2008) [9] found that most of the robot users for both sex are almost equal in percentage and are educated with technical background and they are mostly young. These robots are also considered to be working in public places of organizations where robot-human interaction is even more critical. Thus, their design and function plays a vital role to help the situation [10]. Moreover, shaping of the relationship with robots at home change over time and goes through several stages like pre-adoption, adoption, adaptation, and use retention [11] and modification to the environment to support the navigation of the robot is largely accepted [12].

This project shows the implementation of the above concepts into a mobile platform that can mop a room following a pattern (not in a random path). It shows the fact that the consumption of both water and energy can be reduced if the robot is properly programmed. Saving resources is coherent with engineering practices as it is good for the environment and economy. The dimensions and user-friendliness of the mobile platform are taken into consideration.

2. Methodology

The goal of this research is to develop a user-friendly moping robot. Thus, the activities needed are to enhance desired functionality in the robot blending of different mechanisms is desirable. It helps to achieve an optimum design. The robot mechanism designing moves forward considering the various aspects of machine design, performance, and viability. Actuators, links are designed at this stage. For the locomotion, cleaning, avoiding obstacles, and safety of the robot, optimized motion control was achieved through designing, analysing, and modifying the algorithm. In order to have proper locomotion and functionality a controller was designed according to the algorithm, considering as many exceptional and accidental cases as possible. Finally, the robot was fabricated to with proper functionality, which is consistent with the design of the controller.

3. Development

The development of the robot started with two main goals. Firstly, the product should be hassle free, and secondly, it should be economical. With these aims in front, the materials are chosen and design was made.

3.1. Hardware

The main body of the robot is cylindrical in shape with a diameter of 375mm and is made of tin sheet of thickness 1mm at the sides and top and bottom are made of acrylic plates of thickness 3mm. The bottom was reinforced by a $375\times75\times4mm^3$ aluminium plate to sustain the weights of the equipment inside. The robot is about 250mm high from the ground.

Two individually motor powered wheels accomplish the movement of the robot. There is one supporting wheel at front of the robot. The front wheel also can adjust the front height of the robot using a dc motor-screw attachment. The front height determines the amount of pressure applied on the floor by the mopping roller. The mopping roller is powered by a dc motor with speed reduction gear set and the whole set is clamped beneath the base platform of the robot. The roller is equipped with a cloth to wipe away the dirt and water from the floor. A hot water nozzle is placed at the front of the robot at a height of 193mm from the ground that spray water from the water tank with the help of a centrifugal pump. A complete perspective view of the 'HotBot' is shown in Figure 1.

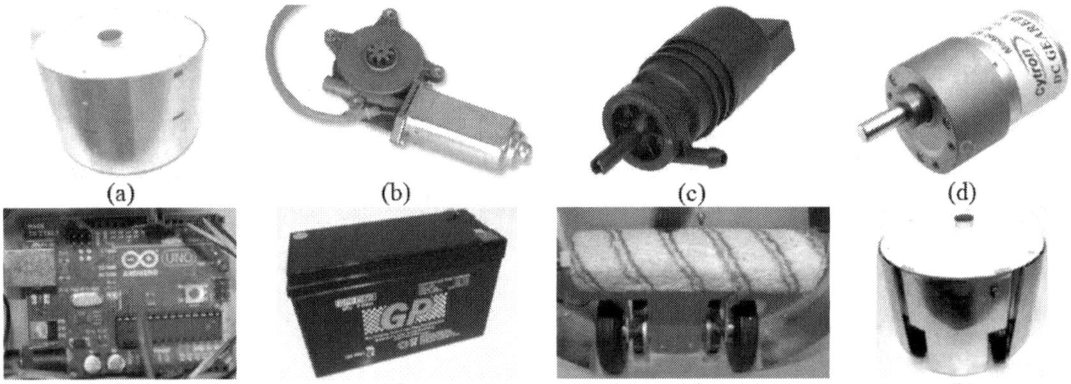

Figure 1: Hardware of the Robot: (a) front perspective view of 'HotBot', (b) wheel motor, (c) pump, (d) roller motor, (e) controller circuit, (f) battery, (g) roller, (h) rear perspective view of 'HotBot'.

The energy source of the robot is a maintenance free sealed lead acid battery with a rating of 7.2Ah and voltage of 12V, as all the electrical equipment used inside require 12V to run. A microcontroller is used to run the control program and to control the operation sequence. A summarized configuration is shown in Table 1.

Table 1. Specifications of the Hardware.

	Specification	Use
Wheel Motor	12 volt, 85 rpm, 2.95 N.m torque.	Used to move the robot.
Water Pump	12 volt, 2200 ml/min delivery.	Used to spray the water.
Roller Motor	12 volt, 26 rpm, 2.1 N.m torque.	Used to rotate the cleaning roller.
Battery	12 volt, 7.2 Ah, 2.1 initial current.	Used as the main source of energy.

3.2. Software

Behavioral approaches are used to autonomously operate and control the mopping robot. 'HotBot' control software is designed using the MPLAB IDE. The algorithm shown figure 2 relates the actors responsible for the autonomous movement with the robot peripheral elements by means of the sequence of events performed by each actor. The algorithm is implemented for studying the robot structure by attributes and relationships among. Finally, the algorithm is generated to visualize where the different system components are distributed among the hardware devices.

Table 2. The working principles of Behavioral approaches are discussed with the following algorithms.

	Use of the function
Moving ( ):	It is the behavior for the moving of the robot from one place to another. By moving around, the robot is used the roller to wipe the place.
Avoidance ( ):	It is the behavior to avoid obstacles such as walls, objects, etc. It uses sensors to detect obstacles and keeping data into the mop model.
Timer ( ):	The behavior bounds the mopping time.
ObstacleDir( ):	It is the behavior to search for obstacle direction for determining to find the moving way.
DistanceMeasure( ):	It is the behavior to spray water for cleaning the place by detecting no obstacle.

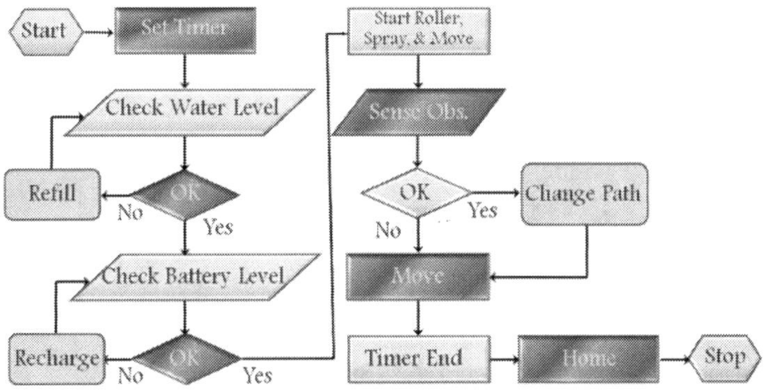

Figure 2: Flow-chart of working sequence.

4. Features

'HotBot' needs list attention during its operation. Starting from a particular home position it continues throughout the room or space to be cleaned and cleans sequentially. It can detect obstacles as well as falls, stairs, or holes and avoid accordingly. After being started 'HotBot' initially checks all it prerequisite conditions and then starts working. If any of the parameters, such as water level, battery charge level, water temperature etc., is in abnormal state it gives signal associated with that parameter. Using the hot water spray system it can remove sticky dirt with less amount of water that helps keeping the floor almost dry after the process is over. A user can chose between the optimum mood and the economy mode to operate. The optimum mode is for usual operation and gives fine cleaning. On the other hand, economy mode can be used for light dirt and can be used to save energy.

5. Results and Discussions

At the beginning, the user needs to plug in the robot to start the heating of water. After reaching to the optimum temperature, the heater will stop automatically and give a signal to unplug the robot. If the

IOP Conf. Series: Materials Science and Engineering **53** (2013) 012008 doi:10.1088/1757-899X/53/1/012008

water level is low, it will give corresponding signal to the user to refill the water pot. Then the user needs to press the start button and the set the time for the robot to work and the user is done. Afterwards the robot will check its initial conditions and start operating according to the instruction given in the controller. At the end of the process, i.e. when the timer ends the robot will go back to its initial position. The robot uses simple program and equipments (optical sensors) to avoid obstacles for cleaning. According to the information provided for the capacity of the robot to clean, the user has to set the timer and thus the robot will continue cleaning. It will maintain a continuous communication with the base sensor during this time and will understand its position. In case of obstacle, it just moves to the sensor that shows no obstacle or turns around.

From the inspection, the robot can clean sticky dirt from the floor. In this operation robot uses hot water to clean dirt. Figure 3 shows the implementation of the mopping application algorithm with 'HotBot'. The place shown in the Figure 3 is very dirty. However, at key points, the robot does show the behavioural approaches to clean the place and perform obstacle avoidance navigation. The robot starts its navigation by spraying hot water from its start position as shown in Figure 3(a). Figure 3(b) shows that the robot performs its navigation and keeps cleaning the place, shown in Figure 3(c). The robot measures distance once it finds any obstacle as shown in Figure 3(d) and it applies obstacle avoidance algorithm to perform smooth navigation, Figure 3(e). At this moment of obstacle avoidance robot does not perform water spraying and roller movement. Once it is free from obstacle, it again starts cleaning. Finally, robot come to its home position as shown in Figure 3(f) once it run out of time that is set by the user. In this operation, hot water is not only helps to clean the sticky dirt but also it consumes more energy than the normal operation. It consumes approximately 42W and with the energy source supplied can run for 1.25 hours. The robot has an average speed of 8.5 m/min and the mopping roller has a length of 21 cm. Thus, the robot is capable of wiping approximately 14 m^2 (as a standard room) in 10 minutes.

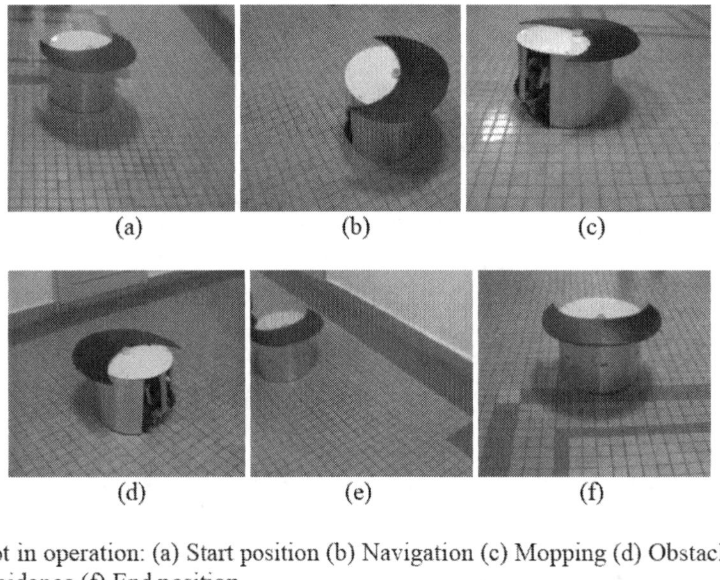

(a) (b) (c)

(d) (e) (f)

Figure 3: Robot in operation: (a) Start position (b) Navigation (c) Mopping (d) Obstacle measurement (e) Obstacle avoidance (f) End position.

6. Conclusion

The technology that helps us in our daily life with a minimum human assistance is one of the most expected and useful one. 'HotBot' gives almost the output as expected. From visual observations, it served with adequate efficiency on several occasions. It cleans heavy stains, leaves almost dry floor, and can do it quick. On the other hand, 'HotBot' has several safety equipments to avoid obstacles, holes, and stairs, to prevent accident or malfunction at low water level, low battery, or low water temperature. In such cases, it gives necessary signals and stops the function that may cause harm to it. It is capable of working in power and economy moods.

Despite of having several positive features, it requires more modification for more optimistic outcome. The cleanliness was not measured based on any scientific method but visual inspection. The structure can be made more attractive for better acceptance at home or office environments. Heating system is dependent on AC power supply and unfavorable for battery, especially for long operation and cost effectiveness. The mopping roller replacement is not convenient once it is worn out. However, the 'HotBot' is capable of proving itself as a good cleaning servant.

References

[1] Prassler, E., Ritter, A., Schaeffer, C., & Fiorini, P. (2000). A short history of cleaning robots. *Autonomous Robots*, *9*(3), 211-226.

[2] Hofner, C., & Schmidt, G. (1995). Path planning and guidance techniques for an autonomous mobile cleaning robot. *Robotics and Autonomous Systems*, *14*(2), 199-212.

[3] Horn, J., & Schmidt, G. (1995). Continuous localization of a mobile robot based on 3D-laser-range-data, predicted sensor images, and dead-reckoning.*Robotics and Autonomous Systems*, *14*(2), 99-118.

[4] Magin, G., Ruβ, A., Burschka, D., & Färber, G. (1995). A dynamic 3D environmental model with real-time access functions for use in autonomous mobile robots. *Robotics and Autonomous Systems*, *14*(2), 119-131.

[5] Forlizzi, J., & DiSalvo, C. (2006, March). Service robots in the domestic environment: a study of the roomba vacuum in the home. In *Proceedings of the 1st ACM SIGCHI/SIGART conference on Human-robot interaction* (pp. 258-265). ACM.

[6] Jansri, A., Klomkarn, K., & Sooraksa, P. (2004, November). On comparison of attractors for chaotic mobile robots. In *Industrial Electronics Society, 2004. IECON 2004. 30th Annual Conference of IEEE* (Vol. 3, pp. 2536-2541). IEEE.

[7] Martins-Filho, L. S., & Macau, E. E. (2007). Patrol mobile robots and chaotic trajectories. *Mathematical Problems in Engineering*, *2007*.

[8] Forlizzi, J., & DiSalvo, C. (2006, March). Service robots in the domestic environment: a study of the roomba vacuum in the home. In *Proceedings of the 1st ACM SIGCHI/SIGART conference on Human-robot interaction* (pp. 258-265). ACM.

[9] Sung, J. Y., Grinter, R. E., Christensen, H. I., & Guo, L. (2008, March). Housewives or technophiles?: understanding domestic robot owners. In *Human-Robot Interaction (HRI), 2008 3rd ACM/IEEE International Conference on* (pp. 129-136). IEEE.

[10] Mutlu, B., & Forlizzi, J. (2008, March). Robots in organizations: the role of workflow, social, and environmental factors in human-robot interaction. In*Human-Robot Interaction (HRI), 2008 3rd ACM/IEEE International Conference on* (pp. 287-294). IEEE.

[11] Sung, J., Grinter, R. E., & Christensen, H. I. (2010). Domestic robot ecology.*International Journal of Social Robotics*, *2*(4), 417-429.

[12] Gutmann, J. S., Culp, K., Munich, M. E., & Pirjanian, P. (2012, May). The social impact of a systematic floor cleaner. In *Advanced Robotics and its Social Impacts (ARSO), 2012 IEEE Workshop on* (pp. 50-53). IEEE.

Fuzzy modeling and control of rotary inverted pendulum system using LQR technique

M A Fairus[1], Z Mohamed[2] and M N Ahmad[2]
[1]Faculty of Electrical Engineering, Universiti Teknikal Malaysia Melaka, Melaka, Malaysia
[2]Faculty of Electrical Engineering, Universiti Teknologi Malaysia, Johor, Malaysia

E-mail: mfairus@utem.edu.my

Abstract. Rotary inverted pendulum (RIP) system is a nonlinear, non-minimum phase, unstable and underactuated system. Controlling such system can be a challenge and is considered a benchmark in control theory problem. Prior to designing a controller, equations that represent the behaviour of the RIP system must be developed as accurately as possible without compromising the complexity of the equations. Through Takagi-Sugeno (T-S) fuzzy modeling technique, the nonlinear system model is then transformed into several local linear time-invariant models which are then blended together to reproduce, or approximate, the nonlinear system model within local region. A parallel distributed compensation (PDC) based fuzzy controller using linear quadratic regulator (LQR) technique is designed to control the RIP system. The results show that the designed controller able to balance the RIP system.

1. Introduction

The Rotary Inverted Pendulum (RIP) as in figure 1 is inherently nonlinear, non-minimum phase, unstable and yet controllable system. Since it was introduced by Professor Katsuhisa Furuta, Tokyo Institute of Technology [1] this system has been a benchmark for the control community as a testbed system for linear and nonlinear control law verification. As an underactuated system, the only actuator that is used to control the two degree of freedom of the system is typically a dc motor. By controlling the torque of the dc motor, the angle of both arm and pendulum can be positioned as required.

Modeling the RIP is the first crucial task before any control technique can be implemented. Among the well-known techniques are the Euler-Lagrange technique which has extensively discussed in [2, 3, 4, 5, 8, 14] and the Newton-Euler technique in [6, 8]. The resulting equations which can then be used for controller design, involve nonlinear elements in order to closely represent the behaviour of the real system. In linear control methods, the linearization process will took place and the designed controller able to perform well in a small range of operation however under large range of operation this controller will prone to be unstable due to the nonlinearities in the plant. Thus, researchers are tends to focus on nonlinear control methods which has the capability to manage the nonlinearities exist in a plant. However, most nonlinear control will involve complex mathematical solution which could demand higher processing resource when implemented.

The complexity has forced researchers to venture into developing methods that can simplify the design process of nonlinear control. One of the methods is fuzzy modeling which used the concept of fuzzy set theory where the model is develop based on either the input-output data

Figure 1. Rotary inverted pendulum by TeraSoft

or the original mathematical model of the system. The fuzzy model proposed by Takagi and Sugeno [7] is described by fuzzy IF-THEN rules where the local dynamics of each fuzzy rule are expressed by a linear system models. These linear system models will be fuzzy "blending" to form the overall Takagi-Sugeno (T-S) fuzzy model which approximates the nonlinear system.

In this paper, the model is obtained from [13] which is based on the Euler-Lagrange technique and from here the T-S fuzzy model under sector nonlinearity approach is then developed. A parallel distributed compensation (PDC) based fuzzy controller using LQR technique is designed to control the RIP system. This paper will focus on the stabilisation controller while the swing-up controller is based on [13].

This paper is organized as follows: Section 2 describes the construction of T-S fuzzy model. Section 3 provides the controller designed based on the PDC. In Section 4, stability analysis of the closed-loop system is explained. Experimental results are given in Section 5. Finally, the concluding remarks are provided in Section 6.

2. Takagi-Sugeno Fuzzy Model of RIP

In this section the fuzzy model of the RIP is discussed. The mathematical model of the RIP as stated in [13] is derived according to Euler-Lagrange equations. The physical parameters of the system are define in table 1. The following is the mentioned model based on figure 2.

$$M(q)\ddot{q} + V_m(q,\dot{q})\dot{q} + G(q) = \begin{bmatrix} \tau_l \\ 0 \end{bmatrix} \tag{1}$$

where:

$$\begin{aligned}
q &= \begin{bmatrix} \theta_1 \\ \theta_2 \end{bmatrix} \\
M(q) &= \begin{bmatrix} M_{11} & m_2 l_1 c_2 \cos\theta_2 \\ m_2 l_1 c_2 \cos\theta_2 & J_2 + m_2 c_2^2 \end{bmatrix} \\
V_m(q,\dot{q}) &= \begin{bmatrix} \frac{1}{2} m_2 l_1^2 \dot{\theta}_2 \sin(2\theta_2) & V_{m12} \\ -\frac{1}{2} m_2 c_2^2 \dot{\theta}_1 \sin(2\theta_2) & 0 \end{bmatrix} \\
G(q) &= \begin{bmatrix} 0 \\ -m_2 c_2 g \sin\theta_2 \end{bmatrix} \\
\tau_l &= \frac{k_b}{R_m} e - \frac{k_b^2}{R_m} \dot{\theta}_1 \\
M_{11} &= J_1 + m_2 l_1^2 + m_2 c_2^2 \sin^2\theta_2 \\
V_{m12} &= -m_2 l_1 c_2 \dot{\theta}_2 \sin\theta_2 + \frac{1}{2} m_2 c_2^2 \dot{\theta}_1 \sin(2\theta_2)
\end{aligned}$$

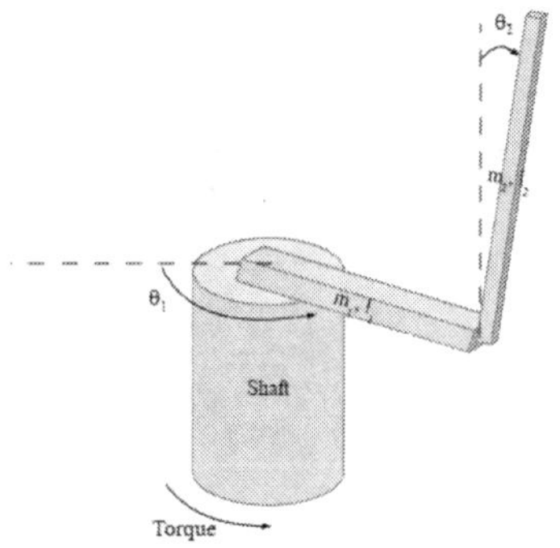

Figure 2. Schematic diagram of rotary inverted pendulum

Table 1. The Mechanical and Electrical System Parameters [13]

Symbol	Physical Quantity	Numerical Value
m_1	Mass of arm	0.056 kg
l_1	Length of arm	0.16 m
c_1	Distance to arm center of mass	0.08 m
J_1	Inertia of arm	0.00215058 kg.m^2
m_2	Mass of pendulum	0.022 kg
l_2	Length of pendulum	0.16 m
c_2	Distance to pendulum center of mass	0.08 m
J_2	Inertia of pendulum	0.00018773 kg.m^2
R_m	Armature resistance	2.5604 Ω
K_b	Back-emf constant	0.01826 V.s/rad
K_t	Torque constant	0.01826 N.m/A
θ_1	Angular displacement of arm	-
$\dot{\theta}_1$	Angular velocity of arm	-
θ_2	Angular displacement of pendulum	-
$\dot{\theta}_2$	Angular velocity of pendulum	-
τ_l	Applied torque	-
e	Control voltage	-

Defining the state variables as

$$[x_1\ x_2\ x_3\ x_4]^T = \left[\theta_1\ \theta_2\ \dot{\theta}_1\ \dot{\theta}_2\right]^T$$

(1) can be simplified and rearranged into more familiar form

$$\dot{x}_1 = x_3$$

$$\dot{x}_2 = x_4$$

$$\begin{aligned}\dot{x}_3 = &\left[\frac{-P_3P_6\cos x_2 \sin x_2}{\left(P_4\left(P_1+P_2\right)-\left(P_2P_4+P_3^2\right)\cos^2 x_2\right)x_2}\right]x_2 \\ &+\left[\frac{-P_4P_7-P_2P_3x_3\sin x_2\cos^2 x_2}{P_4\left(P_1+P_2\right)-\left(P_2P_4+P_3^2\right)\cos^2 x_2}\right]x_3 \\ &+\left[\frac{P_3P_4x_4\sin x_2-P_4\left(P_2+P_5\right)x_3\sin x_2\cos x_2}{P_4\left(P_1+P_2\right)-\left(P_2P_4+P_3^2\right)\cos^2 x_2}\right]x_4 \\ &+\left[\frac{P_4P_8}{P_4\left(P_1+P_2\right)-\left(P_2P_4+P_3^2\right)\cos^2 x_2}\right]e \qquad (2)\end{aligned}$$

$$\begin{aligned}\dot{x}_4 = &\left[\frac{P_6\left(P_1+P_2P_9\right)\sin x_2}{\left(P_4\left(P_1+P_2\right)-\left(P_2P_4+P_3^2\right)\cos^2 x_2\right)x_2}\right]x_2 \\ &+\left[\frac{P_3P_7\cos x_2+P_2\left(P_1+P_2P_9\right)x_3\sin x_2\cos x_2}{P_4\left(P_1+P_2\right)-\left(P_2P_4+P_3^2\right)\cos^2 x_2}\right]x_3 \\ &+\left[\frac{-P_3^2x_4\sin x_2\cos x_2+P_{10}x_3\sin x_2\cos^2 x_2}{P_4\left(P_1+P_2\right)-\left(P_2P_4+P_3^2\right)\cos^2 x_2}\right]x_4 \\ &+\left[\frac{-P_3P_8\cos x_2}{P_4\left(P_1+P_2\right)-\left(P_2P_4+P_3^2\right)\cos^2 x_2}\right]e \qquad (3)\end{aligned}$$

where

$$\begin{array}{ll} P_1 = m_2l_1^2+J_1 & P_6 = m_2c_2g \\ P_2 = m_2c_2^2 & P_7 = k_b^2/R_m \\ P_3 = m_2l_1c_2 & P_8 = k_t/R_m \\ P_4 = m_2c_2^2+J_2 & P_9 = 1-\cos^2 x_2 \\ P_5 = m_2l_1^2 & P_{10} = P_3\left(P_2+P_5\right) \end{array}$$

From (2) and (3), the equations are clearly contain nonlinearities. To develop the T-S fuzzy model, these nonlinearities are chosen as the premise variables, $z_j(t)$ ($j = 1, 2, \ldots, p$ where p is the number of premise variables). In local sector nonlinearity approach, the number of rules that represent the model can be relate through 2^p. Thus, in order to reduce the number of rules the following are chosen (the variables are chosen by referring to [10, 11]):

$$\begin{array}{ll} z_1 = \sin x_2/x_2 & z_3 = x_3\sin x_2\cos x_2 \\ z_2 = \cos x_2 & z_4 = x_4\sin x_2 \end{array}$$

Thus, (2) and (3) can be written as

$$\begin{aligned}\dot{x}_3 = &\left[\frac{-P_3P_6z_1z_2}{\left(P_4\left(P_1+P_2\right)-\left(P_2P_4+P_3^2\right)z_2^2\right)}\right]x_2+\left[\frac{-P_4P_7-P_2P_3z_2z_3}{P_4\left(P_1+P_2\right)-\left(P_2P_4+P_3^2\right)z_2^2}\right]x_3 \\ &+\left[\frac{P_3P_4z_4-P_4\left(P_2+P_5\right)z_3}{P_4\left(P_1+P_2\right)-\left(P_2P_4+P_3^2\right)z_2^2}\right]x_4+\left[\frac{P_4P_8}{P_4\left(P_1+P_2\right)-\left(P_2P_4+P_3^2\right)z_2^2}\right]e \qquad (4)\end{aligned}$$

$$\begin{aligned}\dot{x}_4 = &\left[\frac{P_6\left(P_1+P_2\left(1-z_2^2\right)\right)z_1}{\left(P_4\left(P_1+P_2\right)-\left(P_2P_4+P_3^2\right)z_2^2\right)}\right]x_2+\left[\frac{P_3P_7z_2+P_2\left(P_1+P_2\left(1-z_2^2\right)\right)z_3}{P_4\left(P_1+P_2\right)-\left(P_2P_4+P_3^2\right)z_2^2}\right]x_3 \\ &+\left[\frac{-P_3^2z_2z_4+P_3\left(P_2+P_5\right)z_2z_3}{P_4\left(P_1+P_2\right)-\left(P_2P_4+P_3^2\right)z_2^2}\right]x_4+\left[\frac{-P_3P_8z_2}{P_4\left(P_1+P_2\right)-\left(P_2P_4+P_3^2\right)z_2^2}\right]e \qquad (5)\end{aligned}$$

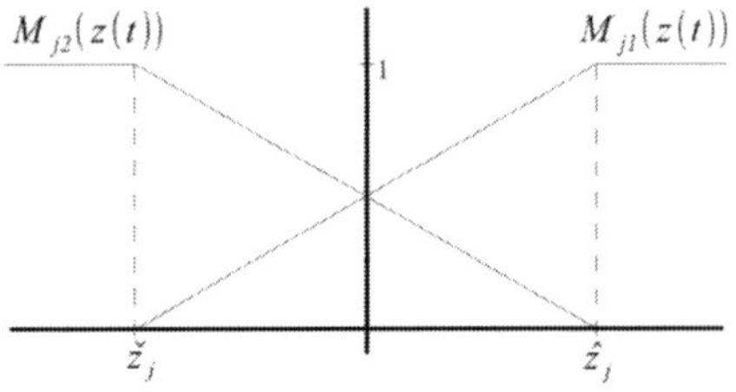

Figure 3. Membership function of M_{j1} and M_{j2}

Through local sector nonlinearities approach, the fuzzy model can approximate the nonlinear system within $x_2 \in [-a \quad a]$, $x_3 \in [-b \quad b]$ and $x_4 \in [-c \quad c]$. Thus, the maximum and minimum value of the premise variables can be define as,

$$\begin{array}{ll} \max\limits_{x_2} z_1 \equiv \hat{z}_1 & \min\limits_{x_2} z_1 \equiv \check{z}_1 \\ \max\limits_{x_2} z_2 \equiv \hat{z}_2 & \min\limits_{x_2} z_2 \equiv \check{z}_2 \\ \max\limits_{x_2,x_3} z_3 \equiv \hat{z}_3 & \min\limits_{x_2,x_3} z_3 \equiv \check{z}_3 \\ \max\limits_{x_2,x_4} z_4 \equiv \hat{z}_4 & \min\limits_{x_2,x_4} z_4 \equiv \check{z}_4 \end{array}$$

From the maximum and minimum values, $z_1(t)$, $z_2(t)$, $z_3(t)$ and $z_4(t)$ can be represented in terms of membership functions $M_{j1}(z_j(t))$ and $M_{j2}(z_j(t))$

$$\begin{aligned} z_1(t) &= M_{11}(z_1(t)) \cdot \hat{z}_1 + M_{12}(z_1(t)) \cdot \check{z}_1 \\ z_2(t) &= M_{21}(z_2(t)) \cdot \hat{z}_2 + M_{22}(z_2(t)) \cdot \check{z}_1 \\ z_3(t) &= M_{31}(z_3(t)) \cdot \hat{z}_3 + M_{32}(z_3(t)) \cdot \check{z}_3 \\ z_4(t) &= M_{41}(z_4(t)) \cdot \hat{z}_4 + M_{42}(z_4(t)) \cdot \check{z}_4 \end{aligned}$$

where

$$M_{j1}(z_j(t)) + M_{j2}(z_j(t)) = 1$$

Therefore the membership functions can be calculated as

$$M_{j1}(z_j(t)) = \frac{z_j(t) - \hat{z}_j}{\hat{z}_j - \check{z}_j} \qquad M_{j2}(z_j(t)) = \frac{\hat{z}_j - z_j(t)}{\hat{z}_j - \check{z}_j}$$

Figure 3 shows the membership function of M_{j1} and M_{j2}.

According to [9], RIP system can be represented by the following T-S fuzzy model:

Rule 1:
IF $z_1(t)$ is ${}^1M_{12}$ and $z_2(t)$ is ${}^1M_{22}$ and $z_3(t)$ is ${}^1M_{32}$ and $z_4(t)$ is ${}^1M_{42}$ THEN $\dot{\mathbf{x}}(\mathbf{t}) = {}^1A\mathbf{x}(\mathbf{t}) + {}^1B\mathbf{u}(\mathbf{t})$

Rule 2:
IF $z_1(t)$ is ${}^2M_{12}$ and $z_2(t)$ is ${}^2M_{22}$ and $z_3(t)$ is ${}^2M_{32}$ and $z_4(t)$ is ${}^2M_{41}$ THEN $\dot{\mathbf{x}}(\mathbf{t}) =$

IOP Conf. Series: Materials Science and Engineering **53** (2013) 012009 doi:10.1088/1757-899X/53/1/012009

${}^{2}A\mathbf{x}(\mathbf{t}) + {}^{2}B\mathbf{u}(\mathbf{t})$

⋮

Rule 16:
IF $z_1(t)$ is ${}^{16}M_{11}$ and $z_2(t)$ is ${}^{16}M_{21}$ and $z_3(t)$ is ${}^{16}M_{31}$ and $z_4(t)$ is ${}^{16}M_{41}$ THEN $\dot{\mathbf{x}}(\mathbf{t}) = {}^{16}A\mathbf{x}(\mathbf{t}) + {}^{16}B\mathbf{u}(\mathbf{t})$

where

$$ {}^{i}A = \begin{bmatrix} 0 & 0 & 1 & 0 \\ 0 & 0 & 0 & 1 \\ 0 & {}^{i}A_{32} & {}^{i}A_{33} & {}^{i}A_{34} \\ 0 & {}^{i}A_{42} & {}^{i}A_{43} & {}^{i}A_{44} \end{bmatrix} \qquad {}^{i}B = \begin{bmatrix} 0 \\ 0 \\ {}^{i}B_3 \\ {}^{i}B_4 \end{bmatrix} $$

and $i = 1, 2, \ldots, 16$, $\mathbf{x}(\mathbf{t}) \in \Re^n$ is the state vector and $\mathbf{u}(\mathbf{t}) \in \Re^m$ is the input vector.

By choosing $a = \pi/4$, $b = 0.1$ and $c = 6$ thus $\overset{\vee}{z_1} = 0.9003$, $\overset{\wedge}{z_1} = 1$, $\overset{\vee}{z_2} = 0.7071$, $\overset{\wedge}{z_2} = 1$, $\overset{\vee}{z_3} = -0.0707$, $\overset{\wedge}{z_3} = 0.0707$, $\overset{\vee}{z_4} = -4.2426$ and $\overset{\wedge}{z_4} = 4.2426$. From these values, the state equation for each of the rules can be determined as follows,

$$ {}^{1}A = \begin{bmatrix} 0 & 0 & 1 & 0 \\ 0 & 0 & 0 & 1 \\ 0 & -3.53 & -0.05 & -0.43 \\ 0 & 49.39 & -0.002 & 0.26 \end{bmatrix} \qquad {}^{2}A = \begin{bmatrix} 0 & 0 & 1 & 0 \\ 0 & 0 & 0 & 1 \\ 0 & -3.53 & -0.05 & 0.47 \\ 0 & 49.39 & -0.002 & -0.28 \end{bmatrix} $$

$$ {}^{3}A = \begin{bmatrix} 0 & 0 & 1 & 0 \\ 0 & 0 & 0 & 1 \\ 0 & -3.53 & -0.05 & -0.47 \\ 0 & 49.39 & 0.06 & 0.28 \end{bmatrix} \qquad {}^{4}A = \begin{bmatrix} 0 & 0 & 1 & 0 \\ 0 & 0 & 0 & 1 \\ 0 & -3.53 & -0.05 & 0.43 \\ 0 & 49.39 & 0.06 & -0.26 \end{bmatrix} $$

$$ {}^{5}A = \begin{bmatrix} 0 & 0 & 1 & 0 \\ 0 & 0 & 0 & 1 \\ 0 & -5.38 & -0.05 & -0.46 \\ 0 & 51.86 & 0.01 & 0.40 \end{bmatrix} \qquad {}^{6}A = \begin{bmatrix} 0 & 0 & 1 & 0 \\ 0 & 0 & 0 & 1 \\ 0 & -5.38 & -0.05 & 0.50 \\ 0 & 51.86 & 0.01 & -0.43 \end{bmatrix} $$

$$ {}^{7}A = \begin{bmatrix} 0 & 0 & 1 & 0 \\ 0 & 0 & 0 & 1 \\ 0 & -5.38 & -0.06 & -0.50 \\ 0 & 51.86 & 0.08 & 0.43 \end{bmatrix} \qquad {}^{8}A = \begin{bmatrix} 0 & 0 & 1 & 0 \\ 0 & 0 & 0 & 1 \\ 0 & -5.38 & -0.06 & 0.46 \\ 0 & 51.86 & 0.08 & -0.40 \end{bmatrix} $$

$$ {}^{9}A = \begin{bmatrix} 0 & 0 & 1 & 0 \\ 0 & 0 & 0 & 1 \\ 0 & -3.92 & -0.05 & -0.43 \\ 0 & 54.88 & -0.002 & 0.26 \end{bmatrix} \qquad {}^{10}A = \begin{bmatrix} 0 & 0 & 1 & 0 \\ 0 & 0 & 0 & 1 \\ 0 & -3.92 & -0.05 & 0.47 \\ 0 & 54.88 & -0.002 & -0.28 \end{bmatrix} $$

$$ {}^{11}A = \begin{bmatrix} 0 & 0 & 1 & 0 \\ 0 & 0 & 0 & 1 \\ 0 & -3.92 & -0.05 & -0.47 \\ 0 & 54.88 & 0.06 & 0.28 \end{bmatrix} \qquad {}^{12}A = \begin{bmatrix} 0 & 0 & 1 & 0 \\ 0 & 0 & 0 & 1 \\ 0 & -3.92 & -0.05 & 0.43 \\ 0 & 54.88 & 0.06 & -0.26 \end{bmatrix} $$

$$ {}^{13}A = \begin{bmatrix} 0 & 0 & 1 & 0 \\ 0 & 0 & 0 & 1 \\ 0 & -5.98 & -0.05 & -0.46 \\ 0 & 57.62 & 0.01 & 0.40 \end{bmatrix} \qquad {}^{14}A = \begin{bmatrix} 0 & 0 & 1 & 0 \\ 0 & 0 & 0 & 1 \\ 0 & -5.98 & -0.05 & 0.50 \\ 0 & 57.63 & 0.01 & -0.43 \end{bmatrix} $$

$$ {}^{15}A = \begin{bmatrix} 0 & 0 & 1 & 0 \\ 0 & 0 & 0 & 1 \\ 0 & -5.98 & -0.06 & -0.50 \\ 0 & 57.63 & 0.08 & 0.43 \end{bmatrix} \qquad {}^{16}A = \begin{bmatrix} 0 & 0 & 1 & 0 \\ 0 & 0 & 0 & 1 \\ 0 & -5.98 & -0.06 & 0.46 \\ 0 & 57.63 & 0.08 & -0.40 \end{bmatrix} $$

$$^{1}B = \begin{bmatrix} 0 \\ 0 \\ 2.68 \\ -1.63 \end{bmatrix} \quad ^{2}B = \begin{bmatrix} 0 \\ 0 \\ 2.68 \\ -1.63 \end{bmatrix} \quad ^{3}B = \begin{bmatrix} 0 \\ 0 \\ 2.68 \\ -1.63 \end{bmatrix} \quad ^{4}B = \begin{bmatrix} 0 \\ 0 \\ 2.68 \\ -1.63 \end{bmatrix}$$

$$^{5}B = \begin{bmatrix} 0 \\ 0 \\ 2.89 \\ -2.48 \end{bmatrix} \quad ^{6}B = \begin{bmatrix} 0 \\ 0 \\ 2.89 \\ -2.48 \end{bmatrix} \quad ^{7}B = \begin{bmatrix} 0 \\ 0 \\ 2.89 \\ -2.48 \end{bmatrix} \quad ^{8}B = \begin{bmatrix} 0 \\ 0 \\ 2.89 \\ -2.48 \end{bmatrix}$$

$$^{9}B = \begin{bmatrix} 0 \\ 0 \\ 2.68 \\ -1.63 \end{bmatrix} \quad ^{10}B = \begin{bmatrix} 0 \\ 0 \\ 2.68 \\ -1.63 \end{bmatrix} \quad ^{11}B = \begin{bmatrix} 0 \\ 0 \\ 2.68 \\ -1.63 \end{bmatrix} \quad ^{12}B = \begin{bmatrix} 0 \\ 0 \\ 2.68 \\ -1.63 \end{bmatrix}$$

$$^{13}B = \begin{bmatrix} 0 \\ 0 \\ 2.89 \\ -2.48 \end{bmatrix} \quad ^{14}B = \begin{bmatrix} 0 \\ 0 \\ 2.89 \\ -2.48 \end{bmatrix} \quad ^{15}B = \begin{bmatrix} 0 \\ 0 \\ 2.89 \\ -2.48 \end{bmatrix} \quad ^{16}B = \begin{bmatrix} 0 \\ 0 \\ 2.89 \\ -2.48 \end{bmatrix}$$

Given a pair of $(\mathbf{x}(\mathbf{t}), \mathbf{u}(\mathbf{t}))$, the final output of the fuzzy system can be calculated through the following defuzzification process

$$\dot{\mathbf{x}}(t) = \frac{\sum_{i=1}^{16} {}^{i}w(z(t)) \left\{ {}^{i}A\mathbf{x}(\mathbf{t}) + {}^{i}B\mathbf{u}(\mathbf{t}) \right\}}{\sum_{i=1}^{16} {}^{i}w(z(t))} \tag{6}$$

where

$$\begin{aligned}
{}^{1}w(z(t)) &= {}^{1}M_{12} \times {}^{1}M_{22} \times {}^{1}M_{32} \times {}^{1}M_{42} \\
{}^{2}w(z(t)) &= {}^{2}M_{12} \times {}^{2}M_{22} \times {}^{2}M_{32} \times {}^{2}M_{41} \\
{}^{3}w(z(t)) &= {}^{3}M_{12} \times {}^{3}M_{22} \times {}^{3}M_{31} \times {}^{3}M_{42} \\
{}^{4}w(z(t)) &= {}^{4}M_{12} \times {}^{4}M_{22} \times {}^{4}M_{31} \times {}^{4}M_{41} \\
{}^{5}w(z(t)) &= {}^{5}M_{12} \times {}^{5}M_{21} \times {}^{5}M_{32} \times {}^{5}M_{42} \\
{}^{6}w(z(t)) &= {}^{6}M_{12} \times {}^{6}M_{21} \times {}^{6}M_{32} \times {}^{6}M_{41} \\
{}^{7}w(z(t)) &= {}^{7}M_{12} \times {}^{7}M_{21} \times {}^{7}M_{31} \times {}^{7}M_{42} \\
{}^{8}w(z(t)) &= {}^{8}M_{12} \times {}^{8}M_{21} \times {}^{8}M_{31} \times {}^{8}M_{41} \\
{}^{9}w(z(t)) &= {}^{9}M_{11} \times {}^{9}M_{22} \times {}^{9}M_{32} \times {}^{9}M_{42} \\
{}^{10}w(z(t)) &= {}^{10}M_{11} \times {}^{10}M_{22} \times {}^{10}M_{32} \times {}^{10}M_{41} \\
{}^{11}w(z(t)) &= {}^{11}M_{11} \times {}^{11}M_{22} \times {}^{11}M_{31} \times {}^{11}M_{42} \\
{}^{12}w(z(t)) &= {}^{12}M_{11} \times {}^{12}M_{22} \times {}^{12}M_{31} \times {}^{12}M_{41} \\
{}^{13}w(z(t)) &= {}^{13}M_{11} \times {}^{13}M_{21} \times {}^{13}M_{32} \times {}^{13}M_{42} \\
{}^{14}w(z(t)) &= {}^{14}M_{11} \times {}^{14}M_{21} \times {}^{14}M_{32} \times {}^{14}M_{41} \\
{}^{15}w(z(t)) &= {}^{15}M_{11} \times {}^{15}M_{21} \times {}^{15}M_{31} \times {}^{15}M_{42} \\
{}^{16}w(z(t)) &= {}^{16}M_{11} \times {}^{16}M_{21} \times {}^{16}M_{31} \times {}^{16}M_{41}
\end{aligned}$$

3. Controller Design via Parallel Distributed Compensation (PDC) Fuzzy Controllers

In PDC fuzzy controllers design, each control rule is designed by the same premise variables as the system's T-S fuzzy model and implement the linear state feedback control laws in the consequent parts:

Rule 1:
IF $z_1(t)$ is ${}^1M_{12}$ and $z_2(t)$ is ${}^1M_{22}$ and $z_3(t)$ is ${}^1M_{32}$ and $z_4(t)$ is ${}^1M_{42}$ THEN $\mathbf{u(t)} = -{}^1K\mathbf{x}(\mathbf{t})$

Rule 2:
IF $z_1(t)$ is ${}^2M_{12}$ and $z_2(t)$ is ${}^2M_{22}$ and $z_3(t)$ is ${}^2M_{32}$ and $z_4(t)$ is ${}^2M_{41}$ THEN $\mathbf{u(t)} = -{}^2K\mathbf{x}(\mathbf{t})$

$\vdots$

Rule 16:
IF $z_1(t)$ is ${}^{16}M_{11}$ and $z_2(t)$ is ${}^{16}M_{21}$ and $z_3(t)$ is ${}^{16}M_{31}$ and $z_4(t)$ is ${}^{16}M_{41}$ THEN $\mathbf{u(t)} = -{}^{16}K\mathbf{x}(\mathbf{t})$

where ${}^iK$ is the feedback matrix of each rule. The final model of this controller is expressed by

$$\mathbf{u}(\mathbf{t}) = \frac{-\sum_{i=1}^{16} {}^iw(z(t))\,{}^iK\mathbf{x}(\mathbf{t})}{\sum_{i=1}^{16} {}^iw(z(t))} \tag{7}$$

substituting (7) into (6) the closed-loop T-S model can be represented by,

$$\dot{\mathbf{x}}(t) = \frac{\sum_{\grave{i}=1}^{16}\sum_{\acute{i}=1}^{16} {}^{\grave{i}}w(z(t))\,{}^{\acute{i}}w(z(t))\left\{{}^{\grave{i}}A - {}^{\grave{i}}B\,{}^{\acute{i}}K\right\}\mathbf{x}(\mathbf{t})}{\sum_{\grave{i}=1}^{16}\sum_{\acute{i}=1}^{16} {}^{\grave{i}}w(z(t))\,{}^{\acute{i}}w(z(t))} \tag{8}$$

where ${}^iw(z(t))$ is as previously discussed. To apply the state feedback control law in PDC fuzzy controller, each of the feedback matrix must be designed for each local linear model then later fuzzy "blending" through (7). In this paper, the feedback gains are calculated based on the LQR technique.

4. Stability Analysis of the Closed-Loop System

The stability of (8) is ensured through a sufficient quadratic stability condition derived in [12].

Theorem: The closed-loop T-S fuzzy model is asymptotically stable in the large if there exists a common positive definite matrix P such that the following two conditions are satisfied:

$$ {}^{\grave{i}\grave{i}}G^TP + P\,{}^{\grave{i}\grave{i}}G < 0$$

$$\left(\frac{{}^{\grave{i}\acute{i}}G + {}^{\hat{i}\grave{i}}G}{2}\right)^T P + P\left(\frac{{}^{\grave{i}\acute{i}}G + {}^{\hat{i}\grave{i}}G}{2}\right) < 0$$

where

$$\begin{aligned} {}^{\grave{i}\grave{i}}G &= {}^{\grave{i}}A - {}^{\grave{i}}B\,{}^{\grave{i}}K, \\ {}^{\grave{i}\acute{i}}G &= {}^{\grave{i}}A - {}^{\grave{i}}B\,{}^{\acute{i}}K, \\ {}^{\hat{i}\grave{i}}G &= {}^{\acute{i}}A - {}^{\acute{i}}B\,{}^{\grave{i}}K \quad \text{and} \end{aligned}$$

$$\grave{i} < \acute{i}\text{s.t.} \left(\frac{{}^{\grave{i}}w(z(t))}{\sum_{\grave{i}=1}^{16} {}^{\grave{i}}w(z(t))}\right) \cap \left(\frac{{}^{\acute{i}}w(z(t))}{\sum_{\acute{i}=1}^{16} {}^{\acute{i}}w(z(t))}\right) \neq 0$$

The above problems can be solved through Matlab Linear Matrix Inequalities (LMI)-toolbox.

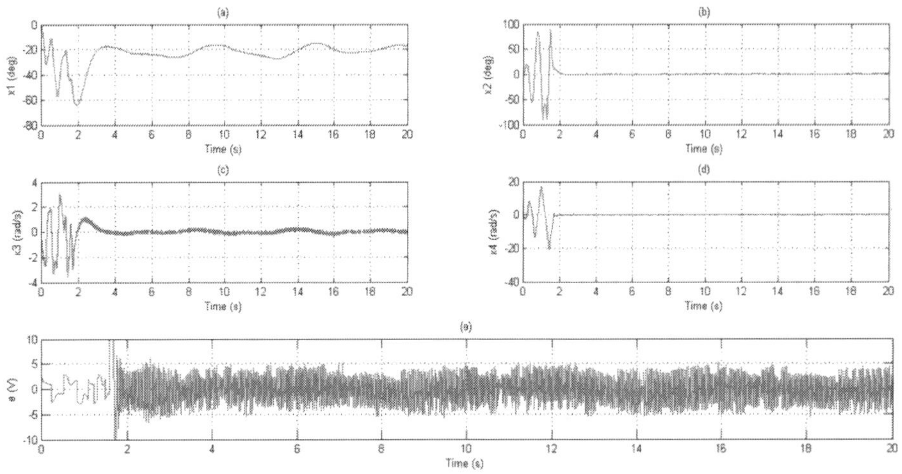

Figure 4. Response of rotary inverted pendulum a) Arm's position b) Pendulum's angle c) Arm's velocity d) Pendulum's velocity e) Control signal

5. Experimental Results

In this section, the feedback gains are calculated based on the LQR technique. By minimizing a quadratic cost function of the form,

$$J = \int_0^\infty \left[\mathbf{x}(\mathbf{t})^T Q \, \mathbf{x}(\mathbf{t}) + \mathbf{u}(\mathbf{t})^T R \, \mathbf{u}(\mathbf{t}) \right] dt$$

with

$$Q = \begin{bmatrix} 10 & 0 & 0 & 0 \\ 0 & 20 & 0 & 0 \\ 0 & 0 & 10 & 0 \\ 0 & 0 & 0 & 1 \end{bmatrix} \text{ and } R = 1$$

the feedback gain can be calculated by Matlab command, `lqr`.

$$\begin{aligned}
{}^{1}K &= [\,-3.162 \quad -149.4 \quad -4.366 \quad -21.72\,] \\
{}^{2}K &= [\,-3.162 \quad -149.4 \quad -4.366 \quad -21.39\,] \\
{}^{3}K &= [\,-3.162 \quad -151.6 \quad -4.560 \quad -22.05\,] \\
{}^{4}K &= [\,-3.162 \quad -151.6 \quad -4.559 \quad -21.72\,] \\
{}^{5}K &= [\,-3.162 \quad -104.1 \quad -4.379 \quad -15.05\,] \\
{}^{6}K &= [\,-3.162 \quad -104.1 \quad -4.379 \quad -14.72\,] \\
{}^{7}K &= [\,-3.162 \quad -105.2 \quad -4.513 \quad -15.22\,] \\
{}^{8}K &= [\,-3.162 \quad -105.2 \quad -4.514 \quad -14.89\,] \\
{}^{9}K &= [\,-3.162 \quad -160.4 \quad -4.322 \quad -22.11\,] \\
{}^{10}K &= [\,-3.162 \quad -160.4 \quad -4.322 \quad -22.78\,]
\end{aligned}$$

$$
\begin{aligned}
{}^{11}K &= \begin{bmatrix} -3.162 & -162.6 & -4.508 & -22.43 \end{bmatrix} \\
{}^{12}K &= \begin{bmatrix} -3.162 & -162.6 & -4.509 & -22.09 \end{bmatrix} \\
{}^{13}K &= \begin{bmatrix} -3.162 & -111.7 & -4.334 & -15.31 \end{bmatrix} \\
{}^{14}K &= \begin{bmatrix} -3.162 & -111.7 & -4.334 & -14.97 \end{bmatrix} \\
{}^{15}K &= \begin{bmatrix} -3.162 & -112.8 & -4.464 & -15.47 \end{bmatrix} \\
{}^{16}K &= \begin{bmatrix} -3.162 & -112.8 & -4.464 & -15.14 \end{bmatrix}
\end{aligned}
$$

By selecting P as

$$
P = \begin{bmatrix} 0.0007 & 0.0046 & 0.0002 & 0.0004 \\ 0.0046 & 0.1065 & 0.0037 & 0.0091 \\ 0.0002 & 0.0037 & 0.0003 & 0.0007 \\ 0.0004 & 0.0091 & 0.0007 & 0.0018 \end{bmatrix} \times 10^{-10}
$$

the stability condition of the above *Theorem* is satisfied. Figure 4 shows the arm's position, pendulum's angle, arm's velocity, pendulum's velocity and control signal of the RIP.

6. Conclusion

In this paper, the T-S fuzzy model of the RIP has been presented. The developed model is then verified through experiment by using state feedback control laws via PDC to stabilise the RIP. Stability conditions of this system can be handled as a LMI feasibility problem which can be solved using Matlab LMI-toolbox. Experimental results demonstrate the effectiveness of this approach.

Acknowledgment

The authors would like to express appreciation to Faculty of Electrical Engineering, Universiti Teknikal Malaysia Melaka, Malaysia for the equipment involved in this paper.

References

[1] Furuta K, Yamakita M and Kobayashi S 1991 *Proc. of Int. Conf. on Industrial Electronics, Control and Instrumentation (Kobe)* vol 3 p 2193
[2] Jian Z and Yongpeng Z 2011 *30th Chinese Control Conf. (Yantai)* p 493
[3] Anvar S M M, Hassanzadeh I and Alizadeh G 2010 *Int. Conf. on Control Automation and Systems (Gyeonggi-do)* p 1952
[4] Ahangar-Asr H, Teshnehlab M, Mansouri M and Pazoki A R 2011 *Int. Conf. on Electrical and Control Engineering (Yichang)* p 5656
[5] Hou Y, Zhang H and Mei K 2011 *Int. Conf. on Electrical and Control Engineering (Yichang)* p 2804
[6] Alt B, Hartung C and Svaricek F 2011 *19th Mediterranean Conf. on Control and Automation (Corfu)* p 1472
[7] Takagi T and Sugeno M 1985 *IEEE Trans. on Systems, Man and Cybernetics* **15** 116
[8] Cazzolato B S and Prime Z 2011 *J. of Control Science and Engineering* **2011** 8
[9] Tanaka K and Wang H O 2001 *Fuzzy Control Systems Design and Analysis: A Linear Matrix Inequality Approach* (New York: Wiley-Interscience)
[10] Taniguchi T, Tanaka K, Ohtake H and Wang H O 2001 *IEEE Trans. on Fuzzy Systems* **9** 525
[11] Wen H H and Jyh H C 2007 *IEEE Trans. on Systems, Man and Cybernetics* A **37** 329
[12] Tanaka K and Sugeno M 1992 *Fuzzy Sets and Systems* **45** 135
[13] TeraSoft Inc. 2010 *Electro-Mechanical Engineering Control System User's Manual* (Taipei City: Terasoft)
[14] Akhtaruzzaman M and Shafie A A 2010 *Proc. of the IEEE Int. Conf. on Mechatronics and Automation (Xi'an)* p 1342

IOP Conf. Series: Materials Science and Engineering **53** (2013) 012010 doi:10.1088/1757-899X/53/1/012010

Assessment on tracking error performance of Cascade P/PI, NPID and N-Cascade controller for precise positioning of xy table ballscrew drive system

L. Abdullah[1], **Z. Jamaludin, N.A. Rafan, J. Jamaludin, T.H. Chiew**
Centre of Excellence, Advanced Manufacturing Centre, Faculty of Manufacturing Engineering, Universiti Teknikal Malaysia Melaka (UTeM), Hang Tuah Jaya, 76100 Durian Tunggal, Melaka, Malaysia

E-mail: lokman@utem.edu.my

Abstract. At present, positioning plants in machine tools are looking for high degree of accuracy and robustness attributes for the purpose of compensating various disturbance forces. The objective of this paper is to assess the tracking performance of Cascade P/PI, Nonlinear PID (NPID) and Nonlinear cascade (N-Cascade) controller with the existence of disturbance forces in the form of cutting forces. Cutting force characteristics at different cutting parameters; such as spindle speed rotations is analysed using Fast Fourier Transform. The tracking performance of a Nonlinear cascade controller in presence of these cutting forces is compared with NPID controller and Cascade P/PI controller. Robustness of these controllers in compensating different cutting characteristics is compared based on reduction in the amplitudes of cutting force harmonics using Fast Fourier Transform. It is found that the N-cascade controller performs better than both NPID controller and Cascade P/PI controller. The average percentage error reduction between N-cascade controller and Cascade P/PI controller is about 65 % whereas the average percentage error reduction between cascade controller and NPID controller is about 82 % at spindle speed of 3000 rpm spindle speed rotation. The finalized design of N-cascade controller could be utilized further for machining application such as milling process. The implementation of N-cascade in machine tools applications will increase the quality of the end product and the productivity in industry by saving the machining time. It is suggested that the range of the spindle speed could be made wider to accommodate the needs for high speed machining. .

1. Introduction

High tracking accuracy and precision are two vital components required in the manufacturing process. A good example of machining application is milling operation where a work piece is fed past a rotating cylindrical tool with multiple cutting edges. Both of these components are paramount important because it will lead to high-quality end product that will be delivered to customer. In an interrupted cutting operation, the teeth of a milling cutter enter and exit the work piece at each revolution. Therefore, it is highly critical that the tool material and the cutter geometry are chosen

[1] Corresponding Author: Faculty of Manufacturing Engineering, Universiti Teknikal Malaysia Melaka (UTeM) 76100 Durian Tunggal, Melaka, Malaysia. Tel.: +6-06-331-6973; Fax: +6-06-331-6411

carefully to withstand the cycles of impact cutting forces and thermal shock that might result from these physical interactions [1]. These forces are natural consequences of the cutting process and could not be avoided. For performance purposes, the ability of the system to withstand these forces and its impact will determine the standard and quality of the end product to be manufactured.

Knowledge on characteristics of these cutting forces is essential in designing the appropriate technique for its compensation [2-5]. The cutting force characteristics are influenced by the cutting parameters such as feed rate, depth of cut, and spindle speed. Variations in these cutting conditions will affect behaviour of the cutting forces in terms of its magnitudes and its harmonics content. Failure to realise this phenomenon could reduce the quality of the finished product as the cutting forces may cause vibration of the structure thus leading to a poor surface roughness measurement. Hence, an efficient and reliable compensation technique is desired in order to improve the tracking performance in machine tools applications. Previous researches on several compensation methods and approaches are discovered and have shown promising results; for example, Inverse Model Based Disturbance Observer [6], classical cascade controller [7] and Repetitive Controller [8,9], and based on the familiar PID control. Result based on PID controller [5] shows that the tracking error can be trim down until millimeter point only whereas results based on [8,9] show that the tracking error is reduced up to micron meter level via direct drive xy table. Furthermore, it is found that most of the research work were based on dedicated cutting forces and did not consider changing of the spindle speed. As a result, there is a need to conduct a research based on various spindle speed. In this research work, the controller designed is based on this needs in which the spindle speed is varied.

2. Methods

Figure 1 shows the ball-screw driven XY milling table of an XYZ-Stage produced by Googol Tech. The XY milling table consists of two axes namely; x and y axes, driven by two Panasonic MSMD 022G1U A.C. servo motors respectively. Both axes are equipped with an incremental encoder for positioning measurement. The resolution of the encoder is 2500 pulse/rev with screw pitch of 5 mm and multiple frequencies of 4. In other words, the resolution of the encoder is 0.0005 mm/pulse. Two limit switches are located in the near end of both axes. The total mass of x-axis is 36.8 kg while the total mass of the y-axis is 23.4 kg. The XY milling table consists of two axes namely; x and y axes, driven by two Panasonic MSMD 022G1U A.C. servo motors respectively. Both axes are equipped with an incremental encoder for positioning measurement. The resolution of the encoder is 2500 pulse/rev with screw pitch of 5 mm and multiple frequencies of 4. In other words, the resolution of the encoder is 0.0005 mm/pulse. Two limit switches are located in the near end of both axes. The total mass of x-axis is 36.8 kg while the total mass of the y-axis is 23.4 kg.

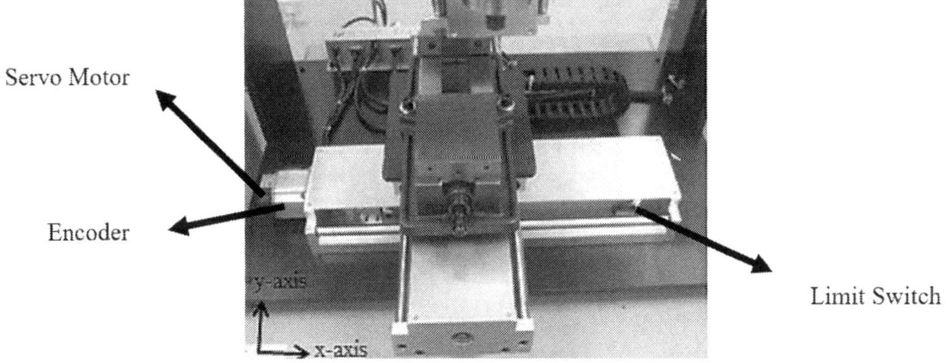

Figure 1. Googoltech xy milling table ballscrew drive system

Figure 2 shows the milling process during the collection of disturbance cutting force data. The aluminium block is placed on top of the Kistler Dynamometer. The diameter of the milling cutter is 10mm. The depth of cut is 0.5mm with the feed rate of 502 mm/min.

Figure 2. Milling operation performed for the purpose of collection of cutting force data

Figure 3 shows a schematic diagram of the overall system. The XY milling table is linked to a servo amplifier which is then connected to a DS1104 DSP board. A personal computer, equipped with ControlDesk and MATLAB software is linked to the DSP board to apply control design and data collection.

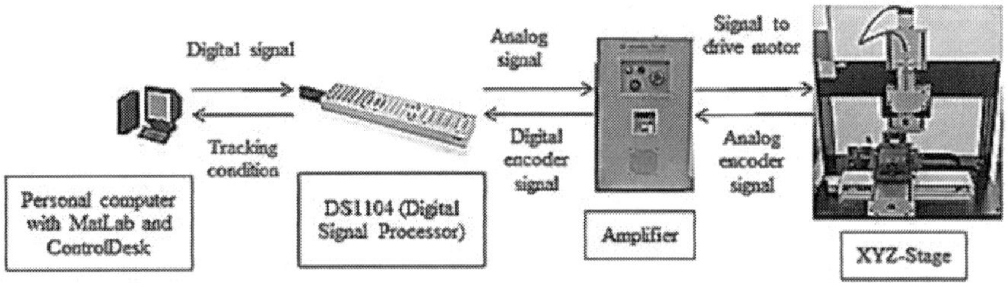

Figure 3. Schematic diagram of the overall system

3. Controller design

Three types of controller namely; Cascade Proportional/Proportional Integral (P/PI) controller, Nonlinear PID (NPID) controller and Nonlinear Cascade (N-cascade) controller is chosen to control the position of the system. The system identification process has been carried out in [10-12]. In general, the plant is an XY table system driven by ballscrew drive. The transfer function of the plant is as follows:

$$\frac{Y}{U} = \frac{69380}{S^2 + 144.8S + 166.3} \tag{1}$$

where Y is the position of the table in millimeter and U is the input voltage signal in volt.

3.1. Design of controller 1(Cascade P/PI)

Cascade P/PI controller belongs to the conventional group of controller and this control structure is widely applied in most machine control algorithm. The cascade P/PI control consists of a PI controlled inner velocity loop with a P controller in the outer position loop. Figure 4 shows the basic structure of a cascade P/PI controller [13].

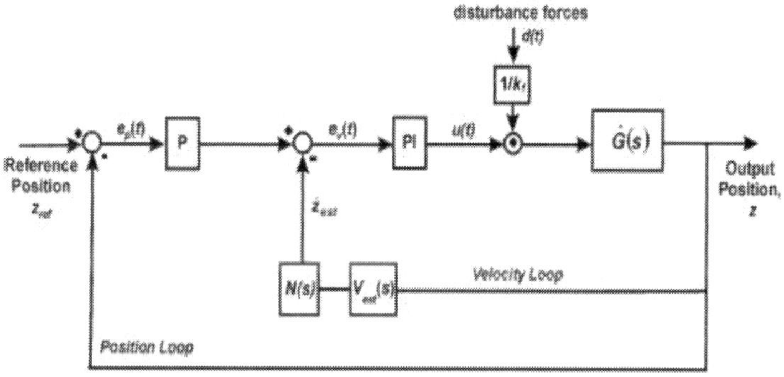

Figure 4. Basic structure of a cascade P/PI controller

The velocity PI controller is first designed before the P (position) controller. Figure 9 and 10 show the Bode open loop transfer function of the speed loop and the Nyquist plot of the position open loop respectively. Figure 11 shows the Bode diagram of the position loop sensitivity transfer function that indicates a system bandwidth of about 41.2 Hz. To conclude, Table 2 illustrates the finalized values of K_p, K_i and K_v for the controlled system.

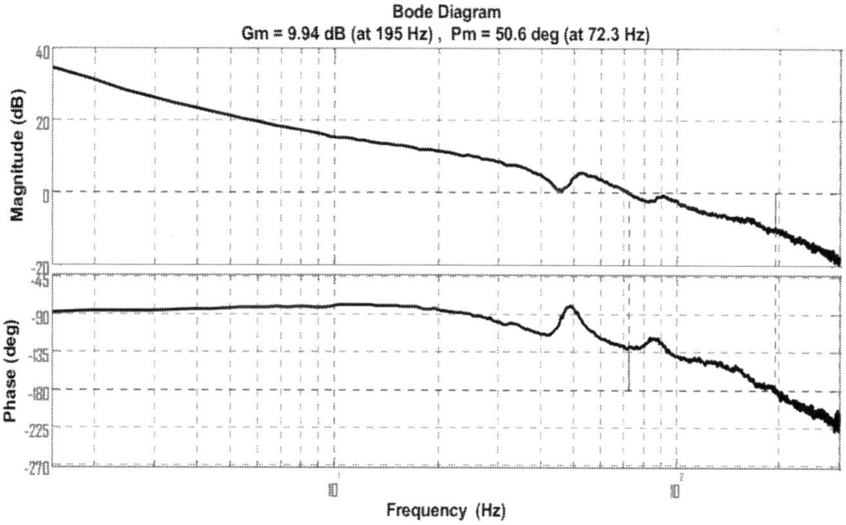

Figure 5. Bode plot of speed open loop transfer function using cascade P/PI

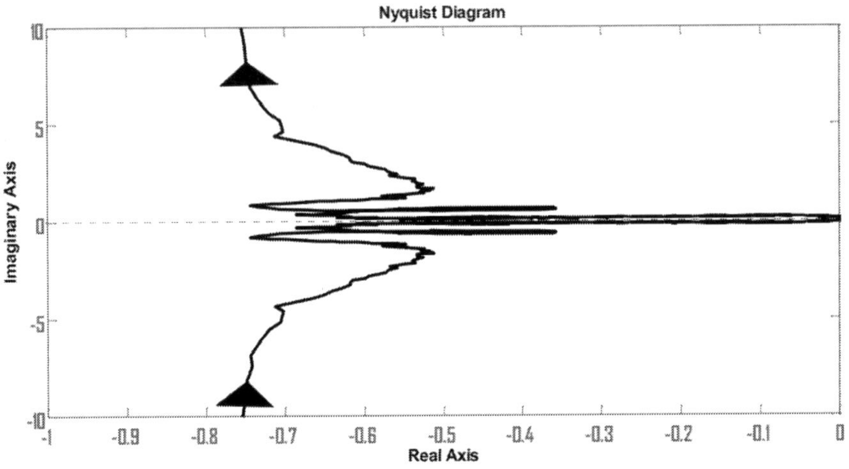

Figure 6. Nyquist plot of open loop position transfer function via cascade P/PI

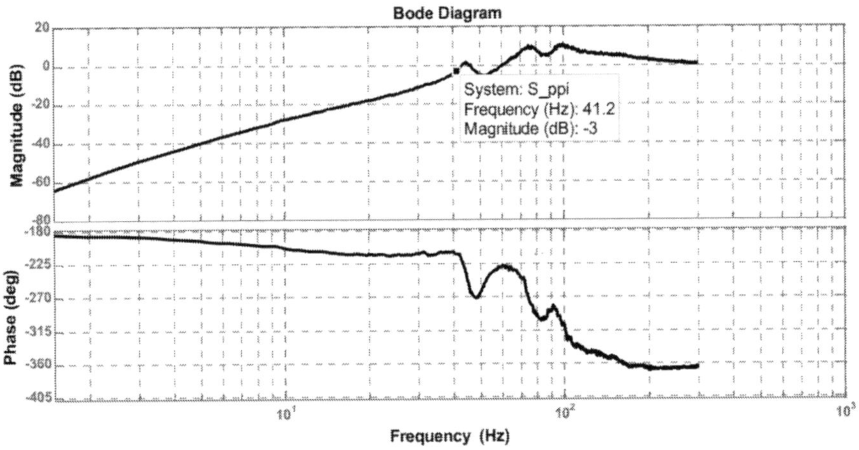

Figure 7. Bode plot of position sensitivity transfer function via cascade P/PI

Table 2: List of the control parameters of the cascade P/PI controller

Parameters	Values
K_p (velocity loop)	0.0326 [volts.s]
K_i (velocity loop)	0.0015 [volts.[illegible]
K_v (position loop)	$282s^{-1}$
Force - voltage converter	1/1623.49 N/V

3.2. Design of controller 2[Nonlinear PID(NPID) controller]

In general the suggested nonlinear PID (NPID) controller is developed by a sector bounded nonlinear gain, K(e) that is integrated in series (cascade) with PID controller. Figure 8 illustrates the block diagram of the NPID control system. The parameters of the traditional PID controller are acquired based on previous work [14] which is based on the desired gain and phase margin. The self tuned gain adjustment, K(e) behave as a nonlinear function of error, e(t) which is bounded in the sector $0 \leq K(e) \leq K(e_{max})$ as pointed out in equation 2 and equation 3. These can be classified as the range of available choice for the nonlinear gain, K(e). The output formed from this nonlinear function is acknowledged as scaled error, f_e and the equation of f_e is presented in equation 4. Alternatively, the overall equation of NPID controller can be viewed as in equation 5.

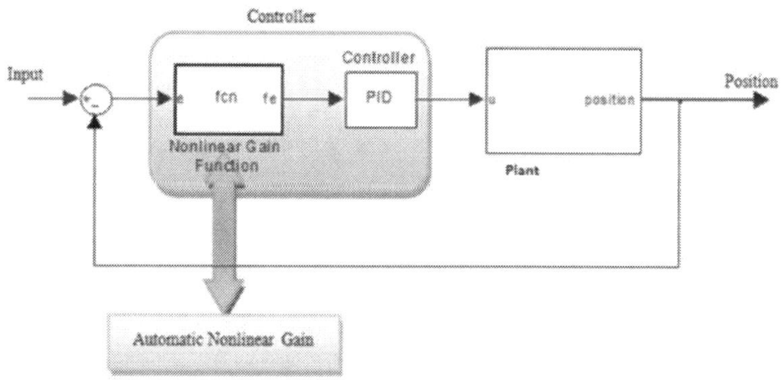

Figure 8. Block diagram of the NPID control system [15]

Nonlinear Gain, $K(e) = \dfrac{\exp(\propto e) + \exp(-\propto e)}{2}$ (2)

Error, $e = \begin{cases} e \quad ; \text{ if } & |e| \leq e_{max} \\ e_{max}*\text{sign}(e); \text{ else } & |e| > e_{max} \end{cases}$ (3)

$$K(emax) = -\frac{1}{\|G(j\omega)\|}$$

Scaled Error, $f_e = K(e)*e(t)$ (4)

General transfer function of NPID, $G_{NPID}(s) = [K_p*f_e]+[K_i/S*fe]+[K_dS*f_e]$ (5)

Table 3. List of the control parameters of NPID controller

Parameters	**Values**	
Range of error, E_{max}	3.0	[mm]
Range of variation of Nonlinear Gain, α	0.5	
Proportional Gain, K_p	1.32	[volts/mm]
Integral Gain, K_i	0.000825	[volts/mm.s]
Derivative Gain, K_d	0.006805	[volts. s/mm]

IOP Conf. Series: Materials Science and Engineering **53** (2013) 012010 doi:10.1088/1757-899X/53/1/012010

3.3. Design of controller 3[Nonlinear cascade (N-Cascade) controller]

Basically, the proposed nonlinear cascade controller is constructed by a combination of two basic function block; namely nonlinear function and cascade P/PI function block. By combining those two function blocks, it is expected that this controller could inherit the plus point of Nonlinear function in NPID controller and the robustness characteristic from cascade P/PI controller. In other words, by having the characteristics of nonlinear function and robustness properties, the controller could have the ability not only to compensate cutting force disturbance at fix value but also at multiple value. It is because the nonlinear algorithm has automatic self-adjusting mechanism in which it will inject higher gain when the system detect higher value of error and in contrast, it will inject lower gain when the system detect lower value of error. Figure 9 portrays the block diagram of nonlinear cascade control structure.

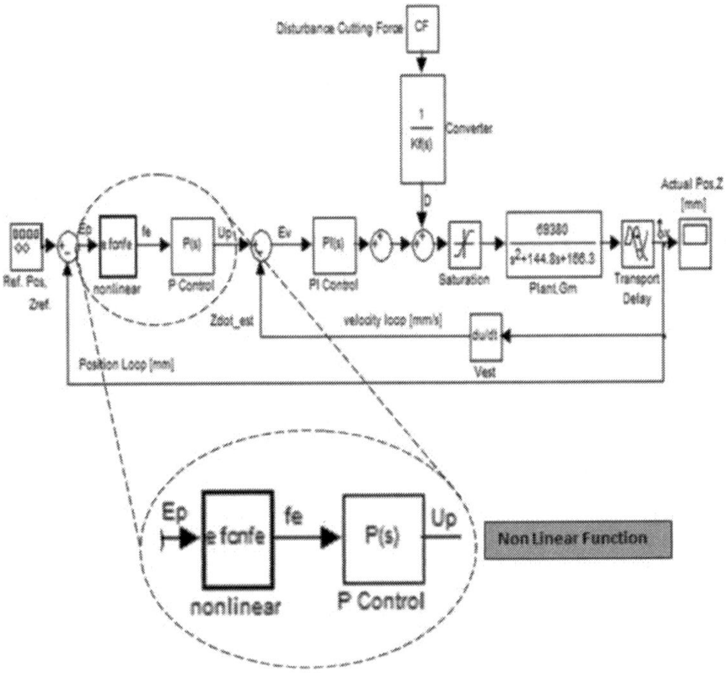

Figure 9. Block diagram of the nonlinear cascade control system

Table 4. List of the control parameters of Nonlinear Cascade controller

Parameters	Values	
Range of error, E_{max}	0.1465	[mm]
Range of variation of Nonlinear Gain, α	2.0	
Proportional Gain, K_v (Position Loop)	450	[s^{-1}]
Proportional Gain, K_p (Velocity Loop)	0.007594	[volts.s]
Integral Gain, K_i (Velocity Loop)	1.0	[volts.[illegible]

4. Result and discussion

The measured cutting forces are injected to the system (as illustrated in figure 4) as disturbances. The tracking performance cascade P/PI, Nonlinear PID (NPID) and Nonlinear Cascade (N-cascade) controller are analysed and compared based on their robustness against variable cutting force disturbance. Analyses on magnitudes of the harmonic components of the error signal is conducted in order to quantify the compensation performance of these controllers against external cutting forces. Table 5 compares the harmonic amplitudes of the position error signal recorded between cascade P/PI, Nonlinear PID (NPID) and Nonlinear Cascade (N-cascade) at varying spindle speed rotations. As expected, the N-cascade controller produces improved performance compared to the other two controllers. The results appeared to concur as N-cascade controller successfully inherit the attributes of NPID controller and cascade P/PI controller.

Table 5. FFT Analyses on harmonic components of the position errors

Spindle speed [rpm]	Harmonic Frequencies [Hz]	Amplitude of Tracking Error [μm] Cascade P/PI	NPID	N-Cascade	Error Reduction [%] Cascade P/PI with NPID	N-Cascade with NPID	N-Cascade with Cascade P/PI
1000	2.0	0.03495	0.4264	0.04247	91.80	90.03	-21.51
	3.5	0.13260	0.7743	0.04212	82.87	94.56	68.23
	5.0	0.08883	0.5321	0.04697	83.31	91.17	47.12
	6.3	0.09024	0.5446	0.05025	83.43	90.77	44.32
	7.7	0.10050	0.6331	0.06248	84.13	99.01	37.83
2000	2.0	0.04617	0.3975	0.03692	88.38	90.71	20.03
	3.5	0.05869	0.6987	0.03747	91.60	94.64	36.16
	5.0	0.07327	0.5718	0.03528	87.19	93.83	51.84
	6.3	0.05534	0.6676	0.03773	91.71	94.35	31.82
	7.7	0.05068	0.5319	0.05635	90.47	89.40	-11.18
3000	2.0	0.07913	0.2349	0.02098	66.31	91.07	73.49
	3.5	0.07545	0.7151	0.03752	89.45	94.75	50.27
	5.0	0.16180	0.4540	0.05117	64.36	88.73	68.37
	6.3	0.19010	0.2831	0.07471	32.85	73.61	60.69
	7.7	0.28960	0.2048	0.07318	-41.40	64.27	74.73

Referring to table 5, it is observed that nonlinear cascade controller demonstrates better performance compared to the other two controllers. Figure 10 ,11 and 12 show the spectrum analysis (Fast Fourier Transform, FFT of the position errors using Cascade P/PI, NPID and N-cascade at 1000rpm spindle speed rotations. Results show an improved performance using Nonlinear cascade compared to both NPID and cascade P/PI controllers for all three cases which are 1000 rpm, 2000rpm and 3000 rpm spindle speed rotations. For instance, for the case of 3000 rpm, the amplitude tracking error drops from 0.2349 μm at 2.0 Hz of NPID controller to 0.07913 μm of cascade P/PI controller. This result demonstrates an improvement of about 66.31 percent. The amplitude tracking error reduced further to 0.02098 μm when nonlinear cascade controller is applied. As a whole, the average percentage error reduction between N-cascade controller and Cascade P/PI controller is about 65 % whereas the average percentage error reduction between N-cascade controller and NPID controller is about 82 % at spindle speed of 3000 rpm. In addition, it is detected that there are cases in which the percentage of error reduction for N-cascade showing negative percentage. It simply means that at that particular frequency, the error is increasing instead of decreasing. In control system theory, this phenomenon is called water bed effect. It means when at certain point the error is being push to reduce, the other point will be likely to increase. Results also show that nonlinear cascade controller is able to compensate and reduce tracking error components generated from various different cutting force disturbances.

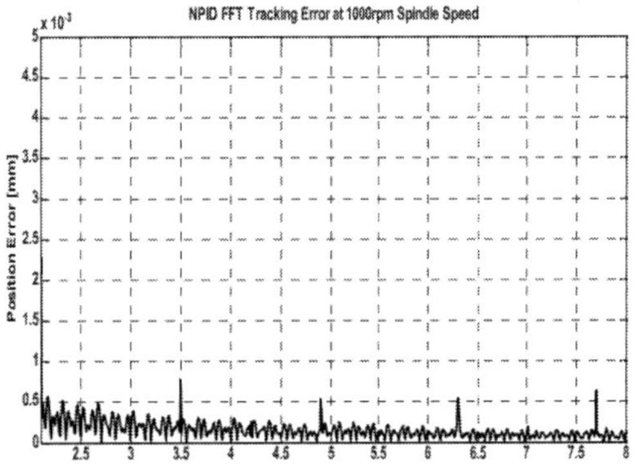

Figure. 10. FFT Error via NPID

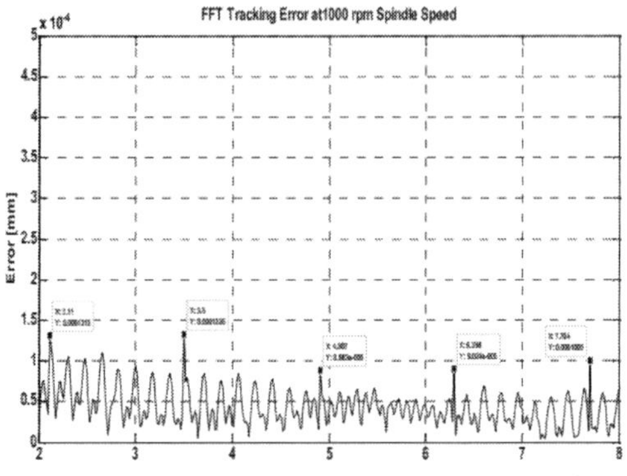

Figure 11. FFT Error via Cascade P/PI

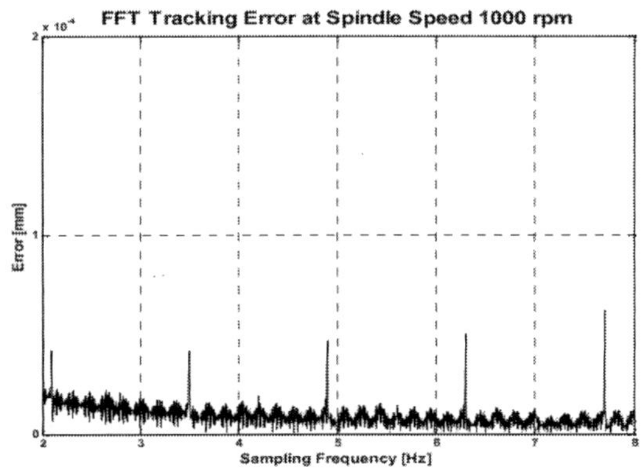

Figure 12. FFT Error via Nonlinear cascade

5. Conclusion

It can be concluded that the performance of Nonlinear Cascade controller outweigh both NPID and cascade P/PI controller. The maximum FFT tracking errors of Nonlinear Cascade is approximately about 0.07471 micrometer, while for the case of NPID and cascade P/PI controllers, the maximum FFT tracking errors are 0.7743 and 0.28960 micrometer respectively. Analyses on the harmonics content of the position tracking errors have shown the supremacy of Nonlinear cascade (N-cascade) controller. The advantages of the implementation of N-cascade in machine tools applications are obvious in which it will increase the quality of the end product and the productivity in industry by saving the machining time.

Acknowledgement

This research is financially supported by the Short Term Grant Scheme (PJP), Centre for Research and Innovation Management with reference no. PJP/2012/FKP(57A)/S01064 and the Faculty of Manufacturing Engineering, Universiti Teknikal Malaysia Melaka (UTeM).

6. References

[1] Kalpakjian, S., and S. Schmid, 2006. *Manufacturing Processes for Engineering*, (3rd ed.). Addison Wesley : Longman.
[2] Jamaludin, Z., H.V. Brussel, G. Pipeleers and J. Swevers, 2008. Accurate motion control of XY high-speed linear drives using friction model feedforward and cutting forces estimation. *CIRP Annals – Manufacturing Technology*, 57(1) : 403-406.
[3] Pritschow, G. and W. Phillip, 1992. Research on the efficiency of feedforward controllers on machine direct drives. *CIRP Annals – Manufacturing Technology*, 41(1) : 411-415.
[4] Skogestad, S., and I. Posletthwaite, 2005. *Multivariable Feedback Control - Analysis and Design*, Chichester, West Sessex, England, John Wiley & Sons.
[5] Wang, J., 2004. *Robust tracking controller design with application to the motion control of an x-y feed table for high-speed machining"*, PhD Thesis, Faculty of Engineering, Department of Mechanical, Division of Production, Machine Design and Automation (P.M.A), Katholieke Universiteit Leuven, Belgium.
[6] Jamaludin, Z., H.V. Brussel and J. Swevers, 2008. Quadrant glitch compensation using friction model-based feedforward and an inverse-model-based disturbance observer, *Proc.Int. Conf.on Advanced Motion Control*, Centro Santa Chiarra Hotel, Trento, Italy. 26 - 28 March 2008.
[7] Jamaludin, Z., H.V. Brussel and J. Swevers, 2007. Classical cascade and sliding mode control tracking performances for a x-y feed table of a high-speed machine tools, *Int. J. of Precision Technology*, 1(1) : 65-74.
[8] Hara, S., Y. Yamamoto, T. Omata and M. Nakano, 1988. Repetitive control system: a new type servo system for periodic exogenous signals, *IEEE Trans. on Automatic Control*, 33(7) : 659-668.
[9] Pipeleers,G., B. Demeulenaere, J. DeSchutter, and J. Swevers , 2008. Robust High-Order Repetitive Control: Optimal Performance Trade-Offs, *Automatica*, 44(9) : 2628-2634.
[10] Abdullah, L., Z. Jamaludin, T.H. Chiew, N.A. Rafan and M.S. Syed Mohamed, 2012. System identification of xy table ballscrew drive using parametric and non parametric frequency domain estimation via deterministic approach . *Procedia Engineering*, 41(1) : 567-574.
[11] Pintelon, R. and J. Schoukens, 2001. *System identification – A frequency domain approach*, New York: IEEE Press.
[12] Pintelon,R., P. Guillaume, Y. Rolain, J. Schoukens, and H. Van Hamme, 1994. Parametric Identification of Transfer Functions in the Frequency Domain – A Survey. *IEEE Trans.on Automatic Control*, 39(11) : 2245-2260.
[13] Abdullah, L., Z. Jamaludin, Q. Ahsan, J. Jamaludin, N.A. Rafan, T.H. Chiew, K. Jusoff and M. Yusoff, "Evaluation on tracking performance of PID, Gain Scheduling and Classical Cascade P/PI controller on XY Table Ballscrew Drive System". *World Appl. Sci. J.*, Vol. 21, pp. 01-10, 2013
[14] Seraji, H., " A new class of nonlinear PID controllers with robotic applications". *Journal of Robotic System*, Vol. 15, No. 3, pp. 161-181, 1998
[15] Rahmat, M.F., Sy Najib Sy Salim, N.H. Sunar, Ahmad Athif Mohd Faudzi, Zool Hilmi Ismail and K. Huda, "Identification and non-linear control strategy for industrial pneumatic actuator". *Int. J. of the Physical Sci.*, Vol. 7, No. 17, pp. 2566-2579, 2012.

Optimal Performance of a Nonlinear Gantry Crane System via Priority-based Fitness Scheme in Binary PSO Algorithm

Hazriq Izzuan Jaafar[1a], Nursabillilah Mohd Ali[1,2b], Z. Mohamed[3c], Nur Asmiza Selamat[1d], Amar Faiz Zainal Abidin[3e], J.J. Jamian[3f], Anuar Mohamed Kassim[1g]

[1]Faculty of Electrical Engineering, Universiti Teknikal Malaysia Melaka, 76100 Durian Tunggal, Melaka, Malaysia.
[2]Autonomous Agent Resarch Group (AA-RG), Faculty of Engineering, International Islamic University Malaysia, P.O. Box 10, 50728 Kuala Lumpur, Malaysia.
[3]Faculty of Electrical Engineering, Universiti Teknologi Malaysia, 81310 Skudai, Johor, Malaysia.

[a]hazriq@utem.edu.my, [b]nursabillilah@utem.edu.my, [c]zahar@fke.utm.my, [d]nurasmiza@utem.edu.my, [e]amarfaiz@fke.utm.my, [f]jasrul@fke.utm.my, [g]anuar@utem.edu.my

Abstract. This paper presents development of an optimal PID and PD controllers for controlling the nonlinear gantry crane system. The proposed Binary Particle Swarm Optimization (BPSO) algorithm that uses Priority-based Fitness Scheme is adopted in obtaining five optimal controller gains. The optimal gains are tested on a control structure that combines PID and PD controllers to examine system responses including trolley displacement and payload oscillation. The dynamic model of gantry crane system is derived using Lagrange equation. Simulation is conducted within Matlab environment to verify the performance of system in terms of settling time (Ts), steady state error (SSE) and overshoot (OS). This proposed technique demonstrates that implementation of Priority-based Fitness Scheme in BPSO is effective and able to move the trolley as fast as possible to the various desired position.

1. Introduction

Limitations of manpower makes gantry crane system are always required for handling heavy load system. It aims to facilitate the delivery of heavy loads with accurate position. However, the crane acceleration is required for motion and induces undesirable load swing [1]. If the speed of trolley is increased, the oscillation of payload become larger and causes the payload difficult to settle down. Thus, position control is difficult and requires a relatively long time for unloading. In order to attain positional accuracy of the gantry crane, a control mechanism that account for positioning of the trolley and oscillation of the payload are required.

Several control techniques have been proposed previously for controlling the gantry crane system. However, PID control schemes based on the classical control theory have been widely used for various industrial control systems for a long time [2]. PID controller is seen as a good prospect due to simple

[1a] Corresponding author. Tel: +06-555 2358; Fax: +06-555 2222
E-mail address: hazriq@utem.edu.my

structure and robust performances in a wide range of operating conditions. Thus, PID controller is chosen and used for this work. However, it has some difficulties in tuning the PID parameters. Therefore, various investigations on intelligent techniques have been conducted to optimize PID parameters. For instance, Genetic Algorithm (GA), Artificial Bee Colony (ABC) algorithm, Ant Colony Algorithm (ACA) and many more was proposed to optimize the parameter of the controller in designing of a nonlinear system. It has flexible and adaptive characteristic in order to find the PID parameters. Nevertheless, since Binary Particle Swarm Optimization (BPSO) is well known compared to the other of some optimization method, thus it is being chosen to be implemented for this work.

2. Nonlinear Model of a Gantry Crane System

Figure 1 shows a schematic diagram of a gantry crane system that is considered in this work. The parameters of m_1, m_2, l, x, θ, T and F are payload mass, trolley mass, cable length, horizontal position of trolley, swing angle, torque and driving force respectively. Nonlinear model of the gantry crane system is modeled based on [3]. Some assumptions have been made to minimize the difficulties of modeling such as cable of trolley and hanged load are assumed to be rigid and massless. The system parameters are shown in Table 1.

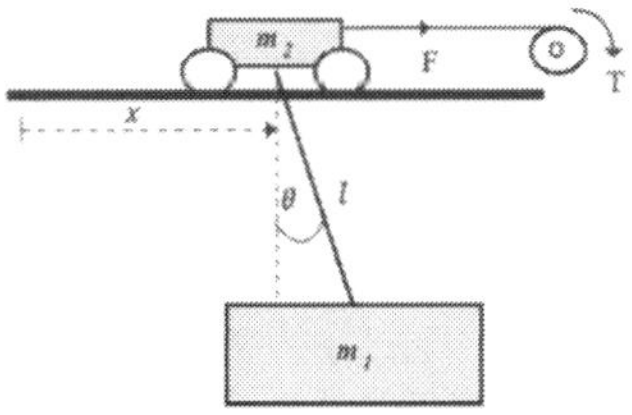

Figure 1. Schematic diagram of a Gantry Crane System [3]

Table 1. System parameters

Parameters	Value (Unit)	Parameters	Value (Unit)
Payload mass *(m1)*	1.0 kg	Resistance *(R)*	2.6 Ω
Trolley mass (m_2)	5.0 kg	Torque constant (K_T)	0.007 Nm/A
Cable length *(l)*	0.5 m	Electric constant (K_E)	0.007 Vs/rad
Gravitational (g)	9.81 m/s^2	Radius of pulley (r_P)	0.02 m
Damping coefficient *(B)*	12.32 Ns/m	Gear ratio *(z)*	15.0

3. Modeling of a Gantry Crane System

Several methods can be used to model the gantry crane system. From the investigations, it is found that the Lagrange's equation is more suitable to derive the mathematical expression for modeling the system. The GCS has two independent generalized coordinates namely trolley displacement, x and payload oscillation, θ. The standard form for Lagrange's equation is given as:

$$\frac{d}{dt}\left[\frac{\partial L}{\partial \dot{q}_i}\right] - \frac{\partial L}{\partial q_i} = Q_i \tag{1}$$

where L, Q_i and q_i represent Lagragian function, nonconservative generalized forces and independent generalized coordinate. The Lagragian function can be written as:

$$L = T - P \tag{2}$$

with T and P are respectively kinetic and potential energies. This relationship involved in calculating on how it is related to be more flexible coordinates. Kinetic and potential energies can be derived as:

$$L = \frac{1}{2}\left(m_1\dot{x}^2 + m_2\dot{x}^2 + m_1 l^2 \dot{\theta}^2\right) + m_1 \dot{x}\dot{\theta} l \cos\theta + m_1 g l \cos\theta \quad (3)$$

Solving for equation (1) yields nonlinear differential equations as:

$$(m_1 + m_2)\ddot{x} + m_1 l\ddot{\theta} \cos\theta - m_1 l\dot{\theta}^2 \sin\theta + B\dot{x} = F \quad (4)$$

$$m_1 l^2 \ddot{\theta} + m_1 l\ddot{x} \cos\theta + m_1 g l \sin\theta = 0 \quad (5)$$

Since the dynamic DC motor is included in this gantry crane model, differential equations with their effects is derived. By considering the dynamic of DC motor, a complete nonlinear differential equation of the gantry crane can be obtained as equation (6) and (7) where V is an input voltage.

$$V = \left[\frac{RBr_p}{K_T z} + \frac{K_E z}{r_p}\right] + \left[\frac{Rr_p}{K_T z}\right](m_1 l)\left[\ddot{\theta}\cos\theta - \dot{\theta}^2 \sin\theta\right] + \left[\frac{Rr_p}{K_T z}\right]\ddot{x} \quad (6)$$

$$m_1 l^2 \ddot{\theta} + m_1 l\ddot{x} \cos\theta + m_1 g l \sin\theta = 0 \quad (7)$$

Thus, PID and PD controllers are implemented for this nonlinear gantry crane as shown in figure 3.

4. Priority-based Fitness Binary Particle Swarm Optimization (PFBPSO)

The basic PSO is developed by Kennedy and Eberhart in 1995 [4]. It based on behaviors of fish schooling and bird flocking in order to search and move to the food with certain speed and position. It has been applied successfully and applied easily to solve various function optimization problems especially for nonlinear models. In 1997, Binary PSO (BPSO) has been introduced to solve discrete optimization problem [5]. Applications of BPSO can be seen in many engineering problems, such as routing in VLSI, computational biology, job scheduling and agriculture.

In this research, a new method of Priority-based Fitness Binary Particle Swarm Optimization (PFBPSO) is proposed for tuning of PID and PD parameters. In this work, settling time (Ts) is set as highest priority, followed by steady state error (SSE) and overshoot (OS). Figure 2 illustrated the PFBPSO process where the P_{BEST} and G_{BEST} are updated according to the priority. The particles find for the local best, P_{BEST} and subsequently global best, G_{BEST} for each iteration in order to search for optimal solution. Each particle is assessed by fitness function. Thus, all particles try to replicate their historical success and in the same time try to follow the success of the best agent. It means that the P_{BEST} and G_{BEST} are updated if the particle has a minimum fitness value compared to the current P_{BEST} and G_{BEST} value. Nevertheless, only particles that within the range of the system's constraint is accepted. The new velocity can be calculated and as in equation (8).

$$v^{i+1} = \omega v^i + c_1 r_1\left(P_{BEST} - x^i\right) + c_2 r_2\left(G_{BEST} - x^i\right) \quad (8)$$

where r_1 and r_2 represent random function values [0,1] while c_1 is cognitive component and c_2 is social component. Next, new particles are updated using equation (9) based on the sigmoid concept which is probability of the normal distribution. The all five parameters are obtained based on binary numbers (either 0 or 1) and then converted into decimal number that represents K_P, K_I, K_D, K_{PS} and K_{DS}.

$$sigmoid = \begin{cases} 1, rand < \dfrac{1}{1 + e^{-v^{i+1}}} \\ 0, rand \geq \dfrac{1}{1 + e^{-v^{i+1}}} \end{cases} \quad (9)$$

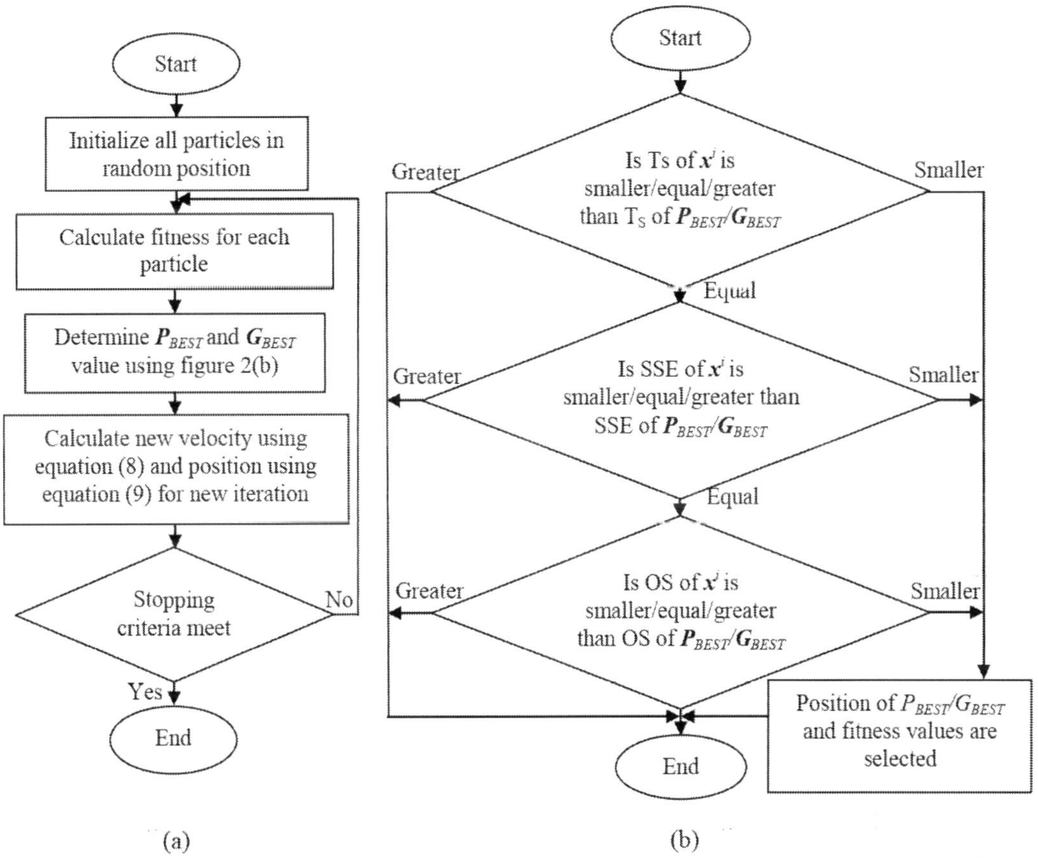

Figure 2. The flow chart of PFBPSO in determining gantry crane system parameters
(a) Process of PFBPSO algorithm (b) Updated rules for P_{BEST} and G_{BEST}

5. Implementation, Results and Discussion

In this work, a control structure that combines PID and PD controllers as shown in figure 3 is proposed. For this structure, PID controller is used for the positioning control meanwhile the PD controller is used for adjusting the payload oscillation. Therefore, PFBPSO algorithm is designed to tune and find all of these five optimal parameters of controllers.

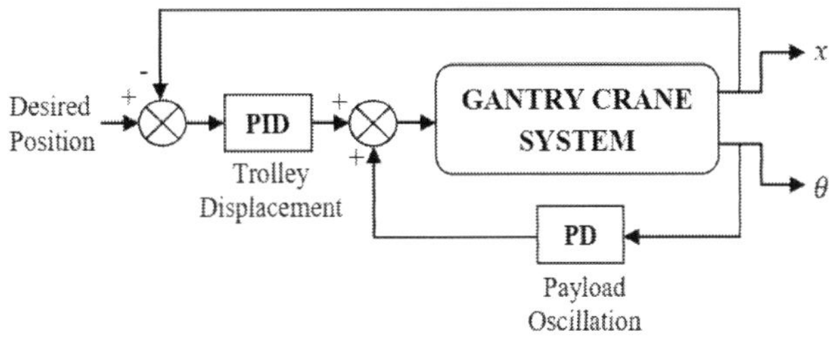

Figure 3. Control structure with five controller gains (PID and PD)

Simulation exercises are conducted with Intel Core i5-2450M Processor, 2.5GHz, 6GB RAM, Microsoft Window 7 and MATLAB as a simulation platform. The gantry crane system model with nonlinear differential equations in equation (6) and (7) are designed via Simulink. With an input voltage, two system responses namely trolley displacement and payload oscillation are examined. In this study, 20 particles are considered with 100 iterations. As default values, c_1 and c_2 are set as 2. The initial value of ω is 0.9 and linearly decreased to 0.4 at some stage in iteration. Table 2 shows the optimal PID and PD parameters are obtained using the PFBPSO algorithm.

Table 2. Optimal PID and PD parameters using PFBPSO

Controller Gains	Binary (12 Bits)	Parameters
K_P	011011.001100	27.188
K_I	000000.010010	0.005
K_D	001011.011100	11.438
K_{PS}	100000.110011	32.797
K_{DS}	000001.100000	1.500

The control structure in figure 3 is then simulated with the PFBPSO tuned controller parameters. Figure 4(a) and 4(b) shows the trolley displacement and payload oscillation responses respectively with payload of 1.0 kg and desired position at 1.0 m. It is noted with the proposed algorithm, zero SSE with low OS (1.691 %) and T_S (1.743 s) of displacement response is achieved. Moreover, low payload oscillation is observed. Table 3 summarizes the performances of the gantry crane system that obtained with the controller.

Table 3. Control performances with payload 1.0 kg and desired Position 1.0 m

	Trolley Displacement			Payload Oscillation	
	Steady State Error, SSE (m)	Overshoot, OS (%)	Settling Time, Ts (s)	Max. Angle, ϴ (rad)	One Cycle Oscillation, T (s)
Performances	0.000	1.691	1.743	0.249	2.038

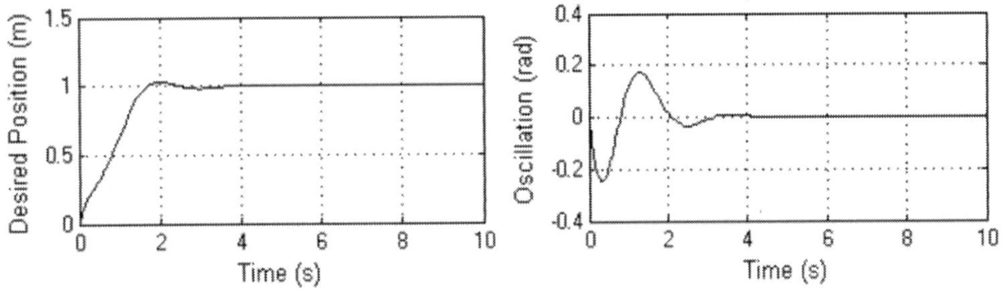

(a) (b)

Figure 4. Performance of (a) trolley displacement (b) payload oscillation with payload 1.0 kg and desired Position 1.0 m

Subsequently, it is desirable to examine the controller's performance under various desired positions. Figure 5 shows the system responses with desired positions at 1.0 m, 0.8 m and 0.2 m. It is shown that the system response successfully track desired positions. In the PFBPSO algorithm, overshoot, settling times and payload oscillation are affected with various desired positions. Therefore, SSE is achieved for all conditions. Table 4 summarizes simulation results with various desired positions.

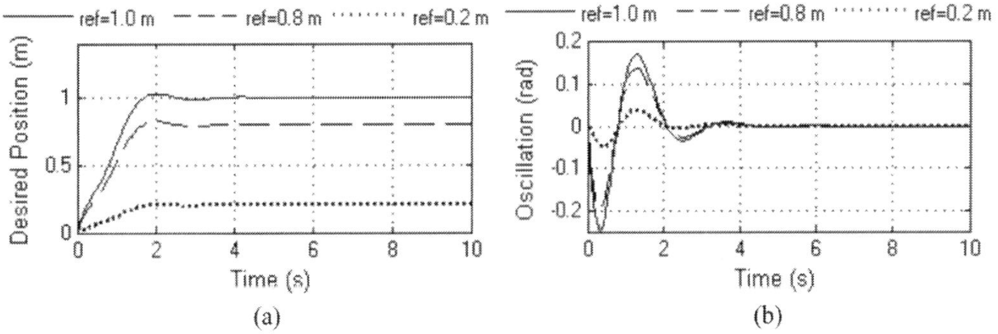

Figure 5. System response with various desired positions at 1.0 m, 0.8 m and 0.2 m
(a) trolley displacement (b) payload oscillation

Table 4. Control performances with payload 1.0 kg and various desired position

	Trolley Displacement			Payload Oscillation	
Various Desired Position	Steady State Error, SSE (m)	Overshoot, OS (%)	Settling Time, Ts (s)	Max. Angle, ϴ (rad)	One Cycle Oscillation, T (s)
1.0 m	0.000	1.691	1.743	0.249	2.038
0.8 m	0.000	1.702	1.721	0.214	2.029
0.2 m	0.000	1.711	1.699	0.193	2.021

Conclusion

This paper has presented design of an optimal PID controller for control of a gantry crane system. Nonlinear differential equations of the system has been derived and used for verification of control algorithm. In this work, the PFBPSO algorithm has been proposed in order to find all of these five optimal controller gains. The optimal gains have been tested based on a control structure that combines PID and PD controllers. System responses including trolley displacement and payload oscillation have been examined. Simulation results have shown that the controller is effective to move the trolley as fast as possible to the various desired position with low payload oscillation based on priority-based fitness scheme.

References

[1] Butler H, Honderd G, Van Amerongen J 1991 "Model Reference Adaptive Control of a Gantry Crane Scale Model", *IEEE Control Systems Magazine* **11** 57-62

[2] K J Astrom, T Hagglund 2006 "Advanced PID Control", *ISA-The Instrumentation, Systems and Automation Society.*

[3] M I Solihin, Wahyudi, M A S Kamal and A Legowo 2008 "Optimal PID Controller Tuning of Automatic Gantry Crane using PSO Algorithm," *Proceeding of the 5th Int. Symposium on Mechatronics and its Applications (Amman, Jordan, 27–29 May 2008)* pp 1-5

[4] Kennedy J and Eberhart R 1995 "Particle Swarm Optimization", *Proceedings of the 1995 IEEE International Conference on Neural Networks (Perth, WA, 27 –30 November 1995)* pp 1942-1948

[5] J Kennedy and R Eberhart 1997 "A discrete binary version of the particle swarm algorithm," *Proc. of IEEE International Conference on Systems, Man, and Cybernetics (Orlando, FL, 12–15 October 1997)* pp 4104-4108

Development of a Novel Locomotion Algorithm for Snake Robot

Raisuddin Khan[1], Md. Masum Billah, Mitsuru Watanabe and A.A. Shafie
Department of Mechatronics Engineering, Faculty of Engineering
International Islamic University Malaysia (IIUM), 53100 Kuala Lumpur, Malaysia

E-mail: raisuddin@iium.edu.my

Abstract. A novel algorithm for snake robot locomotion is developed and analyzed in this paper. Serpentine is one of the renowned locomotion for snake robot in disaster recovery mission to overcome narrow space navigation. Several locomotion for snake navigation, such as concertina or rectilinear may be suitable for narrow spaces, but is highly inefficient if the same type of locomotion is used even in open spaces resulting friction reduction which make difficulties for snake movement. A novel locomotion algorithm has been proposed based on the modification of the multi-link snake robot, the modifications include alterations to the snake segments as well elements that mimic scales on the underside of the snake body. Snake robot can be able to navigate in the narrow space using this developed locomotion algorithm. The developed algorithm surmount the others locomotion limitation in narrow space navigation.

1. Introduction

The calamitous earthquake (Sendai Earthquake) that ruined chaos Japan in 2011, tsunami hit Indonesia several times in 2012 and the terrorist attacks on the World Trade Centers in 2001 are clear sign that we are not prepared for disaster recovery at all. In all cases the infrastructure could not withstand the fury of nature, even in the case of WTC the NYPD was not prepared for such gigantic task of disaster recovery. The conventional reaction to such disaster is not adequate; a new paradigm shift is needed to address such calamities utilizing all resources at hand. Disaster recovery is defined to be the emergency response function which deals with the collapse of manmade structures. In any disaster either man made or due to Mother nature, the elementary tasks at hand are: (i) reach the affected hazardous field (ii) find and get information about victims, and (iii) rescue as many of them as possible. It is possible for robot to reach any hazardous field unlike who have limited mobility in such missions. Nowadays legged and wheeled robots are involved in such mission, but limbless robot like snake robot bring a great advantages to overcome navigation over the very narrow space in disaster area. Few researches have been conducted on snake robot locomotion unlike narrow space consideration [1-5]. Having developed the active chord mechanism to model the movement of lateral undulation [1], Hirose went on to conduct further research on the same type of locomotion [2,3,4].

[1] Raisuddin Khan

Numerous studies based on Hirose's work have also cropped up during the years. Initially Hirose used passive wheels on his snake robots, and Saito's research [5] looked as achieving the same locomotion without any such wheels, with the body of the robot in direct contact with the ground. Other variations include the application of the same type of locomotion to different surfaces such as a sloped surface [6] or uneven surfaces [7]. In recent years, Hirose teamed up with other Japanese researchers to develop a 3-D version his active chord mechanism [8]. However, all these works are principally based on one type of locomotion: lateral undulation. This is not to say that other modes of locomotion have not been studied. Many years back, Burdick developed a model for the side-winding movement [9], while somewhat more in the recent past a robot was developed based on rectilinear motion [10]. Even different gaits not found in natural snakes have been examined by the likes of Chen [11], where a movement known as lateral rolling is studied. Interestingly enough, studies on concertina locomotion is surprisingly absent. Coming back to the application of snake movements, the advantages of the serpentine movement has been abundantly demonstrated. The movement is efficient and can be utilized in various environments. There is, however, a limitation. As Hirose evaluated, the snake is only able to propel forward when it assumes the serpenoid curve. Unlike a simple sinusoidal curve, the curvature of this curve changes sinusoidally over its length. Thus if the snake is to travel along a certain axis, then it must displace its body both above and below this axis to form the curve. The problem arises here, if the minimum perpendicular displacement is not maintained, the snake will not move forward. Even with the increase of links, the minimum perpendicular displacement will be much greater than the width of the body. The immediate approach to solving the issue of narrow space may be to use a different type of locomotion. Two possible candidates come forth, the rectilinear and the concertina, since sidewinding also requires the body to undergo a curve of even greater amplitude. The movement of lateral undulation and rectilinear motion are completely different, and it would be unwise to try and implement both sets of movement in one robot. The concertina too has issues when considered in practical terms. Sinus-lifting is used in this type of motion, which means the robot must lift part of its body above the ground. To bypass this vertical degree of freedom, perhaps mechanisms could be added where friction could be controlled. Though it may be possible to design such a structure, it would hamper the ability of the robot to undertake lateral undulation. A robot developed exclusively for concertina motion would end up being highly inefficient in environs where space is not an issue. To overcome the limitations of serpenoid motion, a completely different type of locomotion is introduced in this paper [12]. The directional friction element of the snake is in fact provided by the scales on the underside of the body. The scales are arranged in a manner such that friction is low in the forward direction while high in reverse and lateral directions. Using this idea and adding additional elements to the generic design of the serpentine robot, a strategy for a novel type of locomotion is proposed. The strategy is evaluated through kinematics and kinetics studies.

2. Materials And Method

2.1. Snake Robot Mechanism

In Fig. 1 mechanical skeleton of a conceived snake robot is shown where each link consists of a solid cross. The legs of the cross, connected in a chain fashion are called centre legs, while the other legs are called transverse legs. Edges of the alternate transverse legs have got snake scale like protrusions that give low frictional resistance during forward motion and high resistance during backward motion. In figure 1 link 1, 3 have got such snake scales. Link 2 acts as a connector between the alternate links with scales. The transverse legs of link 2 are used to make the link symmetric and to reduce complexity of calculating its mass moments of inertia.

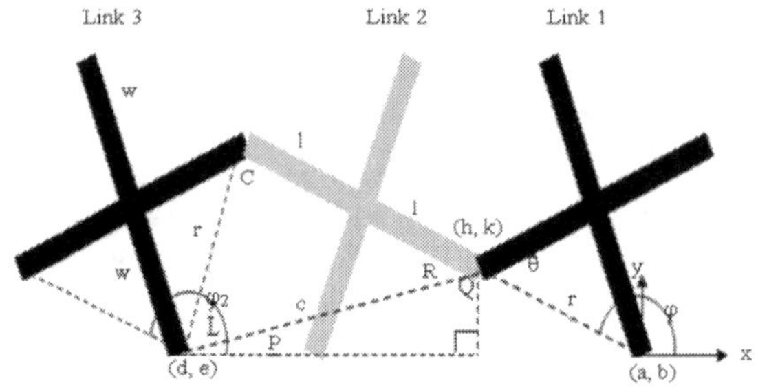

Figure 1. Figure with Coordinate System for a Five-Link snake Structure

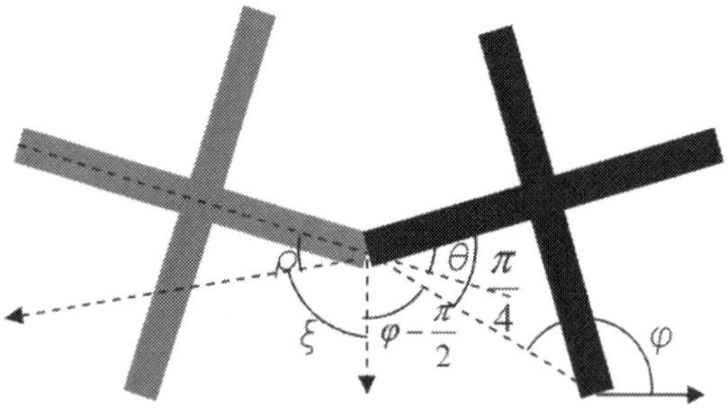

Figure 2. Exaggerated view of a general geometry at a joint

2.2. Kinematics for snake robot with scales underneath

In Fig.1, Link 1 and Link 3 rotate in arcs about point (a, b) and (h, k) respectively. Link 2 on the other hand does not move in a defined arc about a fixed point, and hence must be free to move. To facilitate this unimpeded movement, no frictional appendages are added.

Moving on with the kinematics, following the geometric relations in Figure 1, the coordinates of (h, k) would be

$$(r\cos\varphi + a, r\sin\varphi + b) \tag{1}$$

The length of segment c connecting (d, e) and (h, k) would be given by the expression

$$c = \sqrt{(h-d)^2 + (k-e)^2} \tag{2}$$

Using the cosine rule on the triangle formed between the points (d, e), (h, k), and the joint between Link 2 and 3, the three angles ρ, γ, and λ can be determined as

$$\rho = \cos^{-1}\left(\frac{4l^2 + c^2 - r^2}{4lc}\right) \tag{3}$$

$$\gamma = \cos^{-1}\left(\frac{4l^2 + r^2 - c^2}{4lr}\right) \tag{4}$$

$$\lambda = \cos^{-1}\left(\frac{r^2 + c^2 - 4l^2}{2rc}\right) \tag{5}$$

Drawing a right triangle with its hypotenuse as segment c, two more angles are obtained:

$$\xi = \tan^{-1}\left(\frac{h-d}{k-e}\right) \quad (6)$$

$$\psi = \tan^{-1}\left(\frac{k-e}{h-d}\right) \quad (7)$$

From all this, θ, the angle between the extension of the lengthwise axis of Link 1 and the central member of Link 2, is therefore evaluated by adding the angles concentric to point (h,k). It would be important to note that though this angle appears to simply be $\frac{\pi}{4}$ in the initial configuration, as Link 1 rotates about (a,b) in the clockwise direction.

Angle θ can then be evaluated by adding the following angles:

$$\rho + \xi + \left(\varphi - \frac{\pi}{2}\right) + \frac{\pi}{4} \quad (8)$$

and then subtracting π from the result

$$\rho + \xi + \left(\varphi - \frac{\pi}{2}\right) + \frac{\pi}{4} - \pi = \theta \quad (9)$$

Which rearranged would give the final equation of

$$\theta = \varphi + \rho + \xi - \frac{5\pi}{4} \quad (10)$$

Similarly, referring back to Figure 1, φ2, the Link 3 equivalent to the angle φ for Link 1, is

$$\varphi_2 = \lambda + \psi + \frac{\pi}{2} \quad (11)$$

The angles for following links can then be calculated using the same method on φ_2 as was illustrated here with φ. Thus from the first angle φ, all other positions and angles between links can be determined. However, this still constitutes as inverse kinematics, since the angle that is actuated in the model is θ, and the resulting configuration and position of the structure is denoted by φ. Thus to anticipate the position of the members from the actuated angle θ, a forward kinematics needs to be developed. However, though the forward kinematics is not readily apparent, the maximum angle for θ is evident. As mentioned earlier, for the links to avoid singular positions and keep moving forward, at maximum deflection, the segment r of Link 1 must be in line with, i.e. parallel to, the lengthwise member 2l of Link 2. Keeping this in mind, it becomes clear that the range is

$$-\frac{\pi}{4} \le \theta \le \frac{\pi}{4} \quad (12)$$

2.3. Tables for Actuators value calculated from equations

Table I lists the necessary angles that need to be tracked. The table really consists of six rows as the seventh row is just a repeat of row one. What are deducible from the table are the varied differences in angles from one row to the next. Take the difference from row one to two for example: θ_1 has increased by 42° while θ_4 decreases by only 7°. All the angles in each row must be met before continuing on to the next row, and the angles must traverse their respective displacements and arrive at the next configuration at the same time. Thus, in the example, θ_1 must deflect 42° in the same amount of time θ_4 moves just 7°. It becomes clear that the speeds of the servo motors must also be controlled.

Table 1. Angular positions for gait sequence

Sequence Number	θ_1 (°)	θ_2 (°)	θ_3 (°)	θ_4 (°)
1	-45	45	-45	45

2	-3	19	-30	38
3	35	-25	13	3
4	45	-45	45	-45
5	1	-17	28	-37
6	-36	27	-16	0
7	-45	45	-45	45

For the purpose of controlling the servo motors using the microcontroller, rather than building the board and interface in the lab, a servo controller board as well as a microcontroller interface is procured from the same source as the servo motors themselves. The servo controller is able to control both position and speed, and uses values from 0 to 1454 to control the position which spans a range of 180°, and uses values of 0 to 63 for the speed, where 63 is the maximum controlled speed, and 1 is the slowest speed achievable. Using both the data sheet provided by the source as well as experimental data taken from the completed prototype, the following conversion tables are deduced (table 2 and table 3).

Table 2. Simple conversion table for servo motor position

Angular Displacement (°)	Controller Value
-45	420
0	811
45	1202

Table 3. Conversion table for angular velocity of servo motor

Angular Velocity (°s^{-1})	2	13	26	43	50	64	82	95
Controller Value	1	10	20	30	40	50	60	63

For both the conversion tables, all the middle values are determined through linear interpolation. One of the necessary preparations is that of controlling the speed, as mentioned earlier the structure is required to traverse each sequence in the same amount of time. Thus, firstly the differences in angular displacement from the first position to the next must be determined. Applying this to table 1 the results obtained are shown in table 4:

Table 4. Angular displacement from one row to the next

Difference	θ_1 (°)	θ_2 (°)	θ_3 (°)	θ_4 (°)
From Row 1 to 2	42	-26	15	-7
From Row 2 to 3	38	-44	43	-35
From Row 3 to 4	10	-20	32	-48
From Row 4	-44	28	-17	8

to 5				
From Row 5 to 6	-37	44	-44	37
From Row 6 to 7	-9	18	-29	45

From each row of the table above, then, the maximum absolute value is identified, which in the case of "From Row 1 to 2" would be 42°, and in the case of "From Row 4 to 5" this is 44°. This maximum absolute value is then assigned the maximum velocity of 95°s^{-1}. The other values in the same row are assigned proportionally slower velocities such that if those very displacements are traversed according to the assigned speeds then all the angles would arrive at their respective positions in the same amount of time.

Continuing with the example from the first row, since 26° is 1.62 times smaller than the maximum of 42°, its speed would also be 1.62 times slower at 59°s^{-1}. The corresponding speeds can then be expressed as shown in table 5:

Table 5. Associated angular velocities for the sequence

Sequence Number	Angular Velocity (°s^{-1})			
	θ_1	θ_2	θ_3	θ_4
1	95	59	34	16
2	82	95	93	76
3	20	40	63	95
4	95	60	37	17
5	80	95	95	80
6	19	38	61	95

3. Results and Discussions

As displayed in the figure 4 actuating the first joint creates a moment on the first link. This moment then translates into the ends of the cross-members, and the side that tends to move backwards would "dig in" and thus the whole link would rotate about that point. In the diagram, the "dug in" ends are marked with arrows. The joint between Link 2 and 3 is actuated in the opposite direction so as to neutralize the effect of the moment on Link 3 from Link 1. The positions of the angles are also coordinated such that Links 3, 4, and 5 remain in line and only move forward along the axis of the lengthwise members.

The movement is clearly a stop-and-go movement, and not continuous nor as efficient as the serpentine locomotion. Nevertheless, even the concertina is stop-and-go, and as for narrow spaces, this gait hardly occupies more space then the width of its body.

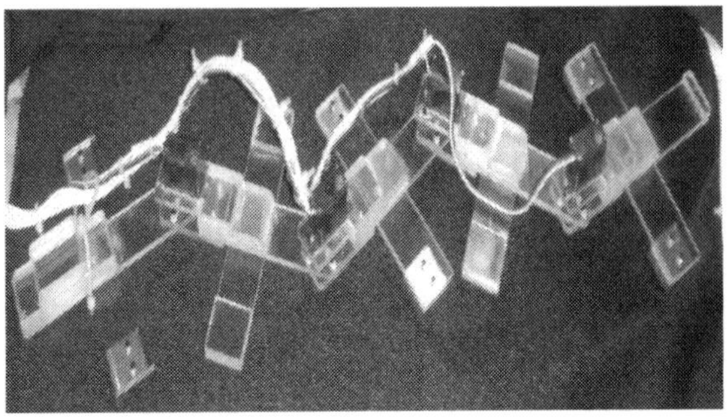

Figure 4. Snake robot navigation (top view)

4. Conclusion

A novel gait for snake robots with scales underneath has been successfully implemented using stand alone hardware. This algorithm requires less transverse movement of the snake body compared to the traditional algorithms based on serpenoid curve. Traversing narrow spaces would be possible for snake robots using this new gait.

References

[1]. Hirose S and Umetani Y 1976 *Kinematic control of active cord mechanism with tactile sensors* in Proc. 2nd RoMAnSy Symp Warsaw pp. 241-252.

[2]. Hirose S 1993 *Biologically Inspired Robots: Snake-Like Locomotors and Manipulators* Oxford University Press.

[3]. Hirose S and Morishima A 1990 *Design and control of a mobile robot with an articulated body* Int. J. Rob. Res. 9(2) pp. 99-114.

[4]. Hirose S and Mori M 2004 *Biologically inspired snake-like robots* in Proc. IEEE Int. Conf. Robotics and Biomimetics pp. 1-7.

[5]. Saito M, Fukaya M and Iwasaki T 2002 *Serpentine locomotion with robotic snakes* IEEE Contr. Syst. Mag. 22(1) pp. 64-81.

[6]. Ma S and Tadokoro N 2006 *Analysis of creeping locomotion of a snake-like robot on a slope* Autonomous Robots pp. 15-23.

[7]. Chernousko F L 2000 *Snake-like motions of multibody systems over a rough plane* in Proc. 2nd Int. Conf. Control of Oscillations and Chaos pp. 321-326.

[8]. Yamada H and Hirose S 2006 *Development of practical 3-dimensional active cord mechanism: ACM-R4* Journal of Robotics and Mechatronics 18(3) pp. 1-7.

[9]. Burdick J W, Radford J, and Chirikjian G S 1993 *A 'sidewinding' locomotion gait for hyperredundant robots* in Proc. IEEE Int. Conf. Robotics and Automation pp.101-106.

[10]. Liu C Y and Liao W H 2004 *A snake robot using shape memory alloys* in Proc. IEEE International Conference on Robotics and Biomimetics ROBIO.

[11]. Chen L., Wang Y, Ma S, and Li B *Studies on lateral rolling locomotion of a snake robot* in Proc. IEEE Int. Conf. Robotics and Automation pp.5070-5074.

[12]. Khan M R, Watanabe M, Shafie A A 2010 *Kinematics Model of Snake Robot Considering Snake Scale* American Journal of Applied Sciences 7 (5): 669-674.

Neutron Radiation Effect On 2N2222 And NTE 123 NPN Silicon Bipolar Junction Transistors

Myo Min Oo[1], N K A Md Rashid[1], J Abdul Karim[2], M R Mohamed Zin[2], N F Hasbullah[1]
[1]Department of Electrical and Computer Engineering, Department of Mechatronics Engineering International Islamic University Malaysia, Kuala Lumpur
[2]Nuclear Power Sector, Malaysian Nuclear Agency

E-mail: nfadzlinh@iium.edu.my

Abstract. This paper examines neutron radiation with PTS (Pneumatic Transfer System) effect on silicon NPN bipolar junction transistors (2N2222 and NTE 123) and analysis of the transistors in terms of electrical characterization such as current gain after neutron radiation. The key parameters are measured with Keithley 4200SCS. Experiment results show that the current gain degradation of the transistors is very sensitive to neutron radiation. The neutron radiation can cause displacement damage in the bulk layer of the transistor structure. The current degradation is believed to be governed by increasing recombination current between the base and emitter depletion region.

1. Introduction

Bipolar junction transistors (BJTs) have important applications on the modern electronics applications such as BiCMOS (Bipolar Complementary Metal Oxide Semiconductor) circuits. In additional, they are also used in the space system [1] and aircraft, spacecraft, military weapons and radiation environment. There are two kinds of neutron sources: reactor source and large accelerator-based on neutron source. Bombardment of neutrons on the semiconductor devices such as BJTs will increase the base current while degrading the current gain. Current gain degradation induced by radiation is one of the important characteristics of BJTs. Increase base current (I_B), resulting from neutron radiation exposure, has been determined to be a primarily cause of BJTs current gain degradation [2]. Typically, specific parameter degradation after radiation, such as current gain, is measured as a function of the particle influence. This degradation is due to a displacement effects in the Si bulk [3].

This study will investigate the degradation of electrical characteristics (2N2222 and NTE 123) under neutron radiation. It will focus on the radiation induced changes in the base current and in the changes of current gain in both models.

2. Experimental Details

The commercial silicon NPN BJTs of NTE123 (TO-39 metal casing) and 2N2222 (TO-18 metal casing), used as samples in this study, are audio amplifier and switch. These devices are manufactured by Multicop and NTE electronics respectively. The irradiation facility used in this study is Trigger Mark 2 at Agency Nuclear Malaysia (ANM) with pneumatic transfer system (PTS). The BJTs were prepared and labeled accordingly before the bombardment. All samples are inserted in the capped polyethylene vials before radiation and then they are exposed to radiation. The duration of irradiation is 1, 3 and 5 minutes. During irradiation, all transistors are unbiased.

All measurements of the samples were carried out in the dark using the Keithley 4200 semiconductor characterization system (SCS). The BJTs are placed on the metal 8101-4TRX test fixture to measure its current-voltage (I-V) characteristics. They were measured at room temperature before and after irradiation.

3. Results And Discussion

3.1 Current Gain

The main performance of the bipolar junction is the current gain (β) where:

$$\beta = I_C/I_B \qquad (1)$$

The current gain depends on both current of base and collector. Figures 1 and 2 show current gain (Beta) versus collector current (I_C) for silicon NPN 2N222 and NTE123 bipolar transistors irradiated for 1, 3 and 5 minutes with neutron radiation.

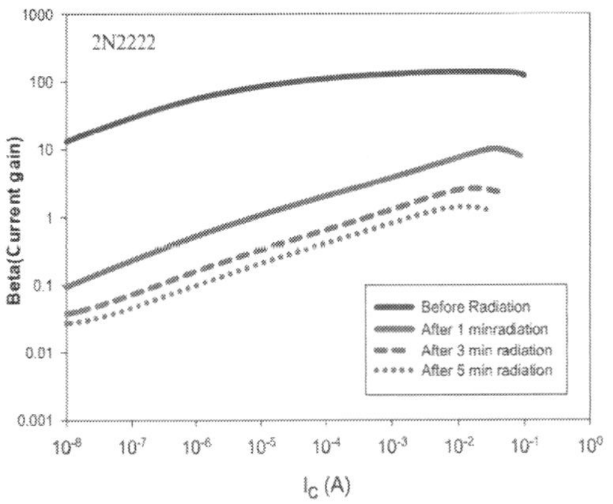

Figure 1 Plot of Current gain (Beta) versus collector current (I_C) at different irradiation time for Si NPN 2N222

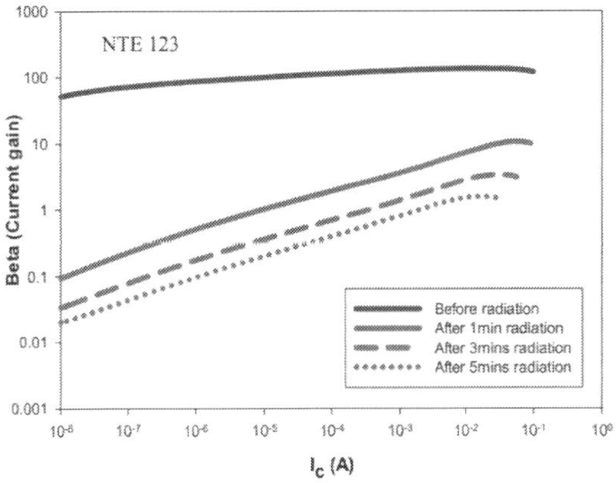

Figure 2 Plot of Current gain (Beta) versus collector current (I_C) at different irradiation time for Si NPN NTE 123

Before radiation, the gain for both types of transistor is practically constant throughout the range of the collector current as shown in Figure 1 and 2. The current gain was dramatically decreased after neutron radiation. However, at large collector current ($3x10^{-2}$A in Figure 1 and 2), there is a noticeable drop in the current. We believe that the drop is due to the high level injection or series resistance of collector [4]. The current gain of silicon 2N2222 and NTE 123 before and after radiation which is taken at constant I_C value is described in Table1.

Table 1 Variation of Current Gain of 2N2222 and NTE 123 before and after radiation at I_C=10^{-5}A and 10^{-2}A

	2N2222 Current Gain (β)				NTE 123 Current Gain (β)			
Collector Current (I_C)	Before radiation	After 1 minute radiation	After 3 minutes radiation	After 5 minutes radiation	Before radiation	After 1 minute radiation	After 3 minutes radiation	After 5 minutes radiation
$1x10^{-5}$ A	86.6	1.09	0.33	0.21	102	1.04	0.36	0.19
$1x10^{-2}$ A	138	8.17	2.54	1.39	133	7.74	2.96	1.53

IOP Conf. Series: Materials Science and Engineering **53** (2013) 012013 doi:10.1088/1757-899X/53/1/012013

Figure 3 and 4 show negligible changes in collector current before and after radiation. The collector current has not significantly changed before and after radiation for both types of transistors: 2N2222 and NTE123.

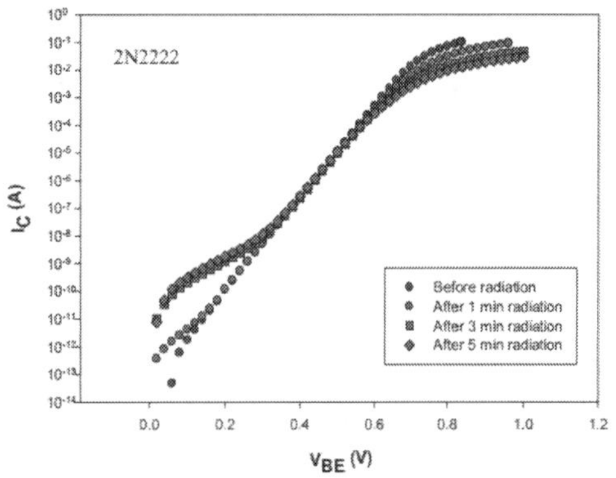

Figure 3 Plot of collector current (I_C) versus base-emitter voltage (V_{BE}) at different irradiation time for Si NPN 2N2222

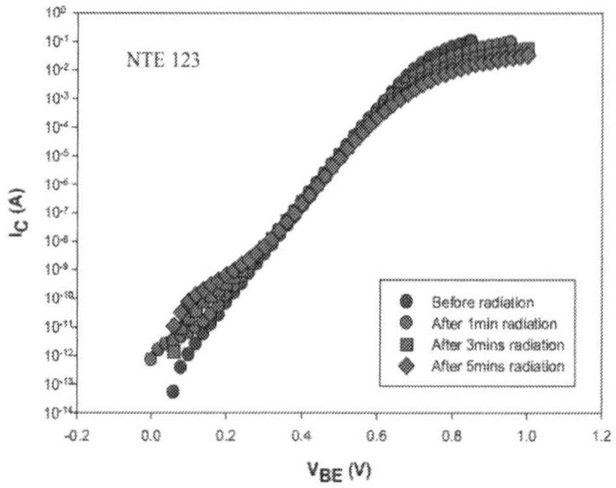

Figure 4 Plot of collector current (I_C) versus base-emitter voltage (V_{BE}) at different irradiation time for Si NPN NTE 123

Hence, the possible factor to explain the degradation of current gain is the base current.

3.1 Base Current

Figure 5 and 6 show the plot of base current (I_B) against base-emitter voltage (V_{BE}) for the silicon NPN 2N2222 and NTE123 bipolar transistors irradiated for 1, 3 and 5 minutes with neutron radiation. The base current can be seen to increase after radiation, especially at lower V_{BE}.

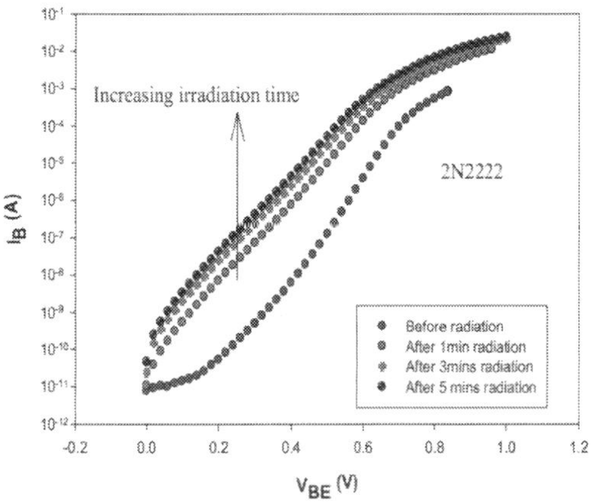

Figure 5 Plot of base current (I_B) versus base-emitter voltage (V_{BE}) at different irradiation time for Si NPN 2N2222

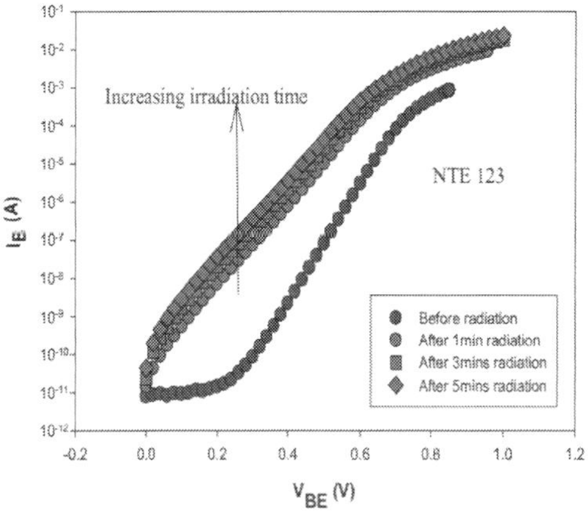

Figure 6 Plot of base current (I_B) versus base-emitter voltage (V_{BE}) at different irradiation time for Si NPN NTE 123

The base current taken at constant V_{BE} before and after radiation is stated in the following Table 2. From Table 2, it can be seen that the base current increased significantly after radiation.

Table 2 Variation of base current of 2N2222 and NTE 123 before and after radiation at V_{BE}=0.4V and 0.7V

Base-emitter voltage (V_{BE})	2N2222 I_B (A)				NTE 123 I_B (A)			
	Before radiation	After 1minute radiation	After 3 minute radiation	After 5 minutes radiation	Before radiation	After 1 minute radiation	After 3 minutes radiation	After 5 minutes radiation
0.4 V	$6.14x10^{-9}$	$7.49x10^{-7}$	$2.44x10^{-6}$	$4.28x10^{-6}$	$2.18x10^{-9}$	$7.70x10^{-7}$	$2.17x10^{-6}$	$3.68x10^{-6}$
0.7 V	$9.39x10^{-5}$	$9.35x10^{-4}$	$1.78x10^{-3}$	$2.29x10^{-3}$	$8.09x10^{-5}$	$9.57x10^{-4}$	$1.68x10^{-3}$	$2.16x10^{-3}$

The neutron can induce the ionization in the oxide layer and displacement damage in the Si bulk, both of which will cause the current gain degradation of NPN BJTs. The displacement damage could produce stable vacancy defect that are effective the centers of recombination and trapping, leading to a decrease minority carrier lifetime. The degradation of minority carrier lifetime makes the semiconductor device properties decreased such as current gain [5]. The displacement damage which is occurred by neutron radiation in the base-emitter junction space charge region and neutral base region also gives rise to an increased recombination base current. The base current increases significantly while the collector current is not affected by radiation [6]. The ionization damage could cause the interface traps and net positive charge in the oxide overlying the emitter-base junction, leading to an increase in the base current. However, ionizing damage is also a factor for failure mechanism.

The excess base current contributed to the recombination base current in the emitter-base depletion region is defined as

$$\Delta I_B = I_B - I_{Bo}\ , \qquad (2)$$

where I_{Bo} is the pre-radiation base current and I_B is the post- radiation base current [7-9].

The recombination current increases in ionizing damage because of the two interacting effects: 1)spreading of the emitter-base depletion layer and 2)increased surface recombination velocity (v_{sr}) which is proportional to the formation of the generation-recombination centers at Si/SiO_2 interface covering EB junction [10].

After radiation, the depleted surface area of NPN transistors increases because of the buildup of the positive trapped charge in the oxide [11]. Consequently, the increasing of depleted surface causes an increase in the surface recombination current. The more recombination current is in the device, the more base current flows at a given base-emitter voltage. As long as the oxide charge increases, the depletion layer moves out further into the base until the entire base region becomes depleted.

The surface recombination velocity increases because of the increasing of the interface traps at the silicon-silicon dioxide interface. These traps do not contribute to the surface current while they are in the un-depleted surfaces. However, they become the effective recombination centers when they break into the depleted layers [12]. Consequently, there is a relation between the positive trapped oxide charge and the interface traps. As the positive charge increases, the depletion region increases in which more recombination centers are exposed due to interface traps. Therefore, as long as the oxide charge and interface traps increase, base current also increases. Hence, the current gain degrades when the base current increases.

4. Conclusion

The current gain of silicon NPN BJTs 2N2222 and NTE 123 is found to increase linearity after neutron radiation. The results in this paper presented that both types of silicon transistors show a significant degradation of the current gain and also an increase in the base current after neutron bombardment. This is believed due to an increase in the base recombination current where generation recombination is occurred as the result of the displacement damage in the bulk layer of the transistor structure.

Acknowledgments

Authors would like to thank the cooperation of staffs from the Nuclear Power Division and Malaysia Nuclear Agency in providing the facilities.

5. References

[1] Claeys C and Simoen E 2002 *Radiation Effects in Advanced Semiconductor Materials and Devices*: Springer)

[2] Enlow E W, Pease R L, Combs W, Schrimpf R D and Nowlin R N 1991 Response of advanced bipolar processes to ionizing radiation *Nuclear Science, IEEE Transactions on* **38** 1342-51

[3] Srour J R, Marshall C J and Marshall P W 2003 Review of displacement damage effects in silicon devices *Nuclear Science, IEEE Transactions on* **50** 653-70

[4] Neudeck G W 1989 *The bipolar junction transistor*: Addison-Wesley)

[5] Xingji L, Chaoming L, Erming R, Hongbin G and Jianqun Y 2012 Simultaneous and Sequential Radiation Effects on NPN Transistors Induced by Protons and Electrons *Nuclear Science, IEEE Transactions on* **59** 625-33

[6] Vuppala S, Li C, Zwicknagl P and Subramanian S 2003 Neutron, proton, and electron irradiation effects in InGaP/GaAs single heterojunction bipolar transistors *Nuclear Science, IEEE Transactions on* **50** 1846-51

[7] Kosier S L, Shrimpf R D, Nowlin R N, Fleetwood D M, DeLaus M, Pease R L, Combs W E, Wei A and Chai F 1993 Charge separation for bipolar transistors *Nuclear Science, IEEE Transactions on* **40** 1276-85

[8] Hart A R, Smyth J B, Van Lint V A J, Snowden D P and Leadon R E 1978 Hardness Assurance Considerations for Long-Term Ionizing Radiation Effects on Bipolar Structures *Nuclear Science, IEEE Transactions on* **25** 1502-7

[9] Kosier S L, Wei A, Schrimpf R D, Fleetwood D M, DeLaus M D, Pease R L and Combs W E 1995 Physically based comparison of hot-carrier-induced and ionizing-radiation-induced degradation in BJTs *Electron Devices, IEEE Transactions on* **42** 436-44

[10] Schrimpf R D 1996 Recent advances in understanding total-dose effects in bipolar transistors *Nuclear Science, IEEE Transactions on* **43** 787-96

[11] Johnston A H, Swift G M and Rax B G 1994 Total dose effects in conventional bipolar transistors and linear integrated circuits *Nuclear Science, IEEE Transactions on* **41** 2427-36
[12] Nowlin R N, Enlow E W, Schrimpf R D and Combs W E 1992 Trends in the total-dose response of modern bipolar transistors *Nuclear Science, IEEE Transactions on* **39** 2026-35

IOP Conf. Series: Materials Science and Engineering **53** (2013) 012014 doi:10.1088/1757-899X/53/1/012014

Simplified Heat Generation Model for Lithium ion battery used in Electric Vehicle

Nur Hazima Faezaa Ismail [1], **Siti Fauziah Toha** [2], **Nor Aziah Mohd Azubir** [1], **Nizam Hanis Md Ishak** [1], **Mohd Khair Hassan** [3], **Babul Salam KSM Ibrahim** [4].

[1]nurhazimafaezaa@yahoo.com, [2]tsfauziah@iium.edu.my

Abstract. It is known that temperature variations inside a battery may greatly affect its performance, life, and reliability. In an effort to gain a better understanding of the heat generation in Lithium ion batteries, a simple heat generation models were constructed in order to predict the thermal behaviour of a battery pack. The Lithium ion battery presents in this paper is Lithium Iron Phosphate ($LiFePO_4$). The results show that the model can be viewed as an acceptable approximation for the variation of the battery pack temperature at a continuous discharge current from data provided by the manufacturer and literature.

1. Introduction

Electric and hybrid vehicle may present the most promising solution for pollution problems caused by the emission of conventional internal combustion engine. These technologies will highly depend on battery packs. The role of battery as the power source in electric drive vehicle is important. Due to the high energy density, Lithium-ion batteries have gained much attention as a viable candidate to increase vehicle range and performance of electric drive vehicle application. However, there is a variety of thermal limitations for Lithium-ion battery. Most common thermal issues related to Lithium-ion batteries are capacity/power fade, selfdischarge, thermal runaway, pack electrical imbalance and cold temperature performance [1]. The battery positive electrode and electrolyte chemistry is said to have an impact of capacity/power fading. Batteries also generate heat during charge and discharge due to enthalpy changes, electrochemical polarization and resistive heating inside the cell [2, 3]. Temperature variation inside the batteries can lead to uneven temperature distribution which creates uneven charge/discharge behaviour within the pack [4]. Worst cases, thermal runaway may occur in a cell when heats are not properly control and possible to cause fire and explosion. For these reason, the need for battery thermal management is vital for electric and hybrid vehicles in order to keep the vehicle at its optimum performance. This paper provides an introduction on the theory and objectives of battery thermal management. The simplified heat generation model was developed to estimate the heat generation inside the battery will be discussed.

2. Basic information of Lithium-ion battery

[1] Research Department, Perusahaan Otomobil Nasional Sdn Bhd.
[2] Department of Mechatronics, International Islamic University Malaysia.
[3] Department of Electrical and Electronic, University Putra Malaysia.
[4] Department of Mechatronics and Robotic Engineering, University Tun Hussein Onn Malaysia.

2.1.1. Battery structure

The basic structure of a battery consists of five major components as shown in Figure 1. There are electrodes (anode and cathode), a separator that prevents electron flow between the electrodes, a current collector on the outside of each electrode, and a case or enclosure. Electrons enter and leave the current collectors via current collector tabs, which connect the battery to the external circuit that contains the load to be powered or the source for battery charging.

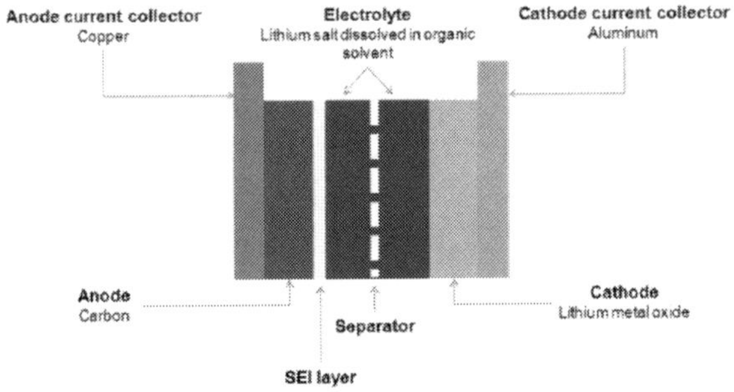

Figure 1.Battery basic structure

2.1.2. Chemical reactions

The Lithium-ion battery discussed in this paper is Lithium Iron Phosphate (LiFePO4). Generally the anode material for LiFePO4 is a form of carbon while the cathode material is LiFePO4 material. During charge or discharge, the reactions occurring in the two electrodes for discharging can be written as in equation (1) and for charging as in equation (2)

$$LiFePO_4 - xLi+ - xe - xFePO_4 + (1\text{-}x)\ LiFePO_4 \quad (1)$$

$$FePO_4 + xLi+ + xe - xLiFePO_4 + (1\text{-}x)\ FePO_4 \quad (2)$$

During the Lithium-ion battery discharge, Lithium ions (Li+) are released from the anode and travel through electrolyte toward the cathode. When the Lithium ions (Li+) reach the cathode, they are quickly incorporated into the cathode material.

2.1.3. Heat generation inside a battery

It is important to understand how heat generated inside a battery. Heat is produced in batteries from two sources; electrochemical operation and Joule heating [5-7]. Reference [8] found a way to calculate the battery heat using a thermodynamic energy balance and cited frequently in the literature in its simplified form is shown in equation 3 below where the first term is the heat generation due to Joule heating and the second term is the heat generation due to entropy changes.

$$q = I(U - V) - I(T\frac{dU}{dT}) \quad (3)$$

3. Cell modelling

To derive the model it relies heavily on the fundamental of heat transfer. Deriving model for heat transfer is inherently challenging due to the complexity of battery chemistry and composition. Thus, many assumptions and simplification must be made. The Lithium-ion battery considered in this study is a $LiFeYPO_4$ (WB-LYP40AHA).Their main properties can be found in Table 1.

Table 1. Battery specification

Type	Nominal Capacity (Ah)	Nominal Voltage (V)	Upper cut-off voltage (V)	Lower cut-off voltage (V)	Operating Temperature (T)
$LiFeYPO_4$	40	3.7	4	2.8	-45°C ~ 85°C

The simulation work for this study was done based on PROTON's electric drive vehicle demonstrator battery's pack. The battery pack comprises of 12 modules in series. 4 modules were arranged 10 cells in series while the rest of the 8 modules were arranged 9 cells in series. As such, there will be a total of 112 cells.

3.1.1. Governing equation
The heat transfer inside the battery can be divided into three parts, i.e., the heat generated by the cell's internal resistance, the change of entropy that occur in the cell components during discharge and heat transferred to ambient conditions by convection [3] and could be written as equation (4)

$$mc_{cell}\frac{dT_{cell}}{dt} = I^2 R + T_{cell}\,\Delta S\,\frac{I}{nF} + Ah(T_{cell} - T_{amb}) \tag{4}$$

Use of this equation assumes that the cell temperature is uniform. Hence a lumped capacitance model must be assumed in order to apply the equation correctly. In heat transfer theory, the validity of this assumption through the use of the Biot number. It is a dimensionless number that compares heat transfer occurring inside a body and at its surface using equation (5)

$$B_i = \frac{hL_c}{k} \tag{5}$$

The characteristic length was calculated with the manufacturer's cell specification. For thermal conductivity and heat transfer coefficient, typical values were taken from literature [9], gave k = 3 W/mK.

3.1.2. Parameter values

Typical values from literature were used for the value of the entropy change [10]. ΔS= -30 J/molK. ΔS can be obtained in the function of state of charge in simplified as a second order polynomial as in equation (6). Table 2 describes the model parameters use.

$$\Delta S = -179(SOC^2) + 274.29(SOC) - 116 \tag{6}$$

Table 2. Model parameters

Parameter	Unit	Value
Mass	kg	1.5
Specific heat capacity	J/kg.K	1350
Internal resistance	Ω	0.02
Entropy change	J/molK	-30
Surface area	m^2	0.21228
Heat transfer coefficient	W/m^2K	18.9

4. Results and discussion

The output data for the Matlab model is shown in Figure 2 for 3A discharge current. Assumed within the single cell Matlab model was that the cell transferred heat from its entire surface and it was in an environment where the ambient conditions were constant. The discharge process was also treated as being 100% efficient irrespective of the discharge current used.

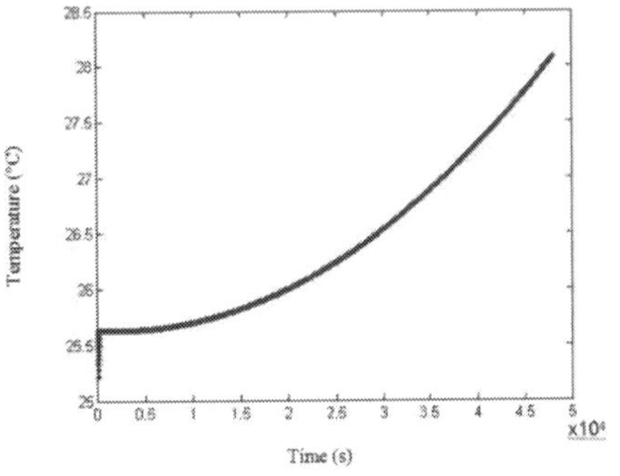

Figure 2. Cell temperature for 3A discharge current in Matlab simulation

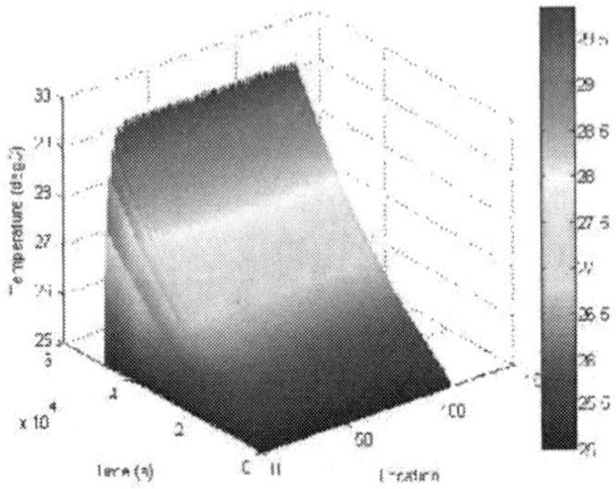

Figure 3. Matlab results for modelling the battery pack

This single cell model was then used to construct a second Matlab model for a battery module. Whilst each of the two battery modules was contain 112 cells in series, symmetry allows for only the first 56 to be modelled. The result of this simulation is shown in Figure 3, where the peak cell temperature for the central cell at the end of the 3A discharge is 29.8°C based on an initial ambient temperature of 25°C.

5. Conclusion

The simplified battery model may not be accurate enough to represent the thermal behaviour of Lithium ion battery. There is a great scope for improvement and future work in this area of battery heat generation modelling. If an experiment investigating the cell temperature can be performed in advanced, the Matlab model for the single cell could have been altered to include the actual data to generate acceptable results.

6. References

[1] Bandhauer T M, Garimella S and Fuller T F 2011 A Critical Review of Thermal Issues in Lithium-ion Batteries *J. Electrochem. Soc.* **158** R1-R25
[2] Sato N 2011 Thermal behaviour analysis of lithium-ion batteries for electric and hybrid vehicles *J. Power Sources* **99** 70-77
[3] Onda K, Ohshima T, Nakayama M, Fukuda K and Araki T 2006 Thermal behavior of small lithium-ion battery during rapid charge and discharge cycles *J. Power Sources* **158** 535-542
[4] Pesaran A A 2001 *Advanced Automotive Battery Conference* Las Vegas,Nevada
[5] Benger R, Wenzl H, Beck H, Jiang M, Ohms D and Schaedlich G 2009 Electrochemical thermal modeling of lithium-ion cells for use in HEV or EV application *World Electric Vehicle Journal* **3** Stavenger, Norway
[6] Jeon D H and Baek S M 2011 Thermal modeling of cylindrical lithium ion battery during discharge cycle *Energy Conversion and Management* **52** 2973-81
[7] Wang F and Li M 2010 Thermal performance analysis of the Lithium-ion Batteries International Conference on Parallel and Distributed Computing, Applications and Technologies (PDCAT) (Wuhan, 8-11 Dec. 2010) 483-486
[8] Bernadi D, Powlikowski E and Newman J 1985 A general energy balance for battery systems *J. Electrochem. Soc.* **132** 5-12

[9] Khateeb S A, Amiruddin S, Farid M, Selman J R and Al-Hallaj S 2005 Thermal management of Li-ion battery with phase change material for electric scooters: experimental validation *J. Power Sources* **142** 345-353
[10] Villano P, Carewska M and Passerini S 2003 Specific heat capacity of lithium polymer battery components *Thermochemica Acta* **402** 219-224

Acknowledgments
The authors acknowledge the Ministry of Higher Education Malaysia for the financial support under the Exploratory Research Grant Scheme (EFGS): ERGS12-033- 0033.

Model building of thermoelectric generator exposed to dynamic transient sources

A Md Yusop[1], R Mohamed and A Ayob
Department of Electrical, Electronic and Systems Engineering, Faculty of Engineering and Built Environment, Universiti Kebangsaan Malaysia, 43600 UKM Bangi, Selangor, Malaysia

E-mail: azdiana@siswa.ukm.edu.my

Abstract. This paper presents the modeling of thermal and power generation behavior of a thermoelectric generator (TEG) exposed to transient sources. Most of the previous research concerned the analysis for steady-state behavior which only involves constant temperature value. However, in practice, the temperature of the TEG input fluctuates with time. Therefore this research will look into a focal point on transient heat sources that is being supplied to the hot junction with natural convection cooling process at the cold junction for single and multiple configuration of TEG. The model obtained the data from existing experiments with predicted various conditions of temperature, heat gradient, internal resistance and current attribute of TEG. Transient analysis on single TEG has shown that the value of Seebeck coefficient, thermal conductivity and figure-of-merit vary with the value of cold side temperature. When the ratio between the load and the internal resistance increases, the voltage increases. By considering the multiple TEGs, the matched voltage shows different values when the number of cascaded TEGs is varied. The simulation results have proven that the variation in the number of cascaded TEGs can be used to determine the output power characteristics of a TEG.

1. Introduction

The pressure to reduce our dependence on fossil fuel in producing electricity is getting greater day by day. Thus, the need for alternative power generation has attracted wide research interest. In the present days, there are many solutions to alternative power generation; for example solar energy, hydro energy, nuclear energy and wind energy. Each of these types of energy has its own inadequacy. Although solar energy is the most commonly used type of renewable energy, it needs extra effort on designing a method of storing energy for later use since its limitation is due to the sunlight availability. Nuclear sources are able to supply sufficient amount of energy to produce electricity but its hazardous waste is very harmful to the environment. This paper will discuss the use of thermoelectric device to overcome the limitations of current alternative energy sources.

Thermoelectric generator (TEG) is a solid state device which transforms thermal energy into electrical energy when there is thermal gradient between its two legs. Due to the absence of moving parts when converting thermal energy into electrical energy, such device can operate for a long period of time with free maintenance. Recently, this device has become a favorable solution in energy conversion since it produces no waste matter during the conversion process. Nuwayhid et.al has used

[1] To whom any correspondence should be addressed.

TEG with domestic woodstove to convert the waste heat of the woodstove to beneficial power to be used in rural places where electric power is subjected to long-drawn-out interruptions [1].

This paper is aimed to conduct a transient and steady state analysis for a single TEG configuration. Previous studies have shown that when it comes to multiple TEGs, researches are mainly focused on the steady state analysis. In [2] the author converts some of the condenser wasted energy to electricity by using Bi_2Te_3 as thermoelectric material and also focuses on the steady state analysis. According to Nguyen and Pochiraju in [3], although there are a number of papers discussing TEG modules, including analytical and numerical steady state models, fully coupled and complete transient analyses of TEG are often not been discussed. Crane [4] also discussed on the transient operating model for the application of segmented thermoelectric elements. An examination on the behavior of a TEG has been studied by Carmo et.al [5] and Hsu et.al [6] but still solitude on single TEG.

To ensure a reliable model, equivalent circuit model of the TEG had been developed using PSPICE software [7]-[10]. TCAD software has been used in [11] to model the TEG using its three dimensional properties. For the design controller, MATLAB/SIMULINK software is a choice because of its ability to model the controller using control principle theory. In addition, the control strategy can be built by using built-in control blocks combined with the use of SimPowerSystem tools for the Maximum Power Point Tracking (MPPT) circuit development. Tsai and Lin [12] has used MATLAB for modeling the thermoelectric module which focused on the modeling of both thermoelectric cooler (TEC) and TEG.

2. Thermoelectric module principle

The basic unit of TEG consists of an array of p-type and n-type thermocouples which is connected electrically in series and thermally in parallel. This structure can be used to convert thermal energy into electricity when heat is applied to one side of the TEG. At this time, electrons in the n-type semiconductor and the holes in the p-type semiconductor will move away from the heating source. This will increase the electrical current produced by the movement of the electrons and holes. From figure 1 it can be seen when the heat flows across the junction, electrical current will be created. A TEG can be connected either in series to increase the output voltage or in parallel to increase the output current.

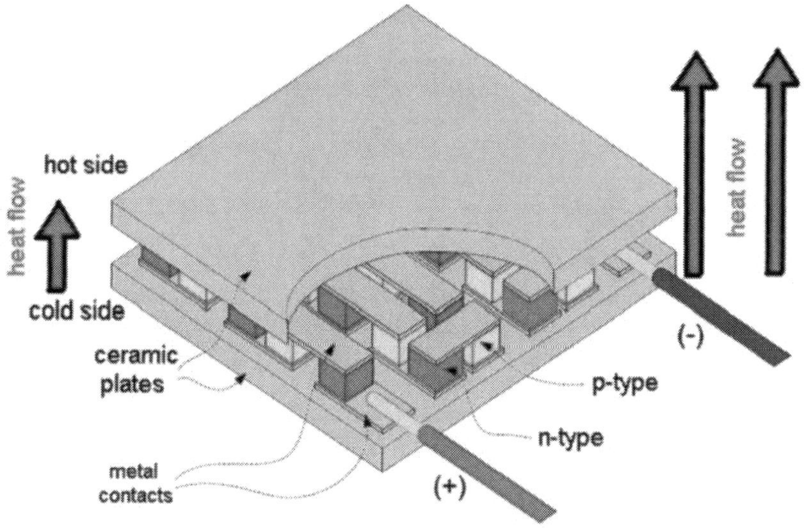

Figure 1. TEG electrical current production [13].

There are several effects that are taken into account in order to obtain the heat modelling of the TEG including thermal conduction, Joule, Peltier, Seebeck and Thomson effects. This research will only focus on thermal conduction, Joule, Peltier and Seebeck effect to encounter the extrinsic temperature of the TEG which includes hot temperature, T_h and cold temperature, T_c. In this TEG modelling, the additional Thomson effect which is described by the derivative of the Seebeck effect in time function is neglected since the value is too small.

TEG has high thermal conductivity to ensure the heat will be distributed equally at both end of the leg. The thermal conductivity follows the Fourier process with its heat transfer, Q_{tc} described by

$$Q_{tc} = -\Delta T \kappa_{tc} \tag{1}$$

where κ_{tc} is the thermal conductivity and ΔT is the difference between the hot side and cold side temperature. Joule effect is generated internally when electrical current, I is flowing through thermoelectric leg. This effect is on both hot side and cold side with the same amount of energy as

$$Q_{joule} = I^2 R \tag{2}$$

where R is the electrical resistance. The Peltier effect is the heating effect when electrical current is flowing through two dissimilar junctions and the total heat transfer is represented by

$$Q_{peltier} = \alpha \Delta T I \tag{3}$$

Where α is the Seebeck coefficient of the TEG. A phenomenon in which a temperature difference between two dissimilar electrical conductors or semiconductors produces a voltage difference between its two junctions is known as Seebeck effect. Seebeck coefficient is also defined as

$$\alpha = \frac{V}{\Delta T} \tag{4}$$

When designing the TEG for a certain condition of thermal flow and temperature, it is possible to obtain the maximum output power. This will happen when load resistance equals to the internal resistance of the TEG. Equation (5) and (6) shows energy balance equations for steady state analysis at the hot and cold junction of the TEG.

$$Q_h = \alpha T_h I - \kappa_{tc} \Delta T - 0.5 I^2 R \tag{5}$$

$$Q_c = \alpha T_c I - \kappa_{tc} \Delta T + 0.5 I^2 R \tag{6}$$

A good TEG must have a large Seebeck coefficient together with low electrical resistance and low thermal conductivity. The term that relates these three parameters is known as figure-of-merit, Z, and is expressed as

$$Z = \frac{\alpha^2}{R \kappa_{tc}} \tag{7}$$

3. TEG's electrical parameter determination

The following parameters are used to specify TEG characteristics: T_h, the hot side temperature; T_c, the cold side temperatures; W_m, the power at the load resistance, R_L matched to the internal resistance, R where (R_L=R) and V_m, the matched load voltage. The electrical resistance and Seebeck coefficient value is defined as

$$R = R_L = \frac{V_m^{\;2}}{W_m} \tag{8}$$

and

$$\alpha = \frac{2V_m}{\Delta T} \tag{9}$$

The load resistance can be varied in proportion to its internal resistance with the following relationship

$$R_L = mR \tag{10}$$

where m is the ratio between the load resistance and internal resistance. From equation (4), the current expression can be defined as

$$I = \frac{\alpha \Delta T}{(1+m)R} \tag{11}$$

The equation will reflect onto the relationship between the internal and the load resistance. As the current varies with the value of m, voltage also varies

The most important part before designing any controller or MPPT circuit, is to maximize the output power at matched load, R_L=R. The maximum current of TEG is the short-circuit current at zero load voltage, V_L=0 which referred as

$$I_{short-ckt} = 2I_m = \frac{2W_m}{V_m} \tag{12}$$

Based on Ohm's Law and the resulting equations of (11) and (12), the TEG voltage can be obtained as

$$V = -R(I - I_{short-ckt}) \tag{13}$$

4. Model building and simulation analysis

The steady-state analysis is done to verify the electrical parameters of the TEG model. The specifications of the TEG module (TEP1-12656-0.6) are listed in table 1. By applying equations (1) to (7), one may obtain the model parameters for steady state condition of the followings; α=0.031 V/K, R=1.2 Ω, κ_{tc}=20.85 W/K and Z=3.869x10^{-5} K^{-1}.

Table 1. Specifications of TEG module (TEP1-12656-0.6).

Specifications	Values
Hot side temperature (°C)	300
Cold side temperature (°C)	30
Open circuit voltage (V)	8.4
Matched load resistance (Ω)	1.2
Matched load output voltage (V)	4.2
Matched load output current (A)	3.4
Matched load output power (W)	14.6
Heat flow across the module (W)	≈365
Heat flow density (Wcm^{-2})	≈11.6

4.1. Transient analysis of a single TEG

Cold side temperature data from [3] is used for the transient analysis of both single TEG and multiple TEGs arrangement. The hot side temperature is fixed at 115 °C. Figure 2 shows the variation of the temperature on its cold side. The cold side temperature is gradually increased from 25 °C to 75 °C and is maintain at that value until the end of the simulation time.

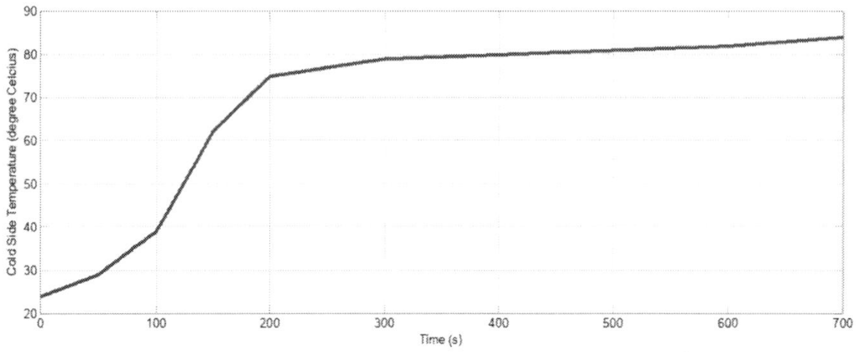

Figure 2. Temperature of the cold side of the TEG.

The TEG model is implemented using MATLAB/SIMULINK as shown in figure 3. The inputs of the model are the cold side temperature and the hot side temperature. For the transient analysis, the value of α, κ_{tc} and Z vary with the value of T_c in which the effect is shown in figure 4, 5 and 6.

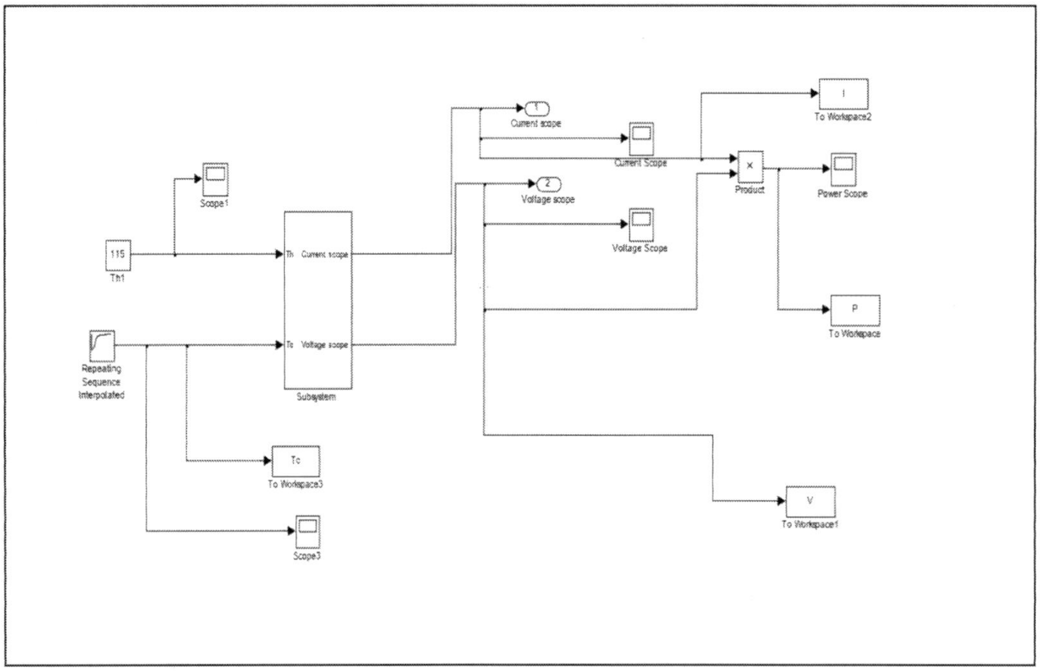

Figure 3. TEG block model as applied in MATLAB.

From the figure, the waveform behaviour of α, κ_{tc} and Z follows the nature of the cold side temperature. As the cold side temperature increase, the value of α, κ_{tc} and Z will also increase since all these values are proportional to the temperature gradient of the TEG.

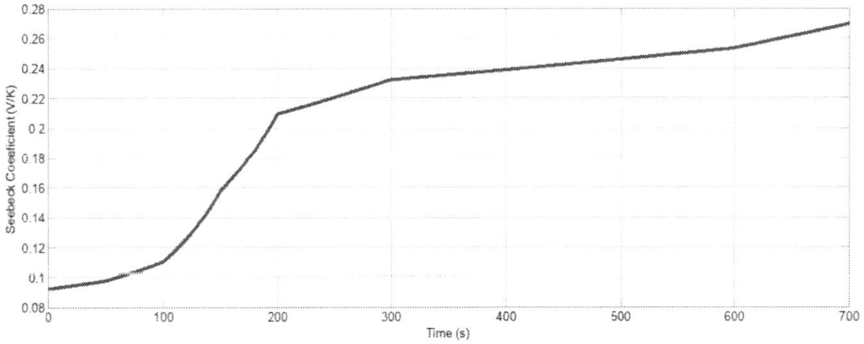

Figure 4. Seebeck coefficient of the TEG.

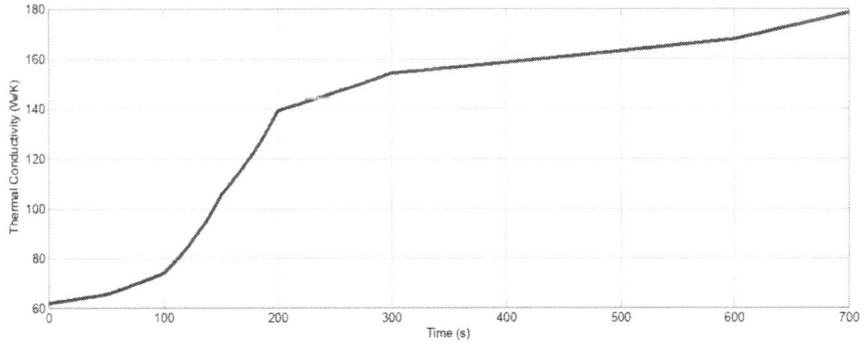

Figure 5. Thermal conductivity of the TEG.

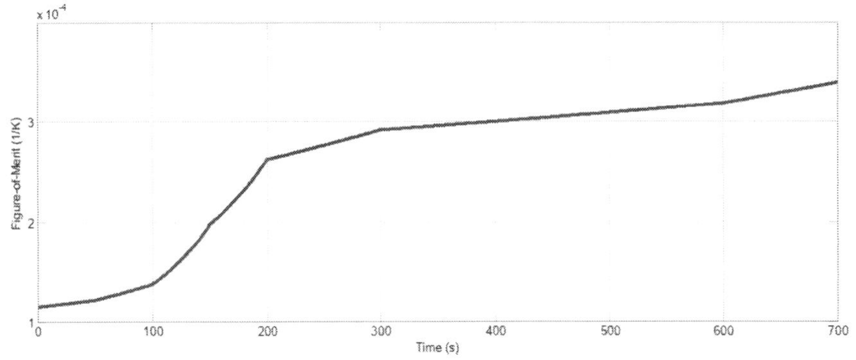

Figure 6. Figure-of-merit of the TEG.

The PI and VI characteristics of the single TEG with transient sources is shown in figure 7. This result reveals that the maximum power is 14.6 W at matched load current of 3.4 A and matched load voltage of 4.2 V. The PI and VI characteristics of the TEG are still the same even when the hot side temperature varies. This results show that the converter of the MPPT circuit must follow the voltage, current and power attribute of the selected TEG.

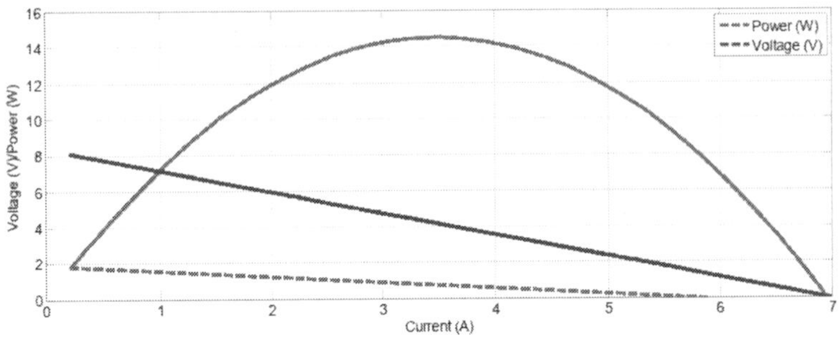

Figure 7. PI and VI characteristics of TEP1-12656-0.6.

An analysis has been carried out to see the effect of the voltage when the ratio between the load and internal resistance is varied. Figure 8 shows the voltage at four load resistance values: *m2=m*, *m2=2m*, *m2=1.5m* and *m2=0.5m* where *m* is the actual ratio and *m2* is the new ratio. Simulation result shows that the voltage increases as the ratio increases.

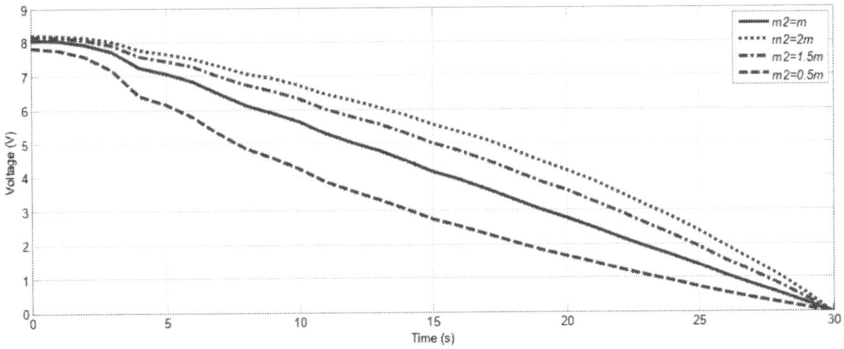

Figure 8. TEG voltage at various load and internal resistance ratios.

In terms of current attributes, when the ratio between load and internal resistance increases, the current decreases. This fact can be clearly seen in figure 9 where the analysis is done at four load resistance values: *m2=m*, *m2=2m*, *m2=1.5m* and *m2=0.5m* where *m* is the actual ratio and *m2* is the new ratio.

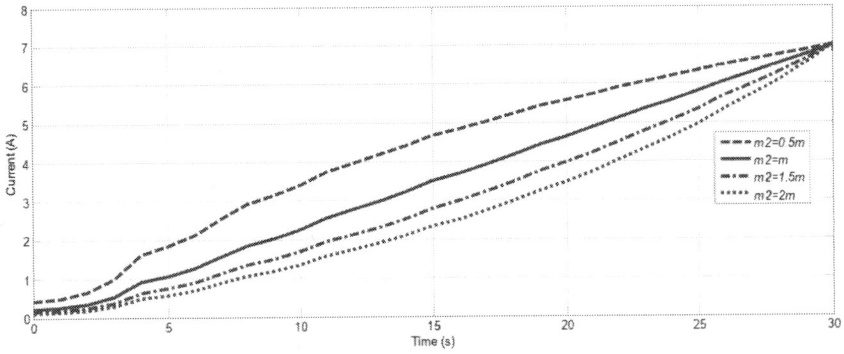

Figure 9. TEG current at various condition of load and internal resistance ratios.

4.2 Transient analysis of a multiple TEG

This section will describe the characteristics of the multiple TEGs where three and six TEGs are connected in series. The matched voltage depicted from figure 10 shows a diverse value when different numbers of TEGs are connected electrically in series and thermally in parallel. In contrast, the matched current in figure 11 gives the same value as the single TEG. When the TEGs are cascaded, it will sum up all the single TEG voltage that can be expressed as

$$V_i = i \times V \tag{14}$$

where i is the number of cascaded TEGs. Since power is directly proportional to the voltage, the total power will be the maximum power of the single TEG multiplies by the factor of i. For TEG (TEP1-12656-0.6), the total power will be 43.8 W and if there are six TEGs cascaded together the total power can be increased up to 87.6 W. These results are shown in figure 10.

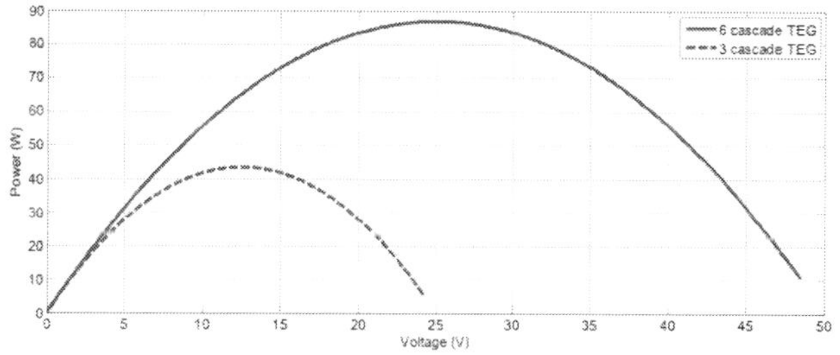

Figure 10. PV characteristics of multiple TEGs.

By considering the same number of cascaded TEGs, the PI characteristics of both cases show the same manners as can be seen from figure 11. With this validation, it is proven that cascaded TEGs will not give any effect to the current value. The only parameter that can be adjusted is the output power since it reflects to the number of cascaded TEGs.

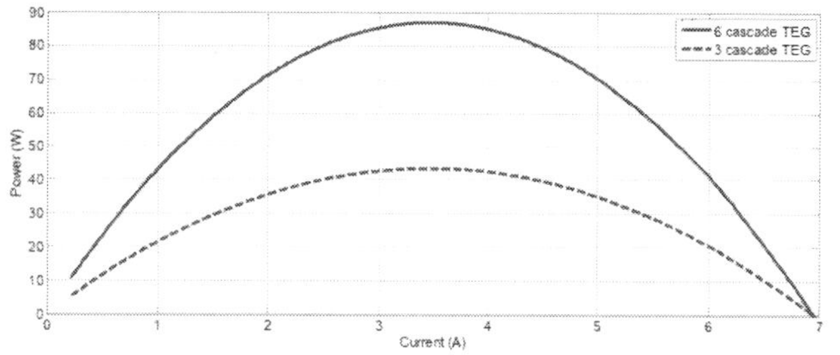

Figure 11. PI characteristics of multiple TEGs.

5. Conclusion

A formulation of the TEG behavior which includes thermal behavior and electrical properties has been developed using MATLAB/SIMULINK. This model is used to determine the output power characteristics of both single and multiple TEGs. The output characteristics are mainly used for transient analysis where the cold side temperature is varied with the simulation time.

The analysis with various conditions of temperature, heat gradient, internal resistance and current attribute of TEG also has been discussed. Simulation model using available TEG parameters has been proven to be the same as the provided parameters by the manufacturer. The results show that the prediction with various input parameters was able to determine the output power characteristics of a TEG.

For power condition, the steady state voltage is half of the peak voltage of the TEG. The hot side temperature of the TEG must be controlled to hold the temperature gradient at this optimal point. In this case, a suitable controller need to be accurately designed to maintain the output power behavior as outlined in this paper.

Acknowledgements
The authors would like to thank the Department of Electrical, Electronic and Systems Engineering, Faculty of Engineering and Built Environment, Universiti Kebangsaan Malaysia (UKM), the Universiti Teknikal Malaysia Melaka (UTeM) and the Ministry of Higher Education for moral, operational and financial support throughout the project.

References

[1] Nuwayhid R Y, Shihadeh A and Ghaddar N 2005 Development and testing of a domestic woodstove thermoelectric generator with natural convection cooling *J. Energy Conversion and Management* **46** 1631-43
[2] Rafiee M, Siadatan A, Afjei E and Abadi E Z A 2012 Improving the efficiency of thermal power plant using thermoelectric material *4th Int. Conf. on Intelligent and Advanced Systems*
[3] Nguyen N Q and Pochiraju K V 2013 Behavior of thermoelectric generators exposed to transient heat sources *J. Applied Thermal Engineering* **51** 1-9
[4] Crane D T 2011 An introduction to system-level, steady-state and transient modeling and optimization of high-power-density thermoelectric generator devices made of segmented thermoelectric elements *J. Elec Materi* **40** 561-9
[5] Carmo J P, Antunes J, Silva M F, Ribeiro J F, Goncalves L M and Correia J H 2011 Characterization of thermoelectric generators by measuring the load-dependence behavior *J. Measurement* **44** 2194-9
[6] Hsu C T, Huang G Y, Chu H S, Yu B and Yao D J 2011 An effective Seebeck coefficient obtained by experimental results of a thermoelectric generator module *J. Applied Energy* **88** 5173-9
[7] Chavez J A, Ortega J A, Salazar J, Turo A and Garcia M J 2000 SPICE model of thermoelectric elements including thermal effects. *Proc. Conf. on Instrumentation and Measurement Technology Conference*
[8] Mitrani D, Tome J A, Salazar J, Turo A, Garcia M J and Chavez J A 2005 Methodology for extracting thermoelectric module parameters *Instrumentation and Measurement IEEE Transactions* **54** 1548-52
[9] Lineykin S and Ben-Yaakov S 2007 Modeling and analysis of thermoelectric modules *Industry applications IEEE Transactions* **43** 505-12
[10] Mitrani D, Salazar J, Turo A, Garcia M J and Chavez J A 2007 Lumped and distributed parameter SPICE models of TE devices considering temperature dependent material properties *13th Int. Workshop on Thermal Investigation of ICs and Systems* (17-19 September 2007)
[11] Gould C A, Shammas N Y A, Grainger S and Taylor I 2011 Thermoelectric power generation: Properties, application and novel TCAD simulation *Proc. of Power Electronics and Applications* (30 August 2011-1 September 2011)
[12] Tsai H L and Lin J M 2010 Model building and simulation of thermoelectric module using Matlab/Simulink *J. Elec Materi* **39** 2105-11
[13] Silva M F. 2010 Thin-films for thermoelectric application *Msc Thesis on Micro/Nanotechnologies* University of Minho
[14] Liang G, Zhou J and Huang X 2011 Analytical model of parallel thermoelectric generator *J. Applied Energy* **88** 5193-9
[15] Kim S 2013 Analysis and modeling of effective temperature differences and electrical parameters of thermoelectric generators *J. Applied Energy* **102** 1458-63

IOP Conf. Series: Materials Science and Engineering **53** (2013) 012016 doi:10.1088/1757-899X/53/1/012016

Speech Analysis Based On Image Information from Lip Movement

Kamil S. TALHA[1], Khairunizam WAN, S.K.Za'ba, Zuradzman Mohamad Razlan and Shahriman A.B
Advanced Intelligent Computing and Sustainability Research Group
School Of Mechatronics Engineering, Universiti Malaysia Perlis (UniMAP), 02600 Arau, Malaysia

Email: kamilsyahid89@gmail.com, khairunizam@unimap.edu.my

Abstract. Deaf and hard of hearing people often have problems being able to understand and lip read other people. Usually deaf and hard of hearing people feel left out of conversation and sometimes they are actually ignored by other people. There are a variety of ways hearing-impaired person can communicate and gain accsss to the information. Communication support includes both technical and human aids. Human aids include interpreters, lip-readers and note-takers. Interpreters translate the Sign Language and must therefore be qualified. In this paper, vision system is used to track movements of the lip. In the experiment, the proposed system succesfully can differentiate 11 type of phonemes and then classified it to the respective viseme group. By using the proposed system the hearing-impaired persons could practise pronaunciations by themselve without support from the instructor.

1. Introduction

Speech recognition is not purely auditory. When a listener can see the speaker, visual information is used in the speech recognition process. The contribution of this visual information to overall speech recognition was first reported by McGurck and MacDonald [1]. Speech command based systems are useful as a natural interface for users to interact and control computers. Such systems provide more flexibility as compared to the conventional interfaces such as keyboard and mouse. However, most of these systems are based on audio signals and are sensitive to signal strength, ambient noise and acoustic conditions [3]. To overcome this limitation, speech data that is orthogonal to the audio signals such as visual speech information can be used. The systems that combine the audio and visual modalities to identify utterances are known as audio-visual speech recognition (AVSR) system. Visual speech recognition (VSR) system refers to the systems which utilizes the visual information of the movement of the speech articulators such as the lips, teeth and somehow tongue of the speaker. The advantages are that such a system is not sensitive to ambient noise and change in acoustic conditions, does not require the user to make a sound, and provides the user with a natural feel of speech and dexterity of the mouth. This paper is structured as follows: Section 2 addresses the related researches to the approaches, applications and problems of recognizing the human gesture. Section 3 describes the configuration of the system and describes the proposed algorithm for the classification of motion patterns. Section 4 presents the results of the classification and the article is concluded with the summary in section 5.

[1] To whom any correspondence should be address

2. Literature Review

Lip-Reading has been practised over centuries for teaching deaf and dumb to speak and communicate effectively with the other people. The automatic lipreading (or multimodal speech processing in general) quickly becomes a mainstream part of the speech related research. In the recent years some prototype systems have already been presented or announced. The need for lipreading in human computer interaction is no longer questioned as the speech recognition based only on audio signal hits its limits.

2.1. Lip reading

Lip reading, also known as lipreading or speechreading, is a technique of understanding speech by visually interpreting the movements of the lips, face and tongue with information provided by the context, language, and any residual hearing. Each speech sound (phoneme) has a particular facial and mouth position (viseme), although many phonemes share the same viseme and thus are impossible to distinguish from visual information alone. Thus a speechreader must use cues from the environment and a knowledge of what is likely to be said.

Several experiments show that lip-reading can be efficiently applied in limited-vocabulary speech recognition [4,5], recognition of speech uttered by speech impaired [6] and also in case of continuous speech signal [7]. Techniques developed for automatic lip-reading find their way also in the world of computer generated facial animation and multimodal speech synthesis [8]. The lip movements and other visually distinguishable changes in articulatory system are represented by different researchers in multitude of possible geometric and non-geometric models [9,10]. According to Conrad (1979)[11], the capacity for lipreading seems to be determined by the person's degree of hearing and the levels of intelligence and of speaking.

However, the studies that have focused on establishing the relation between the degree of hearing and the level of lipreading have not reached unanimous conclusions. Some have observed that hearing people are usually better lipreaders than deaf people and, therefore, they have concluded that the more the loss of hearing, the more difficult lipreading will be. However, there are 43 phonemes in the English language, while there exist only 28 different mouth shapes that separate them [2]. For example, 'd' and 't', or 'f' and 'v' produce the same mouth shape. Therefore, the art of lip reading for humans is context sensitive: it consists not only in visually recognising mouth shapes, but also mentally recognising key elements to predict the word, as well as further recognising key words to predict the sentence.

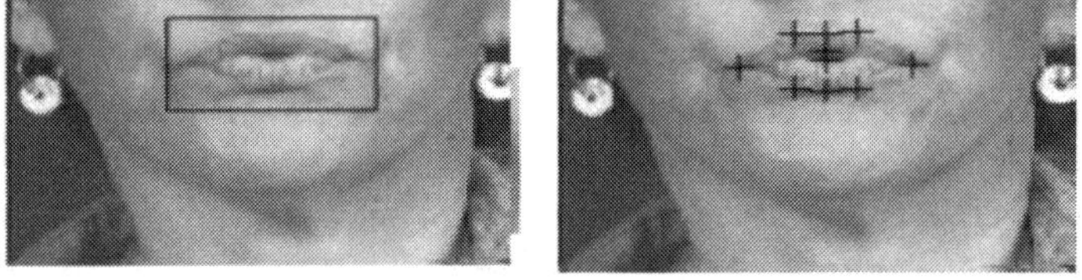

Figure 1. Lipreading researched by Leon J. M. Rothkrantz (2006)

Table 1. Fourteen Visemes

Visemes Number	Corresponding phonemes	Vowel/ Consonant
1	/p/, /b/, **/m/**	Consonant
2	/f/, **/v/**	Consonant
3	**/th/**, /d/	Consonant
4	**/t/**, /d/	Consonant
5	**/g/**, /k/	Consonant
6	**/ch/**, /j/, /sh/	Consonant
7	**/s/**, /z/	Consonant
8	**/n/**, /l/	Consonant
9	**/r/**	Consonant
10	**/A/**	Vowel
11	**/E/**	Vowel
12	**/I/**	Vowel
13	**/O/**	Vowel
14	**/U/**	Vowel

3. Methodologies

The system has three stages: preprocessing, tracking and developing database. Figure 2 shows the block diagram of the proposed method. Videos for different movement of lips are captured and Figure 5 shows some of the sample images explained. During the preprocessing, the movie frames are converted into indexed image format. Median filter is applied on these images and the unwanted noises are removed. For each image frame, the images are converted into binary.

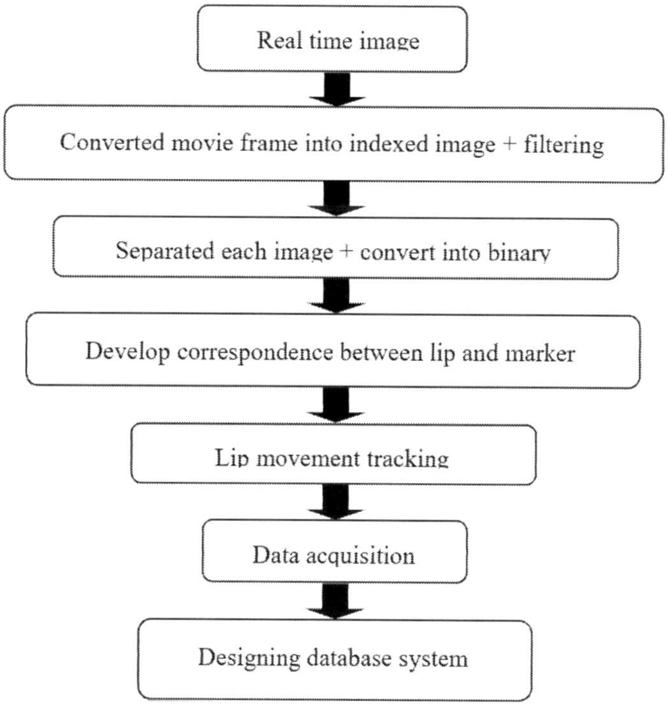

Figure 2. The overall process of the system

3.1. Pre-processing

3.1.1. RGB colour on image
A pixel is a single point in a graphic image. Each pixels of an image contains 3 dimension of colour which is Red, Green and Blue (RGB). The RGB colour is added together in various ways to reproduce a broad array of colours. A RGB decimal value is range from 0 to 255 where will be represented by (0, 0, 0) to (255, 255, 255).
In this project, red marker will be use and will be extracting from other color on an image. The marker will be placed on lip which is 3 on the upper lip and another 3 on the lower lip. The movement of the lip will represent the word that will be spoken. Extracting must be done precisely to eliminate any noise and also to eliminate background colour that can be same colour as the marker.

3.1.2. Intensity image (grayscale image)
This is the equivalent to a "gray scale image" and this is the image we will mostly work with in this course. It represents an image as a matrix where every element has a value corresponding to how bright/dark the pixel at the corresponding position should be colour. There are two ways to represent the number that represents the brightness of the pixel: The double class (or data type). This assigns a floating number ("a number with decimals") between 0 and 1 to each pixel. The value 0 corresponds to black and the value 1 corresponds to white. The other class is called uint8 which assigns an integer between 0 and 255 to represent the brightness of a pixel. The value 0 corresponds to black and 255 to white. The class uint8 only requires roughly 1/8 of the storage compared to the class double. On the other hand, many mathematical functions can only be applied to the double class.

3.1.3. Thresholding
Thresholding is the simplest method of image segmentation. From a grayscale image, thresholding can be used to create binary images. During the thresholding process, individual pixels in an image are marked as "object" pixels if their value is greater than some threshold value (assuming an object to be brighter than the background) and as "background" pixels otherwise. This convention is known as *threshold above*. Variants include *threshold below*, which is opposite of threshold above; *threshold inside*, where a pixel is labeled "object" if its value is between two thresholds; and *threshold outside*, which is the opposite of threshold inside (Shapiro, et al. 2001:83). Typically, an object pixel is given a value of "1" while a background pixel is given a value of "0." Finally, a binary image is created by coloring each pixel white or black, depending on a pixel's labels.

3.1.4. Binary image
A binary image is a digital image that has only two possible values for each pixel. Typically the two colors used for a binary image are black and white though any two colors can be used. The color used for the object(s) in the image is the foreground color while the rest of the image is the background color. In the document scanning industry this is often referred to as bi-tonal.

Binary images are also called *bi-level* or *two-level*. This means that each pixel is stored as a single bit (0 or 1). The names *black-and-white*, *B&W*, monochrome or monochromatic are often used for this concept, but may also designate any images that have only one sample per pixel, such as grayscale images.

3.2. Lip tracking
On this project, the system must track a lip in real-time condition where a real-world process is simulated at a rate that matched that of the real process. In other word, processing the image must be done as the image been captured by camera. considering that camera can captured images at a rate of 30 frames per

second (fps), several image from this frames will be grab to be process. This can be done by using MatLab command below;

```
data=getsnapshot(vid);  where
vid=source of
```

After frame has been grabbed and save, image processing can be done by extracting red colour. To extract the desired colour, each pixels on the image will be examined whether it achieve desired RGB value. For instant, if red colours want to be extract, every pixel that contain RGB decimal value from (0, 0, 0) to (255, 0, 0) will be selected. These selected pixels than will be convert into grayscale colour according to their intensity of red colour. Grayscale is also known as black and white, which are composed exclusively of shades of gray, varying from black at the weakest intensity represented by 0, to white at the strongest represented by 255. This method can be explained through this mathematical equation;

$$B_{xy} = A_{xy} \qquad \text{where; } B = \text{gray intensity}, \quad x,y = \text{coordinate of pixel}$$

This mathematical equation can be simplify by using MatLab command below:

```
diff_im = imsubtract(data(:,:,1),
rgb2gray(data));
```

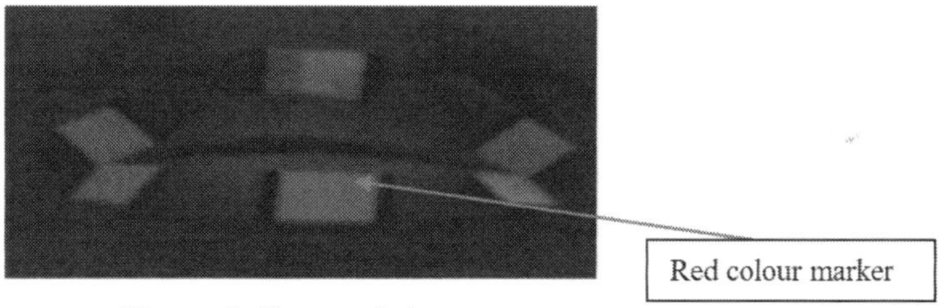

Figure 3. Gray scale image

3.2.1. Red colour extraction

Once the grayscale has been determined, the grayscale image is converted into binary image. By setting factor of binary conversion into certain value, it will remove noise and also background image. Below is the mathematical equation for this method;

$$C_{ij} = \begin{cases} 1, & B_{ij} \geq 200 \\ 0, & \text{else} \end{cases} \qquad \text{where C = binary conversion}$$

This equation can be simplified by using the MatLab command below;

```
diff_im = im2bw(diff_im,0.18);
```

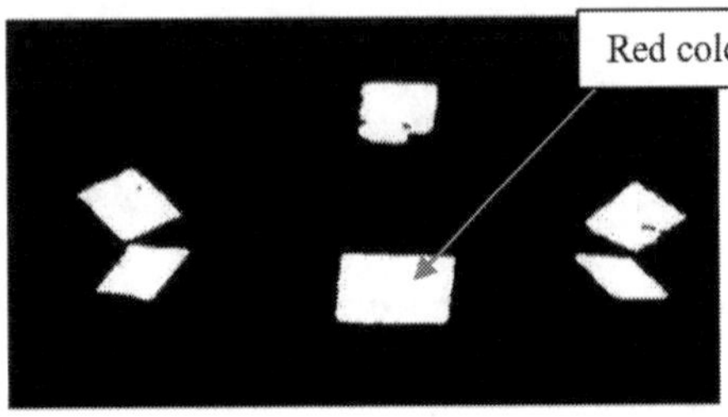

Figure 4. Binary image

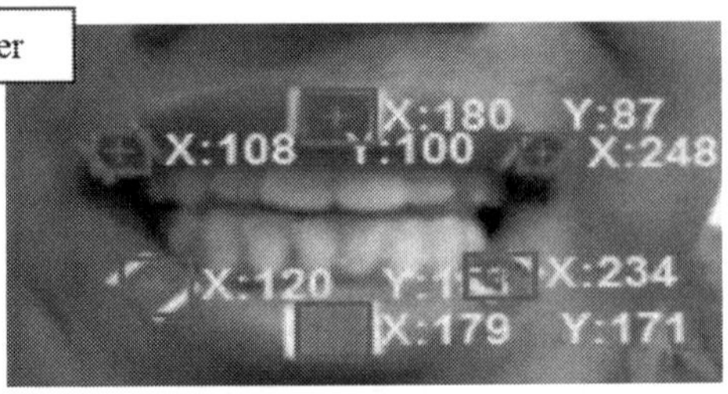

Figure 5. Labeling the image with red rectangular and centroid

3.2.2. Blob analysis

A blob (binary large object) is an area of touching pixels with the same logical state. All pixels in an image that belong to a blob are in a foreground state. All other pixels are in a background state. In a binary image, pixels in the background have values equal to zero while every nonzero pixel is part of a binary object. After the system able to track the red marker, it will now labeling or draw a virtual graphic on the image for further extraction. On this project, red rectangle will be draw on perimeter of red marker. This method also will indicate that the system able to detect red marker when red rectangle is present. MatLab command for this method is;

```
bw = bwlabel(diff_im, 8);
stats = regionprops(bw,
'BoundingBox', 'Centroid');
```

In MatLab command above, there are two image properties analysis that been use which are BoundingBox and Centroid. BoundingBox is the bounding boxes around the blobs in a binary image. This function extracts the coordinates and dimension values of each blob from stats.BoundingBox structure one by one and draws the rectangle around them. Centroid is function return X and Y coordinates.

4. Experiment

4.1. Subject

In this project, only one subject had been tested to doing the corresponding phoneme. Before start the test, on the subject lip need to put a red color marker. After that, subject need to sit in front of the laptop and the lips must be approximately in front of the webcam. The webcam need to focusing to on the lips so that the system may determine the color marker. There are 11 phoneme had been tested in this experiment. The result from the phoneme then we determine it pattern to show which phoneme is in the same group and the difference pattern to the other group.

Table 2 - Five visemes tested

VISEMES	PHONEMES
1	p /b / m
2	f / v
3	t / d
4	g / k
5	s / z

The above table is the five visemes that been tested in this project. From this five viseme tested it may determine the corresponding phoneme group.

4.2. *Testing*

When performing the test, 5 to 10 time of test for every phoneme conducted. The value from the phoneme was calculated to get it average value. First a comparison among the test of same phoneme will be conducted. After had satisfied, the average value of the phoneme then been compare to the others phoneme.

4.3. *Experimental result*

The following result is from the experiment. The result divided to 4 type of viseme and every viseme group contain it corresponding phoneme. In this result, it shows the X-coordinate and Y-coordinate graph. For more detail and to show it pattern more clearly, for every graph had been include with staked line graph. Stacked line graph is the type of graph that categorizes the entire phoneme by alphabetical sequence.

4.3.1. *Viseme 1*

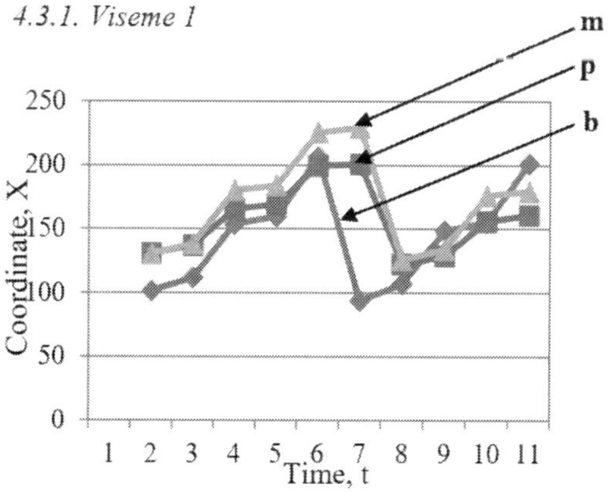

Figure 6. Trajectories of lip for p/b/m

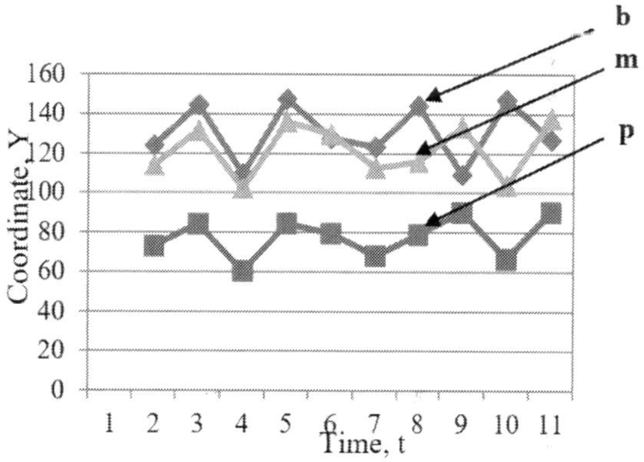

Figure 7. Trajectories of lip for p/b/m

4.3.2. *Viseme 2*

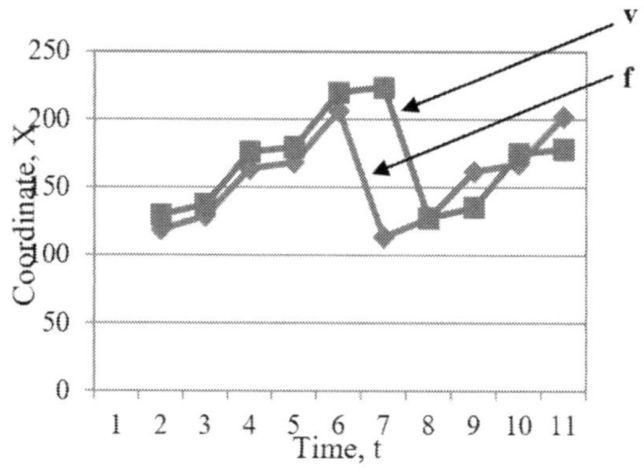

Figure 8. Trajectories of lip for f/v

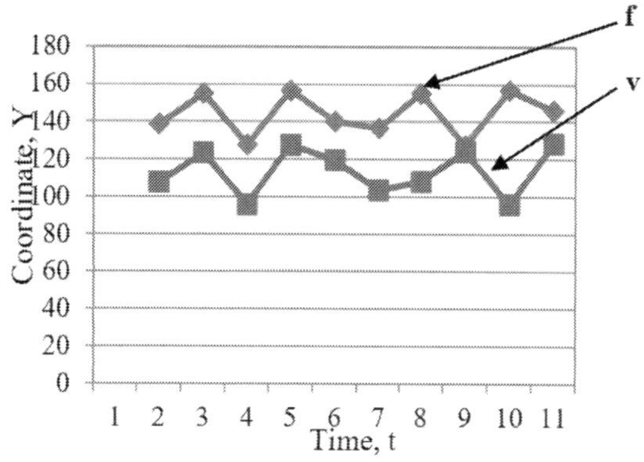

Figure 9. Trajectories of lip for f/v

IOP Conf. Series: Materials Science and Engineering **53** (2013) 012016 doi:10.1088/1757-899X/53/1/012016

5. Conclusion

Overall for this project, the concept of image or video analysis had been taken seriously in determining the success of this project. From starting of this project, many concepts to analyze of image had been done. Several techniques such as colour marker detection, optical flow analysis and snake tracking been taken look. After making a research, the colour marker detection is much easier and much more reliable to determine the movement of the lip. The other technique more reliable to be use for others application and for different situation.

From the research and experiment that had been done, the result may be use to make a more reliable lip reading system. Factor like intensity of light in the test environment need to be consider. Besides that, when operating with the system that need more space of memory, our device of the laptop been use in this experiment need to be consider more. From the test had been done, after making several of test, the laptop may have a problem to processing the image. It may come out the result with an error or sometime the laptop itself will automatically shut down because of the problem of memory dump.

For determining the successful of the project, a lot of researches need to be done and a lot of tests need to be working on. When facing with the signal processing concept, it needs a lot of understanding of the concept and need to analyze a lot of mathematical problem to determine our flow of the project.

6. References

[1] D. McGurck and J. MacDonald. Hearing lips and seeing voices. *Nature*, 264, December 1976.

[2] A. Greenwald. *Lipreading made easy*. Alexander Graham Bell Association for the Deaf, 1984.

[3] T. Chen, "Audiovisual speech processing. Lip reading and lip synchronization," *IEEE Signal Processing Mag*., vol. **18**, January 2001, pp. 9-21.

[4] S. Lucey, S. Srindharan and V. Chandran (2001): "An investigation of HMM classifier combination strategies for improved audio-visual speech recognition," In: [*Dalsgaard et al. (2001) Dastard, Lindberg and Benner*], pp. 1185-1188.

[5] J. C. Wojdel and L. J. M. Rothkrantz (2001): "Using aerial and geometric features in automatic lipreading," In: *[Dalsgaard et al.(2001)Dalsgaard, Lindberg and Benner]*, pp. 2463-2466.

[6] G.Potamianos and C.Neti (2001): "Automatic speechreading of impaired speech," In: *[Massaro et al.(2001)Massaro, Light and Geraci]*, pp. 177-182.

[7] P. Wiggers J. C.Wojdel and L. J. M. Rothkrantz (2002): "Medium vocabulary continuous audiovisual speech recognition, " In: *[Hansen and Pellom(2002)]*, pp. 1921-1924.

[8] T.Chen, (January 2001): "Audiovisual speech processing," In *IEEE Signal Processing Magazine*, pp.9-21.

[9] M. E.Hennecke, D. G. Stork and K. V.Prasad(1996): "Speechreading by humans and machines," In: *[Stork and Hennecke(1996)]*, pp. 331-349.

[10] G.Potamianos, C. Neti, G. Gravier et. al(2003): "Recent advances in the automatic recognition of audiovisual speech," In *the Proceedings of IEEE*, Volume **91**, No. 9, pp. 1306-1388.

[11] R.Conrad(1979): "The deaf school child: Language and cognitive functions," London*: Harper & Row*.

Performance analysis of robust road sign identification

Nursabillilah M Ali[1,3], Y M Mustafah[2,4] and N K A M Rashid[2,5]

[1]Department of Mechatronic Engineering, Faculty of Electrical Engineering, Universiti Teknikal Malaysia Melaka, Hang Tuah Jaya, Durian Tunggal, Melaka
[2]Department of Mechatronic Engineering, Faculty of Engineering, International Islamic University Malaysia, Gombak, Kuala Lumpur

E-mail: [3]nursabillilah@utem.edu.my, [4]yasir@iium.edu.my, [5]nahrul@iium.edu.my

Abstract. This study describes performance analysis of a robust system for road sign identification that incorporated two stages of different algorithms. The proposed algorithms consist of HSV color filtering and PCA techniques respectively in detection and recognition stages. The proposed algorithms are able to detect the three standard types of colored images namely Red, Yellow and Blue. The hypothesis of the study is that road sign images can be used to detect and identify signs that are involved with the existence of occlusions and rotational changes. PCA is known as feature extraction technique that reduces dimensional size. The sign image can be easily recognized and identified by the PCA method as is has been used in many application areas. Based on the experimental result, it shows that the HSV is robust in road sign detection with minimum of 88% and 77% successful rate for non-partial and partial occlusions images. For successful recognition rates using PCA can be achieved in the range of 94-98%. The occurrences of all classes are recognized successfully is between 5% and 10% level of occlusions.

1. Introduction

In recent years, road sign identification system has attracted numerous research works with the possibility of using in autonomous or driver assistance system (ADAS). Research in road sign identification with occlusion, however, is still lacking. Many existing techniques that have been developed produce inaccuracy when dealing with occlusions. Even though the occurrences of road signs with the presence of occlusions are small, yet it is a problem that needs to be addressed.

Road sign gives vital information about the situation and conditions of the road. There are varieties of road signs utilized on the road and they are different in color and shape to represent road sign such as restrictions, prohibitions, warnings and obligations. The study aims to evaluate a system performance of the road sign identification system that able to detect and identify road signs in order to increase its accuracy and robustness. The identification of road signs can be divided into two main processes which are addressed as detection and recognition stages [1-3]. In detection stage, the candidates of possible road sign are extracted. Then, in the recognition stage, the specific type of road sign of the extracted candidates is recognized.

[1] Corresponding author. Tel.: +606- 555 2215; fax: +606- 555 2222
E-mail address: nursabillilah@utem.edu.my

IOP Conf. Series: Materials Science and Engineering **53** (2013) 012017 doi:10.1088/1757-899X/53/1/012017

2. Literature review

According to Pazhoumand Dar and Yaghobi [4], road signs information can be illustrated using three features; shapes, color and pictogram. Khan et al. [5] stated that road sign detection becomes an important task for intelligent and autonomous vehicles in order to make the traffic regulation and guiding work smoothly. Road sign is normally detected based on their color or shape. Color model is the most popular approach in color segmentation technique. Amongst the existing techniques, the popular technique that has been widely used is Red, Green and Blue (RGB) color space. Hue, Saturation and Intensity (HSI) or Hue Saturation Luminance (HSL) has been used to segment the sign images and also has been recommended by other researchers [4].

The initial research on recognition of road sign has started in 1984. Basically PCA is widely used in face recognition. PCA is introduced in 1991. The principal component actually generates *eigenfaces* resulting from *eigenvectors* of the face image. Besides that, PCA also has been used in handling road sign of the partial occlusion situation. Implementing a robust road sign recognition system is not easy. A lot of conditions should be considered such as illumination conditions, orientation and condition of the road sign as well as occlusions on the road signs. Several important and the most dominant methods are artificial neural network (ANN), principal component analysis (PCA) and hidden Markov models (HMM).

3. Overview of the system

In this study, the focus is on road signs used in Malaysia. Sign identification is based on color and shape. A total of 12 types of road signs were used to form 403 images for detection and 600 images for recognition. These images were taken at night and daytime involving illumination and rotational changes as well as partial occlusions. The road sign was detected when the algorithm conducts the searching for the pixel data in the image based on its colors. Only colors of red, blue and yellow road signs are considered in the research. The algorithm is also able to detect more than one road sign in the frame.

3.1 Methodology

The dataset images consisted of 12 road sign classes. They are *Uneven Road, No Entry, No-U-Turn, Right Junction, Give Way, Keep Both Side, Stop Sign, Keep Left, Left Junction, Speed Limit 30km/hr, Speed Limit 50km/hr* and *Speed Limit 60km/hr* signs. These signs are chosen because they are the most common traffic signs in Malaysia. The images size used in the study are with a resolution of 500 x 667 pixels. The color space conversion is used where the images segmentation of the detection process is implemented using HSV color algorithm. A pixel is marked based on the hue, saturation and value ranges.

The recognition process using PCA is divided into two steps; training and testing processes. Based on the flow process of road signs recognition using PCA, the system starts with creating a training dataset folder. In order to make use the dataset that have been created, several pre-processing techniques are required so that the dataset can be easily performed in recognition stage.

To recognize road signs using the PCA technique, we have to produce a matrix for training and testing images. After we find the mean from training database image, we then calculate the covariance matrix. Then, we calculate the *eigenvector* and *eigenvalue* of the images respectively. Based on the eigenvector value of covariance matrix, we produce *eigensign* value which is then projected to sign space. Before we extract the feature vector, the input of training images are resized in two dimensional pixels and reshape to one dimensional vector. Then, the difference of values between the feature vector and the mean are computed. Finally, based on the minimum value of *Eucledian distance* that is produced from the difference of value of projected space, the system will be able to recognize the road signs. The flow chart of road sign recognition is illustrated in Figure 1.

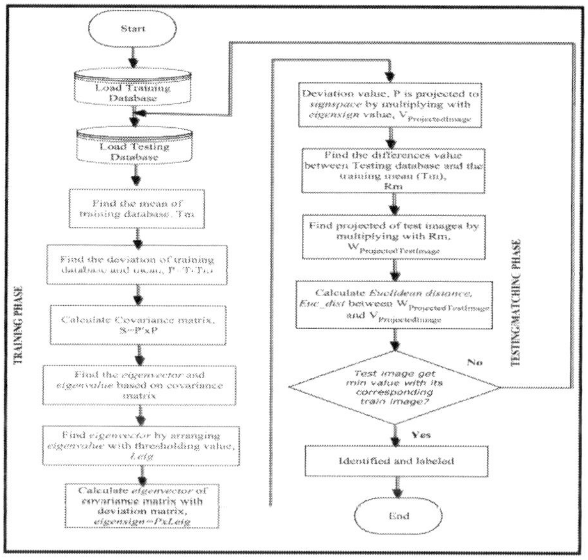

Figure 1. The workflow of road sign recognition using PCA

3.1.1 Pre-processing process of road sign image data set

Once the dataset is ready to be used, they will be converted from grayscale to binary image so that the processing step in feature extraction technique would be faster and easier. Figure 2 illustrates the technique of the image of *no-u-turn* sign is cropped, resized and converted to grayscale color using IrfanView program. Similarly, the process to resize the detected sign and convert to grayscale color using Matlab is shown in Figure 3. Once the image is resized in resolution size of 40 by 40 and converted to grayscale color, it is subsequently put in a testing folder. We did the similar process for training folder. The images in these folders consequently are used for training and testing process.

The database consists of road sign images in gray scale color with each image consists of resolution of 40 by 40 dimensions. Training data is set to 30 for each class and 20 signs for each class is set for testing data. Hence, 360 images are employed for training, while 240 images are for testing of the trained recognition system.

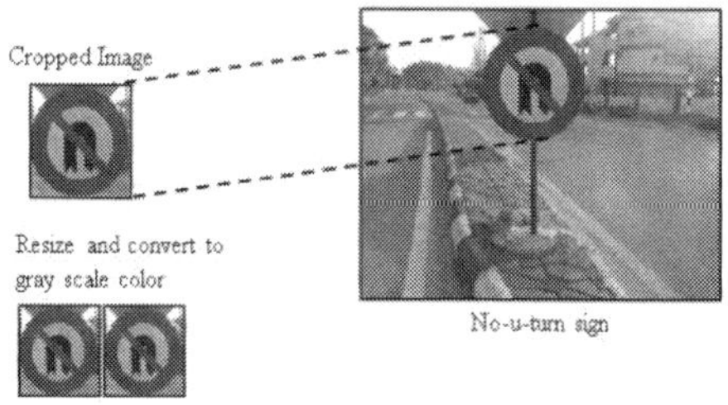

Figure 2. Image is manually cropped using IrfanView program.

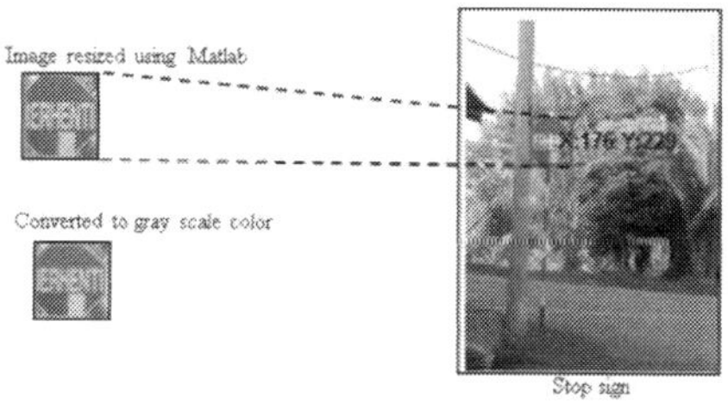

Figure 3. Examples of partially occluded sign image is automatically resize and convert to gray scale color in detection stage.

3.1.2 Partial occlusion and rotational changes signs

In order to evaluate the performance of the system in partial occlusions, testing dataset comprised partial occlusion as well as rotational changes are utilized. The different percentage of occlusion in dataset is important as it will be used to see how robust the system is towards the severity of occlusions. Therefore, the percentage of partial occlusion (O_p) is calculated as follows:

$$O_p = \frac{a_{op}}{t_{dm}} x100\% \qquad (1)$$

where a_{op} is the number of partial occlusion coverage on the candidate signs and t_{dm} is the total dimension of the pixels that represented as the whole of the candidate signs. The partial occlusion on each sign is set to maximum range of 44% coverage. The rotational changes on each signs are simply determined by setting it manually using IrfanView program. Figure 4 shows 20 test images for the *No-Entry* sign with their respective rotational degrees and partial occlusion values. The signs dataset will be subsequently used in recognizing road sign using PCA techniques.

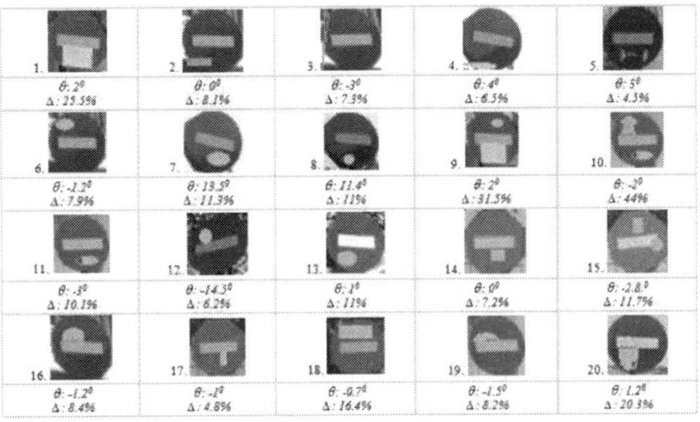

Figure 4. Maximum degree of occlusions for recognition of different road signs. {θ = degree of rotation, Δ = percentage of occlusion}

4. Result

4.1 Evaluation of Accuracy of Detection Stage

Table 1 shows the percentage of detection performance based on HSV color segmentations. Images are split into two parts; 131 images with occluded signs and 272 images are non-occluded signs. The detection accuracy is computed using the following equation:

$$a_d = \frac{b_{sd}}{b_{td}} x100\% \qquad (2)$$

where b_{sd} is number of successful detected and b_{td} is the total number of detected images.

Based on Table 1, HSV produced minimum of 77% successful detection for occluded images. It shows very high detection accuracy with minimum rates of 88% on non-occluded images. This is due to the fact that HSV color ratio is invariant to illumination changes.

IOP Conf. Series: Materials Science and Engineering **53** (2013) 012017 doi:10.1088/1757-899X/53/1/012017

Table 1. Accuracy performance of HSV color space

Sign	Partial occlusion sign(131)	Non-occluded signs with rotational changes(272)	Detected sign		Percentage (%)	
			Partial occlusion sign	Non-occluded sign	Partial occlusion sign	Non-occluded sign
Red	55	95	43	88	78.2	92.6
Blue	35	40	27	35	77.1	87.5
Yellow	41	137	33	135	80.5	98.5

4.2 Evaluation of accuracy of recognition stage

The result of the experiment carried out in the research is computed based on the percentage of accurately recognized road sign. The correctly recognized road sign is calculated by dividing the overall number of testing road sign database images and times by 100. Hence, the percentage of correctly recognized, s_r (%) is computed as below:

$$s_r = \frac{y_r}{y_{tr}} x100\% \quad (3)$$

where y_r is number of correctly recognized sign and y_{tr} is the total number of testing signs.

From the experiment conducted in this study, the percentage of road sign identification is described in Table 2. The result in the Table 2 indicates that PCA produces a better detection rate on every type of road signs tested. With a very low false rate, we can also say that PCA based recognition system is more robust towards occlusion. It shows that the technique using PCA has a high accuracy with a minimum of 94% classification rate of road signs.

4.3 Robustness to partial occlusion

The percentage of occlusion coverage on each sign are indicated into six ranges namely; 5%, 10%, 20%, 30%, 40% and more than 50% occlusions. Graph in Figure 5 illustrated the overall occurrences performance based on PCA techniques in recognizing occlusion signs.
As can be seen from the summary of recognition rates (%) in figure 5, 12 classes are successfully recognized with 5% and 10% occlusion. It is also produced strong recognition rates with at least 6 classes of signs can be recognized when the percentage of occlusion increased gradually. However the system falsely recognized road signs with mostly more than 50% occlusions mark and more than 16 degrees rotation.

Table 2. Accuracy performance of PCA technique

Road sign		Percentage of accuracy rate	
Shape	Partially occluded road sign	**Success**	**False**
Circular	No-entry	95	5
	No-u-turn	96	4
	Keep both side	97	3
	Keep left	97	3
	Speed limit 30km/hr	95	5
	Speed limit 50km/hr	94	6
	Speed limit 60km/hr	94	6

Inverted triangular	Yield/Give away	97	3
Diamond	Uneven road	98	2
	Right junction	95	5
	Left junction	96	4
	Stop	97	3

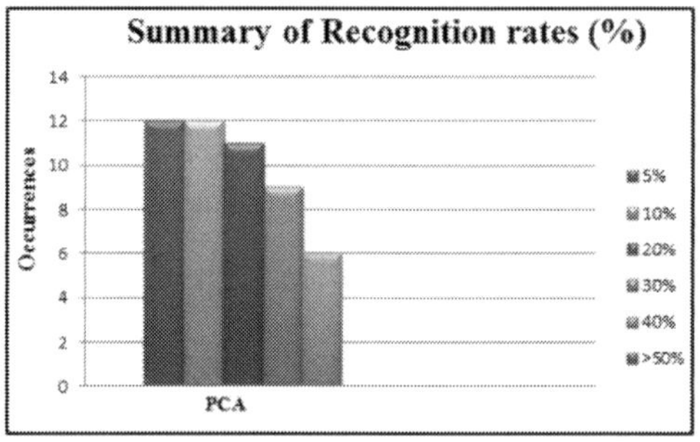

Figure 5. Summary of recognition rates using the PCA technique

4.4 Recognition processing time

PCA is faster in the recognition process and widely used in other applications. Therefore, processing speed using PCA is suitable to be used in offline and real time basis. Real time implementation needs at least 30 frames per second of processing time. In this work, it is observed that the recognition processing time per frame using PCA is *0.0238s*. Hence, the average processing time for 30 frames is only *0.714s* which is comfortably less than *1s*.

5. Conclusion

To sum up, a technique towards road sign detection and recognition using both HSV and PCA was presented in this paper. It shows that these algorithms able to identify the 12 road sign classes successfully. From the experiments conducted with dataset consisting of partial occlusions and rotations variants, we verified the performance of the proposed approach. It can be concluded that the percentage of accuracy using HSV robust and invariant to lighting conditions and PCA approachable to identify occluded images.

6. References

[1] Y. Y. K. Nguwi, A. Z., "A Study on Automatic Recognition of Road Signs," in *Cybernetics and Intelligent Systems, 2006 IEEE Conference on*, 2006, pp. 1-6.

[2] M. Sajjad Hossain, Mahmudul Hasan, M., Ameer Ali, M., Humayun Kabir, Md, & Shawkat Ali, A. B. M., "Automatic detection and recognition of traffic signs," in *IEEE Conference on Robotics Automation and Mechatronics (RAM)*, 2010, pp. 286-291.

[3] K. Wen-Jia and L. Chien-Chung, "Two-Stage Road Sign Detection and Recognition," in *Multimedia and Expo, 2007 IEEE International Conference on*, 2007, pp. 1427-1430.

[4] H. Pazhoumand-Dar and M. Yaghobi, "DTBSVMs: A new approach for road sign recognition," in *Proceedings - 2nd International Conference on Computational Intelligence, Communication Systems and Networks, CICSyN 2010* 2010, pp. 314-319.

[5] J. F. Khan, Bhuiyan, S. M. A., & Adhami, R. R., "Distortion invariant road sign detection," in *Proceedings - International Conference on Image Processing, ICIP* 2009, pp. 841-844.

IOP Conf. Series: Materials Science and Engineering **53** (2013) 012018 doi:10.1088/1757-899X/53/1/012018

Hierarchical Self Organizing Map for Novelty Detection using Mobile Robot with Robust Sensor

M N A H Sha'abani[1,4], M F Miskon[2] and H Sakidin[3]

[1,2,3]CeRIA Laboratory, Faculty of Electrical Engineering, Universiti Teknikal Malaysia Melaka, Hang Tuah Jaya, 76100, Durian Tunggal, Melaka, Malaysia.

E-mail: m011010006@student.utem.edu.my

Abstract. This paper presents a novelty detection method based on Self Organizing Map neural network using a mobile robot. Based on hierarchical neural network, the network is divided into three networks; position, orientation and sensor measurement network. A simulation was done to demonstrate and validate the proposed method using MobileSim. Three cases of abnormal events; new, missing and shifted objects are employed for performance evaluation. The result of detection was then filtered for false positive detection. The result shows that the inspection produced less than 2% false positive detection at high sensitivity settings.

1. Introduction

Novelty detection is a process of identifying abnormalities in an environment. It has been widely used in several fields [1-3]. Recently, there are interests of novelty detection on mobile robot such in [4-6], which generally errands as surveillance and inspection application. The problems of training a robot for novelty detection are sensor error and the inconsistency localization of robot during training and inspection. This will affect the detection performance.

S.Marsland et. al [4] using habituation Self Organizing Map (SOM) to detect novel objects by measuring the novelty level of the same measurements. This work does not concern on locating the novel object. J. Fleischer and S. Marsland [6] then overcome the problem by using landmark categorization. Then, A. Gopalakrishnan et al. extracting salient features to detect landmark for novel position estimation [5]. However, these works are not locating novelty with respect to the robot location.

M.F. Miskon and A. Russel [7] using Flexible Region Map to compare the current and normal measurements based on habituating self organizing map. In this work, the consideration of robot heading for training and inspection is only for 0, 90, 180 and 270 degrees with a 10 degree of tolerance. The result shows that the robot was able to locate the novelty. However, by fixing the heading angle of robot for measuring data, it seems that the robot will lose information for any heading angle that is out of the angle range.

This paper presents a framework of the novelty detection mechanism that learns measurement data and associate the data to the robot pose. The work focuses on developing a mobile robot to act as an inspection agent in an indoor environment. The objective is to make the robot autonomously adapt to normal perception while considering sensor measurement error and localization error.

[4] Corresponding Author.

This paper is organized as follows. First, the overview of the system is discussed. Then, the proposed novelty detection architecture is presented. This is followed by a detailed explanation of the research methodology. The simulation results are given in the next section and finally a conclusion is provided.

2. System Overview

The system is divided into two parts; mobile robot navigation system and novelty detection system. Six sonar sensors are used. The sensor array covers 180 degrees of scan area. The robot uses wall following algorithm to imitate a patrolling task and particle filter method for localization. A Repetitive Observation Strategy (ROS) [8] is used for observation where detection of anomaly is confirmed only if a repeated observation occurs. Data processing composed of novelty detection mechanism, clustering and false positive filter. Figure 1 shows the overview of the hierarchical novelty detector system.

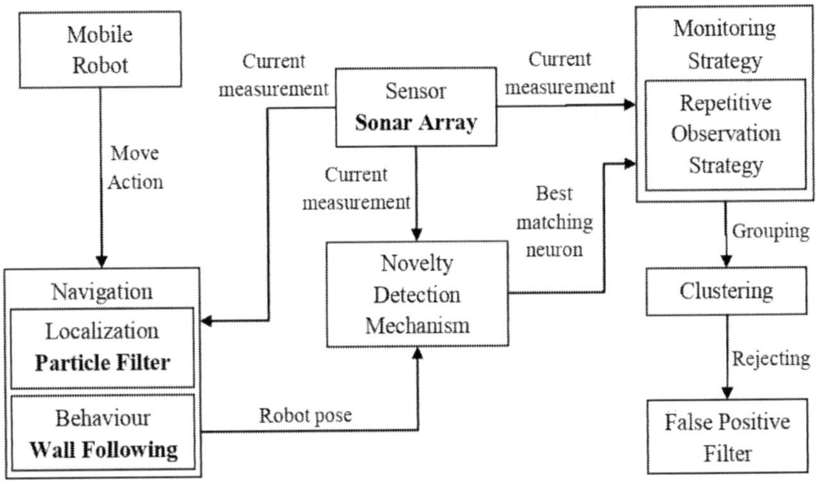

Figure 1. Overall system consisting of mobile robot and novelty detection system.

3. Novelty Detection Mechanism

The novelty detection mechanism consists of hierarchical Self Organizing Map and statistical parameters; mean and variance. The architecture of novelty detection mechanism is shown in figure 2.

3.1. *Self Organizing Map*

SOM is an unsupervised neural network which has a capability in competitive learning. The importance of competitive learning in this network is to determine which neuron represents the robot pose at a certain place. The weight will keep updating and the network will keep growing if required until the training is stopped. Equation (1) shows how the neuron's weight is updated,

$$w_{ij}(t+1) = w_{ij}(t) + \alpha(t)h_j(t)[v_i(t) - w_{ij}(t)] \tag{1}$$

where w is the weight vector for a neuron, α is the learning rate, h is the neighbourhood function and v is the input vector. The neuron's weight vector is updated by three processes; competitive, cooperative and adaptive process [9]. In a competitive process, the neuron with the closest weight vector, w to the current input v will be chosen as a winner and updated. The cooperative process allowed the neighbouring neuron of the winning neuron also to be updated but with fewer amounts based on the neighbourhood function, h. The adaptability of the network to the input pattern is based on the value of learning rate, α. The learning process is stopped when there are no significant changes in the network.

3.2. Statistical Parameters

The error of sensor measurement is depending on the distance measurement and angle of detection. To handle this problem, the statistical parameters of the measurement at a specific robot pose were calculated. The mean and variance is calculated using a single pass algorithm as shown in equation (2) and equation (3),

$$u_n = u_{n-1} + \frac{x_n - u_{n-1}}{n} \tag{2}$$

$$s_n^{\ 2} = \frac{(n-2)s_{n-1}^{\ 2} + (x_n - u_n)(x_n - u_{n-1})}{n-1}, \qquad n > 1 \tag{3}$$

where x_n is the new input, u_n and $s_n^{\ 2}$ is the new mean and variance of total number of samples, n.

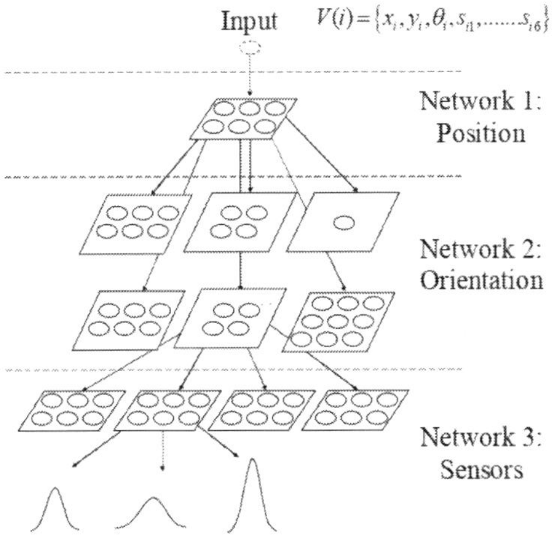

Figure 2. The hierarchical novelty detection mechanism consists of three networks where each network has their own role in the determination of normalities.

4. Methodology

The algorithm was tested using MobileSim simulator. MobileSim is a mobile robot simulator with ARIA platform. Figure 3 shows the environment setup and route plan for the robot to perform inspection. The robot was tested in three cases of novelty; presence of new objects, missing object and shifted object. Each of the case was trained by 10 times of complete rounds.

4.1. Data collection

By referring to figure 3, the robot was initially positioned at point 'Start' and run through the environment using wall following behaviour. As the implementation of ROS, data are collected for every 100 millimetres of robot movement until the robot reaches at point 'End'. After each complete round, the robot was positioned again at the starting point.

4.2. Training network

The collected data was then fed into the network. A batch of data for each measurement $V(i) = \{x_i, y_i, \theta_i, s_{i1}, \ldots\ldots s_{ik}\}$, is separated into three parts; robot position, (x_i, y_i), orientation, θ_i

and sensor measurements, $(s_{i1},.....,s_{ik})$ where k is the number of sensors used. The initial learning rate, α was set to 0.2 for the first network and 0.01 for the second network. The initial radius of neighbourhood function, h for each network are 1000 mm and 1 degree. The network is trained within 10 times of complete round using offline learning.

4.3. Inspection

Initially, the robot was positioned at point 'Start'. The inspection was done by moving the robot along the same path of training route. Along the inspection, the robot was trying to identify any changes of measurement based on the Euclidean distance method.

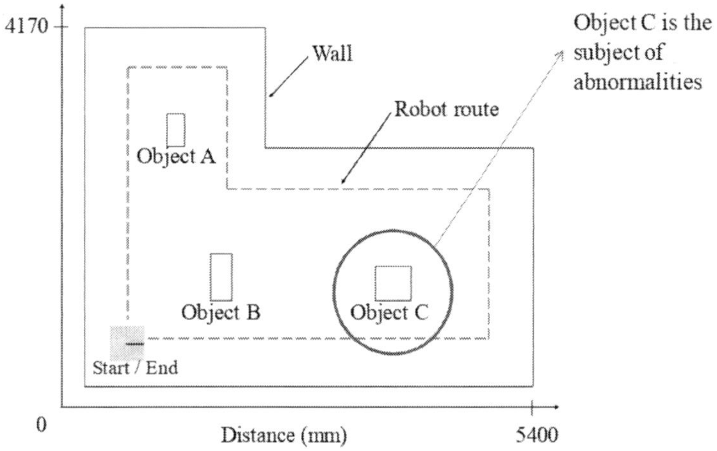

Figure 3. The environment of mobile robot is setup in an L shape with a fixed route (dash line) of mobile robots for training and inspection. Object C is the main subject to be changed while inspecting.

4.4. Anomaly detection

The identification of anomaly measurement is depending on the difference of mean and current measurement at specific pose. The first layer of network fined the best match of neuron with the current position of the robot. Then, the second layer is searching the best match of neuron for orientation. After the robot pose was confirmed to be matched from the previous training, the third layer was acting as the anomaly detector by comparing the current measurement for each sensor with its average measurement during training. If the difference is larger than the threshold value, it is classified as an anomaly. If the pose of the robot does not match to any neuron in the network, the current measurement is also classified as an anomaly.

4.5. Clustering

Clustering is a process of separating and grouping of unlabelled data set based on similar features. The similarity measure of anomaly is the distance between two anomaly points. By using equation (4), the distance can be expressed as

$$r = \sqrt{\sum_{i=1}^{n} |x_i - y_i|^2} \tag{4}$$

where r is the Euclidean distance between two anomaly points. The similarity measure can be set using a threshold value called maximum allowable distance, d_{max}. This threshold was set by referring the resolution of measurement taken (100mm/input data). Any anomaly point that has distance lower than this threshold from its neighbour is grouped in the same group.

4.6. Filtering

Filtering is a process of rejecting untrusted detection. In this research, filtering process is done based on Singleton Filter method [8]. Any cluster containing only one anomaly point was simply rejected.

4.7. Environmental setup
By referring figure 3, it is mentioned that the object C is the subject of abnormalities. The summarization of environmental setup for the simulation is shown in table 1.

Table 1. Environmental setup.

Novelties [a]	Condition	
	Normal	Abnormal
Missing	Present	Absent
Exist	Absent	Present
Shifted	Present	Shifted

[a] The novelty is referred to the object C.

4.8. Parameters setup
The sensitivity setting of anomaly identification was varied from low to high by adjusting the different threshold of current and normal measurement. The purpose is to observe the relationship between true positive rate and false positive rate of the system, thus the performance of the system.

4.9. Performance parameters
Based on the Receiver Operating Characteristics (ROC) curve, there are four possibilities of detection outcomes; true positive, false positive, true negative and false negative detection. True positive detection occurs when the system was correctly identified an anomaly. If a normal measurement was detected by the system as an anomaly, it is a false positive detection. True negative is a classification for a normal detection, whereas false negative is a wrong classification of an abnormal as a normal. For false detection classification, a tolerance boundary of true positive region is created by considering the sensor and localization error. About 100mm of boundary around the surface of novel object was created as shown in figure 4.

5. Results
A sample of simulation results is shown in figure 4. The analysis of novelty detection performance is done after the robot finished a complete round of inspection.

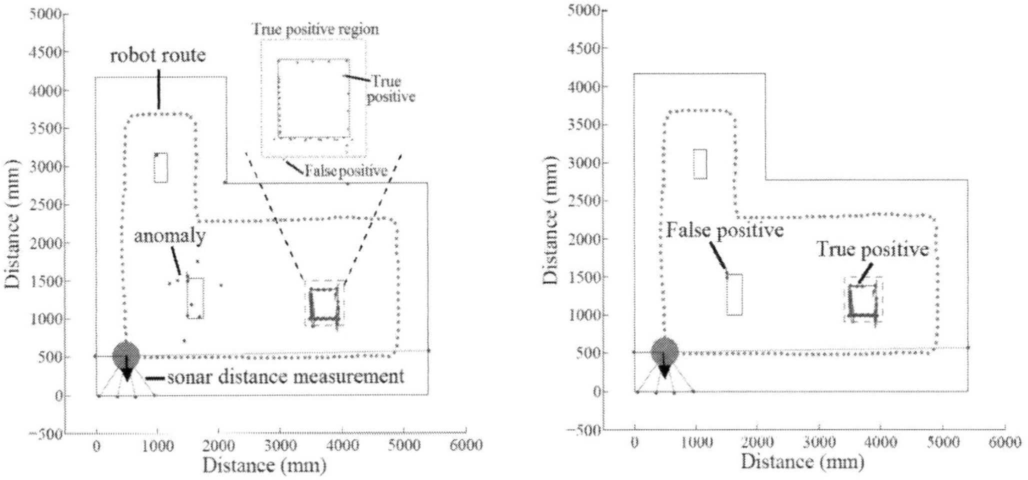

(a) Non-filtered (b) Filter

Figure 4. Sample result of simulation for a present of new object at 0.75 sensitivity settings. The figure shows the result for before and after using Singleton Filter. As seen in figure, any single clustered member was rejected. Even some of the single cluster in the true positive region also rejected, the true positive rate is still significant to the performance of novelty detection.

By referring to figure 5, it can be seen that the performance for missing and exist cases are almost the same, so as the accuracy. For both cases, there are no significant changes on true positive rate at a false positive rate higher than 0.1 but the accuracy of the novelty detection performance reduces as the increment false positive rate. This means that, the higher the sensitivity settings, the occurrence of false positive detection also higher, thus producing high false alarm.

However, the result for shifted object case is unacceptable because the true positive rate and false positive rate are changing closely at the same rate. One of the reasons for this occurrence is because only small changes occurred in the environment. Only the best performance of this case may occur at the centre of the curve (0.5, 0.5). However, by considering the accuracy of detection, the result is unacceptable.

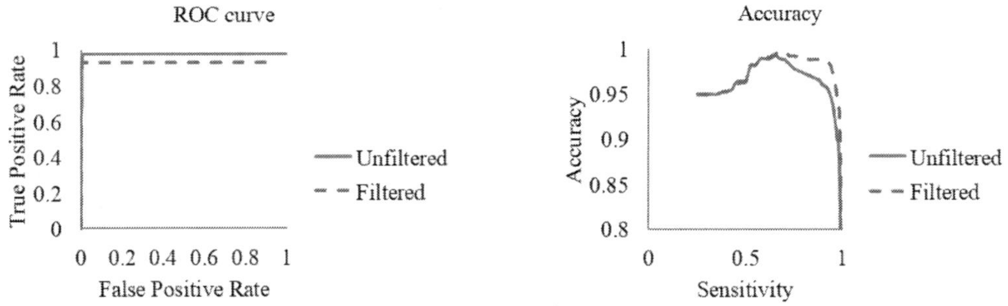

(a) Missing object.

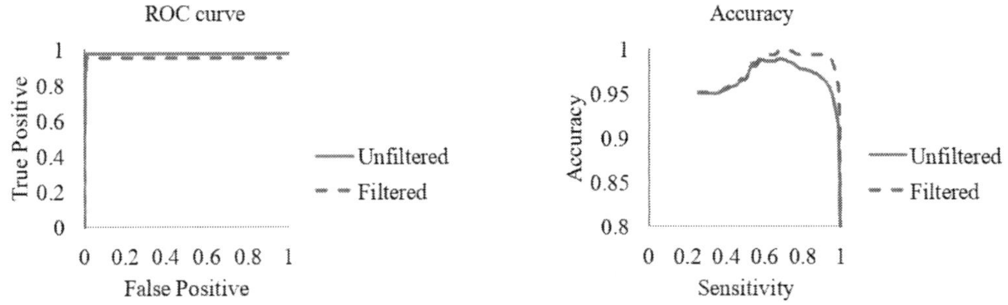

(b) Exists object.

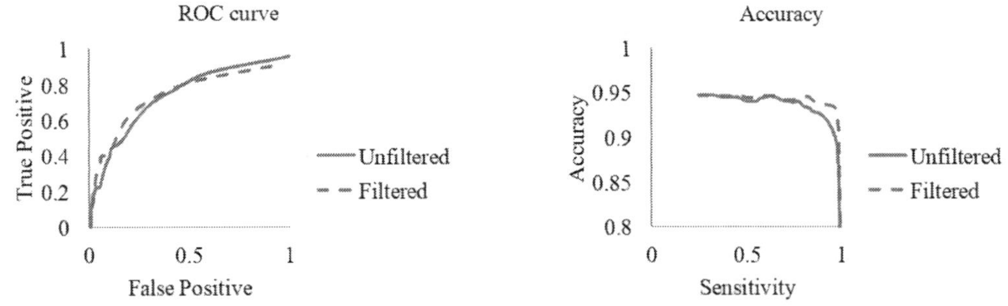

(c) Shifted object

Figure 5. Comparison of performance for three cases of novelties.

Table 2 shows the summarization result for all three cases of novelties. Note that the value of true positive rate and false positive rate for unfiltered and filtered are taken at the beginning of the higher true positive rate.

Table 2. Unfiltered and filtered results using the Singleton Filter of missing exists and shifted of novelties.

	Unfiltered		Filtered		Percentage reduce (%)	
Novelties	TP rate	FP rate	TP rate	FP rate	TP rate	FP rate
Missing	0.97619	0.01033	0.92857	0.0000	4	1
Exist	0.97619	0.01119	0.95238	0.0000	2	1.1
Shifted	The result of this case is unacceptable.					

6. Summary and Conclusion

Novelty detection aims a performance of high true positive rate and low false positive rate of detection. However, it is a trivial task to implement it using a mobile robot. Thus, this paper proposed a method that handling both localization and perception error based on hierarchical Self Organizing Map and statistical approach. Based on the simulation results, it can be concluded that the mechanism is able to detect changes in its environment. For missing and exist cases, a high true positive detection was produced at more than 97% with a lower false positive detection (less than 2%). Furthermore, the implementation of Singleton Cluster reduces the false positive rate by 4% and true positive rate by 1%. Missing a few single clusters in true positive region does not affect the overall performance of detection. For future work, the algorithm will be implemented using an Amigobot mobile robot.

7. References

[1] Tarassenko L, Hayton P, Cerneaz N and Brady M 1995 *Int. Conf. on Art. NN (Cambridge)* vol 4 (Cambridge: British Computer Society) p 442-447
[2] Hayton P, Schölkopf B, Tarassenko L and Anuzis P 2001 *Adv. in Neural Information Processing Sys.* **13** 946-952
[3] Manikopoulos C and Papavassiliou S 2002 *Comm. Magazine IEEE* vol 40 (IEEE: Communications Society of Institute of Electrical and Electronics Engineers) p 76-82
[4] Marsland S, Nehmzow U and Shapiro J 2000 *Int. Conf. on Sim. of Adaptive Behaviour (Paris)* vol 6 p 189-198
[5] Gopalakrishnan A, Greene S and Sekmen A 2005 *IEEE Int. Workshop on Robot and Human Interactive Comm. (Nashville)* vol 14 (USA: IEEE) p 48-53
[6] Fleischer J and Marsland S 2002 *Proc. Int. Conf. on Sim. Adaptive Behaviour (Edinburgh)* vol 7 (Massachusetts: MIT Press) p 151
[7] Miskon MF and Russell AR 2009 *Int. Conf. on Intell. Robotics and Appl. (Singapore)* vol 2 (Germany:Springer-Verlag) p 846-855
[8] Miskon MF and Russell AR 2008 *Austalasia Conference on Robotics and Automation (Canberra)*
[9] Haykin S 2009 *Neural networks and learning machines* (New York: Prentice Hall)
[10] Fawcett T 2006 *Pattern Recognition Letters* **27** 861-874

Robust Features Of Surface Electromyography Signal

M I Sabri[1,4],M F Miskon[2],M R Yaacob[3]

[1,2,3]Faculty of Electrical Engineering, Universiti Teknikal Malaysia Melaka (UTeM)

E-mail:ihsansabriutem@gmail.com

Abstract. Nowadays, application of robotics in human life has been explored widely. Robotics exoskeleton system are one of drastically areas in recent robotic research that shows mimic impact in human life. These system have been developed significantly to be used for human power augmentation, robotics rehabilitation, human power assist, and haptic interaction in virtual reality. This paper focus on solving challenges in problem using neural signals and extracting human intent. Commonly, surface electromyography signal (sEMG) are used in order to control human intent for application exoskeleton robot. But the problem lies on difficulty of pattern recognition of the sEMG features due to high noises which are electrode and cable motion artifact, electrode noise, dermic noise, alternating current power line interface, and other noise came from electronic instrument. The main objective in this paper is to study the best features of electromyography in term of time domain (statistical analysis) and frequency domain (Fast Fourier Transform).The secondary objectives is to map the relationship between torque and best features of muscle unit activation potential (MaxPS and RMS) of biceps brachii. This project scope use primary data of 2 male sample subject which using same dominant hand (right handed), age between 20- 27 years old, muscle diameter 32cm to 35cm and using single channel muscle (biceps brachii muscle). The experiment conduct 2 times repeated task of contraction and relaxation of biceps brachii when lifting different load from no load to 3kg with ascending 1kg The result shows that Fast Fourier Transform maximum power spectrum (MaxPS) has less error than mean value of reading compare to root mean square (RMS) value. Thus, Fast Fourier Transform maximum power spectrum (MaxPS) show the linear relationship against torque experience by elbow joint to lift different load. As the conclusion, the best features is MaxPS because it has the lowest error than other features and show the linear relationship with torque experience by elbow joint to lift different load.

1. Introduction

Nowadays, application of robotics in human life has been explored widely. Robotics exoskeleton system are one areas in recent robotic research that has high impact in human life. These system have been developed significantly to be used for human power augmentation, robotics rehabilitation, human power assist, and haptic interaction in virtual reality. This paper focus on solving challenges in problem using neural signals and extracting human intent. Commonly, surface electromyography signal (sEMG) are used in order to control human intent for application exoskeleton robot[1-5]. But the problem lies on the difficulty of pattern recognition of the sEMG features due to high noises which

[4] Corresponding Author.

are electrode and cable motion artifact, electrode noise, dermic noise, alternating current power line interface, and other noise came from electronic instrument [1,2].For sure unlike the robots used in industry, the robotic exoskeleton systems should be designed with special consideration since they directly interact with human user which is well known as anthropomorphic design. This paper presents a study to find the best features for electromyography signal. Then connected with relationship between best features muscle unit activation potential against torque produced by elbow joint for lifting different load. In this paper propose the novel technique to get the best features by using error to mean percentage.

2. Research Methodology

The purpose of this experiment is to acquire data for recording sEMG signals. The sEMG signal was recorded from biceps brachii muscles (refer in SENIAM group muscle position) of two healthy male (normal BMI and age range from 20-27 years old) by a Z03-002 EMG Preamplifier gain 300 with consist of medical grade stainless steel surface electrode (Motion Lab System, Inc) on the dominant hand. Sampling frequency was set at 1kHz using NI USB-6009 14 bit,48ks/s multifunction input/output (National Instruments). The NI USB-6009 attach with Z03-002 EMG Preamplifier connected to the PC by type-B cable as shown in figure 1

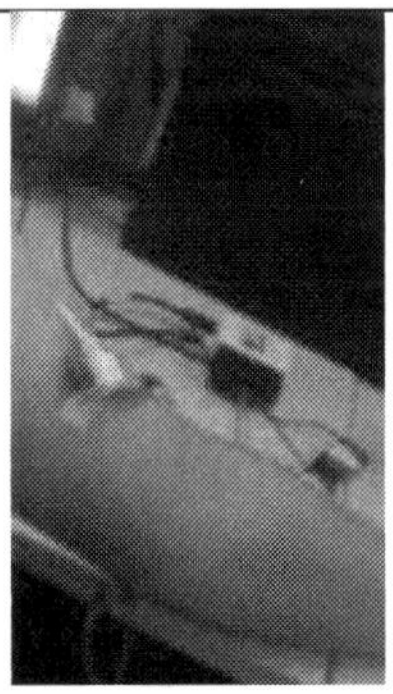

Figure 1. Connection for the data acquisition hardware

2.1 Skin Preparation Procedure

When deal with acquisition of EMG signal need to bear in mind that noise are the main disturbance that can effected the informative raw EMG signal. Thus, for minimizing the error, skin preparation need to be done in order to acquire the informative raw EMG signal. Firstly, skin must be cleaned by shaved the hair to reduce impedance on the skin and clean again with alcohol prep to remove the death skin. After that, measured the skin impedance by using multi meter to ensure that the skin impedance is less than 10kΩ [3]. If the impedance reading high than 10kΩ then the skin need to be done skin preparation procedure until reach less than 10kΩ in order acquire an informative EMG signal.

2.2 Experimental Protocol and EMG Data Gathering

In this experiments, 2 healthy subjects were asked to perform the lifting exercises during the flexion-extension tasks (muscles isometric contraction) . A range of dynamic contraction was from a full extension (0° , forearm in the vertical position) to a full flexion (approximately 150°) with 10s in duration (2s start lifting, 4s holding full flexion position, 2s falling down). Four different loads, consisting of 0,1,2, and 3 kg, were experience by all subject while done this experiment. For this task, there are 2 trials in each load experience by all subject. The experiment setup is shown in figure 2.

Figure 2. Show the step lifting task for acquire Raw EMG signal of biceps brachii muscles

2.3 Extraction features

From the raw EMG data acquire by undergoes experimental procedure the signal can be extracted in term of three main domain: (1) time domain , (2)frequency domain and (3)time-frequency domain[7,9] But commonly the raw EMG signal extracted in term of two different type domain which is time domain and frequency domain. In time domain features, raw EMG signal extracted into five features which are sum absolute value (IEMG), Mean absolute value (MAV), Root Mean Square (RMS) value ,standard deviation (STD), and Variance (VAR). In addition, in frequency domain by using Fast Fourier Transform (FFT) had been extracted into five features which are Mean value, Absolute median value, Absolute Root Mean Square, Maximum Power Spectrum (MaxPS) value and Minimum Power Spectrum (MinPS) value[7,9].

2.3.1 Time domain Features Extraction

Generally, time domain features is the most simple analysis of sEMG signal. In this analysis basically can be extracted the sEMG signal in term of sum absolute value (IEMG), Mean absolute value (MAV), Root Mean Square (RMS) value ,standard deviation (STD), and Variance (VAR).All this features commonly used to learn a pattern of sEMG signal in the machine learning system[6].

2.3.1.1 Integrated EMG

Integrated EMG (IEMG) is calculated as the sum of absolute values in the sEMG signal amplitude. Generally, IEMG is act as an onset index to detect the muscles activity that used for control command of assistive control device. It is related to the sEMG signal sequence firing point, which can be expressed as

$$IEMG = \sum_{n=1}^{N} |x_n| \quad (1)$$

Where N denotes the length of the signal and x_n represents the sEMG signal in segment.

2.3.1.2 Mean Absolute Value

Mean Absolute Value (MAV) is equivalent to average rectified value (ARV). It can be calculated using the moving average of full-wave rectified EMG. On the others word, it is calculated by taking the average of the absolute value of sEMG signal. It is an easy way for detection of muscle contraction levels and it is a popular features used in myoelectric control application. It is represent as

$$MAV = \frac{1}{N}\sum_{n=1}^{N}|x_n| \qquad (2)$$

2.3.1.3 Root Mean Square

Root mean square (RMS) is modeled as amplitude modulated Gaussian random process whose RMS is related to the constant force and non-fatiguing contraction[6]. It is relates to standard deviation, which can be defined as

$$RMS = \sqrt{\frac{1}{N}\sum_{n=1}^{N} {x_n}^2} \qquad (3)$$

2.3.1.4 Standard deviation of EMG

Standard deviation is commonly used to measure confidence in statistical conclusion of sEMG signal. A low standard deviation indicate that the data points tend to be very close to mean and vice versa if a high standard deviation value. Generally, the standard deviation of experimental data, and only effects that fall far outside the range of standard deviation which is consider statistically significant.

$$STD = \sqrt{\frac{1}{N-1}\sum_{n=1}^{N} {x_n}^2} \qquad (4)$$

2.3.1.5 Variance of EMG

Variance of EMG (VAR) uses the power of the sEMG signal as feature. Generally, the variance is the mean value of square of deviation of that variable. However, mean of EMG signal is close to zero. It can be shown as

$$VAR = \frac{1}{N-1}\sum_{n=1}^{N} {x_n}^2 \qquad (5)$$

2.3.2 Frequency Domain

In frequency domain, commonly by generate power spectrum of Fast Fourier Transform (FFT) as the tool to extract sEMG signal in term of five features component which is Mean value , Absolute median value, Absolute Root Mean Square, Maximum Power Spectrum (MaxPS) value and Minimum Power Spectrum (MinPS) value[7,9].

2.3.2.1 Absolute Mean value

Mean value is the most common extraction features technique that used had been used for pattern recognition of sEMG purpose. Generally can be expressed as

$$AMV = abs(\frac{1}{N}\sum_{n=1}^{N} x_n) \qquad (6)$$

2.3.2.2 Absolute median value

Absolute median value is a robust estimator dispersion that used to estimate the dominant frequency for sEMG signal. In other word can be define as

$$AMDV = abs\ (median\ of\ EMG\ value) \tag{7}$$

2.3.2.3 Absolute root mean square
Absolute root mean square is basically the absolute value of root mean square of the sEMG signal that had been FFT. It can be represent as

$$ARMS = abs(\sqrt{\frac{1}{N}\sum_{n=1}^{N} x_n{}^2}) \tag{8}$$

2.3.2.4 Maximum Power Spectrum (MaxPS)
Maximum Power Spectrum (MaxPS) is maximum amplitude frequency spectrum via a Fourier Transform of the sEMG signal. For this project performed spectrum analysis of the entire signal. It can be simplified as

$$\begin{aligned} z &= abs(Y) \\ MaxPS &= \max(z) \end{aligned} \tag{9}$$

Where Y is power spectrum of sEMG signal and z is absolute value of Y.

2.3.2.5 Minimum Power Spectrum (MinPS)
Minimum Power Spectrum (MinPS) is minimum amplitude frequency spectrum via a Fourier Transform of sEMG signal.

$$\begin{aligned} z &= abs\ (Y) \\ MinPS &= \min(z) \end{aligned} \tag{10}$$

Where Y is power spectrum of sEMG signal and z is absolute value of Y.

2.4 Testing the robustness of features
After extracted all the features, the next to test the robustness of the features. In this paper present the technique without using machine learning to indicate the best features [9] as shown in equation 11 and If the error to mean percentage less than 6% thus the features choose as the best features for sMEG signal [9].

$$EMP(\%) = \left|\frac{x_i(actual) - \mu(mean)}{\mu(mean)}\right| \times 100\% \tag{11}$$

3. Result and Discussion
The result show the examples of signal acquire from experiment. This graph show that from first of samples data until 18866 sample data, subject was doing isometric contraction of elbow flexor. Then, from 18867 sample data until 25000 sample data subject was relaxing of elbow flexor. During the isometric contraction of elbow flexor show that the amplitude of voltage increase dramatically until reach the full contraction of biceps muscle then will decrease drop when muscle start to relax until full relax from any others muscles activity.

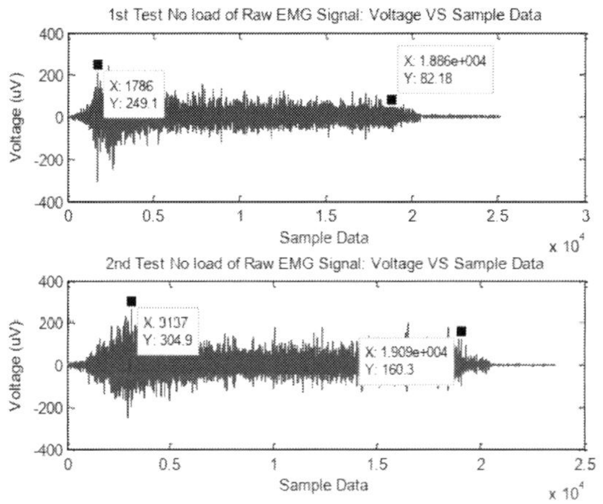

Figure 3. Show example of sEMG signal acquire from same experiment setup. Noted that the reading of sEMG voltage give different amplitude voltage value for each repeated task. Thus need to be get the mean value in order to calculate the error to mean percentage.

Table 1: Error to mean percentage for time domain statistical analysis of raw sEMG signal show that IEMG is the lowest value which is less than 10% of overall error to mean percentage compare with RMS, STD, MAV,and VAR

Error to mean percentage(%)	Sample subject A				Sample subject B			
	0kg task	1kg task	2kg task	3kg task	0kg task	1kg task	2kg task	3kg task
IEMG	5.52	4.50	8.16	8.62	9.45	8.85	5.82	7.29
MAV	2.23	1.28	4.74	5.47	12.37	44.06	4.98	13.74
RMS	1.27	2.93	3.60	6.66	9.83	30.44	2.75	11.33
STD	1.85	2.93	3.60	6.66	9.98	30.44	2.76	11.33
VAR	4.37	5.82	7.25	13.25	19.47	55.72	5.52	22.37

Table 2: Error to mean percentage for frequency domain analysis of power spectrum sEMG signal show that the lowest error to mean percentage is MaxPS (less than 10%) followed by ARMS,AMDV,MinPS and lastly AMV.

Error to mean percentage(%)	Sample subject A				Sample subject B			
	0kg task	1kg task	2kg task	3kg task	0kg task	1kg task	2kg task	3kg task
AMV	36.26	57.93	15.13	42.59	40.22	30.92	35.22	37.10
AMDV	9.67	11.57	23.77	21.88	31.37	43.57	32.96	31.83
ARMS	4.56	5.83	9.54	4.91	19.57	31.33	4.95	34.66
MaxPS	3.68	4.52	0.43	8.83	11.35	6.80	3.28	3.27
MinPS	15.36	16.27	40.88	44.31	50.10	29.36	19.83	4.92

By refer to the table 1 and table 2 can be conclude that the best features is MaxPS because overall error to mean percentage record 5.27% error compare to others features.

The graph directly map from 2 sample subject data acquired in this experiment shown that the relationship between MaxPS features and torque experience by elbow joint to lift hand. Figure below shows linear relationship between MaxPS and torque experienced by elbow joint while was doing isometric contraction task using different load.

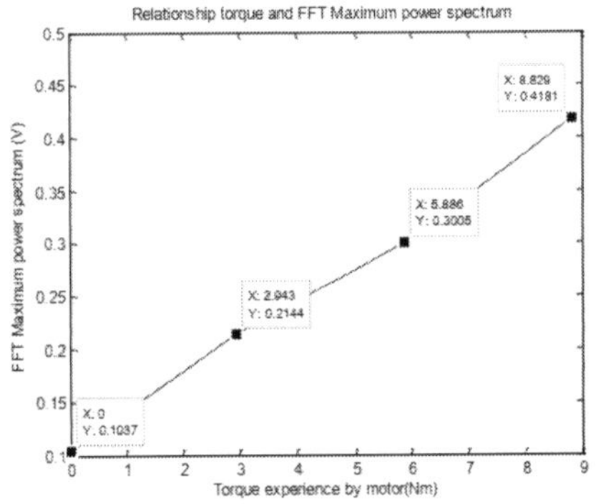

(a)

(b)

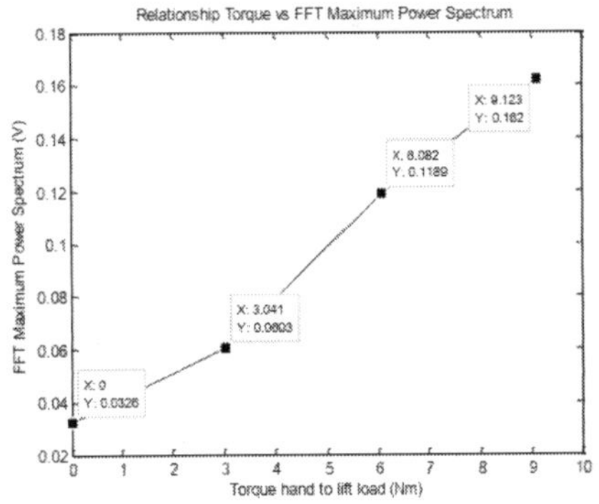

Figure 4 (a). The graph plot from data acquire by subject sample A. The graph show the relationship between MaxPS and torque experience by elbow joint for isometric contraction task using different load.

Figure 4 (b). The graph plot from data acquire by subject sample B. The graph show the relationship between MaxPS and torque experience by elbow joint for isometric contraction task using different load.

Figure 4 (a) shows that maximum amplitude of MaxPS value is 0.4181 compare to 0.162. This is because the difference in muscle strength (due to diameter of biceps brachii). It can be predicted that if the person have the larger diameter of biceps brachii, it will show bigger value activation voltage of muscle during isometric contraction[8].

CONCLUSION

As the conclusion, the best features is Maximum Power Spectrum (MaxPS) of Fast Fourier Transform because it has the lowest error than other features which is 5.17% and show the linear relationship with torque experience by elbow joint during isometric contraction.

References

[1] R M S and S Chatterji August 2012 Trends and Challenges in EMG based Control Scheme of Exoskeleton Robots –A Review *Int. J. of Sc. & Eng. Research* **3** 1-8
[2] Q C Ding, A B Xiong, X G Zhao, and J D Han 2011 A novel EMG-driven state space model for estimation of continous joint movements *IEEE Int. Conf.* 2891-2897
[3] T An-Chih, L Jer-Junn, and L Ta-Te 2012 A modified multi-channel EMG feature for upper limb motion pattern recognition *IEEE Ann. Int. Conf.* 3596-3599
[4] T Beravs, P Rebersek, D Novak, J Podobnik, and M Munih 2011 *Humanoid 11th IEEE-RAS Int. Conf.* 212-217
[5] Adam B Zoss, H Kazeronni and A Chu 2006 *IEEE/ASME Trans. On Mechatronic* **11** 128-138
[6] E A Clancy, E L Morin and R Merletti 2002 *J. of Electromyography and Kinesiology* **12** 1-116
[7] M Reaz, M Hussain and F Yasin 2006 *Biol. Proc. Online* **8** 11-35
[8] K Peter 2005 The ABC of EMG: *A practical introduction to kinesiological Electromyography* ***1***
[9] C L Angkoon Phinyomark and Pornchai Phukpattaranont 2009 A Novel Feature Extraction for robust EMG Pttern Recognition *J. of Computing* **1** 71-80

IOP Conf. Series: Materials Science and Engineering **53** (2013) 012020 doi:10.1088/1757-899X/53/1/012020

An investigation of interference coordination in heterogeneous network for LTE-Advanced systems

M. K. Hasan[1], A.F. Ismail, Aisha-Hassan A. H., Khaizuran Abdullah, H.A.M. Ramli
Department of Electrical and Computer Engineering, Faculty of Engineering, International Islamic University Malaysia, Kuala Lumpur.

E-mail: hasankamrul@ieee.org

Abstract. The novel "femtocell" in Heterogeneous Network (HetNet) for LTE-Advanced (LTE-A) set-up will allow Malaysian wireless telecommunication operators (Maxis, Celcom, Digi, U-Mobile, P1, YTL and etc2.) to extend connectivity coverage where access would otherwise be limited or unavailable, particularly indoors of large building complexes. A femtocell is a small-sized cellular base station that encompasses all the functionality of a typical station. It therefore allows a simpler and self-contained deployment including private residences. For the Malaysian service providers, the main attractions of femtocell usage are the improvements to both coverage and capacity. The operators can provide a better service to end-users in turn reduce much of the agitations and complaints. There will be opportunity for new services at reduced cost. In addition, the operator not only benefits from the improved capacity and coverage but also can reduce both capital expenditure and operating expense i.e. alternative to brand new base station or macrocell installation. Interference is a key issue associated with femtocell development. There are a large number of issues associated with interference all of which need to be investigated, identified, quantified and solved. This is to ensure that the deployment of any femtocells will take place successfully. Among the most critical challenges in femtocell deployment is the interference between femtocell-to-macrocell and femtocell-to-femtocell in HetNets. In this paper, all proposed methods and algorithms will be investigated in the OFDMA femtocell system considering HetNet scenarios for LTE-A.

1. Introduction

Nowadays, the wireless subscriber increased exponentially day by day wireless industry challenged with an increasing demand for ubiquitous wireless coverage and larger data rates. To support this highly demand for data traffic, Third Generation Partnership Project (3GPP) Long Term Evolution (LTE) Release 8 is tried experienced by the cellular operators. The release 8 standard offers significant advantages with respect to its prototype, High Speed Packet Access (HSPA), such as higher spectral efficiency, lower latency due to its flat all internet Protocol (IP) architecture, and larger throughputs [1]. However, the performance of Release 8 does not meet the International Mobile Telecommunications (IMT) Advanced requirements for the fourth generation of mobile networks defined by the International Telecommunication Union (ITU). Accordingly, to meet such necessities which is for downlink data rates of up to 100 Mb/s and 1 Gb/s for mobile and nomadic users,

[1] Mohammad Kamrul Hasan, email: hasankamrul@ieee.org
Department of Electrical and Computer Engineering,
Faculty of Engineering, International Islamic University Malaysia

IOP Conf. Series: Materials Science and Engineering **53** (2013) 012020 doi:10.1088/1757-899X/53/1/012020

respectively, LTE Release 10 is now under standardization. LTE-A femtocells (HeNodeBs) are viewed as a promising option for mobile operators to improve coverage and provide high-data-rate services in a cost-effective manner by reducing the macro-eNodeB traffic load and offloading it over public broadband connections to the core network. Though HeNodeB technology reduces the cost but at the same time introduced complexity of deploy higher-capacity links to the eNodeB. Furthermore, Deploying large number of HeNodeBs in the indoor/outdoor environment is also a critical problem for the synchronization. HeNodeBs synchronization and control is centralized using IEEE1588. Since HeNodeBs depends on third party broadband operators for the control purpose which creating many problems. Synchronization is one of the principle concerns because of high traffic and limited broadband services. HeNodeBs synchronization is very important in order to avoid the interferences. The co-channel deployment in macro-eNodeB and HeNodeBs could increase the capacity of the network manifold through high spatial frequency reuse. However co-channel deployment in macro-eNodeB and HeNodeBs results interference in the network the ultimate is total system performance degradation and this interference becomes a key challenge in HetNet. The key challenges for LTE HetNets include backhaul for the small cells and effective use of interference cancellation so that the various overlapping cells do not interfere with one another.

This paper is organized as follows: section 2 presents some heterogeneous network architecture for LTE-A system. Section 3 presents related works and section 4 presents the Performance Analysis of Self-organized Approaches. Conclusion is given in Section 5.

2. Heterogeneous Network Architecture for LTE-A System

A Heterogeneous Network (HetNet) is the result of an operation approach consisting of two or more cellular layers [2]. Consequently the resulting network comprises a various mix of base-stations types such as macro-eNodeB, micro- eNodeB, Pico- eNodeB and more in recent time's HeNodeBs (femtocells). In HetNet among the low-power nodes, HeNodeB and Pico-eNodeBs will play a key role. As HeNodeB, Pico-eNodeBs are low-power nodes that are typically deployed by operators within the coverage areas of eNodeB for capacity enhancement and coverage extension. In figure 1 HetNet network scenario is decomposed. Pico-eNodeBs usually have the same back-haul and access features as macro-eNodeBs. However, Pico-eNodeBs have lower transmit powers ranging from 23 to 30 dBm, and serve tens of User Equipment's (UEs) within a coverage range of up to 300 m. The roll-outs of Pico-eNodeBs are expected to offload eNodeBs and increase network capacity. However, they also face technical challenges arising from the large difference in Down Link (DL) transmit powers among eNodeB (≈46 dBm) and pico BSs (≈30 dBm) [3]. However, HetNets are not a completely new concept; there is much unexploited potential branching from new network topologies. LTE-Advanced includes features that improve the support for heterogeneous deployments such as eenhanced Inter-Cell Interference Coordination (eICIC).
To get rid of these issues and offer a noteworthy network performance leap, HetNets have been initiated in the LTE-standardization. LTE-A is a flat network based on packet only RAN architecture where macro-eNodeBs are interlinked through X2 interfaces. The basic principles for LTE-A are as bellows:

1. OFDMA (Orthogonal frequency division multiple access) in downlink.
2. SC-FDMA (Single carrier frequency division multiple access) in uplink
3. Scalable spectrum use from 1.4MHz to 20MHz.
4. Localized or distributed resource allocation for frequency selective or frequency diverse scheduling
5. Support for spatial multiplexing (MIMO/MU-MIMO)
6. Frequency and time division duplex for paired and unpaired spectrum

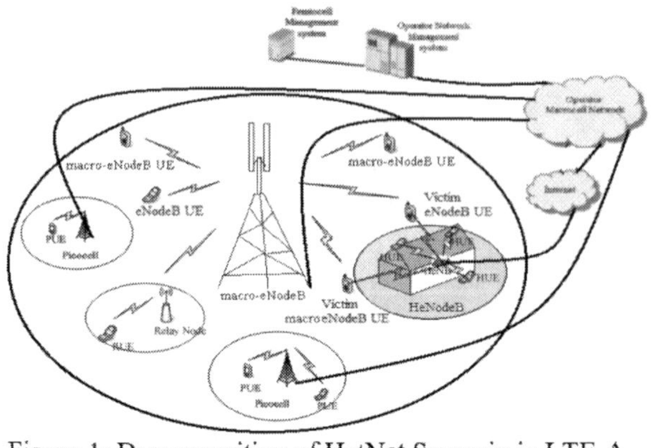

Figure 1: Decomposition of HetNet Scenario in LTE-A

2.1. OFDMA for Downlink

The OFDM signal used in LTE includes a maximum of 2048 different sub-carriers with a spacing of 15 kHz. Although it is obligatory that the mobiles to be accomplished for receiving all 2048 sub-carriers. Here not all are required to be transmitted by the base station; rather it is only required to support the transmission of 72 sub-carriers. Like this way all mobiles will be able to communicate with any base station. Inside the OFDM signal it is likely to select between three types of modulation which are QPSK (= 4QAM) 2 bits per symbol, 16QAM 4 bits per symbol, and 64QAM 6 bits per symbol. The precise format is selected based on the predominant circumstances. QPSK, the lower forms of modulation, do not necessitate such a big signal to noise ratio but are not capable of sending the data in a faster rate. The larger order modulation format can be used only when a satisfactory signal to noise ratio exists.

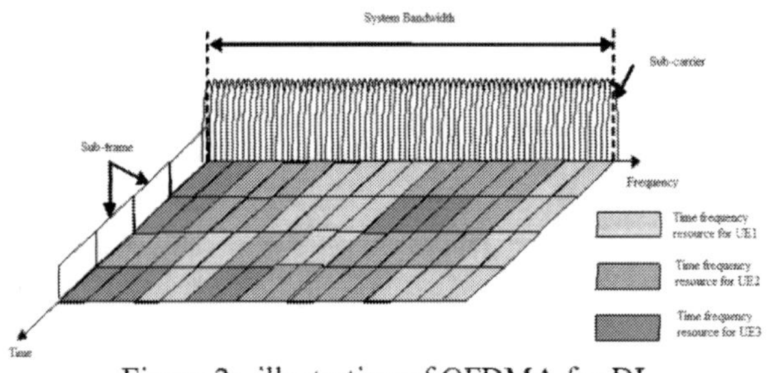

Figure 2: illustration of OFDMA for DL

In case of downlink, the subcarriers are divided into resource blocks which empower the system to be capable of arranging the data across standard numbers of subcarriers compartment wise. Resource blocks comprise of 12 subcarriers, one slot in the time frame irrespective of the general LTE-A femtocell signal bandwidth (see in figure 2). It can be understood that dissimilar LTE signal bandwidths will have diverse numbers of resource blocks.

2.2 SC-DMA for Uplink

For the LTE uplink, a different perception uses of the access technique while still using OFDMA technology, the implementation is called Single Carrier Frequency Division Multiple Access (SC-FDMA). A major parameter that has an effect on all mobiles is that of battery life. Despite the fact that the performance of battery is being upgraded continuously, it is still crucial to assure that the mobiles use as little battery power as possible. With the RF power amplifier that transmits the radio frequency signal through the antenna to the base station being the maximum power item inside the mobile, it is essential that it functions as competent mode as possible. It can be meaningfully affected by the procedure of radio frequency modulation and signal format. Signals containing a large peak to average ratio and necessitate linear amplification do not lend themselves to the usage of efficient RF power amplifiers. Consequently the implication of a transmission mode has a continuous power level while in function. However, OFDM contains a high peak to average ratio. While this is not a difficulty for the base station where power is an imprecise problem, it is unsuitable for the mobile. Thus, LTE makes use of a modulation method addressed as SC-FDMA- Single Carrier Frequency Division Multiplex which is hybrid format and integrates the low peak to average ratio, where offered by single-carrier systems along with the multipath interference resilience as well as flexible subcarrier frequency allocation offered by OFDM.

3. Related works

Nowadays, different industrial challenges towards large deployment of HeNodeBs have been deliberated and researcher has considered it for achieving the solution. The interference is vastly occurred in macro-eNodeB-to-macro-eNodeB, HeNodeB-to-HeNodeB, and macro-eNodeB-to-HeNodeB which eventually worsen system performance. With the increasing number of cells the number of users at cell edges suffers from low throughput caused by interference [4]. However, interference for LTE-A systems is regarded as a key challenge in heterogeneous multi-cell networks where HeNodeBs makes use of identical licensed frequency spectrum with macro-eNodeB. Consequently, Interference is a noteworthy issue connected with HeNodeB within HetNets. Several key issues regarding the interference should be studied for ensuring that the deployment of any HeNodeBs in HetNet will occur effectively. The query increases from the circumstance that HeNodeBs will apply the spectrum formerly allocated for cellular telecommunications. The HeNodeBs will be deployed where it can be termed an ad-hoc fashion; devoid of the network planning which is considered for the deployment of cellular telecommunications base stations in general. Therefore interference will be increasing to a greater extent. Certain difficulties will arise with the main network causing unexpected performance level degradation by both the HeNodeB-UEs and people who may be collaborating through the principal cellular network. Consequently, the level of performance trims down which is notified as a challenging concern for the telecom operators with the intention of managing the interference from the HetNets HeNodeBs-HeNodeBs and HeNodeBs-macro-eNodeB. Nevertheless, for enhancing the overall performance of the network an appropriate interference management is needed in OFDMA for the HetNet of LTE-A systems. Accordingly, an actual investigation is needed and the difficulties should be encountered for enhancing the performance of the HetNet. A number of requirements and parameters to be used for HeNodeB self-organization have been recognized in [5]. The uplink interference in two-tier HeNodeB networks was evaluated in [5], illustrating that tier-based open access can lessen the interference and offer an advancement in the network-wide area spectral efficiency – the feasible number of HeNodeBs and macro-eNodeB UEs per cell-site. Nevertheless, the evaluation of a self-organizing technique is still required and extra requirements and parameters are needed to investigate. Identical conclusions were established in different simulation-centric studies accomplished by the 3GPP RAN 4 group [6-8]. Downlink network capacities under open and closed access were explored in [7]; possible combinations of HeNodeBs and macro-eNodeBs under the restriction of network interference were scrutinized in [6]. Different developments were detailed in [8] for comparing HeNodeB open and closed access. All these simulations illustrate that with adaptive open access, the interference in two-tier networks is diminished and the deployment of co-channel HeNodeBs becomes feasible. Nevertheless, as HeNodeBs are mounted and paid for by their owners, it is required to assess their damage of HeNodeB resources in open access. It is significant that the advantages of reduced interference are not undermined by the loss of HeNodeB resources, namely over-the-air (OTA) and backhaul capacity.

4. Performance Analysis of Self-organized Approaches

Macro-eNodeB networks require complex and expensive manual planning and configuration. The main functionalities of SON for integrated HeNodeB and macro-eNodeB networks are self-configuration, self-optimization, and self-healing. The self-configuration function includes smart frequency allocation among HeNodeB neighbour networks, self-optimization feature includes optimization of transmission power HeNodeBs neighbour networks, maintenance of adjacent cell list, coverage control, and robust mobility management; and self-healing feature includes automatic detection and resolution of most failures. In hybrid solution SON logic is divided between network management system and network elements. However, decentralized approaches also need to be more concern for SON. In OFDMA based technology, an option for the femtocell is to select those subcarriers not being in use by the macro-eNodeB network [9]. In order to solve the problem for better system performance of the self-organization of Orthogonal Frequency Division Multiple Access (OFDMA) HeNodeBs needs more focus on dynamically air interface and tune its sub-channel allocation to reduce inter-cell interference and enhance system capacity [10]. SON permit HeNodeBs to associate themselves into the network of the operator and learn about their neighbouring cells, interference and tune their parameters (power, frequency) consequently [11].If HeNodeBs and macro-eNodeBs share the similar spectrum, interference difficultyrises [12]. In [13], a power control method for pilot and data channels is introduced in UMTS networks that assure a continuous coverage HeNodeB radius. Every HeNodeB sets its power to a value that is almost equivalent to the power received from the nearest macro-eNodeB at a target HeNodeB radius. However, with the purpose of fully benefit from LTE-A HetNet deployments, the main challenges for comprises interference. A number of self-organized interference management techniques and methods are discussed in the following subsections

4.1 Power Control Technique

In [4], the authors projected interference management techniques for both downlink and uplink of HeNodeBs operating based on HSPA. Authors suggested HeNodeB carrier selection and HeNodeB DL transmit power self-calibration techniques for interference management techniques for downlink. Uplink interference management was proposed using adaptive attenuation and limiting the transmitting power at the HeNodeB and as well as HeNodeB UEs. Figure 3 demonstrated that 10 units of macro-UE and 12 of HeNodeB UE per macro-eNodeB cell exist with the presence of HeNodeBs. There are 22 (10 + 12) UE units per cell operated by the macro when HeNodeBs are not present.

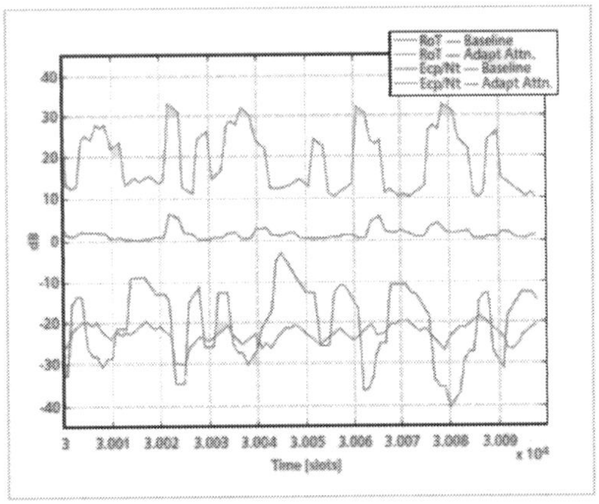

Figure 3: HSPA+ HeNodeB UL performance with strong interference [4].

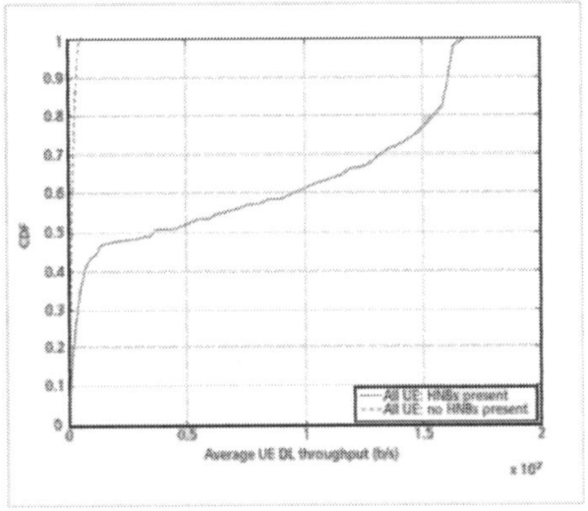

Figure 4: HSPA+ DL user throughput distributions on f1 [4].

Additionally, in figure 3 illustrates that the adaptive UL attenuation algorithm assures stable UL operation and better user experience with the existence of strong explode interference. Figure 4 illustrates the HSPA+ throughput cumulative distributed operations (CDFs) on the shared frequency with and without HeNodeB deployment for the DL and UL, correspondingly. 10 units of macro-eNodeB UE and 12 of HeNodeB UE per macro-eNodeB exist with the presence of HeNodeB UEs. Besides, 22 (10+ 12) UE units per cell served by the macro-eNodeB. Besides, it can be perceived from Figure 4 that when HeNodeB deployed significant capacity gains achieved which is advantageous for both macro-eNodeB and HeNodeB.

4.2 Resource Partitioning Technique

Co-tier interference is addressed by the undesirable signals received by UEs from co-channel LPNs. Co-tier refers to interfering signal coming from the similar network tier. As open access LPNs, namely Pico-eNodeBs, relay nodes, are planned by operators; the co-tier interference is less strong between them. ICIC techniques standardized by 3GPP in LTE/LTE-A can handle this interference effectually. The most severe co-tier interference occurs in HeNodeBs, deployment which typically functions in close access mode. HeNodeBs are usually deployed by end users and has lacking of powerful backhauls, henceforth fast reacting ICIC techniques are inaccessible [4]. On the contrary, co-tier interference is produced by HeNodeBs owing to the low isolation of walls or windows. Nevertheless, OFDMA HeNodeBs are capable of avoiding co-tier interference by assigning frequency resources appropriately among user in a greater time scale or by self-organizing methods. Nevertheless, cross-tier interference can be escaped by initiating a new interference coordination model in macro-eNodeB–Pico-eNodeB and HeNodeB heterogeneous networks.

In [15], interference avoidance is thoroughly analysed while using a time-hopped Code Division Multiple Access (CDMA) physical layer as well as sectorial antennas. As HeNodeBs are conceived to be user-deployed, it is of great interest that HeNodeBs is able toarrange themselves for dimishing interference towards the macro-eNodeB and selecting optimal resource allocation for transmissions. For instance, when OFDMA based technology is considered, an option for the HeNodeB is to choose those subcarriers not being in use by the macro-eNodeB network. For avoiding incessant collision within nearby OFDMA HeNodeBs, Long Term Evolution Advanced (LTE-A) system has implemented orthogonal frequency division multiplexing (OFDM) technique as the air-interface technology, the key target is to provide large wireless data rates for growing user demands. In [16], the necessities of LTE-A system are mentioned for supporting HeNodeB networks through enhancing the system capacity. Generally HeNodeB networks provide voice, data, video along with abundant wireless services within a small range of indoor coverage or inadequate geographical zone. These are raised by different companies as an important research field in the IMT-Advanced standardization exertions. Henceforth, it is indispensable to analyse the interference coordination schemes to improve the capacity of the HeNodeB networks. The inter-cell interference should be considered and the intra-cell interference can be avoided as the orthogonal frequency resource assigned in OFDM based cellular network. For reducing inter-cell interference, frequency reuse has been expansively regarded for wireless systems.

4.3 Hardware Centric Approaches

Authors in [17] presented hardware centric approaches for interference cancellation. The techniques employed either at MAC or physical layer to control or mitigate interference. Practically, in the cellular system the downlink and uplink characteristics are very different to increase the capacity of cellular systems. In the downlink each receiver only needs to decode a single desired signal from the K of intra-cell signals, while suppressing other cell interference from a few dominant neighbour cells as shown in figure 5. As all K users' signals originate at the base station, the link is synchronous and the K-1 intra-cell interferers can be orthogonalized at the base station transmitter [17]. However, some orthogonality is lost in the channel.

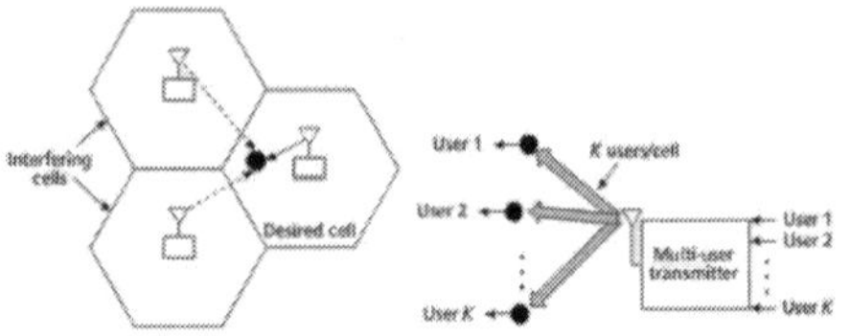

Figure 5: Downlink interference scenario [17]

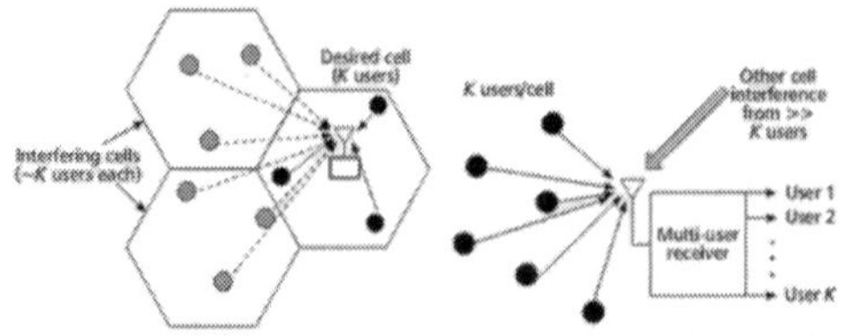

Figure 6: Uplink interference scenarios [17]

Conversely, in the uplink the base station receiver must decode all K desired users while suppressing other cell interference from many independent sources, as shown in figure 6.

4.4 Frequency scheduling Techniques

Authors in [18], a different type of interference challenges also observed where inter-carrier-interference (ICI) analysed. It is essential to manage the received interference by avoiding the use of those frequency resources at the HeNodeB network and exploited the spectrum resources competently. To mitigate the CSG HeNodeB deployment CCI problem between the HeNodeBs and macro-eNodeB UEs for both in UL and DL authors proposed an interference handling framework for near-by macro-eNodeB UEs and HeNodeBs. HeNodeB senses the spectrum during UL, identifies the close-by users through UL scheduling information, and avoids using their spectrum specified in DL scheduling information. However, this scheme utilizes the result of spectrum sensing in terms of energy detection with respect to the distance as availability of scheduling information. Authors in [18] proposed an interference-aware radio resource management scheme where receivers inform about their throughput, interference, and signal levels by means of broadcast messages tied to data reception. The proposed scheduler is structurally related to the non-interference aware Proportional Fair (PF) scheduler. They propose a signaling structure for TDD systems that enables distributed interference-aware scheduling. In which, the receivers transmit a small broadcast interference report after reception of data, which allows other transmitters to become active on the corresponding resources only in the case when it is beneficial for the overall system performance. Furthermore, authors outlined the signaling implementation and distinguished the overhead caused. The performance of interference-aware scheduler are estimated in numerical instances, it's also compared to both PF scheduler and the global optimum transmission schedule obtained by a centralized scheduler having full system wide information. However, in this proposed algorithm still there have research issues. For instance, the challenge stands that of the scalability with the amount of users per cell. In case the amount of UEs per cell grows, it becomes necessary to either time-domain multiplex the UEs over frames or subdivides the resource units to finer granularity. This would mean that the persistence would be partially lost or increased overhead.

4.5 Frequency Reuse Technique

In [19], an adaptive fractional frequency reuse (AFFR) scheme was proposed typically and implemented in planning cell coverage where SINR is calculated in accordance with the received signal power and interference power level. Furthermore, throughput attained through mapping the calculated SINR in accordance with the ideal link-adaptation based LTE link-level capacity. Single Input Single Output (SISO) system capacity is estimated through the following equation 1.

$$\left[S = \{ BW_{eff} * \log_2 \left(1 + \frac{\overset{0 SINR < SINR_{min}}{SINR}}{\underset{5.SINR > SINR_{max}\ 4}{SINR_{eff}}} \right) SINR_{min} \le SINR_{max} \right] \quad (1)$$

where Sis signified as estimated spectral efficiency in bps/Hz, which is upper limited according to the hard spectral efficiency given by 64QAM with coding rate 4/5; BW_{eff} adjusts for the system bandwidth efficiency of LTE and $SINR_{eff}$ adjusts for the SINR implementation efficiency of LTE. However, this scheme was not appropriate for the flexible deployment scenario as the HeNodeB networks.

In [24], the traditional frequency reuse strategy was proposed in on cellular clustering, which decreases inter-cell interference at the cost of both average cell throughput and spectrum efficiency. Originated from the idea of classical reuse clustering, fractional frequency reuse [25], soft frequency reuse [19] and partial frequency reuse [20] assigned the cell-edge users with advanced reuse factor than centre users in order to acquire an efficient factor larger than 1. A distributed utility based signal to interference plus noise ratio (SINR) adaptation in HeNodeBs was developed in [20] for reducing cross-tier interference at the macro-eNodeBs from co-channel HeNodeBs. An interference avoidance using a time hopped Code Division Multiple Access (CDMA) physical layer and sectorial antennas is investigated in [21]. Nevertheless these methodologies are typically founded upon wideband code division multiple Access (WCDMA) networks, and it is difficult to minimize interference through sub-channel allocation, since these are sophisticated features of current Orthogonal Frequency Division Multiple Access (OFDMA) systems. Furthermore in [21], the authors investigated a fixed frequency reuse scheme (FFR), which is competent for local area scenarios in the LTE-A system; though the restrictions on the frequency bandwidth reduce the spectrum efficiency.

In [22], authors projected the interference coordination between HeNodeBs in LTE-proposing CA-based interference coordination. In the Downlink LTE-A network with HeNodeB authors presumed few HUEs with very little or even no mobility linked to the HeNodeB in the small coverage area per cell for simplicity. Additionally, an analytical technique prolonged for the cases with more than one UE per cell [22]. For finding out the optimization issue author extended three equations. In the downlink transmission with N users and K carriers in the network are considered. The transmitting power P_T is remained identical in each carrier per cell and for the resource allocation among the users a binary matrix is also assumed as below, where $a_{k,n} = 1$ is denoted as that carrier k is assigned to user n, or else$a_{k,n} = 0$. $A = \{a_{k,n} | a_{k,n} \in \{0,1\}\} K \times N$. Henceforth, the achievable rate on carrier k in HeNodeB n is expressed by the following equation 2 [53].

$$\eta_{k,n} = Wlog_2\left(1 + \frac{L_{n,n}P_T}{\sum_{j=1,j\neq n}^{N} L_{n,j}a_{k,j}P_T + \sigma^2 N}\right), \quad (2)$$

$$\text{for } 1 \leq k \leq K, 1 \leq n \leq N$$

where W is the carrier bandwidth $L_{n,j}$ is the path loss (PL) from the jth user's serving HeNodeB to UE n, and $\sigma^2 N$ is the noise power of the additive white Gaussian noise (AWGN). The key objective is to find A to interference coordination difficulty namely an objective function is optimized. Typically, the subsequent optimization difficulties with different objectives are required to be solved for interference coordination. Furthermore, authors in [23] further added that in order to coordinate the interference problem the maximization of throughput and maximization of the proportional fair with different objectives are required. To attain the maximum possible system spectrum efficiency throughput maximization is essential which can be formulated as equation 3.

$$\max_{A} \sum_{n=1}^{N} \sum_{k=1}^{K} a_{k,n} \eta_{k,n} \quad (3)$$

For attaining the maximum system throughput while assuring proportional fairness among HeNodeBs, the total of the logarithmic average cell throughput should be maximized as in equation 4.

$$\max_{A} \sum_{n=1}^{N} \log\left(\sum_{k=1}^{K} a_{k,n} \eta_{k,n}\right) \quad (4)$$

Furthermore, author's also organized two steps based carrier aggregation (CA). The first step is founded on the measurement at HeNodeB of the inter-cell interference. Every HeNodeB is allocated to a carrier that is on-overlapping with its interfering HeNodeBs or involves the least interference. To assure the fairness between HeNodeBs, a single carrier is allocated to each HeNodeBs.

In view of the second step, the carriers already used by a HeNodeB are shared to other HeNodeBs for enhancing the spectrum efficiency of the network. It is based on the utility function calculated at HeNodeB UE in equation 5.

$$\xi_n^{(k)} = \frac{r_{n,n}}{\sum_{m\epsilon I_n^{(k)}} r_{n,m}} \qquad (5)$$

Where $r_{n,n}$ and $r_{n,m}$represent the RSRP measured at HeNodeB UE n and $m$from HeNodeB n and m, correspondingly, and $I_n^{(k)}$ embodies the set of interfering HeNodeBs of HeNodeB n which uses carrier *k*. If SIR is more than a specific threshold HeNodeB n decides to use the carrier *k*. It then sends a frequency reuse request to those HeNodeBs in $I_n^{(k)}$ which are currently using carrier k. The frequency reuse request message carries a utility value, which is determined by the different objectives. Regarding the maximum throughput or maximum proportional fair objective, the utility can be calculated by either of the formulas given as

$$U_{k,n}^{A} = \frac{g_n^{(k)}}{\left\|I_n^{(k)}\right\|} \qquad (6)$$

$$U_{k,n}^{B} = \frac{g_n^{(k)}}{\tilde{\eta}_n \cdot \left\|I_n^{(k)}\right\|} \qquad (7)$$

$$\tilde{U}_{k,m}^{A} = g_m^{(k)} \qquad (8)$$

$$\tilde{U}_{k,m}^{B} = \frac{g_m^{(k)}}{\tilde{\eta}_m} \qquad (9)$$

$g_m^{(k)}$symbolizes the throughput gain of HUE n if carrier k is granted to it (i.e. the throughput of HeNodeB UE *n* on carrier *k* which is the function of $\xi_n^{(k)}$.$\tilde{\eta}_n$is current throughput of HeNodeB *n* without using carrier *k*, and$\left\|I_n^{(k)}\right\|$signifies the size of set$I_n^{(k)}$. The equation 6 considers maximizing the sum throughput of all the HeNodeBs and equation 7 for fairness issues.

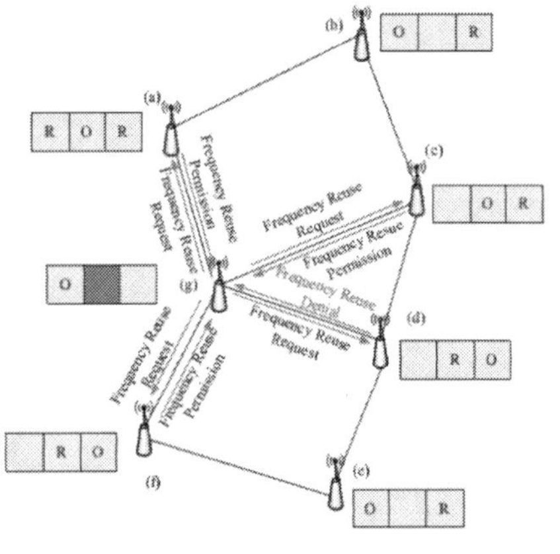

Figure 7: Adaptive Frequency Reuse process [23]

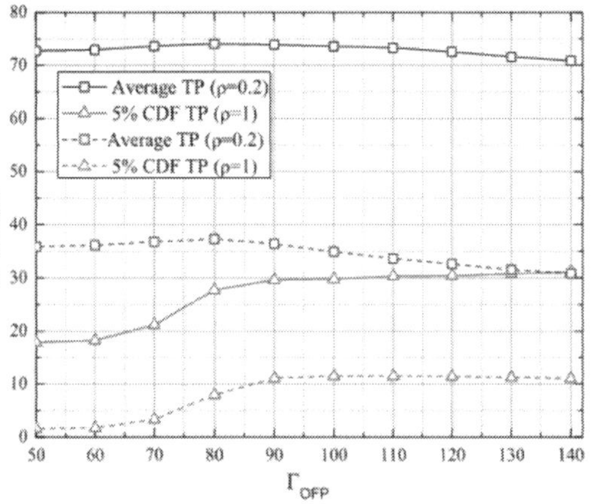

Figure 8: Throughput performances with different threshold in OFP [23]

An instance of the AFR process is illustrated in figure 7, where three carriers are assumed to be present in the network. In figure 7, "O" symbolizes for orthogonal carriers while "R" signifies for the reuse carrier. Once HeNodeB g is powered on and has selected its orthogonal carrier, it scrutinized whether it is able to reuse the other carriers with its neighbours. Here, in this case, second carrier is the candidate reuse carrier of HeNodeB g is assumed. Afterwards, a "Frequency Reuse Request" signalling is sent to other HeNodeBs by HeNodeB g belonging to the set $I_n^{(2)}$ which are using the second carrier. In the request signalling message, HeNodeB g sends the value of its $U_{2,g}^{A}$or $U_{2,g}^{B}$.After receiving the request signalling, the HeNodeBs in the set $I_n^{(2)}$HeNodeB *a, c, d* and *f*, calculate the i$\tilde{U}_{2,m}^{A}$ or $\tilde{U}_{2,m}^{B}$, $(m = a, c, d, f)$ values. When $\tilde{U}_{2,m}^{A} < U_{2,g}^{A}$, $\tilde{U}_{2,m}^{B} < U_{2,g}^{B}$, HeNodeB m will send a "Frequency Reuse Permission" signalling to HeNodeB g; otherwise, it will send a "Frequency Reuse Denial" signalling. HeNodeB g cannot use the second carrier until all the HeNodeBs feedback "Frequency Reuse Permission" signalling messages.

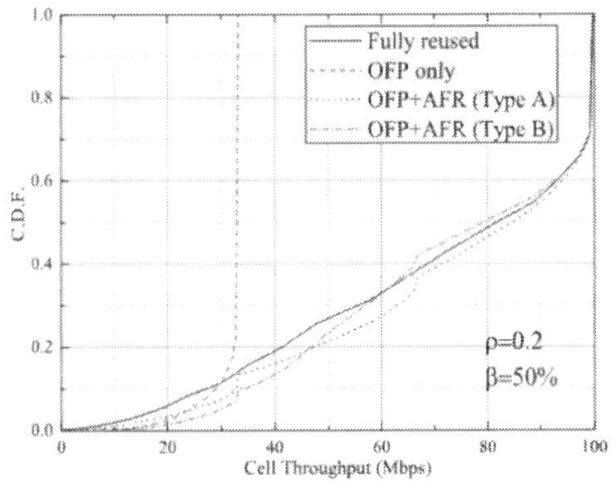

Figure 8: CDF of cell throughput in the network

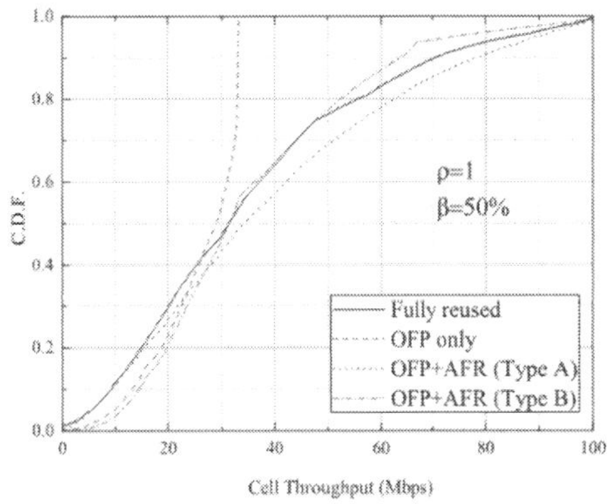

Figure 9: CDF of cell throughput [23]

In figure 7, the average throughput (TP) performances of the network along with developed channel allocation method are detailed while using numerous values of threshold in the first step, i.e. OFP. More average throughput can be achieved because carriers are reused by femtocells more often with the lesser value of the threshold. The CDF performances of the networks with or without the suggested interference coordination arrangement are compared in figure 8 and figure 9, where the deployment ratio $\rho$is 0.2 and 1, respectively.

5. Conclusion

With the intention of improving the coverage and capacity for the Malaysian Telecommunication service HeNodeB deployment is the critical challenge since all operators established their network with macrocell, microcell, radio heads, and relay station. In this paper, we investigated the HeNodeB deployment issues in heterogeneous network in LTE-A systems. Additionally in terms of HeNodeB adoption, the architecture for HetNets in LTE-A is also highlighted. Deploying HeNodeB in HetNets inter-cell interference was found one of the key issues. For interference mitigation a number of self-organizing approaches are evaluated with the performance analysis. From performance analysis it can be conclude that to increase the system throughput inter-cell coordination is highly recommended.

References

[1]. 3GPP, TR 36.814 (V9.0.0), "Further advancements for E-UTRA physical layer aspects," Mar. 2010.

[2]. A. Khandekar, N. Bhushan, J. Tingfang, and V. Vanghi, "LTE-advanced: Heterogeneous networks," in Wireless Conference (EW), 2010 European, pp. 978–982, 2010.

[3]. A. Khandekar, N. Bhushan, J. Tingfang, and V. Vanghi, *LTE- Advanced: Heterogeneous networks*, European Wireless 2010, pp. 978-982, Apr. 2010.

[4]. Y. Mehmet, M. Farhad, N. Sanjiv, P. Akhilesh, J. Nick, R. Balaji and R. Andy, Interference Management and Performance Analysis of UMTS/HSPA+ Femtocells, *IEEE Communications Magazine, IEEE Journal and Magazine,* 47(9),pp. 102 – 109, doi:10.1109/MCOM.2009.5277462

[5]. V.H. MacDonald, The cellular concept. *Bell System Technical Journal,*vo1.58. 1979.

[6]. K. Begain, G. I. Rozsa, A. P fening, and M.Telek. Performance analysis of GSM networks with intelligent underlay-overlay. IEEE ISCC 2002.

[7]. 3GPP RI-050507. Soft Frequency Reuse Schemefor UTRAN LTE. Ruawei. 2005.

[8]. 3GPP RI-060291. OFDMA Downlink Inter-Cell Interference Mitigation.Nokia. 2006.

[9]. A. Saleh, A. Rustako, and R. Roman, "Distributed Antennas for Indoor Radio Communications," IEEE Trans. Commun., vol. 35, no. 12, Dec. 1987, pp. 1245–51.

[10].C.-L. I, L. J. Greenstein, and R. D. Gitlin, "A Microcell/ Macrocell Cellular Architecture for Low- and High- Mobility Wireless Users," IEEE JSAC, vol. 11, no. 6, Aug. 1993, pp. 885–91.

[11].SOCRATES Project. Use cases for self-organising networks, 2008.

[12].E. Juan, M. Jan, "Analysis of Macro - Femtocell Interference and Implications for Spectrum Allocation." IEEE, International Symposium on Personal, Indoor and Mobile Radio Communications PIMRC 2009, September 2009.

[13].H. Claussen, L. T. W. Ho, and L. G. Samuel. An overview of the femtocell concept. Bell Labs Technical Journal - Wiley, 3(1):221–245, May 2008.

[14].H. Holma and A. Toskala, LTE for UMTS Evolution to LTE-Advanced. John Wiley & Sons, 2 ed., 2011.

[15].H. Andersson, R.S. Karlsson, P. Larsson, P. Wikstrom, "Improving system performance in a WCDMA FDD network using indoor pico base stations," IEEE VTC2002-fall,Sept-2002.

[16].V. Chandrasekhar, J. Andrews, A. Gatherer, "Femtocell Networks: a Survey," IEEE Communications Magazine, Sept. 2008.

[17].W. Yuanye, and I. P. Klaus, *Performance Analysis of Enhanced Inter-cell Interference Coordination in LTE-Advanced Heterogeneous Networks,* IEEE 75th Vehicular Technology Conference (VTC Spring), pp. 1-5, Jul. 2012, doi:: 10.1109/VETECS.2012.6240233

[18].Sahin ME., Guvenc .I, Moo-Ryong J., Arslan H., Handling CCI and ICI in OFDMA femtocell networks through frequency scheduling. IEEE Trans Consumer Electron 2009;55(4):1936–44.

[19].3GPP RI-050507. Soft Frequency Reuse Scheme for UTRAN LTE.Ruawei. 2005.

[20]. 3GPP RI-060291. OFDMA Downlink Inter-Cell Interference Mitigation. Nokia. 2006.

[21].M. Rahman, H. Yanikomeroglu and W. Wong. Interference Avoidance with Dynamic Inter-Cell Coordination for Downlink LTE System. IEEE WCNC2009.

[22].Andrews JG. Interference cancellation for cellular systems: a contemporary over-view. IEEE Communication Magazine 2005 Apr.

[23].Z. Jun, T. Hui, T. Peng, H. Yao, G. Liqi, Dynamic Frequency Reservation Scheme for Interference Coordination in LTE-Advanced Heterogeneous Networks, IEEE 75th Vehicular Technology Conference (VTC Spring), pp. 1-5, Ma. 2012, doi: 10.1109/VETECS.2012.6239915.

[24].V.H. MacDonald, The cellular concept. *Bell System Technical Journal,* vo1.58. 1979.

[25].K. Begain, G. I. Rozsa, A. Pfening, and M.Telek. Performance analysis of GSM networks with intelligent underlay-overlay. IEEE ISCC 2002.

IOP Conf. Series: Materials Science and Engineering **53** (2013) 012021 doi:10.1088/1757-899X/53/1/012021

Investigation of Three Dimensional Empirical Indoor Path Loss Models for Femtocell Networks

A Zyoud[1], J Chebil, M H Habaebi, Md R Islam, and A K Lwas
Department of Electrical and Computer Engineering, Faculty of Engineering, International Islamic University Malaysia (IIUM), 53100 Gombak, Kuala Lumpur, Malaysia

[1]E-mail: al7 z@yahoo.com

Abstract. Femtocell which is a home access point that installed by end consumers inside their houses is an important parameter and a promising technology in future wireless networks. It is proposed as a solution to the indoor propagation problems and to increase the indoor bandwidth. However, many challenges need to be addressed before deployment of femtocell. One of these challenges is the indoor propagation. Most of the available models are for long range communication like macro and micro cellular networks. Models for femtocell networks where the effects of walls and floors are appeared are necessary. In this paper eight different models of indoor propagation were studied and compared with measured data. Measurements were conducted in three story building. Four different scenarios with different numbers of penetrated walls and floors were considered. The results were statistical analysed by calculating the average error and the error standard deviation.

1. Introduction

Femtocells provides a scalable and cost efficient solution to overcome the dramatically increase in the demand of indoor wireless bandwidth. As the deployment of femtocells increases [1], it is necessary and very important to predict the effects of the new deployment on existing and adjacent networks. Moreover, it is important for energy consumption and frequency reuse techniques to calculate the minimum required power to transmit from a given femtocell at a given frequency, and in the same time provides acceptable coverage to the served users. During the signal propagation between the transmitter and the receiver, the signal strength degrades due to the interaction with the surrounding environment which is called path loss. Therefore, predicting the path loss is important in order to evaluate the wireless communication system. Moreover, path loss is a significant parameter that really curbs and reduces the efficiency and the coverage of the cellular networks.

In literature different types of propagation models were proposed. Most of the proposed models are for outdoor environment and for long rang networks like macro and microcellular networks. According to Saunders [2], these models are not accurate enough to be applied in small cells like femtocell. In short range networks it is more accurate to consider more specific details in addition to the direct path between the base station and the user equipment (UE). In this case the number of floors and walls in addition to the furniture are severely curbing the signal strength. Some site specific models were proposed in the literature [3, 4]. These models are based on simulation, ray tracing and

[1] To whom any correspondence should be addressed.

IOP Conf. Series: Materials Science and Engineering **53** (2013) 012021 doi:10.1088/1757-899X/53/1/012021

finite difference time-domain technique. However, site specific models are memory and time consumption. Moreover, these models are not suitable for other scenarios with different topology. Therefore, general site models which are based real measurement and experiment are still widely used [5-11]. Recently researches discussed the indoor to outdoor propagation [12-14]. However, indoor to indoor propagation is more significant. Letourneux *et al* [15] and Lee et al [16] applied the Cost231 multi walls multi floors model for indoor path loss calculations.

In this paper eight different models of indoor propagation were studied and compared with measured data. Measurements were conducted in three story building. Four different scenarios with different numbers of penetrated walls and floors were considered. This paper is organized as follows: the open literature indoor propagation models are presented in the next section. Experimental setup is explained in Section 3. Section 4 shows a comparison between models and measurements. Finally, a conclusion is given in Section 5.

2. Indoor Path Loss Propagation Models

In the last decades many of indoor path loss measurements were carried out in different indoor environments, such as houses, corridors and offices In this section, Indoor propagation models that are proposed in literature are presented, these models could be studied and discussed from different points of view such as slope, line of sight (LOS) or non line of sight (NLOS), one dimension or three dimensions. Some of these models considered the reference point at one meter for the indoor propagation and others are based on the free space losses (FSL). Next subsections discuss the available models: Cost231, multi walls multi floors (MWMF), ITU-R P1238, IEEE802.11, WINNER II, and ITU-R M2135.

2.1. Cost231 Models

This model was proposed based on real measurements conducted in European cities at two different frequencies 900 and 1800 MHz, then the model were scaled for other frequencies. Three different models for path loss were presented [5]. The Cost231 one slope model where the path loss is calculated as:

$$PL = L_0 + 10\, n \log(d) \tag{1}$$

where L_0 is the path loss (dB) at the reference distance which is one meter for indoor scenarios. It is calculated as in (2), n is the power decay index, and d is the distance between access point and UE in meter.

$$L_0 = 20 \log\left(\frac{4\pi}{\lambda}\right) \tag{2}$$

The second model is Cost231 linear attenuation model which is expressed as:

$$PL = L_{FSL} + \alpha d \tag{3}$$

This model indicates that there is a linear relation between the path loss and the access point-UE distance. L_{FSL} is the free space losses (dB) as in (4), α is the attenuation coefficient (dB/m), and d is the distance between the access point and UE (m).

$$L_{FSL} = 32.4418 + 20 \log(d) + 20 \log(f) \tag{4}$$

where f is the frequency in GHz and d is the distance in meter.

The previous two models are not considering the numbers of penetrated walls and floors between the transmitter and the receiver as in the real scenarios. Therefore the Cost231 multi walls multi floors model (Cost231 MWMF) was proposed as follows:

$$PL = L_{FSL} + L_c + \sum_{i=1}^{I} K_{wi} L_{wi} + K_f^{\left[\frac{k_f+2}{k_f+1} - b\right]} L_f \tag{5}$$

In this model, the attenuation due to walls and floors is added to the free space loss, L_C is a constant loss which determines from measurements results and it is close to zero [5], k_{wi} and k_f are the number traversed walls and floors respectively, L_f and L_{wi} are the losses of penetrated walls and floors respectively, and b is an empirical parameter.

2.2. The Multi Wall Multi Floor Model (MWMF)

This model is firstly proposed based on ray tracing at 5.2 GHz. Then it was verified using open literature results at different frequencies [6]. The path loss is expressed as

$$L_{MWF} = L_0 + 10\, n \log(d) + \sum_{i=1}^{I} \sum_{k=1}^{K_{wi}} L_{wik} + \sum_{j=1}^{J} \sum_{k=1}^{K_{fj}} L_{fjk} \quad (6)$$

The difference between this model and the Cost231 model is the second term in the MWMF model where n is determined between 1.96 and 2.03while it is 4 in the Cost231 one slope model and 2 in the Cost231 MWMF model. Comparison between these models where n is 4 was presented in [17]. In addition, The MWMF model shows a nonlinear relation between the traversed walls or floors and the penetration loss. L_{wik} and L_{fjk} are the loss of kth wall and floor that has been traversed (dB), K_{wi} and K_{fj} are number of traversed walls and floors, I is the number of wall categories, and J is the number of floor categories.

2.3. ITUR P1238-7 Model

The ITUR P1238 model was proposed for a wide range of frequency, 900 MHz to 100 GHz [7]. Path loss is calculated as follows:

$$PL = 20 \log(f) + N \log(d) + L_f(n) - 28 \quad (7)$$

where N is the distance power loss coefficient, f is the frequency (MHz), d is the distance in meter between the access point and UE (d >1 m), L_f is the floor penetration loss factor (dB), and n is the number of floors separate between the access point and UE. Since this model was proposed as a solution to the frequency reuse between floors, only number of floors is considered.

2.4. IEEE 802.11n Model

All the previous models could be defined as one slope models since only one value for n is considered. This model defined a breakpoint (BP) distance where the path loss before this BP follows the free space loss L_{FS} (slope of 2) as in (4) and after the BP the slope is increased up to 3.5 as [8]:

$$PL = L_{FSL}(Bp) + 10\, n \log(\frac{d}{d_{Bp}}) \quad (8)$$

The BP for the indoor scenario is 5 m.

2.5. WINNER II D112 V1.2 Model

The WINNER II channel mode was proposed for different scenarios of radio propagation such as indoor propagation, indoor to outdoor propagation, outdoor to indoor propagation, and outdoor propagation scenarios [9]. The model was applied at 2 and 5 GHz, then it was extended over frequency range 2-6 GHz. The model is divided to two parts, LOS part is presented as:

$$PL = 18.7 \log(d) + 46.8 + 20 \log(f_c / 5) \quad (9)$$

The second part is NLOS is expressed as:

$$PL = 20 \log(d) + 46.4 + 20 \log(f_c / 5) + 12\, n_W + F_L \quad (10)$$

$$F_L = 17 + 4(n_f - 1) \quad (11)$$

where f_c is the frequency in GHz, d is the distance between the access point and the UE in meter, n_w is number of walls, n_f is number of floors. Equations (10) and (11) show that the relation of penetrated walls or floors and path loss is constant and linear. It is increased by 12 dB for each wall and 4 dB for each floor.

2.6. ITU-R M.2135-1/ 3GPP TR 36.814 Model

WIINER II model was modified and accepted by the ITU-R M.2135 [10] and the 3GPP for indoor femtocell scenario [11]. It was modified based on real measurement carried out in China [18-23]. The model is proposed for LOS as in (12) and NLOS as in (13).

$$PL = 16.9 \log(d) + 32.8 + 20 \log(f_c) \quad (12)$$

$$PL = 43.3 \log(d) + 11.5 + 20 \log(f_c) \quad (13)$$

All previous models could be applied for three dimensional indoor environments. Next section is presenting the experimental setup where the measurements were carried out.

3. Experimental Setup

The measurements were conducted in three story building. The received power was measured continuously with one meter separation between two consecutive measurements. The experiment equipment in Figure 1 is including TP-LIINK modem router and Dell Inspiron laptop in addition to free open source networks scanner which called "inSSIDer" that measure the received signal strength and display the information as graphs and numbers.

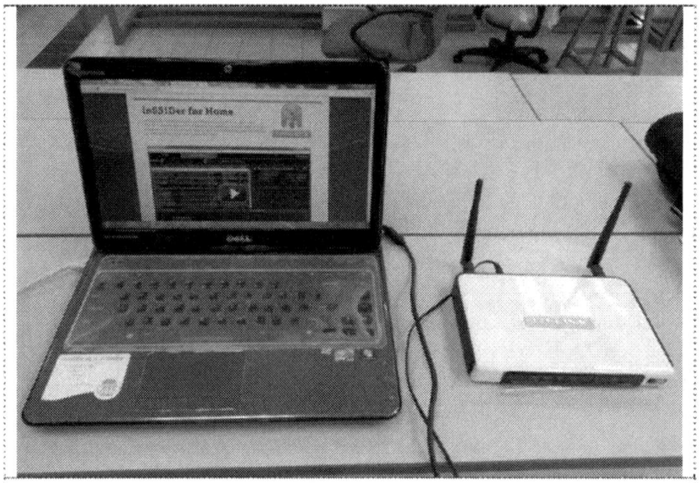

Figure 1. Experiment equipments.

The wifi access point was used instead of the femtocell since the femtocell works on liciance spectrum and it is very difficult to order and deploy. LTE networks in Malaysia operate at 2.6 GHz. Therefore, there is no significant difference between using wifi at 2.4 GHz and the femtocell access point.

The building consists of three floors with 4.3m ceiling height. The first floor is including three rooms as in Figure 2, while the second and the third floors have different dimensions and layout as in Figure 3. The access point was fixed in the first floor one meter above the ground, centered between the left and right walls, and one meter distance from the rare wall. Then using the laptop, the received signal strength was measured and converted to the path loss as:

$$PL = P_t + G_t + G_r - P_r \quad (14)$$

where P_t is the access point transmission power which is 19 dBm, P_r is the received power at the UE terminal, G_t is the transmitter gain which is 3 dBi, and G_r is the receiver gain which is 1 dBi.

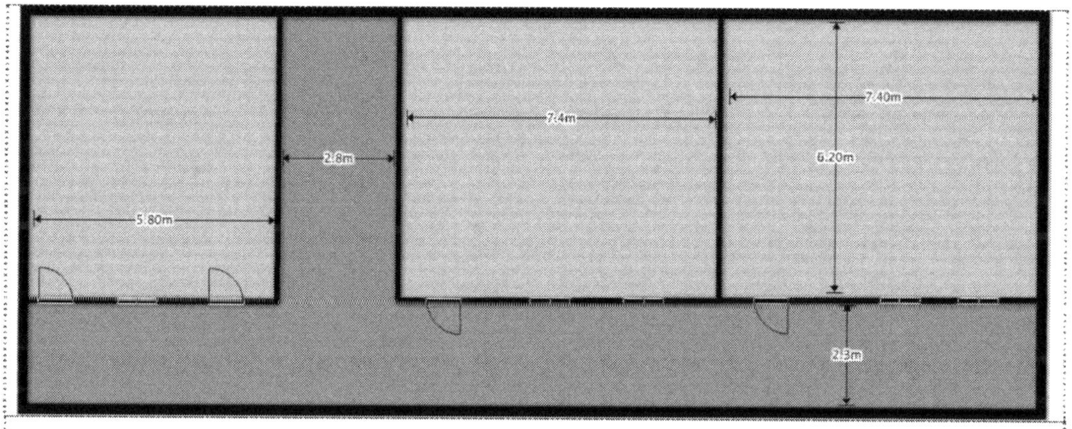

Figure 2. Layout of the first floor

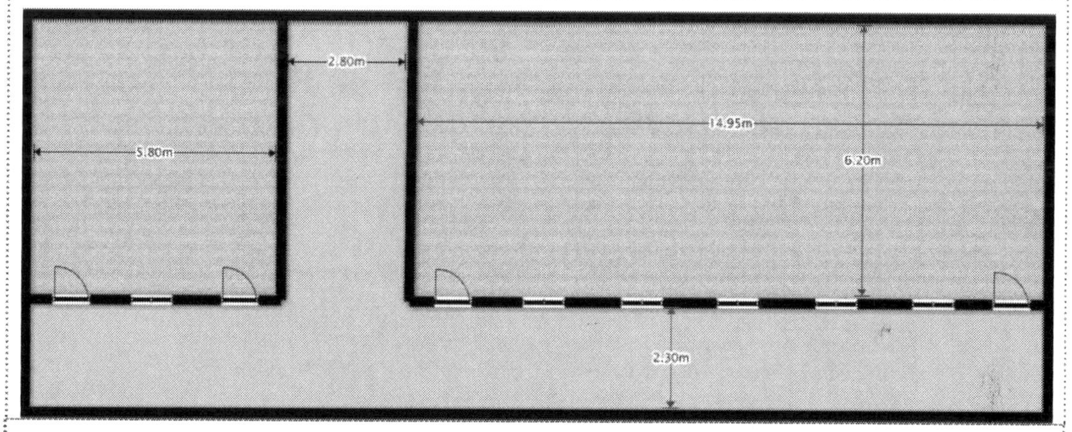

Figure 3. Layout of the second and the third floors

At each location 10 measurements were taken. Then the mean value of measurements at each point was compared with results obtained with the empirical models: Cost231, MWMF, ITU-R P1238, IEEE802.11, WINNER II, and ITU-R M213. After that the measurements were statistically compared with the empirical models as shown in next section.

4. Results and Discussions

The first scenario where is the access point and the UE are in the same room and LOS connection occurs between the transmitter and the receiver. The maximum distance is the length of the room which is 7.4m. The access point was one meter next to closest wall. The path loss exponent n for the Cost231 MWMF model is 4 [5] and for the MWMF model is 2 [6]. The distance power loss coefficient N in the ITUR model equals to 28 [7]. The value of the attenuation coefficient α in Equation (3) is 0.62 [5].

Results of measurements as well as predictions of path loss are given in Figure 4. It is noticed that the measured data is higher than all models at the reference distance. Then it is close to the one slope model at d equals 3 m. After d equals 5 m, the Cost231 one slope, ITUR 1238 and Cost231 linear models are overestimating the measured data. The IEEE802.11 model is close to the measured data while the rest models are underestimating the measured path loss.

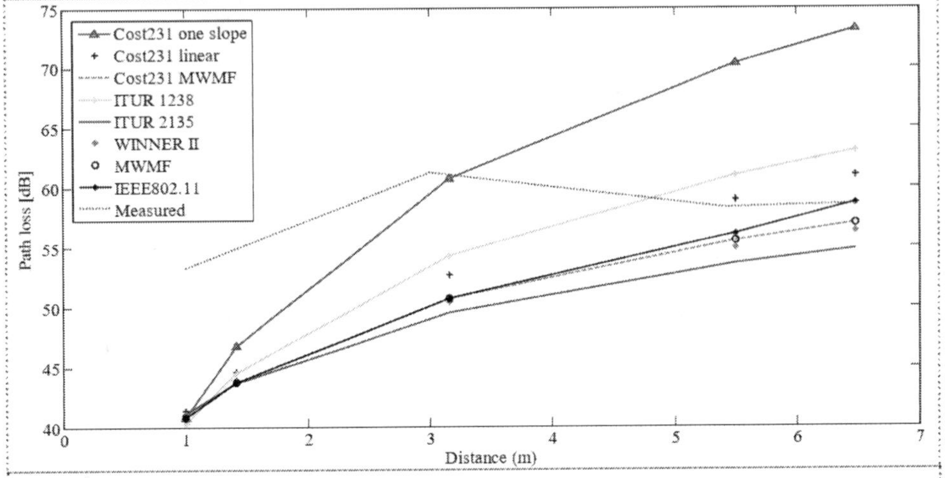

Figure 4. Comparison of measurement and empirical models for the first scenario

The results were statistically proved as in Table 1. Prediction error standard deviation for Cost231 linear model is the lowest between other models, which means it is the closest statistically. The positive mean for the Cost231 one slope model shows that the model is overestimating the average measured data by 1.41 dB. The rest of the models underestimate the average of the measured data by different values. The worst model in this scenario is the Cost231 one slope model since it has the highest error standard deviation with 4.685 dB.

Table 1. Statistical analysis of the first scenario

Model	Mean	Standard deviation	Error standard deviation
Cost231 one slope model	1.41	10.57	2.11
Cost231 linear model	-5.22	7.70	1.54
Cost231 MWMF model	-7.39	8.59	1.72
ITUR 1238 model	-4.31	8.01	1.60
ITUR 2135 model	-8.40	9.11	1.82
WINNER II model	-7.56	8.58	1.72
MWMF model	-7.39	8.59	1.72
IEEE802.11 Model	-6.93	8.53	1.71

In the second scenario, the access point and the UE are in the same floor where the UE is moving from LOS connection to NLOS connection with maximum three walls. The maximum distance is the length of the floor which is 23.85m. n, N and α are same as in the first scenario. L_w for Cost231 MWMF model is 6.9 dB [5]. L_{w11}, and L_{w12} in the MWMF model are 16 and 14 dB respectively [6]. Figure 5 shows the results for the measured data and the prediction models. The effects of walls in this scenario are clear enough in the figure. The path loss at the reference distance still 13 dB higher than the predicted models as in the first scenario. Linear, ITUR 1238, IEEE802.11 and ITUR 2135 models are close to each other but none of them is fitting the measured data. The closest model to the data is the WINNER II model.

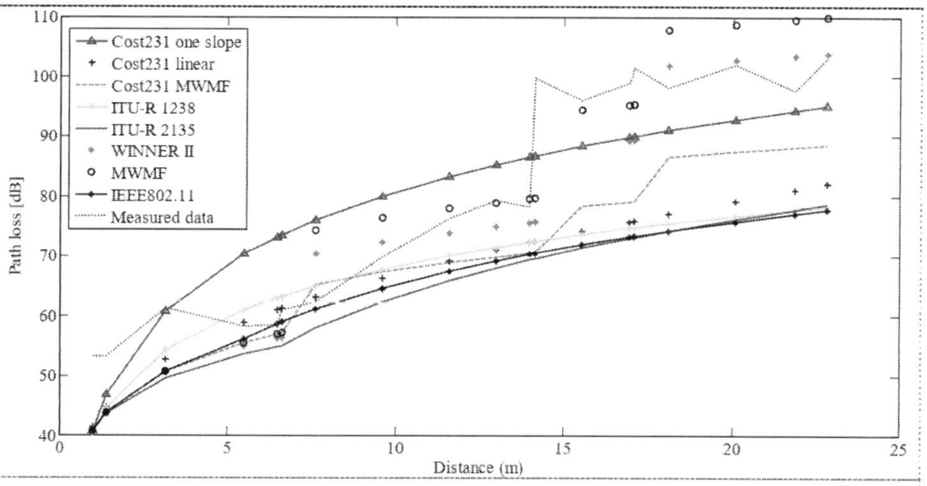

Figure 5. Comparison of measurement and empirical models for the second scenario

Table 2 shows the statistical data for the measurements and the empirical models. The mean values indicate that all models underestimate the average of the measurements. The error standard deviation reading proved that the closest model to the measurement is the MWMF model followed by WINNER II model and Cost231 one slope model.

Table 2. Statistical analysis of the second scenario

Model	Mean	Standard deviation	Error standard deviation
Cost231 one slope model	-0.22	9.77	0.51
Cost231 linear model	-12.07	15.49	0.82
Cost231 MWMF model	-10.74	13.19	0.69
ITUR 1238 model	-12.21	16.52	0.87
ITUR 2135 model	-15.47	17.88	0.94
WINNER II model	-4.40	8.63	0.45
MWMF model	-0.83	8.45	0.44
IEEE802.11 Model	-14.38	17.55	0.92

In the third scenario, the access point is located in the first floor one meter above the ground as in the first scenario while the UE is moving in the second floor. One floor and some cases one wall are considered in the path loss calculations. The following parameters are considered: n for one floor is 4 [5], N is 28, α is 0.62, L_f in Cost231 MWMF model is 18.3 dB, and L_{f11} in the MWMF model is 19 dB [6].

Figure 6 shows that five of the models are close to each other but below the measured data. Two models are close to the data which are Cost231 one slope model and MWMF model. Lastly, the linear model is far and overestimates the measurement. In this scenario, the minimum distance between the transmitter and the receiver is 4.3 m. Therefore the difference between the measurement and the predicted results at the reference distance did not rise.

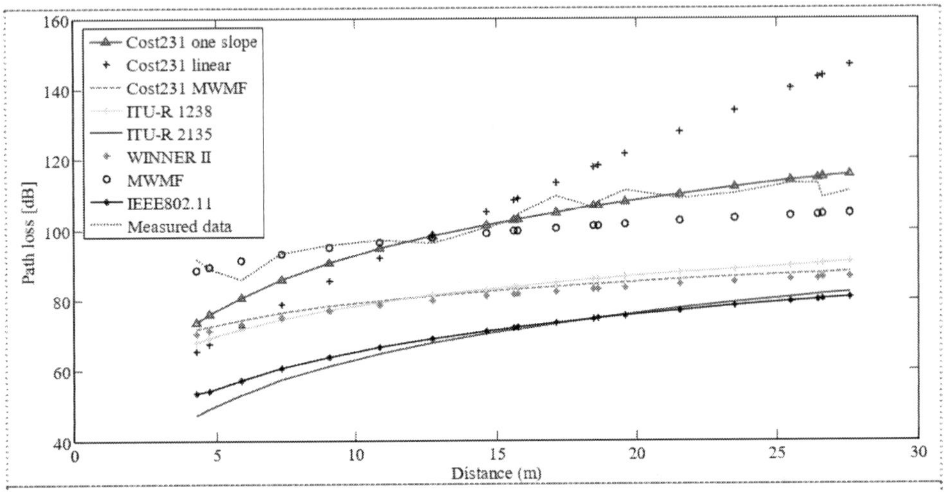

Figure 6. Comparison of measurement and empirical models for the third scenario

The third scenario statistical analysis results are presented in Table 3. It is confirmed that the only model that overestimates the measurements is the linear model while the rest are underestimation. Based on the error standard deviation, the closest model to the measurement is the MWMF model followed by the Cost231 one slope model. The worst predicted model for this scenario is the ITUR 2135 model.

Table 3. Statistical analysis of the third scenario

Model	Mean	Standard deviation	Error standard deviation
Cost231 one slope model	-2.08	5.97	0.30
Cost231 linear model	6.64	18.94	0.95
Cost231 MWMF model	-20.68	21.06	1.05
ITUR 1238 model	-20.20	20.40	1.02
ITUR 2135 model	-33.05	33.28	1.66
WINNER II model	-22.00	22.36	1.12
MWMF model	-3.98	5.63	0.28
IEEE802.11 Model	-32.10	32.21	1.61

The last scenario considered the effects of traversed floors on path loss between the access that located in the first floor and the UE that moving in the second and the third floors. The parameters that considered in this scenario are as follows: n for two floors is 5.2 [5], N is 28, α is 2.8, L_f in Cost231 MWMF model is 18.3 dB, L_{fl1} and L_{fl2} in the MWMF model are 19 and 15.2 dB respectively [7].

The comparison of the measurements and the predicted models is presented in Figure 7. It is shown in the figure that all the predicted models are below the measurements. The mean values in Table 4 are confirming this observation. It also noticed that the Cost231 and the MWMF model are fitting each other and have the same behaviour with the measurement. In addition, they are the closest models to the measurement as observed from Figure 7 and Table 4.

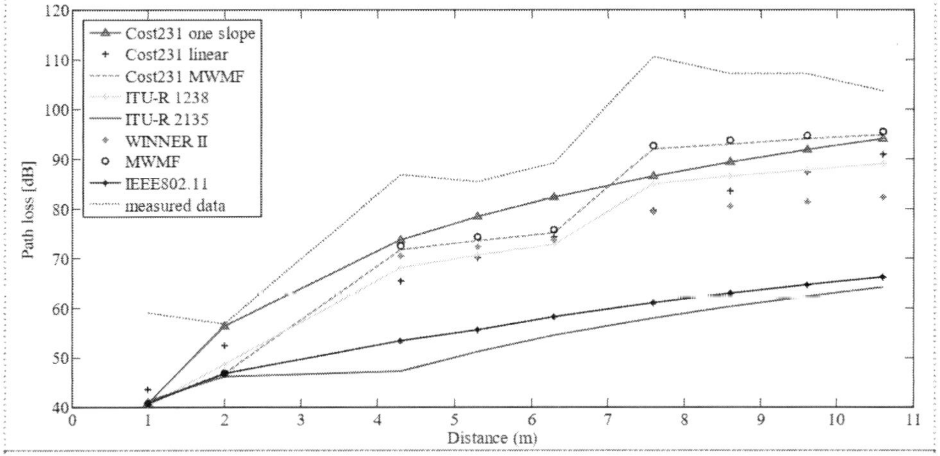

Figure 7. Comparison of measurement and empirical models for the fourth scenario

Table 4. Statistical analysis of the fourth scenario

Model	Mean	Standard deviation	Error standard deviation
Cost231 one slope model	-12.49	14.22	1.58
Cost231 linear model	-17.62	18.97	2.11
Cost231 MWMF model	-13.83	14.17	1.57
ITUR 1238 model	-17.46	18.04	2.00
ITUR 2135 model	-35.66	37.88	4.21
WINNER II model	-19.80	20.86	2.32
MWMF model	-13.29	13.66	1.52
IEEE802.11 Model	-32.91	34.97	3.89

In brief, none of the empirical models predicted the measurement in the four scenarios. It is observed that the models which considered the traversed walls and floors are the closest to the measured data. Figure 7 shows one more interesting thing, where the path loss for the same type of floor is not constant with the number of traversed floors. The path loss is 30 dB for the first floor and 21.5 dB for the second floor. The same thing was observed in [6].

5. Conclusion

This paper compares eight indoor path loss models with measured data for different scenarios at 2.4 GHz. Measurements were carried out in three story building. For LOS scenario as in the first scenario, none of the models are predicting the measured data. This is very clear for the very short distances which close to reference distance. The empirical models seem not working properly for the short range connection. For other scenarios, the closest model to the data is the MWMF model. On the other hand, the worst model is ITUR M2135.

6. References

[1] Small cell forum, "Public access small cell market to hit US$16 billion in 2016", small cell forum Press Releases, 26th February 2013.

[2] Saunders, S. R., *Antennas and Propagation for Wireless Communication Systems*, Wiley, 1999.

[3] M. Hassan-Ali and K. Pahlavan "A new statistical model for site-specific indoor radio propagation prediction based on geometric optics and geometric probability", *Wireless Communications, IEEE Transactions on*, vol. **1**, no. 1, pp.112 -124, 2002.

[4] C.-P. Lim , J. L. Volakis , K. Sertel , R. W. Kindt and A. Anastasopoulos "Indoor propagation models based on rigorous methods for site-specific multipath environments", *Antennas and Propagation, IEEE Transactions on*, vol. **54**, no. 6, pp.1718 -1725, 2006.

[5] "Digital mobile radio towards future generation systems, final report," COST Action 231, European Communities, EUR 18957, Tech. Rep., 1999.

[6] M. Lott and I. Forkel, "A multi-wall-and-floor model for indoor radio propagation," inProc. VTC 2001 Spring Vehicular Technology Conf. IEEE VTS 53rd, vol. **1**, 2001, pp. 464 – 468.

[7] Report ITU-R P.1238-7, "Propagation data and prediction methods for the planning of indoor radiocommunications systems and radio local area networks in the frequency range of 900 MHz to 100 GHz", 2012.

[8] V., Erceg, et al, "Tgn channel models," IEEE P802.11Wireless LANs, Tech. Rep., 2004.

[9] IST-4-027756, "WINNER II D1.1.2 V1.2", WINNER II Channel Models, Part I Channel Models, 2008.

[10] Report ITU-R M.2135-1, "Guidelines for evaluation of radio interface technologies for IMT-Advanced", 2009.

[11] "Further advancements for e-utra (physical layer aspects)," 3GPP TR 36.814 v9.0.0, Tech. Rep., March 2010.

[12] Corre Y, Stephan J, Lostanlen Y: Indoor-to-outdoor path-loss models for femtocell predictions. In Proceedings of IEEE 22nd Symposium on Personal, Indoor and Mobile Radio Communications (PIMRC). Toronto, Canada; 2011:824–828.

[13] Rose, D. M., Jansen, T. and Kürner, T. Indoor to Outdoor Propagation Measuring and Modeling of Femto Cells in LTE Networks at 800 and 2600 MHz. Presented at IEEE Global Communications Conference (GLOBECOM). Houston, Texas, 5th-9th December 2011.

[14] Valcarce A. and Zhang J. "Empirical indoor-to-outdoor propagation model for residential areas at 0.9–3.5 GHz", *Antennas and Wireless Propagation.IEEE Letter*, vol. **9**, pp.682 -685 2010.

[15] Letourneux, F., Corre, Y., Suteau, E., Lostanlen, Y. "3D coverage analysis of LTE urban heterogeneous networks with dense femtocell deployments", *EURASIP Journal on Wireless Communications and Networking* (October), 2012.

[16] Lee J., Bae S., Kwon Y., and Chung M., "Interference analysis for femtocell deployment in OFDMA systems based on fractional frequency reuse," *Communications Letters, IEEE,* vol. **15**, no. 4, pp. 425–427, April 2011.

[17] Zyoud A., Chebil J., Habaebi M. H., Islam Md. R. And Zeki A. M., "Comparison of Empirical Indoor Propagation Models for 4G Wireless Networks at 2.6 GHz", to be presented at International Conference on Control, Engineering & Information Technology, Tunisia, June 2013.

[18] Dong, W., Zhang, J., Gao, X., Zhang, P. and Wu, Y. "Cluster Identification and Properties of Outdoor Wideband MIMO Channel", Vehicular Technology Conference. VTC-2007.

[19] Fujii, T. "Path-loss Prediction Formula in Mobile Communication -an Expansion of "SAKAGAMI" Path-loss Prediction Formula". Trans. IEICE, Japan, J86-B, 10, p. 2264-2267, 2003.

[20] Lu, Y., Zhang, J., Gao, X., Zhang, P. and Wu, Y. "Outdoor-Indoor Propagation Characteristics of Peer-to-Peer System at 5.25 GHz", Vehicular Technology Conference, 2007.

[21] Xu, D., Zhang, J., Gao, X., Zhang, P. and Wu, Y. "Indoor Office Propagation Measurements and Path Loss Models at 5.25 GHz", Vehicular Technology Conference, 2007.

[22] Zhang, J., Gao, X., Zhang, P. and Yin, X. "Propagation Characteristics of Wideband MIMO Channel in Hotspot Areas at 5.25 GHz", Personal, Indoor and Mobile Radio Communications, 2007.

[23] Zhang, J., Dong. D., Liang. Y., Nie. X., Gao. X., Zhang, Y., Huang, CH. and Liu, G. "Propagation Characteristics of Wideband MIMO Channel in Urban Micro- and Macrocell", IEEE PIMRC 2008, Invited paper.

A comparative signaling cost analysis of Macro Mobility scheme in NEMO (MM-NEMO) with mobility management protocol

Shayla Islam [1], Aisha H. Abdalla, Mohamed H. Habaebi, Suhaimi A. Latif, Wan H. Hassan, Mohammad.K. Hasan, H.A.M. Ramli, Othman O. Khalifa
Department of Electrical and Computer Engineering, Faculty of Engineering, International Islamic University Malaysia, Kuala Lumpur.

E-mail: iium19612@hotmail.com

Abstract. NEMO BSP is an upgraded addition to Mobile IPv6 (MIPv6). As MIPv6 and its enhancements (i.e. HMIPv6) possess some limitations like higher handoff latency, packet loss, NEMO BSP also faces all these shortcomings by inheritance. Network Mobility (NEMO) is involved to handle the movement of Mobile Router (MR) and it's Mobile Network Nodes (MNNs) during handoff. Hence it is essential to upgrade the performance of mobility management protocol to obtain continuous session connectivity with lower delay and packet loss in NEMO environment. The completion of handoff process in NEMO BSP usually takes longer period since MR needs to register its single primary care of address (CoA) with home network that may cause performance degradation of the applications running on Mobile Network Nodes. Moreover, when a change in point of attachment of the mobile network is accompanied by a sudden burst of signaling messages, "Signaling Storm" occurs which eventually results in temporary congestion, packet delays or even packet loss. This effect is particularly significant for wireless environment where a wireless link is not as steady as a wired link since bandwidth is relatively limited in wireless link. Hence, providing continuous Internet connection without any interruption through applying multihoming technique and route optimization mechanism in NEMO are becoming the center of attention to the current researchers. In this paper, we propose a handoff cost model to compare the signaling cost of MM-NEMO with NEMO Basic Support Protocol (NEMO BSP) and HMIPv6.The numerical results shows that the signaling cost for the MM-NEMO scheme is about 69.6 % less than the NEMO-BSP and HMIPv6.

Keyword: *MM-NEMO; NEMO; MR; MNN; CoA;*

1. Introduction

In order to achieve the continuous internet connectivity at anytime in any place, Internet Engineering Task Force (IETF) has designed solutions to overcome the inefficiency of current IP addressing to support host based mobility. However, Mobile IPv6 is not able to handle the mobility of an entire network properly, since mobile network introduces much more complex mobility scenarios than host mobility [1-4]. Hence, the NEMO Basic Support Protocol (NEMO BSP) has been proposed by the Network Mobility (NEMO) working group. Simplicity is the most important feature of this protocol since it is a logical extension of the MIPv6 operation. The main purpose of NEMO BSP is to provide seamless connectivity of the whole mobile

[1] Shayla Islam, email: iium19612@hotmail.com
Department of Electrical and Computer Engineering,
Faculty of Engineering, International Islamic University Malaysia

network. There are mainly two major entities in NEMO which are - Mobile Routers (MRs) and Mobile Network Nodes (MNNs). For real time applications in NEMO, continuous Internet connectivity during handoff is generally expected which includes both features i.e. smooth (no or very little packet loss) as well as fast (low delay) handoff. But in accordance with NEMO Basic Support Protocol (NEMO BSP), only one primary Care of Address (CoA) of Mobile Router (MR) can be registered with HA, which affects the handoff performance resulting packet loss and delay. Moreover, there are some other mobility issues which include multi-homed mobile networks, sub optimal routing, and route optimization as well as security issues [5-10].

This paper is organized as follows: section 2 presents some background on handoff delay in NEMO environment. Section 3 presents the proposed handoff cost model. Section 4 evaluates the signaling costs of the proposed scheme with numerical analysis through comparison with NEMO BSP and HMIPv6. Conclusion is given in section 5 (shown in figure 1).

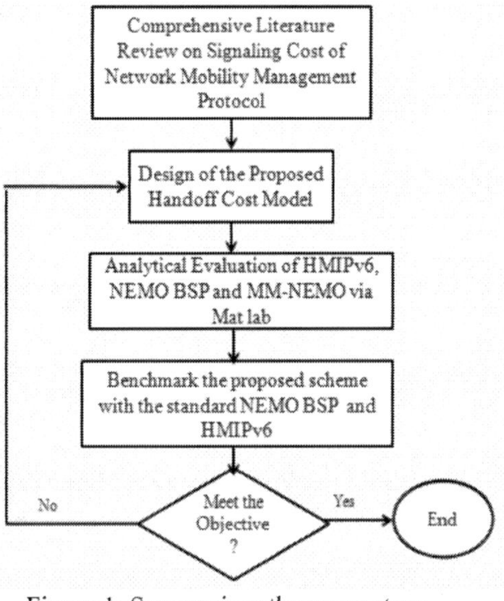

Figure 1: Summarizes the paper steps

2. Related Work

It is possible to use Hierarchical MIPv6 (HMIPv6) and fast MIPv6 (FMIPv6) jointly as Fast Hierarchical MIPv6 (enhancement of MIPv6) to reduce signaling overhead and packet loss. Normally, the Mobility Anchor Point (MAP) is placed at a comprehensive point above the New Access Router (NAR) and Previous AR (PAR) to integrate HMIPv6 and FMIPv6. However, the forwarding of packets would be inefficient between PAR and NAR because data packets will traverse the MAP–PAR link two times before reaching at the NAR. Therefore, in HMIPv6 architecture, the MN exchanges all the signaling messages with MAP, not with PAR to establish the tunnel for the handoff. Introducing FHMIPv6 [2], [4] in IPv6 network is more convenient. However, combining FHMIPv6 with mobile network is still an open issue since it needs to take care of both MR and MNN during handoff from home network to visiting network [2]. As MIPv6 and its enhancements (i.e. FHMIPv6) possess limitations like higher handoff latency, packet loss, NEMO BSP (standard for mobile network) also faces all these shortcomings by inheritance [4]. NEMO BSP is a solution for persevering session continuity by means of bidirectional tunneling between Home Agent (HA) and a mobile network whereas NEMO Extended Support (NEMO ES) that is a result to provide the essential optimization between uninformed Mobile Networks Nodes (MNN) and correspondent Nodes (CN) [4]. NEMO BSP operates in the IP layer and inherits the benefits of Mobile IPv6 [1] by extending the binding mechanism of the ancestor.

In [2], handoff procedure of mobile router (MR) in NEMO Basic Support Protocol (BSP) is almost similar with the mobile node (MN) in MIPv6. Disruption time and packet loss ratio are the critical performance issues during handoff in NEMO network. The parameters that are used to measure the performance analysis are mainly: total handover delay which is depends on movement detection time, care of address configuration delay and registration delay. In figure 2, MR detects its movement by receiving Router Advertisement (RA). Here, RA interval is a random value between minimum router interval (MinRtrInterval) and maximum router interval (MaxRtrInterval) where Layer 2 handoff is transparent to network layer (Layer 3). For layer2 handoff, MR waits for RA to perform Layer 3 handover that is known as movement detection (MD). In [6], [1], MinRtrInterval and MaxRtrInterval are set 30 ms and 70 ms. MR configures CoA after receiving RA according to prefix acquired from RA message received. It should wait for a random time to avoid multiple MRs or MNs configured the same CoA performing duplicate address detection (DAD) at the same time. T_{RED} is the time to complete the registration between the MR and its Home Agent (HA).

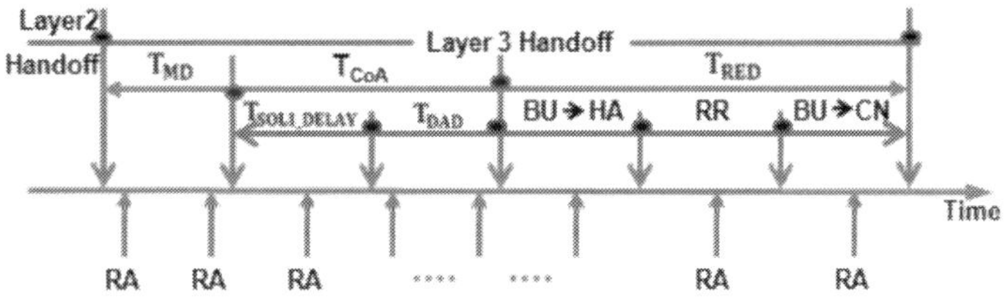

Figure 2: Handoff delay of NEMO BSP [6]

Therefore, the total handoff delay of MR in NEMO can be calculated as

$$T_{HD} = E\,(T_{MD}) + E(T_{DAD}) + T_{RED} \qquad (1)$$

Therefore, NEMO handoff delay is longer than 1.5 s, which is not applicable for real time applications [6].

3. Proposed Handoff Cost Model

Usually, NEMO network is considered as mobility in a car or vehicular network. Hence, for this type of network, advance preparation mechanism work very well. Therefore, the main idea of the proposed handoff cost model is to apply the fast handoff mechanism for the Serving MR (SMR) with its Local Fixed Node (LFN) in NEMO network to acquire seamless handoff. This method is performed in layer 3 by taking the information from layer 2. For simplicity, the following assumption has been made in the proposed cost model [13].

a) It is assumed that under the Serving MR (SMR) the local nodes are fixed.

b) In addition, during handoff signaling message for Local Fixed Nodes (LFNs) is completely handled by the Serving MR.

c) The Serving MR does not need to register with the Home Agent of Mobile Router (HAMR) until it moves from the current Mobility Anchor Point (CMAP) to New MAP (NMAP) as it is assumed that HMIPv6 is supported by network. Hence, the Serving MR only needs to update current location to CMAP through Current Mobile Router (CMR) during local movement.

d) Moreover, Serving MR is communicating with one Corresponding Node (CN). Additionally, Binding Update (BU) refreshment cost is not considered.

In the proposed system model, it is assumed that the hop distances of A, B, F, G are one where as E,C,D,H,I (as shown in figure 3) are considered as five hops because of Internet link connection [11-12]. According to fast handoff procedure in MM-NEMO scheme, after Layer 2 trigger the Current Mobility Anchor Point (CMAP) create New Regional Care of Address (NRCoA) with New Link Care of Address (NLCoA) instead of the Serving MR. This is based on the Mobile Network Prefix (MNP) of the new link. A temporary bi-directional tunnel is established between the Current MAP (CMAP) and the New MR (NMR) as well. This tunnel allows the CMAP to transfer the data packets to the Serving MR's new address as well as buffer these data packets at the new location of the Serving MR. By applying this mechanism, it is possible to minimize packet loss and delay in NEMO network. For performance analysis in NEMO network, a widely known simple mobility model i.e. random waypoint model [13] has been used to find out the residence time of the SMR. In random-based mobility model, the Serving MR (SMR) move liberally on a random basis where destination, speed, and direction all these factors are selected autonomously and independently of other routers. Different values for system parameters are used in our proposed model for numerical analysis are listed in Table 1 [11-13].

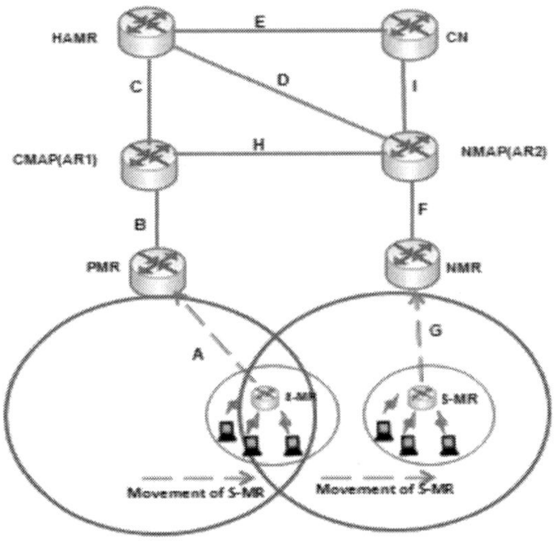

Figure 3: Proposed MM-NEMO Cost Model in NEMO

3.1 Analysis of Location Update Cost

In this section, a signaling cost model is developed to analyze the performance of the MM-NEMO, NEMO BSP, and HMIPv6. In MM-NEMO, extra signaling cost is needed to decrease the packet loss during handoff. When a MR with local fixed node enters into a NMAP domain, the total signaling cost can be calculated as:

$$C_{SIGNALING\ DATA} = C_{SIGNALING\_BASIC} + C_{SIGNALING\_FNA} + C_{SIGNALING\_LBU/LBAck} + C_{SIGNALING\_BA\_SMR\_HA} \quad (2)$$

In equation (2), $C_{SIGNALING\_BASIC}$ is a signaling cost incurred during the time $T_{SIGNALING\_BASIC}$, $C_{SIGNALING\_FNA}$ is a signaling cost incurred during the time $T_{SIGNALING\_FNA}$, $C_{SIGNALING\_LBU/LBAck}$ is a signaling cost incurred during the time $T_{SIGNALING\_LBU/LBAck}$ and $C_{SIGNALING\_BA\_SMR\_HA}$ is a signaling cost incurred during the time $T_{SIGNALING\_BA\_SMR\_HA}$ (shown in figure 4). In the proposed cost model, it is assumed that, HA of SMR send the data packets at a mean rate α as well as the SMR moves from one subnet to another at a mean rate β.

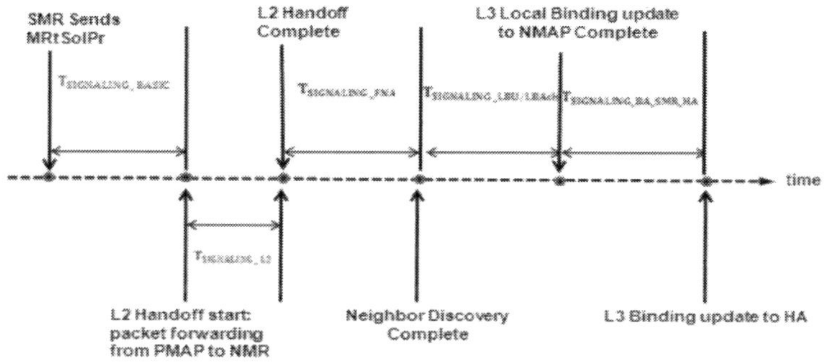

Figure 4: Timing diagram of the MM-NEMO scheme

3.2 *The Cost for Transmitting Data Packets*

Packet delivery cost $C_{DATA\ PACkET}$ is caused when a HA_SMR send packets to a SMR. Therefore, Packet delivery cost is defined as the total of the packet tunneling cost (C_{tunnel}) and the packet loss cost (C_{LOSS}). So, the packet delivery cost is calculated as follows:

$$C_{DATA\ PACkET} = P_{SUCCESS} \cdot C_{TUNNEL} + \lambda . P_{FAIL} . C_{LOSS} \quad (3)$$

Where, $P_{SUCCESS}$ is the probability for MR to perform fast handover successfully in Predictive mode and P_{FAIL} is the probability of failure. λ represent the increasing rate while HA_SMR retransmits the packets to SMR. $P_{SUCCESS} \cdot C_{TUNNEL}$ is a packet delivery cost when the handover is success, $P_{FAIL} . C_{LOSS}$ is a Packet delivery cost when the handover is fail. C_{Tunnel} is the packet tunneling cost for the SMR_HA to send packets through the tunnel between CMAP and NMR.

On the other hand, the packet delivery cost of HMIPv6 based NEMO environment can be expressed as:

$$\frac{C_{LOSS}}{C_{ORIGINAL\_PACKET}} = \lambda * P_{FAIL} * \alpha * C_{CMAP\_LOSS} * (T_{L2} + T_{IP} + \frac{T_{SIGNALING\_LBU}}{L_{BAck}} k + T_{SIGNALING\_BU\_SMR\_HA}) \quad (4)$$

4. Numerical Analysis

The system parameters for numerical analysis are shown in table 1. Numerical results are obtained using MATLAB 7.0.4. The following observations are made: In figure 5 and figure 6, the variation of location update cost with average cell residence time and velocity are illustrated. Here, the longer an S-MR remains in a current cell, the lower the location update cost. It shows that, location update cost for the proposed scheme is about 69.6 % less than that of the NEMO-BSP. It is observed that the location update cost increases with velocity since cell residence time is inversely proportional to velocity. In this case, it results same amount of cost reduction from NEMO-BSP when the number of S-MR is 4. The variation of total cost of MM-NEMO, NEMO BSP, and HMIPV6 with cell residence time is illustrated in figure 7. Since the cell residence time increases, both the total cost of MM-NEMO scheme and NEMO-BSP decrease while HMIPV6 increases linearly. When the time is less than 5 second the total cost of NEMO-BSP is higher than that of HMIPV6 while for MM-NEMO scheme, it is minimized to 2 seconds which is explained in figure 8. In figure 9, it is observed that the cost ratio of NEMO-BSP and HMIPV6 is higher than that of the MM-NEMO and HMIPV6 schemes. When the cell residence time approaches to 2 seconds the cost ratio for the MM-NEMO scheme goes below unity at the same time this ratio is obtained in 5 seconds for NEMO-BSP.

Table 1. System Parameters

Parameter	Value
$T_{SIGNALING\_BASIC}$	43.253 msec
T_{L2}	55 msec
T_{FNA}	5.726 msec
$T_{LBU/LBAck}$	11.452 msec
T_{BU_HA}	25.018 msec
T_{RR}	38.584 msec
T_{BU_CN}	25.018 msec
$c\ d\ f\ g$	1 hop
e	2 hop
h $i\ j\ k\ l$	5 hop

The influence of the number of Serving Mobile Router (S-MR) on packet delivery cost is rising exponentially which is represented in figure 10. By analyzing the figure 9, it is summarized that when the number of S-MR is below 8, the cost ratio is decreasing which shows better performance for MM-NEMO. Furthermore, the proposed MM-NEMO scheme reduces the total cost by 72% over NEMO-BSP when the number of S-MR is 4.The significance of this figure is that if the time is set at 16 seconds the cost for proposed MM-NEMo will be reduce by 83% over the NEMO-BSP at any number of S-MR.

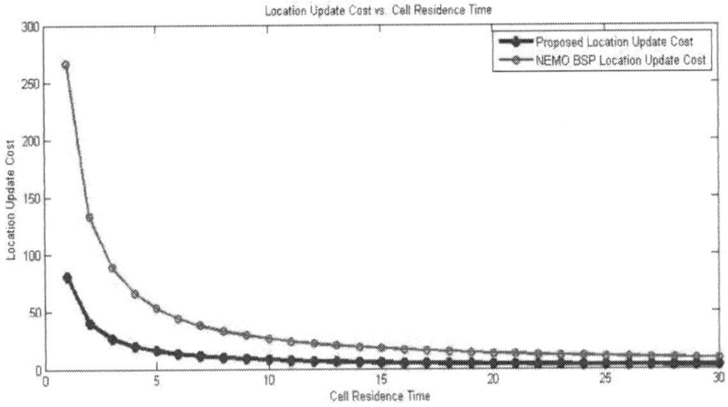

Figure 5: Location Update Cost vs. Cell Residence Time

IOP Conf. Series: Materials Science and Engineering **53** (2013) 012022 doi:10.1088/1757-899X/53/1/012022

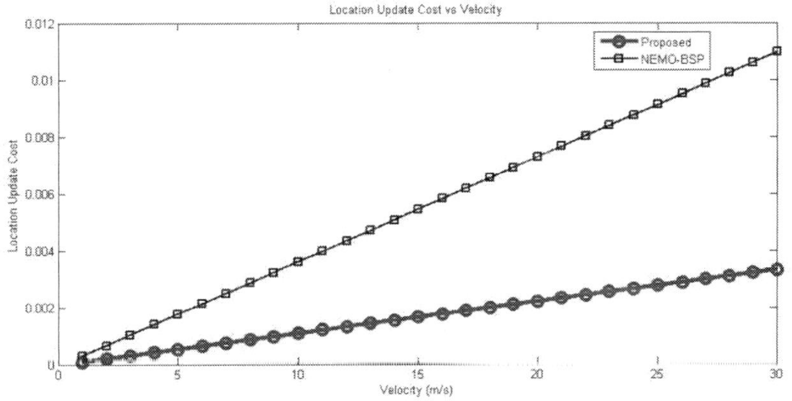

Figure 6: Location Update Cost vs. Velocity

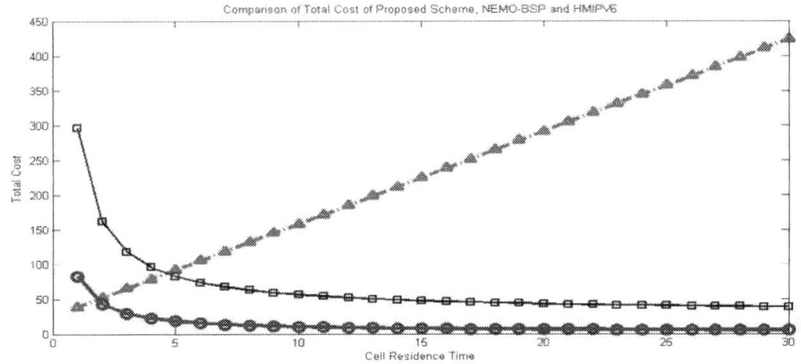

Figure 7: Comparison of Total Cost (MM-NEMO, NEMO BSP, and HMIPV6)

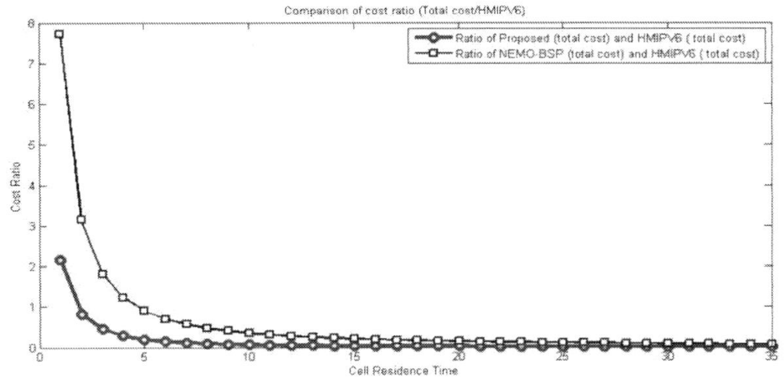

Figure 8: Ratio Comparison of MM-NEMO and NEMO-BSP with HMIPV6

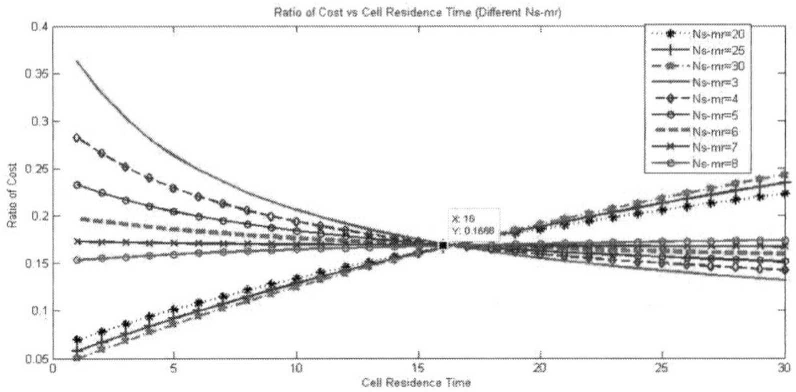

Figure 9: Ratio of Cost vs Cell Residence Time (different N_{S-MR})

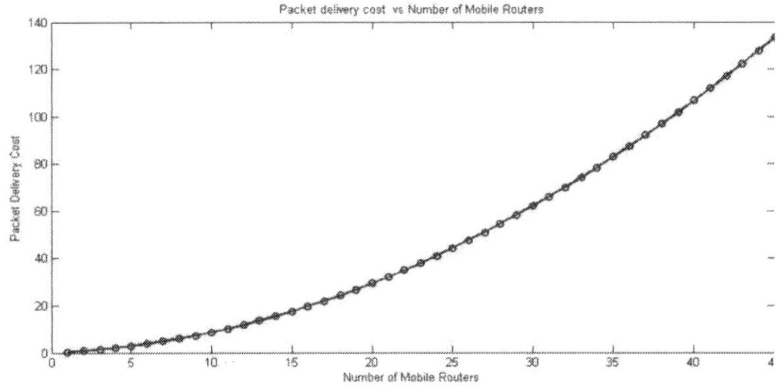

Figure 10: Packet Delivery cost vs. Number of S-MR

5. Conclusion

In this paper, we propose the macro mobility architecture that supports fast handover between the MAP domains by adopting the FHMIPv6 (fast HMIPv6) to improve handover between MAP domains in NEMO environment. Thus, the proposed macro mobility scheme can reduce the signaling cost almost 69.6 % less than NEMO BSP as well as the handoff delay which is incurred when the serving MR (S-MR) with fixed nodes moves between the MAP domains. We observe that the proposed scheme can provide better performance than NEMO BSP that is essential for seamless/uninterrupted handoff. However in order to more precise evaluation, need to consider visiting and local mobile node in the proposed scheme is another requirement as a future work.

References

[1] V. Devarapalli, R. Wakikawa, A. Petrescu, & P. Thubert, Network Mobility (NEMO) basic support protocol. RFC 3963, 2005.

[2] S. Yoo, S. Choi, & D. Su, Analysis of Fast Handover Mechanisms for Hierarchical Mobile IPv6 Network Mobility. *Wireless Personal Communication.* 48, 2009, pp: 215–238, doi: 10.1007/s11277-008-9518-x.

[3] X. Chen, H. Zhang, Y. Chang, & H. Chao, Experimentation, and performance analysis of multi-interfaced mobile router scheme. *Simulation Modelling Practice and Theory, pp.* 407–415. 2010, doi: dx.doi.org/10.1016/j. simpat. 2009. 09. 005.

[4] C. Ng, P. Thubert, M. Watari, & F. Zhao, Network mobility route optimization problem statement. RFC 4888, 2007.

[5] Z. Slimane, M. Feham, & A. Abdelmalek, A Seamless and Transparent MN-Proxy based Mobility Support For (n, n, 1) Multihomed NEMO Model. *International Journal of Computer Science and Network Security (IJCSNS),* vol.10(4), Apr. 2010.

[6] H. Petander, E. Perera, KC. Lan, & A. Seneviratne, Measuring and improving the performance of network mobility management in ipv6 networks. *IEEE Journal on Selected Areas in Communications*, 2006, 24(9), pp. 1671–1681.

[7] K. Romain, M. Julien& N. Thomas. Multiple Mobile Routers in NEMO: How Neighbor Discovery Can Assist
Default Router Selection. IEEE 19th International Symposium on Personal Indoor and Mobile Radio Communications (PIMRC'08), 2008, pp.1-6, doi: 10.1109/PIMRC.2008.4699613

[8] K. Romain, M. Julien& N. Thomas. Multihoming in IPv6 Mobile Networks: Progress, Challenges, and Solutions. *IEEE Communications Magazine,* 2013, pp.128-135.

[9] L. Hai&L. Houda. Hybrid Handover Optimization for Multiple MobileRouters-based Multihomed NEMO Networks. , IEEE International Conference on Pervasive Services, 2007, pp. 136 – 144, doi: 10.1109/PERSER.2007.4283903.

[10] S. Zohra, F. Mohamed &A. Abdelhafid(2012).Seamless Infrastructure independent Multi Homed NEMO Handoff Using Effective and Timely IEEE 802.21 MIH triggers. International Journal of Wireless & Mobile Networks (IJWMN), Vol. 4(3), pp. 119-139, doi: 10.5121/ijwmn.2012.4308.

[11] L. Zhang & S. Pierre (2008). Evaluating the performance of Fast Handover for Hierarchical MIPv6 in cellular Networks. *Journal of networks*, vol.3 (6), doi: 10.4304/jnw.3.6.36-43.

[12] J. HeeYoung & K. Seokjoo (2004). Fast Handover Support in Hierarchical Mobile IPV6. International Conference on Advanced Communication Technology, doi: 10.1109/ICACT.2004.

[13] A. Reaz, M. Atiquzzaman, & W. Ivancic, Signaling Cost Analysis of SINEMO: Seamless End-to-End Network Mobility. *ACM/IEEE international workshop on Mobility in the evolving internet architecture (MobiArc'06)*, 2006, pp. 37-42, doi: 10.1145/1186699.1186712.

Human Classification Based on Gestural Motions by Using Components of PCA

Azri A. AZIZ[1], Khairunizam WAN[1], S.K. Za'aba[1], Shahriman A.B[1], Nazrul H. ADNAN[2], Asyekin H[1]. and Zuradzman M. R[1].
[1]Advanced Intelligent Computing and Sustainability Research Group
School Of Mechatronics Engineering, Universiti Malaysia Perlis (UniMAP), 02600 Arau, Perlis, Malaysia.
[2]Bahagian Sumber Manusia, Tingkat 17 & 18, Ibu Pejabat MARA Jalan Raja Laut,50609 Kuala Lumpur, MALAYSIA.

Email: azriaziz@unimap.edu.my ; khairunizam@unimap.edu.my

Abstract. Lately, a study of human capabilities with the aim to be integrated into machine is the famous topic to be discussed. Moreover, human are bless with special abilities that they can hear, see, sense, speak, think and understand each other. Giving such abilities to machine for improvement of human life is researcher's aim for better quality of life in the future. This research was concentrating on human gesture, specifically arm motions for differencing the individuality which lead to the development of the hand gesture database. We try to differentiate the human physical characteristic based on hand gesture represented by arm trajectories. Subjects are selected from different type of the body sizes, and then acquired data undergo resampling process. The results discuss the classification of human based on arm trajectories by using Principle Component Analysis (PCA).

1. Introduction

The understanding of human physical characteristic specifically to the hand gesture, which can be used as a replacement existing tool to interact with a machine mechanism such as computer, is an interactive alternative in communication area [1, 2]. Recently, human can interact with computer without physical interaction such in computer game application like kinect [3] and changing television program using gesture product by Samsung Smart Interaction TV. All of the non-physical contact interaction can be done through gesture implementations. Human gesture is one of the non-verbal communications for transmitting and receiving information. It is such a way that technology been introduce inside the modern device.

Human gestures are influence by their individuality factors, which are physical body characteristic and emotional state. During the situation of angry or sad, human feeling are revealing by their actions. This action can produce the difference in the gesture performance. Moreover, the result of gesture will vary although the same gesture performs by the same person. The factor of the body size contributes to

[1] khairuniam@unimap.edu.my

the differences in the performance of gestures. Then, in the research the body size of the performer will be considered in the development of the gesture database.

In this paper, motion of arm trajectories are achieved by using an optical motion captures system (MoCap). Hence, the motion features are extracted from three-dimensional (3D) movement of hand. In classification of the motion features, the distribution of motion data once undergoes resampling process. Principle Component-Analysis (PCA) approach is used to classify motion data.

This paper is structured as follows: Section 2 addresses the related researches to the research approaches, applications and problems of recognizing the human gesture. Section 3 describes the methodologies and describes the proposed PCA algorithm for the classification of human gesture based on motion patterns. Section 4 presents the experiment setup and results of the classification and the article is concluded with the summary in section 5.

2. Related Research

Many researchers relate the human motion to generate and integrate the movement in the robot. Researcher from Hanyang University of Soul, Korea developed humanoid robot that have such ability of human arm motion [4]. Furthermore, the development of dataglove also related to the gesture understanding of hand and figure in controlling machine [5]. Additionally, Okumoto et al. uses the implementation of augmented reality to presenting tactile sensation for tactoglove which can bring the fill on sense although just wearing a glove [6].

K. Wan et al. prove that the use of human gesture especially human upper body motion for recognizing the dynamic human gesture and give good performance in classify the various gesture pattern [7]. In addition, hand gesture recognition suit for discover human gesture because the representation of gesture mostly concentrate on human upper body [8-10]. As such, sign language communication also involves mostly the movement of hand as well as arm movement. In real application of machine vision toward gesture, the most suitable gesture is using hand trajectories.

Many approaches in attaining hand motion data, which are vision system, accelerometer sensor [12] and dataglove [2]. The suitability and user-friendly factors are the main rules for researchers to determine the most favorable way in their experiment approaches. The differences will create the better results and by comparing between one to another is importance in the process for improvement in the future.

3. Methodologies

The Human arm is the most significant body part in performing gesture while support by others like head movements, facial expressions and arm movements. **Figure 1** illustrates the flow of the methodologies in processing the arm motion data. In overall, system has three stages and the first is acquisition of raw data, then resampling of motion data and, follows by the classification of motion data by using PCA. In the series of experiment, there are several geometrical gesture involved. **Figures 2** shows 10 geometrical gestures used in the experiments. For the acquisition of motion data, motion capture system (MoCap) consist of five high-speed cameras are placed around the subject to record the gestures. Every gesture are captured and stored in term of .csv format for all three dimension x, y and z. These data represent the motion coordinate in time frame format. **Figure 3** shows the rectangle gesture performs by the subject.

Huge amount of tabulated raw data are collected and store in excel format. Furthermore, raw data undergo preprocessing stage to eliminate the noise and extract the features of the movement. The preprocessing process are called 'resampling process' [7][11] that generate 30 resample points and also known as 30 features point of the gesture movements. Resampling process eliminate the error in time base data then allow it be represented in term of distance based. Every features point represents the gesture characteristic at that certain resample point. Each and every resample point goes through classification process by using (PCA). PCA allows the tabulated data to be grouped according to characteristic of the features point.

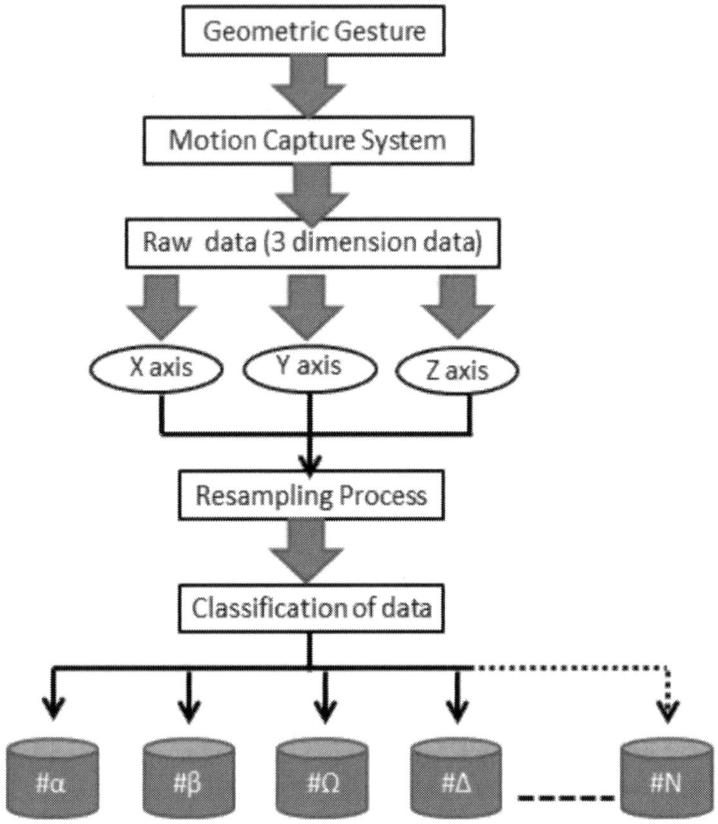

Figure 1. Flow of the proposed methods

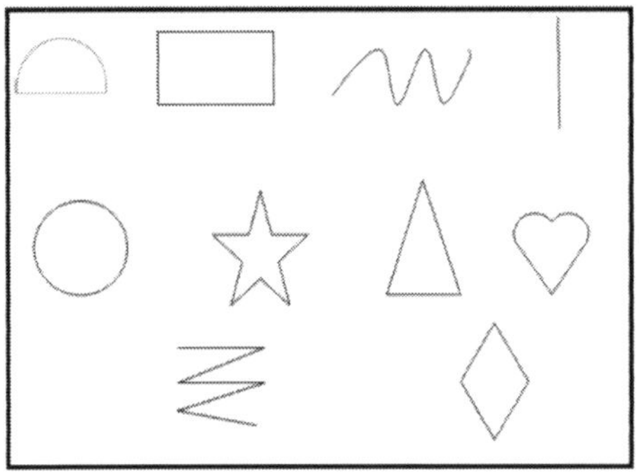

Figure 2. Geometrical gesture used in experiment.

Figure 3. Subject performs the rectangle gesture.

3.1. Principle Component Analysis (PCA)

PCA is one of the method in reduce dimensionality of the numerous data and eliminate the factor of redundancy of the identical data value [12]. It has been found in various applications such as data analysis, monitoring process and rectification of data [14]. The coordinate representation of the new data is calculated by changing the coordinate of the initial coordinate. It is the matter of computing the meaningful basis to filter out the noise and reveal the vital data. Moreover, it is also the revolutions of liner multispectral space (measurement space) into a space of Eigengesture (features space) [5]. The initial data Y are adjusted to be Y_n by subtracting the mean.

$$Y_n = Y - \bar{y} \tag{1}$$

Then, series of result yield from equation (1) are arranged in calculating the covariant of the data set. There are more than single covariance measurement can be calculated and the calculation is depending on the dimension of the data.

$$Cov(Y,U) = \frac{\sum_{a=1}^{m}(y_a - \bar{y})(U_a - \bar{u})}{(m-1)} \tag{2}$$

Equation (2) shows the covariant formula in determining the two dimensional data. $\bar{y}$ is the average value of Y. For data that arrange in $m \times n$ matrix, the possible covariance value between all the different dimensions is,

$$K^{m\,x\,n} = \left(d_{i,j}\,, d_{i,j} = cov\left(Eim_i, Eim_j\right)\right) \tag{3}$$

where $K^{m\,x\,n}$ is a matrix with m rows and n columns then Eim_x is the xth dimension.
Then, to come through PCA process, the determination of eigenvalue and eigenvector called Eigengesture is required. Eigengesture refers to the vector analysis of the eigenvector and it also known as feature vector. Both *Eigengesture* and Eigenvalue can be determined as:

$$Ag = \lambda g \tag{4}$$

Can be simplified as:

$$(A - \lambda I)g = 0 \tag{5}$$

g represent an *Eigengesture* of data A and λ is the eigenvalue of the equation which both are calculated using Jacobian method [13]. The number λ is an eigenvalue of A if and only if $A - \lambda I$ is singular, so the determinant of the *Eigengesture* is:

$$det(A - \lambda I) = 0 \tag{6}$$

Deriving the new data set from the PCA analysis as new representation of data is calculated. In other hand, this new data set can be plot in 2 dimensional (2D) and three dimensional (3D) plotting considering the first, second and third principle component.

$$FinalDataPCA = Eigengesture^T \times OriginData \tag{7}$$

where $Eigengesture$ is the matrix yield from the eigenvector calculation.

4. Experiments

4.1. Experiment Setup

The system consist of five high-speed cameras with an image resolution 640 x 480 pixels and were able to capture 200 frames per second. Optical motion capture system has an ability to capture motion in real time trajectories [15] then image captured were in term of frame per second. The marker was placed in the features point of the body of a performer which the finger. Markers were made of reflective material that can allow the cameras to tract the movement.

The environment parameters must be controlled such as less of light intensity around the system to allow the system to work. **Figure 4** shows the flow of the system configuration of motion detection and development of gesture data. 12 subjects will perform the experiment and they had been given briefing and watch some example video in performing the gesture experiment. They also fill the form that consist data of weight, high and their arm's distance. Then, for every gesture, they need to repeat 10 times. The purpose of repetition is to reduce error and enhance the result. Furthermore, the collected data will ultimately be classified. Classification process will determine data substantial enough of classify group of people according to be body size effect. The space of experimental data collection shows in **Figure 5**.

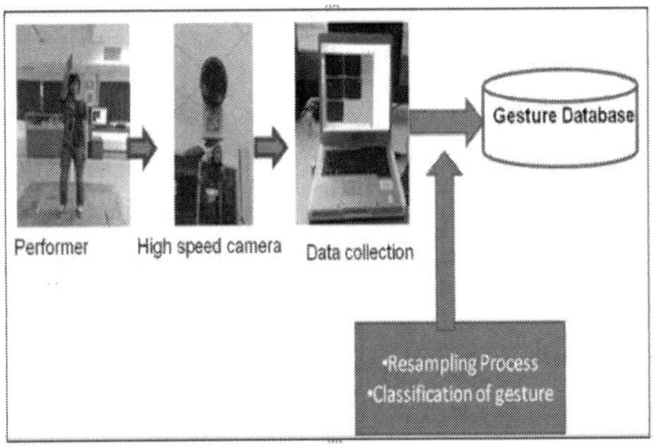

Figure 4. Experimental setup

Figure 5. An optical motion capture system.

IOP Conf. Series: Materials Science and Engineering **53** (2013) 012023 doi:10.1088/1757-899X/53/1/012023

4.2. Experiment Result and Discussion

The subjects need to perform gesture 10 times repetitively to obtain precise data value. The results of 10 trials were shown in **Figure 6**. Every three dimension (*x*, *y*, and *z*) data for each subject was resampled [11] then revealed 30 resample points. This resample data signify by **Figure 7** which only data from one subject.

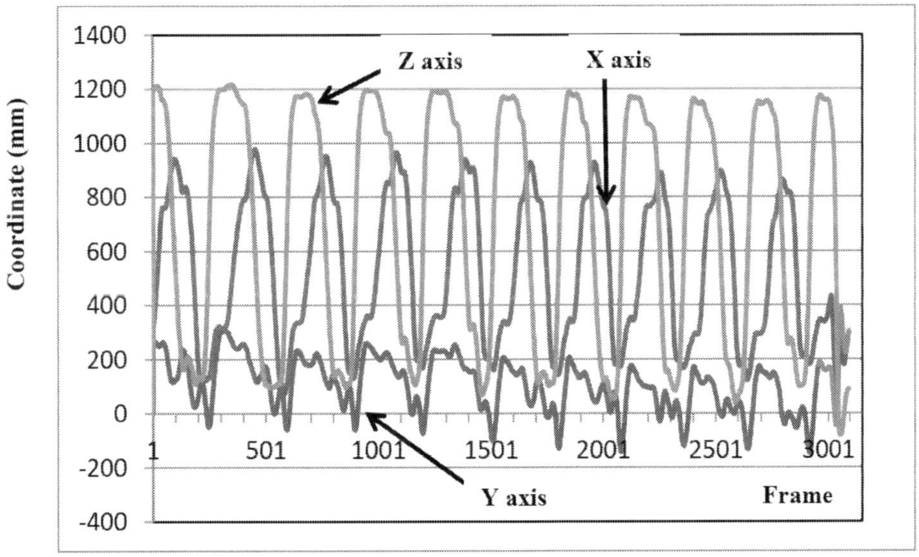

Figure 6. Raw data for gesture rectangle performed 10 trials continuously.

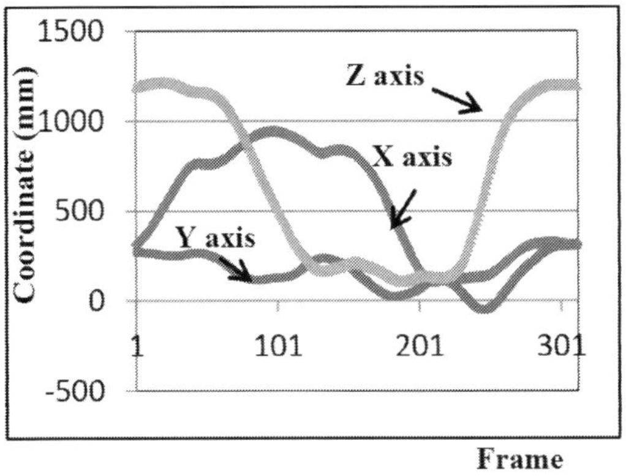

Figure 7. Single trial for the gesture rectangle.

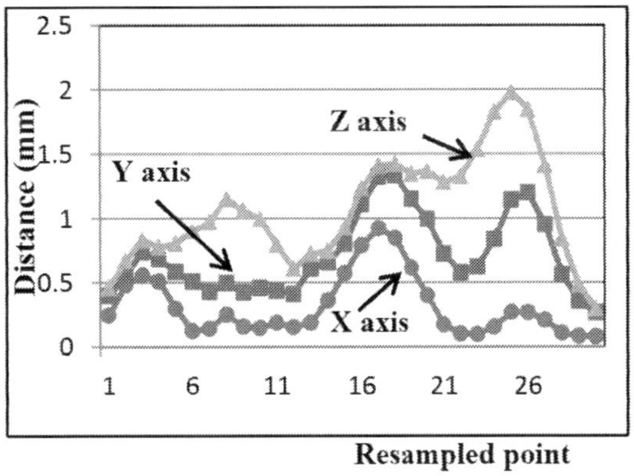

Figure 8. 30 resampled data for the gesture rectangle.

Meanwhile, **Figure 8** shows data after resampling process that make the analysis easier compare to the raw data. 30 features of distance based data from 12 difference subjects were analyzed by using PCA and resulting the classification of data into several groups as shown in **Figures 9 and 10**. There are three dimensional plotting represent by first, second and third principle components. Analysis of each features point explicates the number of group that can be classified on every features point. **Table 1** shows the classification group. It explains that, at every resampled point, how many group data could be possibly classified.

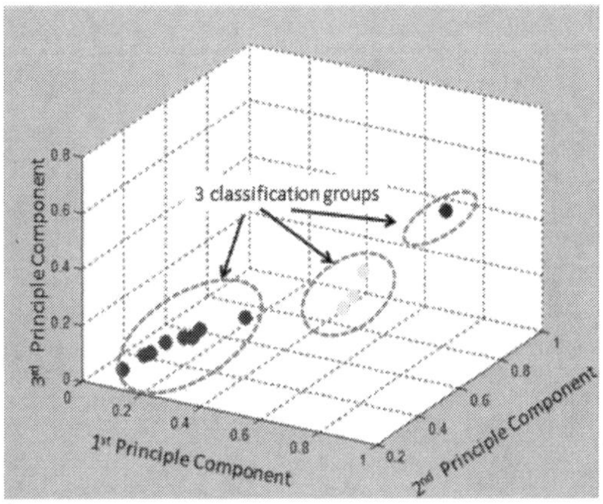

Figure 9. Three classification groups 'data for resampled point #14

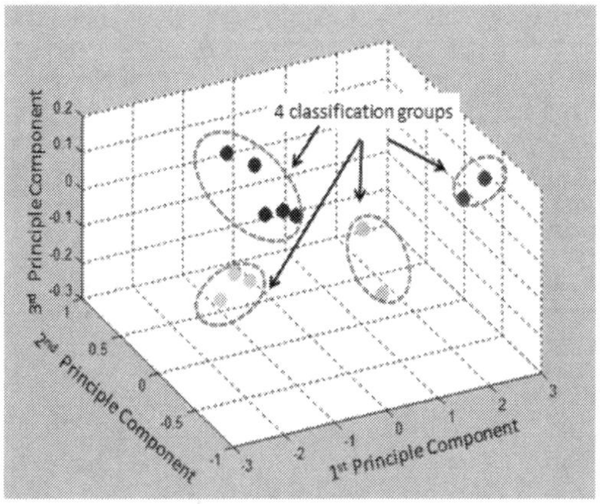

Figure 10. Four classification groups' data for resampled point #2

Gesture rectangle performed by 12 subjects

Table 1: Number of Classified Group

Resampled point	Number of Classified group	Resample point	Number of Classified group
#1	1	#16	2
#2	4	#17	3
#3	4	#18	4
#4	5	#19	4
#5	4	#20	4
#6	3	#21	4
#7	3	#22	4
#8	4	#23	4
#9	4	#24	4
#10	4	#25	4
#11	2	#26	4
#12	4	#27	4
#13	4	#28	4
#14	3	#29	4
#15	4	#30	4

4.3. Discussion.

Based on **table 1**, a further analysis was done to determine the most number of classified groups. From the analysis, four classes group of data was the highest appeared in many resampled points. **Table 2** shows the calculated percentage of the classification. For four groups of classification, there was 73.33% of classification rate. Then three classes group performed 13.33%.

Table 2: Percentage of Group's Classification

Gesture rectangle performed by 12 subjects	
Number of Classification group	**% of classification rate**
1	3.33
2	6.66
3	13.33
4	73.33
5	3.33

5. Conclusion

PCA in this research is used to group human based on their gestural motions. The acquired motion data must primarily be preprocessed to eliminate the noises and errors. In clustering motion data produced by 12 subjects, several groups of them possibly be classified and the best number of group is 4 with the group classification rate 73.33%.

6. Acknowledgment

Special thanks to all members of UNIMAP Advanced Intelligent Computing and Sustainability Research Group and School Of Mechatronics Engineering, Universiti Malaysia Perlis (UNIMAP), 02600 Arau, Perlis, Malaysia for providing the research equipment and internal foundation. This work is also supported by the fundamental research grant scheme (FRGS) awarded by the Ministry of Higher Education to Universiti Malaysia Perlis (FRGS 9003-00313).

7. References

[1] Nazrul H. ADNAN, Khairunizam WAN, Shahriman A.B., M. Hazwan ALI, M. Nasir Ayob and Azri A. AZIZ, 2012, "Development of Low Cost "GloveMAP" Based on Fingertip Bending Tracking Techniques for Virtual Interaction," *International Journal of Mechanical & Mechatronics Engineering IJMME-IJENS* **12 04** 41-51.

[2] Nazrul H. ADNAN, Khairunizam WAN, Shariman AB, Juliana A. Abu Bakar, Azri A. AZIZ 2013 "PCA-based Finger Movement and Grasping Classification using Data Glove 'Glove MAP'," *International Journal of Innovative Technology and Exploring Engineering (IJITEE)* **2**, **-3** 66-71.

[3] Smisek J., Jancosek, M., Pajdla, T. (2011): "3D with Kinect", *IEEE International Conference on Computer Vision Workshops (ICCV Workshops)*, 1154-1160.

[4] Hsin-Yu Liu, Wen-June Wang 2011 "A Course in Simulation and Demonstration of Humanoid Robot Motion" *IEEE transaction on Education*, **54- 2** 255-262.

[5] Nazrul H. ADNAN, Khairunizam WAN ,Shariman AB , Juliana A. Abu Bakar 2013: "PRINCIPAL COMPONENT ANALYSIS FOR THE CLASSIFICATION OF FINGERS MOVEMENT DATA USING DATAGLOVE 'GLOVEMAP'," *International Journal of Computer Engineering and technology (IJCET)* **4- 2** 79-93.

[6] Nazrul H.ADNAN, Khairunizam WAN, Shahriman AB, SKZa'ba, ShafrizaNisha BASAH, Zuradzman M. Razlan, HazryDesa, M. NasirAyob, Rudzuan M. Nor and Azri A.Aziz 2012 "Measurement of the Flexible Bending Force of the Index and Middle Fingers for Virtual Interaction",*International Symposium on Robotics and Intelligent Sensors*, Procedia Engineering **41** 388 –394.

[7] Khairunizam Wan and Hideyuki Sawada, 2008 "Dynamic Gesture Recognition Based on the Probabilistic Distribution of Arm Trajectory, *Proceedings of 2008 IEEE International Conference on Mechatronics and Automation*, 426-431.

[8] Shuying Zhao, Wenjun Tan, Chengdong Wu, et al 2009 "A Novel Interactive Method of Virtual Reality System Based on Hand Gesture Recognition," *CCDC 2009.IEEE*, 5879-5882.

[9] Kye Kyung Kim, Keun Chang Kwak and Su Young Chi 2006 "Gesture Analysis for Human-Robot Interaction," *ICACT 2006.*

[10] J.S. Kim, C.S. Lee, K.J. Song et al. 1997 "Real-time hand gesture recognition for avatar motion control, " In the Proceedings of *HCI'97.*

[11] Azri A. AZIZ , Khairunizam WAN , SK Zaaba, Shahriman A.B , Nazrul H. ADNAN , Rudzuan M. Nor , M. Nasir Ayob ,A.H. Ismail and M. Fadhil Ramly,2012 " Development of Gesture Database for an Adaptive Gesture Recognition System", *International Journal of Electrical & Computer Sciences IJECS-IJENS* **12 -04** 38-44.

[12] R. Valavi and H. Ghassemian, 2004 "A fusion Approach of Multi-Sensor Remote Sensing Data Based on Retina Model",*Proceedings of 12th Iranian Conference on Electric Engineering.*

[13] Yu Kinoshita, Daisuke Takeda, Akinori Sasaki, Hiroshi Hashimoto, Chiharu Ishii, 2006 "Archive and Instruction of Hand Motion: Analysis and Evaluation of Hand Motion", IEEE, 3030-3035.

[14] Jae-Kuk Lee, Kyung-Hun Kim, Tae-Young Kim, and Won-Ho Choi, 2003Non-Linear Principle Component Analysis Using LocalProbability, *Proceedings of the 7th Korea-Russia International Symposium, KORUS* 103-107.

[15] Khairunizam Wan, Azri Abdul Aziz, Shahriman Ab, Siti Khadijah Zaaba, Zuwairie Ibrahim, Zulkifli Md.Yusof, Ismail Ibrahim, Jameel Abdulla Ahmed Mukred, and N. Mokhtar 2012" Probability Distribution of Arm Trajectory for Motion Estimation and Gesture Recognition ," *Adv. Sci. Lett.* **13** 534-539.

A Comparative Study of Small Voltage Rectification Circuits for Implanted Devices

Samnan Haider[11], Hasmah Mansor[1], Sheroz khan[1], Atika Arshad[1], Mohammed M. Shobaki[1] and Rumana Tasnim[1]
[1]*Department of Electrical and Computer Engineering*
International Islamic University
53100 Kuala Lumpur, Malaysia
Email: syedsamnanhaider855@gmail.com

E-mail: syedsamnanhaider855@gmail.com

Abstract: Biomedical implants have acquired an invulnerable and significant importance over the recent years due to their low voltage requirements and trifling dimensions. Biomedical implants are being widely employed for the continuous monitoring of intended parameters as well as stimulation of target organs in patient's body. Being low power devices, the continuous powering without using battery dependent sources necessitates the alternative sources such as energy harvesting from the surroundings. Harvested output needs to be processed before using it for powering purposes, particularly in harsh and challenging premises. Even after processing, it needs to be made robust against the contiguous hazards so that maximum power is transferred to the load. A rectifier is employed to make this harvested energy usable. This paper compares the rectifier circuits, comparing their simulation output in terms of regulation, stability and power transferred. The rectification of small voltages is usually confronted by the threshold voltage drops that affect the output causing an effective drop in the power transfer efficiency and other parameters related. The solution to this problem has been suggested comparing the previous approaches with some modifications. Obtained results have been plotted in terms of ripple factor comparing them with the analytical calculations in order to show the role of capacitance in the reduction of ripple factor.

1. Introduction

Undoubtedly, technological development has bequeathed the innovations to all fields of science but has imparted the biomedical domain with bamboozling novelty [1]. This origination not only whirls around the miniaturization of implantable devices but also spurs towards the development of sensing networks in order to collect the data from subject and process it for monitoring purposes [2-3]. Automated sensing networks have attained an invulnerable importance in biomedical implanted devices as they provide the facility to monitor the intended parameters [4].

Biomedical implanted devices help in performing a number of tasks efficiently such as the collection of data from different body parts for monitoring purposes and to rejuvenate the lost functionality of organs by stimulating the muscular tissues [5-6]. The stimulation of tissues by

[1] To whom any correspondence should be addressed.

electrical means in biomedical implants has always been a field of interest, as this technique has remained helpful in restoring organs' lost functionality in the aftermath of injuries received in accidental situations [7]. However, it has been taken under serious consideration recently, for using the electrical stimulation as a permanent source for impaired tissue operations [8].

Although biomedical implants perform efficiently in fulfilling the intended tasks, but it is quite difficult to meet the power requirements for these devices. Batteries cannot be used for this because of longevity issues and the unacceptable sizes. Therefore energy harvesting is considered as an effective alternative that replaces the batteries fulfilling the energy requirements. Energy obtained as a result of energy harvesting process is not readily usable because of its too small amplitude therefore it needs to be processed by using some amplification and regulatory circuits that bring it to an operating level. Mainly, it suffers of continuity and stability and issues related to power transfer efficiency, which can be overcome by implementing different circuits proposed recently by contemporary researches.

A contemporary research claims to enhance the power transfer efficiency by using a model of Printed Spiral Coil PSC applying an algorithm to make this PSC operate at a frequency of 13.56MHz [11]. The suggested algorithm improves the efficiency and stabilizes the output as well. Class-E oscillator [12] has been implemented in order to enhance the power transfer efficiency. The paper endorses to use the same link for powering as well as for communication purposes at the same time avoiding the switches and connectors. This technique not only ameliorates the power transfer efficiency but regulates the output as well. A neoteric approach in [13] proposed a full wave CMOS rectifier for non-invasive powering of low voltage biomedical implants. Bootstrapped capacitors are used for the reduction of threshold voltage drops to overcome the start-up voltage of selected transistor. This technique improves the power efficiency minimizing the voltage drop.

This paper focuses on the rectification and regulation aspects of the low power output produced as a result of energy harvesting. It compares a number of rectification circuits proposed by contemporary researches on the basis of output power and regulation of output voltage making such small voltage usable by bringing it to an operational level.

2. Methodology

Harvesting energy using thermal and other sources like solar, vibrational etc. produces an output of minimal value. Fig. 1 exhibits the basic concept behind energy harvesting using two sources only i.e. thermal and vibrational. The output produced as a result of energy harvesting cannot be directly utilized to power up the buried implant, so it needs to be amplified, regulated and stabilized. In the recent past, many researches have proposed visible solutions to enhance the power received. This paper is a comparative study of a number of approaches based on rectifiers used in different topologies, in terms of all the tasks quoted above.

Circuits including Full wave bridge rectifier, Gate cross coupled, fully cross coupled rectifier and bootstrapping capacitor based rectifier have been simulated for an input of 3V, f= 50Hz. Simulation plots for input and output currents, voltages and power have been obtained comparing them in the light of regulation and power transferred consequently in order to know that how much power is transferred over a small distance using CMOS transistors. Circuits have been compared on the basis of ripple factor and hence categorized by calculating the values for each circuit. The role of a capacitor filter in reduction of the ripple factor has been shown via simulations by connecting a capacitor of a value equal to 100uF at the output.

3. Results and Discussion

Generally, the betrothed applications for rectifier are low voltage energy harvesters with very low output voltage amplitudes of around 500 mV generally. Usually, these rectifiers are used in biomedical implants in order to rectify the harvested output and regulate it to a usable level. Dealing with such a small voltage demands an efficient circuit in terms of power conversion i.e. lesser threshold voltage drops. Recent reported researches have considered this issue vitally and many solutions have been proposed to for the minimal threshold voltage drop resulting in the enhanced output. This section is based on the simulation results of all the circuits mentioned in section II chronologically.

Fig. 1 represents a conventional full wave bridge rectifier circuit with 4 diodes connected in such a way that D1 and D3 conduct during the positive half cycle whereas the other pair D2 and D4 conduct in the negative half cycle. It rectifies the input voltage and brings the negative half above zero. Currently, a 3V input is applied to this circuit and the rectified output v/s input voltage is shown in fig. 2.

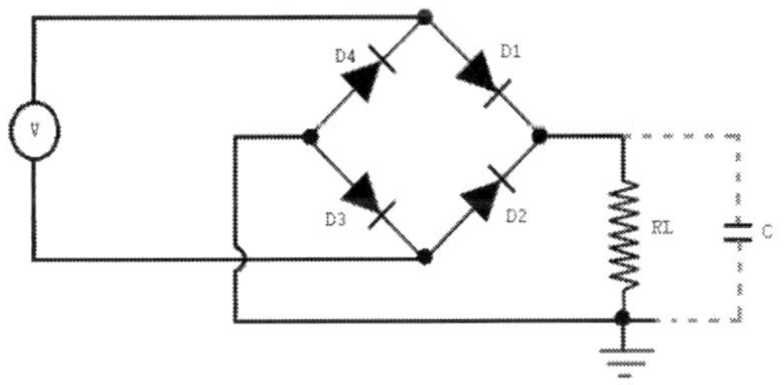

Fig. 1: Full wave bridge Rectifier

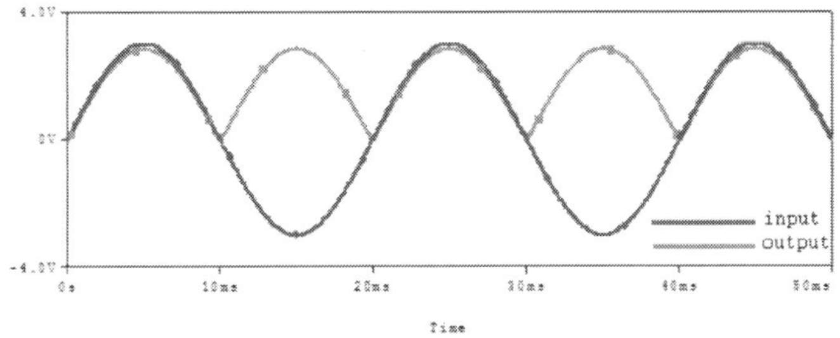

Fig. 2: Full wave rectified output

A full wave rectified output approx. equal to 2.8V is produced as a result of rectification process. The loss is because of threshold voltage drops in diodes. This drop affects the output power as well that can be observed in Fig. 3. An output power of 12mW is obtained for an input of 15mW. Also the output voltage constitutes of a vigilant AC part which is not suitable for powering small voltage devices such as biomedical implants. So a capacitor is added in parallel to the load and the output becomes more regulated with less ripples and smaller ripple factor.

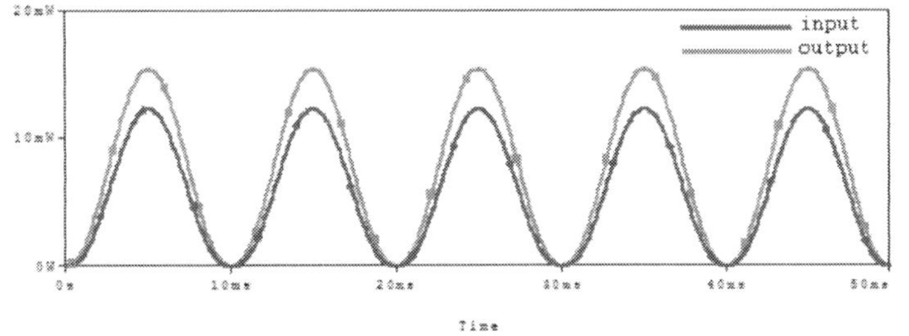

Fig. 3: Input power v/s Output power

Adding a capacitor regulates the output and brings it near to DC which is shown in Fig. 4, hence such regulated voltage can be used directly to power up the biomedical devices.

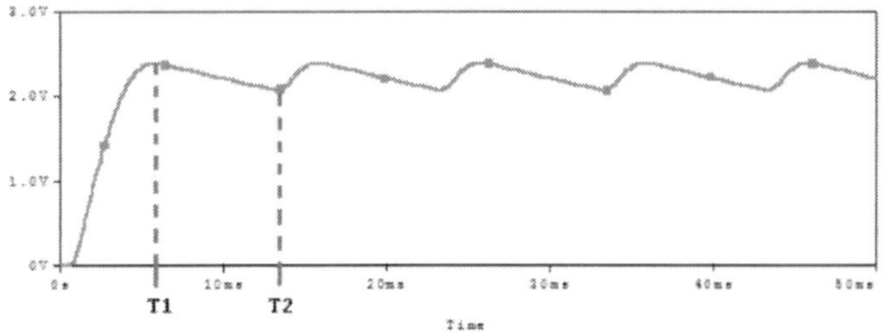

Fig. 4: Output voltage with a capacitor filter

Fig. 4 shows the result in case of a capacitor connected in parallel to the load. The output is more regulated and near to DC voltage as compared to the Fig. 2. Preemptively, looking at the waveform gives an idea that the ripple factor has been reduced as well. This can be vindicated by a very simple calculation. First of all, the ripple factor for Fig. 2 needs to be calculated. The peak amplitude for the input voltage is 3V. The output DC voltage is **Vavg=2Vm/π= 6/π=1.90V**. The peak-to-peak ripple voltage is the difference between the maximum and the minimum in the *vo* waveform that means **Vp-p=Vm-0=3V.** The percentage ripple factor equals **Vp-p/Vavg*100= 157%** for frequency of 50Hz. The ripple factor for the waveform in Fig. 9 can be calculated as under

As the output Dc voltage is equal to Vavg =2Vm= 5V in this case.

The load current I_L is given by I_L=Vavg/R_L=5/500=0.01A.

The capacitor discharges between T1 and T2 which is 20msec for a frequency of 50Hz so the charge supplied by the capacitor to the output resistor during this period is given as Eq.1.

$$\Delta Q = I_L * T = 0.01 \times 20 \times 10^{-3} = 0.0002 \text{ Coulomb} \quad (1)$$

Peak to peak ripple voltage is **Vp-p=** ΔQ **/ C= 2V**. This is almost 80% of the value of ripple factor for Fig. 2 so it has reduced almost more than 50% compared to the case when capacitor is not used. A small capacitor of value equal to 100uF has reduced the ripple factor to the half of the original value and it will go down some more if the capacitance value is increased.

A gate cross coupled rectifier is shown in Fig. 5 that has a pair of switch connected NMOS transistors at the input and a pair of diode connected PMOS transistors at the output. The working principle of this rectifier is quiet simple. The switch connected PMOS helps to overcome the threshold voltage drop during each half cycle. The rectified output is shown in Fig. 6 where it has been compared with the input voltage as well to show the drop caused due to the threshold voltage.

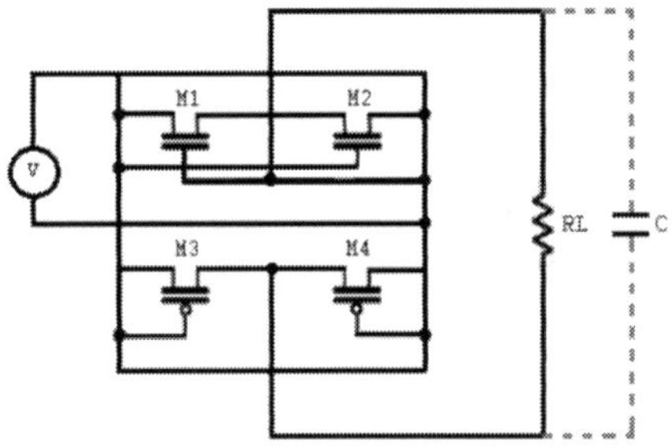

Fig. 5: Gate cross coupled rectifier [13]

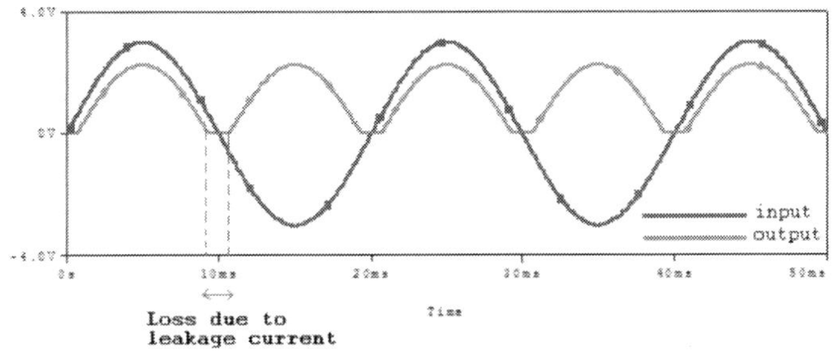

Fig. 6: Input voltage v/s output voltage

An output of 2.6 V approx. is produced after the rectification process but it contains a delay as the active area is reduced as compared to the input waveform. This is because of leakage current and it can be eradicated by using a comparator proposed in [9-10] . The output contains a noticeable AC part that can be supressed easily by using the same capacitor as added in the previous cases. Adding a capacitor caused a visible drop in the ripple factor in this circuit as well. Output for this case is shown in Fig. 7. The percentage drop can be calculated by using the same method and for these waveforms, it has been dropped from 157% to 82% approx. for a value of C=100uF. It can be reduced even more by using a higher value of capacitance.

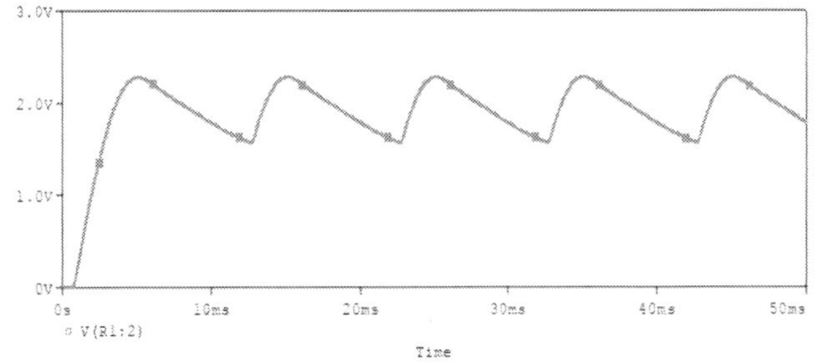

Fig. 7: Output voltage with a capacitor filter

This threshold voltage drop due to the diode connected transistors causes the power to decrease as well which is shown in Fig. 13 where an output power of 10mW is produced against an input of 13mW.

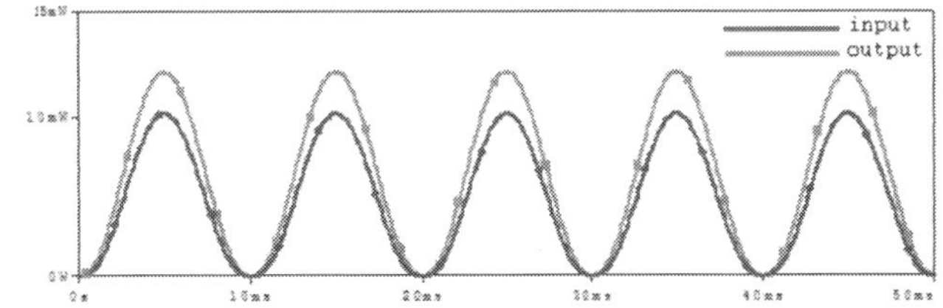

Fig. 8: Input power v/s output power

Fig. 9 shows another topology of full wave rectifier with two pairs of transistors i.e a pair of NMOS switches at the input and a pair of PMOS switches at the output. This circuit overcomes the threshold voltage drop to a higher level as compared to the previous configuration because the threshold drop is replaced by switches during each half cycle. The resulting output after rectification is shown in comparison with the input voltage in Fig. 10. A subsequent Fig. 11 shows the plots for input and output power as well to show the efficiency of the circuit. Output power is almost 16mW in this case. The output voltage in this case also needs to be regulated in order to suppress the AC portion in it, hence it is passed through a capacitor filter for a value of C=100uF and the resulting output is shown in Fig. 12.

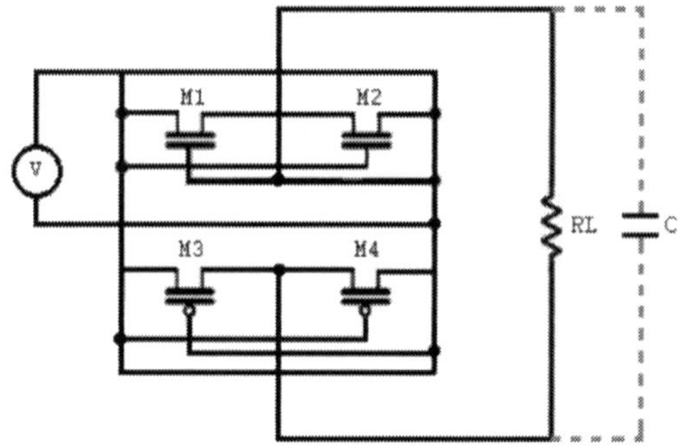

Fig. 9: Fully cross coupled rectifier [13]

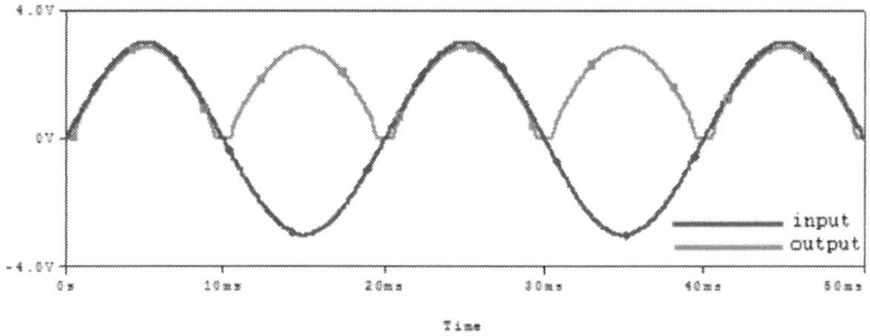

Fig. 10: Input voltage v/s output voltage

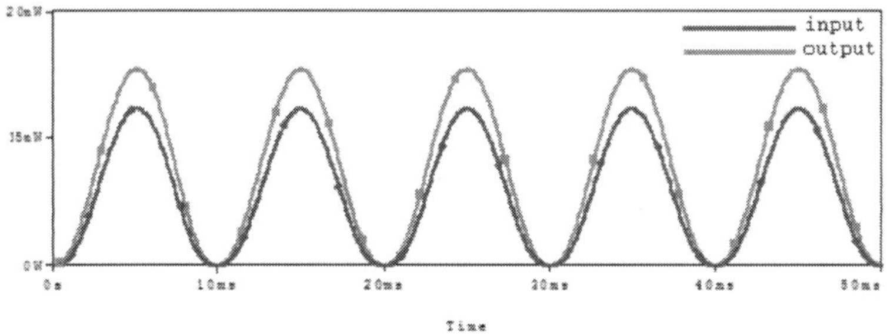

Fig. 11: Input power v/s output power

After adding a capacitor in parallel to the load resistor, the output becomes more regulated and it is averaged to a value that equals approx. 2.5 V. The ripple factor is reduced as well and by calculation it comes out to be 80% of the value for ripple factor in the normal case without using a capacitor.

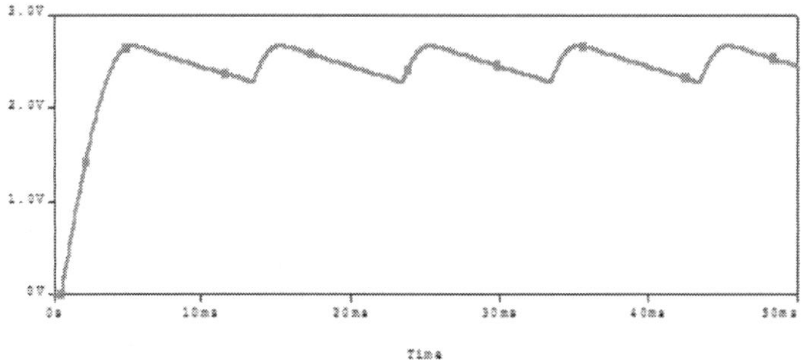

Fig.12: Output voltage with a capacitor filter

Fig. 13 shows a contemporary design of a full wave rectifier with functionality based upon the dynamic bulk biasing and bootstrapping capacitors. The dynamic bulk solves the problem of leakage current through the bulk whereas small bootstrapping capacitors not only reduce the effective threshold voltage of the main pass transistor; hence this circuit proves to be the most efficient amongst all the configurations discussed in this paper. Simulations results for input voltage v/s output voltage have been shown in Fig. 14 where an output voltage equal to 2.9 V approx. is produced by this circuit. Plots for input v/s output power are shown in Fig. 14; this circuit produces an output power equal to 17mW which is almost the highest in all the discussed configurations. Similar to the other cases, the output voltage contains an AC portion which is regulated by using the same value of capacitor as used for the other circuits. Fig. 16 shows the output for this case. The output produced is enhanced and more ripple free compared to all other techniques.

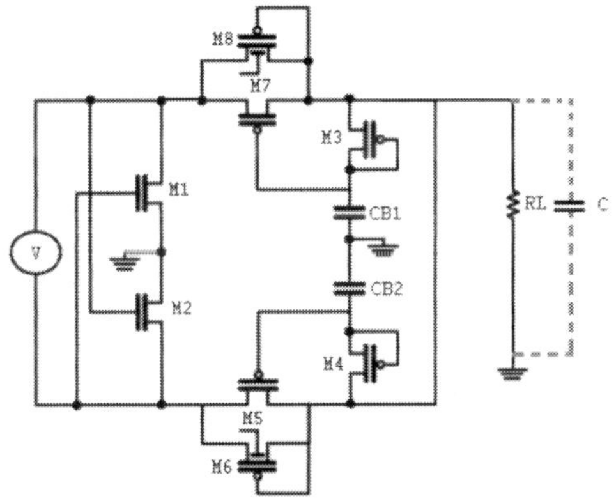

Fig. 13: Bootstrapping capacitors based rectifier [13]

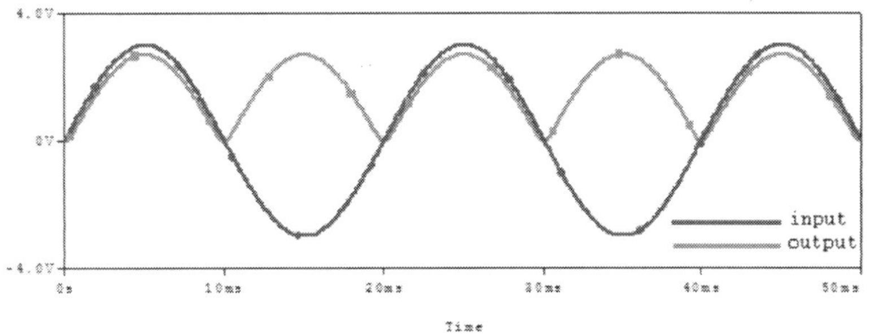

Fig. 14: Input voltage v/s output voltage

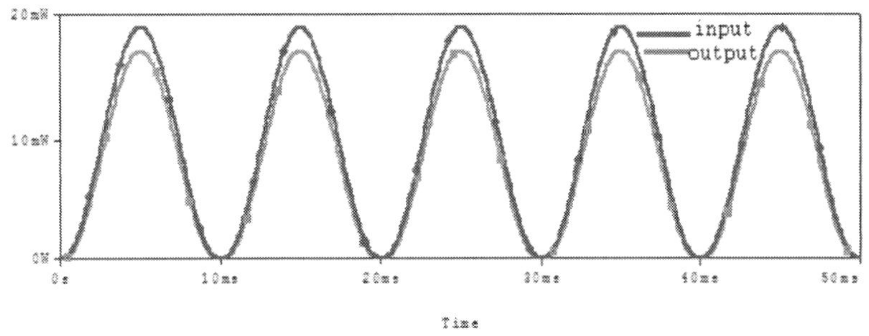

Fig. 15: Input power v/s output power

The capacitor filter reduces the ripples to a higher extent and makes this voltage regulated, bringing it closer to DC. An average voltage equal to 2.8V is produced in this case and the ripple factor has reduced to almost 85% compared to the original rectified voltage waveform.

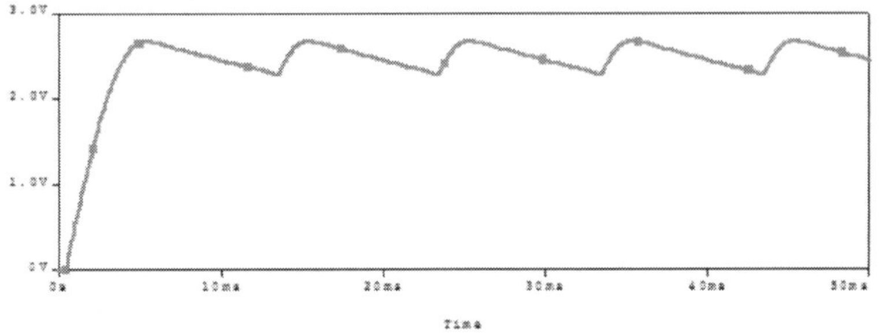

Fig. 16: Output voltage with a capacitor filter

4. Conclusion

Low power devices require a regulated input for their proper functioning. Energy harvesting is preferred over batteries due to the low voltage requirements but the output of energy harvesting system is so small that it needs to be regulated to usable level. Rectifiers are used to fulfil this job usually. Although these circuits rectify the output but it contains AC ripples that can interlude the operation of the device, so a solution has been proposed to cope against this problem. The output becomes more regulated if a capacitor is connected in parallel to the load resistor. The larger the value of capacitor, the more regulated output is. This paper compared the contemporary rectifier approaches in terms of output voltage, output power and the ripple factor. All above mentioned parameters have been plotted and compared. The ripple factor for each case has been calculated using analytical approach as well. Results show that the last circuit with bootstrapping capacitors performs best in terms of achieving all the intended parameters.

References

[1] Gosselin, B., Ayoub, A.E., Roy, J.-F., Sawan, M., Lepore, F., Chaudhuri, A. and Guitton, D., "A Mixed-Signal Multichip Neural Recording Interface with Bandwidth Reduction," *IEEE Trans. Biomedical Circuits and Systems,* 2009, vol. 3, no. 3, pp. 129–141.

[2] Weir, R.F., Troyk, P.R., DeMichele, G.A., Kerns, D.A., Schorsch, J.F. and Maas, H., "Implantable Myoelectric Sensors (Imess) for Intramuscular Electromyogram Recording," *IEEE Trans. Biomedical Engineering,* 2009, vol. 56, no. 1, pp. 159–171.

[3] Eric Y. C., Arthur L. C. and Pedro P. I., "A Miniature-Implantable RF-Wireless Active Glaucoma Intraocular Pressure Monitor," *IEEE Trans. Biomedical Circuits and Systems,* 2010, vol. 4, no. 6, pp. 340- 349.

[4] Uei-Ming., J., and Maysam G., "Optimization of Data Coils in a Multiband Wireless Link for Neuroprosthetic Implantable Devices," *IEEE Trans. Biomedical Circuits and Systems,* 2010, vol. 4, no. 5, pp. 301-310.

[5] Kelly, S.K., Wyatt, J.L., "A Power-Efficient Neural Tissue Stimulator with Energy Recovery," *IEEE Trans. Biomedical Circuits and System,* 2011, vol. 5, no. 1, pp. 20–29.

[6] Grace, E.A., Rajan, S.E. and Asis, A.A.C., "Performance Evaluation of Different Rectifiers for Piezoelectric Energy Harvesting Applications," in Proc. *IEEE International Conference on Recent Advancements in Electrical, Electronics and Control Engineering*, 2011, pp. 248 – 252.

[7] Qingyun Ma., Haider, M.R. and Massoud, Y., "A Low-Loss Rectifier Unit for Inductive-Powering of Biomedical Implants," in *Proc. IEEE 19th International Conference on VLSI and System-on-Chip,* 2011, pp. 86-89.

[8] Ramadass, Y.K. and Chandrakasan, A.P., "A Battery-Less Thermoelectric Energy Harvesting Interface Circuit with 35 mV Startup Voltage," *IEEE Journal of Solid-State Circuits,* 2011, vol. 46, no. 1, pp. 333 - 341.

[9] Peters, Christian, Handwerker, Jonas, Maurath, Dominic and Manoli, Yiannos, "An ultra-low-voltage active rectifier for energy harvesting applications," in *Proc. of IEEE Int. symp. Circuits and Systems,* 2010, pp. 889 - 892.

[10] Niu, D., Zhangcai Huang, Minglu Jiang and Inoue, Y., "A Sub-O.3V CMOS Rectifier for Energy Harvesting Applications," in *Proc. IEEE Int. Symp. Circuits and Systems,* 2011, pp. 1-4.

[11] Wei Wu and Qiang Fang, "Design and Simulation of Printed Spiral Coil used in Wireless Power Transmission Systems for Implant Medical Devices," in *Proc. IEEE Int. Conference on Medicine and Biology Society*, 2011, pp. 4018-4021.

[12] Troyk, P.R. and DeMichele, G.A. , "Inductively-Coupled Power and Data Link for Neural Prostheses using a Class-E Oscillator and FSK Modulation," in *Proc. IEEE Int. Conference on Medicine and Biology Society*, 2003, pp. 3376 – 3379.

[13] Hashemi, S.S., Sawan, M. and Savaria, Y., "A High-Efficiency Low-Voltage CMOS Rectifier for Harvesting Energy in Implantable Devices," *IEEE Trans. Biomedical Circuits and Systems*, 2012, vol. 6, no. 4. pp. 326 – 335.

A Fusion of Sensors Information for Autonomous Driving Control of an Electric Vehicle (EV)

Hasri Haris[2,3], Khairunizam WAN[1], D Hazry[1] and Zuradzman M Razlan[1]
[1]Center of Excellence UAS,
[2]Advanced Computing and Sustainable Research Group,
Universiti Malaysia Perlis, Malaysia.
[3]Bahagian Sumber Manusia,
Majlis Amanah Rakyat (MARA), Kuala Lumpur, Malaysia.

*E-mail: khairunizam@unimap.edu.my

Abstract. The study uses the environment of the road as input variables for the main system to control steering wheel, brake and acceleration pedals. A camera is installed on the roof of the Electric Vehicles (EV) and is used to obtain image information of the road. On the other hand, users or drivers do not have to directly contact with the main system because it will autonomously control the devices by using fuzzy information of the road conditions. A fuzzy information means in the preliminary experiments, reasoning of the various environments will be done by using fuzzy approach. At the end of the study, several existing algorithms for controlling motors and image processing technique could be combined into an algorithm that could be used to move EV without assist from human.

1. Introduction

Autonomous driving is increasingly attracting public interest due to various research projects over the past years [1-7]. Usually, conventional cars are converted with significant effort and many different sensors are placed on the roof. The advance of electro-mobility provides the chance for completely new vehicle concepts. By breaking away from classic approaches, it is possible to consider and integrate autonomous driving into the vehicle architecture with respect to Information Technology (IT) and sensor network systems, energy management and design. These kinds of cars are the upgrade version of the EV.

Recently a lot of EVs and related vehicles such as a hybrid car have been developed to solve environment and energy problems caused by the use of an internal combustion engine vehicle. Developing such vehicles for solving the environment and energy problems is a great idea. Currently, many researches publish technical papers in journals, which are related to autonomous EV. In their researches, steering wheel, brake and acceleration pedals are control by using computers [8-10]. On the other hand, users and drivers do not have a direct contact with them. A touch panel is installed in the EV and it serves a user Graphical User Interface (GUI) for users and drivers interact with controlling devices. Unfortunately, based on current outcomes more effort should be done for making

[1]khairunizam@unimap.edu.my

sure that autonomous EV could move with safety. Although mechanism of mechanical could be used to solve safety and reliability issues of an autonomous EV, the computational approach is also very important. The computational approach for example the algorithm for controlling motor device, the capacity of data transmission device, image processing technique and etc. [11-13].

2. Literature Review

The focus of current research towards electric, hybrid electrics, and fuel cell vehicles has been on increasing energy efficiency and reducing emissions. Future vehicles will include electric drive-train components that must be capable of performing conventional anti-lock braking, traction control, and active yaw control safety functions. From the viewpoint of electric and control engineering, Hybrid electric vehicles (HEVs) have evident advantages over conventional internal combustion engine vehicles (ICVs). Firstly, torque generation is very quick and accurate, for both accelerating and decelerating. This should be the essential advantage. In Hybrid electric vehicles, motor and traction control system (TCS) should be integrated into Hybrid Traction Control System (HTCS), since a motor can either accelerate or decelerate the wheel. Its performance should be advanced one, if we can fully utilize the fast torque response of motor. Secondly, output torque is easily comprehensible. There exists little uncertainty in driving or braking torque inputted by motor, compared to that of combustion engine or hydraulic brake.

In recent years fuzzy logic control techniques have been applied to a wide range of systems. Many electronic control systems in the automotive industry such as automatic transmissions, engine control and traction control systems are currently being pursued. These electronically controlled automotive systems realize superior characteristics through the use of fuzzy logic based control rather than traditional control algorithms. Fuzzy Logic Control is a type of control, which is based on Fuzzy set theory and reasoning. David Elting and Mohammed Fennich in their research told that automotive systems realize superior characteristics through the use of fuzzy logic controllers [14] especially in nonlinear cases. The brake system is a challenging control problem because the vehicle-brake dynamics are highly nonlinear with uncertain time-varying parameters [15]. Fuzzy controllers have the benefit of not requiring a mathematical model of the plant [16], while still being highly robust [17]. Also, certain fuzzy control designs can be implemented that have the ability to learn [18] or to adapt [19] themselves to improve its performance. Because of these features, fuzzy controllers have been successfully implemented in the automotive field for controlling both wheel dynamics and vehicle dynamics.

3. Methodologies

In this study, a pure differential drive of the electric vehicle is considered as shown in figure 1 and it is assumed that the posture and the orientation of the vehicle are known at each instant. By referring to the figure, L is the base width of the vehicle and R is the distance between the centres of the wheel. The orthonormal inertial basis is $\{x, 0, y\}$, which is called the world coordinate system that is fixed to cartesian workspace. The centre of gravity of the mobile vehicle is the point C, and the basis $\{M_1, C, M_2\}$ is attached to the EV, which is called the local coordinate system. M_1 refers to longitudinal direction, while M_2 is the vertical direction, and $OC=xI_1+yI_2$. The location of EV could be determined by three variables, which are the spatial positions x and y of the reference point C, and its orientation. The vector Ps describing the EV posture and is defined as in equation (1).

$$p_s(t) = \begin{bmatrix} x(t) \\ y(t) \\ \theta_c(t) \end{bmatrix} \tag{1}$$

Where $t_0<t<t_n$ and t_0 is the initial moment of motion and t_n is a final moment of motion. For example, in the perspective of brake control of the EV, two degree of freedom is considered, which is velocity V_C and angular velocity θ_c as shown in figure 2. The vehicle kinematic associated with the Jacobean matrix and the velocity vector is defined in equation (2).

$$\begin{bmatrix} \dot{x}_c \\ \dot{y}_c \\ \dot{\theta}_c \end{bmatrix} = \begin{bmatrix} \cos\theta_c & 0 \\ \sin\theta_c & 0 \\ 0 & 1 \end{bmatrix} \begin{bmatrix} v_c \\ \omega_c \end{bmatrix} \qquad (2)$$

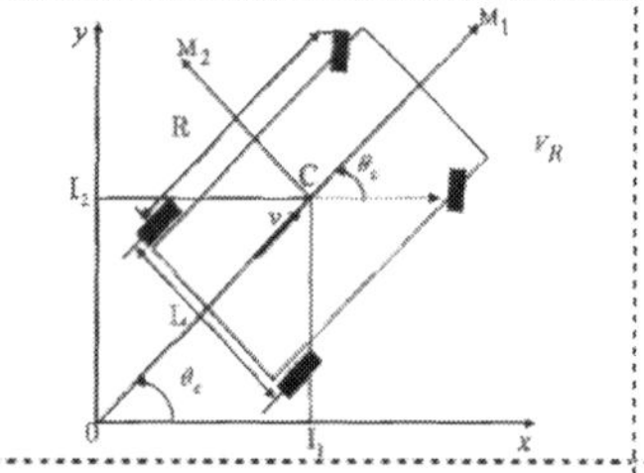

Figure 1. A pure differential drive of the EV model

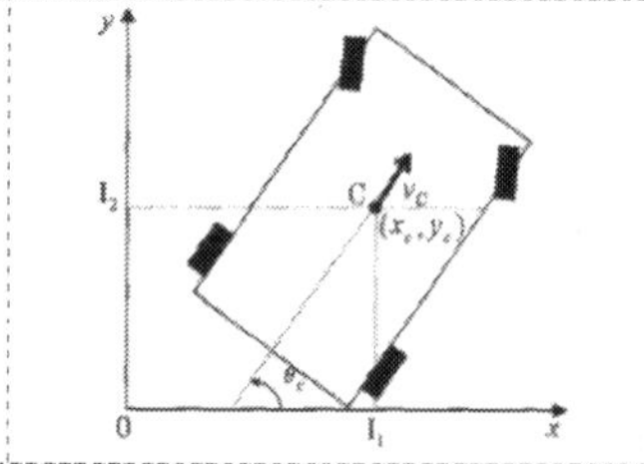

Figure 2. Two degree of freedom of EV model.

Besides brake and acceleration pedals, a steering wheel also needs to be considered in an autonomous EV. Directions of the EV depend on the current degree of the steering wheel. By extracting a rear wheels as shown in figure 2 as the driving wheel, figure. 3 give the description of the steering wheel. In the autonomously control of the EV, the input images from the camera are used to control the degree of the steering wheel. On the other hand, β(t) and α(t) must be adjusted for the EV to move straight in a straight road, and for the EV to turn left or right in the curvature condition.

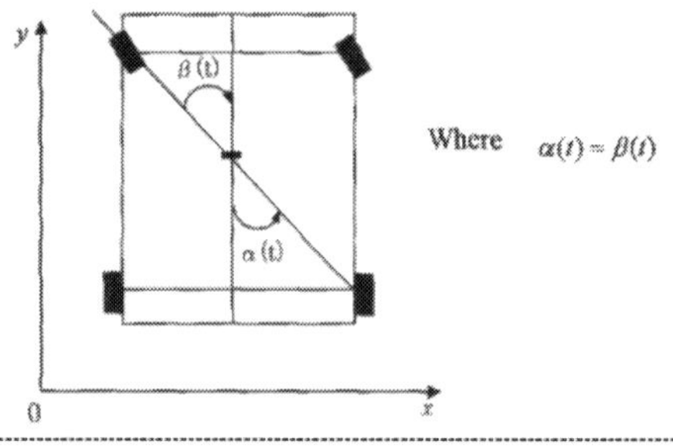

Figure 3. Description of the steering wheel.

Figure 4 shows an electrical configuration of the proposed EV. Two personal computers will be installed in the EV. The first PC will be employed for the motor control included I/0 board, motor control board. The second PC is for image processing included image process board. These two PC is connected through Ethernet hub or a wireless data transmission device. The on-board computer via mouse and touch panel is installed to operate steering wheel, gear shift, brake and acceleration pedals. On the other hand, the computer must control the vehicle brake pedal without having direct access to the steering mechanism, drive motor, and the transmission of the vehicle. The computer must be able to control the vehicle brake in the same manner as a human being does. The accelerator and the brake use stepping motor and the steering wheel part uses AC servo motor.

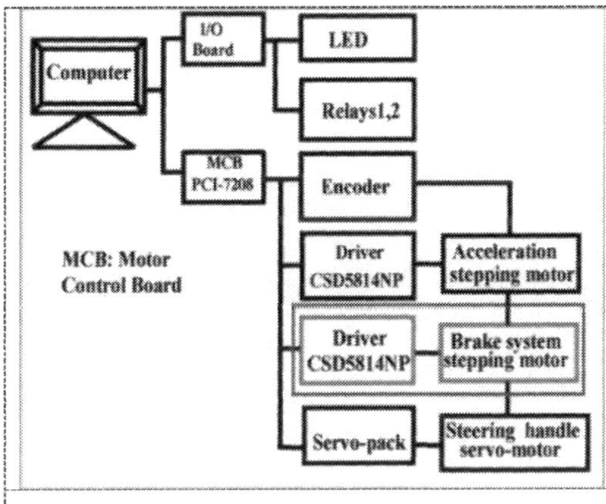

Figure 4 Electrical configuration of the proposed EV.

4. Results and discussions

This paper presents the design and implementation of an autonomous Electric Vehicle (EV) with intelligent driving control to provide the driver assistance as well as unmanned driver. It is an automatic guided vehicle and able to move automatically along the tracks in a given region. For the purpose of prototyping, a buggy car will be re-designed and several sensors are installed. The camera will be installed to the EV as a vision system and connects to the personal computer (PC) for the processing of image information. Image processing algorithm will be employed for the detection of the signage and the centre of gravity (COG) of the road. Another PC will be installed for controlling motors to operate acceleration pedal, brake pedal and steering wheel. Information from several sensors is fused to move the EV intelligently without control by the human.

Figure 5. Buggy Car.

In the experiment, the EV will follow the line in the middle of the road, and the line drawn in both left and right of the road. A CCD camera is used to capture images of the road while EV moves. The fuzzy information of the image captures is transferred to the second PC to control stepping and servo motors. The control mechanisms are stepping motors for brake and acceleration pedals, and a servo motor for the steering wheel. Figure 5, shows the image at time t captured by the CCD camera. The CCD camera is controlled by the first PC, and the image information is sent to the second camera for the main system to make a decision and produces command to the motor control board. For the purpose of reducing a computational cost, the image information contains the centre of gravity of the line and the values of θ. The image information captured by the CCD camera is processed by using image processing technique. The centre of the gravity of the line and the angle of the current location of the EV are sent to the second PC. Here, a predictive algorithm is needed to move the EV to the correct position at the time t+1. A Kalman filter technique is proposed to predict the future location of the EV. To move toward correct location, steering wheel, brake and acceleration pedals will be controlled.

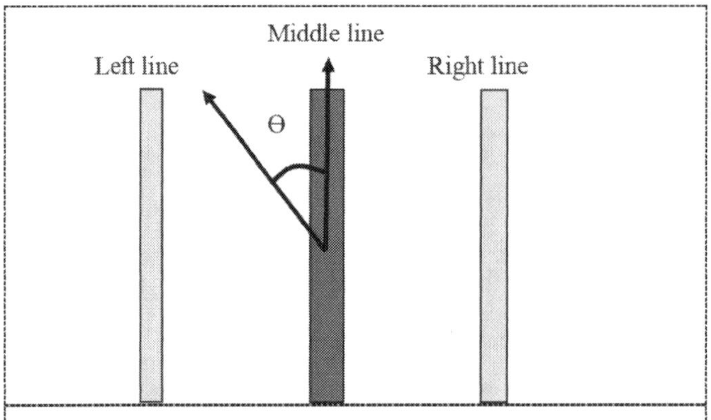

Figure 5. The road condition with the middle and side lines. Image information captures by a CCD camera.

5. Conclusions

The research of the electric vehicle (EV) in the proposal is developed in two fields. One is the control of the accelerator and the brake, the other is the controls of the steering wheel. Both of them are interfaced with stepping and servo motors with controllers. Needing the control of the accelerator and the brake is combined with the control of steering wheel is running at a constant speed. This becomes a necessary, indispensable technology in doing the driving support and actually running on the road. Therefore, we will set a speed control as a basic research to combine the control of the accelerator and the brake with the steering control by an image processing technique. The control of the steering wheel is done by referring to input images captured by using a CCD camera equipped on the EV. In this paper we produce the theory and conceptual of further experiment on Autonomous Driving Control of an electric vehicle EV. Hopefully after the future research and experiment, the results will appear as what we discuss on discussion and expected results.

Acknowledgements
Special thanks to all members of UniMAP Advanced Intelligent Computing and Sustainability Research Group, COEUAS and School Of Mechatronics Engineering, Universiti Malaysia Perlis, Malaysia for providing the research equipment and internal foundation. This work is also supported by the fundamental research grant scheme (FRGS) awarded by the Ministry of Higher Education to Universiti Malaysia Perlis (FRGS 9003-00313).

References

[1] P E Dumont, A Aitouche and M Bayart 2004 *IEEE Vehicle Power and Propulsion.* pp.198 - 203.

[2] Hammouri, M Kinnaert and E H El Yaagoubi 1999 *IEEE Transactions on Automatic Control* **44**: 1879-1884.

[3] M A Djeziri, A Aitouche and B O Bouamama 2007 *IEEE Conference on Decision and Control* pp. 2578-2583.

[4] K Bouibed, A Aitouche and M Bayart 2009 *IEEE Journal of Energy and Power Engineering* **3**: 10-18.
[5] M A Djeziri, R Merzouki and B Ould Bouamama 2009 *IEEE Transaction on Vehicular Technology* **58**: 4710-4719.
[6] Han-Shue Tan, Rajesh Rajamani and Wei-Bin Zlhang 1998 *Proceedings of the American Control.*
[7] Systems Control Technology Inc. 1994 *California PATH Research Paper.*
[8] S R Cikanek and K E Bailey 2002 *Proceedings of the American Control Conference Anchorage.*
[9] H X Li and S Guan 2001 *IEEE Control Systems Magazine* **21**
[10] Q CHEN, B QIU and Q XIE 2005 *Tsinghua University Press.*
[11] P G Howlett, P J Pudney and V Xuan 2009 *Journal of Automatica* **45**: 2692-2698.
[12] General definitions of highspeed 2009 *International Union of Railways.*
[13] Masamichi Ogasa 2010 *IEEJ Trans. on Electrical and Electronic Engineering* **5**: 304-311.
[14] S K Mazumdar and C C Lin 2005 *Proceedings of the IEEE International Conference on Neural Networks.*
[15] G F Mauer, G F Gissinger and Y Chamaillard 2004 *SAE International Congress & Exposition.*
[16] E David, F Mohammed, K Robert and H. Bert Fuzzy *Intel Corporation, Automotive Operation.*
[17] G F Mauer 1995 *IEEE Transactions on Fuzzy Systems* **3**.
[18] S T Lin and A K Huang 1998 *IEEE Transactions on Industrial Electronics* **45**
[19] J R Layne, K M Passino and S Yurkovich 1993 *IEEE Transactions on Control Systems Technology* **1**:

IOP Conf. Series: Materials Science and Engineering 53 (2013) 012026 doi:10.1088/1757-899X/53/1/012026

Optimization of wireless power transmission for two port and three port inductive link

Samnan Haider[11], Hasmah Mansor[1], Sheroz khan[1], Atika Arshad[1], Mohammed M. Shobaki[1] and Rumana Tasnim[1]
[1]*Department of Electrical and Computer Engineering*
International Islamic University
53100 Kuala Lumpur, Malaysia
E-mail: syedsamnanhaider855@gmail.com

Abstract: Recent developments have shown that the use of buried electronic devices or body implants has been becoming prevalent. Such low power devices are being powered up through non-contact means utilizing inductive coupling from external powering source. Inductive coupling not only solves the issue of energy availability but helps collecting the sensed data that can be archived or used for subsequent monitoring purposes. This paper analyses the performance of two-port and three-port inductive links in terms of power sent, power received and power transfer efficiency. All the above mentioned parameters have been plotted using analytical approach and obtaining simulation where required. The effect of mutual coupling has been studied in detail for both systems and demonstrated by plotting the power transfer efficiency for different values of the coefficient of coupling (k) using MATLAB. Results show that power transfer efficiency depends highly upon the value of k.

1. Introduction

Technological advancement has bestowed its innovation and novelty upon all the scientific fields. The portability in electronic devices is one of those and it has acquired a distinct significance in today's world [1]. The constant and unremitting energy availability to these low power devices is a challenge still. Especially in the case of using as a source of energy, batteries that may drain out after specific duration and need to be replaced. In some applications, the replacement of batteries may not be a costly exercise, while in some other applications it may be costly as well as time consuming and may be even impossible due to inaccessibility concerns; hence a constant voltage supply from batteries is becoming more important than ever [2]. Therefore, interfacing and coupling of an energy harvesting system for charging up of batteries in such devices may significantly improve their energy availability besides reducing maintenance and operational cost and relieving concerns of occasional replacements for batteries [3].

Inductive coupling is considered to be an efficient technique for transfer of power wirelessly. The basic working principle for inductive coupling states that the external power source and the implants to be powered are to be placed in the close vicinity to each other [4]. T[1]here is no need to expose or connect any of the systems to the other physically. The primary coil is excited by the external power

[1] To whom any correspondence should be addressed.

source to store the energy. The contiguity and the nearness of coils play a vital role in the generation of low power electrical field that inter-connects them and hence allows the transfer of electricity between the two systems. The energy sharing between the systems becomes possible only if they have the same amount of power.

The most appealing and alluring use of inductive coupling is in the field of biomedical implants where it is not only be used for powering these devices but also for gathering of data in telemetry purposes. Powering refers to deliver the effective amount of energy to the implanted device from the external source in order to operate it and telemetrically insinuate to the transfer of intuited data from the buried devices or implanted device to an peripheral device and vice versa [5]. Fig. 1 below shows an arrangement how an inductive coupling can be used for powering and telemetry purposes.

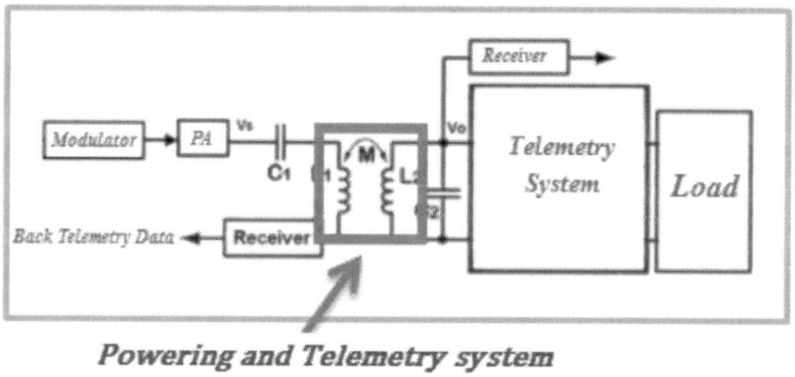

Fig. 1: Simple model for inductive coupling to be used for powering and telemetry

Inductive coupling makes the biomedical implants possible to operate by powering it according to the requirements and also in monitoring the subject. Due to which these implanted devices help in the real time monitoring of human body providing an exclusive prospect for the timely diagnosis and treatment of many diseases by collecting the data from the internal parts of human body [6].

Initially, the transfer of energy using implanted devices has been quite clumsy and a knotty procedure because the wires as an interfacing medium used to be pierced through the skin. But this method soon proved to be useless as it not only hindered the patient's movements but also had an astounding chance of infection. As explained earlier that a battery also cannot be inserted to power up the implanted device due to its size and draining out reasons. The total size of the implant increases eventually even if the battery is small enough, therefore restraining the possible implant sites. Furthermore, lifetime of the batteries is limited and even the rechargeable batteries have limited number of recharge cycles before becoming completely unusable.

In recent research works, many techniques have been proposed to power up the implanted devices and to collect the data sensed from them. One of the most expedient approaches is the use of energy harvesting method to power up the device using thermal [7], light [8] and piezoelectric sources [9]. Most of these techniques make use of the external environmental resources of temperature, wind, water, sunlight as energy reservoir. But all these sources are not available in the case of biomedical implanted devices. In the case of thermal energy harvesting, the working environment is restricted so in case of unavailability of thermal sources, therefore it requires an advance arrangement for the alternative powering source such as vibrational source in order to complement the process and hence keep the device running. Using solar energy harvesting also suffers from similar problems regarding the limitation of working environment. The use of implanted and external antennae to wirelessly transmit energy to the implant appears to be a suitable alternative. However, parameters such as human safety, power transfer efficiency and simplicity of electronics imply that the operational frequency must be in the frequency range between 1 and 20MHz [10]. Due to this constraint, the size of the implanted antenna, which is closely related to the electromagnetic radiation frequency, becomes too large to be implanted. But if inductive coupling is used for both powering an telemetry purposes, it proves to be the most efficient method as it has got several preferential advantages including reliable continuity of supply as compared to that from environmental parameters such as temperature, light etc. Being advantageous from many aspects, inductive coupling has some disadvantages too such as

- More heat generation as compared to other techniques
- Low efficiency
- Emission of harmful magnetic radiations
- More power consumption

A typical research proposes an OP-AMP based inductive coupling circuit to scrutinize the magnetic coupling of two coils that can be used to power up the implanted devices in the microwatt range [11]. The circuit for the powering system is espoused with a transmission coil and a monitoring device and an OP-AMP circuit connected with an inductive sensor. Furthermore, the paper includes some important parameters that need to be taken into consideration while designing an inductive power link. The inductive coupling has been further elaborated in [12] by proposing a system consisting of a reader coil inductively coupled to the sensing circuit, a front end analogue circuit, and a digital signal processing (DSP) unit. The results have been shown in the form of analytical derivations and simulation of a typical telemetric circuit. The simulation is shown to be perfectly matched with the analytical results.

This paper compares two coils and three-coil energy transfer system based upon the phenomenon of inductive coupling in terms of power transfer efficiency. Simulation results show that it is no more a monotonically decreasing function of the distance between the coils if it is tuned to a resonant frequency that can be calculated easily. This is the point where the system shows the maximum performance at a relatively larger operating distance by tuning it to certain required operating frequencies.

The paper is organized such that Section II presents the analysis of telemetry systems containing the subsections (a) and (b) for the modelling and simulation results for two coils and three coils inductive coupling system respectively. Section III presents the discussion and comparisons and section IV concludes with the conclusion.

2. Analysis of telemetry system

The power received by the reader side mainly depends upon the separation between the coils or the mutual coupling. Two coils are said to be linked inductively, when the primary coil excited by the external source is able to have a magnetic field induced in the secondary coil, producing a voltage across its terminals as a result.

Mutual coupling, playing a vital role in overall performance of the whole system, primarily depends on the dimensions and positioning of the two coils. Analytically the mutual coupling can be expressed as

$$M = \frac{\mu_0 N_1 R_1^2 N_2 R_2^2 \pi}{\sqrt{(R_1^2 + x^2)^3}} \tag{1}$$

Where M is mutual coupling coefficient
μ_0= permeability of air ($4\pi \times 10^{-7}$ H•m^{-1})
N_1 and N_2 are the turns of primary and secondary coils respectively
R_1= radius of the coil in the source loop
R_2= radius of the coil in the load loop
x= separation between the coil

Assuming N_1=N_2=1000 turns and radius of the coil R_1=3mm and R_2=1mm, varying 'x' between 1mm to 0.1 meter, Equation (1) above is plotted for M as a function of distance as shown in Fig. 2.

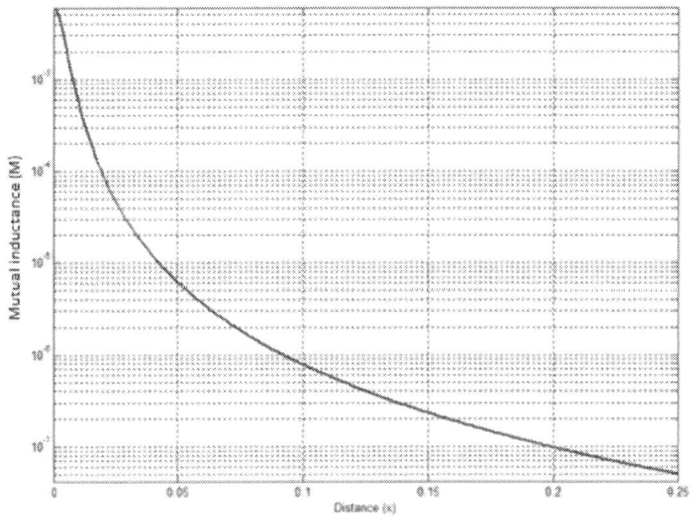

Fig.2: Mutual inductance as a function of distance '*x*'

It can be observed that mutual inductance M decreases remarkably as the separation between the coils increases. So for higher values of mutual inductance, coils need to be placed near to each other as this determines the value of K, the coupling coefficient between the coils given as

$$K = \frac{M}{\sqrt{(L1L2)}} \qquad (2)$$

2.1 Modeling and plots for two coils inductive coupling system

As shown in the Fig. 3, the circuit consists of two parts each with a coil separated by a certain distance from the other having a mutual coupling $j\omega M$ between them. One of the coils, connected to the source is the primary coil whereas the other side including load is the secondary coil.

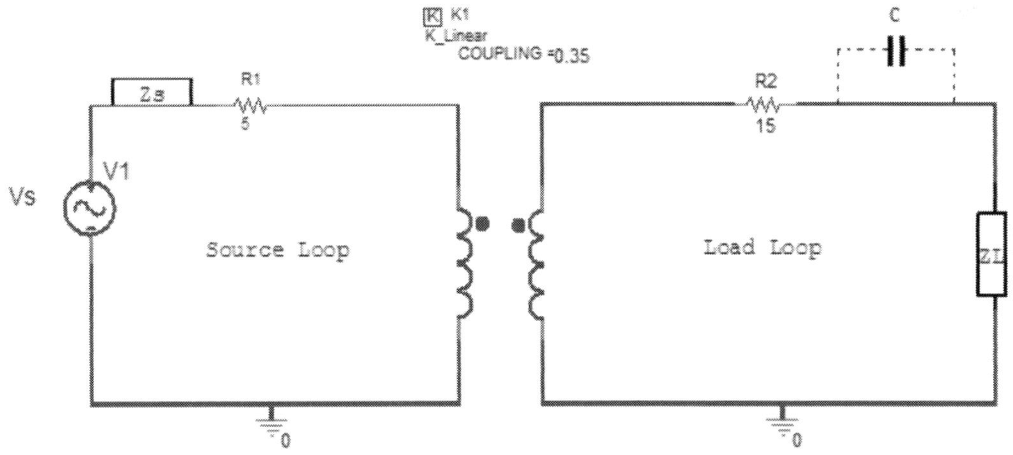

Fig.3: Two coils inductive coupling system

The circuit shown above follows the principle of inductive coupling for transferring the power wirelessly. The currents for each loop can be calculated by applying KVL that comes out to be

$$I_1 = \frac{\begin{vmatrix} V_S & -j\omega M \\ 0 & Z_{22} \end{vmatrix}}{\begin{vmatrix} Z_{11} & -j\omega M \\ -j\omega M & Z_{22} \end{vmatrix}} = \frac{V_S Z_{22}}{Z_{11}Z_{22} + \omega^2 M^2} \tag{3}$$

$$I_2 = \frac{\begin{vmatrix} Z_{11} & V_S \\ -j\omega M & 0 \end{vmatrix}}{\begin{vmatrix} Z_{11} & -j\omega M \\ -j\omega M & Z_{22} \end{vmatrix}} = \frac{j\omega M V_S}{Z_{11}Z_{22} + w^2 M^2} \tag{4}$$

The power sent from the source loop equal the source voltage times the current in the source loop that can be expressed mathematically as

$$P_{sent} = |I_1|\, V_s \tag{5}$$

Whereas the received power for the load/ reader loop is given by

$$P_{recivied} = |I_2|^2 R_l \tag{6}$$

For the parameters used in the circuit in Fig. 3, the power transfer efficiency, η, (as a ratio of power received to the power sent) is as shown in Equation (7).

$$P_{eff} = \frac{P_{recieved}}{P_{sent}} * 100 \tag{7}$$

Substituting the values of each parametric value, final expression for the power transfer efficiency comes out as shown in Equation (9).

$$P_{eff} = \frac{|I_2|^2 R_l}{|I_1|\, V_s} * 100 \tag{8}$$

$$P_{eff} = \frac{w^2 M^2\, R_l}{|Z_{22}|\, |Z_{22}Z_{11} + w^2 M^2|} * 100 \tag{9}$$

Equations (5) and (6) are simulated for the component values of circuit shown in Fig. 3 and the resulting plots are shown in Fig. 4. It can be observed that power received reaches maximum at a frequency of 10 kHz approximately. Maximum power can be transferred if the circuit is tuned to this frequency in order to make this circuit to perform more efficiently.

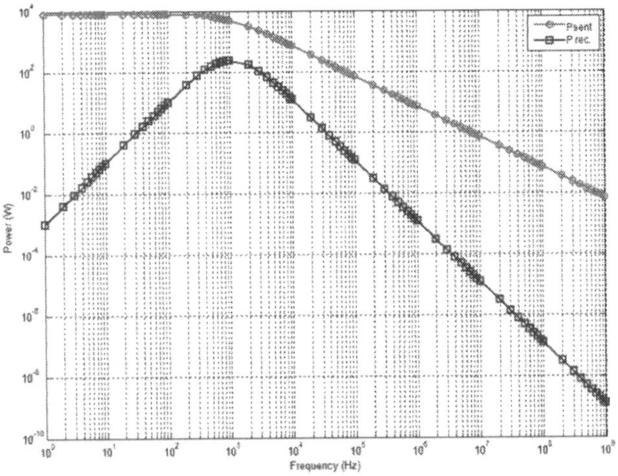

Fig. 4: Power sent, power received as a function of Frequency

Similarly, as shown in which is shown in Fig. 5, equation (9) can be plotted to show the power transfer efficiency for a fixed resistive load, R_L=15Ω. It can be seen in the plot that the power transfer efficiency reaches the maximum value at the same value of resonant frequency. Both power transfer efficiency and resonant frequency can be enhanced by changing the value of '*k*' and load resistance.

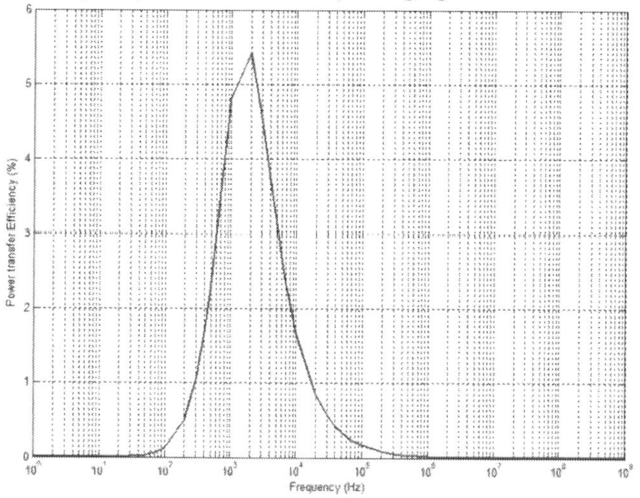

Fig. 5: Power Transfer Efficiency as a function of frequency

2.1.1 Effect of load variations

An increase in the value of load resistance reduces the impedance of secondary coil because they are inversely proportional to each other according to Equation (10). This decrease in impedance increases the power transfer efficiency subsequently. To check the effect of load variations on the power transfer efficiency, Equation (9) is plotted keeping all the parameters constant except the value of R_L varying it from 0Ω to 1000 Ω (Fig. 6). Results show that the power transfer efficiency increases almost linearly with a higher slope until it reaches the maximum at R_L=110 Ω and then start decreasing. In Fig. 7, the load resistance and frequency both are made variable and the plots show that power transfer efficiency increases with R_L until 100 Ω where it achieves the maximum. After that it becomes constant and only the resonant frequency keeps increasing, even if the load is varied after that.

$$Z = (R_1 j\omega L_1)\left(\frac{\omega^2 M^2}{R_2 + j\left(2\pi L_2 + \frac{1}{\omega C_2}\right)}\right) \quad (10)$$

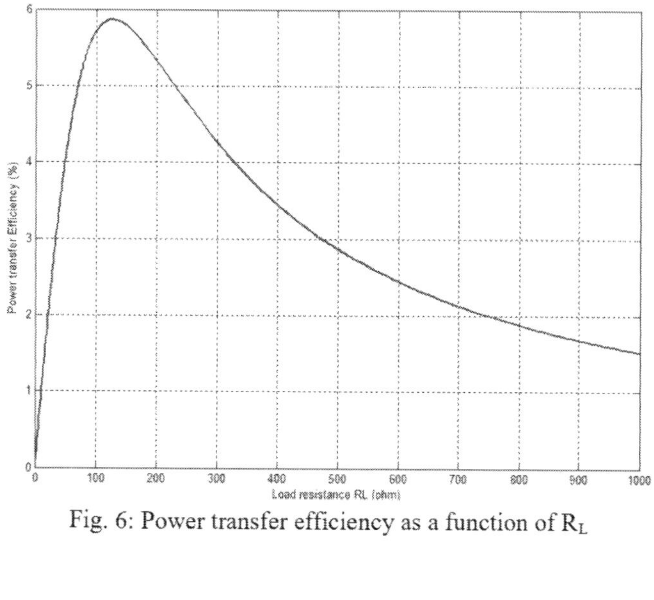

Fig. 6: Power transfer efficiency as a function of R_L

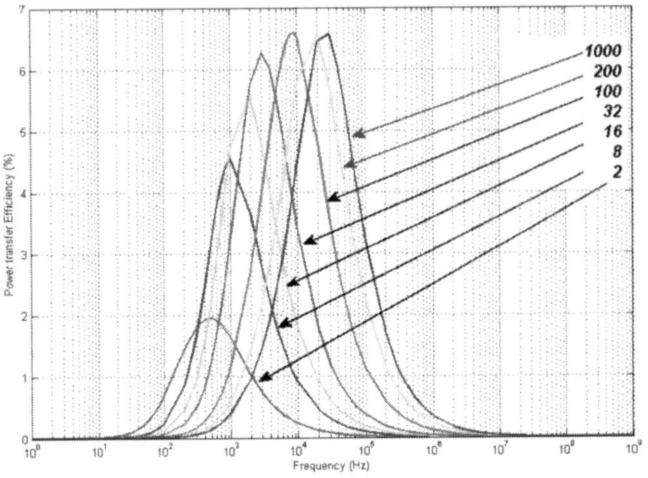

Fig. 6: The power transfer efficiency as a function of frequency with different value of R_L

2.1.2 Effect of changing coefficient of coupling '*K*'

As explained earlier in the literature, that the value of coupling coefficient plays an important role in determining the efficiency of an inductively coupled circuit. According to Equation (1) the mutual coupling depends upon the separation distance between the coils. Fig. 2 shows that an increase in the distance between the coils reduces the mutual inductance visibly that eventually reduces the mutual coupling. Fig 7 illustrates the effect of changing value of mutual coupling between the coils and it can be noticed here that the circuit achieves higher power transfer efficiency for a larger value of '*K*'. If the coils are placed near to each other, it makes the input impedance independent of frequency changes but it increases the power transfer efficiency effectively. We can conclude that when the distance between the coupled coils is smaller, the value of M is higher and hence the value of '*K*', so the higher power transfer efficiency is achieved.

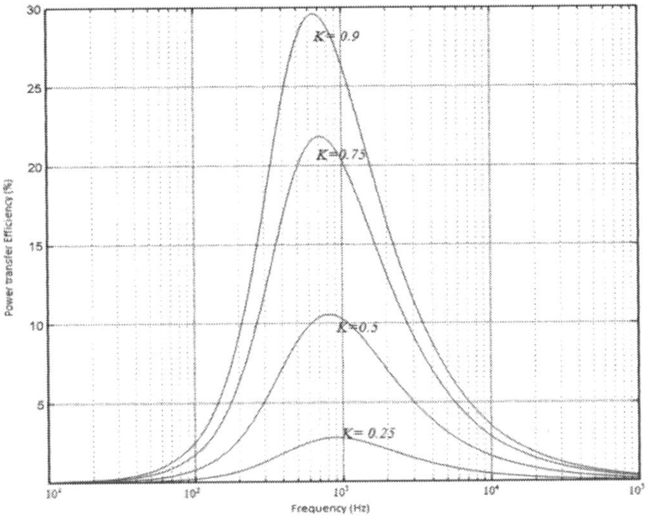

Fig. 7: Power transfer efficiency as a function of frequency for different values of '*K*'

2.1.3 Effect of added capacitance

If a capacitor is added in the circuit as shown in Fig. 2 with dotted lines, it increases the power transfer efficiency immensely because of the fact explained in Equation (10). As the impedance is inversely proportional to the value of capacitance and resistance so an insertion of capacitance reduces the impedance that increases the power transfer efficiency. The capacitive transducers are preferred over the resistive transducers because of their enhanced immunity to the temperature changes. The effect of addition of a capacitor is shown in Fig. 8, where it can be observed that the efficiency has increased up to 95% for C= 1ηF but the bandwidth has reduced as well. So it demands to tune the circuit carefully because of this shorter bandwidth.

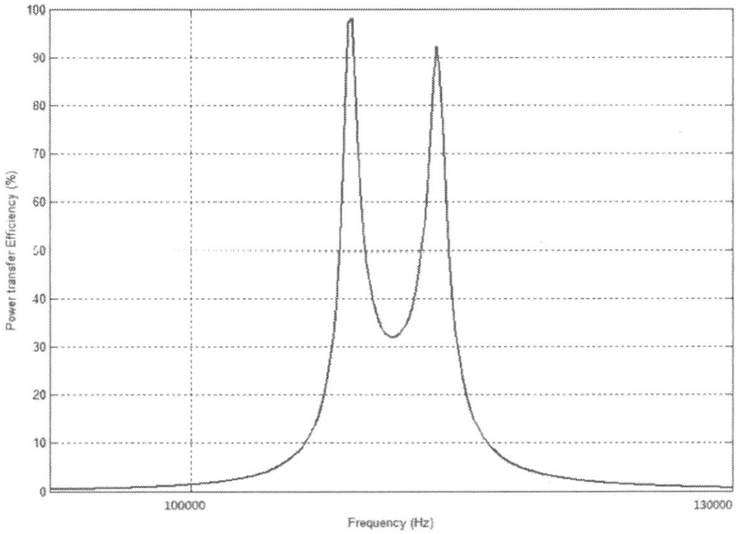

Fig. 8: The power transfer efficiency as a function of frequency in the presence of capacitor

The dip in the middle is because of the smaller value of capacitance and it can be removed easily by using a capacitor with higher value but the higher value of capacitance will cause some more convergence in the bandwidth of resonant frequency.

2.2 Modelling and plots for three coils inductive coupling system

Figure 9 shows a three coils inductive coil energy transferring system. It consists of a source loop containing the primary coil, a Hub loop containing a coil that is acting as a hub to link the other two inductively and a load loop containing the secondary coil. Two coupling coefficients have been used to couple three coils that can be mathematically termed as $j\omega M_1$ *and* $j\omega M_2$. For transferring the power to the load loop, it has to go through three coils i.e. primary to the secondary.

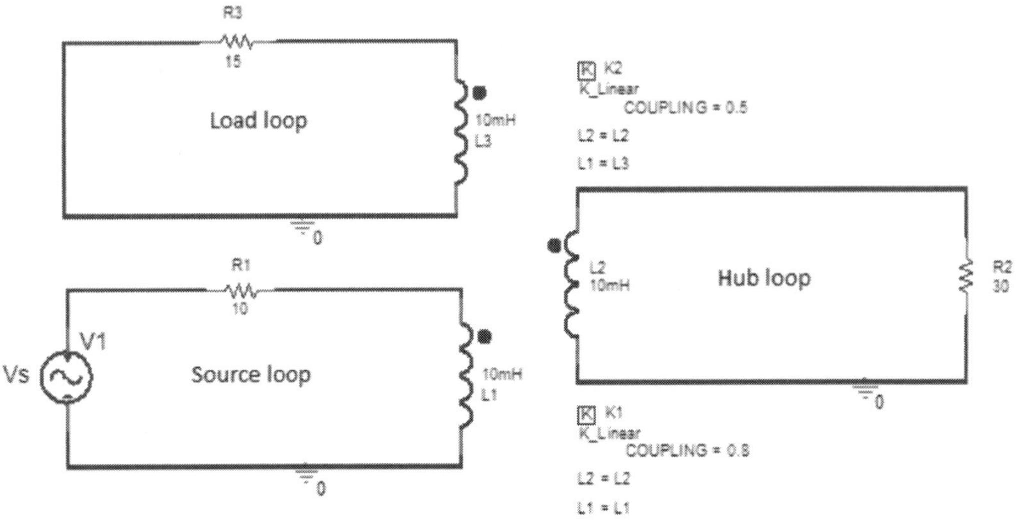

Fig. 9: Three coils inductive coupling system

Current for each loop can be calculated by applying KVL as done for the two coils system. Assuming the impedance of source loop as Z_{11}, hub loop as Z_{22} and load loop as Z_{33}, the currents for each loop can be expressed by following equations.

$$I_1 = \frac{V_s(Z_{22}Z_{33}+\omega^2M^2)}{Z_{11}(Z_{22}Z_{33}+\omega^2M_2^2)+\omega^2M_1^2Z_{33}} \quad (11)$$

$$I_2 = \frac{V_s(j\omega M_1 Z_{33})}{Z_{11}(Z_{22}Z_{33}+\omega^2M_2^2)+\omega^2M_1^2Z_{33}} \quad (12)$$

$$I_3 = \frac{V_s(\omega^2M_1M_2)}{Z_{11}(Z_{22}Z_{33}+\omega^2M_2^2)+\omega^2M_1^2Z_{33}} \quad (13)$$

Similarly, the power sent and power received by the load loop can be expressed as follows:

$$P_{sent} = |I_1|\,V_s \quad (14)$$

$$P_{receivied} = |I_3|^2R_l \quad (15)$$

The power transfer efficiency is given by Equation (7) where substituting the values for power sent and power received we get the following expression.

$$P_{eff} = \frac{V_s(\omega^4M_1^2M_2^2)}{(Z_{22}Z_{33}+\omega^2M_2^2)} \quad (16)$$

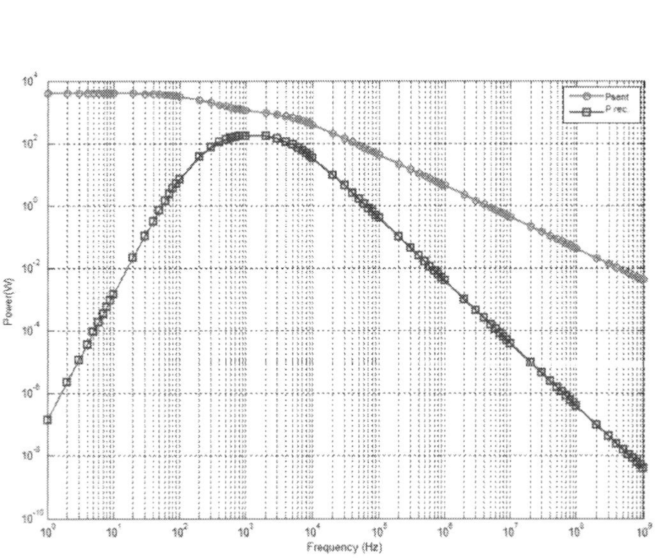

Fig. 10: Power sent, Power received as a function of frequency

Equations (14) and (15) are simulated for the component values as shown in Fig. 9 and the resulting plots are shown in Fig. 10 where it can be observed that power received reaches maximum at resonance frequency of 10^3 Hz approx. this means that at this value of frequency, maximum power is being transferred so for the efficient operation and maximum efficiency the circuit needs to be tuned to this value if the same component values are used.

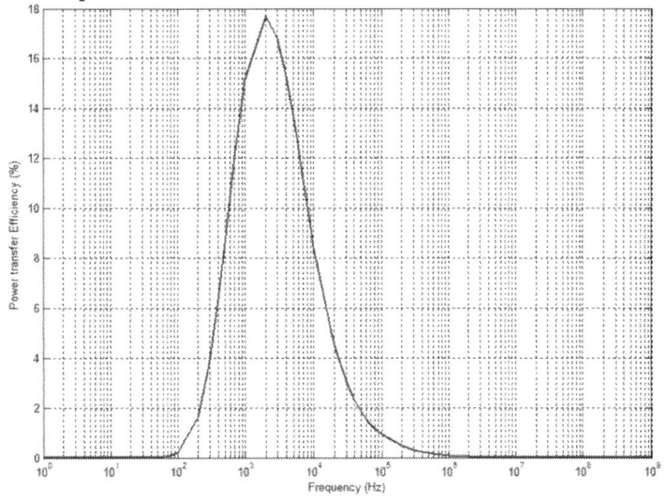

Fig. 11: Power transfer efficiency as a function of frequency

Fig. 11 shows a plot for power transfer efficiency as a function of frequency which has been plotted using Equation (16). It can be observed that the system shows maximum power transfer efficiency at the resonant frequency of 10 kHz. So it can be concluded that for the component values used for the simulation, this circuit needs to be tuned to the resonant frequency for the maximum power transfer efficiency. It must be kept in mind that the value of *'K'* and load resistance has the same effects on the performance of circuit as well similar for the two coils inductive system.

IOP Conf. Series: Materials Science and Engineering **53** (2013) 012026 doi:10.1088/1757-899X/53/1/012026

3. Discussion and comparison

Equations for currents, power sent, power received and power transfer efficiency for two coils as well as three coils inductive coupling system have been derived using analytical approach. All these equations have been plotted for fixed values of components shown in Fig. 3 and Fig. 9 for both the circuits using MATLAB. The effect of load variations and coupling coefficient *'K'* is studied for fixed as well as variable frequency. The plots obtained as a result show an evident dependency of the performance of whole system on all these factors. Specially, it depends upon the mutual coupling which refers to the separation between the coils. The smaller the gap between the coils, the more efficient the circuit performs. The power transfer reaches a maximum value at a certain frequency which can be referred to as a resonant frequency for that system and it should be tuned to that frequency to get the maximum output.

4. Conclusion

Inductive coupling can be employed to power up the biomedical implants and buried devices normally in the very low voltage requirement. The source for generation can be the human body itself where energy will be harvested using thermal and piezoelectric transducers. This paper has presented the modelling results for an inductively coupled system using two coils and three coils for wireless transfer of power over small gaps. The effect of mutual coupling and load variations, upon the power transfer efficiency of the system has been studied in detail and illustrated with analytical plots using MATLAB. Results show that the coils placed in close vicinity to each other with a smaller gap separation, perform better for transferring the power. A resonant frequency for both the systems has been obtained at which the maximum power transfer takes place. If the circuit is regulated to this value, then the power transfer efficiency is no more a decreasing function of the changes in frequency, distance and load.

References

[1] A. Sanni, A. Vilches and C. Toumazou, "Inductive and Ultrasonic Multi-Tier Interface for Low-Power, Deeply Implantable Medical Devices," *IEEE Trans. Biomed. Circuits Systems*, vol.6, no. 4, pp. 297-308, August 2012.

[2] Z. G. Wan, Y. K. Tan, and C. Yue, "Review on Energy Harvesting and Energy Management for Sustainable Wireless Sensor Networks", *in Proc. IEEE Intern. Conf. Communication Technology*, pp. 362-367, Sept. 2011.

[3] A.P. Sample, D.A. Meyer, J.R. Smith, "Analysis, Experimental Results, and Range Adaptation of Magnetically Coupled Resonators for Wireless Power Transfer", IEEE Trans. Industrial Electronics, vol. 58, no. 2, pp.544-554, February 2011.

[4] R. Harrison, P. Watkins, R. Kier, R. Lovejoy, D. Black, R. Normann, F. Solzbacher, "A Low-Power Integrated Circuit for a Wireless100-Electrode Neural Recording System," in *Proc. IEEE Intern. Conf. Solid- State Circuits,* pp.2258-2267, February 2006.

[5] Ghovanloo, Maysam and Atluri, Suresh, "An Integrated Full-Wave CMOS Rectifier with Built-In Telemetry for RFID and Implantable Biomedical Applications", *IEEE Trans. on Circuits and Systems,* vol. 55, no. 10, pp. 3328-3334, November 2008.

[6] Hunter H. Wu, Aaron Gilchrist, Kylee D. Sealy, and Daniel Bronson, "A High Efficiency 5 kW Inductive Charger for EVs Using Dual Side Control," *IEEE Trans. Industrial Informatics*, vol. 8, no. 3, pp. 585-595, August 2012.

[7] D.C. Hoang, Y.K. Tan, H.B. Chng and S.K. Panda, "Thermal Energy Harvesting from Human Warmth for Wireless Body Area Network in Medical Healthcare System", in Proc. *IEEE Intern. Conf. on Power Electronics Drive Systems,* pp. 1277- 1282, November 2009.

[8] S. Ayazian, V. A. Akhavan, E. Soenen and A. Hassibi, "A Photovoltaic-Driven and Energy-Autonomous CMOS Implantable Sensor", *IEEE Trans. Biomed. Circuits and Systems*, vol. 6, no. 4, AUGUST 2012.

[9] E. A. Grace, S. E. Rajan, A. A. Clarence, "Performance Evaluation of Different Rectifiers for Piezoelectric Energy Harvesting Applications", in Proc. *IEEE Intern. Conf. on Recent Advancements in Electrical, Electronics and Control Engineering,* pp. 248-252, December 2011.

[10] Shuenn-Yuh Lee, Chih-Jen Cheng and Ming-Chun Liang, "A Low-Power Bidirectional Telemetry Device with a Near-Field Charging Feature for a Cardiac Micro stimulator," *IEEE Trans. On Biomed. Circuits and Systems*, vol. 5, no. 4, pp. 357-367, August 2011.

[11] Atika Arshad, Sheroz Khan, AHM Zahirul Alam and Rumana Tasnim, "Investigation of Inductive Coupling Approach for Non-Contact Bidirectional Transfer of Power and Signal," in Proc. *Intern. Conf. on Computer and Communication Engineering*, pp. 570- 573, July 2012.

[12] R. Nopper, R. Has, and L. Reindl, "A Wireless Sensor Readout System—Circuit Concept, Simulation, and Accuracy," *IEEE Trans. on Instrumentation and Measurement,* vol. 60, no. 8, August 2011.

IOP Conf. Series: Materials Science and Engineering **53** (2013) 012027 doi:10.1088/1757-899X/53/1/012027

High Quality Acquisition of Surface Electromyography – Conditioning Circuit Design

Mohammed M. Shobaki[1], Noreha Abdul Malik, Sheroz Khan, Anis Nurashikin, Samnan Haider, Sofiane Larbani, Atika Arshad and Rumana Tasnim[1]
Department of Electrical and Computer Engineering
International Islamic University
53100 Kuala Lumpur, Malaysia
Email: m.shobaki@gmail.com

Abstract: The acquisition of Surface Electromyography (SEMG) signals is used for many applications including the diagnosis of neuromuscular diseases, and prosthesis control. The diagnostic quality of the SEMG signal is highly dependent on the conditioning circuit of the SEMG acquisition system. This paper presents the design of an SEMG conditioning circuit that can guarantee to collect high quality signal with high SNR such that it is immune to environmental noise. The conditioning circuit consists of four stages; consisting of an instrumentation amplifier that is used with a gain of around 250; 4th order band pass filter in the 20-500Hz frequency range as the two initial stages. The third stage is an amplifier with adjustable gain using a variable resistance; the gain could be changed from 1000 to 50000. In the final stage the signal is translated to meet the input requirements of data acquisition device or the ADC. Acquisition of accurate signals allows it to be analyzed for extracting the required characteristic features for medical and clinical applications. According to the experimental results, the value of SNR for collected signal is 52.4 dB which is higher than the commercial system, the power spectrum density (PSD) graph is also presented and it shows that the filter has eliminated the noise below 20 Hz.

1. Introduction

The acquisition of SEMG signals during contractions is used for many applications including the diagnosis of neuromuscular diseases, prosthesis control as such signals could prove to be a diagnostic tool for estimating muscle fatigue or for evaluating patients with neuromuscular diseases, low backache, and disorders of motor control [1-5].

SEMG acquisition quality is influenced by different parameters including inter-electrode distance (IED), EMG electrodes position and electrodes configurations. The effect of variation such parameters is studied in [6-7]. Moreover, different methods of crosstalk quantification and reduction are discussed in [8] and suggestions for better electrode locations are presented in [9] to face up the cross talk issue.

SEMG signal quality can be represented by calculating signal to noise ratio (SNR), to increase the SNR, the measurement system must include special electronics for the signal filtering and processing. Surface EMG measurement system with high SNR is presented in [10], however, a notch filter used in the system can delete some useful components from EMG signals.

Some EMG measurement systems in previous mentioned works suffers from noise interference related issues either due to non-use of low pass filter or of sampling the signals with rate less than the needed, these probably the factors lead to reduce accuracy of the results. Further, using notch filters

[1] To whom any correspondence should be addressed.

could not remove unwanted frequency components alone without washing some useful frequency component along side, once again at the cost of poor quality signals.

The SEMG signals have a very low amplitude and low frequency contents. The quality of such signals is highly dependent on the conditioning circuit design. This paper presents a design of SEMG conditioning circuit that can guarantee the high quality signal acquisition and immune the environmental noise. The SNR is calculated and compared to the other works, and the PSD graph for the SEMG signal is presented.

2. The SEMG measurement system

As shown in Fig. 1 the SEMG system consists of surface electrodes, leads, conditioning circuit, data acquisition device (DAD) and a computer.

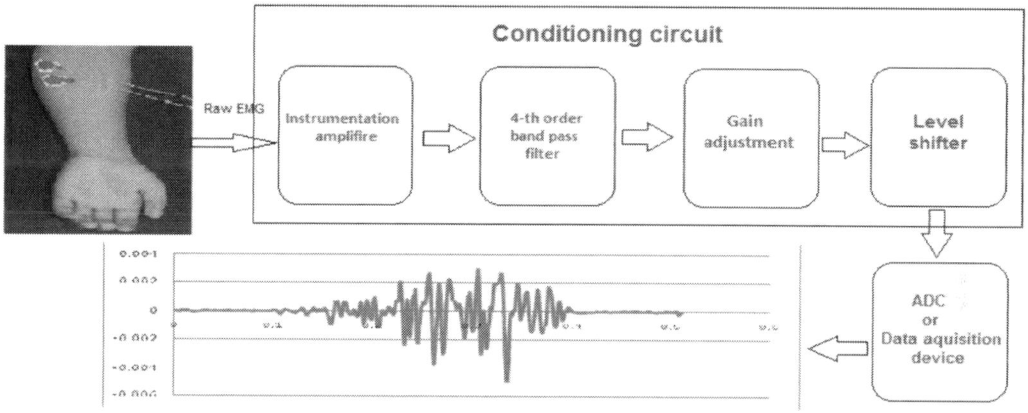

Figure 1: SEMG measurements system.

The electrodes attached to biceps muscle and connected to the conditioning circuit using the leads. In order to get the signal for analysis and storing we need to convert it to a digital form using DAD, then the signal can be transferred to the computer using USB connection.

3. The conditioning circuit

The goals of the signal conditioning are noise reduction, signal strengthening, removing unwanted signal component and signal translation fulfill the requirements of the next stage for further processing.

3.1 Amplifier:

A differential amplifier is used with bipolar SEMG electrodes. Since the signals have low amplitude a large gain is required. As shown in Fig.2, we choose the INA 2128 instrumentation amplifier due to its high CMRR at high gain, the CMRR of the INA2128 is about 120 dB which is sufficient for the design. Moreover, the effect of 50Hz power supply interference will be removed. This amplifier has a high input impedance to limit the attenuation from electrode-skin interference. Also, it shows that the high CMRR can improve the signal to noise ratio as in equation 1 as below:

$$U = K(V_{e1} - V_{e2}) + \frac{K}{CMRR} V_{common} \qquad (1)$$

$$R_G = \frac{50K}{G-1} \qquad (2)$$

To calculate the resistance for a certain gain, Equation 2 can be used, where G is the gain. For this stage of amplification a gain of 250 is sufficient, and the value of RG = 200 Ω.

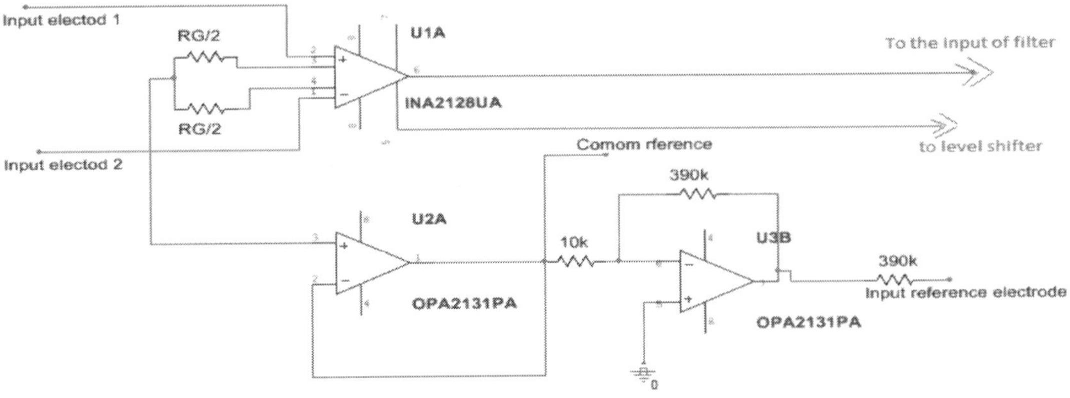

Figure 2: Instrumentation amplifier.

The reference electrode used is mono-polar and this is to add stability during measurement of signals to correct these signals level. The body reference circuit used is taken from the INA129P instrumentation amplifier datasheet as suggested.

3.2 Band pass active filter (4-th order)

The band pass filter input is the output of the instrumentation amplifier. It's used to eliminate the unwanted frequency component that is outside the typical SEMG signal frequency range 20-500Hz. High-pass filtering is used to remove movement artifacts and DC components (<10Hz). And low pass filtering is desirable to remove high frequency content above 500Hz to avoid signal aliasing, the filter structure is shown in fig.3.

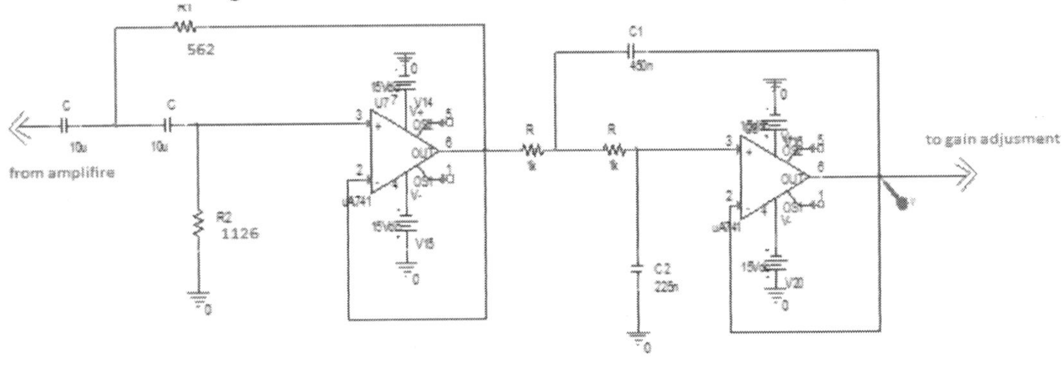

Figure 3: 4th order band pass active filter.

To find high cut-off frequency fc1 and fc2 we assume that R= 1k and solving the equations 4, 5, 6 and 7 for C1, C2, R1 and R2. The frequency response for the filter is given in Equation 3, and the response shown in Fig. 4 by Pspice simulation.

$$H(s) = \frac{\frac{1}{R^2C_1C_2}}{\frac{1}{R^2C_1C_2} + (\frac{2}{RC_1})s + s^2} * \frac{s^2}{\frac{1}{C^2R_1R_2} + (\frac{2}{CR_2})s + s^2} \quad (3)$$

$$w_{c1} = \sqrt{\frac{1}{R^2C_1C_2}} \quad (4)$$

$$Q = \frac{1}{2}\sqrt{\frac{C_1}{C_2}} \qquad (5)$$

$$W_{c2} = \sqrt{\frac{1}{C^2 R_1 R_2}} \qquad (6)$$

$$Q = \frac{1}{2}\sqrt{\frac{R_1}{R_2}} \qquad (7)$$

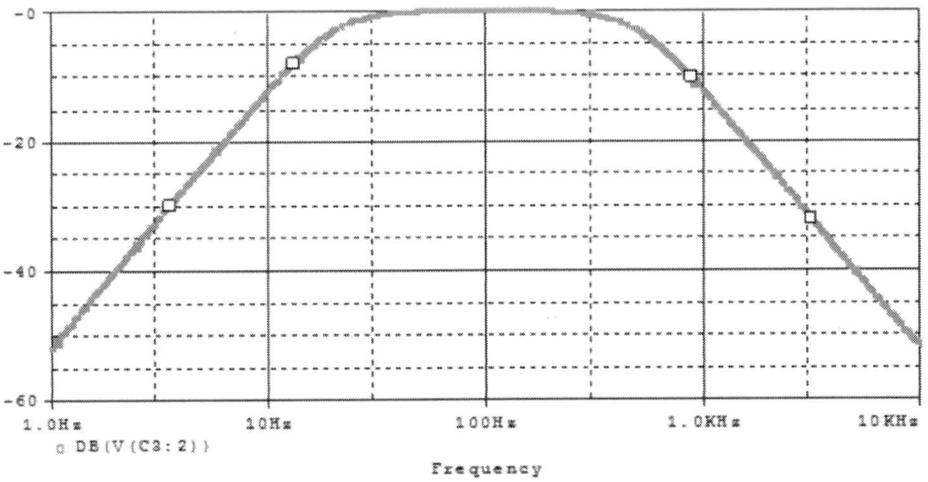

Figure 4: the frequency response of the proposed filter (20-500Hz) simulation result.

3.3 Gain adjustment and signal translation.

Since we have used a 250 gain for the instrumentation amplifier to avoid the saturation, a second stage of amplifier is needed to bring the voltage level up to meet the analogue to digital converter or data acquisition device requirement. The amplifier circuit is shown in fig. 5.

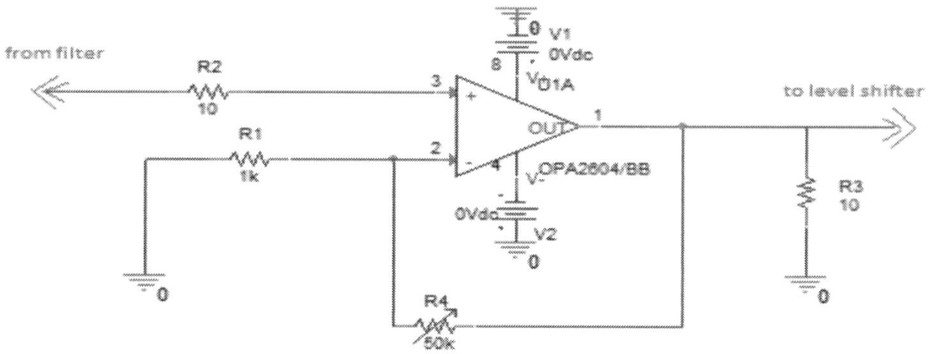

Figure 5: Gain adjustment circuit for sensitivity corrections.

We use Equation 8 to calculate the gain, when we made changes of R4 values from 0- 50k the gain changed from 1 to 51 at this stage.

$$G = 1 + \frac{R_4}{R_1} \qquad (8)$$

The output of gain adjustment will be ±10 V single ended. Interfacing this output to the ADC requires attenuation and a level shift, the circuit shown in figure converts the ±10 V single ended to +5V signal can connected to ADC in differential configuration for more noise rejection.

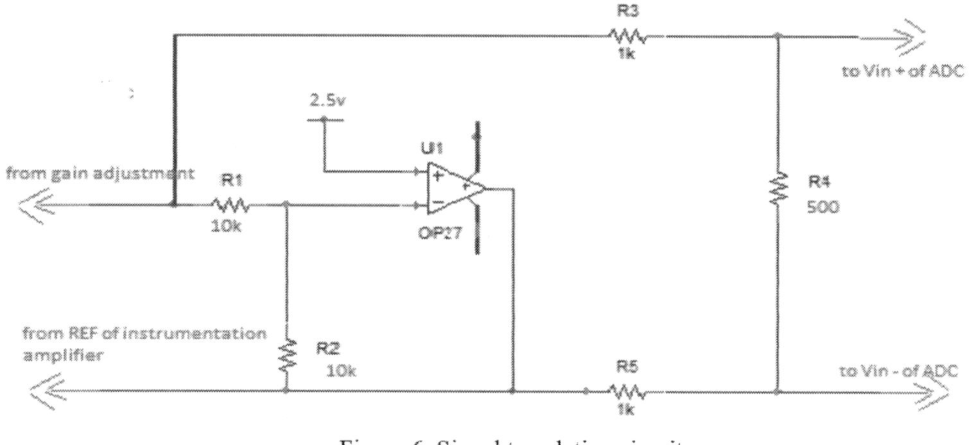

Figure 6: Signal translation circuit

4. Experimental Procedures

The conditioning circuit is used to collect signals form biceps muscle, it is connected to the PicoLog 1012 device and the signal is recorded using PicoLog recorder software, the sampling rate used is 2 kHz. The duration for collected signal is 10s, during the 10s two isometric contractions is recorded. Signal-to-noise ratio is defined as the power ratio between a signal (meaningful information) and the background noise (unwanted signal):

$$SNR = \frac{P_{signal}}{P_{noise}} \qquad (9)$$

Which could be written using amplitude ratios as:

$$SNR_{dB} = 20\, log_{10}\left(\frac{A_{signal}}{A_{noise}}\right) \qquad (10)$$

5. Results and Discussion

The raw EMG data is imported to excel and the plots shown in fig. 7. While the PSD graph is shown in Fig. 8. The plots shows that the filter has successfully eliminate the signal components below 20Hz. These components mostly generated by cables movement or electrode displacement.

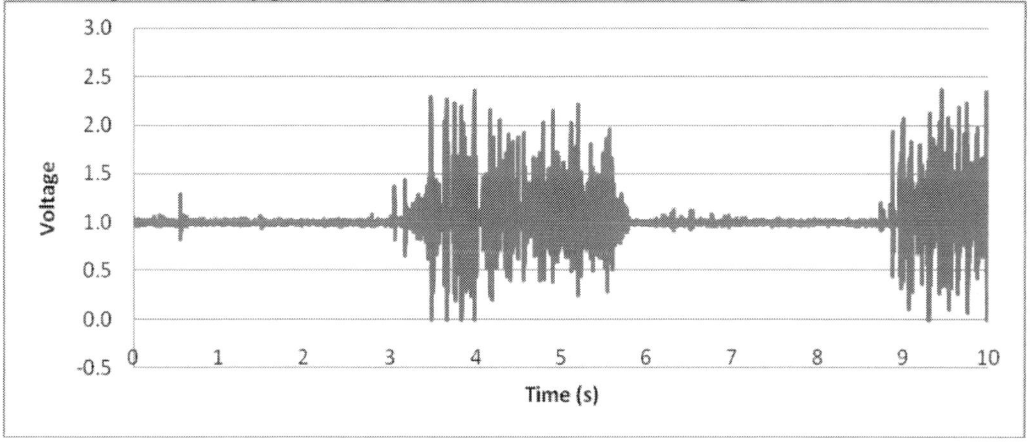

Figure 7: Raw EMG signal during isometric contraction of biceps muscles.

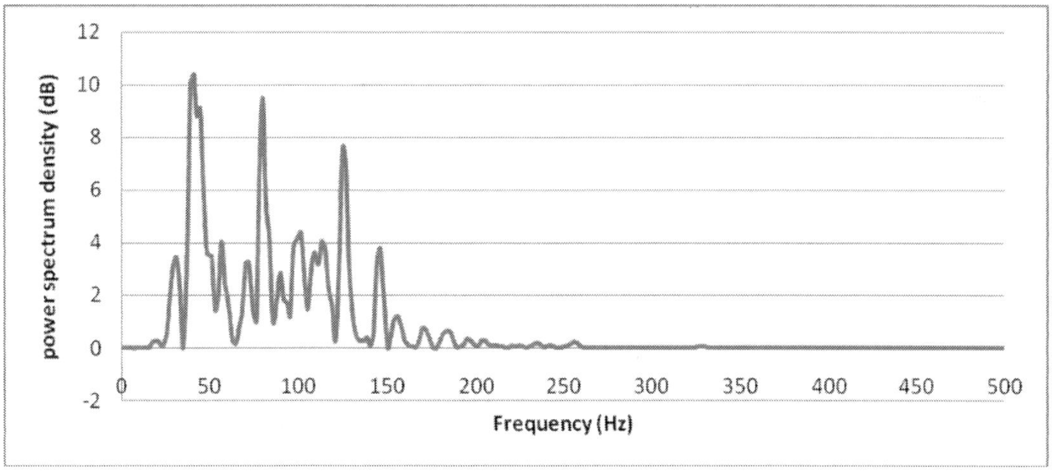

Figure 8: PSD for the collected SEMG signal.

The SNR ratio can be calculated using the amplitude relation as in equation 10. The SNR calculated for this study is 52.39 dB which is better than the SNR for commercial system (42dB) that is calculated according to results presented in [10], and it is still less than the SNR for the system proposed in the same paper (59dB). However, the system in [10] uses a notch filter which may lead to lose some useful frequency components.

6. Conclusion

The characteristics for the medical and clinical application can be obtained by acquisition of accurate signals. The design of an SEMG conditioning circuit presented in this paper guarantees the collection of high quality signal with high SNR that is immune to environmental aberrations. Experimental results show that the value of SNR for collected signal is 52.4 dB which is higher than the commercial system. The power spectrum density (PSD) graph is presented as well which shows that the filter has the capability to eliminate the noise below 20 Hz.

Acknowledgment

We would like to acknowledge and extend my heartfelt gratitude to Research Management Centre IIUM who has made the improvement of this research possible by granting us Research Endowment Fund (Type B).

References

[1] Jie Liu and Ping Zhou, Senior Member, IEEE, "A Novel Myoelectric Pattern Recognition Strategy for Hand Function Restoration After Incomplete Cervical Spinal Cord Injury," IEEE TRANSACTIONS ON NEURAL SYSTEMS AND REHABILITATION ENGINEERING, VOL. 21, NO. 1, JANUARY 2013.

[2] Cara E. Stepp, James T. Heaton, Rebecca G. Rolland, and Robert E. Hillman, "Neck and Face Surface Electromyography for Prosthetic Voice Control After Total Laryngectomy". IEEE TRANSACTIONS ON NEURAL SYSTEMS AND REHABILITATION ENGINEERING, VOL. 17, NO. 2, APRIL 2009.

[3] N.Yozbatiran, J. Berliner, M. K. O'Malley, A. U. Pehlivan, Z. Kadivar, C. Boake, and G. E. Francisco, "Robotic training and clinical assessment of upper extremity movements after spinal cord injury; A single case report," J. Rehabil. Med., vol. 44, pp. 186–188, Feb. 2012.

[4] Edoardo Fiorucci, Giovanni Bucci, RuggeroCattaneo, and Annalisa Monaco, "The Measurement of Surface Electromyographic Signal in Rest Position for the Correct Prescription of Eyeglasses" IEEE TRANSACTIONS ON INSTRUMENTATION AND MEASUREMENT, VOL. 61, NO. 2, FEBRUARY 2012.

[5] Anders Fougner, Member, IEEE, Øyvind Stavdahl, Member, IEEE, Peter J. Kyberd, Yves G. Losier, and Philip A. Parker, Senior Member, IEEE, "Control of Upper Limb Prostheses: Terminology and Proportional Myoelectric Control—A Review," IEEE TRANSACTIONS ON NEURAL SYSTEMS AND REHABILITATION ENGINEERING, VOL. 20, NO. 5, SEPTEMBER 2012

[6] D K Kumar, A Melaku, "Electrode Distance and Magnitude of SEMG" , Proceedings of the Second Joint EMBS BMES Conference Houston, TX, USA October 23-26, 2002.

[7] A. Rainoldi, G. Melchiorri, I. Caruso, "A method for positioning electrodes during surface EMG recordings in lower limb muscles", Journal of Neuroscience Methods 134 (2004) 37–43.

[8] D. Farina1, R. Merletti, B. Indino, T. Graven-Nielsen, "Surface EMG Crosstalk Evaluated from Experimental Recordings and Simulated Signals Reflections on Crosstalk Interpretation, Quantification and Reduction" Methods Inf Med 2004; 43: 30-5.

[9] Isabel C.N. Saccoa, Aline A. Gomes, Mitie E. Otuzi, Denise Pripas, Andrea N. Onodera, "A method for better positioning bipolar electrodes for lower limb EMG recordings during dynamic contractions", Journal of Neuroscience Methods 180 (2009) 133–137.

[10] Wonkeun Youn; Jung Kim, "Development of a compact-size and wireless surface EMG measurement system," ICCAS-SICE, 2009 , vol., no., pp.1625,1628, 18-21 Aug. 2009

IOP Conf. Series: Materials Science and Engineering **53** (2013) 012028 doi:10.1088/1757-899X/53/1/012028

Performance characterization of inductive coupling system

Mohammed M. Shobaki[1], Noreha Abdul Malik, Sheroz Khan, Anis Nurashikin, Samnan Haider, Atika Arshad, Rumana Tasnim and Foughalia Aissa[1]
Department of Electrical and Computer Engineering
International Islamic University
53100 Kuala Lumpur, Malaysia
Email: m.shobaki@gmail.com

Abstract: The use of biomedical implants has been on the rise over the recent past years and this use is expected to increase exponentially over the comings years. Such devices are becoming the most feasible interface to monitor the betrothed parameters. Biomedical implants usually require a low input voltage which is provided making use of inductive coupling. Inductive coupling not only solves the powering issue but also helps to collect the data through non-invasive means. Power transfer efficiency of an inductive link system depends upon factors like mutual coupling, separation between the coils and most importantly the shape of the input voltage. The power transferred or signal received as a result of coil separation under optimum size conditions are reported in contemporary works. This paper comes is testing the effect of different shaped input voltages on the performance of inductive coupling system in terms of power transfer efficiency. Circuits have been tested for three different types of input waveforms of sine, square and ramp. Comparison of plots shows that the shape of input plays a major role on the entire performance of inductive coupling system.

1. Introduction

Inductive coupling based power transfer has several advantages over other methods. It is immune to environmental factors such as dust, water, or chemicals. Inductive coupling transfers electrical power between two isolated coils placed in closed vicinity to each other. The contiguity and the nearness of coils play a vital role in the generation of low power electrical field that inter-connects them and hence allows the transfer of electricity between the two systems. Powering the implanted devices and collecting data the sensed from such devices is being studied recently. Different techniques have been proposed for these purposes. Energy harvesting from different sources from the external environment such as thermal, light and piezoelectric is used which is not suitable in the case of biomedical implanted devices like using solar energy harvesting faces problems regarding the limitation of working environment[1-3]. So, transmitting energy wirelessly to the implant devices using implanted and external antenna is considered to be a suitable solution to all these issues [4].

Inductive coupling has several advantages over the others for powering and telemetry purposes, such as the implanted devices can be operated continuously because external source keeps providing the power and its independent environmental factors such as temperature, light etc. A telemetry chip using inductive coupling supplies power to and transmits data from an implantable sensor. Such sensors offer several advantages to study and monitor the human body. The telemetry system used two coils to transmit both power and data. The distance between coils is up to 25 mm [5]. Inductive power transfer for transferring the power wirelessly is being used in different applications. For example, Inductive coupled charger used for charging of mobile phones or other handled devices. Biomedical

[1] To whom any correspondence should be addressed.

application for this technique is to power up and collect data from implanted devices such as cochlear and retinal implants and inductive coupling is used to fulfil this. [7].
These previous papers studied the performance of the telemetry system using a sinusoidal power supply only; this work will focus on studying the performance by simulating different input wave forms as a voltage source including sinusoidal, square and triangular inputs. Moreover, this work will study the effect of the capacitance at the load side on the power transfer efficiency.

2. Analysis of the telemetry System

A two coils inductive coupling circuit is shown in Fig. 1 and in order to find out the power transfer efficiency, the values of I1 and I2 need to be evaluated. There are two methods to do so i.e. Applying KVL and VNA, or by reflecting the source or load loop impedance. Both ways have been illustrated below.

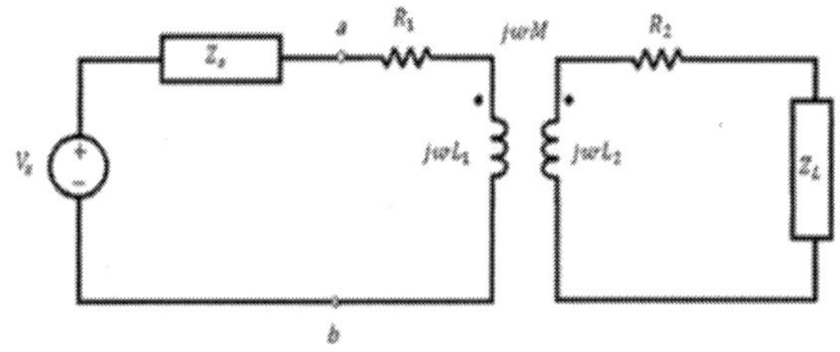

Fig. 1: Two loop inductive coil system

Applying KVL on the source loop

$$V_s = (Z_s + R_1 + jwL_1)I_1 - jwMI_2 \quad (1)$$

For the load loop

$$0 = -jwMI_1 + (R_2 + jwL_2 + Z_L)\, I_2 \quad (2)$$

We assume that

$$(Z_s + R_1 + jwL_1) = Z_{11} \quad (3)$$

$$(R_2 + jwL_2 + Z_L) = Z_{22} \quad (4)$$

Putting these in Eq. 1 and Eq. 2 we get

$$V_s = Z_{11}I_1 - jwMI_2 \quad (5)$$

$$0 = -jwMI_1 + Z_{22}\, I_2 \quad (6)$$

$$I_1 = \frac{\begin{vmatrix} V_S & -jwM \\ 0 & Z_{22} \end{vmatrix}}{\begin{vmatrix} Z_{11} & -jwM \\ -jwM & Z_{22} \end{vmatrix}} = \frac{V_S Z_{22}}{Z_{11}Z_{22} + w^2 M^2} \quad (7)$$

$$I_2 = \frac{\begin{vmatrix} Z_{11} & V_S \\ -jwM & 0 \end{vmatrix}}{\begin{vmatrix} Z_{11} & -jwM \\ -jwM & Z_{22} \end{vmatrix}} = \frac{jwMV_S}{Z_{11}Z_{22} + w^2M^2} \tag{8}$$

Where

$$I_2 = \frac{jwM}{Z_{22}} I_1 \tag{9}$$

$$P_{sent} = |I_1|\, V_s \tag{10}$$

$$P_{recivied} = |I_2|^2 R_l \tag{11}$$

The power efficiency is given by:

$$P_{eff} = \frac{P_{recieved}}{P_{sent}} * 100 \tag{12}$$

$$P_{eff} = \frac{|I_2|^2 R_l}{|I_1|\, V_s} * 100 \tag{13}$$

Substitute equation 9 in 13 we get:

$$P_{eff} = \frac{|jwM|^2 |I_1|\, R_l}{|Z_{22}|\, V_s} * 100 \tag{14}$$

Substitute the value of I_1 in equation 7we get the following:

$$P_{eff} = \frac{w^2M^2\ R_l}{|Z_{22}|\ |Z_{22}Z_{11} + w^2M^2|} * 100 \tag{15}$$

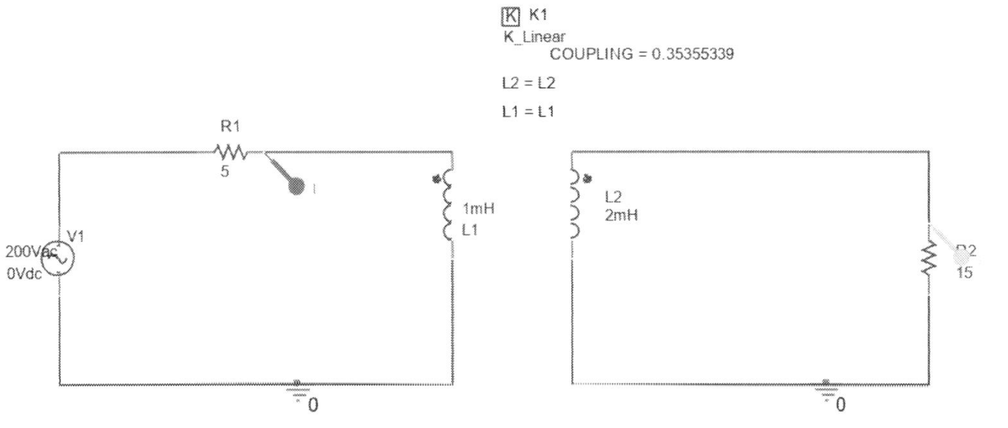

Figure 2: Inductive coupling system

For the circuit shown in fig. 2 assume that R1, R2, L1, L2, M, and Vs have the values 5, 15, 1mH, 4mH, 1mH and 200volts respectively. The terms Z11 and Z22 as a function of w are given by:

$$Z_{11=}\ 10 + 0.001\mathrm{w}\ j$$

$$Z_{22=}\ 30 + 0.004\mathrm{w}\ j$$

The values of the power sent from the source loop, the power received by the load and the power efficiency is calculated according to the equations for the circuit shown in figure, these value plotted and shown in fig.3 (a) and 4 (a). The simulation results for these parameters are shown in fig.3 (b) and 4 (b).

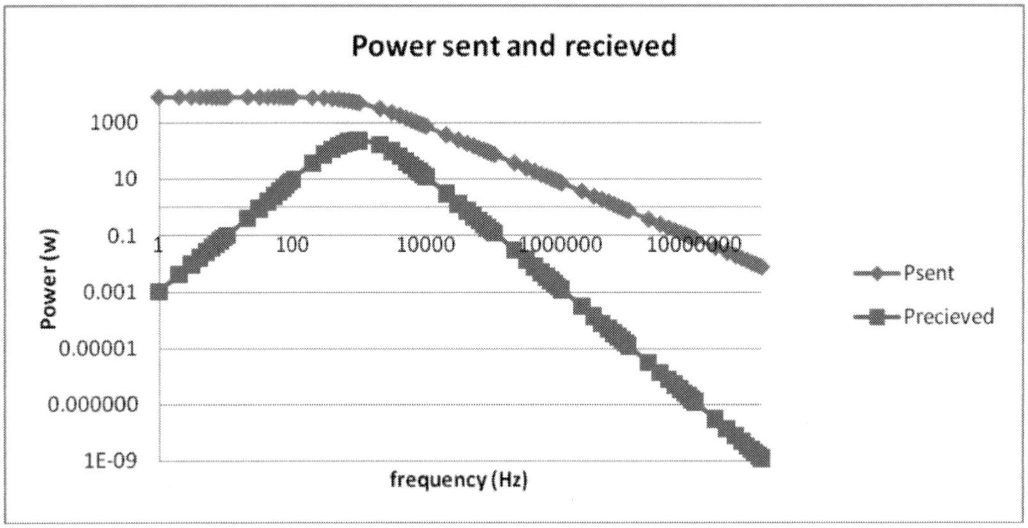

Figure 3 (a): Analytical plots for power sent and received as a function of frequency, both x and y axis in logarithmic scales.

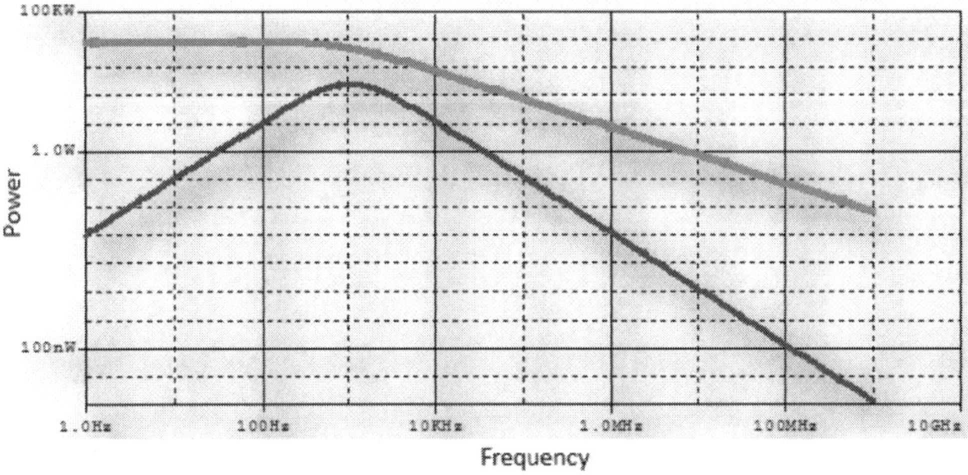

Figure 3 (b): Simulation plots for power sent and Power received as a function of frequency, Red line is the power received while the green line is power sent, both x and y axis in logarithmic scales.

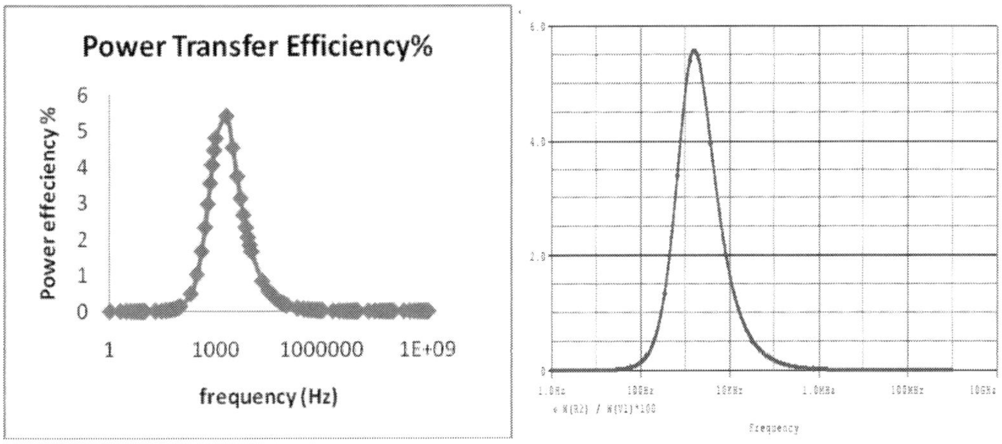

Figure 4 (a): Analytical plots for Power Transfer Efficiency% as a function of frequency (b): Simulation Power Transfer Efficiency% as a function of frequency.

2.1 Different input wave forms

We used three differnt shaps of voltage sources in simulation of the inductive coupling system using Pspice. The wave forms shown in Fig. 5 are of the same amplitude and the corresponding plots for power transfer efficiency are shown in Fig 6. Power recieved at the load side has been shown in Fig 7 when all three different supplies are used.

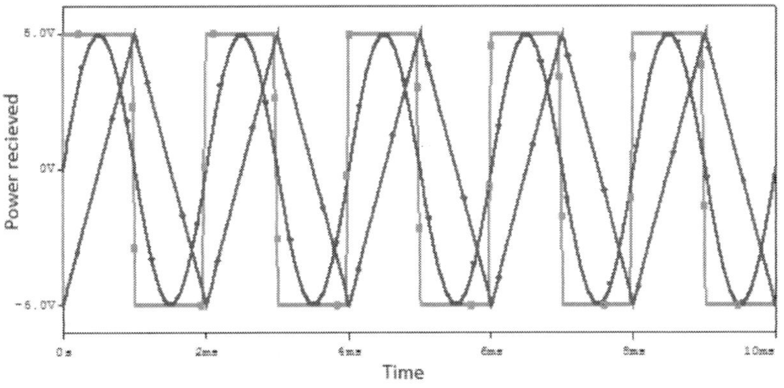

Figure 5: Different voltage sources.

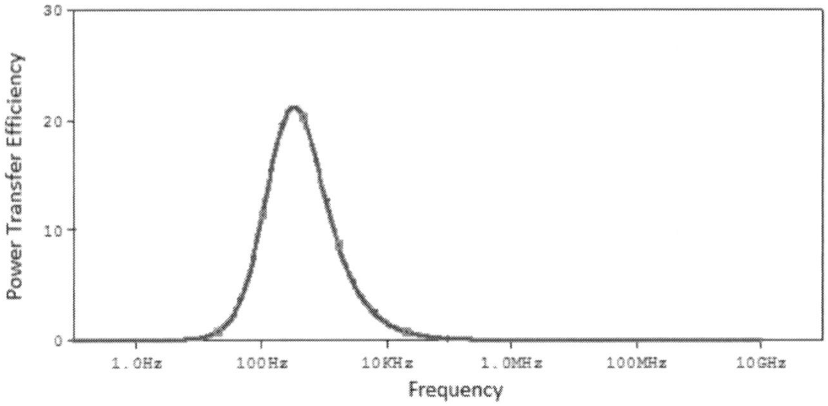

Figure 6: Power Transfer Efficiency for the systems

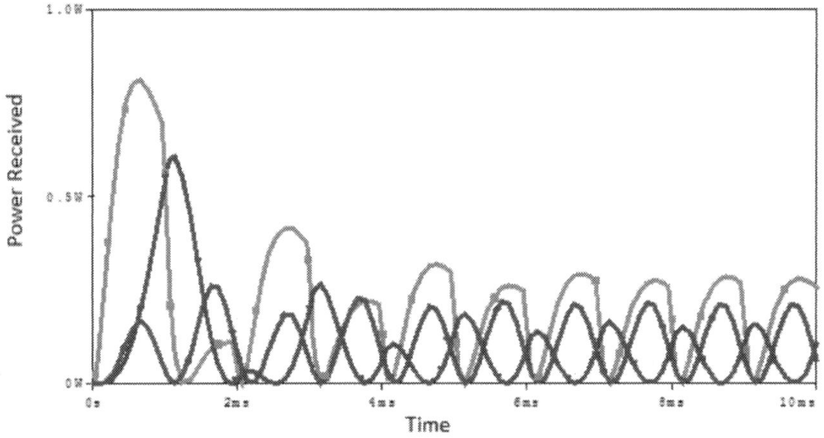

Figure 7: Power received using different voltage sources, square wave: green, sinusoidal wave: red, triangular wave: blue.

3. Discussion and comparison

Comparison between simulation results and the analytical plots for power sent, power received and power transfer efficiency in Fig. 3 and Fig. 4 shows that that both exactly match. The power transfer efficiency in fig 4 have a maximum value at 1.6 kHz frequency, which can be referred to as the resonant frequency for this circuit .So the telemetry system needs to be tuned to this frequency for maximum power transfer in this case.

According to the simulation results shown in fig 6, the shape of the voltage source has no effect on the value of the power transfer efficiency. And this is because the PSPICE simulation deals with the three shapes as an AC source during the AC analysis. However, the results for the transient analysis in fig. 7 shows clear differences in terms of power received for the different shapes. The highest power is received in the case of square wave voltage source.

4. Conclusions

The telemetry system based on coupled coils structure is studied using three shapes of voltage sources. The currents, impedances and the power equations are derived for both circuits. The efficiency of these circuits is studied as a function of frequency, using analytical and simulation plots which are both exactly matched. The simulation results show that there is no effect of the source shape on the power efficiency in the AC analysis while it shows differences in the power received in the transient simulation, the highest power could be received to the load is in the case of using a square wave power supply.

Acknowledgment

We would like to acknowledge and extend my heartfelt gratitude to Research Management Centre IIUM who have made the improvement of this research possible by granting us Research Endowment Fund (Type B).

References

[1] D.C. Hoang, Y.K. Tan, H.B. Chng and S.K. Panda, "Thermal Energy Harvesting From Human Warmth For Wireless Body Area Network In Medical Healthcare System", International Conference On Power Electronics And Drive Systems, PEDS 2009.

[2] Sahar Ayazian, Student Member, IEEE, Vahid A. Akhavan, Eric Soenen, Member, IEEE and Arjang Hassibi, Member, IEEE, "A Photovoltaic-Driven and Energy-Autonomous CMOS Implantable Sensor", IEEE TRANSACTIONS ON BIOMEDICALCIRCUITS AND SYSTEMS,VOL.6,NO. 4, AUGUST 2012.

[3] E. Annie Grace, Dr. S. Edward Rajan, A. Allwyn Clarence Asis, "Performance Evaluation of Different Rectifiers for Piezoelectric Energy Harvesting Applications", IEEE International Conference On Recent Advancements In Electrical, Electronics And Control Engineering, 2011.

[4] Ghovanloo, Maysam and Atluri, Suresh, "An Integrated Full-Wave CMOS Rectifier with Built-In Telemetry for RFID and Implantable Biomedical Applications", IEEE Transactions on Circuits and Systems, Vol. 55, No. 10, Page No. 3328-3334 (November 2008)

[5] Christian Sauer, Student Member, IEEE, Milutin Stanac'evic', Student Member, IEEE,Gert Cauwenberghs, Senior Member, IEEE, and Nitish Thakor, Fellow, IEEE,Power Harvesting and Telemetry in CMOS for Implanted Devices," IEEE TRANSACTIONS ON CIRCUITS AND SYSTEMS—I: REGULAR PAPERS, VOL. 52, NO. 12, DECEMBER 2005

[6] Klaus Finkenzeller "RFID Handbook: Fundamentals and Applications in Contactless Smart Cards and Identification"Second Edition, Copyright □ 2003 John Wiley & Sons, Ltd.

[7] C. Koch, W. Mokwa, M. Goertz, P. Walter, "First Results of a Study on a completely implanted retinal prosthesis in blind humans," in Sensors IEEE 2008, Oct 2008, pp. 1237 - 1240.

Electrical performances of commercial GaN and GaAs based optoelectronics under neutron irradiation

D Ahmad Fauzi[1], N K A Md Rashid[1], J Abdul Karim[2], M R Mohamed Zin[2], N F Hasbullah[1], O A Sheik Fareed [1]

[1]Department of Electrical and Computer Engineering, Department of Mechatronics Engineering International Islamic University Malaysia, Kuala Lumpur
[2]Nuclear Power Sector, Malaysian Nuclear Agency

E-mail: dhiyauddin.fauzi@gmail.com, nfadzlinh@iium.edu.my

Abstract. This paper aims to demonstrate the effects of displacement damage caused by high energetic neutron particle towards the electrical performances of gallium arsenide (GaAs) and gallium nitride (GaN) *p-n* based diodes. The investigations are carried out through current-voltage (I-V) and capacitance-voltage (C-V) measurements using Keithley 4200 SCS. Two different commercial optoelectronics diodes; GaN on SiC light emitting diode (LED) and GaAs infrared emitting diode (IRED) were radiated with neutron using pneumatic transfer system (PTS) in the PUSPATI TRIGA Mark II research reactor under total neutron flux of 1×10^{12} neutron/cm^2.s. Following the neutron exposure for 1 ,3 and 5 minutes, the I-V forward bias and reverse bias leakage current increase for GaAs IREDs, but minimal changes were observed in the GaN LEDs. The C-V measurements revealed that the capacitance and carrier concentration of GaAs IREDs decrease with increasing radiation flux.

1. Introduction

Optoelectronics diodes are extensively used in many electronics devices for various applications including devices that require robust performance in radiation environment. Due to economics and time constraints, commercial diodes are often chosen for the ground and space operations. Devices used in the radiation harsh environments where commercial diodes are applicable includes; radiation detectors and optical link readouts used in CERN and ATLAS tracking systems, sensors and detectors mounted on search and rescue robots used in nuclear facilities, optocouplers and solar cells in satellites and space station on-board circuitries[1, 2]. These off-the-shelf devices need to be tested to comply with specific requirements of radiation tolerance since the use of these devices in radiation harsh environment will cause degradation in the performance which leads to permanent failure. Two significant effects of radiation on semiconductor devices known so far are the displacement damage(DD); damage that is caused by high energy particles and the ionisation damage (ID); damage that is caused by energetic charged particles. In semiconductor material physics, DD is associated with the creation of vacancy-interstitial pair caused by the displacement of atom from its lattice site when collided with heavy particles (i.e:neutron particles). ID is the damage instigated from the accumulation of charges discharge into a device by charged particles (i.e:x-ray,gamma ray). Other possible damages caused by ID are single event upset (SEU) and single latch-up upset (SLEU) which are able to alter the state of a device through the flow of free charges in the material [3].

In this paper, we report studies on the application of diode-based devices in a nuclear environment, where high neutron flux is continuously generated in the nuclear reactor. The DD effect is highly significant in the nuclear reaction compared to the secondary effect, ID induced by the gamma ray product. However, one must not overlook at the effect that can be caused by ID as the gamma-ray product is considerably generated through the decay of radioactive fission products and radiative capture process which, if accumulated, can cause crucial damage [4].

Reports from past years have proven that gallium-nitride (GaN)-based devices are highly radiation tolerant compared to gallium arsenide (GaAs)-based devices to some magnitude. This paper is presented to investigate the changes in electrical performances of both diodes subjected to thermal neutron radiation with the analysis of both displacement damage as well as ionisation damage taken into consideration. A critical way to observe the impact of neutron radiation besides output power and peak wavelength measurements is through current-voltage (I-V) and capacitance-voltage (C-V) characterisation [5]. These methods are proven to provide subtle information on the contact resistance as well as changes in electrical performances occur in the junction.

2. Experimental Details

Two commercial optoelectronics diodes; GaAs grown by liquid phase epitaxy (LPE) IREDs (CQY37N) and GaN on SiC technology grown by metal-organic chemical vapor deposition (MOCVD) LEDs (TLHB5800) from Vishay were used in this investigation. The GaAs IRED, have a current rating of 100 mA while the GaN LED have a less maximum rating of 20 mA. Mostly, the devices can only be allowed to perform under its maximum rating to prevent from device-wearout. The schematic cross section diagrams of the devices are provided in Figure 1 (a) and (b) respectively.

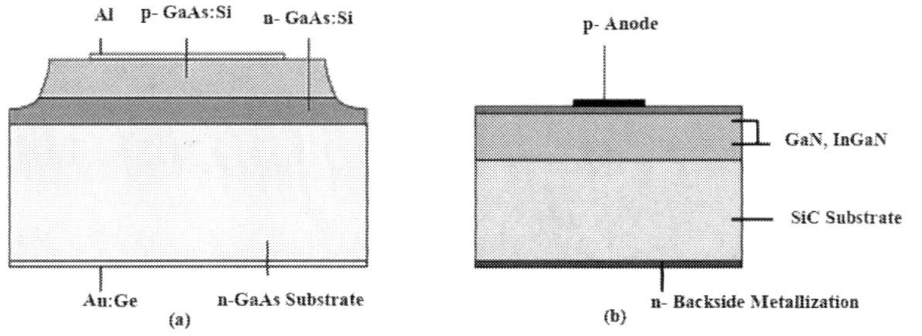

Figure 1. Schematic Cross Section Diagram of (a) GaAs IRED and (b) GaN LED [6, 7]

The diodes were exposed to thermal neutrons (E<0.5 eV) by means of pneumatic transfer system (PTS) operated at a power rating of 750kW in the PUSPATI TRIGA Mark II research reactor, Malaysian Nuclear Agency (ANM). The PTS in PUSPATI TRIGA Mark II is designed to transport the samples to ring G20 of the reactor core which has a standard neutron flux of 1×10^{12} neutron/cm^2.s. Three samples for each model with technically the same pre-irradiation I-V and C-V characteristics were used to achieve reproducibility and consistency in readings. All samples were encapsulated in polyethylene (PE) vials and radiated in the PTS for 1 minute, 3 minutes and 5 minutes of maximum corresponding neutron fluence; 6×10^{13}, 1.8×10^{14} and 3×10^{14} neutron/cm^2. The pre and post-irradiation I-V and C-V measurements were performed in a room temperature using Keithley 4200 semiconductor characterization system (SCS) via Keithley's Interactive Testing Environment (KITE) software. To ensure no deviation in the photocurrent results, samples were placed in Keithley 8101-4TRX during measurement. The wearout factor; that is the defects caused in normal operation [8] is neglected as the study is primarily concerning on the changes in the electrical performance due to the alteration of atomic structure. Furthermore, the limited space of the PTS's transportation pipeline disallowed such experimentation to be conducted.

3. Results and Discussion

3.1 The Current-voltage (I-V) Characteristics

The room-temperature forward and reverse biased *I-V* characteristics for both GaAs and GaN based diodes were plotted in a semi-log *I* versus *V* with current compliance set to a constant value of 0.01 A (to prevent the diode from self-heating). The forward-bias *I-V* results of pre and post neutron irradiation for GaAs and GaN diodes are presented in Figure 2(a) and (b) respectively. The forward leakage current of GaAs IREDs increased substantially after being irradiated with 6×10^{13}, 1.8×10^{14} and 3×10^{14} neutron/cm^2 of neutron fluence. In general, the maximum corresponding current increment observed is in 3 orders of magnitude. The relative percentage leakage current increments for each irradiation time at specified voltages are tabulated in Table 1. The difference in the relative percentage leakage current increments between 1.8×10^{14} to 3×10^{14} n/cm^2 of neutron fluence was determined to be marginal. In Figure 2(b), the forward leakage current of the GaN LEDs exhibited no change under the same irradiation conditions. After additional 1 and 5 minutes, similar trends were also observed. Small reductions in the leakage current for both devices were noted over higher reverse-bias voltage which is attributed by the ohmic effect [9].

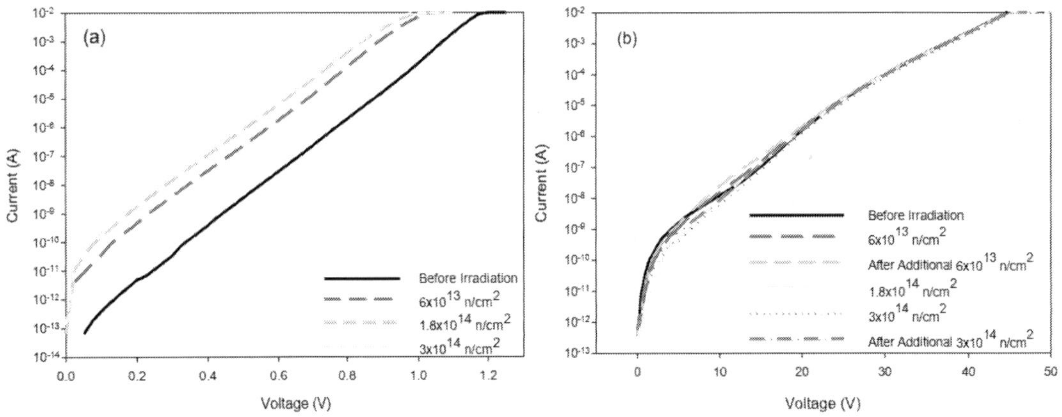

Figure 2. Forward Bias Dark I-V Characteristics of (a) GaAs IRED and (b) GaN LED

Table 1. Relative Percentage Leakage Current Increment in GaAs FB I-V Characteristics

Applied Voltage	Irradiation Period		
	1 min	3 min	5 min
0.4	117.9%	417.5%	572.4%
0.8	58.6%	213.6%	276.2%
1.2	71.2%	162.2%	184.4%

The difference in radiation hardness between both devices can be perceived in Figure 3(a) and 3(b); an extrapolation of reverse bias *I-V* characteristics of the GaAs and GaN based diodes. In reverse-bias, the leakage current of GaAs IREDs, were observed to increase with increasing radiation fluence. After a maximum neutron exposure 3×10^{14} n/cm^2, the reverse breakdown voltage is found to increase from 34V to 43.6V. The reverse-bias characteristics of GaN LEDs indicated no sign of degradation even though the samples were exposed to another additional neutron fluence of 6×10^{13} and $3\text{x}10^{14}$ n/cm^2.

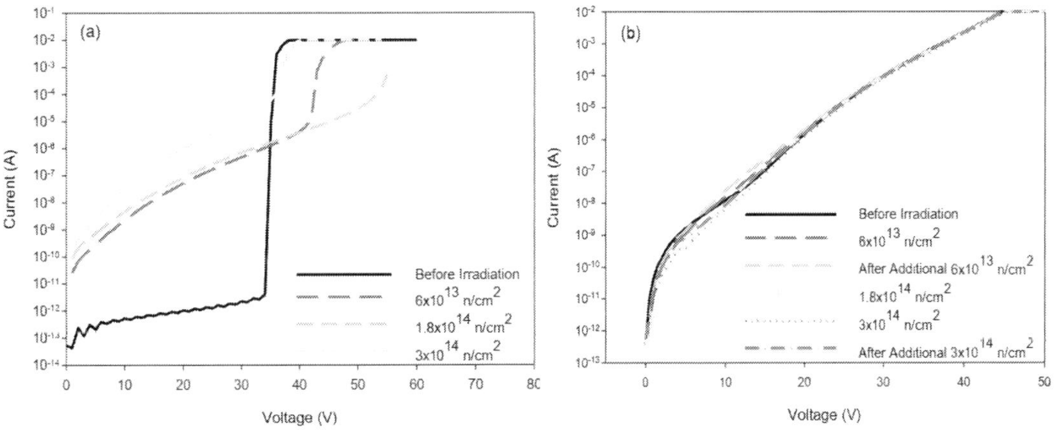

Figure 3. Reverse Bias Dark I-V Characteristics of (a) GaAs IRED and (b) GaN LED

The ideality factor, series resistance as well as saturation current of the diodes were modelled by the ideal Shockley equation given as follows [10, 11];

$$I = I_S\left[\exp\left(\frac{qV}{nkT}\right) - 1\right] \quad (1)$$

With the existence of series resistance, R_S the real diode model can be further expressed as [12];

$$V = \left(\frac{nkT}{q}\right)\ln\left(\frac{I}{I_S}\right) + IR_S \quad (2)$$

where n is the ideality factor, k is the Boltzman constant, T is the measured diode temperature, q is the electron charge and I_S is the saturation current. The ideality factor is included to account for the imperfection of the junction in a commercial diode. Essentially, it is used as an indicator to determine the dominant current mechanism contributing to the generated current inside the diode. The diffusion is said to be the dominant mechanism if $n = 1$ (ideal) and generation-recombination (GR) is the dominant mechanism if $n = 2$. It may occur that both mechanisms operates at the same time if n is in the range of 1 to 2 [13].

The increment in the leakage current of GaAs diode is known to be predominantly caused by the formation of non-radiative generation-recombination (GR) centers or traps in the forbidden band gap [14, 15]. Due to displacement defects, these induced acceptor-like traps lead to the increment rate of GR of holes and electrons, hence, increment of both radiative and non-radiative recombination current [16, 17]. The minority carrier lifetime however, decreases to increasing number of traps which can be quantify as;

$$\tau_0^{-1} = \tau_R^{-1} + \tau_{NR}^{-1} \quad (3)$$

where τ_0 is the initial total lifetime, τ_R and τ_{NR} is the associated radiative and non-radiative lifetime respectively [18]. The traps density was observed to vary proportionally with irradiation time. Due to that, a reduction in the total carrier lifetime is expected relative to the exposure time.

On top of that, the increase in the rate of GR process might be contributed by the transition of additional vacancy-interstitial pairs; caused by preceding high energy interstitial atom or primary knock-on atoms (PKAs) [3]. In elastic collision between high energetic neutrons with the material's atomic structure, further vacancy-interstitial pair can be created by the preceding interstitial atom. This atom can cause damage similar to DD if high kinetic energy is being transferred by the neutron particles. Considering the secondary effects of ionizing damage, the changes in the dark current of a diode may also be resulted from gamma induced electron-hole pair [17]. Based on the minor differences in the relative percentage current increment between 3 and 5 minutes, we can deduced that the induced vacancies led to a self-annealing process which cures the defects to minimal as reported by [19, 20].

The ideality factor, the saturation current and the series resistance were obtained by fitting the derived equation (2) onto the linear region of the experimental *I-V* data. The ideality factor of GaAs IRED increases from 1.76 to 1.9 with respective to irradiation fluence of 6×10^{13} to 3×10^{14} n/cm^2. This suggests that trap-assisted GR mechanism is responsible for the carrier transport in the depletion region as described above. The trap formed at the mid band is assumed to be shallow; otherwise deep trap would lead to reduction in magnitude of the leakage current. The ohmic effect mentioned in the previous discussion can be explained by the increment in the series resistances of the GaAs diode. The results were recorded to have increased from 0.5Ω to 3Ω after a maximum exposure fluence of 3×10^{14} n/cm^2.

Since there are fairly no changes in GaN diode's leakage current, neither the ideality factor nor the series resistance were observed to have changed. This is due to the GaN's wide bandgap energy, higher thermal stability and small thermal neutron capture cross section of nitride; 1.91 compared to arsenic; 4.3[21]. The ideality factor and series resistance however were measured to have high values of 5.3 and 50Ω respectively. High value of n and R_s indicates that high formation of defect density in the device layer is formed, which may be contributed by the excessive metal- *p*-type contact resistance of GaN [22, 23]. Due to high ideality factor, the recombination process in GaN LED could not be simply modelled by SRH recombination statistics.

One of the valid reasons in the increment of the resistance in both devices is due to the effects of neutron transmutation doping (NTD). NTD occur when neutron particles undergo inelastic interaction with the material, and consequently change the type of the material from n-type to p-type and vice versa [3].

3.1 The Capacitance-voltage (C-V) Characteristics

The room temperature *C-V* analysis of the samples was realized to obtain the approximation values of the doping profile and width of the depletion region. Results were obtained at a frequency of 500 kHz and supply voltage V_{rms} of 100mV. Assuming the diodes are abrupt junction, the *C-V* characterisation is carried out using the expression below [11];

$$C = \frac{A}{2}\sqrt{\frac{2q\varepsilon_0\varepsilon_s N_B}{(V_o - V_a)}} \tag{4}$$

where C is the capacitance, A is the active area, q is the charge, ε_0 and ε_s are the vacuum and material's relative permittivity respectively, N_B is the doping concentration and V_a is the applied voltage.

The width of the depletion region is defined as;

$$W = \frac{\varepsilon_0\varepsilon_s A}{C} \tag{5}$$

and the carrier concentration at specified width is given by;

$$N_d(W) = \frac{2}{q\varepsilon_0\varepsilon_s A^2 \left|\frac{d}{dV_a}\left(1/C^2\right)\right|} \tag{6}$$

Since the active area for both GaAs and GaN devices are undisclosed by the manufacturers, we assumed it to be equivalent to 0.09 mm^2 (minimum active area for such commercial devices). The relative doping concentration versus width of the depletion region is represented as in Figure 4. After a maximum exposure to neutron fluence of 3×10^{14} n/cm^2, the capacitance of GaAs diode was observed to decrease. As illustrated in Figure 4(a), the decrement in capacitance signifies that the doping concentration has also decreased. The decrement is interpreted as the effects of carrier removal. The formation of deep traps in the near mid-band causes low-energy carriers to be trapped, which reduces the concentration of the free carriers. Reduction in the carrier concentration causes the series resistance of the diode and the width of the depletion region to increase [3, 24]. Hence, this explains the fact that the GR process of the carriers can only be achieved through trap-assisted emission as band-to-band emission is barely impossible to occur at wider bandgap. As for GaN LEDs, the capacitance as a function of voltage is constant after similar maximum neutron fluence of 3×10^{14} n/cm^2. Hence, there is relatively no significant change observed in the doping profile of the diode depicted in Figure 4(b).

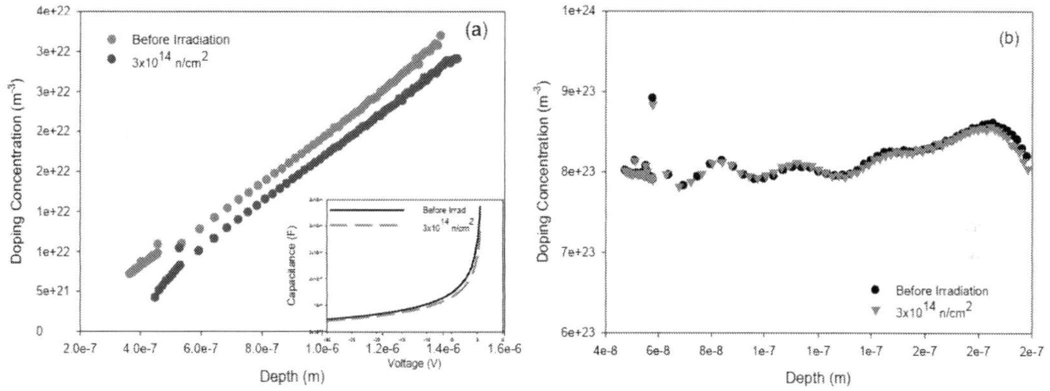

Figure 4. The Doping Profile of (a) GaAs IRED (inset is the C-V Charateristics) and (b) GaN LED.

4. Conclusion

The effects of thermal neutron towards the electrical characteristics of commercial GaAs and GaN optoelectronic diodes have been studied in this paper. The results demonstrated that GaAs IREDs suffered degradation resulted from neutron's displacement damage and gamma's ionisation damage. The defects are indicated by the increment in the values of FB and RB leakage currents, ideality factor, series resistance as well as the saturation current. Further investigation reveals that the DD induced defects caused the value of the carrier concentrations to decrease whilst the depletion region increase with respective irradiation time. However, the GaN based LEDs showed no signs of degradation towards the thermal neutron under the same experimental conditions. Hence, this would approved that GaN based devices are more radiation tolerant than its competitors.

Acknowledgement
The authors would like to acknowledge Mr. Shaiful Rizaidie Mohd. Yakin from the nuclear power division as well as the health physics group for their incessant help and cooperation. We would also like to thank International Islamic University Malaysia (FRGS12-081-0230) for financing the research work.

References

[1] Cochran D J, Buchner S P, Dakai C, Kim H S, LaBel K A, Oldham T R, Campola M J, O'Bryan M V, Ladbury R L, Marshall C, Sanders A B and Xapsos M A 2009 Total Ionizing Dose and Displacement Damage Compendium of Candidate Spacecraft Electronics for NASA .In: *Radiation Effects Data Workshop, 2009 IEEE,* pp 25-31

[2] Moll M 2006 Radiation tolerant semiconductor sensors for tracking detectors Nuclear Instruments and Methods in Physics Research, Section A: Accelerators, Spectrometers, Detectors and Associated Equipment 565 202-11

[3] Claeys C and Simoen E 2002 Radiation Effects in Advanced Semiconductor Materials and Devices: Springer(

[4] Lamarsh J R 2002 *Introduction to nuclear reactor theory*: American Nuclear Society(

[5] Poyai A, Simoen E, Claeys C, Hayama K, Kobayashi K and Ohyama H 2002 Radiation effects on the current-voltage and capacitance-voltage characteristics of advanced p-n junction diodes surrounded by shallow trench isolation *Nuclear Instruments and Methods in Physics Research, Section B: Beam Interactions with Materials and Atoms* 186 409-13

[6] Telefunken V 2000 Physics of Optoelectronic Devices. Vishay GmbH) pp 1-7

[7] Telefunken V 2008 Physics and Technology. Vishay GmbH) pp 1-8

[8] Johnston A H and Miyahira T F 2000 Characterization of proton damage in light-emitting diodes *IEEE Transactions on Nuclear Science* 47 2500-7

[9] Pillai V R V, Khamari S K, Dixit V K, Ganguli T, Kher S and Oak S M 2012 Effect of γ-ray irradiation on breakdown voltage, ideality factor, dark current and series resistance of GaAs p-i-n diode *Nuclear Instruments and Methods in Physics Research, Section A: Accelerators, Spectrometers, Detectors and Associated Equipment* 685 41-5

[10] Sze S M and Ng K K 2006 *Physics of Semiconductor Devices*: John Wiley & Sons(

[11] Streetman B G and Banerjee S K 2009 *Solid State Electronic Devices*: Pearson Prentice-Hall(

[12] Umana-Membreno G A, Dell J M, Parish G, Nener B D, Faraone L and Mishra U K 2003 60Co Gamma Irradiation Effects on n-GaN Schottky Diodes *IEEE Transactions on Electron Devices* 50 2326-35

[13] Che Omar N I, Alang Md Rashid N K, Abdullah J, Abdul Karim J and Hasbullah N F 2012 Effects on the forward bias characteristics of neutron irradiated Si and GaAs diodes *Research Journal of Applied Sciences, Engineering and Technology* 4 5079-83

[14] Beringer J, Borer K, Mommsen R K, Nickerson R B, Weidberg A R, Monnier E, Hou H Q and Lear K L 1999 Radiation hardness and lifetime studies of LEDs and VCSELs for the optical readout of the ATLAS SCT *Nuclear Instruments and Methods in Physics Research, Section A: Accelerators, Spectrometers, Detectors and Associated Equipment* 435 375-92

[15] Linhart V, Bém P, Götz M, Honusek M, Mareš J J, Slavíček T, Sopko B and Šimečková E 2006 Radiation damage study of GaAs detectors irradiated by fast neutrons *Nuclear Instruments and Methods in Physics Research, Section A: Accelerators, Spectrometers, Detectors and Associated Equipment* 563 66-9

[16] Khanna S M, Rejeb C, Jorio A, Parenteau M, Carlone C and Gerdes Jr J W 1993 Electron and neutron radiation-induced order effect in gallium arsenide *IEEE Transactions on Nuclear Science* 40 1350-9

[17] Srour J R and McGarrity J M 1988 Radiation effects on microelectronics in space *Proceedings of the IEEE* 76 1443-69

[18] Stewart Share A S E, Roland A. Polimadei 1974 Properties of Compensated GaAs Light-Emitting Diodes Following 6oCo Irradiation *Solid State Electronics* 18 471-80

[19] Lischka H, Henschel H, Kohn O, Lennartz W and Schmidt H U 1993 Radiation effects in light emitting diodes, laser diodes, photodiodes, and optocouplers. In: *Radiation and its Effects on Components and Systems, 1993.,RADECS 93., Second European Conference on,* pp 226-31

[20] Kim H-Y, Kim J, Ren F and Jang S 2010 Effect of neutron irradiation on electrical and optical properties of InGaN/GaN light-emitting diodes *Journal of Vacuum Science & Technology B: Microelectronics and Nanometer Structures* 28 27-9

IOP Conf. Series: Materials Science and Engineering **53** (2013) 012029 doi:10.1088/1757-899X/53/1/012029

[21] Li C and Subramanian S 2003 Neutron Irradiation Effects in GaN-Based Blue LEDs *IEEE Transactions on Nuclear Science* 50 1998-2002

[22] Jungjin P, Taewook K, Daeyoung W, Joong-Kon S, Jong-Ho L, Park B-G and Hyungcheol S 2011 Non-ideal characteristic analysis of GaN-based light-emitting diode using current-voltage (I–V) and low-frequency noise experiment. In: *Physical and Failure Analysis of Integrated Circuits (IPFA), 2011 18th IEEE International Symposium on the,* pp 1-4

[23] Shah J M, Li Y L, Gessmann T and Schubert E F 2003 Experimental analysis and theoretical model for anomalously high ideality factors (n≫2.0) in AlGaN/GaN p-n junction diodes *Journal of Applied Physics* 94 2627-30

[24] Williams J G, Patel J U, Ougouag A M and Yang S Y 1991 Carrier removal and changes in electrical properties of neutron irradiated GaAs *Journal of Applied Physics* 70 4931-7

A qualitative review for wireless health monitoring system

Atika Arshad[1], **Ahmad Fadzil Ismail, Sheroz Khan, AHM Zahirul Alam, Rumana Tasnim, Syed Samnan Haider, Mohammed M. Shobaki and Zeeshan Shahid**
Department of Electrical and Computer Engineering
International Islamic University
53100 Kuala Lumpur, Malaysia

E-mail: atikaarshad@hotmail.com

Abstract. A proliferating interest has been being observed over the past years in accurate wireless system development in order to monitor incessant human activities in health care centres. Furthermore because of the swelling number of elderly population and the inadequate number of competent staffs for nursing homes there is a big market petition for health care monitoring system. In order to detect human researchers developed different methods namely which include Field Identification technique, Visual Sensor Network, radar detection, e-mobile techniques and so on. An all-encompassing overview of the non-wired human detection application advancement is presented in this paper. Inductive links are used for human detection application while wiring an electronic system has become impractical in recent times. Keeping in mind the shortcomings, an Inductive Intelligent Sensor (IIS) has been proposed as a novel human monitoring system for future implementation. The proposed sensor works towards exploring the signature signals of human body movement and size. This proposed sensor is fundamentally based on inductive loop that senses the presence and a passing human resulting an inductive change.

1. Introduction

Generally the phrase wireless health monitoring system is used for describing a type of patient care system which involves monitoring of a patient's condition via wireless technology located at a healthcare facility which is remote with respect to the location of the patient, allowing overcoming inconveniences of services or devices. An advantage of having wireless monitoring systems is that they are inexpensive to install because extensive wiring is no longer required between sensors and the data acquisition system. Wireless communication can precisely unknot a patient from a monitoring entity, which is particularly valuable to supervise for long time. Researchers have been incessantly exploring that wireless sensors, as an exciting technology are able to process the human response and monitor data for signs of irregularly human actions. Moreover, wireless sensors have limitations that require novel system architectures and modes of operation. This paper is intended to serve as an overview of the several approaches available to be used in health monitoring systems.

[1] Atika Arshad, Email: atikaarshad@hotmail.com
Department of Electrical and Computer Engineering International Islamic University

2. State of the current art

Recent surveys reveal that elderly individuals or aged people have the strongest impact on long-term healthcare public investments [1-4]. Therefore, current health care system developed should be able to provide efficient, high quality and inexpensive health care services to elderly people by reducing mobility, primarily enabling the elderly to live independently like entering/leaving a room, or taking medication on time, utilization of cooking, washing, and toilet facilities thus allowing elderly people to remain engaged in their normal activities of daily life.
Nowadays, there are several sensors widely used for automated human activity monitoring and detection such as Global Positioning System (GPS), cellular modem, infrared (IR) sensors and image processing, etc.

Previously GPS and cameras were one of the most general way of tracking human activities. Satellite is being used in GPS to give information required to monitor the patient, for instance longitude/latitude coordinates, speed, time and direction. Nevertheless this way of tracking human activities is considered quite undependable as GPS are generally inadequate by dreadful weather, landscapes, and buildings (e.g., indoor) [5].

Cellular modem or the mobile monitoring system allows cellular phone to access the network. The phones will be able to transmit data where the mobile base station is located however the mobile phones cannot communicate when they are unable to connect to the base station [6].
The Smart Medical Home technology makes use of infrared sensors and video cameras, this technology provides medical assistance to patient and health care expert, however IR is sensitive to bad weather conditions and ambient light hence conflicts in identifying and monitor the subjects outside of the home [6]. Image processing usually necessitates extorting required information from data offered by an imaging sensor which gets its information from TV camera or an infrared camera. However, the inadequacy of these cameras is that they require overcoming detection artifacts caused by shadows, and reflections [7-8].

Another method for human detection was published in [9]. The approach used was based on Near Field Imaging (NFI). This technology is able to detect falls i.e. a person lying on the floor, electrodes made of thin films are positioned under the floor surface and these electrodes will produce electric field thus detecting objects near the floor, such as human body. A paper on remote monitoring pacemaker system has been presented [10] employing a near-field coupling technique. A recent work in [11] presents the findings that demonstrate the use and application of magnetic sensors in automobiles where a deployed sensor can detect seat occupancy of a travelling passenger. The detection of vacant or occupied status of seats in vehicles can improve an important stimulus for the timely operation of car airbag safety system. Effective and timely measures can certainly - prevent passengers from receiving fatal injuries during accidents.

Another similar concept has been reported in [12], where the authors have applied an inductive transducer method. The distinctive impedance peaks were characterized by passive component in high frequency range under known DC biasing condition. The impedance behaviour of 30Ω and 220Ω resistor and 250μH and 47μH (chosen as the components under test) were detected for a wide range of frequency under different biasing conditions. The circuit element inductance has been proved in showing variations as a function of frequency. The main reason for characterization of resistive elements was to enable the designer to select proper component and deviation or drift in functional specifications and also to achieve optimal circuit performance.

In the most recent reported works [13-14], the authors presented the design idea of a road condition monitoring system (vehicle detection) using inductive sensor loop-based intrusive transducer buried under the surface of roads. The experimental results showed that classifying the sizes of heterogeneous traffic was possible besides merely counting the passing by vehicles. The basic principle used here is that a vehicle passing over the sensor brings about a change in the inductance of the transducer loop buried under the road.

Moreover by using the concept of the works reported by [11-14] the proposed work aims at deriving signature signals using multiple inductive loop shaped coils used as inductive sensory transducers for the purpose of detection and identifying wave shapes with features representative of different human body sizes and passing gait. Most of the existing sensing technologies provide measurement with its own shortcomings, affecting the results of the measurement process. Among all the sensing techniques, inductive loop sensors had demonstrated a better reliability.

A general review of existing, relevant technology and products was then undertaken to determine the extent to which these criteria were already being met and to identify the sorts of issues and technologies that might require special consideration in order to develop a proof-of-concept system that would fully satisfy the criteria.

The review indicated that the identified criteria were not being met with existing products and highlighted the techniques of sensing and transmission of information as major areas requiring attention in the development of a system.

3. Proposed architecture of health care system

Researchers have been exploring various novel forms of detectors and health care systems in order to provide more precise and detailed information. In this paper, a novel method of monitoring health care system is proposed. The proposed sensing technique produces changes in the electromagnetic field which is due to the presence of the conductive material. Here, a monitoring system will make use of inductive loops that can collect unique signature waves from inductive loop shaped coils (as in inductive transducers). The produced wave shapes include features representing target body's conditions. Different size of humans or different human postures can be associated with distinctive signature waveforms.

Fig.1 presents a proposed human monitoring sensor model which is typically a passive sensor assembly modeled as an LC circuit [15] where the sensor acts as a changing capacitive, inductive or resistive element. The inductively coupled transceiver and passive sensor system can be modeled as the primary and secondary sides of an air-core transformer respectively. An inductive link is formed by a loosely coupled transformer consisting of two inductor coils L_1 and L_2 placed relatively close to one another with currents I_1 and I_2 and having primary and secondary side with their effective series resistance R_1 and R_2.

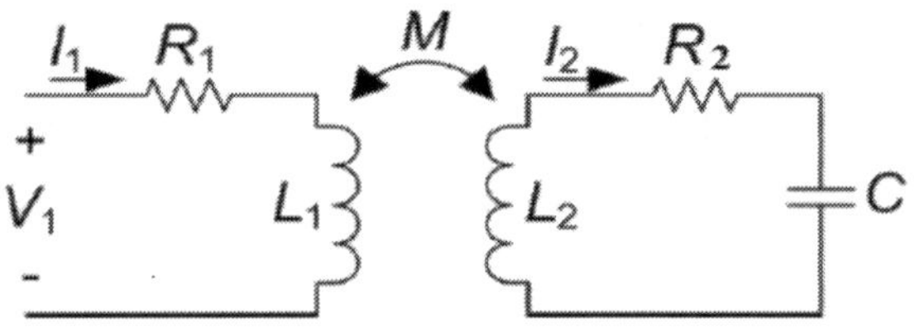

Fig.1 A schematic illustrating human monitoring wireless sensory system

The external or the primary coil is driven by an alternating current to create an electromagnetic field with its magnitude dependent on the dimensions of the coil, the drive current and the frequency of operation. The magnetic flux generated by the primary coil will be linked to the internal or the secondary coil thus inducing a current through this secondary coil which is proportional to the rate of change of the flux and the number of turns in the secondary coil (Faraday's law of electromagnetic induction). The proportion of energy captured by the secondary coil can be represented by the coupling factor, k. This value (dimensionless and always between 0 and 1) is an important factor in the operation of any inductively coupled system. The inductance impedance of the transformer model is

$$V_1 = j\omega L_1 I_1 + j\omega M I_2$$

$$0 = j\omega L_2 I_2 + j\omega M I_1 \tag{1}$$

where the mutual inductance, M is between loop and loop is given by . The transfer function of the proposed monitoring system is derived in Equation (2).

$$H(s) = \frac{V_0}{V_{in}} = R_1 + j2\pi f L_1 + \frac{(2\pi f)^2 M^2}{R_2 + j(2\pi L_2 - \frac{1}{2\pi f C_2})} \tag{2}$$

The detection is carried out as a function of frequency, the signal detected at the secondary coil is simply a function of the electromagnetic coupling and the impedance of the primary coil. Here the LC circuit identifies the resonant frequency of the inhabitant humans, hence the resonant frequency of the LC circuit changes when a change is brought in either inductance or capacitance. Hence in the proposed work the detection principle is the change introduced in the inductance of the sensor which will shift the resonant frequency. The shift in frequency can be given by the formula in Equation (3).

$$f = \frac{1}{2\pi\sqrt{LC}} \tag{3}$$

The inductance and the capacitance values are chosen to resonate by detection, that is a change in human movement will bring about the change in it's inductance.

4. Results

The sensor portion of the system consists of an LC circuit. The LC circuit has an inductor with a fixed value of capacitor. The detection principle in this work is the shift brought about in the resonant frequency when there are changes in inductance of the employed sensor. The change in resonant are assumed to be under the following scenarios:

- Case 1: assume that there is no object present in the surrounding area; mathematically this is represented by an inductance of 100ηH.
- Case 2: assume that there is a metallic object present in the surrounding area, shown by an inductance of 100µH.
- Case 3: assume a human subject present in the area be represented by inductance value of 100mH.The power transfer reaches a maximum value at a certain frequency which can be

referred to as a resonant frequency for that system and it should be tuned to that frequency to get the maximum output.

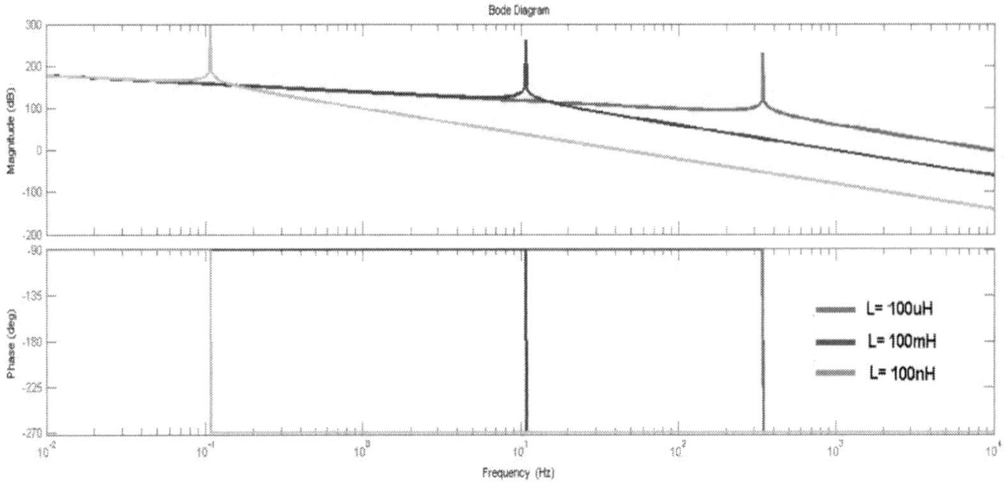

Fig.2 Magnitude and phase response for the resonance frequency plots of three different assumed scenarios

Using the scenario stated earlier, the values for the sensor's inductance are L= 100nH, L= 100μH and 100mH with a capacitor having a value of 15pF. The calculated resonant frequency of the LC circuit is 41.09 kHz, 1.30 kHz and 129.95 kHz respectively. This is within the low frequency range for biomedical implanted devices. From the assumed results, it is obvious that the distinctive impedance waveforms can be represented as the change in human movement for real time biomedical application.

5. Conclusion

The presented health care monitoring technologies have carried out the detection via wireless means. The modeling and design of few schemes have been discussed in these works. In accordance with the reviewed works, it is apparent that the IIS monitoring system will be able to provide a better solution in term of reliability that ensures interoperability with external health information systems. The system will be able to obtain feedback reports which can be used to inform patients about their daily monitored activities.

References

[1]. "Trends in Aged Care Services: some implications", *Productivity Commission Research Paper*, September 2008

[2]. J. Laditka, "Providing behavioral incentives for improved health in aging and medicare cost control: A policy proposal for universal medical savings accounts," *Journal of Health and Social Policy*, vol.13, no.4, pp.75-90, 2001.

[3]. C. Wakabayashi, "Effects of Immigration and Age on Health of Older People in the United States," *Journal of Applied Gerontology*, vol.29, pp. 697-719, December 201

[4]. Population Division, Department of Economic and Social Affairs, United Nations, "World Population Ageing 1950-2050".

[5]. S.M. Verulkar, M. Limkar, "Real Time Health Monitoring Using GPRS Technology", *International Journal of Computer Science and Network (IJCSN)*, vol. 1, no. 3, June 2012.

[6]. C. NI Scanaill, S. Carew, P. Barralon, N. Noury, D. Lyons, and G. M. Lyons, "A Review of Approaches to Mobility Telemonitoring of the Elderly in Their Living Environment", *Annals of Biomedical Engineering*, vol. 34, no. 4, pp. 547–563, April 2006.

[7]. M. Belshaw, B. Taati, J. Snoek, A. Mihailidis, "Towards a single sensor passive solution for automated fall detection", *IEEE 2011 Annual International Conference on Engineering in Medicine and Biology Society, EMBC*, pp. 1773 – 1776, 2011.

[8]. M. Elsayed, A. Alsebai, A. Salaheldin, N. Gayar, M. Helw, "Ambient and wearable sensing for gait classification in pervasive healthcare environments", *IEEE International Conference on e-Health Networking Applications and Services (Healthcom)*, pp. 240 – 245, 2012.

[9]. U. Maurer, A. Rowe, A. Smailagic and P. Siewiorek, "eWatch: a Wearable Sensor and Notification Platform," *In Proceedings of International Workshop on BSN, Wearable and Implantable Body Sensor Networks*, pp. 144-145, 2006.

[10]. Shuenn-Yuh Lee, Chih-Jen Cheng and Ming-Chun Liang, "A Low-Power Bidirectional Telemetry Device With a Near-Field Charging Feature for a Cardiac Microstimulator", *IEEE Transactions on Biomedical Circuits and Systems*, vol. 5, no.4, pp. 357 – 367, August 2011.

[11]. B. George, H. Zangl, T. Bretterklieber and G. Brasseur, "A combined Inductive-Capacitive proximity Sensor for seat occupancy detection", *IEEE Transactions on Instrumentation and Measurement*, vol. 59, no.5, pp.1463-1470, May, 2010.

[12]. Chang, Weng-Yew, Kye-Yak See, and Bo Hu, "Characterization of Component Under DC Biasing Condition Using an Inductive Coupling Approach", *IEEE Transactions on Instrumentation and Measurement*, vol. 59, no.8, pp. 2109-2114, August, 2010.

[13]. S. Sheik Mohammed Ali, B. George, L. Vanajakshi and J. Venkatraman, "A Multiple Inductive Loop Vehicle Detection System for Heterogeneous and Lane-Less Traffic," *IEEE Transactions on Instrumentation and Measurement*, vol. 61, no.5, pp.1353-1361, May, 2012.

[14]. S.S.M. Ali, B. George and L. Vanajakshi, "A magnetically coupled inductive loop sensing system for less-lane disciplined traffic," *IEEE International Instrumentation on Measurement Technology Conference (I2MTC)*, pp. 827- 832, May, 2012.

[15]. R. Nopper, R. Has, and L. Reindl, "A Wireless Sensor Readout System—Circuit Concept, Simulation, and Accuracy," *IEEE Trans. on Instrumentation and Measurement,* vol. 60, no. 8, August 2011.

IOP Conf. Series: Materials Science and Engineering **53** (2013) 012031 doi:10.1088/1757-899X/53/1/012031

Transforming human gait for signature signals characterization

Atika Arshad[1], Ahmad Fadzil Ismail, Sheroz Khan, AHM Zahirul Alam, Rumana Tasnim, Syed Samnan Haider, Mohammed M. Shobaki and Zeeshan Shahid
Department of Electrical and Computer Engineering
International Islamic University
53100 Kuala Lumpur, Malaysia

E-mail: atikaarshad@hotmail.com

Abstract. An integrated wireless inductive sensor is reported based on a system for monitoring human movement and body size. The proposed system senses the presence of human beings using electromagnetic field by making use of basic inductive coupling approach, hence analysing the performance of human monitoring. The amalgamation of the integrated system proposed will help in providing better services to the elderly people resided in healthcare centres. The developed sensing system is of low cost, flexible, robust, and easily implantable and capable of inductive sensing through marking signature waveforms as a result of human movements.

1. Introduction

Non-contact sensor systems are an emerging technology consisting of miniaturize, low-power, and low-cost devices. This technology has the potential to have enormous impact on many aspects of signal monitoring system. An inductive coupling technique is a very reliable method for supplying power to remote inaccessible region such as biomedical implants, structure-embedded micro-sensors, or safety monitoring devices in harsh environments conditions, since inductive coupling gives higher level of power transfer efficiency when compared with target devices powered directly from solar light, thermal energy and so on. Energy harvesting using light sources restricts the working environment of electronic systems. Such systems cannot work normally in low light or dirty conditions, whereas thermal energy can be converted to electrical energy by the seebeck effect however the working environment for thermo-powered systems is also limited [1]. Also, energy harvesting using light can be practical for autonomous implantable biomedical sensors [2], but the energy harvested is in low power (μW range). However, inductive coupling from an external power source is preferred because it performs better compared to the different methodologies of energy harvesting mechanisms where the devices are powered directly from the energy source; such as piezoelectric, thermoelectric, photovoltaic, or ultrasound. Inductive coupling provides higher power

[1] Atika Arshad, Email: atikaarshad@hotmail.com
Department of Electrical and Computer Engineering International Islamic University

transfer efficiency and reliable control. The power received could be in the milli-watt ranges with a reasonable distance while the transmission could be electronically controlled [3].

Several non-contact techniques for the detection of remote sensors are cited. In [4], a research was conducted using passive telemetry technology for measuring the intraocular pressure (IOP) of an eye in a continuous manner. The sensor proposed is implanted between the eyeball and the orbital bones where it can be fixed and also where it will be an easy access for surgery. The powering source system is an external coil wound around an eyeglass producing magnetic field which is magnetically coupled by the IOP sensor allowing the changes in impedance to be measured with response to variations of pressure on the sensor. The basic idea proposed in this system is to use an implanted resonant LC circuit magnetically coupled to an external coil embedded in the glasses. The capacitor of the LC resonant circuit is the pressure sensing device. The aim was to measure the resonance frequency of the sensor, which depends only on the value of its inductance and capacitance. Since the capacitance of the sensing device is related to the exerted pressure, the IOP will be obtained from the measured resonance frequency by varying the capacitance value, hence simulating for different pressures.

Recently a research showed an innovative technology to charge vehicle batteries without using the traditional method of gasoline. This novel technique is based on inductive power transfer interface (IPT) technique that allows a bi-directional and contactless power interface for the electric vehicle (EV) to be wirelessly charged or discharged by loose magnetic coupling [5].

Moreover, inductive loop sensors are currently being used to detect vehicle under a heterogeneous and less-lane-disciplined traffic [6-7] for traffic controlling and management system. These inductive loop sensors can not only detect the size of vehicles but also enables accurate counting of the number of vehicles in peak time. These types of sensor are basically embedded under the surface of the road.

Another paper [8] reports the use of wireless, passive sensor based on an inductive-capacitive resonant circuit for monitoring the growth of bacteria. The sensor is a planar, printed or photolithographically defined resonant circuit, the response of which is remotely detected by a loop antenna measuring the increase in complex permittivity of the bacteria concentration level thus indicating if a food product is safe for human consumption. The sensor comprises of an inductor-capacitor (LC) resonant circuit, which is placed within the biological medium of interest.

Furthermore the same authors proposed another practical wireless telemetric sensor for monitoring humidity levels in the packaged food [9]. The sensor developed is made of a planar inductor and capacitor printed on a paper substrate. The taste quality of the packaged food can be indirectly determined by measuring the change in the sensor's resonant frequency, which is proportional to the environment moisture content. To monitor food quality, the sensor is embedded inside the food package by adhering it to the package's inner wall; its response is remotely detected through a coil connected to a sensor reader. As food quality degrades due to increasing humidity inside the package, the paper substrate absorbs water vapour, changing the capacitor's capacitance and the sensor's resonant frequency.

Among the different sensor strategies, different groups have proposed the use of inductive-based devices, where sensing is determined by the changes in the inductance of a coil. Moreover by using the concept of the work reported in the literature this paper proposes a novel wireless inductive sensor based on a system for monitoring the signature signals of human movement and body size. The potential of the inductive device for the development of the sensor is strongly conditioned to the availability of a suitable signal conditioning circuit. One of the most used approaches is based on the measurement of the resonant frequency of an LC circuit containing the inductive sensor. In this framework, this paper analyzes the viability of a signal conditioning circuit sensor for monitoring purpose.

This paper is organized as follows: Section II describes the system architecture. The theoretical derivations are given in Section III, while the analytical and simulation results under assumed working conditions of the system proposed are given in Section IV. Conclusion of the outcomes for the proposed system, together with some ideas for expansion of this work is at the end.

2. System Architecture

The sensor architecture for remote sensing is shown in Fig.1. The sensor is endowed with computation and communication capabilities collaborating with each other in order to transfer the information of interest from desirable sources (at places such as control room, diagnostic facilities) to intended target devices. Such target devices may be employed to humans being diagnosed. Telemetric systems such as these could be used for remote sensing online, continuous, and real time telemedicine monitoring. Such systems are beneficial not only in the detection of abrupt changes of patients' health, but also in the assessment of the right dose at the right of medication. This type of monitoring allows patients in elderly homes to remain engaged in the normal activities of daily life while treating their complications by either staying at home or close to places and premises providing specialized medical services.

The sensor architecture is composed of three sections, as depicted in Fig.1. The general requirements at each section of this architecture are presented as follows.

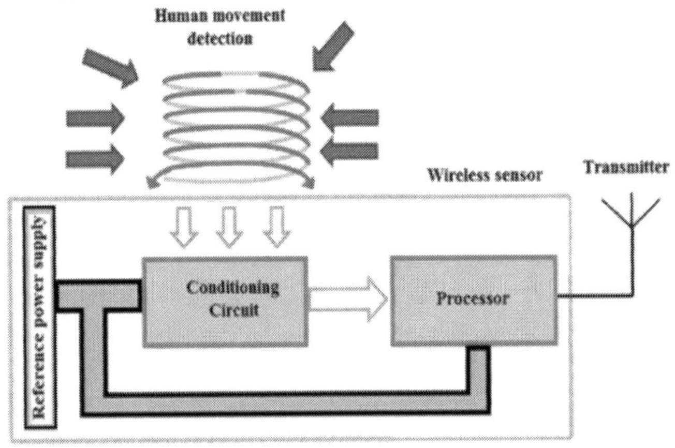

Fig.1 Sensor detection architecture

Using non-contact sensor technology, this telemetric link consists of a sensor (conditioning circuit) at the point of signal origin which collects data about the environment showing the status of the monitored inhabitant (presence of human, presence of object, and presence of no object/human). Hence the sensor can use wireless link to collect data about the status of the inhabitant remotely and perform appropriate actions when necessary that is by sending data to the second part which is the processor. The analysis and filtration of the data into information is done by the processor which will assist in suitable decision making; like the monitored person's nature is checked for deviations from their "normal" behaviour. The normal behaviour is able to signify worsening conditions of elderly patient's health. Lastly the transmitter will send the data to a computer for displaying the acquired data on a computer before a medical expert.

In this paper, a wireless inductive sensor is proposed for monitoring of human body size and movement in a closed room. The proposed sensor consists of a core-less, hollow and multi-turned coil that is capable of storing energy in its magnetic field when a current passes through its turns. The sensor is proposed to be employed under the floor of a patient's room in such a way that a passing human brings about changes in inductance. The change in inductance produces an impedance spectrum which is detected by measuring the impedance across the terminals of the loop.

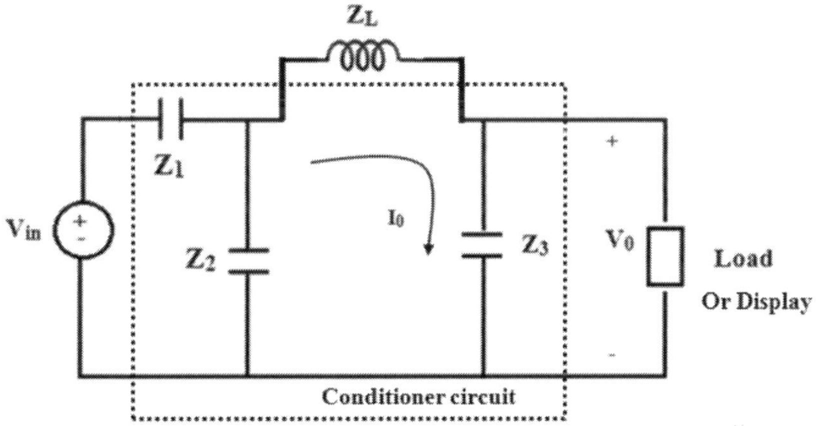

Fig.2 Circuit schematic of detection system

Fig.2 represents a given sensing system; relevant analysis is made for obtaining transfer function. The magnitude and phase of the same are plotted to be compared with similar plots obtained from PSpice simulation of the same circuit. Here Z_L is the sensor inductance (with a parallel capacitance and a load). The difference between the signals V_0 and V_{in} is analyzed versus the inductance of the sensor Z_L. The transfer function of the proposed circuit is derived as

$$H(s)=\frac{V_0}{V_{in}}=\frac{C_1C_2C_3}{s^3(LC_1C_2{}^2C_3{}^3+LC_3{}^3C_2{}^3)+s(C_2C_3{}^2C_1+C_2{}^2C_3{}^3+C_3{}^2C_2{}^3)} \quad (1)$$

The results can ultimately be used for identification of gaps and proximity of humans moving around the coil under question.

IOP Conf. Series: Materials Science and Engineering **53** (2013) 012031 doi:10.1088/1757-899X/53/1/012031

3. Results

The circuit in Fig.2 is applied to the characterization on the status of the inhabitant of human using inductive coils. The analysis of the proposed system will allow the demonstration of the ability of this simple circuit for the high-sensitivity detection of the inductive changes from integrated inductive coils.

The derived transfer function in equation (1) is plotted to show that it peaks up to a certain frequency. This frequency depends upon the inductive element in the circuit. The sensor portion of the system consists of an LC circuit. The LC circuit has an inductor with a fixed value of capacitor. In the specific sensor proposed, in this investigation the magnet wire will be assumed to be non-magnetic and with no core, hence the permeability can then be said to be equal to that of free space. The change of inductance is measured for three inductance value 100mH, 150mH and 200mH. Fig.3 indicates the graph obtained analytically by plotting magnitudes and the phase of the transfer function as a function of frequencies reflecting equation (1), whereas Fig.4 presents the simulation results. There is similarity between the analytical derivations and the simulated results obtained; both are on accordance to each other.

Moreover from the results obtained, it is obvious that the distinctive impedance waveforms can be represented as the change in human movement for real time biomedical application.

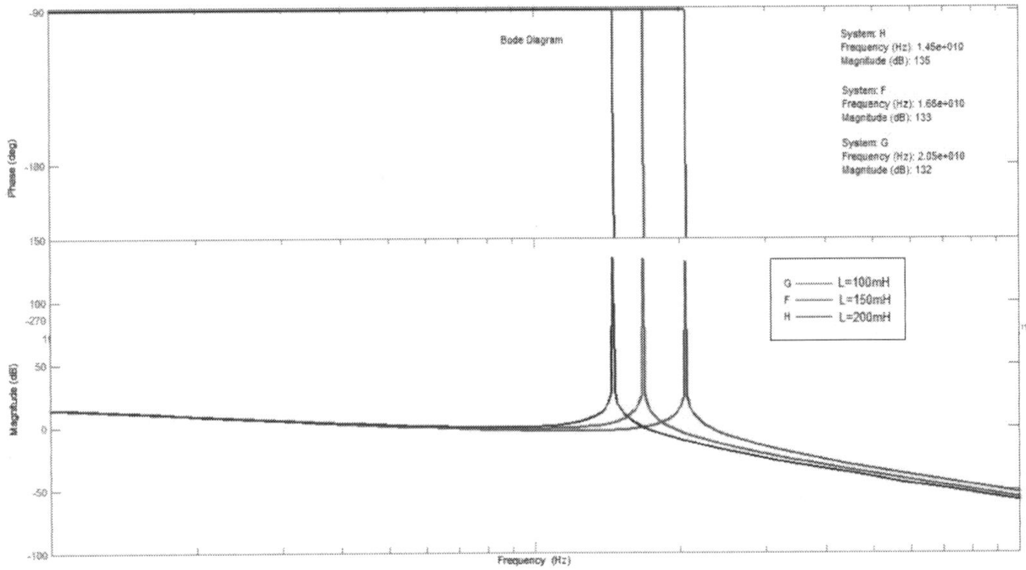

Fig.3 Magnitude and phase response for the resonance frequency plots for the change in inductance via analytical derivation

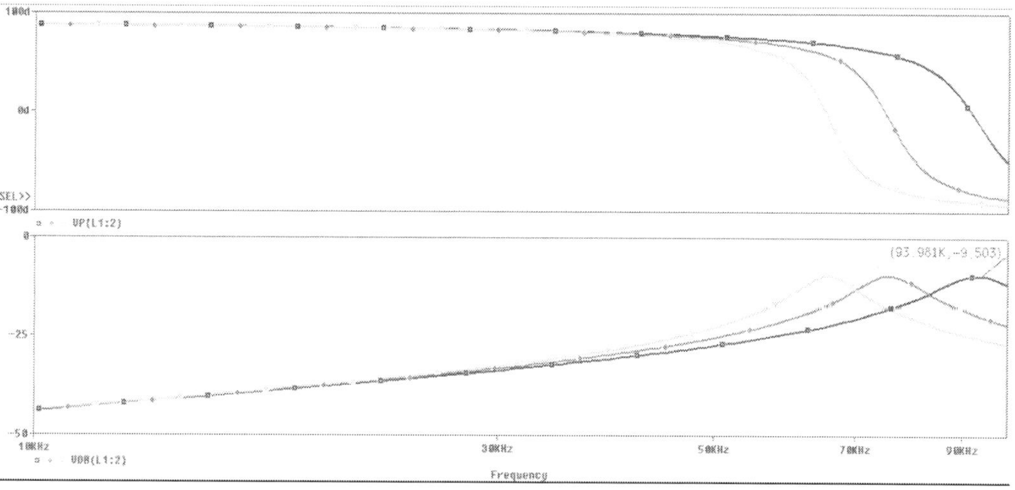

Fig.4 Magnitude and phase response for the resonance frequency plots for the change in inductance via Pspice

4. Conclusion

This work builds a model of the inductive loops on signal acquisition to monitor and analyse different postures of human body under varied condition (change in inductance value). This analysis has allowed demonstration of the ability of a very simple circuit for the high-sensitivity detection of the inductive changes from integrated coils. The resulting signals are very susceptible to ambient noise sources, and decontamination can be carried out by adaptive filtering methods. To develop such adaptive filter, robustness and an intelligent filtering system is required for obtaining a well-defined detection signal which will be an additional room for this research work.

References

[1]. Yen Kheng Tan, "Sustainable Energy Harvesting Technologies – Past, Present and Future", Published by InTech, Chapter 2, p.25, Dec 2011.

[2]. S. Ayazian, V.A. Akhavan, E.Soenen and A. Hassibi, "A Photovoltaic-Driven and Energy-Autonomous CMOS Implantable Sensor", *IEEE Transactions On Biomedical Circuits And Systems*, Vol. 6, No. 4, pp 337-343, August 2012.

[3]. Young-Sik Seo, Minh Quoc Nguyen, Zachariah Hughes, Smitha Rao and J.-C. Chiao, "Wireless Power Transfer by Inductive Coupling for Implantable Batteryless Stimulators", *International Microwave Symposium*, Montreal, Canada, June 17-22, 2012.

[4]. S. Lizon-Martinez, R. Giannetti, J.L. Rodriguez-Marrero and B. Tellini, "Design of a System for Continuous Intraocular Pressure Monitoring," *IEEE Transactions on Instrumentation and Measurement*, Vol. 54, No. 4, August 2005

[5]. U.K. Madawala, D.J. Thrimawithana, "A Bidirectional Inductive Power Interface for Electric Vehicles in V2G Systems," *IEEE Transactions on Industrial Electronics*, 58(10), pp. 4789-4796, October, 2011.

[6]. S.S.M. Ali, B. George,L. Vanajakshi and J. Venkatraman, "A Multiple Inductive Loop Vehicle Detection System for Heterogeneous and Lane-Less Traffic", IEEE Transactions on Instrumentation and Measurement, May, 2012, 61(5), pp.1353-1361.

[7]. S.S.M. Ali, B. George, and L. Vanajakshi, "A magnetically coupled inductive loop sensing system for less-lane disciplined traffic", *IEEE International Instrumentation on Measurement Technology Conference (I2MTC),* May, 2012, pp. 827- 832.

[8]. K. G. Ong. J. S. Bitler, C. A. Grimes, L. G. Puckett and L. G. Bachas, "Remote query resonant circuit sensors for monitoring of bacteria growth: application to food quality control," *Sensors 2002*, 2, 219-232.

[9]. Ee Lim Tan, Wen Ni Ng, Ranyuan Shao, Brandon D. Pereles and Keat Ghee Ong, "A wireless, passive sensor for quantifying packaged food quality," *Sensors 2007*, 7, 1747-1756

IOP Conf. Series: Materials Science and Engineering **53** (2013) 012032 doi:10.1088/1757-899X/53/1/012032

Evaluation of QoS supported in Network Mobility NEMO environments

L F Hussien[1*], A H Abdalla[1], M H Habaebi[1], O O Khalifa[1] and W H Hassan[2]

[1]Department of Electrical and Computer Engineering, Faculty of Engineering International Islamic University Malaysia, 50728 Kuala Lumpur, Malaysia

[2]Malaysia Japan International Institute of Technology, Universiti Teknologi Malaysia, 54100 Kuala Lumpur, Malaysia

[*]E-mail: lolo_cts1@yahoo.com

Abstract. Network mobility basic support (NEMO BS) protocol is an entire network, roaming as a unit which changes its point of attachment to the Internet and consequently its reachability in the network topology. NEMO BS doesn't provide QoS guarantees to its users same as traditional Internet IP and Mobile IPv6 as well. Typically, all the users will have same level of services without considering about their application requirements. This poses a problem to real-time applications that required QoS guarantees. To gain more effective control of the network, incorporated QoS is needed. Within QoS-enabled network the traffic flow can be distributed to various priorities. Also, the network bandwidth and resources can be allocated to different applications and users. Internet Engineering Task Force (IETF) working group has proposed several QoS solutions for static network such as IntServ, DiffServ and MPLS. These QoS solutions are designed in the context of a static environment (i.e. fixed hosts and networks). However, they are not fully adapted to mobile environments. They essentially demands to be extended and adjusted to meet up various challenges involved in mobile environments. With existing QoS mechanisms many proposals have been developed to provide QoS for individual mobile nodes (i.e. host mobility). In contrary, research based on the movement of the whole mobile network in IPv6 is still undertaking by the IETF working groups (i.e. network mobility). Few researches have been done in the area of providing QoS for roaming networks. Therefore, this paper aims to review and investigate (previous /and current) related works that have been developed to provide QoS in mobile network. Consequently, a new proposed scheme will be introduced to enhance QoS within NEMO environment, achieving by which seamless mobility to users of mobile network node (MNN).

1. Introduction

The emergence and success of cellular communication networks indicate that there are wide users' demands to mobility. Actually, such networks are developed to grant not only the traditional voice

[1*] **Corresponding Author:** Loay F. Hussien, Department of Electrical and Computer Engineering, Faculty of Engineering, International Islamic University Malaysia, 50728 Kuala Lumpur, Malaysia, lolo_cts1@yahoo.com

service but also data services as well. Internet Protocol (IP) seems to be the base technology of future networks to provide all kind of services to both (fixed and mobile) users via different access technologies. Nowadays, there are multiple technologies existing such as Wireless Local Area Network (WLANs), Bluetooth, Satellite communications, Microwaves, Wireless Metropolitan Area Networks (WiMAX) and cellular networks. However, IP was not designed taking into account the mobility of both users and their own terminals. In few years, the Internet Engineering Task Force (IETF) has developed protocols such as Mobile IPv4 (MIP) [1] and Mobile IPv6 (MIPv6) [2] for supporting seamless connectivity to mobile hosts. Mobile IPv4 protocol was more defined in (RFC 3344) and the updates were added in (RFC 4721). The protocol Mobile IPv6 was examined well in (RFC 3775 and RFC 6275).

Nevertheless, these protocols do not support the movement of a complete network that changes its point of attachment from fixed infrastructure, while still maintaining sessions of every device belong to the network. For the time being, we are witnessing the emergence of mobile networks, namely a set of hosts that move collectively as a whole unit for instance: ships, submarines, buses, trains and aircrafts as illustrated in Figure 1. The IETF has created a working group known as: Network Mobility (NEMO) (RFC 3963), with the aim of extending existing host mobility solutions (i.e. an individual mobile device) to enable the movement of an entire mobile network in IPv6 [3].

Figure 1. NEMO applications usage scenarios

Quality of Service (QoS) can be characterized in more than one way. In the field of computer networking it is the ability of the network element (e.g. application, host or router) to provide some level of assurance for consistent network data delivery. In other words, QoS is a set of technologies that enables network administrators to manage the effects of congestion on applications traffic flows by using network resource optimally rather than conditionally adding extra capacity.

The various QoS characteristics are divided in two groups, technology-based and user-based QoS parameters. Technology-based parameters contain superior performance in terms of delay, response time, jitter, data rate and loss rate. On the other hand, user-based QoS parameters are more likely subjective. They include categories such as perceived QoS, the visual quality of a streaming video, cost per unit time or per unit of data and the security. As an example, a user browsing the web and watching public news broadcast would be more interested in the quality of the picture rather than its security. A user who is remotely connecting to a corporate network would be most interested in the security of the connection and less interested in costs.

Within a few past decades, QoS is certainly not supported over the IP-based networks. Working on QoS support in IP networks, has led to three distinct approaches namely, Integrated Services (IntServ) [4], Differentiated Service (DiffServ) [5] and Multiprotocol Label Switching (MPLS) [6]. Unfortunately, these approaches were initially designed for static networks without mobility in-mind. Thus, they are not fully adapted to mobile environments yet. In fact, it is anticipated that more mobile users will be connected to the Internet rather than PCs users. These mobile users are interested to get similar QoS in mobile terminals as in fixed terminals (i.e. wired networks) in order to run real-time applications properly. Above all QoS models, the most promising approach due to its simplicity and scalability advantages, is DiffServ. Therefore, integrating QoS with mobility support seems to be needed to fulfil the necessity of users.

The essential goal of the NEMO working group is to specify a solution to provide continuous Internet connectivity to nodes in a mobile network at all times while the mobile network changes its point of attachment [3]. NEMO networks might lead to frequent handovers while mobile networks roam among different Access Routers (ARs) and still staying connected to the fixed network. Quality of Service degradation or enforced service termination may arise when there are insufficient resources required for handover requests in NEMO networks. In fact, users inside a NEMO require QoS guarantee while accessing the Internet via a mobile. Therefore, resources should be ideally provisioned before the handover process to eliminate or minimize the packet drops due to insufficient resources.

The rest of the paper is structured as follows. Section 2 & 3 will explore literature review and the related works that gather the integration of QoS within NEMO area. Section 4 will deliberate the open issues. After that, Section 5 will present the proposed scheme. Finally, the conclusion and future work are given in Section 6.

2. Literature review

NEMO basic support protocol introduces several advantages, such as reduced signalling, increased manageability, reduced power consumption and conservation of bandwidth when it is compared with individual host mobility. However, it inherits all the drawbacks of Mobile IPv6-based, such as inefficient routing path, single point bottleneck, increased handover latency and packet loss rates, increased packet overhead and so on so forth. To overcome these drawbacks of NEMO, one must perform an extensive research works investigation. Therefore, a new yet novel break-through with a new proposed scheme could be achieved.

Beside NEMO BS protocol basic support, the IETF has developed and extended two more complex versions: nested network mobility and multi-homed network mobility.

2.1. Nested network mobility(NEMO)

A Mobile networks can encompass a very complex hierarchy, e.g. individual mobile networks operating within a larger mobile network, the visiting mobile nodes (VMNs) in a mobile network is actually a mobile router that has its own mobile network nodes, and so on. Thus, NEMO model that supports the connectivity of one entire mobile network through another mobile network resulting in topologies referred to as Nested NEMO networks. In other words, Nested mobile networks occur when the VMN with different home address from a MR is attached to the mobile network, or a mobile network is attached to another mobile network. The Nested NEMO scenarios quickly become extremely inefficient and are not suitable for real life deployment solutions. The inefficient routing model that occurs in Nested NEMO Networks is commonly known as "pinball routing problem". Each mobile router (MR) and mobile network node (MNN) has its own home agent (HA). So, if the correspondent nodes (CNs) desire to send data to the leaf MNN which is located at the bottom of the nested mobile network, the packets have to travel to all home agents that are mapped to the MRs of the nested mobile networks. Therefore, the mobile network will become more complicated. The process of the nested mobile network is depicted in Figure 2.

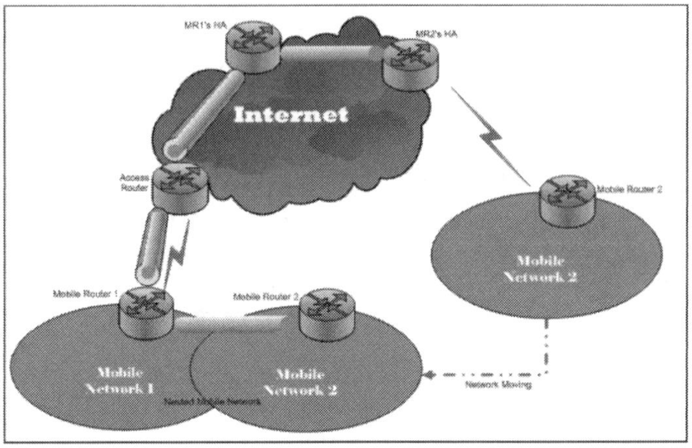

Figure 2. Nested mobile network NEMO

2.2. Multihomed network mobility(NEMO)
In the NEMO terminology, Multihomed network mobility consists of:

- Multiple egress interfaces on a MR as shown in Figure 3. Egress interface is an interface that connected to access router from the MR, where the packets from the mobile MNNs routed their ways out), or

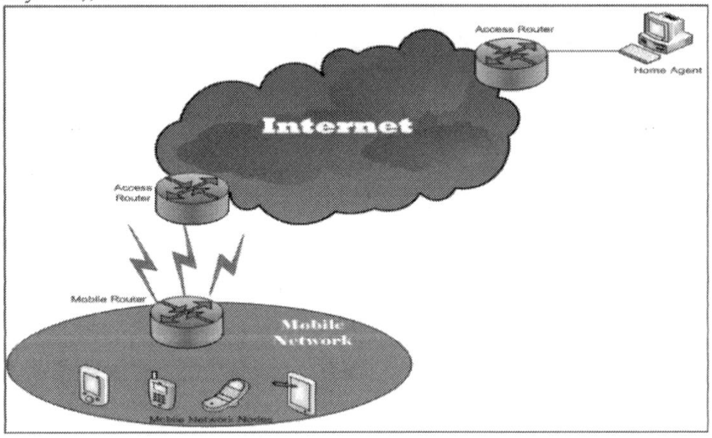

Figure 3. Multihomed NEMO multiple egress interfaces

- Multiple MRs in a mobile network as illustrated in Figure 4, by which the mobile network is simultaneously connected to the Internet via more than one mobile router, or

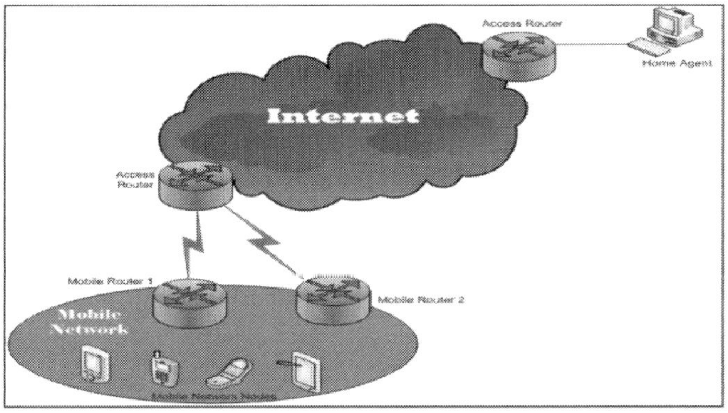

Figure 4. Multihomed NEMO multiple mobile routers

- Multiple global prefixes are available on the mobile network or mobile networks associated with multiple HAs as shown in Figure 5. This situation exists when the MNN/ MR has different home addresses.

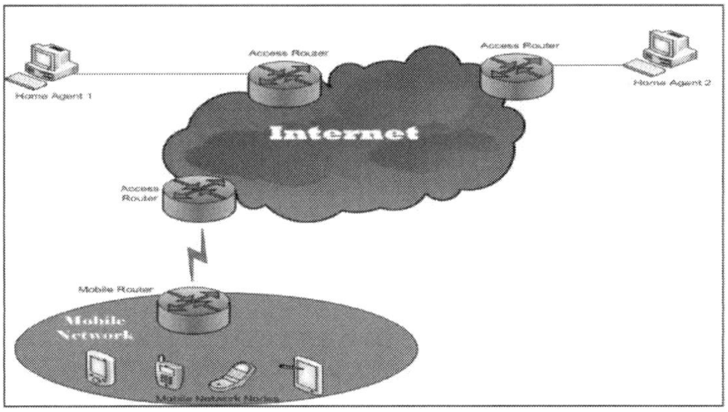

Figure 5. Multihomed NEMO multiple HAs

Multihomed NEMO yields many benefits such as:

1. Fault-Tolerance/Redundancy.
2. Load-Sharing.
3. Policy-Routing.

3. Related work

Several aggregation policies have been proposed in the literature by [7], [8] to control and manage the buffering duration at the MR. All of these use a static parameter upon which to make the aggregation decision. However, they might handle the cost inefficiency when the request rate unpredictably fluctuates over time. In the work done in [9], [10], a dynamic cost-driven QoS aggregation policy was proposed and analyzed. It is able to cost-efficiently operate under burst QoS requests, while at the same time not disadvantaging the user with long waiting time. The process of handover happens when the mobile network roams to a new point of attachment. Resources might not be enough in the mobile network due to loss of signal, QoS degradation or service termination [11]. Authors in this paper [12] defined an IntServ and DiffServ model to propose an alternative approach which combines the advantages of IntServ and DiffServ protocols to provide QoS guarantees in NEMO networks. They studied the handover process in handling the resource reservation, where the mobile node searches a good level of QoS before joining the mobile network. This work proposes a protocol that could support RSVP signaling, called NEMO Reservation (NEMOR). The protocol reserved the resources according to aggregated flows. The implementation comprised two phases: the MR and HA, and the HA and CN. However, the NEMOR protocol was written as an Internet draft and has not been extended for further work. No performance evaluation study was conducted in order to appreciate the benefits and gauge their associated overhead.

The binding cache of Home Agent (HA) only contains home address of mobile router. The addresses of mobile network nodes are not binded with current CoA of mobile router. As a result, packets cannot be tunneled to the mobile router correctly. To solve this issue, the idea of Prefix Scope Binding Update (PSBU) [13] is proposed to support advanced network mobility in IPv6. It suggests that an MR advertise its mobility through both the classic Mobile IPv6 binding update and PSBU. A PSBU binds the mobile network prefix with the COA of the MR rather than HoA with COA of the MN. With this approach, upon reception of a packet from MR's HA, the MR will send a PSBU to the CN which enables the CN to bind the mobile network prefix to the MR's COA. Thus, all the subsequent packets addressed to the nodes in the mobile network will be sent toward the MR's COA to achieve route optimization.

The work in [14] has proposed the use of two new Mobility Option (MO) headers to trade priority information between a MR and its respective HA. A negotiation signal is exchanged between the MR and HA. This signal is a mobility option header carried in Mobile IPv6. The former MO header is a priority negotiation request that holds the priority and the address information for the destination. The latter header is a priority negotiation reply that includes a return value that is either a simple success or failure. If the MR and the HA obtain a request for a priority from their ingress network, they set off the MO header for the negotiation signal and send it to their respective MR or HA. The technique splits up the traffic between the highest and normal priority.

In NEMO BS, the binding update traffics are minimized at the cost of tunneling overhead. To reduce the tunneling overhead, an adaptive NEMO support protocol based on hierarchical mobile IPv6 (HMIPv6) was proposed in [15]. The general idea is a trade-off between adaptive binding update traffic and tunneling overhead. An adaptive binding update strategy was adopted based on the session to mobility ratio (SMR). This SMR with a predefined threshold was compared with different binding update procedures. An optimal threshold for adaptive binding updates was also derived in [16].

Network mobility could be split up into two domains, a wireless domain and wired domain. Authors in [17] have proposed a two-level aggregation-based QoS architecture to offer QoS support in NEMO. The architecture is divided into two levels, a node level and a network level. The QoS requirements for each flow are assembled at the mobile network nodes, while the MR at the network level gathers the QoS requests and aggregates them into a single Service Level Specification (SLS) [18] request for the entire NEMO subnet. The MNNs forward the resource requests for several flows or applications and allocate the resources to these flows. Moreover, they also proposed a universal signaling protocol to exchange the SLS between MNNs and the MR, and MR and the visited networks. The SLS is presented to carry QoS information for traffic aggregations. Another QoS aggregation approach has been proposed in [19] to provide for signaling control between the MR and the access network. Three policies were proposed which are temporal, cardinal and resource-threshold. The paper in [20] proposed a reasonable solution for a scheduling algorithm in network mobility. The authors assessed the performance of priority scheduling and fair scheduling. They proposed a scheduling algorithm Adaptive Rotating Priority Queue (ARPQP) that has exposed QoS guarantees for the higher priorities and maintains the reasonable throughput for the lower priorities.

Traffic scheduling and shaping mechanisms were utilized based on DiffServ model in [21] to constrain the traffic sources. The approach is to reduce the 802.11b wireless channel of saturation. Traffic is grouped from a high priority to a low priority. The EF (Expedited forwarding) class has the higher priority than AF (Assured Forwarding) and BE. Worst-case fair weighted fair queuing (WF2Q) mechanism is configured to distribute the bandwidth for AF and BE traffic. The technique concentrates on constraining the traffic rate rather than allocating the bandwidth for each class. In the same way, an admission control is required to control the amount of EF traffic to avoid the wireless channel being saturated. When there is no EF traffic transmitted over the wireless channel, the bandwidth is given to the AF and BE traffic. However, BE traffic carries on utilizing the channel bandwidth with the AF traffic.

A couple of attempts have been done to analyze the signaling cost for mobility protocols. This framework [22] performed cost analysis of Mobile IP to minimize the signaling cost while introducing a novel regional location management scheme. The research work in [23] analyzed the signaling costs of SIGMA [45] and HMIPv6. In [24] an analytical model for the performance and cost analysis of IPv6-based mobility protocols were presented. Authors of [25] performed the signaling cost analysis of Seamless IP diversity based Network Mobility (SINEMO) [26]. Though, these analyses overlooked all possible costs (e.g. costs for sending query message by CN, securing location updates, obtaining IP address by MH, etc.), they failed to consider computing the signaling costs on various mobility entities. The work in [27] analyzes the total cost (including data delivery cost) of various mobility entities of NEMO basic protocol and figures out how those costs are affected by various network parameters, such as network size, mobility rate, traffic rate, and data volume. However, more optimizations are needed.

4. Open issues

From the previous related works that have been conducted, obviously there are a lot of opportunities to improve Quality of Service in network mobility. Some proposals have succeeded to tackle some issues in network mobility. However, certain research areas are still having open issues and they need to be highlighted:-

- Performance Issues (Delay, Packet loss): The handover process results in increased latency due to the multiple levels of indirection involved. The chances of packet loss are also more significant as a result of increase in latency. Research is required to adapt mechanisms such as Fast handover, Hierarchical handover or DiffServ to support QoS in mobile networks.
- Increase in Packet Size: Each added level of nested mobility requires an additional tunnel encapsulation, and these extra IPv6 headers increase the packet size and the associated overheads. Solutions that optimize the routes and reduce the levels of indirection are needed.
- Sub-Optimal Routing: the NEMO basic protocol uses bidirectional tunnelling between the MR and HA to achieve global connectivity to nodes within the mobile network. In case of nested NEMO network, several tunnelling occurs and this leads to several complexities. One of the significant problems regarding the nested NEMO is the pinball routing problem. Packets will flow via the Home Agents of all the Mobile Routers on their paths within the mobile network. When the mobile network moves to a new location, the new location has to be informed to the home agent of the mobile network. This binding update and binding acknowledgement has to pass through several MR-HA tunnels. This process is called as pinball routing. This is highly undesirable because it would incur delays, more bandwidth usage and increase the packet size at each level of nested mobility.
- Multi-Homing Issues: Support for multi-homing in a network mobility environment is crucial since if a MR fails to maintain session continuity this would affect the session preservation of the entire network. Also multi-homing support would enhance the load sharing and fault tolerant capabilities of mobile networks. However, NEMO Basic Support protocol neither prevents nor explicitly specifies mechanisms to handle multi-homing issues. Providing QoS for mobile network nodes by exploiting the redundancy provided in multi-homed mobile networks is a research area that needs further research.
- Scalability Issues: how to grow mobile network (i.e. by increasing the number of MNNs, MRs, Nested NEMO or Multihomed NEMO) while maintaining the performance is a challenge need to be taken under consideration.

- Signalling cost Issues: the cost analysis of network mobility and the underlying mobility management entities have become essential to avoid their performance degradation. However, there have been a few comprehensive cost analyses of NEMO protocol that considers all possible costs. Thus, developing analytical framework to measure network mobility costs and comparative analysis of the schemes based on the network mobility costs require more investigations.

5. The proposed scheme

The most important intention of studying QoS support in mobile environment lies in two aspects: One concentrates on how the mobility node affects end-to-end QoS guarantees. While the second concentrates on how to apply the existing QoS technologies in wired networks to wireless networks, namely how to append mobility support to these solutions and how these solutions suit the wireless link characteristics [28].

The topology depicted below in Figure 6, it based on an IPv6 network with mobility support and DiffServ model composite within the network to offer privilege QoS guaranteed service. The whole network literally moves as single unit.

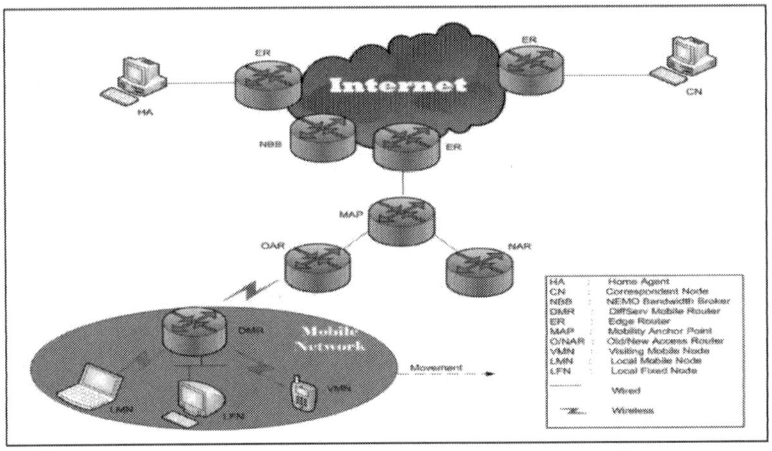

Figure 6. The proposed network topology

This topology allows the nodes to simply connect as if they were on a fixed point of attachment, hiding all the complexity of handling mobility into three devices called DiffServ Mobile Router (DMR), Home Agent (HA) and MAP (mobility anchor point). The HA is a fixed device responsible for transmitting data from correspondent nodes to mobile nodes through the DMR. The DMR is located in a mobile environment. It can utilize any available access technology (e.g. 3G, 4G, WLAN or WiFi) to ensure proper routing functions between the mobile nodes and the Internet. Furthermore, it the DMR can also use more than one access technologies at the same time for load sharing and facilitating the vertical handover to the MNNs.

The proposed framework integrates the existing QoS over IP architecture with NEMO BS protocol. The aim is to suit the needs of both QoS guaranteed and mobility in communication. Taking the advantages of Differentiated Service (DiffServ) approach into a count, our focus is to propose the necessary modification in NEMO to ensure that the seamless mobility with the required QoS parameters can be achieved. Differentiated Service (DiffServ) has been proposed by Internet Engineering Task Force (IETF) to extend the Internet to be a QoS-capable, efficient and scalable network supportive. The Differentiated Services (DS) network architecture provides QoS guarantees in a scalable and least complex manner. Service diferentiation is desirable to accommodate heterogeneous application requirements and user expections. It interconnects heterogeneous wire-line/wireless networks with the Internet backbone to provide end-to-end QoS and seamless roaming to mobile users. It also premits differentiated pricing of Internet service.

The proposed scheme assumes that the DMR in the proposed architecture has the functionally of the Edge Router (ER). So, it will be empower to implement the police. On the other hand, the NBB acts as bandwidth broker to manage and monitor the network resource. By allocating resources to forwarding classes and controlling the amount of traffic for these classes, the framework will create different levels of services and resource assurance but not absolute bandwidth guarantees or delay bounds for individual flows.

Precisely, NEMO Bandwidth Broker is responsible for mapping packects to one of forwarding classes supported in the network. It must ensure that the traffic conforms to SLA for particular users. The SLA specifies packet classification, re-marking rules, traffic profiles and actions to traffic streams which are In/Out-of-profile. The TCA between the domains is derived (explicitly or implicitly) from this SLA. The boundary nodes ranslate the TCA into traffic profile for each user connect to. Once the packects pass the boundary nodes into interior of the network, resource allocation is preformed based on forwarding class.

Mobile network node (MNN) defines its requirements using SLA to request resources. In contrary, NBB agent has to perform admission control task by accepting or rejecting bandwidth requests. It maintains a database of parameters, in accordance with which reservations are made and then the DSCP for those services are going to be assigned. The negotiations for bandwidth allocation will initially occur between the MNN and NBB. After which, if the NBB accepts the QoS request to grant the resources, it will configure the edge routers and DMR to help optimizing the existing resources (i.e. controlling the network load) as shown in Figure 7.

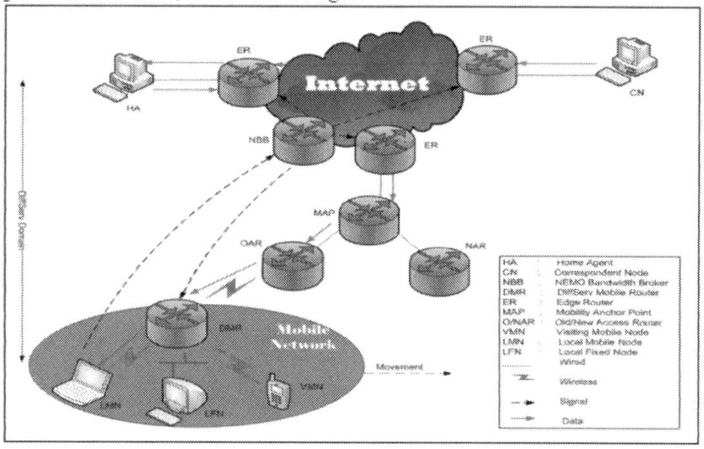

Figure 7. The Generic signalling and data flow in the proposed scheme

As we know that Network Mobility is an extension of Mobile IPv6 (MIPv6). In conjunction with MIPv6, the MN performs binding update to HA/CNs regardless of its movements to other subnets. This induces unnecessary signaling overhead and latency. Therefore, the proposed scheme makes the use of F-HMIPv6 protocol. This is due to the fact that F-HMIPv6 protocol intends to combine the Fast Handovers for Mobile IPv6 (FMIPv6) protocol with the Hierarchical Mobile IPv6 Mobility Management protocol (HMIPv6). It means that the fast handover mechanism will be deployed over the HMIPv6 networks using the F-HMIPv6 protocol. Therefore, the protocol provides the advantages of both schemes (i.e., a seamless handover scheme with less signalling overhead and lower handover latencies). Hierarchical topology hides the network mobility form internet. Moreover, the overall

handover latency achieved by FMIPv6 will be further reduced because of local location updating in HMIPv6, while in the original FMIPv6, the Home Agent (HA) and CNs are usually far away.
Priority Queue algorithm (PRI) has been chosen as a scheduler at all of edge and core routers to provide service differentiation by classifying the arriving data to different priority classes. The PRI scheduler at the output link should ensure that EF always receives a better QoS than AF and while the AF class receives better service than BE class. It also has the advantage of being simple and easy to implement for scheduling of the traffic at each output link in a DiffServ network.

6. Conclusions

In conclusion, Quality of Service is a set of service requirements (performance guarantees) to be met by the network while transporting a flow. Obviously, mobile users need to achieve QoS similar to wired users. However, the current QoS approaches are designed without mobility in mind. Therefore, there is a need to come up with a set of protocols to ensure that mobile users along with their terminals can obtain feasible required QoS. This paper reviewed (the previous /and current) works that have been adopted different techniques to provide QoS in mobile network context. Then, the main fundamental open issues are presented and discussed in general. A new proposed scheme is introduced in this paper to optimize QoS in NEMO environment. Currently, Network Simulator NS-2 is been used to carry out the proposed scheme. In future work, we will further carry out the proposed scheme using analytical analysis to generate the signaling cost. Moreover, we plan to extent the model to investigate a detailed analysis of different techniques with certain performance metrics to check how the proposed scheme is working in a complex network such as nested or multihomed network mobility.

References

[1] C. Perkins, ed. Oct. 1996 IPv4 Mobility Support RFC 2002
[2] D. Johnson, C. Perkins, J. Arkko June 2004 Mobility Support in IPv6 RFC3775 http://www.ietf.org/rfc/rfc3775.txt
[3] T. Ernst 2005 Network Mobility Support Goals and Requirements IETF Internet draft <draft-ietf-nemo-requirements-05>
[4] Braden, R., Clark, D. Shenker, S. June1994 Integrated Services in the Internet Architecture: an Overview Internet Engineering Task Force Request for Comments (RFC) 1633
[5] Blake, S., Black, D., Carlson, M., Davies, E., Wang, Z., Weiss, W. Dec. 1998 An Architecture for Differentiated Services Internet Engineering Task Force Request for Comments (RFC) RFC 2475
[6] Rosen. E., Viswanathan, A., Callon, R. January 2001 Multiprotocol Label Switching Architecture Request for Comments (proposed standard) 3031 Internet Engineering Task Force
[7] M. A. Malik, S. S. Kanhere, M. Hassan, and B. Benatallah 2004 On-Board RSVP: An Extension of RSVP to Support Real-Time Services in On-Board IP Networks *ser. Springer LNCS* vol. 3326 *A. Sen et al. Eds. New York: Springer-Verlag* pp. 264–275
[8] M. A. Malik, S. S. Kanhere, and M. Hassan 2005 Aggregation policies over RSVP tunnels *in Proc. 62nd IEEE VTC—Fall* vol. 2 pp. 1249–1253
[9] G. Kamel, A. Mihailovic, P. Pangalos, and A. H. Aghvami Nov. 2007 Cost-optimal QoS aggregation for network mobility *in Proc. IEEE GLOBECOM* pp. 5006–5010
[10] G. Kamel, A. Mihailovic, and A. H. Aghvami Feb. 2008 Case analysis of a costoptimal QoS aggregation policy for network mobility *IEEE Commun. Lett, vol. 12, no. 2*, pp. 130–132
[11] Yan Xuying et al. June 2007 NSIS Based Resource Reservation for NEMO *Control Conference* Hainan China

[12] M. Tlais and H. Labiod. June 2005 Resource reservation for nemo networks *International Conference on Wireless Networks Communications and Mobile Computing* volume 1 pp. 232-237

[13] P. Thubert, and M. Molteni October 2002 IPv6 Reverse Routing Header and Its Application to Mobile Networks draft-thubert-nemo-reverse-routing-header-01.txt

[14] Han Sunyoung. 2006 QoS Negotiation MO Headers for NEMO Home Network Model Network Mobility Group Technical Report draft-han-nemo-qos-mo-homenet-00.txt

[15] Pack S, Kwon T, Choi Y, Park E. 2009 An adaptive network mobility support protocol in hierarchical mobile ipv6 networks *IEEE Transactions on Vehicular Technology*

[16] Yaning Wang, Linghang Fan, Nadeem Akthar, Kar Ann Chew, and Rahim Tafazolli 2005 An Aggregation-based QoS Architecture for Network Mobility *4th IST Mobile Summit Conference*

[17] Badr Benmammar, Zeina Jrad, and Francine Krief 2009 QoS Management in Mobile IP Networks using a Terminal Assistant *International Journal of Network Management* 19:1-24

[18] George Kamel, Mihailovic Andrej, and A.Hamid Aghvami September 2009 A Cost-Optimal QoS Aggregation Policy for Network Mobility: Analysis and Performance Comparisons *IEEE Transactions on Vehicular Technology* 58(7)

[19] Yaning Wang, Linghang Fan, Dan He, and Rahim Tafazolli 2008 Performance Comparison of Scheduling Algorithms in Network Mobility Environment *Computer Communications Journal, 31(9):pp.*1727 - 1738

[20] Martin Heusse, Paul Starzetz, Franck Rousseau, Gilles Berger-Sabbatel, Andrzej Duda 2003 Bandwidth Allocation for DiffServ based Quality of Service over 802.11

[21] J. Xie and I. Akyildiz, July 2002 A novel distributed dynamic location management scheme for minimizing signaling costs in Mobile IP *IEEE Transactions on Mobile Computing* vol. 1 no. 3 pp. 163–175

[22] S. Fu and M. Atiquzzaman Nov 2005 Signaling cost and performance of SIGMA: A seamless handover scheme for data networks *Wireless Communication and Mobile Computing* vol. 5 no. 7 pp. 825–845

[23] A. S. Reaz, P. K. Chowdhury, and M. Atiquzzaman, Dec. 2006 Signaling cost analysis of SINEMO: Seamless End-to-End Network Mobility *in First ACM/IEEE International Workshop on Mobility in the Evolving Internet Architecture* San Francisco CA

[24] C. Makaya and S. Pierre March 2008 An analytical framework for performance evaluation of IPv6-based mobility management protocols *IEEE Transactions on Wireless Communications* vol. 7 no. 3 pp. 972–983

[25] S. Fu and M. Atiquzzaman 2006 SIGMA: A Transport Layer Handover Protocol for Mobile Terrestrial and Space Networks *e-Business and Telecommunication Networks, Springer* pp. 41–52

[26] P. Chowdhury, M. Atiquzzaman, and W. Ivancic July 2006 SINEMO: An IP-diversity based approach for network mobility in space *in Second IEEE International Conference on Space Mission Challenges for Information Technology (SMC-IT)*

[27] Md. Shohrab Hossain and Mohammed Atiquzzaman 2011Cost and Scalability Analysis of Mobility Management Entities of NEMO

[28] LIU Chu-da,LIU Yi, Sun Haitao, ZHU Guo jun2004 QoS Provisioning in Mobile IPv6 China

Inductive Loops for Sensing Position as Signature Signals

Sofiane Larbani[1], Noreha Abdul Malik, Anis Norashikin Nordin, Sheroz Khan, Mohammad Shobaki
Department of Electrical and Computer Engineering, Kulliyyah of Engineering
INTERNATIONAL ISLAMIC UNIVERSITY MALAYSIA

E-mail: sofilarbani@gmail.com

Abstract. In this paper, an inductive sensing technique made of a special shaped inductive loop is proposed. The inductive loop has an inner turn fitted within an outer turn, making a total inductance value 100μH. This loop is made to be shown with balanced response using three capacitance values of 0.068μF each when a sinusoidal voltage source of 5V peak-to-peak is applied. The variation of the relative permeability of the inductance of the inductive loop (ΔL) results in a variation of the overall inductance value ($L+\Delta L$), that causes the output signal to change in term of shape and amplitude for variation of total inductance sweep over a given period of time. As a result of change in inductance value (1μH) there is a correspondence increase of 300mV. Theoretical derivations have showed in close agreement with the simulation plots obtained using Multisim software.

I. Introduction

Detection and sensing procedures are becoming more important as the demand for better accuracy for more comfort and security has increased. The required sensing techniques are diversified in nature and in the basic principle in producing electro-chemical signals in view of issues related to linearity, better sensitivity, and improved accuracy. Such issues not considered significant some time ago, have been becoming of serious research concerns [1-4].

All these techniques depend on the surface and other similar features of the target entity being sensed. This makes every type of application different in demand, and more demanding in the principle of technique being adopted. For sensing of relative position, tracking of objects and sensing traffic flow of moving vehicles, or sensing the shape and size of objects in question, are used electronic detection techniques in contrast to the mechanism being employed for sensing of constituent elements in soil or fluid or for that matter the constituents in human blood. Broadly can be divided into contact and non-contact categories, sensors are taking up new shapes and sizes including those buried in concrete structures or surface layer.

Inductive loops has been used for sensing of targets such as vehicles [5], traffic control and vehicle classification [6]. However detection of non-metallic object can be done knowing that there is a small change of the inductance value as small as it is with an adequate sensing system like the circuit presented in this paper can achieve the goal of identification of signature signals as a detection technique.

In this paper and inductive loop based sensing technique is suggested, primarily meant for sensing through non-contact means. The signal response is conditioned through using a simple circuit ultimately producing very promising results, good linearity and sensitivity over a reasonable sensing range of interest.

---

[1] sofilarbani@gmail.com

II. Theoretical Details and Schematics:

The basic sensing principle used in this work is as shown in Figure 1. The inductive changes on the loop inductance value (L1) are directly represented by variation in the output voltage. The input applied is a periodic source signal of 5Volts amplitude with a frequency of 65 kHz. The capacitors used are all 68μF. The output signal is measured across capacitor, C_3.

The variation of the inductance value is represented here as a variation of the variable inductor L_1, the other components' values remain unchanged, as well as the input frequency and amplitude. Figure 1 is the schematic diagram of circuit used for this sensing technique.

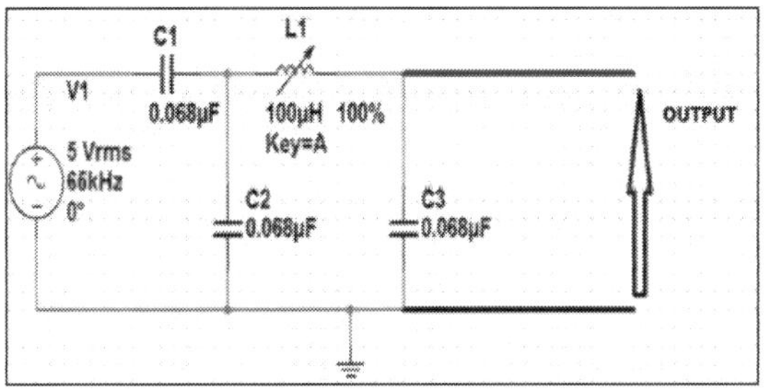

Figure 1: Circuit schematic

The variable inductor is representing a loop, multiple loop's shape can be used at this point rectangular (widely used for the traffic detection), and in our case it is preferable to use the modified loop as illustrated in Figure 2, instead of the conventional rectangular shape.

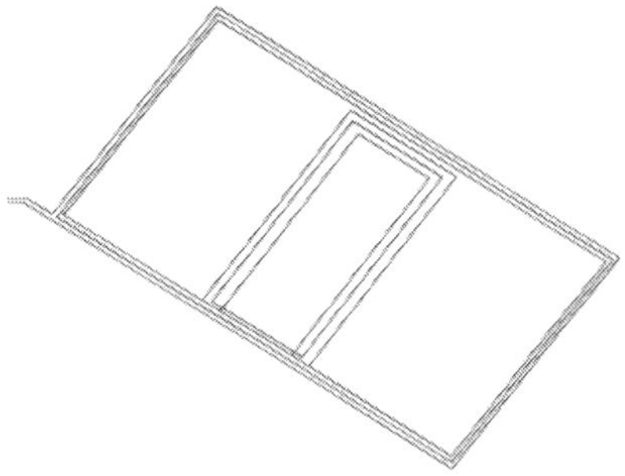

Figure 2: The modified inductive loop [3]

This loop has been proven to be more sensitive [3] as it allows detecting very small variation of inductance values as a result of bodies approaching or receding in front leading to make the loop assuming different relative permeability values.

IOP Conf. Series: Materials Science and Engineering **53** (2013) 012033 doi:10.1088/1757-899X/53/1/012033

III. Analytical Model

The circuit schematic shown in Figure 3 is used for making deriving of the transfer function of the circuit.

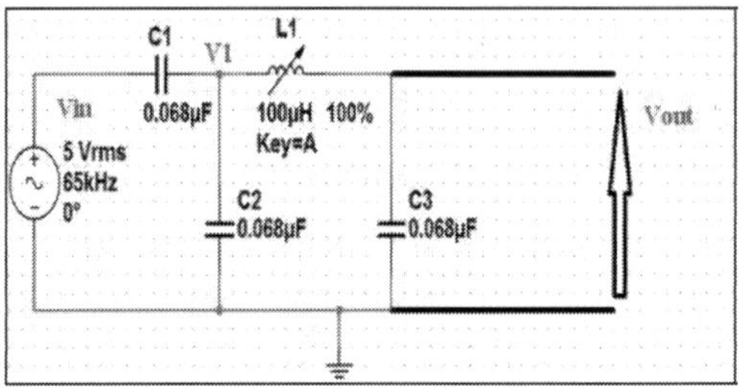

Figure 3: Circuit Schematic used for derivations.

The transfer function H(s) is calculated as follow:

$$V1 - \frac{[(ZL1 + Zc3) \backslash\backslash Zc2]}{[(ZL1 + Zc3) \backslash\backslash Zc2] + Zc1} \text{ x Vin} \tag{1}$$

$$H(S) = \frac{1}{1 + S^2 L1c3} \text{x} \frac{1}{2 + \frac{1}{(1 + S^2 L1C3)}} \tag{2}$$

By substituting the L1=100µH and C3=0.068µH, equation (2) becomes:

$$H(S) = \frac{1}{1 + (S^2 6.8x10^{-12})} \text{x} \frac{1}{2 + \frac{1}{1 + (S^2 6.8x10^{-12})}} \tag{3}$$

$$H(S) = \frac{1}{1 + (S^2 6.8x10^{-12})} \text{x} \frac{1 + (S^2 6.8x10^{-12})}{3 + [2x(S^2 6.8x10^{-12})]} \tag{4}$$

Simplifying equation (4), H(s) becomes:

$$H(S) = \frac{1}{3 + [2x(S^2 6.8x10^{-12})]} \tag{5}$$

Replacing L_1 by inductance of 100µH gives us transfer function as given in (6)

$$H(S) = \frac{1}{3 + (13.6x10^{-12} x S^2)} \tag{6}$$

IV. Simulation Results and Discussion

Using Multisim, the simulation of the circuit under consideration is carried out while Matlab is used to plot the analytical derivations including the transfer function.

A- Multisim based plots and analysis:

The simulation results shown in Figure 4 till Figure 6 are for selected values of loop inductance.

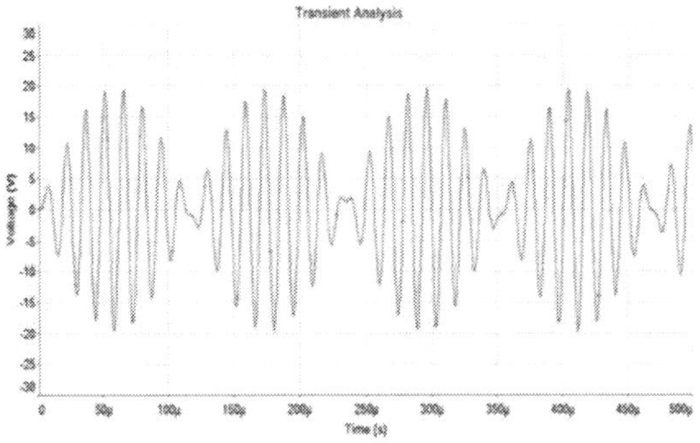

Figure 4: Multisim plot of the output signal at 100μH

Figure 4 shows the output signal when the inductance value of the loop is equal to 100μH. In this figure, it can be seen that the shape is quite periodic and have maximum voltage amplitude of 17.99 V with a periodicity of the signal at about 140μs.

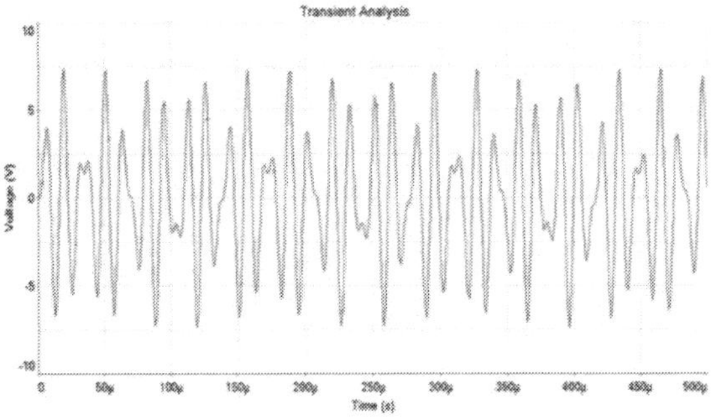

Figure 5: Multisim plot of the output signal at L=60μH

Here again the periodicity of the signal is very clear and the signal make a cycle in every 140μs, it is the same as in Figure 4 with an inductive value of the loop at 100μH. The maximum value of the amplitude of the voltage is equal to 6.99 V, showing a drop from 17.99 V of the output signal amplitude at the inductance value of 100μH.

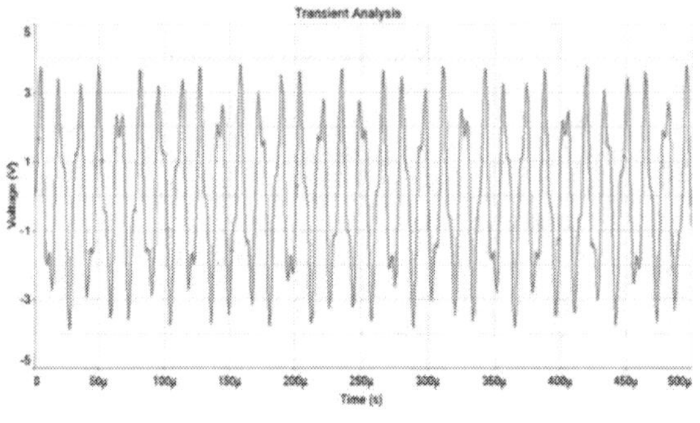

Figure 6: Multisim plot of the output signal at L=20μH

Figure 6 shows that the shape of the output signal has changed, as well as the output voltage value that has dropped again as the inductance value is reduced. The maximum voltage value of the output is equal to 3.85V in this output result where the inductance value is equal to 20μH. The periodicity cycle time is equal to 140μs as the two previous figures (Figure 4 and 5).

The periodicity and the repeatability of the shape of the signal in the three figures (Figure 4,5 and 6) can be used as shape signature of the inductive change that happens at the inductive loop. It can easily see that the voltage value of the output is dropping as the inductance value decreases; the linearity of this decrease is linear at certain ranges of the inductive change.

B- Bode Plots of the phase and frequency under both Multisim and Matlab Softwares

In the following figures Bode magnitude and phase plots are obtained using Multisim simulation software:

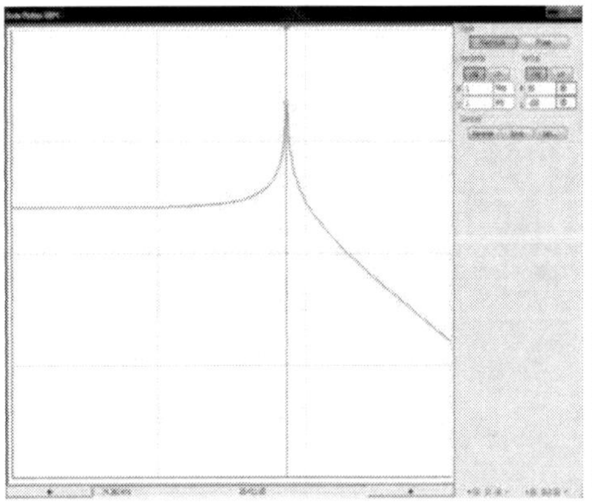

Figure 7: (a) Bode plot Multisim at L=100μH

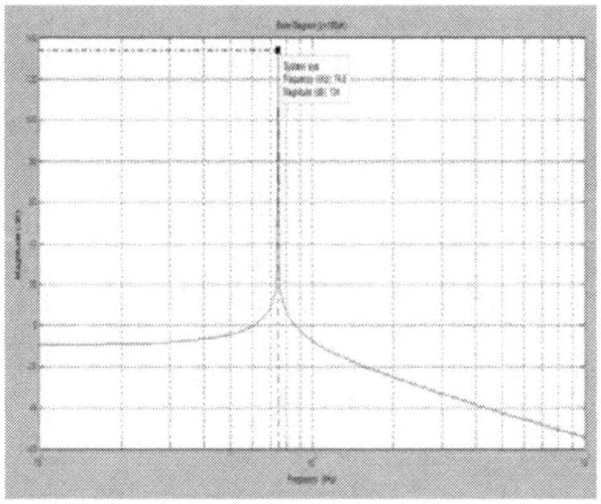

Figure 7: (b) Bode plot using Matlab at L=100 μH

The plots shown in Figure 7: (a) and (b) represent Bode magnitude and phase plots of the circuit under Multisim simulation software, the obtained values are 74.36 kHz for the resonance frequency value and 25.41 dB of magnitude value for the considered inductance value of 100μH while the phase is changing from 0 -180 ° at the resonance frequency.

To verify the reliability of the transfer function that has been derived in the analytical model section, and also for more accuracy of the results, Matlab based plots are being presented in the next figure.

The following figures are Matlab based plots of the transfer function of the system calculated in the analytical model section at 100μH of inductance value.

$$\mathbf{H(S)} = \frac{1}{1+(1.36\times10^{-12}\times S^2)} \quad \textbf{(7)}$$

The resonance frequency value that has been obtained from the figure is equal to 74.8 kHz, which is almost the same as the simulation run under Multisim that gives a value of 74.36; the difference is due to the non-ideality of the components simulated as well as the known accuracy of the mathematical tool (Matlab) for calculation.

The phase shifting from 0 to -180 degrees remain the same at the resonance frequency as it was under Multisim simulation.

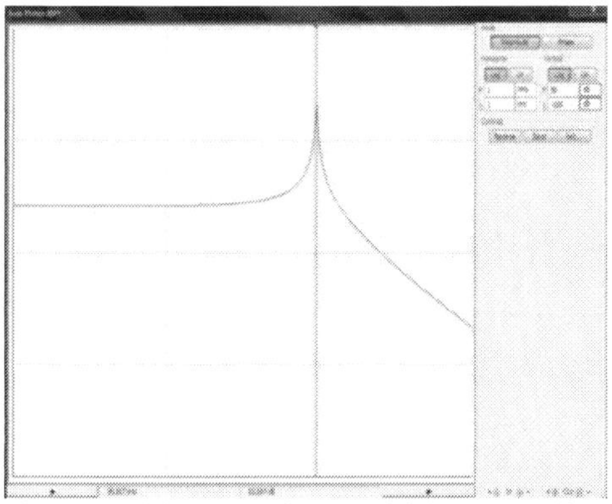

Figure 8: (a) Bode plot (Multisim at L=60μH)

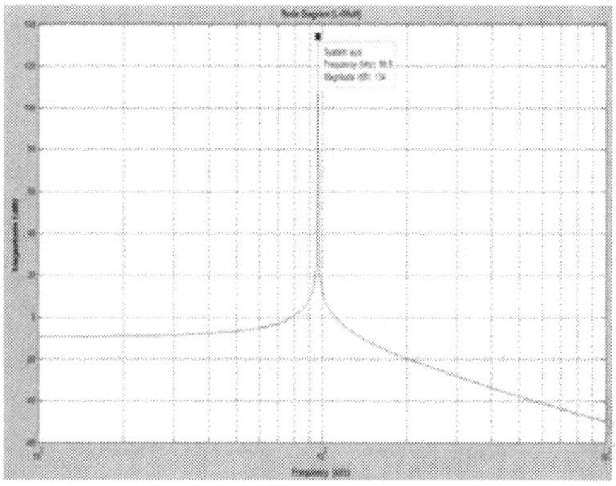

Figure 8: (b) Bode Magnitude plot (Matlab at L=60μH)

Here again the match of the Multisim simulation is showing almost the same result (95.80 kHz), and under Matlab plot of the equivalent circuit mathematically calculated and then plotted the value of the resonance frequency is equal to (96.5 kHz). This small difference is again due to the accuracy of Matlab and the non-ideality of the component used when under Multisim Software.

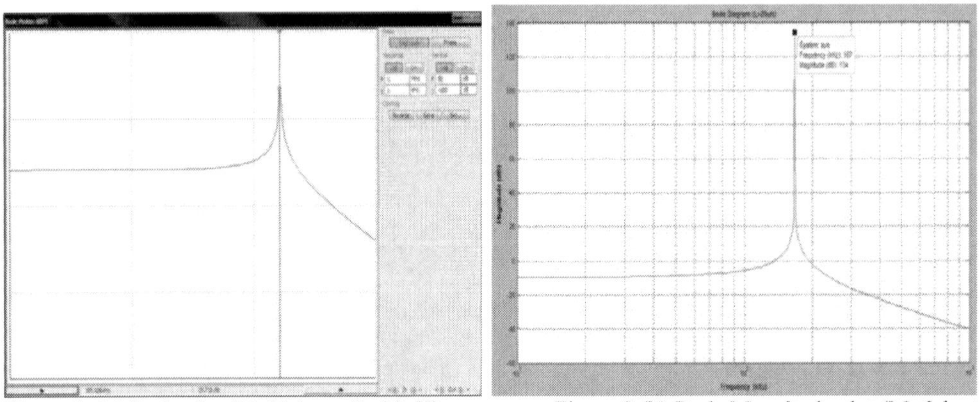

Figure 9: (a) Bode plot (Multisim at L=20μH) Figure 9 (b) Bode Magnitude plot (Matlab at L=20μH)

Accurate Matlab Plots that match the Multisim is none than a great validation of the simulation based results obtained; the analytical derived model is then tangible and can be used for the detection and the shape recognition.

Table 1: Parameters of the output signal Vs Inductive Change

Inductance (μH)	**20**	**40**	**60**	**80**	**100**	**120**	**140**	**160**
Voltage (V)	**3.85**	**4.88**	**6.96**	**10.49**	**17.99**	**49.82**	**80.08**	**23.20**
Resonance Frequency (kHz)	**166.29**	**117.85**	**95.69**	**83.29**	**74.02**	**67.81**	**62.97**	**59.03**
Bode Magnitude (dB)	**26.80**	**27.73**	**23.98**	**30.21**	**25.37**	**24.70**	**40.39**	**31.79**

Table 1 represents the values obtained when simulating the circuit. These very same values are used to plot the different parameters versus the inductive change using Matlab.

C- Linearity of the output parameters:

All the following plots will be plotted based on the simulation's results obtained when sweeping the inductance value from 20μ to 160μH, and are illustrated in the following table:

From the table 1, the voltage variation for 1μH is not less than 300mV in the following values interval of the inductance which is 80μH-160μH.

We begin with the study of the frequency linearity; Figure 10 shows resonance frequency versus the inductive change of the loop.

The linearity of the resonance frequency versus the inductive change will be discussed; as a matter of fact this system is quite linear at certain values of the inductance.

It is clear that the interval 60μH-160μH, this figure has been plotted using Matlab Software; a graphic representation of a polynomial of the first degree that we got by approximating the 6 last points of the data (60μH, 80μH… 160μH).

This polynomial of the first degree order is given as:

$$\mathbf{F = f(L) = -0.3578L + 113.1612} \quad (8)$$

This equation gives the resonance frequency of the system in kHz for μH change of the inductance value of the loop. The following plot represents the Bode Magnitude versus the inductive change.

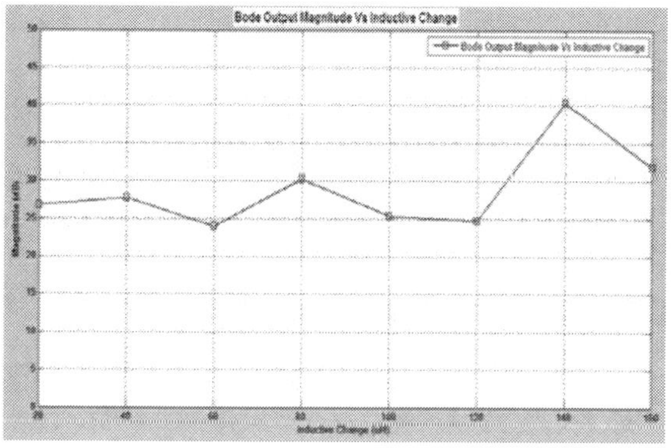

Figure10. Bode plot Magnitude versus Inductive Change

Figure 10 represents the variation of the Bode magnitude by respect to the variation of the inductance value of the loop. In fact, the variation cannot be linear because of two factors: the circuit reaction and the accuracy of the software used. The plot shows a non-linear behavior in the overall interval of 20μH-160μH.

The output bode magnitude cannot be used because of the non-linearity of it, however the resonance frequency and voltage consist very good parameters for the detection in the inductive variation range.

V. Conclusion

In this paper an inductive loop for sensing position as signature signals has been presented. The simplicity of the circuit makes it very easy to be implemented. Theoretical analysis is carried out for deriving the transfer function of the circuit and the required parameters come to validate the simulation results. The linearity of the resonance frequency and the shape formed make the key of the recognition of the signature signals with respect to the inductive change of the loop. The system is linear with the range of 20μH-160μH. Furthermore, the system is very sensitive as it can permit the detection of less than 1μH as presented in the result and discussion section. The sensitivity of the circuit is established for a change of 1μH results leading to produce 300mV in the output voltage. The future work will be writing programs and algorithm using PC microcontrollers-based platforms for recognition of these signature signals according to the shape and positions of the target objects and bodies.

Acknowledgment

The authors would like to thank the Ministry of Higher Education Malaysia for funding this research under the Research Acculturation Grant Scheme (RAGS) RAGS12-014-0014.

IOP Conf. Series: Materials Science and Engineering **53** (2013) 012033 doi:10.1088/1757-899X/53/1/012033

Appendix A

We know that:

$$Vout = \frac{Zc3}{Zc3 + ZL1} x\, V1 \qquad (9)$$

We replace V1 By its value then we'll have

$$Vout = \frac{Zc3}{Zc3 + ZL1} x \frac{[(ZL1 + Zc3)\setminus\setminus Zc2]}{[(ZL1 + Zc3)\setminus\setminus Zc2] + Zc1} xVin \qquad (10)$$

We know:

$$H(t) = \frac{Vout}{Vin} \qquad (11)$$

Then:

$$H(t) = \frac{Zc3}{Zc3 + ZL1} x \frac{[(ZL1 + Zc3)\setminus\setminus Zc2]}{[(ZL1 + Zc3)\setminus\setminus Zc2] + Zc1} \qquad (12)$$

$$H(t) = \frac{1}{1 + \frac{ZL1}{Zc3}} \quad x \quad \frac{1}{1 + \frac{Zc1}{[(ZL1 + Zc3)\setminus\setminus Zc2]}} \qquad (13)$$

Let's calculate the term:

$\frac{Zc1}{[(ZL1+Zc3)\setminus\setminus Zc2]}$, First we calculate $[(ZL1 + Zc3)\setminus\setminus Zc2]$

$$[(ZL1 + Zc3)\setminus\setminus Zc2] = \frac{(ZL1 + Zc3)\,x\,Zc2}{ZL1 + Zc3 + Zc2} \qquad (14)$$

Then we have the inverse

$\frac{ZL1 + Zc3 + Zc2}{(ZL1 + Zc3)\,x\,Zc2}$ That we are going to replace in the H (t) equation finally we get finally:

$$H(t) = \frac{1}{1 + \frac{ZL1}{Zc3}} \quad x \quad \frac{1}{1 + \frac{Zc1x(ZL1 + Zc3 + Zc2)}{(ZL1 + Zc3)\,x\,Zc2}} \qquad (15)$$

$$H(t) = \frac{1}{1 + \frac{ZL1}{Zc3}} \quad x \quad \frac{1}{1 + [\frac{Zc1}{Zc2} x \frac{(ZL1 + Zc3 + Zc2)}{(ZL1 + Zc3)}]} \qquad (16)$$

Here is more simplified form of H (t):

$$H(t) = \frac{1}{1 + \frac{ZL1}{Zc3}} \quad x \quad \frac{1}{1 + [\frac{Zc1}{Zc2} x\, (1 + \frac{Zc2}{(ZL1 + Zc3)})]} \qquad (17)$$

By replacing ZL1 by SL1, Zc1 by 1/ (Sc1), Zc2 by 1/ (Sc2), and Zc3 by 1/ (Sc3) we get H(S):

IOP Conf. Series: Materials Science and Engineering 53 (2013) 012033 doi:10.1088/1757-899X/53/1/012033

$$H(S) - \frac{1}{1+\ S^2L1c3} \times \frac{1}{1+[\frac{c2}{c1}x\,(1+\frac{\frac{1}{Sc2}}{\left(SL1+\frac{1}{Sc3}\right)})]} \tag{18}$$

$$H(S) = \frac{1}{1+\ S^2L1c3} \times \frac{1}{1+\left[\frac{c2}{c1}x\left(1+\frac{\frac{1}{Sc2}}{\frac{1+S^2L1C3}{Sc3}}\right)\right]} \tag{19}$$

$$H(S) = \frac{1}{1+S^2L1c3} x \frac{1}{1+\left[\frac{c2}{c1}x\left(1+\frac{c2}{c3x(1+S^2L1C3)}\right)\right]} \tag{20}$$

Hence

$$H(S) = \frac{1}{1+S^2L1c3} x \frac{1}{1+\left[\frac{c2}{c1}x\left(1+\left[\frac{c2}{c3}x\frac{1}{(1+S^2L1C3)}\right]\right)\right]} \tag{21}$$

The final form of the transfer function:

$$H(S) = \frac{\frac{1}{L1c3}}{\frac{1}{L1c3}+S^2} x \frac{1}{1+\left[\frac{c2}{c1}x\left(1+\left[\frac{c2}{c3}x\frac{\frac{1}{L1c3}}{\left(\frac{1}{L1c3}+S^2\right)}\right]\right)\right]} \tag{22}$$

From last time we had the following transfer function:

$$H(S) = \frac{\frac{1}{L1c3}}{\frac{1}{L1c3}+S^2} x \frac{1}{1+\left[\frac{c2}{c1}x\left(1+\left[\frac{c2}{c3}x\frac{\frac{1}{L1c3}}{\left(\frac{1}{L1c3}+S^2\right)}\right]\right)\right]} \tag{23}$$

But for easier calculation the use of the transfer function that we obtained before the last simplification which is like follow:

$$H(S) = \frac{1}{1+S^2L1c3} x \frac{1}{1+\left[\frac{c2}{c1}x\left(1+\left[\frac{c2}{c3}x\frac{1}{(1+S^2L1C3)}\right]\right)\right]} \tag{24}$$

As we know that in our circuit c1=c2=c3=0.068μF and that L1 is a variable because it is susceptible to change of inductance value.

A lot of terms will be reduced to 1.

$$H(S) = \frac{1}{1 + S^2 L1c3} x \frac{1}{1 + \left[1x\left(1 + \left[1x\frac{1}{(1 + S^2 L1C3)}\right]\right)\right]} \tag{25}$$

References

[1] Janusz Gajda, Ryszard Sroka, Marek Stencel, " A Vehicle Classification Based On Inductive Loop Detectors", IEEE TRANSACTIONS ON INSTRUMENTATION AND MEASUREMENT, MAY 21-23 2001

[2] Boby George, Hubert Zangle, Thomas Betterklieber, George Brasseur, "A Combined Inductive-Capacitive Proximity Sensor For Seat Occupancy Detection", IEEE TRANSACTIONS ON INSTRUMENTATION AND MEASUREMENT, VOL. 59, NO. 5, MAY 2010

[3] S. Sheik Mohammed Ali, Boby George, Lelitha Vanajakshi, and Jayashankar Venkatraman, "A Multiple Inductive Loop Vehicle Detection System for Heterogeneous and Lane-Less Traffic", IEEE TRANSACTIONS ON INSTRUMENTATION AND MEASUREMENT, VOL. 61, NO. 5, MAY 2012

[4] S. Sheik Mohammed Ali, Boby George and Lelitha Vanajakshi, "A Magnetically Coupled Inductive Loop Sensing System for Less-lane Disciplined Traffic", IEEE TRANSACTIONS ON INSTRUMENTATION AND MEASUREMENT, VOL. 61, NO. 5, MAY 2012

[5] Michael P. Perkins, Member, IEEE, Mike M. Ong, Ron D. Speer, Member, IEEE, and Charles G. Brown, Jr., Member, IEEE , "Analysis of a Small Loop Antenna With Inductive Coupling to Nearby Loops," IEEE TRANSACTIONS ON ELECTROMAGNETIC COMPATIBILITY, VOL. 53, NO. 4, NOVEMBER 2011

[6] Satish Chandra Bera, Nirupama Mandal, and Rajan Sarkar, "Study of a Pressure Transmitter Using an Improved Inductance Bridge Network and Bourdon Tube as Transducer", IEEE TRANSACTIONS ON INSTRUMENTATION AND MEASUREMENT, VOL. 60, NO. 4, APRIL 2011

Experimental evaluation of agricultural biomass flow sensing behaviour using capacitive technique

Khalid. A S Al Khateeb[1], RumanaTasnimAnika, Sheroz Khan, MusseMohamud, AtikaArshad, Khalid Hasan, Syed SamnamHaider, Mohammed M. Shobaki
Department of Electrical and Computer Engineering
International Islamic University
53100 Kuala Lumpur, Malaysia

Email: rumana_anika@yahoo.com

Abstract. To enhance industry control quality level as well as uphold enterprise economic benefit precise sensing and measurement of biomass flow is a major concern among researchers worldwide. Keeping in mind the shortcomings of existing sensing technologies this paper has developed a capacitive sensing method by making use of aop amp based bridge circuit along with particularly designed sensing electrodes. The objective of this work is fulfilled via experimental validation through a prototype hardware implementation of a flow sensing set up. The experimental results have specified the measurement system which is able to sense flow variation as a change of dielectric permittivity of different biomass materials under room condition. Moreover, the obtained results have revealed distinctive features clearly signifying the shapes and physical characteristics of electrodes, locations of the mounted electrodes on test pipe wall, dielectric permittivity and characteristics of test biomass materials.

1. Introduction

Processes relating biomass materials are of emergent concern, although handling and conveying biomass particles are quite challenging and complex because of its unusual physical properties. Over the recent past years, researchers have been researching and reporting works on using various techniques for biomass or gas-solid flow measurement. The wide-spreading applications of agricultural biomass (e.g., grains, flour, and corn) are found not only in the food processing industries but also in chemical, pharmaceutical industries and power plants. A comprehensive overview has been conducted by some researchers on agricultural biomass particles flow measurement techniques [1]. When biomass particles are subjected to multiphase flow, it faces distinctive challenges since these are particularly anisotropic. Also atypical features like large mean particle sizes, extreme shape including flakes, chips, fibres, wide size distributions, general heterogeneity of biomass particles make the sensing and mass flow measurement challenging. Among all the sensing techniques, capacitive technique has proven to be much more reliable than the other techniques in term of sensitivity and cost namely electrostatic, optical, ultrasonic, microwave etc. In capacitive technique, the measured capacitance can be used to visualize the phases' concentration ratios [2]. In [3] the researchers developed a computer processing system including a measurement model unit for measuring agricultural powder flow rate based on the

[1] RumanaTasnim, Email: rumana_anika@yahoo.com
Department of Electrical and Computer Engineering International Islamic University

measurement values of the total pressure drop of gas-solid two-phase flow and the concentration of powder. Differential pressure transducer, a capacitance concentration sensor, a pressure transducer, and a temperature transducer were used in this technique. Some researchers developed capacitive sensing techniques implying novel measurement circuits with innovative electrode shapes. For past few years this technique has been being used vastly in industrial sectors in order to monitor the flow of dry agricultural powders [2].

Moreover, a number of capacitance measurement techniques were developed earlier and further implemented in Integrated Circuit (IC) which have been reported [5-6] to make the conversion of capacitance-to-digital (CDC). An innovative capacitive sensor [7] with ring and concave type electrode has been proposed for void fraction measurement and flow pattern identification. In [8] the authors developed a multi-electrode sensing structure (11 ring shaped) for enhancing the sensitivity to dielectric permittivity variations. In [9] the authors conducted a brief overview on the developed capacitive measurement techniques over recent years for the application of flow measurement in process instrumentation. Also a capacitance based sensor to measure industrial process parameters and a capacitance-to-digital conversion system were developed in this paper. Such technologically advanced multi-purpose capacitance-to-digital conversion hardware can work under several carrier frequencies. In order to validate the real-time utility of flow sensor, different electrode geometries with the application of potential electrode excitation modes are implied by some authors in [10]. Here, the sensor electronic is comprised of a capacitance to digital converter Application Specific Integrated Circuit (ASIC) and a Digital Signal Processor (DSP) running on C. Linux Operating System. A good level of sensitivity is not maintained due to material changes inside the measuring volume that contributes to measurement inaccuracy. Most of the existing technologies made use of complex measurement interface circuit model as well as electrode arrangement used for flow piping arrangement where the effect of stray capacitance is a key issue. On top of this, it is more challenging while focusing at the types of electrodes, unusual particle size of biomass, and the accuracy and simplicity of the measurement electronics circuit.

In this paper, the developed sensing system can detect very slight capacitance variations through reducing stray and residual capacitance effect.Two agricultural biomass materials or grains namely: fodder (cattle food) and flour are used as biomass test materials under similar flow and measurement condition in normal room temperature.This paper is organized as follows: Section 2 introduces the methodological foundation of the sensing system architecture along with measurement interface circuit. Experimental approach along with results is presented in Section 3. Section 4 details discussion and finally the paper is concluded in section 5.

2. Sensing System and Measurement Interface

The Capacitive sensing technique for biomass flow works on the basic principle of particulate flow within a capacitive sensing volume through causing changes in the dielectric constant of gas/solids mixtures. Flow sensing system comprises of a custom developed horizontally and vertically connected pipe on which capacitive sensing electrodes are fitted. Five pairs of sensing electrodes are hereby connected on the pipe section with a measurement interface circuit which is based on an adjusted bridge network. The block diagram of sensing system is presented below in Fig 1.

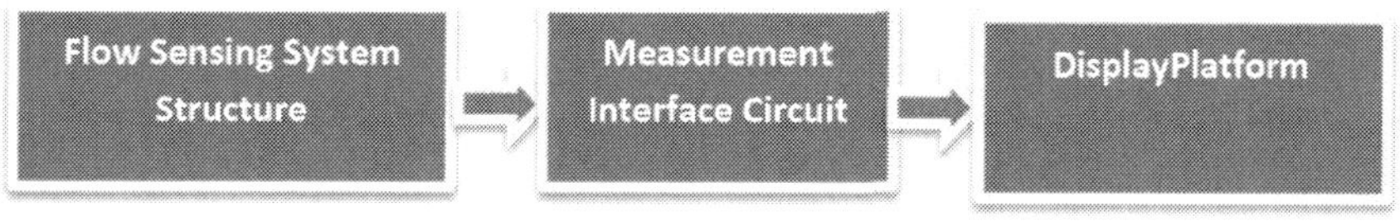

Fig. 1Block Diagram of sensing system

The signals obtained at the end of the measurement interface circuit become a source of information for flow sensing of agricultural biomass materials. Figure 2 illustrates the measurement interface circuit.

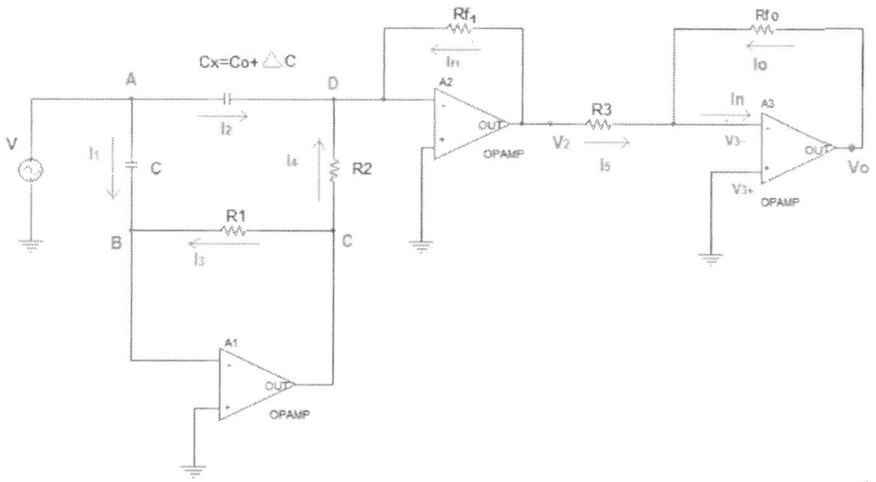

Fig. 2 Measurement Interface Circuit

The measurement interface circuit is built with operational amplifiers. The lower bridge arms' pair is built of resistors where point B remains at virtual ground for better stability of flow sensing by the variable capacitance arm Cx. It diminishes the stray capacitance effect in between the output leads of bridge network. The bridge output will be in voltage form being linearly balanced with capacitance change for different biomass concentration during flow phenomenon. The output voltage derived from measurement interface circuit is,

$$\dot{V}_o = \frac{-j\omega v R_{fo} R_{f1}}{R_3 R_2}(C_x - C_0) \quad (1)$$

The review indicated that the identified criteria were not being met with existing products and highlighted the techniques of sensing and transmission of information as major areas requiring attention in the development of a system.

3. Experimental Detail and Result Analysis

The sensing set up uses an industrial vacuum pump for sucking the fodder and flour materials independently by means of a horizontal and vertically linked polyvinyl chloride 1000 mm pipe with an inside diameter of 25 mm and outside diameter of 27 mm. The following Figure shows the images of test materials used to conduct the experiments.In the test section, five capacitance sensors for the measurement of capacitance are located keeping certain distances through a 1000 mm horizontally and vertically connected pipe. Two pairs of electrodes are mounted which are constructed by mild steel and circular electrode and another three pair are made by semicircular plates. Moreover, the ratio of electrode width W to electrode separation d is made 4:1, the total number of electrodes are K = 10, the electrodes' width are W =12 mm, electrode separation d= 3 mm (As in Figure 2)

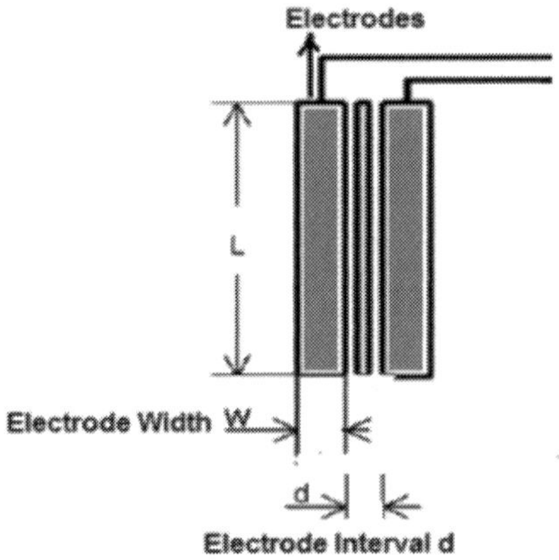

Fig.3Electrode structure

The schematic of custom developed electrode shapes are given in the following Figure.

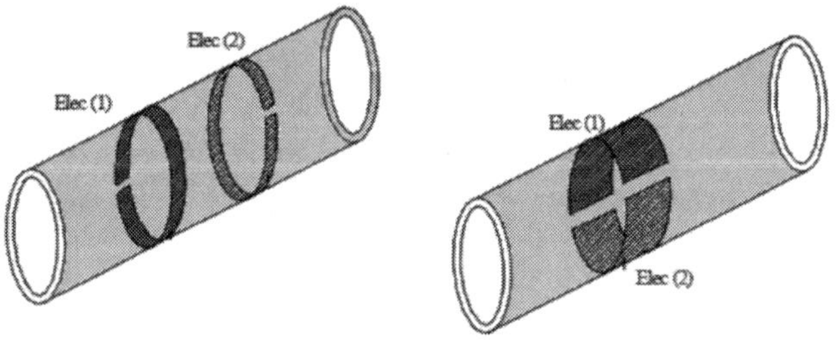

(a) (b)
Fig. 4 Schematic of electrode shapes (a) Circular (b) Semi-circular (Wael et al, 2008)

The calibration is made for bridge circuit to balance the bridge for no flow. During unbalanced condition of bridge biomass flow gets started in flow piping system which is depicted in Figure 4. The flow piping system involves a stunner at the extreme end for making the materials flow through the pipe whereas an industrial suction motor for sucking the materials which get stored in collector bag at the other end.

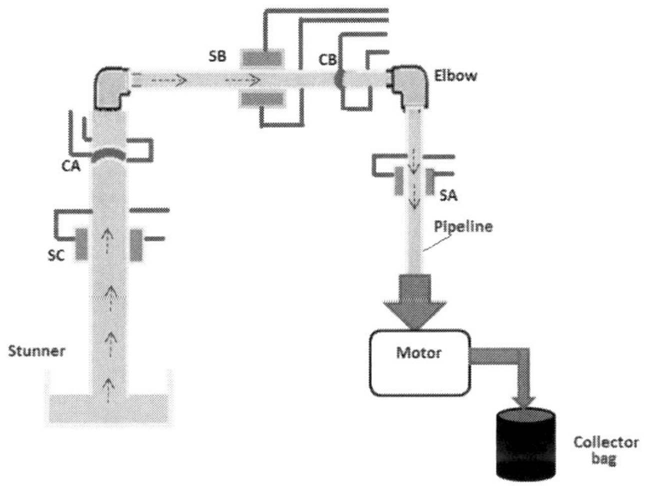

Fig. 5 Flow Sensing Arrangement

Firstly fodder and flour biomass materials are flown at 1000g/min and at 600g/min respectively. The measurements of capacitance changes (in Pico-farad) and voltage outputs (in Volts) are kept in record in every 10 seconds of flow and shown in the following bar graphs. Figure 6 a-e) shows the voltage outputs and capacitance bar graphs. The semicircular electrode SC (mounted close to the container and on vertically upward pipe section) achieved the maximum output voltage with maximum capacitance due to location reason. The SB electrode (horizontally placed) demonstrates the minimum capacitances (65.6-68pF) throughout the whole flow period. Semicircular SA (close to the suction pump and on the vertically downward pipe) illustrates fluctuated and varying voltages while the capacitance sensed showed no obvious fluctuation. The similarity in such pattern shows the effect of identical location on test pipe wall.

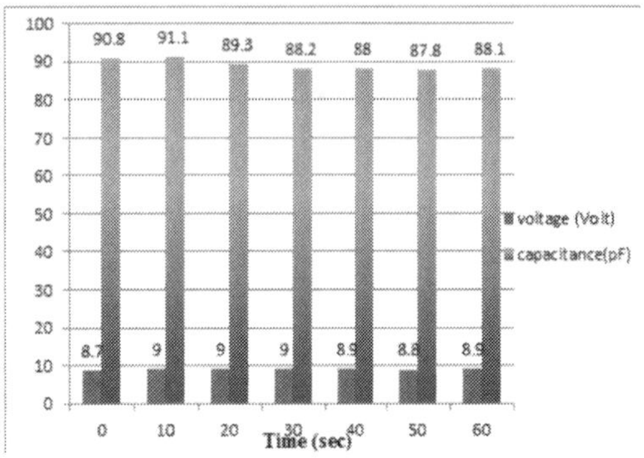

(a)

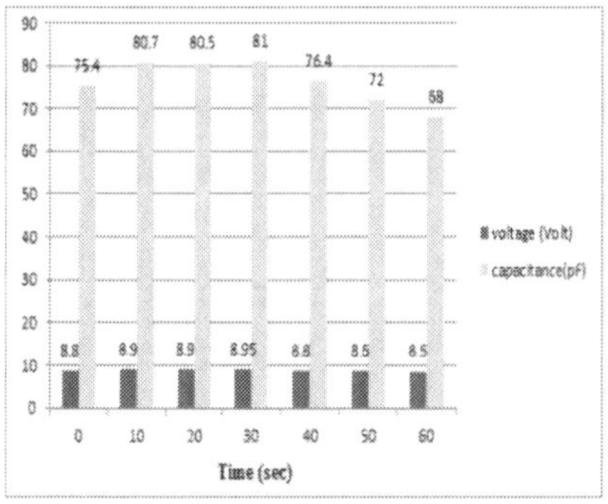

(b)

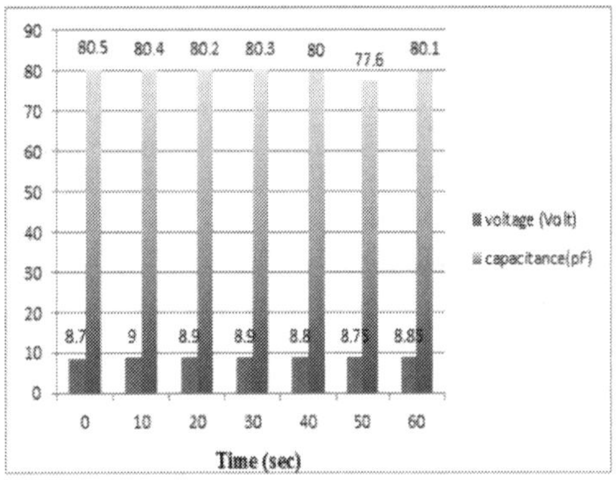

(c)

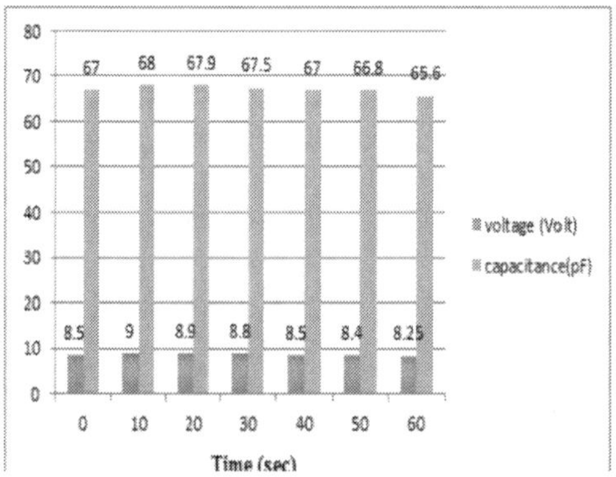

(d)

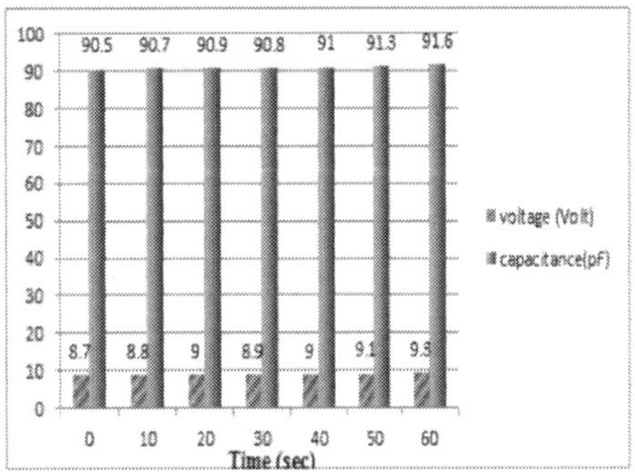

(e)

Fig. 6Bar graph of fodder flow (1000gm/min) for (a) circular CA (b) circular CB (c) semicircular SA (d) semicircular SB (e) semicircular SC

Even though possessing dissimilar physical properties of electrodes, both SB and CB have shown matching behavior in the measured voltages. As a continuation, the other biomass material flour is flown and once again the bar graphs in Figure 7(a)-(e) have pointed up proportional correspondence between the capacitance sensed and output voltage obtained for different electrodes confirming the proper functionality of measurement circuit.

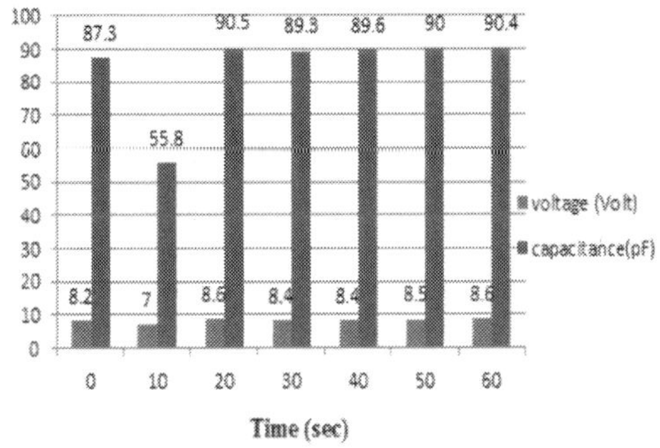

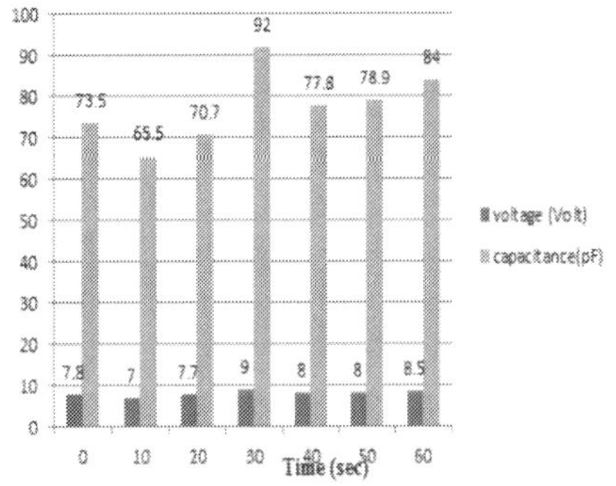

(a) (b)

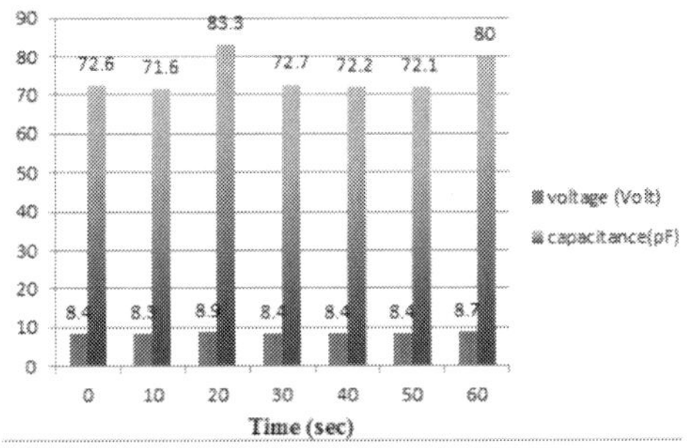

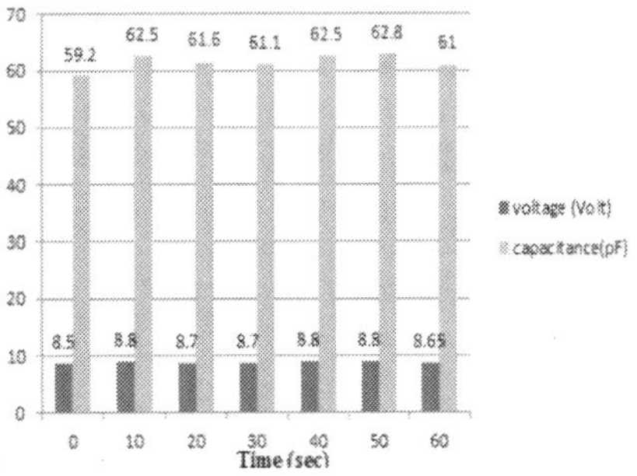

(c) (d)

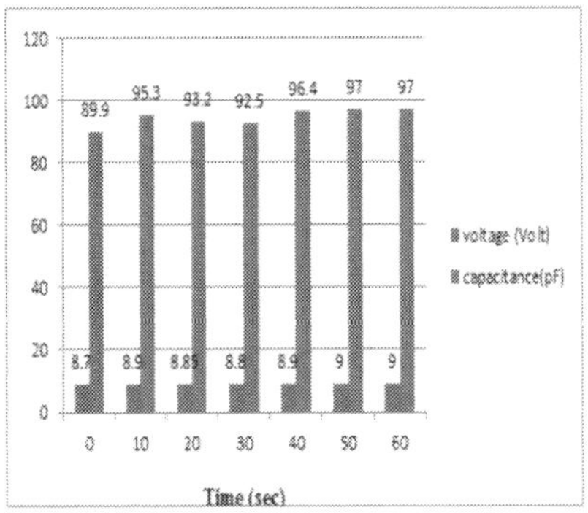

(e)

Fig. 7: Bar graph of flour flow (600gm/min) for (a) circular CA (b) circular CB (c) semicircular SA (d) semicircular SB (e) semicircular SC

The bar graphs proved that the voltage and capacitance graphs showed consistent results when looking at the electrodes' location (same as previous sensing behaviour), types and physical properties. Like fodder the dissimilarity of flour materials has been sensed well by circular electrodes when semicircular ones showed steady response mostly. Because of having bigger diameter (500μm) of fodder, it flows faster than the fine particle Flour (shaped as 100μm). The dielectric permittivity has also proven its effect on the sensed results. Fodder materials having dielectric permittivity of 2.5-3.0 has shown its effect on the capacitance and voltage waveforms which are higher and smooth in comparison with these of flour. Due to the variation in the flow regime in the measurement pipe, the capacitance has been increased and decreased accordingly which has shown its effect on the voltage output. Eventually the final output has been recorded with different biomass flow rate in the flow pipe. In the present scheme, very small capacitance variation has been converted into voltage variation by using a simple low cost operational amplifier based capacitance bridge circuit. The recorded values sensed by the electrodes across the sensors are directly related to the air-biomass phase distribution, shape and physical characteristics of electrodes, locations of the mounted electrodes on test pipe wall and physical characteristic as well as dielectric permittivity of test biomass materials.

4. Discussion

Due to the atypical characteristics of biomass materials, industrial biomass processes have not shown expected outputs in terms of measurement interface circuit used in capacitive sensing techniques. Hence, the prospective of biomass as a means of highly regulated generation of renewable energy source is yet to be perceived fully. This paper has looked into a rare research issue to let the unrealized biomass resources recognized. The positioning of the electrodes has shown noteworthy effect on the measured data. The nearest electrodes from the biomass container have shown larger concentration (both in terms of capacitance and voltage sensed) than the other parts. Moreover as the galvanized iron semi-circular electrodes have better conductivity than that of mild steel circular ones, sensitivity have been shown to be relatively better for semicircular ones. On the whole, the multiphase flow sensing as well as flow sensing of biomass process instrumentation is addressed in this paper through realizing the sensing behaviour of two different agricultural biomass materials' flow; particularly concentrating at the types and design of electrodes, the irregular particle size and the efficiency of measurement electronics.

5. Conclusion

In this paper, the experimental work has been carried out in line with the aspect stated in the research problem, tuning the findings to achieving the research objectives. The developed system is able to detect very minor capacitance variations offsetting stray and residual capacitances' effect. This sensing system has been validated to be used in real time flow measurement by setting up a data processing unit in a Programmable Logic Controller (PLC)-based environment for measuring mass flow rate as well as concentration and velocity of biomass materials. The useful data extracted from real time flow measurement will allow the plant operators enhancing flow control of process instrumentation.

References

[1]. Heping Cui & John R. Grace (2006) Pneumatic conveying of biomass particles: a review, China Particuology, Vol. 4, Nos. 3-4, 183-188

[2]. Sen, S., Das,P. K.,Dutta, P. K.,Maity, B., Chaudhuri, S., Mandal, C.,&Roy,S. K.(2000).PC-based gas-solids two-phase mass flowmeter for pneumatically conveying systems. Flow measurement and Instrument, 11, pp. 205-212.

[3]. Zhiyao Huang, Baoliang Wang &Haiqing Li. (2002).*An intelligent measurement system for powder flow rate measurement in pneumatic conveying system.* IEEE Transactions on Instrumentation and Measurement, 51 (4), pp. 700-703.

[4]. Dyakowski, T., Edwards, R.B., Xie, C.G., & Williams, R.A.(1997). Application of capacitance Tomography to gas-solid flows. Chemical Engineering Science, 52, pp. 2099-2110.

[5]. Brasseur, G.,Brandstatter, B., &Zangl, H. (2003, August).*State of the Art of Robust Capacitive Sensors.* IEEE Transactions on Instrumentation and Measurement, 52 (4), pp. 1261-1265.

[6]. Bretterklieber, T., Zangl, H., Hrach,D., Holler,G.,Hammerschmidt, D., &Motz, M. (2008).Versatile Programmable Integrated Interface for Robust Capacitive Sensors.Elektrotechnik und Informationstechnik, 125 (4),pp.123-128.

[7]. Ahmed, W.H., & Ismail, B.I. (2008, November).*Innovative Techniques for Two-Phase Flow Measurements.* Recent Patents on Electrical Engineering.

[8]. Hrach, D., Zangl, H., Fuchs, A., &Bretterklieber, T. (2008). *A Versatile Prototyping System for Capacitive Sensing.* Sensors and Transducers Journal, 90, pp. 117-127.

[9]. Anton,F., Hubert zang, &Michael J.M.(2009).Thomas bretterklieber, Capacitive sensing in process instrumentation. Metrol. meas. syst., 17(4), pp. 557-568.

[10]. Holler, G., Fuchs, A., &Hrach, D. (2008, November). Flow velocity determination in cryogenic media by means of capacitive sensing. Sensing Technology ICST 2008, pp. 472–476.

IOP Conf. Series: Materials Science and Engineering **53** (2013) 012035 doi:10.1088/1757-899X/53/1/012035

A study of pneumatic conveying of gas-solid flow for industrial application

Khalid A. S. Al-Khateeb[1], **Rumana Tasnim, Sheroz Khan, Musse Mohammod, Atika Arshad, Mohammed M. Shobaki, Samnan Haider, Nazmus Saquib, Tawfilur Rahman**
Department of Electrical and Computer Engineering International Islamic University 53100 Kuala Lumpur, Malaysia

E-mail: rumana_anika@yahoo.com

Abstract. The complicated nature of gas-solids' physical properties have challenged the researchers over past decades who have led their efforts in developing its' flow sensing and measurement methods. The term 'gas-solid flow' signifies dilute- or dense-phase flow with a very little concentration of solids. For conducting such flow measurement, generally velocity profile and volumetric concentration of the flow particles being conveyed are needed to be measured. An important application of gas-solid flow has taken root in the form of biomass flow in pneumatic conveying systems, and its' online measurement has proven to be an exigent research pursuit. Additionally the other applications have been explored in power plants, food, chemical and automobiles industries as well. This paper aims at exploring the evolution of flow measurement methods along with a brief explanation on existing fundamental sensing techniques. Furthermore, the most recent patents developed for such measurements in pneumatic conveying pipelines are scrutinized along with their concomitant pros and cons.

1. Introduction

The advancement of suitable instrumentations for on-line monitoring and measurement in gas-solid two-phase flow has proven to be an exigent task for scientists and researchers at universities and institutes universally. Solids generally appear in the form of powder and grain transported in air which has extensive applications across food processing, agricultural particles (e.g., grains, wheat, corn) [1] flowing, chemical, pharmaceutical industries and also in transporting pulverized coal in fuel lines of thermal power stations. The development of suitable instrumentation for such application frequently poses exigent difficulties particularly in the light of inconsistency of plant operating conditions due to this multifaceted flow medium. The term 'gas-solid mass flow' indicates the amount of conveyed material mass per unit time through a conveying pipe. In order to conduct the flow measurement, generally velocity profile and volumetric concentration of the flow particles being conveyed are required to be measured. In view of that, the sensing techniques should be developed so as to perceive the flow nature of flow materials which have not yet well-researched. The developed technologies have not come within the reach of attaining an excellent level of measurement accuracy and precision due to the lack of research activities toward sensing techniques to perceive the flow behaviour.

[1] Rumana Tasnim, Email: rumana_anika@yahoo.com
Department of Electrical and Computer Engineering International Islamic University

However, the capacitive and electrostatic sensing technique are advantageous and favourable over other measurement techniques namely optical, ultrasonic, microwave techniques etc and these are currently being used numerously in industries due to its' special features like low production cost, simplicity in structure, non-contact, non-invasion, radiation free and maintenance free nature, high precision, good linearity, quick response, non-interference of the flow field and direct electric signal output. This paper has conducted a review on flow sensing techniques used for determining the key areas (e.g., concentration, velocity and other flow parameter) reported. This paper is organized as follows: Section II describes the current state of art and its' employed features. Section III has concentrated on the overall discussion. Finally Section IV presents the conclusion.

2. State of Conventional Arts on Contact Type Metering

In case of gas-solid two phase flow contact and noncontact type measurement techniques have been widely researched over decades. Those contact type gas-solid flow measurement technologies have been being used extensively; particularly in developing countries. Such techniques are namely venturimeter, orifice flow meter, impact plates, Coriolis etc. In [2] the researchers developed a differential-pressure concentration technique by means of uniting a capacitance meter with a venturimeter for measuring fine particles but did not give any experimental validation of this technique. Some other significant development of flow meters include impact plates, the coriolis principle [3] which cannot be mounted on pipelines reliably and straightforwardly for facilitating on-line measurement. Due to high revenue loss of such techniques during maintenance, inconvenient repeatability and non-user friendly features in the aforementioned techniques, a growing trend towards designing and developing innovative non-contact sensing technologies among the researchers have been observed.

3. Contemporary Work on Non-Contact Type Sensing

The noncontact type measurement of mass flow rate (concentration and velocity) of gas-solid flow is generally categorized into four major categories according to the sensing principles: electrical, attenuation, nuclear magnetic resonance and tomographic techniques. Based on these principles capacitive, electrostatic, ultrasonic, optical, radiometric, acoustic sensors have been developed. The electrical techniques namely electrostatic and capacitive methods [4] have been greatly acknowledged as the most reliable gas-solid flow measurement techniques.

3.1. Optical Sensing Technique

Optical sensors are able to sense measurement system of large and small scaled stacks. It has shown advantageous behaviour over some non-optical sensors to some extent in term of electrical passiveness and minimization of electromagnetic interference. The optical flow meters which are based on laser are capable of measuring the actual speed of particles. The working principle facilitates optical laser technology to offer a good level of flow data in the presence of environmental influence like high temperature, pressure and humidity and low flow rates, pipe vibration and acoustic noise. In [5] the researchers invented an optical system anchored in the detection of light mirrored by solid particles for measuring the solid concentration and velocity in dense gas–solid three-phase flow. The following figure illustrates an optical measurement system.

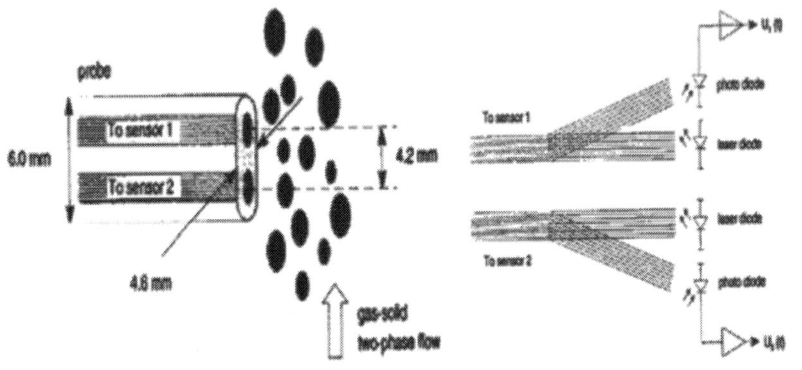

Fig.1 Optical measurement system

In case of the above drawn optical sensors, laser diodes emit light which is guided by the probe tip by half of the glass fibres of multi glass fibre. A part of the reflected light is guided by other half of the glass fibre to photo diodes which are able to convert optical signal into the electrical signal. The succeeding amplification of the photo diode signals send voltage signals as a function of time amendable in order to continue further processing with personal computer. Such optical sensors are usually insensitive and highly expensive while comparing with other traditional measurement technologies.

3.2. Microwave Technique

For last few decades researchers have been developing mass flow measurement scheme through using microwave radiation for pneumatic conveying applications. A number of configurations have been offered for measuring velocity using microwave Doppler technique. The mono-static and the bi-static principles are the most common ones. A transceiver to transmit and receive the microwave energy is used in mono-static technique whereas a separate transmitter and receiver aerial are used in bi-static technique. Furthermore, for facilitating measurements in pneumatic transport systems, an innovative microwave mass flow metering technique [6] has been developed which showed promising and precise measurement of particulate gas-solid flow.

3.3. Ultrasonic Technique

Ultrasonic Doppler flow sensor is capable of sensing gas and gas-solid particle flow. Ultrasonic flow sensors are used commercially while measuring gas and solid flow namely coal flow in biomass-coal fired power plant. Researchers and scientists have accomplished significant development of this technique worldwide and encountered some measurement difficulties namely high attenuation, high particle impingement noise, acoustic impedance mismatch [7]. With the increase in velocity, some more difficulties may show up to be addressed. The highly turbulent fields makes the measurement of received ultrasound signals complex as the sound is scattered by the small eddies in the turbulent flow field. Ultrasonic sensors work only in specific conditions (low particle number concentration). It introduces some measurement problems with sound distribution in the pipe material to measure specially the hydrocarbon gasses with wide range of turn down ratio and a very high or low flow considering the high accuracy.

3.4. Electrostatic Technique
For the measurement of pulverized coal flow in coal fired power plants mostly electrostatic sensors [8-13] are used. Basically the coal particles produce electrostatic charge on their surfaces which is a result of interactions among each other, the pipe wall and also the conveying air. The mass flow rate of the conveyed particles is usually computed through uniting the particle velocity and volumetric concentration. Cross-correlation technique determines particle velocity. Electrostatic sensing has been considered as an effective technique to determine the key parameters of gas–solid flow due to its less cost; uncomplicated structure, and robustness. It uses a metal electrode placed in the particle flow to detect the passing charges and infer flow parameters. Mostly researchers have put forward their effort toward developing circular shaped electrodes which are mounted around the pipeline. In [9] researchers aimed to focus on last 50 years' flow measurement methods' succeeding development and historical advancement review on electrostatic technique which presents numerical evaluation as well as theoretical basics of these techniques. In [11] researchers have evaluated the performance of both circular and rod electrostatic sensors. Usually three types of electrodes are commonly used namely pin, quarter ring and ring shapes. In [13], the authors used pin shape with variation of size for circular and rectangular shapes. Very recently some authors [14] have presented flow velocity measuring electrostatic sensing technique which is another type of capacitive sensing technique. The authors have developed an innovative multichannel instrumentation system with circular and arc-shaped electrostatic sensor arrays in order to measure the online continuous measurement of "mean" and "local" characteristics of blended biomass flow.

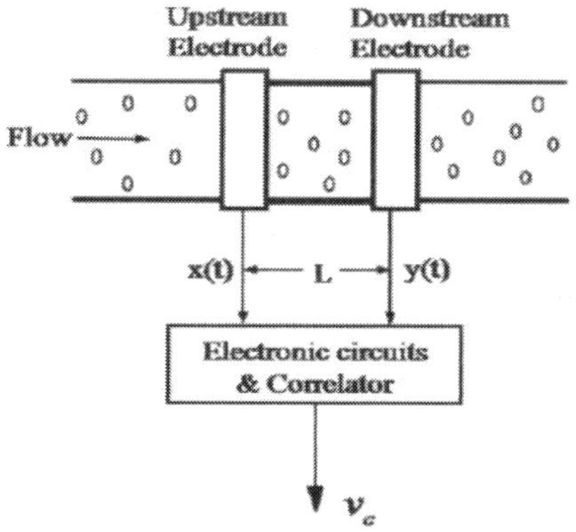

Fig.2 System structure of velocity measurement based on a pair of electrostatic sensors [14]

Several experimental tests were conducted on a pneumatic conveying test rig under various flow conditions. The sensing technique appeared to suffer from nonlinearity and have shown an adverse effect of relying on the physical properties of the particles for which ultimately the magnitude of charge signal got affected.

3.5. Capacitive Sensing Technique
Capacitive sensing technique is extensively used in the industry due to its advantageous features over other hazardous techniques in term of high sensitivity, reliability, simplicity and inexpensive nature. Capacitive sensors use electrodes that produce an electric field which gets excited to a voltage. The produced electric field thereby influences the electric charges on or the voltage of the electrodes, depending on the electrical properties (dielectric constant) of the surrounding materials. A measurement circuit is connected to an influenced electrode which can therefore provide information on the sensor surroundings. The signals obtained at the output of the sensing devices become a source

of information for flow-measurement of particulates. Thus the power plant operators will be notified about the flow parameters like particle velocity, concentration of biomass-coal or blended flow and accordingly take initiatives for detecting fuel supply problems and improving fuel delivery. Capacitive measurement technique for gas-solid flow is founded on the elementary principle of gas-solid flow within a capacitive sensing volume having changes according to the effective dielectric constant of gas/solids mixtures passing through. The measured capacitance is obtained for determining the phases' concentration ratios [15]. Researchers have invented quite a number of capacitive sensing techniques which involves novel measurement circuits with innovative electrode shapes. For past few years this technique is being used in industrial sectors in order to monitor the flow of dry powders [16-17]. In recent times, an instrumentation system based on capacitance sensing was invented for facilitating the measurement of concentration of pulverized fuel in a laboratory scale pneumatic conveying system with facilities for calibration provided [18]. The capacitance sensor has made the use of source grid sensing electrodes to measure the concentration of flow. The surrounding of a capacitive sensor has unfavourable influence on measurement. Some researchers have come forward with innovative electrode design [19-20] and tested those as a versatile hardware platform to facilitate the measurement in a real-life industrial sector. For converting capacitance-to-digital (CDC), a number of capacitance measurement methods were implemented in an Integrated Circuit (IC) [21-23]. Researchers analysed the effect of electrically conductive contamination, moisture and dew which can stay in contact with the sensor [21]. Another promising capacitive sensing technique on the basis of stainless steel source grid sensing electrodes to increase high and homogeneous capacitance sensitivity over the pipe cross section was developed by some researchers for measuring pulverized fuel concentration in a laboratory scale pneumatic conveying system along with the calibration facility [18]. Also a novel interface circuit for the capacitive sensor has been used in this paper to eradicate baseline drift and make the system immune to stray capacitance. The proposed capacitance sensors are connected to a capacitance measuring circuit through a shielded twisted-pair cable. This circuit uses differential capacitance configuration using an active sensor and the dummy sensor which is not susceptible to capacitance changes because of temperature.

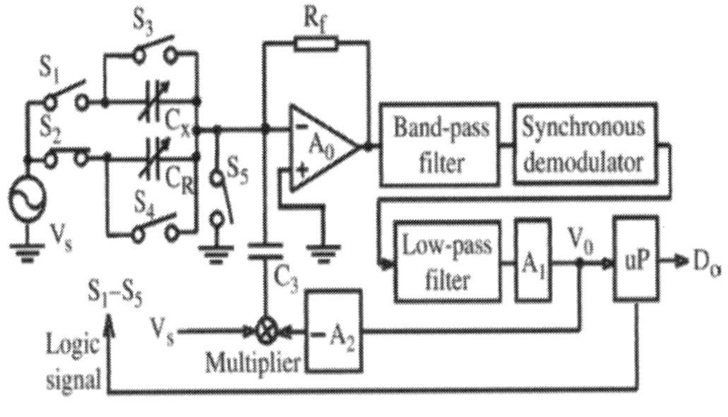

Fig.3 Interface circuit of concentration sensor (Hu et al., 2006)

The excitation source Vs gives a stable and high frequency carrier to let the current flow through active sensor capacitance C1 and dummy sensor capacitance C2 change with the variation in capacitance. The change in concentration of flow is amplitude modulated through capacitance variation. This modulated signal passes through a current amplifier and a band-pass filter before reaching a synchronous demodulator, a low-pass filter and a gain amplifier A1. The output of A1 is a

DC output voltage that passes through the feedback loop. A good level of linearity within +/- 5% at various temperatures or moisture content and capacitance sensitivity of 5% over the entire pipe cross section was obtained. Another research paper [24] focused on gas-slug two-flow measurement technique through non-intrusive capacitive sensing technique in which a signal is applied to transmitter electrode and received from receiver electrode. The probability density function (PDF) and power spectral density (PSD) plots of the received signal are obtained at different point down the same pipe, which become a kind of signature plots for the type and thickness of the slug-flow within the pipe. Relaxation oscillation based circuit is used for measurement of small capacitive changes as a result of concentration flow of pulverized coal, making use of the interfacing circuit which is a modified version of the circuit given in [25] where it is used for converting a resistive change into duty-cycle. However, in [26] while developing capacitance concentration measurement circuit for pulverized coal measurement, the changes in Cx are transformed into DC; where the two diodes are making the V1 and V2 independently available making the former portion of the circuit without affecting the latter potion, and vice versa. Once again bridge circuit made by R1, R2 and Cx and C0 is used with DC output. This circuit is based upon differential structure and relaxation oscillator and the electrodes chosen for the experiments are source grid sensing electrode [27].

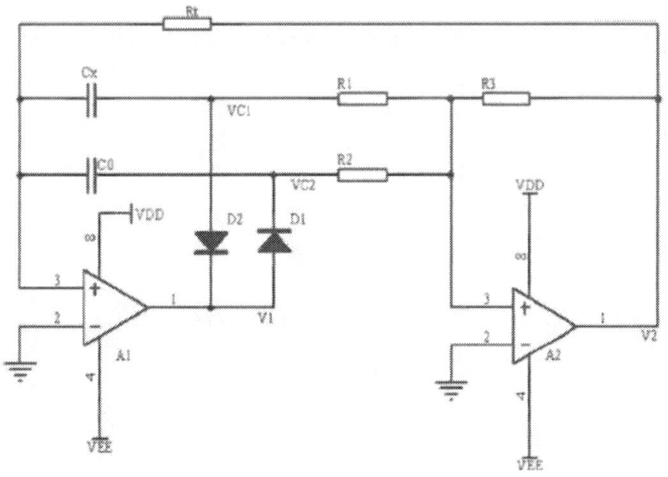

Fig.4 Schematic diagram of the capacitive sensor interface circuit [27]

The experimental results showed the linearity and resolution of the measurement system to be 7.5% and 0.005% respectively under laboratory conditions which is expected to be high while applied in real time flow application. On the whole, the issues concerning capacitive techniques which should be significantly more researched are designing of electrode structure and optimization, pipeline design, complicated and unreliable capacitance measurement circuits (stray capacitance effect, baseline drift, and long cable capacitance), effects of unusual particle properties on sensing phenomenon for pneumatic conveying of biomass.

4. Discussion

The significance and magnitude of measuring an accurate flow rate of gas-solid flow in pipelines has been ever-increasing in the industrialized regions. The transportation of solid biomass by air is very irregular and exigent. Research and studies in the area of flow measuring sensors have been becoming a rehabilitated interest among researchers these days. In this paper, some issues have been highlighted on concentration and velocity measurement of biomass particles. Further research work should be carried out for realizing multiphase flow of biomass materials to improve biomass process instrumentation and utilize biomass resources effectively in such complicated research area.

5. Conclusion

In this paper a descriptive scenario of various sensory systems for gas-solid flow sensing techniques is presented. The pictorial diagrams are supported by describing the principle upon which the workings of the reported sensors are based; the relevant theory of the same is given. Then discussed techniques have been highlighted from recently reported works which have got application utility in industrial environment. This paper after having detailed account of contact and non-contact type measuring techniques highlights the aspects of those measurement methods which are not covered up to mark in some recent papers, setting up the stage for embarking on the scope of novel research areas.

References

[1]. Heping Cui & John R. Grace (2006) Pneumatic conveying of biomass particles: a review, China Particuology, Vol. 4, Nos. 3-4, 183-188, 1998. p. 133–7.

[2]. Jin F, Xia J, Liu R, Wang S. Mathematical models of gas/solid two-phase mass flow-rate measurement using differential pressure concentration method. On Line Flow Measurement of Particulate Solids. UK: University of Greenwich; 1998. p. 133–7.

[3]. Malmgren, M. (1998).Online measurement system for particulate solids using Coriolis principle. University of Greenwich, pp. 126–32.

[4]. Yan, Y. (1996). Mass flow measurement of bulk solids in pneumatic pipelines. Measurement Science and Technology, 7(12), pp.1687–706.

[5]. Nieuwland, J.J., Meijer, R., Kuipers, J.A., & Swaaij, W. (1996).Measurement of solids concentration and axial solids velocity in gas–solids three-phase flows. Powder Technology, 87, pp.127–39.

[6]. Green, R. G., & Thorn, R. (1998).Sensor systems for lightly loaded pneumatic conveyors. Powder Technology, 95, pp. 79-92.

[7]. Williams, R.A., Xie, C.G., Dickin, F.J., Simons, S.J.R. & Beck, M.S. (1991).Multi-phase flow measurements in powder processing. Powder Technology, 66, pp. 203-224.

[8]. Carter, R.M &Yong Yan. (2004). On-line measurement of particle size distribution and mass flow rate of particles using combined imaging and electrostatic sensors. IEEE Instrumentation and Measurement Technology Conference, 2, pp. 864- 867.

[9]. Gajewski, J.B. (2008, September). Electrostatic nonintrusive method for measuring the electric charge, mass flow rate, and velocity of particulates in the two phase gas-solid pipe flows—Its only or as many as 50 years of historical evolution. IEEE Journals & Magazines, 44 (5) pp. 1418-1430.

[10]. Xu, C., Wang, S., Tang, G., Yang, D., & Zhou, B. (2007).Sensing characteristics of electrostatic inductive sensor for flow parameters measurement of pneumatically conveyed particles. Journal of Electrostatics, 65 (9), pp. 582– 592.

[11]. Shao, J., Krabicka, J. & Yan, Y. (2007).Comparative study of electrostatic sensors with circular and probe electrodes for velocity measurement of pulverised coal. Journal of Science, 28 (11), pp. 1921–1926.

[12]. Jiaqing Shao, Krabicka, J., & Yong Yan. (2010, May).Velocity Measurement of Pneumatically Conveyed Particles Using Intrusive Electrostatic Sensors. IEEE Transactions on Instrumentation and Measurement, 59 (5), pp. 1477–1484.

[13]. Ghazali, M.R., Ibrahim, W.I., & Rahmat, M. F. (2011).Electrostatic Sensor in Circular and Rectangular Shape Electrode for Solid Dry Pneumatic Conveyor System. International Conference on Circuits, System and Simulation IPCSIT, 7.

[14]. Xiangchen Qian & Yong Yan. (2012, May).Flow Measurement of Biomass and Blended Biomass Fuels in Pneumatic Conveying Pipelines Using Electrostatic Sensor-Arrays. IEEE Transactions on Instrumentation and Measurement, 61 (5), pp. 1343- 1352.

[15]. Sen, S., Das, P. K., Dutta, P. K., Maity, B., Chaudhuri, S., Mandal, C., & Roy, S. K. (2000).PC-based gas-solids two-phase mass flowmeter for pneumatically conveying systems. Flow measurement and Instrument, 11, pp. 205-212.

[16]. Dyakowski, T., Edwards, R.B., Xie, C.G., &Williams, R.A. (1997). Application of capacitance Tomography to gas-solid flows. Chemical Engineering Science, 52, pp. 2099-2110.

[17]. McKee, S.L., Dyakowski, T., Williams, R.A., Bell, T.A., & Allen, T. (1995).Solids flow imaging and attrition studies in a pneumatic conveyor. Powder Technology, 82, pp. 105-113.

[18]. Hu, H.L., Xu, T.M., Hui, S. E., & Zhou, Q.L. (2006).A novel capacitive system for the concentration measurement of pneumatically conveyed pulverized fuel at power stations. Flow measurement and Instrumentation, 17, pp.87-92.

[19]. Ahmed, W.H., & Ismail, B.I. (2008, November).Innovative Techniques for Two-Phase Flow Measurements. Recent Patents on Electrical Engineering.

[20]. Holler, G., Fuchs, A., & Hrach, D. (2008, November). Flow velocity determination in cryogenic media by means of capacitive sensing. Sensing Technology ICST 2008, pp. 472-476.

[21]. Brasseur, G., Brandstatter, B., & Zangl, H. (2003, August).State of the Art of Robust Capacitive Sensors. IEEE Transactions on Instrumentation and Measurement, 52 (4), pp. 1261-1265.

[22]. Bretterklieber, T., Zangl, H., Hrach, D., Holler, G., Hammerschmidt, D., & Motz, M. (2008).Versatile Programmable Integrated Interface for Robust Capacitive Sensors. Elektrotechnik and Informationstechnik, 125 (4), pp.123-128.

[23]. Zangl, H., Bretterklieber, T., & Holler, G. (2008).On the Design of a Fully Differential Capacitive Sensor Front-End Interface Circuit. Microelectronics Conference ME2008, pp. 132-136.

[24]. Emerson, R., & Leonardo, G. (2010). Characterization of slug flows in horizontal piping by signal analysis from a capacitive probe. Flow Measurement and Instrumentation, 21, pp. 347-355.

[25]. Mochizuki, K., Masuda, T., & Watanabe, K. (1998).An interface circuit for high-accuracy signal processing of differential capacitance transducers. IEEE Transactions on Instrumentation and Measurement, 47 (4), pp. 823-827.

[26]. Chen xia, huhonglit, liuzhihong, (2009, October). A Pulverized Coal Concentration Measurement System Based On Capacitance Sensor. IEEE Technology and Innovation Conference 2009 (ITIC 2009), pp.1-4.

[27]. Cheng-Ta Chiang & Yu-Chung Huang. (2006, December). A Semi cylindrical Capacitive Sensor With Interface Circuit Used for Flow Rate Measurement. IEEE Sensors Journal, 6 (6), pp. 1564-1570.

An Examination of Application of Artificial Neural Network in Cognitive Radios

H. Bello Salau[1], E. N. Onwuka[2] and A. M. Aibinu[3]

[1, 2, 3]Department of Telecommunication Engineering, Federal University of Technology, P.M.B. 65, Minna, Niger State, Nigeria.

E-mail: {[1]habeeb.salau, [2]onwukaliz and [3]abiodun.aibinu}@futminna.edu.ng

Abstract. Recent advancement in software radio technology has led to the development of smart device known as cognitive radio. This type of radio fuses powerful techniques taken from artificial intelligence, game theory, wideband/multiple antenna techniques, information theory and statistical signal processing to create an outstanding dynamic behavior. This cognitive radio is utilized in achieving diverse set of applications such as spectrum sensing, radio parameter adaptation and signal classification. This paper contributes by reviewing different cognitive radio implementation that uses artificial intelligence such as the hidden markov models, metaheuristic algorithm and artificial neural networks (ANNs). Furthermore, different areas of application of ANNs and their performance metrics based approach are also examined.

1. Introduction

Recent advancement in radio technology has led to the development of a smart device known as cognitive radio [1]. This type of radio fuses powerful techniques taken from artificial intelligence, game theory, wideband/multiple antenna techniques, information theory, statistical signal processing and others together to create an outstanding dynamic behavior. This advancement, led to the discovery of a new intelligent wireless dynamic systems for the next generation wireless communication networks.

Basically, cognitive radio (CR) is a type of software radios with additional capabilities and functionalities such as decision-making, learning and environment sensing which enable it to attain the required dynamic performance. CR has the capability of using radio frequency spectrum opportunistically without causing interference or disruption of service to the surrounding systems [1], [2]. This is achieved by analyzing the surrounding radio frequency spectrum with the goal of locating spectrum access opportunities. It uses a smart technique based on cumulative learning to decide on which spectrum access opportunity to utilize [3].

CR is extremely flexible by allowing coexistence with many other communication system, adapt to any channel condition by changing its operation parameters such as channel coding and modulation scheme to achieve the highest possible level of quality of service [4]. CR is considered as a spectrum sharing technology like ultra-wide-band [1] with the main discrepancy being in the method of overlapping of the spectrum signal.

In ultra-wide-band, signal spectrum overlaps with the primary user signal spectrum by transmitting at a very low power levels which barely affect the primary users. CR transmits over a spectrum segments that are temporarily vacant in a flexible opportunistic manner [2]. Consequently, CR may transmit high signal power and therefore have a longer range than an ultra wide band device, provided that it can accurately sense the spectrum segment which is currently unused and the spectral harmful interference in the primary user spectrum bands.

CR emanated from the enhancement of a software radio with cognitive capabilities. Those capabilities are often provided by an intelligent instantiation of a software package called cognitive engine [4]. There are various artificial intelligence (AI) techniques which includes the hidden markov models, metaheuristic algorithm and the ANNs that can be utilized in order to achieve the basic components of a CR namely, awareness, reasoning and learning. Of all the artificial intelligent techniques, this paper focuses on ANNs only. It examines their different areas of applications and performance metrics based approach. The remainder of the paper is organized as follows: Section 2 presents a survey of areas of application of ANNs in CR and their performance metrics based approach. The paper is concluded in Section 3

2. ARTIFICIAL INTELLIGENT TECHNIQUES USED IN COGNITIVE RADIO

In this section, a survey of various AI techniques in different areas of applications, strengths and limitations are reviewed with focus on the ANNs.

2.1.1. Metaheuristic Algorithms

A metaheuristic algorithm is used in CR to search through a solution space while learning and establishing the requisite relationships in computationally hard problems [4]. The technique is good in optimizing parameters and other learning techniques can be used in the training phase. However, the formulation of rule space is difficult if optimization or learning is not restricted. There are various types of metaheuristic algorithms as highlighted below.

I *Tabu search*: This technique can be easily implemented, the performance of the search method is enhanced by using a memory structure called tabu list which ensures that a recent move is not repeated or reversed [4]. Its efficiency is determined by an appropriate parameter selection.

II *Simulated Annealing:* The simulated annealing is a simple approach for global optimization in a large search space. It is easy to implement however, it has a slow convergence rate [4].

III *Ant colony optimization (ACO):* The ACO mimics ant behavior walking around a graph representing the problem to be solved and finding locally productive areas [5]. It is easy to adapt to real time change however, it is not efficient for local searches.

IV *Genetic algorithms (GA):* GA is a particular class of an evolution algorithms used in CR. The appropriate parameter selection is a determinant of the efficiency of the algorithm [4].

Areas of Applications

There are numerous areas of application of metaheuristic algorithm to CR. This includes its utilization in the dynamic configuration of the CR in response to the changing wireless environment [4] and in the adaptation of the radio parameters of a software defined radio to the changing radio environment.

2.1.2. Hidden Markov Model (HMM)

The hidden markov model is a convenient and mathematically tractable statistical model which was developed in order to describe and analyze the dynamic behavior of a complex random phenomenon that can be modeled as a markov process with observable and unobservable states [4]. This technique has the capability of predicting based on experience. Also, it is easily scalable, excellent for classification and can be used to model complex statistical processes. However, it requires a good training sequence and it is computationally complex.

Areas of Application

This includes its utilization in processing signal cyclostationary features for primary signal detection [6]. It can also be used for spectrum sensing as demonstrated in [7]. In [7], an HMM-based dynamic spectrum access algorithm was used in predicting the spectrum occupancy of the licensed radio bands for CR networks.

2.1.3. Artificial Neural Network (ANN)
An ANN is a set of nonlinear functions with adjustable parameters to give a desirable output [4], [8], [9]. There are different types of ANNs which are separated by their network configurations and training methods which facilitate a multitude of application. However, only those that are peculiar to CR are highlighted.

I *A radial basis function networks (RBFNs):* the RBFN has a built-in distance criterion with respect to a center in its hidden layer. This prevents the network from settling into local minimal which is challenge peculiar to perceptron networks [4].

II A multi-layer linear perceptron networks (MLPNs): this network is made up of a layers of neurons where each neuron is a linear combination of the previous output [4], [10], [11]. The size of the network and application determine the performance of the training algorithms.

III A multilayer nonlinear perceptron networks (NPNs): this network which enable the incorporation of non-linearity in the network provides a highly flexible and dynamic results. However, the use of back propagation for training resulted in more processing time needed to achieve a precise result consequently; it is slow to converge during training [4].

Areas of Application

ANN has been applied in spectrum sensing [12], signal classification by utilizing the extracted cyclostationary signal features [13], radio parameter adaptation in CR with three optimization goals viz meeting the BER, maximizing the throughput, and minimizing the transmit power [14], for pattern classification in a pattern-based transmission for CR [4] and in characterizing real time achievable communication performance in CR as proposed by Baldo and Zorzi [15]. This various areas of application are expounded below.

AREAS OF APPLICATION OF ANN IN CR

Feng et al. proposed a statistical mean throughput as evaluation criteria for cognitive user with fixed rate service in spectrum sensing [12]. It was assumed that the state of vacancy and occupancy time of objective license channel are independent of each other with exponential distribution. Furthermore, it was also reported that N objective license channels is being predicted based on the result of machine learning at each time slot. The cognitive user with fixed rate service based on machine learning spectrum sensing has better performance compared to random spectrum sensing. Also, there is decrease in the packet loss rate as the channel time detection threshold increases. However, there is decrease in effective transmission time and difficulty is encountered in determining an appropriate method of increasing the sample size for the statistical mean throughput since it is impossible that the sample size tends to infinity.

In [13], ANN-based signal classifier was developed. It utilizes the extracted cyclostationary signal features, a new method for classification of communication signals based on cyclic spectral analysis and pattern recognition performed by a neural network was proposed [13]. Two cases were studied: Classification with and without prior knowledge of carrier and bandwidth of the signal. Distinct features of each signal type were extracted using cyclic spectral analysis, and a neural network was designed to classify signals based on these features. The combination of the cyclostationary analysis and ANN provides an efficient and reliable signal classification. A multilayer linear perceptron network (MLPNs) with 4 neurons in the hidden layer was used for each signal class during training. Each MLPN was trained with a back propagation algorithm.

The strength of the approach was in the reduction of the online processing time. Also, it provides an efficient and reliable signal classification and is robust to stationary noise. However, difficulty is encountered in comparison due to the practical limitations of the number of modulation type analyzed. Furthermore, a scenario where by there are two or more signals such as BPSK, QPSK, FSK etc being present at a time was not considered.

A technique that can be used in adjusting radio parameter was presented in [14]. The technique, an adaptation mechanism uses GA to optimize the radio parameters on the physical layer was proposed. The GA is used to realize cross layer optimization and a method of adaptive waveform control [4]. It is referred to as the Wireless System Genetic Algorithm (WSGA) [14]. The adaptive component, that uses the GAs lead to the development of a radio defined by a chromosome. The chromosome's genes represent the adjustable parameters in a given radio and by genetically manipulating the chromosome, the GA can find a set of parameters that optimize the radio for the user's current needs. In the WSGA, the radio behavior is interpreted as a set of physical and MAC layer operation parameters defined by traits encapsulated in the genes of a chromosome [14]. Power, frequency, pulse shape, symbol rate and modulation are used to represent chromosome for GA manipulation and are also identified as possible genes in the chromosome definition to allow for future growth through each layer of the radio communication's stack.

The advantage of this technique is that it has a dynamic fitness definition and evaluation where the weighting of each function is adjustable. Also, it has the capability of enabling any fitness function to be used as required by the current link conditions and user requirements. Furthermore, it has the capability of applying constraints to the optimization problem which facilitated the incorporation of regulatory and physical restrictions during chromosome evolution and the WSGA is capable of working with real hardware to accomplish real goals. However, the use of proxim tsunami radios which has a limited number of adaptable parameter for experiment is a demerit. In addition to this, the convergence speed of the system is slow and there is no assurance that the winning member after all the single comparisons in all dimension are the best or non dominated member of the population.

In [15], Baldo and Zorzi proposed multilayered feed forward neural networks as an effective technique for real-time characterization of the communication performance. This provide a general-purpose black box modeling (this consists of analyzing input-output relations of the system under consideration, and trying to build a predictor with the purpose of estimating output values for unknown combinations of the inputs) of the performance as a function of the measurements collected by the CR thus providing good modeling accuracy. Also, it is flexible and as such can be used for runtime modeling in different applications and scenarios and no prior knowledge of the laws driving the real system is required. Furthermore, it avoids all assumptions which are not verifiable in practice and has the ability to account for non-idealities in parameters i.e. tolerance of components, device failures, etc. However, a great challenge is how to effectively determine the values of the weights and biases which will provide the desired approximation for arbitrary continuous functions defined over compact subsets during training. Also, an additional effort and time is needed in training the multilayered feed forward neural network (MFNN) when adding a new variable to a MFNN performance predictor.

Tsagkaris et al. focus on the channel estimation and predictive modeling phases where a benchmarking work that aims at evaluating the applicability of multiple types of NNs in the learning module of the cognitive engine within a cognitive terminal is developed [16]. The proposed neural network (NN)-based learning schemes relax the reasoning process and assist in the optimum decision regarding the radio-configuration settings (mainly PHY and MAC layer) that provide the best QoS for the given problem and user application needs. A potential solution is proposed that assists the cognitive radios in the derivation and enforcement of decisions regarding the selection of desired radio configuration which will optimize its QoS. The NNs was used as the technique for the proposed solution which was motivated by the fact that the NNs have the ability to learn from given examples. Two NN-based learning schemes were set up and tested namely, the 'basic' and the 'extended' one. While the former one aims at building the framework for developing such learning schemes and apply them into future cognitive radio based systems, the latter one stresses that such a learning scheme should be extensible, i.e., flexible in incorporating further information data in the learning process [16].

In the basic NN-based learning scheme, an Elman network is the NN that was selected for the basic scheme which is a two-layer back-propagation, recurrent network, with the addition of a feedback connection from the output of the unique hidden layer to the input layer [16]. This recurrent connection allows the Elman network to both detect and generate time-varying patterns. The NN uses the Tansig function shown in equation (1) for the neurons in its hidden (recurrent) layer, and the Logsig function in equation (2) for the neuron in its output layer respectively.

$$F_K(S_K) = \frac{1}{1+e^{-S_K}} \qquad (1)$$

$$F_K(S_K) = \frac{2}{1+e^{(-2S_K)}} - 1 \qquad (2)$$

where S_K = input neurons from neighbors belonging to different layers and F_K = Activation function. A delay line of five slots was inserted in the input layer, which corresponds to the time window which are slots that was used to represent past experience and knowledge collected by the NN.

The proposed approach has the ability to learn from given example and hence perform better in cognitive tasks. Furthermore, it has a good data rate prediction capability. However, the assumption that the NN-based scheme should be tuned in an arbitrary radio configuration is a drawback of the technique. Also, there may be overtraining or over fitting of the training data as a result of the network not been properly trained which may lead to the network not learning the basic structure of the data but instead learning irrelevant details of the individual cases. In addition to this, a realistic input time series and environment situations are not considered; this may affect the NN-based schemes validity and robustness.

2.1.4. Performance Metrics of ANN Based Approach
The performance metrics of the papers reviewed above is summarized in table (1) below for easy and quick comparisons.

Table 1: Summary of Performance Metrics of ANNs in CR

Paper	Input Parameter	Output Parameter	Usage	Environment and Other Factor Considered	Activation Function	Layers	Performance Metrics
Feng Qia, Ye Zhihuia, Shen Keqinb "Spectrum Environment Machine Learning in Cognitive Radio" (2012)	Alternate time of channel vacancy/ occupancy state	Throughput Packet loss rate	Spectrum Sensing	Number of detected transmission	Threshold Function	Statistical mean throughput	Throughput Packet Loss rate
A. Fehske, J. Gaeddert and J. H. Reed "A New Approach to Signal Classification Using Spectral Correlation and Neural Networks"	Carrier Frequency Symbol Rate SNR	Spectral coherence function Relinbility	Signal Classification	The signal's carrier and bandwidth is known It was assumed that there is no prior knowledge of the signal other than its presence	Linear Function	Multi-layer linear perceptron network	Spectral coherence function of each of the signal type i.e BPSK, QPSK, FSK and MSK Probability of classification for varying SNR Probability distribution for confidence levels
Thomas W. Rondeau, Bin Le, Christian J. Rieser, and Charles W. Bostian "Cognitive Radios with Genetic Algorithms: Intelligent Control of Software Defined Radios"	BER Frequency Power	Spectral Occupancy Throughput Interference	Radio Parameter Adaptation	Incorporation of regulatory and physical restrictions during chromosome evolution	It utilizes the dynamic fitness definition and evaluation where the weight of each function is adjustable and any fitness function may or may not be used depending on the current link condition and user requirement	Multi objective genetic algorithm	Interference avoidance Throughput Spectral Occupancy
Nicola Baldo and Michele Zorzi "Learning and Adaptation in Cognitive Radios using Neural Networks" (2008)	SNR Received Frames Erroneous Frames Idle Time	Throughput Delay Reliability	Performance characterization of the components of a cognitive radio system	Number of detected transmissions/ Number of users	A Sigmoid Function	Multi layer feed forward neural networks	Throughput
K. Tsagkaris, A. Katidiotis and P. Demestichas "Neural network based learning schemes for CR systems" (2008)	Time series collected by the radio-scene analysis phase	Mean squared error	Channel estimation and predictive modeling phases	Estimation of time zone parameter	Sigmoid Function	Multilayer linear perception	Mean squared error

3. Conclusion

The prospects and the effective utilization of CR for various applications such as spectrum sensing, radio parameter adaptation and signal or pattern classification depends greatly on the choice of the AI techniques being utilized in the design of the CR. This paper reviewed various areas of application of AI techniques in CR particularly the use of ANN. The performance metrics of each of the ANN technique have also been examined in order to understand the tradeoff in the selection of the ANN technique to be used which is critical in successful CR design.

References

[1] J Mitola. "Cognitive Radio: An Integrated Agent Architecture for Software Defined Radio". *Doctor of Technology, Royal Institute of Technology (KTH), Stockholm, Sweden*, 2000.

[2] I.F. Akyildiz, W.Y. Lee, M.C. Vuran, and S. Mohanty. "NeXt Generation Dynamic Spectrum Access/cognitive radio wireless networks: a survey. *Computer Networks Journal (Elsevier)*, vol. 50, pp. 2127-2159, September 2006.

[3] J. Mitola, III, "Cognitive radio for flexible mobile multimedia communications," in Proc. IEEE Int. Workshop MoMuC, San Diego, CA, pp. 3–10.Nov. 15–17, 1999.

[4] A. He, K. K. Bae, T. R. Newman, J. Gaeddert, K. Kim, R. Menon, L. Morales-Tirado, J. J. Neel, Y. Zhao, J. H. Reed, and W. H. Tranter, "A survey of artificial intelligence for cognitive radios," IEEE Trans. Veh. Technol., vol. 59, no. 4, pp. 1578–1592, May 2010.

[5] M. Dorigo and C. Blum, "Ant colony optimization theory: A survey," Theor. Comput. Sci., vol. 344, no. 2/3, pp. 243–278, Nov. 2005.

[6] K. Kim, I. A. Akbar, K. K. Bae, J.-s. Urn, C. M. Spooner, and J. H. Reed, "Cyclostationary approaches to signal detection and classification in cognitive radio," in Proc. 2nd IEEE Int. Symp. New Frontiers DySPAN, Dublin, Ireland, pp. 212–215Apr., 17–20, 2007.

[7] I. A. Akbar and W. H. Tranter, "Dynamic spectrum allocation in cognitive radio using hidden Markov models: Poisson distributed case," in Proc. IEEE SoutheastCon, Richmond, VA, pp. 196–201, March 22–25, 2007.

[8] A.M. Aibinu, M.J.E. Salami, A.A. Shafie. "Artificial neural network based autoregressive modeling technique with application in voice activity detection" Engineering Applications of Artificial Intelligence (Elsevier), Vol. 25, pp. 1265–1276, June 2012.

[9] A. M. Aibinu, A. A. Shafie, and M. J. E. Salami. "Performance Analysis of ANN based YCbCr Skin Detection Algorithm" In Proceedings of Elsevier International Symposium on Robotics and Intelligent Sensors (IRIS 2012) Procedia Engineering Vol. 41 PP. 1183 - 1189, 2012

[10] A. M. Aibinu, M. J. E. Salami, and A. A. Shafie. "Determination of Complex-Valued ParametricModel Coefficients Using Artificial Neural Network Technique" Advances in Artificial Neural Systems Journal, Volume 2010, Article ID 984381, March 2010.

[11] A.M. Aibinu, M.J.E. Salami, A.A. Shafie. "A novel signal diagnosis technique using pseudo complex-valued autoregressive technique" Expert Systems with Applications Journal, Vol. 38, PP 9063–9069, 2011.

[12] Q. Feng, Y. Zhihui and S. Keqin, " Spectrum Environment Machine Learning in Cognitive Radio", International workshop on Information and Electronics Engineering (IWIEE Elsevier), pp 4181-4185, 2012.

[13] A. Fehske, J. Gaeddert, and J. H. Reed, "A new approach to signal classification using spectral correlation and neural networks," in Proc. 1st IEEE Int. Symp. New Frontiers DySPAN, Baltimore, MD, Nov. 8–11, 2005, pp. 144–150.

[14] T. W. Rondeau, B. Le, C. J. Rieser, and C. W. Bostian, "Cognitive radios with genetic algorithms: Intelligent control of software defined radios," in Proc. Forum Tech. Conf. SDR, Phoenix, AZ, pp. C-3–C-8, Nov. 15–18, 2004.

[15] N. Baldo and M. Zorzi, "Learning and adaptation in cognitive radios using neural networks," in Proc. 5th IEEE CCNC, Las Vegas, NV, pp. 998–1003, Jan. 10–12, 2008.

[16] K. Tsagkaris, A. Katidiotis, P. Demestichas, "Neural network-based learning schemes for cognitive radio systems", Computer Communications, Volume 31, Issue 14, Pages 3394-3404, 5 September 2008.

How to model wireless mesh networks topology

M L Sanni[2], A A Hashim[1,2], F Anwar[2], G S M Ahmed[3] and S Ali[3]
[2]Department of Electrical and Computer Engineering,
[3]Department of Science,
Faculty of Engineering, International Islamic University, Malaysia

E-mail: aisha@iium.edu.my

Abstract. The specification of network connectivity model or topology is the beginning of design and analysis in Computer Network researches. Wireless Mesh Networks is an autonomic network that is dynamically self-organised, self-configured while the mesh nodes establish automatic connectivity with the adjacent nodes in the relay network of wireless backbone routers. Researches in Wireless Mesh Networks range from node deployment to internetworking issues with sensor, Internet and cellular networks. These researches require modelling of relationships and interactions among nodes including technical characteristics of the links while satisfying the architectural requirements of the physical network. However, the existing topology generators model geographic topologies which constitute different architectures, thus may not be suitable in Wireless Mesh Networks scenarios. The existing methods of topology generation are explored, analysed and parameters for their characterisation are identified. Furthermore, an algorithm for the design of Wireless Mesh Networks topology based on square grid model is proposed in this paper. The performance of the topology generated is also evaluated. This research is particularly important in the generation of a close-to-real topology for ensuring relevance of design to the intended network and validity of results obtained in Wireless Mesh Networks researches.

1. Introduction

The growing demand for Wireless Mesh Networks (WMN) is majorly motivated by its provision of low cost broadband access, facilitating real-time web-based applications for mobile users. WMN is a self- healing, self-configuring and self-organised communication network with fixed wireless Mesh Routers (MR) and static or mobile Mesh Clients (MC). And being an infrastructure wireless ad hoc, it has the capability of service delivery coordinated by an important node referred to as Internet gateway (IGW). With IGW, WMN is able to internetwork with other networks such as Internet, Cellular and Sensor networks [1]. MC can generate and serve as sink for traffic in the network, while MR is only responsible for multihop relaying of traffic in the backbone of WMN. The multihop provided by the MR extends connection beyond the adjacent nodes while IGW provides the bridge between WMN and outside network as shown in figure 1. MC is usually an IP enabled devices like PDAs, mobile phones, laptops and desktops.

WMN offers low deployment cost for mobile Internet users, hence an indispensable access network for delivery of multimedia applications such as e-learning, games, disaster recovery operation and teleconferencing. Consequently, WMN researches are motivated by the need to design efficient

[1] A A Hashim.

protocols for the delivery of these applications which are based on certain algorithm design objectives and constraints. An important starting point in the design of such network protocols is the network topology design for modelling of the simulated scenario to capture the network components characteristics and interactions for design. Furthermore, topology design marks the template for the evaluation of new protocol via simulation. However, the validity of simulation highly depends on the simulation model employed [2].

Topology generation is the design of representation of the network characteristics and connectivity which are important basic research task to be accomplished in the development and/or analysis of networks and protocols. Node connectivity in WMN is typically ensured among adjacent nodes especially if the nodes are within the transmission range of one another. Apart from 1-hop connectivity, nodes also connect via intermediate relay network, however, with a bound on the number of hop possible without signal level degeneration.

Graph theory is usually the basic tool used in modelling the network topology, thus several graph models have been proposed. Furthermore, a significant parameter in selecting or designing topology generators is the quantification of its adherence or conformity to the technical characteristics of nodes interactions and existence of links between nodes while satisfying the architecture requirements of the physical network. The importance of generating close to real network topology is identified [3] - [5]. Pure random models, Waxman model and its variants are usually implemented at the core of random, hierarchical and degree-related synthetic topology generators such as GT-ITM, Inet, BRITE and Tiers [6]. However, these synthetic topology generators based on geographic mathematical models of certain probability may not adequately capture the inherent topological properties of multihop infrastructure ad hoc like WMN [2]. And topology generation of a close to real network characteristics ensures relevance of the proposed protocol and validity of evaluation results obtained in such researches [7].

In this paper, an algorithm based on mathematical and interference model (protocol model) is proposed for the generation of WMN topology based on grid structure to serve as the starting point in the design of network protocols researches for WMN. The proposed algorithm models the node placements, link establishments and node distribution with a view to obtaining realistic network connectivity model as a premise to the validity of proposed protocols and its observed performance. The remaining sections of the paper is organised as follows. Section 2 is a review of existing synthetic topology models and generators. The algorithm for the design of the proposed WMN network topology generation is described in Section 3. Section 4 is the performance evaluation of the proposed algorithm while Section 5 concludes the paper.

2. Related works

Multihop network design consists of node, deployment and mobility, radio, wireless signal propagation and traffic pattern model designs [2]. Two of the sub models (wireless signal propagation and traffic pattern) are handled by measurement. However, most research efforts in the design of node models are not verified by real measurement, furthermore, they are not motivated by the structure of real networks. In uniform node placement model, the placement area L^2 and number of nodes n are selected with a placement probability of n/L^2. The chain node placement model implements equidistant internodes distances among the network nodes placed on a line. The grid node placement model uses the intersection points of cells as the node location points, in this case, the basic node degree are typically 2, or 3 or 4 depending on the location in the rectangular or square grid structure.

In the existing topology designs, node placements and interactions are based on random model which establishes links between randomly placed nodes based on certain probability. The probability of link between the nodes and maximum distance between any two nodes are expresses in [8] as:

$$P(u,v) = \alpha \exp[-d/\beta L] \quad (1)$$

where $0<\alpha, \beta\leq 1$ are Waxman parameters, d is the Euclidean distance between nodes u and v and L is the maximum distance between any two nodes. In addition randomly generated number in the range between 0 and L replaces d in one of the variants of Waxman model. Another variation includes specification of average node degree requirements by the introduction of a scaling factor kε/n on P(u,v) for controlling the number of edges placed between nodes. Also [3] employs the probability:

$$P(u,v)=\alpha \exp[-d/L-d] \qquad (2)$$

Three hierarchical design levels of Internet topology - LAN, MAN (by minimum spanning tree) and WAN are presented in [9]. The initial connected random graph generated is a transit domain, while each of the transit domains is connected with stubs. The backbone / access hierarchical topology of transit and stub domains is generated in GT-ITM [10].

Inet [11] is another synthetic Internet topology generator based on random network generation. It builds spanning tree for all nodes with node degree greater than 1 with uniform probability of d/D for node degree *d* and the sum of node degrees of other nodes with at least one free degree, D. Node connectivity process is based on some heuristics. Inet topology generators still fall short in capturing some properties of Internet such as maximum clique size and clustering coefficient. Even in cases of uniform node placement, [12] - [13] achieved 10.8 and 13.78 average node degree respectively at connectivity probability of 0.99, however, high node degree implies connected network with many links for all pairs, hence certain impacting factors cannot be studied in such networks. The poor performance and lack of robustness to failure of some synthetic topology generators are highlighted in [14]. In [4], the tasks of human designer in a network topology design are simulated in the design of IGen, additionally, the procedures are automated to produce IGen topology based on certain heuristics and hence it is not based on probabilistic method. IGen groups the nodes into different clusters based on their different Euclidean distances and/or traffic demand however, node connectivity model and determination of existence or otherwise of link or edges are not considered; and further researches are also required on link capacity assignment.

BRITE [15] implements preferential attachment based on Waxman or incremental node growth model while in NPART [2], the node placement/ topology design algorithm permits the network to grow, thus the placement area is not predefined. Link-level topology generator that captures both the subnet and degree distributions of the measured network is proposed in [7]. The node centric model assumes predefined node degree in the topology generation, and then the links are merged to obtain the required subnet distribution. Furthermore, in the subnet approach, the nodes in different subnet generated are combined in such a way to satisfy the node degree specification. However the work did not address network parameters such as the average node shortest path and graph density.

From the foregoing, the topology generation of GT-ITM, NPART and BRITE considers only graph properties, while the consideration of technical and economic constraints are also important as random, hierarchical, degree-related and other graph properties as proposed in the design of network topologies [4].

3. The proposed topology generation model

3.1. Motivation

The main motivation for the proposed topology generator in this paper is the connectivity issue in WMN which arose when Waxman model of random graph is simulated for WMN. The topology obtained is as shown in figure 2. However, the existing topology generators based on Waxman model depicts a geographic and probability based topology which may not to be suitable in this case, given the following observations from figure 2:

- Waxman model creates links with far away nodes while the nearest nodes are left unconnected.
- The topology generated by Waxman may contain unconnected nodes however, connectivity cannot be taken for granted, and it has to be ensured.
- The topology generated by Waxman may contain unconnected nodes however, connectivity cannot be taken for granted, and it has to be ensured.

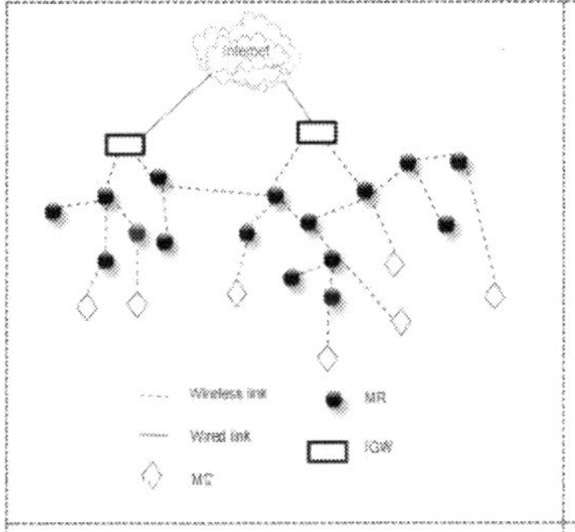

Figure 1. Wireless Mesh Network architecture.

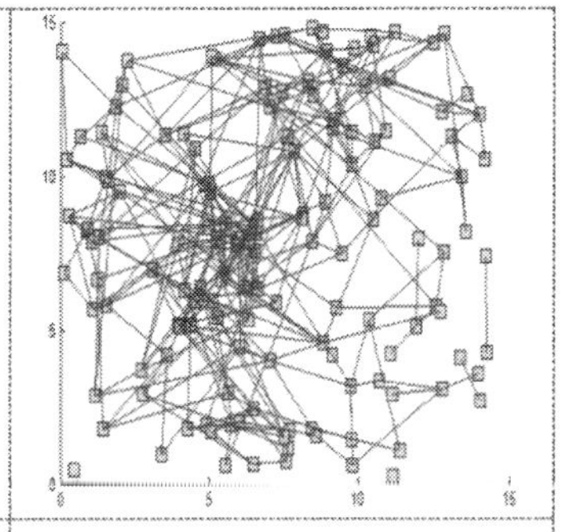

Figure 2. Waxman topology generation.

- The fan-out of the mesh node should form the upper bound of the node degree and maximum hop count in WMN.
- Inter node placement distance should reflect close to real router placement, thus too close/far nodes may not be realistic.

Furthermore, for network connectivity, [16] highlights that probability of each node communicating with another node is informed by their geometric distance, however, this is usually a function of nodes transmission range, multipath fading or shadowing. Moreover, NPART topology generator is based on similar focus however link establishment model differs. Hence the topology design problem in WMN addressed in this paper is to design placement of nodes on a network area such that the characteristics of WMN which determines the nature of inter node communication is simulated given that a connected topology is guaranteed.

3.2. Algorithm description and operation

The following assumptions are made:

- l should be less than T_r so as to isolate unrealistic node degree and cases when connectivity is difficult to achieve.
- l should be a factor of L in order to obtain equal square sections in the square grid structure
- All wireless nodes have same transmission power hence equal T_r
- To ensure connectivity, each node has at least one adjacent node

The algorithm generates random nodes on a square grid placement area to simulate WMN topology. The topology model proposed is represented mathematically with predetermined placement area and the establishment of links is achieved based on the interference model (protocol model) [17]. The protocol model states that a packet transmitted by a node at location X_i is received by another node placed at X_j if

$$\begin{aligned} &|X_i - X_j| < Tr \\ &|X_k - X_j| > (1+\Delta)Tr \end{aligned} \tag{3}$$

For every node placed at location X_k, with a simultaneous transmission in the same channel and where Δ is the guard zone between adjacent nodes to prevent interference for nodes within each other's T_r. The randomly generated nodes are independently and uniformly distributed over given network area. 1-hop nodes neighbour list (adjacent nodes) is generated based on the parameter relations specified in the protocol model using the transmission range of wireless nodes, thus the connectivity matrices is created with 0 for no link and 1 if link exists between the two nodes (figure 3) .The following is the procedure implemented for the random node placement and node connectivity based on the protocol model for WMN topology design:

1. Input L, l, and T_r
2. Initialise all variables
3. Compute the network topology of N number of squares for WMN radio placement area
4. Create network grid for the total number of vertical v_k and horizontal h_k lines contributing to the number of square sections in the area.
5. Compute number of nodes r, as a function of predefined area as $r = \eta N$ where η is the percentage of the total network square sections to be assigned a WMN radio node
6. Generate vector of random integers to depict nodes' positions and adjacent nodes list given the protocol model.
7. Generate connectivity matrix
8. Generate node degree for the topology and check the connectivity status of the network

4. Results and discussions

The wide difference in the average diameter of PROTOPO and Leipzig is a reflection of existence of bridge nodes and delay in transmission in Leipzig (figure 4).

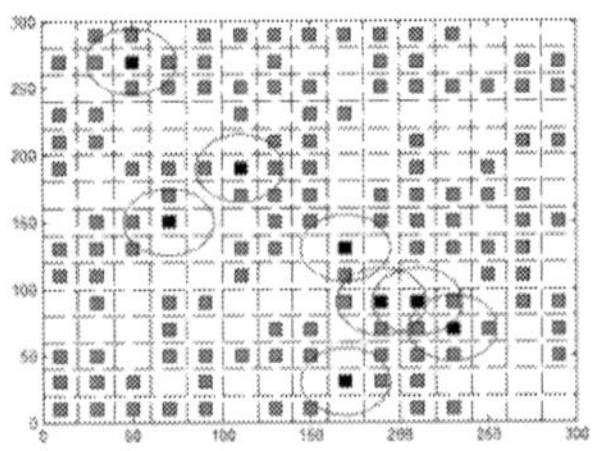

Figure 3. PROTOPO topology with illustration of node coverage areas.

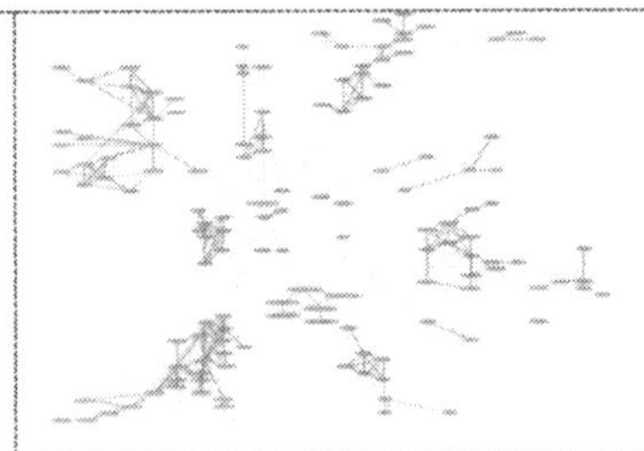

Figure 4. Leipzig topology www.Leipzig.freifunk.net/.

Table 1 shows the analysis of PROTOPO and Leipzig based on 11 topology samples ranging from 143 to 177 nodes, with bridge nodes inclusive while the number of generations of samples in PROTOPO achieves 95% level of confidence.

Table 1. Comparison of PROTOPO with Leipzig topology.

Topology	PROTOPO	Leipzig
Average number of links	415	277
Average node degree	5.33	3.57
Network diameter	43.01	4.19
Bridges	0	14.27%

The number of adjacent nodes in Leipzig samples is lower than that observed in PROTOPO due to the emphasis placed on the connectivity in the model of PROTOPO (figure 5).

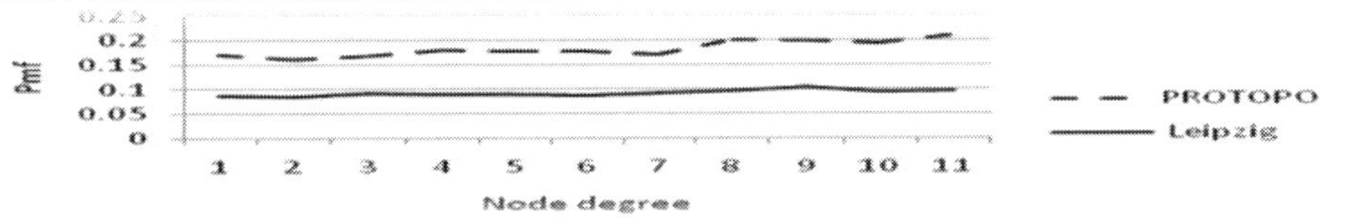

Figure 5. Comparison of node degree distributions

The average share of disconnected nodes is 14.27% in Leipzig, PROTOPO could not capture this parameter because the algorithm usually test the topology generated for connectivity however, and its adjacent list can consists of a single link. Thus PROTOPO makes provisions to avoid scenarios of disconnected networks which in reality may be the ideal except if such nodes exist to serve as connecting hub for new joins.

5. Conclusion and future works

The extent of practical and realistic objectives and constraints considered by network topology designers determines the extent of acceptability and validity of design and evaluation of protocols based on these topology models. Thus, a practical flexible network topology model based on the protocol model depicting the criteria for existence or otherwise of network links among the communicating nodes and the generation of connected networks is presented. This tool can serve as the basis for path planning optimisation in real time for multicast routing, route for autonomous system and gathering information from data servers at varying locations. At present, only topology characteristics can be sampled, further research work is also required to optimise PROTOPO.

6. References

[1] Akyildiz I, Wang X and Wang 2005 *Elsevier Computer Network* **47** 445-87
[2] Milik B and Malek M *Proc. 2nd Int. Conf. on Simu. Tools and Tech for Comm., Networks and Systems*, Italy, March 2-6, 2009
[3] Zegura E W Calvert K L and Bhattacharjee *Proc 15th IEEE INFOCOM* March 24-28 1996 **2** pp 594-602
[4] Quoitin B Schrieck V François P and Bonaventure O *Proc. 21st Int. Conf. on Teletraffic Congress*, Sept. 15-17, 2009 pp 1-8
[5] Kojic N, Reljin I and Reljin B 2012 *Sensors* **12** 7548-57
[6] Medina A, Lakhina A Matta I and Byers J *Tech. Rep.*Boston University, Boston, MA, 2000
[7] Akgun M B and Gunes M H *Proc. 31st ICDCSW* 140-5, June 20-24, 2011
[8] Waxman B M 1988 *IEEE J. Select. Areas Comm.* **6** 1617-22
[9] Doar M and Leslie I 1993 *Proc. IEEE INFOCOM* 82–9
[10] Thomas M and Zegura E W 1994 *College of Computing, Tech. Rep.*
[11] Winick J and Jamin S 2002 *University of Michigan, Tech. Rep*
[12] Bettstetter C *Proc. 3rd ACM Intl Symp. on Mobile ad hoc Networking and Computing*, Switzerland 2002
[13] Li X, Wan P, Wang Y and Yi C 2003 *Proc. 4th ACM Int. Symp. on Mobile ad hoc Networking and Computing*, Maryland, USA
[14] Li L, Alderson D, Willinger W and Doyle J *ACM SIGCOMM*, 2004
[15] Boston University, http://www.cs.bu.edu/brite/
[16] Abou El Saoud M and Mahmoud S *Proc. IEEE Symp on Comp. and Comm.* pp 404-10, July 6-9, 2008
[17] Gupta P and Kumar P R 2000 *IEEE Transactions on Information Theory* **46** 388-404
[18] Leipziger freifunk-community,www.Leipzig.freifunk.net/
[19] Milik B and Malek M 2007 *IEEE Communication Lett.*, **11** 580-2

Lower-Limb Rehabilitation Robot Design

E M Bouhabba[1], A A Shafie[1], M R Khan[1], and K Ariffin[2]
[1]Departement of Mechatronics Engineering, Faculty of Engineering, International Islamic University Malaysia
[2]Department of Orthopaedics, Traumatoligy and Rehabilitation. Faculty of Medicine, International Islamic University Malaysia

E-mail: b_mehdi1983@yahoo.fr

Abstract. It is a general assumption that robotics will play an important role in therapy activities within rehabilitation treatment. In the last decade, the interest in the field has grown exponentially mainly due to the initial success of the early systems and the growing demand caused by increasing numbers of stroke patients and their associate rehabilitation costs. As a result, robot therapy systems have been developed worldwide for training of both the upper and lower extremities. This paper investigates and proposes a lower-limb rehabilitation robot that is used to help patients with lower-limb paralysis to improve and resume physical functions. The proposed rehabilitation robot features three rotary joints forced by electric motors providing linear motions. The paper covers mechanism design and optimization, kinematics analysis, trajectory planning, wearable sensors, and the control system design. The design and control system demonstrate that the proposed rehabilitation robot is safe and reliable with the effective design and better kinematic performance..

1. Introduction

A rehabilitation robot is a robot with the application of robotic technology to the rehabilitation needs of people with disabilities as well as the growing elderly population [1]. An intelligent rehabilitation robot should be a human adaptive mechatronics (HAM) system. The HAM main aim is to introduce to a human-machine system a capability; that the machine can measure, or understand, the level of the skill of its human operator. Another HAM aim is to develop an integrated theory and technology which can adapt the HAM machine to the human according to their skill, in order to maximize not only the total performance of the human-machine system, but also to help improve the skill of the human [2]. This paper researches a human and machine interaction system – a rehabilitation robot to assist and support people with a paralysis problem. Thus, the research conducted in this paper belongs to a core part of research into HAM. The lower-limb rehabilitation robot aims to help people, who suffer paralysis caused by disease or lower-limb movement disorder following an accident, to improve and resume limb functions. According to World Health Organization (WHO) statistics, the world population of people over 60 years old will double by 2050, and the number of handicapped people from disease will increase. Thus, the requirement of the rehabilitation robots is in demand [3].

[1] El Mehdi Bouhabba, b_mehdi1983@yahoo.fr
Department of Mechatronics Engineering
Faculty of Engineering , International Islamic University Malaysia

In recent years, the research on rehabilitation robots has become a hot topic, and several rehabilitation robots have been developed. The Lokomat, developed by Hocoma AG (Volketswil Switzerland), was the first driven gait orthosis that helped to improve the walking movements of patients who are gait-impaired. Colombo et al. [4, 5] gave a detailed description of the device. Veneman et al. [6, 7] developed a gait rehabilitation device called LOPES which combined a freely translatable and twodimensional (2D)-actuated pelvis segment with a leg exoskeleton containing three actuated rotational joints (two at the hip and one at the knee). Cullell et al. [8] designed a knee-ankle-foot orthosis based on biomechanical data to restore the human gait, providing stability during the standing phase. Pennycott et al. [9] studied a body-weight supported robotic-assisted device; they presented a method of estimating and controlling the rate of active work done during supported stepping, and conducted the experimental validation through ambulation of three subjects with a spinal cord injury. Yaskawa Electric Corporation of Japan made a lower-limb rehabilitation robot, TEM LX2 type D [10], which was mainly used for patients with lowerlimb disease in an acute phase. Its main purpose was to enable the patients to restore their physical functions and walking ability. Saito et al. [11] developed an orthosis externally powered by a bi-articular muscle mechanism with a bilateral-servo actuator. Hunt et al. [12] investigated feedback control strategies for integration of electric motor assist and functional electrical stimulation (FES) for paraplegic cycling, and provided indicative results from one paraplegic subject where a series of feedback-control tests illustrated accurate control of cycling cadence, leg power control, and external disturbance rejection. Chen et al. [13] proposed a new prototype of lower-limb rehabilitation robots, and the prototype might give considerable protection to patients. Yeh and Yu [14] developed a lower-limb mobile training robot that provided adequate motion functions, such as the passive and resistive motions of multiple joints for increasing training performances and for shortening training programs. Beyl et al. [15] reported on the development of a gait rehabilitation exoskeleton with a knee joint powered by pleated pneumatic artificial muscles. Simon et al. [16] built a lower-limb robotic device that used a novel control strategy for increasing force symmetry during bilateral lower-limb extensions. Costa et al. [17] reported on the design and control of a new 'human friendly' orthosis (exoskeleton), powered by high power pneumatic muscle actuators (pMAs). Willems et al. [18] used a Biodex System 3 Dynamometer made by Biodex Medical Systems Inc. to determine the isokinetic peak torque and peak torque/body-weight values for reciprocal concentric and eccentric eversion–inversion movements of the ankle. Tsoi and Xie [19] designed a parallel ankle rehabilitation robot and investigated the potential advantage of using a variable impedance controller to perform ankle rehabilitation exercises. Saglia et al. [20] presented a redundantly actuated parallel mechanism for the ankle rehabilitation. With the advance of computer network technology and remote operation technology, remote operating robots began to be used in telemedicine, remote surgery and remote rehabilitation [21–23]. Girone et al. [24] proposed a stewart platform based system for the ankle tele-rehabilitation.

Although researchers have developed many kinds of rehabilitation robots, there are still few intelligent rehabilitation robots that can carry out targeted rehabilitation training from the motion intention of patients. In this paper, a lower-limb rehabilitation robot used for patients with paraplegia is proposed. The rehabilitation robot is an exoskeleton rehabilitation robot with multi-joints which can realize the single joint movement of the lower-limb. In contrast to the rehabilitation robots adopting the suspension system and treadmill to realize walking training, the rehabilitation robot is mainly used for paraplegic patients whose lower-limbs cannot move. Therefore, the robot mechanisms are installed in a chair so that the patient can be trained while sitting in the chair, which is more comfortable than standing on a treadmill or suspending body using a weight-support system, and the more reasonable for the clinical rehabilitation of the paraplegia patients. Moreover, the proposed rehabilitation robot has the following design features: (1) three rotary joints; (2) wearable sensors force sensors, which can be installed easily, are used to measure the force creating by muscles, which can greatly reduce the manufacturing cost; and (3) the design takes fully into account the human lower-limb data, and the length of the connecting link can be adjusted over a certain range according to the size of the patient. This paper is organized as follows. In section 2, the mechanism design and sensors are detailed. In section 3, modeling of sensors. In section 4, the control system and design. Finally, conclusions are presented in section 5.

2. Mechanism design and sensors

2.1. Determination of degrees of freedom

The movement of a human lower limb includes: the vast scale forward swing, small-scale back and side swing of the thigh; the vast scale back swing of the knee joint; and the foot rotation around the ankle joint in the vertical plane and horizontal plane [25].Since the movement of the human lower limb in the sagittal plane is the basic movement form, the rehabilitation robot is designed as three degrees of freedom that can realize the movement of the hip joint, the knee joint, and the ankle joint in the sagittal plane.

According to the movement features of the human lower limb, a rehabilitation robot with three-planed links is designed and shown in figure 1. Three links are connected by three revolute joints. The revolute joint o_1 connects with link 1 and the base; the revolute joint o_2 connects with link 1 and link 2; and the revolute joint o_3 connects with link 2 and link 3.

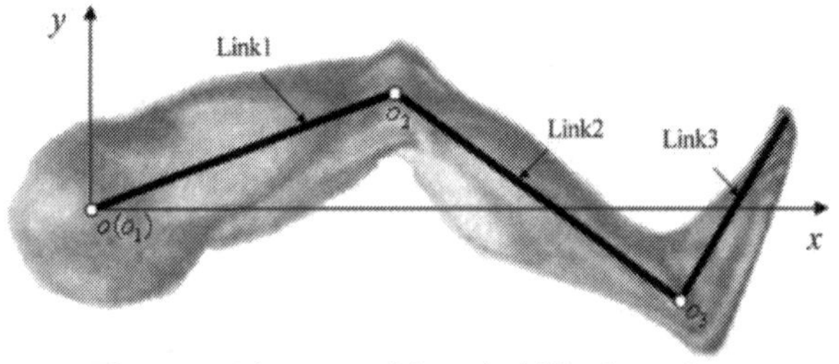

Figure 1. Diagram of the rehabilitation robot.

2.2. Determination of link length and joint rotation range

Since the size of the lower limb differs for each patient, the length of the three links should be designed to meet the requirements of a height range from 1500mm to 1900mm. Since the lower limb length of a male is greater than that of a female, the data of a male is taken as the maximum of height, and the data of a female as the minimum of height. From reference [26], the dimension proportions of thigh, calf, and ankle relative to height are 0.471, 0.265, and 0.045, respectively for a male, and are 0.466, 0.261, and 0.042, respectively for a female. Therefore, the maximum and minimum length of the every link can be calculated as shown in table 1. In the actual design, the design length of every link should be greater than the theoretical length.

Table 1. Dimension of links

Links	Theory length (mm)		Design length (mm)	
	Minimum	Maximum	Minimum	Maximum
Thigh	306	420	300	470
Calf	328	410	320	430
Ankle	100	200	90	220

The joint rotation range of the rehabilitation robot should be in accordance with the motion range of the lower limb when the patient is trained sitting on a chair. Therefore, it has to be determined based on the movement angle and the movement type of the lower limb joint [25]. When the patient

IOP Conf. Series: Materials Science and Engineering **53** (2013) 012038 doi:10.1088/1757-899X/53/1/012038

sits on the chair, the motion ranges of the hip joint, the knee joint and the ankle joint are 60^0, 110^0, and 70^0, respectively, which are taken as the joint rotation ranges of the rehabilitation robot.

2.3. Calculation of Joint Moments

Different types of sensors can measure force and movement of human body segments, but joint moments are difficult to directly measure by sensors as it relates to many other parameters of human body. However, joint moment is one of the most important parameters in evaluating the motor function of human limbs, and it can reflect accurately the effectiveness of rehabilitation activities. Therefore, in this paper the joint moments of hip, knee, and ankle are calculated based on the data measured by the sensor system.

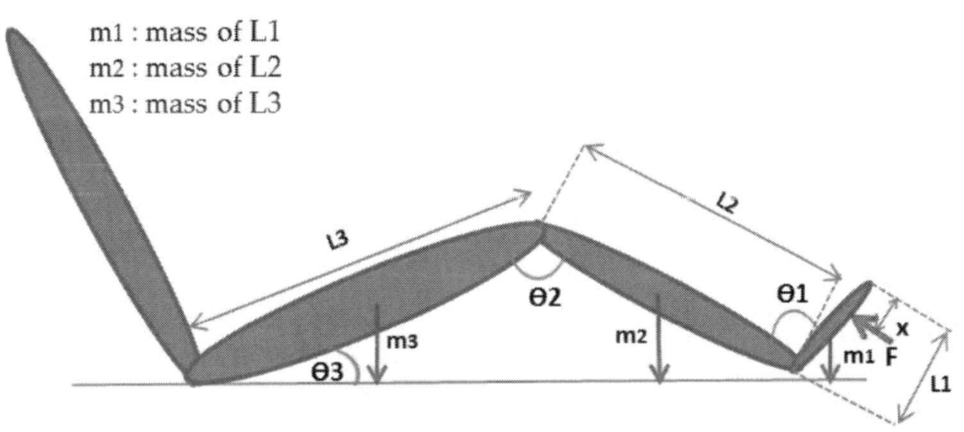

Figure 2. A diagram illustrating dynamic.

As shown in figure 2, the magnitude and location of F can be acquired from GRF and COP data measured by the force plate. θ_1, θ_2 *and* θ_3 can be acquired from the angle data measured by the three wearable motion sensors; L1, L2, and L3 can be acquired by the lower limb lengths of users; and m1, m2, and m3 can be estimated by the weight of users and the mass percentage law of human body. Moreover, the parameters p, q, and w are introduced to indicate the mass centre locations of foot, and shank. As all the data are real-time measured by the sensor system of the rehabilitation robot, the joint moments can be real-time calculated in the computer control program. Finally the joint moments (M) of ankle, knee and hips can be represented by the equation:

$$M_{ankle} = Fx_1 - pL_1m_1g \quad (1)$$

$$M_{knee} = F[L_2\cos\theta_1 - x_1] - qL_2\cos\theta_1\, m_2g - [L_2\cos\theta_1 - pL_1]m_1g \quad (2)$$

$$M_{knee} = F[L_3\cos\theta_2 - L_2\cos\theta_1 + x_1] - wL_3\cos\theta_2\, m_3g - [L_3\cos\theta_2 - qL_2\cos\theta_1]m_2g - [L_3\cos\theta_2 - L_2\cos\theta_1 + pL_1]m_1g \quad (3)$$

3. Modeling of sensors

3.1. Modeling of IPMC sensor

An ionic polymer-metal composite (IPMC) consists of a polyelectrolyte membrane containing cat ions with a solvent and metal electrodes chemically plated on both surfaces of the membrane. The present study investigated an IPMC sensor model using a RC circuit where the electric components were related to the physical parameters of the IPMC. A charge model that describes a linear kinematic relationship between the charge distributions and the applied bending angles was developed, and the time derivative of the charge model was implemented to the circuit model as a current source. The parameters were estimated by minimizing the error between the real output signal and the simulated output signal from the model. The effectiveness of the model was also tested using the inverse model that reproduces the input bending angles. Finally, the IPMC sensor was mounted on the surface of an upper limb to monitor the knee flexion and extension using the inverse model that was realized in an analog electric circuit using operational amplifiers [23]. Figure 3 shows the IPMC sensor used to measure the angular rotation of human shank and trunk.

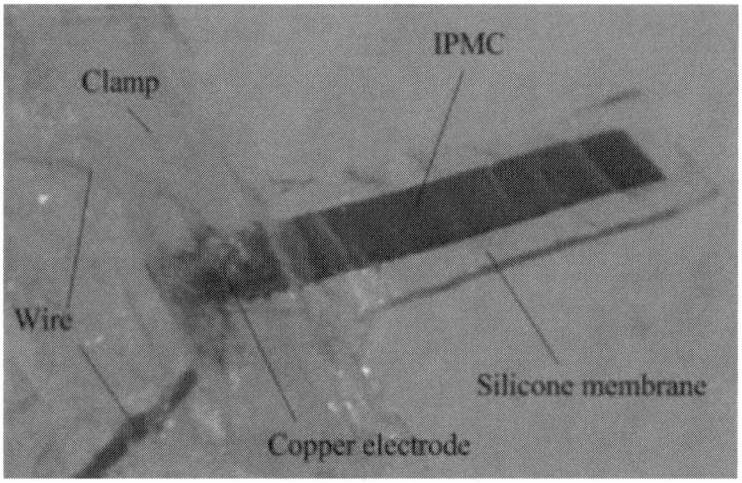

Figure 3. The image of coated IPMC sensor [23].

The main contribution of this paper is the development of a practical model of IPMC sensor which is easily applicable to real world. Another contribution is that the performance of the IPMC sensor is demonstrated in a real biomedical application for lower limbs Robot.
The relation between the current (I) and generation and the bending angle (o) can be obtained as

$$I = K\frac{d\theta}{dt} \tag{4}$$

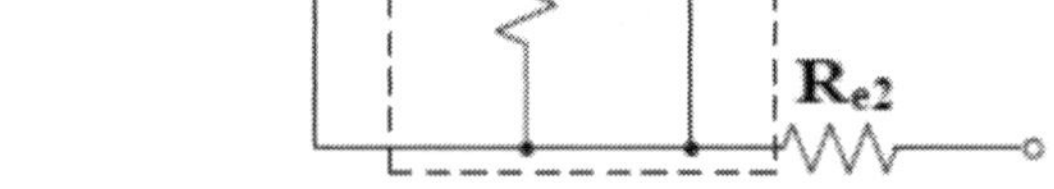

Figure 4 Circuit model of the IPMC sensor [23].

Figure 4 shows the circuit model of IPMC sensor that has the current source. The circuit model is connected to the measurement device through the resistances (R_{e1} and R_{e2}) acting as the resistances along the surface of the IPMC. Since the polymer membrane is sandwiched between the two electrodes, the IPMC has a capacitance (C), as shown in figure 4. R_c represents an ion diffusion resistance, and R represents the resistance of polymer [23].
Because the input buffer resistance of DAQ is much greater than the surface resistances of
IPMC, the effects of R_{e1} and R_{e2} are ignored for the simplification of the model. Then, the output voltage (Vo) is written by

$$V_0 = IZ_T \tag{5}$$

Where Laplace domain expression of Z_T is

$$Z_T = \frac{sRR_cC+R}{s(R+R_c)C+1} \tag{6}$$

After assuming all the initial conditions are zero, the governing differential equation between the input bending angle (θ) and the output voltage (Vo) is represented by

$$V_0' = \frac{1}{(R+R_c)C}V_0 = \frac{KRR_c}{R+R_c}\theta'' + \frac{KR}{(R+R_c)C}\theta' \tag{7}$$

3.2. Modeling of force plat sensor
Force plates are commonly used in biomechanics laboratories to measure ground forces involved in the motion of human subjects. A force plate is simply a metal plate with one or more sensors attached to give an electrical output proportional to the force on the plate. The sensor can either be a strain gauge or a piezoelectric element. Figure 5 shows the architecture of a force plat sensor.

If a force is applied to the center of the plate, each piezo will share the load equally and generate the same charge. An off-center force results in an unequal sharing of the load. In commercial force plates the four piezo outputs are monitored separately and processed electronically to provide measurements of the total force as well as the line of action of the force. In the present case, all four piezos were connected directly in parallel. Each piezo generates a charge proportional to the local force on the piezo. When all four piezos are connected in parallel, the output signal is proportional to the total charge and is hence proportional to the total force on the plate. Consequently, the same signal is generated regardless of whether one stands in the middle of the plate or on one corner, provided all four piezos are closely matched in sensitivity.

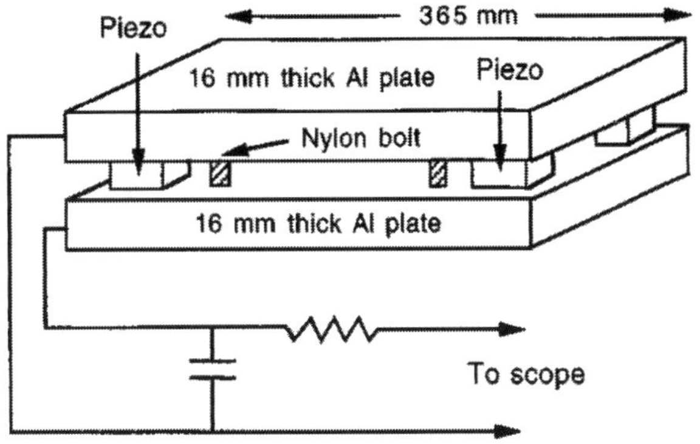

Figure 5. Architecture of force plate

The figure 6 and 7 shows where we place force plate sensor based on located forces in the foot. Two load cells are set at the location of 1and 2 of the force plate. Based on this structure, load (F) can be calculated by summation. Furthermore, the data measured by the force plate can be real-time utilized in the control program

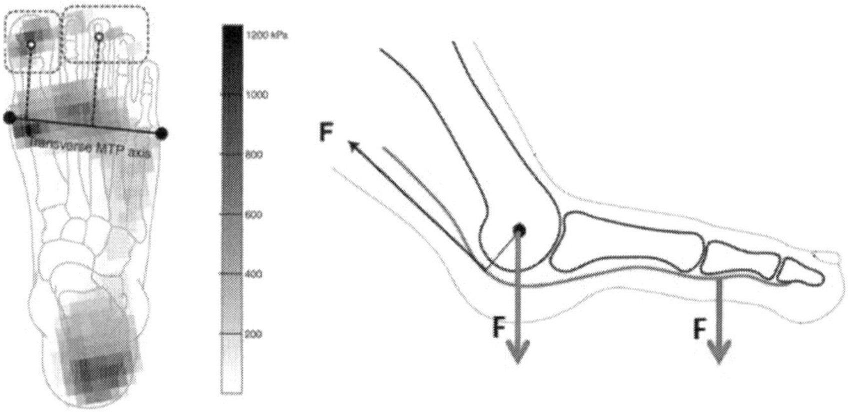

Figure 6. Located forces in the feet

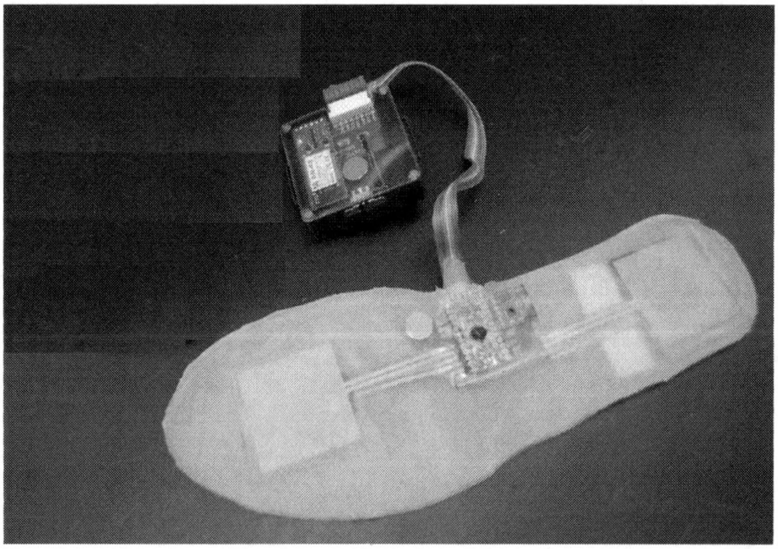

Figure 7. Force plate sensor placed in foot

4. Lower limbs robot design and control system

4.1. Lower Limbs Robot Design

Lower Limbs Robot Design (LLRD) is designed to be comfortable and safe for the patient, and adjustable. LLRD have wearable sensors which controls two types of motors, the first one creates a movement on the knee and second on the ankle, as shown in figure 8. Figure 9 shows the relationship between the sensors and the motors in terms of control system.

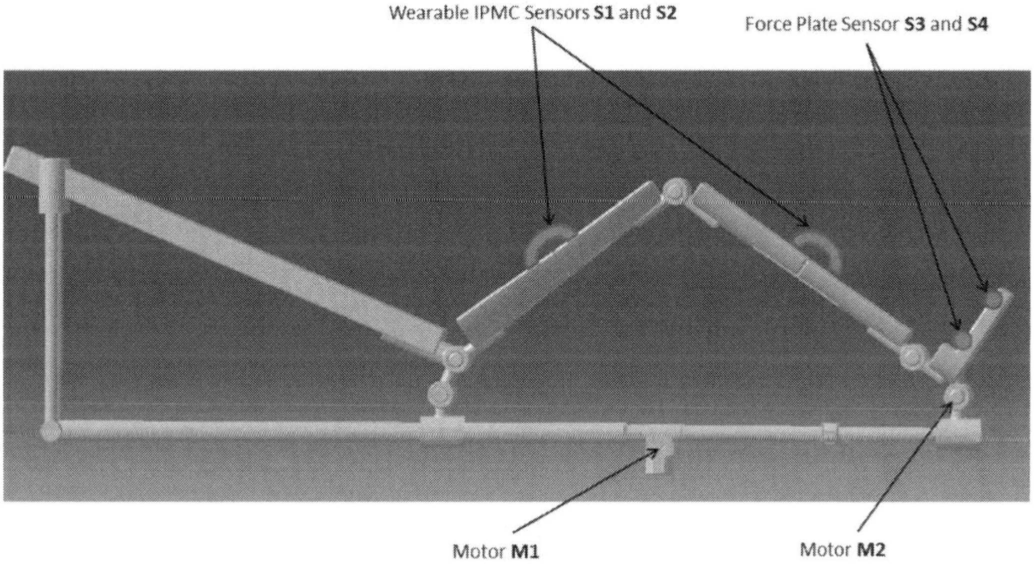

Figure 8. Lower Limbs Robot Design.

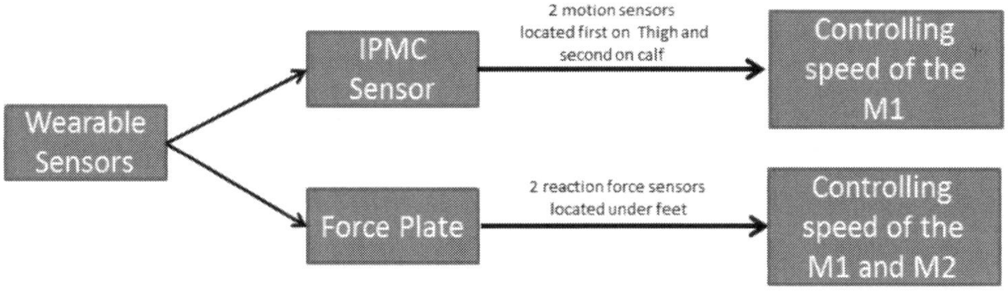

Figure 9. Control the motors by wearable sensors.

4.2. Control system

To meet the needs of active, passive and assisted training, the system adopts the variable servo control modes that include position control, speed control, and current control. In active training, the method of current control is used, so that the active training of every joint can be carried out. In passive training, the method of speed and position control is used, so that the hip, knee, and ankle can be exercised according to the foot trajectory planning. In assisted training, the method of current control, based on the torque feedback and the electro-myographic signal (EMG), is used. Training effects evaluation is based on the feedback of surfaces EMG, motion measurement and force/torque measurement

5. Conclusion

A lower-limb rehabilitation robot with the function of active training, passive training, and assisted training used for a patient with a lower-limb paralysis has been investigated and developed in this paper. The research conducted is based on the HAM concept. The developed rehabilitation robot has a mechanism where three rotary joints consist of three crank rocker mechanisms with an identical module. This feature makes the rehabilitation robot flexible enough to adapt to the variety of user needs. The values of this paper are that it covered the mechanism design, mechanism optimization, kinematics analysis, trajectory planning, motion simulation, control system, and experimental research of the rehabilitation robot. The experimental results have demonstrated that the proposed rehabilitation robot is effective, safe, and reliable and has a potential application. The mechanism developed in this paper provides a good supporting case for the HAM framework.

6. Acknowledgement

This work was supported by the Research Centre Management, Islamic International University Malaysia (Grant No. 2009AA04Z201).

References

[1] Tejima, N. Rehabilitation robotics: a review. Advanced Robotics, 2000, 14(7), 551–564.
[2] Yu, H. Overview of human adaptive mechatronics. In Proceedings of the 9th WSEAS International Conference on Mathematics and computers in business and economics, Bucharest, Romania, June, 2008, 152–157
[3] Zhu, Y., Zhen, R., Liu, H., and Meng, S. The research of rehabilitation robots BCI technology based on EEG. In Proceedings of the 2008 International Conference on Computer science and software engineering, Wuhan, People's Republic of China, 12–14 December 2008, pp. 1107–1110.
[4] Colombo, G., Wirz, M., and Dietz, V. Driven gait orthosis for improvement of locomotor training in paraplegic patients. Spinal Cord, 2001, 39(5), 252– 255.
[5] Colombo, G., Matthias, J., Reinhard, S., and Volker, D. Treadmill training of paraplegic patients using a robotic orthosis. J. Rehabilitation Res. Develop., 2000, 37(6), 693–700.
[6] Veneman, J. F., Kruidhof, R., Hekman, E. E. G., Ekkelenkamp, R., Asseldonk van, E. H. F., and Kooij van der, H. Design and evaluation of the LOPES exoskeleton robot for interactive gait rehabilitation. IEEE Trans. Neural Syst. Rehabil. Engng, 2007, 15(3), 379 386.
[7] Veneman, J. F, Ekkelenkamp, R., Kruidhof, R., Helm van der, F. C. T., and Kooij van der, H. A series elastic- and bowden-cable-based actuation system for use as torque actuator in exoskeletontype robots. Int. J. Robotics Res., 2006, 25(3), 261– 281.
[8] Cullell, A., Moreno, J. C., Rocon, E., Forner Cordero, A., and Pons, J. L. Biologically based design of an actuator system for a knee-ankle-foot orthosis. Mech. Machine Theory, 2009, 44(4), 860– 872.
[9] Pennycott, A., Hunt, K. J., Jack, L. P., Perretb, C., and Kakebeeke, T. H. Estimation and volitional feedback control of active work rate during robotassisted gait. Contr. Engng Practice, 2009, 17(2), 322–328.
[10] Miyamura, T., Murai, S., and Tomisaki, H. Therapeutic Exercise Machine LX2. Yaskawa Technical Review, 2003, 67(3), 224–228.
[11] Saito, Y., Kikuchi, K., Negoto, H., Oshima, T., and Haneyoshi, T. Development of externally powered lower limb orthosis with bilateral-servo actuator. In Proceedings of the International Conference on Rehabilitation robotics, Chicago, Illinois, 28 June–1 July 2005, pp. 394–399.
[12] Hunt, K. J., Stone, B., Negard, N. O., Schauer, T., Fraser, M. H., Cathcart, A. J., Ferrario, C., Ward, S. A., and Grant, S. Control strategies for integration of electric motor assist and functional electrical stimulation in paraplegic cycling: utility for exercise testing and mobile cycling. IEEE Trans. Neural Syst. Rehabil. Engng, 2004, 12(1), 89–101.

[13] Chen, S., Wang, Y., Li, S., Wang, G., Huang, Y., and Mao, X. Lower limb rehabilitation robot. SME/ IFToMM International Conference on Reconfigurable mechanisms and robots, London, UK, 22–24 June 2009, pp. 439–443.

[14] Yeh, S. and Yu, H. Design and implementation of a lower-limb mobile training robot. IEEE International Conference on Systems, man. and cybernetics, Singapore, 12–15 October 2008, pp. 3459–3464.

[15] Beyl, P., Damme, M. V., Ham, R. V., Vanderborght, B., and Lefeber, D. Design and control of a lower limb exoskeleton for robot-assisted gait training. Appl. Bionics Biomech., 2009, 6(2), 229–243.

[16] Simon, A. M., Gillespie, R. B., and Ferris, D. P. Symmetry-based resistance as a novel means of lower limb rehabilitation. J. Biomech., 2007, 40(6), 1286–1292.

[17] Costa, N., Bezdicek, M., Brown, M., Gray, J. O., Caldwell, D. G., and Hutchins, S. Joint motion control of a powered lower limb orthosis for rehabilitation. Intl J. Autom. Comput., 2006, 3(3), 271–281.

[18] Willems, T., Witvrouw, E., Verstuyft, J., Vaes, P., and De Clercq, D. Proprioception and muscle strength in subjects with a history of ankle sprains and chronic instability. J. Athletic Training, 2002, 37(4), 487–493.

[19] Tsoi, Y. H. and Xie, S. Q. Impedance control of ankle rehabilitation robot. Proceedings of the 2008 IEEE International Conference on Robotics and biomimetics, Bangkok, Thailand, 14–17 December 2008, pp. 840–845.

[20] Saglia, J. A., Tsagarakis, N. G., Dai, J. S., and Caldwell, D. G. A high-performance redundantly actuated parallel mechanism for ankle rehabilitation. Int. J. Robotics Res., 2009, 28(9), 1216–1227.

[21] Popescu, V. G., Burdea, G. C., Bouzit, M., and Hentz, V. R. A virtual reality-based telerehabilitation system with force feedback. IEEE Trans. Inf. Technol. Biomed., 2000, 4(1), 45–51.

[22] Butner, E. S. and Ghodoussi, M. Transforming a surgical robot for human telesurgery. IEEE Trans. Robotics Autom., 2003, 19(5), 818–824. 23 Jack, D., Boian, R., Merians, A. S., Tremaine, M., Burdea, G. C., Adamovich, S. V., Recce, M., and Poizner, H. Virtual reality-enhanced stroke rehabilitation. IEEE Trans. Neural Syst. Rehabil. Engng, 2001, 9(3), 308–318.

[23] Kiwon, P. Bonghwan, L. Hyung,M. Choi, K. Gunyong, H. Byun, G and Hyung, K . IPMC Based Biosensor for the detection of Biceps Brachii Muscle Movements . International Journal of Electrochemical Science, pp. 4098-4109.

[24] Girone, M., Burdea, G., Bouzit, M., Popescu, V., and Deutsch, J. E. A stewart platform-based system for ankle telerehabilitation. Autonomous Robots, 2001, 10(2), 203–212.

[25] Zhang, M. and Li, X. J. A design of the mechanism for lower limbs passive rehabilitation and kinematics analysis. In Proceedings of the Fifth International Conference on Frontier of computer science and technology (FCST), Changchun, China, 18–22 August 2010, pp. 603–607.

[26] Zhang, Q. and Gu, L. Human engineering. Beijing Institute of Technology Press, Beijing, June 2009.

Hybrid pid and pso-based control for electric power assist steering system for electric vehicle

R A Hanifah[1], S F Toha[2*] and S Ahmad[3]
Department of Mechatronics, Faculty of Engineering, International Islamic University Malaysia, Kuala Lumpur, Malaysia

E-mail:[1]rabiatul.hanifah@gmail.com, [2]tsfauziah@iium.edu.my, [3]salmiah@iium.edu.my

Abstract: Electric power assist steering (EPAS) system provides an important significance in enhancing the driving performance of a vehicle with its energy-conserving features. This paper presents a hybrid PID (Proportional-Integral-Derivative) and particle swarm optimization (PSO) based control scheme to minimize energy consumption for EPAS. This single objective optimization scheme is realized using the PSO technique in searching for best gain parameters of the PID controller. The fast tuning feature of this optimum PID controller produced high-quality solutions. Simulation results show the performance and effectiveness of the hybrid PSO-PID based controller as opposed to the conventional PID controller.

Keywords: Electric Power Assist Steering (EPAS); C-type EPAS; PID; Particle Swarm Optimization (PSO); electric vehicle.

1. Introduction

In electric vehicle technology, batery served as a power supply to drive the vehicle. Hence, it has been a major concern in EV technology in terms of battery energy capacity to support the long-range operation [1]. As the result of this limitation give an opportunity to electric assist power steering (EPAS) system by replacing mechanical steering system. Mechanical steering system or hydraulic power assist steering (HPAS) draws a constant energy supply from battery to maintain the pressure in hydraulic pump [2]. Whereas the EPAS system improves the energy efficiency due to its on-demand system feature which is only operating when the steering wheel is turned [3].

It offers an additional advantages than hydraulic power steering as it can reduce the steering torque, provide various steering feel and improve return-to-center performance of a steering wheel when it is steered [4]. The EPAS system becomes an alternative to the automotive manufacturer in providing the convenience of steering assist without adding value to the engine cost and fuel consumption. Only approximately one-

* Corresponding author. Tel: +603 6196 5705

twentieth of energy is consumed by the EPAS system compared to the conventional HPAS system and this feature make EPAS a very suitable candidate in EV steering system [5].

The electric motor is attached to the steering rack or column via gear mechanism and sensors are located on the input shaft in EPAS system. It uses electromechanical actuation where the sensors determine the driver's torque, steering angle and speed and direction of the steering wheel. The sensors together with the vehicle velocity are fed into the ECU. The resulting demand from the ECU process is used to excite the circuitry of the motor and finally giving an output to the rack [6].

In the research on EPAS system, the primary goals are to reduce the steering torque exerted by the driver and also to improve the steering performance. In line with these goals, many control strategies have been applied and implemented in the EPAS system such as a reference model, fuzzy logic, PID, H_2 and H_∞ [7-11].

2.0 PID Controller

Proportional-Integral-Derivative (PID) control technique continues to provide the simplest and effective solutions in industrial control system application since its introduction in 1936. The PID controller has been widely used in the industry because of its simple structure and robust performance as well as low cost and cheap maintenance. [12, 13]. Three individual values of proportional, integral and derivative parameters involved in the PID controller calculation. The proportional value determines the reaction of the current error, the integral value determines the reaction based on the sum of recent errors, and the derivative value determines the reaction based on the rate at which the error has been changing the weighted sum of these three actions is used to adjust the process via the final control element [14].

The PID controller tuning aim to determine its proportional, integral and derivative parameters that fit into the system to assure the closed loop system performance specification and robust performance of control loop over a wide range of operation are met [14]. In practice, it is difficult to ensure the system meets all the desirable qualities simply because the behavior of the parameters. For example, in achieving a better transient response of the system, its response under disturbance condition will be slow.

In the industrial plants, it has been difficult to tune the PID controller gains properly because of the uncertainties encountered in control system both in the environment and within the system [15]. Resulting from this, the conventional PID controller cannot provide solutions to all control problems because of the complexity, inadequate dynamic definition and time-variant with delays and non-linearity [13]. To improve the conventional PID, many optimization methods have been developed such as auto tuning and intelligent PID controller.

3.0 Particle Swarm Optimization (PSO)

Particle Swarm Optimization is a population based evolutionary algorithm which originally developed by J. Kennedy and R. C. Eberhart on 1995 [16]. It was developed through simulation of a simplified social system, and has been found to be robust in solving continuous nonlinear optimization problems [12]. Using this algorithm, the solution obtained is a high-quality solution which is generated within the shorter calculation time and can achieve stable convergence characteristic.

This method is developed from research on swarm such as bird flocking and fish schooling. It has features of easy implemented technique, stable characteristic convergence and good computational efficiency[17]. Instead of using evolutionary operators such as mutation and crossover to manipulate the particle, a flock of particles is put into the d-dimensional search space for a d-variable optimization problem. Each particle flies within the search space with velocities and positions which is dynamically adjusted according to its

IOP Conf. Series: Materials Science and Engineering **53** (2013) 012039 doi:10.1088/1757-899X/53/1/012039

own flying experience and its neighbors' flying experience [12, 14]. A fitness function is used in determining the optimal solution. At each iteration, the fitness of each particle is evaluated. Each particle keeps track of its own best position and its best fitness in the problem space it has achieved so far. This is called particle best or pbest. The algorithm used for updating the particle velocity and position at each iteration is in equation (1) and (2) below:

$$v_{id_{new}} = (w \times v_{id}) + C_1\left(rand_1(p_{id} - x_{id})\right) + C_2\left(rand_2\left(p_{gd} - x_{id}\right)\right) \tag{1}$$

$$x_{id_{new}} = x_{id} + v_{id_{new}} \tag{2}$$

PSO also tracked another best value associated with the location of the best fitness particle which is called as global best or gbest. These two best values influence how the solution change or particle movement in the search space. PSO is faster in finding solutions than evolutionary computation methods because it produces random initial population and generates new population based on current cost [18].

4.0 System Modeling

There are four types of EPAS system which are column-type (C-type), pinion type (P-type), rack type (R-type) and dual pinion type. This paper uses C-type EPAS which consist of a torque sensor, an electric motor, a reduction gear, a column and a rack-pinion mechanism. Figure 1 shows the schematic diagram of EPAS system.

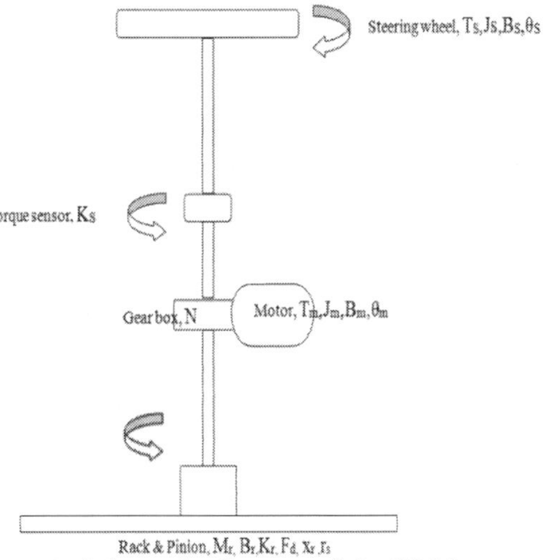

Figure 1. Schematic diagram of the EPAS system

The mathematical model of the EPAS system can be described as equations shown below,

$$J_s\ddot{\theta}_s + B_s\dot{\theta}_s + K_s\theta_s = T_s + K_s\left(\frac{x_r}{r_s}\right) \tag{3}$$

$$J_m\ddot{\theta}_m + B_m\dot{\theta}_m + K_m\,\theta_m = T_m + GK_m\left(\frac{x_r}{r_s}\right) \tag{4}$$

$$M_r\ddot{x}_r + B_r\dot{x}_r + K_r x_r = \frac{K_s\theta_s}{r_s} + \frac{GK_m\theta_m}{r_s} - F_d \tag{5}$$

where Js is the steering column moment of inertia; Bs is the steering column viscous damping; Ks is the steering column stiffness; Jm is the motor moment of inertia; Bm is the motor viscous damping; Km is the motor stiffness; G is the motor gear ratio; Mr is the mass of the rack; Br is the rack viscous damping; Kr is the tire spring rate; Fd is the road random force; xr is the rack position; rs is the pinion radius; Ts is the steering wheel torque from the driver and Tm is the electromagnetic torque provided by electric motor.

According to the mathematical model, a Simulink diagram is designed. PID controller is used in the system to help minimizing the current and hence reducing its power consumption . The PID is known for its satisfying impact to the system performance, however it needs some tuning optimization to ensure maximum performance. Hence, the hybrid PID and PSO are applied in the system to improve the tuning of the PID controller. The mechanism is illustrated in the Figure 2.

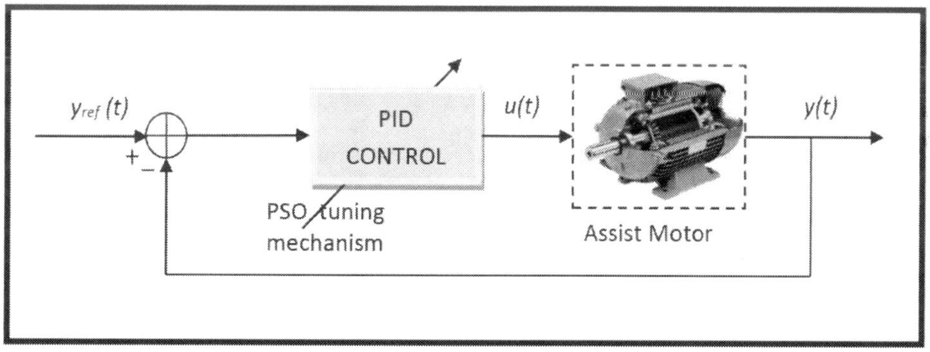

Figure 2. PID controller with PSO tuning of the EPAS subsystem

5.0 Simulation and Result

From the mathematical model obtained, a Simulink diagram is constructed. The system utilizes the PID controller to reduce the power consumption of EPAS system. The hybrid PSO-PID controller is then applied to the system to find the optimal tuning performance of the PID controller. This approach is opposed to the trial and error technique used in conventional PID tuning.

Preliminary test is conducted to validate all parameters of PID controller and justify the potential of reducing the EPAS system power consumption. Figure 3 and Figure 4 show the performance of PID controller with random signal as the input and sinusoidal input signal respectively.

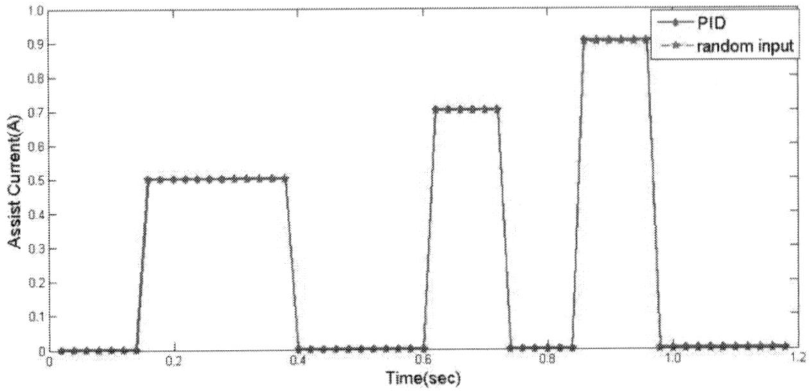

Figure 3. Controller performance with random input signal.

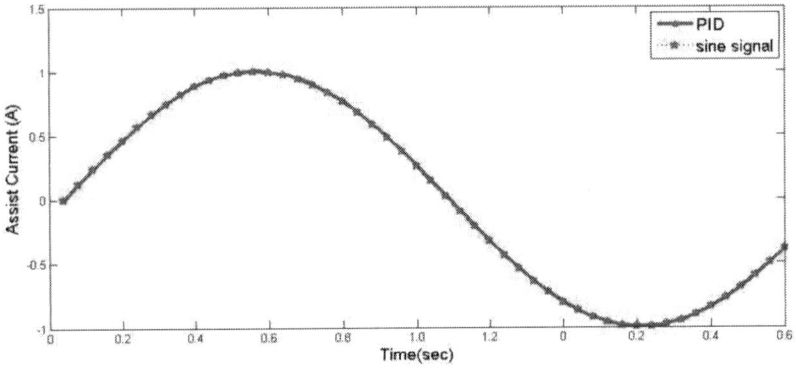

Figure 4. Controller performance with sinusoidal input signal.

From Figure 3 and Figure 4, it shows that the PID controller provides a good control signal and able to control current accordingly. The test was simulated under a constant speed of 30km/h and at random driver's torque. It shows that the PID controller was able to control the assist current supplied to the motor as per requirement. However, the PID controller is tuned by trial and error which is time consuming and not the perfect optimal tuning.

Thus, to reduce the time consumption in parameters tuning of the PID hybrid PSO-PID is applied. In PSO, a swarm of particles is defined and each particle will be evaluated for its fitness at each iteration within the swarm. Particle with best fitness value is called local best and the best among the local best is called global best. This global best value is the optimum solution for the tuning. Parameters used in PSO algorithm are tabulated in Table 1. Figure 5 illustrates the controller performance for PID and hybrid PSO-PID controllers. Hybrid PID_PSO shows a better performance than PID controller. Simulation on power consumption shows that PSO-PID based controller produced a smooth power compared to PID controller. Figure 6 illustrates the comparison.

Table 1. Parameters of PSO algorithm

Parameters	Value
Number of iterations	50
Number of variables	3 (Kp,Ki,Kd)
Swarm	70
Inertia weight, w	Wmax = 0.9, Wmin = 0.4

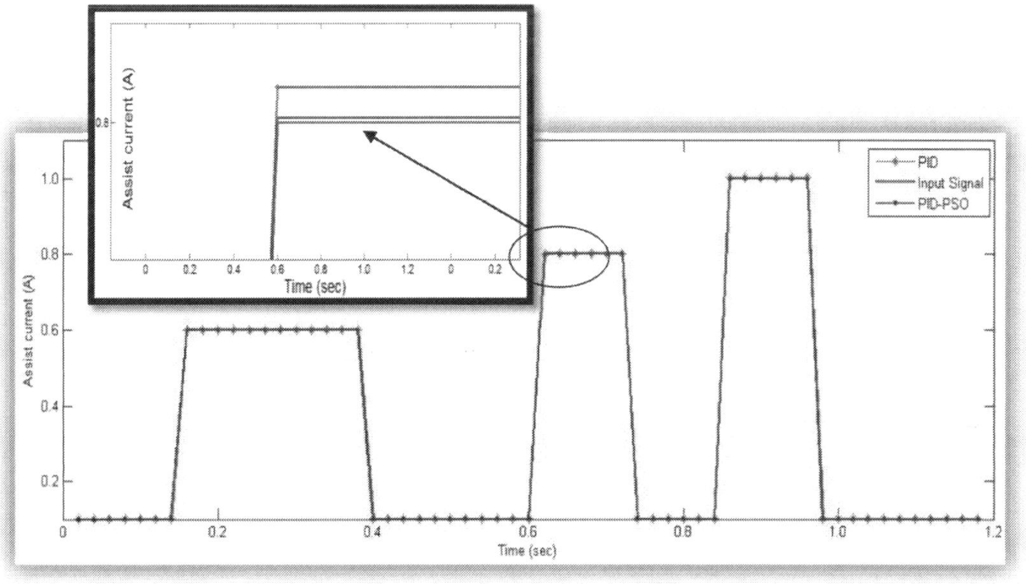

Figure 5. Controller performance of PID controller and PSO-PID controller

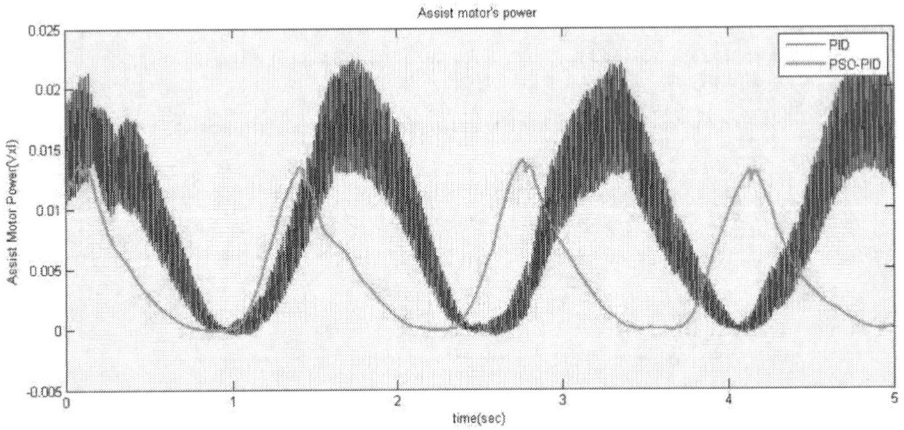

Figure 6. Power consumption comparison of PID controller and PSO-PID controller

6.0 Conclusion

The simulation done for PID and PSO-PID based shows an improvement in terms of assist current control and also power consumption of the assist motor. The result shows that PSO-PID controller able to control the current almost accurately and produce a smooth motor power. However, further improvement need to be carried out in the case of dynamic vehicle velocity. The PSO-PID based controller shows a potential in optimizing energy efficiency of EPAS system.

References

[1] Cunningham, J.S., *An Analysis of Battery Electric Vehicle Production Projections*, in *Department of Mechanical Engineering*. 2009, Massachusetts Institute of Technology. p. 30.

[2] Hassan, M.K., et al., *Optimal Design of Electric Power Assisted Steering System (EPAS) Using GA-PID Method.* Procedia Engineering, 2012. **41**(0): p. 614-621.

[3] Tsung-Hsien, H., et al. *Design of control logic and compensation strategy for electric power steering systems*. in *VPPC '08. IEEE Vehicle Power and Propulsion Conference, 2008.*

[4] Chih-Jung, Y., et al. *Development of a Test Bench for Tuning and Validating Electric Power Steering Control Method*. in *VPPC 2007. IEEE Vehicle Power and Propulsion Conference, 2007.*

[5] Hao-bin, J., et al. *Low-pass filter based automotive EPS controller and comparative full-vehicle tests*. in *Control and Decision Conference, 2008. CCDC 2008. Chinese*. 2008.

[6] Bhangu, B. and C. Bingham, *GA-tuning of Nonlinear Observers for Sensorless Control of Automotive Power Steering IPMSMs*, in *Vehicle Power and Propulsion 2005 IEEE Conference*. 2005. p. 772- 779.

[7] Wei, J.S., et al. *The Integrated Control of Electronic Suspension and Steering System Based on Fuzzy Logic Method in Vehicle*. in *FSKD 2007. Fourth International Conference on Fuzzy Systems and Knowledge Discovery, 2007.*

[8] Zhang, H., et al. *Modeling and characteristic curves of electric power steering system*. in *2009. PEDS 2009. International Conference on Power Electronics and Drive Systems*, 2009.

[9] Marouf, A., et al. *Control of an Electric Power Assisted Steering system using reference model.* in *2011 50th IEEE Conference on Decision and Control and European Control Conference (CDC-ECC),* 2011.
[10] Huaiquan, Z. and L. Min. *Fuzzy Neural Network PID Control for Electric Power Steering System.* in *2007 IEEE International Conference on Automation and Logistics,* 2007.
[11] Xiang, C., et al., *A Generic Model-Based Advanced Control of Electric Power-Assisted Steering Systems.* IEEE Transactions on Control Systems Technology, 2008. **16**(6): p. 1289-1300.
[12] Zwe-Lee, G., *A particle swarm optimization approach for optimum design of PID controller in AVR system.* IEEE Transactions on Energy Conversion, 2004. **19**(2): p. 384-391.
[13] Kumar, V., Rana, K.P.S., & Gupta, V. , *Real-Time Performance Evaluation of a Fuzzy PI + Fuzzy PD Controller for Liquid-Level Process.* International Journal of Intelligent Control and Systems, 2008. **13**(2): p. 89-96.
[14] Nagaraj, B. and N. Murugananth. *A comparative study of PID controller tuning using GA, EP, PSO and ACO.* in *2010 IEEE International Conference on Communication Control and Computing Technologies (ICCCCT),* 2010.
[15] Zhicheng, X. *A novel robust PID controller design method.* in *2010 International Conference on.Computer Application and System Modeling (ICCASM),* 2010.
[16] Xu, Z. *A novel robust PID controller design method.* in *2010 International Conference on* IEEE *Computer Application and System Modeling (ICCASM),* 2010.
[17] Eberhart, R.C. and Y. Shi. *Comparison between genetic algorithms and particle swarm optimization.* in *Evolutionary Programming VII.* 1998: Springer.
[18] Toha, S.F. and M.O. Tokhi, *PID and inverse-model-based control of a twin rotor system.* Robotica, 2011. **29**(06): p. 929-938.

IOP Conf. Series: Materials Science and Engineering **53** (2013) 012040 doi:10.1088/1757-899X/53/1/012040

Testing of Lightweight Fuel Cell Vehicles System at Low Speeds with Energy Efficiency Analysis

Muhammad Rizuwan B. Mustaffa, Wan Ahmad Najmi B. Wan Mohamed

Alternative Energy Research Centre (AERC), Faculty of Mechanical Engineering, Universiti Teknologi MARA (UiTM), Malaysia

Email : rizuwan.mustaffa@yahoo.com

Abstract. A fuel cell vehicle power train mini test bench was developed which consists of a 1 kW open cathode hydrogen fuel cell, electric motor, wheel, gearing system, DC/DC converter and vehicle control system (VCS). Energy efficiency identification and energy flow evaluation is a useful tool in identifying a detail performance of each component and sub-systems in a fuel cell vehicle system configuration. Three artificial traction loads was simulated at 30 kg, 40 kg and 50 kg force on a single wheel drive configuration. The wheel speed range reported here covers from idle to 16 km/h (low speed range) as a preliminary input in the research work frame. The test result shows that the system efficiency is 84.5 percent when the energy flow is considered from the fuel cell to the wheel and 279 watts of electrical power was produced by the fuel cell during that time. Dynamic system responses was also identified as the load increases beyond the motor traction capabilities where the losses at the converter and motor controller increased significantly as it tries to meet the motor traction power demands. This work is currently being further expanded within the work frame of developing a road-worthy fuel cell vehicle.

1. Introduction

Fuel cell vehicles (FCV) are widely perceived as a long-term technology in transportation with the promise of clean emission exhaust and efficient vehicle system [1]. By electrochemical reaction, fuel cells (FC) generates electricity and releases water and heat as by-products, with the advantage of having a fuel (hydrogen) that can be produced from many sources. The commercialization of electric vehicle (EV) in stand-alone and hybrid versions has set the stage for FCV as the next big step for automotive industries to achieve sustainable mobility while accounting for CO_2 gas emission and sustainable energy sources [2].

FCV consist of a fuel cell system that is integrated with other major component such as the vehicle and power controller modules, DC/DC converter and electric traction drive motor. The electrical power produced by the fuel cells is converted into mechanical power by traction motors and between that the DC/DC converter is required to supply constant voltage to motor. The performances of fuel cell system, DC/DC converter and traction motors contributes to the overall fuel efficiency; thus, a good system design traces the energy flow within the system and an intelligent energy management is implemented based on the energy efficiency for optimum fuel consumption.

[1] No17, Jalan Pahlawan 3, Taman Pahlawan, 42500, Selangor, Malaysia.

Interest in FCV research and technology are growing among academic institutions. Previously, the now defunct Formula Zero competition that promotes hydrogen as vehicle propulsion source has sparked the interest among the universities and research institute to develop FCVs. The Foundation for the Development of New Hydrogen Technologies in Aragon has developed a single seated FCV that is able to run at a maximum speed of 150 km/h [3]. Then, the University of Tainan developed the light electric vehicle (LEV) which was powered by a 6 kW PEM fuel cell and able to drive up to 60 km/h maximum speed. FCV development is not limited to car-type vehicles. At Birmingham campus a group of researchers has converted an electric battery scooter into a fuel cell powered scooter with a specific range around 9 km/kWh and maximum range of 15 miles.

Honda, as a major automotive manufacturer, also plays a key role in promoting FCVs. FCX Clarity is a car model that revolutionizes the automotive industry. With a maximum power of 100 kW, the FCX Clarity is also capable of giving a smooth, powerful and continuous extendable acceleration feel. In terms of fuel economy, it has been enhanced up to 26 percent from the previous FCV version with the energy efficiency at 60 percent and is scheduled for commercial release in 2015 [6].

Performance study in FCV system is a required step in order to develop a highly efficient propulsion system. In [7], the study of power train performances for FCV concluded that the energy consumption is highly related to the acting forces on the vehicle and an efficiency mapping of the traction motor was developed as an energy management strategy. Detail energy flow analysis and variable traction loads is also a required step for integration with the overall energy mapping [8][9], as it will bring a big impact for vehicle fuel consumption.

Automotive propulsion system testing is traditionally performed on a dynamometer system where an artificial load is applied to the system and the behavior of the propulsion is recorded in adapting to the required load. Due to the non-existent of a commercial test bench product for FCV analysis, a mini FCV test bench was self-developed with a single wheel-motor configuration. The traction loads are mechanically applied on the wheel to simulate different driving conditions. This FCV test bench is applied here and the paper reports the energy flow analysis of a lightweight FCV propulsion system at low speeds as reference for future works. The energy analysis acquired represents part of an overall energy management strategy for UiTM's self-developed three-wheel FCV from the basis of first-order energy flow mapping.

2. FCV Description

The lightweight FCV propulsion system was developed based on the Shell Eco-Marathon Asia competition requirement. The details of the FCV power train configuration and test bench is explained briefly in subsequent sections.

2.1. FCV Configuration

FCV power plant is based on the fuel cell stand-alone system where the power is totally supplied by the fuel cell only without any support from any energy storage devices. The power train consist of a 1 kW Polymer Electrolyte Membrane (PEM) fuel cell, co-developed with G-Energy Technologies (M), and an electric motor.

2.1.1. Fuel cell stack

The 65 cell FC stack is an open cathode design where the air supply acts as both the reactant and cooling agent. The maximum rated power is 1 kW (33V at 30A) and was developed using Membrane Electrode Assemblies developed locally by G-Energy Technologies (M). The graphite bipolar plates integrate parallel flow channels for hydrogen flow and straight flow channels for air flow.

2.1.2. Vehicle control system

All the subsystem is controlled by an *Ecomat*mobile basic controller from IFM Electronic. The *Ecomat*mobile operates at 24 volt and provides a central vehicle system controller where all the input from the fuel cell stack is measured and monitored before distributed to the subsystems. It is also capable of operating as a data acquisition system.

2.1.3. DC/DC Converter

The DC/DC converter used in this work is a step down converter which converts the dynamic input voltage from the fuel cell around 65 V - 36 V to 24 V. The converter manufacturer is Zahn Electronics Inc. With efficiency around 97 percent, it contributes highly to an efficient propulsion system.

2.1.4. Electric Propulsion System

The propulsion system consists of an electric motor, motor controller and gearing system. The propulsion motor is a DC motor manufactured by Midwest Motion Product (MMP), with a peak torque of 0.6 Nm and the maximum speed is 4530 rpm. It is operated at 24 V and weighs around 2 kg.

2.2. Test Bench

The mini FCV test bench was developed for lab-scale testing to analyze the behaviour of a simple FCV propulsion system configuration and sub-components performance. Figure 1 shows the mini FCV test bench schematic. The hydrogen is fed by a compressed hydrogen tank to the fuel cell and a hydrogen flow meter records the hydrogen flow. The fuel cell converts the hydrogen energy into electrical energy. The electricity produced goes through the DC/DC converter to maintain the voltage at 24 V. The stabilized electrical power is then supplied to the electric motor based on the load demand. Power trains convert the electricity to mechanical movement. At every component of FCV, current sensors are installed so that the amount of electrical energy usage at each sub-component level can be recorded for energy flow analysis.

The artificial mechanical load is developed based on the inertia concept where the load is provided by the weights that represents the inertia of actual vehicles at various driving conditions. Other than that, the test bench is also able to simulate the rolling resistance acting on the wheel.

For this study, the experimental methodology is as follows:

- Three loads are given at 30 kg, 40 kg and 50 kg. The loadings represent the vehicle weight received by one wheel, for example a load of 30 kg represents 120 kg weight of vehicle (30 kg for each wheel).
- The driving conditions are based on starting speed (3 km/h) and low speed cruising (4,5,6,8,10,12,14 and 16 km/h).
- Each experiment at different loads is repeated 3 times for data reliability.
- The recorded data are hydrogen consumed, fuel cell voltage and current, motor current and controller current.

3. Result and Discussion

The performances analyses are conducted from 3 to 16 km/h except for loading at 50 kg. At this high traction load, the electric motor does not have enough torque (with a fixed transmission gear ratio of 12) to overcome the inertia acting on the wheel. The 50 kg load is a representation of the forces acting on the wheel when a 90 kg vehicle climbs a 3° slope.

3.1.1. Fuel Cell Stack Performances

The performance of the fuel cell stack power over vehicle speed is shown in figure 2. Fuel cell power generation increases with vehicle speed due to the higher demands of the electrical motor at high speeds. At the starting condition the power consumed for the three different vehicle loads are nearly similar. When the vehicle starts to move, the fuel cell produces the power proportional to the applied artificial loads as higher loads increases the rolling resistance and inertia of the vehicle. The power generation is highly dependent on the type of motor used; here, the motor power demand is linear to the increase in speed.

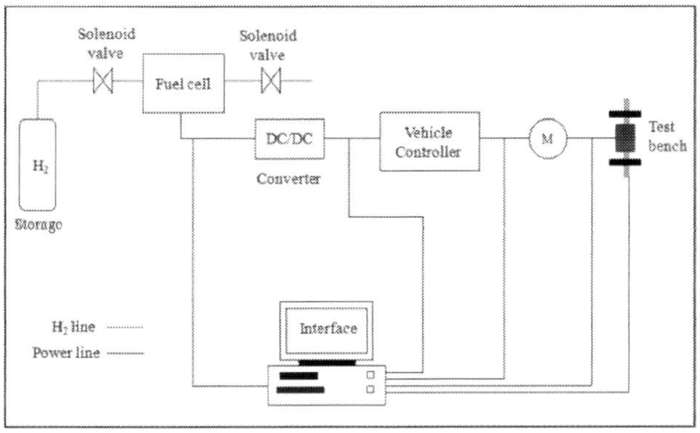

Figure 1. Test bench diagram for the fuel cell vehicle system.

Consequently, a fuel cell operation generates thermal energy as its by-product through the irreversibilities within the electrochemical reaction. Figure 3 traces the changes in stack temperature relative to the load and speed variations. The increase of stack temperature is around 14% to 16% and the highest increase is at the highest load where the power demand is higher. At these low operating speeds, the fuel cell cooling mechanism is still capable to handle the generated heat and maintain a low stack temperature (max 27°C) at all times.

3.1.2. Fuel Consumption

Fuel consumption during the testing is recorded using the hydrogen flow meter (before and after load is applied) and presented in Figure 4. Theoretically, a fuel cell hydrogen consumption is a direct conversion of the electrical current produced. In this case, the current is the load demand by the motor which varies with the artificial load given and the vehicle speed requirements. The results show that the measured hydrogen consumption closely follows the theory where the consumption increases linearly with the speed. For example, 2.8 litres per minutes of hydrogen is needed at 16 km/h for the traction load of 30 kg and increases by 200% as the traction load increases by 10 kg.

3.1.3. FCV System Efficiency

Figure 5 shows an example of the energy efficiency mapping (illustrated using the Sankey diagram) of the FCV propulsion system during cruising at the maximum speed, excluding the fuel cell stack itself. Here, the energy flow is traced downstream of the stack as the stack operating parameters in this study was not critically measured. The most efficient energy flow from the three different load settings can be identified from the efficiency comparison for the transmission system, motor controller, vehicle control system (VCS) and DC/DC converter.

At 40 kg and 16 km/h, the major energy loss from the system is contributed by the DC/DC converter at 10.6%. The VCS and motor controller consumes around 2.8% and 1.9% respectively from the total energy. Therefore, approximately 84.5% of electrical energy generated by the fuel cell is successfully transmitted to the wheel for propulsion.

Figure 6 shows the comparison of efficiency mapping between the three conditions tested at its maximum experimental speeds. The 30 kg and 40 kg traction loads has near similar efficiency values for each sub-systems compared to the condition at 50 kg load. Efficiency mapping at 50 kg load shows that much higher loss occurs at the DC/DC converter (46.1%) and motor controller (40.9%). This proves the dynamic relationship between the sub-system components at different operating conditions and is an important element in the development of the power management strategies of a FCV.

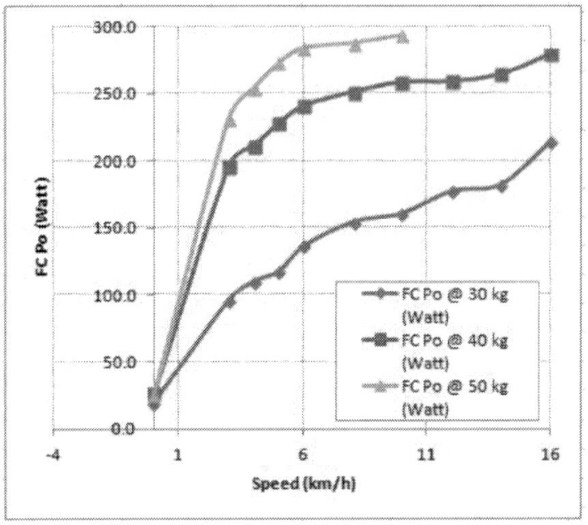

Figure 2. Performances of fuel cell power (watt) over vehicles speed (km/h)

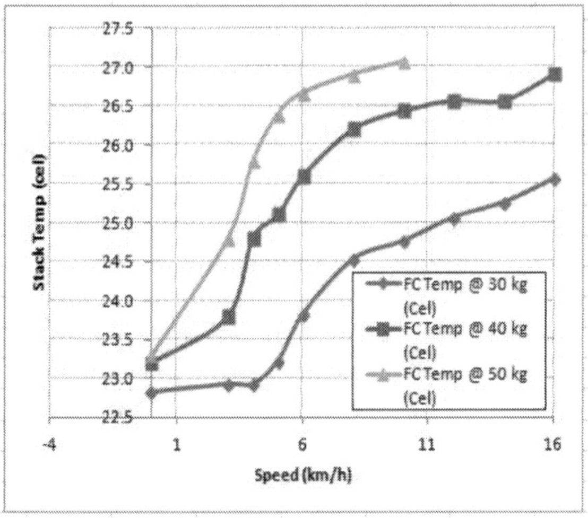

Figure 3. Performances of fuel cell temperature (°C) over vehicle speed (km/h)

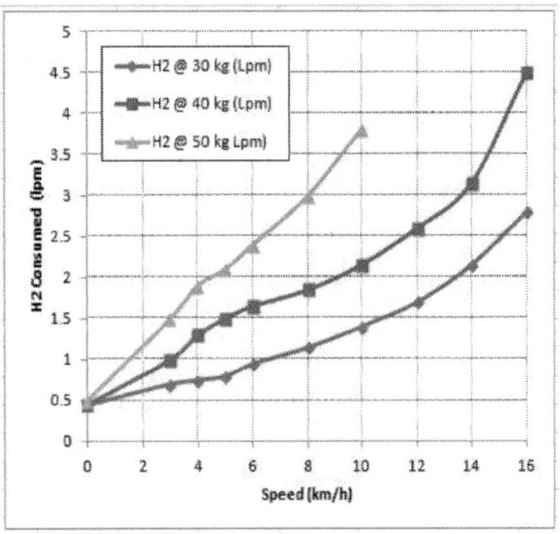

Figure 4. Hydrogen consumption rate (in Lpm) at variable vehicle speed and traction load

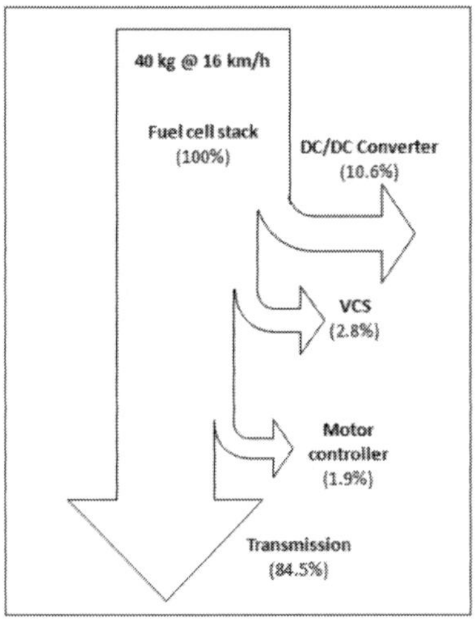

Figure 5. Example of energy efficiency mapping using the Sankey diagram method at 16km/h with traction load of 40 kg .

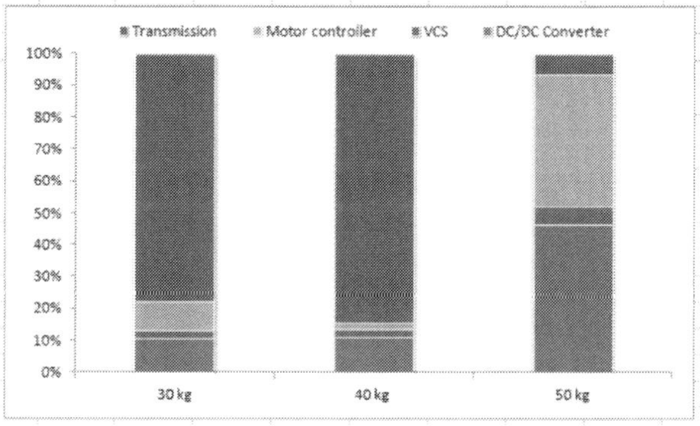

Figure 6. Comparison of efficiency map for three experiments tested at the maximum speed

4. Conclusions

A mini test bench was developed for a lightweight FCV design performance and the energy efficiency mapping was reported for low speed drive conditions. The fuel cell acts accordingly with the required load demand from the electrical motor and a linear temperature increase was noted as the power generation increase. The energy flow mapping based on the Sankey diagram clearly provides a good tool in tracing the energy losses as the generated energy flows from the fuel cell to the wheel. However, limitations in fuel cell operating data measurement limit this analysis from conducting a comprehensive well-to-wheel energy analysis. The energy efficiency comparison indicates that the sub-systems react sensitively to the selection of the motor for the right driving condition. In this case, it was seen that the inability of the electrical motor to adequately supply the required torque at 50 kg traction loads (equivalent to a 3° climb) increases the energy losses significantly at the converter and motor controller. This dynamic system response will be an avenue for future study to improve the energy management strategy for an FCV.

Acknowledgments

The authors would like to acknowledge Excellence Fund 600-RMI/ST/DANA/5/3(477/2011) from Research Management Institute (RMI) of UiTM for funding this research. The authors are grateful of the support provided.

References

[1] Chan CC 2007 The State of Art of Electric, Hybrid, and Fuel Cell Vehicles *Proceeding of IEEE* **95** 704-718
[2] Aso S, Kizaki M and Nonobe Y 2007 *Power Conversion Conference* (Nagoya/IEEE) p 1606-1611
[3] Mora J, Romero L and Ruperez M 2011 *Int. J. Hyg Energ* **36** 8008-8016
[4] Hwang J J and Chang W R 2012 *J Power Sources* **207** 111-119
[5] Shang J L and Pollet B G 2010 *Int J Hyg Energ* **35** 12709-12715
[6] Matsunaga M, Fukushima T and Ojima K 2009 Powertrain System of Honda FCX Clarity Fuel Cell Vehicle *World Electric Vehicle Journal 3* 1-10
[7] Wasselynck G, Auvity B, Oliver J C, Trichet D, Josset C, and Maindru P 2012 *Energy* **38** 414-424
[8] Saxe M, Folkesson A and Alvtors P 2008 *Energy* **33** 689-711
[9] Ouyang M, Xu F, Li J, Lu L, Gao D and Xie Q 2006 *J Power Sources* **163** 467-479

Edge detection techniques for iris recognition system

U T Tania[1], S M A Motakabber[2] and M I Ibrahimy[3]
Department of Electrical and Computer Engineering, International Islamic University Malaysia, 53100 Kuala Lumpur, Malaysia

Email: tandria_17@yahoo.com, amotakabber@iium.edu.my, ibrahimy@iium.edu.my.

Abstract. Nowadays security and authentication are the major parts of our daily life. Iris is one of the most reliable organ or part of human body which can be used for identification and authentication purpose. To develop an iris authentication algorithm for personal identification, this paper examines two edge detection techniques for iris recognition system. Between the Sobel and the Canny edge detection techniques, the experimental result shows that the Canny's technique has better ability to detect points in a digital image where image gray level changes even at slow rate.

1. Introduction

Iris recognition system mainly includes eye image capturing, image pre-processing and edge detection through iris region segmentation, feature extraction and pattern matching. Among them edge detection is one of the major part in iris recognition system. Edge detection technique makes pupil boundary detection accurately and easier. A circular edge detection method is used to look for a circle in the image which has maximum gray level difference with its neighbour [1]. Hough transform uses different approaches which are computationally complex [2]. A texture based method first determines high contrast parts of boundary then detects outer boundary and eyelids [3]. However, its each step is more challenging than previous ones [4]. The segmentation method first uses wavelet transform and then different integral operator is used for localizing the iris [5]. A new noise removing approach is introduced based on the fusion of edge and region information. In this case, whole procedure includes three steps namely, rough localization and normalization, edge information extraction based on phase congruency, and the infusion of edge and region information [6]. Another segmentation method is introduced based on integro-differential operators and Hough transforms [7]. Some researchers mainly focused on the iris image representation and feature matching without introducing a new method for segmentation [8, 9].

2. Methodology

Edge detection is divided into three main steps: image pre-processing, feature extraction of iris image and template matching. Image pre-processing consists image conversion from RGB image to gray image, edge detection, localization of iris in a given eye image, filtering etc. Generally edge detection aims at identifying points in a digital image where image brightness changes sharply. The points at which image brightness changes sharply are typically organized into a set of curved line segments termed edges. In this study the Canny edge detection and the Sobel edge detection techniques are considered. In Sobel technique, an operator is selected for each point of the image and it generates the corresponding gradient vector. The main principle of this operator is to convolve the image with a small, separable, and integer filtered values of horizontal and vertical directions of image points. The operator calculates the gradient of the image intensity at each point. For high frequency variations in the image supposed to be easier to find out the gradient of the image. The Sobel edge detection, the

Email: tandria_17@yahoo.com Cell: +60162245097

operator uses two 3×3 matrix kernels which are convolved with the original image to calculate the approximations of the derivatives. If the matrix A is defined as a source image, and the W_x and the W_y are the two images which contain the horizontal and vertical derivative of the approximations respectively then the relation can be represented as follows:

$$W_x = \begin{bmatrix} +1 & 0 & -1 \\ +2 & 0 & -2 \\ +1 & 0 & -1 \end{bmatrix} *\mathrm{A} \text{ and } W_y = \begin{bmatrix} +1 & +2 & +1 \\ 0 & 0 & 0 \\ -1 & -2 & -1 \end{bmatrix} *\mathrm{A} \qquad (1)$$

Where * here denotes the 2-dimensional convolution operation. Since the Sobel kernels can be decomposed as the products of an averaging and a differentiation kernel, they compute the gradient with smoothing. The x-coordinate is defined as increasing in the "right"-direction, and the y-coordinate is defined as increasing in the "down"-direction. Each point of the image, the resulting gradient approximations can be combined to give the gradient magnitude and gradient's direction by using Equation (2) and Equation (3).

$$W = \sqrt{W_x^2 + W_y^2} \qquad (2)$$

$$\theta = \tan^{-1}\left({W_y}/{W_x} \right) \qquad (3)$$

Canny edge detection algorithm runs in several steps. First in smoothing step, the operators blur the image to remove noise. Then in finding gradients step when operator detects the large magnitude of gradient of image it marks the edges. In non-maximum suppression step the operator only look for local maxima and marked it as edges. Then the operator applies threshold to determine potential edge. In final step edges are determined by suppressing all edges that are not connected to strong edge. All the edge detection method applied by using MATLAB image processing tool [10]. Original image collected from CASIA database [11].

3. Result and Discussion

Figure 1 shows the results obtained by applying the Sobel edge detection and the Canny edge detection techniques.

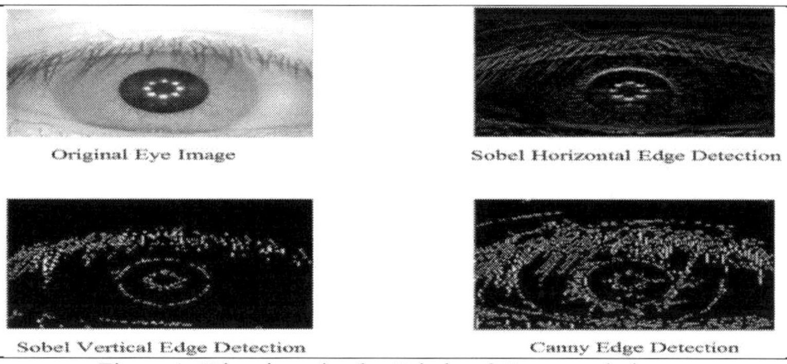

Figure 1: Edge detection by Sobel and Canny techniques.

The result shows how rapidly and smoothly the image changes. The Canny operator is optimum even for noisy images. This operator fills the gap between strong and weak edges of the image. Compared to other edge detection techniques, this operator is less fooled by specious noise. Generally horizontal edge detection is more readable than vertical edge detection. In Sobel edge detection method, it looks for both horizontal and vertical edges independently then represents them together. The Canny edge detection technique is more efficient to detect both the slow variation of gray level as well as strong variation of gray level of the image.

4. Conclusion

It is observed that the Canny edge detection technique has better performance compared to the Sobel edge detection technique. Unless the preconditions are particularly suitable, it is difficult to find an edge detector that performs significantly better than the Canny edge detector. However Sobel edge detection technique has the ability to detect both horizontal and vertical edges individually that presented in combined form, thus computation is relatively cost-effective.

References

[1] Daugman J 2004 How iris recognition works, Circuits and Systems for Video Technology IEEE Transactions on, **14** 21.

[2] Wildes R P 1997 Iris recognition: an emerging biometric technology, Proceedings of the IEEE, **85** 1348

[3] Ma L, Tan T, Wang Y and Zhang D 2003 Personal identification based on iris texture analysis, Pattern Analysis and Machine Intelligence IEEE Transactions on, **25** 1519

[4] Barzegar N, and Moin M S 2008 A new approach for iris localization in iris recognition systems, In Computer Systems and Applications, AICCSA , IEEE/ACS International Conference on 516

[5] Sun Z, Wang Y, Tan T and Cui J 2005 Improving iris recognition accuracy via cascaded classifiers, Systems, Man, and Cybernetics, Part C: Applications and Reviews, IEEE Transactions on, **35** 435

[6] Huang J, Ma L, Tan T and Wang Y 2003 Learning based enhancement model of iris, In Proceedings of the 14th British Machine Vision Conference (BMVC'03) 153

[7] Tisse C L, Martin L, Torres L and Robert M 2002 Person identification technique using human iris recognition, In Proc. of Vision Interface

[8] Boles W W and Boashash B 1998 A human identification technique using images of the iris and wavelet transform, Signal Processing, IEEE Transactions on, **46** 1185

[9] Lim S, Lee K, Byeon O and Kim T 2001 Efficient iris recognition through improvement of feature vector and classifier, *ETRI journal*, **23** 61

[10] MATLAB6.5 Image processing toolbox tutorial http://www.cs.otago.ac.nz/cosc451/Resources/matlab_ipt_tutorial.pdf, [Last viewed March 3, 2013]

[11] Center for biometrics and security research, http://www.cbsr.ia.ac.cn/english/index.asp [Last viewed June 17, 2013]

IOP Conf. Series: Materials Science and Engineering **53** (2013) 012042 doi:10.1088/1757-899X/53/1/012042

Mechanical design and simulation of two-wheeled wheelchair using solidworks

T M Altalmas [1], S Ahmad [2 3], A Aula [1], R Akmeliawati [2] and S N Sidek [2]
[1] Intelligent Mechatronics Research Unit, Department of Mechatronics Engineering, International Islamic university Malaysia, Gombak, Selangor 50728,Malaysia.
[2] Department of Mechatronics Engineering, International Islamic university Malaysia, Gombak, Selangor 50728, Malaysia

Abstract. This article is presented a new design of two-wheeled wheelchair that can balance on two wheels to make it suitable in the narrow areas, especially in the domestic environments; it has the ability to extend the height of the chair to help the user to act independently in the life for example, in the library to pick and put books on the shelves. The 3D model has been built up using SolidWorks Software. Nowadays, SolidWorks environment is considered as a powerful tool that is helping designer to design products and attain its performance before physical prototype stage. SolidWorks simulation model has been employed to test the frame of the wheelchair under the weight of the human body and the upper part of the wheelchair. The static analysis has been done on the frame using steel and aluminium; however the aluminium material has been selected due to its light weight

1. Introduction

In the recent years, the use of wheelchairs has increased tremendously due to the growing number of aged and disabled people [1]. Many researchers have been done for the sake of helping aged and disabled people to have independent life instead of waiting the help of others. Some researches is related to refine the environment around the wheelchair users to make it suitable for them, others try to produce products to make the life of disabled people easier. There are a lot of wheelchairs types; each type has its specifications, size, dimensions and functions. Two-wheeled wheelchairs one of the types that has been discovered in the last few years. Two-wheeled wheelchairs give the users the ability to move in narrow areas easily. In addition to that the ability to extend the height of the seat helps the users to act independently during dally activity, for example when they want to collect books from the shelf in the library.

Many commercial designs have appeared in the recent years. iBot is a powered wheelchair with a lot of abilities working on 4 wheels , balancing on 2 wheels, climbing stair and giving users the ability to join eye-to-eye conversation with people but unfortunately the company has stopped producing it[2]. Genny is another two wheeled wheelchair, it can balance on 2 wheels but it does not has the ability to lift the seat [3].

International Organization for Standardization (ISO) has done standards that are related to the working area of wheelchairs. These standards can help the architectures to design the houses to be suitable for the users of wheelchairs, ISO 7176-5:2008 standard was focused in the determination of

[3] Corresponding author Email: salmiah@iium.edu.my
Tel: +603 6196 4506

dimensions, mass and manoeuvring space [4]. Regarding to ISO 7176-5 maximum occupied length of manual and powered wheelchair has to be 1300mm, occupied width of manual wheelchair has to be 800mm but electrical type has to be 700mm, and ISO stated that the maximum turning diameter has to be 2000mm for manual wheelchair and 2000 to 2800 in powered wheelchair .

Nowadays, designing such mechanical design has become easier than previous because many software have been found out to help engineering in designing and testing the products. SolidWorks is considered one of the powerful tools to design and simulate the mechanical designs and products [5]. Using SolidWorks, engineers can shorten the required designing time, and optimising and validating the design to reach the optimal design as possible. SolidWorks simulation another important tool is included in SolidWorks software. It can help in implementing many tests on the design to ensure the safety of this design and to test the design under the external forces.R Sam et. al have designed and simulated a robotic systems in the SolidWorks environment to reduce the designing time and test the parts of the robot to reach the optimal design[6].L Zhao et al have done the virtual design of bike. Simulation of the bike has been done to select the best material for the frame [7].

This paper presents the 3D design of two-wheeled wheelchair design using SolidWorks. Which is the design process is following the systematic steps for designing using CAD software. Finite element analysis has implemented on the frame of the wheelchair to test it under the external loads and select the optimal material for making the weight is suitable.

2. SolidWorks Overview

SolidWorks is an outstanding 3D design tools. It gives designers the ability to sketch out ideas, experiment with features and dimensions and produce models and detailed drawing. 3D design approach is used in SolidWorks. It is start from initial sketch on papers then going to SolidWorks environment to build the parts from the initial sketch after that combining the parts together in one assembly to come out with the final design. 2D drawing is the final step before sending the design to the workshop to build the real prototype. Figure 1 shows the four important sections of SolidWorks [6].

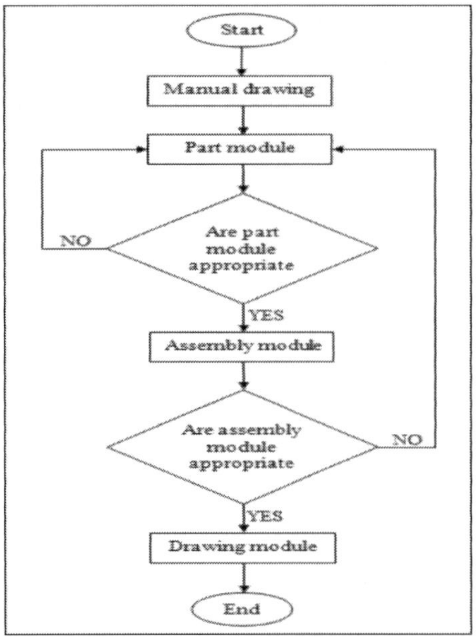

Figure 1. Systematic steps in designing using SolidWorks

IOP Conf. Series: Materials Science and Engineering **53** (2013) 012042 doi:10.1088/1757-899X/53/1/012042

3. Three Dimensional Design of two-wheeled wheelchair

Designing steps are following the flow chart in figure1. First of all, deciding the dimensions of each part to come out with suitable design and to achieve the ISO- standard values. Second step is modelling each part individually. The wheelchair is divided into three parts; rear wheels, the frame and casters. In addition to the parts there are two subassemblies, the lifting mechanism - lifting mechanism- and the chair with human body. Figure2 shows the parts and subassemblies of the design.

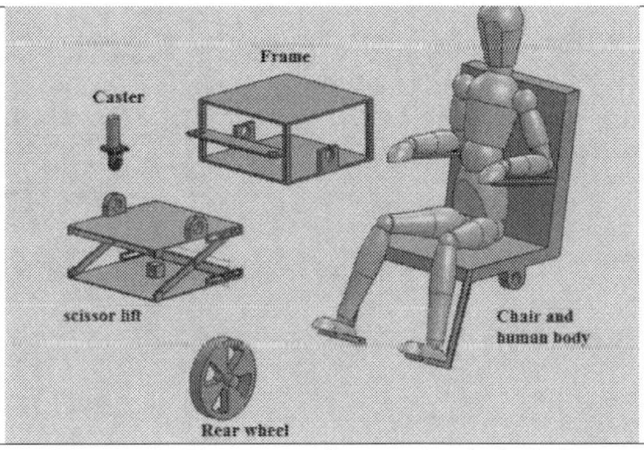

Figure 2. The parts of two-wheeled chairs

Next step is to combine all parts and subassemblies together to build the final design. In this step bottom-up technique is used, which it starts by building the parts then integrate these parts together using mates. Mates create the geometric relations, such as coincident, perpendicular and tangent. These mates help in restricting the movement of the parts. Figure3 shows the full design after combining the parts together.

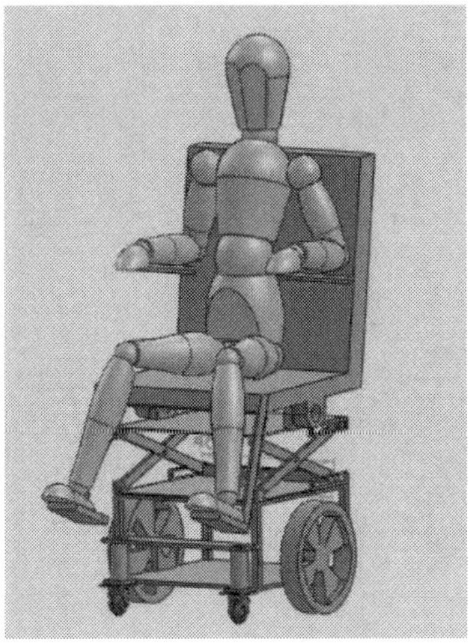

Figure 3. Full design of two-wheeled wheelchair

4. Working area and dimensions:

International Organization for Standardization (ISO) has set the standard dimensions and turning radius of powered wheelchairs. Whereas the maximum occupied length and width should be 1300mm and 700mm respectively. While the maximum turning radius should be 2800 mm. . The proposed

IOP Conf. Series: Materials Science and Engineering **53** (2013) 012042 doi:10.1088/1757-899X/53/1/012042

design achieves the ISO-7176-5 standards of dimensions and turning radius, where it is occupied length and width are 904.34mm and 660.65 respectively these dimensions with the ability to work on two wheels give this design the ability to work in narrow areas. Figure 4 shows the occupied length, occupied width and turning radius of two-wheeled design.

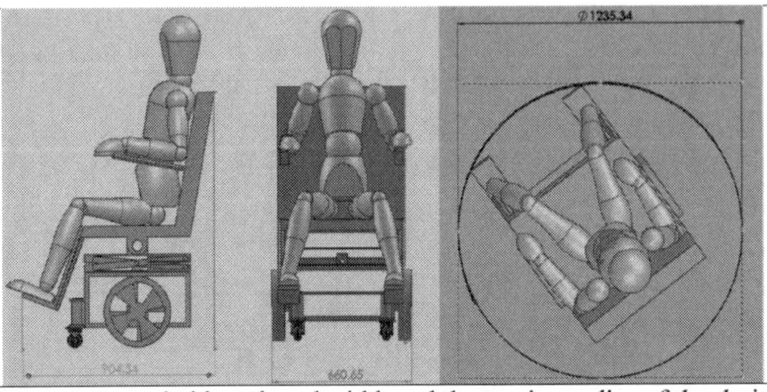

Figure 4. Occupied length and width and the turning radius of the design

5. Finite element analysis:

The static analysis was employed on the frame of the wheelchair to check its ability to stand under the weight of the human body, check the deformation under this load and the safety factor for two selected materials. Table1 shows the material that is selected to build up the frame. The steel frame has mass of 47891.30 grams, while the aluminium one has almost third of this mass.

Table 1. Specification of used materials

	Steel AISI1020	Aluminium alloy 1350
Elastic Modulus (N/m^2)	2e+0011	6.9e+0010
Density(Kg/m^3)	7900	2700
Yield strength (N/m^2)	351571000	27574200

The frame was analyzed through static analysis test. 980 N external force has been applied at the upper side of the frame this external load considered as 100 KG is up to the frame. The static test has been done using steel and aluminum.

5.1 static analyses for steel frame:

Figure5. Shows that the steel frame is pass the test successfully, due to the many criteria. First of all from the stress graph it is clear that the maximum stress is less than the yield strength, so the part will not broke under this external load. Secondly, the maximum deformation is very small, it is 2.964e-1mm. Finally, for safety factor it is clear from the graph that all the graph has the blue colour and there is not any red area which this is meant that the part is pass the test successfully.

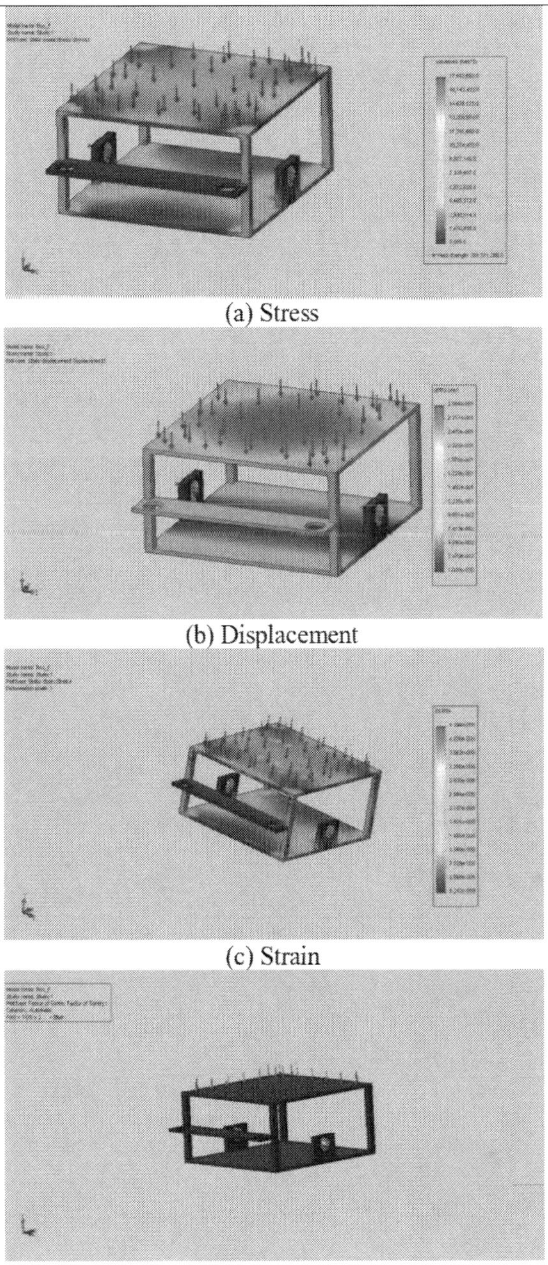

(a) Stress

(b) Displacement

(c) Strain

(d) Factor of safety

Figure 5. The result of static analysis for steel frame

5.2 static analyses for aluminum frame:

Figure6 shows that the aluminium frame is pass the test successfully, due to the many criteria. Firstly from the stress graph it is clear that the maximum stress is less than the yield strength, so the part will not broke under this external load. Secondly, the maximum deformation is very small, it is 8.409e-1mm, but it is almost three times the value of steel. Finally, for safety factor it is clear from the graph that all the graph has the blue colour and there is not any red area which this is meant that the part is pass the test successfully.

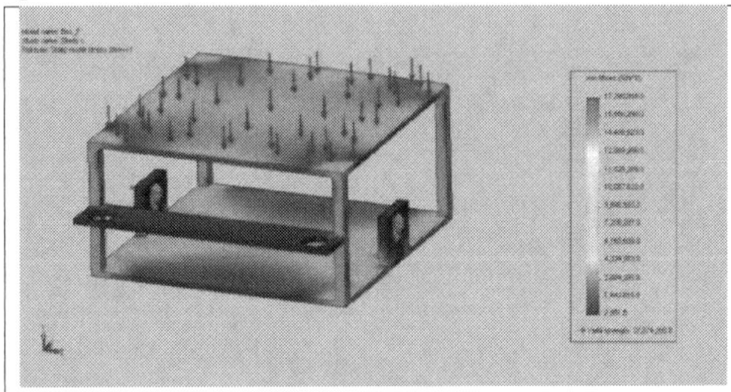

(a)Stress

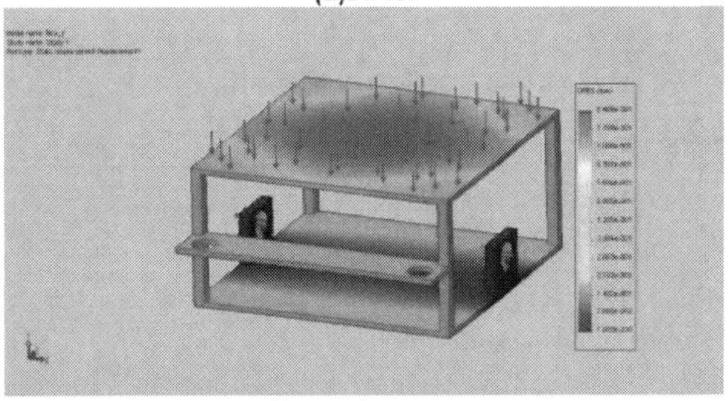

(b) Displacement

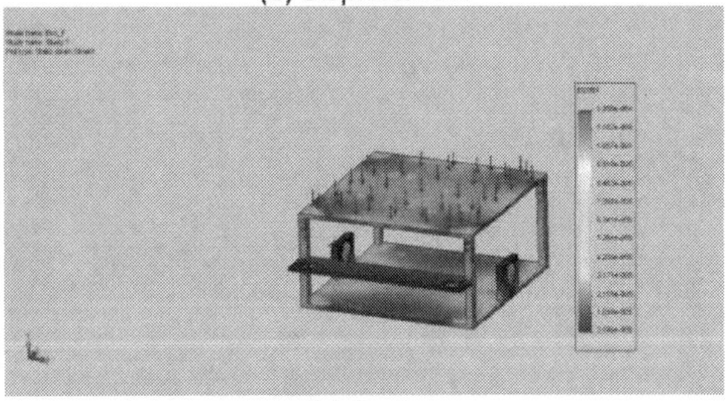

(c) Strain

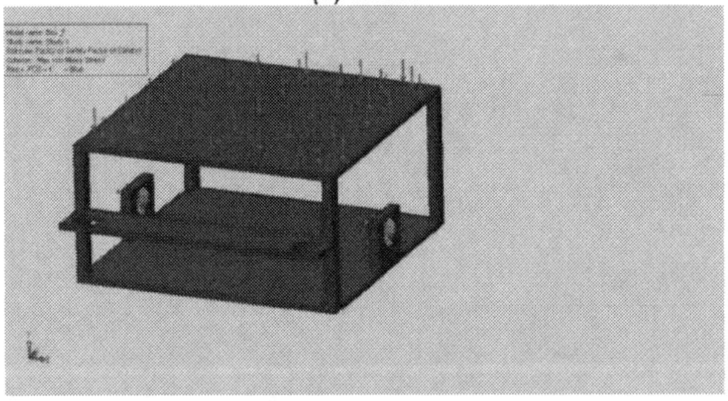

(d) Factor of safety

Figure 6. The result of static analysis for aluminum frame

5.3 material selecting

The previous graphs show that there are differences between the using of steel and aluminium. Regarding to stress analysis the stress of both materials is almost same. It is 17610692N/m^2 and 17290958N/m^2 for steel and aluminium respectively. For the displacement graph it is clear that the aluminium material presented more distortion than steel, it is deform three times more than the steel. Going to the factor of safety graph it is evident that both of the materials don't have any red areas, so both material success in the analysis. But aluminium steel is selected because of the Light weight.

6. Conclusion

In this paper a bottom-up assemblies is used to build the 3D model of the two-wheeled wheelchair. Starting from building the parts, then integrate it together in one assembly to reach the full 3D model. SolidWorks simulation has been used to ensure the capability of the frame to Stand under the influence of an external force – human body weight - . Finite element analysis has been done using steel and aluminium to build up the frame, both material success in the test, but aluminium has been selected because of its light mass.

References

[1] http://www.aoa.gov/aoaroot/aging_statistics/index.aspx
[2] http://www.hizook.com/blog/2009/02/11/ibot-discontinued-unfortunate-disabled-perhaps-budding-robotics-opportunity
[3] http://www.gennymobility.com/Genny/WhyChooseGenny.aspx
[4] F Wien and I Johan, Working area of wheelchairs, Space requirements of wheeled mobility international workshop, 9-11 october,2003,NY,USA
[5] https://www.solidworks.com/
[6] R Sam, K Arrifin and N Buniyamin, Simulation of Pick and Place Robotics System Using Solidworks Softmotion, International Conference on System Engineering and Technology, September 11-12, 2012, Bandung, Indonesia
[7] L Zhao and S Zhao, Virtual Design of Bike Based on SolidWorks, International Conference on Electronic and Mechanical Engineering and Information Technology (EMEIT),12-14 Aug. 2011, Libr., Handan Coll., Handan, China.

A Framework for Developing Real-Time OLAP algorithm using Multi-core processing and GPU: *Heterogeneous Computing*

H I Alzeini[1,4], Sh A Hameed[2] and M H Habaebi[3]
Electrical and Computer Department, IIUM, PO: 53100, Jalan Gombak, Kuala-Lumpur, Malaysia

E-mail: [1] h.alzeini@hotmail.com, [2] shihab@iium.edu.my and [3] habaebi@iium.edu.my

Abstract. The overwhelmingly increasing amount of stored data has spurred researchers seeking different methods in order to optimally take advantage of it which mostly have faced a response time problem as a result of this enormous size of data. Most of solutions have suggested materialization as a favourite solution. However, such a solution cannot attain Real-Time answers anyhow. In this paper we propose a framework illustrating the barriers and suggested solutions in the way of achieving Real-Time OLAP answers that are significantly used in decision support systems and data warehouses.

1. Introduction

This Online analytical processing (OLAP) [1] is a category of software technology that enables analysts, managers and executives to gain insight into data through fast, consistent, interactive access to a wide variety of possible views of information that has been transformed from raw data to reflect the real dimensionality of the enterprise as understood by the user. OLAP functionality is characterized by dynamic multidimensional analysis of consolidated enterprise data supporting end user analytical and navigational activities including calculations and modelling applied across dimensions, through hierarchies and/or across members, trend analysis over sequential time periods, slicing subsets for on-screen viewing, drill-down to deeper levels of consolidation, rotation to new dimensional comparisons in the viewing area …etc. OLAP works with data warehouses that have sizes of not only terabytes but also petabytes which results in slower response time to answer queries. Consequently, processors have to tackle billions of rows. The issue has been intensively addressed and we would say that the most of solutions have relentlessly gone to materialization direction. Materialization by definition means pre-fetching data and pre-computing prospective queries – which are usually expected – load these answers with their features into the main memory. Subsequently, any coming query will find the answer waiting for it. This technique has been widely used and improved as it implies other problem such as the main memory occupation and excluding latest data. The latter issue has caused a significant debate, that is, by materializing result the system will present static answers that often do not take into account the latest updates of the warehouse. Warehouse usually is being updated periodically (e.g. daily, weekly, monthly … etc.). Therefore, pre-fetching and pre-

[4] To whom any correspondence should be addressed.

computing are performed periodically either which would necessarily lead to some answers that do not include current time interval updates (non Real-Time answers). Thus, the Real-Time OLAP answers issues is about if the latest updates are included in the answers or not, rather than the absolute answer's time values – which varies dramatically according to several factors (e.g. query size, answer size, processing capabilities, distribution issues … etc.). Other minor solutions have been suggested will be elaborated in the following sections. However, all the materialization solutions cannot meet the Real-Time requirements which motivated other researchers to focus on different direction recently; how to optimally exploit processing facilities in order to function them to immediately answer queries. Such an approach, of course, needs powerful processing capabilities to process millions, and sometimes billions, of rows in few seconds to answer each query instantaneously. Recent suggestions included multicore processing, distributed processing as well as GPU utilizing.
The following sections of this paper are organized as follow: related work has been presented and briefly criticized. Then, the possible solution has been presented. Afterwards, our chosen suggestion has been shown in addition to the expected outcomes. Finally, a summary and future research directions.

2. Materialization

We can categorize the literature (to optimize materialization objective) into two main areas: how to optimize materialized views (sub-cube) construction and how to optimize materialized views distribution over multi-nodes systems.

2.1. Sub-Cube Construction

A dimension-oriented storage was one of the suggested solutions that have been proposed to minimize the cube aggregation time [2]. Z. Jing-hua et al have presented that the traditional row-oriented storage restricting the performance, accordingly; a new parallel aggregation mechanism utilizing MapReduce jobs [3] and HBase mechanism have been implemented and a performance improvement has been demonstrated. However, using HBase is restricted to several factors, such as the size of data must be enormous enough to achieve the efficiency (e.g. hundreds of millions or billions row) otherwise (e.g. few thousand/million rows) the traditional RDBMS is better choice due to the fact that HBase is efficient if we have 5 or more nodes at least. Furthermore, in order to use HBase we should tradeoff between extra features of RDBMS (e.g. transaction, typed columns, secondary indexes … etc.) and Hbase features. Third and most importantly; RDBMS cannot simply be ported to HBase by just replacing a JDBC driver, the operation require a full redesign cycle [5]. In spite of these limitations, one can benefit from [2] approach in parallelism part, using MapReduce [3] can be very helpful in developing a distributed computing system.

2.2. Distributed OLAP

Many researches have investigated views materialization over distributed processing systems as one of the potential solution aiming at reducing processing time consuming by OLAP queries. One of the resent researches for S. Sen et al [6] has focused on presenting new architecture to maintain materialized views for cloud computing environment. Considering the subtle differences between them; the proposed architecture is more pertained to distributed systems rather than the cloud computing.

A different infrastructure has been introduced by W. Li et al [7] with different utilization (distributed databases). In this paper there is single frontend database server and multiple backend databases. The materialized views (MQTs) have been used and the main contribution is the Data Placement Advisor (DPA - is a replacement of MQTA that had been designed to deal with standalone database server in which MQTs are placed on the same server with the base tables) that recommend MQTs and allocation mechanisms in order to minimize the response time of the queries. The new DPA build recommendations according to the analytic results of the MQTA and recommend whether to place MQTs at the backend or at the frontend based on the response time. A significant approach has been proposed in [8] defining the policies of OLAP queries caching in internet proxies, the solution suggests that we can utilize the LAN proxy by locating the materialized views in the proxy server in order to achieve the same purpose; reducing the response time for OLAP queries from scattered private LANs. The research suggests a mechanism to estimate access costs in order to decide whether the query will be satisfied locally (at the proxy) or at the central DW. An analytical cost model has been derived in order to quantify the benefits, a cache algorithm (VHOW) has been developed and the simulation results assured the schema efficiency.

J. Loureiro , O. Belo [9] have introduced an evolutionary approach of selection and allocation materialized view (sub cubes) over a distributed nodes. The solution utilizes a linear cost model, such that, it takes into account the communication costs and nodes processing power (e.g. aggregation, integration ... etc.). The results have shown the superiority of genetic algorithms (GA) against the greedy algorithm in terms of cube selection and allocation as well as the scalability. Moreover, M. Lawrence et al [10] have proposed new query processing method over distributed nodes, the schema essentially rely on the Fragment Aggregation and Recombination (FAR) approach, in which the queries are divided into sub queries and sent to different nodes in order to be answered. The Site Brokers are the responsible servers for cache indexing, organization and coordination of resources within each site, the indexes, in turn, contain the data cube lattice structure as well as the R-Tree for each view indexing cached fragments. Implementation and evaluation have shown 50% - 60% response time reduction for realistic network parameters. However, the schema is lacked of mathematical models that might help designers in terms of which appropriate number of nodes should they utilize, and what is the best hierarchy in order to attain lowest response time. L. Wang has presented the most recent approach [11] in which followed mostly the same four layered structure that has been presented in [10]. Since the approach is specified to serve real estate investment systems, which narrows the assumptions; many simplifications have been assumed, such as the application service procedures are the same on each server and most critically the task assigning to be static which is mostly unpractical.

Among the two OLAP standards; F. Dehne and et al [12] have shown that MOLAP is suffering from scalability problems such that representing extremely sparse spaces is not easily adapted. On the other hand, ROLAP facing speed indexing problem for OLAP queries (performance problem). Accordingly, new solution has been proposed (RCUBE) based on maintaining the ROLAP scalability advantage with improving query indexing speed. The new approach uses packed R-trees, round robin distributed disk striping and Hilbert curve based (previous approaches used XYZ data format – also called lowX or nearest-X – which its response time deteriorate rapidly when relevant point are dispersed over the data set) for data ordering in order to achieve the minimum communication volume among p processors that process a balanced load and without a need for a shared disk in addition to maintain the scalability in terms of number of processors, data size and dimensions. The proposed solution considers that the queries are passed directly to each processor and the intermediate results remain distributed for further processing, previous RCUBE approaches considered global R-tree on the front-end. Furthermore, one of the significant improvements that have been presented; a threshold (α) such that if the number of records in the result set is below the threshold, the partial result set are sent to one processor to be sorted, otherwise, will be sent to multiple processors, the α value can be set as a physical characteristics of the parallel machine.

The scheme has been implemented in C++, STL and MPI, and tested on a 17 nodes Beowulf cluster (a front-end and 16 compute nodes) with 1.8 GHz Intel Xeon processors, 1 GB RAM per node, 2X40 GB 7200 RPM IDE disk drives per node, the operating system was Linux Redhat 7.2 gcc 2.95.3 and all

IOP Conf. Series: Materials Science and Engineering **53** (2013) 012043 doi:10.1088/1757-899X/53/1/012043

nodes are connected via 100 MB Ethernet switch. The results have shown that the index construction speed is near to the optimum, figure 1 shows some sample results:

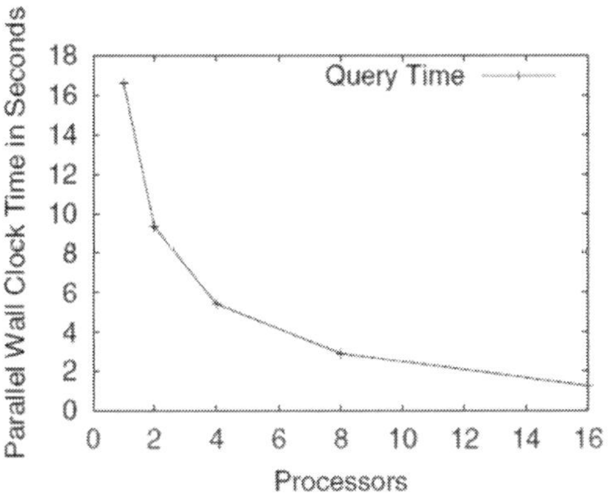

Figure 1 (a). Distributed query resolution [12].

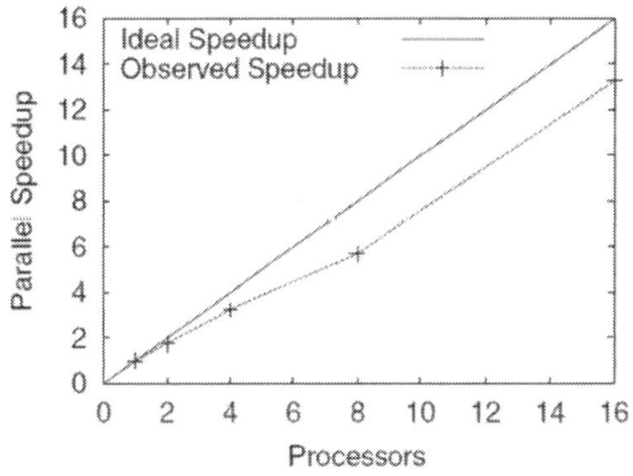

Figure 1 (b). Corresponding speedup [12].

Another almost similar approach has been presented by J. You [13]. Namely, HDW solution, in which, GFS has been utilized as infrastructure for partitioning, load balance and failure tolerance purposes, Bigtable for distributed Cubetable management purposes and MapReduce for task parallelism purposes (The three components: GFS, Bigtable, MapReduce [3] pertain to Google distributed file system's infrastructure). Furthermore, MPI has been used for communication. The schema has been implemented in C++ and the conducted experiments in an 18-node PC cluster and 36 total cores, 18 GB RAM and 540 GB disk volumes, the experiments have shown that 60 million rows (1.37 GB text file) needed less than 5 minutes to be constructed. In addition, if the number of rows increases (e.g. 100 million) the construction time will not exceed 7 minutes. For the 36 nodes; the speedup is almost linear.

Regarding dataset distribution, there are two suggestions, first: data replication, second, centered dataset, the last suggestion has been indicated as undesired since it causes bottleneck in the central server. However, replication is considered as storage wasting choice also. An improvement has been suggested to be partial replication. In this regard there are many of what so called replica algorithms such as [14], in which, a new algorithm (OLAP4RS) has been presented based on pre-processing framework solution instead of post processing solutions (start processing the replica selection after creation request received).

In fact, in addition to the aforementioned examples; there are a significant number of researches that have addressed sequential and parallel cube construction methods (e.g. 15, 16, 17, and 18). However, all these approaches are following the static cube construction proposed by Gray et al [1]. The major disadvantage of such cubes is that the system can be only updated periodically. Therefore, the latest information might not be bundled with the relatively old ones in order to answer some queries. Another shortcoming is the massive space of memory that is needed due to materialized views storing in such structures as well as data duplicating.

3. Materialization Enhancements

The last issue – memory space - has been addressed by a considerable number of researches that suggest two solutions: compression and view selection.

3.1. Compression

The compression has been early suggested as a potential solution for enormous storage size. In the literature we can find a scalable and efficient parallel algorithm has been introduced for Closed Cube Computation – a lossless cube compression method – [19]. Utilizing MapReduce [3] framework for parallelism; the results have indicated significant reduction of storage size when the number of the closed cubes goes up. [21] Also has utilized MapReduce paradigm [3] for parallel closed cubes. Another example [20] by Z. Wang, Y, Xu have demonstrated the minimal condensed cubes challenges and discussed alternative computation algorithm (MinBST) that outperform previous ones. [22] has presented compression algorithm based on the balanced tree bitmap compression mechanism and has shown that the structure compression time can be logarithmic which outperforms the traditional bitmap compression techniques (linear time).

However, such a solution cannot be applied to the materialized views due to the fact that with more compression and storage size saving, there is relatively longer time needed for computation and answering the queries. Therefore, we can easily find in [20], [21], [22] the queries answer's time tends to rise which make the compression solution, as a result, inefficient for Real-Time purposes.

3.2. View Selection

Furthermore, some researches have focused on view selection issue, many algorithms have been suggested to choose the most suitable views to be loaded into the main memory due to the fact that we have enormous amount of data views and limited main memory size. Graph OLAP [23] is a suggested solution for views materialization issue by considering particular constrains in order to decide to choose partial or full materialization. A dynamic tuning of aggregation tables (materialized views) has been addressed as an online solution [24]. Aggregation tables tuning represent the stage of views selection in which an optimized selected views can be only loaded into the main memory to enhance the response performance. Recently the Genetic Algorithm (GA) has been introduced [25], [26] as one of potential solutions in this regard. CSSAA_VSP clonal based algorithm [27] has been proposed also to solve views selection issue. Remarkably, according to the empirical results, the algorithm has outperformed GA algorithm in terms of speed and solution quality, it has been proved as feasible and efficient algorithm. S. Sen et al [28] also have presented SelectPath algorithm that chooses the path which require generation of smallest number of cuboids among others. The approach utilizes Galois connection method. A numerical example has been given, yet the algorithm had not been investigated. Additionally, Galois connection has been used also in [29] to develop an algorithm utilizing dynamic data structure aiming at reducing time consuming and space occupation over time.

4. Real-Time OLAP

In order to bundle the latest updates with OLAP answers; researchers have focused on hardware processing capabilities. Three approaches have been addressed as follow:

4.1. Multi-Core Processing Solution

Authors F. Dehne and H. Zaboli [30] have presented a novel Parallel Real-time OLAP that is exploiting Multi-core processors systems. On contrast to the most introduced solutions in this regard (except rare instances such as [24] which utilize dynamic data cube); this approach does not rely upon views materialization or any static structure or pre-computed aggregations. Therefore, in order to compensate the performance degradation resulted from ignoring materialization solution; the authors have suggested taking advantage of the revolutionary hardware developments, specifically, multi-core processors that have the ability to handle several operations at the same time. In more details, the solution developed a parallel DC-tree for multi-core architectures as the sequential DC-tree [31] is the only fully dynamic data structure has been published so far. DC-tree is an extension of X-tree [32] and R-tree [33]. The new structure has been used with two main operations proposing two algorithms: 1) PARALLEL_OLAP_INSERT and 2) PARALLEL_OLAP_QUERY. The tests (using TPC_DS as benchmark) have demonstrated a significant speedup in terms of response time and throughput. A different approach to achieve Near Real-Time has been proposed by P. Mazur et al [34]. The solution suggests a universal database analysis repository model that contains all descriptive information about previous queries such that no need to repeat the whole calculation operation to find an answer, instead, small update can be applied to previous results to answer the query faster. Same concept has been portrayed also in [35] called the minimum incremental maintenance of materialized view, in which updating views occurring periodically and incrementally to avoid re-compute the entire view.

4.2. GPU Solution

I3DC (Interactive Three-Dimensional Cubes) is a substantial solution based entirely on GPU [36], the approach utilizes a new blending-as-aggregation (BAA) algorithm that maps OLAP aggregations to GPU mechanisms of rendering. Without the need of pre-computation and translation to integers by CPU; the new systems can build and visualize data cubes for tens of millions of records in milliseconds. The only concern of this approach is that show the results in chromatically graphical manner (in order to save up time of translation to integers and to be completely compatible with GPU structure and operations). Therefore, the output will be limited to the color range that the human eyes can recognize. However, the application that has been presented uses cubes of up to three dimensions, which is trivial number comparing to the real world datasets that can have much larger number of dimensions. Visualizing bigger numbers will definitely result in bigger number of colors which would cause difficulties for human eyes to distinguish between them.

4.3. Multi-Core Processing + GPU Solution

Exploiting the revolutionary hardware capabilities, M. Malik et al [4] have proposed three significant performance improvements: algorithm of string dictionary system used to translate literal data into integers for GPU processing purpose (translation performance evaluation based on TPC-DS fact tables and has shown 7% performance degrading due the translation time). Second, parallel version of OLAP processing using the CPU that has been implemented in OpenMP as well as a performance model needed by the scheduling algorithm in order to decide to which resource (CPU or GPU) should different queries be assigned. Third, the scheduling algorithm that supports text-to-integer translation and multiple CPU and GPU partitions. Each of the two resources has two main tasks; CPU processes OLAP queries that are not costing as well as performing text-to-integer translation. GPU in turns builds the cubes from traditional tables that are stored in GPU memory in addition to performing queries that are costly for CPU.

GPU has six partitions that are used to process queries from fact tables, while CPU has two partitions, figure 2 - [4]. The same idea (exploiting GPU to serve OLAP purposes) has been addressed also by C. Weyerhaeuser et al [38] with Business Intelligence Accelerator (BIA: highly distributed analytical engine supports OLAP). The results and comparison have shown substantial gain over CPU-based implementation although that the final recommendation was not to use GPU for BIA for particular reasons (e.g. GPU has smaller memory, data transfer from main memory to GPU memory has a cost … etc.)

5. Proposed Solution

Whatever improvements have been applied to materialization approach; none of them could overcome the Real-Time problem. Hence, our suggested approach is based on the concept of enhancing the hardware processing utilization. In particular, take advantage of both CPU and GPU. GPU significantly outperform CPU. However, using only GPU has shortcoming that the resultant answers will not be in numerical manner, rather, it might be multidimensional colourful graphs, shapes … etc. Hence, combining both CPU and GPU is a must in order to assign translation tasks to the CPU and processing tasks to the GPU. The proposed solution is essentially based on [4]. The algorithm will be based on TCP-DS benchmark by modifying the code of DsGen software which is open-source code. The software allows us to generate data with different scales from 100GB to 300TB. Afterward we can apply queries and compare the results in terms of response time. Our major contribution will be using this task partitions to achieve real-time OLAP, the following diagram shows the solution approach. Figure 2 shows the block diagrams of the suggested GPU and CPU partitions [4].

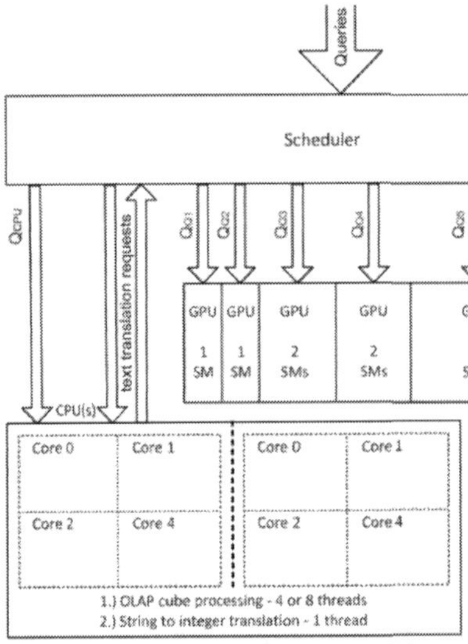

Figure 2. The block diagram of the GPU and CPU [4].

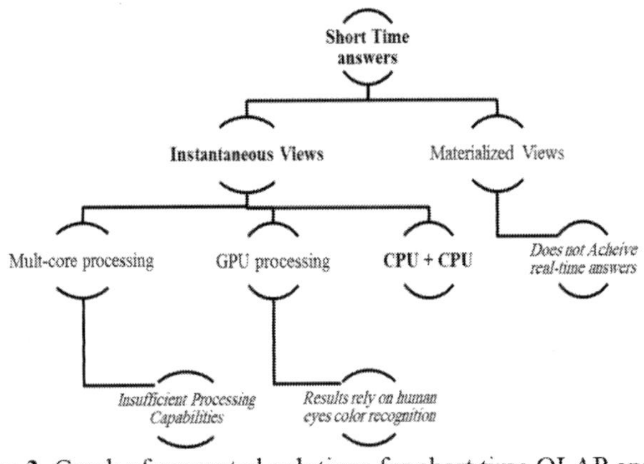

Figure 3. Graph of presented solutions for short time OLAP answers.

6. Expected Results

By applying the aforementioned approach we expect to achieve real-time answers by ignoring materialization approach and utilizing dynamic cubes rather than static ones. Furthermore, the increasing time needed to handle queries and processing them from the scratch will be compensated by exploiting the revolutionary processing capabilities that GPU offer. However, GPU by itself cannot present results in numerical manner, thus, including the multi-core processor is essential in order to accomplish two tasks, scheduling and translating data from literal to integer and vice versa. The approach tremendously affects the e-business [37] sector in addition to healthcare and artificial intelligence fields

7. Conclusion

In this framework paper essential approaches suggested by researches in order to optimize OLAP performance have been introduced on two levels: materialized answers and instantaneous answers. We have shown why materialized views cannot achieve real-time answer as well as why instantaneous answers are problematic issue. Hence, the powerful processing capabilities that are needed for instantaneous answers have been handled by GPU + CPU hybrid systems approach. Furthermore, an approach which utilizes a distributed algorithm presented in [12] and a partitioning algorithm presented in [4] in order to achieve real-time answers have been suggested. Future trends might implement the solution on Beowulf cluster and compare the enhancement in addition to developing the partitioning algorithm.

Acknowledgments
This research work has been funded by MOHE Malaysia under research grant fund (RFGS 13-028-0269).

References

[1] Gray J, Chaudhuri S, Bosworth A, Layman A, Reichart D, Venkatrao M, Pellow F, and Pirahesh H, 1997 Data Cube: A Relational Aggregation Operator Generalizing Group-By, Cross-Tab, and Sub-Totals, Data Min. Know. Disc., vol. 1, pp. 29–53,
[2] Jing-hua Z, Ai-mei S and Ai-bo S, May, 2012 OLAP Aggregation Based on Dimension-oriented Storage, 2012 IEEE 26th International Parallel and Distributed Processing Symposium Workshops & PhD Forum.
[3] Dean J and Ghemawat S, 2004 MapReduce: Simplified data processing on large clusters, OSDI '04, pp. 137–150,. [Online]. Available: http://www.usenix.org/events/osdi04/tech/dean.html
[4] Malik M, Riha L, Shea C and El-Ghazawi T, May, 2012 Task Scheduling for GPU Accelerated Hybrid OLAP Systems with Multi-core Support and Text-to-Integer Translation, Parallel and Distributed Processing Symposium Workshops & PhD Forum (IPDPSW), 2012 IEEE 26th International.
[5] http://hbase.apache.org/book/architecture.html#arch.overview, accessed on 19/1/2013.
[6] Sen S, Datta D and Chaki N, Septemper, 2012 An Architecture to Maintain Materialized View in Cloud Computing Environment for OLAP Processing, Computing Sciences (ICCS), 2012 International Conference.
[7] Li W, Zilio D, Batra V, Subramanian M, Zuzarte C and Narang I, April, 2007 Load Balancing for Multi-tiered Database Systems through Autonomic Placement of Materialized Views, 22nd International Conference on Data Engineering (ICDE'06),.
[8] Loukopoulps T and Ahmad I, October, 2006 Policies for Caching OLAP Queries in Internet Proxies, Parallel and Distributed Systems, IEEE Transactions.
[9] Jorge L and Orlando B, December, 2006 An Evolutionary Approach to the Selection and Allocation of Distributed Cubes, Database Engineering and Applications Symposium, 2006. IDEAS '06. 10th International.
[10] Lawrence M, Dehne F and Rau-Chaplin A, March, 2007 Implementing OLAP Query Fragment Aggregation and Recombination for the OLAP Enabled Grid, Parallel and Distributed Processing Symposium, 2007. IPDPS.
[11] Wang L, Aug, 2011 Design and implementation of a distributed OLAP system, Artificial Intelligence, Management Science and Electronic Commerce (AIMSEC), 2011 2nd International Conference.

[12] Dehne F, Eavis T and Rau-Chaplin A, May, 2003 Parallel multi-dimensional ROLAP indexing, Cluster Computing and the Grid, 2003. Proceedings. CCGrid 2003. 3rd IEEE/ACM International Symposium.

[13] You J, Xi J, Zhang C and Guo G, October, 2008 HDW: A High Performance Large Scale Data Warehouse, Computer and Computational Sciences, 2008. IMSCCS '08. International Multisymposiums.

[14] Jiang J, Xu G, Wei X and Zhou R, August, 2008 A Positive Preprocessing Framework for Mapping Traditional Replica Selection Algorithms, Computing, Communication, Control, and Management, 2008. CCCM '08. ISECS International Colloquium.

[15] Chen Y, Dehne F, Eavis T and Rau-Chaplin A, Jan. 2008 PnP: sequential, external memory, and parallel iceberg cube computation, Distributed and Parallel Databases, vol. 23, no. 2, pp. 99–126. [Online]. Available: http: //www.springerlink.com/index/10.1007/s10619-007-7023-y.

[16] Dehne F, Eavis T and Hambrusch S, 2002 Parallelizing the data cube, Distributed and Parallel Databases, vol.11, pp.181–201. [Online].Available:http://www.springerlink.com/index/BGN4YJUMUBPELXK0.pdf.

[17] Guo-Liang Z, Hong C, Cui-Ping L, Shan W and Tao Z, 2010 Parallel Data Cube Computation on Graphic Processing Units, Chines Journal of Computers, vol. 33, no. 10, pp. 1788–1798,. [Online]. Available: http: //cjc.ict.ac.cn/eng/qwjse/view.asp?id=3197.

[18] You J, Xi J, Zhang P and Chen H, May 2008 A Parallel Algorithm for Closed Cube Computation, IEEE/ACIS International Conference on Computer and Information Science, pp. 95–99. [Online]. Available: http://ieeexplore.ieee.org/lpdocs/epic03/wrapper.htm?arnumber=4529804

[19] You J, Xi J and Zhang P, May, 2008 A Parallel Algorithm for Closed Cube Computation, Computer and Information Science. ICIS 08. Seventh IEEE/ACIS International Conference.

[20] Sergey K and Yury K, March, 2009 Applying Map-Reduce Paradigm for Parallel Closed Cube Computation, Advances in Databases, Knowledge, and Data Applications, 2009. DBKDA '09. First International Conference.

[21] Wang Z and Xu Y January, 2008 Minimal Condensed Cube: Data Organization, Fast Computation, and Incremental Update," Internet Computing in Science and Engineering, 2008. ICICSE '08. International Conference.

[22] Zalaket J, April, 2009 Speed up the Search in Bitmap Based Compressed Sparse Arrays, Information Management and Engineering, 2009. ICIME '09. International Conference.

[23] Chen Ch, Yan X, Zhu F, Han J and Yu Ph, 2008 Graph OLAP: Towards Online Analytical Processing on Graphs, ICDM 2008: 103-112.

[24] Hose K, Klan D, and K.-U S, April, 2009 Online Tuning of Aggregation Tables for OLAP, Data Engineering, 2009. ICDE '09. IEEE 25th International Conference.

[25] Talebian S.H and Kareem S.A, May, 2009 Using Genetic Algorithm to Select Materialized Views Subject to Dual Constraints, 2009 International Conference on Signal Processing Systems.

[26] Talebian S.H and Kareem S.A, May, 2010 A Lexicographic Ordering Genetic Algorithm for Solving Multi-objective View Selection Problem, Computer Research and Development, 2010 Second International Conference.

[27] Yuhang Z, Qi K and Wei Y, May, 2010 Materialized view selection algorithm—CSSA_VSP, Computational Intelligence and Natural Computing Proceedings (CINC), 2010 Second International Conference.

[28] Sen S, Chaki N and Cortesi A, November, 2009 Optimal Space and Time Complexity Analysis on the Lattice of Cuboids Using Galois Connections for Data Warehousing, Computer Sciences and Convergence Information Technology, 2009. ICCIT '09. Fourth International Conference.

[29] Sen S, Chaki N, February, 2011 Efficient Traversal in Data Warehouse Based on Concept Hierarchy Using Galois Connections, Emerging Applications of Information Technology (EAIT), 2011 Second International Conference.

[30] Dehne F, Zaboli H, 2012 Parallel Real-Time OLAP on Multi-core Processors, CCGRID '12 Proceedings of the 2012 12th IEEE/ACM International Symposium on Cluster, Cloud and Grid Computing (ccgrid 2012),Pages 588-594.

[31] Ester M, Kohlhammer J and Kriegel H-P, 2000 The DC-tree: a fully dynamic index structure for data warehouses, 16th International Conference on Data Engineering (ICDE), pp. 379–388. [Online]. Available: http://ieeexplore.ieee.org/lpdocs/epic03/wrapper.htm?arnumber=839438.

[32] Berchtold S, Keim D and Kriegel H.-P, Oct. 1996 The X-tree: An index structure for high-dimensional data, VLDB, pp. 28–39.

[33] Guttman A, 1984 R-trees: a dynamic index structure for spatial searching, ACM SIGMOD, pp. 47–57. [Online]. Available: http://portal.acm.org/citation.cfm?id=602266.

[34] Mazur P, Murlewski J, Kaminski M, Sakowicz B and Napieralski A, 2008 Universal database analysis repository model for use in near real-time database warehouse environment, Modern Problems of Radio Engineering, Telecommunications and Computer Science, 2008 Proceedings of International Conference.

[35] Zhou L, Shi Q and Geng H, March, 2010 The minimum incremental maintenance of materialized views in data warehouse, Informatics in Control, Automation and Robotics (CAR), 2010 2nd International Asia Conference.

[36] Yang K, Li Y, Luo Q, P.V. S and Shi J April, 2009 I3DC: Interactive Three-Dimensional Cubes, Data Engineering, 2009. ICDE '09. IEEE 25th International Conference

[37] Yadav S, Ahmad K and Shekar J, 2011 Analysis of web mining application and beneficial areas, IIUM Engineering Journal, Vol. 12, No. 2, pp. 185-195.

[38] Weyerhaeuser C, Mindnich T, Faerber F and Lehner W, December, 2008 Exploiting Graphic Card Processor Technology to Accelerate Data Mining Queries in SAP NetWeaver BIA, Data Mining Workshops, 2008. ICDMW '08. IEEE International Conference, December, 2008.

IOP Conf. Series: Materials Science and Engineering **53** (2013) 012044 doi:10.1088/1757-899X/53/1/012044

Performance analysis of multi-radio routing protocol in cognitive radio ad hoc networks under different path failure rate

Z Che-aron[1], A H Abdalla[1], K Abdullah[1] and W H Hassan[2]

[1]Department of Electrical and Computer Engineering, International Islamic University Malaysia (IIUM), Jalan Gombak, 53100 Kuala Lumpur, Malaysia
[2]Malaysia-Japan International Institute of Technology (MJIIT), Universiti Teknologi Malaysia (UTM), Jalan Semarak, 54100 Kuala Lumpur, Malaysia

E-mail: one_zamree@hotmail.com

Abstract. In recent years, Cognitive Radio (CR) technology has largely attracted significant studies and research. Cognitive Radio Ad Hoc Network (CRAHN) is an emerging self-organized, multi-hop, wireless network which allows unlicensed users to opportunistically access available licensed spectrum bands for data communication under an intelligent and cautious manner. However, in CRAHNs, a lot of failures can easily occur during data transmission caused by PU (Primary User) activity, topology change, node fault, or link degradation. In this paper, an attempt has been made to evaluate the performance of the Multi-Radio Link-Quality Source Routing (MR-LQSR) protocol in CRAHNs under different path failure rate. In the MR-LQSR protocol, the Weighted Cumulative Expected Transmission Time (WCETT) is used as the routing metric. The simulations are carried out using the NS-2 simulator. The protocol performance is evaluated with respect to performance metrics like average throughput, packet loss, average end-to-end delay and average jitter. From the simulation results, it is observed that the number of path failures depends on the PUs number and mobility rate of SUs (Secondary Users). Moreover, the protocol performance is greatly affected when the path failure rate is high, leading to major service outages.

1. Introduction

Traditional fixed spectrum allocation policies have been proposed to allow each wireless service to use certain frequency bands [1]. With the widespread usage of wireless devices, the spectrum bands become rapidly congested. Moreover, the tremendous growth in low-cost wireless applications that utilizes the unlicensed spectrum bands (e.g. ISM bands) leads to the increasing congestion and scarcity of unlicensed radio spectrum resources.

In addition, the experiments from the Federal Communication Commission (FCC) [2] reveal that the spectrum usage is concentrated on certain portions of the spectrum while a significant amount of the spectrum remains unutilized. Consequently, the emerging field of Cognitive Radio Networks (CRNs) [3, 4] is geared towards addressing the problems of spectrum scarcity and spectrum underutilization by enabling the unlicensed users, known as Secondary Users (SUs), to opportunistically utilize the vacant licensed spectrum, such as frequencies licensed for television

[1] Corresponding author: Zamree Che-aron

broadcasting, in such a manner as not to cause harmful interference to the licensed users (or Primary Users (PUs)), i.e. the SUs need to immediately vacate the licensed band being used once a PU is detected.

Various routing protocols have been proposed in order to provide efficient routing performance for mobile ad hoc networks [5, 6]. However, with the distinctive characteristics of Cognitive Radio Ad Hoc Networks (CRAHNs) (see Figure 1) that are wireless, multi-hop, self-organized, dynamically topology changing and spectrum availability varying networks [7], the routing protocol must provide multi-radio communication support so that the SUs are able to access multiple radio spectrum bands (both unlicensed and licensed bands) for data delivery.

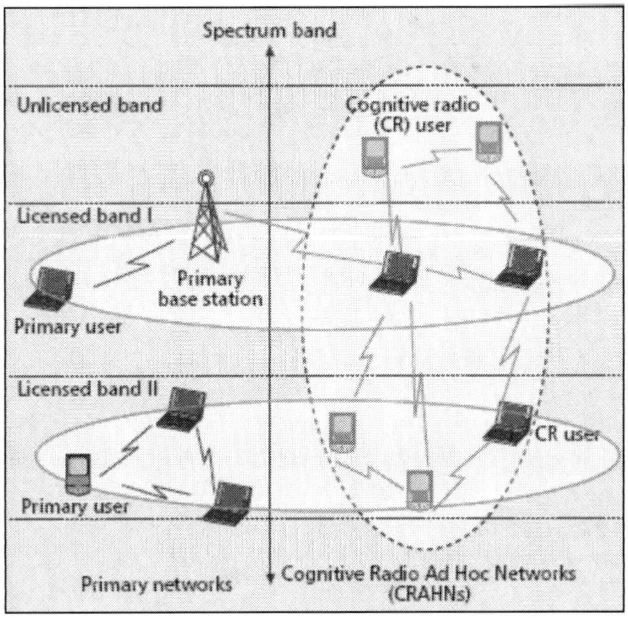

Figure 1. Cognitive radio ad hoc network architecture [7].

In this paper, we evaluate the performance of the Multi-Radio Link-Quality Source Routing (MR-LQSR) protocol [8] in CRAHNs based on different path failure rate by means of simulation using NS-2 [9]. Our evaluation metrics include average throughput, packet loss, average end-to-end delay and average jitter.

The remainder of the paper is organized as follows. We briefly review the MR-LQSR protocol in section 2. The simulation model and parameters are described in section 3. Finally, we present the simulation results and discussion in section 4 followed by the conclusion in section 5.

2. Multi-Radio Link-Quality Source Routing (MR-LQSR) protocol

In [8], the MR-LQSR protocol was proposed by Draves *et al.* that provides multi-radio communication support for multi-hop wireless ad hoc networks. The MR-LQSR is an extension of the Link-Quality Source Routing (LQSR) protocol [10] which is a source-routed link-state protocol derived from DSR (Dynamic Source Routing) [11]. The new routing metric called WCETT (Weighted Cumulative Expected Transmission Time) was presented to provide better route selection by taking into account for not only the link loss rate and bandwidth but also the interference among links that use the same spectrum channel as well as the channel diversity.

The protocol is mainly composed of four components: 1) discovering the neighboring nodes, 2) assigning the weight to links connected with the neighboring nodes, 3) broadcasting the links' weight information to other nodes in the networks, and 4) selecting the optimal path towards the desired destination based on the link weights. For the first and the third component, the MR-LQSR does not require any modification to the DSR's corresponding components. However, the MR-LQSR utilizes the WCETT metric to select a transmission path, while the DSR assigns equal weight to all links in the network and chooses the shortest path for data delivery.

In MR-LQSR protocol, each link is assigned a weight which is equal to the expected time it takes to transmit a fixed-size packet on that link called Expected Transmission Time (ETT). It is defined as:

$$ETT = ETX * \frac{S}{B} \tag{1}$$

where S is the packet size , B is the link bandwidth, and ETX [12] is the expected number of transmissions needed to send a unicast packet on a link, which can be calculated as follows:

$$ETX = \sum_{K=1}^{\infty} K \times S(K) = \frac{1}{(1-P_f)(1-P_r)} \tag{2}$$

where P_f and P_r is the packet loss rate of forward link and reverse link respectively, $S(K)$ means the total packet loss rate of the link, and K is the number of retransmissions.

For a path consisting of n links with assuming that the system has a total of k channels, the definition of WCETT routing metric can be estimated as follows:

$$WCETT = [(1-\beta) * \sum_{i=1}^{n} ETT_i] + [\beta * \max_{1 \le j \le k} X_j] \tag{3}$$

where β is a tunable parameter subject to $0 \le \beta \le 1$ and X_j is the sum of ETT of links using channel j which can be defined as:

$$X_j = \sum_{\text{All links using channel } j} ETT_j \tag{4}$$

For the MR-LQSR protocol, the source node always selects a path with the lowest value of WCETT for data transmission (low WCETT value implies better routes). However, the different values of β can impact the protocol performance. The WCETT uses β as the weight given to the channel-diversity component. When the value of β is high, the protocol always selects paths with more channel diversity. On the other hand, with low value of β, the shorter paths are more preferred.

3. Simulation environment

In this section, we evaluate the protocol performance of the MR-LQSR (at $\beta = 0.6$) in CRAHNs with different number of path failures from 0 to 6 times. The network simulations have been done via NS-2 simulator with Cognitive Radio Cognitive Network (CRCN) patch [13]. The simulations are performed in multi-hop network topology where 25 movable SUs are deployed in an area of 1000 x 1000 m^2. The source-destination node pair is selected to create UDP connections. Each UDP connection transmits CBR traffic with 512 byte packets at packet interval of 0.03 seconds. The IEEE 802.11 standard is used as MAC protocol at bandwidth of 2 Mbps. The Two-Ray Ground model [14] is selected as the wireless propagation type. The simulation time is set to 100 seconds. Table 1 summarizes the simulation parameters.

In addition, we change the number of PUs and the mobility rate of SUs in order to produce different number of path failures occurred during simulation time because, from the simulation results, we observe that the number of path failures varies according to the PUs number and SUs' mobility rate. By increasing the number of PUs or mobility rate of SUs, the number of path failures is higher. During the simulation time, SUs must immediately vacate the channel which overlaps a PU's transmission frequency in order not to cause a harmful interference to the PUs, resulting in a path breakage. Moreover, the increased mobility rate of SUs can make the network topology change more frequent, easily leading to a path failure.

Table 1. Simulation parameters.

Parameters	Values
Number of movable SUs	25
Simulation area	1000 x 1000 m^2
Transport layer protocol	UDP
Data packet size	512 bytes
Packet interval	0.03 seconds
MAC layer	IEEE 802.11
Link bandwidth	2 Mbps
Propagation model	Two-Ray Ground
Traffic type	CBR
Simulation time	100 seconds
Interface queue type	DropTail/PriQueue

4. Simulation results and discussion

The simulation results of the MR-LQSR protocol in CRAHNs under different path failure rate are generated in the NS2 output trace files and we analyze the results by using the NS2 Visual Trace Analyzer [15]. The results are displayed in the form of line graphs based on the performance metrics including average throughput, packet loss, average end-to-end delay and average jitter.

In Figure 2, the performance behavior of the MR-LQSR in term of average throughput versus the number of path failures is shown. It is observed that the results of average throughput decrease rapidly when the number of path failures increases. This behavior can be justified because the efficient route recovery algorithm is not provided by the routing protocol. When a transmission path failure is detected, the data traffic is still delivered through the failed path until the source node is informed with a Route ERRor (RERR) packet and finds a new path. As a result, a lot of data packets are dropped in the network, as shown in Figure 3 which exhibits the result of the number of dropped data packets against the number of path failures.

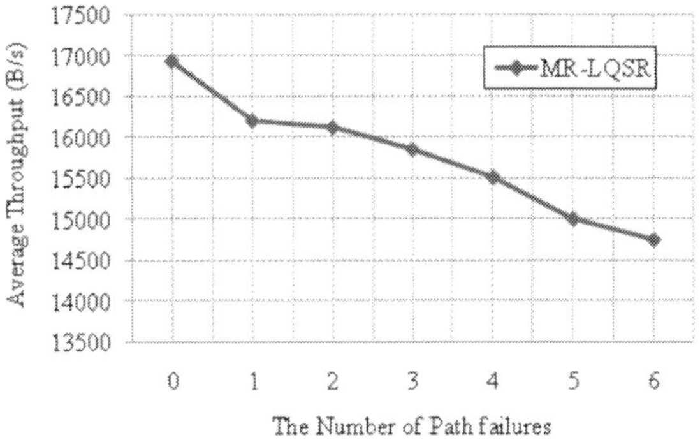

Figure 2. Simulation results of average throughput.

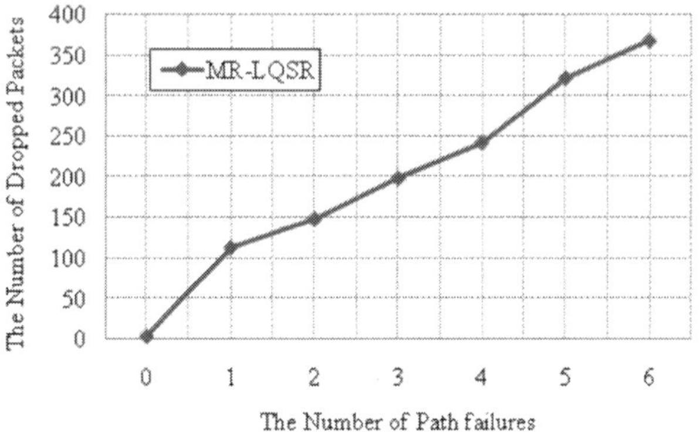

Figure 3. Simulation results of packet loss.

Figure 4 displays the performance behavior of the protocol in term of average end-to-end delay with the number of path failures. From the graph, with increased number of path failures, it is clear that the protocol produces longer average end-to-end delay. With lack of efficient route recovery algorithm, when a transmission path is failed either because of PUs activity or SUs mobility, some data packets are buffered until a new path is discovered, leading to long packet latency. Furthermore, in case a transmission path is broken due to node mobility, the new path which has longer path length is always created for transmitting the next data flows, resulting in higher end-to-end delay.

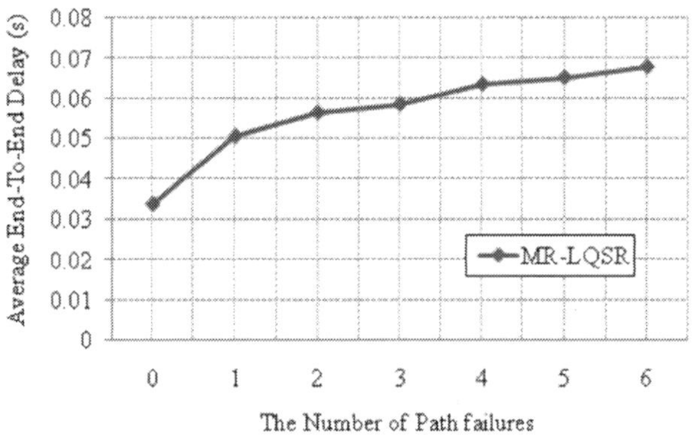

Figure 4. Simulation results of average end-to-end delay.

In addition, the results of average jitter against the increased number of path failures, as shown in Figure 5, exhibit a similar trend to the results of average end-to-end delay. The jitter is defined as a measure of the variability over time of the data packet latency across a network. The average jitter is a significant metric in an assessment of network performance, especially, for a real-time application. A system with lower jitter provides better QoS (Quality of Service). From the graph, it is observed that the results of average jitter rise dramatically when the number of path failures is increased. Due to lack of efficient route recovery mechanism, a path breakage can cause a significant interruption of data transmission, leading to high average jitter.

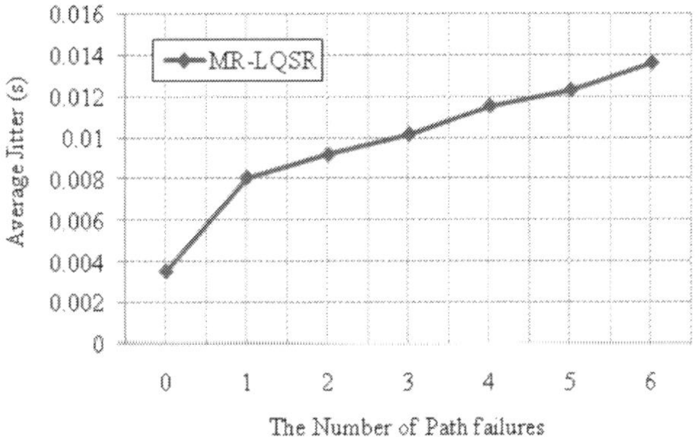

Figure 5. Simulation results of average jitter.

5. Conclusion

Cognitive Radio (CR) technology holds great promises for addressing the problems of spectrum scarcity and spectrum underutilization in order to meet the increasing demand for radio spectrum. As the special network characteristics, CRAHNs have received increasing research attention in recent years. Since such networks do not require any centralized administration or established infrastructure, they can be deployed for a variety of military and civilian applications.

In this paper, the MR-LQSR routing protocol has been reviewed briefly. Additionally, we have simulated the protocol and evaluated the performance in CRAHNs under different path failure rate, using NS-2 simulator. The considered performance metrics are average throughput, packet loss, average end-to-end delay and average jitter. From the simulation results, since the protocol lacks the efficient route recovery algorithm which can quickly repair the broken paths, when the number of path failure is increased that causes significant interruptions of data transmission, the results of average end-to-end delay, average jitter and packet loss rise rapidly. On the other hand, the throughput results decrease dramatically, when the path failure rate is high.

For further research, we plan to investigate and study on how to improve the protocol performance in CRAHN under high path failure rate. We believe that providing the efficient route recovery approach which can cope with a large number of path failures is the right way to further enhance the protocol.

6. References

[1] McHenry M 2003 Spectrum white space measurements *New America Foundation's Broadband Forum*, http://www.newamerica.net/files/nafmigration/archive/Doc_File_185_1.pdf

[2] Federal Communications Commission (FCC) 2003 Notice of proposed rule making and order, ET Docket No. 03-222

[3] Haykin S 2005 Cognitive radio: brain-empowered wireless communications *IEEE Journal on Selected Areas in Communications* **23** 201-220

[4] Mitola J 2000 Cognitive radio --- an integrated agent architecture for software defined radio, PhD Dissertation, Department of Teleinformatics, Royal Institute of Technology (KTH)

[5] Gupta N and Gupta R 2010 Routing protocols in mobile ad-hoc networks: an overview *Proc. Int. Conf. on Emerging Trends in Robotics and Communication Technologies (Chennai)* (New York: IEEE) pp 173-177

[6] Bakht H 2011 Survey of routing protocols for mobile ad-hoc network *International Journal of Information and Communication Technology Research* **1** 258-270

[7] Akyildiz I F, Lee W Y and Chowdhury K R 2009 CRAHNs: cognitive radio ad hoc networks *Ad Hoc Networks* **7** 810-836

[8] Draves R, Padhye J and Zill B 2004 Routing in multi-radio, multi-hop wireless mesh networks *Proc. 10th Int. Conf. on Mobile Computing and Networking (Philadelphia)* (New York: ACM) pp 114-128
[9] Information Sciences Institute (ISI) 1989 The network simulator (NS-2), http://www.isi.edu/nsnam/ns/
[10] Draves R, Padhye J and Zill B 2004 The architecture of the link quality source routing protocol, Microsoft Research Technical Report MSR-TR-2004-57
[11] Johnson D B, Maltz D A and Broch J 2001 DSR: the dynamic source routing protocol for multihop wireless ad hoc networks *Ad Hoc Networking* ed Perkins C E (Massachusetts: Addison-Wesley Longman Publishing) pp 139-172
[12] De-Couto D S J, Aguayo D, Bicket J and Morris R 2005 A high-throughput path metric for multi-hop wireless routing *Wireless Networks* **11** 419-434
[13] Zhong J and Li J 2009 Cognitive radio cognitive network simulator, http://stuweb.ee.mtu.edu/~ljialian/
[14] Eltahir I K 2007 The impact of different radio propagation models for mobile ad hoc networks (MANET) in urban area environment *Proc. 2nd Int. Conf. on Wireless Broadband and Ultra Wideband Communications (*Sydney*)* (New York: IEEE) pp 30-38
[15] Rocha F 2012 NS2 visual trace analyzer, http://nsvisualtraceanalyzer.wordpress.com/

Acknowledgments
We would like to acknowledge the Research Management Centre (RMC) at International Islamic University Malaysia (IIUM) for their support in this research.

IOP Conf. Series: Materials Science and Engineering **53** (2013) 012045 doi:10.1088/1757-899X/53/1/012045

Rain Attenuation Analysis using Synthetic Storm Technique in Malaysia

A K Lwas, Md R Islam[1], J Chebil, M H Habaebi, A F Ismail, A Zyoud and H Dao
Kulliyyah of Engineering International Islamic University Malaysia
Jalan Gombak, 53100 Kuala Lumpur, Malaysia

E-mail: rafiq@iium.edu.my; alilawas@yahoo.com

Abstract Generated rain attenuation time series plays an important role for investigating the rain fade characteristics in the lack of real fade measurements. A suitable conversion technique can be applied to measured rain rate time series to produce rain attenuation data and be utilized to understand the rain fade characteristics. This paper focuses on applicability of synthetic storm technique (SST) to convert measured rain rate data to rain attenuation time series. Its performance is assessed for time series generation over a tropical location Kuala Lumpur, in Malaysia. From preliminary analysis, it is found that SST gives satisfactory results to estimate the rain attenuation time series from the rain rate measurements over this region.

1. Introduction

In satellite communications which operate at frequencies above 10 GHz, the signal will be subjected to various propagation impairments such as rain attenuation, cloud attenuation, tropospheric scintillation, water vapour attenuation, and rain and ice depolarization. Rain attenuation is the most significant parameter among propagation impairments [1-3]. It is one of the most fundamental limitations to the performance of satellite communication links in the microwave region, causing large variations in the received signal power, with little predictability and many sudden changes. In tropical areas, where the problem is very severe [4], lack of real data especially in higher frequencies makes the generation of rain attenuation time series a requirement for investigating the rain fade characteristics.

By applying the synthetic storm technique, rain attenuation time series can be generated at any frequency and polarization. This work focuses on applicability of SST by comparing rain attenuation with that predicted by SST from rain rate both measured concurrently. The measurements were conducted over Kuala Lumpur (3.25°N, 101.7°E) where located in heavy rain zone by receiving signal at 10.982 GHz (Ku-band) from MEASAT3 (91.5_E) on 77.4_ elevation angle. The measurements have been carried out for one month period. Rain rate time series is obtained from a collocated tipping bucket rain gauge data which is time-synchronized with attenuation data logging system. Section two of the paper provides brief details on the SST. Section three includes the experimental setup details. Results are introduced and analyzed in section four. Section five concludes the paper contribution.

2. Synthetic Storm Technique

A summary of the fundamentals of the SST as developed by [5] is: converting a rain rate time series, which is measured using a rain gauge at a point for 1 minute and the average value is taken, to a rain rate space series along a line, using an estimate of the storm translation speed v to transform time to distance. Mathematically, this is carried out by convolution.

The Synthetic Storm Technique is a physical-mathematical radio propagation method that can be used to generate reliable rain attenuation time series by converting a rain rate time-series at a specified

[1] Md R Islam: rafiq@iium.edu.my

site into a rain attenuation time-series. This process requires knowledge about the length of the signal path through the rain cell, the rain cell velocity and the rain rate at the site under investigation [5]. Moreover, by applying the SST can generate rain attenuation time series at any frequency and polarization, and for any slant path above approximately 10°, as long as the hypothesis of isotropy of the rainfall spatial field holds, in the long term. The SST is very useful for designing communications satellite systems and improve its performance [6].For calculating rain attenuation using SST over a satellite path, following are the basic assumptions which are explained in Figure. 1 [6-8].

During rain, the vertical structure of the troposphere becomes two layers. A is the rain layer and B is the melting layer as shown in Figure. 1. The signal attenuation in case of satellite path is obtained from specific attenuation at a point using following expression:

$$A(x) = K_A \int_0^{L_A} R^{\alpha_A}(x_0 + \Delta x_0, \xi) d\xi + K_B r^{\alpha_B} \int_{L_A}^{L_B} R^{\alpha_B}(x_0, \xi)\, d\xi \qquad \text{(1-a)}$$

where $A(x)$ is the attenuation at specific point and ξ is the distance measured along the satellite path. where, k and α depend on the electromagnetic wave frequency and polarization, raindrop size distribution. They are given by the [9] for water temperature of 20° C and Parson's law drop size distribution and for 0° C, are given by [10]. The relation between rain rate of layer A and layer B can describe by

$$R_B = r\, R_A \quad \text{with } r = 3.134 \qquad \text{(1-b)}$$

According to (ITU-R P.839-3, 2001) the height of the upper limit of the layer B above sea level is

$$H_B = h_0 + 0.36 \text{ km} \qquad \text{(1-c)}$$

where h_0 is the mean $0^\circ C$ isotherm height above mean sea level. H_A is the height above sea level of the upper limit of layer A. It is given as

$$H_A = H_B - h \text{ km} \qquad \text{(1-d)}$$

where h is a thickness of the melting layer. The radio path lengths L_A and L_B are given as:

$$L_A = (H_A - H_S) / \sin(\theta) \qquad \text{(1-e)}$$

and

$$L_B = (H_B - H_S) / \sin(\theta) \qquad \text{(1-f)}$$

where 'θ' is the link elevation angle and H_S is the height of the Earth station above sea level.

$$\Delta x_0 = \Delta L \cos(\theta) = h/\tan(\theta) \qquad \text{(1-g)}$$

and

$$\Delta L = L_B - L_A = h / \sin(\theta) \qquad \text{(1-h)}$$

Corresponding to time series of rain rate, an attenuation time series is generated for every observation stations and an attenuation data set is derived.

According to (Matricciani, 1996), the below equation was derived by applying the Fourier transforms theory and some assumptions to the Equation (1). Attenuation time series

$$A(t) = K_A R^{\alpha_A}(t) L_A + r^{\alpha_B} K_B R^{\alpha_B}(t)(L_B - L_A) \qquad (2)$$

where R(t) is rain rate time series.

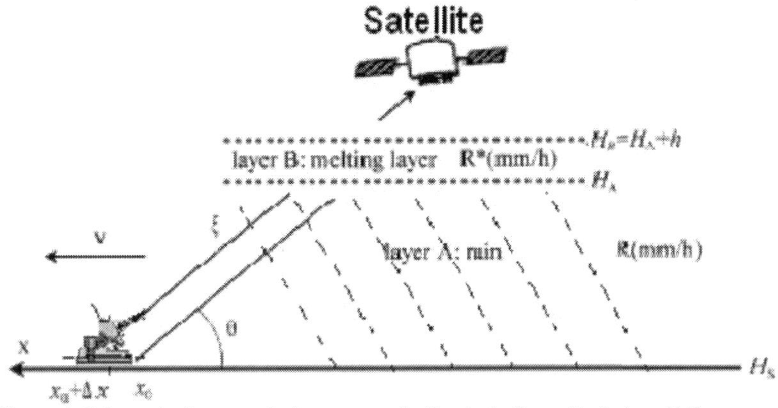

Figure 1. Schematic diagram of rain structure for Synthetic Storm Technique [11].

3. Experimental Setup

Vertically polarized Ku-band satellite signal level from MEASAT3 (geostationary at 91.5°E) has been continuously monitored at rooftop of block E2 (3.25°N, 101.7°E) at Kulliyyah of Engineering, International Islamic University Malaysia (IIUM) for 1 month from 1st April 2012 to 30th April 2012. At Ku-band, 10.982 GHz of QPSK modulated signal level has been received by an offset feed parabolic antenna of 0.6m diameter size. The elevation angle of the earth-satellite path is 77.4°. The logging data was sampling rate of 0.1s time interval and subsequently process to 1s. A 0.2 mm tipping-bucket rain gauge is integrated in the system and logged every 10s time interval to measure rainfall rate and determine rainy events as shown in Figure. 2 [13].

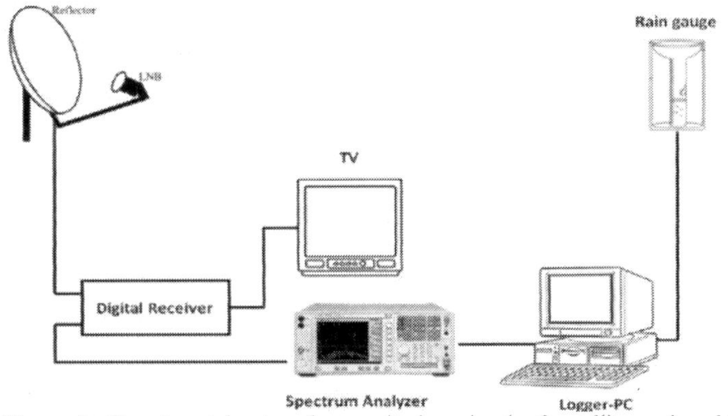

Figure 2. Experimental setup for monitoring signal of satellite and real-time raining at IIUM, Gombak.

4. Results Analysis

In April 2012, all rain events have been observed nine events have been recorded. These rain events have been converted to rain attenuation time series using SST by Equation (2). According to SST the rain attenuation time series due to rainy layer and melting layer has been calculated separately. The advection velocity v is assumed as originally suggested in [5]. The radio path lengths L_A and L_B are calculated according to Equations (1-e and 1-f) respectively. The specific attenuation of melting region (B) is constructed according to [10] and the values for coefficients k and α are taken to be 0.016354 and 1.1562 respectively. The k and α values are related to rainy layer (A) and are selected from [12] to be 0.015184 and 1.2144 respectively.

Figures 3 and 4 give examples of two measured rain rate time series and its corresponding rain attenuation time series for two events on 9/4/2012 and 13/4/2012 respectively. Both events, the actual rain rate time series and the rain attenuation time series by SST demonstrate strong correlation throughout the event. The predicted rain attenuation time series also follows similar pattern observed in rain rate time series.

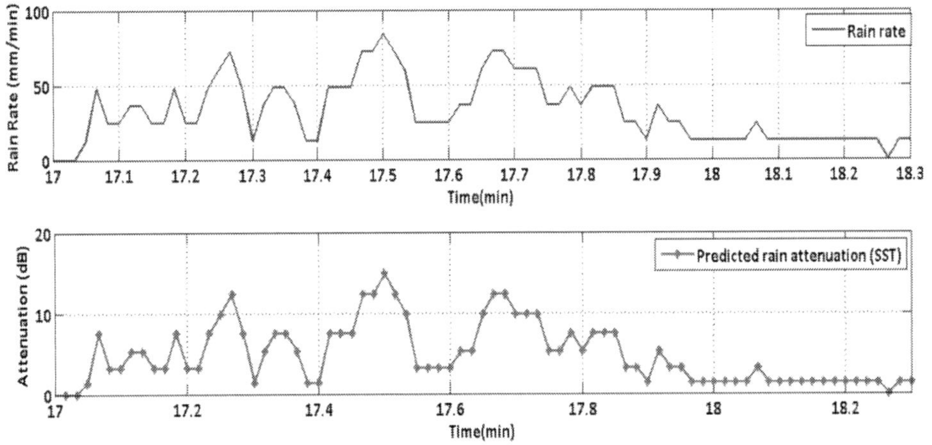

Figure 3. Rain rate and rain attenuation time series converted by SST for a measured rainy event on 9/4/2012.

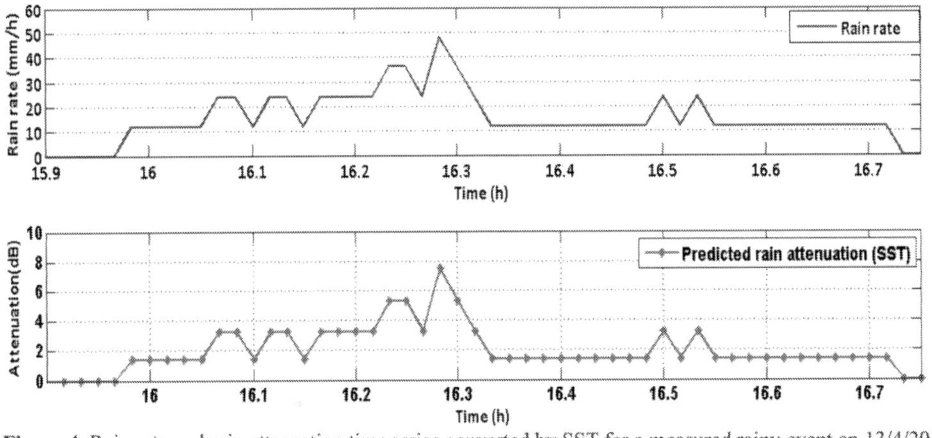

Figure 4. Rain rate and rain attenuation time series converted by SST for a measured rainy event on 13/4/2012.

The rain attenuation time series obtained from measured one minute rain intensity data using SST was compared against the measured one minute rain attenuation for same period for two events on 9/4/2012 and 13/4/2012. Figures 5 and 6 show comparison between predicted rain attenuation time series using SST and measured rain attenuation time series. The rain attenuation time series by SST are almost leading on measured rain attenuation time series in both events. This may occur because of little resynchronization between devices of measurement. There are many peaks observed in Figure 5. The highest peak the value of rain attenuation is close to 15 dB by SST but the measurement is about

13 dB. The value of rain attenuation peak observed in Figure 6 is about 7.5 dB by SST but the measurement is close to 5.5 dB.

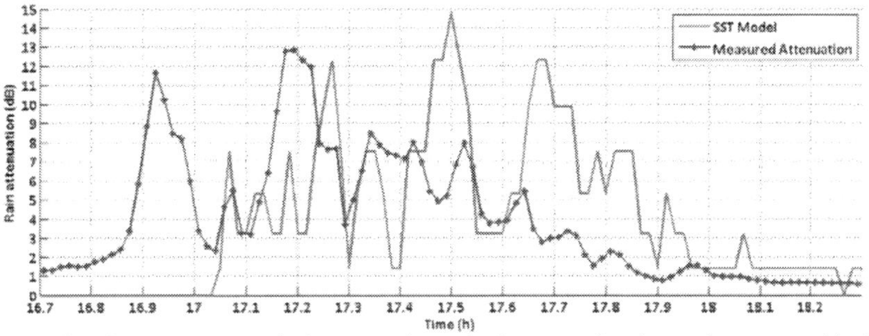

Figure 5. Comparison between measured rain attenuation and rain attenuation time series converted by SST for a rainy event on 9/4/2012.

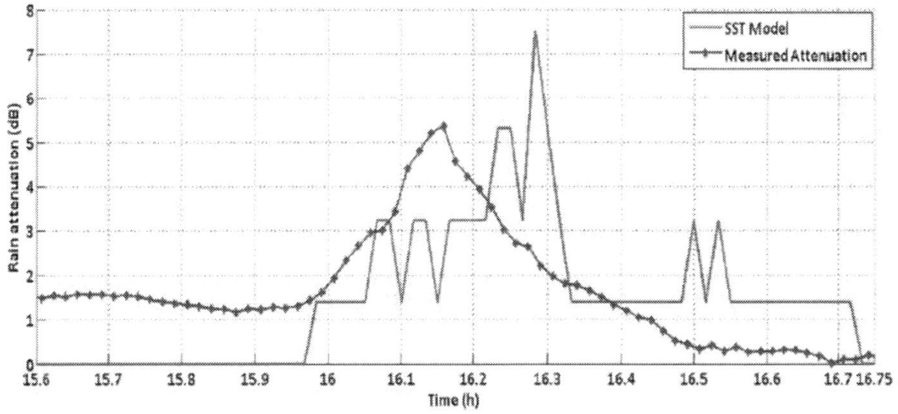

Figure 6. Comparison between measured rain attenuation and rain attenuation time series converted by SST for a rainy event on 13/4/2012.

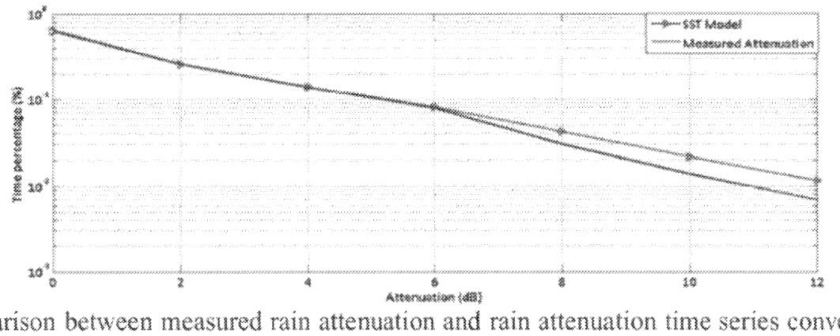

Figure 7. Comparison between measured rain attenuation and rain attenuation time series converted by SST for month of April.

Cumulative distribution of one month percentage exceedance of rain attenuation using SST and measured rain attenuation are calculated. Therefore a comparison between them is done and is shown in Figure 7. It is seen from the plot that measured rain attenuation exceedances match well with predicted rain attenuation values before 6 dB while after 6 dB SST is overestimated.

5. Conclusion

This paper focuses on the applicability of SST method in Malaysia which has been established as an accurate method in Europe. SST is used to generate rain attenuation time series by using rain rate data measured for one month in Malaysia. Converted attenuation by using SST is compared with actual rain attenuation at 10.982 GHz recorded at Satellite Lab, IIUM, Kuala Lumpur from MEASAT3. It is found that SST derived attenuation is consistent well with the measured attenuation in lower levels but it is overestimated in higher levels. However this result is preliminary, a long term measurement is recommended to conclude and validate the above findings.

6. References

[1] Matricciani E 2006 Time Diversity As A Rain Attenuation Countermeasure In Satellite Links In The 10-100 Ghz Frequency Bands Department of Electronic and Information Milan Polytechnic ESA SP-626 Milan.

[2] Sridhar M, Raju K P and Rao Ch. S 2012 Estimation of Rain Attenuation based on ITU-R Model in Guntur (A.P): *ACEEE Int. J. on Communications* Vol. **03** No. 03 India.

[3] Nakajo R, Maekawa Y 2012 Characteristics of Rain Attenuation Time Variation in Ka Band Satellite Communications for the kind of Rain Types in Each Season: Proceedings of ISAP2012 Nagoya: Japan.

[4] Lam Y H, Luini L, Din J, Capsoni C and Panagopoulos A D 2012 Investigation of Rain Attenuation in Equatorial Kuala Lumpur: *IEEE Antennas and Wireless Propagation Letters IEEE* Vol. **11** pp. 1002-1005.

[5] Matricciani E 1996 Physical-mathematical model of the dynamics of rain attenuation based on rain rate time series and a two-layer vertical structure of precipitation: *Radio Science* Vol. **31** No. 2 pp. 281–295.

[6] Matricciani E, Riva C and Castanet L 2006 Performance of the synthetic storm technique in a low elevation 5 slant path at 44.5 GHz in the French Pyrénées: in Europ.Conf.Antennas Propag. (EuCAP 2006) Nice: France.

[7] Panagopoulos A. D, Arapoglou P D, MChatzarakis G E, Kanellopoulos J D and Cottis P G 2004 A new formula for the prediction of the site diversity improvement factor: Int. J. *Infrared Millimeter Waves* Vol. **25** No. 12 pp. 1781–1789.

[8] Sánchez-Lago I, Fontán F P, Mariño P and Fiebig U C (2007) Validation of the Synthetic Storm Technique as Part of a Time Series Generator for Satellite Links *IEEE Antennas and Wireless Propagation Letters*, Vol. **6** pp. 372-375.

[9] ITU-R P.838-3 2005 Specific attenuation model for rain for use in prediction methods ITU-R Recommendations Geneva.

[10] Maggiori D, 1981 Computed transmission through rain in the 1–400 GHz frequency rangefor spherical and elliptical drops and any polarization *Alta Frequenza*, Vol. **50** pp. 262–273

[11] Shukla K A, Das S and Roy B 2009 Rain Attenuation Measurements using Synthetic Storm Technique over Ahmedabad, International Conference on Computers and Devices for Communication India.

[12] ITU-R P.839-3 2001 Rain height model for prediction methods: Recommendation P. Series Fascicle Radio wave propagation International Telecommunication Union Geneva.

[13] Dao H, Islam MD R and Al-khateeb K A S 2012 Analysis of Rain Fade Duration Over Satellite-Earth Path at Ku-Band in Tropics IEEE Computer and Communication Engineering (ICCCE, Kuala Lumpur) Malaysia.

IOP Conf. Series: Materials Science and Engineering **53** (2013) 012046 doi:10.1088/1757-899X/53/1/012046

Performance Analysis of IIUM Wireless Campus Network

Suhaimi Abd Latif[1], Mosharrof H. Masud, Farhat Anwar
Department of Electrical and Computer Engineering
International Islamic University Malaysia
Kuala Lumpur, Malaysia

E-mail: suhaimie@iium.edu.my, masuduia@yahoo.com, farhat@iium.edu.my

Abstract. International Islamic University Malaysia (IIUM) is one of the leading universities in the world in terms of quality of education that has been achieved due to providing numerous facilities including wireless services to every enrolled student. The quality of this wireless service is controlled and monitored by Information Technology Division (ITD), an ISO standardized organization under the university. This paper aims to investigate the constraints of wireless campus network of IIUM. It evaluates the performance of the IIUM wireless campus network in terms of delay, throughput and jitter. QualNet 5.2 simulator tool has employed to measure these performances of IIUM wireless campus network. The observation from the simulation result could be one of the influencing factors in improving wireless services for ITD and further improvement.

1. Introduction

The Internet is one of the most powerful, cost effective and easy resources in the educational institutions especially at tertiary level for state-of-the-art research activities. Good connectivity with the Internet is not luxury to a university, in fact is a basic necessity of today's modern education. Today's Internet is the combinations of multiple set of technologies like WiFi, WiMAX, GSM. The next generation wireless technology systems are being devised with the vision of heterogeneity where a mobile user will be connected with multiple wireless networks like GPRS/UMTS, IEEE 802.11x, WiMAX and satellite etc. [1]. The Internet Provides vast amount of information to the users to enhance research activities by the students and lecturers. Therefore, a university needs to be connected with high speed connections to provide satisfactory services to its users. However, educational institutions face budgetary restraints and limited resource that makes a complicated decision to allot bandwidth in proper way to the campus users. The demand for bandwidth is increasing every year due to rising number of new student's enrolment with multiple devices like, laptops, smart phones, etc. Hence, it is important to allocate bandwidth in proper way among campus users.

It has been observed that available bandwidth of a university is generally not enough to meet rising demand to support optimal usage. Management of the bandwidth is a generic term that describes different techniques, technologies, tools and policies employed by an organization to provide optimum usage of its resources [2]. It is noted that bandwidth management procedures can be classified into three broad categorized including techniques and technologies, organizational management policies and monitoring. Some commonly used techniques are [3]:

[1]Suhaimi Abd Latif, Department of Electrical and Computer Engineering, International Islamic University Malaysia, Kuala Lumpur, Malaysia, suhaimie@iium.edu.my.

- Data compression
- Local caching
- Application prioritization
- Distributed content
- Blocking unauthorized traffic and
- Good practice among users of the Internet

IEEE 802.11g has been deployed in IIUM to provide wireless services. The Information Technology Department (ITD) of IIUM has set some rules and regulations for its users that reduces burden on the overall network and improves the quality of wireless facilities. For example, ITD has set restriction on social sites (facebook, twiter, youtube) during office time (8am to 6pm) so that essential activities can be performed smoothly. It also limits the download capacity for an individual user and given an equal priority to all users. The ITD is trying to increase the quality of wireless services by taking many managerial polices. The analysis of technical aspects also needs to be considered. Therefore, this paper aims to investigate the performance of IIUM wireless campus network in terms of throughput, delay and jitter.

The rest of the paper organized as follows. Section II addresses the issues of WiFi in details. The background of IIUM wireless campus network has mentioned in section III. Simulation scenario has drawn and highlighted in section IV. Performance analysis is discussed in section V. Finally, section VI concludes this paper.

2. Wireless Network (WiFi)

Wireless Fidelity (WiFi) is a popular technology that has been used for data transmission from last decade for small coverage area. Data transmission can be performed either connected with an Access Point (AP) at broadband speed or in ad hoc mode. In 1997, Institute of Electrical and Electronic Engineer (IEEE) drafted 802.11 standard for Wireless Local Area Network (WLAN) that is known as IEEE 802.11. This protocol was capable to transmit only 2 Mbps which is too slow for many applications and therefore, it was not manufactured at that time. Later on some versions have been released namely IEEE 802.11a and IEEE802.11b that work independently. In 1999, some network hardware companies have manufactured IEEE 802.11a that operating at 5 GHz with transmission speed of 54Mbps. However, it was not well accepted due to its heavy power consumption and high cost.

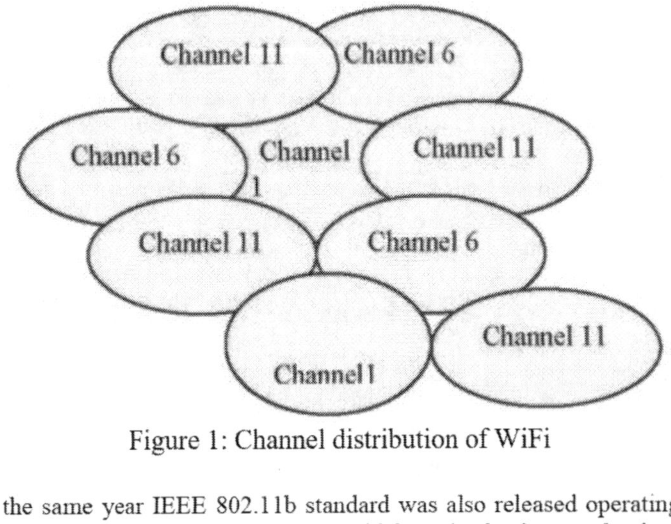

Figure 1: Channel distribution of WiFi

Subsequently, in the same year IEEE 802.11b standard was also released operating at 2.4 GHz with transmission speed 11Mbps that has been accepted widely. It is also known that both 11a and 11b are the extension of original IEEE 802.11 and IEEE 802.11b released before IEEE 802.11a in same year and gained popularity. IEEE 802.11a is mostly used in business networks and IEEE 802.11b is suits with home networks but they are incompatible with each other.

WiFi has 11 channels with 3 non-overlapping that uses direct-sequence spread-spectrum depicted in figure 1 [7]. Channel 1, 6 and 11 are typically used for large coverage of area. However, this protocol is criticized for lacking of security. IEEE 802.11g is the extension of IEEE 802.11b that released in 2003 and operates at 2.4 GHz with transmission speed at 54Mbps. This standard is backward compatible with IEEE 802.11b. Frequency Division Multiplexing (FDM) technology is used for this protocol. Finally, the newest IEEE 802.11n has been released in 2009 for high data transmission. MIMO technology has utilized to maximize transmission speed that is up to 600 Mbps using 2.4 GHz. A comparison of different versions' of WiFi is listed in table 1. The performance of WLAN, particularly delay and jitter has been addressed and showed an improvement for interactive multimedia transmission [8,9].

Table 1. Comparison of wireless technologies

Technology	**Data Rate**	**Tx-Range Indoor/Outdoor**	**Frequency Band**	**Coding Technique**
IEEE 802.11a	54Mbps	35/120m	5.5 GHz	OFDM
IEEE 802.11b	11 Mbps	38/140m	2.4 GHz	FDM
IEEE 802.11g	54Mbps	38/140m	2.4 GHz	OFDM
IEEE 802.11n	100Mbps	70/250m	2.4 GHz	MIMO

2.1 Radio Channel

Wi-Fi protocol uses spread spectrum techniques in the 2.4 GHz band that is unlicensed in most countries and known as the Industrial, Scientific, and Medical (ISM) band. Wi-Fi utilizes Direct Sequence Spread Spectrum DSSS (802.11), Complementary Code Keying (CCK, 802.11b), or OFDM modulation (802.11a/g/n) with 14 Radio Frequency (RF) channels (11 available in US, 13 in Europe, and just 1 in Japan) and 22 MHz bandwidth [10].

2.2 Network Size

The basic cell of IEEE 802.11 is known as Basic Service Set (BSS) that is a set of mobile or fixed stations. If a Mobile Device (MD) moves from the BSS coverage area, it can not communicate with other BSS directly. The coverage area of a BSS is typically 38 meter and 140 meter for indoor and outdoor respectively for IEEE 802.11b/g and the range is listed in table 1.

2.3 Security

IEEE 802.11 uses the RC4 stream cipher for encryption and the CRC-32 checksum for integrity. WiFi Protected Access (WPA) and WAP2 encryption mechanisms are employed to WiFi technology for further enhancement that includes IEEE 802.11i standard which are AES block cipher and CCM.

2.4 Mobility

Mobility is one of the most important issues that has been supported by WiFi. WiFi provides high speed data transmission with low mobility as its coverage area is very limited. To support seamless mobility for the MD, the coverage area must be extended to the certain level.

3. IIUM Wireless Network

International Islamic University Malaysia (IIUM) is one of the unique universities in the world in terms of internationalization, Islamization and comprehensive excellence of teaching and learning. It has four campuses including Gombak - the main campus, Kuantan campus, Petaling Jaya campus and Kuala Lumpur campus. The wireless network of IIUM has been maintained and monitored by ITD in all these campuses. The wireless network has been covered in academic and residential areas where there are approximately thirty thousand users are being benefited from wireless facilities. The IIUM wireless network has implemented in three phases in 2003, 2009 and 2013. In first phase, it covers the entire academic area at Gombak Campus, Kuantan campus and Kuala Lumpur campus which includes; administrative office areas, class rooms, meeting rooms, conference rooms, cafeteria etc.[11]. In second and third phases, the wireless coverage has been increased to the full campus including academic and residential areas. For example, the main campus Gombak has more than 700 acres and most of the places have covered by the wireless network.

This research only discusses on the performance of wireless network of Gombak campus particularly for a small coverage area. There are two wireless controller are connected with the Internet and these two controllers connect all the APs with wire. IEEE 802.llg technology has deployed to provide IIUM wireless services. One AP can support concurrently maximum 50 nodes either static or mobile. Considering the population density of a specific location, numbers of APs have been deployed in those areas. In this research, some nodes are considered static and some are considered mobile.

4. Simulation Setup

QualNetT is one of the network simulators that can predict and act like real network and the performance of a specific network can be measured through simulation or emulation. Wired, wireless and combined network platforms can be developed using this simulator. QualNet is developed by Scalable Network Technologies; Inc. [12]. QualNet comes with different libraries such as Wi-Fi, WiMAX, CDMA, GSM etc. This experiment uses WiFi libraries. It is a quite simple and easy to design as it is Graphical User Interface (GUI) based simulator. In this experiment, a small wireless network of IIUM has been designed for the analysis. Three parameters, namely; throughput, delay and jitter have been considered for the analysis.

Table 2. Simulation Parameters

Parameters	Values
Physical layer technology	IEEE 802.11
Data rates bits/sec	54 Mbps
Area in meters	100m X 100m
Nodes	4 APs, 16 static nodes and 20 mobile nodes
Simulation time	1000 seconds
Traffic type	CBR
Packet size	512 Bytes
Mobility model	Random waypoint
Mobility speed	Minimum 5m/s and maximum 30m/s

There are 4 nodes are considered as APs those are directly connected with the wireless controller. 16 nodes are static and 20 nodes are mobile moves at the speed from 5m/s to 30 m/s based on random way point model. The Constant Bit Rate (CBR) traffic is employed to with packet size 512 bytes in a 100m X 100m area with 54MBps data rate. The simulation scenario has shown in figure 2.

IOP Conf. Series: Materials Science and Engineering **53** (2013) 012046 doi:10.1088/1757-899X/53/1/012046

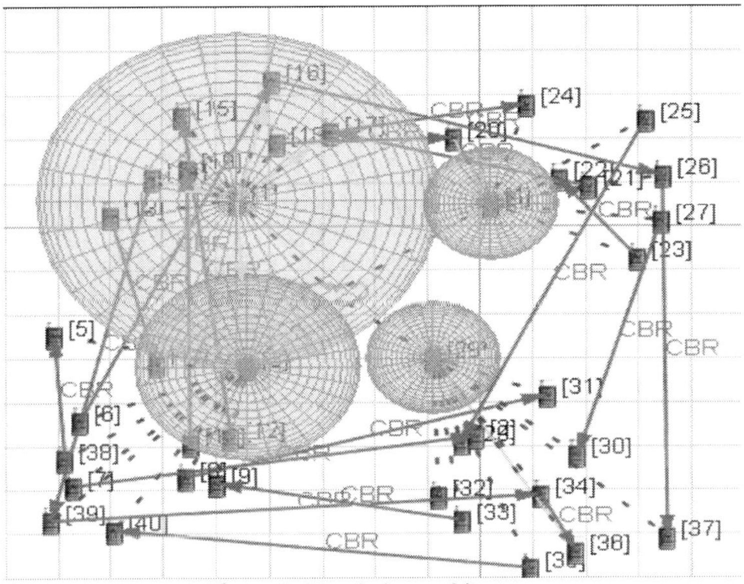

Figure 2: Simulation architecture

5. Performance Analysis

5.1. Average delay

A packet experienced of end-to-end delay due to queuing and different routing paths travelling that takes time to reach to the desired destination. The end-to-end delay occurred by the packets for each flow the individual packet delay are summed and the average is computed [13]. The delay is usually measured in seconds and delay calculation is given in equation 1 and 2.

Average end-to-end delay = (Total of Transmission Delays of All Received Packets) / (Number of Packets Received) (1)

Where,

Transmission Delay of a Packet= (Time Packet Received. at Server – Time Packet Transmitted at Client) (2)

here 'time' is in seconds.

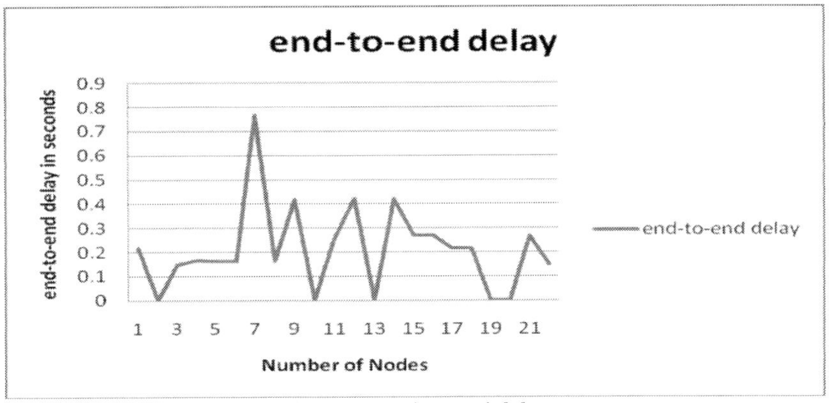

Figure 3: End-to-end delay

According to the simulation result that has been depicted in figure 3, it can be said that the maximum end to-end delay occurred 0.76 seconds for 7th node and minimum is 0 for some nodes. The average delay is 0.22 seconds. This delay seems too high for many data transmission particularly for real time data transmission.

5.2. Throughput

Throughput is the average rate of successful message delivery over a communication channel. Data may be delivered over a physical or logical link, measured in bits per second. The system throughput or average throughput is the sum of the data rates that are delivered to all the terminals in a network [14]. Throughput is usually measured in bits per second (bit/s or bps), and sometimes in data packets per second or data packets per time slot and the calculation is written in equation 3.

Throughput = [(Total Bytes Sent * 8) / (Time Last Packet Sent - Time First Packet Sent)] (3)
where 'time' is in seconds.

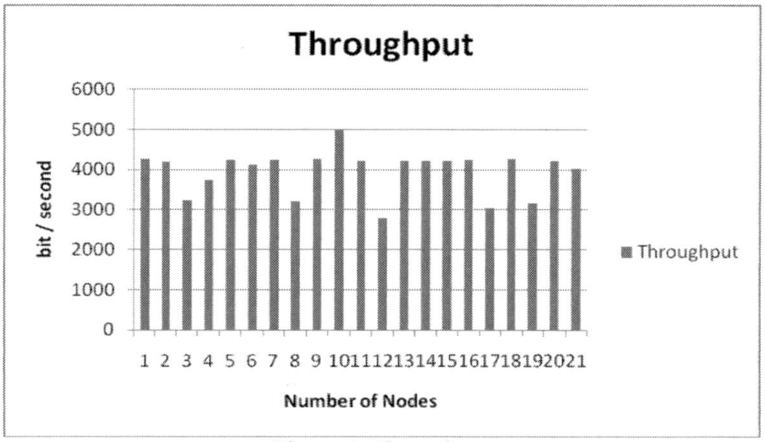

Figure 4: Throughput

The simulation result shows in figure 4 that the average throughput is achieved around 4000 bps when CBR traffic is transmitted in the designed network.

5.3. Average Jitter

Jitter is a variation in packet transit delay caused by queuing, contention and serialization effects on the path through the network. In general, higher levels of jitter are more likely to occur on either slow or heavily congested links. The usual causes include connection timeouts, connection time lags, data traffic congestion, and interference. Simply put, this jitter is an undesirable output of system flaws and interruptions. Thus when jitters occur, computer monitors and computer processors may malfunction, files may get lost, downloaded audio files may acquire noise, Internet phone calls may get interrupted, suffer time lags or get disconnected. The jitter calculation is provided in equation 4 and 5.

Average Jitter = (Total Packet Jitter for all Received Packets) /
(Number of Packets Received - 1) (4)

Where,

Packet Jitter = (Transmission Delay of Current Packet – Transmission. Delay of Previous Packet) (5)

Jitter can be calculated only if at least two packets have been received.

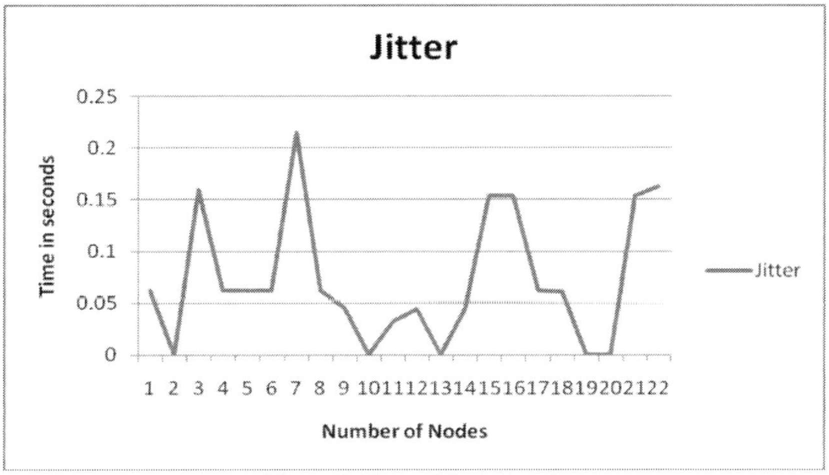

Figure 5: Jitter of the network

According to the simulation result, the maximum jitter sows the 0.21 seconds and minimum is 0 for some nodes. The average jitter is approximately 0.08 seconds.

6. Conclusion

This paper discusses the basics of WiFi and its applications. The performance of IIUM campus network has been analysed in terms of throughput, delay and jitter. The average throughput is 4000 bps, average delay is 0.22 seconds and jitter is 0.08 seconds. It can be said from above simulation results that the delay and jitter is quite high for data transmission. This research only considers a small coverage area of IIUM while the large coverage will be analysed in future work. The real time applications are not considered for this research that will be considered in further research.

7. Acknowledgement

The authors would like to thank to the Research Management Centre of IIUM for funding this research work under the Research Endowment Fund B – EDW B12-351-0829 Scheme and ITD for their information sharing.

8. References

[1] Mosharrof H. Masud, F. Anwar, S. M. S. Bari, A primary interface selection policy in heterogeneous networks based on QoS, *4th Int. Conf. on Computer and Communication Engineering (ICCCE'12)*, 3-5 July, 2012, Kuala Lumpur, Malaysia.

[2]. Lockias Chitanana, Bandwidth management in universities in Zimbabwe: towards a responsible user base through effective policy implementation, *International Journal of Education and Development using Information and Communication Technology (IJEDICT)*, 2012, Vol. **8**, Issue 2, pp. 62-76.

[3]. Linda Deneen, Bandwidth management tools, strategies and issues, University of Minnesota Duluth, 2002)

[4]. Vivaldi I., Improved handover routing scheme in hierarchical mobile IPv6 networks, A Masters Thesis, School of Graduate Studies, University Putra Malaysia (UPM), 2003.

[5]. Kandaraj Piamrat, Adlen Ksentini, Jean-Marie Bonnin, César Viho, Radio resource management in emerging heterogeneous wireless networks, *Computer Communications* **34** (2011) 1066–1076

[6]. Kandaraj Piamrat, C´esar Viho, Adlen Ksentini, Jean-Marie Bonnin, Resource management in mobile heterogeneous networks: state of the art and challenges, *Research Report of Institut National De Recherche En Informatique Et En Automa Tique*, 2008.
[7]. Ms. Shilpa Jindal Ms. Alka Jindal Dr. Neena Gupta, Grouping Wi-MAX, 3G and Wi-Fi for wireless broadband, *The First IEEE and IFIP International Conference in Central Asia on Internet,* 2005.
[8]. Latif, S. A.; Rashid, M.A.; Alam, F., "Effects of Varying Superframe Duration on Jitter in Interactive Multimedia Traffic Transmission over WLANs," *6th IEEE/ACIS International Conference on Computer and Information Science, 2007. ICIS 2007.* vol., no., pp.257,260, 11-13 July 2007.
[9]. Latif, S. A.; Rashid, M.A.; Alam, F., "Profiling Delay and Throughput Characteristics of Interactive Multimedia Traffic over WLANs Using OPNET," *21st International Conference on Advanced Information Networking and Applications Workshops, 2007, AINAW '07.* vol.2, no., pp.929,933, 21-23 May 2007
[10]. Jin-Shyan Lee, Yu-Wei Su, and Chung-Chou Shen, A comparative study of wireless protocols: Bluetooth, UWB, ZigBee, and Wi-Fi, *The 33rd Annual Conference of the IEEE Industrial Electronics Society (IECON),* Nov. 5-8, 2007, Taipei, Taiwan.
[11]. http://www.iium.edu.my/wireless/project%20info.html, accessed on 14 May, 2013
[12]. QualNet 5.2User's Guide", Scalable network technologies, Inc.,
[13]. Lea Skorin -Kapov, Darko Huljenic, Dario Mikic, Danko Vilendecic, Analysis of end - to -end QoS for networked virtual reality services in UMTS, *IEEE Communications Magazine*, April 2004
[14]. Sabyasachi Chatterjee, Siladitya Sen, Suchismita De, Effects on throughput for different data rates of IEEE 802.11b standard on Wi-Fi network using QualNet simulator, *National Conf. on Materials*, Devices and Circuits in Communication Technology (MDCCT 2012).

Review paper of gateway selection schemes for MANET of NEMO (MANEMO)

Z Mahmood, A Hashim, O Khalifa, F Anwar and S Hameed
Department of Computer and Information Engineering, Faculty of Engineering, International Islamic University Malaysia, Gombak, Malaysia

E-mail: zainab48@hotmail.com

Abstract. The fast growth of Internet applications brings with it new challenges for researchers to provide new solutions that guarantee better Internet access for mobile hosts and networks. The globally reachable, Home-Agent based, infrastructure Network Mobility (NEMO) and the local, multi-hop, and infrastructure-less Mobile Ad hoc Network (MANET) developed by Internet Engineering Task Force (IETF) support different topologies of the mobile networks. A new architecture was proposed by combining both topologies to obtain Mobile Ad Hoc NEMO (MANEMO). However, the integration of NEMO and MANET introduces many challenges such as network loops, sub-optimal route, redundant tunnel problem, absence of communication without Home Agent reachability, and exit router selection when multiple Exit Routers to the Internet exist. This paper aims to review the different proposed models that could be used to implement the gateway selection mechanism and it highlights the strengths as well as the limitations of these approaches.

1. Introduction

The wide use of the Internet and the need of continuous connection to the wireless network opened up new areas for researchers to solve the challenges of the new era of communication. Mobility is still one of the main fields of studies to keep the ongoing connections when the device moves from one network to another. The Mobile IPv6 [1] protocol was designed to enable a single node's mobility using the same IPv6 address in different locations. However, by time, the increasing use of Internet required the same features for the entire network of IPv6's nodes.

1.1. Network Mobility (NEMO)

The new IETF NEtwork MObility Basic Support (NEMO BS) protocol [2] uses Mobile IPv6 to provide mobility for a network as a whole and this includes access networks scenarios on public transport like trains, buses, planes, and ships. With NEMO BS, the network's devices do not require any additional software or protocol supporting mobility; they are connected to a Mobile Router (MR) which deals with different access networks while the network is on the move. Since a NEMO MR can transparently perform its mobility process on behalf of any IPv6 enabled device and at the same time it behaves as an IPv6 device when they connect to a network, this means that NEMO MRs can connect with other NEMO MRs and access the Internet through them. When this situation occurs, the resulting network is known as a Nested NEMO network [3]. A highly suboptimal structure can occur with the possibility of getting loops and the packets may pass through a highly inefficient route via each of the

HAs of the NEMOs located in the path from source to correspondent node; this is known as Pinball Routing Problem and it is depicted in figure 1.

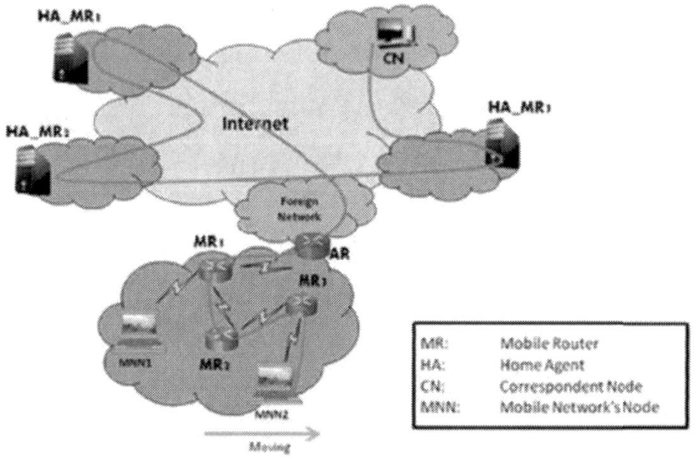

Figure 1. The pinball routing problem in nested NEMO

1.2. Mobile Ad Hoc Network (MANET)

On the other hand, the Mobile Ad Hoc Networks (MANET) [4] supports mobility (Mobile Host and Mobile Router) by deploying optimized routing protocols specifically designed to operate between mobile devices to mainly support networking scenarios that have no former infrastructure [5]. MANET protocols were intended to maintain inter-connection between the devices of the MANET network and were improved later to connect any node on the Internet. Figure 2 shows a mobile ad hoc network (MANET).

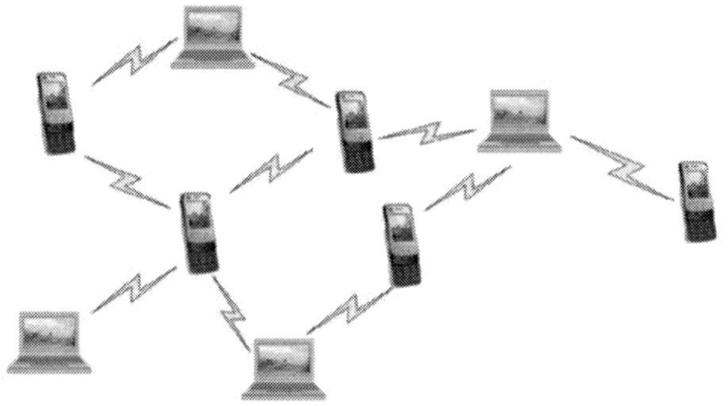

Figure 2. Mobile Ad Hoc Network (MANET)

1.3. MANET for NEMO (MANEMO)

The combination of the Mobile Ad Hoc Networks (MANETs) and Network Mobility (NEMO), which is known as (MANEMO), can form global reachable, scalable, and optimized network topologies. Hence, the optimized, multihop routing provided by MANET protocols can be utilized in order to solve the inefficient routing problems experienced within Nested NEMO network configurations.

Also, the globally reachable Home Agent-based properties of NEMO can be leveraged to provide MANET networks with permanent reachability in the Internet without flooding their changing routing information into the infrastructure [6]. As in NEMO BS model, the mobile entity is considered to be a Mobile Router (MR) that manages the mobility of the entire network over its Egress interface (i.e. its connection to the Internet) and presents its Ingress interface to IP devices as a normal, static IP connection [7].

However, the MANEMO came out with new challenges, which are specified in [8]. Mainly these challenges are: Sub-optimal route and Redundant tunnels, disconnection when the HA cannot be reached, Exit Router Selection which occurs when a mobile node obtains multiple Exit Routers toward the Internet, and Network Loop that occurs when multiple mobile routers are nested and suddenly the Exit Router loses connectivity to the Internet and connects to the mobile network of a lower hierarchical MR.

Furthermore, since MANEMO is enabling a MR to install one or more default routes towards the Internet and multiple Exit Routers may present in the network, there is a need for additional information to be made available to Mobile Routers to help them in selecting the gateway which ensures the best performance and load balance of the whole network while reducing network loops. (Arkko et al., 2005)

Many RFCs and Internet drafts of IETF are discussing MANEMO from different perspectives as in [8], which outline the problems and the approach rationale of MANEMO. While [9] described the MANEMO architecture including possible topology configuration and addressing assignment, [10] defines the MANEMO problem statement as several issues exist.

2. Models of Gateway Selection in MANEMO

According to [11], there is no standard method that can ensure optimum gateway selection that maximizes Internet connection performance and robustness. Any proposed solution should perform four tasks in order to achieve its design goals, these tasks being: capability measurement, capability information dissemination, optimum route selection and route enforcement. The proposed model informs each mobile node about the performance of each Internet gateway and enables these nodes to select their default gateway from the available gateways. The traffic is then routed from and to the mobile router so that it passes through the selected gateway. However, the evaluation of the model needs to be done. Table (1) concludes the solution models as in [11].

Table 1. Solution models for gateway selection in MANEMO

Solution Model	*Suggested Modifications*	*Limitation*	*Challenges*
Tree discovery protocol gateway advertisement	Using the Tree Information Option (TIO) of TD in nested NEMO to broadcast gateways' metrics down the tree	It can only maintain a single exit route to the Internet in all cases. The MR changes its default gateway only if it lose the connection with the current gateway	Can be used in NCM, however no gateway criteria included yet
OLSR protocol gateway advertisement	Using extended OLSR's Topology Control packets (TC) and Host and Network Association messages	Can support multiple exit routes but restricts to MANEMO using OLSR as MANET's routing protocol	Can be used in certain MCM scenarios
MANET independent gateway advertisement	Extending IPv6 NDP messages to advertise gateways inside the MANET	Not describing the gateways' capabilities	Can be further extended to append more options

Intelligent gateway selection in MANEMO	Extending the NEMO BS to support intelligent exit route selection and select one gateway as default and other as backup	The selected two gateways may not be the 'best' gateways for the MR	Depends on TDP to advertise the gateways' capabilities
Simultaneous gateway utilization	Using multihomed NEMO when each MR has multiple CoA-HoA bi-directional tunnels	Should be MANET of multihomed NEMOs	The same way used by a MR to distribute traffic on each bi-directional tunnel can solve the multiple exit route selection of MANEMO

[12] proposed a multi-dimensional protocol to support the process of gateway discovery in mobile hotspots, and to help in selecting the best path able to satisfy the user's requirements and to guarantee a target end-to-end service quality. It was pointed out that the mobile nodes need to indicate their service preferences, discover available gateways, and select the best gateway for network connectivity and service provisioning. [12] combines functionality of a well known protocol to describe multimedia sessions characteristics such as SDPng (Session Description Protocol Next Generation), and a highly popular routing protocol in ad hoc networks, that is Ad hoc On-Demand Distance Vector routing (AODV) [13], to provide quality of service support and Internet gateway discovery. However, the MNs requirements are very different from the MRs considered in NEMO or MANEMO. Furthermore, the cross layering involved in this proposed solution may lead to a complicated code that is hard to maintain, and it also can stifle proliferation since every update may require a complete redesign and replacement. Moreover, cross-layer design creates interactions, some intended, and others unintended [14].

[15] proposed Mobile Gateway supporting policy based routing. It manages both NEMO connectivity and MANET connectivity and switches path to a destination according to application requirements and network environment. The decision is made with preference information and three parameters such as bandwidth, round trip time, and hop count. These parameters are measured during communications and route exchange of a MANET routing protocol.

It is important for the mobile nodes to detect efficiently available Internet gateways providing access to the Internet. Internet gateway discovery time and handover delay have strong influence on packet delay and throughput. Therefore, [16] focuses on proposed technical solutions on Internet gateway discovery for MANET. Some existing proposals do not consider multiple gateways and thus they lack mechanisms for load balancing and scalability. The proactive gateway discovery consumes excessive network resources due to the frequent advertisement flooding. Though many researchers have proposed proactive gateway discovery solutions, the solutions lacks load balancing, hence there is a need to propose a gateway discovery protocol that takes path load into consideration. The solicitation is a broadcast, which consumes network resources, just as the flooding of advertisement does. In case of hybrid gateway discovery approaches, a significant disadvantage is that in the proposed approaches some roaming nodes beyond the TTL scope of the Internet gateway may never receive gateway advertisements. Moreover, some mobile nodes that have registered with the gateway cannot receive the latest advertisement broadcast to update registrations with their home agents [16]. A summary of the existing gateway discovery and/or selection schemes is shown in table 3.

Table 2. Summary of Features of Existing Proposals as in [16]

Proposed Architecture/ Protocol	*Gateway Discovery Approach*	*Mobile IP Support*	*Ad Hoc Routing Protocol*	*Multiple Gateways support*	*Parameters for Gateway Selection*	*Modification*	*Performance Evaluation*
Bin et al.	Reactive	YES	AODV	YES	---	---	Simulation using NS 2
Ratanchand	Hybrid	YES	AODV	YES	---	---	Simulation

ani et al.							
Rosenschon	Proactive	---	AODV	YES	Hop count	Enhanced AODV's HELLO packets to interact with link layer	Simulation using NS 2
Sun et al.	Proactive, Reactive	YES	AODV	YES	Hop count when not hearing anything from the GW for more than one beacon interval		Simulation
Jonsson et al.	Proactive, Reactive	YES	AODV	YES	Hop count, The MN change to a new GW if it's at least two hop nearer than the current one	MIPMANET	Simulation using NS 2
Lee et al.	Hybrid (RMD based)	YES	DSR	YES	Hop count		QualNet Simulator
Tseng et al.	Proactive, Reactive	YES	DSDV	YES	Default hop count then measure: Minimum Load-Index (MLI) GWs broadcast the to the one with maximum bandwidth	Using NAT server	Testbed
Ghassemian	Proactive, Reactive	-	AODV	YES	Hop count		Simulation
Ruiz	Hybrid (MBC)	NO	AODV	NO	---	---	Analytical model and simulation using NS 2
Hossam El-Moshrify et al.	Proactive, Reactive, Hybrid	NO	AODV	YES			Simulation using NS 2
Rafi U Zaman et al.	Reactive	-	AODV	YES	Aggregate Interface Queue Length	Modified AODV	Simulation using NS 2
Ammari et al.	Reactive	YES	DSDV	YES	Select a closest (hop count) and a least loaded (number of nodes being served)		Simulation using NS 2
Khan et al.	Proactive	YES	DSDV	YES	Hop count, if same, the host with the latest routing update time	Extended DSDV (Eff-DSDV)	NCTUns network simulator

2.1. Load Balancing in the Network

One of the things to be considered in any gateway selection model for MANEMO is load balancing. [17] proposed a multi mobile router selection method in order to improve the bandwidth, network coverage and reliability of NEMO. Instead of using a single mobile router in a mobile network to carry out all operations related to mobility and packet forwarding, a dynamic load sharing mechanism between all available mobile routers is supported. This solution mainly relies on Neighbor Discovery. However, the primary results show that the neighbor unreachability detection may not match latency requirements of time-sensitive communication since it takes a long time.

[18] presented a review of the gateway load balancing solutions under common parameters. Table 3 summarizes these strategies.

Table 3. Summary of load balancing solutions as in [18]

Load Balancing Strategy	*MANET Routing Protocol*	*Gateway Discovery*	*Load Balancing Mechanism*	*Load Balancing Parameters*
Zhou et al.	DSDV	Proactive + Reactive	Gateway selection	Distance , Registered Nodes, Queue Length
Trivino-Cabrera et al	AODV	Proactive	Gateway Discovery	Bandwidth
Hsu et al	DSDV	Reactive	Gateway Discovery and selection	Traffic Load, Bandwidth
Shin et al	Reactive	Reactive	Gateway Discovery	Packet Queue Size
Le-Trung et al	Proactive	Proactive	Gateway selection	Distance, Traffic Load

[19] proposed two algorithms for load balancing for the nested NEMO topology to reconstruct the Internet-based MANET using a load-information in the route advertisement (RA) message. The Mobile Router (MR) is the point of Internet attachment for the MNNs, and it is likely to be a potential bottleneck because of its limited wireless link capacity and all packets in/out the mobile network must pass through it. However, since the NEMO protocol is a hierarchical construction the ARs and MRs could be the overloaded points with the most traffic load. [19] proposed a dynamic load-balancing scheme to balance the load between ARs and MRs by adding some information in BU and RA. Using Lowest-ID (LID) algorithm and Highest-Degree algorithm (HD) can achieve this goal. In LID, each node has a unique ID. Every node periodically broadcasts its ID through a 'Hello' message. By receiving the message, the nodes can determine the neighbor ID. The lowest-ID node in the same neighborhood is selected as a cluster head (CH); nodes which can 'listen' to two or more CHs become cluster gateways (CG), while other nodes are cluster member (CM). In HD, the highest degree node in a neighborhood is selected as a CH. The degree value represents the number of neighbors. In contrast to LID, HD uses location information for the cluster composition. Like LID, a node periodically broadcasts its degree through a 'Hello' message. Adopting two MANET algorithms could partially improve the throughput of the NEMO. However, the end-to-end delay that the packet lost should be tested as well. Furthermore, the results in the combined NEMO and MANET may not be as good as when they are separated.

2.2. Neighbor Discovery Protocols

The IPv6 nodes on the same link use Neighbor Discovery to discover each other's presence, to determine each other's link-layer addresses, to find routers and to maintain reachability information about the paths to active neighbors as described in IETF RFC 2461 [20]. The Neighbor Discovery Protocol (NDP) for IPv6 defined in IETF RFC 4861 [21], (obsoletes: RFC 2461), can be used to distribute the information needed for the gateway selection process. When a packet is forwarded for which the corresponding Neighbor Cache Entry is missing or incomplete, it notifies NDP to take appropriate action. NDP picks up requests from the shared queue and performs any necessary discovery action. In many implementations the NDP is also responsible for responding to router solicitation messages and Neighbor Unreachability Detection.

However, many extensions have been done to the NDP in order to match with different technologies, as in [22] which described the modifications to IPv6's neighbor discovery protocol needed to avoid vulnerabilities when a network is scanned. Another extension to IPv6 Neighbor Discovery Protocol is defined in [23]; it allows management of IPv6 traffic offloading to IPv4 and moving IPv4 traffic away from a specific interface. Current Internet multi-interfaced hosts generally prefer IPv6 addresses over IPv4 addresses when performing source and destination address selections.

To use the NDP for IPv6 in different networks it needs to be extended to match their specification. In [24], Neighbor Discovery Optimization for sensor networks is discussed. IPv6 Neighbor Discovery was not designed for non-transitive wireless links. The traditional IPv6 link concept and heavy use of multicast make the protocol inefficient and sometimes impractical in a low power and lossy network.

NEMO specifications needed a new Neighbor Discovery Protocol option and modified operation of MR and MN to support route optimization. These modifications are described in RFC 3963 [2] and [25]. On the other hand, RFC 4861 [21] extends the Neighbor Discovery Protocol in order to carry information and metrics which will help a Mobile Router select its Attachment Router(s) in an autonomous fashion and provides generic rules which guarantee that the interaction of different selection processes will not create loops.

For MANET, RFC 6130 [26] describes a 1-hop and symmetric 2-hop neighborhood discovery protocol (NHDP) for Mobile Ad Hoc Networks (MANETs). This protocol uses a local exchange of HELLO messages so that each router can determine the presence of, and connectivity to, its 1-hop and symmetric 2-hop neighbors.

3. Conclusion

As the use of Internet is getting wider, the need of continuous connection to the wireless network opened new areas for researchers to solve the challenges of the new era of communication. Network Mobility (NEMO) enables continuous connectivity of a whole network with the Internet while the infrastructure-less Mobile Ad Hoc Network (MANET) enables the direct connection between mobile nodes. Merging the two technologies came out with a new efficient, scalable architecture named Mobile Ad Hoc NEMO (MANET). Although no complete image of the MANEMO architecture is found at present, many researches have been done to solve some of the MANEMO challenges. The models and protocols used for NEMO or MANET may have to be extended if used for the MANEMO. This research has reviewed some of the existing works regarding gateway selection models which may satisfy the requirement of MANEMOs. The future of MANEMO at a large scale relies on the ability to provide advanced mobility support that can coexist with the currently deployed protocols.

4. References

[1] Johnson D, Perkins C and Arkko J June 2004 Mobility Support in IPv6 *RFC* 3775
[2] Devarapalli V, Wakikawa R, Petrescu A and Thubert P 2005 Network Mobility (NEMO) Basic Support Protocol *RFC* 3963
[3] Ernst T and Lach H July 2007 Network Mobility Support Terminology *RFC* 4885
[4] Broch J, Maltz D, Johnson D, Hu Y and Jetcheva J 1998 A Performance Comparison of Multi-Hop Wireless Ad Hoc Network Routing Protocols *Proc. of the 4th Annual ACM/IEEE Int. Conf. on Mobile Computing and Networking* pp 85—97

[5] McCarthy B, Edwards C and Dunmore M 2008 Using NEMO to Extend the Functionality of MANETs *IEEE Int. Conf. on Communications Workshops 2008* ICC (Workshops '08. IEEE) pp 455-60

[6] Investigations in Mobile IP 2009 *The Progression from Mobile IPv6 to MANEMO* (Retrieved on: 30 April 2012) from http://ebookbrowse.com/c4-the-progression-from-ipv6-to-manemo-pdf-d182189108

[7] McCarthy B, Edwards C and Dunmore M 2007 Advances in MANEMO: Definition of the Problem Domain and the Design of a NEMO-Centric Approach *Proc. of the 2007 Int. Symp. on Applications and the Internet Workshops (SAINTW'07)* (IEEE computer society)

[8] Wakikawa R February 2007 *MANEMO Problem Statement* Internet draft <draft-wakikawa-manemo-problem-statement-00>

[9] Wakikawa R July 2007 *MANEMO Topology and Addressing Architecture* Internet draft<draft-wakikawa-manemoarch-00>

[10] McCarthy B, Edwards C and Dunmore M 2007 Network transparency in a mountain rescue domain *Internet Research* **17** 465 – 78

[11] Ahmed M, McCarthy B and Edwards C 2010 Gateway Selection in Mobile Ad-hoc Networks *The 11th Annual PostGraduate Symp. on the Convergence of Telecommunications, Networking and Broadcasting*

[12] Iera A, Molinaro A, Polito S, Ruggeri G and Vilasi A. 2006 Gateway Discovery and Selection in Mobile Hotspots *Int. Conf. on Wireless Broadband and Ultra Wideband Communications (Auswireless)*

[13] Perkins C, Belding-Royer E and Das S 2003 Ad hoc On-Demand Distance Vector (AODV) Routing *IETF RFC* 3561

[14] Kawadia V and Kumar P *2005* A cautionary perspective on cross-layer design *IEEE Wireless Communications* **12** 3-11

[15] Okada K, Wakikawa R and Murai J 2006 MANET and NEMO Converged Communication *AINTEC 2006* (*LNCS 4311*) pp 235–51

[16] Rakesh K, Sarje A and Misra M 2010 Review Strategies and Analysis of Mobile Ad Hoc Network-Internet Integration Solutions *IJCSI* **7** 18-28

[17] Kuntz R, Montavont J and Noel T 2008 Multiple mobile routers in NEMO: How Neighbor Discovery can assist default router selection *IEEE PIMRC 2008* pp 1-6

[18] Rafi Z, Khan K, Razzaq M and Reddy V 2009 *A Simulation Based Comparison of Gateway Load Balancing Strategies in Integrated Internet-MANET* (17th Int. Conf. on Advanced Computing and Communication ADCOM'09) *IISC* (Bangalore: IEEE Computer Society) pp 270 –2

[19] Li Long-Sheng, Lee Gwo-Chuan and Kang Li-Keng 2009 Balanced Topology NEMO Construction for the Internet-Based MANET *IJCNS* **2** 895-902

[20] Narten T, Nordmark E and Simpson W 1998 *Neighbor Discovery for IP Version 6 (IPv6)* IETF *RFC* 2461

[21] Narten T, Nordmark E, Simpson W and Soliman H 2007 *Neighbor Discovery for IP Version 6 (IPv6) RFC* 4861

[22] Herberg U, Cole R and Chakeres I January 2012 *Definition of Managed Objects for the Neighborhood Discovery Protocol* Internet draft <draft-ietf-manet-nhdp-mib-11>

[23] Korhonen J, Savolainen T and Ding Y March 2012 *Controlling Traffic Offloading Using Neighbor Discovery Protocol* Internet draft <draft-korhonen-mif-ra-offload-04.txt>

[24] Shelby Z, Chakrabarti S and Nordmark E 2011 *Neighbor Discovery Optimization for Low Power and Lossy Networks (6LoWPAN)* Internet draft <draft-ietf-6lowpan-nd-18>

[25] Kyeong J, Jae H, Jung S and Hyoung J February 2004 *Route Optimization for Mobile Nodes in Mobile Network based on Prefix Delegation* Internet draft <draft-leekj-nemo-ro-pd-02.txt>

[26] Clausen T, Dearlove C and Dean J 2011 Mobile Ad Hoc Network (MANET) Neighborhood Discovery Protocol (NHDP) *RFC* 6130

Novel design of a self powered and self sensing magneto-rheological damper

Mohammad Meftahul Ferdaus[1, a], M M Rashid[2, b], M.M.I. Bhuiyan[3, c], Asan Gani Bin Abdul Muthalif[4], M R Hasan[5]

[12345]Department of Mechatronics Engineering
International Islamic University Malaysia

[a]ferdaus57@gmail.com, [b]mahbub@iium.edu.my, [c]moinul@iium.edu.my

Abstract Magneto-rheological (MR) dampers are semi-active control devices and use MR fluids. Magneto-rheological dampers have successful applications in mechatronics engineering, civil engineering and numerous areas of engineering. At present, traditional MR damper systems, require a isolated power supply and dynamic sensor. This paper presents the achievability and accuracy of a self- powered and self-sensing magneto-rheological damper using harvested energy from the vibration and shock environment in which it is deployed and another important part of this paper is the increased yield stress of the Magneto rheological Fluids. Magneto rheological fluids using replacement of glass beads for Magnetic Particles to surge yield stress is implemented here. Clearly this shows better result on yield stress, viscosity, and settling rate. Also permanent magnet generator (PMG) is designed and attached to a MR damper. For evaluating the self-powered MR damper's vibration mitigating capacity, an Engine Mount System using the MR damper is simulated. The ideal stiffness of the PMG for the Engine Mount System (EMS) is calculated by numerical study. The vibration mitigating performance of the EMS employing the self-powered & self sensing MR damper is theoretically calculated and evaluated in the frequency domain.

1. Introduction

Magneto-rheological fluids or MR fluids is a sort of smart material. By changing external magnetic field strength the rheological properties of MR fluids can be varied. Such fluids consist of micron sized particles, made of magnetic material, suspended in a carrier fluid, normally a type of oil. In the absence of an applied field, the MR fluids exhibit Newtonian-like behavior. However, in the presence of an applied magnetic field, the iron particles acquire a dipole moment aligned with the external field which causes the particles to form linear chains parallel to the field [3]. Recently, magneto-rheological (MR) fluid based dampers, which are a most promising class of semi-active energy absorbers, have been widely applied to control and minimize undesirable vibration and shock for numerous systems. Such as different types of vehicle suspension system systems, and civil structures. Magneto-rheological actuators were proposed, as well as magneto-rheological energy absorbers for shock applications [2]. MR dampers possess main performance advantages, including continuously controllable force, quick response, and low power consumption, which can be easily improved in

[b]Corresponding Author: H/P: +60163132236, E-mail: mahbub@iium.edu.my

design procedures. MR dampers can be analyzed using available models and are amenable to innovative design concepts such as increasing the viscosity of the MR fluids by changing the properties of the fluid [7]. To practically construct vibration and shock mitigation systems using MR dampers, either a power supply or a current amplifier is required to activate the electromagnetic coils in the MR dampers to supply magnetic field to the MR fluid [4]. However, if mechanical energy resulting from vibration and shock can be converted into electrical energy in order to provide power to the MR dampers, substantial reductions in system volume, weight, cost, and maintenance can be realized. MR dampers are typically composed of a piston rod, a piston head, as well as hydraulic and pneumatic reservoirs separated by a floating piston or diaphragm. Inside the hydraulic cylinder, the piston rod is attached to the piston head, which contains the magnetic circuit [1]. When the piston rod assembly moves, the fluid flows through an annular gap in the piston head. Current supplied to the coil in the piston head creates a magnetic field in the gap and increases the yield stress of the MR fluid in the annular gap. This increase in yield stress changes the velocity profile of the MR fluid in the gap and raises the pressure drop down the length of the piston head. In this way, MR dampers can produce controllable field-dependent yield force, in addition to passive velocity dependent viscous damping force.

This study focuses on the development of a self-powered and self-sensing MR damper operated using the energy harvested from its operating environment, and sedimentation improved new MR fluid's impact on this damper [8]. For using the new fluid the off state viscosity has increased by 22% and the sedimentation rate decreased slightly [6]. To explain the self-power generation idea, an energy harvesting device that is a permanent magnet generator is considered and added to a MR damper. The generator comprises of a stator, a permanent magnet, and a spring and operates as an energy garnering dynamic vibration absorber of which resonance frequency is matched with a target system whose vibration is to be suppressed. In addition, the resonance frequency of the target system is typically chosen to be much less than the disturbance spectrum in order to take advantage of the mount's low transmissibility at high frequency. Generally speaking, vibration isolation systems require high damping near resonance and lower damping above resonance in order to achieve good vibration isolation performance. Therefore, when MR dampers are used for vibration isolation purposes, it would be ideal if the MR damper force can be maximized near resonance and minimized above resonance see Fig. 3. If the energy-harvesting dynamic vibration absorber is itself a single-degree-of-freedom (SDOF) system whose natural frequency is matched to that of the target system, then the displacement of the permanent magnet will be greater near the resonance of the target system. Above resonance, the displacement of the permanent magnet is substantially reduced. Thus, for a given dynamic vibration absorber vibration level, the current produced is injected into the magnetic circuit, thereby commanding a damping level proportional to the current produced by the dynamic vibration absorber. Thus, the energy-harvesting dynamic vibration absorber acts as a semi-active control algorithm that produces high current and commands high damping near resonance of the target system, while producing low current and command lower damping above resonance. A key benefit of this approach is that the MR damper no longer needs a sensor to measure motion as a feedback signal to a microprocessor that calculates a control input based on modern control algorithms.

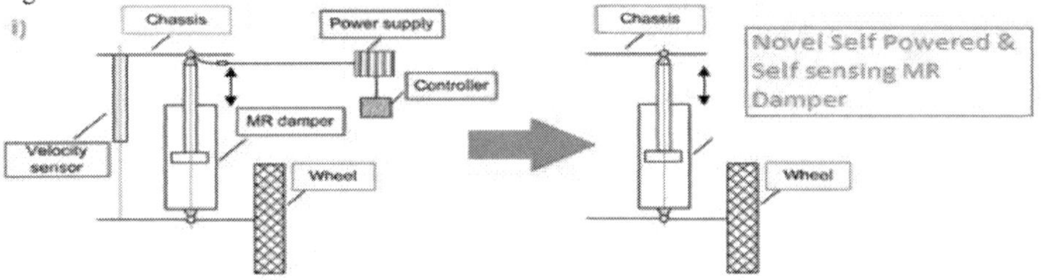

Figure 1: i) Conventional MR Damper ii) Novel Self Powered & Self sensing MR Damper

Many professor, researchers such as Russell et al, Nehl et al, Wang and his associates, Or et al, Jung and his associates, Choi and Wereley, Jung et al, Bogdan works on Self powered magneto-rheological dampers. While the previous studies on Magneto-Rheological dampers with self-sensing or power generation functions have presented some favorable results, but this is a integration of the sensing and power generation ability within the same device [5].

2.Self-powered & self sensing MR damper

Figure below presents a schematic diagram of a self-powered and Self Sensing MR damper. The hydraulic cylinder contains the piston rod which is attached to the piston head. At one side of the piston head, the PMG is configured.

The motion of the permanent magnet as a result of the piston rod motion generates electromotive force emf in the stator and the strength of the emf is proportional to the velocity of the permanent magnet. So velocity information can be easily calculated from here. The emf induced voltage or current is directly provided to activate the coil winding in the piston head so as to produce a magnetic field into the MR fluid inside the MR valve of the self-powered MR damper.

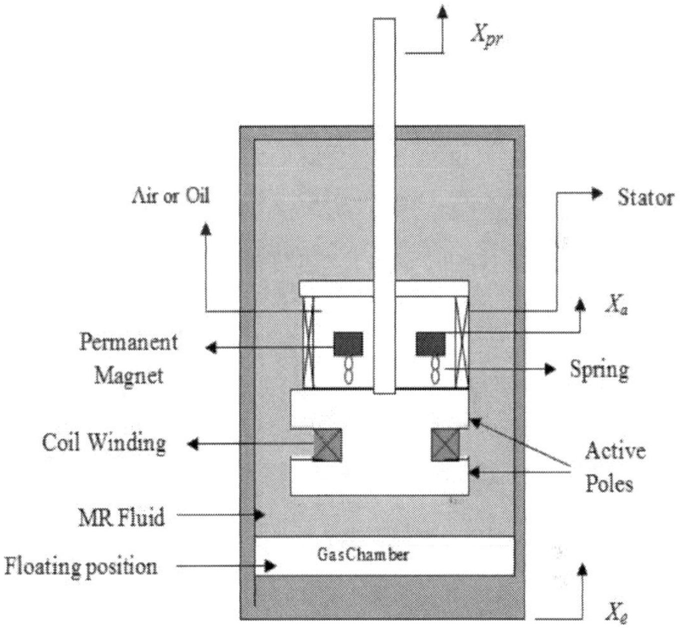

Figure 2: Schematic Diagram new proposed MR damper

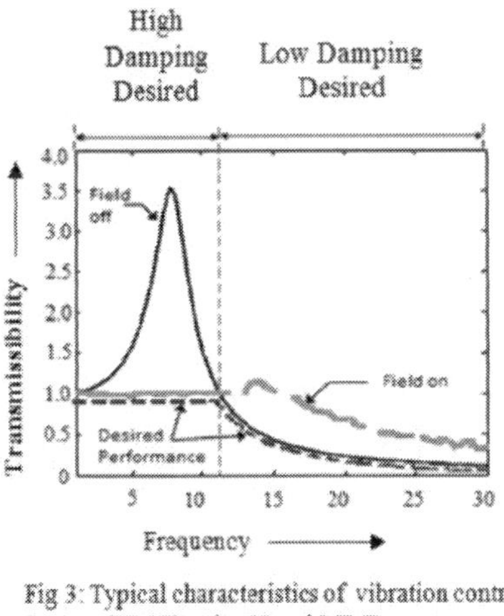

Fig 3: Typical characteristics of vibration controlling System (SDOF) using Novel MR Damper

Electromotive Force, E_{emf} generated by the motion of the permanent magnet in the PMG is give

$$E_{emf} = -N_m \frac{d\phi}{dt} = -N_m \alpha B_m 2\Pi r_m (\dot{x}_{pr} - \dot{x}_{pm})$$

Here, N_m is the turn number of the coil in the stator affected by the permanent magnet at a time, ϕ is the magnetic flux, B_m is the magnetic density of the permanent magnet, r_m is the radius of the permanent magnet, $\dot{x}_{pr}$ is the velocity of the piston rod, $\dot{x}_{pm}$ is the velocity of the permanent magnet, and α is the empirical correction factor for the effective magnetic density of the permanent magnet since there is an air gap clearance between the permanent magnet and the stator. In this study, the design parameters were chosen to be N_m=120 turns, α = 0.75, B_m =1.3T, and r_m=15 mm.

If the time constant, equal to inductances/resistances, of the coils is assumed to be small, the current I produced by the energy harvesting PMG is given by

$$I = \frac{E_{emf}}{R_{st} + R_{cl}}$$

Here, R_{st} is the resistance of the stator and R_{cl} is the resistance of the coil winding. In this study, $R_{st} + R_{cl} = 3$ ohm.

The force of the self-powered MR damper is given by

$$F_{nMR} = F_{Passive} + F_{Semi} + F_{gas}$$

Where

$$F_{passive} = A_p^2\left(\frac{6\eta L}{\pi r d^3} + \frac{6\eta L_c}{\pi r_c d_c^3} + \frac{6\eta L_s}{\pi r_s d_s^3}\right)(\dot{x}_{pr} - \dot{x}_e) + \left(\frac{2\eta\pi r L}{d} + \frac{2\eta\pi r_c L_c}{d_c} + \frac{2\eta\pi r_s L_s}{d_s}\right)(\dot{x}_{pr} - \dot{x}_e)$$

And $F_{semi} = (2A_p \frac{L}{d} + 2\pi r L)\tau_y(H)\operatorname{sgn}(\dot{x}_{pr} - \dot{x}_e)$

$$F_{gas} = K_{gas}(x_s - x_e)$$

Here, A_p is the effective piston area, A_r is the piston rod area, and x_e is the displacement of the hydraulic cylinder. d, d_c, and d_s are the gaps of the active pole, the coil winding, and the stator, respectively. In addition, r, r_c, and r_s are the radii of the active pole, the coil winding, and the stator, respectively. The dimensions of L, L_c, and L_s are the lengths of the active pole, the coil winding, and the stator, respectively. η is the fluid viscosity and $\tau_y(H)$ is the yield shear stress of the MR fluid Lord Corporation, Cary, NC, MRF-40 and is assumed to be a function of the magnetic field strength H as follows :

$$\tau_y(H) = 1.93H^{1.73} \text{ (Pa)}$$

Where

$$H = \left|\frac{N_c I}{2d}\right|$$

Here, N_c is the turn number of the coil winding in the piston head and K_{gas} is the stiffness due to the gas pressure in the gas chamber and is given by

$$K_{gas} = \frac{n A_r^2 P_0}{V_0}$$

Here, V_0 and P_0 are the initial volume and pressure of the gas chamber, respectively and n is the specific heat ratio of the gas. In this study, the design parameters were chosen to be A_p= 1700 mm^2, A_r=79 mm^2, L=20 mm, L_c= 15 mm, L_s=30 mm, r=24 mm, r_c=21 mm, r_s=21 mm, d=1 mm, d_C=4 mm, d_S=4 mm, η =0.18 Pa s, and N_C=150 turns.

To emulate a practical force response of the self-powered MR damper, we consider the time response of the semi-active damper force F_{semi} as follows:

$$\dot{F}^*_{semi} = \frac{-F^*_{semi} + F_{semi}}{\tau}$$

Here, F^*_{semi} is the emulated semi-active damper force and τ is the time constant of the self-powered MR damper. In this study, the time constant was chosen to be τ = 5 ms. During a computer simulation, the semi-active damper force F_{semi} will be replaced by the emulated semi-active damper force F^*_{semi} .

3. Ems with new proposed MR damper

The self-powered & self-sensing MR damper's vibration controlling performance can be theoretically evaluated by considering an EMS (SDOF) with the MR damper as shown in the figure below:

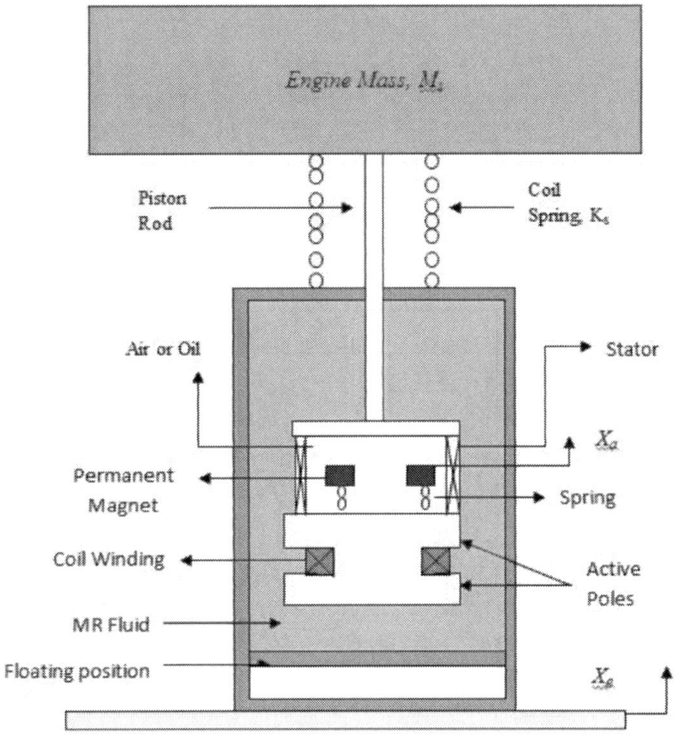

Figure 4: An EMS (SDOF) attach with the self-powered & Self Sensing MR damper

The governing equation of motion for the EMS attach with the new proposed MR damper is as follows:

$$M_s\ddot{X}_{pr} = -(K_s + K_{gas})(x_{pr} - x_e) - F_{passive} - F^*_{semi} + K_a(x_{pm} - x_{pr}) + C_a(\dot{x}_{pm} - \dot{x}_{pr})$$

$$M_a\ddot{x}_{pm} = -K_a(x_{pm} - x_{pr}) - C_a(\dot{x}_{pm} - \dot{x}_{pr})$$

Here, M_s is the engine mass, M_a is the mass of the permanent magnet in the energy-harvesting PMG, K_s is the stiffness of the coil spring, K_a and C_a are the stiffness and the damping of the spring in the energy-harvesting PMG, respectively, x is the displacement of the engine mass, and x_e is the excitation displacement. In this study, representative parameters were chosen to be M_s =60 Kg, M_a = 0.4 Kg, Ks + K_{gas} = 150 KN/m, and C_a =4 Ns /m.

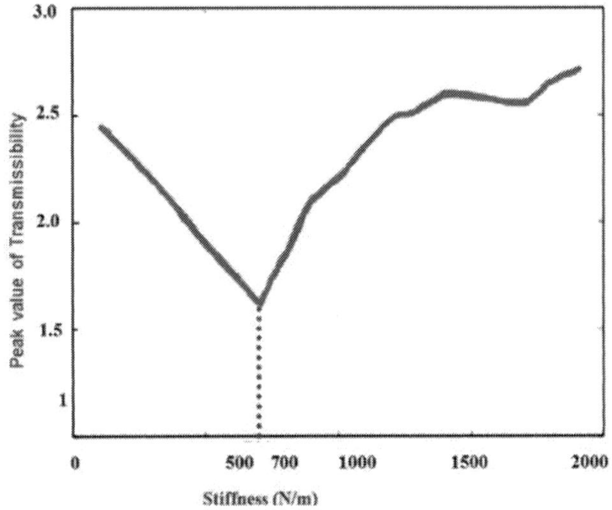

Fig. 5: Maximum peak value of the transmissibility of the EMS versus the stiffness of the permanent magnet generator.

Because the resonance of the EMS is 8.0 Hz, an initial choice for the stiffness of the permanent magnet generator was chosen to be K_a = 1000 N / m so as to match the resonance of the permanent magnet generator with that of the EMS. However, in order to find an optimal stiffness, we plot the maxi- mum peak value of the transmissibility $|x_{pr} / x_e|$ of the EMS versus the stiffness of the PMG in the self-powered MR damper in Fig. 8. From this plot, it was deter- mined that the optimal stiffness value of the permanent magnet generator for the EMS was K_a = 700 N / m. This optimal stiffness, K_a = 700 N / m, suggests that the resonance of the permanent magnet generator should be slightly below that of the EMS. In this study, the optimal stiffness, K_a =700 N / m, of the PMG for the EMS is used to estimate the vibration controlling capability of the self-sensing & powered MR damper.

IOP Conf. Series: Materials Science and Engineering **53** (2013) 012048 doi:10.1088/1757-899X/53/1/012048

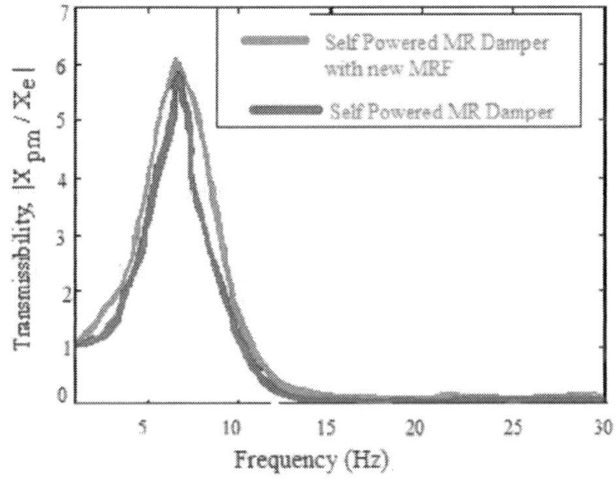

Figure: B

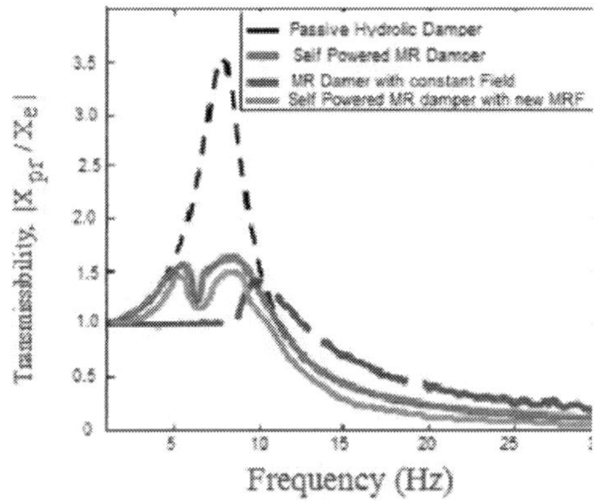

Figure: A

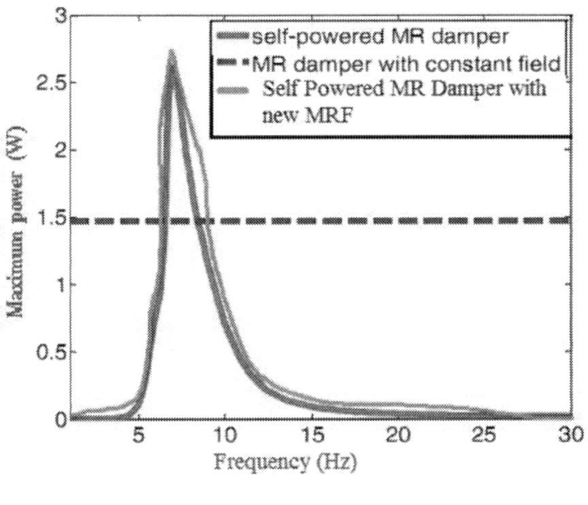

Figure: D

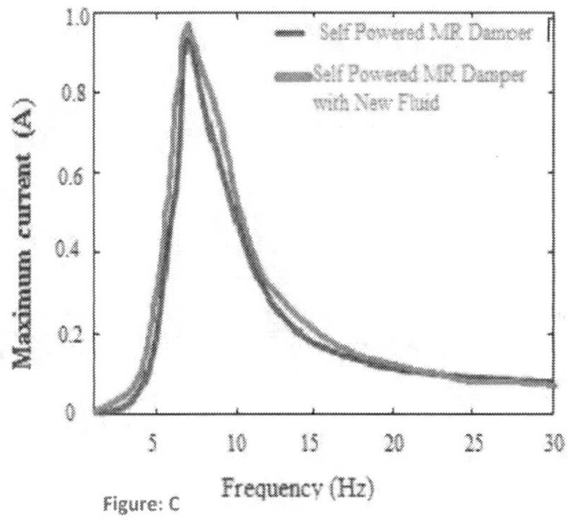

Figure: C

Fig. 6: Simulated transmissibility of the vibration controlling system with the self-powered & sensing MR damper filled by MRF-40 & New MRF-37. A. transmissibility of the engine mass x_{pr} / x_e. B. transmissibility of the energy-harvesting DVA x_{pm} / x_e. C. maximum current and D. maximum power

IOP Conf. Series: Materials Science and Engineering **53** (2013) 012048 doi:10.1088/1757-899X/53/1/012048

4. Performance Evaluation

Figure 6 presents the simulated transmissibility of the EMS with the self-powered MR damper. In this case, a sinusoidal excitation displacement with constant amplitude of 1.0 mm was used. Note that the key benefit of the semi-active control system is to reduce the effects of resonance in the engine mass, without increasing high frequency transmissibility. In Fig 6, the dash-dotted line represents the passive hydraulic damper that has the same geometric dimension as the self-powered MR damper without the energy-harvesting PMG, the dashed line rep- resents the MR damper with constant field that has a MR valve in the piston head and is applied at the current of 0.7 A from an external power source, and the red solid line represents the self- powered MR damper and green solid line represents self powered & self sensing MR damper with new fluid . In Fig. 6 A, the MR damper with constant field can mitigate resonance, but increases transmissibility above resonance. The self-powered MR damper also substantially reduces transmissibility of the target system at resonance, but it does not increase transmissibility above the crossover frequency. This results from the fact that the energy-harvesting PMG produces a current proportional to the motion of the target system mass, as shown in Figs. 6 B and 6 C. The maximum power of the self- powered MR damper is 2.6 W (see Fig. 6 D, and the power consumption of the MR damper with the constant current of 0.7 A is 1.5 W. Overall, the self-powered MR damper shows good vibration isolation performance over the simulated excitation frequency and self-powered & self-sensing MR damper with new MRF shows much better vibration controlling performance with more control ability.

5. Conclusions

In this study, the achievability and accuracy of a self-powered & sensing MR damper operated using energy harvested from the operating environment were addressed. A PMG was theoretically constructed and added to a MR damper and also experimented with two new MRF with a little mixing nonmagnetic particle. The dynamic equation for the self-powered & sensing MR damper was derived theoretically. In order to theoretically evaluate the vibration controlling capability of the self-powered & sensing MR damper, a SDOF EMS using the MR damper was constructed and its governing equation of motion was theoretically derived. In order to find the optimal stiffness of the PMG, a parametric study was conducted. This parametric study indicated that the optimal stiffness was slightly lower than the stiffness required to match the PMG and engine mount frequencies. The isolation performance of the EMS with the self-powered MR damper was calculated in the frequency domain. It was observed through this theoretical analysis that the self-powered MR damper can provide good vibration isolation performance using neither a sensor nor a control processor/algorithm. And on the other side replacing the ordinary MRF with new MRF with glass beads presented a significant enhancement in yield force, off-state viscosity etc. With this power generation and self-sensing features, this new MR damper could bring great benefits for the MR damper systems such as saving energy, size and weight reduction, lower expenses, and less preservation.

6. References

[1] Y.-T. Choi and N. M. Wereley, "Self-Powered Magnetorheological Dampers," *Journal of Vibration and Acoustics*, vol. **131**, no. 4, p. 044501, 2009.

[2] M. Zapateiro, F. Pozo, H. R. Karimi, S. Member, and N. Luo, "Semiactive Control Methodologies for Suspension Control With Magnetorheological Dampers," no. April 2012, pp. 370–380, 2011.

[3] Z. Parlak, T. Engin, and İ. Çallı, "Optimal design of MR damper via finite element analyses of fluid dynamic and magnetic field," *Mechatronics*, vol. **22**, no. 6, pp. 890–903, Sep. 2012.

[4] M. M. Rashid, N. A. Rahim, S. Member, M. A. Hussain, M. A. Rahman, and L. Fellow, "Analysis and Experimental Study of Magnetorheological-Based Damper for Semiactive Suspension System Using Fuzzy Hybrids," no. April 2011, pp. 1051–1059, 2010.

[5] C. Chen and W.-H. Liao, "A self-sensing magnetorheological damper with power generation," *Smart Materials and Structures*, vol. **21**, no. 2, p. 025014, Feb. 2012.
[6] L. A. Powell, N. M. Wereley, J. Ulicny, and A. R. Testing, "Magnetorheological Fluids Employing Substitution of Nonmagnetic for Magnetic Particles to Increase Yield Stress," vol. **2202885**, no. November, pp. 3764–3767, 2012.
[7] Y. D. Liu, B. J. Park, F. F. Fang, H. J. Choi, and W.-S. Ahn, "Iron oxide/MCM-41 mesoporous nanocomposites and their magnetorheology," *Colloid and Polymer Science*, Mar. 2013.
[8] H. H. Sim, S. H. Kwon, and H. J. Choi, "Xanthan gum-coated soft magnetic carbonyl iron composite particles and their magnetorheology," *Colloid and Polymer Science*, vol. **291**, no. 4, pp. 963–969, Sep. 2012.

IOP Conf. Series: Materials Science and Engineering 53 (2013) 012049 doi:10.1088/1757-899X/53/1/012049

Design and Development of a Simulator for Modelling Carbon Nanotube

S Farhana[1], Z Alam, S Motakabber and S Khan
Department of Electrical and Computer Engineering
Faculty of Engineering, International Islamic University Malaysia

E-mail: farhana.engineer@ieee.org

Abstract. In this paper, a Matlab simulator is developed for modelling of carbon Nanotubes (CNTs). First electronics structure is derived and discussed to facilitate the analysis and modelling of the simulator. Then a calculation is done for electron energy dispersion relations for single-wall zigzag CNTs with different tube indexes. From the analysis of energy dispersion relation, it is derived an analytical models for CNT parameters such as the mobility and the intrinsic carrier concentration. Once obtain the CNT parameters, and then import them to the developed simulator. The simulator is based on the semiconductor equations and quantum effects. Finally the simulator produces the result for optimum modelling of CNT for electronic device application.

1. Introduction

As roadmap, investigators are exploring new paradigms for electronic devices. Carbon nanotubes (CNTs) are being explored as a structure that may play leading role in future electronic systems [1-4]. Carbon nanotubes (CNTs) are sheets of graphite which formed into cylinders. A nanotube with one layer of carbon atoms is single-wall carbon nanotube (SWCNT), and a CNT with multiple layers of carbon atoms is multi-wall carbon nanotube (MWCNT). SWCNT has the ability to act as a conductor (metal) and as a semiconductor as well [5,6]. However, a self-detecting cantilevered CNT-gated FET consisting of a single-walled CNT (SWNT) channel with a multi-walled CNT (MWNT) cantilever functioning as a gate. With this device, the resonance of an MWNT cantilever was detected for the first time as the change in drain current toward the future application of CNT cantilevers to the mass measurement of biological molecules. Theory indicates that CNTs can be metallic or semiconducting according to the fundamental tube indexes (lm) with bandgap of the semiconducting tube depending on the CNT diameter [7].

Low-dimensional carbon, such as single-walled carbon nanotubes (SWNTs) [8-9], graphene [10,11] and hydrogen passivity diamond surface [12,13] has remarkable electronic properties for high frequency electronics. These unusual electronic properties originate from the unique crystal and electronic band structure which provides extremely high carrier mobilities [14,15]. Owing to the high charge mobility of SWNTs and low geometric capacitance, carriers on the tubes can respond to an external electric field in a picoseconds time scale [16]. This very fast response time opens new prospects for high frequency device applications. These drawbacks make single-tube devices

[1] Soheli Farhana.

impractical for high frequency electronics. Dense arrays of SWNTs, however, increase the total gate capacitance and the contribution of the fringe capacitance becomes smaller.
This paper describes the modelling of the simulator for designing the electronic structure, energy dispersion relation and intrinsic carrier concentration in the following sections.

2. CNT Electronic Structure

Carbon nanotubes (CNTs) which have unique electronic properties are being considered for various next generation device applications like field effect transistors or sensing elements. Due to the tiny radius of carbon nanotubes, the quantization of wavevectors to be quantized in the circumferential direction. Moreover, the thinness of the nanotube's cylindrical shell obviously yields an even shorter length of confinement in the radial direction, thus making the material virtually one-dimensional as far as electron transport is concerned.

2.1. Reciprocal lattice

The primitive cell of a carbon nanotube can be described in the unit vectors:

$$R_1 = \frac{a}{2}\left(\sqrt{3}\hat{x} + \hat{y}\right) \text{ and } R_2 = \frac{a}{2}\left(\sqrt{3}\hat{x} - \hat{y}\right) \tag{1}$$

where, a = 2.49 Å is the carbon to carbon atom distance between two carbon atoms.
The reciprocal lattice vectors are of the form [15,16]:

$$b_1 = \frac{2\pi}{a}\left(\frac{1}{\sqrt{3}}\hat{x} + \hat{y}\right) \text{ and } b_1 = \frac{2\pi}{a}\left(\frac{1}{\sqrt{3}}\hat{x} - \hat{y}\right)$$

2.2. Electronics properties

Armchair tubes: Simple zone folding predicts that if the difference of the n, m: chiral indices is divisible by 3, and then the (n, m) nanotube is metallic. We shall refer to these nanotubes as "zone folding metallic", or shortly, ZF-M tubes.
Using the following equations to obtain calculate different nanotube.
The chiral vector of the nanotube is defined in (2),

$$C_h = na_1 + ma_2 \tag{2}$$

Length of the chiral vector is the peripheral length of the nanotube:

$$L = |C_h| = a\sqrt{n^2 + nm + m^2} \tag{3}$$

The tube diameter is then given by,

$$d_t = \frac{L}{\pi} = \frac{a}{\pi}\sqrt{n^2 + nm + m^2} \tag{4}$$

The bandgap of a single wall nanotube (SWNT) is given by,

$$E_g = 2\gamma_0 a_{cc}/d_t \tag{5}$$

$$d = \gcd(n, m) \tag{6}$$

if (n-m) is divisible by 3d then $dr = 3d$ otherwise, $dr = d$.

$$N_{at} = 4 * \frac{n2 + m2 + nm}{dr} \tag{7}$$

3. CNT Simulator

3.1. Energy Dispersion Relation

To account for the CNT-related quantum effects in the CNT-MOSFET channel, we need to determine the band-structure of the CNTs. Due to confinement, the band structure splits into a system of subbands when graphene is wrapped into a CNT. Each of the subbands has a characteristic effective mass, mobility, and band energy minima. We determine the energy levels of CNTs by applying zone-folding methods to graphene [14]. From the two-dimensional (2-D) graphene band diagram, we cut 1-D slices, whose numbers and locations are set by the fundamental tube indexes. The resulting band structure is given in figure 1. We use the bandgap of the CNT and the electron affinity of the graphite to obtain the electron affinities of CNTs. We then calculate CNT affinities by subtracting half the bandgap value of the lowest subband of the CNT from the electron affinity of graphite.
In order to examine the conductivity properties of the nanotube, it is necessary to derive its k relation. This is done by starting from the equivalent relation of a two dimensional graphene lattice. The energy dispersion for carbon nanotubes can be calculated from the electronic structure of graphene. The energy dispersion of graphene is:

$$E_{2D}(K) = \pm V_{pp\pi}\{3 + 2Cos(KR_1) + 2Cos(KR_2) + 2Cos[K(R_1 - R_2)]\}^{1/2} \quad (8)$$

where Vpp, is the nearest neighbour transfer integral, R_1 and R_2 are the unit cell vectors of a carbon nanotube given in equation (8). The 3d view of energy dispersion relation [17] is shown in figure 1.

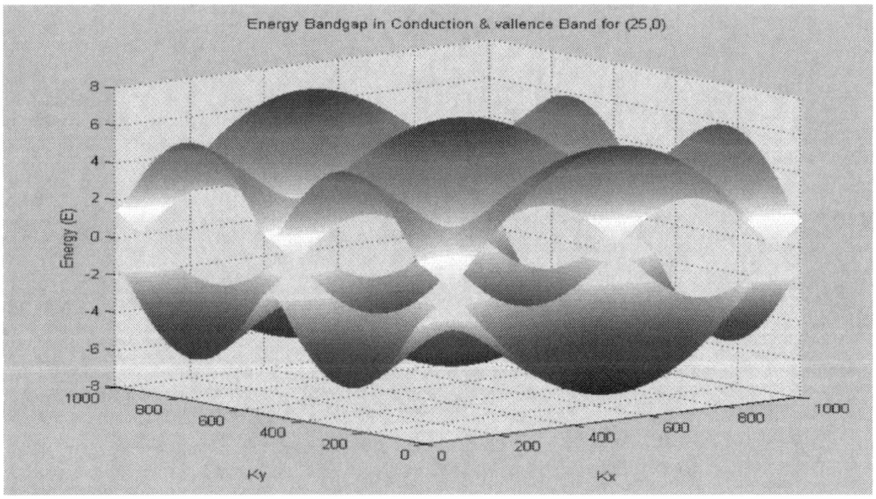

Figure 1. Plot of the Energy dispersion relation between valence and conduction band

3.2. Density Of States

The allowed k-vectors in momentum space depend on vectors K_1 and K_2 as shown in figure 2. It is possible to represent the area in momentum space for a single state as, $A_P^{1-state} = h^2|K_1||K_2|/2$ and a differential area as, $A_P = h^2|K_1|dk$, where dk is in the direction of K_2 and h is the Planck's constant divided by 27w. Therefore, the density of states per unit energy can be defined as follows [18]:

$$D(E)dE = 2\frac{dA}{A_P^{1-State}} = \frac{4}{h^2|K_1||K_2|}h^2|K_1|\frac{dk}{dE} = \frac{2|T|}{\pi}\left(\frac{dE}{dk}\right)^{-1} \quad (9)$$

Where,

$$K_1 = \frac{(2n + m)b_1 + (2m + n)b_2}{Nd_R} \text{ and } K_2 = \frac{mb_1 - nb_2}{N}$$

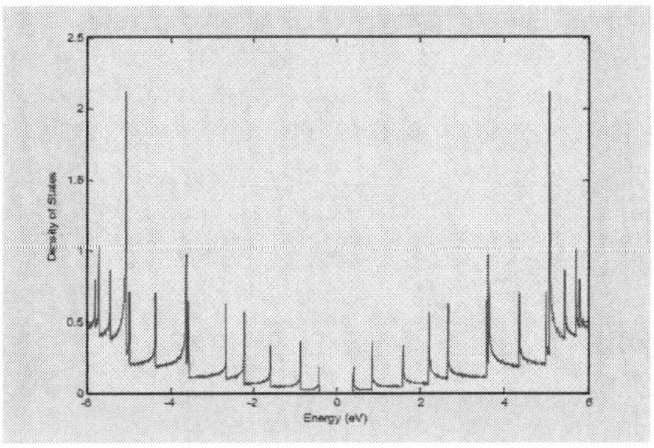

Figure 2. Density of states plot for a CNT

3.3. CNT Intrinsic Carrier Concentration

With the density of states calculated, it is possible to calculate the carrier concentration of carbon nanotubes. The carrier concentration in any semiconductor is given by [19],

$$n_{cnt} = \int_{E_c}^{\infty} D(E)f(E)dE \tag{10}$$

where D(E) is the density of states, f(E) is the Fermi level, and E, is the conduction band minima.
Using equations (9), the carrier concentration becomes,

$$n_{cnt} = \sum_{i}^{\text{All Bands}} \left[\frac{4}{\pi V_{pp\pi} a\sqrt{3}} \cdot \int_{E_c}^{\infty} E(E^2 - E_{Cmin_i}^2)^{-1/2} + \left(1 + e^{\frac{E-E_F}{KT}}\right)^{-1} dE \right] \tag{11}$$

The equation (11) can be simplified using mathematical manipulations to arrive at the following expression:

$$n_{cnt} = \frac{4}{\pi V_{pp\pi} a\sqrt{3}} \int_0^{\infty} (E' + E_c)(E^2 + 2E_c E')^{-1/2} \left(1 + e^{\frac{E'-E_F+E_c}{KT}}\right)^{-1} dE' \tag{12}$$

The limits of integration have been changed by replacing the variable E in equation (8) with (E, - E), where E, is the lowest conduction energy. Furthermore, the summation has also been dropped as the Fermi function becomes negligible for conduction energy-minimums beyond the first band.
The integral in equation (12) is still very difficult to solve analytically, nevertheless, by letting x = E/KT and $\eta = \frac{E_F - E_c}{KT}$,
equation (9) becomes,

$$n_{cnt} = \frac{4\sqrt{KT}}{\pi V_{pp\pi} a\sqrt{3}} \int_0^{\infty} (KTx + E_c)[x(KTx + 2E_c)]^{-1/2} (1 + e^{x-\eta})^{-1} dx \tag{13}$$

To investigate the effects of embedding a CNT into a MOSFET, we developed a novel device simulator. One of the fundamental quantities required by our CNT-MOSFET solver is the CNT intrinsic carrier concentration is shown in figure 3. Thus we develop a methodology to obtain the intrinsic concentrations of different tubes. We start from the parabolic energy dispersion approximation. The density of states for each subband is zero for energies less than the energy minimum of that particular subband, and becomes the following for energies greater than the subband energy minimum.

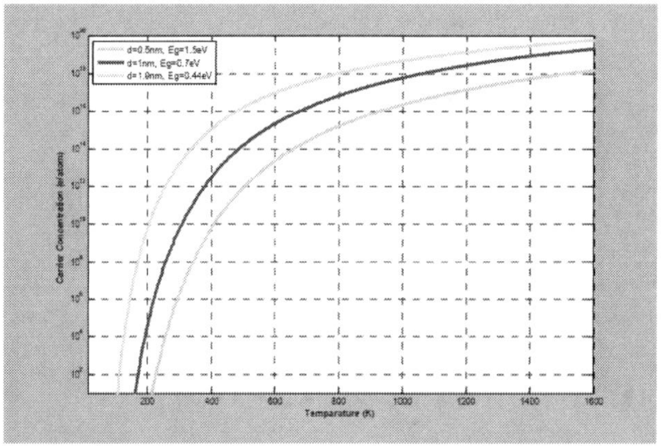

Figure 3. CNT intrinsic carrier concentration

4. Conclusion

A Matlab simulator is developed for the analysis of carbon Nanotubes (CNTs). A quantum method is derived to develop this simulator. The result shows that this simulator is capable to perform the modelling of CNT by exploring its properties such as energy dispersion relation, density of states and CNT intrinsic carrier concentration. Thus, we concluded that CNT employing higher mobility, lower bandgap carbon Nanotubes appear to exhibit improved capabilities and therefore may represent a new paradigm for devices in the 21st century.

References

[1] Lundstrom M 2003 *A top-down look at bottom-up electronics* in Symp. VLSI Tech. Dig. pp. 5–8.

[2] Wong H S P 2002 *Field effect transistors from silicon MOSFETs to carbon nanotube FETs* in Proc. MIEL vol. 1pp. 103–107.

[3] Guo J, Datta S and Lundstrom M 2002 *Assessment of silicon MOS and carbon nanotube FET performance limits using a general theory of ballistic transistors* in IEDM Tech. Dig. pp. 29.3.1–29.3.4.

[4] Avouris P 2002 Molecular electronics with carbon nanotubes Acc. Chem. Res. vol. 35 pp. 1026–34.

[5] Navi K, Rashtian M, Hashemipour O, Khatir A, Keshavarzian P 2010 *High speed capacitor-inverter based carbon nanotube full adder* Nanoscale Research Letters 859:862.

[6] Navi K, Rad R, Moaiyeri M, Momeni A 2010 *A low-voltage and energyefficient full adder cell based on carbon nanotube technology* Nano Micro Letters 114:120.

[7] Naoyuki M, Takayuki A and Seiji A 2012 *A Carbon Nanotube Field-Effect Transistor with a Cantilevered Carbon Nanotube Gate* Applied Physics Express 5.

[8] Pennington G and Goldsman N 2003 *Semiclassical transport and phonon scattering on electrons in semiconducting carbon nanotubes Phys. Rev. B, Condens Matter vol. 68 pp. 45* 426-1–45 426-11.

[9] Pennington G and Goldsman N 2003 *Monte Carlo study of electron transport in a carbon nanotube* IEICE Trans. Elect. vol. E86-C, pp. 372–8.

[10] Lin Y M, Dimitrakopoulos C, Jenkins K A, Farmer D B,Chiu H Y, Grill A and Avouris P 2010 *100 GHz transistors from wafer-scale epitaxial graphene Science* 327 662.

[11] Xia F N, Mueller T, Lin Y M, Valdes-Garcia A and Avouris P 2009 *Ultrafast graphene photodetector Nature Nanotechnol* 4 839–43.

[12] Kasu M, Ueda K, Ye H, Yamauchi Y, Sasaki S and Makimoto T 2006 *High RF output power for H-terminated diamond FETs Diamond Relat.* Mater. 15 783–6.

[13] Ueda K, Kasu M, Yamauchi Y, Makimoto T, Schwitters M,Twitchen D J, Scarsbrook G A and Coe S E 2006 *Diamond FET using high-quality polycrystalline diamond with f(T) of 45 GHz and f(max) of 120 GHz* IEEE Electron Device Lett. 27 570–2.

[14] Zhou X J, Park J Y, Huang S M, Liu J and McEuen P L 2005 *Band structure, phonon scattering, and the performance limit of single-walled carbonnanotube transistors Phys* Rev. Lett. 95146805.

[15] Jimenez D, Cartoixa X, Miranda E, Sune J, Chaves F A and Roche S 2007 *A simple drain current model for Schottky-barrier carbon nanotube field effecttransistors* Nanotechnology 18 025201.

[16] Zhong Z H, Gabor N M, Sharping J E, Gaeta A L and McEuen P L 2008 *Terahertz time-domain measurement of ballistic electron resonance in a single-walled carbon nanotube* Nature Nanotechnol. 3 201–5.

[17] Farhana, S., Alam, A. Z., Motakabber, S., & Khan, S. 2013 *Analysis of CNT electronics structure to design CNTFET* in Nanoelectronics Conference (INEC), 2013 IEEE 5th International (pp. 329-332). IEEE.

[18] Griffiths D. 1995 *Introduction to Quantum Mechanics* Prentice Hall, Inc., Upper Saddle River, NJ.

[19] Shur M. 1990 *Physics of Semiconductor Devices*, Prentice Hall.

IOP Conf. Series: Materials Science and Engineering **53** (2013) 012050 doi:10.1088/1757-899X/53/1/012050

μ-Patterning of Carbon Nanotube (CNT) forest for MEMS applications

T.Saleh[1]
Department of Mechatronics Engineering, International Islamic University Malaysia, Kuala Lumpur 50728, Malaysia
E-mail: tanveers@iium.edu.my

Abstract. This paper proposes three new approaches for micro patterning of CNT forest in order to make it useful for MEMS based applications. The first two techniques are based on micro electro discharge machining (μ-EDM). However, the biggest problem associated with μ–EDM is the spark gap which limits the structural resolution of the fabricated pattern. In order to overcome this challenge the first technique proposed in this paper is reverse μ–EDMing of CNT forest where the CNTs are used as cathode instead of tungsten tool. This dramatically reduces discharge voltage hence the spark gap. In the second method Sulphur Hexafluoride (SF_6) SF_6 was used as dielectric instead of air which has three time higher dielectric strength than air. This helps to reduce spark gap further. This research work also discusses the experimental results when SF_6 was used as dielectric medium for reverse EDMing CNT forest. It was observed that at too low voltage (~10V) air gives lower spark gap than SF_6, however at moderately high voltage (~25V) SF_6 performs better. Finally, the third approach for patterning CNT forest described in this paper is mechanical bending of CNTs. In this method patterning of CNT forest is carried out by moving a rotating cylindrical μ–tool (3000RPM) in X,Y and Z direction. The Z movement of the tool is controlled in step mode to provide the overall depth of the μ–structures with 1μm/step. In XY plane the tool moves continuously at 1mm/min speed. The movement of the tool on the CNT forest causes the CNTs to be bent and flattened in the direction of the tool motion hence the patterns are formed on bare CNT forest. The most significant observation made from the processed CNT forest is the visible optical reflection from bent and flattened area. Typically, CNT forest is known to be the darkest material on earth. However, this new processing technique causes the CNT surface to reflect light like mirror. A detail comparison between all proposed techniques (mechanical and reverse μ–EDM) for patterning CNT forest is also included in this paper.

1. Introduction

Vertically aligned Carbon Nano Tubes (CNTs) which are commonly referred to as CNT forests is drawing attention by the researchers in the field of micro-electro-mechanical systems (MEMS) because of its very attractive mechanical, electrical, optical, and thermal properties **[1-5]** . Patterning of CNT forest is very important for MEMS based application and this is currently carried out through selective growth of the forests by chemical vapor deposition (CVD) on pre-patterned catalyst on the

[1] To whom any correspondence should be addressed.

substrate. This method has limitation in producing true 3-D structure. Free-form and three dimensional micro-patterning of CNT forest by micro-electro-discharge machining (μEDM) was first proposed by Takahata group from UBC, Canada **[6, 7]**. Dry air (mixture of N_2 and O_2) was used as the dielectric medium in the process as liquid dielectric causes damage to the forest because of capillary action. The optimal machining condition was observed at the discharge voltage of 30 V. This technique was further investigated to understand possible removal mechanism of CNTs by μEDM **[7, 8]**. These studies suggest the removal of CNTs during μEDM is carried out by oxygen plasma etching rather by direct thermal removal (evaporation and melting) process. The machining tolerance which is mostly governed by the spark gap was observed to be in a range of 10 μm; significantly larger than typical values (of one to a few μm) involved in standard μEDM for metal erosion. In order to achieve high tolerance on a patterned CNT forest by μ–EDM spark gap is a big challenge that needs to be addressed. In this paper two approaches shall be discussed in order to achieve higher tolerance on patterned CNT forest by μ–EDM.

In typical EDM (including μEDM), the workpiece and the electrode are generally arranged to be the positive and the negative, respectively, as this polarity is helpful for efficient EDM operation. CNTs are known to significantly enhance an applied electric field and thus have excellent electron emission properties because of their extremely small tips with nanometer radii and high aspect ratios. Previous report suggests that CNT cathodes can reduce gas breakdown voltage while enhancing discharge current compared to tungsten cathodes **[9]**. This suggests that the use of reverse polarity in μEDM of CNT forests could be effective in lowering the discharge voltage which will be helpful to pattern CNT forest at lower energy thus to reduce the spark gap. This hypothesis is explored experimentally in this present study by carrying out reverse-polarity μEDM of pure CNT forests in dry air.

Dielectric strength plays an important role in μEDM. Sulphur hexafluoride (SF_6) has a dielectric strength much higher than that of N_2 (by a factor of ~3). This suggests that in SF_6 ambient the gap between the machining electrode and workpiece needs to be much smaller to initiate breakdown for a given electric field strength. This hypothesis suggests that use of SF_6 as the dielectric medium may further reduce the spark gap which will eventually increase the resolution of the patterning of CNT forest by μEDM. This was verified experimentally in this study and discussed in the second part of the paper.

Finally, another approach for patterning CNT forest described in this paper is mechanical bending of CNTs. Patterning of CNT forest is carried out by moving a rotating cylindrical μ–tool (3000RPM) in X,Y and Z direction. The movement of the tool on the CNT forest causes the CNTs to be bent and flattened in the direction of the tool motion hence the patterns are formed on bare CNT forest without any spark gap which is normal μEDM based patterning method.

2. Experimental Setup

CNT forest samples are first grown on highly doped silicon (Si) substrate (<100> n-type) by a method called chemical vapour deposition (CVD). All the experiments related to patterning of CNT forest samples using various methods were carried out using a servo-controlled 3-axis μEDM system with a 0.1-μm positioning resolution. The discharge pulses were generated with a relaxation-type resistor-capacitor (*R-C*) circuit **[10]** (for μ-EDM based experiments).

The experimental set-up for reverse μ-EDM in air is illustrated in Fig. 1. As shown, a current probe (CT-l, Tektronix, USA) was used to monitor pulses of the discharge current in real time. The discharge current data were captured from the oscilloscope using a GPIB interface and stored in a computer for subsequent analysis. A series of μEDM experiments at both normal and reverse polarities were performed for various energy levels that were controlled by varying the discharge voltage in a range of 10-60 V with a fixed capacitance of 10 pF in the *R-C* circuit. The electrode (tungsten, diameter of 40-93 μm) was rotated at 3000 rpm during all the machining experiments. The X-Y scanning rate and the electrode feed rate in the Z direction were set to 1 mm/min and 10 μm/min, respectively.

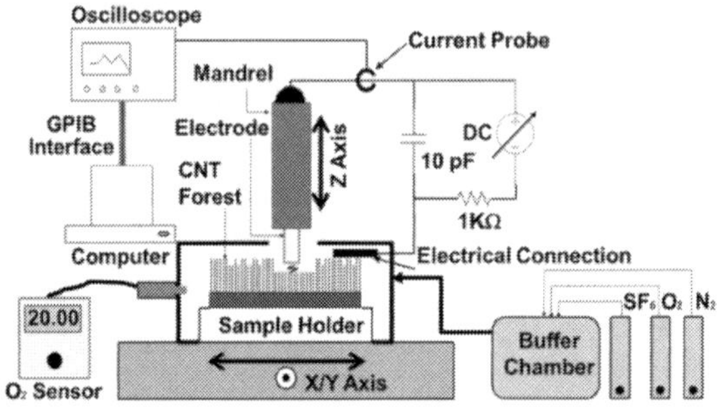

Figure1: Set-up used for characterization of reverse-polarity μEDM of pure CNT forests in air and SF_6 environment.

Same setup was used (Fig. 1) for SF_6 based μEDMing of CNT forest. According to the illustration first O_2 was mixed with either N_2/SF_6 inside a buffer chamber, and the mixed gas was introduced to the machining chamber, where the O_2 concentration was measured using an oxygen sensor (VN202, Vandagraph Co., UK). The flow rates of O_2 and N_2/SF_6 was adjusted so that the O_2 concentration reached the target value and was stabilized in the machining chamber for at least 5 minutes prior to machining. A series of μEDM experiments (with SF_6) were performed using cylindrical tungsten electrodes rotating at 3000 rpm. The μEDM conditions are summarized in Table 1. A 10-pF capacitor was used in the discharge circuit for all the experiments in this study.

Table 1 – μEDM conditions used for CNT machining experiments.

Parameters		Values used
Machining voltage (V)		25 or 10
Capacitance (pF)		10
Electrode feed rate during EDM (mm/min)		Z: 0.03, X-Y: 1
Ambient	O_2 % in SF_6 (for the SF_6-O_2 system)	50, 20, 10, or 0
	O_2 % in N_2 (for the N_2-O_2 system)	20 or 0

Finally Fig.2 illustrates the concept of mechanical processing of the CNT forest that is locally bent by scanning a microscopic metallic tool; a cylindrical piece of tungsten (diameter ≤ 300 μm) rotating at a speed of 3,000 rpm is moved along the X, Y, and Z directions to carry out the process. This experiment was also performed on the same machine as mentioned earlier. However, in this case the EDM power was turned off during patterning. The movement in the Z direction was performed in the step mode with a step size of 1 μm, and the back-and-forth scanning speed in X and Y directions was 1 mm/min.

Figure 2: Concept of patterning CNT forest by mechanical bending of CNTs.

3. Results and Discussions:

3.1 Reverse polarity μEDM in air:

In this section results related to reverse polarity μEDMing of CNT forest (in air medium) have been discussed. At first discharge pulses generated with both reverse and normal polarities were characterized to observe the differences between the two conditions. Fig.3 shows the comparison of the average peak current of the pulses (n = 400) measured at different voltage levels. It can be understood from the comparison that the peak currents with the reverse polarity are higher than those with the normal polarity, and that the difference in the peak current at lower voltages is significant.

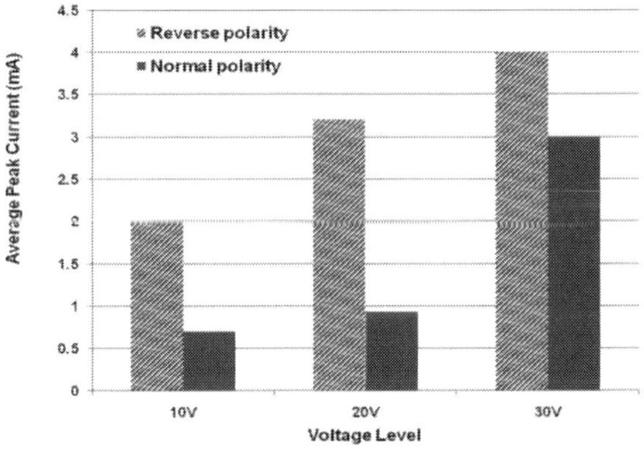

Figure 3: Average peak current of discharge pulses measured at different voltages (with 10-pF capacitance).

A possible reason behind the higher discharge current in the reverse-polarity case that uses the forest as the cathode could be enhanced field-emission (FE) from CNTs **[11]**; the forest cathode can more easily emit electrons from the nano-scale tips of the CNTs compared to a conventional metallic cathode (such as the tungsten tip) at a given voltage.

The previous studies reported that the lowest value of the optimal voltage with the normal polarity was around 30V, and that further lowering of the voltage led to mechanical grinding of the CNTs **[6]**. To evaluate the effect of the reverse polarity at low voltage, square patterning with both polarities were carried out at 10 V and 10 pF. The results in Figs. 4(a) and 4(b) indicate that the reverse-polarity process produced sharper, smoother, and cleaner microstructures compared to the normal-polarity case. Moreover, the patterns created with the reverse polarity exhibited a narrower width in the machined grooves compared to those with the normal polarity even though identical electrode and machining conditions were used. From the dimensions shown in Fig. 4, as well as the diameter (93 μm) of the electrode used, the discharge gap clearance is calculated to be 7.5 μm for the normal polarity, whereas for the reverse polarity it is 2.5 μm, which is 3× smaller (and shows a ~4× improvement over the previous result reported in **[6]**). This means that reverse μEDM enables much tighter tolerances and higher precision in CNT forest patterning.

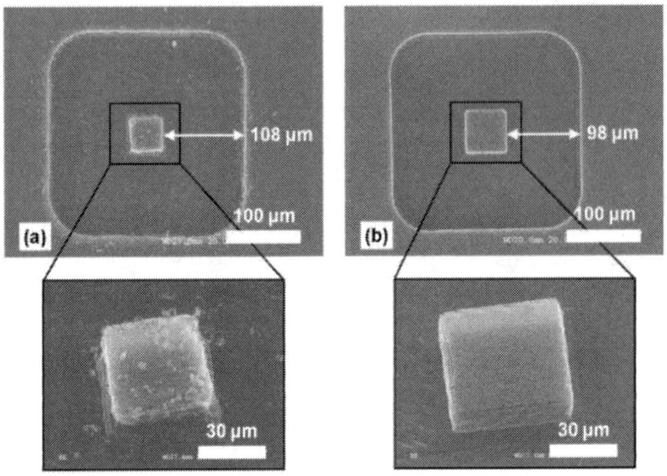

Figure 4: SEM images show micropatterns machined in a CNT forest using 10 V for a depth of 40 µm with (a) normal polarity and (b) reverse polarity.

Machining stability is another important aspect of µEDM processing of CNT forests. In the process, electrode feeding is feedback controlled so that when a short circuit is detected, the Z stage retracts the electrode upward until the circuit is opened and then moves the electrode downward to resume its feeding and the machining process. The process becomes unstable and slow if many short circuits occur during the process because of frequent up/down motion of the Z axis. It was observed that reverse-polarity machining at low voltages was consistently more stable than normal-polarity machining for CNT forests. This tendency can be seen in Fig. 5 that plots the electrode position on the Z axis during

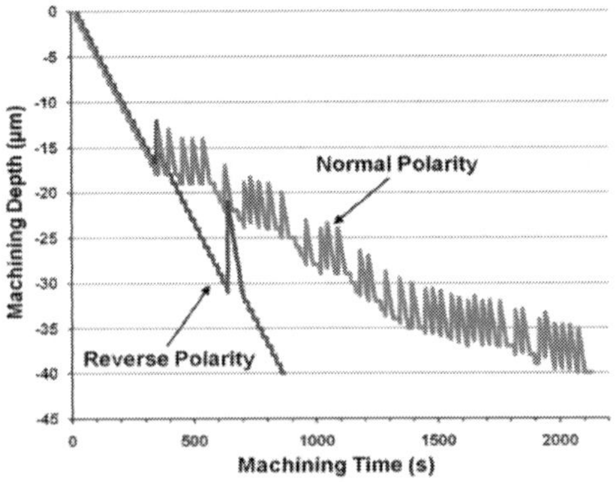

Figure 5: Electrode positions on the Z axis tracked in real time during machining with normal and reverse polarities, both using the same machining condition.

each of the machining processes conducted under the same conditions except for the voltage polarity. It is clear from the graph that the normal-polarity case produced frequent short circuits that led to ripples in the electrode motion, whereas the reverse-polarity case resulted in much more stable and faster electrode feeding or removal of the CNTs. For this particular machining condition, the total machining time with the reverse polarity was approximately 60 % shorter than the time with the normal polarity.

The effectiveness of low-energy reverse μEDM in high-aspect-ratio micromachining of CNT forests was evaluated by patterning conical and pyramid like microstructuers at the same EDM condition (10V and 10 pF) using a tapered cylindrical electrode as shown in Fig. 6.

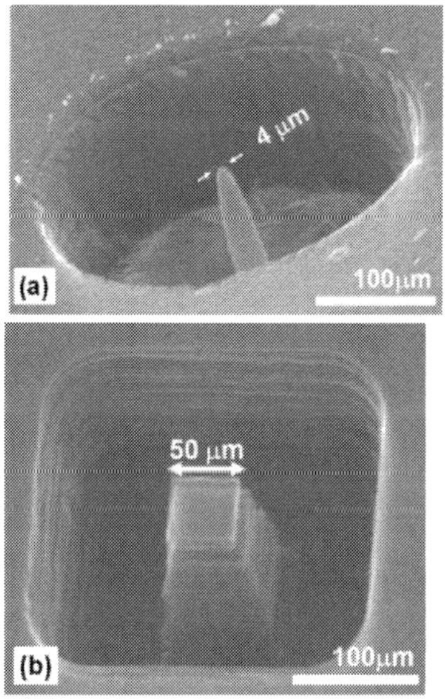

Figure.6: SEM images of high-aspect-ratio microstructures patterned using reverse μEDM at 10V by scanning a cylindrical electrodes along a circular or square orbit: (a) A needle-like microstructure with an aspect ratio of ~25; (b) a deep pyramid structure.

3.2 Reverse polarity μEDM of CNT forest in SF_6 medium:

In order to further enhance the resolution of the micro structures patterned by EDM on CNT forest gaseous medium with high dielectric strength (SF_6) was used. In this section experimental result using SF_6 ambient shall be discussed. To study the role of O_2 in the SF_6 case, the μEDM process was characterized with O_2 concentrations of 10%, 20%, and 50%. The results machined at 25 V shown in Fig. 7 suggest that the structural and surface quality improved with increasing O_2 concentration up to 20% (Fig. 7(a) and (b)), and that the structures became distorted (e.g., the top surface of the centre post as seen in Fig. 7(c), possibly due to sparks propagating and etching portions of it) again when the concentration was further increased to 50%. The structure machined at 10 V and 20% O_2 in SF_6 shown in Fig. 7(d) indicates deteriorated structural quality compared to Fig. 7(b), the 25-V case under the same ambient. The results obtained at the same voltage levels, 25 V and 10 V, with 20% O_2 in N_2 ambient are shown in Fig. 7(e) and (f), respectively. These suggest that, in contrast to the SF_6-O_2 ambient cases, the 10-V condition led to higher machining quality than the 25-V condition in the N_2-O_2 ambient; this result is consistent with the previous findings reported in [6]. However, a comparison between Fig. 7(b) and (f), representing the conditions that provided the highest machining quality for the SF_6 and N_2 environments, respectively, suggests that SF_6 results in sharper corners (without extended portions at the corner of the centre post) than N_2, and the sidewall of the resultant structure is smoother (free from extended layers) for SF_6 ambient. Possible sources of the different optimal voltage levels for the SF_6 and N_2 environments (25 V and 10 V, respectively) found above will be discussed later.

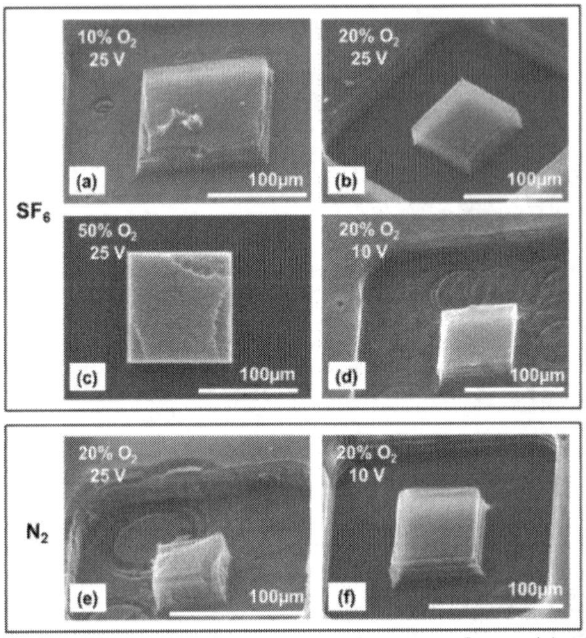

Figure 7: SEM images of the microstructures machined in a CNT forest: (a) 10% O_2 in SF_6 at 25 V; (b) 20% O_2 in SF_6 at 25 V; (c) 50% O_2 in SF_6 at 25 V; (d) 20% O_2 in SF_6 at 10 V; (e) 20% O_2 in N_2 at 25 V; (f) 20% O_2 in N_2 at 10 V.

Fig. 8 shows, in SF_6 (at 20% O_2), processing at 25 V resulted in the minimum discharge gap and the highest machining quality; however, this is not the case for N_2 (at the same O_2 concentration), in which processing at 10 V led to the minimum gap and the highest machining quality. In the SF_6 environment, 25 V was a suitable voltage level to produce spark discharge pulses and perform smooth machining (Fig. 7(b)); however, 10 V may have made the discharge gap too small because of the high-dielectric ambient and thus caused physical touching and mechanical rubbing between the rotating electrode and the forest surface (which may have caused the circular marks on the bottom of the structure shown in Fig. 7(d) due to non-ideal mechanical/positioning instability in the μEDM system used, deteriorating the processed structure/surfaces. This undesired mechanical contact may have also occurred on the sidewalls of the patterned structures and caused slight bending or displacement of the CNTs on the walls, which may be the probable cause of the larger gap compared with the N_2 environment case at 10 V (Fig. 7(d)).

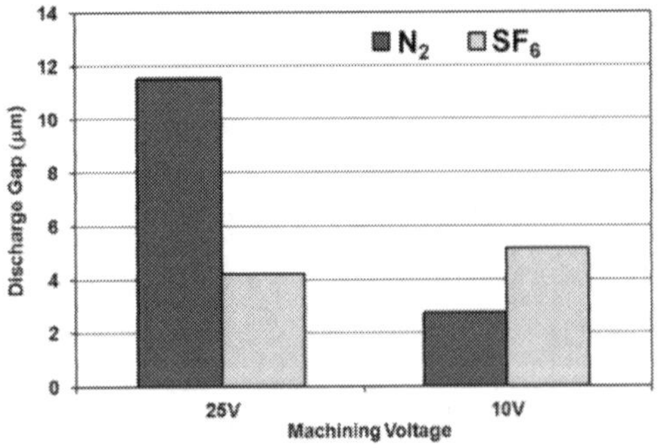

Figure 8: Measured discharge gaps resulted from the SF_6 and N_2 environments (with 20% O_2) and two different discharge voltages.

With a discharge voltage of 25 V and an O_2 concentration of 20%, the use of SF_6 was revealed to give better results compared to N_2, the conventional ambient medium; the discharge gap could be reduced to 4.2 μm. At a lower voltage level (10 V), an N_2 ambient with 20% O_2 produced an even smaller discharge gap, although the occurrence of short circuits slowed the removal process somewhat.

3.2 Patterning of CNT forest by mechanical bending:

Finally to eliminate completely the spark gap problem associated with EDM based patterning method efforts were made to pattern bare CNT forest by mechanical bending of CNTs. The process is described in the section 2 earlier. In this method the spark gap was zero as no spark erosion was taken place during the process. However another significant observation made from the patterned CNT forest is the reflectivity from bent and flattened CNT forest. Typically, CNT forest is known to be the darkest material on earth. **[12]**. However, when mechanical processing technique is applied on the CNT forest it starts to reflect light like mirror as shown in Fig. 9(a). The reflectivity of the surface was measured and quantified and the result is shown in Fig. 9(b). The result shows the bent region of the CNT forest has optical reflectivity in a range of 10% to 15% from visible to infrared spectrum of light, which is two orders of magnitude higher than bare CNT forest.

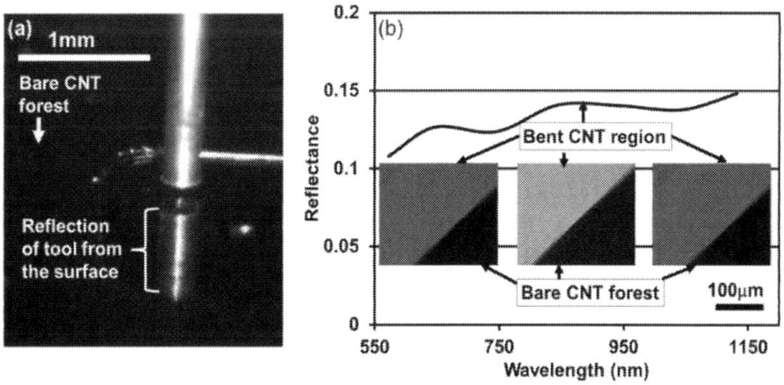

Figure 9: (a) Optical image showing the reflected image of the tool on the bent-CNT surface. (b) Measured normal-incidence spectral reflectance of the surface (averaged) for visible-near infra-red light accompanied by images of red, green and blue light reflection from the processed surface.

The possible reason behind the observed reflectivity from the bent CNT surface is stated as follows. The CNTs of the grown forest are essentially Multi-walled CNTs (MWCNT), hence metallic. When they are bent and flattened by the moving tool a nice smooth surface texture is created with surface roughness S_a as low as 5nm (shown in Fig. 10). This surface roughness is of optical quality. Therefore, the processed metallic CNT surface starts to reflect light.

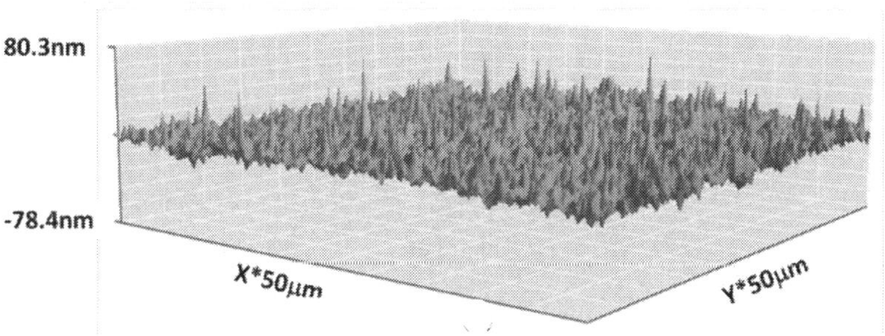

Figure 10: AFM result of the surface shows high quality surface finish with S_a 5nm and S_y 0.1 mm.

4. Conclusion:

Post patterning of CNT forest is necessary in order to use it for MEMS based applications. In this paper three methods for micro patterning of CNT forest have been discussed. First two methods are μEDM based technology. It was shown for μEDM based patterning method, reverse EDM with high dielectric environment (SF_6) can reduce spark gap at a voltage level of 25V however if the voltage is reduced further down (~10V) then reverse EDM in air proved to produce low spark gap with significantly long machining time. Mechanical patterning method can eliminate spark gap problem completely with high quality surface roughness (Sa = 5nm). Though, this method flattens the protruded CNTs on the patterned path. Therefore, it may not be suitable where protrusion of CNT in patterned zone is necessary.

References

[1] M. M. J. Treacy, T. W. Ebbesen and J. M. Gibson, Nature 381, 678 (1996).
[2] J. P. Lu, Phys. Rev. Lett. 79, 1297 (1997).
[3] K. Hsieh, T. Tsai, D. Wan, H. Chen, N. Tai, ACS Nano 4, 1327 (2010).
[4] Z. Yang, L. Ci, J. A. Bur, S. Lin, and P. M. Ajayan, Nano Lett. 8, 446 (2008).
[5] N. Hamada, S. Sawada, and A. Oshiyama, Phys. Rev. Lett. 68, 1579 (1992).
[6] W. Khalid , M. S. Mohamed Ali , M. Dahmardeh, Y. Choi, P. Yaghoobi, A. Nojeh , K. Takahata, Diam. Relat. Mater. 19, 1405 (2010).
[7] M. Dahmardeh, W. Khalid, M.S. Mohamed Ali, Y. Choi, P. Yaghoobi, A. Nojeh, K. Takahata, 24th IEEE Int'l Conf. Micro Electro Mechanical Systems (MEMS 2011), Cancun, Mexico, 2011, pp. 272-275.
[8] M. Dahmardeh, A. Nojeh, and K. Takahata, J. Appl. Phys. 109, 093308 (2011).
[9] B. Liang, A. Ogino and M. Nagatsu, J. Phys. D. Appl. Phys. 43, 275202 (2010).
[10] M.P. Jahan, Y.S. Wong, and M. Rahman, J. Mater. Process. Tech. 209, 1706 (2009).
[11] S. Fan, M. G. Chapline, N. R. Franklin, T. W. Tombler, A. M. Cassell, H. Dai, Science 283, 512 (1999).
[12] Z. P. Yang, L. Ci, J. A. Bur, S. Y. Lin, and P. M. Ajayan, Nano Lett. **8**, 446 (2008).

IOP Conf. Series: Materials Science and Engineering **53** (2013) 012051 doi:10.1088/1757-899X/53/1/012051

Fuzzy Logic Controller Design for A Robot Grasping System with Different Membership Functions

Hamzah Ahmad[1], Saifudin Razali[2], and Mohd Rusllim Mohamed[3]
[1,2,3]FKEE, University Malaysia Pahang, Pekan Campus, 26600 Pekan, Pahang, MALAYSIA
E-mail: hamzah@ump.edu.my

Abstract. This paper investigates the effects of the membership function to the object grasping for a three fingered gripper system. The performance of three famously used membership functions is compared to identify their behavior in lifting a defined object shape. MATLAB Simulink and SimMechanics toolboxes are used to examine the performance. Our preliminary results proposed that the Gaussian membership function surpassed the two other membership functions; triangular and trapezoid memberships especially in the context of firmer grasping and less time consumption during operations. Therefore, Gaussian membership function could be the best solution when time consumption and firmer grasp are considered.

1. Introduction

Robot grasping system has several issues to be solved such as the position control, the force control, the hybrid control, and the grasping control[1]. To ensure better performance and efficient grasping, artificial intelligent techniques are always the best choice to be included in the grasping system. The method is selected as it could reacts and decides effectively as what human will do during picking or lifting an object.

Since 1990's, Fuzzy Logic has been proposed to provide alternative solutions for grasping purposes which analyzed a number of identified issues[2, 3, 4, 5]. The earliest contribution in determining the Fuzzy Logic in robot grasping system was proposed by R.Alberto et al[6]. They discussed the performance of the system in two cases;known and unknown conditions. Based on their findings, Fuzzy Logic can guarantee better results as they mimics the human behavior very well during the operation. Their results was supported and demonstrated by others that suggested fuzzy logic technique could provide good results by determining initial pattern of objects to be lifted[3]. N.I. Glossas et al.[7] proposed the fuzzy logic grasping system by using a subtractive clustering algorithm for their two fingers gripper. By considering three inputs and an output, they adjust the gripper motion through tactile sensors through very minimum rules. Their experimental results expressed that the fuzzy logic with clustering algorithm perform well than any other fuzzy based system.

Not limited to the standalone technique, the control system for the robot grasping system has also combined two different methods. For example, L.Minzhuo et al.[8] demonstrated that a robot grasping system in a space exploration can achieve better results when a number of heterogenous sensors are attached to the gripper with designated control strategies when Fuzzy Logic and Artificial Neural Network are hybridized. Other than this paper, there are also a number of books that analyze the robot grasping system. A.M Zaki et al.[9] studied the performance of fuzzy logic technique and ANFIS system for an intelligent gripper. Triangular and trapezoid memberships are used for the system to investigate the behavior of the gripper. Some guidelines on building the gripper was also proposed and they mainly analyzed the gripper system that used two fingers gripper. They claimed that the time consumption can be decrease as well as improved the slip displacement of the gripper.

This paper attempts to examine the performance of Fuzzy Logic Control to lift known objects considering an object shape. Our work extends the analysis performed by [3] to understand the effects of membership function to the grasping system. Besides, the work considers three fingers type gripper compared to the literatures presented above. Until now, in our best knowledge the analysis is still unavailable even though the membership function is known to be one of the factors that leads the decision. This is probably due to the triangular membership function is easier and has faster computation time when compares to other membership types. In contrast to these claims, our work has identified that the computation time also can be faster for gaussian membership. Therefore other membership function should also examine to understand how they behave in different conditions. It was also claimed that the triangular membership is lacking of linguistic interpretation in comparison to gaussian approach[10]. Hence, the gaussian membership could provide better results in pursuing better grasping for the system by incorporating the human-like behavior in the system in the sense of decision making.

To demonstrate the results, three different cases of different fuzzy sets effects are observed to identify the Fuzzy Logic Controller influence in controlling the gripper. The analysis through MATLAB Simulink and SimMechanics Toolboxes virtually identifies the best selections of fuzzy sets design to be incorporated into the grasping system when a cube shape and rectangle shape objects are lifted. The selections are based on the time consumed during operation. Analysis through SimMechanics is still unavailable which considerably makes this investigation become interesting as designer could design the system with exact parameters before realizing the design. Hence, the design will improve the performance with less expenditure.

2. Fuzzy Logic Controller Design

In our system, we divided the analysis into two different parts. The first part will only consider the object shape as our only input and the output is the angle orientation while the second part analyzes the performance when two inputs are fed into the system. These cases are organized in order to demonstrate easier on how actually the fuzzy sets membership affects the performance of the system. The fuzzy logic is designed based on three stages as listed below;

- Fuzzification

  This stage defines the fuzziness of the object size. Three fuzzy set boundaries are designed for the object shapes; small, medium and large and ranges between $0cm$ to $10cm$. For the output, the thumb finger angle has also three fuzzy set; small, medium and large which ranges between $20°$ to $50°$. This range was identified in correspond to the object size. Based on these memberships, the fuzzy rule are constructed. The fuzzification is shown in table 1. The simplest fuzzification is considered to easily shown the impact of the fuzzy condition to the robot grasping system.

Table 1. Fuzzy Logic : Fuzzy sets Input

Cube size	Angle movements
Small	Large
Medium	Medium
Large	Small

- Rule Evaluations
  This step calculates the fuzziness of the system. By using the Mamdani technique in Fuzzy Logic toolbox in Simulink, the rules are evaluated to identify the best position of robot fingers to be allocated. It is expected that, based on the evaluations, the robot can grasp effectively without dropping the object. Notice that, the rules are constructed to control the robot finger angle to move in the specified value. Only three rules are considered as shown in table 2. The fuzzy set membership for the inputs which are the cube dimensions or sizes is classified into three categories as mentioned in above. Similar settings are done for the other two fingers.

Table 2. Fuzzy Rule for the thumb finger

IF cube is small THEN angle movement is large
IF cube is medium THEN angle movement is medium
IF cube is large THEN angle movement is small

- Defuzzification
  This process determines the output of the system. The results must be able to identify the best gripping angle that the robot can perform. The simulink and fuzzy logic toolbox are used to simulate the performance of our design.

The work only focusses on a three fingers robot grasping system that has a thumb finger and two other fingers. Each of the finger has two degree of freedom(d.o.f) which we would like to examine based on the proposed fuzzy membership function. As explained above the fuzzy logic steps, the system attempts to understand the performance of different fuzzy sets membership in selecting the best method for use in the grasping system. Even though the fuzzy rules are seems very simple, but it provide a general pictures that by choosing appropriate membership function, the result would become better especially when a best grasping is considered. In addition, remark that the works do not examine in details about the force being applied when lifting the object using the gripper to examine the robustness of the system.

3. Preliminary Results and Discussion

In this first session, an object with a cube shape is only considered as an input for analysis purposes of the fuzzy controller design. The three fingers gripper is designed through MATLAB Simulink and SimMechanics toolboxes with specific weights and dimensions. The model developed by SimMechanics is shown in fig.1 stating the initial position of the gripper and fig.2 for the final position of the gripper. The initial position describes the position of the gripper before grasping the object. To grasp the given object, the gripper then moves as shown in fig.2. At the same time, each finger is configured to pick the objects with pre-described force which is priorly defined to avoid any damage to the object. The gripper is also has been programmed to be at least able to lift the object. Remark that the gripper and objects are located in a known position before task operation. This work does not considers any sensors input for the system to operate but only depends on the pre-determined size of the cube size or dimension. Besides, the gripper is expected to be able to grasp the object with suitable force without damaging the object.

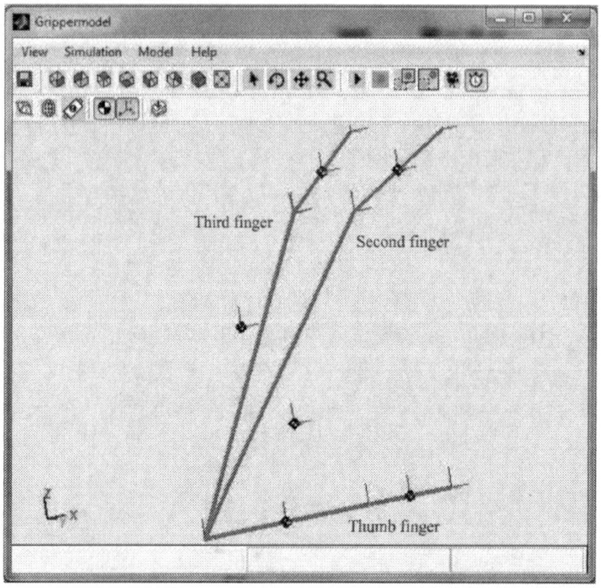

Figure 1. Gripper initial position

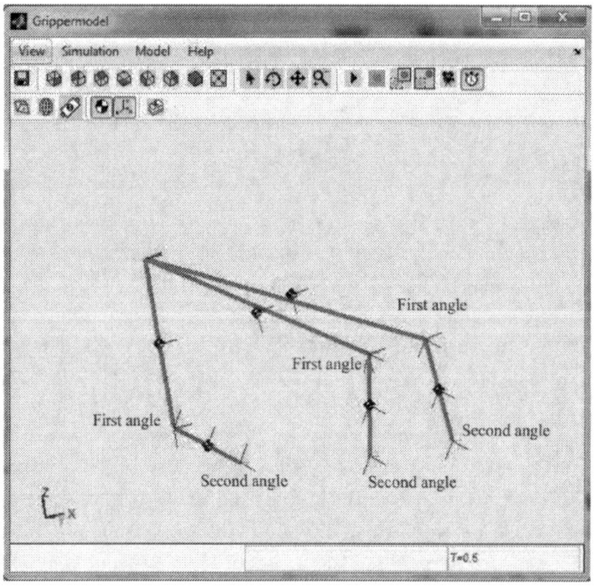

Figure 2. Gripper final position

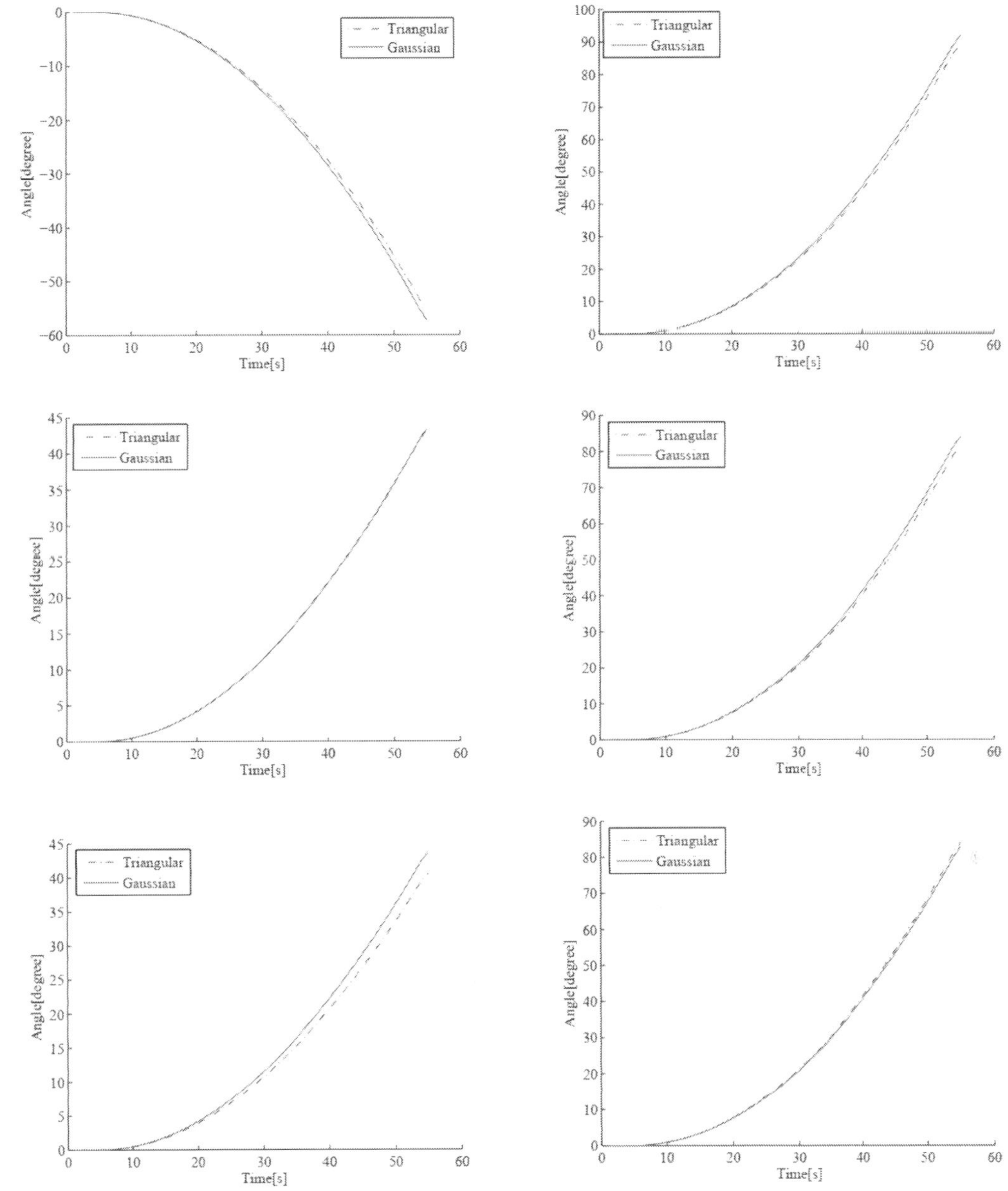

Figure 3. The performance comparison between gaussian membership and triangular membership fuzzy sets for robot thumb and fingers. The first upper two figures demonstrate the thumb fingers for first(left) and second angle(right). This is then followed by the other two fingers with each of it angle position

Figure 3 describes the angle movements for each fingers for the three fingered gripper when the gripper attempts to grasp a cube. The angle movements for the gaussian membership shows slightly bigger movements compared to the triangular membership. In other words, the gripper has better and firm grasp to the object if the gaussian membership is applied.

One of the other factors to be examined is the time required for the grasping system to operate. This issue is very significant in pursuing lower cost and increasing the products lifting especially in manufacturing industries that employs the gripper system. Our investigation has find that the average times taken for the robot to grasp the objects are shown in Table 1. The time consumption is calculated based on the average results of simulation after examining the results for five different simulation outputs.

Table 3. Comparison of time consumption for different fuzzy membership types

Fuzzy Membership functions	Gaussian	Triangular	Trapezium
Time consumed[s]	2.637038	2.9293148	2.8996832

The works also explored the conditions when trapezoid membership function is used for fuzzy set membership. Our findings are still consistent with the our previous results which illustrated that the gaussian membership is the best solution even if the trapezoid function is referred.

To support and further evaluates the performance, system with two inputs was also taken into account for verifications. Other than the cube dimension as input, the analysis is also looking into the effects of material types whether soft or hard type. The fuzzy rules are created to demonstrate how the gripper will moves if both of these inputs are feed into the system. The fuzzy rules are listed as below.

- IF the cube is small AND material type is soft, THEN theta is large
- IF the cube is small AND material type is hard, THEN theta is very large
- IF the cube is medium AND material type is soft, THEN theta is medium
- IF the cube is medium AND material type is hard, THEN theta is large
- IF the cube is large AND material type is soft, THEN theta is small
- IF the cube is large AND material type is hard, THEN theta is medium

The surface viewer for both fuzzy membership types are shown in fig.4 below. From the simulations, the angle movements and final position for thumb finger do not show too much differences. However, it is recognized that the second finger has exhibit better pictures of the performance as presented in fig.5. Both angle 1 and angle 2 on the second finger for gaussian membership function shows better grasping when a hard cube is lifted. It is also expected that even if the fuzzy sets for the input is increased especially for the material type, then the merely same results should be achieved. The same outcomes is also possible for the number of fuzzy sets in input 1; cube dimension. It can be concluded by referring to these figures that the gaussian membership offers better solution for researcher in robot grasping system in terms of reducing operation time and better decision of the fingers movement when collecting a cube size objects. The main reason to this results could be due to the interpretation of linguistic variable as been explained in the previous section that the triangular membership function cannot appropriately simulates the variable. Furthermore, the gaussian membership function is somehow very near to the human behavior in making decision. Even though many researcher has used the triangular membership for their system, now they should also consider the gaussian membership in pursuing their expected outcomes.

IOP Conf. Series: Materials Science and Engineering 53 (2013) 012051 doi:10.1088/1757-899X/53/1/012051

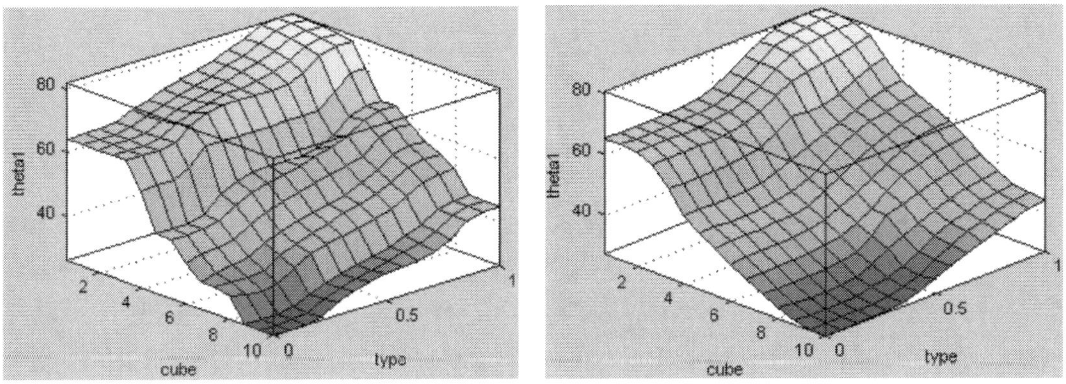

Figure 4. The surface viewer for both triangular membership(left) and gaussian membership(right)

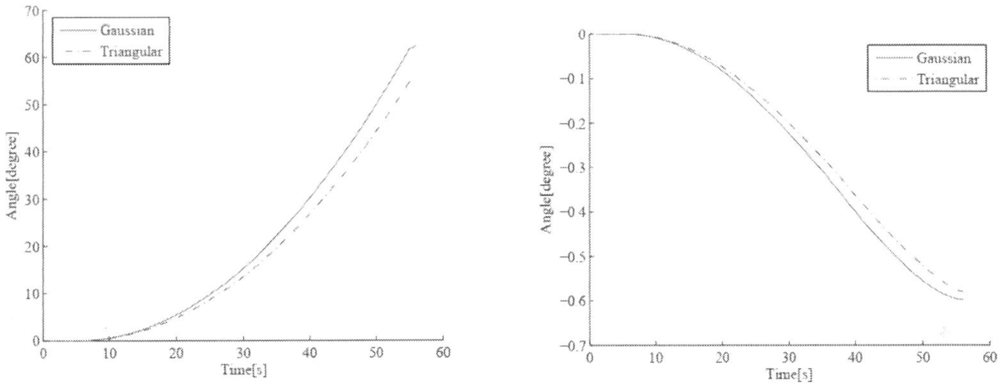

Figure 5. The performance comparison between gaussian membership and triangular membership fuzzy sets for second finger. The left side figure show the movements for first angle and the right side figure illustrate the second angle of the finger

4. Conclusion

This paper discussed the performance evaluation between fuzzy set membership function in a robot grasping system by using fuzzy logic controller. A fuzzy logic based robot grasping system that has three fingers are used for evaluations which consists of a thumb finger and other two normal fingers. There are two types of memberships that was mainly analyzed which are the triangular membership and the gaussian membership. The investigation falls into two aspects, which first determines the performance when only one single input is considered and then proceed to a case when two inputs of the system. Both cases suggested that gaussian membership can provide better results in comparison to the triangular membership function or the trapezoid membership function. Nevertheless, further analysis must be carried in analyzing the actual force exhibit by each fingers especially when fragile objects needs to be lifted.

4.1. Acknowledgments

Authors would like to thank UMP for the continuous supports on this research.

References

[1] S.R. Khuntia, K.B. Mohanty, S.Panda and C.Ardil, "A Comparative Study of P-I, I-P, Fuzzy and Neuro-Fuzzy Controllers for Speed Control of DC Motor Drive",International Journal of Electrical and Computer Engineering, 2010.

[2] Bayro-Corrochano, E., "Object Manipulation using Fuzzy Logic and Geometric Algebra", 18th International Conference on Pattern Recognition, 2006. ICPR 2006, pp. 1120 - 1123.

[3] Aydin, K.K, "Fuzzy logic, grasp preshaping for robot hands", Third International Symposium on Uncertainty Modeling and Analysis, 1995, and Annual Conference of the North American Fuzzy Information Processing Society. Proceedings of ISUMA - NAFIPS '95, pp. 520 - 523.

[4] Palm, R.,"Learning of Grasp Behaviors for an Artificial Hand by Time Clustering and Takagi-Sugeno Modeling", IEEE International Conference on Fuzzy Systems, 2006 , pp. 291 - 298.

[5] Glossas, N.I.,"A cluster based fuzzy controller for grasp and lift fragile objects", 18th Mediterranean Conference on Control & Automation (MED), 2010, pp. 1139 - 1144.

[6] R.Alberto, W.Xia, "Fuzzy Logic In Robot Grasping Control",IEEE/RSJ International Workshop on Intelligent Robots and Systems IROS 1991, pp. 1632 -1637.

[7] L. Minzhou Luo, L.Wei, S.Bingyu,M.Tao, "Autonomous Grasping of a Space Robot Multisensory Gripper",2006 IEEE/RSJ International Conference on Intelligent Robots and Systems, pp. 1014 - 1019.

[8] N.I. Glossas , N.A. Aspragathos,"A Cluster Based Fuzzy Controller for Grasp and Lift Fragile Objects", 18th Mediterranean Conference on Control & Automation, 2010, pp. 1139 - 1144.

[9] A.M. Zaki, O.A. Mahgoub, A.M. El-Shafei and A.M. Soliman, "Control of Efficient Intelligent Robotic Gripper Using Fuzzy Inference System" Book chapter 5 of Fuzzy Inference System - Theory and Applications, Intech Open, 2012, pp. 85 - 112.

[10] Vincent O. S. Olunloyo, Abayomi M. Ajofoyinbo, Oye Ibidapo-Obe, "On Development of Fuzzy Controller: The Case of Gaussian and Triangular Membership Functions", Journal of Signal and Information Processing, Vol.2, pp. 257-265, 2011.

Parameter Estimation of a Closed Loop Coupled Tank Time Varying System using Recursive Methods

Siti Nora Basir[1,3,5], Hazlina Selamat[2], Hanafiah Yussof[1], Nur Ismarrubie Zahari[4,6], Syamimi Shamsuddin[1]

[1]Centre of Excellence for Humanoid Robots and Biosensing, Faculty of Mechanical Engineering, Universiti Teknologi Mara, Selangor, Malaysia

[2]Faculty of Electrical Engineering, Universiti Teknologi Malaysia, Johor, Malaysia

[3]Faculty of Electrical & Electronic Engineering, Universiti Tun Hussein Onn Malaysia, Johor, Malaysia

[4]Faculty of Engineering, Universiti Putra Malaysia, Selangor, Malaysia

E-mail: sitinora.basir@gmail.com

Abstract. This project investigates the direct identification of closed loop plant using discrete-time approach. The uses of Recursive Least Squares (RLS), Recursive Instrumental Variable (RIV) and Recursive Instrumental Variable with Centre-Of-Triangle (RIV + COT) in the parameter estimation of closed loop time varying system have been considered. The algorithms were applied in a coupled tank system that employs covariance resetting technique where the time of parameter changes occur is unknown. The performances of all the parameter estimation methods, RLS, RIV and RIV + COT were compared. The estimation of the system whose output was corrupted with white and coloured noises were investigated. Covariance resetting technique successfully executed when the parameters change. RIV + COT gives better estimates than RLS and RIV in terms of convergence and maximum overshoot.

1. Introduction

Since the past decades, the development in closed loop system identification has shown much improvement in several important aspects. There are many conditions in which identification in open loop is difficult and not practical. This includes the validation, robustness, stability; which is unstable in open loop, and also the significant drift of the plant [1]. One more disadvantage of open loop identification is that the process has to be taken out of commission while the test is being carried out [2]. It will make the identification process more complicated especially in industrial production processes, where performing identification experiments in closed loop identification are important due to safety or production restrictions reasons [3].

Because the restrictions gain by open loop identification, the experimental data can only be obtained by closed-loop system identification. But this approach faces some problems that open-loop identification does not. The most fundamental problem in closed loop identification is the correlation between the immeasurable noise and the plant input. Whenever the feedback controller is not

[5] Corresponding author

[6] A part of this project is supported by the Ministry of Higher Education Malaysia (MOHE) under the Fundamental Research Grant Scheme (FRGS).

identically zero, then the input and the noise will be correlated [4]. This is the reason why several methods that work in open loop be unsuccessful when applied to closed-loop data.

Closed loop identification can be categorized into three main approaches, direct, indirect and joint input output [5]. The more frequently used is the direct method [4]. This method disregards any potential feedback and identifies the open-loop system using measurements of the closed-loop process input and output. It has the advantages over the other two closed loop identification method in the optimal statistical properties, the simplicity, and the general applicability [6].

A time varying system has parameters that vary to time, where the variation needs to be tracked. Parameter estimation is a process of determining the system transfer function coefficients relating to different signal in certain system. It is fundamental when the system model is difficult to obtain [7]. To get the estimation of the coefficient, control input and output data frequently used as inputs to the parameter estimator. This is typically carried out by deriving a parametric model of the plant and then applying an algorithm to estimate the unknown parameters.

In order to track time-varying model parameters, recursive identification is frequently applied. This enables an update of the parameters as new measurements become accessible and one does not have to wait the computation of data to do the identification process [8].

2. System Description

There are numerous numbers of controlled systems in this world that need to be estimated in such a way that the controller of the system can be designed. The control of liquid level in a tank and the flow of liquid between tanks are one of a good example of system in the process industries.

Coupled-Tank CTS 001 [9] in Figure 1 is a computer-controlled coupled tank system that is used for liquid level control.

From the input output experimental data, the model structure of the plant will be determined. It will be in a parametric form where the parameters of the system are derived such as in differential equation figure. ARX form will be used in this project since it is the simplest form of parametric module [10] that represents both deterministic and the stochastic part of the system.
Given the coupled tank transfer function is

$$G(z^{-1}) = \frac{Bz^{-1}}{Az^{-1}} = \frac{b_1 z^{-1} + b_2 z^{-2}}{1 + a_1 z^{-1} + a_2 z^{-2}} \tag{1}$$

where

$\theta \quad = [a_1\, a_2\, b_1\, b_2] = [\text{-}1.238\ 0.2404\text{-}0.025\ 0.02486\,]\text{for } 0 < t < 500$

$\quad\quad = [a_1\, a_2\, b_1\, b_2] = [\text{-}0.05566\text{-}0..4333\text{-}0.0075\ 0.0127\,]\text{for } 500 < t < 1000$

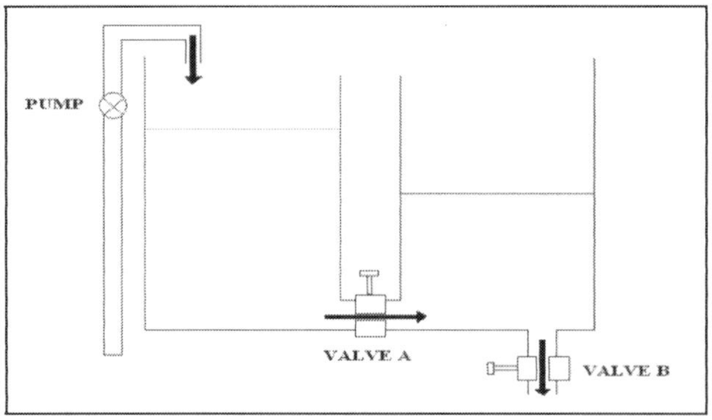

Figure 1: Coupled-Tank CTS 001

3. Recursive Identification Methods

There are two ways of parameter estimation which is online and offline ways. In offline identification, the accumulation of data used to estimate the parameters. All the data are available to be analyzed. On the other hand, online identification is done recursively in time. It means that the parameter estimates is updated recursively within the time limit, which one does not have to wait the group of data to be estimated. In this project, the identification using online approach will be used to estimate the parameter within a given model structure.

The recursive least squares method is one of the most popular adaptive filters [11, 12]. As compared to the least-mean-square (LMS) algorithm, the RLS offers a superior convergence rate, especially for highly correlated input signals. However there would be an increase in the computational complexity. Recursive least squares method has been widely used and it has several advantages such as easy numerical solution and fast parameter convergence [13] and with those advantages it is feasible to apply the recursive least squares methods to the direct closed-loop identification approach [14]. The stochastic approximations method has excellent computational characteristics, but slow convergence, which requires recycling of data to achieve sufficient models. It is also subject to considerable bias in the presence of noise and poor reliability. The recursive instrumental variables (RIV) approach provides unbiased estimation with fast convergence, but lower reliability than achieved by recursive least squares method [15].

3.1. Recursive Least Square

Equations [5] below show how to compute the estimate parameters using RLS identification method.

$$\hat{\theta} = \hat{\theta}(t-1) + K(t)\varepsilon(t) \tag{2}$$

$$K(t) = \frac{(P(t-1)\varphi(t)}{(\lambda + \varphi^T(t)P(t-1)\varphi(t)} \tag{3}$$

$$\varepsilon(t) = y(t) - \varphi^T\,\theta(t-1) \tag{4}$$

$$P(t) = P(t-1) - \frac{P(t-1)\varphi(t)\varphi^T(t)P(t-1)}{\lambda + \varphi^T(t)P(t-1)\varphi(t)} \tag{5}$$

Equation (2) shows that current estimates are computed by adding some adjustment to the previous estimates using residual error, $\varepsilon(t)$. K(t) is a least squares weighting matrix while $\varphi(t)$ is a linear regressor. P(t) is a matrix proportional to the variance of the previous estimates. λ is the forgetting factor that determines how old data are discarded. No data will be forgotten if $\lambda = 1$ and the smaller is its value, the faster old data are discarded.

3.2. Recursive Instrumental Variable

A basic Instrumental Variable was proposed [5] to overcome the drawback of least squares method, which is the estimates are consistent only under restrictive conditions. By doing some modification, the instruments, $z(t)$ will be used instead of $\varphi(t)$. The instruments vector $z(t)$ composed of delayed samples of the reference signal.

$$\hat{\theta} = \hat{\theta}(t-1) + K(t)\varepsilon(t) \tag{6}$$

$$K(t) = \frac{(P(t-1)z(t)}{(\lambda + \varphi^T(t)P(t-1)z(t)} \tag{7}$$

$$\varepsilon(t) = y(t) - \varphi^T \hat{\theta}(t-1) \tag{8}$$

$$P(t) = P(t-1) - \frac{P(t-1)z(t)\varphi^T(t)P(t-1)}{\lambda + \varphi^T(t)P(t-1)z(t)} \tag{9}$$

$$z(t) = [u(t-1)u(t-2)u(t-3)u(t-4)] \tag{10}$$

It can be seen from the equations (6) to (9) that they are very similar to RLS algorithm but $\varphi(t)$ have been changed to $z(t)$ while $\varphi^T(t)$ remains.

3.3. Recursive Instrumental Variable + Centre of Triangle [16]

This method is the modification of Recursive Instrumental Variable method where the new estimation will takes the prior estimates. The theory is by taking the average of three estimations consequently, the estimation overshoot will be reduced and steady state estimation will be obtained.

$$\hat{\theta}(t)_{new_{cot}} = \frac{1}{3}\left(\hat{\theta}(t) + \hat{\theta}(t-1) + \hat{\theta}(t-2)\right) \tag{11}$$

It is hard to produce full convergence theory for time varying case as compared to linear time invariant system [17]. Practically the least square algorithm has fast initial convergence rate, but the algorithm gain reduces radically when the covariance matrix gets small after a few iterations. So to get the strength back, the covariance matrix is reset at various times, which will give the energy back to the algorithm and is useful in maintaining an overall fast convergence rate. This is particularly useful when tracking time varying parameters. The time to reset the covariance matrix is when a significant parameter change has occurred [18].

4. Covariance Resetting Technique

The tracking ability of the parameter estimator depends on the covariance element but its strength is decays within time. To overcome this problem, covariance resetting approach will be employed to revitalize the strength of the tracking variations of parameters. However it is not easy to employ the resetting technique since the time of parameter changing is unknown. One way to detect the changes of parameter is to measure the residual error inside the recursive algorithm. Without the execution of the covariance resetting technique, the residual error will have a sudden change. Covariance resetting technique is executed whenever the algorithm of recursive identification methods perceives changes. One way of ensure the changes is to monitor the residual error movements. Before the covariance resetting technique takes place, the residual error, Equation (8) will face some irregular movements when the parameter varies. These irregularities then used to execute covariance resetting algorithm, Equation (12). Figure 2 shows the irregularities of residual error after the parameter changes at t = 500.

Given the algorithm with covariance resetting technique as follows:

$$P(t) = \begin{cases} P(t-1) - \dfrac{P(t-1)\varphi(t)\varphi^T(t)P(t-1)}{\lambda + \varphi^T(t)P(t-1)\varphi(t)}, |\varepsilon| < 0.02 \\ \sigma^2 I, |\varepsilon| \geq 0.02 \end{cases} \tag{12}$$

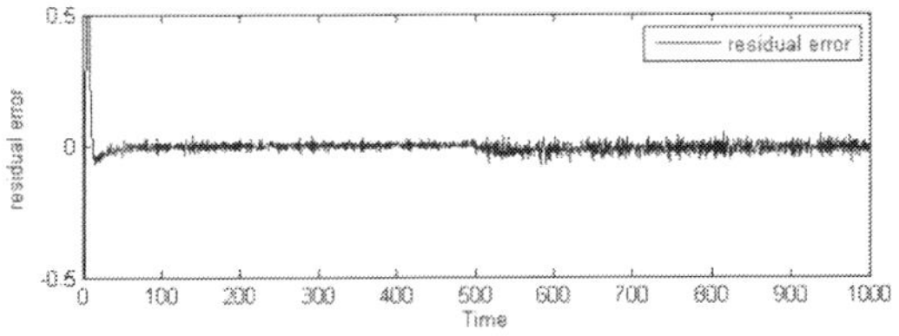

(a)

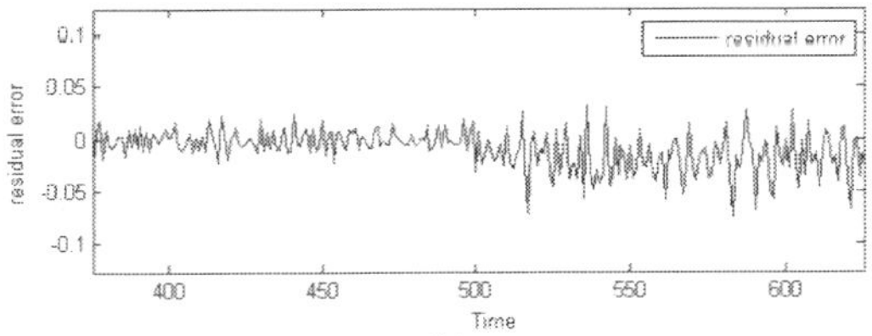

(b)

Figure 2: The occurrence of residual error changes after t = 500 a) residual error b) residual error (zoomed in)

5. Simulation

The main purpose of this study is to compare the performance of three parameter estimation techniques, Recursive Least Squares (RLS), Recursive Instrumental Variable (RIV) and Recursive Instrumental Variable with Centre-Of-Triangle (RIV + COT). Figure 3 shows parameter estimation before the covariance resetting technique is applied.

The estimation still track the previous value without able to detect the changes. This situation happened because the ability of the estimation to track the parameter is decays within time. It is why the covariance resetting should be employed whenever the changes are detected to give those algorithms energy to keep on tracking the parameter.

5.1. Comparison of Parameter Estimation using White Noise as Disturbance

Figure 4 shows the comparison of estimation using three different parameter estimator, Recursive Least Squares (RLS), Recursive Instrumental Variable (RIV) and Recursive Instrumental Variable with Centre-Of-Triangle (RIV + COT) using white noise as disturbance to the coupled tank system.

The parameters estimated using RLS did not converge to its true value and give biased estimation, especially for parameter a_1 and a_2, while for parameter b_1 and b_2, the parameter converge in slow responses. These actions occur before and after the changes of parameters take place. For RIV estimator, the parameters converge to its true value faster than RLS method and also give unbiased estimation. However RIV method has its drawback which is high overshoot of the estimation while its algorithm tracking the parameter changes. The RIV + COT though gives smaller overshoot then RIV method.

Obviously shown that for estimated parameter a_1 and a_2, RLS method have no overshoot, while the other two methods produced overshoot for the estimation. For parameter b_1 and b_2, the RIV + COT gave lower maximum overshoot than RIV method while RLS gave the lowest maximum overshoot. For final value of estimation, RIV and RIV + COT gave better estimates than RLS.

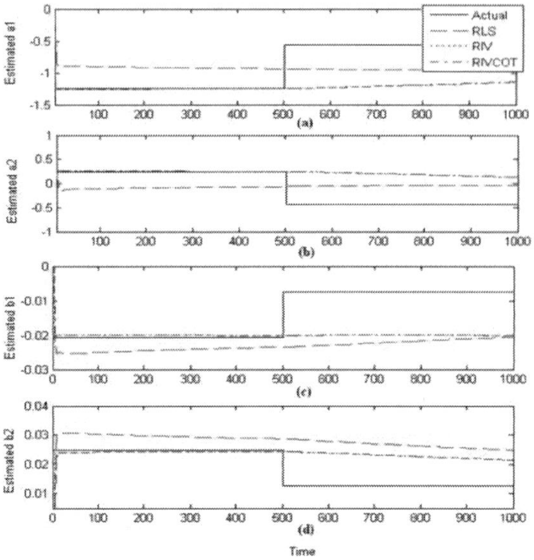

Figure 3: Estimation without covariance resetting employment, a) parameter a_1, b) parameter a_2, c) parameter b_1, d) parameter b_2

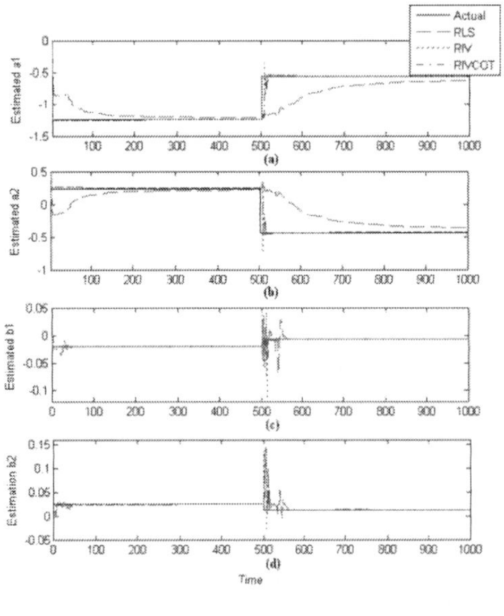

Figure 4: Parameter estimation using white noise as disturbance, a) parameter a_1, b) parameter a_2, c) parameter b_1, d) parameter b_2

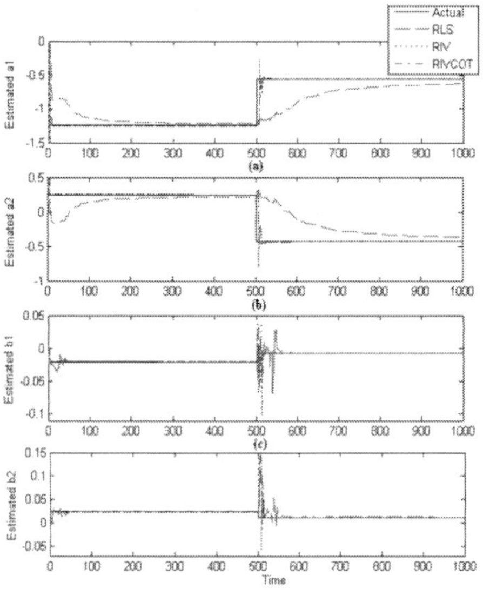

Figure 5: Parameter estimation using coloured noise as disturbance, a) parameter a1, b) parameter a2, c) parameter b1, d) parameter b2

IOP Conf. Series: Materials Science and Engineering **53** (2013) 012052 doi:10.1088/1757-899X/53/1/012052

Table 1 presents the maximum overshoot and final value of estimation produced for the parameter estimation using three recursive identification methods.

Table 1: Parameter estimation using white noise as disturbance

Parameter	Method	Maximum overshoot	Final Value Actual	Estimated	Error percentage
a_1	RLS	-	-0.5566	-0.6257	12.4147%
	RIV	-0.3499		-0.5586	0.3593%
	RIVCOT	-0.5315		-0.5586	0.3593%
a_2	RLS	-	-0.4333	-0.3635	16.1089%
	RIV	-0.7037		-0.4313	0.4616%
	RIVCOT	-0.4592		-0.4313	0.4616%
b_1	RLS	-0.0699	-0.0075	-0.0074	1.3333%
	RIV	-0.1139		-0.0075	0%
	RIVCOT	-0.0538		-0.0075	0%
b_2	RLS	0.1178	0.0127	0.0130	2.3622%
	RIV	0.1468		0.0127	0%
	RIVCOT	0.1449		0.0127	0%

Table 2: Parameter estimation using coloured noise as disturbance

Parameter	Method	Maximum overshoot	Final Value Actual	Estimated	Error percentage
a_1	RLS	-	-0.5566	-0.6255	12.3787%
	RIV	-0.2709		-0.5582	0.2875%
	RIVCOT	-0.5322		-0.5582	0.2875%
a_2	RLS	-	-0.4333	-0.3637	16.0628%
	RIV	-0.8078		-0.4317	0.3693%
	RIVCOT	-0.4583		-0.4317	0.3693%
b_1	RLS	-0.07	-0.0075	-0.0074	1.3333%
	RIV	-0.1015		-0.0075	0%
	RIVCOT	-0.0588		-0.0075	0%
b_2	RLS	0.1179	0.0127	0.0130	2.3622%
	RIV	0.1472		0.0127	0%
	RIVCOT	0.1445		0.0127	0%

5.2 Comparison of Parameter Estimation using Coloured Noise as Disturbance

Figure 5 shows the estimation results using three different parameter estimator, RLS, RIV and RIVCOT using coloured noise as disturbance.

Table 2 presents the maximum overshoot and final value of estimation produced for the parameter estimation using three recursive identification methods. Both white and coloured noise as disturbance show almost identical results, where conduct to another analysis to differentiate the effect of each of the noise.

5.3 Analysis on Different Noise Power

Further analyses were made to get a clearer difference between white and coloured noise as disturbance by using higher and lower noise power. The noise power represents the height of power spectrum density of the white noise. The estimation method to be considered is only RIV + COT method. Figure 6 shows that all of the parameters that have to be estimated, a_1, a_2, b_1 and b2 using both types of noises with 1×10^{-5} g^2/Hz noise power converges almost at the same time, with coloured noise is faster than white noise with white noise gave higher overshoot. Figure 7 shows the result using lower noise power which is 1×10^{-8} g^2/Hz. All of the parameters that have been estimated converge almost at the same time with coloured noise converges faster than white noise but gave higher overshoot of estimation.

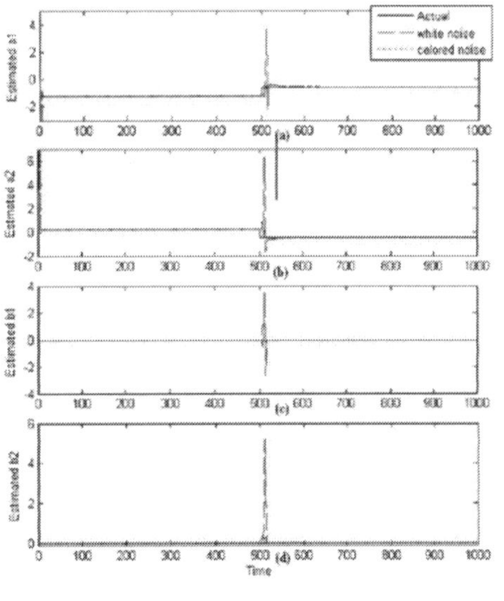

Figure 6: Convergence comparison between white and coloured noise with 1×10^{-5} g^2/Hz noise power, a) parameter a_1, b) parameter a_2, c) parameter b_1, d) parameter b_2

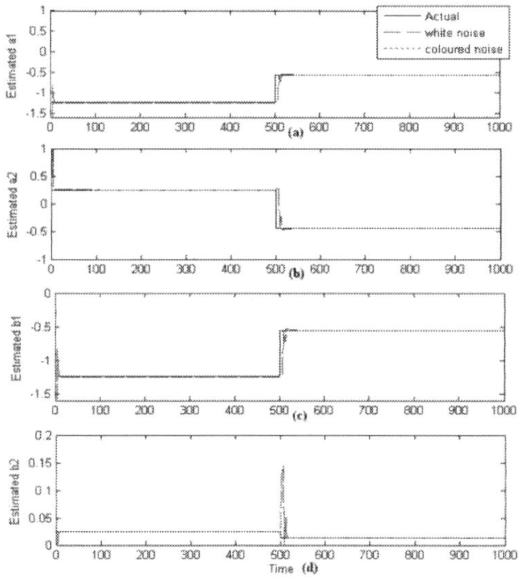

Figure 7: Convergence comparison between white and coloured noise with $1 \times 10^{-8} g^2$/Hz noise power, a) parameter a_1, b) parameter a_2, c) parameter b_1, d) parameter b_2

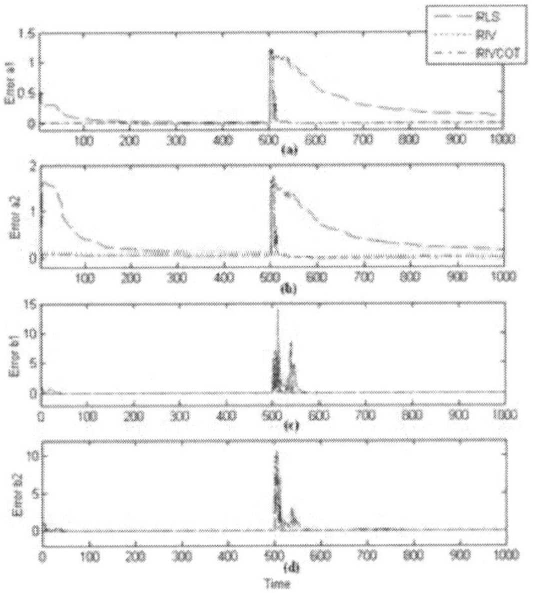

Figure 8: δ values for RLS, RIV and RIV + COT using white noise a) parameter a_1, b) parameter a_2, c) parameter b_1, d) parameter b_2

5.4 Error Analysis

From Figure 8, it is obviously shown that for all parameters, RLS method gives more error, δ in the estimation especially for parameter a_1 and a_2, while smaller value of δ given by RIV and RIV + COT methods.

6. Conclusion

The execution of covariance resetting technique is fundamental when recursive methods are applied to a time varying system estimation. An algorithm has been proposed to execute the covariance resetting technique when the time of the parameter changing is unknown. The results show that the covariance resetting technique is successfully executed when the parameters changed for every method. The parameters estimated using RLS did not converge to its true value and give biased estimation for both before and after the changes of parameters take place. The parameters converge to its true value for RIV and RIV+COT cases faster than RLS method and also gives unbiased estimation but then produced higher overshoot of the estimation when the parameter changes. RIV + COT methods though give smaller overshoot then RIV method. It is also shown that for final value of estimation, RIV and RIV + COT gave better estimates than RLS for all parameters that have been estimated. The estimation using higher power of noise gave higher overshoot than the lower power of noise. Based on the work undertaken in this project, the following suggestions for future works are recommended. Improve the estimation by using other controller such self tuning controller or fuzzy controller and improve the covariance resetting technique using statistical methods.

7. Acknowledgement

The authors would like to thank Ministry Of Higher Education Malaysia, Universiti Tun Hussein Onn Malaysia, Universiti Teknologi Malaysia, Universiti Teknologi Mara and Universiti Putra Malaysia for their support. A part of this project is supported by the Ministry of Higher Education Malaysia (MOHE) under the Fundamental Research Grant Scheme (FRGS).

8. References

[1] Landau I. D. and Karimi, A. (1997). Recursive algorithms for identification in closed loop: a unified approach & evaluation. *Automatica.* **33** 1499 -1523.
[2] Kealy, T. and O'Dwyer, A. (2002). Comparison of open and closed loop process identification techniques in the time domain. *Proc. of the 3rd Wismarer Automatisierungs Symposium.* **1** 3 - 4.
[3] Gilson, M., and Van den Hof, P. (2005). Instrumental variable methods for closed-loop system identification. *Automatica.* **41** 241 - 249.
[4] Forssell, U. and Ljung, L (1999). Closed-Loop identification revisited. *Automatica.* **35** 1215–1241.
[5] Söderström, T. and Stoica, P. (1989). *System Identification.* Englewood Cliffs, NJ: Prentice-Hall.
[6] Zhou, Y., Han, A., Yan, S. and Chen, X. (2006). A fast method for online closed loop system identification. *International Journal of Advance Manufacturing Technology.* **31** 78 - 84.
[7] Razap, A. N. A. and Selamat, H. (2009). Parameter estimation of a time-varying system operating in closed-loop. *The Second International Conference on Control, Instrumentation and Mechatronic Engineering (CIM09).*
[8] Waller, M. and Saxen, H. (2000). Estimating the degree of time variance in a parametric model. *Automatica* **36** 619 - 625.
[9] Coupled-Tank Liquid Level Computer-Controlled Laboratory Teaching Package: Experimental and Operational (Service) Manual; Augmented Innovation Sdn. Bhd., Kuala Lumpur, Malaysa
[10] Taib, M. N., Adnan, R., and Rahiman, M. H. F. (2007). *Practical System Identification.* Faculty of Electrical Engineering, UiTM Shah Alam
[11] Haykin, S. (2002). *Adaptive Filter Theory*, 4th ed. Englewood Cliffs, NJ: Prentice-Hall,
[12] Benesty, J. and Huang (2003). *Adaptive Signal Processing-Applications to Real-World Problems.* Springer-Verlag.
[13] Hang, C. C., Lee, T. H., and Weng, K. H. (1993). Adaptive control. instrument society of America, Research Triangle Park, NC.
[14] Eker, I. Vural, M. (2003). Experimental online identification of a three-mass mechanical system. *Proc. of the 2003 IEEE Conference on Control Applications.* **1** 60 - 65.
[15] Robertson, D. G. and Lee, J. H. (2002). On the use of constraints in least squares estimation and control. *Automatica* **38** 1113 - 1123.
[16] Razap, A. N. A. A., Selamat, H., Yusof, R., & Sam, Y. M. (2011). Improved recursive instrumental variable parameter estimator in the self-tuning control of a couple tank system. *Int. Journal of Innovative Computing, Information and Control.* **7** 5699 – 5715
[17] Goodwin G.C. and Elliot H. (1983). Deterministic convergence of a self tuning regulator with covariance resetting. *IEEE Proceeding* **130**.
[18] Goodwin G.C. and Kwai S.S. (1984). *Adaptive Filtering Prediction and Control.* Prentice Hall,Inc., Eaglewood Cliffs, New Jersey.

Analysis of rain fade duration models for Earth-to-satellite path based on data measured in Malaysia

Hassan Dao, Islam Md Rafiqul[1] and Khalid A. S. Al-Khateeb

Electrical and Computer Engineering Department, Kulliyyah of Engineering
International Islamic University Malaysia (IIUM)
Kuala Lumpur, Malaysia

E-mail: rafiq@iium.edu.my;sun.spu@gmail.com

Abstract. Statistical analysis of rain fade duration is crucial information for system engineer to design and plan a fade mitigation technique (FMT) for the satellite communication system. An investigation is carried out based on data measured of one year period in Kuala Lumpur, Malaysia from satellite path of MEASAT3. This paper presents statistical analysis of measured fade duration on high elevation angle (77.4°) in Ku-band compared to three prediction models of fade duration. It is found that none of the models could predict measured fade duration distribution accurately.

1. Introduction

Duration statistical knowledge such as probability of duration for rain fades at a given attenuation, what time and how frequent it occurs and when it may be next happened is important key to get insight into the performance availability and quality of service. Service provider must consider such aspects to implement an appropriate forward error correction (FEC), choice of modulations and the range of adaptive power control (APC) during severe rain fade events in the overall design of their satellite communication systems, especially for the low-margin and Ku-band and higher frequencies systems [1].

The number of events for low attenuations and short durations mainly depends on scintillation and receiver noise amplitude. In literatures, different values are usually proposed to identify and separate fade events into short and long durations between 10 and 30s [2]. The number of events of filtered and unfiltered data of this link was already shown in [3]. In this case, duration above 30s was effective for separating long duration caused by attenuation from short duration due to fast fluctuations of scintillation.

Many works have been studied and published on fade duration statistics both satellite and terrestrial path links whereas it is no model for interfade duration. All models fundamentally developed a cumulative distribution based on power law, lognormal and two lognormal distributions.

This paper is presented statistical analysis of fade duration based on measured data compared to three prediction models of cumulative distribution function (CDF), namely ITU-R, Timothy and Cheffana-Ayama models [4-6]. Testing method for comparing error of fade duration predictions were also calculated and presented in this paper.

[1] Islam Md Rafiqul: rafiq@iium.edu.my

2. Measurement setup and data processing

The measurement was monitored the MEASAT3 signal at 10.982 GHz (Ku-band), vertical polarization with QPSK modulation and was sampled for time interval length of 0.1s and subsequently processed to 1s time interval length. The signal was collected for 12 months during 1st September 2011 to 31st August 2012 located in the International Islamic University Malaysia (IIUM), Gombak, Kuala Lumpur. Tipping bucket rain gauge of 0.2mm was synchronized with time of attenuation logging system to determine raining events. Details of the experimental setup are available in [7-8].

3. Fade duration distribution models

The statistics of fade duration distribution is analysed from unfiltered measured data of 1s time interval length. The data points for calculating the number of events of fade duration *N(d>D | a>A)* and probability of occurrence *P(d>D | a>A)* are preferably used corresponding ITU-R P 311-12 [9], namely 1, 10, 30, 60, 120, 180, 300, 600, 900, 1200, 1500, 1800, 2400 and 3600 s, respectively. The number of events *N(d>D | a>A)* statistics and calculation method of probability of occurrence *P(d>D | a>A)* are already presented in [3].

3.1 ITU-R model. ITU-R model is able to calculate fade duration statistics including effect of gases, clouds, rain and scintillation for earth-satellite path links. The model presents long-term of fade duration follows a lognormal distribution and short-term of fade duration follows a power-law distribution [4].

Calculation of *P(d>D | a>A)* of duration *d* longer than *D* given that attenuation *a* is greater than *A* as (1-2).

$$\text{For } 1 \le D \le D_t \qquad P(d > D|a > A) = D^{-\gamma} \tag{1}$$

$$\text{For } D > D_t \qquad P(d > D|a > A) = D_t^{-\gamma} \cdot \frac{Q\left(\dfrac{\ln(D) - \ln(D_2)}{\sigma}\right)}{Q\left(\dfrac{\ln(D_t) - \ln(D_2)}{\sigma}\right)} \tag{2}$$

Where: exponent γ of the power-law distribution of the fraction of fading time due to fades of short duration and D_t, is boundary between short and long fade durations. Details of the ITU-R model are available in [4].

3.2 Timothy model. The model [1, 5] developed by normalizing a lognormal distribution that depends on the average fade duration $\overline{D}$. If fractions of fade events exceeding the normalised fade duration, the fade duration distribution approximation is given by (3).

$$P(D_f \ge x) = \frac{1}{2} erfc\left(\frac{\ln x - \ln \overline{x}}{\sigma\sqrt{2}}\right) \tag{3.1}$$

$$= \frac{1}{x\sigma\sqrt{2\pi}} \int_x^{\infty} \exp\left[-\frac{1}{2}\left(\frac{\ln x - \ln \overline{x}}{\sigma}\right)^2\right] dx \tag{3.2}$$

Where $x = D/\overline{D}$ is the normalised fade duration, $\overline{D}$ is the average fade duration for a particular threshold, σ is the standard deviation of $\ln(D/\overline{D})$, *erfc* is the complementary of error function, and $\overline{\ln x}$ is the mean of $\ln x$.

IOP Conf. Series: Materials Science and Engineering **53** (2013) 012053 doi:10.1088/1757-899X/53/1/012053

3.3 Cheffena-Ayama model. The model [6] was modelled and tested the database created covering Ku, Ka, and Q/V frequency bands, elevation angles between 14°-89° in various climatic conditions such as tropical, subtropical and temperate. The model was developed the prediction model of a cumulative distribution function (CDF) of probability of occurrence of fade duration, by the sum of two lognormal functions. The conditional fade duration distribution is given by (4).

$$P(d > D \mid a > A) = \alpha \frac{Q\left(\frac{\ln(D/m_s)}{\sigma_s}\right)}{Q\left(\frac{\ln(1/m_s)}{\sigma_s}\right)} + (1-\alpha)\frac{Q\left(\frac{\ln(D/m_r)}{\sigma_r}\right)}{Q\left(\frac{\ln(1/m_r)}{\sigma_r}\right)} \quad (4)$$

Where: D ≥ 1 (s) and Q is the standard cumulative distribution function for a normal distributed variable as same as equation in [4].

$$\alpha = 0.8881 - 0.3168\beta + 0.1636\exp\left(-\frac{A}{2.61}\right) \quad (5)$$

A is the attenuation threshold (dB), β is the rain convectivity parameter obtained from recommendation ITU-R P.837 [10], f is frequency (10-50 GHz) and ε is elevation angle (5°-90°).

The mean m_s and standard deviation σ_s of the first lognormal function are defined as

$$m_s = 0.3636 - 2.0411\cdot 10^{-6} f^3 + 11117\exp(-\varepsilon) \quad (6)$$

$$\sigma_s = 1.6462 + 29.8038\exp\left(-\frac{f}{3.5}\right) - 1.3671\cdot 10^{-6} A^3 \quad (7)$$

Similarly to the second lognormal function, mean m_r and standard deviation σ_r are expressed as

$$m_r = 686.59 - 173.51\log(f) \quad (8)$$

$$\sigma_r = 0.621 + 4.3516\cdot 10^{-3} f^{1.5} + 3.3637A^{-2} \quad (9)$$

4. Comparison of fade duration distribution models and measurement

Probability of occurrence $P(d>D \mid a>A)$ of measured data compared to three prediction models is presented in figure 1-3. The ITU-R model predicts the measured statistics quite well for short duration (<30s) except 1dB attenuation as shown in figure 1. However, the model underestimated the measured data for long duration (>30s) that mainly caused by rain attenuation.

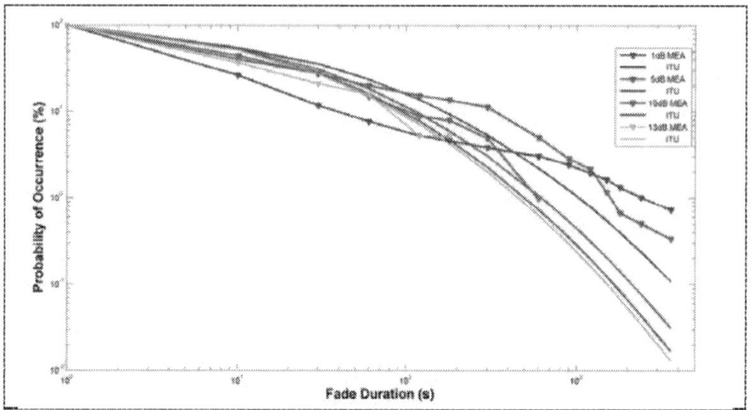

Figure 1. Probability of occurrences P(d>D I a>A)comparison between ITU-R model (ITU) and measured data (MEA) for given attenuation thresholds.

Figure 2 depicts probability of occurrence *P(d>D I a>A)* comparison between Timothy model and measured data for given attenuation thresholds. The Figure clearly shows that the model overestimates both short and long durations for all attenuation thresholds.

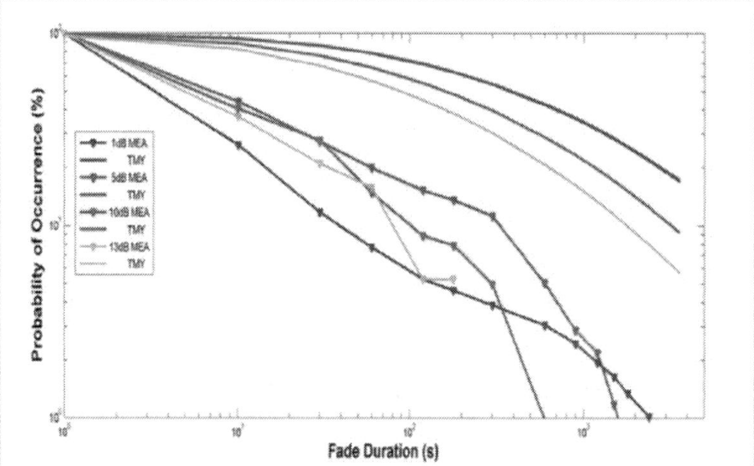

Figure 2. Probability of occurrence *P(d>D I a>A)* comparison between Timothy model (TMY) and measured data (MEA) for given attenuation thresholds.

Cheffena-Ayama model and measured data statistics for given attenuation thresholds is presented in figure 3. In general, the model predicted similar trend to measured data statistics for given attenuation thresholds. However, the model relatively overestimated the measured data for all attenuation thresholds. The overestimation was much lower than Timothy model as shown in figure 2.

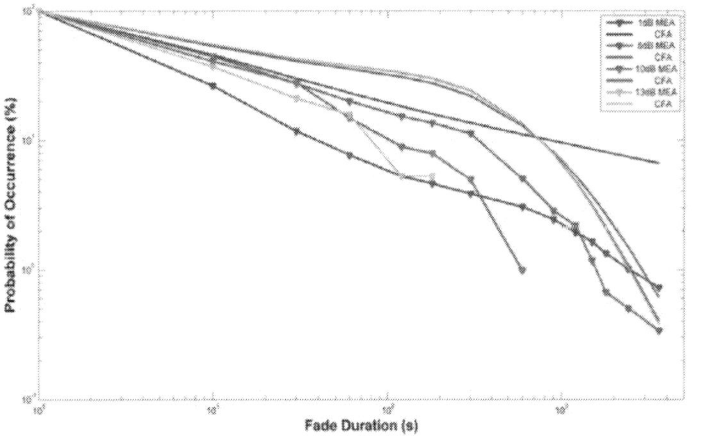

Figure 3. Probability of occurrence $P(d>D \mid a>A)$ comparison between Cheffena-Ayama model (CFA) and measured data (MEA) for given attenuation thresholds.

In order to compare with prediction models, measured data are tabulated for fixed individual duration D in the range of 1 to 3600 s and for fixed attenuation threshold A, namely 1, 5, 10 and 13dB. The test variable is defined by the logarithm ratio of the predicted probability to the measured probability as proposed by ITU-R P.311 [9]. For testing methods of the probability of occurrence $P(d>D \mid a>A)$, the test variable is defined as the natural logarithm of the ratio of predicted probability $P_p(d>D \mid a>A)$ to measured probability $P_m(d>D \mid a>A)$, for each A and D.

The mean, standard deviation and root mean square (RMS) for each attenuation threshold A and each duration D are then calculated the error of different models.

Table 1: Prediction test variables of $P(d>D \mid a>A)$ with different models for given attenuations

Models	A (dB)s	1	5	10	13
ITU-R	Mean	0.369	-0.089	0.030	0.110
	Std	0.096	0.064	0.048	0.017
	RMS	0.161	0.021	0.041	0.063
Timothy	Mean	1.472	1.124	0.892	0.747
	Std	0.085	0.052	-0.303	-0.466
	RMS	0.763	0.656	0.512	0.414
Cheffena-Ayama	Mean	0.284	0.113	0.278	0.301
	Std	0.0001	-0.009	-0.120	-0.172
	RMS	0.069	0.033	0.084	0.104

Three distribution prediction models of fade duration were presented and compared with measured data. Table 1 shows calculation results of test variables of probability of occurrence $P(d>D \mid a>A)$ by mean, standard deviation (Std) and root mean square (RMS) for three models with different attenuations. Calculation details are proposed by ITU-R P.311 [9]. In general, the ITU-R model predicts better than other models followed by Cheffena-Ayama and Timothy, respectively.

5. Conclusion

Rain fade duration data were measured for one year in Kuala Lumpur, Malaysia from satellite path of MEASAT3. The probability distributions of fade duration were found almost similar in short and long durations. Three distribution prediction models of fade duration were presented and compared with measured data. It was found that none of the models predicted measured fade duration distribution accurately. ITU-R predicted the measured data statistics well for short duration except 1dB attenuation whereas long duration, the model underestimated the measured data. Prediction by Timothy was much overestimated for both short and long duration. Cheffana-Ayama model predicted similar trend to the measured data statistics and overestimated for both short and long durations. However, the overestimation by Cheffana-Ayama was much lower than Timothy model.

References

[1] Pan, Q. W., & Allnutt, J. E. 2004. 12-GHz Fade Durations and Intervals in the Tropics. *IEEE Transactions on Antennas and Propagation,* Vol 52(3), p 693-701.

[2] García-del-Pino, P., Riera, J. M., & Benarroch, A. 2011 Fade and interfade duration statistics on an Earth-space link at 50 GHz. *IET Microwave Antennas Propagation,* Vol 5(7), p 790-794.

[3] Dao, H., Islam, M. R., Al-Khateeb, K., & Ismail, A. F. 2012. Analysis of rain fade duration over satellite-earth path at Ku-Band in tropics. *International Conference on International Conference on Computer and Communication Engineering (ICCCE) 2012*, p 953 - 956.

[4] ITU-R P.1623-1 2005 Prediction method of fade dynamics on Earth-space paths, *International Telecommunication Union (ITU).*

[5] Timothy, K. I., Ong, J. T., & Choo, E. B. L. 2000 Fade and nonfade duration statistics for earth-space satellite link in Ku-band. *Electronics Letters,* Vol 36, p 894-895.

[6] Cheffena M., Amaya C. 2008 Prediction Model of Fade Duration Statistics for Satellite Links Between 10–50 GHz, IEEE Antennas and Wireless Propagation Letters, Vol **7**, p 260 - 263.

[7] Dao H, Islam M.R., Al-Khateeb K., 2011 Modification of ITU-R Rain Fade Slope Prediction Model Based on Satellite Data Measured at High Elevation Angle, *IIUM Engineering Journal (IIUMEJ),* Vol 12(5), p 53-59.

[8] Dao, H., Islam, M. R., & Al-Khateeb, K. A. S. 2013. Rain Fade Slope Model in Satellite Path Based on Data Measured in Heavy Rain Zone. *IEEE Antennas and Wireless Propagation Letters,* , Vol 12, p 50-53

[9] ITU-R P.311-12, 2005 Acquisition, presentation and analysis of data in studies of tropospheric propagation, *International Telecommunication Union (ITU).*

[10] ITU-R-P.837-6. 2012. Characteristics of precipitation for propagation modelling. *International Telecommunication Union (ITU).*

A Survey of Bandwidth Utilization: Case Study of Federal University of Technology Minna.

A. Haq[1], H Bello Salau[2], A. M. Aibinu[3] and E. N. Onwuka[4]

[1]Department of Computer Engineering, [2,3,4]Department of Telecommunication Engineering, Federal University of Technology, P.M.B. 65, Minna, Niger State, Nigeria.

E-mail: {[1]arifa, [2]habeeb.salau, [3]abiodun.aibinu and [4]onwukaliz}@futminna.edu.ng

Abstract. The effective utilization of the limited scarce bandwidth resources allocated by a spectrum regulator usually the Nigerian Communications Commissions (NCC) in Nigeria universities is paramount in maximizing the usage of the expensive scarce bandwidth resources. A significant bandwidth allocation is needed in order to meet up with the challenges of the day and the task of networking, communicating and reaching the word in our universities. Therefore, bandwidth management becomes necessary and essential. This paper contributes in that direction by surveying the bandwidth utilization at Federal University of Technology Minna campus with the hope of proffering a general solution that can be adopted in Nigerian universities for effective bandwidth management. Also, factors that hindered the development of most Nigeria universities are also examined. The federal university of technology minna which is structured like all other universities in the country in terms of the bandwidth requirement was choosing as a case study for this research work. Furthermore, some policies which can be adopted in order to effectively manage the scarce bandwidth resources in Nigerian universities are also proposed.

1. Introduction

Internet connectivity and access to networked information resources are in increasing demand and very paramount in carrying out meaningful research in any educational institution and research institute [1]. In view of this, Nigerian universities are not an exception. The Nigeria University is an establishment where a seat of higher learning is housed, including administrative and living quarters as well as facilities for research and teaching [1]. Internet access is made possible through allocated frequency (bandwidth) usually by spectrum regulators; the Nigerian Communications Commissions (NCC) in Nigeria. Bandwidth is simply used to refer to the capacity of a communication media to transfer data from source to destination [2], [3]. The bandwidth management describes the various techniques, technologies, tools and policies employed in an organization to enable the most efficient utilization of bandwidth resources.

The Internet connectivity supports numerous applications which consume a considerable proportion of the limited allocated bandwidth. This application includes the distribution of email spam and viruses, music and video downloading by music lovers and it's used in pornography viewing. It also serves as a gateway for computer hackers to practice their skills. Furthermore, it facilitates cross-border movement by all kinds of software, agents, worms and spiders harvesting and indexing content or simply tracking usage and offering upgrades [2], [4], [5]. These numerous applications consumed a considerable proportion of the limited allocated bandwidth in our higher educational institutions in Nigeria. Therefore, it slows down the response times and degrades internet network performance [1], [2]. This makes the internet access usage hectic in carrying out any meaningful research work in the universities. Hence, it becomes necessary to address this great challenge of

[1] A. Haq

ineffective bandwidth utilization and management. Therefore, a novel artificial based automated traffic control technique is proposed in this paper to address this (FUTM) that is choosing as a case study for this research work.

An effective bandwidth management and optimization technique is promising in improving the quantity and quality of service of the allocated bandwidth for research and other educational purposes in the institution. It also aims at forestalling network congestion. Furthermore, this paper presents various factors that hindered the development of most Nigeria universities. In addition to this some policies which can be adopted in order to effectively manage the scarce bandwidth resources in Nigerian universities are proffered. The rest of the paper is structure as follow; section 2 presents a concise survey of some bandwidth management technique. Section 3 presents the methodology used in this study. The results analysis and discussion is presented in section 4. The paper is concluded in section 5

2. Concise review of Bandwidth Management Techniques

In this section, a brief and detailed survey of various approaches that have been adopted in managing the scarce and limited allocated bandwidth is presented.

Maduka et.al presents an evidence-based policy-making approach where bandwidth is being optimized in a university network. A network event logs from a campus network was used as a research data [6]. This approach was able to minimize and saves cost. It also improves the network performance of the campus network and an optimum bandwidth management is achieved through this approach. However, it is tedious and difficult to get data from the network logs of the University of Ibadan used as the case study for the research work due to irregular power supply, and lack of up to date equipment that can be used to monitor the network.

Artificial intelligent based techniques for bandwidth management which include traffic shaping, scheduling algorithm is presented in [7]. Also another form of intelligent based method for utilization and bandwidth management is presented in [8], [9]. However, this technique which was effective in improving the network performance in the scenario considered for this study is not realistic in the Nigerian universities setting which is considered in this study due to shortage of power supply.

A considerable research work has also been carried out in using public policy techniques for bandwidth management. This includes: bandwidth prioritization schemes which are often trailed by heavy debates relating to net neutrality and privacy issues as presented in [10], [11]. Another approach for bandwidth management is proposed in [12] which use the promotion of local content, peering and interconnection caching. An economic measure was also used for bandwidth management as carried out in [13]. This economic measure includes users being asked to pay for what they use by metering traffic.

Irrespective of the approach being adopted for bandwidth management, it is essential to clearly understand the characteristics of user activities on the network under consideration which is FUTM network (FUTnet) in this research work. The proper analysis of user activity, bandwidth management need not always be overly technical in order to achieve improved network performance.

3. Research Methodology

A network monitoring was done by logging data of the traffic and network congestion on the routers for a period of three weeks. These logs were collected using a Multi-Router Traffic Grapher (MRTG) across the main router of the FUTnet campus network. The actual bandwidth usage is measured across **nodes on the main router. The** research ports were selected based on the stability of electric power supply to the ports and the capability of the port equipment. The MRTG which is a powerful tool that can combines data collection with graphical trending to monitor the traffic load on a network link is utilized. This help presents the data in a graphical format and the amount of traffic in each section considered in this study plays a vital role in determining and predicting the likely nodes where congestion might occur in future. The collected logs were analyzed at the end of the monitored period

[1] A. Haq

to reveal hard facts that become useful in evidence based policy-making. The MRTG can present the graph obtained for the bandwidth in and out of any simple network management protocol (SNMP)-enabled network device which include servers, routers, firewalls and switches.

4. Results Analysis and Discussions

The results obtained from the MRTG placed on the main router to monitor traffic congestion and network bandwidth usage for a period of three weeks is presented in this section.

4.1. *Plot of Average Graphs of Port Gio/32*: An average five minute plot of graphs for port Gio/32 allocated to staff quarters as obtained on the 12th April is presented in Figure (1) below. The bandwidth in bit/sec was plotted on the vertical axis against the days of the weeks on the horizontal axis. It is observed that the peak network consumption at the staff quarters was calculated to be about 3Mb/s from the data obtained from the MRTG. This bandwidth usage is quite high for a residential place like the staff quarters at the period where by most staff will be in their various offices on the campus thus causing unnecessary traffic concentration and slowing the network down for research and other academic work purposes. Furthermore, a plot of 30 minutes graph is shown in Figure (2) on weekly bases for the staff quarters. It is seen from the plot that the average bandwidth weekly consumption is about 3.2Mb/s which is considerable high at the period where most staff were supposed to be in their various offices. It was also seen from Figure (2) that the bandwidth usage from Tuesday to Friday was very minimal and negligible at about 0Mb/s due to power outage during that period which is one of the greatest predicaments being faced in our institution and most institutions in the country in general.

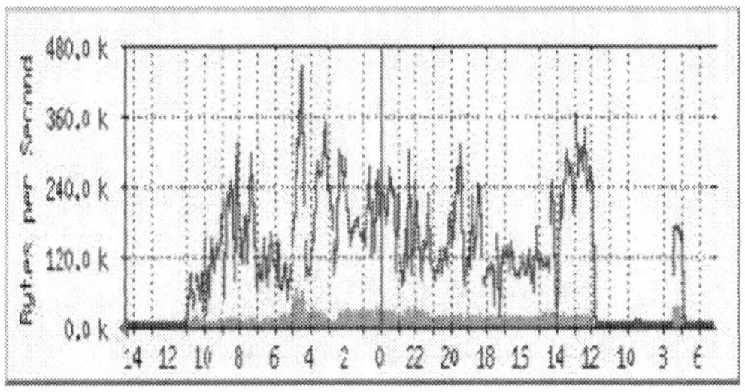

Figure 1: A 5 minute average graphs for port Gio/32 at 12th of April 2013

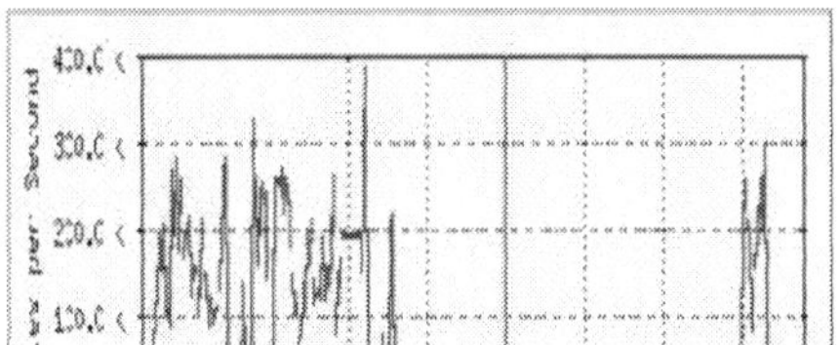

Figure 2: A 30 minute average graphs for port Gio/32 at 12th of April 2013

4.2. *Plot of Average Graphs of Port Gio/41:* The Gio/41 port is the main campus truck of the campus network infact, it is the backbone of the university network. An average five minute plot of graphs for the port Gio/41 as obtained on the 12th April is presented in Figure (3) below. The bandwidth in bit/sec was plotted on the vertical axis against the days of the weeks on the horizontal

[1] A. Haq

axis. It is observed that the peak network consumption at the main port was calculated to be about 13Mb/s from the data obtained from the MRTG. This bandwidth usage is quite high however, it is not out of place because it is expected to be at the peak period at those hours when most staffs are in their offices and students are on campus. Though it is worth noting that if during the holiday periods when this study is carried out that most students are not on campus and the bandwidth consumption is high and closed to the maximum allocated. One may imagine what the situation will be when the school is fully in section thus there is need for developing methods that can check this excessive bandwidth usage without interfering with the progress and speed of utilizing it for research work and other educational purposes. Furthermore, a plot of 30 minutes graph is shown in Figure (4) on weekly bases for main port. It is seen from the plot that the average bandwidth weekly consumption is about 12Mb/s which is considerable high at the period where most staff and students are on holiday. It was also seen from Figure (4) that the bandwidth usage from Tuesday to Friday was very minimal and negligible at about 0Mb/s due to power failure during that period which is one of the greatest challenge militating against most universities in Nigeria.

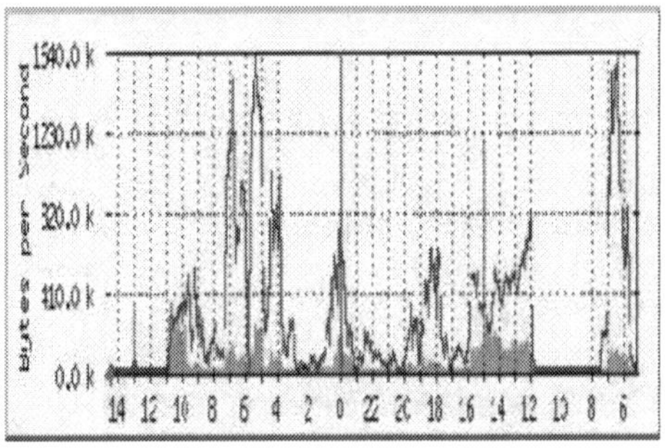

Figure 3: A 5 minute average graphs for port Gio/41 at 12th of April 2013

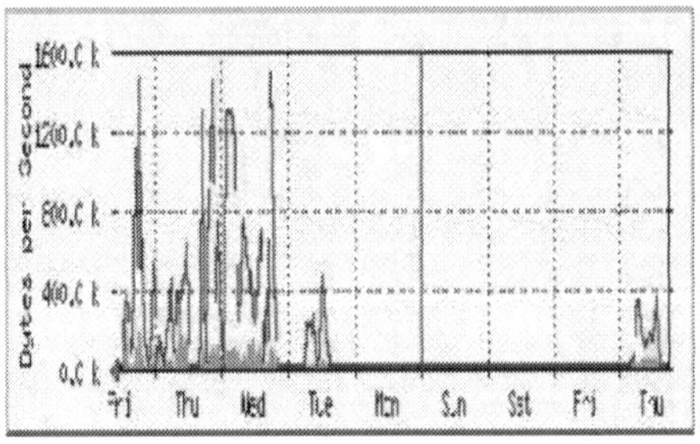

Figure 4: A 30 minute average graphs for port Gio/41 at 12th of April,2013

5. Conclusion and Recommendation

This paper presents an analysis of bandwidth utilization at FUTM with the hope of generalizing for other universities in Nigeria since they all have similar structures in terms of network design, staff and student ratio. It was shown in the paper that a considerable amount of the allocated bandwidth is still being used in places like staff quarters during the off peak hours when most staff are suppose to be in their offices and similarly on campus during the peak hours. This ineffective utilization of bandwidth tends to degrade the quality and output of staff research activities as well as serious minded students who may need the internet access for educational purposes. As a way of addressing this challenge, this paper recommended that some public policy that will serve as a check and balances to the bandwidth usage should be enacted so that higher priority are giving to informative and research article site while high bandwidth consuming

[1] A. Haq

video and audio downloads should be blocked during the peak periods. It is also recommended that an artificial based automated traffic control should be used in the traffic control and bandwidth measurements.

References

[1] A. A. Ojedokun, "Internet access and usage by students of the University of Botswana", African Journal of Library and Archives information Science, Vol. 11, No. 2, pp 93-103, 2001.
[2] F. O. Olabude, "Utilization of Internet sources for Research by Information professionals in Sub-Saharan Africa", African Journal of Library Archives and Information Science, Vol. 17, No. 1, pp 53- 54, 2007.
[3] S. Igun, "Users and Internet skills: A report from Delta State University Abraka, Nigeria" Electronic Journal of Academic and Special Librarianship, 2005.
[4] S. Kim, and P. K. Varshney "Adaptive online bandwidth allocation and reservation for QoS" 2005
[5] R. Flickenger, "How To Accelerate Your Internet". A practical guide to Bandwidth Management and optimization using Open Source Software. INASP/ICTP. October 2006
[6] K. A. Maduka, "Bandwidth management: the public policy approach in a university campus network", Internet governance research paper, Diplo Foundation and European Union, 2012
[7] A. Brzezinski, "Scheduling algorithms for throughput maximisation in data networks". Boston:
[8] MIT Department of Electrical Engineering and Computer Science. pp 226, 2007
[9] S. J. Bauer, "Congestion on the Internet: operator responses, economic analysis, and improving the network architecture". Boston, MIT, Dept of Electrical Engineering and Computer Science, pp156, 2008
[10] M. Allman, V. Paxson, and E. Blanton " TCP Congestion Control". RFC 5681, IETF, September 2009
[11] D. C. Nunziato, "Virtual freedom: net neutrality and free speech in the Internet age". Stanford, Calif. Stanford Law Books. pp 194, 2009
[12] E. D. Cohen, " Mass surveillance and state control", the total information awareness project. 1st ed., New York: Palgrave Macmillan, pp 252, 2010.
[13] AAISOA and CIPACO , "Synthesis of an e-discussion". Dakar, Panos Institute West Africa, pp5, 2005.

[1] A. Haq

IOP Conf. Series: Materials Science and Engineering **53** (2013) 012055 doi:10.1088/1757-899X/53/1/012055

Scorpion image segmentation system

E Joseph[1], A M Aibinu[2], B A Sadiq[3], H Bello Salau[4], M J E Salami[5]
[1,3] Department of Computer Engineering, Federal university of Technology, P.M.B 65 Minna, Nigeria.
[2,4] Department of Telecommunication Engineering, Federal university of Technology, P.M.B 65 Minna, Nigeria.
[5]Department of Mechatronic Engineering, International Islamic University P.O Box 10, 53100, Gombak, Malaysia.

E-mail: elijoe3yk@yahoo.com

Abstract. Death as a result of scorpion sting has been a major public health problem in developing countries. Despite the high rate of death as a result of scorpion sting, little report exists in literature of intelligent device and system for automatic detection of scorpion. This paper proposed a digital image processing approach based on the floresencing characteristics of Scorpion under Ultra-violet (UV) light for automatic detection and identification of scorpion. The acquired UV-based images undergo pre-processing to equalize uneven illumination and colour space channel separation. The extracted channels are then segmented into two non-overlapping classes. It has been observed that simple thresholding of the green channel of the acquired RGB UV-based image is sufficient for segmenting Scorpion from other background components in the acquired image. Two approaches to image segmentation have also been proposed in this work, namely, the simple average segmentation technique and K-means image segmentation. The proposed algorithm has been tested on over 40 UV scorpion images obtained from different part of the world and results obtained show an average accuracy of 97.7% in correctly classifying the pixel into two non-overlapping clusters. The proposed [1]system will eliminate the problem associated with some of the existing manual approaches presently in use for scorpion detection.

1. Introduction

Scorpion envenomation has been a major public health problem around the world. Despite the high rate of death as a result of scorpion sting, the approaches of detection of this poisonous arthropod's have been of the medieval approach [1-6]. Available approaches to scorpion detection are crude and time consuming as it sometimes involves the presence of an investigator. Thus, new approach of detection using image processing is proposed in this work.

Scorpions are arthropods which are poisonous and found in several habitats, they are of the class Arachnida and of the order scorpiones [6]. They are nocturnal and feed at night on a variety of insects; scorpions sometimes used their venom for defence and predation [1].

Scorpions cuticle fluoresces a brilliant cyan-green under ultraviolet (UV) light. Scorpion investigators use this unique characteristic to sample these arthropods at night. Two molecules at their

[1] Corresponding editor: E Joseph elijoe3yk@yahoo.com, Department of Computer Engineering, Federal university of Technology, P.M.B 65 Minna, Nigeria.

cuticles, beta-carboline and 4-methyl-7-hydroxycoumarin, is responsible for the fluorescence [4]. Their exoskeleton Fluoresces green when they are exposed to Ultraviolet light [4]. Figure 1a show scorpion under ultraviolet light while figure 1b show scorpion under natural light. The rest of this paper is organised as follows, section 2 is a review of scorpion detection technique, section 3 is the methodology, section 4 is the result and discussion of the results, section 5 of the work is the conclusion while the last part section 6 is acknowledgement.

(a)

(b)

Figure 1: (a) Scorpion under UV light. (b) Scorpion in Natural environment.

2. Review of scorpion detection technique

Scorpion detection technique sometimes referred to as scorpion sampling technique [8, 10] can be categorise into two, namely the unaided detection method and the aided detection method. The former method is further subdivided into Rock rolling, location burrowing detection technique, and peeling loose back of tree detection technique. The aided detection technique is subdivided into the use of ultraviolet flashlight and touch, and the use of pitfall traps [7, 10].

Rock rolling detection technique involves opening up rock and checking the presence of this arthropod's in there shelter under the rock [8]. The location of burrow detection method involves locating appropriate shelter under burrows and digging up the burrow for scorpion acquisition [10]. Figure 2a show scorpion burrow while Figure 2b show loose bark of tree where scorpions use as shelter.

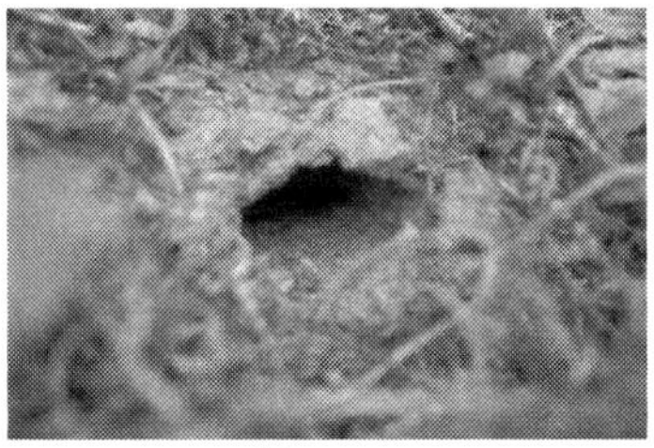
(a)

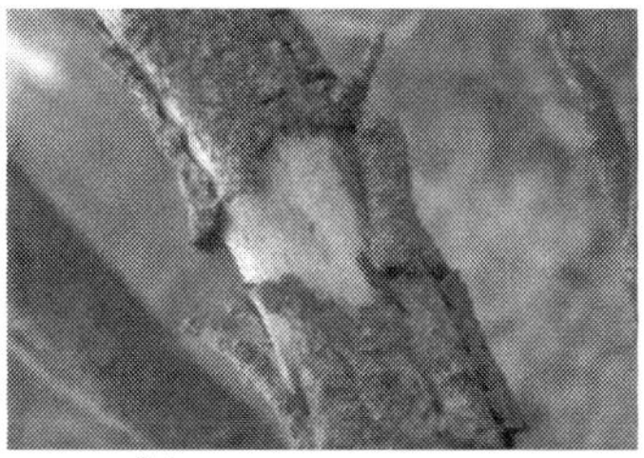
(b)

Figure 2: (a) Scorpion Burrow. (b) Loose bark of tree.

The Peeling loose back of tree detection method involves peeling back of trees in the day time to search for scorpions [3]. The aforementioned methods are crude and time consuming, unreliable, and dangerous. Furthermore, if proper care is not taking the person carrying out the search might get sting by scorpion. The finding that scorpions florescence under ultraviolet wavelengths of black light has led to the use of ultraviolet light in the detection and study of scorpion [5], in this detection technique ultraviolet touch are used by investigator, who searches an area in the night for possibility of locating scorpion [8,9], Figure 3a show an investigator using UV light for scorpion detection.The use of UV is also time consuming and crude and involve the presence of a person for possible detection of this arthropods.

The pitfall detection technique involves setting pitfall traps for scorpion, by digging up ground surface for trap setting, covering and checking for possibility of catching scorpion; Figure 3b shows a pitfall trap.

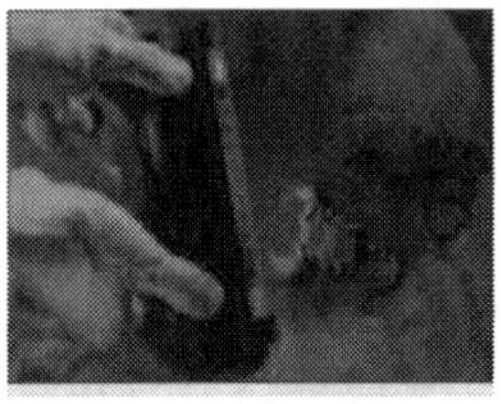
(a)

(b)

Figure 3: (a) Scorpion fluorescence under UV light. (b) A pitfall trap.

Similarly this technique is not suitable for some environments and it is time consuming [8,10]. Hence the use of image processing technique has been proposed in this work as alternative detection technique.

One of the key features of the proposed technique is the colour segmentation approach. This involves analyzing the captured image using the RGB information. Hence detailed review of colour segmentation is undertaking in this section. In Colour-based segmentation of berlin edema from fundus image, using K-means clustering algorithm [11], the input image was sharpened using Laplacian filter mask of 3*3 matrix, then the image was pre-processed with Laplacian filter mask, K-means clustering was used for pixel based segmentation of the image, fifteen fundus images (768X576) resolution was used [11].

The detection of colour image, using saturation invariant in color image and hue invariant to detect edges and corner of the image was reported in [12], The dichromatic reflection model for colour mode analyses was use in getting the hue invariant, and the saturation invariant of the colour to detect the edges of the image. Extension of Haris detector in colour space to detect the corners of the colour image using invariant of hue and saturation [12]. Colour images were collected from Google image [12].

In lip tracking using local region based approach to tracking, [13] divide the local region based approach into lip colour extraction for the first lip frame and tracking of the lip frame after the colour extraction, the tracking algorithm uses a lip contour extraction for the first lip frame and a lip tracking in the second phase an illumination equalization was used to reduce the lighting asymmetry of the lip, it uses a 16point geometric deformable model to model lip shapes [13]. Two hundred (200) frontal face image from the CVL face database and five hundred (500) from the laboratory database was used in evaluating the performance of the system. The review of segmentation of images with multi-objectives using weighted formula approach and pareto approach was carried out in [14], approaches to multi-objective segmentation and image clustering was compare, using images from UCI repository and the brain web.

Improved thresholding-based segmentation under natural environment was proposed in [15]. Detailed investigation and comparison of three different techniques was done in the work. It was concluded that TsTN outperform other techniques due to uneven illumination in the acquired images. The use of color information for recognition and sorting of fruit was proposed in [16]. It was shown that the use of any of the three techniques proposed in the work is sufficient for identifying and classifying apple, Banana and Mango from an acquired image [16].

3. Methodology

The main stages used in the design of the algorithm is made up of two framework namely, image pre-processing and image segmentation. A General block diagram of this framework is shown in Figure 4.

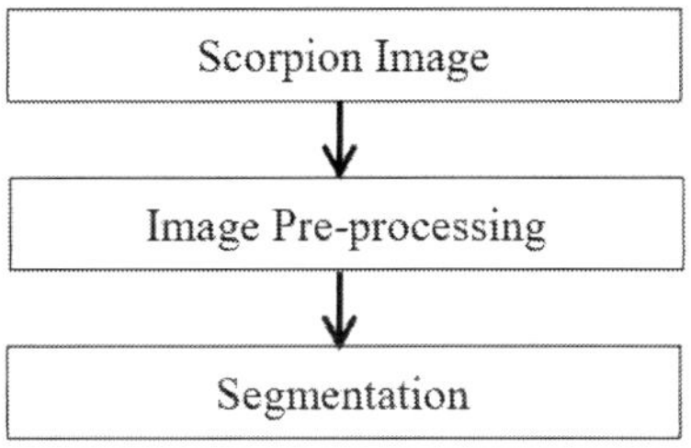

Figure 4: General block diagram of the framework

Based on Figure 4, two new approaches for Scorpion segmentation have been proposed in this work. The first method involve the block diagram shown in Figure 5. In this approach, the image pre-processing stage involves conversion of the image from RGB to YCbCr colour space to see which of the channels of the colour space provides a better quality image for segmentation.

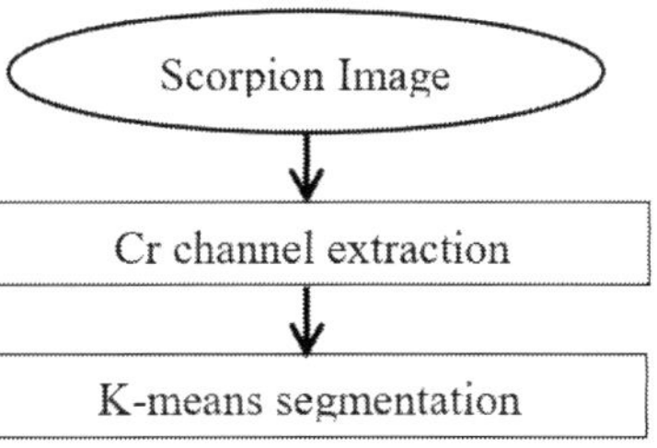

Figure 5: Block diagram of the K-means Segmentation approach

The input image was converted to the YCbCr colour image using the mathematical relation shown in equation (1):

$$\begin{bmatrix} Y \\ Cb \\ Cr \end{bmatrix} = \begin{bmatrix} 16 \\ 128 \\ 128 \end{bmatrix} + \begin{bmatrix} 0.25678824 & 0.50412941 & 0.09790588 \\ -0.1482229 & -0.29099279 & 0.43921569 \\ 0.43921569 & -0.36778831 & -0.07142737 \end{bmatrix} \begin{bmatrix} R \\ G \\ B \end{bmatrix} \quad (1)$$

The K-means image segmentation is carried out using the mathematical relationship given as

$$L = \sum_{j=1}^{k} \sum_{i=1}^{n} a_{ij} \left\| x_i - \mu_j \right\|^2 \quad (2)$$

Where the given data points $n = (x_1, x_2 \dots\dots\dots . x_n)$ and K is the number of clusters $K = (\mu_1, \mu_2, \dots\dots\dots . \mu_K)$.

The second technique involves the use of the block diagram shown in Figure 6. The green channel of the RGB scorpion image is extracted for further segmentation using simple averaging technique.

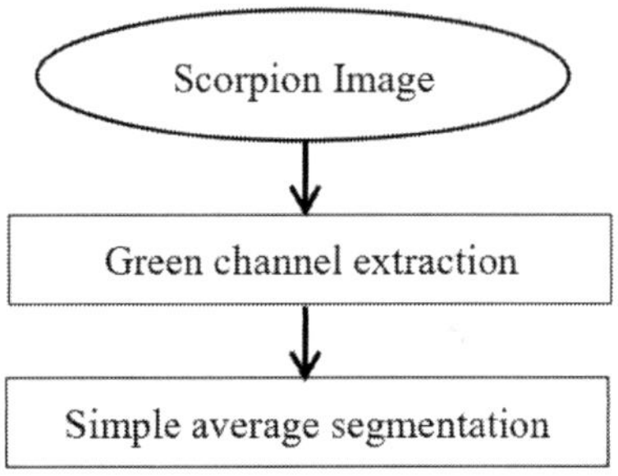

Figure 6: Block diagram of the simple Average Segmentation approach

The simple average thresholding invole the use of (3).

$$\bar{X} = \frac{\sum_i^m k_i}{m} \qquad (3)$$

Where $\bar{X}$ is the mean of the image, m is the total intensity value of foreground image, and k is the total intensity value of the foreground image.

$$g(x,y) = \begin{cases} 1 & if\ f(x,y) > T \\ 0 & if\ f(x,y) \leq T \end{cases} \qquad (4)$$

Where $g(x,y)$ and $f(x,y)$ are pixel value of binary image and the pixel value of the green channel image.

The green channel of the RGB image used for thresholding is expressed in equation (5) below

$$f(x,y) = I(x,y,2) \qquad (5)$$

$$where\ I(x,y,2) is\ the\ green\ channel\ of\ the\ original$$

this section we discuss how to format the title, authors and affiliations. Please follow these instructions as carefully as possible so all articles within a conference have the same style to the title page. This paragraph follows a section title so it should not be indented.

4. Results and discussions

Performance evaluations of the proposed methods were done using the receiver operating characteristics (ROC). Pixels will be classified into True Positive (TP), False Positive (FP), True Negative (TN) and False Negative (FN).

False positive are regarded as background pixels incorrectly classified as foreground pixels while TN are background pixels correctly classified as background pixels [17-19]. Furthermore, FN are foreground pixels that are incorrectly classified as background [17,18]. Similarly TP are foreground pixel that are correctly classified as foreground.

Performance evaluation of the system was done using true positive rate, true negative rate, false positive rate and false negative rate. Other performance metrics include:

TP: True Positive Detection; TN: True Negative Detection
FP: False Positive Detection; FN: False Negative Detection
MP: Missed Point; PR: Precision
FPR: False Positive Rate; TPR: True Positive Rate

SN: Sensitivity; Acc: Accuracy
SP: Specificity; ToC: Time of Completion
The TOC is a measure of the time taken for complete segmentation. Other metrics are defined as

$$PR = \frac{TP}{TP + FP}$$

$$FPR = \frac{FP}{N}$$

$$TPR = \frac{TP}{P}$$

$$SN = \frac{TP}{TP + FN}$$

$$SP = \frac{TN}{FP + TN}$$

$$Accuracy(ACC) = \frac{TP + TN}{P + N}$$

The ground truth segmented images was done using adobe Photoshop cs5 extended version 12.0.2x 32. The images were imported into the workspace and the quick selection tool was used to select the whole area of the image for segmentation. The result obtained from the application of K-means segmentation technique and simple average segmentation technique, are as shown in Table 1-2 below.

Table 1
The receiver operating characteristics (ROC) of the K-means segmentation technique.

Details	ROC of the Cr channel				
No	**ACC**	**PR**	**TPR**	**FPR**	**Toc**
Im1	0.9463	1.0000	0.9593	0	0.2537
Im2	0.9421	1.0000	0.8408	0	1.5554
Im3	0.9729	1.0000	0.9347	0	1.6154
Im4	0.9615	1.0000	0.9104	0	1.4102
Im5	0.9751	1.0000	0.9091	0	0.1237
Im6	0.9655	1.0000	0.9245	0	1.3812
Im7	0.9761	0.9978	0.9865	0.0005	1.3930
Im8	0.9566	1.0000	0.8784	0	2.0971
Im9	0.9864	1.0000	0.9920	0	1.5990
Im10	0.9826	1.0000	0.9913	0	1.6382

Table 2
The receiver operating characteristics (ROC) of the simple average segmentation technique.

Details	ROC of the Green channel				
No	**ACC**	**PR**	**TPR**	**FPR**	**Toc**
Im1	0.9658	1.0000	0.8270	0	0.0032
Im2	0.9513	1.0000	0.7445	0	0.0027
Im3	.9802	1.0000	0.8239	0	0.0029
Im4	0.9686	1.0000	0.8182	0	0.0030
Im5	0.9968	1.0000	0.8798	0	0.0002
Im6	0.9744	1.0000	0.8430	0	0.0027
Im7	0.9898	0.9996	0.9467	0.0001	0.0028

Im8	0.9627	1.0000	0.7775	0	0.0029
Im9	0.9940	1.0000	0.9575	0	0.0028
Im10	0.9884	1.0000	0.9233	0	0.0032

Results obtained as shown in Table 1-2. It shows that the accuracy value obtained in the Simple average segmentation technique of the green channel is better than that obtained from the K-means segmentation of the Chroma red channel. Furthermore, it was observed that the precision of the simple average segmentation technique is better than that obtained from the K-means segmentation technique. Similarly, it was observed that the time of completion of execution of the simple average segmentation technique is faster than K-means segmentation technique. Samples of resulting images are shown in Appendix I

5. Conclusions

Scorpion stings which are vital health issue that requires a considerable attention in minimizing the pain inflicted on victims and possibly avert death. A possible way of achieving this is the capability of being able to detect the presence of these arthropods earlier before it stings its victim. This paper proposed the use of digital image processing approach for the detection of scorpion. It has been shown that the simple averaging of the Green channel of the acquired RGB image is sufficient for segmenting the image from the background. Furthermore, the risk associated with the use of ultraviolet touch detection technique and pitfall trap which involves the presence of an investigator and also time consuming is being eliminated by the novel digital image processing approach proposed in this paper.

Acknowledgement

The authors will like thank Arnourd Quanjer (http://www.wildlifephotography.nl/gallery/scorpions-under-uv-light) for the scorpion images, Dr. Victor Fet, for advice on where to get UV touch for sampling scorpion, and Jan ove rein, Norman Larsen, Jonathan Leeming for their advice on techniques of sampling scorpions.

Appendix A.

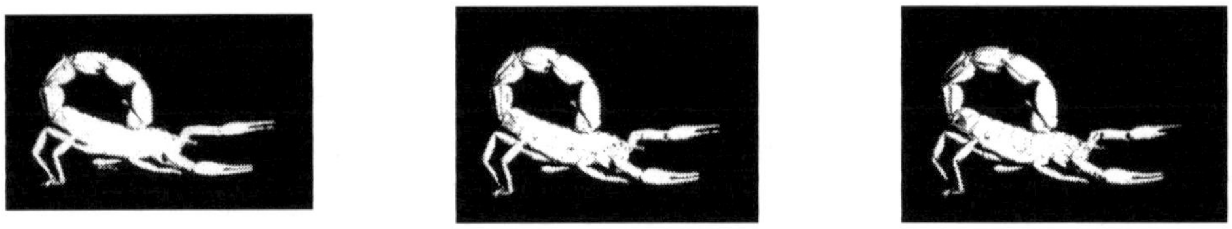

a b c

Figure 7: (a) Ground truth (b) K-means Segmentation (c) Simple Average Segmentation

References

[1]. As Chowel, G., Hyman, J. M., Diaz-Duenas, P., & Hengartner, N. W. (2005). Predicting Scorpion sting incidence in an endemic region using climatological variables. *International Journal of Environmental Health Research*, 425-435.

[2]. Bradley, R. A., & Brody, A. J. (1984). Relative abundance of three vaejovid scorpion across a habitat gradient. *J. Arachnol, 11*, 437-440.

[3]. Druce, D., Hamer, M., & Slotow, R. (2004). Sampling strategies for millipedes (Diplopoda) centipedes (Chilopoda) and scorpions (Scorpionida) in savanna habitats. *Africa Zoology, 39*(2), 293-304.

[4]. Gaffin, D. D., Bumm, L. A., Taylor, M. S., Popokina, N. V., & Mann, S. (2012). Scorpion fluorescence and reaction to light. *Animal Behaviour, 83*, 429-436.

[5]. Lourenco, W. R. (2012). Fluorescence in scorpions under UV light; can chaerilids be a possible exception? *Comptes Rendus Biologies, 335*, 731-734.

[6]. Ozkan, O., Yagmur, E. A., & Ark, M. (2011). A newly describe scorpion species, leiurus abdullahbayrami (Scorpion Buthidae) and the lethal potency and in vivo effect of its venom. *The Journal of Venomous Animals and Toxins Including Tropical Diseases*, 414-421.

[7]. Raz, S., Sion, R., Tomas, P., Adam , H., Hagay , K., Dan , Z., et al. (2009). Scorpion Biodiversity and Interslope Divergence at "Evolution Canyon", Lower Nahal Oren Microsite, Mt. Carmel, Israel. *Peer-Reviewed open access Journal*, e5214.

[8]. Shehab, A. H., Amr, S. Z., & Lindsell, A. J. (2011). Ecology and biology of scorpions in Palmyra, Syria. *Turk J Zool, 35*(3), 333-341.

[9]. Volschenk S. Erich. (2005). A new technique for examining the surface morphosculture of scorpion. *The Journal of Arachnology*, 820-825.

[10]. Williams, S. C. (1968). Method of Sampling Scorpion Population. *Proceeding of the California Academy of Science*, (pp. 221-230). California.

[11]. Datta, S., & Monisha, C. (2012). Application of colour based segmentation approach to detect Berlin's Edema. *Proceedia technology* (pp. 600-603). Elsevier Ltd.

[12]. Chu, J., Miao, J., Zhang, G., & Wang, L. (2013). Edge and Corner Detection by Colour invariant. *Optics and Laser Technology* (pp. 756-762). Elsevier Ltd.

[13]. Cheung, Y.-M., Liu, X., & You, X. (2012). A Local region based aroach to lip tracking. *Pattern Recognition, 45*, 3336-3347.

[14]. Bong, C.-w., & Rajeswari, M. (2012). Multi-objective nature-inspire clustering and classification for image segmentation. *Applied soft computing, 11*, 3271-3282.

[15]. Abdullah, S. L., Hambali, H. A., & Jamil, N. (2012). Segmentation of natural images using an imroved thresholding-based technique. *Procedia Engineering, 41*, 938-944.

[16]. Aibinu, A., Salami, M. J., Shafie, A. A., Hazali, N., & Termidzi, N. (2011). Automatic Fruit Identification System using Hybrid Technique. *IEEE Computer Society*, 217-221.

[17]. Aibinu, A. M., Iqbal, M. I., Salami, M. J., & Nilsson, M. (2010). Vascular intersection detection in retina fundus images using a new hybrid approach. *Computers in Bilogy and medicine, 40*, 81-89.

[18]. Aibinu, A. M., Salami, M. J., & Shafie, A. A. (2011). A novel signal diagnosis technique using pseudo complex-valued autoregressive technique. *Expert Systems with Applications, 38*, 9063-9069.

[19]. Sivabalakrishnan, M., & Manjula, D. (2010). Human Tracking and segmentation using colour space conversion. *International journal of computer science issues., 7*(5), 284-288.

Artificial Intelligence based technique for BTS placement

C O Alenoghena[1,4], J O Emagbetere[2] and A M Aibinu[3,5]
[1]*Dept. of Telecommunications Engineering, Federal University of Tech. Minna, Nigeria.*
[2]*Dept. of Electrical/Electronic Engineering University of Benin, Nigeria.*
[3]*Dept. of Telecommunications Engineering, Federal University of Techn.[1] Minna, Nigeria.*
[4]alenca2@yahoo.com
[5]maibinu@gmail.com

Abstract. The increase of the base transceiver station (BTS) in most urban areas can be traced to the drive by network providers to meet demand for coverage and capacity. In traditional network planning, the final decision of BTS placement is taken by a team of radio planners, this decision is not fool proof against regulatory requirements. In this paper, an intelligent based algorithm for optimal BTS site placement has been proposed. The proposed technique takes into consideration neighbour and regulation considerations objectively while determining cell site. The application will lead to a quantitatively unbiased evaluated decision making process in BTS placement. An experimental data of a 2km by 3km territory was simulated for testing the new algorithm, results obtained show a 100% performance of the neighbour constrained algorithm in BTS placement optimization. Results on the application of GA with neighbourhood constraint indicate that the choices of location can be unbiased and optimization of facility placement for network design can be carried out.

1. Introduction

The deployment of Global System for Mobile (GSM) service in developing countries has been massive with an unprecedented market growth[9], this phenomenon is expected after long years of experiencing poor communication services. The fast growth rate of the GSM customer base has brought about a deployment of base transceiver stations (BTS) to meet coverage and capacity demands[2]. The dramatic increase on the number of BTS is attributed to the increase of subscribers, and advancement in technology (3G, UMTS) which opens room for advanced services and use of multimedia facilities in hand held phones. The telecommunication industry in Nigeria requires the construction of more mast and towers to improve teledensity which is presently only 50%[2].

The Base Station System (BSS) of the GSM network is the equipment located at a cell site. The BTS is responsible for maintaining connections to the mobile station (MS) within its cell. The BTS for a given geographical area have to be well positioned for maximum coverage and minimal interference which are the indexes for measuring the quality of any mobile service [5]. Unlike the mobile station that is handheld, the BTS is a conspicuous structure that has been the cause of much agitation from several quarters on the need for RF safety in base stations. The BTS contains several transceivers (TRXs) for transmitting and receiving information, the antennas for receiving

[1] C. O. Alenoghena, Dept. of Telecommunications Engineering, Federal University of Tech. Minna, Nigeria, alenca2@yahoo.com

information to/from one or more transceiver are also connected to the BTS. These antennas have to be mounted high to avoid interference and maximum coverage, necessitating the need for mast and towers and even the use of roof tops of high buildings, figure1 shows the picture of a BTS.

Figure1: Picture of a BTS

The presence of a BTS is always felt by the surrounding structures, human and material, irrespective of the fact that the BTS may be stand alone on a space with the minimum space as specified by regulations, the facility may be said to be optimally placed if the neighbours are not complaining. The process of cell planning is conducted in an ad-hoc fashion after manually inspecting maps depicting the propagation properties of the service area. There is a need therefore for a more efficient and automated cell planning model that taking compliance to regulation into consideration.

In this work an intelligent based technique for optimal site selection for facility placement, here referred to as the neighbourhood constraint based genetic algorithm. Genetic algorithm is an optimization tool capable of giving optimum results in situation where there are many conflicting options an requiring optimal balance. It has also proved to be a viable tool for search problems that are presented as N-P hard. This is so as Genetic algorithm have the capability to search large spaces efficiently without the need for derivative information. The concept of neighbourhood constraints is born out of a need to quantitatively evaluate the process of decision making for the best placement of facilities in a wireless telecommunication network, this case the Base transceiver station (BTS). Its implementation will lead to a situation where placement choice would be made based on realistic, unbiased evaluation with all variables and options taken into consideration and adherence to regulating bodies. The use of the neighbor constraint tool will lead to a saving of time that will be spent on manual site verification; it will also result in a reduction of cost that will hitherto be spent on site verification and subsequently automation of planning process leading to improved efficiency.

The remaining part of this paper is divided into three sections; section two is on literature review, a review of traditional network planning process and optimizing cell planning process are presented. Section three discusses the new methodology and implementation while results conclusion and references are presented in section four.

2. Literature review

In this section, detail review of the processes involve in telecommunication network facility placement is discussed. The issues discussed are basically traditional planning process, optimization algorithm for cell planning and use of genetic algorithm in facility placement. Detail discussions are subsequently presented.

2.1 *Traditional network planning:*

In cellular telecommunication provision, cell planning is done to ensure coverage and avoid interference in a network [1]. Deciding where to place the base station of a cellular network is a very

IOP Conf. Series: Materials Science and Engineering **53** (2013) 012056 doi:10.1088/1757-899X/53/1/012056

important issue that is usually decided during the process of cell planning. One of the essential tools used in cell planning is the MapInfo which is a windows based software used for site location analysis. The selection process is typically based on location demographics, nearby facilities depicted in a geography map. Their limitation is that physical inspection and final decision of proposed sites will have to be taken by a team of radio planners. Manual investigations and tests are usually carried out to ascertain if the potential sites meet legal, civil and environmental requirements. This manual exercise is rigorous, inefficient and could be prone to errors as they are based on personnel experience[10]. Figure 2. is a chart showing a typical cell planning process[3].

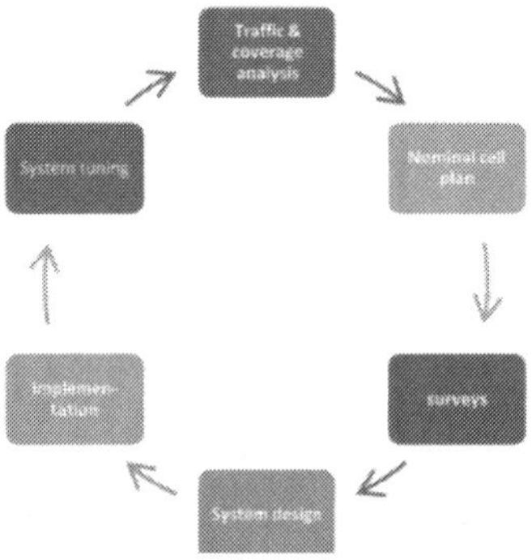

Figure 2: Traditional cell planning Process

2.2 Optimizing Cell Planning Algorithms.

Research have been ongoing on the optimization of cell planning process, the design issues in wireless network are very complex, involving resource allocation, interference cancellation and signal detection[7]. Most of these problems have large search space which makes them non-deterministic polynomial (NP)-hard hence solutions using analytical approaches have proved to be intractable and tedious [8]. These reasons have made many researchers to propose various techniques of heuristics methods for solving these problems. Finding the most optimal locations to place a base station which encompasses the antenna and other transmission facilities poses one of the most challenging design problems in the planning and deployment of a communications network [11]. The use of artificial intelligence have been undergoing much research and genetic algorithm has been found to be useful for solving this type of NP-hard problem [7,8,11,12].

3. Methodology

In this section, detailed explanation of Genetic algorithm and the neighbourhood constraint concept, and its implementation on a generated territory is presented.

The choice of locations for placement will be determined using an optimization tool referred to as Genetic algorithm GA. GA is a form of evolutionary algorithm (EA), which are algorithms that obtain their solutions based on the principle of natural evolution, such as adaptation and survival of the fittest. It is capable of giving optimum results in situation where there are many conflicting options and has proved to be a viable tool for search problems and also in problems that are presented as N-P hard.

GA utilizes a clearly defined procedure for solving problems with a finite time for termination. The major issues addressed in the implementation of GA's include, parameters coding, fitness function development and strategy for chromosome selection.

Basic steps taken in the implementation of Genetic algorithm begins with defining the optimization variables, its constitution and cost. It ends by testing for convergence.

3.1 Defining the neighbour constraint:

The neighbourhood constraint process intends to place values (weights) on the neighbours surrounding a choice position, these are them evaluated and the position with the least weight or optimal value is chosen, it is a minimization algorithm. Figure 3 is a typical neighbourhood map depicting the various neighbours of the BTS as $L_1 - L_8$, the summation of the various weight ($\sum L$) gives the optimization equation. An open space as neighbour is assigned the value zero. In applying GA to optimization placement selection, the neighbour evaluation and criteria will be used selecting the optimal site.

L1	L2	L3
L4	BTS	L5
L6	L7	L8

Figure 3: A 3 X 3 neighbor map

$$F(L) = \sum L_i$$
$$= L_1 + L_2 + L_3 + L_4 + L_5 + L_6 + L_7 + L \quad (1)$$

3.2 Assigning weights to neighbour:

The International Telecommunication Union (ITU) and bodies' in-charge of telecommunications regulations in various countries have guidelines bordering on the building of towers and mast. In these guidelines minimum setbacks of towers from existing structures are specified in meters. In computing weights for the BTS neighbours a bar chart is generated depicting the minimum distances and the corresponding weights. Table 1 below shows the allocation of weights which is proportional to the specified standard clearance required by law.

Table 1: Minimum setbacks for towers and mast/neighbor weights

S/N	Constraint list	Min. Distance (m)	weight (L)	Legend
1	Residential	20 x 20	2	
2	Commercial	12 x 12	1.2	
3	Industrial	12 x 12	1.2	
4	Civic/cultural	20 x 20	2	
5	Educational	150	15	
6	Health /clinics	150	15	
7	Highways	50m	5	
8	Open space		0	

3.3 Problem Representation

A decision on mode and form of representing the system information is taken prior to starting optimization, for this work the relevant parameters will be encoded using binary numbers, for ease of implementation and best performance. The BTS are encoded using two coordinates X-Y of the territory which form the chromosomes. The number of bits to be used for representation was determined using equation (2) [6].

$$2^{mj-1} < (bj - aj)X10^p \leq 2^{mj} - 1 \qquad (2)$$

Where;

m is the required binary bits
a is the lower limit of the decimal range
b is the upper limit
p is the required precision in decimal points.

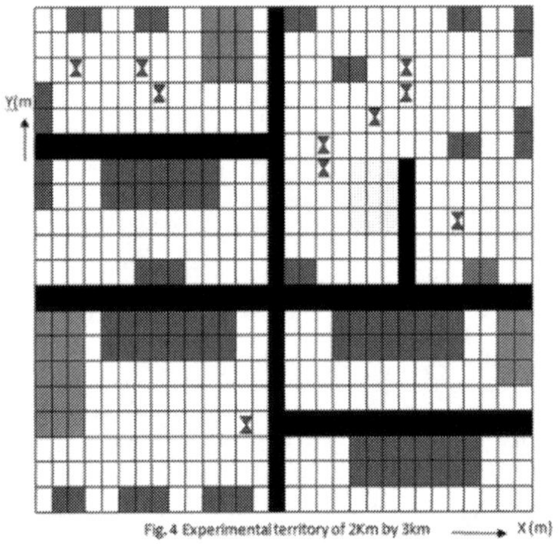

Fig. 4 Experimental territory of 2Km by 3km

4. Results

The experimental data is a 2km by 3km territory developed to test the new algorithm. The experimental data is a typical suburban area with existing infrastructures as represented in figure 4. A data base of the territory was generated and fed into the algorithm. The programme for the genetic algorithm was written in matlab 7.5

The following parameters setting were used for optimization: Initial Population =50; Generations=30; Probability of crossover =0.25; Mutation Rate=0.01 ; Precision=1

Ten rounds of iteration were run and the results obtained are shown in table 2. The algorithm in each round selects the optimal location for the BTS based on optimization criteria. The coordinates of the optimal position selected by the algorithm is given (row and column). The 'Type' column depicts the existing structure on selected site. The neighbor weighting is also shown. Selected optimal sites are plotted on the territory map as purple colored asterisk (♦).

Table2: Results of GA optimization

RUNS	LIKELY LOCATIONS			Neighbour weight
	Row	*Column*	*Type*	(∑ L)
1	14	18	Open space	0
2	16	21	"	0
3	18	7	"	0
4	17	23	"	0
5	17	8	"	0
6	18	23	'	0
7	12	26	"	0
8	15	18	"	0
9	18	3	"	0
10	4	13	"	0

4.1 Discussion of results:

Ten rounds of iteration were carried out; in each round the algorithm selected the optimal location For this experiment, the sites selected by the algorithm from the experimental data were all open spaces with minimal neighbor weighting of zero. This showed a 100% performance of the neighbor constrained algorithm in placement. Selection of sites is in adherence to regulation and consideration of neighbor which is unbiased and efficient.

The results reveal a concentration of the selected location on the upper half of the territory which is sparsely populated with less buildings. 90 % of sites selected are located in the upper half of the territory with only 10% in the lower half of the territory. Each round of iteration took a few second approximately 30 seconds, which is clear indication of the time saving to be derived from its use. The algorithm through the several iterations there was a clear avoidance of BTS placement on roads, and algorithm to seek out optimally regions with less neighbor.

4.2 Conclusion

In this paper, the application of GA with neighbourhood constraint has been presented. Results indicate that the choices of location are unbiased and optimization of facility placement for network design can be carried out. Further work will involve the implementation of this new algorithm in a real territory and combining with other objectives of coverage optimization and interference minimization. The application will lead to a quantitatively evaluated decision making process that is unbiased, fast and automated for the placement of Base transceiver station (BTS).

5. References

[1] Mishra A 2004 *Fundamentals of cellular network planning and optimisation* (England: John Wiley & sons) pp 21-31

[2] Association of licensed telecommunications operators of Nigeria, 2012 *NESREA's draft on environmental regulations for the telecommunications industry* , assessed on line on 13th August 2012

[3] Ericson comapny, 2004 *Cell planning principles,* Ericson manual (En/LZT 123 3314 R3A) pp 1-10

[4] Government of Ghana, *Guidelines for the deployment of communications towers* http://www.nca.org.gh/128/120/Guidelines.html assessed on August 13th 2012 pg. 41-43

[5] Hanif D S, Chandra M P and Theodore S R 1996 Optimal location of transmitters for micro-cellular radio communication system design, IEEE *journal on selected areas in communications,* **14**, no 4, 662-73

[6] Jamaluddin H 2011 *lecture notes on genetic algorithm*, faculty of mechanical Engineering Unuversiti Teknologi Malaysia. (unpublished)

[7] Munyaneza J, Kurien A and Van Wyk B 2008 *Optimization of antenna placement in 3G networks using genetic algorithms* Proc IEEE third international conference on broadband communication, information technology and biomedical applications (Pretoria, South Africa, 23-26 Nov. 2008) pp. 30 – 37

[8] Zimmermann J, Hons R and Mu¨hlenbein H 2003 *ENCON*: an evolutionary algorithm for the antenna placement problem, Elsevier science ltd, *computers & Industrial Engineering* 209-226.

[9] Eugene J 2011 *Telecommunications business in Nigeria: how far?*" International communications lecture organized by the Nigeria-Swedish chamber of commerce (Victoria Island, Lagos, April 8th 2011)

[10] Nagy L and Farkas L 2000, *Indoor base station location optimization using genetic Algorithm,* The 11th IEEE international symposium on personal, indoor and mobile radio communication **2** 843-46.

[11] Rambally R S and Maharajh A 2009 "Cell planning using genetic algorithm and tabu search" Proc IEEE second International conference on applications of digital information and web technologies, (London, ICADIWT '09.) pp 640-645

[12] Chio Y S, Kim K S and Kim N 2008 The displacement of base station in mobile communication with genetic approach EURASIP *Journal on Wireless communications and networking,* article ID **580761**.

IOP Conf. Series: Materials Science and Engineering 53 (2013) 012057 doi:10.1088/1757-899X/53/1/012057

[1]Model for simulating scorpion substrate vibration and detection system

B A Sadiq[1], A M Aibinu[2], E Joseph[3], H B Salau[2] and M J E Salami[4]

[1] Department of Electrical Engineering, Niger State Polytechnic P.M.B. 01,Zungeru, Nigeria,

[2] Department of Telecommunication Engineering federal University of Technology, P.M.B. 65 Minna, Nigeria,

[3] Department of computer Engineering, federal University of Technology, P.M.B. 65 Minna, Nigeria,

[4] Department of Mechatronics Engineering, International Islamic University P.O. Box 10, 53100, Gombak, Malaysia.

E-mail: balaasadiq@gmail.com

Abstract. Scorpion stings are vital health issues which requires prompt attention to minimize the pain inflicted on victims and avert death. A possible solution in averting the sting is the capability of detecting its presence earlier before it stings. Scorpion like other arthropods have a specific kind of movement pattern called substrate vibration, which generates a specific signal that is used in recognizing and locating mates and preys. This paper aims at developing an intelligent scorpion detection system using vibration frequency detection technique. A six step model for simulating scorpion substrate vibration and detection has been proposed. The surrounding vibrating signal is acquired and passed through a band pass filter. The resulting signal is model using autoregressive modeling technique.. Resulting co-efficients are further analyzed for activity detection. The frequency response of scorpion activities for mating behaviour was simulated, detected analysed using MATLAB environment. The resulting coefficients was also compared and analysed. Results obtained shows that the proposed technique is appropriate for model and simulating scorpion substrate vibration and detection system.

1. Introduction

Scorpions are ancient animals that have been in existence since 425 – 450 million years ago-according fossil records [1,2]. They evolved from an amphibious ancestor, and occur in many habitats ranging from deciduous forest, motane pine forests, intertidal zones, rain forest, caves, grasslands and savannahs [3,4]. Like the insects and spiders, scorpions belong to the phylum Arthropoda and like the spiders they belong to the Arachnida but belong to a different order, Scorpiones [2]

Scorpions are mostly nocturnal and hunt insects, spiders, centipedes and other arachnids including scorpions [3,4]. Figure 1a shows a Scorpion feeding on grasshopper. Bigger scorpions sometimes feed on vertebrate like small lizards, snakes and mice [4]. During the day they can be found under rocks, bark, cow-pats and rock crevices, often they are attracted to insect activity around lights. Many people leave lights on during the night and these attract insects that in turn attract scorpions [2].

---

[1] To whom any correspondence should be addressed.

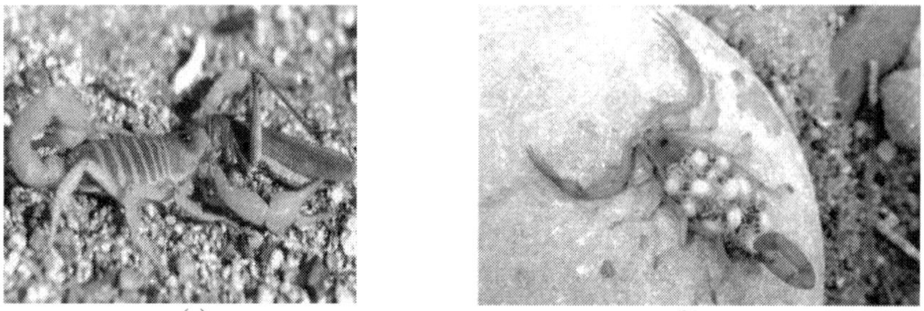

(a) (b)

Figure 1: (a) Scorpion feeding on a grasshopper. [5]
(b) Female scorpion with young on her back. [4]

In most species (not all) the male initiates the courting. The male grips the female pedipalps (chelae) and leads her in a mating dance that usually lasts about 30 to 60 minutes, but can vary from 5 minutes to 2 days and they can cover a distance of about 25 metres. Scorpions are viviparous (carry the eggs inside the reproductive tract and give birth to live young). After a 3 to 18 month gestation period, 1-105 live young are born, the average numbering approximately 26. The female, in most species, forms a basket with her first or first and second pairs of legs to catch the newborn at birth [2]. They then climb up her legs onto her back where they will moult for the first time as shown in figure 1b [4].

Envenomings by scorpion stings is an important but yet neglected health issue in many parts of the world, particularly in the extreme Northern and Southern parts of Africa; the Middle East; Southern states of USA; Mexico; some parts of South America, and the Indian sub-continent. Scorpion stings are very lethal in young children [6]. The venoms release autonomic nervous system mediators causing myocardial damage, cardiac arrhythmias, pulmonary oedema, shock, paralysis, muscle spasms and pancreatitis [6]. Despite the high rate of death associated with scorpion stings, little or no work has been reported which involve the use of intelligent approach in the detection of this arthropod's [7]. It is for this reason; this paper proposes an automatic and intelligent approach for the detection of scorpion.

The remaining part of this paper is organized as follows. Section II dwells on the review of various scorpion detection methods as well as the performance analysis of each of the methods. In section III, the scorpion substrate vibration was discussed and analysed. An overview of the development of the scorpion substrate vibration and detection modelling was presented in section IV. Section V present the result obtained. Section IV; and the discussion of the result. Section VI contains the conclusion of the paper.

2. Review of scorpion detection techniques

The earlier Scorpion detection methods involved locating areas suspected to be hiding places for scorpions and digging up such places in search of scorpions. This method is called the burrowing detection technique [8]. Another method is the rolling of rocks and checking for the presence of the scorpion [8]. Also detection can be done by peeling the tree-bark in search of the scorpion [8]. Furthermore scorpion can be detected using the pitfall traps method, in which hollow traps are set for scorpion to fall into and consequently captured [8]. All the aforementioned methods of detection, prove to be dangerous, time consuming, and also exposes the detector to the danger of being sting.

A more scientific method was proposed, based on the fact that scorpions fluorescence in the presence of ultraviolent (UV) light was proposed in [9]. The UV light is used by searcher usually at night for possible detection of scorpion. However, this method although more scientific, is only effective at night. Another scientific method was proposed by [7], which was based on image processing technique by using colour segmentation approach.

Table 1: Performance Evaluation of Scorpion Detection Methods

S/N	Detection Method	Advantages	Disadvantages
1	Burrowing Detection	It is cost effective, as the tool require are not expensive.	It is dangerous. It is tedious. It is time consuming. It is destructive as it involve digging the earth.
2	Rock Rolling	It is cost effective, as it requires simple or no tool at all.	It is dangerous. It is tedious. It is time consuming.
3	Bark Peeling	It is cost effective, as the tool require are not expensive.	It is dangerous. It is tedious. It is time consuming. It is destructive as it involved destruction of the tree.
4	Pitfall Trap	It is relatively less dangerous. It is also cost effective.	It relies on human intervention, as the detector has to manually set the trap and monitor it. It is also time consuming.
5	UV light Method	It is relatively less dangerous.	It relies on human intervention, as it is not intelligent enough to detect the scorpion automatically. It is more effective at night and not efficient during the day.
6	Image processing	There is absolute no chance of getting sting by the scorpion. It involved little or no human intervention.	It is more effective at night and not efficient during the day, as it also required the use of UV light. It is relatively expensive to deploy.

3. Scorpion substrate vibration

Insect of the order scorpiones (mantophasmatodia) to which scorpion belong to, are known to have a specific kind of movement pattern called substrate vibration [10]. Signal usually exhibited in recognition and locating their mates. This is achieved by making abdominal contact with the substrate in which signal are produced which generates compression and transverse waves, which propagate through the material. The signal propagation depends on the resonance properties as well as the characteristic of the substrate [10]. Sand scorpions receive two kinds of waves: longitudinal sound waves and surface waves also known as Rayleigh waves. It is believed that the mechanoreceptors respond to Rayleigh waves to detect the direction of a vibration source and longitudinal vibrations to estimate the distance [11].

Substrate vibration is detected by mechanoreceptors located in the scolopidia organ (i.e the legs) of the scorpion. Scorpion has five scolopidia organs in each leg of the male and female: femoral chordotonal organ subgenual organ, tibia distal organ, tibio-tarsal scolopidial organ and tarso-pretarsal scolopidial organ [10].These organs were found to be very sensitive vibration receptors (especially the femoral chordotonal and the subgenual organs) for certain range of frequency vibration [10].

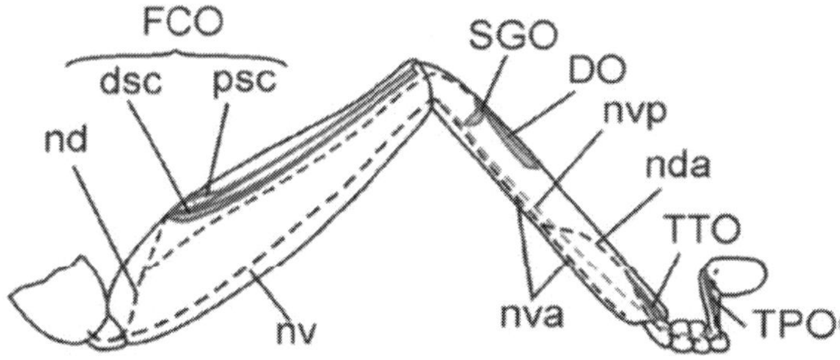

Figure 2: Schematic drawing of the leg scolopidial organs of Arthropods. [10]

The femoral chordotonal is responsible for detection of very low frequencies below 100Hz, while the subgenual organ is responsible for detection of frequencies between 100Hz-160Hz. It is therefore conclude that scorpions response best to frequencies below 120Hz [10].

(a) (b)

Figure 3: (a) Internal structure of femoral chordotonal organ (FCO) [10]
(b) Internal structure of subgenual organ (SGO) and distal organ. [10]

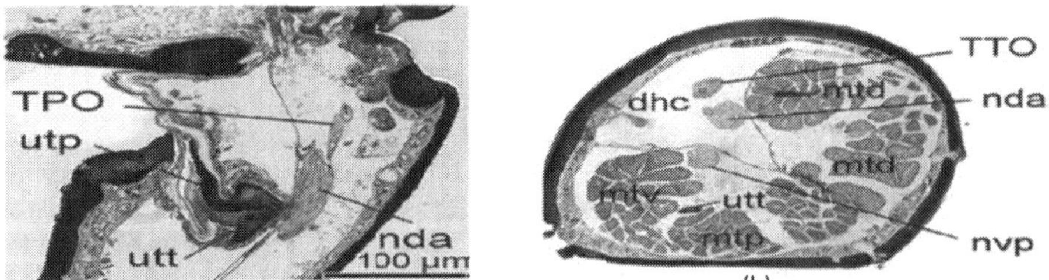

(a) (b)

Figure 4: (b) Internal structure of tarso-prestarsal organ (TPO) [10
(b) Internal sctructure of tibio-tarsal organ (TTO). [10]

The vibrating signal in female scorpion consist of individual pulses repeated at equal interval, while that of the male scorpion consist of repeated trains of pulses, the frequency distribution of the percussive call depends on the resonant properties of the material the scorpion is tapping on [12].Once a duet is established, the female becomes stationary and wait for the arrival of the male scorpion.

In scorpions a Duet is accomplished when a female within the range of communication respond to the male vibration signal using it own vibration signals and as a result become stationary [12]. When this is accomplished, the male began a localised search by orientation towards the stationary female until it locates it.

There are basically two types of Duet: 'two-way' duet and 'three-way' duet. In two-way the male call and then the female answer while in three-way duet the male call, female answer and then the male reply [12]

4. Model for scorpion substrate vibration and detection

The proposed model for simulating scorpion vibration technique is as shown in Figure 3.1. Detailed explanation of each unit is subsequently presented herewith.

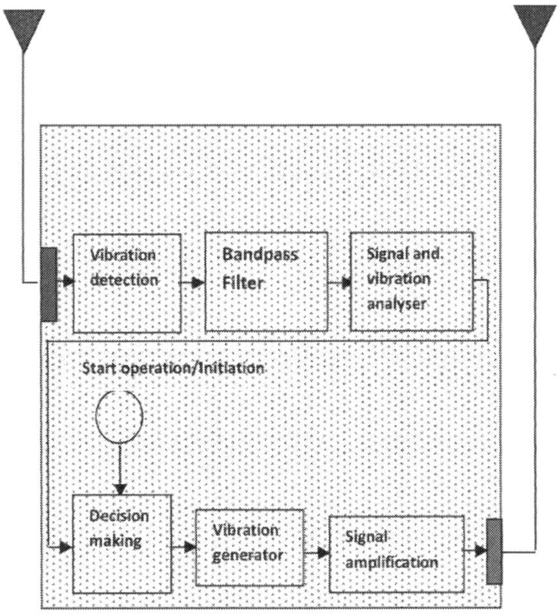

Figure 5: Block diagram of the proposed system.

4.1 Vibration detection module

The vibration signal propagated through the substrate is received via the receiving antenna; and passed unto the vibration detection module. Vibration sensors are used in this module. The receiving antenna is carefully selected as it has been observed that the choice of antenna improves system performance and at the same time reduce the cost.

4.2 Band pass Filter

A band-pass filter is a circuit that passes frequencies within a certain range and rejects (attenuates) frequencies outside that range [14]. A simple cascade of a low pass filter in series with a high pass filter has been used in selecting appropriate frequencies in this work. Figure 3.2 shows the simple block diagram of the Bandpass filter used in this work.

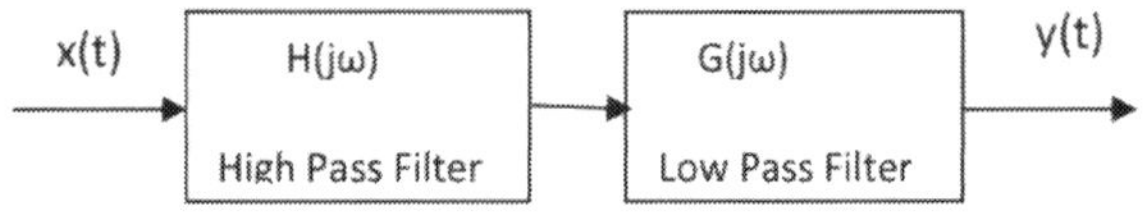

Figure 6: High and low pass filters combined to produce a band pass filter

4.3 Signal and vibration analysing module

In this stage, artificial neural network (ANN) based autoregressive (AR) modelling technique proposed in [15] is proposed for the analysis and detection of the frequency contained in the filtered signal. Detailed analysis of the technique is as contained in [15-17]. The filtered signals are windowed, and formatted for model coefficient determination using ANN based AR technique. A fixed model order of two have been proposed for this work. The received filtered signal is assumed to be an AR signal generated by the difference equation given as

$$y(n) = -a_1 y(n-1) - a_2 y(n-2) + w_n \qquad (1)$$

where the AR model coefficients are $a_1$and a_2, and w_n is a white Gaussian-distributed noise with zero mean and variance σ^2. As proposed in [15,18], the AR model coefficients are to be subjected to the following constraints.

$$-(1 + a_2) < a_1 < 0 \qquad (2)$$

$$| a_1(1 - a_2) | \leq | 4a_2 | \qquad (3)$$

$$a_1^2 - 4a_2 < 0 \qquad (4)$$

ANN based AR technique is applied in the determination of the coefficients and the computation of the signal frequency is computed using

$$\cos(2\pi f_0) = -\frac{a_1(1+a_2)}{4a_2} \qquad (5)$$

4.4 Decision making module

Here decision will be taking based on the on the previously analysed frequency (f_o), as to whether the frequency of the vibrating scorpion is for food or mating purposes for example.

TABLE 2: Likely Frequency and Activities of Scorpion.

S/N	SORPION ACTIVITIES	FREQUENCE RANGE (f_O)
1	Prey detection	4-17Hz
2	Mating Activity	40-80Hz

4.5 *Start operation/initiator*
This initiates the commencement of scorpion activities which are related to vibration movement of the Scorpion. These activities include prey detection and location of a partner for mating purpose.

4.6. Vibration generator Module
The vibration module generates the required vibration signal in accordance to intended activity.

4.7 Signal amplifier
The generated signal is amplified and conditioned by the signal amplifier before transmission.

5. Results and discussions
Typical result obtained using the proposed model will be reported in this section. A 60Hz signal for mating activity is sampled at the rate of 300Hz by the receiving scorpion module in a two way duet communication. The received was firstly filtered for noise and unwanted signal removal using a Bandpass filter. The resulting filtered signal was then analyzed using ANN based AR model technique. Comparison was done with the use of Yule Walker equation. Figure 7 shows the results obtained using both ANN based AR modeling technique and Yule Walker AR technique. Furthermore, Table 3 shows the results of the resulting coefficients for the two techniques.

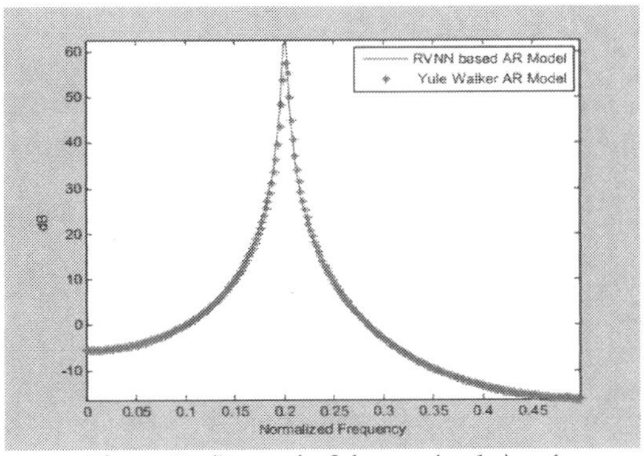

Figure 7: Spectral of the received signal

Table 3: Coefficients of the Received Scorpion Signal

AR Coefficients	Actual Values	ANN based AR technique	Yule Walker Method
a_0	1.0000	1.0000	1.0000
a_1	-0.6180	-0.5877	-0.5843
a_2	1.0000	0.9581	0.9537

6. Conclusions

The method of detecting scorpion using the vibrating pattern discussed in this paper, will definitely be a better method of detecting scorpion; since it can be used to effectively detect this dangerous animal before it could harm the victim.

Acknowledgement
We wish to express our propound gratitude to our colleagues who have stood for us during the course of this research. Our propound gratitude also go wives for their understanding during the compilation of this research work.

References

[1] Bradley, R. A., & Brody, A. J. (1984). Relative abundance of three vaejovid scorpion across a habitat gradient. *J. Arachnol, 11*, 437-440.

[2] Eastwood, E. B. 1977. Notes on the scorpion fauna of the Cape. Part 2. The *Parabuthus capensis* (Ehrenberg) species -group; remarks on taxonomy and bionomics (Arachnida, Scorpionida, Buthidae). *Annals of the South African Museum* 73: 199-214.

[3] Chowel G., Hyman J. M., Diaz-Duenas P., & Hengartner N. W. (2005). Predicting Scorpion sting incidence in an endemic region using climatological variables. *International Journal of Environmental Health Research*, 425-435.

[4] Dawn H.G. and Carl O. (2001).scorpions Arizona Poison Control and Drug Information Center - 1 (800) 222-1222).

[5] URL http://www.ntnu.no/ub/scorpion-files/p_cavimanus3.jpg

[6] Warrell D., Gutierrez J-M, and Padilla A., (2007). Rabies and envonomings: a neglected public health issue: Report of consultative world health organization, Geneva. Pp 12.

[7] Elijah J., sadiq B.A., Aibinu M. A.& Salami M. J. E. (2013). An itelligent Scorpion detection system. Pp 1-2

[8] Shehab, A. H., Amr, S. Z., & Lindsell, A. J. (2011). Ecology and biology of scorpions in Palmyra, Syria. *Turk J Zool, 35*(3), 333-341.

[9] Raz, S., Sion, R., Tomas, P., Adam , H., Hagay , K., Dan , Z., et al. (2009). Scorpion Biodiversity and Interslope Divergence at "Evolution Canyon", Lower Nahal Oren Microsite, Mt. Carmel, Israel. *Peer-Reviewed open access Journal*, e5214.

[10] Eberhard M. J., Lang D., Metscher B., Pass G., Picker M. D. and Wolf H. (2010). Structure and sensory physiology of the leg Scolopidial organ in mentatophasmotodea and their role in vibration communication. Journal of arthropod structure and development. Vol. 39; Pg 230-241.

[11] DaeEun K. (2006). Neural Network Mechanism for the Orientation Behavior of Sand Scorpions Towards Prey: IEEE transactions on neural networks, Vol. 17, N0. 4. Pg 1070-1076.

[12] Kenneth W. S. And John B. S. (2006). Insect sound communication: physiology, Behaviour Ecology and Evolution. Pp. 179-186.

[13] M. Vetterli, P. Marziliano, and T. Blu, "Sampling signals with finite rate of innovation," IEEE Trans. Signal Processing, vol. 50, no. 6, pp. 1417-1428, June 2002.

[14] Bowick C. (1997). RF Circuit Design. Newnes is an imprint of Elsevier Science. Pp 47-49.

[15] A.M. Aibinu, M.J.E. Salami and A. A. Shafie, "Artificial neural network based autoregressive modeling technique with application in voice activity detection", Engineering Applications of Artificial Intelligence 25 (2012) 1265–1276.

[16] A. M. Aibinu, A. A. Shafie and M. J. E. Salami, Performance Analysis of ANN based YCbCr Skin Detection Algorithm, International Symposium on Robotics and Intelligent Sensors 2012 (IRIS 2012), Procedia Engineering, Vol. 41, August 2012, ISSN 1877-7058.2012.

[17] A. M. Aibinu, M.J.E. Salami and A. A. Shafie, "*A novel signal diagnosis technique using pseudo complex-valued autoregressive technique.* Expert Systems with Application", 38 (8). pp. 9063-9069. ISSN 0957-4174, 2011

[18] S. Haykin,and S. Kesler, "Prediction error filtering and maximum entropy spectral estimation", In: Non-Linear Methods of Spectral Analysis, Springer-Verlag, New York, pp. 16–32 (Chapter 2), 1979.

Human Identification at a Distance Using Body Shape Information

N K A M Rashid[1], M F Yahya and A A Shafie
Department of Mechatronics Engineering, Faculty of Engineering,
International Islamic University Malaysia, 53100, Kuala Lumpur, Malaysia

E-mail: nahrul@iium.edu.my

Abstract. Shape of human body is unique from one person to another. This paper presents an intelligent system approach for human identification at a distance using human body shape information. The body features used are the head, shoulder, and trunk. Image processing techniques for detection of these body features were developed in this work. Then, the features are recognized using fuzzy logic approach and used as inputs to a recognition system based on a multilayer neural network. The developed system is only applicable for recognizing a person from its frontal view and specifically constrained to male gender to simplify the algorithm. In this research, the accuracy for human identification using the proposed method is 77.5%. Thus, it is proved that human can be identified at a distance using body shape information.

1. Introduction

Human identification is critical in today's society for enhancing public safety and privacy protection. Many methods published so far for human identification at a distance focused on face and gait like [1], [2], and [3]. However, these methods have their own vulnerabilities such as face is prone to fake facial images as demonstrated in [4] and gait is susceptible to occlusion as in [5]. For that reason, developing a new algorithm by utilizing information other than the face or gait is highly anticipated. One of the available information if seen from a far is the shape of human body. Basically, the shape of human body offers distinctive features not only at the upper part of the body which is the shape of the head but also the slant of the shoulder and the physique of the trunk. These distinctive features when combined form a set of body features which are unique from one person to another. In this paper, an intelligent system approach by utilizing fuzzy logic and multilayer neural network is developed to recognize human at a distance based from their body shape information.

2. Methodology

The development of the proposed human identification system consists of three stages. At the initial stage, the process is concerned about extracting body features data from human subjects (i.e. data of the head, shoulder, and trunk). In the second stage, based on the acquired data, fuzzy logic is utilized in a way such that each body feature is classified into its respective classes. The classes are determined based on the shape of the head, the slant of the shoulder, and the physique of the trunk. Then, in the final stage, the information gained from the fuzzy logic implementation is fed into an artificial neural network. This is so that the network can learn and recognize the pattern of the information which belongs to that of a specific human subject.

[1] To whom any correspondence should be addressed.

2.1. Human subject body features data acquisition

In order for a system to perform human recognition, it needs a priori data about the human subject. The priori data is collected based on the body features of human specifically the head, shoulder, and trunk. The head, shoulder, and trunk are detected using a developed set of image processing techniques. Figure 1 shows the original image, the detected shape of body features and their corresponding parameters which will be processed further for recognition purposes. Parameters H_w, H_h, H_{bp}, and H_{wp} in figure 1 (b) symbolize the width of the head, the height of the head, the accumulation of black pixels, and the accumulation of white pixels in the bounding box which encloses the head respectively. In figure 1 (c), parameter γ_s represents the topmost angle of the marked triangle and figure 1 (d) shows T_w and T_h which are the parameters which represent the width and the height of the trunk respectively.

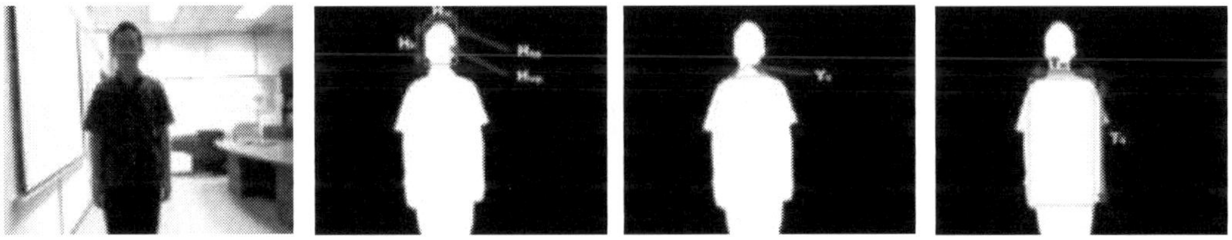

(a) (b) (c) (d)

Figure 1. Images used to obtain the priori data of a human subject:
(a) original image (b) head (c) shoulder (d) trunk

2.2. Body features classification using fuzzy logic

Since human have ambiguous body shape, the accurate mathematical models for classifying their features are difficult to establish. Fortunately, fuzzy logic provides an alternative approach for modeling problems which are imprecisely defined by mathematical models. In this research, the overall application of fuzzy logic concepts is utilized using a fuzzy inference system (FIS). A FIS defines a nonlinear mapping of the input data vector of body features into a scalar output (the type of body feature). Three FISs are developed for classifying the shape of the head, the slant of the shoulder, and the physique of the trunk. Additionally, triangular membership function is utilized for all of the developed membership functions.

2.2.1. The shape of the head. From cropping and observing several processed images of human head, its shape can be classified either as oval vertical, square, circle, or oval horizontal as shown in figure 2.

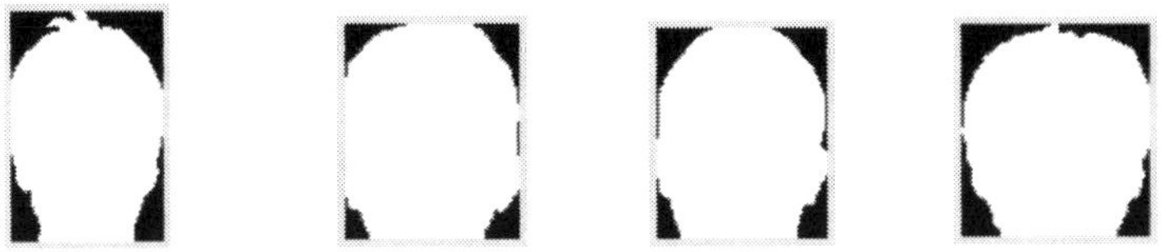

(a) (b) (c) (d)

Figure 2. The shape of the head: (a) oval vertical (b) square (c) circle (d) oval horizontal

From the figure, there are two parameters which can be used to determine the different type of head which are the aspect ratio and the deviation of its shape to the bounding box which encloses the head. The head aspect ratio R_h is defined as

$$R_h = \frac{H_w}{H_h} \tag{1}$$

and the head shape deviation to the bounding box is defined as

$$D_h = \frac{H_{bp}}{H_w} \tag{2}$$

For both of the parameters R_h and D_h, their linguistic variables are {*Small*, *Medium*, *Big*}. Then, the rules are as tabulated in table 1.

Table 1. FIS rules for head classification

Deviation \ Ratio	*Small*	*Medium*	*Big*
Small	*Oval_Vertical*	*Square*	*Oval_Horizontal*
Medium	*Oval_Vertical*	*Circle*	*Oval_Horizontal*
Big	*Oval_Vertical*	*Circle*	*Oval_Horizontal*

Output of the FIS is a defuzzification value from 1 to 4 where 1, 2, 3, and 4 represent oval vertical, square, circle, and oval horizontal respectively.

2.2.2. The slant of the shoulder. Shoulder slant can be either one of the three classes of very concave down, concave down, or straight as shown in figure 3.

(a) (b) (c)

Figure 3. The slant of the shoulder: (a) very concave down (b) concave down (c) straight

The slant of the shoulder is determined by making use of the constructed triangular geometry using three points between the neck and shoulder of a person. From basic triangular geometry as shown in figure 4, the angle $\gamma_{P_{st}}$ plays an important role to classify shoulder to its respective classes.

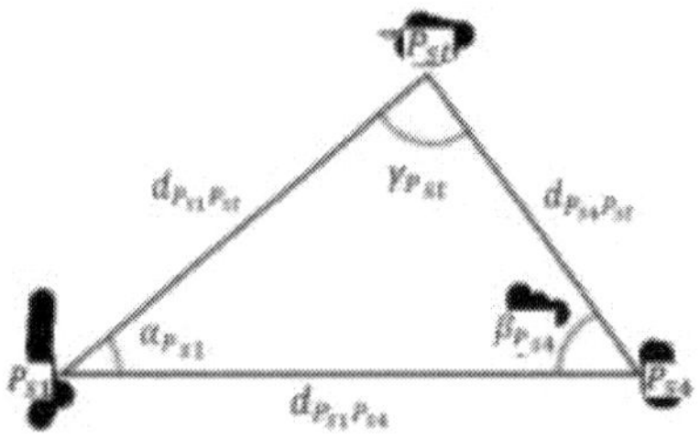

Figure 4. Triangular geometry

The angle γ at P_{st} of the triangle by using the law of cosine is found to be

$$\gamma_{P_{st}} = \cos^{-1}\left(\frac{{d_{P_{s1}P_{st}}}^2 + {d_{P_{s4}P_{st}}}^2 - {d_{P_{s1}P_{s4}}}^2}{2(d_{P_{s1}P_{st}})(d_{P_{s4}P_{st}})}\right) \tag{3}$$

The shoulder angle is defined as

$$\gamma_s = \gamma_{P_{st}} \tag{4}$$

For the input, let γ_s indicates the shoulder slant angle with linguistic variable be {*Small*, *Medium*, *Big*}. For the output membership function (the shoulder type), it has linguistic variable set to {*Very_Concave_Down*, *Concave_Down*, *Straight*}. The rules are stated as

R_1: If Angle is *Small* Then Shoulder Type is *Very_Concave_Down*
R_2: If Angle is *Medium* Then Shoulder Type is *Concave_Down*
R_3: If Angle is *Big* Then Shoulder Type is *Straight*

Output of the FIS is a defuzzification value from 1 to 3 where 1, 2, and 3 represent very concave down, concave down, and straight respectively.

2.2.3. The physique of the trunk. In this research, it is discovered that the physique of the trunk can be classified into five classes of very thin, thin, slim, fat, and very fat as depicted in figure 5.

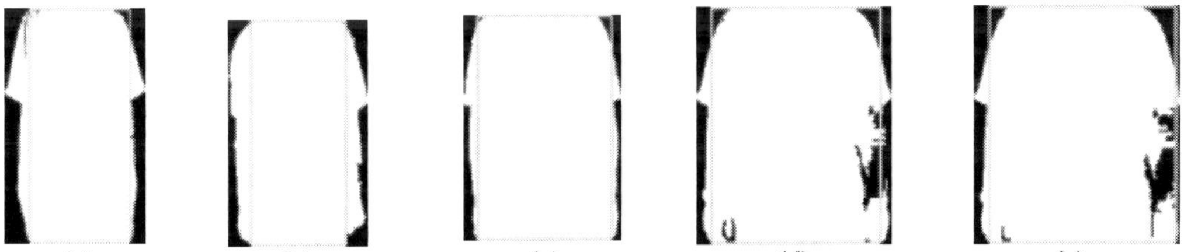
(a) (b) (c) (d) (e)

Figure 5. The physique of the trunk: (a) very thin (b) thin (c) slim (d) fat (e) very fat

The physique of the trunk can be determined by using the aspect ratio of the bounding box which encloses the trunk. The trunk aspect ratio is defined as

$$T_r = \frac{T_w}{T_h} \tag{5}$$

The input to the FIS is defined as T_r with its subsequent linguistic variable be {*Very_Small*, *Small*, *Medium*, *Big*, *Very_Big*}. The output membership function (the trunk type) has linguistic variable term set to {*Very_Thin*, *Thin*, *Slim*, *Fat*, *Very_Fat*}. The rules statement for classifying the type of trunk are stated as

R_1: If Aspect Ratio is *Very_Small* Then Trunk Type is *Very_Thin*
R_2: If Aspect Ratio is *Small* Then Trunk Type is *Thin*
R_3: If Aspect Ratio is *Medium* Then Trunk Type is *Slim*
R_4: If Aspect Ratio is *Big* Then Trunk Type is *Fat*
R_5: If Aspect Ratio is *Very_Big* Then Trunk Type is *Very_Fat*

Output of the FIS is a defuzzification value from 1 to 5 where 1, 2, 3, 4, and 5 represent very thin, thin, slim, fat, and very fat respectively.

2.3. Training of artificial neural network

In this research, since there are multiple inputs of body features information and multiple outputs of Boolean algebra sequence (assigned to each of the human subjects initially), a multilayer network architecture is best suited to perform mapping of inputs to outputs. Figure 6 shows an architectural design of the network. The network basically consists of three layers; one input layer L_1, one hidden layer L_2, and one output layer L_3.

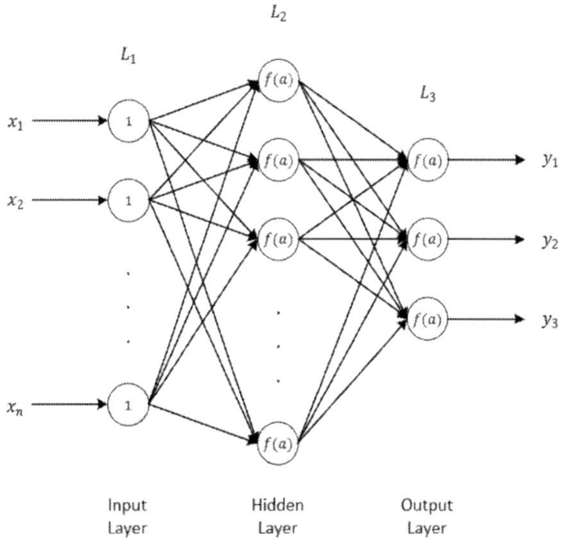

Figure 6. A three-layer feed-forward network

$x_1, x_2, \dots, x_n$ signify the input (from the defuzzification values of the head, shouder, and trunk) to the multilayer neural network. Each neuron uses sigmoid activation function because it is smooth and bounded function. The output of the network is $y_3\, y_2\, y_1$ or combination of Boolean algebra set to 000, 001, 010,…,111. Backpropagation learning algorithm is used to train the network since the inputs (body features information) are non-linearly separable. During training, it is found out that the optimum network have 0.01 mean-squared error with 7078 epochs for which the network is set to have 10 hidden neurons. This network is selected to undergone a test for accuracy of the developed human identification system.

3. Results and Discussion

The human identification accuracy is defined as

$$hi_a = \frac{n_{si}}{n_{ti}} \times 100\% \tag{1}$$

where n_{si} is the number of successful identification and n_{ti} is the total number of identification. Before the accuracy test can be performed, the analysis requires experimental setup. A total of 20 video sequences each for 8 people are recorded, where one sequence last for approximately 15 seconds. During the recording, each person was instructed to go towards the camera from multiple directions and stop at a distance of about 2 meters in front of the camera. Then, the images from the video are captured to go through the body features data acquisition stage and afterwards the body features classification using fuzzy logic stage. The initial 10 processed datasets are used as continuous input to train a neural network and another 10 processed datasets are used to test the network. Table 2 shows the tabulated result for successful identification of the human subject.

Table 2. Successful human identification

Human subject	Successful identification (out of 10)
HS001	7
HS002	7
HS003	9
HS004	8
HS005	8
HS006	8
HS007	7
HS008	8

Based on table 2, the final calculated accuracy is found to be 77.5%.

4. Summary

In this paper, an intelligent system approach for human identification at a distance using the body shape information is proposed. The system utilizes three body characteristics of human: the shape of the head, the slant of the shoulder, and the physique of the trunk. The shape of the head can be categorized into four different classes: oval vertical, square, circle, and oval horizontal. The slant of the shoulder can be either one of the class of very concave down, concave down, or straight. The physique of the trunk classes are very thin, thin, slim, fat, and very fat. By using FIS, those classes can be represented as linguistic variables which may yield a unique single value representing each class of body feature. These values are then fed into an artificial neural network to generalize the pattern of given inputs during training process. For testing purposes, the response outputs of neural network are converted into combination of Boolean algebra values to represent or identify specific human subject. During testing of the proposed system, the successful identification rate of 77.5% is achieved. Thus, it is concluded that human can be recognized from a distance using body shape information.

References

[1] Bhanu B and Zhou X 2006 Feature fusion of face and gait for human recognition at a distance in video *Int. Conf. on Pattern Recognition* vol 4 pp 529-532

[2] Almohammad Manhal Saleh, Salama Gouda Ismail and Mahmoud Tarek Ahmed 2012 Human identification system based on feature level fusion using face and gait biometrics *Int. Conf. on Engineering and Technology (ICET)* pp 1-5

[3] Kale Amit, Roychowdhury A K and Chellappa R 2004 Fusion of gait and face for human identification *Proc. IEEE Conf. on Acoustics, Speech, and Signal Processing* vol 5 pp 901-904

[4] Määttä J, Hadid A and Pietikainen M 2012 Face spoofing detection from single images using texture and local shape analysis *IET Biometrics* vol 1 pp 3-10

[5] Hofmann M, Wolf D and Rigoll G 2001 Identification and reconstruction of complete gait cycles for person identification in crowded scenes *Proc. Int. Conf. on Computer Vision Theory and Applications (VISAPP)*

Optimizing POF/PCF based optical switch for indoor LAN

M M I Bhuiyan[1, 4], M M Rashid[1, 5], Sayem Ahmed[1, 6], M Bhuiyan[2] and M Kajihara[3]

[1] Department of Mechatronics Engineering, International Islamic University Malaysia, Jalan Gombak, 53100 Kuala Lumpur, Malaysia
[2] SEGi University, Faculty of Medicine, No. 9, Jalan Teknologi, Taman Sains Selangor, Kota Damansara PJU 5, 47810 Petaling Jaya, Selangor, Malaysia
[3] Department of Materials Science and Engineering, Tokyo Institute of Technology, Yokohama 226-8502, Japan
E-mail: [4]moinul@iium.edu.my, [5]mahbub@iium.edu.my, [6]sayem7746@gmail.com

Abstract. For indoor local area network (LAN) the Polymer optical fiber (POF) is mostly appropriate, because of its large core diameter and flexible material. A 1×2 optical switch for indoor LAN using POF and a shape memory alloy (SMA) coil actuator with magnetic latches was successfully fabricated and tested. To achieve switching by the movement of a POF, large displacement is necessary because the core diameter is large (e.g., 0.486mm). A SMA coil actuator is used for large displacement and a magnetic latching system is used for fixing the position of the shifted POF. The insertion loss is 0.40 to 0.50dB and crosstalk is more than 50dB without index-matching oil. Switching speed is less than 1s at a driving current of 80mA. A cycling test was performed 1.4 million times. Polymer clad fiber optical (PCF) switch also fabricated and tasted.

***Keywords**: Optical switch, Shape memory alloy, Plastic optical fibre, Indoor LAN, Magnetic latch*

1. INTRODUCTION

For office and home environment Polymer optical fiber (POF) is more suitable than the quartz optical fibre to construct LAN (Local Area Network) [1-3], because of its flexibility and ease of connection by relatively large core diameter. Demand of treating huge information in home or office is increasing and it is predicted that the huge information from the outdoors will be transmitted to various information terminals and AV equipment using an optical fibre through information consents in the near future (Figure 1).

In order to transmit the information from the outdoor to each information terminal, optical switches are needed in every bifurcation.

[1] M M I Bhuiyan

IOP Conf. Series: Materials Science and Engineering **53** (2013) 012059 doi:10.1088/1757-899X/53/1/012059

Small core Silica optical fibers are widely used for communication systems. However, such small core, only 0.01 mm optical fibers (Figure 1(b) shows the fiber diameter comparison) require high precision alignment between two fiber ends, and which is the main problem in indoor LAN connection area. In this research we approached POF/PCF (although higher optical attenuation than silica fiber but larger diameter) to overcome the problem. Because of larger core diameter, POF/PCF is selected for the LAN switch. A large displacement is required for switching. SMA [4-6] coil is preferred for actuation and power consumption is also considered.

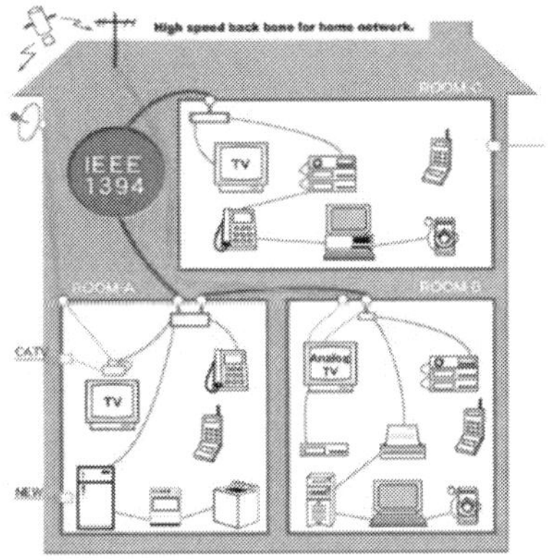

(a)

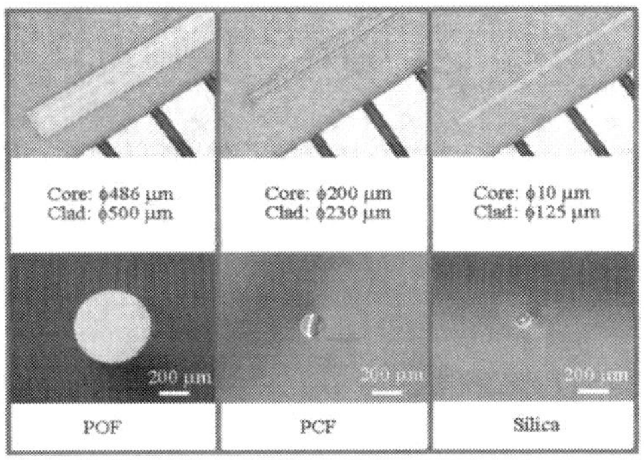

(b)

Figure 1: (a) Concept of future home network (b) Large diameter optical fibre (POF/ PCF) and conventional silica fibre.

A magnetic latching system [7] is introduced in this device for holding the position of the shifted POF, and therefore a continuous current supply is not necessary. There are some optical fibre switches, which use solenoid coils and permanent magnets, which move and latch fibre magnetically. Such switches are for silica optical fibre and required small displacement of the fibre.

2. Structure and Principle of Switch

Structure of the optical switch is shown in Figure. 2(a) and a circuit diagram is shown in Figure. 2(b). There is a POF fixed to a SMA coil in the right side of the figure and two POF(s) are fixed in the left side. One of the fibre ends of the left side POF(s) and the fibre end of the right side POF are faced each other. If switch No.1 is turned on (Figure. 2 (b)), electrical current is applied to upper half of the SMA coil and the upper half of SMA is heated above a certain transition temperature by direct heating and contracts. POF, which is attached to the SMA moves towers upper side and the light is passed into the upper left side POF. If No.2 switch (Figure. 2(b)) is turned on similarly, the same change will be occurred in the lower SMA and the POF is moved lower side. For switching by movement of a POF, large displacement is needed because of large core diameter (in this case, Core diameter: 0.486mm, Outer diameter: 0.5mm) and this device needs displacement of 0.5mm. It is relatively difficult to get displacement of 0.5mm by micro actuator for example, electrostatic actuator, piezoelectric actuator etc. We chose SMA coil actuator for large displacement.

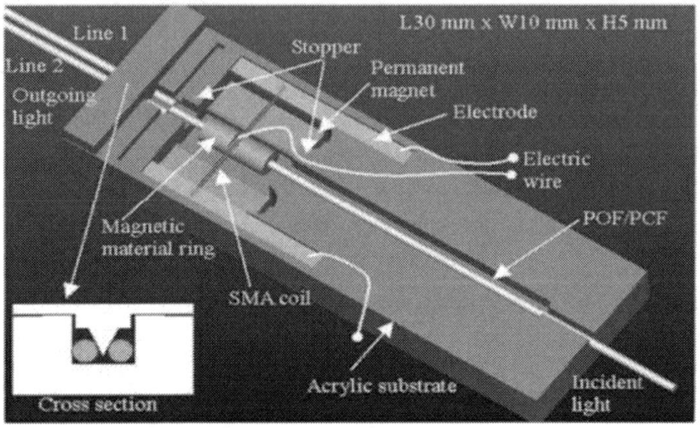

(a): Structure of switch

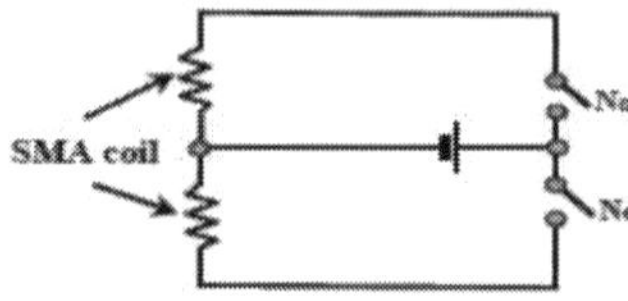

(b): SMA drive circuit

Figure 2: Structure of switch and SMA drive circuit

As shown in Figure. 2(a), two pairs of stoppers are located both sides of the right side POF. The stoppers enable its fibre end and left side POF end (in Figure.2) to face correctly each other. However, the current must be applied continuously to maintain the displaced position in the case of using SMA actuator [1]. To solve this problem the magnetic material ring (nickel) is attached to the right side POF, and two permanent magnets are placed at the both side of the POF. This magnetic latch mechanism holds the position of the POF even if driving current is turned off. Moreover, this magnetic latch prevents upward displacement of the POF by pulling it down by embedding two permanent magnets in the substrate.

Acrylic substrate, which is carrying on all the parts, is fabricated using high-speed milling. The size of the whole switch is L25mm×W15mm×H5mm. The size of the permanent magnet was L4mm×W4mm×H2mm. Surface magnetic flux density was 320mT and magnetic force was 0.35kg. Ni-Ti SMA coil micro actuator was used. Its external diameter was 0.2mm and wire diameter was 0.05mm. External diameter, wall thickness and length of Ni magnetic material ring were 1.5mm, 0.5mm and 4.0mm respectively.

3. Fabrication

3.1 The processing method and result

In order to make precise alignment of the POF, it is necessary to fabricate with sufficient accuracy. The high-speed milling machine is preferred for cutting the acryl substrate. Before the processing, appropriate tool and processing conditions are chosen and checked using simulator as shown in Figure. 3.

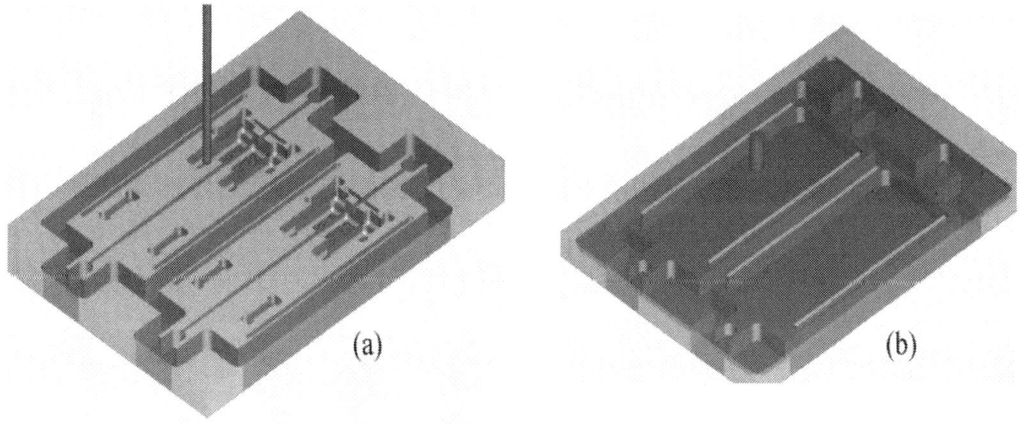

Figure 3: Process simulation of Optical Switch Structure

Three types of switches were fabricated using the POF of 0.5mm, 0.75mm and 1.0mm diameter respectively. Three types of acrylic substrates fabricated using high-speed milling machine are shown in Figure 4. Acrylic substrate was cooled by water to avoid deformation during high-speed milling. Firstly, a switch using 0.5-mm-diameter PMMA POF (Core diameter: 0.486mm, Outer diameter: 0.50mm) was fabricated and its performance was evaluated.

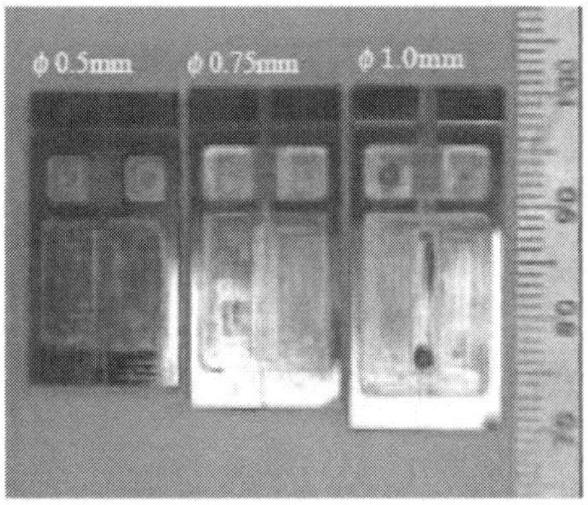

Figure 4: Three types of acrylic substrates

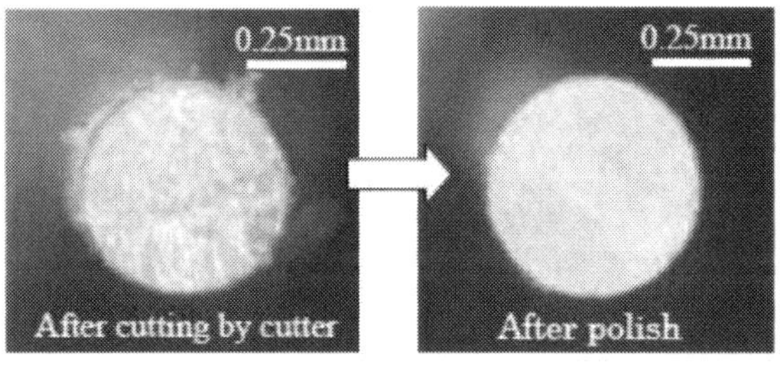

Figure 5: The optical images of POF end

3.2 Processing of POF End

A parallel POF end with a smooth surface is beneficial in reducing insertion loss in an optical switch. In this study, a mechanical polishing technique [8] was used for POF to obtain the parallel end face with a lower roughness reproducibly. It is reported that the mechanical polishing technique could form a smooth surface with lowest loss compared to other techniques such as hot plate, hot cutter technique. Figure 5 compares the end face polished by a fibre polisher (Seikoh Giken SEP-7002) to that produced by a cutter. The insertion loss will be reduced further by alignment of POF(s) using the magnetic latch and the stopper (Figure. 2(a)), gap control between the fibre ends (Figure. 2(a)) and precise fixation of POF(s).

IOP Conf. Series: Materials Science and Engineering **53** (2013) 012059 doi:10.1088/1757-899X/53/1/012059

3.3 Assembly Process

The above-mentioned parts are assembled on the processed acrylic substrate as the following.

1. Cu electrodes for driving of the SMA were attached to the substrate with an adhesive.
2. The middle of the SMA coil was fixed to a metallic ring and connected to the electric cables.
3. Two permanent magnets were fixed to each predetermined position.
4. The right side POF was inserted into the metallic ring and fixed using the adhesive.
5. The fibres in the left side were placed in the grove of the substrate and properly fixed by pressing from upper side using a V-shaped convex. (See Figure.6).
6. Gap between the POF ends was adjusted under a microscope.
7. Finally, the magnetic material ring was attached around

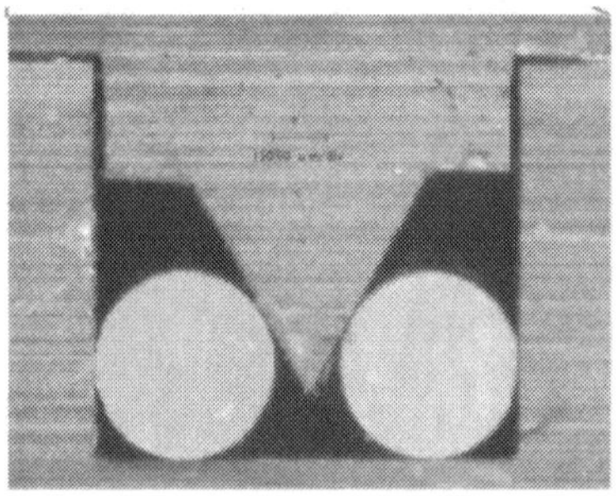

Figure 6: Fixation of left side POF(s)

4. Results

4.1 Characteristics of SMA actuator

Comparison between coupling loss using index-matching oil and coupling loss without index-matching oil is shown in Figure 7. In this case mechanically polished ends were used. The insertion loss of the fabricated switch was 0.50 dB or less, namely, a minimum value of 0.40 dB to a maximum value of 0.50 dB when index-matching oil was not used. When index-matching oil was used, insertion loss was less than 0.10 dB, namely, a minimum value of 0.06 dB to a maximum value of 0.09 dB.

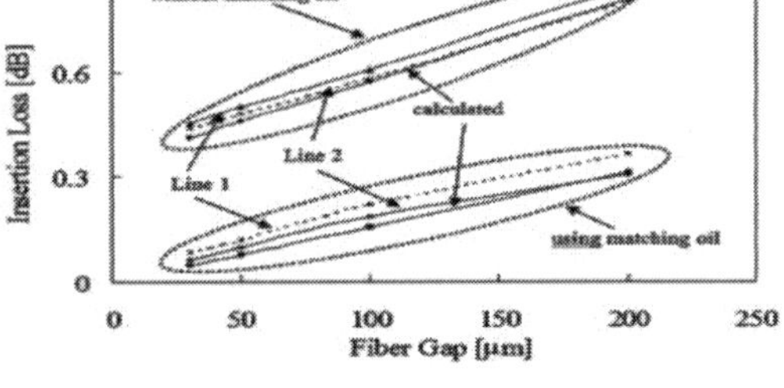

Figure 7: Required release power from magnetic latch.

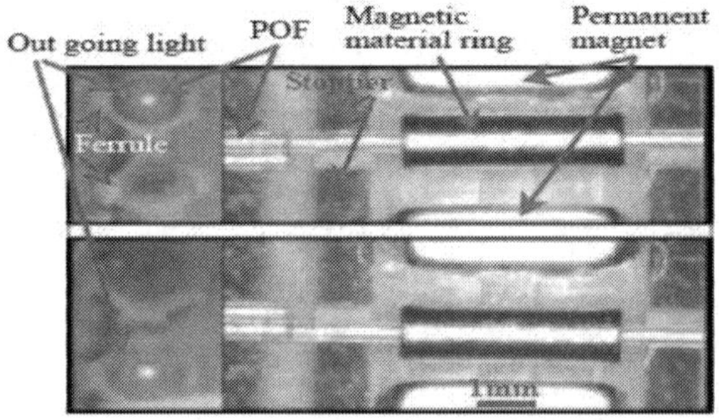

Figure 8: Switching condition

4.2 Evaluation of Switching Speed

650nm laser diode was used light source. The light intensity, which came out from the output side, was measured as shown in Figure.10. In this measurement, index-matching liquid was not used. To measure coupling loss, intensity of outgoing light from the switch (Figure 9) was compared to that of reference POF (Figure 9). The entire pass lengths are 300mm. Insertion loss was less than 1[dB] (minimum value of 0.62[dB] to maximum of 0.74[dB]) (Figure.10). Almost insertion loss of conventional optical switch for communication is less than 1 [dB]. To reduce insertion loss further, antireflective-coated fibre end could be effective. This would also reduce back reflections. Index matching liquid is not preferable, because their surface tension prevents switching motion and the liquid have to be sealed in the device.

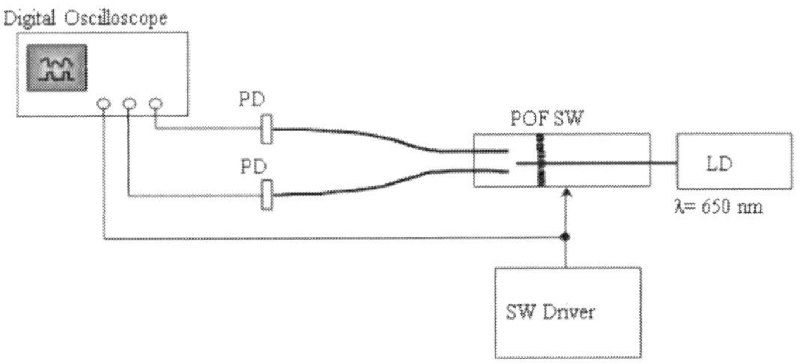

Figure 9: Measurement method

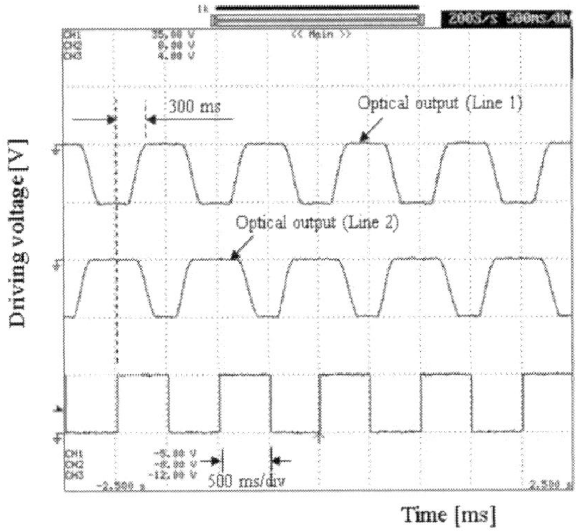

Figure 10: Measurement of coupling loss

5. Conclusion

Optical switch (1×2) using POF and SMA coil actuator with magnetic latch has been successfully fabricated and actuated. The continuous switching operation was confirmed and during this test switching speed was less than 1 second at driving current of 80mA. This speed is sufficient for indoor LAN.

The loss of the fabricated switch was less than 1 [dB] without using index-matching liquid. This value is almost equivalent to the conventional quartz fibre switch. This seems to be the firstly realized optical switch using large diameter (0.5mm) optical fibre.

Other large external diameter (for example, 0.5mm) plastic clad fibre (PCF) can be used for this device. To fabricate the substrate at low cost, injection moulding can be applied instead of high-speed milling machine.

Fabricated 1×2 switch can produce a good result at stand-by application. In future as the insertion loss is very low, by adding several 1×2 optical switches, it is possible to use at 1×8 optical switch application. To fabricate the substrate at low cost, injection molding can be applied instead of high-speed milling machining. We got new information by doing this research work, which will help to fabricate polymer optical switch with low insertion loss in future.

Acknowledgment

A part of this work was performed in Venture Business Laboratory in Tohoku University, Japan.

References

[1] Large, M. C J; Blacket, D.; Bunge, C. -A, "*Microstructured Polymer Optical Fibers Compared to Conventional POF: Novel Properties and Applications*," *Sensors Journal, IEEE* , vol.10, no.7, pp.1213,1217, July 2010 doi: 10.1109/JSEN.2010.2041056

[2] Tao Geng; YuZhuo Wang and HaiBin Wu, "Research on vision-based waveguide alignment of optical fiber," *Measurement, Information and Control (MIC), 2012 International Conference on* , vol.1, no., pp.286,289, 18-20 May 2012

[3] Horino M , Sato K , Hayashi Y, Mita M and Nishiyama T, "*Plane Type Fiber Optic Switches*," The Institute of Electronics, Information and Communication Engineers Paper Magazine Vol.J83-C (2000), pp.681-688 (in Japanese)

[4] Bhuiyan M. M I; Haga Y and Esashi M., "Design and characteristics of large displacement optical fiber switch," *Quantum Electronics, IEEE Journal of* , vol.41, no.2, pp.242,249, Feb. 2005
doi: 10.1109/JQE.2004.839681

[5] Adelaide Nespoli, Stefano Besseghini, Simone Pittaccio, Elena Villa and Stefano Viscuso, "*The high potential of shape memory alloys in developing miniature mechanical devices: A review on shape memory alloy mini-actuators, Sensors and Actuators*". A: Physical, Volume 158, Issue 1, March 2010, Pages 149-160, ISSN 0924-4247, 10.1016/j.sna.2009.12.020.

[6] Hoffman M, Member, IEEE, Kopka P and Voges E, "*All-Silicon Bistable Micromechanical Fiber Switch Based on Advanced Bulk Micromachining*," IEEE Journal of Selected Topics in Quantum

[7] Bachman M, Yang Zhang, Minfeng Wang; Li, G., "*High-Power Magnetically Actuated Microswitches Fabricated in Laminates*," *Electron Device Letters, IEEE* , vol.33, no.9, pp.1309,1311, Sept. 2012 doi: 10.1109/LED.2012.2206553

[8] Bilro L , Alberto N , L. M. Sá , Pinto J. L and Nogueira R N , "*Side-polished plastic optical fibre as refractive index, cure and viscosity sensor.*" Proc. SPIE 8001, International Conference on Applications of Optics and Photonics, 800142 (July 26, 2011); doi:10.1117/12.892099

Design and fabrication stable LNF contact for future IC application

M M I Bhuiyan[2, 4], M M Rashid[1, 5], Sayem Ahmed[1, 6], M Bhuiyan[2] and M Kajihara[3]

[1] Department of Mechatronics Engineering, International Islamic University Malaysia, Jalan Gombak, 53100 Kuala Lumpur, Malaysia
[2] SEGi University, Faculty of Medicine, No. 9, Jalan Teknologi, Taman Sains Selangor, Kota Damansara PJU 5, 47810 Petaling Jaya, Selangor, Malaysia
[3] Department of Materials Science and Engineering, Tokyo Institute of Technology, Yokohama 226-8502, Japan
E-mail: [4] moinul@iium.edu.my, [5] mahbub@iium.edu.my, [6] sayem7746@gmail.com

Abstract. Enable the design of a small contact spring for applications requiring high density, high speed and high durability. A low normal force (LNF) contact spring with high performance is fabricated using a unique combined MEMS photo resist lithography and electro fine forming (EFF) technology. Reducing a total contact material cost of a connector, a high-Hertz stress with LNF contact will be a key technology in the future. Only radius R 5μm tip with 0.1N force contact provides an excellent electrical performance which is much sharper than conventional contact. 0.30million cycle's durability test was passed at 300μm displacement and the contact resistance was ≦50mΩ.

Keywords*: LNF contact, high-Hertz stress, electro fine forming, NiCo, photolithography*

1. Introduction

In present situation of electrical industry, miniature, light weight and high performance electronic devices production has been accelerated [1-2]. In order to increase the performance of a device the number of I/O pin is increasing. The size of those devices getting smaller. After introducing the Micro-Electro-Mechanical Systems (MEMS) it becomes possible to develop micro sized element for devices. On the other hand, miniature, light and high functional electronic devices have been developed using high functional LISs called SOC (System-on-Chip) [3-5].

Pogo pin is using in test socket area as shown in Figure 1. However, its contact force is approximately 0.5N. In this case 9806N/20,000pin is required for one board testing. This value is very

[1] M M I Bhuiyan

high for the board and IC contact also. Sometime IC board or contact receives big damage and it is economically not acceptable. The user's requirement is the load need to decrease at least half. If we can reduce this total load by 1/3 with high performance, that will be an excellent innovation. Therefore, an immediate solution is required.

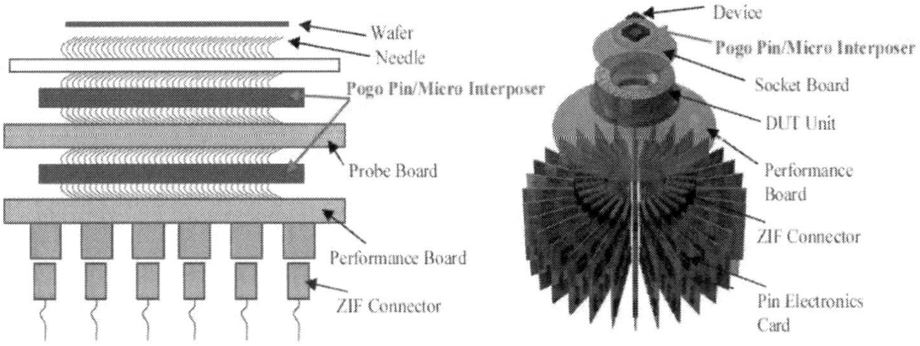

Figure 1. IC test socket application

The commercial stamping process has size limitation that R≥200μm=low Hertz-stress and it can't meet the user's requirements. Here is our innovation that, if it is possible to fabricate a very high Hertz-stress contact with LNF contact then it could make a stable contact resistance and this technology will help to reduce the material cost drastically.

We developed an excellent material Ni-Co to decrease active resistance of contact spring. Its durability performance is good. Although a suitable contact with appropriate R is the Key factor for LNF, it has not been optimized yet. We tried to optimize an appropriate R using several size R (5 to 200μm) by the same LNF contact which will be described in this paper. To fabricate this miniature contact, a MEMS technology is used instead of the existing pressing and injection moulding method because it is not compatible with the high-density packaging required in the advanced IT market. This research also focuses on a fabrication of micro interposer by combining UV thick resist photolithography and Ni-Co fine electroforming technology. This fabrication method is expected to develop high precision micro interposer and high-density packaging in mass production and is very essential to realize micro order devices.

The business value of this contact will be utilized on the Test Interface Connection area. For example, test sockets for IC-devices, test board I/F for IC-tester, and High speed board to board for I/O device. Furthermore, on a Medical Equipment area like 3D transducer interconnection for ultra-sound, X-Ray sensor I/O for CT scanner is also necessary.

2. HERTZIAN CONTACT MECHANISM & DESIGN

A small spring with high reliability has to be developed to satisfy mentioned requirements which are focused in our research. Since sufficient contact pressure cannot be taken when the conventional material is used, although terminal size is made into a micro level, a reliable spring is not made. The commercial material Vickers hardness, as for example; Brass is 100–175, Phosphor Bronze is 150–200 are very low, where we needed the hardness more than double. Also a high Young's modulus material should be chosen for this research. Considered all of mentioned problems, NiCo material is selected for our miniature size contact.

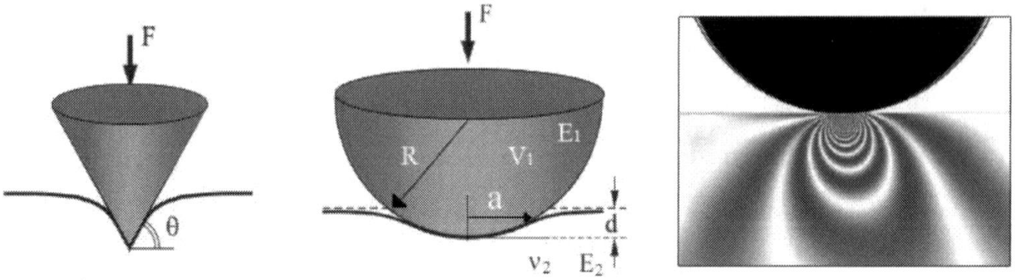

Figure 2: The geometry of Hertzian contact

Figure 2 shows a sphere (radius R, elastic contacts v_2, E_2) pressed by a load F into a flat substrate (with elastic constants v_1, E_1). The load is supported over a circular area whose radius, a, is given by:

$$= 0.721\sqrt[3]{FE * R_D} \tag{1}$$

Where,

$$E^* = \frac{1-v_1{}^2}{E_1} + \frac{1-v_2{}^2}{E_2} \tag{2}$$

The peak pressure under the sphere, σ_0, is given by:

$$\sigma_0 = 1.5\ \frac{F}{\pi a^2} \tag{3}$$

The stress field outside the contact zone has a tensile radial stress near to the surface which rapidly decreases, and soon becomes compressive, with depth [9].

Here R_D: Diameter of the sphere, F: Contact force, E_1, E_2: Young's Modulus, V_1, V_2: Poisson's ratio. σ_0: Hertzian stress.

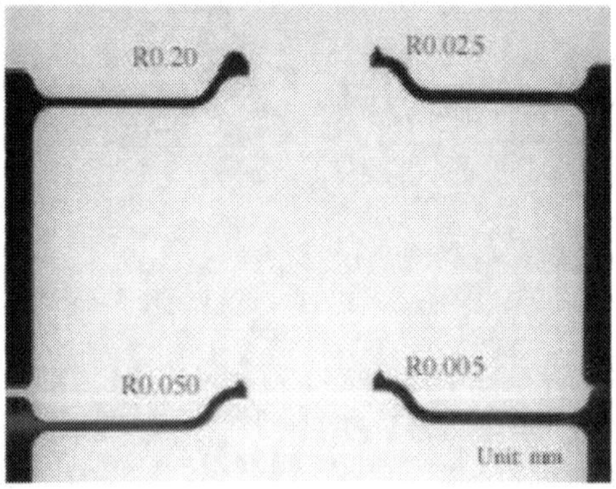

Figure 3: Different R in equal diameter structure

In order to increase Hertzian stress [9-12] of contact point of the terminal have to be lower radius (sharp) to stick to the upper terminal. To investigate an appropriate radius R, we designed four types of radiuses such as 0.005, 0.025, 0.050, and 0.2mm with same size of the contact structure as shown in Figure 3.

In our case, the values of all parameters are, R_D: 10μm, F: 0.16N, E1, E2: 180, 110GPa respectively, V_1, V_2: 0.31, 0.33 respectively. Calculating by using the equation (3), a compression stress of 19.14GPa is obtained. Using another larger diameter R_D=50, 100, 400μm are calculated by same equation, then the compression stress are 6.54, 4.12, 1.64GPa respectively. We can understand here, a small contact point obtains high contact stress and a steady contact resistance consequently. In this research, it verifies what kind of phenomenon occurs in a contact surface by using contacts with different R sizes.

3. FABRICATION:

The high aspect ratio micro interposer increases the contact resistance. Consequently, it improves the reliability of electrical signal transmission. A fabrication process of the micro interposer combining UV thick resist photolithography [11] and Ni-Co electro fine forming (EFF) technology were adopted to satisfy the above-mentioned specification. In order to make the terminal thicker, UV thick negative type photo resist is necessary. In this research we tried to fabricate a low resistance contact using a new process for a Ni-Co/Au/Ni-Co sandwich type structure as shown in Figure 4. The previous contact resistance was 200mΩ. The target of our contact is less than 50mΩ. Therefore, a thick contact and an Au sandwich type contact are decided to fabricate.

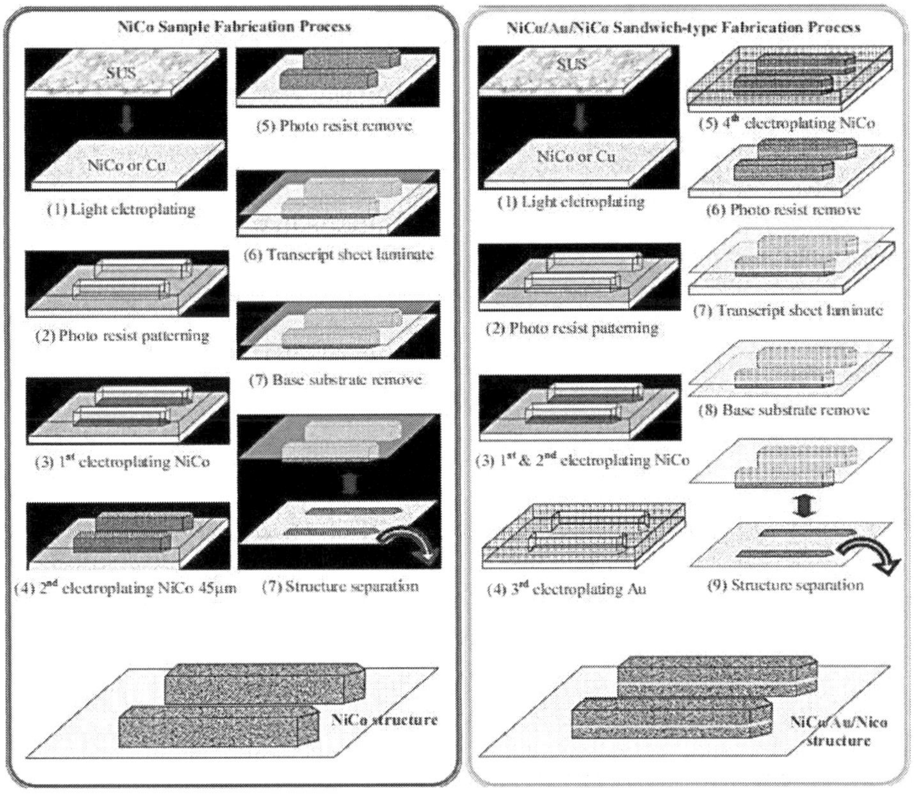

Figure 4: Fabrication process

At first, I would like to explain here a basic type Ni-Co contact fabrication process. First, a Ni-Co bright electroplating (t=15μm) process is done on the base substrate (SUS) (1). A high sticking power is not necessary with this substrate. Second a UV thick photo resist is formed as a mould for the terminals (2). Then, the 1st electroplating (t=4Oμm) of Ni-Co or Cu is done in the mould (3). Au sandwich type structure will be electroplated here if necessary (4). In our case, 1μm Au layer is added for the sandwich contact. The 2nd electroplating (Ni-Co) is done for the structure in the mould, and the thickness is 45μm (5). Here the balance of sticking power of the 1st and 2nd electroplating is very important in order to remove the structure from the 1st electroplating. A neutral sticking power is suitable for here. The photo resist is removed (6). The transcript seat lamination is done to the top of the structure (7). Base structure removed from bright electroplates (8). Lastly, the structure is removed from the 1st electroplates (9).

3.1. Fabricated Structures

Figure 5 shows SEM (scanning electron microscope) photographs of the structures which are formed by the Ni-Co electroforming. The thickness of fabricated structures was measured by laser beam 3D microscope. The thickness was approximately 46μm (design 45μm), radius r=5μm (design 5μm). The edge surface was little bit rough but remained an acceptable level. Also a taper angle about 2% of the cross section of the photo resist as a mould for the terminals was observed. The current density of Ni-Co electroplating was 0.5A/mm^2 constant.

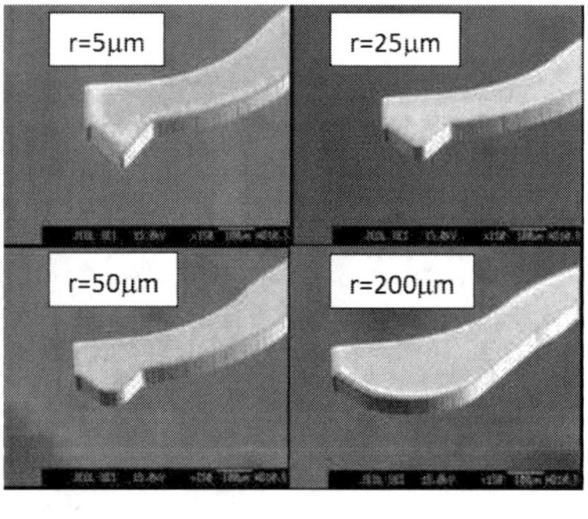

Figure 5: SEM photographs of fabricated structure.

4. Contact Force vs. Contact Resistance

The contact force and contact resistance is measured in different scenario to compare them. Figure 6 illustrate that the contact force is almost stable and same for all of the contacts where the maximum value was same around 0.16N. For each type of contact structure we measured n=5 samples. The lowest radius contact resistance, which is R 5μm, was the mostly stable compared to others. On the other hand, the largest R200μm was the very unstable. Also the R25μm and 50μm was unstable compared to the smaller R5μm contact.

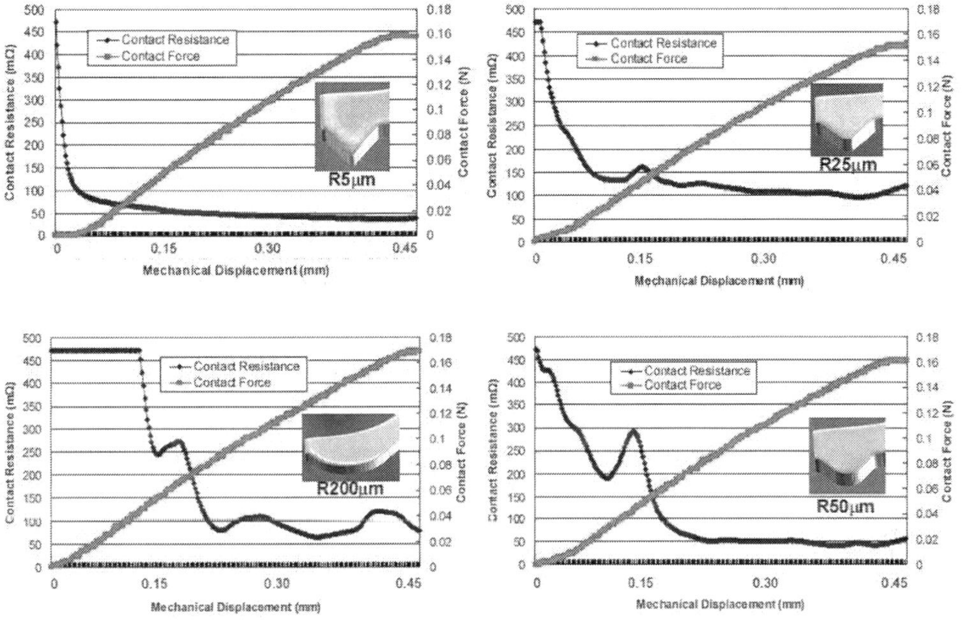

Figure 6: Compression between contact Force and Resistance with different R

Similarly we check these contact springs with Au sandwich type structure and measured by the same process. However, the result was almost alike. For the Au sandwich type the Au layer was only 1μm where the total contact spring thickness is 45μm. To find out the actual result further examination required.

5. Conclusion

In this research we figured out how to provide a stable contact resistance by LNF contact with very sharp contact like R=5μm with high Hertz stress will be needed than the larger contact R=25 to 200μm. The evaluation result proved that the capability of low normal force (LNF) which is 0.06N-0.15N at 0.45mm mechanical displacement by the contact point 5μm radius, the contact resistance was less than 50mΩ. This value is except conductor resistance. After finish the test, the surface condition of both parts (tip of contact and against electro pad) is analysed by SEM and mapping process. In the beginning of contact displacement the entering amount of R=5μm contact to the opposite electro pad was much higher than R=200μm at the same LNF. As a result, the braking ability of R=5μm contact of surface oxide film is much better than the larger one. For this reason, the contact resistance was stable.

Furthermore, we have done a durability test for 0.3million cycles of contact at 0.3mm mechanical displacement. The data of contact resistance was very stable, which is less than 50mΩ. This value is also except conductor resistance. There was no damaged of contact during the durability test.

In order to optimize an appropriate contact R with high Hertz stress for LNF contact, new technology photo lithography (MEMS) and NiCo electro fine forming technology is utilized to fabricate as a minimum R was only 5μm and thickness was only 45μm which is very difficult to fabricate by an existing commercial stamping method. As a result of using this process, we have fabricated successfully in these type of contact.

IOP Conf. Series: Materials Science and Engineering **53** (2013) 012060 doi:10.1088/1757-899X/53/1/012060

Also, to realize a high durability capable contact with high yield strength (more than 180Gpa), a suitable content Percentage of Ni and Co and electroforming current density is checked by some experiments. We have determined by the experiment that in order to fabricate an excellent contact, the ratio of Ni-Co is 8:2 and current density 0.5A/mm2 is necessary. The Vickers hardness rate was more than 500Hv and it is 1.5 times larger from the commercial contact like phosphor bronze.

6. Acknowledgements

This work was partially developed in cooperation with Kyushu Hitachi Maxell Co. Ltd. The author would like to thank Mr. Tsuyoshi Tawara of Kyushu Hitachi Maxell Co. Ltd., Japan for providing valuable manufacturing data and comments on this development. Specially, the author would like to thank Mr. Eiichiro Takemasa who gave me a support to continue this project.

References

[1] Akhtar Razul Razali, Yi Qin, "*A Review on Micro-manufacturing, Micro-forming and their Key Issues*", Procedia Engineering, Volume 53, 2013, Pages 665-672, ISSN 1877-7058, 10.1016/j.proeng.2013.02.086.

[2] Markets Survey Outlook for Electronic Components, http://www.sangyo-joho.co.jp/

[3] Carvalho de, Bernardi M, Sanchez P, Reorda E, M.S., "*An Enhanced Strategy for Functional Stress Pattern Generation for System-on-Chip Reliability Characterization,*" *Microprocessor Test and Verification (MTV), 2010 11th International Workshop on*, vol., no., pp.29,34, 13-15 Dec. 2010

[4] Bhuiyan M M I, Unno T, Yokoyama Y, Toriyama T and Sugiyama S, "*Micro Connector Fabricated by Micro Process Technology*," MHS2000, pp. 83-88, Oct. 2000, Japan

[5] Unno T, Toriyama T, Bhuiyan M M I, Yokoyama Y and Sugiyama S、"*Micro Connector for High Packaging Density Fabricated by Using UV Thick Photoresist*," The Transactions of The Institute of Electrical Engineers of Japan, The Inst. of Electrical Engineers of Japan, Vol. 122-E, No., 5, pp. 249-255, May. 2002

[6] Michel F and Ehrfeld W, "*Mechatronic Micro Devices*," Proceedings of the 1999 International Symposium on Micromechatronics and Human science, pp. 27-32, 1999

[7] Warren P D, "*Determining the Fracture Toughness of Brittle Material by Hertzian Identation,*" Journal of the European Ceramic Society, vol. 15, pp. 201-207, 1995

[8] Paduture N P, "*Hertzian Contacts*," Elsevier Science Encyclopedia of Materials Science and Technology, pp. 3750-3752. 2001

[9] Petit F, et all. "*Fracture Toughness and Residual Stress Measurements in Tempered Glass by Hertzian Indentation,*"Acta materialia 55, pp. 2765-2774, 2007

[10] P. D. Warren, et all. "*Mechanics of Hertzian Cracking*," Tribology International, Vol. 28, No. 6, pp. 357-362, 1995

[11] Hirai Y, Nakai Y, Sugano K, Tsuchiya T and Tabata O, "*Embedded double-layered microchannel fabrication for microfluidic devices using developer permeability of negative thick-film resists,*" Procedia Engineering, Volume 5, 2010, Pages 854-857, ISSN 1877-7058, 10.1016/j.proeng.2010.09.243.

IOP Conf. Series: Materials Science and Engineering **53** (2013) 012061 doi:10.1088/1757-899X/53/1/012061

New consumer load prototype for electricity theft monitoring

A I Abdullateef[1,3,4], M J E Salami[1], M A Musse[2], M A Onasanya[3] and M I Alebiosu[3]
[1] Department of Mechatronics Engineering, International Islamic University Malaysia, Malaysia
[2] Electrical and Computer Engineering Department, International Islamic University Malaysia, Malaysia
[3] Electrical Engineering Department Yaba College of Technology, Yaba, Lagos, Nigeria

E-mail:isqeel@hotmail.co.uk

Abstract. Illegal connection which is direct connection to the distribution feeder and tampering of energy meter has been identified as a major process through which nefarious consumers steal electricity on low voltage distribution system. This has contributed enormously to the revenue losses incurred by the power and energy providers. A Consumer Load Prototype (CLP) is constructed and proposed in this study in order to understand the best possible pattern through which the stealing process is effected in real life power consumption. The construction of consumer load prototype will facilitate real time simulation and data collection for the monitoring and detection of electricity theft on low voltage distribution system. The prototype involves electrical design and construction of consumer loads with application of various standard regulations from Institution of Engineering and Technology (IET), formerly known as Institution of Electrical Engineers (IEE). LABVIEW platform was used for data acquisition and the data shows a good representation of the connected loads. The prototype will assist researchers and power utilities, currently facing challenges in getting real time data for the study and monitoring of electricity theft. The simulation of electricity theft in real time is one of the contributions of this prototype. Similarly, the power and energy community including students will appreciate the practical approach which the prototype provides for real time information rather than software simulation which has hitherto been used in the study of electricity theft.

1. Introduction

One of the major challenges facing power providers worldwide is electricity theft which is the practice of using electricity from the utility company without the company's authorization or consent. Electricity theft, which could be in form of billing irregularities, meter tampering, and illegal connection and unpaid bills is commonly carried out at the consumer end, [1]. The prevalence of electricity theft in developing countries range from 20 to 30% losses in the distribution network [2], and [3] reported a wider range of 10 - 40%. Recent worldwide report indicates an increase in theft records as well as numbers of offenders include residential customers. American Electric Power reported 27% increase in electricity theft between January and February 2009 as compared to previous year [4]. Similarly, investigation of 14,000 customers by PECO Energy Company, (USA), in early 2008, showed 30% electricity pilfering [4].

[4] Ayodele Isqeel Abdullateef

Furthermore, the consequence of this menace, in terms of revenue loss, is well evident in developed countries such as the USA and Canada which recorded huge revenue loss, amounting to $6 billion and $100 million, respectively, to electricity theft in 2010 [5-6]. Similarly, fast developing countries, such as Malaysia, reported a revenue loss ranging from RM150million to RM500 million, between 2010 and 2011, [7-8]. In addition, losses amounting to US$32 million due to electricity theft was recorded by Jamaica Public Service Company (JPS) in 2012 [9] while losses recorded in South Africa ranged between R2.5 billion and R3.6 billion also in 2012 [10-11].

Billing irregularities, non-payment by the consumers, illegal connections, meter tampering and metering errors, have been identified as the widespread operations under electricity theft [12]. The most prominent being illegal connection and meter tampering and various ways of tampering with energy meter are elucidated in [13]. Basically, electricity theft often results to overloading of the generation units and this adversely affects the utility company in terms of revenue losses [14]. Besides, electricity theft poses health hazard as well as safety concerns leading to death and maiming of personnel [15-18]. Hence, the need to prevent and minimize these adverse impacts of electricity theft has attracted the interest of this study.

Various schemes have been designed to solve the problem [19-20] however, recent reports indicate that the menace has not been managed properly, because the prevalence of the electricity theft is mushrooming at alarming rate and the perpetrators are becoming sophisticated as technology advances [21]. Consequently, the detection and minimization of electricity theft remains one of the major challenges in the distribution network, even as several methods proposed for their amelioration are either not implementable in real time or yet to be applied in real situations. This paper discusses the problem of electricity theft and elaborates on the proposed prototype for the study of electricity theft in real time.

2. Literature review

Detailed review of several techniques proposed and developed by various researchers, over the years, for the detection and estimation of electricity theft can be found in [22]. In [23-25], automatic meter reading system incorporated with tampering detection and various communication media such as Global System for Mobile Communications (GSM) as well as Zigbee, to track electricity theft have been proposed. [2, 26-29] employed artificial intelligent system such as Support Vector Machine (SVM) to classify electricity theft based on the energy consumption pattern of the consumer and pre-select suspected consumers for inspection.

In addition, the power line impedance technique proposed in [30], involves disconnection of subscribers and that a low voltage signal of 2V at 150Hz would be transmitted to the network to detect impedance of the network. Comparisons were made with the installed impedance values and the difference indicates the theft location with respect to the location of legitimate consumer. On the other hand, [31] proposed Central Observer Meter (COM) to monitor and identify the perpetrators. The method proposed in [32] uses two energy meters to track illegal connection.

Meter tampering detection based on monitoring the live current, neutral current and the voltage at the meter input to detect various tampering have been proposed in [33]and the changes in their values depict electricity theft. [34] and [35] proposed the injection of unwanted harmonics into the distribution network in order to cause damage to the appliances of the suspected illegal users and application of smart resistance incorporated in Smart Meter as a mode of detecting illegal electricity usage, respectively.

3. The prototype design consideration

Consumer Load Prototype (CLP) in this study represents the loads of a typical domestic consumer connected to power distribution network in real life situation. The design of a consumer load prototype involves the architectural plan of consumer's building which depicts necessary information about the layout of the building and the electrical design with lighting outlets, socket outlets and switches symbols (Figure 1). This facilitates the calculation and selection of the materials needed for the final construction.

The level of illumination, type of illumination, height of fitting above the floor and the size of the room(s) were considered for number of lighting outlets. The control switches, for controlling the light, are located on the inner wall of the room, adjacent to the door. The socket outlets for fixed equipment are located at anticipated positions while the sockets for portable equipment were adequately provided.

The regulations and standards of the Institution of Engineering and Technology-IET and Wiring Regulations (BS 7671) were used to govern the design and installation of the electrical equipment and wiring. The installation contains efficient and effective earthing system, recommended by IEE regulations [36].

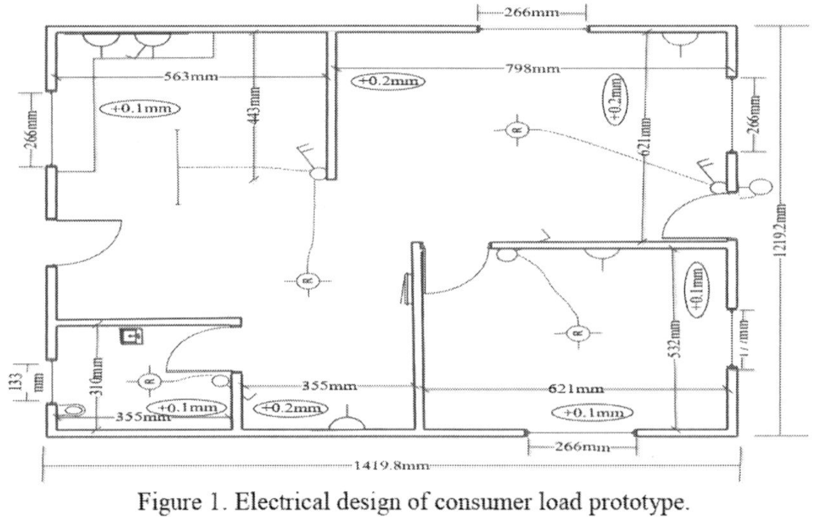

Figure 1. Electrical design of consumer load prototype.

3.1. Assumptions

The following assumptions were considered in the construction of the CLP.

1. A typical consumer house consisting of one bedroom apartment was chosen for the prototype design. This is to minimize the space, construction cost and moreover, most of the household electrical loads can be represented adequately without large deviation from those used by typical domestic consumers.

2. The consumer building is assumed to occupy a 9144mm by 7924mm in area (comprising a bedroom, a living room, dining, toilet, laundry and kitchen as detailed in Table 1).

3. Voltage supply is 240V

Table 1: Dimensions on the Plan and the Prototype.

	Actual house dimension (*mm*)	Plan dimension (*in*)	Prototype dimension (*mm)*
Room 1	4267 x 3657	2.80 x 2.40	621 x 532
Living room	4877 x 4267	3.60 x 2.80	798 x 621
Toilet	2438 x 2133	1.60x 1.40	355 x 310
Kitchen	3962 x 4267	2.54 x 2.80	563 x 621

4. Circuits system of the CLP

The consumer load consists of three types of sub-circuits which include the lighting sub-circuit, heating sub-circuit and power sub-circuit.

4.1. Lighting sub-circuits:

Lighting sub-circuit is usually rated at 5A according to IEE regulation A10. In deciding the number of lighting per circuit, the number of lighting outlet was obtained from the electrical design plan and minimum 100W per outlet is assumed IEE regulation. This means that the highest number of outlets that can be connected per circuit is 11, however, more than one circuit is preferable to prevent total darkness in case of fault. Lighting circuits was wired with 1.0mm or 1.5 mm cable while the cord suspending the pendant fitting is 0.5mm or 0.7mm.

4.2. Power sub-circuit

This is usually in form of socket outlet which could be either in ring circuit or radial circuit and are designed to provide a power source for portable power equipment. The most acceptable rating is 13A as indicated in IEE regulation A30-41. The usual number of sockets allowed in both ring and radial circuits is unlimited depending on the sizes of the room and type of cable used [37]. The kitchen was given a special consideration, due to the fact that loads are expected to be concentrated there. The assumed current for socket outlet is 13A according to IEE regulation A30-A41 and protected with 20A fuse radial connection

4.3. Heating sub-circuit:

Appliances such as air condition, cooker of higher rating and immersion water heater fitted to vessels in excess of 15 litres usually have separate circuit because of their current rating which is in excess of 15A and fused with 30 or 32A fuse (IEE regulation A26). Exception to this rule, as contained in the regulation A28 and A29, it allowed final sub-circuit to supply two cooking appliances if (i) they are in the same room (ii) situated within 2m apart and (iii) the combined rating does not exceed 30A. Otherwise, a separate cooker circuit is recommended and the current demand of the appliances shall be assessed according to the IEE regulation A27.

Table 2. Summary of main cable selection

Sub-circuit	Number of outlets	Actual current (A)	Applying Diversity factor	Main-sub circuits current	Cable size
Lighting	6	3.166	0.66 X 3.1	2.1A	1mm
Radial	4	20	100% of full load of largest circuit 40% of full load of all other circuits	Since it is one circuit =20A	2.5mm
Cooker/AC	1	30	the first 10 A of the cooker load 30% of the remainder 5A if the cooker unit has a socket outlet	17.5A	2.5mm
			Total	= 39.6A	

4.4. Selection of Cables for Sub-Main

Different factors considered in the selection of cable for sub-main circuit includes:

i. Capability to carry full load current
ii. The maximum permissible voltage drop ≤ 2.5% of supply voltage

IOP Conf. Series: Materials Science and Engineering **53** (2013) 012061 doi:10.1088/1757-899X/53/1/012061

iii. Method of insulation
iv. Grouping of the cables
v. Ambient temperature
vi. Class of excess current protection

In the selection of main cable that supply the consumer from the distribution feeder, diversity factor is applied to the sub-circuit as indicated in Table 2. The total current value is 39.6A. The equivalent cable, from [38], to this value is $4mm^2$ and, miniature circuit breaker of 45A rating was chosen as a protective device. The voltage drop which should not be greater than 2.5% of the supply voltage (240V) for the $4mm^2$ cables is 11mV/A/m [38]. The length of the cable is 5m and this leads to the calculation of the main sub-circuit which supply power to the consumer load and the selection of switchgears.

4.5. 4-way Distribution board
The distribution board or consumer unit is where various load sub-circuits and earthing arrangements are connected. It is partitioned into various sections with fuses or circuit breaker of different ratings to isolate the sub-circuit load in the event of a fault. These sections are known as 'ways' hence, the unit derives its name from the number of ways it is enclosed i.e. 4-way distribution board (5A circuit breaker- lighting sub-circuit; 20A circuit breaker- power sub-circuit; 32A circuit breaker-heating sub-circuit; 40A residual current device (additional protection).

4.6. Construction
All the components and wires were connected to form a complete Consumer Load Prototype (CLP). The final outlook of the prototype is shown in Figure 2 and the flow chart reflected in Figure 3.

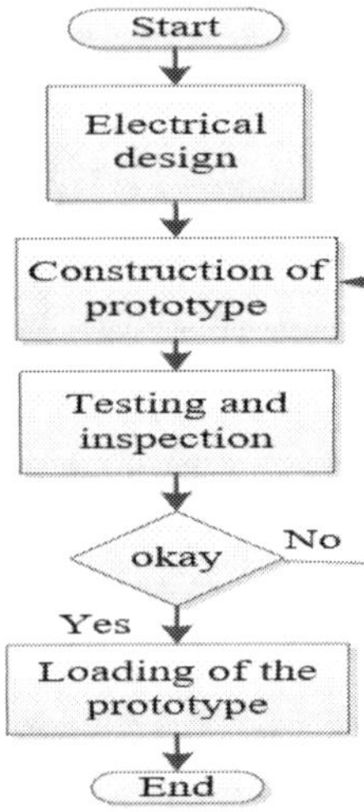

Figure 3: Flow chart for the Prototype development

Figure 2: Final outlook of the Prototype

5. Load monitoring and sensor selection
Current sensors detect the flow of current in a conductor and generate a signal (analog or digital) that is proportional to it. Current sensor is employed in order to monitor the loads connected to the prototype and acquire real time data. Rogowski coils, Current transformer and Hall Effect sensor are the most commonly used sensors for current detection in a conductor. One major advantage of these sensors is that they provide galvanic isolation between the power network, the operator and other instrument connected to them. Their selection is based on accuracy, drift, linearity, phase shift,

integration and cost, however Hall Effect current sensor (ACS785, Allegro MicroSystem Inc., USA) was chosen based on its size and other advantages expressed in [39] having satisfied the selection criteria.

Energy meter (misa SDN BHD, Malaysia) as well as Hall Effect sensors was connected to the prototype, while the meter shows the energy consumption in kWh, the sensors acquired the current data, via the labview hardware and software, which is used for monitoring purpose. Finally, the reading of the energy meter was benchmarked with the output of the sensors.

5.1. Experimental set up

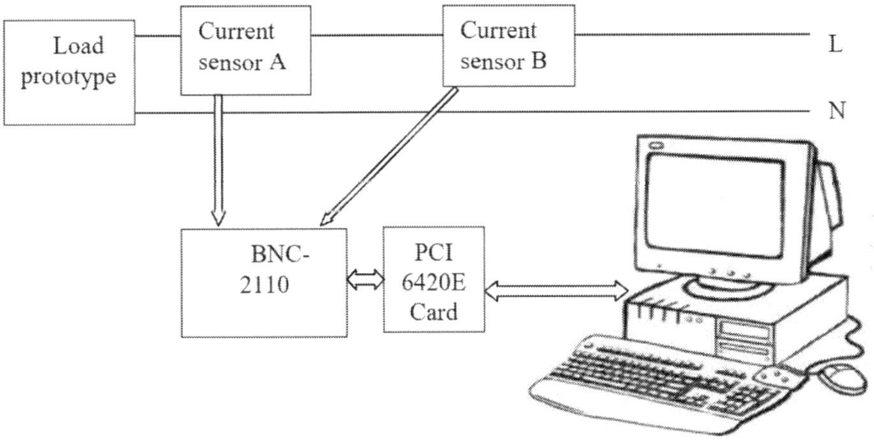

Figure 4. Block diagram of the monitoring system

Real time information was fed directly from the sensors attached to the conductor feeding the prototype both at the prototype terminal and power source (pole). The sensors were connected to (BNC-2110, National Instrument, USA) device which is linked to the PCI 6420E channel in the computer. The software programme was configured to log acquired data directly into the computer hard disk as well as show the monitoring processes in real time Figure 5. Data acquired were stored for further analysis.

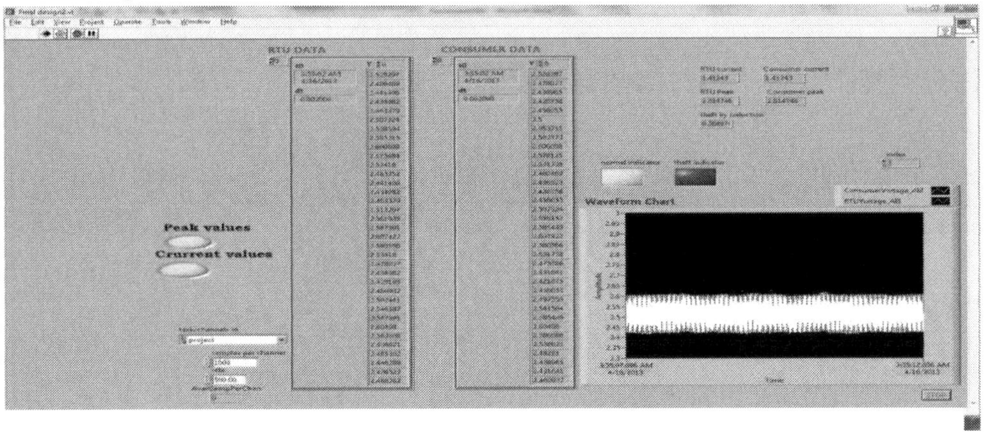

Figure 5. Real time load monitoring system using Labview platform

6. Result and discussion

The prototype was connected to power source and the loads including: refrigerator, incandescence bulb (100W), fluorescence lamp (80W), table fan (40W), and electric kettle (1500W) were connected for 15minutes each accordingly. The energy consumption of these loads was monitored through the energy meter as well as the real time data acquired through the current sensors (ACS 758) connected to the prototype via LABVIEW platform. Figure 6a to 6e show the graph of the individual load. A rapid increase in load consumption pattern was observed whenever the switch was ON, this is due to the effect of the switches contact during this period.

However, this rapid increase is pronounced in the current value of the refrigerator (Figure 6a) when it was switched ON. This was due to the single phase induction motor that drives the compressor and it is the major component in the refrigerator that consumes energy. It takes six to ten times it running current value at starting point [40]. Figure 7 illustrates the graph of the total load connected to the prototype and switch ON at 15 minute interval for about one hour 15miniute. It was observed that the time for electric kettle was shorter and fixed, this is because it is controlled automatically by the control switch which goes OFF when the water boils and corresponds to the time it took to boil 1.5liter of water. These observations suggest that the energy consumed by the loads connected to the prototype representing a typical consumer can be measured, monitored, studied and detected in real time. Furthermore, the consumer behaviour with respect to illegal activities such as electricity theft on the distribution network can be detected based on the consumer load pattern which is acquired through the sensors. Hence this could assists in the development of new algorithm for real time electricity theft detection and monitoring.

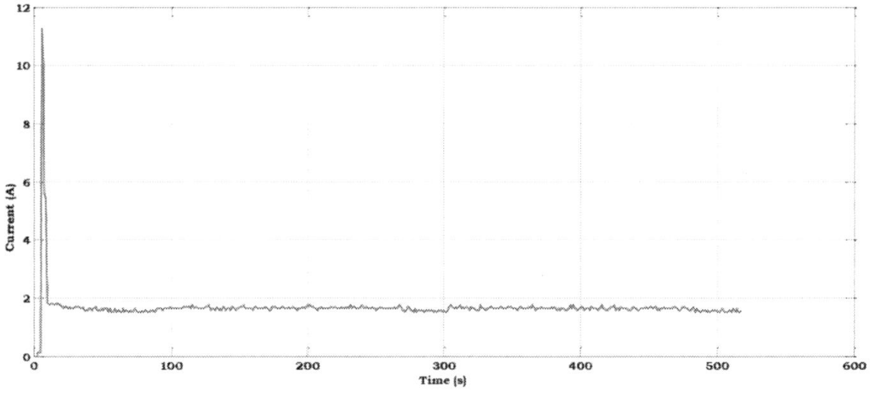

(a)

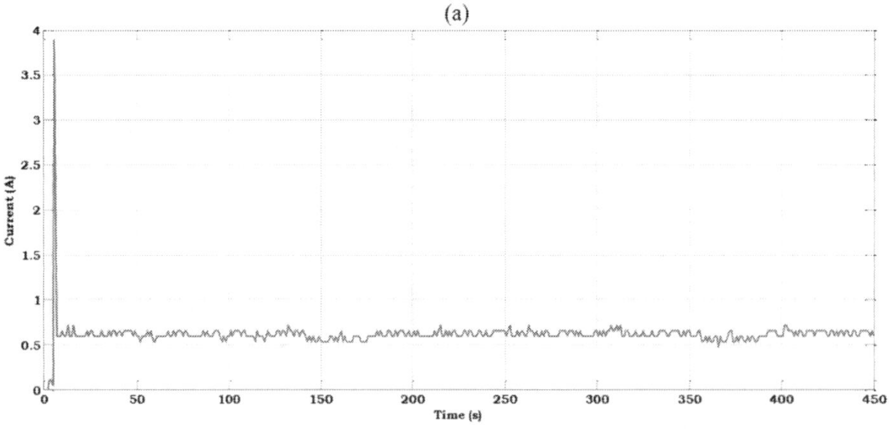

(b)

IOP Conf. Series: Materials Science and Engineering **53** (2013) 012061 doi:10.1088/1757-899X/53/1/012061

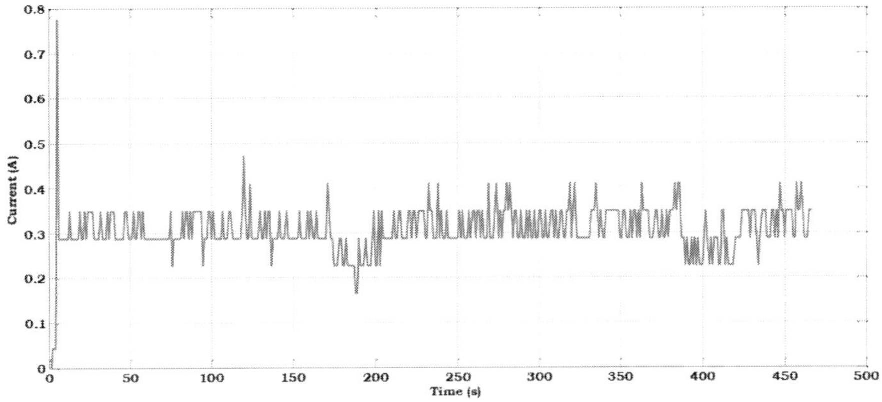

(c)

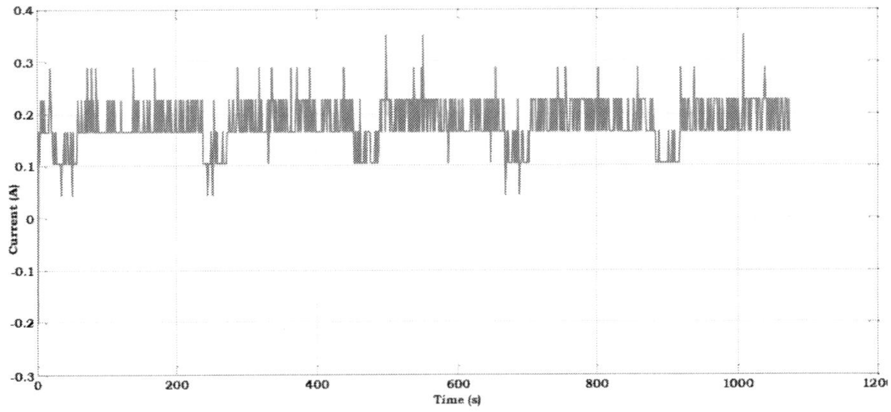

(d)

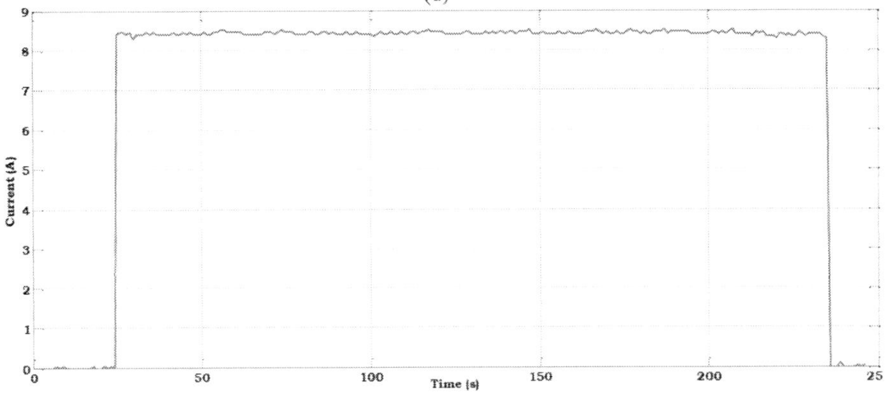

(e)

Figure 6. 15minutes load simulation graph : (a) Refrigerator, (b) Incandescence bulb, (c) Fluorescence lamp, (d) Table fan and (e) Electric kettle graph

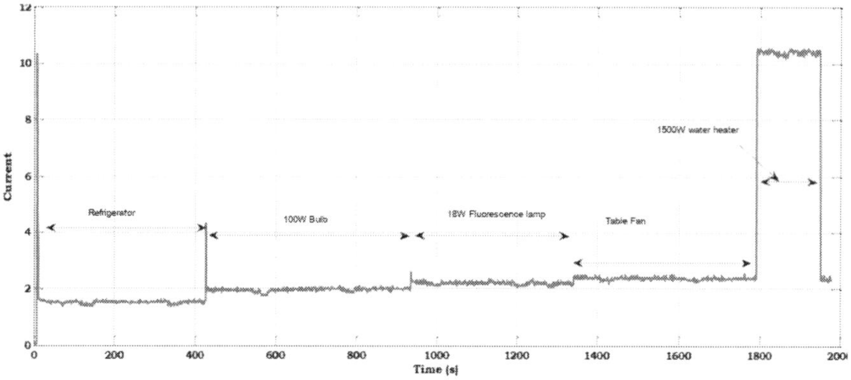

Figure 7. Combined loads at 15minutes interval

7. Conclusion

A CLP has been designed and constructed for the study of electricity theft in this paper. The Prototype which was constructed on a 4ft by 8ft plywood will facilitate real time simulation and data collection for the monitoring of electricity theft on low voltage distribution system. Standard regulations from Institution of Engineering and Technology (IET) were used to standardise the prototype. The ability to generate real time data for the study of electricity theft monitoring is considered as one of the contributions of this prototype.

The prototype will assist researchers and power utilities alike who are facing challenges in getting real time data for the study and monitoring of electricity theft. Similarly, the power and energy community including students will appreciate the practical approach from which the prototype provides real time information rather than software simulation which has hitherto been used in the study of electricity theft. Future work is to analyse the acquired data using advance signal processing and intelligent system techniques so as to study consumers' power consumption and to detect electricity theft.

References

[1] SmithT B, 2004 *Energy Policy*, **32** 2067 – 76
[2] Nizar A H and Dong Z Y 2009 *In Proc. Power System Conf. and Exposition, PSCE, Seattle, WA*,1-10.
[3] Wang L, Devabhaktuni V and Gudi N, *In Proc. North American Power Symposiums* 1-8
[4] http://www.azcentral.com/news/articles/2009/03/16/20090316gan-powertheft16-ON.html?nclick_check=1
[5] http://www.khou.com/news/crime/Electricity-theft-a-dangerous-trend-thats-costing-honest-Houstonians-money-92450694.html
[6] http://www.cbc.ca/news/canada/british-columbia/story/2010/10/08/bc-hydro-grow-op-theftw.html
[7] http://www.bernama.com.my/bernama/v5/newsindex.php?id=617926
[8] Zuhrin Y N, Raman A *The Star Newspaper,* Malaysia. December, 31, 2010.
[9] http://www.stabroeknews.com/2012/archives/11/13/electricity-theft-jumps-in-Jamaica/
[10] http://www.fin24.com/Economy/R4bn-lost-to-electricity-theft-20100323
[11] http://www.environment.co.za/south-africas-energy-crisis/gauteng-has-highest-rate-of-electricity-theft-eskom.html
[12] Nizar A, Dong Z, Devabhaktuni V and Gudi N, 2006 *In Proc. Power and Energy Conf. (Putra Jaya)* pp. 82-87.

[13] DepuruS S S R, Wang L, Devabhaktuni V and Gudi N, 2010 *Proc. IEEE North American Power Symposium(Texas)*
[14] Depuru S S S R, Wang L and Devabhaktuni V 2011 *Energy Policy*, **39** pp 1007-1015.
[15] http://www.accidents.co.za/2012/08/01/man-fights-for-life-after-illegal-connection-lectrocution
[16] http://articles.timesofindia.indiatimes.com/2012-07-21/jaipur/32775988_1_pali-sister-jaipur
[17] http://www.newvision.co.ug/news/633444-six-killed-in-illegal-electricity-connections.html
[18] http://www.newsabahtimes.com.my/nstweb/fullstory/60915
[19] Solomon N and Joseph C, 2011 *J. Theo.and appl. inf.o. tech.* **26** pp112-117.
[20] Rahul A, Sapatarshi D and Naveen A 2003 *Proc. Student conf. on research and devel.(Putrajaya)* pp 15-18
[21] http://www.forbes.com/sites/kensilverstein/2012/08/05/india-and-u-s-share-energy-woes-stealing-electricity/
[22] Abdullatee f A I, Salami M J E, Tijani I B and Onasanya M A 2012 *in Proc. National Conf. of Elect.and Elect. Eng. (Johor, Malaysia)* pp 199-204
[23] Fuxiang G, Wenxin X and Langtao L 2010 *2nd Inter. Conf. on Ind. and info. Syst.(Shenyang, China)* pp 270-273
[24] Tan H G R, Lee C and Mok V 2007 *Inter.Power Eng. Conf.* (Kuala Lumpur) pp. 465-469.
[25] Zheng G, Gao Y and Wang L 2010 *Power and Energy Eng. Conf. Asia-Pacifi* pp 1-6
[26] Nagi J, Yap K S, Tiong S K, Ahmed S K and Mohamad M 2008 *IEEE Trans. Power Delivery* **2** pp 907-912
[27] Nagi J, Yap K S, Tiong S K, Ahmed S K and Nagi F *IEEE Trans. Power Deli.* **2**pp1284-1285.
[28] Nizar A, Dong Z and Wang Y 2008 *IEEE Trans. Power System* **23** pp 946-955
[29] Depuru S S S R, Wang L and Devabhaktuni V *in Proc. Power Syst. Conf. and Expo.*(Toledo) pp. 1-8.
[30] Pasdar and Mirzakuchaki S 2007 *2nd Inter. Workshop on Soft Comp. Appl.*(Tehran) pp163-167
[31] Bandim C, Alves Jr J, Pinto Jr A, Souza F, Loureiro M, Magalhaes C and Galvez-Durand F 2003 *Trans. and Distr. Conf. and Expos.*(Rio de Janeiro)**1**pp 163-168.
[32] CavdarI H 2004 *IEEE Trans. Power Delivery* **19** pp 1663-1667
[33] Naiman S, Kissaka M and Mvungi N 2004 *in AFRICON* **2** pp 821-826,
[34] Bat-Erdene B, NamS Y and KimD H *Future Info.Tech.* pp 214-223
[35] Kadurek P, Blom J, Cobben J and Kling W 2010 *in Innov. Smart Grid Tech. Conf.* (Europe) pp. 1-6
[36] Scaddan B 2011*Electrical installation work* (Newnes)
[37] Linsley T 2008 *Basic Electrical Installation Work* (Routledge)
[38] Scaddan B 2008 *Electric wiring domestic* (Elsevier)
[39] Allegro Microsystems. (2010). *Datasheet ACS758xCB* [Online]. Available: http://www.allegromicro.com/en/Products/Part_Numbers/0758/0758.pdf
[40] Gupta J 2009 *Theory & Performance Of Electrical Machines*, (SK Kataria & Sons.)

Hybrid algorithm for NARX network parameters' determination using differential evolution and genetic algorithm

M J E Salami[1,4], I B Tijani[1, 2], A I Abdullateef[1] and M A Aibinu[3].
[1] Department of Mechatronics Engineering, Faculty of Engineering, IIUM, Malaysia
[2] Electronics Eng. Dept., ADMC, HCT, Abu Dhabi, UAE
[3] Federal University of Technology, Minna, Nigeria

E-mail: momoh@iium.edu.my

Abstract. A hybrid optimization algorithm using Differential Evolution (DE) and Genetic Algorithm (GA) is proposed in this study to address the problem of network parameters determination associated with the Nonlinear Autoregressive with eXogenous inputs Network (NARX-network). The proposed algorithm involves a two level optimization scheme to search for both optimal network architecture and weights. The DE at the upper level is formulated as combinatorial optimization to search for the network architecture while the associated network weights that minimize the prediction error is provided by the GA at the lower level. The performance of the algorithm is evaluated on identification of a laboratory rotary motion system. The system identification results show the effectiveness of the proposed algorithm for nonparametric model development.

1. Introduction

Non parametric modeling approach based Nonlinear Autoregressive with eXogenous inputs Network (NARX-network) has been a successful technique for time series prediction, filtering, system modeling and identification in almost all fields of studies, [1], [2]. However, one of the major challenges in the application of this NARX network is in the choice of appropriate network architecture in terms of number of input and output tapped delays, TD, and number of hidden neurons (H) for a single hidden layer network, [3]. The decision has been based majorly on iterative experimentation (trial and error basis). This problem has initiated the use of evolutionary optimization algorithm in neural network design for optimal/sub-optimal performances [3], [4], and by extension to NARX network. Many research works have been reported on network (MLN) architecture optimization using simulation annealing and tabu search, [5], using GA, [4], and DE, [6],and recently a multi-objective Differential Evolution (MODE) was proposed for network architecture optimization, [7]

In this paper, a hybrid optimization algorithm involving DE and GA is proposed to determine optimal NARX network architectures and weight. The motivation is to explore the optimization strength of both DE and GA to search simultaneously for both model architecture and weight.

The rest of the paper is organized as follows. A brief overview of both NARX network is given in Section 2 followed by the basic description of DE and GA algorithms in Section 3. The proposed

[4] Prof. Momoh Jimoh Eyiomika Salami

hybrid algorithm is presented in Section 4. The application to system identification of a practical system is reported in Section 5 while the results and discussion are given in Section 6. The paper is concluded in Section 7.

2. NARX network overview

NARX network generally describes a discrete nonlinear system as linear combination of its past output values as well as those of the input, [8]:

$$\hat{y}(k) = f[y(k-1), y(k-2), \ldots, y(k-n+1);\ \delta(k), \delta(k-1), \ldots, \delta(k-m+1)] \qquad (1)$$

where $\delta(k)$ and y(k) represent input and output of the system as time step k, n and m are the input and output orders respectively, while the function $f[\bullet]$ is a nonlinear mapping function which is given by the Multilayer Perceptron (MLP) network.
As shown in Figure 1, the input-output mapping implemented with two layers MLP is expressed as:

$$\hat{y}(k) = \Gamma(u(k), y(k)) = \Gamma\left[[W^L \Pi [W^I(\delta^*, y^*)]]\right] \qquad (2)$$

where $\Gamma(\bullet)$ is mapping function implemented by the MLP, $\Pi(\bullet)$ is nonlinear activation function usually a sigmoid function presents in both hidden and output layers, W^I and W^L represent the connection weights for input and hidden layers respectively.

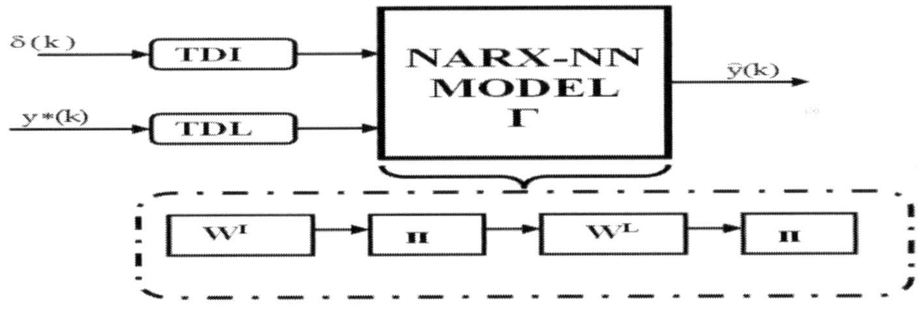

Figure 1. Architecture of series-parallel NARX-Network, [7]

A general configuration of NARX network for multi-inputs and multi-outputs (MIMO) systems with two inputs, two outputs and one hidden layer with three nodes together with biases is shown in Figure 2.

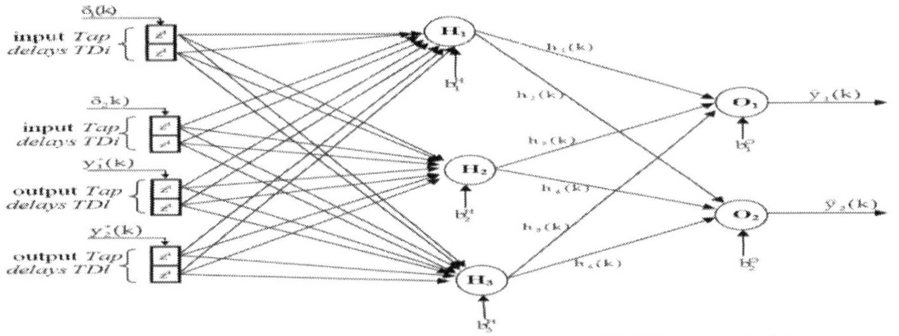

Figure 2 A three layer, two inputs, two outputs NARX network [7]

Given n as number of inputs, m as number of outputs with O nodes, assuming a single layer network with H hidden nodes, the network parameters consist of: (i) input states $\delta = [\delta_1, \delta_2, \ldots, \delta_\Phi]$ where Φ is total number of the states given as:

$$\Phi = H[(n * TDI) + (m * TDL)] \tag{3}$$

TDI and TDL are number of input and out time delays (z^{-1}), respectively.

(ii) input biases, $b^H = [b_1, b_2, \ldots, b_H]$; (iii) input weights W^I:connection weight between input state, j and hidden layer nodes, i, expressed as $w_{i,j}$. Note that total number of this weights is equal to Φ; (iv) layer weights, W^L: connection weight between the hidden layer nodes i and output nodes κ expressed as $w_{\kappa,i}$. The total of this connection weight is given by $H * O$; and (v) output biases $b^O = [b_1, b_2, \ldots, b_O]$.

Thus, the total network parameters N_p to be adjusted during identification process is given as sum of total input connection weights, input biases, total hidden layer connection weight and output biases:

$$N_p = \Phi + H * O + b^H + b^O \tag{4}$$

The output of the hidden layer is given by:

$$h_i = \Pi\left[\sum_{j=1}^{N} w_{i,j} x_j + b_i^H\right] \tag{5}$$

while the estimated model output is:

$$y_\kappa = \Pi\left[\sum_{i=1}^{H} w_{\kappa,i} h_i + b_i^O\right] \tag{6}$$

The network learning process generally involves determination of the network parameters W^I and W^L that minimizes the prediction error between the actual and predicted output given as:

$$V_N = \frac{1}{2N} \sum_{k=1}^{N} \varepsilon^T \varepsilon \tag{7}$$

where ε is prediction error expressed as $\varepsilon = y(k) - \hat{y}(k)$

Conventionally, the training process is carried out with back propagation learning algorithm with Levenberg-Marquardt learning function [10].

3. Overview of GA and DE algorithm

GA and DE are among the prominent evolutionary optimization algorithms with vast practical applications as reported in the literatures, [10], [11]. The GA was originally proposed by John Holland, [12], and represents the pioneer evolutionary approach to practically optimization problem. It mimics biological process of natural selection mechanism in which stronger individuals have higher probability to win in a competing environment. DE on the other hand, was latter proposed by Storn and Price, [13]. Unlike GA which uses genetic operator in evolving new solutions/generation, DE uses differential operator to evolve new solutions/generation. Also, DE uses real values to represent the solution candidates while GA originally represents the parameters in form of chromosomes (a binary encoding of the actual parameters). Generally, both are population based algorithm and shared similar major concepts of: population initialization, crossover, mutation and selection, [14].

Initialization involves generation of initial population using a random process for the optimization process. In GA, the individuals are encoded as a string of bits (0 & 1) using the process of encoding. While in DE the random generated real parameters values are used directly.

Mutation and crossover are two consecutive operators for generation of new vector known as child/trial vector. The process begins by selecting two individuals from the current generation, then, crossover and mutation operators are applied to generate child vector. In GA, a crossover point is randomly set and the portions of the two genes beyond this set point are swapped to form new vector (child). This process is controlled by a pre-specified probability rate known as crossover rate usually between 0.6 and 1.0. Then, mutation is applied to each child string to alter each bit randomly with a probability determined by the pre specified mutation rate which is usually between 0.1 and 0.001, [14].

The DE mutation comes first and involves addition of weighted differential of two or more randomly selected vectors to another randomly selected based vector. Though there are several strategies of computing the weighted differential, the basic and common one is "DE/rand/1/bin", [13]. Then a recombination process known as crossover is used to alter randomly the new vector. It ensures that each parameter of the differential mutant vector is accepted into the trial vector with some probability know as crossover constant which is usually between 0 and 1. It determines the fraction of parameter values that are copied from the mutant into the trial or child vector.

Selection operator generally involves comparison of the objective values known as fitness function of both parent and child vector to determine who survives in the next generation

Once the new population is generated, the process of mutation, crossover (recombination) and selection is repeated until the pre-specified termination criterion is satisfied. Based on this routine, the general flowchart of the DE and GA algorithms are shown in Figure 3 (a) and Figure 3(b), respectively.

4. Proposed hybrid DE-GA algorithm

The proposed hybrid algorithm explores the optimization strength of both DE and GA sub optimization algorithms in determining the optimal parameters of NARX network. The algorithm comprises two level optimization processes. The first level (upper level) comprises the DE sub algorithm which searches for the optimal NARX network architecture, while the second level involves GA sub algorithm searches for the optimal network weights that minimizes a given objective function. The flowchart of the proposed algorithm is shown in Figure 4. As shown in Figure 4, the GA sub algorithm is integrated into the DE algorithm to compute the overall objective function.

The problem is formulated as follows: given a set of input and output data pairs ($u(k),y(k)$) of a system to be modeled, the objective of the DE upper level optimization is to search for set of network architecture parameters, φ,

$$\varphi := [TD, H] \tag{8}$$

that minimizes the objective function, $f(\varphi)$ as follows:

$$\min f(\varphi) = 0.5 * (\Gamma(W) + p_c) \tag{9}$$

where p_c is the percentage of network connection used, and $\Gamma(W)$ is the objective function values returns by the GA sub-algorithm.

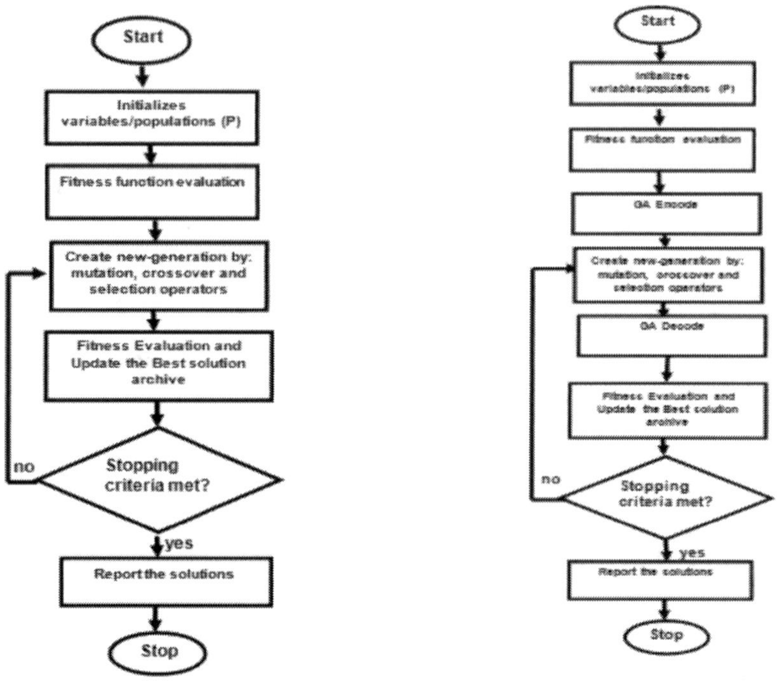

Figure 3 Flowchart of (a) DE algorithm, (b) GA Algorithm

The GA sub-algorithm searches for a set of network weights, W, which minimizes the objective function, $\Gamma(W)$, where $\Gamma(W)$ is defined as prediction error between the actual response and predicted response for a given data, length, *N:*

$$_{min}\Gamma(W) = \frac{1}{2N}\sum_{k=1}^{N}\varepsilon^{T}\varepsilon \tag{10}$$

where $\varepsilon = y(k) - \hat{y}(k)$

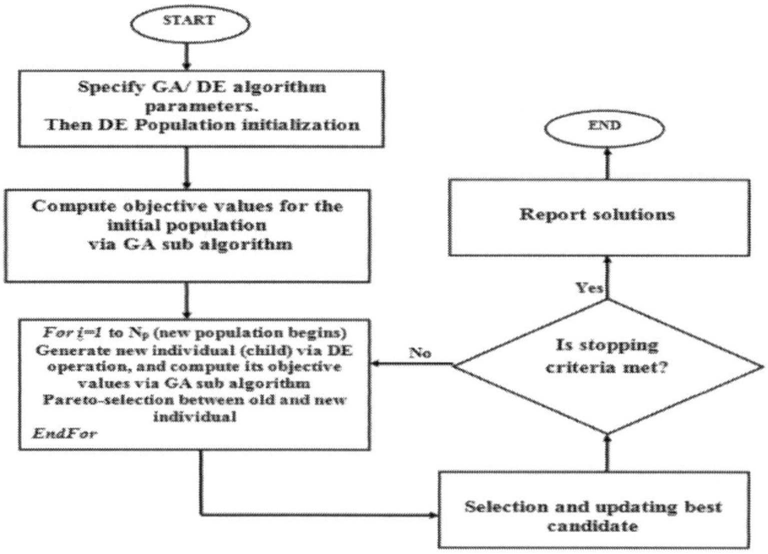

Figure 4 Flowchart of the proposed hybrid DE-GA Optimization algorithm

The major algorithm procedures are summarized as follows:

Step 1: Specifying the optimization parameters for both DE and GA sub algorithms: DE parameters are: crossover constant, *CR*, mutation Constant, *F*, population size, β_{DE}, and maximum generation g^{DE}_{max} while GA main parameters are: probability crossover, p_c , mutation rate, m_c, population size, β_{GA} , and maximum generation size, g^{GA}_{max}

Step 2: Initialize the DE population, P_o, of size, β_{DE}, for $g^{DE} = 0$. The population comprises of individuals, $\varphi^g_i \; \forall i, i = 1,...,\beta_{DE}$, where $g = g^{DE}$ and . Given a specified lower, *L*, and upper, *U*, bound on the network parameters, an individual of the population is initialized using:

$$\varphi^g_i = randint(1,1,[L,U]) \tag{11}$$

where "*randint*" is a function that generates uniformly distributed random integers, within the range specified by the bounds *L* and *U*.

Step 3: Compute vector of objective values, *Fit*, for all the individuals in the population. This is achieved by evaluating the objective function given in equation (1), hence,

$$Fit_i = {}_{min} f(\varphi^g_i) = 0.5*(\Gamma(W_i) + p_{ci}) \quad \forall i, i = 1,...,\beta_{DE} \tag{12}$$

where $f(\varphi^g_i)$ is the objective functions for a given parameter vector φ^g_i.

As earlier stated, the objective functions expressed in equation (5) comprises the accuracy and complexity of the NARX network. The accuracy is specified by the sub-objective values, $\Gamma(W)$ which is the optimal prediction error returns by the GA sub-algorithm. The GA sub-algorithm is called by the DE sub-algorithm at this point to search for optimal network weight, *W* that minimizes $\Gamma(W)$.

For a given network architecture, φ produced by DE sub-algorithm, GA sub algorithm is called to search for corresponding optimal weights that minimizes prediction error given in equation (3). A nested MATLAB function (m-files) incorporating the MATLAB inbuilt GA algorithm, [15] is developed to integrate the GA into the DE process. The total weight for the given, φ is computed using equation (3), and this with other pre-specified parameters as stated in step 1 are passed into the GA sub algorithm.

Step 4: Begins DE optimization process

while g< g^{DE}_{max}

Step 5: Starts DE operation to generate new population, P_{new} and update best candidate

For *i*= 1 to β_{DE}

1. Using the DE process of mutation and cross-over, generate new candidate, v^{g+1}_i called child vector from a randomly selected three individuals, r_1, r_2, r_3, from the previous population such that $r_1, r_2, r_3 \in [1; N_p]$, *where* $r_1 \neq r_2 \neq r_3 \neq i$ as follows:

i. Mutation: compute the mutant vector υ using DE operation:

$$\upsilon_i = \varphi_{r_3} + (\varphi_{r_2} - \varphi_{r_1}), \; F = 1 \tag{13}$$

ii. Perform crossover to generate child vector, v^{g+1}_i using:

Generate a random number $Rand_j = floor(rand(0,1)*d)+1$

for j=1: d

$$v_{i,j}^{g+1} = \begin{cases} \upsilon_{i,j} & if\ (rand(0,1) \leq CR\ \ or\ \ j = Rand_j \\ \varphi_{i,j}^{g} & otherwise \end{cases} \tag{14}$$

EndFor

2. Check for violation of the bounds L and U, and effect correction if needed

For j=1: d

If $v_{i,j}^{g+1} < L \,||\, v_{i,j}^{g+1} > U$

$v_{i,j}^{g+1} = randint(1,1,[L,U])$

EndIf

EndFor

3. Evaluate the objective function, Fit_i^{g+1} of the child vector, $f(v_i^{g+1})$ as described in step 3.

4. Selection process: let the generated child vector, v_i^{g+1} compete with the parent vector φ_i^g as follows

$$\varphi_i^{g+1} = \begin{cases} v_i^g : if\ \ v_i^{g+1}\ \ wins\ \ \varphi_i^g \\ \varphi_i^g : if\ \ v_i^{g+1}\ \ not\, wins\ \ \varphi_i^g \end{cases} \quad (27)$$

where: $v_i^{g+1}\ \ wins\ \ \varphi_i^g\ \ \ if\ only\ and\ if\ \ : f(v_i^g) \leq f(\varphi_i^g)$ (15)

Update new population and Best candidate index:

$P_{new}(i) = \varphi_i^{g+1}$

$iBest_{new} = i : if\ Fit(iBest_{new}) <= Fit(iBest_{old})$

EndFor

Step 6: set $g=g+1$, and go to Step 4

EndWhileLoop

Step 7: Report the solutions: Best Fit and Best parameters set: φ and W

This hybrid algorithm, subsequently refers to as DE-GA-NARX network algorithm, is written in MATLAB which facilitate the integration of the MATLAB inbuilt GA and NARX network algorithm in the hybrid optimization algorithm process.

5. Application to system identification

The hybrid algorithm is used to identify model of a laboratory rotary motion control system shown in Figure 5. As shown in Figure 5, the DC motor driven rotary motion system represents a typical practical problem in system identification. It consists of servo motor driven by an amplifier and position encoder attached to the shaft as the feedback sensor. Although, an approximate lump mathematical model based on the physical laws of motion and Kirchhoff's laws could be derived for the system as reported in [16], the actual model order and parameters of the complete system components are not always known a priori. In the absence of actual system model detailed physical parameters, the NARX network technique provides an effective approach for the system nonlinear model identification.

The system is excited with Pseudo Random Binary Signal (PBRS) to obtain input-output data pairs requires for system identification process. Given this input and output data pair, the proposed algorithm is applied to identify NARX network parameters the optimally modeled the system.

Two set of data were collected for training with data size, NT=503 and validation with data size, NV=120. Sample of the input and measured output data is shown in Figure 6.

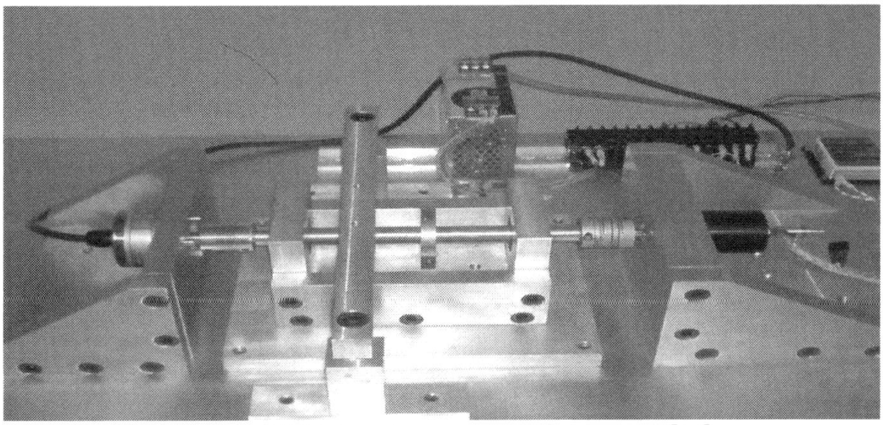

Figure 5 DC-Motor driven rotary motion systems [16].

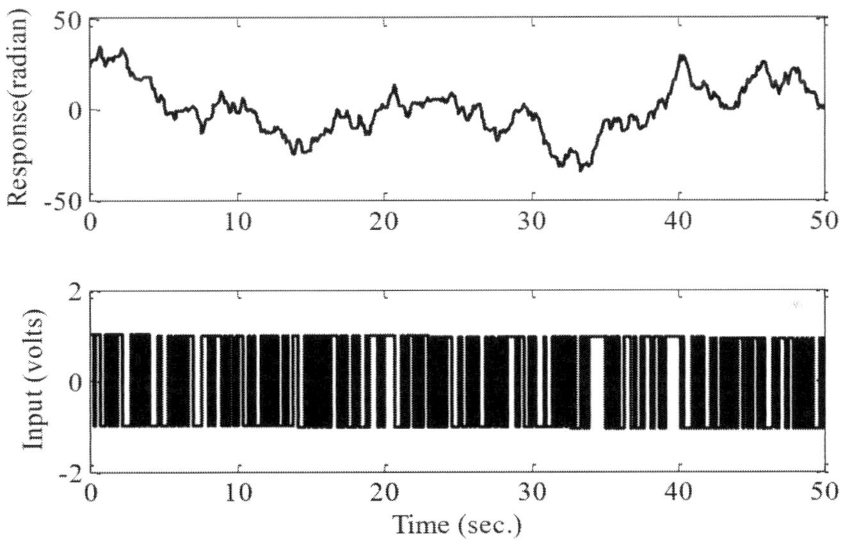

Figure 6 Experimental input and output data pair

6. Results and discussion

The primary optimization parameters are: population size, β_{DE}=30; β_{GA}=20; maximum generation size, g_{max}^{DE}=100, g_{max}^{GA}=20; maximum network size, TD=10, H=10; DE mutation factor, 0.75 and cross over constant, 0.25; GA mutation rate, 0.01, probability cross over, 0.8. Figure 7 shows the comparison of the actual and predicted output with validated data. The results of the optimal network produced by the algorithm are given in Table 1. As shown in Figure 7, and Table 1, the prediction error is considerable minimized for both training and validation data, and yet provides network with less than 50% (17 size) full connection.

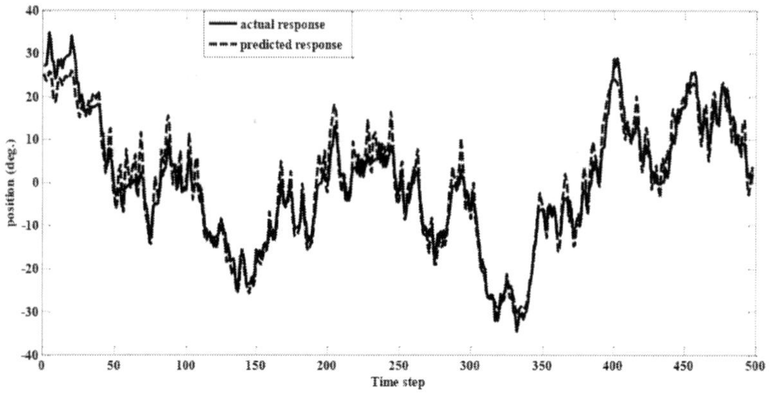

(a)

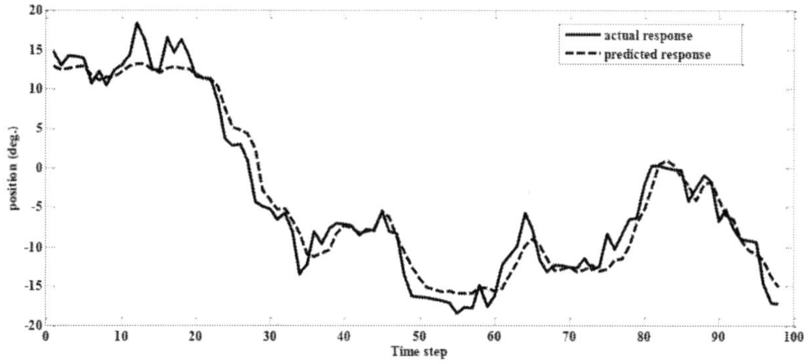

(b)

Figure 7. Measures and predicted plant responses with (a) training and (b) validation data

Table 1: Results of Optimized NARX Network for the plant model

Network size	% connection	Prediction error [1] Training data	Prediction error Validation data
[3,3,2]	7.69	0.0089	0.0135

7. Conclusion

This study has presented an hybrid algorithm for optimal determination of a NARX network parameters. The proposed hybrid algorithm explored the evolutionary optimization strength of both DE and GA to search for both the network architecture and weights which yield optimal model accuracy and least network complexity.

The algorithm is applied to identify model of a rotary motion control system, the results shows the effectiveness of the proposed algorithm in terms of model accuracy with average prediction error of 0.011, while utilizing only 50% of the full network size. Generally, the proposed hybrid algorithm provides automatic network architecture and parameters determination for nonlinear, nonparametric NARX network model. This is expected to reduce the developmental time involves in autonomous helicopter project.

References

[1] Vijaya K M P, Sampath S S, Omkar S N and Ranjan G 2011 J. **83** 283–295.
[2] José M P, Menezes Jr., Guilherme A B 2008 J. *Neurocomputing*, **71**, 3335–3343
[3] Teresa B, Ludermir AY, and Cleber Z 2006 *IEEE Transactions*, **17**
[4] Jinn-Tsong T, Jyh-Horng C, and Tung-Kuan L 2006 *IEEE Transaction on Neural Networks*
[5] Chia-Nan K, *Engineering Applications of Artificial Intelligence* **25** 533–543
[6] Subudhi B, Jena D 2008 *Neural Processing Letters*, **27** 285-296
[7] Tijani I B, Akmeliawati R, Legowo A, Budiyono A 2012 *Under review* (EAAI), Elsevier.
[8] Narendra K S and Parthasarathy K 1990 *IEEE Trans. on neural networks* **1** 4-27
[9] MATLAB, 2009a. Neural Network Toolbox, User Guide. Mathworks
[10] Yoel T, 2012 *Engineering Applications of Artificial Intelligence* **25** 1009–1021
[11] Xin-SheYang, 2010 Wiley, A John Wiley & Sons, Inc., Publication.
[12] Holland J, The University of Michigan Press, Ann Arbor, MI
[13] Storn R, and Price K, 1997 J 341–359 *Kluwer Academic Publishers.*
[14] Man K F, Tang K S, and Kwong S 1996 *IEEE Transactions On Industrial Electronics* **43**
[15] MATLAB, 2009a, User Guide's, Mathworks, 2009
[16] Tijani I B, and Akmeliawati R 2012 J. Elsevier **25**

MEMS and EFF technology based micro connector for future miniature devices

M M I Bhuiyan[1], T Alamgir[2], M Bhuiyan[3], M Kajihara[4]
[1,2]Department of Mechatronics Engineering, International Islamic University Malaysia, Kuala Lumpur 53100, Malaysia
[3]Monash University Sunway Campus, School of Medicine and Health Sciences, Bandar Sunway, Selangor, 46150, Malaysia
[4]Department of Materials Science and Engineering, Tokyo Institute of Technology, Yokohama 226-8502, Japan

E-mail: moinul@iium.edu.my

Abstract. The development of a miniature; size, light and high performance electronic devices; has been accelerated for further development. In commercial stamping method, connector pitch size (radius) is more than 300μm due to its size limitation. Therefore, the stamped contact hertz stress becomes lower and less suitable for fine pitch connector. To overcome this pitch size problem a narrow pitch Board-to-Board (BtoB) interface connectors are in demand for the current commercial design. Therefore, this paper describes a fork type micro connector design with high Hertz-Stress using MEMS and Electro Fine Forming (EFF) fabrication techniques. The connector is designed high aspect ratio and high-density packaging using UV thick resist and electroforming. In this study a newly fabricated micro connector's maximum aspect ratio is 50μm and pitch is 80μm is designed successfully which is most compact fork-type connector in the world. When these connectors are connected, a contact resistance of less than 50mΩ has been attained by using four-point probe technique.

1. INTRODUCTION

The recent developments in miniature according to the size, and performance have been accelerated for the further development and the devices has been developed by high functional large scale integration chip design; called System on chip (SOC) method [2]. In the commercial design industries are using pressing and injection moulding technique and the contact minimum pitch has to be more than the 300μm. In the commercial design therefore it is required to meet this minimum criterion to make a good connector. However, connector manufacture with less than 300μm by this technique it is difficult and not compatible with high-density packaging, which is required in the advanced IT (Information Technology) market [3]. Therefore, high-density micro connectors-with a smaller pitch less than 100 μm are one of promising connecting tools for current electronic devices. At the same time, it is also in need to comprehend high-density wrapping with narrow pitch to meet the aforesaid demand [2-4]. Therefore, this paper focuses on micro connectors' construction by combining UV thick resist photolithography and Ni electroforming [1]. This technique is well-matched to produce a high-

[1] To whom any correspondence should be addressed.

density packaging and high precision micro connectors in mass production. A fork type micro connector is designed including socket/plug with a connector pitch radius of 80μm has been fabricated [2] using UV thick photo-resist and Ni electroforming to facilitate the commercial specifications of high-density packaging and better performance electronic devices. However, this micro connector design was done on elastic properties of bulk Ni at normal temperature, whereby plastic deformation of the socket was induced in a long-lasting temperature humidity testing leads to permanent deformation of Ni. In identifying the newly developed contacts characteristics and performance it is required to perform environmental test is conduct at the temperature range of -40 OC to +85 OC [5] at 30 minutes per steps or one hour cycle.

2. CONNECTOR DESIGN

Figure 1 shows the dimension of the plug and socket of the proposed connector. In using UV thick photo-resist [5] (THB-130N) is used to form the mould of the proposed connector and the connector can be designed with the minimum dimension of thickness 50μm and width 15μm with the minimum gap between to connector is 15μm and the maximum aspect ratio supposed to be 3.3 of the mould for electroforming.

The fork-type micro connector shape in Figure 1 is planned and it can be fabricated using a mono-layer structure and the design is justified in Table 1 with the commercial specification and identified its strength over the commercial demand. The number of photo-masks required for fork-type connectors is less than that of cantilever-type, and the fork-type has been adopted. Once the connectors plug and socket is completely connected than, the connector will be able to transmit stable signal without any interruption. If contact force between the connectors are not strong enough or gaps occurs within the connecting point than, shaky contact resistance occurs and leads to unstable signal transmission. For this reason, it is important to secure the contact force during design process.

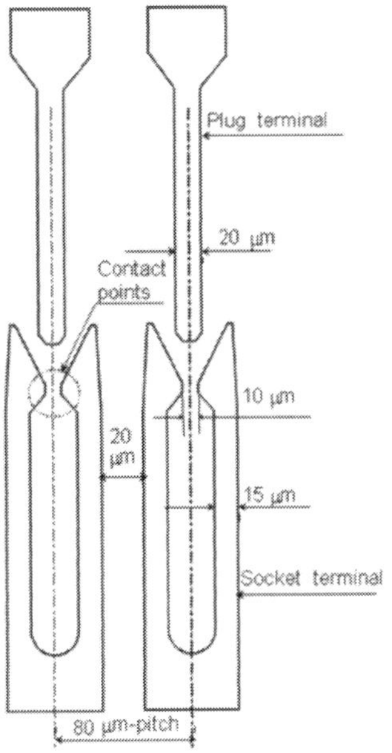

Figure 1: Schematic of socket and plug terminals.

Table 1: Comparison of designed micro connector over the commercial connector specifications

	Commercial connector	Micro connector
Classification	Board to board	Board to board
Structure	Fork-type / Cantilever-type	Fork type
Pitch	300 μm	80 μm
Number of pins	20 □100	100
Allowable current	□ 500 mA	□10 mA
Contact resistance	□ 5 mΩ	□ 50 mΩ
Temperature	Operating temperature: -20□85□ (Including the exothermic temperature 30□) Storage temperature: -40□85□	Operating temperature: -20□85□ (Including the exothermic temperature 30□) Storage temperature: -40□85□
Contact force	□ 290mN /pin	□ 5mN / pin
Packaging density	3pin/ mm	12 pin/ mm
Fabrication technology	Pressing and injection molding	Photolithography Micromachining Technology

Therefore, this study focuses on fabrication of micro connectors by combining electroforming and UV thick resist photolithography. This method makes it possible to realize high-density packaging and high precision micro connectors in mass production.

According to FEM analysis for load–deflection relation of a single cantilever of the socket terminal, the dimensions are identified to meet the specific normal force to make a stable contact. Figure 7 reflects are finite elements modelling. Ni is chosen as the contact material. Therefore, Young's modulus E= 200GPa and poison's Ratio $\upsilon = 0.3$ are used for the analysis [4]. 4.5mN contact force is achieved when the displacement is 5μm.

The schematic view is also provided below in figure 2 and 3 to visualize the designed micro connector in real design.

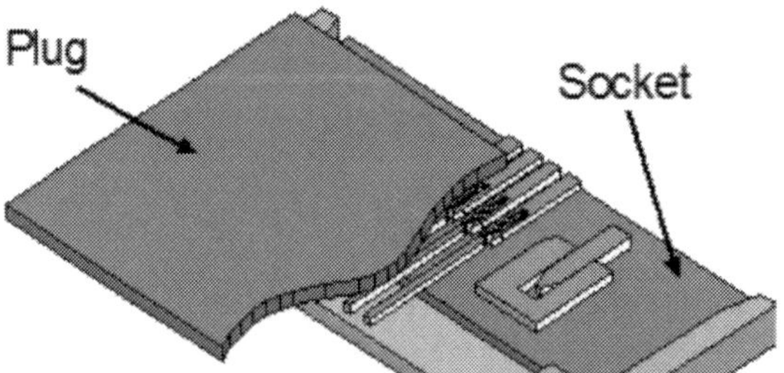

Figure 3: Schematic view of assembled micro connector.

3. FABRICATION

The high aspect ratio micro interposer increases the contact resistance. Consequently, it improves the reliability of electrical signal transmission. A fabrication process of the micro interposer combining UV thick resist photolithography and Ni fine electroforming was adopted to satisfy the above mentioned specification. In order to make the terminal thicker, UV thick negative type photo resist is necessary. In this research we tried to fabricate a low resistance contact using a new process for a Ni sandwich type structure as shown in Figure 4. The previous contact resistance was 200mΩ. The target of our contact is less than 50mΩ. Therefore, a thick contact and an Au sandwich type contact are decided to fabricate and the process is explained in the following Figure 4. Fabricated plug terminals by Ni electroforming are showed in Figure 5 also fabricated socket terminal is displayed in Figure 5.

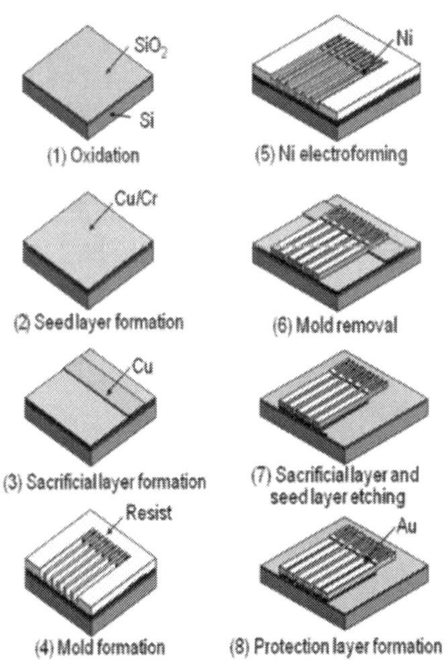

Figure 4: Fabrication Process

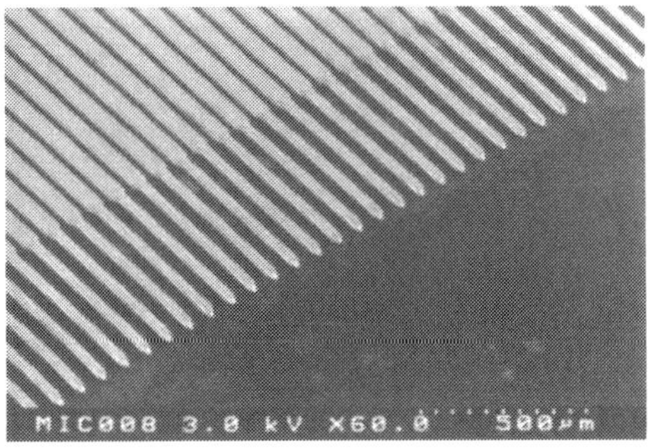

Figure 5: Fabricated plug terminals by Ni electroforming.

IOP Conf. Series: Materials Science and Engineering **53** (2013) 012063 doi:10.1088/1757-899X/53/1/012063

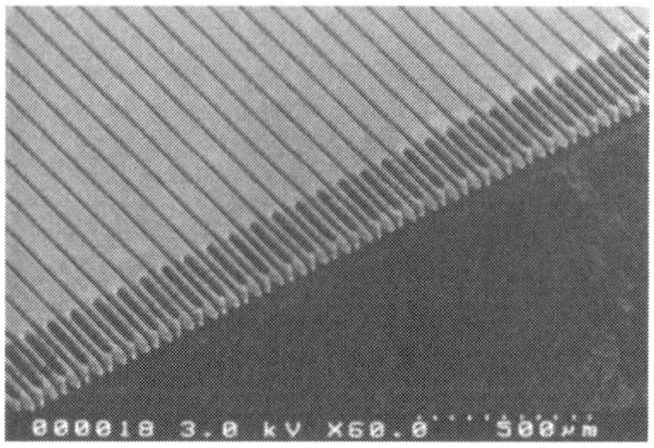

Figure 6: Fabricated socket terminals by Ni electroforming.

4. EXPERIMENTAL RESULTS AND ITS VALIDATION

Connection was confirmed for single pair of plug (Figure 5) and socket (Figure 6) terminals on the prototype connector. The plug terminal was separated from the substrate and was inserted in the socket terminal manually under microscope observation.

Contact resistance between the single plug and the single socket was measured by the four-point probe method (see Figure 9-11). The value of contact resistance was measured at approximately 50 mΩ.

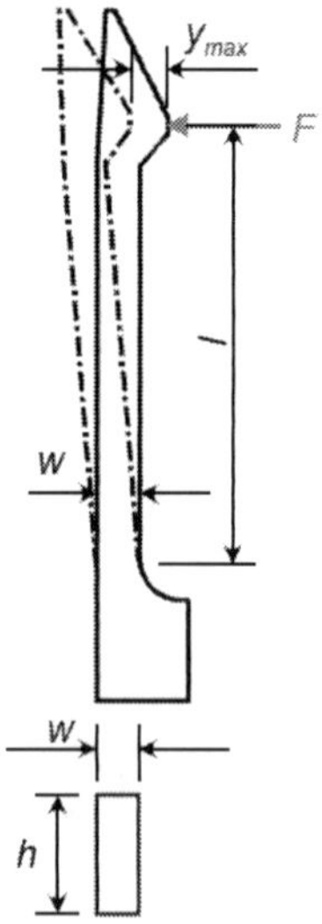

Figure 7: Contact force Measurement experimental setup

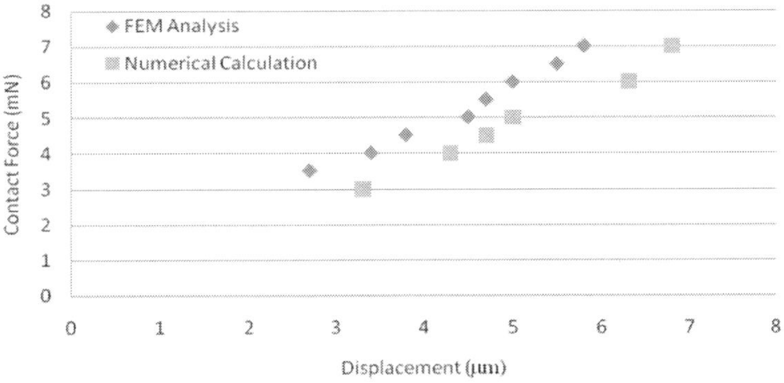

Figure 8: Contact Force (mN) vs Displacement (μm)

$$y_{max} = \frac{Fl^3}{3EI}$$; Cantilever displacement and force relation

$$I = \frac{hW^3}{12}$$; Cross-section equation

Here,

Socket displacement, $y_{max} = 5\mu m$
Contact force, $F \geq 5mN$
Socket length, $l = 200\mu m$
Contact thickness, $h = 50\mu m$
Socket contact width, $w = 15\mu m$
Young Modulus, $E = 207GPa$
Therefore, Force applied on the contact (for 5μm displacement), $F = 5.4mN$

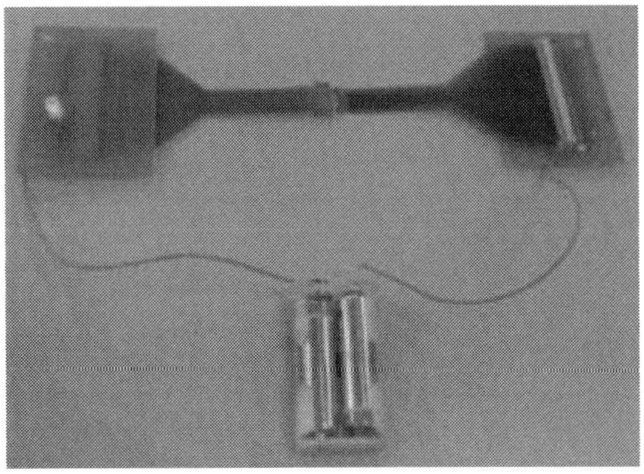

Figure 9: Measurement of contact resistance by four-point probe method

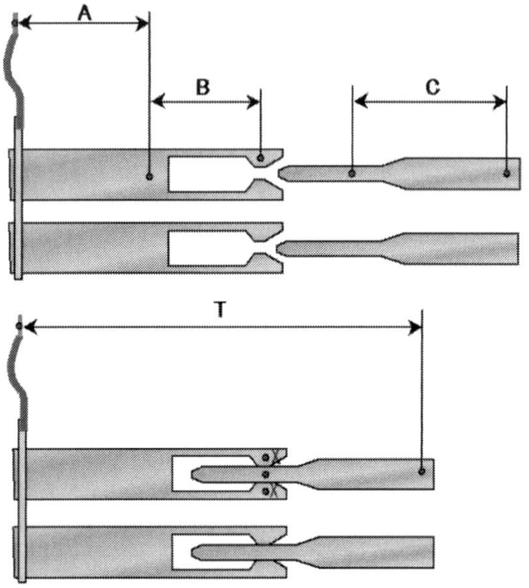

Figure 10: Schematic view of Measurement of contact resistance by four-point method

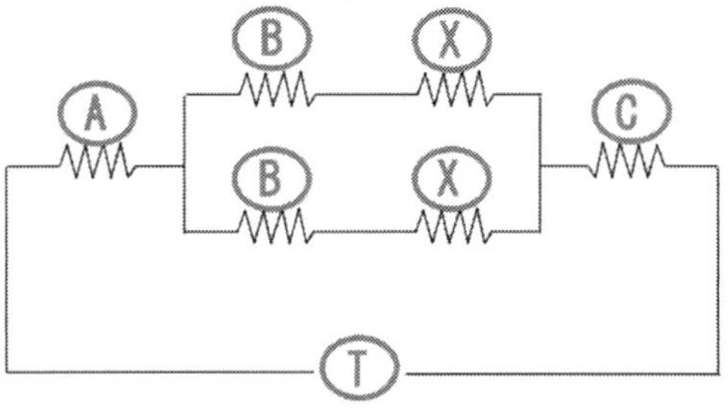

Figure 11: Equivalent circuit diagram of four-point method

Total resistance (RT) = A + (B+X)/2 +C

Steps	1	2	3	4
Temp (°C)	-40	+85	-40	+85
Time (min)	30	30	30	30

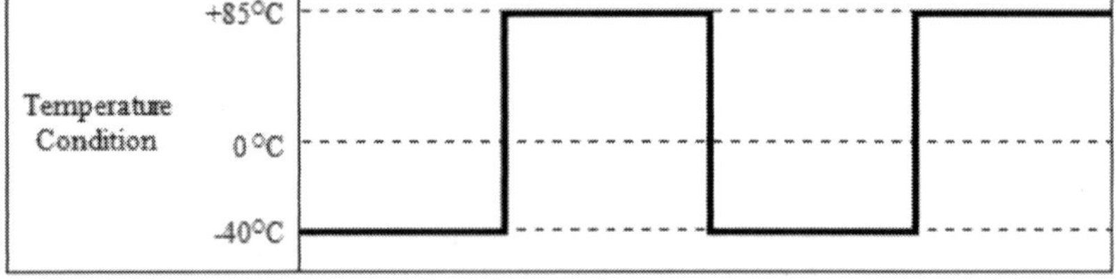

Figure 12: Environmental condition

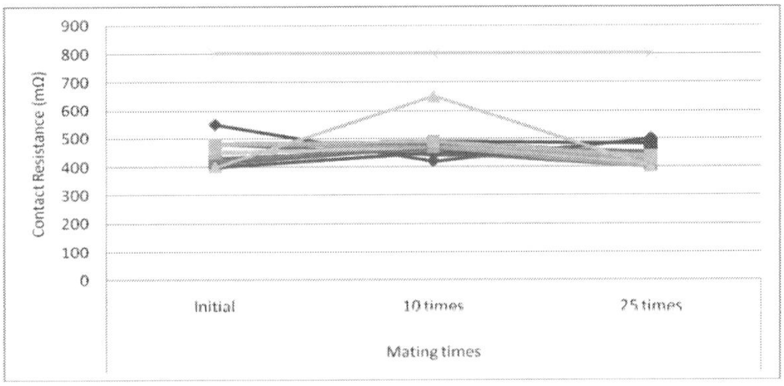

Figure 13: Contact resistance at different mating time

5. FUTURE DEVELOPMENT

To increase contact normal force, the material need to be changed and our future research will be proceed to indentify the suitable material to make more stable contact. On paper [6-7] there are studies conducted on the EFF technology to develop new material NiCo in order to increase the material strength drastically. By using the similar smart material it is possible to come up with more stable contact resistance near future.

CONCLUSION

Due to the recent demand in the industrial production it is required to produce thinner contact which supposed to be more stable and less contact resistance to fulfil the commercial demand. In this research it has been proved that better contact is being produced by MEMS and EFF. The prototype micro connector was fabricated by combining UV thick resist photolithography and Ni electroforming. The size of the terminal of fabricated micro connector thickness was 50 μm, and minimum width 15 μm. Throughout the experimental test we have identified the contact force rage was 2.08 mN to 3.76 mN, and the average value was 3.11 mN. At the same time the contact resistance was approximately 50 mΩ. We also successfully fabricated 100 micro connectors. The maximum aspect ratio of each fabricated micro connector was 3.3 and the terminal pitch is 80 μm. In the future development, connectors with higher aspect ratio will be fabricated to realize commercial micro connectors.

ACKNOWLEDGEMENT

This work was partially developed in cooperation with Ritsumeikan University, Japan. The author would like to thank Ritsumeikan University authority for providing all sort of facilities in order to accomplish is research work.

REFERENCES

[1] M. Bhuiyan, S. Yamada, N. Kurokawa, K. Sakamoto, E. Takemasa, "Development of Low Normal Force Contact using Electro Fine Forming and Lithography Technology," *Tyco Electronics Technical Conference 2008*, Oct. 2008, China

[2] M.Ishii et al.: "Connector Advanced Technology'99", Japan Advanced Technology (1999) (in Japanese).

[3] F.Michel and W.Ehrfeld: "Mechatronic Micro Devices", *Proceedings of the 1999 International Symposium on Micromechatronics and Human Science*, 27(1999).

[4] D.C.Miller, W.Zhang and V.M.Bright: "Microrelay Packaging Technology Using Flip-Chip Assembly", *Proceedings of MEMS 2000*, 265(2000).

[5] T. Unno, et al. "Research on Manufacture of the Micro Connector using a UV-LIGA Process," PhD. Thesis of Ritsumeikan University (2004) (in Japanese).
[6] Bhuiyan, M. M I; Unno, T.; Yokoyama, Y.; Toriyama, Toshiyuki; Sugiyama, S., "Micro connector fabricated by micro process technology," *Proceedings of 2000 International Symposium on Micromechatronics and Human Science, 2000. MHS 2000.*, vol., no., pp.83,88, 2000
[7] M.Bhuiyan, "Optimization of Low Normal Force Contact Spring," Autumn Conference TE KEC 2009, Japan
[8] M.Bhuiyan, "Development and Optimization of High-Hertz Stress Contact to Reduce Future Contact Material Cost," TE Technical Conference 2010, Oct. 2010, USA

IOP Conf. Series: Materials Science and Engineering 53 (2013) 012064 doi:10.1088/1757-899X/53/1/012064

A Qualitative Approach to Mobile Robot Navigation Using RFID

M. Hossain[1], M. M. Rashid, M.M.I. Bhuiyan, S. Ahmed, and M. Akhtaruzzaman
Department of Mechatronics Engineering
International Islamic University Malaysia, 53100 Kuala Lumpur, Malaysia.

E-mail: mahbuba.hsn@gmail.com, mahbub@iium.edu.my, moinul@iium.edu.my, sayem7746@gmail.com, and akhter900@yahoo.com

Abstract. Radio Frequency Identification (RFID) system allows automatic identification of items with RFID tags using radio-waves. As the RFID tag has its unique identification number, it is also possible to detect a specific region where the RFID tag lies in. Recently it is widely been used in mobile robot navigation, localization, and mapping both in indoor and outdoor environment. This paper represents a navigation strategy for autonomous mobile robot using passive RFID system. Conventional approaches, such as landmark or dead-reckoning with excessive number of sensors, have complexities in establishing the navigation and localization process. The proposed method satisfies less complexity in navigation strategy as well as estimation of not only the position but also the orientation of the autonomous robot. In this research, polar coordinate system is adopted on the navigation surface where RFID tags are places in a grid with constant displacements. This paper also presents the performance comparisons among various grid architectures through simulation to establish a better solution of the navigation system. In addition, some stationary obstacles are introduced in the navigation environment to satisfy the viability of the navigation process of the autonomous mobile robot.

1. Introduction

Navigation and localisation of mobile robot both in indoor and outdoor environment is an important and challenging issue to accomplish. Basically mobile robot navigation requires a few abilities; localization, orientation, ability to move, and a means to determine whether the goal is achieved or not. The objective is to find the most efficient way to reach a destination. Many approaches on mobile robot navigation are aforementioned and can be categorized in two ways, Relative Localization (RL) and Absolute Localization (AL). The RL includes dead reckoning and inertial navigation strategies where the AL includes passive/active RFID, landmarks, map-based, and vision-based localization process [1].

Dead reckoning is one of the conventional approaches where the main strategy is to measure the number of wheel revolution and use relative motion [2][3]. Grid based line following mobile robot combining with dead-reckoning strategy is represented in a research where the robot's task was to perform in a soccer game competition [4]. The problems of dead-reckoning process are uncertainty

[1] Corresponding Author, E-mail: mahbuba.hsn@gmail.com

IOP Conf. Series: Materials Science and Engineering **53** (2013) 012064 doi:10.1088/1757-899X/53/1/012064

caused by the wheel slip and error accumulation. The problem can be solved by using external sensor to revise the accumulated errors by estimating the pose and location through observing the existing landmarks in its environment [5][6][7]. Again problem occurs because of ambient lights and shielding obstacles which cause the landmarks not to be observable. So, instead of using landmarks, many researchers tried to use active RFID tags but some problems still remain in terms of maintenance and battery replacement issues. Not only that, using active RFID tags increase the system complexity because of the determination of its path depending on the signal strength of two or more antennas [8].

A recent research [9] represents Received Signal Strength Indicator (RSSI) measurements and comparison technique between two RFID sensors and Back-propagation Artificial Neural Network (BPANN) for robot localization. Such classification system grows complexity and the learning procedure needs huge datasets. The system also increases power consumption.

Navigation routes can be observed through Global Positioning System (GPS) and using active RFID. But GPS cannot be used in indoor environment because it loses range and strength of signals. In outdoor environment, GPS system is used successfully for mobile robots navigation [10] [11]. Park S. et al. (2010) presented intelligent localization system for mobile robot using Passive RFID by calculating the read-time of RFID system but the study does not introduce any obstacle avoidance mechanism [12]. Some researchers describe the estimation of mobile robot orientation in unknown environment by using 2D range scans [13].

Passive RFID can provide a non-vision based navigation system where the small detecting range could be considered. In this research, a passive RFID reader is mounted on the robot and a number of tags are placed on the navigation surface based on specific grid architecture. There are two main objectives of the research. Firstly, provide a suitable navigation technique for the robot to reach its target location avoiding the obstacles. Secondly, verify the performance of navigation based on various grid architectures. Different grid architectures will provide different resolutions for path planning during navigation. The position and orientation of the robot will be identified based on the information detected from the placed tags on the grid points and the compass. The proposed strategy will also minimize the power consumption of the entire system.

The rest of the paper is organized as follows: RFID system, system design, and grid architecture are presented in Section 2. Section 3 represents the Navigation Algorithm. Section 4 presents simulation and results analysis. Concluding remarks are presented in section 5.

2. System description

2.1. RFID system

RFID system has three basic components, RFID reader (interrogators), RFID tags (transponders), and Firmware. The firmware establishes a communication between the reader and tag to retrieve the tag serial (ID) number. The firmware also maintains the communication with main controller to send the tag ID so that the controller can perform its necessary task using that information. The RFID reader can detect a RFID tag in its wireless range. Depending on the operating principles, a tag can be characterized as active, semi-passive, and passive [15]. A passive RFID tag is least complex and cheapest among the all three types. Passive tag does not have its own power source. It uses the Electromagnetic (EM) field from the reader to power up its internal circuit.

2.2. System architecture and design

For this project, the task of the main controller (PIC) is to accumulate and process necessary information obtained from sensors, such as Compass, RFID tags, Infra Red (IR) sensor, Ultrasonic sensor, etc. to drive the mobile robot navigating in the indoor environment. The RFID system used in this research is passive RFID type. The coil antenna of the reader is mounted under the robot base so that the robot can detect the tag underneath while traveling on the navigation surface. Based on the ID number of a tag as detected on the navigation surface, the robot calculates its location respect to that tag information. As the tag detection range is about 7 cm, the signal does not overlap with other tags. Figure 1 represents the physical system architecture with sensors and actuators distribution around the controller of the mobile robot.

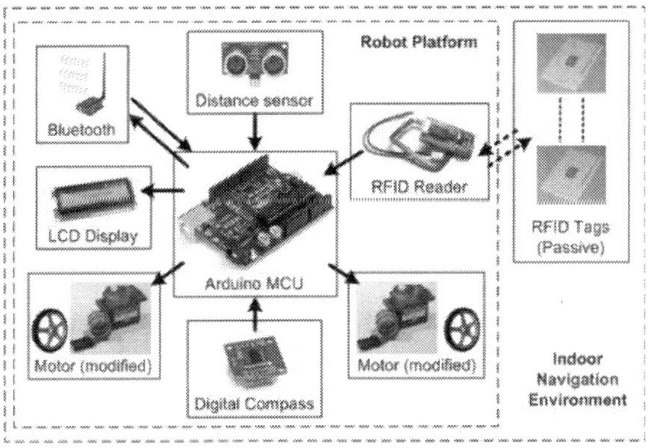

Figure 1. System architecture concept.

Based on the above architecture, the controller unit receives the necessary information about the environment from Distance sensor, Digital compass, and RFID reader. The controller sends the current status and other information to LCD display. Most importantly, controller generates output signals to drive the motors so that the robot can move to its goal following the selected route. The compass helps to identify its orientation and the direction of the target destination based on the angular displacement, θ, from the reference y-axis. The distance sensor helps the robot to detect any obstacle on its navigation route. Furthermore, the RFID reader has the key responsibility which helps the robot to identify its current position, direction and the distance of the goal location. During the navigation task, the RFID reader sense the tags located on the grid points of the navigation surface. If the robot is diverted from its targeted direction line, new direction route is recalculated based on the information taken from RFID and also from compass. The mobile robot controller also has connection with wireless communication device such as Bluetooth to communicate with external device like PC through which several commands could be send to the robot and also the new goals could be assigned.

2.3. System architecture and design

The displacements of the tags on the floor are regular and in square grid patterns as presented in figure 2. This type of structure is considered to have a simplified navigation strategy. In this research, two types of displacement patterns are used, smaller (20cm) and larger (25cm) displacement, between two corresponding tags. Each tag has its own ID so its location in the grid is fixed from other tags. The (x, y) coordinate of each tag is changed to the polar coordinate system represented as (r, θ) where 'r' is the displacement between the goal and the present tag and 'θ' is the angle between the y-axis and the target line. Objective is to reach the target using the shortest possible route.

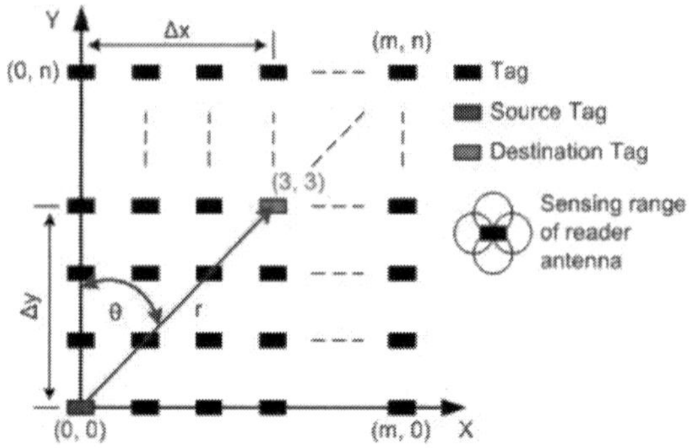

Figure 2. RFID tag displacement and grid architecture of the navigation surface.

3. Navigation algorithm

As the purpose of this navigation is to reach the goal, the basic algorithm can be presented with a flow diagram as shown in figure 3. For this algorithm all the ID of RFID tags are predefined and stored in the MCU with the corresponding coordinate locations. As a result, robot knows all the positions of the tags in its environment as well as the position of the assigned goal position. For the initial condition, the robot is placed on a particular tag. As the process begins, the robot considers the tag as the source position and information are used to calculate (r, θ) and determines the direct distance 'r' which has to travel and the angle 'θ' to fix its orientation by rotational turning. As it detects any new tag on its way, the robot directs itself to the location of the new tag and performs calculations to determine the new (r, θ). Every time the robot detects a new tag, it updates its position and continues to reach the goal.

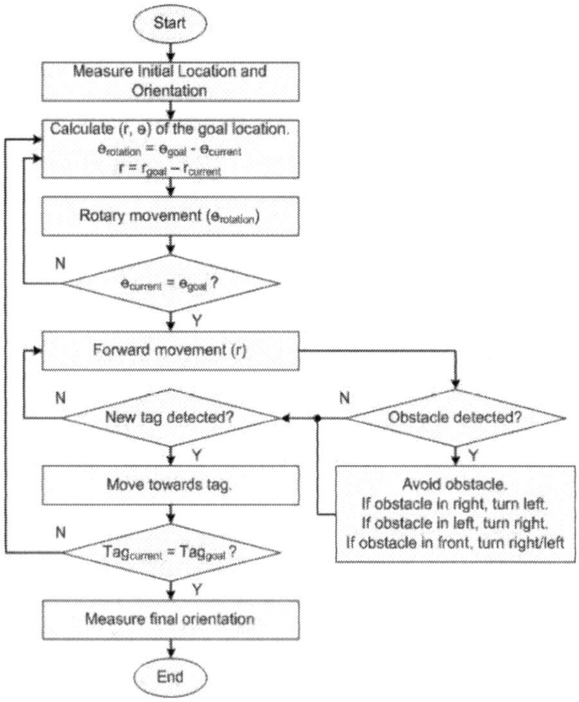

Figure 3. Flow diagram of the navigation algorithm.

During the navigation process, if any obstacle is detected in its way, the robot tries to avoid it by turning counterclockwise or clockwise direction until the obstacle goes out off range. If the obstacle is on the right the robot turns counterclockwise and if the obstacle is on the left, robot turns clockwise. If the obstacle is detected in front, the robot turns clockwise or counterclockwise depending on the direction of the goal. The tags are placed on the navigation surface at a significant distance so that only one tag can be detected in a particular time. If tag detection range is large enough and more than one tag is detected at the same time, the nearest one has to be identified using RSSI technique.

The equations to calculate the distance and the angle can be presented as,

$$r = \sqrt{\Delta x^2 + \Delta y^2} \quad (1)$$

$$\theta = \tan^{-1}[\frac{\Delta x}{\Delta y}] \quad (2)$$

$$\Delta x = x_{tag} - x_{current} \quad (3)$$

$$\Delta y = y_{tag} - y_{current} \quad (4)$$

$r = \mathit{Displacement\ between\ target\ tag\ and\ present\ tag}$
$\theta = \mathit{Angle\ between\ reference\ (y-axis)and\ target\ tag}$
$\Delta x = \mathit{Displacement\ in\ x-axis\ between\ target\ tag\ and\ present\ tag}$
$\Delta y = \mathit{Displacement\ in\ y-axis\ between\ target\ tag\ and\ present\ tag}$

4. Simulation and result analysis

Using the algorithm, some simulations are conducted to validate the performance of the navigation procedure. Two different grid environments are used. Both include square grid structures as described in section 2.3. One environment has spacing (20cm×20cm) and the other with spacing (25cm×25cm). The navigation strategy also tested through simulation both for with and without obstacles.

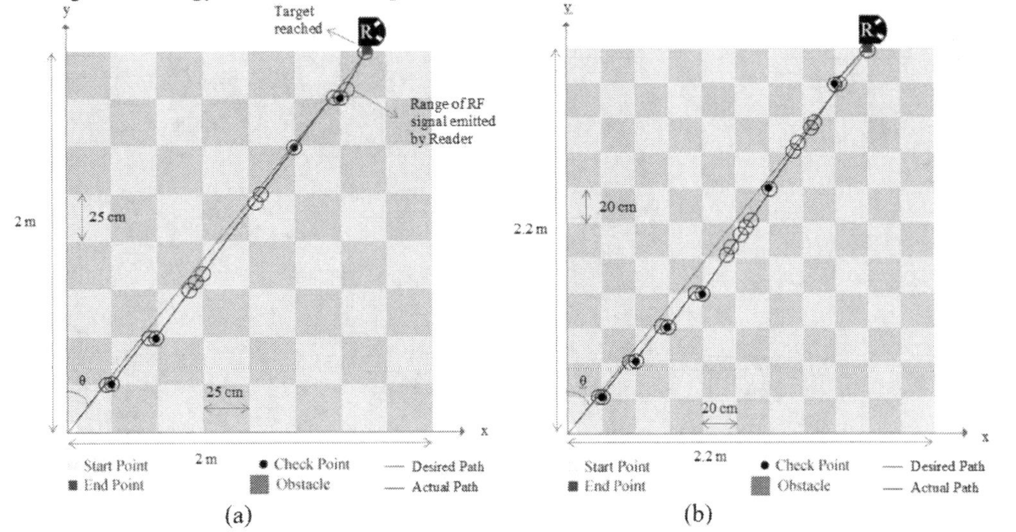

(a) (b)

Figure 4. Robot navigation with tag spacing (a) 25cm×25cm and (b) 20cm×20cm (No obstacles).

Table 1. Errors in angular displacement of the robot for no obstacle (test 1).

IOP Conf. Series: Materials Science and Engineering **53** (2013) 012064 doi:10.1088/1757-899X/53/1/012064

Analysis with Tag Spacing 25cm×25cm			Analysis with Tag Spacing 20cm×20cm		
Update from nearby tag	**Measured angle, θ (⁰)**	**Angle error from Desired Angle (%)**	**Update from nearby tag**	**Measured angle, θ (⁰)**	**Angle error from Desired Angle (%)**
0	39.2	0	**0**	39.3	0
1	38.3	2.295918367	**1**	38.7	1.526717557
2	37	5.612244898	**2**	37.9	3.562340967
3	37.2	5.102040816	**3**	36.9	6.106870229
4	27.5	29.84693878	**4**	35.5	9.669211196
5	Goal	0	**5**	36.9	6.106870229
-	-	-	**6**	45	-14.50381679
-	-	-	**7**	Goal	0

The above two navigation environments, shown in figure 4, are considered as without obstacles. Simulation results present the smooth navigation of the robot with some small deflections in terms of orientations. For both of the cases robot reaches to its destination. In figure 4 (a), where (25cmx25cm) displacement is used, the robot updates its position for four times during the movement while figure 4 (b), where (20cmx20cm) grid displacement is used, presents 6 updates points while navigating from source to destination. So, first grid environment allows the robot to reach its destination faster than the second environment. Table 1 and figure 5 present the errors in angular displacement of the robot from the desired angle during navigation.

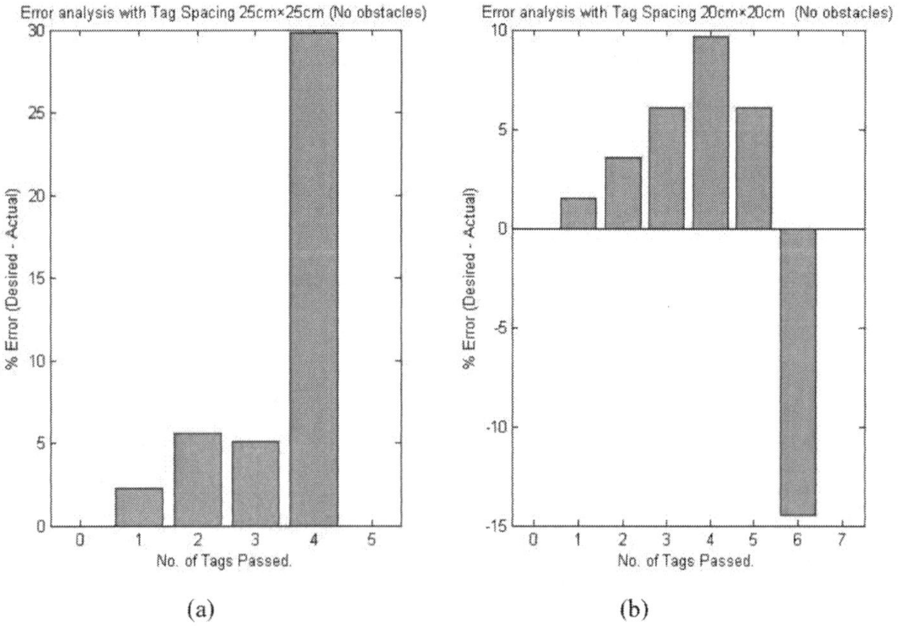

Figure 5. Orientation error during navigation process, (a) with tag spacing 25cm×25cm and (b) tag spacing 20cm×20cm (with no obstacles, test 1).

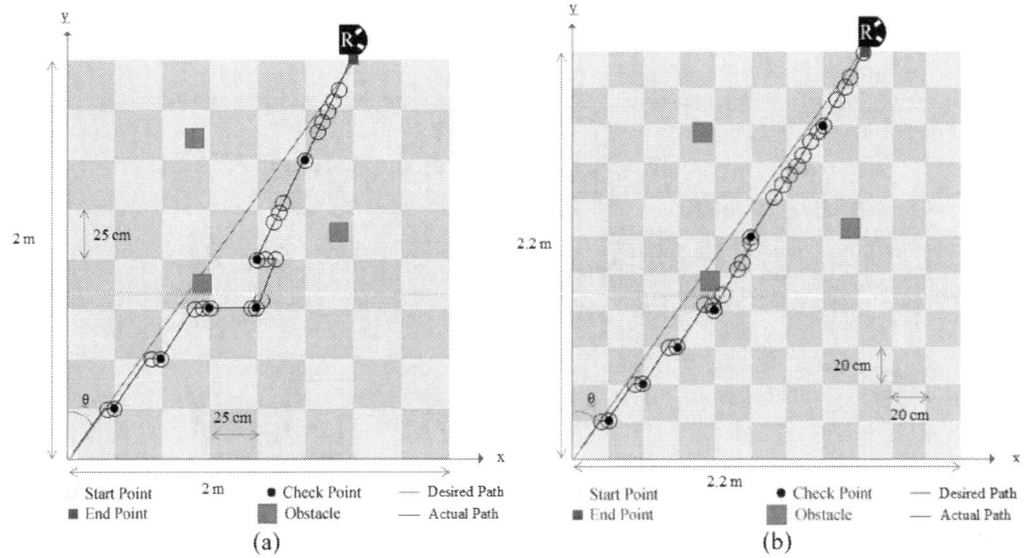

Figure 6. Test case 2; Simulation of the robot navigation (a) with Tags spacing 25cm×25cm and (b) with Tags spacing 20cm×20cm (With obstacles).

Figure 6 shows the result of another simulation test where some obstacles are places. For both of the conditions, the source, destination, and obstacle placements are same so that the performance of the robot navigation for the two different environments can be compared. The result presents that for both of the situations, the numbers of update points are same but for first environment, the robot travels longer distance than the second one. First environment, where (25cm x 25cm) tag displacements are used, also shows that a large angle error occurs at the point when robot meets an obstacle on its way. Table 2 presents the list of angle errors occurred during the navigation process for both of the cases. Figure 7 shows the graphical representation of the angle errors.

The error of the navigation process can be analysed using the position error, orientation error, and time taken to reach the goal. Here only one aspect is considered for the navigation algorithm and that is orientation or angular deflection. Results have shown that the algorithm used is quite simple and straight forward. It also can be observed that without obstacles, the navigation route is smooth having less orientation error. With obstacles, the orientation error increases but it still reaches the goal with less complicacy.

Table 2. Errors in angular displacement of the robot with obstacles (test 2).

Analysis with Tag Spacing 25cm×25cm			**Analysis with Tag Spacing 20cm×20cm**		
Update from nearby tag	**Measured angle, θ (^{0})**	**Angle error from Desired Angle (%)**	**Update from nearby tag**	**Measured angle, θ (^{0})**	**Angle error from Desired Angle (%)**
0	36.9	0	**0**	36	0
1	35.5	3.79403794	**1**	35	2.777777778
2	33.7	8.672086721	**2**	33.7	6.388888889
3 (Obstacle)	90	-143.902439	**3**	32	11.11111111
4	21.8	40.92140921	**4 (Obstacle)**	29.7	17.5

5	26.6	27.91327913	**5**	31	13.88888889
6	26.6	27.91327913	**6**	26.6	26.11111111
7	Goal	0	**7**	Goal	0

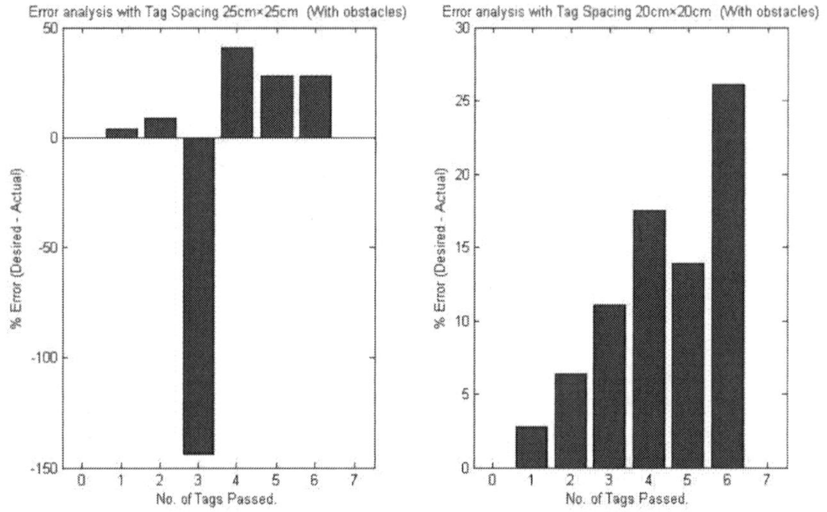

(a) (b)

Figure 7. Graphs showing the orientation error during navigation process with (a) tag spacing 25cm×25cm and (b) tag spacing 20cm×20cm (with obstacles, test 2).

Figure 8 presents another simulation result with different arrangement of obstacles. The result shows that the first environment (25cm x 25cm spacing of tags) causes large angular deflection at the meeting point with obstacle. But both of the cases, robot finds its possible shortest path to reach the goal. Basically, environment with different tag spacing shows different results. Smaller spacing provides more update of its position in the environment compared to larger spacing and therefore better resolution in navigation. But, as there is a better resolution for small spacing environment, there exists frequent updates of positions which increase time in travelling from source to destination. On the other hand, larger tag spacing provides less update points as the gaps among tags are large and the travelling time decreases. For obstacles, smaller tag spacing shows less deviation than the bigger tag spacing environment. Table 3 shows the errors in angular displacement of the robot while travelling from source to destination. Figure 9 represents the graphical view of the errors.

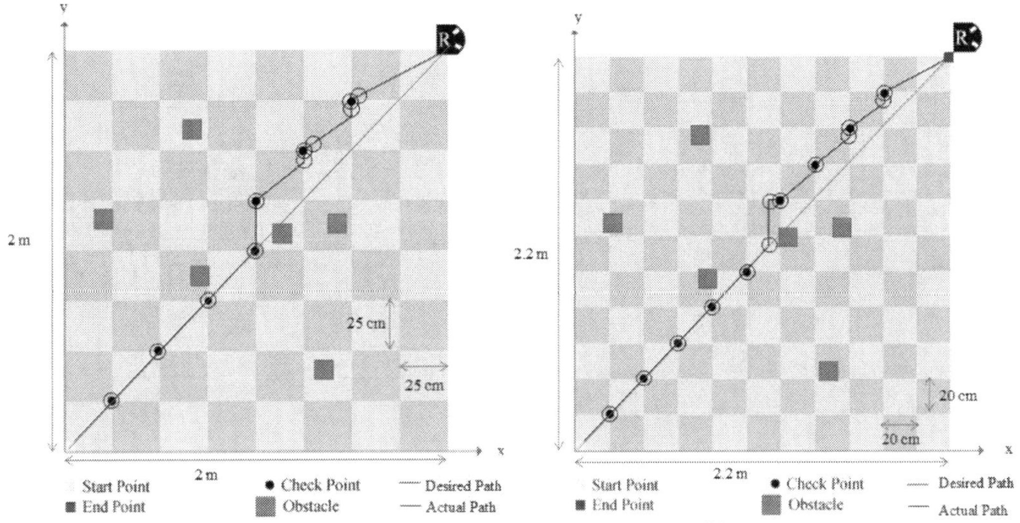

(a) (b)

Figure 8. Test case 3; Simulation of the navigation procedure (a) with tag spacing 25cm×25cm and (b) with tag spacing 20cm×20cm (With obstacles).

Table 3. Errors in angular displacement of the robot with obstacles (test 3).

Analysis with Tag Spacing 25cm×25cm			**Analysis with Tag Spacing 20cm×20cm**		
Update from nearby tag	**Measured angle, θ (°)**	**Angle error from Desired Angle (%)**	**Update from nearby tag**	**Measured angle, θ (°)**	**Angle error from Desired Angle (%)**
0	45	0	0	45	0
1	45	0	1	45	0
2	45	0	2	45	0
3	45	0	3	45	0
4 (Obstacle)	0	100	4	45	0
5	53.1	-18	5	45	0
6	56.3	-25.11111111	**6 (Obstacle)**	51.3	-14
7	63.4	-40.88888889	7	53.1	-18
8	Goal	0	8	56.3	-25.11111111
-	-	-	9	63.4	-40.88888889
-	-	-	10	Goal	0

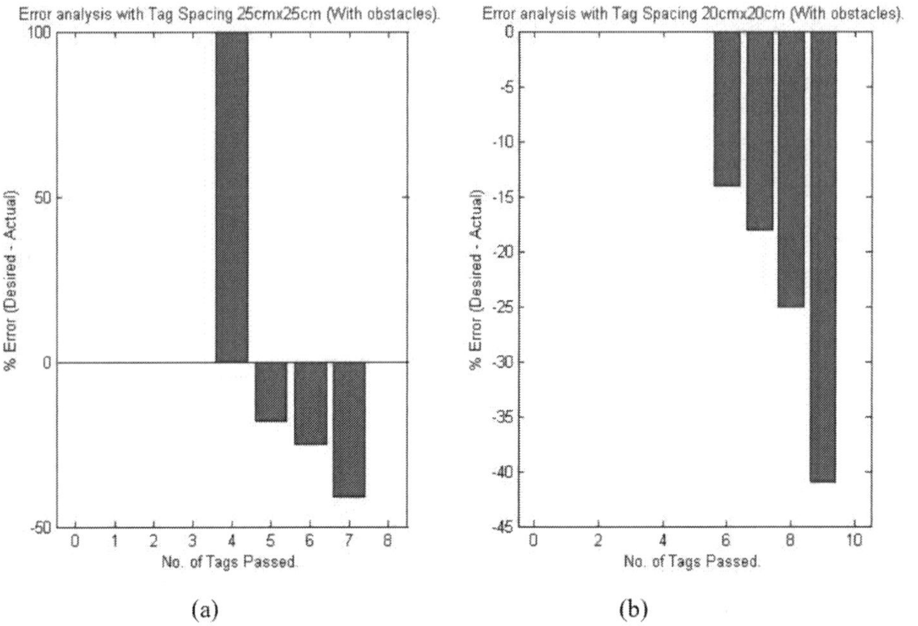

Figure 9. Graphs showing the orientation error during navigation process with (a) tag spacing 25cm×25cm and (b) tag spacing 20cm×20cm (With obstacles, test 3).

For both of the environment shown in this paper, path is easily observed and goal successfully reached. But it is important to keep in mind that the displacement of the tags should not be too small or too large because the smallest displacement will increase frequent and unnecessary updates and largest displacement will increase the possibility of dead-reckoning or sometimes the goal cannot be achieved at all.

5. Conclusion

Based on the simulation and analysis, this method shows effective results in providing good navigation system for mobile robot in indoor environment. The system provides the information for both position and orientation of the robot. No matter what the initial orientation of the robot is, it can orient itself successfully to face to the goal location. As the tags positions are known to the robot and the tag spacing is regular, the robot can perform its navigation tasks only in the same environment. This could be a drawback of the system because the robot will not be able to navigate in unknown environment. Future work will include the experimental investigation of this navigation algorithm to analyze its effectiveness in real environment. With the angle error, the analysis of position error and the travelling time should be taken into consideration to represent a good analytical report of the system. A good hardware with robust controller could lead the system to its level of perfection in the fields of mobile robot navigation.

References

[1] Park S and Hashimoto S 2009 *An approach for mobile robot navigation under randomly distributed passive RFID environment* Proc. of Int. Con. On Mechatronics (ICM 2009) vol. 1 no. 6 pp. 14-17

[2] Park S, Saegusa R and Hashimoto S 2007 *Autonomous navigation of amobile robot based on passive RFID* Int. Symp. On Robot and Human interactive Communication (RO-MAN

2007) vol. 218 no. 223, pp. 26-29

[3] Crowley J L 1989 *Control of Displacements and Rotation in a Robot Vehicle* Proc. of Int. Con. on Robotics and Automation.

[4] Akhtaruzzaman M, Hasan S K and Shafie A A 2009 *Design and Development of an Intelligent Autonomous Mobile Robot for a Soccer Game Competition* Proc. of Int. Con. On Mechanical and Electronics Engineering (ICMEE 2009, World Scientific) pp. 167-171

[5] Tsuji Y, Yagi Y and Yachida M 2001 *Robust Alternative Estimation of an Environment Map and Robot Egomotion Using an Omni directional Vision sensor* Journal of the Robotics Society of Japan vol. 9 no. 1 pp. 59-67

[6] Drocourt C, Delahoche L, Pegard C and Clerentin A 1999 *Mobile Robot Localization Based on and Omni directionalStereoscopic Vision Perception System* Proc. of Int. Con. on Rob. Auto. pp.1329-1334

[7] Betke M and Gurvits K 1997 *Mobile Robot Localization Using Landmarks* IEEE Trans. on Rob. And Auto. vol. 13 no. 2 pp. 251-263

[8] Gueaieb W and Miah M S 2009 *A modular cost-effective mobile robot navigation system using RFID technology* Journal of Communications, USA, 4 (2) pp. 89-95

[9] Nosaiba A S and Khalid A M 2012 *Autonomous mobile robot localization based on RSSI measurements using an RFID sensor and neural network BPANN* Journal of King Saud University - Computer and Information Sciences, Available online 23 October 2012, ISSN 1319-1578, 10.1016/j.jksuci.2012.10.001.

[10] Se-gon R and Hyouk R C 2009 *3-D Tag-Based RFID System for Recognition of Object* IEEE Trans. on Automation Science and Engineering Vol. 6 Iss. 1 pp. 55-65

[11] Shair S, Chandler J H, Gonzalez-Villela V J, Parkin R M and Jackson M R 2008 *The Use of Aerial Images and GPS for Mobile Robot Waypoint Navigation* IEEE/ASME Trans. on Mechatronics vol.13 no.6 pp. 692-699

[12] Park S, Saegusa R and Hashimoto S 2010 *An intelligent localization algorithm using read time of RFID system* Journal of Advanced Engineering Informatics Vol. 24 Iss. 4 pp. 490-497

[13] Lu F and Milios E 1997 *Robot Pose Estimation in Unknown Environments by Matching 2D Range Scans* Journal of Intelligent and Robotic Systems Vol. 18 Iss. 3 pp 249-275

[14] Akhtaruzzaman M, Shafie A A and Rashid M 2012 *Designing an Algorithm for Bioloid Humanoid Navigating in its Indoor Environment* Journal of Mechanical Engineering and Automation Vol. 2(3) pp. 36-44

[15] Chawal V and Ha D S 2007 *An overview of passive RFID* IEEE Applications & Practice pp. 11-17.

Two Dimensional Array Based Overlay Network for Balancing Load of Peer-to-Peer Live Video Streaming

Abdullah Faruq Ibn Ibrahimy, Islam Md. Rafiqul[1], Farhat Anwar and Muhammad Ibn Ibrahimy

Department of Electrical and Computer Engineering,
Kulliyyah of Engineering, International Islamic University Malaysia
Jalan Gombak, 53100 Kuala Lumpur, Malaysia

E-mail: rafiq@iium.edu.my

Abstract. The live video data is streaming usually in a tree-based overlay network or in a mesh-based overlay network. In case of departure of a peer with additional upload bandwidth, the overlay network becomes very vulnerable to churn. In this paper, a two dimensional array-based overlay network is proposed for streaming the live video stream data. As there is always a peer or a live video streaming server to upload the live video stream data, so the overlay network is very stable and very robust to churn. Peers are placed according to their upload and download bandwidth, which enhances the balance of load and performance. The overlay network utilizes the additional upload bandwidth of peers to minimize chunk delivery delay and to maximize balance of load. The procedure, which is used for distributing the additional upload bandwidth of the peers, distributes the additional upload bandwidth to the heterogeneous strength peers in a fair treat distribution approach and to the homogeneous strength peers in a uniform distribution approach. The proposed overlay network has been simulated by Qualnet from Scalable Network Technologies and results are presented in this paper.

1. Introduction

Live video streaming or broadcasting is very expensive through satellite or terrestrial system. On the other hand the huge bandwidth requirement puts lots of load on the video streaming server for streaming video in a server to client network. Peer to peer overlay network can reduce load of the video streaming server by utilizing upload bandwidth of peers to stream video to peers, which is called peer-to-peer video streaming. Any one, who is connected to internet, can provide live video content to any one, who is also connected to internet, with very low cost by peer-to-peer video streaming. Peer-to-Peer video streaming is providing scalability and it is also providing very low cost solution for live video streaming [1]. To ensure optimized utilization of peers upload bandwidth and to ensure a very effective, efficient and stable performance, a highly reliable overlay network, which creates virtual link between servers and peers, is needed.

The live video stream is divided into pieces, called chunk. For live video streaming every chunk has to reach the destination before a specified very sort time, which is playback deadline. Due to

[1] Islam Md Rafiqul: rafiq@iium.edu.my

chunk delivery delay or loss of chunk the stream data cannot be delivered to users before their playback deadline. If the stream data cannot be delivered before their playback deadline then distortion less continuous playback of video cannot be delivered to users. Imbalanced overlay network design is one of the main reason for which chunks get lose or corrupt in a network [8].

The upload bandwidth capacity, which is excluding the upload bandwidth capacity required for uploading the video stream to one peer, is denoted as additional upload bandwidth of a peer. Proposed overlay network distributes chunks to the peers by preferring the peers with higher additional upload bandwidth. If an overlay network utilizes peers additional upload bandwidth to support as many peers as possible and having no support in case of departure of a peer with additional upload bandwidth then this makes the overlay network very vulnerable to churn. But this overlay network always provides support in case of departure of a peer with additional upload bandwidth or without additional upload bandwidth. This makes this overlay network very robust to churn, which enhances the stability of the overlay network.

Instead of considering only upload bandwidth of a peer, download bandwidth of the peer is also considered. Because to upload the video stream at a certain streaming rate the peer has to have a minimum download bandwidth to download the video stream. And peers are arranged according to their upload bandwidths and download bandwidths.

2. Related Works

BitTorrent is one of the most common peer-to-peer file sharing technology. This technology is used by many researchers for live video streaming in a peer-to-peer network [2]. Quality-of-service is one of the main focuses of an overlay network. Fast-Mesh is very much concerned with the quality-of-service [3]. Random peer selection policy is applied in LSONet [4], which is a mesh based overlay network. Some overlay networks like HCPS [5] uses a tracking server or management peer to balance the clusters in it. So the tracking server knows about every single peer. There are also hybrid tree-mesh based overlay network for live video streaming like MultiPeerCast [6]. GradienTv [7] is a mesh based overlay network. It places the video-streaming server at the center of the gradient. PRIME [8] is mesh based overlay network for peer-to-peer live video streaming; it is maximizing the utilization of upload bandwidth of peers while sustaining the quality of content.

3. Proposed Overlay Network

The overlay network puts peers in a two dimensional array. The overlay network has total $((P*n)+c)$ peers. P is number of complete rows and n is number of columns. c is number of peers in the incomplete row. Number of rows depends on number of users and it can be extended to as many as required, which enhances scalability of the overlay network. Number of columns is equivalent to number of live video streams supported by upload bandwidth of streaming server. Number of row is denoted by $1 \leq i \leq P+1$ and number of column is denoted by $1 \leq j \leq n$. $P= ((m/n)+1)$ and m=total number of peers excluding the last row.

The live video streaming server is streaming live video stream data to n peers, which are on first row of the two dimensional array. After that the live video stream data is streaming to the peers of every column in a line topology method. Streaming server and peers are streaming in unicast way. Streaming links from streaming server are unidirectional, but streaming links from peers are bidirectional. Tracking server links with every peer to exchange control signals, which links are also bidirectional. Tracking server receives information from peers about their transmitted and received chunks. And it transmits scheduling policy to every peer about identity of chunks, which will forward and identity of peers, where chunks will forward.

The minimum upload and download bandwidths of each peer are equivalent to video streaming rate. Minimum peer degree of a peer is one and maximum peer degree of a peer depends on upload bandwidth capacity of that peer and the upload bandwidth capacity of the overall system. Each peer $X_{i,j}$ can at least upload or download the stream data to or from peer $X_{i-1,j}$ and peer $X_{i+1,j}$ if they exist. The Fig. 1 shows the arrangement of peers in the proposed overlay network.

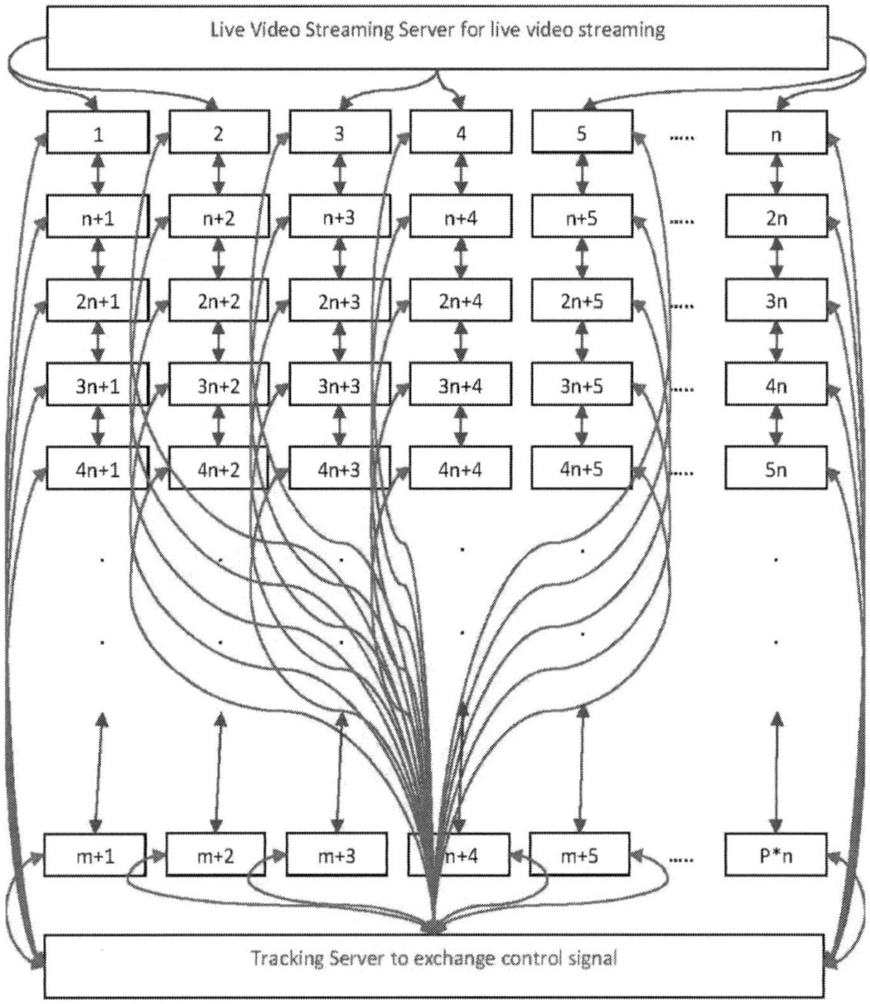

Figure1. Construction of the proposed overlay network.

4. Peer strength measurement

Peers are organized based on their actual upload and download bandwidths. Both upload and download bandwidths are compared to determine the strength of a peer. The one with higher download bandwidth is considered as higher download strength peer if download bandwidth is between the video streaming rate and twice of the video streaming rate. But peers with download bandwidth above twice of the video streaming rate are considered as same download strength peer. And always peer with

IOP Conf. Series: Materials Science and Engineering **53** (2013) 012065 doi:10.1088/1757-899X/53/1/012065

higher upload bandwidth is considered as higher upload strength peer. Peer with higher upload strength and higher download strength is considered as higher strength peer.

Additional upload bandwidth links are distributed to heterogeneous strength peers by preferring the peer with higher strength. And additional upload bandwidth links are distributed to homogeneous strength peers by preferring the peer with longer presence.

5. Simulation

5.1. Setup

The proposed overlay network has been simulated by Qualnet from Scalable Network Technologies. Total number of peer was twenty (excluding the live video streaming server). There were ten peers on every row, so there were two rows of peers and the live video streaming server was streaming directly to the ten peers, which are on the upper row of each column. As the peers on the lower row was receiving packets from the live video streaming server through the peers on the upper row, so we could observe the effect of peer-to-peer packet transmission. We have considered constant-bit-rate coding (CBR) content for the live video streaming service. The propagation delay for each link was 60 milliseconds. The uplink bandwidths of peers are allocated according to the distribution of Table 1 [5]. 320 seconds, 3020 seconds and 30020 seconds simulation time was considered for short duration live video streaming, average duration live video streaming and long duration live video streaming respectively.

Table 1. Distribution of Bandwidth.

Uplink (kbps)	Percentage of peers
128	20
384	40
1000	25
5000	15

5.2. Result and Analysis

Average end-to-end delay, jitter and throughput for different simulation times are plotted in from Fig. 2 to Fig. 10. Columns designated by number 1 to 10 are the number of peer's ID on the lower row of the overlay network. The average end-to-end delay, jitter and throughput variations of peers of different columns on the lower row are very minimum, which confirms that the overlay network is balancing the load of the live video streaming server. With the increase of simulation time, the throughput (bits/s) does not vary much. Also, average end-to-end delay (s)'s variation is not very significant with the change of simulation time. But average jitter (s)'s variation is significant with the increase of simulation time compared to average end-to-end delay (s) and throughput (bits/s). Average jitter (s)'s variation between different columns is reducing as the simulation time is increasing.

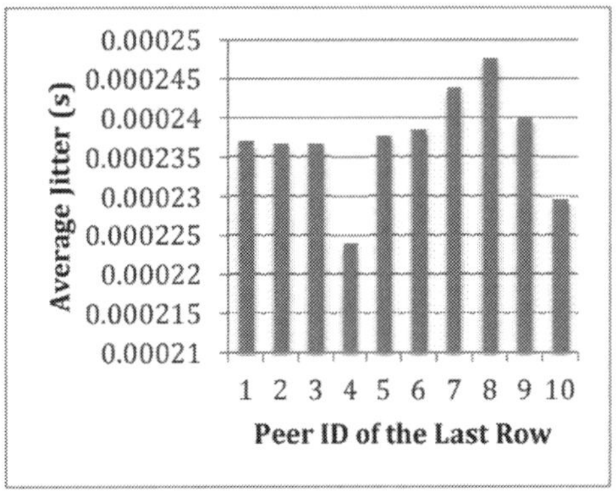

Figure 2. Average Jitter (s) for 320 seconds simulation.

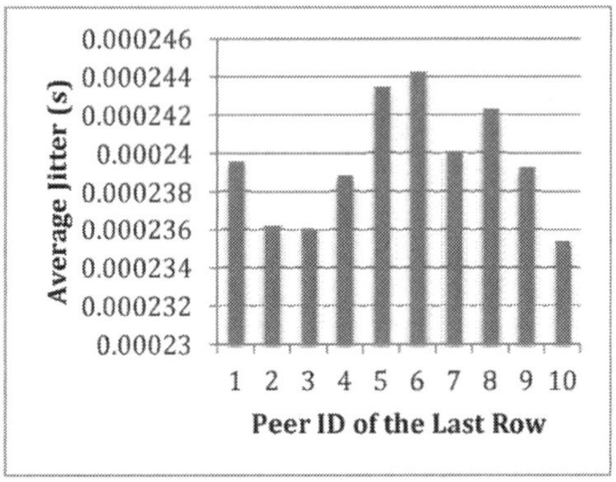

Figure 3. Average Jitter (s) for 3020 seconds simulation.

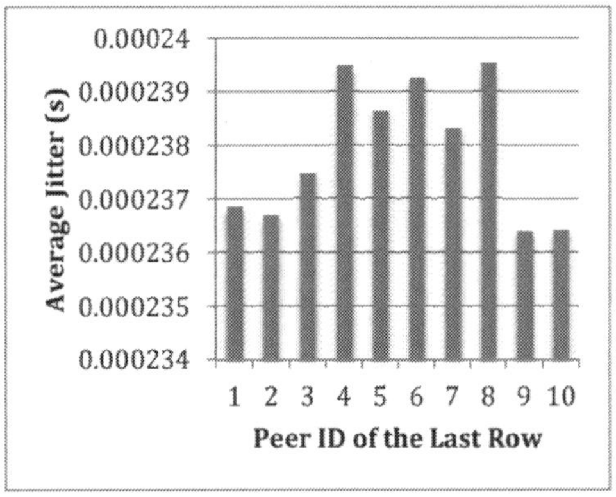

Figure 4. Average Jitter (s) for 30020 seconds simulation.

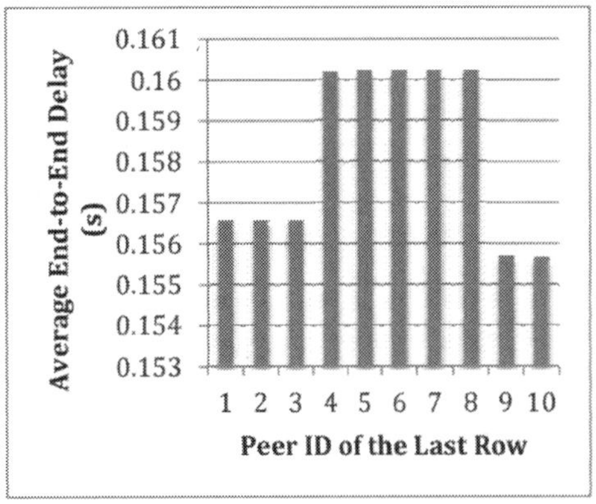

Figure 5. Average End-to-End Delay (s) for 320 seconds simulation.

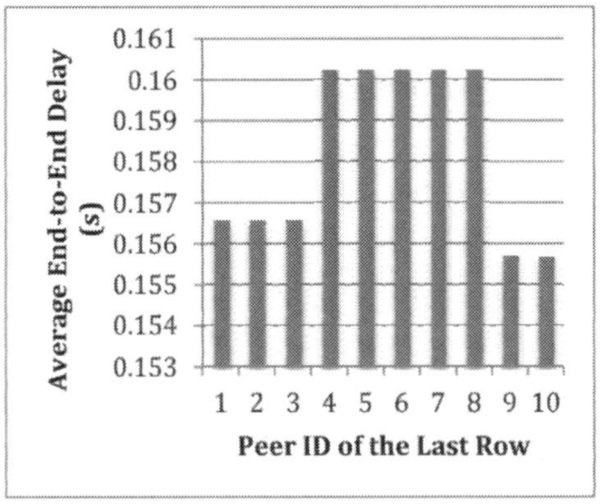

Figure 6. Average End-to-End Delay (s) for 3020 seconds simulation.

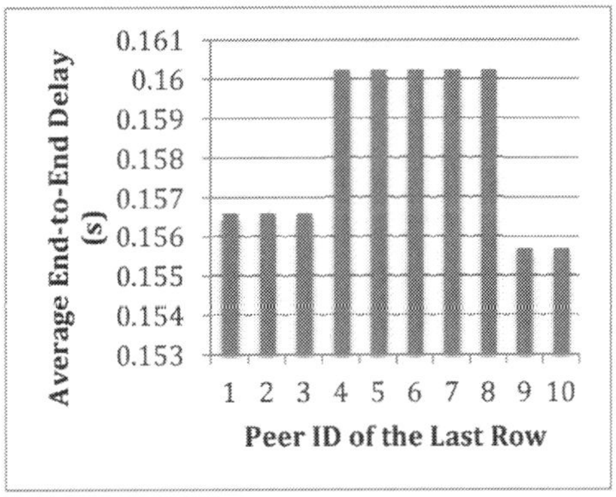

Figure 7. Average End-to-End Delay (s) for 30020 seconds simulation.

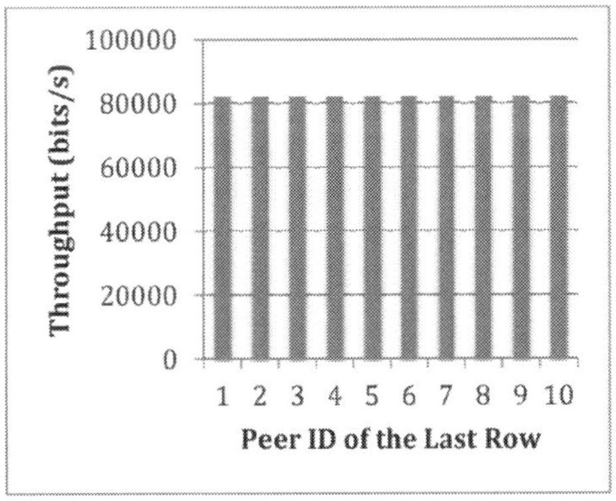

Figure 8. Throughput (bits/s) for 320 seconds simulation.

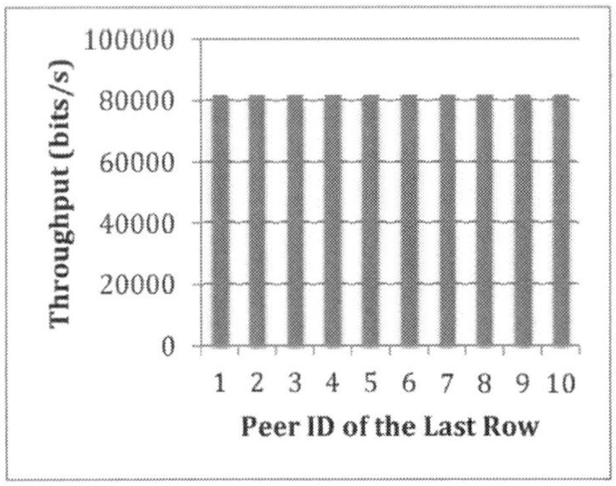

Figure 9. Throughput (bits/s) for 3020 seconds simulation.

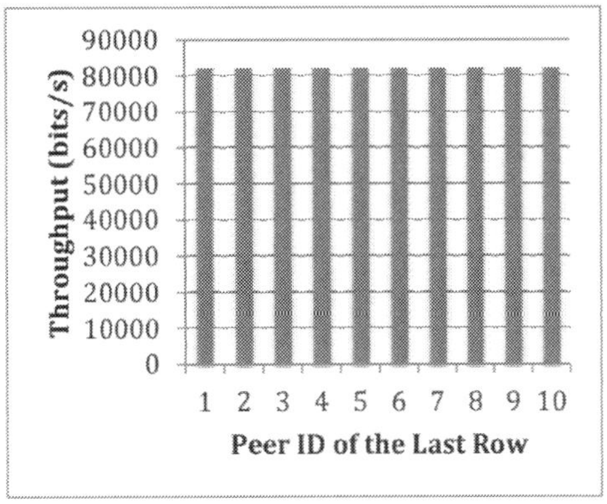

Figure 10. Throughput (bits/s) for 30020 seconds simulation.

Average Jitter varied from 0.000223912 seconds to 0.000247614 seconds for 320 seconds simulation. And it varied from 0.000235382 seconds to 0.000244231 seconds for 3020 seconds simulation. At last it varied from 0.000236399 seconds to 0.000239535 seconds for 30020 seconds simulation.

Average End-to-End Delay varied from 0.155671 seconds to 0.160222 seconds for 320 seconds simulation. And it varied from 0.155679 seconds to 0.160220 seconds for 3020 seconds simulation. At last it varied from 0.155680 seconds to 0.160214 seconds for 30020 seconds simulation.

Throughput is 81933 (bits/s) for 320 seconds simulation. And it is 81921 (bits/s) for 3020 seconds simulation. At last it is 81920 (bits/s) for 30020 seconds simulation.

6. Conclusion

An overlay network has been proposed for both homogenous and heterogeneous strength peers by arranging peers in a two dimensional array to provide better performance. Homogeneous strength peers can be further organized based on their locality and scaling the video can further enhance the scalability of the overlay network. The arrangement of the proposed overlay network provides balanced load on the streaming server. The balancing of load has been observed in simulation by considering 20 nodes over various time durations. Average End-to-End delay, Jitter and Throughput are balanced over the different columns of the overlay network. However the proposed network's performance needs to be tested by increasing the number of nodes to 100 and 1000 and so on.

References

[1] Yi Cui, Baochun Li, and Klara Nahrstedt, "oStream: Asynchronous streaming multicast in application-layer overlay networks", IEEE Journal on Selected Areas in Communications, volume 22, Issue 1, Pages 91 – 106, 2004.

[2] Darshan Purandare, and Ratan Guha, "BEAM: A Peer-to-Peer Framework for Live Media Streaming", Proceedings of the 31st IEEE Conference on Local Computer Networks, Tampa, Florida, Pages 513-514, November 2006.

[3] Dongni Ren, Yui-Tung Hillman Li, and S.-H. Gary Chan, "Fast-Mesh: A Low-Delay High- Bandwidth Mesh for Peer-to-Peer Live Streaming", IEEE Transactions on Multimedia, Volume 11, Issue 8, Pages 1446-1456, December 2009.

[4] Hui Guo, Kwok-Tung Lo, Yi Qian, and Jiang Li, "Peer-to-Peer Live Video Distribution under Heterogeneous Bandwidth Constraints", IEEE Transactions on Parallel and Distributed Systems, Volume 20, Issue 2, Pages 233-245, February 2009.

[5] Yang Guoa, Chao Lianga, and Yong Liub, "Hierarchically Clustered P2P Video Streaming: Design, Implementation, and Evaluation", Computer Networks: The International Journal of Computer and Telecommunications Networking, Volume 56, Issue 15, Pages 3432-3445, October 2012.

[6] ZhiHui Lu, You Li, Jie Wu, ShiYong Zhang, and YiPing Zhong, "MultiPeerCast: A Tree- mesh-hybrid P2P Live Streaming Scheme Design and Implementation based on PeerCast", Proceedings of the 10th IEEE International Conference on High Performance Computing and Communications (HPCC), Dalian, China, Pages 714-719, September 2008.

[7] Amir H. Payberah, Jim Dowling, Fatemeh Rahimian, and Seif Haridi, "gradienTv: Market- Based P2P Live Media Streaming on the Gradient Overlay", Proceedings of the Distributed Applications and Interoperable Systems 10th IFIP WG 6.1 International Conference (DAIS 2010), Amsterdam, Netherlands, Pages 212-225, June 2010.

[8] Nazanin Magharei, and Reza Rejaie, "PRIME: Peer-to-Peer Receiver-Driven Mesh-Based Streaming", Proceeding of the 26th Annual IEEE Conference on Computer Communications (IEEE INFOCOM), Alaska, USA, Pages 1415 – 1423, May 2007.

IOP Conf. Series: Materials Science and Engineering **53** (2013) 012066 doi:10.1088/1757-899X/53/1/012066

Development of Portable Automatic Number Plate Recognition System on Android Mobile Phone

Abdul Mutholib[1], Teddy S. Gunawan[1], Jalel Chebil[1], Mira Kartiwi[2]

[1] ECE Department, Faculty of Engineering International Islamic University Malaysia Kuala Lumpur, Malaysia

[2] Information Systems Department, Faculty of ICT International Islamic University Malaysia Kuala Lumpur, Malaysia

E-mail: tsgunawan@iium.edu.my

Abstract. The Automatic Number Plate Recognition (ANPR) System has performed as the main role in various access control and security, such as: tracking of stolen vehicles, traffic violations (speed trap) and parking management system. In this paper, the portable ANPR implemented on android mobile phone is presented. The main challenges in mobile application are including higher coding efficiency, reduced computational complexity, and improved flexibility. Significance efforts are being explored to find suitable and adaptive algorithm for implementation of ANPR on mobile phone. ANPR system for mobile phone need to be optimize due to its limited CPU and memory resources, its ability for geo-tagging image captured using GPS coordinates and its ability to access online database to store the vehicle's information. In this paper, the design of portable ANPR on android mobile phone will be described as follows. First, the graphical user interface (GUI) for capturing image using built-in camera was developed to acquire vehicle plate number in Malaysia. Second, the preprocessing of raw image was done using contrast enhancement. Next, character segmentation using fixed pitch and an optical character recognition (OCR) using neural network were utilized to extract texts and numbers. Both character segmentation and OCR were using Tesseract library from Google Inc. The proposed portable ANPR algorithm was implemented and simulated using Android SDK on a computer. Based on the experimental results, the proposed system can effectively recognize the license plate number at 90.86%. The required processing time to recognize a license plate is only 2 seconds on average. The result is consider good in comparison with the results obtained from previous system that was processed in a desktop PC with the range of result from 91.59% to 98% recognition rate and 0.284 second to 1.5 seconds recognition time.

1. Introduction

Automatic number plate recognition (ANPR) has played a significant role for traffic law enforcement, access control, automated tolls and automated parking. ANPR system is an image processing technique to identify the vehicles by their number plate. It consists of acquisition of an image, detection the numbers' location in the image and extraction characters to interpret the pixels into numerically readable characters [1]. It became much interest during the last decade along with the improvement of

[1] Corresponding author: Teddy S. Gunawan, tsgunawan@iium.edu.my

digital camera technology and the computational processing [2]. ANPR has also been used in various traffic law enforcement, including speed prosecution, stolen car detection and border monitoring. Moreover, it can be utilized for area management, such as managing of parking lots and toll gate control [3].

Commonly, the ANPR system works in five steps, including image acquisition, image pre-processing, number plate localization, character segmentation to get individual character and finally optical character recognition (OCR) to recognize the individual character with the help of database stored for each and every alphanumeric character [2]. To reduce the computational requirement, number plate localization was not included, instead vehicle plate number image was captured at a specific ROI (region of interest).

There are plenty approaches that can be used to improve the accuracy rate of the image recognition; some common ones are RGB, YCbCr, image binarization, image filter, fuzzy algorithm, genetic algorithms, support vector machines and artificial neural network [4]. In this paper, we focus on minimalizing the computational cost due to the limitation of Android mobile phone device while we keep maintaining a certain degree of recognition accuracy and recognition time. Hence, we proposed the algorithm that suitable for portable ANPR.

The rest of the paper is organized as follows: section 2 presents the ANPR on Android, while Section 3 discusses the hardware requirement and software environment development of the development portable ANPR system and the proposed algorithm. Section 4 discusses the implementation and Section 5 will present the experimental result. Finally Section 6 concludes the paper.

2. ANPR and Android

In the last five years, the biggest improvement in mobile phone development was the introduction of Java-hosted MIDlets. MIDlets are executed on a Java virtual machine, a process that abridgments the core hardware and provide chance for the developers to develop applications which is running on the wide variety of devices that supports the Java run time. Unfortunately, this accessibility comes at the price of restricted access to the device hardware [5]. It was considered normal in the mobile application development for third-party applications to obtain different hardware access and accomplishment rights from those given to native applications written by the phone manufacturers.

Android walk alongside a movement of mobile operating systems which is designed for increasingly powerful mobile hardware, such as, Windows Mobile, iOS Apple and Blackberry. They provide a richer and simplified development environment for mobile applications. However, different with Android, they are built on proprietary operating systems. In some cases it prioritizes the native applications over the application that developed by third parties. It restricts communication between applications and native phone data, and restricts control the distribution of third-party application to their platforms. Android offers an open development environment built on an open-source Linux kernel. Hardware access is available to all applications through the API libraries. In Android, the third-party and native Android applications are constructed with the same APIs and executed on the same run time. Developers can eliminate and replace any native application with a third-party developer alternative, even both the dialer and home screens can be replaced [5].

Android's software architecture as shown on Figure 1 was deliberated to simplify reusable applications' component. The applications can publish their capabilities, and other applications can use those capabilities. Android software development kit (SDK) provides authoring applications with rich functionality. It can handle touch screens, accelerometers, 3D graphics, and GPS as well as collaboration among applications like email messaging, calendars, social networking, and location-based services [6].

ANPR as an intelligent tool that eliminate the need of human intervention requires a high degree of accuracy. It has many variables that need to be optimized in term of producing a high accuracy of the number plate recognition, such as time of day, weather and angles between the cameras and the license plates [3]. Android has the ability to compensate all of the ANPR's variables that can affect the ANPR ability to produce an accurate recognition. An illumination wavelengths have a direct impact on the resolution and detection accuracy.

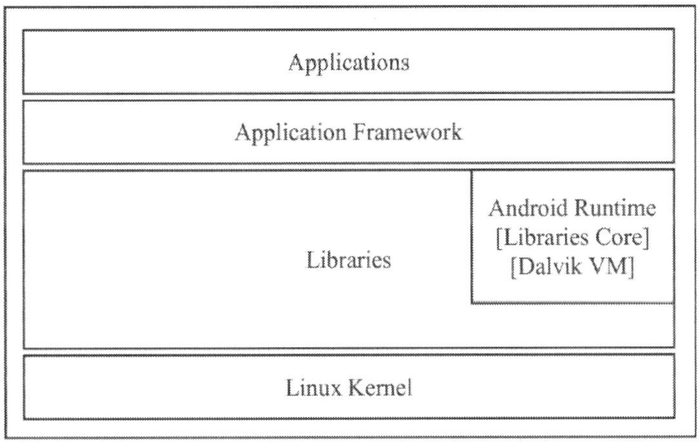

Figure 1. Android Architecture

3. System Design and Proposed Algorithm

In this section, the proposed algorithm in Android mobile phone will be implemented using Eclipse as the IDE (Integrated Development Environment), Android SDK (Software Development Kit), Sun's Java Development Kit (JDK) and the Android Developer Tool (ADT) an Eclipse plug-in. The Eclipse IDE, JDK and ADT are installed in a personal computer as a software environment development. The overall system design of the portable ANPR system can be categorized into two parts, the hardware requirement and software environments developments. Without one of these two parts, we cannot develop any portable ANPR system.

3.1 Hardware Requirement

Hardware is the main part of the portable ANPR system. In this section, we will describe the hardware requirement for constructing and deploying the ANPR system, including personal computer and Android mobile device.

3.1.1 Personal Computer

The PC used in this research is a Toshiba Notebook Satellite M200 released in 2007. The details specification can be seen in Table 1.

IOP Conf. Series: Materials Science and Engineering **53** (2013) 012066 doi:10.1088/1757-899X/53/1/012066

Table 1. Personal Computer Specification Details

Component	Specification
Model	Toshiba Satellite M200
CPU	Intel Core 2 Duo T5450 1.6 GHz
Cache Sizes	2 MB (L2) 64 KB (L1)
System Bus	667 MHz
Memory	3 GB DDR2
Screen size	14.1 inch
LCD technology	XGA
Natural resolution (max)	1280 x 800 pixels
Graphics hardware	Intel 965 Express Chipset Family
Video RAM	384MB

3.1.2 Android Mobile Phone

The Android Mobile Phone used in this research is Samsung GT-S5830 (Galaxy Ace) also known as Galaxy Cooper which is a mid-end smart phone family released by Samsung in February 2010. The details specification can be found in Table 2.

Table 2. Android Mobile Phone Specification Details

Component	Specification
CPU	800 MHz 1 Core ARM v6
GPU	Adreno 200
Chipset	Qualcomm MSM7227
Memory	278.34 MB
ROM	178.39 MB
Network Data	2G / 3G / HSDPA 7.2
OS	Android 2.3.7
Screen size	3.5 inch
LCD technology	TFT 16M colors multi touchscreen
Resolution	HVGA (480 x 320) ~165 ppi pixel density
Camera Resolution	5 MP (2592x1944 pixels)
Digital/Optical Zoom	Up to 2x digital zoom
Shot Mode	Single / Smile / Continuous / Panorama
Photo Effects	Normal / Negative / B&W / Sepia
White Balance	Auto / Daylight / Cloudy / Incandescent / Fluorescent
GPS	AGPS
Wi-Fi	802.11b/g/n
Sensor	Accelerometer, proximity, compass, gravity

3.2 Software Environments Developments

Software controls the hardware resources in this case is due to the processing of the number plate image step by step until producing a set of number plate character. The ANPR system is constructed using Eclipse, JDK 1.7, android SDK and ADT plug-in for Eclipse IDE as the software environments developments. The minimum Android API (Application Programming Interface) that we used for the constructing of the ANPR system is API 10 which can be implemented to Android 2.3 (Ginger Bread) and above.

3.3 Proposed Algorithm

The proposed portable ANPR algorithm can be categorized into four steps as shown in Figure 2.

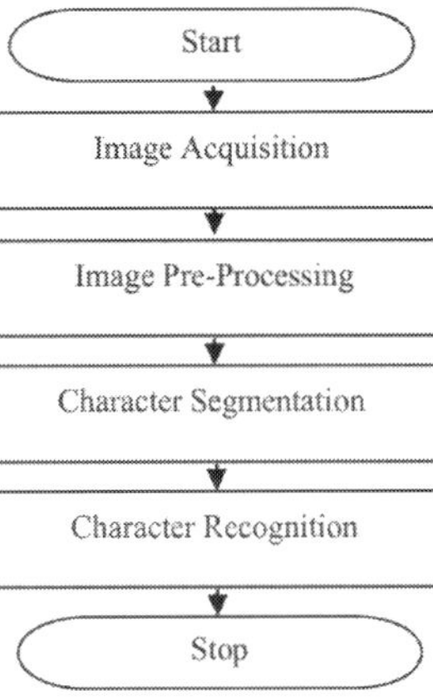

Figure 2. Portable ANPR on Android's flow chart.

3.3.1. Image Acquisition

The first process of the portable ANPR system is very important, because the result of this process will deliver to the next process. In this process, Android camera embedded in the phone was used. It capables to acquire 5 megapixels and can produce image with maximum resolution of 2592x1944 pixels. To reduce the computational requirement, only images with 640x480 pixels (around 80 Kbytes required disk space) were captured. Figure 3 shows how to set the effect of the camera. The camera in Android offers many parameters that also can be used as image pre-processing in ANPR system. One of the parameter is *EFFECT_MONO* that produce image with only black and white. Thus we need have not to develop any function for gray scaling the image and just pass the black white image to next process. The result of acquisition image can be seen in Figure 4.

```
Line Code set camera as mono (black and white)
params.setColorEffect(android.hardware.Camera.Parameters.EFFECT_MONO);
```

Figure 3. Excerpt of Android code to acquire black and white image

Figure 4. Sample of black and white image captured by Android camera

3.3.2. *Image Pre-Processing*

This part is very crucial part of the portable ANPR system. Image pre-processing steps adjusts the image gained from step 1. Furthermore, it enhances the license plate image to ease recognition process. In this procedure, contrast enhancement was applied on the image to alleviate the unwanted effects due to the poor lighting conditions and the similarity between background and text color. A basic enhancement algorithm is given as [7]:

$$G(x,y) = \frac{(S(x,y) - \min(S)) \times 255}{\max(S) - \min(s)} \tag{1}$$

where $G(x,y)$ is the contrast enhanced pixel value, $S(x,y)$ represents the pixel value, min(S) and max(S) are the minimum and maximum grayscale values of $S(x,y)$ respectively. If max(S) – min(S) $\leq$ 80, it means the gray scale image is low in contrast. Gray scale extension will increase the contrast through the calculation to achieve max(S) – min(S) = 255. Figure 5 shows the contrast enhancement image.

Figure 5. Contrast enhanced image

3.3.3. *Character Segmentation*

Character segmentation process is the crucial step for this portable ANPR system. This step associated with the recognition routine. The character segmentation process tries to find the boundaries for characters related with a line of scanned data. Tesseract library offers fixed pitch detection and chopping method. It can determine whether the text line are fixed pitch. If it finds fixed pitch text, Tesseract will chop the words into characters using the pitch, and restricts the chopper and associator on these words for the word recognition step [8]. Figure 6 shows a fixed-pitch word chop into characters.

Figure 6. Character segmentation

3.3.4. *Character Recognition*

The final step of ANPR is character recognition. As the previous step, in this step we used Tesseract as the OCR engine and it is based on the artificial neural network algorithm. The word network in the term artificial neural network mentions to the inter–connections among the neurons in the different layers of the system. As an example, the system has three layers. The first layer has input neurons which send data through synapses to the second layer of neurons, and then through further synapses to the third layer of output neurons. The complex systems will have more extra layers of neurons with some having more layers of input neurons and output neurons. The synapses have parameters called weights that manipulate the data in the calculations. Figure 7 shows the diagram of neuron model.

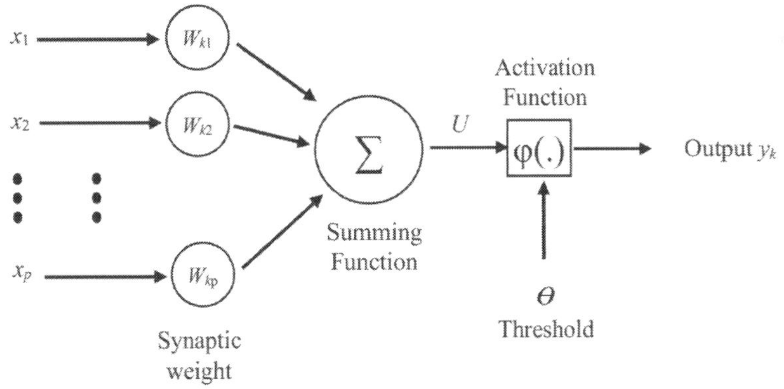

Figure 7. Matematical non-linear model of a neuron

Mathematically, we can define a k neuron with the formulation as the follow.

$$u_k = \sum_{j=1}^{p} w_{kj} x_j \tag{2}$$

and

$$y_k = \varphi(u_k - \theta_k) \tag{3}$$

where x_1, x_2, ..., x_p are the input signals; W_{k1}, W_{k2}, ..., W_{kp} are the synaptic weight from the neuron k; U_k is a linear combiner output; θ_k is a threshold; $\varphi(.)$ is an activation function and y_k is output signal from the neuron [9].

For this ANPR system which has number and character, the ANN's number, there will be 10 neurons for hidden layer and output layer. For ANN's letter, there will be 20 and 26 neurons for hidden layer and output layer respectively. The value of element of each targets are all zero except one element on specific position which represent the number or letter. The output from the network will be two dimensional matrixes with size 26 x n and 10 x n for ANN's letter and ANN's number, respectively.

The value of output will be in range 0 and 1. To be recognize, output should be processed first by converted the highest value to be 1 and other will be 0. Then, the location of element which has value 1 will be founded and the result will represent number or letter. For example, output from ANN's letter which has value 1 at first element will represent the letter A.

Figure 8. Character recognition result

4. Implementation

As the system design has been completed, the system is ready for the construction and implementation. First of all the portable ANPR is constructed under Android platform using android API 10 for android 2.3 Ginger Bread. For the system environment itself Eclipse IDE was used as the coding editor, JDK 1.7 for the java environment, Android SDK and ADT plug-in for Eclipse. Furthermore, several Google's libraries (Tesseract) have been imported in this implementation step. The proposed algorithm contains more than hundred line of code, including the image processing, GPS geo-tagging, sending the result of recognized character to database server and sending the notification to the user. The result of the constructing step can be seen in Figure 9.

Finally, after the construction and implementation of the proposed algorithm for portable ANPR completed, we need to deploy the system into Android mobile phone. This step is performing thru the compiling of the source code in Eclipse IDE with the destination process is Android mobile phone and the ANPR system is automatically installed in Android mobile phone.

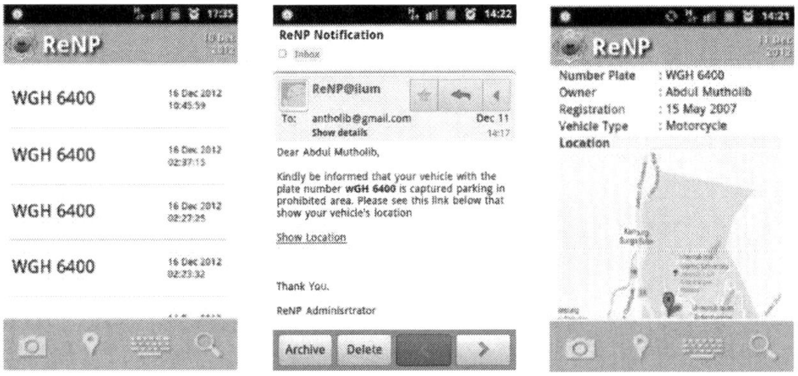

Figure 9. GUI of portable ANPR system

5. Experimental Results and Discussion

Experiments have been performed to evaluate the proposed system and to measure the accuracy of the system. The images for the input to the system are black and white images with the size 640x480 pixels. The test images were taken under various illumination conditions. As shown in Table 3, the recognition accuracy is 90.86% and the average recognition time is 2 second.

Table 3. Recognition rate and recognition time for various number plates.

Number Plate	Accuracy (%)	Time (s)
WGH 6400	88.57%	2.1
WAX 5739	91.43%	2.0
PDG 7628	85.71%	2.1
DAA 9949	94.29%	1.9
WDC 292	85.71%	2.2
DBH 1912	100.00%	2.0
BFP 2569	100.00%	1.8
ABS 1544	94.29%	1.9
WQP 5739	82.86%	2.1
BFD 9917	85.71%	1.9
	90.86%	**2**

The proposed system is different from previous systems in term of algorithm and hardware design. Ying Wen et. al. [10] implemented the ANPR system on desktop PC with 1.8 GHz processor using VC++ 6.0 and achieve 93.54% for recognition accuracy and 0.284 second for recognition time. Ying Wen et. al. used improved Bernsen algorithm connected component for license plate detection method and support vector machine for recognition method. Ankush Roy et. al. [11] achieved 91.59% for recognition accuracy and it was designed using Matlab (R2008a) and implemented using C. Ankush Roy et. al. used bounding box method for segmentation and artificial neural network (ANN) for recognition method. At last Huang et. al. [12] gained 1.5 second for recognition time. They used Otsu method to binarize the image and template matching technology to recognize the license numbers.

IOP Conf. Series: Materials Science and Engineering **53** (2013) 012066 doi:10.1088/1757-899X/53/1/012066

6. Conclusions

This paper has discussed the development of portable ANPR on android mobile phone. The design of portable ANPR have been proposed and implemented on Android mobile phone platform. This system is designed to identify Malaysian number plates. To evaluate its performance, the system is tested over several number plate images. Results show that the accuracy of the proposed system is about 90.86% with the recognition time of 2 second. This proposed system can be easily modified and optimized for multinational car license plates.

7. References

[1] Lakshmi, C.J., et al., *A Novel Approach for Indian License Plate Recognition System*. International Journal Of Advanced Engineering Sciences And Technologies (IJAEST), 2011. 6(1): pp. 010 - 014.

[2] Qadri, M.T. and M. Asif, *Automatic Number Plate Recognition System for Vehicle Identification Using Optical Character Recognition*. International Conference on Education Technology and Computer, 2009: pp. 335 - 338.

[3] Hamey, L.G.C. and C. Priest, *Automatic Number Plate Recognition for Australian Conditions*. Proceedings of the Digital Imaging Computing: Techniques and Applications, 2005: p. 8.

[4] Chen, Z.-X., et al., *Automatic License-Plate Location and Recognition Based on Feature Salience*. IEEE TRANSACTIONS ON VEHICULAR TECHNOLOGY, 2009. 58(7): p. 3781-3785.

[5] Meier, R., *Professional Android 2 Application Development*, 2010, Indianapolis, Indiana: Wiley Publishing, Inc. 399.

[6] Mutholib, A., T.S. Gunawan, and M. Kartiwi. *Design and implementation of automatic number plate recognition on android platform*. in Computer and Communication Engineering (ICCCE), 2012 International Conference on. 2012.

[7] Jian, Y., et al. *Text extraction from images captured via mobile and digital devices*. in Advanced Intelligent Mechatronics, 2009. AIM 2009. IEEE/ASME International Conference on. 2009.

[8] Smith, R. *An Overview of the Tesseract OCR Engine*. in Document Analysis and Recognition, 2007. ICDAR 2007. Ninth International Conference on. 2007.

[9] Gunawan, T.S., O.O. Khalifa, and E. Ambikairajah, *Forward Masking Threshold Estimation Using Neural Networks And Its Application To Parallel Speech Enhancement*. IIUM Engineering Journal, 2010. 11(1): p. 15-26.

[10] Ying, W., et al., *An Algorithm for License Plate Recognition Applied to Intelligent Transportation System*. Intelligent Transportation Systems, IEEE Transactions on, 2011. 12(3): p. 830-845.

[11] Roy, A. and D.P. Ghoshal. *Number Plate Recognition for use in different countries using an improved segmentation*. in Emerging Trends and Applications in Computer Science (NCETACS), 2011 2nd National Conference on. 2011.

[12] Yo-Ping, H., L. Shi-Yong, and C. Wei-Po. *A template-based model for license plate recognition*. in Networking, Sensing and Control, 2004 IEEE International Conference on. 2004.

IOP Conf. Series: Materials Science and Engineering 53 (2013) 012067 doi:10.1088/1757-899X/53/1/012067

An investigation of fatigue phenomenon in the upper limb muscle due to short duration pulses in an FES system.

Jannatul Naeem[1], Amelia Wong Azman[1], Sheroz Khan[1] and Yasir Mohd Mustafah[2]
[1]Department of Electrical and Computer Engineering, International Islamic University of Malaysia, Gombak, 53100, Malaysia
[2]Department of Mechatronics, International Islamic University of Malaysia, Gombak, 53100, Malaysia

E-mail: naeem_088@yahoo.com

Abstract: Functional Electrical Stimulation (FES) is a method of artificially stimulating muscles or nerves in order to result in contraction or relaxation of muscles. Many studies have shown that FES system has helped patients to live a better lives especially those who are suffering from physical mobility. Unfortunately, one of the main limitations of an FES system besides of its high cost is largely due to muscle fatigue. Muscle fatigue will affect the training duration which could delay patients' recovery rate. In this paper, we analyzed the occurrence of this fatigue phenomenon in terms of stimulator parameters such as amplitude, frequency, pulse width and pulse shape. The objective of this investigation is to identify other key features of the FES system parameters in order to prolong the training duration among patients. The experiment has been done on a healthy person for the duration of one minute and later the muscles response will be observed. Resultant muscle response is recorded as force using force resistive sensor. The experimental results show muscles will get fatigue at a different rate as the frequency increases. The experiment also shows that the duty cycle is reciprocal to the resultant force.

1. Introduction

The human nervous system is a complex part of body in which actions are controlled by the human brain. The nervous system is meant to make us respond to involuntary and voluntary actions such as grabbing a glass, restraining body movement due to hazardous circumstances, maintaining balance during walking and many more. The two major parts of the nervous system are Central Nervous System (CNS) and Peripheral Nervous System (PNS). The former is made of spinal cord and brain, while the latter consists of receptors (sensing impulses) and effectors (taking actions in response to impulses) [1]. CNS is the body's main control system which interprets and receives all nerve impulses from body parts and suggests to the muscles what appropriate actions needs to be taken. PNS on the other hand are all the nerve cells outside the brain and spinal cord.

[1] Corresponding Author: naeem_088@yahoo.com

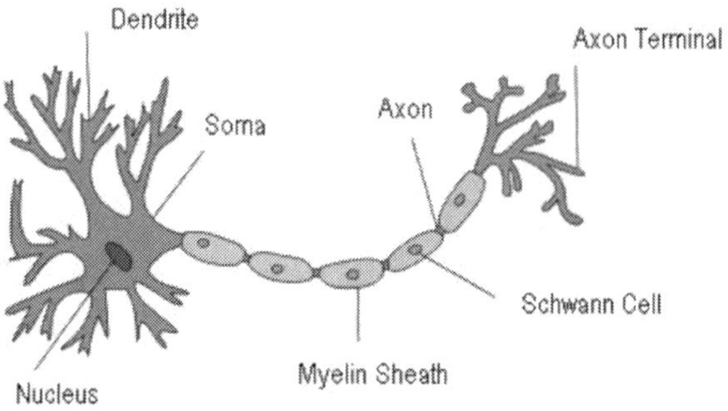

Figure 1. Structure of a Typical Neuron Cell [1].

Figure 1 illustrates a structure of a neuron cell. The human body is constantly responding to changes from the outside environment as well as to changes within the body to maintain equilibrium. An electro-chemical signal or the neuron from the brain passed on to and from the nerve pathways are resulted from the activation of sensory inputs from either internal or external sources. When this happened, a potential difference is created between axon and its surroundings which acted like a tiny battery. As a result of that, a muscle would either contracts or relaxes.

Functional Electrical Stimulation (FES) could interrupt the equilibrium within a human body. This is by artificially introducing short electrical pulses to stimulate the muscles which will result to involuntary movements. These muscle contractions can be adjusted by stimulating one or more muscles that will create torque at the joint of the muscle. When the electrical stimulation is used to enhance the muscle functions such as standing, grasping, rowing and gait pattern, this electrical stimulation is known as the Functional Electrical Stimulation (FES). For this reason, FES has been widely used around the world as part of a rehabilitation program to stimulate the contractions of muscles among patients with paralysis condition. Thus, it is not surprising to see many research works in designing effective FES devices over the years.

In the process of designing an FES system, there are some important parameters that need to be considered. Among the main parameters are the voltage and current level, pulse shape, pulse width and pulse frequency. The amplitude of the current defines the strength of stimulation or how strong is the force generated to the muscle. The stimulation frequency mainly controls twitching of the muscle. As the frequency increases, the stimulated muscle will get fatigue fasters which inadvertently limit the muscle training duration.

In [2], the authors stimulate muscles by inserting modulated pulse at a range of 0-200μs, under a constant current of 20mA fixed at 12Hz frequency. The muscle fatigue is measured using equation (1).

$$\text{Fatigue Index} = \frac{\text{Muscle Tension [Newtons]}}{\text{M}-\text{wave Parameter [Volts,seconds]}} \qquad (1)$$

In the equation above, the muscle tension represent as force muscle output unit in Newton. The M-wave constraint is in quantitative measure that could be articulated in volts, seconds or volts-seconds depending on the choice of M-wave parameter. In selecting M-wave parameter, there are two important criteria. Firstly, when the muscle is either in fatigue or non-fatigue state, the parameters must reflect the degree of muscle unit recruitment during electrical stimulation. Secondly, the parameters should be kept constant in the duration of constant stimulation. However, in this research, only a single waveform has been used.

In another technique elaborated in [3], the strength of the muscle contraction can be changed by modulating the stimulation parameters such as waveforms, intensity, inter pulse interval and frequency of stimulation for signals applied to different electrodes. These fundamental parameters changes are done for an effective FES control system. For a specific stimulus waveform, it depends on the intensity of the physiological effect and the type of electrode used. In restoring motor unit function it is necessary to use a frequency controlled FES system. The frequency of stimuli signal of an FES system is commonly reported as the main cause of rapid muscle fatigue during muscle recruitment. Muscle recruitment is the number of motor unit that is required for a specific function, a process with its utility determined by the stimulation intensity. Muscle contractions induced by FES tend to result in rapid muscle fatigue, which greatly limits activities such as FES-assisted standing and walking [4]. In [4], a study was conducted to show the effects of randomly modulating the parameters of FES signals on muscles activity. However, the study does not provide any conclusion on the effect of signal modulation on muscle fatigue but only suggest that parameters such lowering the stimulation frequency might be a helpful parameter in reducing fatigue.

In [5], M. Sahin and Y. Tie presented the effects of the traditional square wave together with seven different pulse shapes. The waveforms that they have used include rectangular, linear increase and decrease, exponential increase and decrease, Gaussian, and sinusoidal waveform. Using computational stimulation model, they found that non rectangular pulses are more energy efficient than rectangular waveform and the chronaxie time was longer with all the non rectangular waveform. Linearly decreasing ramp provided the best charge injection for all pulse widths when compared with other waveforms. The experimental results also show that Linear, exponential decrease, and Gaussian pulses to be more efficient than the traditional square wave.

There has also been research on proposing closed loop control system towards next generation FES systems [5, 6].

Most of the available FES systems are current controlled and have utilized the square waves pulse shape. Interestingly, recent studies also have shown that not only tradition mono-phasic or bi-phasic square wave can be used in an FES system but also other pulse shape such as the triangular, exponentially increase, exponentially decreased and raising ramp [7]. Different pulse shape has different energy delivering manner therefore the muscle contraction also be different. The pulse width parameter is closely related to the pulse shape. The pulse width defines the effective stimulation time given to the nerve. This paper studies the muscle fatigue with respect to these different stimulation parameters. From our literature review, the current available off-the-shelf FES systems not only are limited in its functionality, they are mostly very expensive and bulky. Hence, another point to be considered is the total cost in designing and developing an FES system.

The use of stimulation parameters in improving stimulation results has been discussed in view of ultimately producing a programmable stimulator implant for minimizing tissue damage [7]. The authors in [7] have studied and analyzed the effects of pulse width and frequency of stimulation signals by measuring the charge delivered and the energy efficiency. A circuit schematic for a portable functional electrical stimulator is suggested in [8] suggesting improvement in gait and muscle strength. Once again, the parameters that were analyzed are limited to pulse amplitude, pulse-width and pulse duration.

Recently, Sabut [9] combined the effects of FES with the conventional rehabilitation program (CRP) to improve better in gait and muscle recovery as well as sub-acute than chronic stroke patient. While the results show significance change in walking velocity among the chronic strokes patients, the FES parameters that is put to the test is only limited to amplitude variations. At the end of 2011, Dongchul Lee [10] patented a multi waveform circuitry for electrical stimulation to be used as therapy level. He described that different shapes have different characteristics of energy delivered to the tissues. Unfortunately, the work did not come with any analysis to suggest that his proposed waveforms will work better for an FES system. To the best of our knowledge, none of the stated researches have experimented on the effect of stimulation parameters on muscle fatigue.

Therefore, one of the main aims of this research is to study the effects of different stimuli parameters on muscle fatigue. The results from this investigation will help physician to make better decision in providing efficient treatment to the patients.

The remaining of this paper will be as the following; Section 2 is detailed the FES circuit that is used in this research followed by the description on the experimental setup in Section 3. Section 4 will discuss on the results and finally the paper will be concluded in Section5.

2. FES Circuit Design

In this work, we have adopted the circuit design proposed by Cheng et al. [11]. In the work, they introduced a 9V battery powered FES system that replaces the use of a transformer with a resonant circuit to produce stimulus pulses. The stimulus parameters are frequency, pulse width and amplitude. Pulse frequency is controlled by the Q1 and Q2 gate duty cycle. Pulse width is controlled by Q2 and Q1 current amplitude ranged by R.

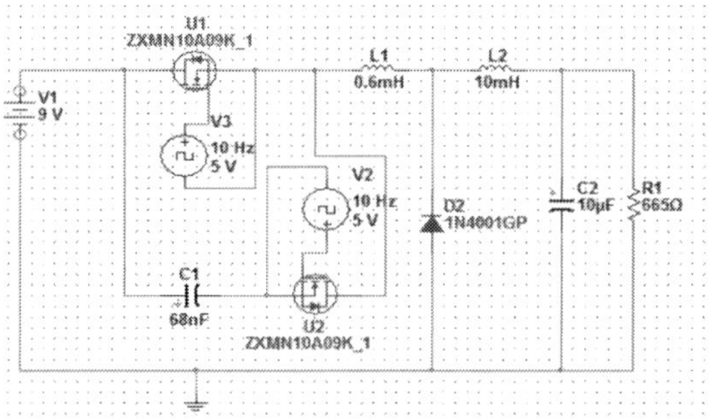

Figure 2. Resonant transformer less FES circuit

Unfortunately, the paper only presented a simulation results that shows that the circuit can achieve up to 100mA. Our work extends this work by implementing and validating the circuit in Figure 2 and to study the effect of varying parameters against muscle fatigue.

3. Experiment Setup

The experiment was on the upper limb muscle of a healthy person. At this stage of the research work, electrodes were placed on the biceps muscle since the biceps have larger muscle group area as shown in Figure 3 to obtained better stimulation outcome. It was noticed that the effectiveness stimulation for muscle activation depends greatly on the placement of the electrodes which would require in depth understanding on the full upper limb anatomy.

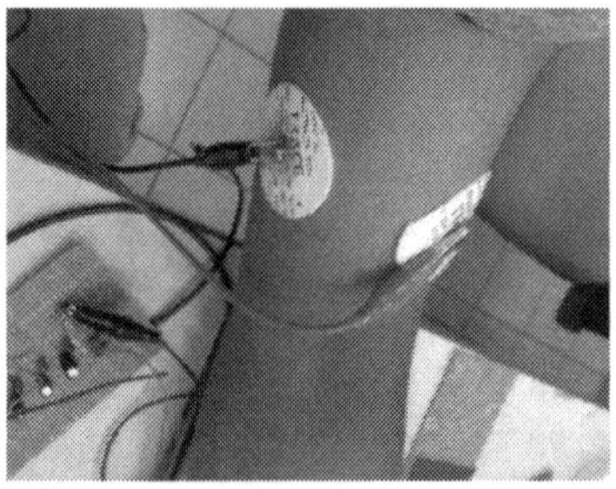

Figure 3. Placement of electrodes on the biceps muscle

In order to measure the muscle fatigue as previously stated by equation 1, the force must first be obtained. To do this, this experiment had utilized the force-sensing resistor (FSR) sensors. An FSR sensor work like a variable resistor. When short electrical pulses from the FES circuit are exerted to the biceps muscle, it was resulted to an involuntary hand motion. This involuntary hand motion was exert force to the surface of the FSR sensors. As the force increases on the calibrated FSR sensors surface, the resistance value of the FSR sensors was decreased. The output from the FSR sensors were then is converted to analogue voltages which were connected to the input pins (pin A0-A5) of an Arduino microcontroller. The voltage change detected from the analogue pins is the equivalent force monitored by the Arduino serial monitor. The following is the pseudocode for the algorithm in the Arduino.

Procedure Muscle fatigue force detection (signal)
Signal ← Square wave with the 100mA amplitude
Force ← resultant muscle response measure in force value by force sensitive resistor sensor
SensorValue ← Analogue value reading from sensor
Weight ← consider the analogue value changes linearly according to sensor output of the graph
Force: = weight* acceleration 9.8m/s^2
while (device on)
weight = analogue value/10
force = weight*9.8
display (force)
end **while**
end **procedure**

Experiments are conducted for one minute on a healthy volunteer. In the first part of the experiment, an initial frequency was set to 1Hz with constant amplitude at 100mA and pulse duty cycle 50 percent. The frequency will later be increased slowly and any changes will be recorded. The second part of the experiment is to vary the duty cycle from 50% to 85%-15%. Variance in the outcome will also be recorded. It needs to be highlighted that the volunteer is given enough time to rest his muscles in between experiments.

4. Result and Discussion

Figure 4 shows a graph of the muscle response with regards to the changes in the stimuli parameters. For the first part of the experiment i.e. varying frequency, there was no activity of muscle contraction observed at 1Hz. The earliest muscle response observed was at the 5Hz frequency. After 1 minute stimulation duration, the muscle begins to show signs of fatigue. As illustrated in Figure 4, the first recorded muscle response was at its highest at 0.38 Newton and slowly decreases over time. As the frequency increases to 10Hz, the first recorded force is at 0.3 Newton. Although there is a slight drop in the force, the experiment shows that the muscle twitched at a faster rate. Unfortunately, because the motor neuron does not have enough time to release the neurotransmitter quickly, hence, the muscles get fatigue at a faster rate as shown in Figure 4.

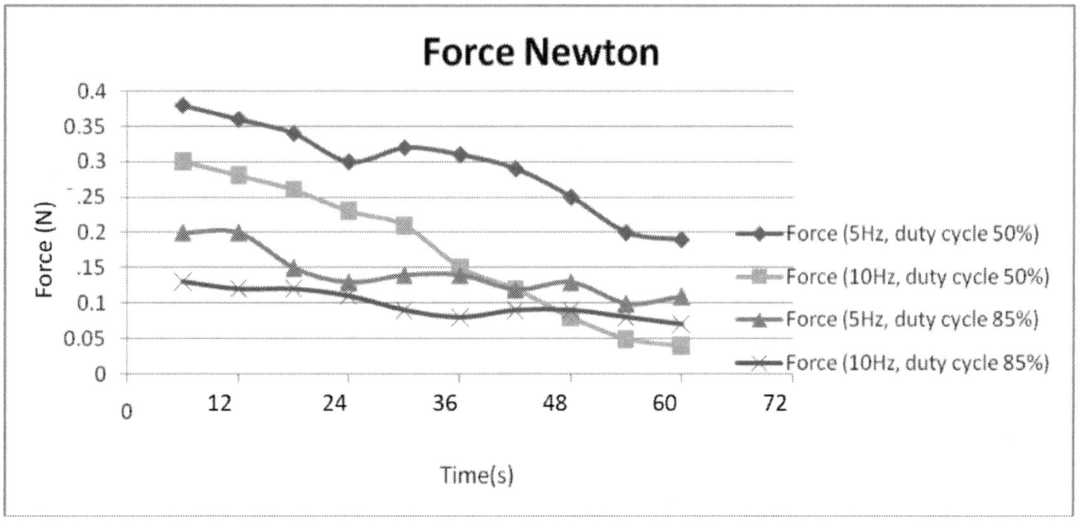

Figure 4. Muscle response with stimuli parameters of current 100mA, Frequency 5Hz, 10Hz and duty cycle 50%, 85%

In the second part of the experiment i.e. changing the duty cycle, the experiment recorded low force at the first reading for both 5Hz and 10Hz as illustrated in Figure 4 at 0.2 Newton and 0.13 Newton respectively. Interestingly, the muscle does not get fatigue quicker when compared with the results from the first part of the experiment. The results also show that there is only a small variation in the force throughout the experiment.

5. Conclusions

This work presents an investigation on the effect of FES parameters that has been linked to muscle fatigue. One of the objectives of this work is to gives clinician a better understanding on the effect of stimuli parameters to increase the training duration when using FES devices among patients. Results shows, frequency and stimulation time greatly affects on the muscle fatigue. Duty cycle pattern of the stimulation has shows active stimulation time of a cycle, which also effect on the muscle fatigue. A transfomerless based FES circuit has been built and tested on a neurological intact person. From this work, we have validated the circuit presented in [11]. The results of this work concluded that the muscle fatigue varies depending on the pulse frequency and the stimulation time and exerted force is related to the pulse amplitude.

6. References

[1] Cantor, S.R.; Cantor, S.G.; "Physiological description of the neuron and the human nervous system," *Frequency Control Symposium, 1995. 49th., Proceedings of the 1995 IEEE International* , vol., no., pp.3-9, 31 May-2 Jun 1995

[2] Heasman, J.M.; Scott, T.R.D.; Vare, V.A.; Flynn, R.Y.; Gschwind, C.R.; Middleton, J.W.; Butkowski, S.B.; , "Detection of fatigue in the isometric electrical activation of paralyzed hand muscles of persons with tetraplegia," *Rehabilitation Engineering, IEEE Transactions on* , vol.**8**, no.3, pp.286-296, Sep 2000

[3] Xu Qi; He Jiping; Wang Yongji; Xu Tao; Huang Jian; , "A review of fundamental mechanisms and techniques in functional electrical stimulation," *Neural Interface and Control, 2005. Proceedings. 2005 First International Conference on*, vol., no., pp. 201-204,26-28 May 2005

[4] Graham, G.M.; Thrasher, T.A.; Popovic, M.R., "The effect of random modulation of functional electrical stimulation parameters on muscle fatigue," *Neural Systems and Rehabilitation Engineering, IEEE Transactions on*, IEEE TRANSACTIONS ON NEURAL SYSTEMS AND REHABILITATION ENGINEERING, VOL. 14, NO. **1**, MARCH 2006, pp.38-45

[5] M. Sahin and Y. Tie, "Non-rectangular waveforms for neural stimulation with practical electrodes," J. Neural Eng., vol. **3**, no. 3, pp.227–233, Sep. 2007

[6] Lynch, C.L.; Popovic, M.R.; , "Functional Electrical Stimulation," *Control Systems, IEEE Magazine*, vol.**28**, no.2, pp.40-50, April 2008

[7] Wongsarnpigoon, A.; Woock, J.P.; Grill, W.M.; , "Efficiency Analysis of Waveform Shape for Electrical Excitation of Nerve Fibers," *Neural Systems and Rehabilitation Engineering, IEE Transactions on*, vol.**18**,no.3,pp.319-328, June2010

[8] Abidin, A.Z.; Agarwal, S.; Chattopadhyay, S.; Poddar, A.; Ganesh, B.; , "A study on portable functional electrical device for small muscles," *Students' Technology Symposium (TechSym), 2010 IEEE*,vol., no. , pp.1-5,3-4, April2010

[9] Sabut, S.K.; Sikdar, C.; Kumar, R.; Mahadevappa, M.; , "Improvement of gait & muscle strength with functional electrical stimulation in sub-acute & chronic stroke patients," *Engineering in Medicine and Biology Society,EMBC, 2011 Annual International Conference of the IEEE* , vol., no., pp.2085-2088, Aug. 302011-Sept.32011

[10] Dongchul Lee; Michael Moffitt; Christopher Ewan Gillespie; Kerry Bradley,"use of stimulation pulse shape to control neural recruitment order and clinical effect". Patent number: US 8036754, Issue date: Oct 11, 2011

[11] Cheng, K. W E; Yan Lu; Kai-yu Tong; Rad, A.B.; Chow, D.H.K.; Sutanto, D., "Development of a circuit for functional electrical stimulation," *Neural Systems and Rehabilitation Engineering, IEEE Transactions on* , vol.**12**, no.1, pp.43,47, March 2004

IOP Conf. Series: Materials Science and Engineering **53** (2013) 012068 doi:10.1088/1757-899X/53/1/012068

Development of a high hertz-stress contact for conventional batch production using a unique scribing technology

M M I Bhuiyan[1], T Alamgir[2], M Bhuiyan[3], M Kajihara[4]
[1,2]Department of Mechatronics Engineering, International Islamic University Malaysia, Kuala Lumpur 53100, Malaysia
[3]Monash University Sunway Campus, School of Medicine and Health Sciences, Bandar Sunway, Selangor, 46150, Malaysia
[4]Department of Materials Science and Engineering, Tokyo Institute of Technology, Yokohama 226-8502, Japan

E-mail: moinul@iium.edu.my

Abstract. Gradually the electronic devices are getting more compact dimension with respect to the width and thickness. As a result, the contacts are becoming thinner and which leads the contact to be loose and unstable contact. In comercial stamping methode, connector tip diameter should be more than 300μm due to its size limitation. Consequently, the connector contact resistance is becoming higher due to weak contact force. To overcome this problem there were few more basic research using MEMS and Electro Fine Forming (EFF) technology to make high Hertz-Stress Contact (5μm) due to the limitation in the commercial stamping process and the result was in satisfactory level. However, since the MEMS and EFF fabrication is costly therefore, a new method is introduced in this paper using the commercial Phosphor Bronze stamping method to reduce the production cost. Moreover, scribing method is used to make tip on the contact. Accordingly, more compact fine pitch contact is successfully fabricated and tested with 5μm High Hertz Stress without using the MEMS and EFF technology. Hence the manufactured contact resistance becomes less than 20mΩ ±5mΩ.

1. INTRODUCTION

Nowadays, developing countries are expanding rapidly. Hence, countries are more depending on the technology and machineries such as the electric machines, electronic devices, and vehicle related devices. Then, according to the research, electronic devices are getting more compact (Figure 1) with time and leads to a high demand on coming up with more compact contacts and in next decade electronics world needs more compact connector for a connector market which is reported by Sangyo-Joho Limited Japan [1-2]. A contact provides an independent connection between two elements of an electronic system without unacceptable signal distortion or power loss. On the other hand, miniature, light and high functional electronic devices have been developed using high functional LISs called SOC (System-on-Chip) [3-5]. The development of a miniature size, light and high performance electronic devices has been accelerated for further development of the highly-networked information society. Thus, this raises the query whether those emerging countries are using currency conventional

[1] To whom any correspondence should be addressed.

technology or not. Subsequently, the answer is "no". They will use the latest technology in a single step. Because of that, miniature connectors with smaller pitches have been required in order to satisfy the specification of these electronic devices. Narrow pitch Board-to-Board (BtoB) interface connectors are also required. It is well understood that less than 50μm pitch connector will be required in the Laptop PC area within next 5-7 years. However, the present commercial connector's minimum pitch is 250μm due to the existing limitation in commercial stamping manufacturing method. Therefore, this research will develop a breakthrough technology to solve current problem for next generation fine pitch connectors. Moreover the future developed connector should be developed in low cost fabrication process.

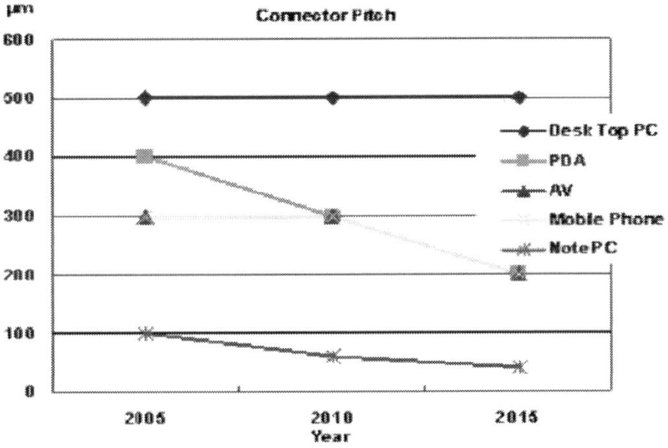

Figure 1: Graphical representation on Contact density vs year

Once a connector is designed than it is required to ensure detachable connection and no unacceptable distorted signal passes through the contact. Therefore, contact normal force is a vital key point to ensure stable connection in maintaining this requirement. Once the contacts are getting more compact, than the contact normal force will be getting smaller due to the reduced size and space limitation. Therefore, the sufficient normal force has to be reduced due to its size limitation. Thus, the electrical connection becomes unstable due to the presence of corrosive layer on the contact surface. The contact force is limited to the size of beam as it is known by general equations in material science. Therefore, it is becoming a challenge to the future development to come up with a stable electrical performance contact with high durability using Low Normal Force (LNF) contact and it is going to be the main target of this research as of (Figure 2) showed relationship between the Contact Pitch versus the Normal force of a contact.

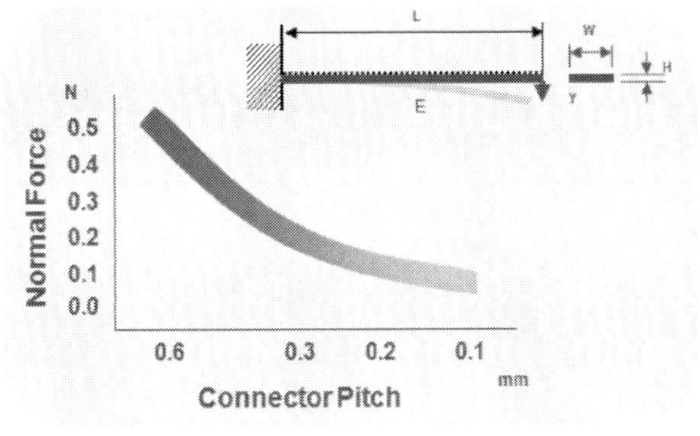

Figure 2: Relation between Connector Pitch vs Normal Force

Here,

$$Spring\ Rate = K = \frac{EWH^3}{4L^3} = \frac{EW}{4}\left(\frac{H}{L}\right)^3$$

$$Elastic\ Range = \frac{2L^2\sigma_{yield}}{3EH} = \frac{2L^2\varepsilon_{yield}}{3H}$$

Extensive studies are performed on the proposed LNF contact with the existing contacts features [13] to overcome the unstable contact dilemma which mainly are due to lack of high hertz stress and reduced normal force (Figure 3). In order to justify this basic idea, it was required to find out some related information on required contact force, and flexible required durability of the contact. Thus the ground research focused in the high Hertz-stress contact properties shown in Figure 4 to realize the prospective solution on the above mentioned problem. Initially, it was an idea that, if a very sharp tip contact is fabricated than the connection will be more stable at the contact point and the current will pass easily.

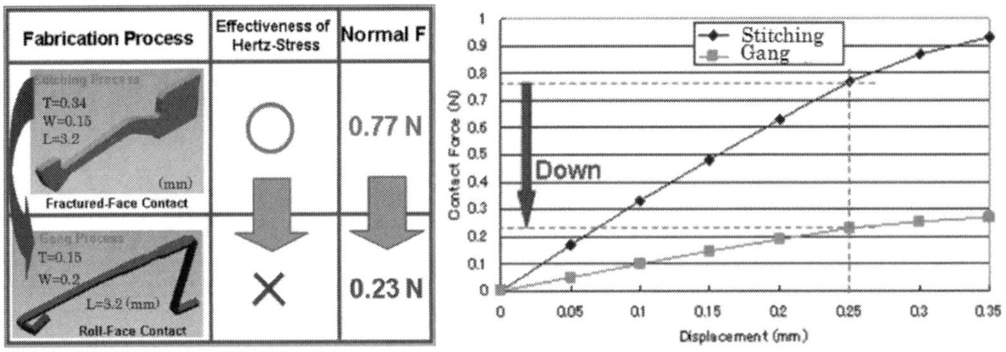

Figure 3: Comparison between with conventional contact with Low Normal Force Contact

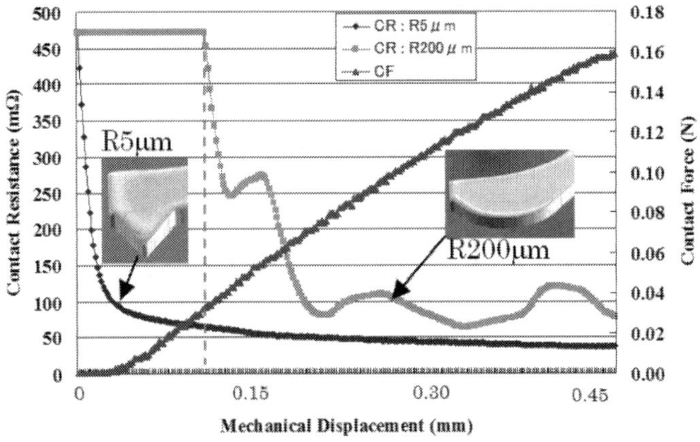

Figure 4: Resistance Comparison between with High vs low Hertz-Stress

According to the past research, we have developed an excellent material Ni-Co for contact spring [5] in MEMS and EFF process and could develop a LNF contact and the contact performance was also in satisfactory level and durability performance was within the acceptable range. Although a suitable contact with appropriate R is the Key factor for LNF and it is yet to be optimized. In past research has tried to optimize an appropriate R using several size (r=5 to 200μm) by the same LNF contact. To fabricate this miniature contact, a MEMS technology is used because the existing pressing and injection moulding method is not compatible with the high-density packaging required in the advanced IT market [3-8, 15]. This research focuses on a fabrication of micro interposer by combining UV thick resist photolithography and Ni-Co fine electroforming technology. This fabrication method is expected to develop high precision micro interposer and high–density packaging in mass production and is very essential to realize micro order devices. The business value of this contact will be utilized on the Test Interface Connection area. For example, test sockets for IC-devices, test board I/F for IC-tester, and High speed board to board for I/O device. Furthermore, on a Medical Equipment area like 3D transducer interconnection for ultra-sound, X-Ray sensor I/O for CT scanner is also necessary.

In order to increase Hertzian stress [9-12] of contact point of the terminal have to be lower radius (sharp) to stick to the upper terminal. A sphere (radius R, elastic contacts υ2, E2) pressed by a load F into a flat substrate (with elastic constants υ1, E1). The load is supported over a circular area whose radius, a, is given by: [9]

However, MEMS and EFF technology is not suitable for industrial batch production since this process is time consuming and economically expensive. Therefore, a cost effective process is main target to meet in order to fulfil the industrial demand and this research is conducted to satisfy the industrial demand.

2. FUNCTIONS

To achieve the industrial demand, a contact has to establish a separable connection between two elements and also should be able to supply undistorted signal through its contact. Both requirements are depending on the connector application and its electrical and environmental requirements. Separable contact gives the flexibility on easy placement [15], easy repair, upgrading, and maintenance or interconnects ability. Undistorted signal flow depends on the resistance within the connecting point. Lower the contact resistance the better signal passes through the contact.

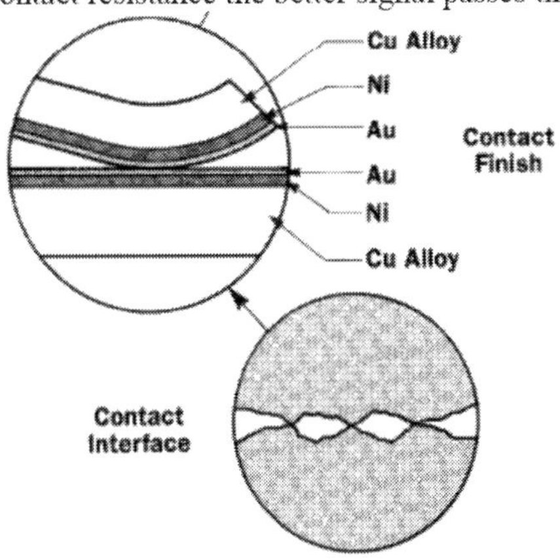

Figure 5: Microscopic view of contact surface

3. MODELLING

A number of possible solutions are gathered and in order to find the best possible design to meet the industrial requirement, an analysis is done through Pugh method and scribing method is chosen out of the all proposed possible design.

Item \ MFG Process	Stamping + MEMS (Photolithography + EFF)	Chemical Etching + Stamping	Scribing	Punching	Conventional Blanking
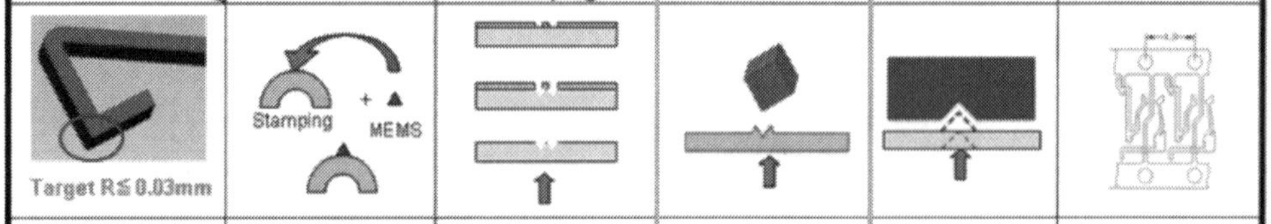					
Present min R	Min 0.005	Min 0.01	Min 0.05	Min 0.1	Min 0.08
Difficulty	△	△	○	×	○
Cost	×	△	○	○	○
Invest	×	×	○	○	△
Overall Rating	2	4	8	5	6

Figure 6: Decision Matrix on the proposed possible designs

In order to make the stable contact in using Low Normal Force material the contact is being modelled from thin Phosphor-Bronze substrate and a dual tip is made on the substrate by scribing method (Figure 7) is used to make a stable contact on the connecting point. Throughout this process a Low Normal Contact Force is being modelled and analyzed its properties in different environment.

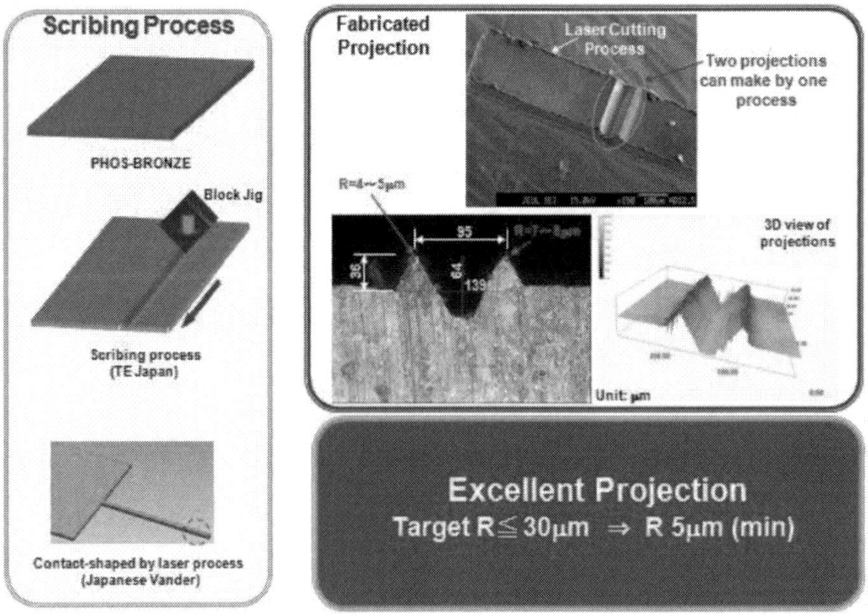

Figure 7: Scribing Process and Tip Structure size Verification

IOP Conf. Series: Materials Science and Engineering **53** (2013) 012068 doi:10.1088/1757-899X/53/1/012068

4. FABRICATION

To facilitate the proposed contact, a micro interposer combining UV thick resist photolithography and Ni-Co fine electroforming fabrication process was adopted to satisfy the above-mentioned specification. SEM (scanning electron microscope) photographs of the structures which are formed by the Ni-Co electroforming. A SEM photograph of the designed contact is shown in Figure 8.

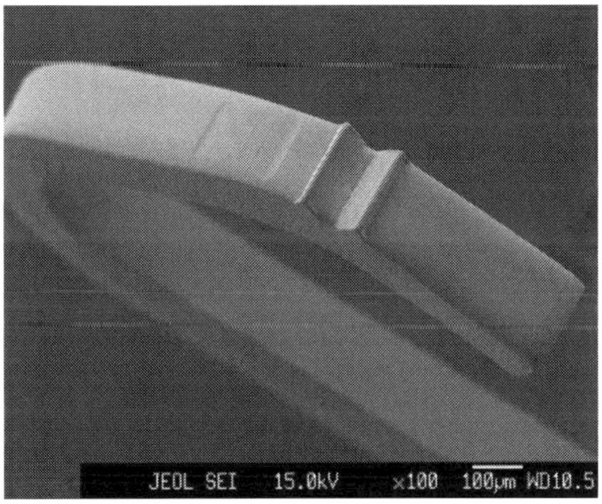

Figure 8: SEM Figure of Dual High Hertz-Stress Contact Tip

5. EXPERIMENTAL RESULTS AND ITS VALIDATION

A number of environmental test are conducted on the contact force versus contact resistance. I the following at Figure 9 shows the initial resistance, resistance at humidity-temperature cycling test and temperature life test. To conduct the environmental test it have ensured the selected samples contact forces were stable for all of the contacts and the average maximum normal force was around 0.16N and the contact resistance was approximately below 30mΩ which reflects a good contact resistance within the acceptable range. For lowest radius r = 5μm contact. We measured n=5 samples for each type of structure. The lowest radius r = 5μm contact resistance was more stable than the largest r = 200μm was checked from the initial data. Also the middle range contact radius r =25μm and 50μm was unstable than the smaller r =5μm.

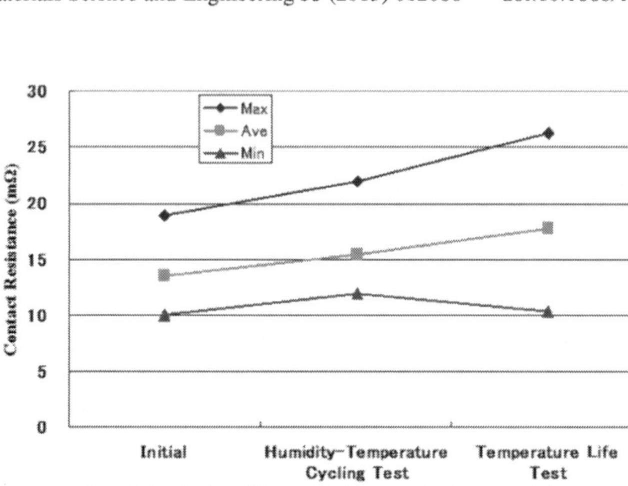

Figure 9: Evaluation of Contact Resistance on Environmental Test

6. FUTURE DEVELOPMENT

We are currently conducting a number of studies on the further development on the design, fabrication, and characterization. In order to go for industrial production system in commercial stamping method we need to do further experiment to replace with this new proposed design.

7. CONCLUSION

Due to the recent demand in the industrial production it is required to produce more compact contact in using conventional batch production. In reality, the similar quality contact is only achievable in using MEMS and EFF technology but these contacts production cost is higher and not suitable for gross production due to slow manufacturing process. Thus, it was a big challenge to come up with the similar type of contact in the commercial production. In this research a no of possible design is studied and it has been indentified scribing technology is the best design to fulfil industrial demand and the contact are within the required criteria which is tested through a number of environmental test. Throughout the test the contact resistance decreased drastically which was the major aim in this research. Consequently, in this basic research we have identified the commercial contact resistance is within the 20mΩ to 30 mΩ.

ACKNOWLEDGEMENTS

This work is partially developed in cooperation with Kyushu Hitachi Maxell Co. Ltd. The author would like to thank Mr. Tsuyoshi Tawara of Kyushu Hitachi Maxell Co. Ltd., Japan for providing valuable manufacturing data and comments on this development. Specially, the Author would like to Mr. Eiichiro Takemasa who gave me a support to continue this project.

REFERENCES

[1] T.Unno, et all. "Research on Manufacture of the Micro Connector using a UV-LIGA Process," PhD. Thesis of Ritsumeikan University (2004) (in Japanese).
[2] Markets Survey Outlook for Electronic Components, "http://www.sangyo-joho.co.jp/"
[3] M. Ishii, et all. "Connector Advanced Technology'99," Japan Advanced Technology (1999) (in Japanese)
[4] M.M.I. Bhuiyan, T. Unno, Y. Yokoyama, T. Toriyama and S. Sugiyama, "Micro Connector Fabricated by Micro Process Technology," MHS2000, pp. 83-88, Oct. 2000, Japan

[5] T. Unno, T. Toriyama, M.M.I. Bhuiyan, Y. Yokoyama and S. Sugiyama, 'Micro Connector for High Packaging Density Fabricated by Using UV Thick Photoresist', The Transactions of The Institute of Electrical Engineers of Japan, The Inst. of Electrical Engineers of Japan, Vol. 122-E, No., 5, pp. 249-255, May. 2002

[6] M. Bhuiyan, S. Yamada, N. Kurokawa, K. Sakamoto, E. Takemasa, "Development of Low Normal Force Contact using Electro Fine Forming and Lithography Technology," Tyco Electronics Technical Conference 2008, Oct. 2008, China

[7] M.Bhuiyan, "Optimization of Low Normal Force Contact Spring," Autumn Conference TE KEC 2009, Japan

[8] M.Bhuiyan, "Development and Optimization of High-Hertz Stress Contact to Reduce Future Contact Material Cost," TE Technical Conference 2010, Oct. 2010, USA

[9] P. D. Warren, "Determining the Fracture Toughness of Brittle Material by Hertzian Identation," Journal of the European Ceramic Society, vol. 15, pp. 201-207, 1995

[10] N. P. Paduture, "Hertzian Contacts," Elsevier Science Encyclopedia of Materials Science and Technology, pp. 3750-3752. 2001

[11] Petit, F., A. C. Sartieaux, Maurice Gonon, and Francis Cambier. "Fracture toughness and residual stress measurements in tempered glass by Hertzian indentation." Acta materialia 55, no. 8: 2765-2774, 2007

[12] Warren, P. D., D. A. Hills, and D. N. Dai. "Mechanics of Hertzian cracking." Tribology international 28.6 (1995): 357-362.

[13] F. Michel and W. Ehrfeld, "Mechatronic Micro Devices," Proceedings of the 1999 International Symposium on Micromechatronics and Human science, pp. 27-32, 1999

[14] D. C. Miller, W Zhang and V. M. Bright, "Microrelay Packaging Technology Using Flip-Chip Assembly," Proceedings of MEMS 2000, pp. 265-270, 2000

[15] H. Okuyama, T. Haga, K. Emura, and H. Takada, "Development of Micro connector with automatic Connecting / Disconnecting Mechanism," Proceedings of MEMS 1999, pp. 257-262, 1999

IOP Conf. Series: Materials Science and Engineering **53** (2013) 012069 doi:10.1088/1757-899X/53/1/012069

Human Gait Recognition And Classification Using Similarity Index for various conditions

Nahid A Makhdoomi[1], **Teddy S Gunawan**[1], **Mohamed H Habaebi**[1]
[1]ECE Department, Faculty of Engineering
International Islamic University Malaysia
Kuala Lumpur, Malaysia

E-mail: nahidameer@yahoo.co.in , tsgunawan@iium.edu.my

Abstract. Gait recognition is usually referred to signify the human identification by the style/way people walk in image sequences. Our aim is to implement the traditional gait recognition algorithm and to show the variation in gait recognition when subject is observed parallel to camera under three conditions- walking normal, carrying a bag and wearing a coat. However in this case, the work devises a novel method for the purpose of similarity computation rather than the traditional recognition where the overall recognition rate of 78.57 percent was obtained.

1. Introduction

With the advent of technology and computers in modern world, security has become one of major concerns. Being humans we always try to ease our access to any system. Traditionally, we used to have passwords, pins and other methods to prevent our data from being misused. As the time passes by and threats grow we try to have unique features that determine the identity of the person and prove the authenticity. All passwords and pins can be shared, so a technique was developed that is based on the unique features of the person, who is allowed to access the system. This was named as the biometrics.

Biometrics is the automated use of physiological or behavioral characteristics to determine or verify identity of a person [1]. Physiological biometrics are related to shape of body parts and are more stable. These include fingerprints, face recognition, iris scan etc. Behavioral biometrics such as signatures, speech, gait, and voice recognition etc are related to behavior of a person and are more prone to changes depending on factors such as aging, injuries, or even mood. Recognition based on the (static) human shape as well as on movement, suggests a richer recognition cue that constitutes our gait. We define gait to be the coordinated, cyclic combination of movements that result in human locomotion [2]. The movements are coordinated in the sense that they must occur with a specific temporal pattern for the gait to occur. Besides other biometrics such as face, eyes and fingerprints, human gait is an important biometric that is used for the identification of people. For every individual, the physical characteristics sets up its gait, so the gait is believed to be as unique to a particular person as eyes or face is.

---

[1] nahidameer@yahoo.co.in

Gait recognition is the process where the features of human motion are automatically obtained/extracted and later these features enable us to authenticate the identity of the person in motion. As like other pattern recognition techniques, gait recognition technique also involves 2 stages: Information is derived from human locomotion in the first stage i.e. feature extraction stage and in the next stage, i.e. the recognition stage, a standard similarity computation technique is used to obtain results for being a match or a mismatch. A unique advantage of gait as a biometric is that it offers potential for recognition at a distance or at low-resolution or when other biometrics might not be perceivable [3].

The rest of the paper is organized as follows: Section 2 presents the Gait Recognition Algorithm, while Section 3 discusses the hardware requirement and software used in the human gait recognition system. Section 4 discusses the implementation and the proposed algorithm and Section 5 provides the experimental results. Finally, Section 6 will end the paper with conclusion and future works.

2. Gait Recognition Algorithm

As a new technology of biometrics, gait recognition has recently gained more and more interests from researchers due to its several attractive properties. The biomechanics' literature makes similar observations "A given person will perform his or her walking pattern in a fairly repeatable and characteristic way, sufficiently unique that it is possible to recognize a person at a distance by their gait" [4]. Gait recognition essentially aims to recognize a person by automatically extracting movement characteristics of the walking person in the video. In fact, it seems that it is the possibility that a subject may not be aware of the surveillance and identification that raises public concerns about gait biometrics.

Based on the work "Simplest Representation Yet for Gait Recognition: Averaged Silhouette" algorithm presented by Z. Liu and S. Sarkar [5], we use it as our base algorithm that presents a robust representation for gait recognition which is compact, easy to construct, and affords efficient matching. It also simply aligns and averages the silhouettes over one gait cycle rather than using time series based representation comprising of a sequence of raw silhouette frames or of features extracted therein. Finally, it bases recognition on the Euclidean distance between these averaged silhouette representations. However in this case, the work will involve a different method for the purpose of similarity computation rather than the traditional recognition technique using the Euclidean distance between the averaged silhouettes.

3. System Design

In this section, we implement the proposed algorithm for human gait recognition. The overall system design of the gait recognition system can be categorized into two parts, the hardware requirement and software environments.

3.1. Hardware Requirement

Hardware is the main part of the gait recognition system. The hardware that is needed to accomplish this task is a personal computer and a video camera with higher resolution quality that will capture the video from which the frames will be extracted. But in this case, we are not using any camera as we will be using an already created database.

3.2. Software

Software that we used to implement this work is MATLAB. MATLAB is a high-level language and interactive environment for numerical computation, visualization, and programming. The MATLAB application is built around the MATLAB language, and most use of MATLAB involves typing MATLAB code into the Command Window (as an interactive mathematical shell), or executing text files containing MATLAB code and functions.

4. Implementation

4.1. Gait Database

In this research, we used the publicly available CASIA gait database: Database B [6]. This database is a large multi-view gait database, consisting of 124 subjects; the data was captured from 11 views. Three variations, namely view angle, clothing and carrying condition changes, are separately considered. The format of the video filename in Dataset B is 'xxx-mm-nn-ttt.avi', where

- xxx: subject id, from 001 to 124.
- *mm: walking status, can be 'nm' (normal), 'cl' (in a coat) or 'bg' (with a bag)*
- *.nn: sequence number.*
- *ttt: view angle, can be '000', '018', ..., '180'.*

4.2. Proposed Algorithm

The research consists of four main phases: The first phase involves extraction of the silhouette of an individual. Calculating the gait period or gait cycle of the individual follows this. Finding the sum of silhouettes is the next step. Finally, similarity score computation and matching process is performed for recognition. The detail of the design is shown in Figure 2.

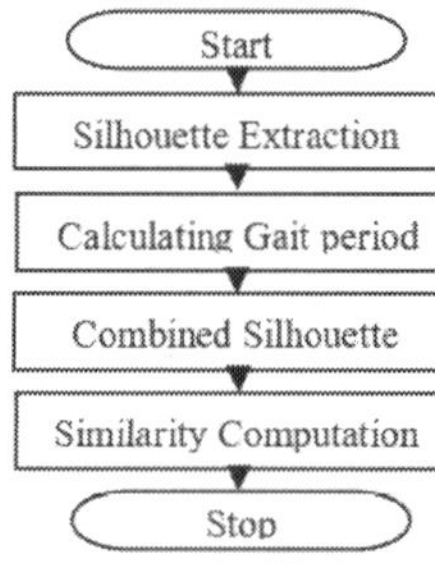

Figure 1: Gait recognition in flow chart.

The formula used for similarity comparisons is:

$$D = \sqrt{\frac{\sum [I_2(x,y) - I_1(x,y)]^2}{\sum [I_1(x,y)]^2}} \quad (1)$$

Any two images when compared using root mean square value are said to be similar if the value falls under the given threshold.

5. Results and Discussions

The preprocessed binary silhouettes were obtained. The sum of silhouettes was calculated as the first step from which we obtained the index of silhouette with maximum sum. This was the silhouette that contained the maximum information because the sum of that silhouette will be highest at the full stride stance. The number of pixels in foreground will reach a maximum when the two legs are farthest apart (full stride stance) and drop to a minimum when the legs overlap (heels together stance). Since the gait is periodic so the information will again be highest at next full stride stance. Now a loop is applied to add all the images between two full stride stances. Besides obtaining the sum of images that fall between two full strides, a counter is run to calculate the number of images that form a gait cycle. Hence, combined silhouette can be obtained for that individual. That silhouette is added to gallery database against that person.

Figure 2. Gait cycle sample images

The images above show a few sample preprocessed silhouettes taken from one gait cycle. The combined silhouette of the images, when calculated is shown below. All the combined images were compared to silhouettes of other people for the purpose of recognition.

Figure 3. A Sample of Combined Silhouette of an individual walking parallel to the image plane.

Any two images when compared using root mean square value are said to be similar if the value falls under the given threshold. Lesser the value more similar is the images. On performing experimentation, results show **0.6** is safe threshold for recognition while some matches were found above the given value and some mismatches were found on the score below the values.

Variation in gait recognition when subject is observed parallel to camera under three conditions - walking normal, carrying a bag and wearing a bag as shown in Table 1 and figure 4.

Table 1. Comparison ratio showing variation in human gait recognition

	Normal Vs Coat	*Normal Vs Bag*	*Bag Vs Coat*
Person1	0.3697	0.2920	0.4384
Person2	0.2785	0.2434	0.3336
Person3	**0.7672**	0.5522	0.5488
Person4	0.3642	0.4897	0.3549
Person5	0.2435	0.4291	0.4873
Person6	0.3342	0.4401	0.3385
Person7	0.5203	0.4317	0.3448
Person8	0.3735	0.4281	0.5418
Person9	0.5920	**0.6566**	0.3527
Person10	**0.6732**	**0.6262**	0.3232
Person11	**0.7823**	**0.8007**	0.3768
Person12	0.2737	0.5245	0.5182
Person13	**0.6802**	**0.7044**	**0.7044**
Person14	0.4536	0.4538	0.2589

71.42% recognition was obtained when normal walking pattern was compared with walking pattern of same person wearing a coat. Again recognition of **71.42%** was obtained when normal walking pattern was compared with walking pattern of same person carrying a bag. **92.85%** recognition was obtained when gait of person wearing a coat and carrying a bag were compared.

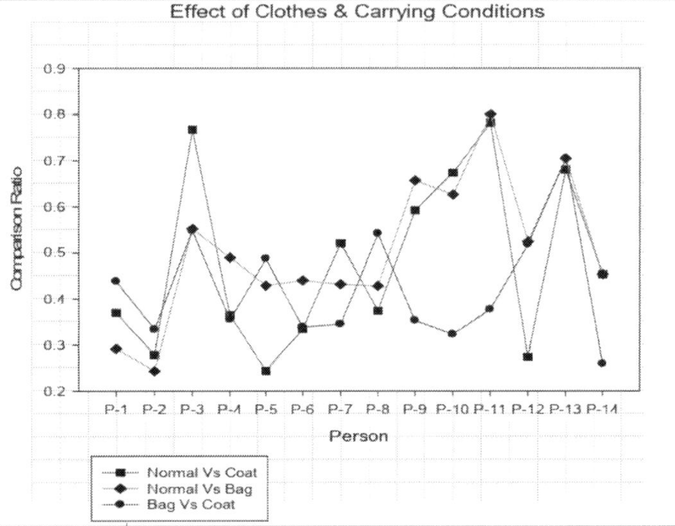

Figure 4. Graph showing comparison ratio in different scenario.

6. Conclusion

Gait Biometrics is an emerging and ever changing field of technology that can be implemented into just about anything that requires a security protocol. Literature survey [7] suggests most of topics have been touched if not explored to the fullest. Most of the results are calculated when the algorithm is implemented in ideal conditions and on particular databases. It is still not fully deployed in practicality because the confounding factors have not yet been fully blotted out. There is a lot of work to be done in different domains such as effect of terrain and slopes, calculation of the gait cycle in such conditions, the effect of bad weather and poor visibility, the effect of shoe type on the gait cycle and the measures that should be take to minimize the effect. One of most exciting works which can be done is that how does age affect the gait and how shall we incorporate the changing factor for each interval of life cycle? We can also try to increase the efficacy of present algorithms that work on multimodal fusion of various techniques. At last we can say that in order to make it stronger faster and more precise a lot of work is still to be done.

7. References

[1] Olufemi Sunday Adeoye, A Survey of Emerging Biometric Technologies. International Journal of Computer Applications (0975 – 8887) Volume 9– No.10, November 2010.

[2] Jeffrey E. Boyd and James J. Little, Biometric Gait Recognition, Biometrics School 2003, LNCS 3161, pp. 19–42, Springer-Verlag Berlin Heidelberg 2005.

[3] S. Yanushkevich, V. Shmerko, A. Stoica, P.Wang, S. Srihari, Introduction to Synthesis in Biometrics, Center of Excellence for Document Analysis and Recognition, State University of New York at Buffalo, Amherst, NY, USA.

[4] Winter D., The Biomechanics and Motor Control of Human Gait, 2nd Ed., Waterloo, 1991.

[5] Z. Liu and S. Sarkar, SimplestRepresentation Yet for Gait Recognition: Averaged Silhouette, Proc. Int'l Conf. Pattern Recognition, vol. 4, pp. 211-214, 2004. Author name / Procedia Computer Science 00 (2010) 000–000.

[6] CASIA Gait Database: http://www.sinobiometrics.com/resources.htm.

[7] Human identification based on Gait, http://asifbanka-ambition.blogspot.com/

IOP Conf. Series: Materials Science and Engineering **53** (2013) 012070 doi:10.1088/1757-899X/53/1/012070

Two degrees of freedom control of a ball and beam system

N N Abdul Aziz[3], M I Yusoff[3], M I Solihin[2,4] and R Akmeliawati[1,2,3]
[2]Intelligent Mechatronics System Research Unit (IMSRU) - International Islamic University Malaysia
[3]Department of Mechatronics Engineering, International Islamic University Malaysia
[4]Faculty of Engineering, Technology and Built Environment. UCSI University, Malaysia

Email: rakmelia@iium.edu.my

Abstract. In this paper, a two-degree of freedom (2DOF) controller is designed for a ball and beam system. The controller is developed based on the algebraic method. The ball and beam system is one of the most popular laboratory experiments for control education. The controller is designed such that the ball can track a square wave with a certain design specification. The advantages of 2DOF controller are the feedback controller takes care of the uncertainty and the feedforward filter ensures the tracking of the reference command. Though the control method is not new, the application of the technique on such unstable system is of interest. Controlling the ball on the beam is a challenging task because of the instability of the system and the fact that the output, i.e. the ball position, increases almost without limit for a fixed input beam angle. The controller needs to regulate the position of the ball by changing the angle of the beam at the pivot point. It is a difficult control task as the ball moves with an acceleration which is proportional to the tilt angle of the beam. Two types of 2DOF controllers are considered; based on dominant pole design and based on integral time absolute error (ITAE) design. The performances of the resulting controllers are compared. Simulations were run over various frequencies. The results indicate the effectiveness of the designed 2DOF controllers in achieving the design specification.

1. Introduction

Two degrees of freedom control is a control technique that can provide pole and zero matching. In this type of controller three conditions must be satisfied such that the desired closed-loop transfer function and the plant transfer function are implementable, i.e. pole-zero excess inequality, retainment of non-minimum phase errors and the Hurwitz requirement of the desired closed-loop transfer function [1].

Typically the structure of the closed-loop system is shown in Figure 1.

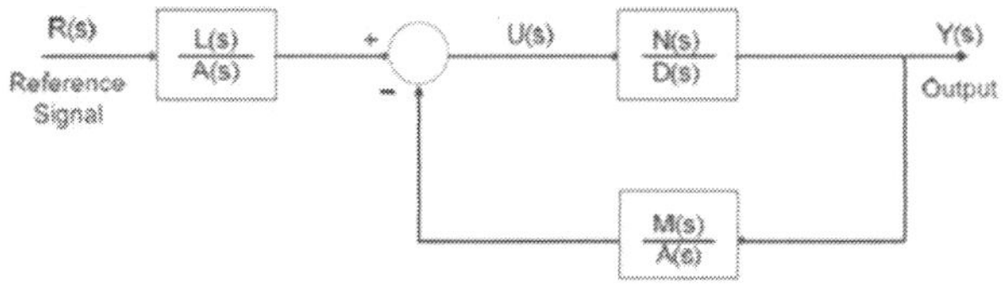

Figure 1 Two degree of freedom controller structure

[1] Corresponding author.

The controllers are described by two transfer functions L(s)/A(s) and M(s)/A(s), in the feedforward and feedback connection, respectively.
This type of controller has been used in applications, such as liquid tank controller [2]. The technique offers the advantage of pole and zero matching between the actual and desired closed-loop transfer function. For this reason, it is also called model matching technique.

The ball on beam balancer system is one of the most enduringly popular and important laboratory models for teaching control systems engineering. The system control job is automatically regulating the position of the ball on the beam by changing the angle of the beam. An example of modeling and controller design for a ball and beam system can be found in [1].
A ball is placed on a beam and is allowed to roll with one degree of freedom along the length of the beam. The model of the ball and beam system is shown in Figure 2. A lever arm is attached to the beam at one end and a servo gear at the other. As the servo gear turns by an angle theta, the lever changes the angle of the beam by alpha. When the angle is changed from the horizontal position, gravity causes the ball to roll along the beam. In this project, the objective is to design a controller to control a ball and beam system.

Controlling the ball on the beam is a challenging task because of the instability of the system and the fact that the output, i.e. the ball position, increases almost without limit for a fixed input beam angle. The controller needs to regulate the position of the ball by changing the angle of the beam at the pivot point. It is a difficult control task as the ball moves with an acceleration which is proportional to the tilt angle of the beam.

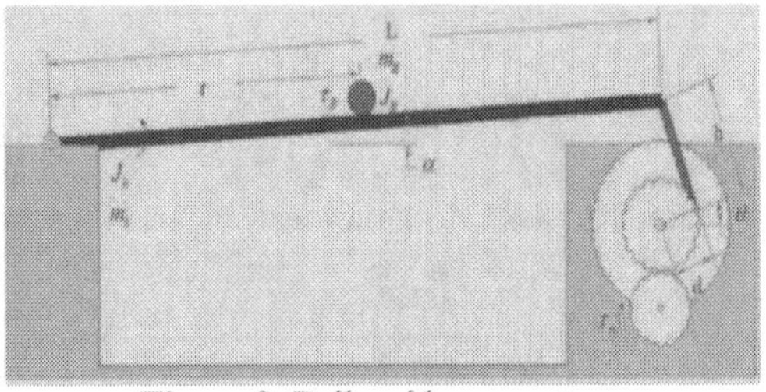
Figure 2. Ball and beam system

The paper is organized as follows. Section 2 provides the mathematical model of the ball and beam system. Section 3 describes the controller design. Section 4 presents the results and discussion. Section 5 concludes the paper

2. Mathematical model
Lagrange method is used to derive the dynamic equations of ball and beam system that is based on energy balance of the system. The derivation in this section is based on [3-4].

2.1. Equations of motion
For deriving the Euler-Lagrange equation, we need to find the kinetic energy and potential energy for the ball and beam system first. The kinetic energy of the system is:

$$T = T_b + T_B, \tag{1}$$

$$T_b = \frac{1}{2} J_b \dot{\alpha}^2 \tag{2}$$

$$T_B = \frac{1}{2}(m_B r^2)\dot{\alpha}^2 + \frac{1}{2}m_B \dot{r}^2 + \frac{1}{2}J_B \omega_B{}^2, \quad (3)$$

$$\dot{r} = R\omega_B, \quad (4)$$

$$\omega_B = \frac{\dot{r}}{R}, \quad (5)$$

$$J_B = \frac{2}{5}m_B R^2. \quad (6)$$

Substituting equation (5) and (6) into equation (3),

$$T_B = \frac{1}{2}(m_B r^2)\dot{\alpha}^2 + \frac{1}{2}m_B \dot{r}^2 + \frac{1}{5}m_B \dot{r}^2 = \frac{1}{2}\left[m_B r^2 \dot{\alpha}^2 + \frac{7}{5}m_B \dot{r}^2\right] \quad (7)$$

Substituting equation (2) and (7) into equation (1),

$$T = \frac{1}{2}J_b \dot{\alpha}^2 + \frac{1}{2}\left[m_B r^2 \dot{\alpha}^2 + \frac{7}{5}m_B \dot{r}^2\right] = \frac{1}{2}\left[(J_b + m_B r^2)\dot{\alpha}^2 + \frac{7}{5}m_B \dot{r}^2\right] \quad (8)$$

The potential energy of the system is exhibited by the rolling ball alone:

$$U = m_B g r \sin\alpha + \frac{L}{2}m_b g \sin\alpha = \left(m_B r + \frac{L}{2}m_b\right) g \sin\alpha \quad (9)$$

where parameters $m_B, m_b, J_b, J_B, R, \omega_B, \dot{r}, g, L, r, \alpha$ are the ball and beam mass, beam moment of inertia, ball moment of inertia, radius of the ball, rotational and radial velocities of the ball, gravity acceleration, length of the beam, linear motion of the beam along the beam and beam angle, respectively.

The Lagrange equation is the difference between kinetic and potential energy that is defined by equation (10).

$$L = T - U. \quad (10)$$

By substituting equation (8) and (9) into equation (10), we will get equation (11).

$$L = \frac{1}{2}\left[(J_b + m_B r^2)\dot{\alpha}^2 + \frac{7}{5}m_B \dot{r}^2\right] - \left(m_B r + \frac{L}{2}m_b\right) g \sin\alpha \quad (11)$$

Since there is no external force on the ball in radial direction, Lagrange's equations of motion are formed as:

$$\frac{d}{dt}\left[\frac{\partial L}{\partial \dot{\alpha}}\right] - \frac{\partial L}{\partial \alpha} = \tau, \quad (12)$$

$$\frac{d}{dt}\left[\frac{\partial L}{\partial \dot{r}}\right] - \frac{\partial L}{\partial r} = 0, \quad (13)$$

where τ is the torque produced by the motor applied at the end of the beam. For equation (12):

$$(J_b + m_B r^2)\ddot{\alpha} + 2m_B r\dot{r}\dot{\alpha} + \left(m_B r + \frac{L}{2}m_b\right) g \cos\alpha = \tau \quad (14)$$

From equation (14), we can obtain

$$\frac{7}{5}\ddot{r} - r\dot{\alpha}^2 + g\sin\alpha = 0. \quad (15)$$

2.2. Linearization around operating-point of the system

Jacobian linearization method is used to find the linear approximation of the dynamic equations. The linear dynamics equation of ball and beam system is presented in state space realization based on equation (16) and (17).

$$\dot{\mathbf{x}}(t) = \mathbf{A}\mathbf{x}(t) + \mathbf{B}\mathbf{u}(t), \quad (16)$$

$$\mathbf{y}(t) = \mathbf{C}\mathbf{x}(t) + \mathbf{D}\mathbf{u}(t), \quad (17)$$

where matrix **A** defines the dynamic behavior of the system, matrix **B** defines the position and behavior of system actuator, matrix **C** defines the relation between states and output of system and matrix **D** is feedforward matrix that is equal to zero in this case.

Jacobian linearization gives the linear dynamic equation around the operating point that is the middle of the beam. Besides, the state space formulation must be derived around operating point. The assumption for state space formulation is to define the operating point of the system. Thus, we used (*) notation to refer to operating point of the system that is as follows.

$$\Delta\dot{\mathbf{x}} = \mathbf{A}\Delta\boldsymbol{x} + \mathbf{B}\Delta u\,, \tag{18}$$

$$\Delta\mathbf{y} = \mathbf{C}\Delta\mathbf{x} + \mathrm{D}\Delta u\,, \tag{19}$$

$$\Delta\mathbf{x} = \begin{bmatrix}\Delta x_1\\ \vdots \\ \Delta x_n\end{bmatrix} = \begin{bmatrix}x_1 - x_1^*\\ \vdots \\ x_n - x_n^*\end{bmatrix}\ , \tag{20}$$

$$\Delta u = u - u^*, \mathrm{and}\, y = y - h(x^*, u^*).$$

The operating point: $x_1^* = 0,\quad x_2^* = \delta,\quad x_3^* = 0,\quad x_4^* = 0.$

Let: $x_1 = \alpha;\quad x_2 = r;\quad x_3 = \dot{\alpha};\ x_4 = \dot{r}$

Thus, equation (14) can be rewritten as

$$(J_b + m_B x_2^2)\dot{x}_3 + 2m_B x_2 x_4 x_3 + \left(m_B x_2 + \frac{L}{2}m_b\right) g\cos x_1 = \tau \tag{21}$$

or

$$\dot{x}_3 = \frac{\tau - 2m_B x_2 x_4 x_3 - \left(m_B x_2 + \frac{L}{2}m_b\right) g\cos x_1}{J_b + m_B x_2^2}\ . \tag{22}$$

Using small angle approximation,

$$\dot{x}_3 = \frac{\tau - 2m_B x_2 x_3 x_4 - m_B x_2 g - \frac{L}{2}m_b g}{J_b + m_B x_2^2}\,. \tag{23}$$

Likewise, equation (15) can be rewritten as

$$\frac{7}{5}\dot{x}_4 - x_2 x_3^2 + g\sin x_1 = 0 \tag{24}$$

Using the small angle approximatinon, i.e. $\sin x_1 \approx x_1$

$$\dot{x}_4 = \frac{5x_2 x_3^2 - 5g x_1}{7}\ , \tag{25}$$

$$\dot{x}_1 = \dot{\alpha} = x_3, \tag{26}$$

$$\dot{x}_2 = \dot{r} = x_4. \tag{27}$$

The dynamic equation of the system is defined as:

$$\dot{x}_1 = x_3 = f_1\ \ , \tag{28}$$

$$\dot{x}_2 = x_4 = f_2 \quad , \tag{29}$$

$$\dot{x}_3 = \frac{\tau - 2m_B x_2 x_3 x_4 - m_B x_2 g - \frac{L}{2} m_b g}{J_b + m_B {x_2}^2} = f_3 \, , \tag{30}$$

$$\dot{x}_4 = \frac{5 x_2 {x_3}^2 - 5 g x_1}{7} = f_4 . \tag{31}$$

The static load required in the operating point is defined by τ^*.

$$\tau^* = g\left(\frac{m_b L}{2} + m_B r\right) . \tag{32}$$

Modeling of DC servomotor can be divided into electrical and mechanical subsystems. The electrical subsystem is based on Kirchhoff's voltage law that is defined by equation (33).

$$V = L_m \dot{I} + R_m I + K_m \dot{\Theta}_m \, , \tag{33}$$

$$\tau_m = K_i I \quad . \tag{34}$$

where V, I, Θ_m are motor voltage, motor current and motor angle. L_m, R_m, K_m, K_i are motor constants. Compared to $R_m I$ and $K_m \dot{\Theta}_m$, the term $L_m \dot{I}$ is very small. In order to simplify the modeling, we neglected the term $L_m \dot{I}$. Thus,

$$V = R_m I + K_m \dot{\Theta}_m \quad . \tag{35}$$

Rearranging equation (35), we will get

$$I = \frac{V - K_m \dot{\Theta}_m}{R_m} \quad . \tag{36}$$

Substituting equation (36) into equation (34), we get

$$\tau_m = K_i \frac{V - K_m \dot{\Theta}_m}{R_m} \quad . \tag{37}$$

The mechanical subsystem is defined in the following equation.

$$\tau_b = \frac{K_g K_i \eta_{total}}{R_m} \frac{L}{d} V - \frac{{K_g}^2 K_i K_m \eta_{total}}{R_m} \frac{L^2}{d^2} \dot{\alpha} . \tag{38}$$

Substituting equation (38) into equation (28), we get

$$f_3 = \dot{x}_3 = \frac{\frac{K_g K_i \eta_{total} L}{R_m} \frac{L}{d} V - \frac{{K_g}^2 K_i K_m \eta_{total} L^2}{R_m} \frac{L^2}{d^2} x_3 - 2 m_B x_2 x_3 x_4 - m_B x_2 g - \frac{L}{2} m_b g}{J_b + m_B {x_2}^2} \tag{39}$$

Based on our assumption, voltage V^* should be defined for the operating point ${x_1}^* = 0, \quad {x_2}^* = \delta,$ ${x_3}^* = 0, \quad {x_4}^* = 0$, which is shown as in the following equations.

$$\frac{\frac{K_g K_i \eta_{total}}{R_m}\frac{L}{d}V - \frac{K_g{}^2 K_i K_m \eta_{total}}{R_m}\frac{L^2}{d^2}x_3 - 2m_B x_2 x_3 x_4 - m_B x_2 g - \frac{L}{2}m_b g}{J_b + m_B x_2{}^2} = 0$$

$$\frac{K_g K_i \eta_{total}}{R_m}\frac{L}{d}V - \frac{K_g{}^2 K_i K_m \eta_{total}}{R_m}\frac{L^2}{d^2}x_3 - 2m_B x_2 x_3 x_4 - m_B x_2 g - \frac{L}{2}m_b g = 0$$

$$V^* = \left(m_B \delta g + \frac{L}{2}m_b g\right)\frac{R_m d}{K_g K_i \eta_{total} L} \tag{40}$$

Thus, the linearization of f_1, f_2, f_3 and f_4 around operating point $x_1{}^* = 0$, $x_2{}^* = \delta$, $x_3{}^* = 0$, $x_4{}^* = 0, \delta = \delta_2$, are as follows.

$$\frac{\partial f_3}{\partial x_2} = \frac{-m_B g(J_b + m_B \delta_2{}^2) - 2m_B \delta_2\left(\frac{K_g K_i \eta_{total}}{R_m}\frac{L}{d}\left(m_B \delta g + \frac{L}{2}m_b g\right)\frac{R_m d}{K_g K_i \eta_{total} L} - m_B \delta_2 g - \frac{L}{2}m_b g\right)}{(J_b + m_B \delta_2{}^2)^2}$$

$$\frac{\partial f_3}{\partial x_2} = -\frac{m_B J_b g + m_B{}^2 g \delta_2{}^2}{(J_b + m_B \delta_2{}^2)^2} \tag{41}$$

$$\frac{\partial f_3}{\partial x_3} = \frac{-\frac{K_g{}^2 K_i K_m \eta_{total}}{R_m}\frac{L^2}{d^2}(J_b + m_B \delta_2{}^2)}{(J_b + m_B \delta_2{}^2)^2} = -\frac{K_g{}^2 K_i K_m \eta_{total}}{R_m(J_b + m_B \delta_2{}^2)}\frac{L^2}{d^2} \tag{42}$$

$$\frac{\partial f_4}{\partial x_1} = \frac{-5g(7)}{7^2} = -\frac{5g}{7} \tag{43}$$

$$\frac{\partial f_3}{\partial V} = \frac{\frac{K_g K_i \eta_{total} L}{R_m \, d}(J_b + m_B \delta_2{}^2)}{(J_b + m_B \delta_2{}^2)^2} = \frac{K_g K_i \eta_{total}}{R_m(J_b + m_B \delta_2{}^2)}\frac{L}{d} \tag{44}$$

The state space matrices and vectors that describe the system are as the following.

$$A = \begin{bmatrix} 0 & 0 & 1 & 0 \\ 0 & 0 & 0 & 1 \\ 0 & -\frac{m_B J_b g + m_B{}^2 g \delta_2{}^2}{(J_b + m_B \delta_2{}^2)^2} & -\frac{K_g{}^2 K_i K_m \eta_{total}}{R_m(J_b + m_B \delta_2{}^2)}\frac{L^2}{d^2} & 0 \\ -\frac{5g}{7} & 0 & 0 & 0 \end{bmatrix},$$

$$B = \begin{bmatrix} 0 \\ 0 \\ \frac{K_g K_i \eta_{total}}{R_m(J_b + m_B \delta_2{}^2)}\frac{L}{d} \\ 0 \end{bmatrix}, \quad C = [0 \quad 1 \quad 0 \quad 0], \text{ and } D = [0]$$

The beam moment of inertia is defined as

$$J_b = \frac{1}{2}m_b L^2 \tag{45}$$

The ball and beam system parameters and DC motor specifications in the present study are shown in Table 1.

Table 1. Ball and beam system parameters

Symbol	Quantity	Value
g	Gravity acceleration	9.8 (m/s^2)
m_B	Ball mass	0.064 (kg)
m_b	Beam mass	0.65 (kg)
R	Ball radius	0.0254 (m)
l	Beam length	0.425 (m)
d	Lever length	0.12 (m)
δ	Equilibrium point of ball position	0.2 (m)
K_m	Back EMF constant	0.00767 (V.sec/rad)
K_i	Torque constant	0.00767 (N.m)
K_g	Gear ratio	14
R_m	Motor resistance	2.6 (Ω)
η_{motor}	Motor efficiency	0.69
$\eta_{gearbox}$	Gearbox efficiency	0.85

The state space matrix representing the ball and beam system is:

$$A = \begin{bmatrix} 0 & 0 & 1 & 0 \\ 0 & 0 & 0 & 1 \\ 0 & -10.2378 & -1.3983 & 0 \\ -7 & 0 & 0 & 0 \end{bmatrix},\ B = \begin{bmatrix} 0 \\ 0 \\ 3.6769 \\ 0 \end{bmatrix},\ C = [0 \quad 1 \quad 0 \quad 0],\quad D = [0].$$

The transfer function of the system is denoted as $G(s)$,

$$G(s) = \frac{Y(s)}{V(s)} = \frac{-25.7382}{s^4 + 1.3893s^3 - 71.6646}.$$

3. Two DOF controller design

In this paper, two degree of freedom controller based on algebraic method is designed to control the position of the ball on the beam such that it tracks a square wave with amplitude of 5cm. Though the control method is not new, the application of the technique on such unstable system is of interest. In achieving the objective, the following design specifications need to be satisfied:

- The settling time is less than 2 seconds,
- The percent overshoot is less than 5 %
- Velocity error constant is larger than 150.

The plant transfer function is defined as:

$$G(s) = \frac{N(s)}{D(s)} = \frac{-25.7382}{s^4 + 1.3893s^3 - 71.6646} \quad .$$

Diophantine equation, $D_0(s) = \overline{D_p(s)}D_p(s) = A(s)D(s) + M(s)N(s)$.
In this case, the order of the plant, $n = 4$. Thus, the order of $\overline{D_p(s)}D_p(s) \geq 2n - 1 = 2(4) - 1 = 7$.

The closed-loop transfer function of the system is 7th order system. Since the damping ratio, $\zeta = 0.7$ and settling time, $T_s = 0.2$sec. So, the natural frequency, $\omega_n = 28.57$rads/sec.

Thus, we obtain the desired characteristics:

$$a_d(s) = s^2 + 2\zeta\omega_n s + \omega_n{}^2 = s^2 + 39.998s + 816.245$$

The dominant poles are, $p = -20 \pm j20.403$
Since this system is 7th order system, we have to add 5 more poles at least 2-6 times faster than dominant poles. So, the rest of poles are: $p \ = [-40 \ \ -60 \ \ -80 \ -100 \ -120]$.
Then, the desired characteristics equation is:

$$\begin{aligned} a_d(s) &= (s^2 + 39.998s + 816.245)(s + 40)(s + 60)(s + 80)(s + 100)(s + 120) \\ &= s^7 + 440.398s^6 + 78975.5s^5 + 7471176s^4 + 405094220s^3 + 1.283948192\times 10^{10}s^2 \\ &\quad +2.29423392\times 10^{11}s + 1.88064\times 10^{12} \end{aligned}$$

Thus, the overall transfer function $G_0(s)$is:

$$G_0(s) = \frac{1.88064\times 10^{12}}{s^7 + 440.398s^6 + 78975.5s^5 + 7471176s^4 + 405094220s^3 + 1.283948192\times 10^{10}s^2 + 2.29423392\times 10^{11}s + 1.88064\times 10^{12}}$$

The necessary condition for $S_m$to have full row ranks is $n - 1 \leq m$, for this case, $m \geq 3$.

Thus, order of $A(s)$ is 3 since the highest order of plant denominator $D(s)$ is 4.

$$S_m C_m = \begin{bmatrix} D_0 & N_0 & 0 & 0 & 0 & 0 & 0 & 0 \\ D_1 & N_1 & D_0 & N_0 & 0 & 0 & 0 & 0 \\ D_2 & N_2 & D_1 & N_1 & D_0 & N_0 & 0 & 0 \\ D_3 & N_3 & D_2 & N_2 & D_1 & N_1 & D_0 & N_0 \\ D_4 & N_4 & D_3 & N_3 & D_2 & N_2 & D_1 & N_1 \\ 0 & 0 & D_4 & N_4 & D_3 & N_3 & D_2 & N_2 \\ 0 & 0 & 0 & 0 & D_4 & N_4 & D_3 & N_3 \\ 0 & 0 & 0 & 0 & 0 & 0 & D_4 & N_4 \end{bmatrix} \begin{bmatrix} A_0 \\ M_0 \\ A_1 \\ M_1 \\ A_2 \\ M_2 \\ A_3 \\ M_3 \end{bmatrix} = \begin{bmatrix} F_0 \\ F_1 \\ F_2 \\ F_3 \\ F_4 \\ F_5 \\ F_6 \\ F_7 \end{bmatrix}$$

$$\begin{bmatrix} -71.6646 & -25.7382 & 0 & 0 & 0 & 0 & 0 & 0 \\ 0 & 0 & -71.6646 & -25.7382 & 0 & 0 & 0 & 0 \\ 0 & 0 & 0 & 0 & -71.6646 & -25.7382 & 0 & 0 \\ 1.3983 & 0 & 0 & 0 & 0 & 0 & -71.6646 & -25.7382 \\ 1 & 0 & 1.3983 & 0 & 0 & 0 & 0 & 0 \\ 0 & 0 & 1 & 0 & 1.3983 & 0 & 0 & 0 \\ 0 & 0 & 0 & 0 & 1 & 0 & 1.3983 & 0 \\ 0 & 0 & 0 & 0 & 0 & 0 & 1 & 0 \end{bmatrix} \begin{bmatrix} A_0 \\ M_0 \\ A_1 \\ M_1 \\ A_2 \\ M_2 \\ A_3 \\ M_3 \end{bmatrix} = \begin{bmatrix} 1.88064\times 10^{12} \\ 2.29423392\times 10^{11} \\ 1.283948192\times 10^{10} \\ 405094220 \\ 7471176 \\ 78975.5 \\ 440.398 \\ 1 \end{bmatrix}$$

From which the coefficients of the controller transfer function are:
$A_0 = 7361602909$, $A_1 = 783616.467$, $A_2 = 438.9997$, $A_3 = 1$, $M_0 = -7.3089 \times 10^{10}$
$M_1 = -8.9139 \times 10^9$, $M_2 = -498.8505 \times 10^6$, and $M_3 = \ -15.3391 \times 10^6$.

Thus, the transfer function of the second controller:

$$C_2(s) = \frac{M(s)}{A(s)} = \frac{-15.3391\times10^6 s^3 - 498.8505\times10^6 s^2 - 8.9139\times10^9 s - 7.3089\times10^{10}}{s^3 + 438.9997s^2 + 78361.6467s + 7361602.909}.$$

Since $L(s)N(s) = N_0(s)$, where $N_0(s) = 1.88064\times 10^{12}$ and $N(s) = -25.7382$,
thus, $L(s) = \frac{N_0(s)}{N(s)} = \frac{1.88064\times10^{12}}{-25.7382} = -7.3068\times 10^{10}$.

The transfer function of the first controller,

$$C_1(s) = \frac{L(s)}{A(s)} = \frac{-7.3068\times 10^{10}}{s^3 + 438.9997s^2 + 78361.6467s + 7361602909}$$

As for satisfying the requirement of the velocity error K_v larger than 150, let $E(s)$ be the error signal, where $E(s) = C_1(s)R(s) - C_2(s)Y(s)$. As $Y(s) = E(s)G(s)$, thus,
$E(s) = C_1(s)R(s) - C_2(s)E(s)G(s)$. Substituting $C_1(s)$ and $C_2(s)$ gives

$$E(s) = \frac{C_1(s)R(s)}{1+C_2(s)G(s)} = \frac{\frac{L(s)}{A(s)}R(s)}{1+\frac{M(s)}{A(s)}G(s)}.$$

Therefore, the steady state error, e_{ss}

$$e_{ss} = \lim_{s\to0} sE(s) = \lim_{s\to0} s\frac{\frac{L(s)}{A(s)}\frac{1}{s}}{1+\frac{M(s)}{A(s)}G(s)},$$

$$= \frac{\frac{-7.3068\times10^{10}}{7\,361602909}}{1+\frac{-7.308\,9\times10^{10}-25.7\,38\,2}{7\,361602909\,-7\,16646}} = 2.7843.$$

From which the uncompensated velocity error constant is

$$K_{vu\ ncompensated} = \frac{1}{e_{ss}} = \frac{1}{2.7843} = 0.359.$$

Since

$$K_{vcompensated} = K_{vu\ ncompensated}\left(\frac{z}{p}\right),$$

and from the specification, $K_{vcompensated}$ must be greater than 150, we choose $K_{vcompensated} = 160$.

$$160 = 0.359\left(\frac{z}{p}\right),$$

$$\frac{z}{p} = 445.68.$$

So, we let $z = 0.001$ which is closed to zero so that it will not give too much effect to the system transient response.

$$p = \frac{0.001}{445.68} = 2.244\times 10^{-6} \approx 0.$$

The additional zero and pole are added to the overall transfer function to satisfy the velocity error requirement. Thus, the third controller is placed just after the summing point and just before the plant block in Figure 1. This extra controller in fact provides an additional integrator to the system, or precisely it is given by the following transfer function.

$$C_3(s) = \frac{s+0.001}{s}.$$

4. Results and discussion

From the simulation result for the 2DOF controller, the values for percent overshoot and settling time satisfy the desired specification which percent overshoot need to be less than 5% and settling time less than 2 sec. From the desired specification, we need to consider the velocity error constant K_v which must be greater than 150. From section 4 which is controller 1 design, we obtained steady state error, e_{ss} of 2.7843 and uncompensated velocity error constant, $K_{v,uncompensated}$ of 0.359. Here, we choose the compensated velocity error constant to be 160. From the calculation, we need to add zero of -0.001 and pole of -2.244×10^{-6}. The value of zero must be nearly to zero so that it will not affect the transient response of the closed loop system. This is proven by performance criteria obtained from Figure 4. The value for settling time remains the same and percent overshoot is increased by 0.03%. Figure 4 and 5 shows that the output of closed loop system tracks the square wave input from +5cm to -5cm. The response of the closed-loop system indicates that the percent of overshoot is 1.93%, settling time is 0.207 second and zero steady state error.

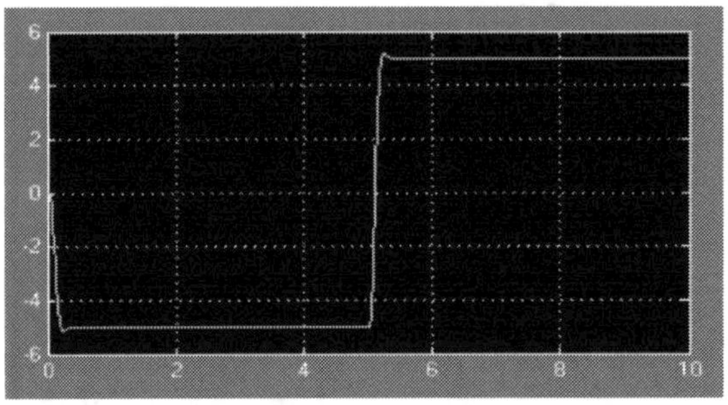

Figure 4. Response for closed loop system with 2 DOF controller without K_v requirement

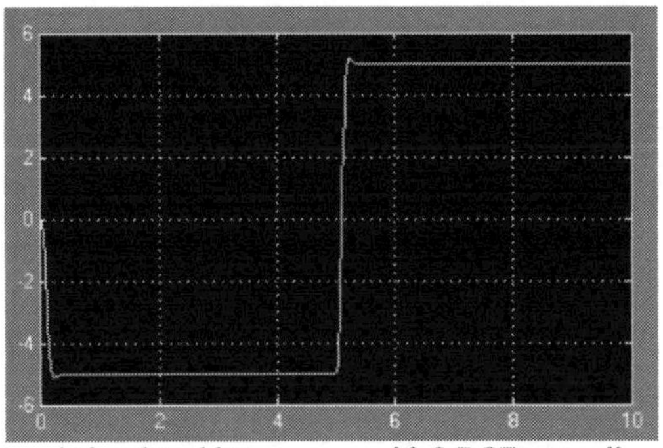

Figure 5. Simulink result for closed loop system with 2 DOF controller with K_v requirement

Comparison between 2DOF and pole placement method

We have designed controller using 2DOF and pole placement methods. For evaluating the performance between pole placement and 2DOF controller, we used different values of natural frequency, ω_n with same value of damping ratio which is 0.707. The optimum coefficient for damping ratio is equal to 0.707. The performance for both controllers is compared based on their percent

IOP Conf. Series: Materials Science and Engineering **53** (2013) 012070 doi:10.1088/1757-899X/53/1/012070

overshoot, settling time and steady state error. The results obtained from both controllers are tabulated in Table 2.

Table 2. Performance result for different value of frequency response

Controller	Frequency response ω_n (rad/sec)	Percent overshoot (%)	Settling time, T_s (sec)	Steady state error, e_{ss}
Pole placement	5.00	4.20	1.270	0
	10.00	3.54	0.670	0
	15.00	2.83	0.465	0
	20.00	2.12	0.355	0
	25.00			
	28.57			
	30.00			
2 DOF	5.00	4.24	1.270	0
	10.00	4.01	0.670	0
	15.00	3.60	0.471	0
	20.00	3.02	0.368	0
	25.00	2.40	0.303	0
	28.57	1.90	0.207	0
	30.00	1.80	0.190	0

From Table 2, for pole placement controller, the percent overshoot, settling time and steady state error for frequency response, ω_n equals to 25 rad/sec, 28.57 rad/sec and 30 rad/sec are left blank (shaded) as the responses become overdamped. As we can see, as the frequency response below 25 rad/sec increases, the values for percent overshoot and settling time are decreasing. As for the steady state error, the values remain 0.

Further performance evaluation of the closed-loop responses is conducted by comparing the 2 DOF controller with dominant pole design and the ITAE design.

Table 3. Performance of 2DOF controller with different poles selection

Controller	Frequency response ω_n (rad/sec)	Percent overshoot (%)	Settling time, T_s (sec)	Steady state error, e_{ss}
Dominant poles	5.00	4.24	1.270	0
	10.00	4.01	0.670	0
	15.00	3.60	0.471	0
	20.00	3.02	0.368	0
	25.00	2.40	0.303	0
	28.57	1.90	0.207	0
	30.00	1.80	0.190	0
Integral of time multiplied by absolute error (ITAE)	5.00	10.90	2.000	0
	10.00	10.90	0.999	0
	15.00	10.90	0.666	0
	20.00	10.90	0.500	0
	25.00	10.90	0.400	0
	28.57	10.90	0.350	0
	30.00	10.90	0.333	0

Table 3 indicates that the dominant pole design provides better responses than the ITAE.

5. Conclusion

In conclusion, the objectives of the research have been achieved. The 2 DOF controller and pole placement approaches have been presented. Evaluation of the performance of the closed loop system has been also presented. The performance of the designed 2DOF controller is comparable to that of the pole placement controller. However, the advantage of using 2DOF architecture is that the controller designer has more freedom in achieving pole and zero matching.

References

[1] Chen, C T 1993 Control System Design: Transfer Function, State-Space and Algebraic Methods, Saunders-HBJ

[2] Gunes N S, Uraz A 2010 Two Degree of Freedom Control of a Liquid TankCanadian Journal on Electrical and Electronics Engineering Vol. 1 No. 3

[3] Keshmiri, M, Jahromi, A. F., Mohebbi, A., Amoozgar, M. H. , Xie, W-F. 2012 Modeling and Control of Ball and Beam System using Model Based and Non-Model Based Control Approached *International Journal on Smart Sensing and Intelligent System*, Vol. 5, No. 1

[4] Control Tutorials for MATLAB and Simulink http://www.library.cmu.edu/ctms/ctms/index.htm

IOP Conf. Series: Materials Science and Engineering **53** (2013) 012071 doi:10.1088/1757-899X/53/1/012071

Development of a tentacle propulsion technique for underwater application

T Alamgir[1], M M Rashid[2], M R Khan[3]
[1,2,3]Department of Mechatronics Engineering, International Islamic University Malaysia, Kuala Lumpur 53100, Malaysia

E-mail: tarikbina@gmail.com

Abstract. As robotic technology matures and more platforms are fielded in unstructured real-world situations, the more new areas of applications are being thought for robotic deployment. After successes in industrial robots, researchers are now trying to explore new robots with biological features of different biological creatures like, snake, bird, and spider for their stunning advantages. Underwater exploration using robots is a new avenue. Research on the tentacle robot for underwater application is a new field of research besides the other research in this arena. There are few researches on this topic are explored and mostly are on biological robot. Besides those researches this paper aims to propose and demonstrate another technique to build a tentacle for propulsion purposes. Therefore, in this paper will discuss more on mathematical development for the propulsion technique and its software verification technique in considering the environmental constrains.

1. INTRODUCTION

As robotic technology developed and more focusing area are explored in unstructured real-world situations, the more new areas of applications are being discover for robotic deployment. After successes in industrial robots, scientist are now trying to explore new robots with biological features of different biological creatures like, snake, bird, spider for their vivid advantages. The various contributions on bio-inspired or bio-mimetic have been increasing dramatically throughout the last decade [1]. Bio-inspired robotics reflects the features and capabilities of natural evolution of a system that could be efficiently replicated or mimicked in a human developed system to design the new technologies and improvement of conventional systems [2]. Underwater exploration using robots is a new avenue [3]. However, fish robot cannot be expected to manipulate objects or undertake maintenance operation underwater, it can only do surveillance activities [4]. But we need multitasking underwater robots, especially one that would be able to propel itself, as well as manipulate objects. A robot that would be able to mimic natural tentacle is a probable solution. In the design and development of tentacle robot, tentacle would play the vital role. Tentacles can be regarded as continuum robot (continuous backbone).Tentacle robot is one of the features in underwater robot created from a sequence of link and joint combinations. The links are rigid members connected by the joints or axes. Whereas the axes are movable component of the tentacle robot that causes relative motion between adjoining links. These joints can be classified in different categories. One of them is

[1] To whom any correspondence should be addressed.

IOP Conf. Series: Materials Science and Engineering **53** (2013) 012071 doi:10.1088/1757-899X/53/1/012071

kinematic analysis which analysis the robot structure topologies such as serial robotic manipulator, parallel manipulator or hybrid robotic manipulator. For any model analysis the analysis can be done in four ways which are: a. Forward Dynamics analysis, b. Inverse dynamics analysis, c. Kinematic analysis and d. trimming analysis methods. For a close loop system the problem is solved mathematically by forward kinematics where individual body mass, acting force, torque and mass moment of inertia are known. The degree of freedom DOF are also predetermined once solving a mechanical problem in Forward dynamics method. For determining the forces/torques system has to be given a set of motions that required applying to the machine can be solved by reverse dynamics. This mode only works with open topology systems whereas the model diagrams is prepared in open loops [1]. The paper focuses on the modelling, Simulation and development studies of an underwater tentacle robot. The works in this paper more specifically identifies the prospects of the tentacle robot and identifies the individual joins position velocity and acceleration needed in the real environment in considering the environmental constrain to the model for simulation.

Since the proposed tentacle is usually an open ended tentacle therefore, an inverse kinematic mathematical and simulation model is designed to identify the required angular position, velocity and acceleration for each joint to acquire the defined location of each link in a particular time.

2. MODEL DESIGN

The kinematics control problem focused upon the computation if the joint positions required in locating the end-effectors' at desired Cartesian position and orientation. The design of kinematics control algorithms involves formulation of kinematic model, inversion of kinematic model and the implementation of inverse kinematics algorithms. However, the most difficult task in the kinematics control algorithm is the formulation of the inverse kinematic model.

A. Inverse kinematics solution (IKS) by mathematical model:

IKS for Two Links Manipulator is simple and the closed from IKS is presented below:

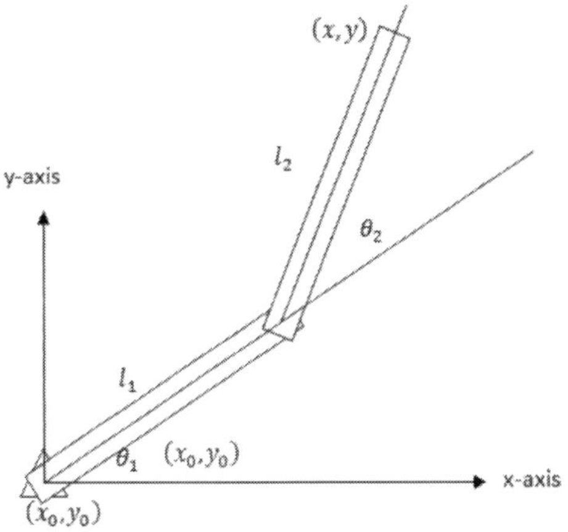

Figure 1. Mathematical model diagram of a two link manipulator.

The forward kinematics equations in terms of the joint variables of the 2-link manipulator are as shown in Figure 1.

$$x = l_1 \text{Cos}\theta_1 + l_2 \text{Cos}(\theta_1 + \theta_2) \quad (1.1)$$

$$y = l_1 \text{Sin}\theta_1 + l_2 \text{Sin}(\theta_1 + \theta_2) \quad (1.2)$$

Solving (1.1) and (1.2) for $\theta_1$and θ_2, the inverse kinematics solution can be obtained, which is shown in the following steps:

$$\mathrm{Cos}\theta_2 = \frac{x^2+y^2-l_1^2-l_2^2}{2l_1l_2} \tag{1.3}$$

$$\mathrm{Sin}\theta_2 = \pm\sqrt{1-\mathrm{Cos}^2\theta_2} \tag{1.4}$$

Here, "+" sign is used for CCW configuration and "_" sign for CW configuration.
By expanding (1.1) and (1.2), the following are obtained,

$$x = (l_1 + l_2\mathrm{Cos}\theta_2)\mathrm{Sin}\theta_1 + l_2\mathrm{Sin}\theta_2\mathrm{Cos}\theta_1$$

$$= K_1\sin\theta_1 + K_1\cos\theta_1 \tag{1.5}$$

$$y - (l_1 + l_2\cos\theta_2)\sin\theta_1 - l_2\sin\theta_2\cos\theta_1$$

$$= K_1\sin\theta_1 - K_1\cos\theta_1 \tag{1.6}$$

Where, $K_1 = (l_1 + l_2\cos\theta_2)$ and $K_2 = l_2\sin\theta_2$
Simultaneous solution of Equations (1.5) and (1.6) gives,

$$\sin\theta_1 = \frac{K_1y-K_2y}{k_1^2+k_2^2} \tag{1.7}$$

$$\cos\theta_2 = \frac{K_1y+K_2y}{k_1^2+k_2^2} \tag{1.8}$$

Thus the joint variable $\theta_1$and θ_2 are obtained from the following two equations:

$$\theta_1 = \mathrm{atan2}(\sin\theta_1, \cos\theta_1) \tag{1.9}$$

$$\theta_2 = \mathrm{atan2}(\sin\theta_2, \cos\theta_2) \tag{1.10}$$

It is showed that the above IKS of the 2-link tentacle input link length, coordinates of base and desired points with sign configuration will provide a unique solution for any IKS problems. However, once the number of links is increased than it raises new complexity in solving that problem. Any Hyper Redundant Robot (HRR) can be considered as a robot consisting of 2-links robot and provided IKS of 2-links can be used repeatedly for the IKS of the HRR. Therefore, this method can be extended to in considering one joint at a time to solve the multilink tentacle desired location and angle. This complexity can be solved more efficiently in making a software simulation and mimic the same sort of response curve for different link of a tentacle.

B. Solidworks model design for the proposed tentacle:

Solidworks, as the mainstream software for the virtual prototype system, combines multi-body dynamics modeling with large displacements as well as multi functioning tools. It also supports data exchange with other geometric modeling software during product development. This software has more powerful geometric modeling function. Due to this a kinematics model of tentacle robot propulsor is designed in Figure 2 and further imported into Matlab through a common data exchange interface with .XML format. Designed tentacle parameters are as below:

IOP Conf. Series: Materials Science and Engineering **53** (2013) 012071 doi:10.1088/1757-899X/53/1/012071

Table 1. Identified Link Parameters of Tentacle Robot Model.

i (joint number)	a_i **(link length)**	m_i **(link weight)**	d_i **(link offset)**	θ_i **(joint angle)**
1	15.89cm	0.587kg	$d_1 = 0$	θ_1
2	15.68cm	0.281kg	$d_2 = 0$	θ_2
3	15.68cm	0.281kg	$d_3 = 0$	θ_3
4	15.68cm	0.281kg	$d_4 = 0$	θ_4
5	15.68cm	0.281kg	$d_5 = 0$	θ_5
6	13.9cm	0.570kg	$d_6 = 0$	θ_6

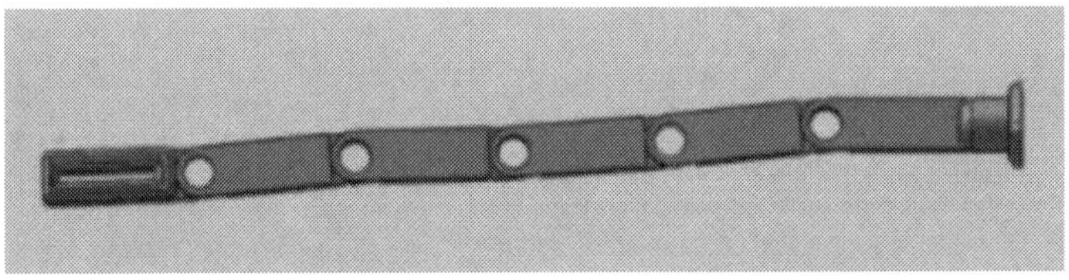

Figure 2. CAD design by using Solidworks software.

C. Sim-Mechanics model for the proposed tentacle:

In computer simulation process this tentacle is analyzed for a single joint to identify the respective angle, velocity and acceleration. Once the tentacle robot propulsor model is designed and exported into physical Modeling STL format than a Sim-Mechanics toolbox Model is generated in Simulink.

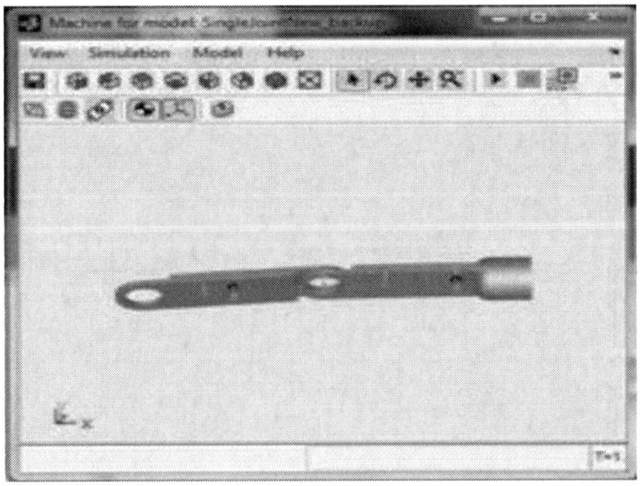

Figure 3. Simulation model diagram in Simulink in Matlab.

D. Simulink Model

After importing the Solidworks physical model .XML file into the Mathlab using Sim-Mechanics, a Simulink model is build. This model is upgraded to provide the real environmental constrain in integrating the drag co-efficient model in the Simulink model to generate the approximate natural movement to extract to the desired torque for individual joint to achieve the desired motion. This

Simulink model is further modified in considering the drag effect in the normal water. The designed Simulink model is provided below.

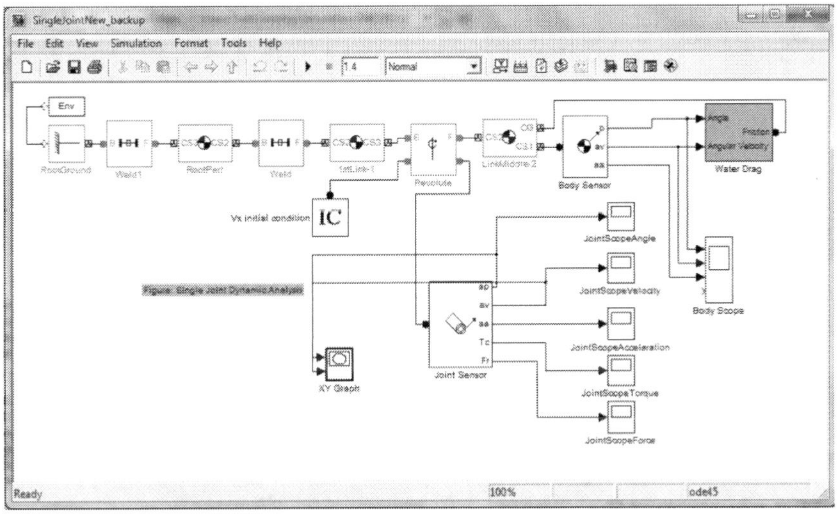

Figure 4. Simulink model diagram of the Designed CAD model for the tentacle.

For the drag equation, the model is modified and designed accordingly to implement the following equation and applied on the link centre of Gravity (CG) point.

$$F_D = \frac{1}{2}\rho\vartheta^2 C_D A$$

$$; \rho = \frac{m}{v}(1000\frac{kg}{m^3})$$

$$; C_D = 1.98$$

$$; A = Facing\ area$$

Therefore, the designed model appears as of Figure 5:

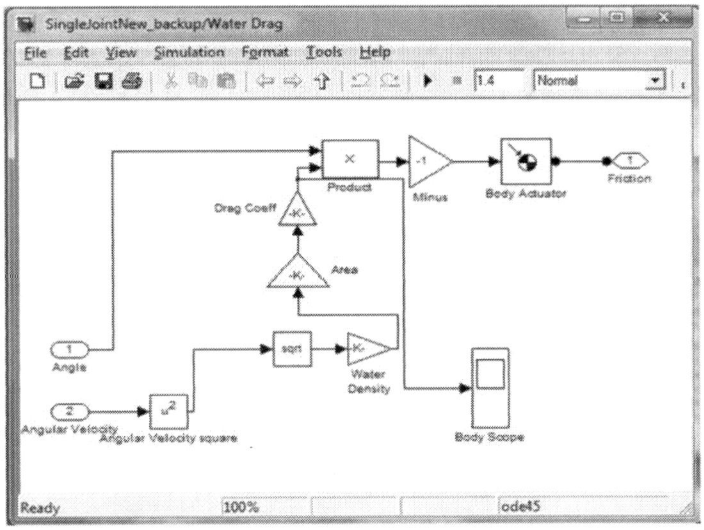

Figure 5. Simulink model diagram for the drag equation of the tentacle.

IOP Conf. Series: Materials Science and Engineering **53** (2013) 012071 doi:10.1088/1757-899X/53/1/012071

3. MODEL ANALYSIS

IKS Response is obtained for the first link joint in considering it acting as a driving link and the first joint as the driving joint. Besides that, rest of the joints will is making the sinusoidal movement from the reaction force acting on individual joint. Therefore this simulation aims to acquire the required torque for the driving joint. In the simulation model we achieved the angular position, velocity and acceleration for individual joints and from the graph we can individual joints diagrams are quite unique to each other. Obtained angular position, velocity and acceleration curves are provided below in Figure 6-9.

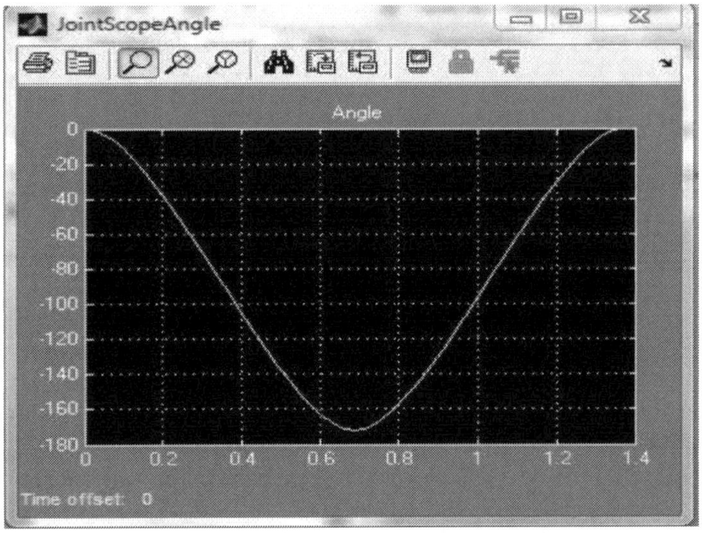

Figure 6. Base joint angular position vs. time response.

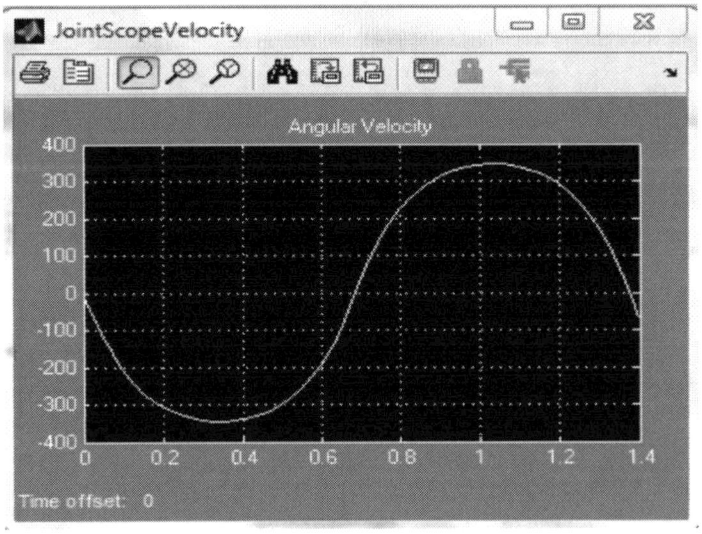

Figure 7. Base joint angular velocity vs. time response.

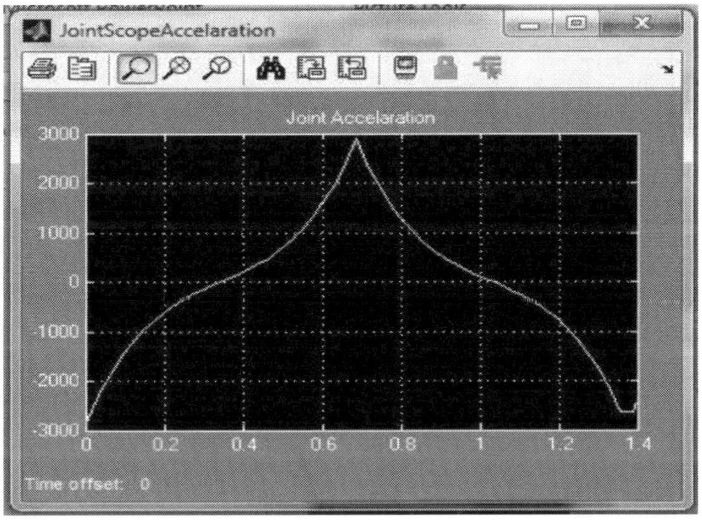

Figure 8. Base joint simulation response on angular acceleration vs. Time.

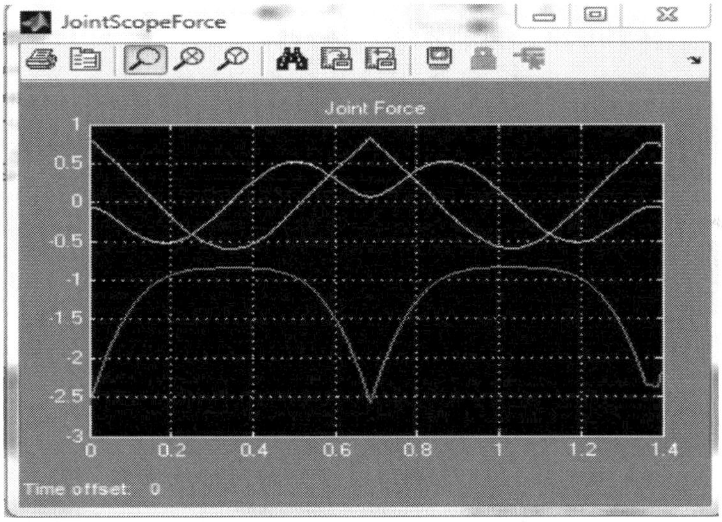

Figure 9. First link simulation response in identifying applied force in the joint.

CONCLUSION AND FUTURE WORKS

This paper describes the modelling method of a Bio-Inspired tentacle robot to achieve the propulsion in underwater. Simulation provides the feasibility of the tentacle robot and through the simulation we achieve the driving joint manipulators angular, velocity and acceleration with respect to the time which can be applied in the real prototype once the prototype designed and simulated to mimic the simulation design. Simulation has been carried out to identify the required torque for the experimental design. This works describes a method and steps to build a simulation model from a CAD model to a simulation model and evaluates the simulation results to justify the acquired results. A real prototype will be designed for further improvement on this study.

REFERENCES

[1] Chowdhury, A.R.; Prasad, B.; Kumar, V.; Kumar, R.; Panda, S.K.; , "Design, modeling and open-loop control of a BCF mode bio-mimetic Robotic Fish," *IEEE International Symposium on Safety, Security, and Rescue Robotics (SSRR), 2011*, vol., no., pp.226-231, 1-5 Nov. 2011

[2] M. S. Triantafyllou, A. H. Techet, and F. S. Hover, "Review of experimental work in biomimetic foils," IEEE J. Ocean. Eng., vol. 29, no. 3, pp. 585–594, Jul. 2004.

[3] P. R. Bandyopadhyay, "Trends in biorobotic autonomous undersea vehicles," IEEE J. Ocean. Eng., vol. 30, no. 1, pp. 109–139, Jan. 2005.

[4] C. Zhou, K. H. Low, "Study and implementation of station-holding performance on a fishrobot in adverse unsteady flow", *International Conference on Intelligent Robots and Systems (IROS)*, Page(s): 928 – 933, 2010.

Towards a Sign Language Synthesizer: *a Bridge to Communication Gap of the Hearing/Speech Impaired Community*

H A Maarif[1], R Akmeliawati[2], T S Gunawan[3], A A Shafie[4]

[1,2,4] Department of Mechatronics Engineering, Faculty of Engineering,
International Islamic University Malaysia, Kuala Lumpur, Malaysia
[3] Department of Electical and Computer Engineering, Faculty of Engineering,
International Islamic University Malaysia, Kuala Lumpur, Malaysia

E-mail: alqodri.maarif@gmail.com

Abstract. Sign language synthesizer is a method to visualize the sign language movement from the spoken language. The sign language (SL) is one of means used by HSI people to communicate to normal people. But, unfortunately the number of people, including the HSI people, who are familiar with sign language is very limited. These cause difficulties in the communication between the normal people and the HSI people. The sign language is not only hand movement but also the face expression. Those two elements have complimentary aspect each other. The hand movement will show the meaning of each signing and the face expression will show the emotion of a person. Generally, Sign language synthesizer will recognize the spoken language by using speech recognition, the grammatical process will involve context free grammar, and 3D synthesizer will take part by involving recorded avatar. This paper will analyze and compare the existing techniques of developing a sign language synthesizer, which leads to IIUM Sign Language Synthesizer.

1. Introduction

To enhance the technology that helps the HSI people, the attempt to provide technology that can result sign language synthesizer is highly required. This due to the need of communication protocol which more mobile and flexible. The HSI people are necessary to be treated as an important group in community. Therefore, a sign language synthesizer is handy solution to this communication gap.

The main idea of SL synthesizer depends on two main components, i.e. speech recognition and signing avatar. Speech recognition is analysed by two common algorithms, i.e. lips reading and voice recognition. Combination between two is an important part of the SL synthesizer. The exact translation from spoken language will be required to understand the meaning of the words. The following process will be the signing avatar. The detail of both main components will be explained in this section. The general audio-visual speech recognition is divided into two parallel stages: the lips reading and the voice recognition.

[1] To whom any correspondence should be addressed.

IOP Conf. Series: Materials Science and Engineering **53** (2013) 012072 doi:10.1088/1757-899X/53/1/012072

The basic idea of signing avatar is based on the recorded signing avatar (Verlinden, Tijsseling, & Frowein, 2002). The signing avatar uses synthetic signing (Elliott, Glauert, Kennaway, & Marshall, 2000; Kennaway, 2002), and recorded signing (Bangham, Cox, Lincoln, Marshall, Tutt, & Wells, 2000).

Signing avatar will involve motion capture or motion vector recording process which records motion information for signing avatar. Motion capture process must involve the sensor device which can send information about position and orientation of each joint in human body. The motion vector information will be saved into animation file format after recorded by using camera or particular sensor. The recorded motion vector (motion capture file) will be transferred into 3D framework synthesizer. 3D framework synthesizer will process the recorded motion capture file into designated animation for signing avatar. Loading and retargeting stage will involve in designing signing avatar ("Mocap: Load And Retarget BVH," 2012).

2. Formatting the title, authors and affiliations

The main idea of SL synthesizer depends on two main components, i.e. speech recognition and signing avatar. Speech recognition is analysed by two common algorithms, i.e. lips reading and voice recognition. Combination between two is an important part of the SL synthesizer. The exact translation from spoken language will be required to understand the meaning of the words. The following process will be the signing avatar. Figure 1 shows the general process of SL synthesizer which involves speech recognition and signing avatar.

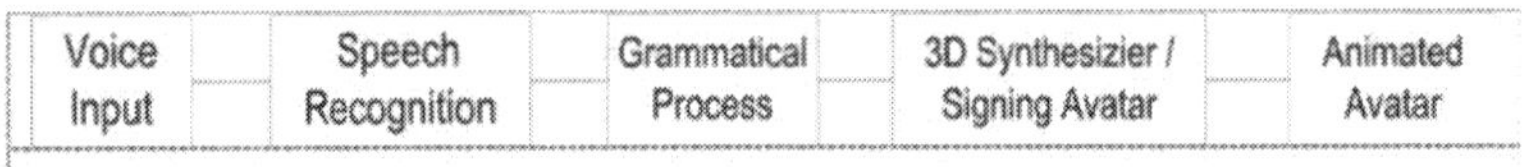

Figure 1. General Procedure of Sign Language Synthesizer

Once the voice has been recognized, the recognized voice will be processed and translated into SL in the form of signing avatar. The animation process will involve the 3D animation which resembles the human motion. This process will analyse the movement as the human body, for example the number of joints and the face expression.

2.1. Speech Recognition

Speech recognition is designed to translate voice into text. The research of speech recognition is conducted by applying audio recognition and video speech information process. From the oldest research [1] until the recent one [2, 3], showed that the audio-video speech recognition system is efficient in recognizing the voice from the spoken language.

The general audio-visual speech recognition is divided into two parallel stages: the lips reading and the voice recognition. Figure 2 shows flow of the audio-visual speech recognition. In this system, voice and image are divided, and then they are processed separately. After passing all processes in each part, integration block combines result of two inputs, and it is finally processed into recognition part, in this diagram; search block is the final process.

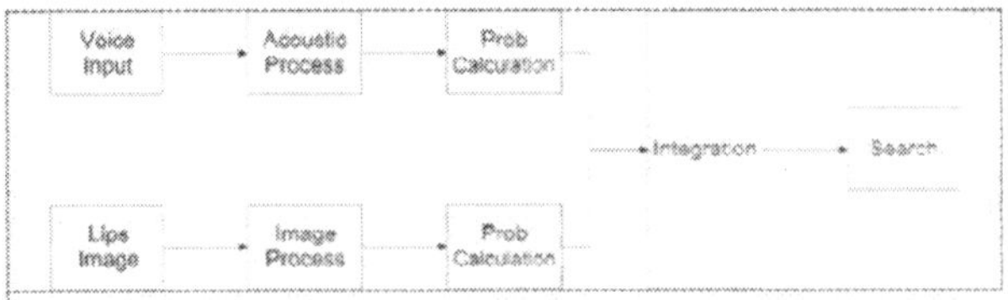

Figure 2. The flow of audio-visual speech recognition

Audio-visual integration solutions can be classified in three different groups: early integration, late integration and finally hybrid integration [4]. In all of these approaches the Hidden Markov Models (HMM) can be used to perform the recognition. In hybrid integration the state emission probabilities from the HMM theory [5] are evaluated independently for the visual and the acoustic channel. However, the Viterbi decoding is performed only once. The integration is made on the emission probabilities level. This kind of feature combination is known as multistream [6].

2.2. Signing Avatar

Signing avatar is based on the recorded signing avatar [7]. It described about the work involved in the realization of a Virtual Human signing the weather forecast on an Internet page. It has an objective to achieve semi-automatic translation from text to SL. With the use of a computer, sentences in a written language are analysed and transposed to a sequence of signs. Then the signs are displayed using computer animation of an avatar.

The signing avatar uses synthetic signing [8, 9], and recorded signing [10]. The general process of signing avatar is the conversion from text to SL; it involves the capturing and animation of signs, and publishes the avatar in a web page. Figure 2 shows the example of signer with motion tracking device.

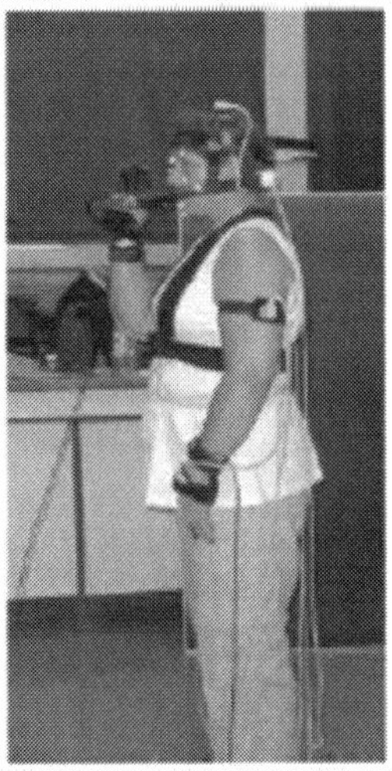

Figure 3. Example of Signer with Motion Tracking Device. [7]

The possible sign in signing avatar recorded [7]. The recording process utilizes the input devices: an optical system for tracking facial expression, a magnetic body suit for posture, and data gloves for hand and finger shapes. With these devices on the body, the signer makes the signs of the word and sentences. The equipment was originally developed for capturing global movements like running or dancing.

For signing, the quality requirements of the capturing are much higher than for previous applications of these techniques. In SL, even the slightest difference in speed, direction or gesture has a bearing upon meaning. But with the latest version, motions, posture, hand shapes and facial expressions are captured at very high resolution. The signs can be easily recognized when the recorded motion data are visualized on the computer. For visualization an avatar program is fed with the recorded data from the three tracking systems. The avatar then makes the same movements as the live person made with the equipment on.

2.2.1. Motion Capture

Signing avatar will involve motion capture or motion vector recording process which records motion information for signing avatar. Motion capture process must involve the sensor device which can send information about position and orientation of each joint in human body. The motion vector information will be saved into animation file format after recorded by using camera or particular sensor.

One of file formats which is widely used in motion capture is .BVH, Biovision Hierarchy. The Biovision Hierarchy (BVH) character animation file format was developed by Biovision, a defunct motion capture services company, to give motion capture data to customers. This format largely displaced an earlier format Biovision providing skeleton hierarchy information as well as motion data [11]

The other motion capture file formats are also possible to be involved in 3D synthesizer. The variety of motion capture file format is based on the development of gesture controllers, haptic systems, motion capture systems, etc., on the one hand, and the need of allowing virtual reality systems to inter-communicate through control data, the question of gesture and motion takes more and more importance.

Motion and gesture file formats are widely used today in many applications that deal with motion and gesture signal. It is the case in domains like motion capture, character animation, gesture analysis, biomechanics, musical gesture interfaces, virtual surgery. Those formats are low-level formats, i.e. formats close to the signal produced by the capture system. A detail of motion capture file format is shown in Table 1.

The recorded motion vector (motion capture file) will be transferred into 3D framework synthesizer. 3D framework synthesizer will process the recorded motion capture file into designated animation for signing avatar. Loading and retargeting stage will involve in designing signing avatar [12]. There are couple software's which can be utilized to process the motion capture files. The open source and commercial software's are available. For instance, 3D Max and Blender represent commercial and open source software, respectively [13].

Table 1. Motion Capture File Format

No	File Format	Developer	Description
1	BVA and BVH	Biovision	Standard representation of movements in the animation of humanoid structures. Widely adopted by the animation community
2	MNM	Autodesk 3D	Renaming the segments of a BVH file.
3	ASK/SDK	Biovision	Variant of the BVH file format. ASK Contain information concerning the skeleton and does not contain any information about the channels of the movement. SDK contain other information concerning the scene.
4	AOA	Adaptive Optics	Describes the captors and their position at each sampling period.
5	ASF/AMC	Acclaim	ASF contains the skeleton description. AMC contains the movement data.
6	BRD	LambSoft	Stocking the data coming from a magnetic motion capture system.
7	HTR and GTR	Motion Analysis	HTR is native format for skeleton. GTR is similar to HTR but less structural element.
8	TRC	Motion Analysis	Raw data from the full body motion capture system.

			Output data coming from face tracker
9	CSM	Character Studio	Importing marker data
10	V/VSK	Vicon Motion System	Marker data. Global segment translation and rotation data. Local rotation data (with root translation data).

3. The Available Sign Language Synthesizer

The project of generating SL in term of 3D animation has been done. There are some projects of SL synthesizer which can create the sign animation from the text [14]. The available SL synthesizer is applied in Italy for Italian Sign Language [15], Australia for Australian Sign Language-AUSLAN [16], United States of America for American Sign Language [17].

Each of the available systems provides a method to perform 3D animation process. ATLAS for Italian SL uses three main components to perform 3D animation, i.e. The Planner, Executor, and Animation engine. Each of the components provides support to show the animation process. Specifically, in Animation engine, it relies on stored data repository which was recorded by using Motion Capture or hand animated. The comparison of the available SL synthesizer system is described in Table 2.

AUSLAN for Australian SL uses Human Modelling System. It involves a parametric human animation technique to provide 3D animation of human body. It consists of Human Modeling Module which provides for input and output of XML definition describing the hierarchical object being modeled, the Rendering Module displays the position of joints of the above kinematic tree graphical, while the while the Model Interpolation Module provides for input, output and high-level control of animations between a set of model key frames [18].

American Sign Language specified the SL synthesizer by allowing animation to adapt visual or gesture language. The adaptation is applied by using hand shape, hand position, palm orientation, and non-manual signal. The system utilizes motion capture which captures the movement of humans in real-time and the data would be saved in order to be used on the 3D animation.

Table 2. **A Comparison of Available SL Synthesizer**

No	Name	Mechanism	Output	Advantages	Limitation	Number Database
1	ATLAS [15]	• AEWLIS structure for linguistic input • Animation track for animation system	Signing Avatar	• Real time signing • Natural Sign	• Only for Italian • Not mobile	Limited for Italian Sign Language
2	Auslan Tuition System [16]	• Human Modeling System • Use Open GL to render	Signing Avatar	• User defined input • Able to update the database • Has two components : Tutorial and Sign Editor Program	• Only for tutorial • Not able to translate in real time • Commercially released • Limited database, based on package	Limited Database Number, based on the released version

3	DePaul ASL Synthesizer [19]	• Motion Capture • Retargeting • Wrinkle Simulation	Signing Avatar	• Real time signing • Natural sign • Face expression	• Not mobile • Time delay due to huge processing stage	N/A (On going project)
4	DictaSign [20]	• Use SiGML • Linguistic Modeling • AnCoLin Annotation Tools • Sign Language Corpora	Signing Avatar	• Multi Sign Language • Natural signing • Non manual signing	• Not mobile	1500 per SL
5	SASL-MT [21]	• Graphical signing avatar • Linguistics aspects • Data building and data entry tools • H-Anim standard	Signing Avatar	• Pluggable Avatar • Reusable Avatar	• Not yet deployed	N/A (On going project)
6	SiSi – IBM [22]	Voice to Sign Translation	Signing Avatar	• Real time translation	• Not mobile • Only for BSL • Commercially released	Unlimited

4. Conclusion

The 3D animation used for Sign Language synthesizer involves motion capture or recorder avatar. It is applied in order to decrease time consuming and complexity. The achievement by proposed systems have shown the usage of motion capture or recorded avatar. It has been a part of 3D animation system for sign language synthesizer.

Motion capture method avatar has shown the benefits of the systems. Some of the advantages of using motion capture are ranging from real time results until less of the amount of work. Real time results can be obtained, since the recorded motion is similar to the real motion from the model. Less amount of work is always obtained with the complexity or length of the performance to the same degree when using traditional techniques.

Unfortunately, motion capture technique is limited into some disadvantages. Firstly, it uses specific hardware and software which are necessary to be owned with some high cost. It can limit into small 3D animation production. The capture system may have specific requirements for the space it is operated in. When problems occur it is sometime easier to reshoot the scene rather than trying to manipulate the data. Only a few systems allow real time viewing of the data to decide if the take needs to be redone. Applying motion to quadruped characters can be difficult. Based on advantages of motion capture, utilization of motion capture method will be more required in the future.

5. References

[1] Petajan, E.D., et al. *An Improved Automatic Lipreading System to Enhance Speech Recognition.* in *Human Factors in Computing System.* 1988: ACM.

[2] Daubias, P., *Modèles A Posteriori De La Forme Et De L'apparence Des Lèvres Pour La Reconnaissance Automatique De La Parole Audiovisuelle.* 2002, l'Université de Maine France.

[3] Goecke, R., *A Stereo Vision Lip Tracking Algorithm and Subsequent Statistical Analyses of the Audio-Video Correlation in Australian English*, in *School of Information Sciences and Engineering.* 2004, The Australian National University Canberra.

[4] Potamianos, G., C.N.G. Gravier, and A. Garg, *Recent advances in the Automatic Recognition of Audio-Visual Speech.* IEEE, 2003. **91**(9).

[5] Rabiner, L.R., *Tutorial on Hidden Markov Models and Selected Applications in Speech Recognition.* IEEE, 1989. **77**(2): p. 257-286.

[6] Dupont, S. and J. Luettin, *Audio-Visual Speech Modeling for Continuous Speech Recognition.* IEEE Transaction on Multimedia, 2000. **2**(3).

[7] Verlinden, M., C. Tijsseling, and H. Frowein, *A Signing Avatar on the WWW*, in *Lecture Notes in Artificial Intelligence.* 2002. p. 169-172.

[8] Elliott, R., et al. *The Development of Language Processing Support for the ViSiCAST Project.* in *4th International ACM SIGCAPH Conference on Assistive Technologies.* 2000. Washington.

[9] Kennaway, R., *Synthetic Animation of Deaf Signing Gestures*, in *Lecture Notes in Artificial Intelligence.* 2002. p. 146-157.

[10] Bangham, J.A., et al. *Signing for the Deaf using Virtual Humans.* in *IEEE Seminar on Speech and Language Processing for Disabled and Elderly People.* 2000. London.

[11] *Biovision Hierarchy.* 2012 [cited 2012 12 December 2012]; Available from: http://en.wikipedia.org/wiki/Biovision_Hierarchy.

[12] *Mocap: Load And Retarget BVH.* 2012 [cited 2012 17 December]; Available from: http://www.makehuman.org/node/286.

[13] *Animation Export and Motion Transfer.* 2012 [cited 2012 14 December]; Available from: http://wiki.ipisoft.com/index.php?title=Animation_export_and_motion_transfer.

[14] Pyfers, L. *Open Sign* 2011; Available from: http://www.opensign.org/.

[15] Atlas. *ATLAS - Automatic Translation into Sign Language.* 2011 [cited 2012 23 March]; Available from: http://www.atlas.polito.it/index.php.

[16] Wong, J.C. *The Auslan Tuition System.* 2004 25 March [cited 2012 22 March]; Available from: http://auslantuition.csse.uwa.edu.au/index.html.

[17] Wolfe, R., J. McDonald, and J. Schnepp. *An Avatar to Depict Sign Language: Building from Reusable Hand Animation.* in *International Workshop on Sign Language Translation and Avatar Technology (SLTAT)* 2011. Federal Ministry of Labour and Social Affairs, Berlin, Germany.

[18] Yeates, S., E.-J. Holden, and R. Owens, *An Animated Auslan Tuition System.* International Journal of Machine Graphics and Vision, 2003. **12**(2): p. 203-214.

[19] *American Sign Language.* 2012 [cited 2012 23 March]; Available from: http://asl.cs.depaul.edu/.

[20] Efthimiou, E. *DictaSign.* 2009 [cited 2012 23 March]; Available from: http://www.dictasign.eu.

[21] Zijl, L.v. and G. Olivrin. *South African Sign Language Assistive Translation.* in *IASTED International Conference on Assistive Technologies.* 2008. Baltimore, USA.

[22] Murph, D. *IBM's SiSi virtually translates speech to sign language.* 2007 [cited 2011 2 November]; Available from: http://www.engadget.com/2007/09/13/ibms-sisi-virtually-translates-speech-to-sign-language/.

Effect of tuber skin on the thermal properties of whole tubers of potato and sweet potato

A A Oluwo[1], R M Khan and M J E Salami.
Mechatronics Engineering Department,
Kulliyyah of Engineering,
International Islamic University Malaysia (IIUM),
Gombak, Selangor, Malaysia.

E-mail: yinkostics2@yahoo.com

Abstract. Temperature-dependent thermal coefficients of mathematical models of the postharvest storage process play an important role in determining the models accuracy. Thermal properties of tubers under storage available in literature are generally of those in semi processed form (skinless) such as those having undergone peeling, dicing and cutting actions. This study investigates the effect of tuber skin on the thermal properties of whole tubers of potato and sweet potato. A direct approach was used to measure the tubers' density and thermal conductivity and thermal diffusivity by the transient heat transfer method. Indirect approach was used to measure the tubers' specific heat. Experimental data were used to develop empirical models of the thermal coefficients as a function of temperature. Results of the study should find great use in the modeling of potato and sweet potato storage process.

Keywords: thermal properties, whole tubers, potato, sweet potato, tuber skin.

1. Introduction

Design of food processing systems, such as postharvest storage and food drying systems involves the proper specification of heating and cooling components of the equipment. Mathematical models are often employed to have a better understanding of the heat and mass transfer processes. One prominent drawback in the model approach is the use of appropriate thermal coefficients. Search through literature shows some works carried out by [1] and [2]. They both provided predictive equations for thermal conductivity of food stuffs. Due to substantial differences in the chemical and physical constituents of the food item, there is a need for each material to have its own predictive equation [3]. The limitation is that Literature search also revealed that thermal coefficients of food stuff are often associated with foods in semi processed form. An example is the abundant data on thermal properties of cuts or sliced pieces of peeled yams, potatoes and sweet potatoes which find use in the design of cooking, frying, or microwaving equipment. However, the postharvest storage process is associated with whole (unpeeled) tubers of food items. Experimental determination of appropriate thermo-physical properties of whole tubers of foods for the storage process will definitely yield better models. Good models will lead to better understanding of the storage process, thus control algorithms for the storage process can be developed. Ultimately, shelf life of stored products will be improved and in turn result in availability of the products outside its season and ensure its price stability.

[1] To whom any correspondence should be addressed.

Attempt was made by [4] to model a potato storage process where due to lack of data on thermal properties of whole tubers of potato, that on mashed potatoes was managed. In determining some selected engineering properties of sweet potato as a function of temperature, [5] employed slab and cylindrical cuts of sweet potato. Investigations on the effects of variation in temperature and geometry on the thermal properties of pieces of yam in slab and cylindrical shapes were carried out by [6] . Experiments to determine the thermal properties of sweet potato and provide models as a function of temperature and moisture content were carried out by [7]. The tubers were peeled and then cut into cylindrical pieces. Measurements of thermal diffusivity of potato, malt bread and wheat flour were carried out by [8]. Potatoes were cooked, mashed and mixed to obtain homogenous products.

This study takes a look at the effect of the tuber skin on the thermal properties of whole tubers under storage with emphasis on potato and sweet potato. This study considered product temperature ranging from 8 to roughly 35°C.

2. Materials and methods

Potato (solanum tuberosum) and sweet potato (Ipomea batatas) were purchased from the Chow kit market, in Kuala Lumpur, Malaysia. Cleaning, involved a simple process of brushing off the soil content from the sample bodies using a soft brush. No form of washing whatsoever was done to any of the tubers. The tubers were left in the whole form as bought from the market.

2.2. *Experimental Procedure*

2.2.1. Density [ρ] of the Tubers. The density of each sample was computed using equation 1. An electronic weighing scale of make and model METTLER Toledo B204-S was employed to determine the mass of the samples in grams. The values were then converted to kilograms. The scale has an accuracy of +_0.0001g. Through the principles of flotation [9], a 1000 ml measuring cylinder initially containing 300 ml of water was used to determine the volume of the samples. The final volumes were noted, the differences calculated and converted to SI units (m^3). The mass (kg) and volume (m^3) data were then plugged into (1).

$$\rho = \frac{mass\ of\ sample(kg)}{volume\ of\ sample\ (m^3)} \tag{1}$$

2.2.2. Thermal Conductivity and Diffusivity [k, α] of the Yam Tuber. The thermal conductivity and diffusivity were measured using a KD2 thermal property analyzer (Decagon devices Inc., Pullman, USA). The device employs the transient method of thermal properties determination. Equation (2) illustrates the working principle of the KD2.

$$T - T_0 = \frac{q\left[\ln(t) - \gamma - \ln\left(\frac{r^2}{4\alpha}\right)\right]}{4\pi k} \tag{2}$$

Where T is final temperature (^{O}C); T_0 is initial temperature before heating (^{O}C); q is heat produced per unit length per unit tie (Wm^{-1}); k is the thermal conductivity of the medium (Wm^{-1O}C^{-1}); t is time (s); γ is Euler's constant (0.5772); r is radial distance from the probe to the point where temperature is being measured (m); and α is thermal diffusivity (m^{2}s^{-1}).

The probe from the KD2 was inserted to the depth of the axis of the sample, then placed inside a temperature/humidity chamber and allowed to equilibrate with the surrounding atmospheric (temperature and humidity) conditions. The temperature/humidity chamber was then preset to the required environmental temperature. After about 30 minutes (after which the sample must have equilibrated with the environment) the KD2 was switched on and after about 25 seconds displays the

IOP Conf. Series: Materials Science and Engineering **53** (2013) 012073 doi:10.1088/1757-899X/53/1/012073

product temperature and the computed thermal conductivity and diffusivity of the sample. Prior to the commencement of the experiment the KD2 was calibrated with an agar solution.

2.2.3. Specific Heat [c_p] of the Tubers. The indirect method (calculation) was used to determine the specific heat capacity -c_p- of the materials. Equation (3) represents the relation used to determine the specific heat capacity after rearranging to make c_p the subject of the formula.

$$c_p = \frac{k}{\rho\alpha} \quad (3)$$

where k is the thermal conductivity ($Wm^{-1}K^{-1}$) of the material.
ρ is density (kg/m^3) of the material.
α is the thermal diffusivity (mm^2/s) of the material.

3. Results and discussion

On inspection of the samples it was discovered there was little variation in terms of skin color and texture and they were free of any skin abrasion. On the basis of these observations one could primarily conclude that the tubers were most likely of the same harvest. Since the sources of both tubers were a fresh vegetables market, estimates of moisture content were made for each material. A moisture content of about 75% was assumed for the potato tubers given the fact that fresh off the farm potato tubers usually have a moisture content of about 80% as stated by [4]. The reason here is that about 5% of the moisture content of the potatoes is assumed to have been lost during transit and initial storage. For the sweet potatoes, a moisture content of about 72% (wet basis) was made by [5].

3.1 Density (ρ)

The density of the materials was determined as earlier described and the results are presented in tables 1and 2. On the average the density of potato was found to be 1072 kg/m^3 with a standard deviation value of ±14.255 kg/m^3, while that of sweet potato was found to be 978 kg/m^3 with a standard deviation of ± 10.840. The results here are in agreement with those of literature [5] and [7].

3.2 Thermal conductivity (k)

The thermal properties for both potato and sweet potato were measured through temperature ranges as shown in tables 3 and 4 respectively. Specifically, that of potato spanned 11.7 to 34.6°C as shown in table 2. Figure 1 shows that the thermal conductivity of whole tubers of potato ranged between values of 0.42 to 0.61W/m°C within the range of study. Between 11.7°C and 29.8°C the thermal conductivity dips from a value of 0.45 W/m°C (11.7°C) to 0.52 W/m°C (25.5°C) then rises to 0.61W/m°C (30°C). The thermal conductivity drops to a value of 0.42 W/m°C at 34.6°C. Though the trend shown here is relatively similar to that observed by both [5] and [6] in their study of thermal properties of cuts of sweet potato and yam respectively, the values in this experiment are about twice theirs'. Equation (4) represents an empirical model of thermal conductivity of potato as a function of temperature.

Table 1. Experimental values of the density of potato.

Sample	Mass (g)	Volume (ml)	Density (kg/m^3)
A	75.1200	70.00	1073
B	121.1462	115.00	1054
C	104.0830	98.00	1062
D	99.3450	92.00	1080
E	79.5485	73.00	1090
Avg			1072 ± 14.255

Table 2. Experimental values of the density of sweet potato

Sample	Mass (g)	Volume (ml)	Density (kg/m^3)
A	135.0563	139.00	972
B	166.8247	170.00	981
C	154.3932	160.00	965
D	99.7649	102.00	978
E	91.4687	92.00	994
Avg			978 ± 10.840

Table 3. Thermal properties of potato as determined by the KD2 Thermal Analyzer and calculated.

	Experimental values					Calculated
S/N	Chamber Temperature (^oC)	Chamber Relative Humidity (%)	Tuber Temperature (^oC)	Thermal Conductivity (W/m^{o}C)	Thermal Diffusivity (x10^{-8}m^2/s)	Specific Heat Capacity (kJ/kg.K)
1	11.4	60.2	11.7	0.45	11.0	3.817
2	18.8	60.0	19.2	0.43	10.0	4.013
3	25.0	60.0	25.5	0.52	12.0	4.044
4	30.2	60.0	29.8	0.61	16.0	3.558
5	35.0	60.0	34.6	0.42	10.0	3.919

Table 4. Thermal properties of sweet potato determined by KD2 Thermal analyzer and calculated.

	Experimental values					Calculated
S/N	Chamber Temperature (^oC)	Chamber Relative Humidity (%)	Tuber Temperature (^oC)	Thermal Conductivity (W/m^{o}C)	Thermal Diffusivity (x10^{-8}m^2/s)	Specific Heat Capacity (kJ/kg.K)
1	7.7	80.0	8.0	0.46	11.0	4.280
2	9.8	80.0	10.2	0.44	10.0	4.504
3	13.8	80.0	14.4	0.50	12.0	4.265
4	18.0	80.0	18.7	0.42	10.0	4.299
5	30.0	80.0	29.7	0.42	10.0	4.299

Figure 2 shows that the thermal conductivity values for sweet potato in this experiment proved also higher than those of [5]. The value remained relatively constant at about 0.44 W/m^{o}C between 8 and 30°C only for a jump to 0.50 W/m^{o}C at about 14.4^{o}C.

Based on the experimental values of k, (4) and (5) represent the empirical models describing thermal conductivity as a function of temperature for the potato and sweet potato tubers respectively.

$$k_p = 0.010T^2 - 0.0225T + 1.854 \qquad (R^2 = 0.938) \qquad (4)$$

$$k_{sp} = 8e-05T^3 + 0.004T^2 + 0.075T + 0.089 \qquad (R^2 = 0.515) \qquad (5)$$

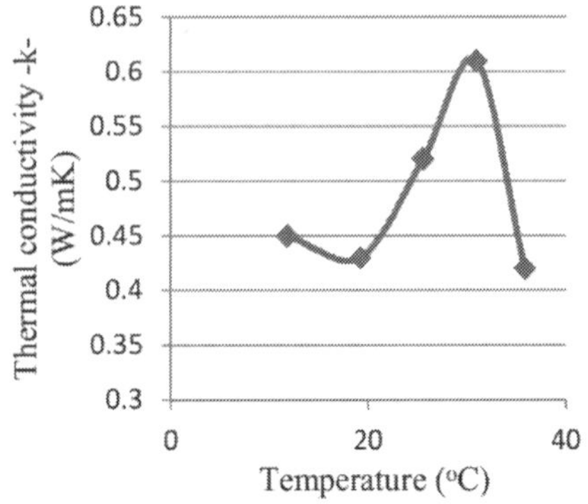

Figure 1. Plot of potato thermal conductivity (k) as a function of temperature.

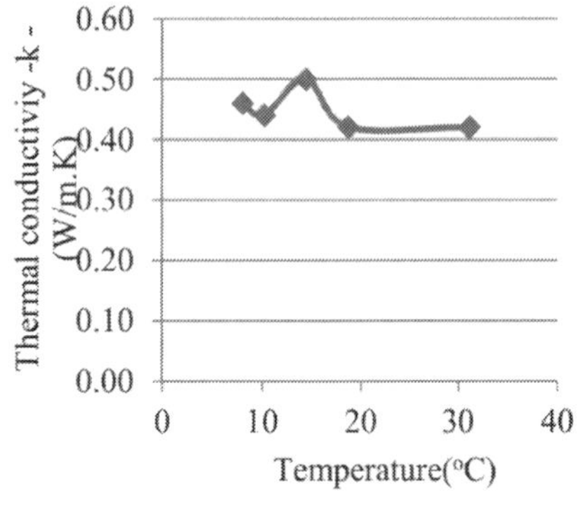

Figure 2. Plot of sweet potato thermal conductivity (k) as a function of temperature.

3.3 Thermal diffusivity (α)

Figure 3 shows that the thermal diffusivity value for potato remains relatively constant at a value of 12.0×10^{-8} through the temperatures of 11.7 to 34.6^{O}C. These values are slightly higher compared to those of [5] and [6] who both recorded on average, values of about $7.0 \times 10^{-8} m^2/s$ for cuts of both sweet potato and yam.

The thermal diffusivities for sweet potato throughout the considered temperature range remained largely at 10.5×10^{-8} m^2/s. This is illustrated in figure 4. In line with that reported for potato, the results are higher than those reported by [5].

Equations (6) and (7) are empirical models describing the thermal diffusivity as a function of temperature for potato and sweet potato respectively.

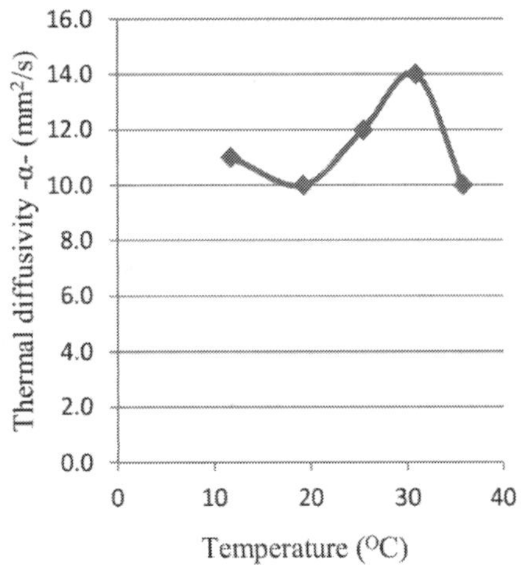

Figure 3. Plot of potato thermal diffusivity (α) as a function of temperature.

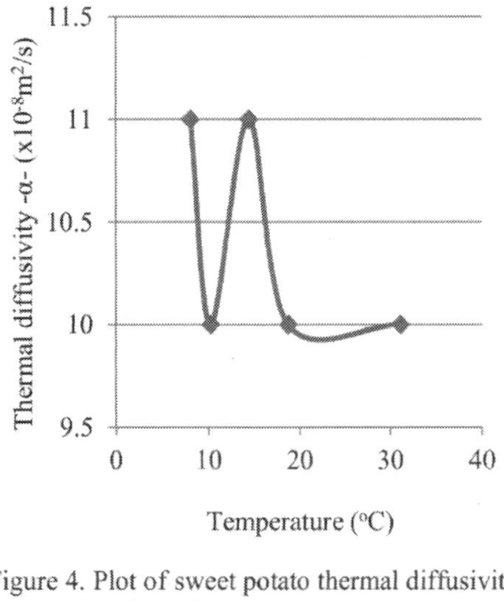

Figure 4. Plot of sweet potato thermal diffusivity (α) as a function of temperature.

$$\alpha_{\text{p}} = -0.003T^3 + 0.251T^2 - 5.291T + 44.37 \quad (R^2 = 0.942) \quad (6)$$
$$\alpha_{\text{sp}} = -0.003T^3 + 1.374T^2 - 13.80T + 59.13 \quad (R^2 = 1.0) \quad (7)$$

3.4 Specific heat (c_p)

The specific heat was determined for both materials using the indirect method. The specific heat for both materials was calculated using (3). Figure 5 indicates that the specific heat of potato rises from a value of 3.817 kJ/kg.K at 11.7°C to 4.044 kJ/kg.K at 25.5°C before dropping to a value of 3.919 kJ/kg.K at 34.6°C. The results are not in agreement with those of [5] and [6] which show a steady and gradual drop in the specific heat for both sweet potato and yam from values of 2.5 kJ/kg.K to about 2.0 kJ/kg.K over the temperature range.

The sweet potato also shows a similar trend to that of the potato with the specific heat peaking at a value of 4.504 kJ/kg.K at a temperature of 10.2°C. This is illustrated in figure 6.

The specific heats in our case also proved to be higher than those of [5] and [6] as was the cases of thermal conductivity and diffusivity.

The empirical models of specific heat as a function of temperature for both materials are represented by (8) and (9).

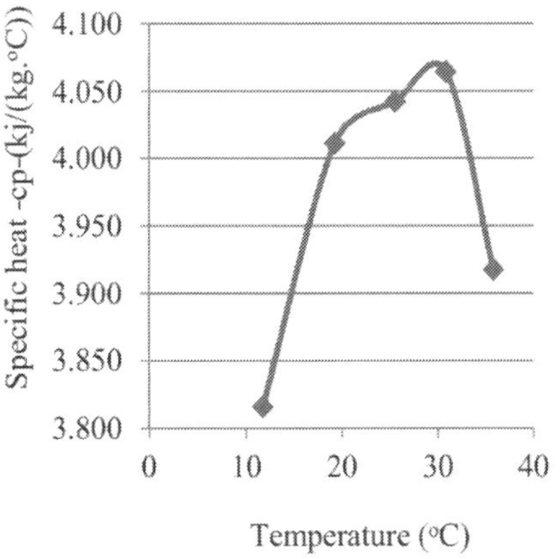

Figure 5. Plot of potato specific heat (c_p) as a function of temperature.

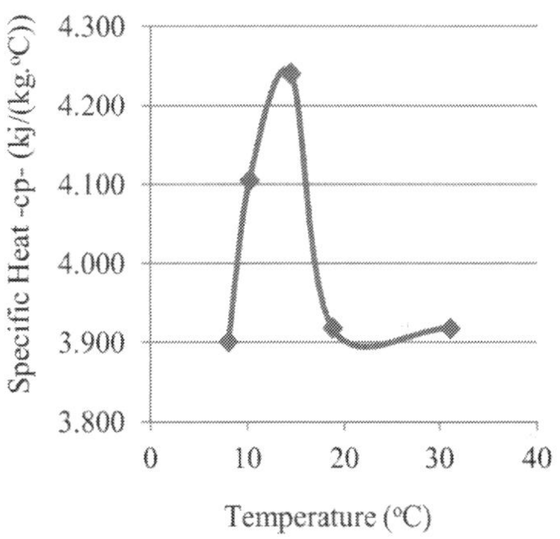

Figure 6. Plot of sweet potato specific heat (c_p) as a function of temperature.

$$c_{\text{pp}} = -3e^{-05}T^3 + 0.025T + 3.487 \quad (R^2 = 0.966) \quad (8)$$
$$c_{\text{psp}} = -0.036T^3 + 0.627T + 0.883 \quad (R^2 = 0.966) \quad (9)$$

4. Conclusion

The effect of the tuber skin on thermal properties of potato and sweet potato over a range of temperatures was studied. Temperature was varied over a range of 8°C to 34.6°C. Two of the thermo-physical properties were measured directly while a third was measured indirectly.

The values of all three thermal properties namely thermal conductivity, diffusivity and specific heat were observed to be higher than those found in literature. The density of the materials, a physical property was found to be in agreement with those of literature.

The high values of the experimental results prove that the tuber skins have an increased effect on the thermal properties. It would seem that the tuber skin create a sort of thermal drag on the food components in whole form. In comparison to tubers in full or semi processed forms, whole tubers will require more energy for processing especially as demanded by a storage (cooling) processes. Variation of the thermal properties was best described using empirical models in the form of polynomial equations of the third order.

5. References

[1] Miles C A, van Beck G and Veerkamp C H 1986 Thermal properties of food *In Engineering properties of foods* ed. Rao M A et al (New York) p 269

[2] Sweat V E 1986 Thermal properties of foods *Engineering Properties of Foods* ed. Rao et al New York p 49

[3] Mohsenin N N 1980 *Thermal Properties of Foods and other Agricultural Materials* CRC PressI Llc

[4] Xu Y and Burfoot D 1999 Simulating the bulk storage of foodstuff *Journal of Food Engineering* **39**(1) pp 23-29

[5] Oke M O, Awonorin S O, Sanni L O, Akanbi C T and Abioye A O 2007 Determination of some selected engineering properties of sweet potato cuts as function of the temperature *J. Food Technol.* **5** (1) p 66-70

[6] Oke M O, Awonorin S O, Oyelade O J, Olajide J O, Ogunlakin G O and Sobukola P O 2009 Some thermo-physical properties of yam cuts of two geometries *African Journal of Biotechnology* **8** (7)

[7] Farinu A and Baik O D 2007 Thermal properties of sweet potato with its moisture content and temperature *International Journal of Food Properties* **10** (4) pp 703-19

[8] Magee T and Bransburg T 1995 Measurement of thermal diffusivity of potato, malt bread and wheat flour *Journal of Food Engineering* **25** (2) pp 223-32

[9] Nwanekezi E and Ukagu J 1999 Determination of engineering properties of some Nigerian fruits and vegetables *Nigeria Food J.* **17** pp 55-60

IOP Conf. Series: Materials Science and Engineering 53 (2013) 012074 doi:10.1088/1757-899X/53/1/012074

Towards Port-Hamiltonian Approach for Modeling and Control of Two-wheeled Wheelchair

A Aula, R Akmeliawati[1], S Ahmad, T M Altalmas, and S N Sidek

Intelligent Mechatronics System Research Unit, Dept. of Mechatronics Engineering International Islamic University Malaysia, Jalan Gombak, 50728 Kuala Lumpur, Malaysia

Email: rakmelia@iium.edu.my

Abstract. This paper introduces the modeling and control design of a two-wheeled wheelchair (TWW) based on structure-preserving port-Hamiltonian concept. In this paper, a model of TWW with features, including space-saving, four to two-wheel transformation, and adjustable seat height is proposed to increased mobility and independence of the user. Then, the mathematical model of a TWW in its balanced mode is derived. The model is based on the total energy in the system. The system is divided into subsystems whereby the interconnections which exist are utilized. The nonlinearity of the model is preserved using port-controlled Hamiltonian (PCH) system and made to advantage. The proposed controlled is designed based on the idea of PCH such that the energy balance in the system can be achieved while stabilizing the system.

1. Introduction

Many researches were conducted to find a better solution for helping the disabled people with different types of disabilities to be independent in their daily activities. Most of these researches on electric powered wheelchair were aimed to increase the comfort, handling and safety of the wheelchair [2-5]. Control algorithms were also embedded to improve its usability. However, existing mechanical designs and control algorithms of electric powered wheelchair do not give the users the full ability to act independently in life.

Though there are few wheelchair models that can adjust the seat to a certain height, only fewer of them that are actually operable in two wheels configuration. This feature increases user's mobility by allowing them to maneuver in narrow spaces. This advantage was already recognized in JOE [6], a two-wheeled mobile robot, which is regarded as a pioneer in this field. This two-wheeled mobile robot, along with other similar robots [7] triggered the research on self-balancing robot, which is inspired by an inverted pendulum concept. These researches were then improved further for wheelchair application. Hence, the dawn of research on two-wheeled wheelchair.

Nakamura and Murakami [8-10] proposed a wheelchair on two wheels. They implemented the so-called disturbance observer (PADO) to balance and steer the wheelchair. A PADO was developed to estimate the disturbance on pitch angle to achieve a robust stabilization. Later, they added a reaction torque observer (RTOB) to realize a power-assist control on yaw direction. In [11-12], a new mechanism that transforms the wheelchair from four wheels to two wheels configuration was

[1] Corresponding author.

proposed. Furthermore, they added an extendable feature to the second link which lifts the user to an eye-level height with a standing person. This feature increases user's confident while having a conversation.

In the last decade, Port-controlled Hamiltonian (PCH) systems have emerged as an interesting class of nonlinear models suitable for a large number of physical applications. This modeling approach originates from the network modeling of energy-conserving lumped-parameter physical systems with independent storage elements. This kind of models encompasses a very large class of physical systems, containing the class of Euler-Lagrange models. The Hamiltonian function, used in this approach, is a good candidate of Lyapunov functions for many physical systems [13]. The name PCH systems refers to two major components of a control paradigm [14]:

- *Port*: the modeling approach is port-based, which successfully composes complex systems by means of *power*-preserving interconnections.
- *Hamiltonian*: the mathematical framework extends the Hamiltonian mechanics, which emphasizes *energy* function Hamiltonian as basic concept for modeling multi-physics system.

Latest applications of PCH systems include various types of control problem, structure-preserved modeling of nonlinear system, and stabilization technique, e.g. [15-19].

In this paper, an electric-powered TWW is modeled using PCH systems. The proposed TWW has a transformable wheel configuration, which turns the four-wheeled wheelchair into a two-wheeled wheelchair. This two-wheeled configuration has similar dynamics to a double-inverted pendulum on two wheels, which is a nonlinear, multi-variable, higher order, and unstable system. The PCH systems approach preserves the nonlinear structure of the system, hence preserving also its nonlinear dynamics [20].

This paper is organized as follows. Section 2 summarizes the concept of Port-controlled Hamiltonian Systems. After developing the simplified model of TWW using MapleSim, the mathematical model is derived in section 3. Section 4 proposes the steps to design a controller based on PCH systems.

2. Port-controlled Hamiltonian Systems

2.1. The Concept of Energy

A dynamical system can be viewed as a set of simpler subsystems that exchange energy, (from theory of passivity, which is compatible with 'interconnection theory' [21]). Energy can serve as a *lingua franca* to facilitate communication among scientists and engineers from different fields. Most engineering applications are mixtures of electrical, mechanical and other domains. In PCH systems-based modeling; an energy-based perspective and the role of interconnection between subsystems provide the basis in modeling physical systems.

Port-controlled Hamiltonian system—sometimes referred also as port-Hamiltonian systems or generalized port-controlled Hamiltonian systems—is first introduced by A.J. van der Schaft [20]. This technique extends the theory of Hamiltonian mechanics. In this technique, the nonlinear dynamics of a system is modeled as the Hamiltonian equations of motion, which are derived from the total energy of the system H, also known as the Hamiltonian.

$$H(q,p)=\frac{1}{2}p^T M^{-1}(q)p+P(q)\ (=\frac{1}{2}\dot{q}^T M(\dot{q})q+P(q)) \tag{2.1}$$

where q is the generalized coordinates, p is the generalized momenta, $p=M(q)\dot{q}$, M is the k x k inertia (or generalized mass) matrix which is symmetric and positive definite for all q.

In PCH systems, any engineering domain (electrical, mechanical or hydraulical, with exception to thermodynamical) is introduced two variables, called *power conjugate variables*, whose product equals power. These variables are labeled as *effort* $\boldsymbol{e}$ and *flow* $\boldsymbol{f}$. For any given system (from any domain), there exists a storage and dissipative element. Within each element, a power port, with $P = e \cdot f$, is then defined. The subdomain classification in PCH framework is based on the Generalized Bond Graph (GBG) framework introduced in [22-23].

When modeling a physical system using this approach, it is viewed as an interconnection between energy storage elements, dissipative (resistive) elements, and the environment. The structure of any storage element is the following:

$$\dot{z}(t) = f_s(t), \quad e_s(t) = \frac{\mathrm{d}E}{\mathrm{d}z}(z(t)), \tag{2.2}$$

where $f_s(t)$ is the flow variable of storage element, $e_s(t)$ is the effort variable of storage element, z is the generalized state variable and $E(z)$ is the stored energy. Figure 1 illustrates this relation.

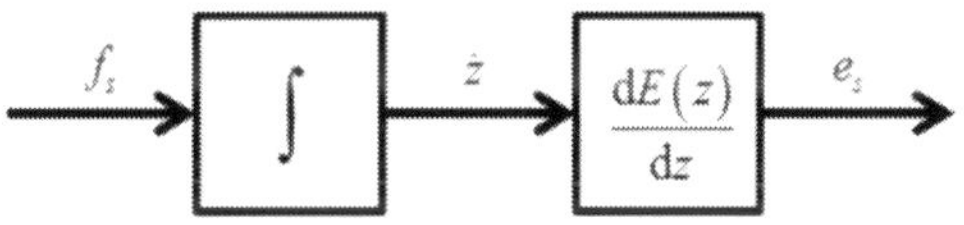

Figure 1. The relationship between effort and flow variables in a storage element.

The change in energy of a storage element is always its external power flow, i.e.

$$\dot{E}(z) = \frac{\mathrm{d}E}{\mathrm{d}z}(z(t))\dot{z} = e_s f_s. \tag{2.3}$$

Thus, by construction, a product of effort and flow implies the integral of power with regards to time yields energy.

For a dissipative element, the energy dissipation $D(x)$ is expressed by direct relation between effort and flow variables. In the linear case,

$$e_R = Rf_R \Rightarrow P_{diss} = e_R f_R = Rf_R^{\,2} \geq 0, \quad R \geq 0 \tag{2.4}$$

where $f_R(t)$ and $e_R(t)$ is the flow and effort variables of dissipative (resistive) element, respectively.In general, the dissipative power $P_{\text{diss}} \geq 0$, otherwise the element would generate energy instead of dissipating it. For the general nonlinear case, the relation can be expressed in the flow-controlled form (equation 2.5) or effort-controlled form (equation 2.6)

$$e_R = \hat{e}_R(f_R), \tag{2.5}$$
$$f_R = \hat{f}_R(e_R). \tag{2.6}$$

The power and energy balance of PCH system with storage and dissipative element are, respectively:

$$e_S^T f_S + e_R^T f_R = 0, \tag{2.7}$$
$$\frac{\mathrm{d}}{\mathrm{dt}} H = e_S^T f_S + e_R^T f_R. \tag{2.8}$$

The Hamiltonian equations of motion satisfies the general energy balance equation,

$$\frac{\mathrm{d}}{\mathrm{d}t}H = \frac{\partial^T H}{\partial q}(q,p)\dot{q} + \frac{\partial^T H}{\partial p}(q,p)\dot{p} = \frac{\partial^T H}{\partial p}(q,p)\tau = \dot{q}^T\tau \tag{2.9}$$

from which the controller will be designed. The designed controller will ensure the stability (with preserving structure of the Hamiltonian system and conservation of energy) of the system in the Lyapunov sense [23].

2.2. PCH-based Modeling

Consider the following nonlinear system in the form

$$\begin{aligned} \dot{z} &= f(z) + g(z)u, \\ y &= h(z). \end{aligned} \tag{2.10}$$

where $f(z)$, $g(z)$, and $h(z)$ are nonlinear functions of the state z, u is the input of the system and y is the output. Equation (2.10) can be rewritten as a general Port-controlled Hamiltonian systems form:

$$\dot{z} = \mathbf{J}(z)\frac{\partial H(z)}{\partial z} + g(z)u, \tag{2.11}$$

$$y = g^T(z)\frac{\partial H(z)}{\partial z}, \tag{2.12}$$

where $z = \begin{bmatrix} q & p \end{bmatrix}^T$ and $\mathbf{J}(z)\frac{\partial H(z)}{\partial z}$ is the internal Hamiltonian dynamics which can be derived based on physical law(s). Equation (2.11) can be rewritten as

$$\dot{q} = \frac{\partial H}{\partial p}(q,p) \tag{2.13}$$

$$\dot{p} = -\frac{\partial H}{\partial q}(q,p) \tag{2.14}$$

where the Hamiltonian H is the total (kinetic and potential) energy as described by equation (2.1). $\mathbf{J}(z)$ is an $n \times n$ *skew-symmetric* matrix.

For a system with dissipation, equation (2.11) and (2.12) are rewritten as:

$$\dot{z} = [\mathbf{J}(z) - \mathbf{R}(z)]\frac{\partial H(z)}{\partial z} + g(z)u, \tag{2.15}$$

$$y = g^T(z)\frac{\partial H(z)}{\partial z}. \tag{2.16}$$

where $\mathbf{R}(z)$ is a positive semi-definite symmetric matrix, which represents the dissipation term corresponds to the internal energy dissipation in the system.

Matrix $\mathbf{J}(z)$ corresponds to the internal power-conserving interconnection structure of the physical systems due to:

1. Basic conservation laws, such as Kirchhoff's laws;
2. Powerless constraints; kinematic constraints;

3. Transformers, gyrators, exchange between different types of energy.

In many examples, the structure matrix $\mathbf{J}(z)$ will additionally satisfy an *integrability* condition.

The main message of this approach is that Port-controlled Hamiltonian system is closer to physical modeling, and capturing more information than just energy-balance of passivity. Three major features of PCH systems are [14]: its scalabilityto very large interconnected multi-physics systems, its ability for incorporating nonlinearities while retaining underlying conservation laws, and its integration of the treatment of both finite-dimensional and infinite-dimensional components.

3. Modeling of Two-wheeled Wheelchair

3.1. 3D Model

Before deriving the mathematical model, a 3D frame-based model of the wheelchair is designed in MapleSim. This step helps in analyzing the interconnection relationship between elements for PCH systems derivation. From [24], the TWW system involves the following mechanism:

1. Transformation from four wheels to two wheels involves lifting the casters (front wheels) using certain mechanism;
2. The lower part of the system, which includes casters, main wheels, motor and motor housing, caster lifting mechanism; is assumed as first link;
3. The second part of the system, located atop the first one has a chair with extendable height; is assumed as second link;
4. While at two wheels, i.e. balancing mode, the system resembles a double-inverted pendulum on two wheels;

Figure 2 shows the 3D model of a TWW in balance mode, i.e. on two wheels. Its similar dynamics with a double-inverted pendulum is shown in figure 3. The definition of each symbol is described in table 1.

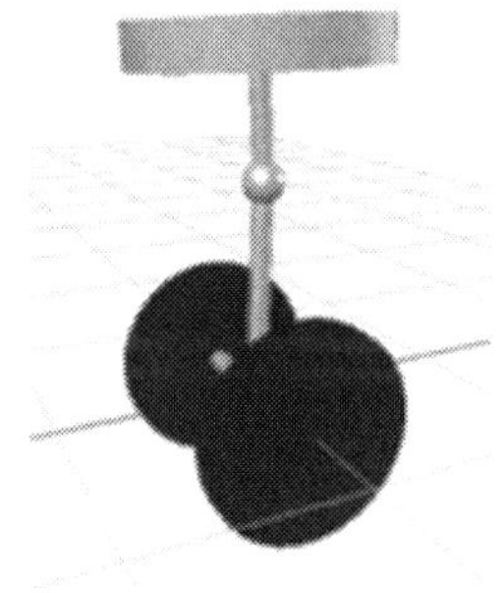

Figure 2. A 3D model developed in MapleSim showing its striking resemblance to a double-inverted pendulum on two wheels when in balance mode.

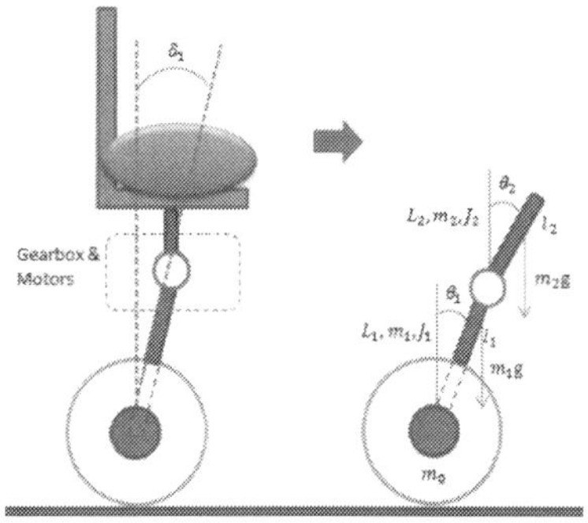

Figure 3. A dynamic analysis of a TWW in balance mode in 2D view reveals its similar characteristics to a double-inverted pendulum on two wheels.

This model is built with these assumptions:

1. The system is divided into three subsystems: base, first link, and second link;
2. The model is constrained to move in x-direction only, i.e. the base only move straight forward and backward without turning and the two links rotate in x-direction only;
3. On the base, there are two identical motors to rotate the two main wheels;
4. On the first link, only the mass of the link itself is considered;
5. On the second link, there are two masses considered, the link's mass itself and the load;

6. The extendable second link is assumed to be fully stretched;
7. The connections between each element in the system are assumed to be rigid and ideal, i.e. no spring-damper characteristics;

Table 1. Nomenclature.

Symbol	Definition
m_0	Mass of the base, including the two wheels
m_1	Mass of the first link
m_2	Total mass of the second link, including the load
L_1	Length of first link
L_2	Total stretched length of second link
θ_1, θ_2	Angular displacement of first and second link, respectively
J_0, J_1, J_2	Moment of inertia of base, first and second link, respectively
g	Gravity constant

3.2. Mathematical Model

Consider the TWW is in balancing mode, i.e. in two wheels as shown in figure 3. The system has two inputs, one torque F applied to the base consisting of synchronous torques from identical left and right motors, and one torque τ from the motor between first and second link, which drives the second link. Table 2 completes the notation described in table 1.

Table 2. Notation used in deriving the mathematical model.

Symbol	Definition
r	Radius of main wheel
x	Linear displacement on x-axis
F	Force acting on the base of the system
τ	Torque acting between first and second link
i	Subsystem; 0 = base, 1 = first link, 2 = second link
p_i	Linear momentum of ith subsystem
K_i	Kinetic energy of ith subsystem
U_i	Potential energy of ith subsystem

As addressed in subsection 3.1, the model is divided it into three subsystems, namely base, first link and second link.

On the base, the kinetic energy is considered from translational dynamic of the two wheels, which is the following:

$$K_0 = 2\left(\frac{1}{2}m_0\dot{x}^2 + \frac{1}{2}J_0\dot{\theta}_0^2\right), \tag{3.1}$$

θ_0 is the rotation angle of each wheel and J_0 is the inertia of the wheel. Since $\dot{\theta} = \frac{\dot{x}}{r}$,

$$K_0 = \left(m_0 + \frac{J_0}{r^2}\right)\dot{x}^2. \tag{3.2}$$

Let $a_0 = \left(m_0 + \frac{J_0}{r^2}\right)$. The potential energy is zero, i.e. $U_0 = 0$. The storage energy on the base, H_0, is the sum of the kinetic and potential energy $K_0 + U_0$. Rewriting H_0 in terms of momentum p_0,

$$H_0(x, p_0) = \frac{1}{4}\frac{p_0^2}{a_0}. \tag{3.3}$$

On the second subsystem, i.e. the first link, the kinetic energy is given as:

$$K_1 = \frac{1}{2}m_1 L_1^2 \dot{\theta}_1^2. \tag{3.4}$$

The potential energy acting on the first link due to gravitational acceleration is:

$$U_1 = -m_1 g L_1 \cos\theta_1. \tag{3.5}$$

Let $a_1 = (m_1 g L_1)$ and $J_1 = m_1 L_1^2$ are constant. The total energy stored in the first link is

$$H_1(\theta_1, p_1) = \frac{1}{2}\frac{p_1^2}{J_1} - a_1\cos\theta_1. \tag{3.6}$$

On the third subsystem, i.e. the second link, the kinetic energy is given as:

$$K_2 = \frac{1}{2}m_2 L_1^2\dot{\theta}_1^2 + \frac{1}{2}m_2 L_2^2\dot{\theta}_2^2 + m_2 L_1 L_2\cos(\theta_1 - \theta_2). \tag{3.7}$$

The potential energy by gravitational force is given by

$$U_2 = -m_2 g(L_1\cos\theta_1 + L_2\cos\theta_2). \tag{3.8}$$

Let $a_{21} = (m_2 L_1^2)$, $a_{22} = (m_2 L_1 L_2)$, $a_{23} = (m_2 g L_1)$, $a_{24} = (m_2 g L_2)$, and $J_2 = m_2 L_2^2$. The total energy stored in the second link is:

$$H_2(\theta_1, \theta_2, p_1, p_2) = \frac{1}{2}\frac{p_1^2}{a_{21}} + \frac{1}{2}\frac{p_2^2}{J_2} - a_{22}\cos(\theta_1 - \theta_2) - a_{23}\cos\theta_1 + a_{24}\cos\theta_2. \tag{3.9}$$

The total energy of the system is the sum of Hamiltonian of base, first link, and second link, i.e.

$$H_{\text{Total}}(\mathbf{q},\mathbf{p}) = \frac{1}{2}\mathbf{p}^T\mathbf{M}^{-1}(\mathbf{q})\mathbf{p} + U(\mathbf{q}), \tag{3.10}$$

where $\mathbf{q}$, $\mathbf{p}$, and $\mathbf{M}$ are the vector of generalized coordinate, vector of generalized momenta, and matrix of inertia, respectively.For our system,

$$\mathbf{q} = [x, \theta_1, \theta_2]^T, \tag{3.11}$$

$$\mathbf{p} = [p_0, p_1, p_2]^T. \tag{3.12}$$

Recall the general Hamilton's equations of motion from equation (2.13) and equation (2.14):

$$\frac{\mathrm{d}\mathbf{q}}{\mathrm{d}t} = \frac{\partial H}{\partial \mathbf{p}} \tag{3.13}$$

$$\frac{\mathrm{d}\mathbf{p}}{\mathrm{d}t} = -\frac{\partial H}{\partial \mathbf{q}} \tag{3.14}$$

Substituting equation (3.11) and (3.12) into equation (3.13) and (3.14) yields

$$\frac{\mathrm{d}x}{\mathrm{d}t} = \frac{\partial H_{\text{Total}}}{\partial p_0} = \frac{1}{2}\frac{p_0}{a_0}, \tag{3.15}$$

$$\frac{\mathrm{d}\theta_1}{\mathrm{d}t} = \frac{\partial H_{\text{Total}}}{\partial p_1} = \left(\frac{1}{J_1} + \frac{1}{a_{21}}\right)p_1, \tag{3.16}$$

$$\frac{\mathrm{d}\theta_2}{\mathrm{d}t} = \frac{\partial H_{\text{Total}}}{\partial p_2} = \frac{p_2}{J_2}, \tag{3.17}$$

$$\frac{\mathrm{d}p_0}{\mathrm{d}t} = -\frac{\partial H_{\text{Total}}}{\partial x} = 0, \tag{3.18}$$

$$\frac{\mathrm{d}p_1}{\mathrm{d}t} = -\frac{\partial H_{\text{Total}}}{\partial \theta_1} = -\left(a_1 \sin\theta_1 + a_{22}\sin(\theta_1 - \theta_2) + a_{23}\sin\theta_1\right), \tag{3.19}$$

$$\frac{\mathrm{d}p_2}{\mathrm{d}t} = -\frac{\partial H_{\text{Total}}}{\partial \theta_2} = -\left(-a_{22}\sin(\theta_1 - \theta_2) - a_{24}\sin\theta_2\right). \tag{3.20}$$

Equation (3.15) implies that the momentum on base, i.e. $p_0 = 2a_0\dot{x}$, is comprised of two momenta from two main wheels. Equation (3.16) implies that the angular momentum of first link, i.e. $p_1 = \dot{\theta}_1\left(\frac{J_1 a_{21}}{J_1 + a_{21}}\right)$, is influenced by the mass of second link. On the second link, however, the angular momentum p_2 is affected only by its own inertia $\dot{\theta}_2 J_2$, as shown by equation (3.17). Equation (3.18) confirms equation (3.2) that the potential energy is zero. Equation (3.19) and equation (3.20) suggest that the two links are affected by downward potential energy due to gravitational acceleration.

From d'Alembert's principles, the interconnections are obtained as follows:

$$F = m_0\ddot{x} \tag{3.21}$$

$$\tau = J_2\ddot{\theta}_2 + \left(a_{22}\sin(\theta_1 - \theta_2) + a_{24}\sin\theta_2\right) \tag{3.22}$$

Therefore, the complete PCH-based model of a two-wheeled system is as follows

$$\dot{\mathbf{q}} = \mathbf{J}\frac{\partial H_{\text{total}}}{\partial \mathbf{q}}(x,\theta_1,\theta_2,p_0,p_1,p_2)+\mathbf{g}\ \mathbf{u}$$
$$\mathbf{y} = \mathbf{g}^T\frac{\partial H_{\text{total}}}{\partial \mathbf{q}}(x,\theta_1,\theta_2,p_0,p_1,p_2) \tag{3.23}$$

where $\mathbf{J} = \begin{bmatrix} 0 & 0 & 0 & 1 & 0 & 0 \\ 0 & 0 & 0 & 0 & 1 & 0 \\ 0 & 0 & 0 & 0 & 0 & 1 \\ -1 & 0 & 0 & 0 & 0 & 0 \\ 0 & -1 & 0 & 0 & 0 & 0 \\ 0 & 0 & -1 & 0 & 0 & 0 \end{bmatrix}$ and $\mathbf{g} = \begin{bmatrix} 1 & 0 \\ 0 & 0 \\ 0 & 0 \\ 0 & 0 \\ 0 & 0 \\ 0 & 1 \end{bmatrix}$.

4. Controller Design

In this section a proposition for controller design using Port-controlled Hamiltonian systems is introduced.

It is well-known that the standard feedback interconnection of two PCH systems is again a PCH system [20]; which can be exploited for stability analysis and control purposes.

The PCH systems use the system's interconnection structure and its Hamiltonian (i.e. its total energy) as the primary means for modeling and control. Theoretically [14], the Hamiltonian of the system assumes its minimum at its stable equilibrium state. So, by introducing dissipation (or damping) in the controller, the energy in the system decreases until the minimum of energy or, equivalently, the desired equilibrium configuration is reached, i.e. the power balance of the (closed-loop) system is then reduced to

$$\frac{\mathrm{d}}{\mathrm{dt}}H_{CL} = \bar{e}_R^T\bar{f}_R \leq 0 \tag{4.1}$$

where $\bar{e}_R^T\bar{f}_R$ refers to the product of dissipative effort and flow of a dissipative element, which is the dissipated power. By standard Lyapunov theory, equation (4.1) implies that the Hamiltonian can be used as Lyapunov function candidate, which is already known in classical Hamiltonian mechanics. However, in some cases, Hamiltonian H cannot be used as a Lyapunov function, i.e. there is no minimum in H. To solve this issue, a Lyapunov function of combination of the total energy H and another *conserved quantity* is proposed.In PCH systems, this methodology is called the Energy-Casimir method [25].

In general, for any PCH systems with dissipation (see equation (2.11)) or without dissipation (see equation 2.15), there exist a conserved quantity C, called Casimir function, as being the solutions of

$$\frac{\partial^T C}{\partial z}(z)\mathbf{J}(z) = 0 \tag{4.2}$$

The time-derivative of C along the PCH system (see equation (4.3)) shows that it remains constant along the trajectory of the PCH system, regardless of the form of the PCH system, thus proofing that it is invariant of the Hamiltonian H.

$$\frac{\mathrm{d}C(z)}{\mathrm{d}t} = \frac{\partial^T C}{\partial z}(z)\dot{z} \tag{4.3}$$

Substituting equation (2.11) to equation (4.3) yields:

$$\frac{dC(z)}{dt} = \frac{\partial^T C}{\partial z}(z)\mathbf{J}(z)\frac{\partial H(z)}{\partial z} + \frac{\partial^T C}{\partial z}(z)g(z)u \tag{4.4}$$

According to equation (4.2), for autonomous system or if $\frac{\partial^T C}{\partial z}(z)g(z)u = 0$, the function $C(z)$ is constant. For a system with dissipation (equation (2.15)), the Casimir functions are obtained from both $\mathbf{J}(z)$ and $\mathbf{R}(z)$, i.e. :

$$\begin{aligned} &\frac{\partial^T C}{\partial z}(z)\mathbf{J}(z) = 0, \\ &\frac{\partial^T C}{\partial z}(z)\mathbf{R}(z) = 0. \end{aligned} \tag{4.5}$$

The existence of Casimir function *C(z)* also hinted that it can be used for stability analysis of the PCH system by using it as a Lyapunov functions [25-27].

Recall a nonlinear PCH plant system with dissipation from equation (2.15) and (2.16):

$$\dot{z} = [\mathbf{J}(z) - \mathbf{R}(z)]\frac{\partial H(z)}{\partial z} + g(z)u, \tag{4.7}$$

$$y = g^T(z)\frac{\partial H(z)}{\partial z}. \tag{4.8}$$

The proposed controller is of the PCH form. Thus, both the controller and the plant are treated as subsystems with the interconnection as shown in figure 4.

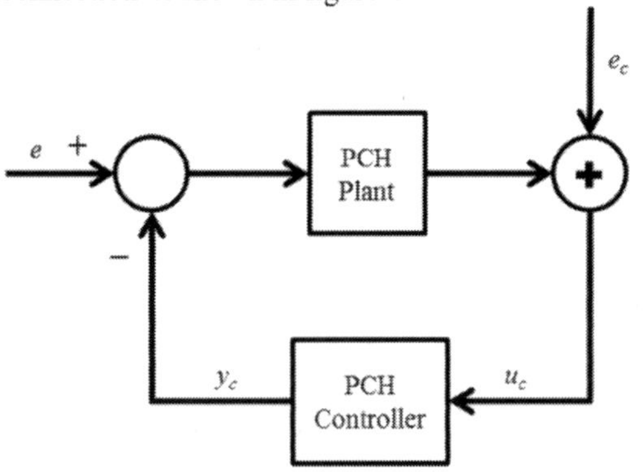

Figure 4. A standard feedback interconnection used by most plant with controller system

with *e* and e_C are external error signals.

Proposition 4.1. Consider the proposed PCH controller system as follows, with the interconnection as shown in figure 4.

$$\dot{\xi} = [\mathbf{J}_C(\xi) - \mathbf{R}_C(\xi)]\frac{\partial H_C(\xi)}{\partial \xi} + g_C(\xi)u_C, \tag{4.5}$$

$$y_C = g_C{}^T(\xi)\frac{\partial H_C(\xi)}{\partial \xi} \tag{4.6}$$

Remark 4.1: The interconnection is defined as follows:

$$\begin{aligned} u &= -y_C + e, \\ u_C &= y + e_C. \end{aligned} \tag{4.7}$$

The closed-loop system is represented by the following

$$\begin{bmatrix} \dot{x} \\ \dot{\xi} \end{bmatrix} = [\mathbf{J}_{\mathrm{CL}}(x,\xi) - \mathbf{R}_{\mathrm{CL}}(x,\xi)] \begin{bmatrix} \dfrac{\partial H_{\mathrm{CL}}(x)}{\partial x} \\ \dfrac{\partial H_{\mathrm{CL}}(\xi)}{\partial \xi} \end{bmatrix} + \begin{bmatrix} g(x) & 0 \\ 0 & g_C(\xi) \end{bmatrix} \begin{bmatrix} e \\ e_C \end{bmatrix}, \tag{4.8}$$

$$\begin{bmatrix} y \\ y_C \end{bmatrix} = \begin{bmatrix} g(x) & 0 \\ 0 & g_C(\xi) \end{bmatrix} \begin{bmatrix} \dfrac{\partial H_{\mathrm{CL}}(x)}{\partial x} \\ \dfrac{\partial H_{\mathrm{CL}}(\xi)}{\partial \xi} \end{bmatrix}, \tag{4.9}$$

which is again a Port-controlled Hamiltonian System.

In the control system, it is the input u that we want to know, as it is the output from the controller, as illustrated by figure 4. To determine H_C, the following steps is suggested:

1. Define state variables of the system, i.e. x;
2. Using the property $\dfrac{\partial^T C}{\partial z}(z)\mathbf{J}(z) = 0$, obtain Casimir function C;
3. Define Lyapunov function V as

$$V = H_p + H_C + C, \tag{4.10}$$

with H_p, H_C are the Hamiltonian of the plant and the controller, respectively.

4. H_C is determined such that $\dot{V} \le 0$, to guarantee stability;
5. From [25], taking the time-derivative of H_C, we can design u.

$$\frac{\mathrm{d}H_C}{\mathrm{dt}} = -u^T y. \tag{4.11}$$

5. Conclusion

In this paper, a model of two-wheeled wheelchair based on Port-controlled Hamiltonian systems was presented. The nonlinearity of the model is preserved using PCH system. Furthermore, a PCH-based controller was also proposed. Due to this, the closed-loop system was also a PCH system, hence preserving the nonlinear structure.

In the future, the derived model and the proposed controller will be implemented into hardware experiment.

Acknowledgments

This work is supported by The Ministry of Science, Technology, and Innovation (MOSTI) Malaysia under project grant SF12-012-0041.

References

[1] Karmarkar A M, Dicianno B E, Cooper R, Collins D M, Matthews J T, Koontz A, Teodorski E

E and Cooper R 2011 Demographic Profile of Older Adults Using Wheeled Mobility Devices *J. Aging Res.* pp 1-11

[2] Ding D and Cooper R A 2005 Electric powered wheelchairs *IEEE Cont. Syst.* **25**(2) pp 22-34

[3] Zhan J 2009 A pressure cushion for control of intelligent wheelchair movements *IEEE 10th Int. Conf. on Computer-Aided Industrial Design and Conceptual Design (26-29 November 2009)* pp 155-157

[4] Kuramatsu T and Murakami T 2010 Force sensorless power-assist control of yaw motion direction for two wheels driven wheelchair *World Automation Congress (Kobe, Japan, 19-23 September. 2010)* pp 1-6

[5] Fan J 2011 Motion control of intelligent wheelchair based on sitting postures *Int. Conf. on Mechatronics and Automation (Beijing, 7-10 August 2011)* pp 301-306

[6] Grasser F, D'Arrigo A, Colombi S and Rufer A C 2002 JOE: a mobile, inverted pendulum *IEEE Trans. Industr. Electr.* **49**(1) pp 107-114

[7] Lee H J, Kim H W and Jung S 2010 Development of a mobile inverted pendulum robot system as a personal transportation vehicle with two driving modes: TransBot *World Automation Congress (Kobe, Japan, 19-23 September 2010)* pp 1-5

[8] Nakamura A and Murakami T 2008 Trajectory Tracking Control of a Two Wheels Driven Wheelchair *Asia Int. Symp. on Mechatronics (Japan, 2008)* pp 255-259

[9] Nakamura A and Murakami T 2009 A stabilization control of two wheels driven wheelchair *IEEE/RSJ Int. Conf. on Intelligent Robots and System (St. Louis, USA, 11-15 October 2009)* p 4863

[10] Kawamura T and Murakami T 2011 Vibration suppression for uprising control of two-wheel driven wheelchair *37th Ann. Conf. on IEEE Industrial Electronics Society (Melbourne, 7-10 November 2011)* p 3323

[11] Goher K, Ahmad S and Tokhi O M 2010 A new configuration of two wheeled vehicles: Towards a more workspace and motion flexibility *4th Ann. IEEE System Conference (San Diego, California, 5-8 April 2010)* pp 524-528

[12] Ahmad S and Tokhi O M 2008 Modelling and Control of a Wheelchair on Two Wheels *2nd Asia Int. Conf. on Modeling & Simulation (Kuala Lumpur,13-15 May 2008)* pp 579-584

[13] Wei A, Wang Y and Hu X 2012 Adaptive robust parallel simultaneous stabilization of two uncertain port-controlled Hamiltonian systems subject to input saturation *Chinese Control Conf. (Hefei, 25-27 July 2012)* pp 727-732

[14] Duindam V, Macchelli A, Stramigioli S, and Bruyninckx H 2009 *Modeling and Control of Complex Physical Systems: The Port-Hamiltonian Approach* (New York: Springer-Verlag)

[15] Ravi B and Dey B 2010 Stabilizing a Flexible Beam on a Cart: A Distributed Port-Hamiltonian Approach *J. Nonlinear Sci.* **20** pp 131-151

[16] Peza-Solis J F,Silva-Navarro G and Castro-Linares R 2010 Control of a rigid-flexible two-link robot using Passivity-based and Strain-feedback approaches *7th Int. Conf. on Electrical Engineering, Computing Science and Automatic Control* (*Mexico City, 8-10 September 2010*) pp 476-481

[17] Yang B, Li H, Kang Z and Jiang H 2012 Hamiltonian-based binocular visual servoing of camera-in-hand robotic systems *Proc. Int. Conf. on Modeling, Identification & Control (Wuhan, China 24-26 June 2012)* pp 388 - 393

[18] Dirksz D A and Scherpen J M A 2012 Adaptive control of port-Hamiltonian systems *Proc. 19th Int. Symp. on Mathematical Theory of Networks and Systems (Budapest 5-7 July 2010)* pp 1503-1508

[19] Dirksz D A and Scherpen J M A 2012 Structure Preserving Adaptive Control of Port-Hamiltonian Systems *IEEE Trans. Auto. Cont.* **57** pp 2880-2885

[20] van der Schaft A J 2000 Port-controlled Hamiltonian Systems: towards a theory for control and design of nonlinear physical systems *J. Society of Instrument and Control Engineers of Japan* **39** pp 91-98

[21] Ortega R, Loria A, Nicklasson P J and Sira-Ramirez H 1998 *Passivity-based Control of Euler-Lagrange Systems* (London: Springer-Verlag)
[22] Breedveld P C 1982 Thermodynamic Bond Graphs and the Problem of Thermal Inertance *J. Franklin Inst.* **314**(1) pp 15-40
[23] van der Schaft A J 1986 Stabilization of Hamiltonian systems *Nonlinear Anal. Theory Methods Appl.* **10**(10)1021-35
[24] Aula A, Akmeliawati R, Ahmad S, Altalmas T and Sidek S N 2013 Integrated Design, Modeling and Analysis of Two-wheeled Wheelchair for Disabled. *Proc. 16th Int. Conf. on Climbing and Walking Robots (Sidney, Australia)* pp 141-152
[25] van der Schaft A J 2004 Port-Hamiltonian systems: network modeling and control of nonlinear physical systems *Advanced dynamics and control of structures and machines* ed H Irschik and K Schlacher (New York: Springer)
[26] Ortega R, van der Schaft A J, Mareels I and Maschke B 2001 Putting Energy Back in Control *IEEE Cont. Sys.* **21**(2) pp 18-33
[27] OrtegaR, van der Schaft A J, Maschke B and Escobar G 2002 Interconnection and damping assignment passivity-based control of port-controlled Hamiltonian systems *Automatica* **38** pp 585-596

Preliminary Analysis of Dust Effects on Microwave Propagation Measured in Sudan

Elfatih A. A. Elsheikh, Islam Md. Rafiqul[1], Jalel Chebil, Saad Bashir and Ahmad F. Ismail
Electrical and Computer Engineering Department, Kulliyyah of Engineering
International Islamic University Malaysia (IIUM)
Jalan Gombak, 53100 Kuala Lumpur, Malaysia

E-mail: rafiq@iium.edu.my

Abstract. Dust storms are meteorological phenomena which occur for a percentage of time in arid and semi arid areas especially at African Sahara and Middle East. Measurements at existing microwave links have showed dust storms can potentially result in serious attenuation in signal level especially at Ku band and higher frequencies with direct impact on telecommunications system performance. Only a limited amount of research has been carried out and the available data was very scarce. Few prediction models have been developed to estimate microwave signal attenuation during the dust storm based on scattering theory and approximation of dust properties. However, real dust storm is a complex phenomenon which is difficult to be described by theoretical physical or mathematical models [5-6]. In this paper, an evaluation of the existing attenuation prediction models has been done based on the measured dust storm properties and measured attenuation in Sudan.

1. Introduction

It has been realized that attenuation is higher at frequency bands above to 10 GHz due to the sensitivity to weather conditions. Consequently, rain, snow, cloud, fog, dust storms and other phenomena cause signal attenuation which can seriously limit the performance of wireless communication systems[1-4]. Microwave signal attenuation due to dust storm is one of those potential problems and technical challenges need to be investigated in order to provide reliable wireless communication services[5-7]. Sand and dust-storms occur in many parts of the world, especially in the Middle East and arid parts of Asia, as well of Southwest USA, in the dray states such as Texas and Arizona.

Dust storms are becoming more frequent in some parts of the world; there is significant correlation between the increased occurrences of dust storm and the climate change phenomenon. From figure 1 it is obvious that the dust storm phenomena is increasing in recent years[8-9].

[1] Islam Md Rafiqul: rafiq@iium.edu.my

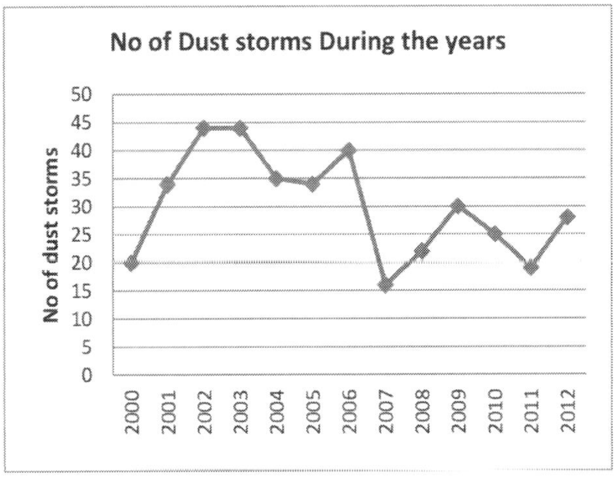

Figure 1: Frequency of Dust storm occurrence in Khartoum (2000-2012) visibilty <1km.

Dust storms affect the microwave propagation and it is severe in high frequencies. It attenuates the signal and degrades the performance of microwave links especially in arid region[10]. Limited research has been conducted in dust storm attenuation area. The scarcity of measured data forces researchers to work for the dust storm prediction modeling. Few theoretical models have been developed based on single scatterer, multiple scattering or mutual interaction phenomena [1-3, 11-13].

2. Measurement setup

Microwave links at 12GHz-13GHz and different lengths have been installed in Khartoum as shown in Fig. 2.and Table 1 All terrestrial links have been monitored and the transmitted and received signal levels have been recorded for one year. The corresponding visibility data have been collected from the metrological department.

Figure 2: Geographical details of the Microwave links

Table 1: Details of the four terrestrial links that are used for the study.

link	Location		Frequency MHZ		Length Km
			Up	down	
JERRAIF SOBA	E 32°36′8.70 E 2°37′51.30"	N 15°36′29.90" N 15°29′39.60"	12765 12905	13031- 13171	12.97
SOBA MODRAAT	E 32°29′56.00 E 2°28′26.72"	N 15°32′41.50" N 15°36′19.10"	13031- 13171	12765- 12905	15.22
MODRAAT Omdurman	E 32°29′56.00 E 2°28′26.72"	N 15°32′41.50" N 15°36′19.10"	12765- 12905	13031- 13171	7.19

Two attenuation prediction models have been evaluated using the analyzed dust particle properties, and based on the measured visibility and attenuation.

3. Dust Storm Prediction Models

Different approaches have been adopted to evaluate microwave signal attenuation due to dust storm in terms of its characteristics such as the number of dust particles per cubic meter, the mass of dust per cubic meter or visibility. Therefore, various investigations showed that the scattering of electromagnetic waves propagating in dust storms is influenced by various factors including incident wave frequency, permittivity, density, geometric scale, distribution of grains, moisture and chemical behavior of dust particles[14].

Two recent attenuation prediction models have been investigated. These models have been based on certain assumptions but rely on some empirical inputs such as particle shape, chemical composition, size and dielectric constant.

3.1 Chu's Modified by Goldhirish Model (1982)

Goldhirish [1] has modified Chu's model by expressing the specific attenuation coefficient as a function of visibility and particle size distribution as follows.

$$A = \frac{5.67\times10^2}{V\times r_e{}^2\times\lambda}\times\frac{\varepsilon''}{[(\varepsilon'+2)^2+\varepsilon''^2]}\sum_i P_i.r_i^3 \ [\mathrm{dB/km}] \quad (1)$$

where
r =particle radius in meters.
λ = wavelength (m).
ε', ε'' =complex dielectric constant of dust particles.
V = visibility (m)
re = Equivalent particle radius in meter.

Although, Chu modified model has attained close agreement between predicted and measured attenuation, the model losses it is reliability when used to predict attenuation above 37GHz [14-15].

3.2 Elshaikh Model (2008)

Elshaikh.Z.[15] has proposed a prediction model based on Mie scattering which can calculate the specific attenuation in microwave wave band with high reliability. The model has related attenuation coefficient to the visibility, particle size distribution.

IOP Conf. Series: Materials Science and Engineering **53** (2013) 012075 doi:10.1088/1757-899X/53/1/012075

$$A = \frac{r_e f}{V}\left(x + y r_e{}^2 f^2 + z r_e{}^3 f^3\right)\ [\text{dB/km}] \quad (2)$$

where

r_e = Equivalent particle radius in meter.
V = Visibility in kilometer.
f = Frequency in GHz.

$$x = \frac{1886 \times \varepsilon''}{(\varepsilon'+2)^2 + \varepsilon''^2} \quad (3)$$

$$y = 137 \times 10^3 \times \varepsilon'' \left\{ \frac{6\left(7\varepsilon'^2 + 7\varepsilon''^2 + 4\varepsilon' - 20\right)}{5[(\varepsilon'+2)^2 + \varepsilon''^2]^2} + \frac{1}{15} + \frac{5}{3\left[(2\varepsilon'+3)^2 + 4\varepsilon''^2\right]} \right\} \quad (4)$$

$$z = 379 \times 10^4 \left\{ \frac{(\varepsilon'-1)^2.(\varepsilon'+2) + [2.(\varepsilon'-1).(\varepsilon'+2) - 9] + \varepsilon''^4}{[(\varepsilon'+2)^2 + \varepsilon''^2]^2} \right\} \quad (5)$$

4. Measurements

In the study conducted by the author, a duststorm was monitored on September, 2008 in Khartoum-Sudan. One of most severe dust storm occurred on 21st of September. Dust storm attenuation measurements recorded by Marconi microwave system which monitoring a microwave network providing cell phone services in Sudan. Figure 3 and 4 show a record of microwave signal level during the storm. On the other hand, Sudan meteorological Authority recorded a visibility less than 50 m during the storm.

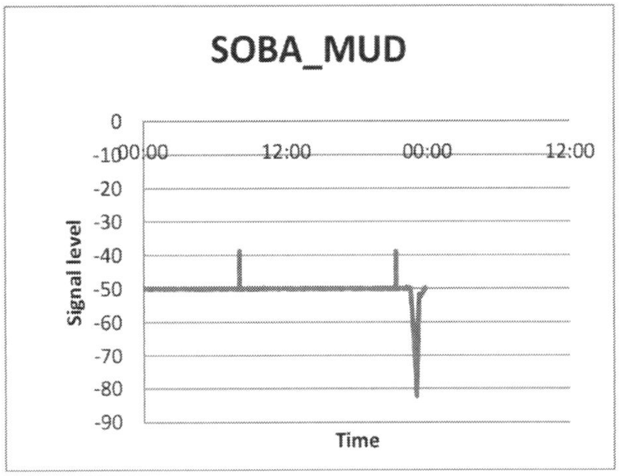

Figure 3: Measured signal level for soba –MUD link on 21st September 2008.

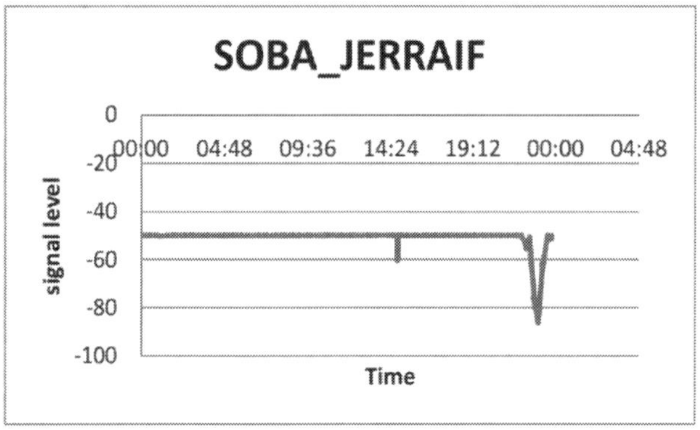

Figure 4: Measured signal levels for soba – JERRAIF link on 21st September 2008.

Figure 3 showed that measured total attenuation for SOBA_MUD link was equal to 32 dB at 13GHz and 15.2 km link. Whereas, Figure 4 showed the measured total attenuation for SOBA_ JERRAIF link equal to 36 dB at13GHz and 13km link length.

5. Predicted attenuation

Dust storm attenuation was predicted using models proposed by Goldhirsh and Zain with the parameters re = 38μm , ε'=4.2 , ε"=1.56, V=50m and f=13GHz. Equations (1) to (5) were used to estimate the attenuations and presented in Table 2. Measured and predicted attenuation are compared in Table 3. It is obvious that the measured attenuation are 20 and 25 dB higher than that predicted by both models for Soba-Mud and Soba-Jerraif links respectively.

Table 2: Predicted attenuation using measured visibility.

Goldhirsh dB/km	Zain dB/km	Length km	Goldhirsh total dB	Zain Total dB
0.72	0.74	15.22	10.8	11.24
0.72	0.74	12.97	9.3	9.6

Table 3: Comparision between measured and predected attenuation

Link	Goldhirsh	Zain	Measured
SOBA_MUD	10.8 dB	11.24 dB	32dB
SOBA_JERRAIF	9.3 dB	9.6 dB	36 dB

6. Conclusion

Dust storms can cause serious attenuation in signal level especially at Ku band and higher frequencies with direct impact on telecommunications system performance. Two operational microwave links at 13 GHz in Sudan were monitored during dust storm and observed more than 30 dB attenuation for about 10 Km lengths. Models proposed by Goldhirsh and Zain were used to predict attenuation on both measurements. It is found that both models underestimate the attenuation measured on both microwave links.

Refences

[1] Goldhirsh, J. 1982. A parameter review and assessment of attenuation and backscatter properties associated with dust storms over desert regions in the frequency range of 1 to 10 GHz., IEEE Transactions on Antennas and Propagation, 30(6): p. 1121-1127.

[2] Ahmed, A.S., A.A. Ali, and M.A. Alhaider. 1987. Airborne Dust Size Analysis for Tropospheric Propagation of Millimetric Waves into Dust Storms. Geoscience and Remote Sensing, IEEE Transactions on. GE-25(5): p. 593-599.

[3] Elshaikh, Z.E.O. and M. Islam. 2009. Mathematical model for the prediction of microwave signal attenuation due to duststorm. Progress In Electromagnetics Research, 6: p. 139-153.

[4] Chen, H.Y. and C.C. Ku, 2012. Calculation of Wave Attenuation in Sand and Dust Storms by the FDTD and Turning Bands Methods at 10–100 GHz. Antennas and Propagation, IEEE Transactions on, 60(6): p. 2951-2960.

[5] Abuhdima, E.M. and I.M. Saleh. Effect of sand and dust storms on GSM coverage signal in southern Libya. 2010: IEEE.

[6] Elabdin, Z., et al. 2008. Duststorm measurements for the prediction of attenuation on microwave signals in Sudan. International Conference on Computer and Communication Engineering, ICCCE 2008.

[7] Md. Rafiqul Islam, Zain Elabdin Omer Elshaikh, Othman O. Khalifa, AHM Zahirul Alam, Sheroz Khan and A.W. Naji. 2010. Prediction Of Signal Attenuation Due To Duststorms Using Mie Scattering. *IIUM Engineering Journal* (ISSN: 1511-788X), Vol. 11, No. 1, pp. 71-87.

[8] Islam, M.R., et al. Proposing a horizontal path adjustment factor for microwave link's attenuation prediction based on the analysis of dust storm behavior. in Communications (MICC), 2011 IEEE 10th Malaysia International Conference on. 2011.

[9] Ghobrial, S., et al., DUST STORMS IN THE SUDAN: INTENSITY AND PARTICLES'CHARACTERISTICS. JIEEE, 1985. 85: p. 326-328.

[10] Xiao-Ying, D., C. Hsing-Yi, and G. Dong-Hui, Microwave and Millimeter-Wave Attenuation in Sand and Dust Storms. Antennas and Wireless Propagation Letters, IEEE, 2011. 10: p. 469-471.

[11] Ruike, Y., W. Zhensen, and Y. Jinguang, The Study of MMW and MW Attenuation Considering Multiple Scattering Effect in Sand and Dust Storms at Slant Paths. International journal of infrared and millimeter waves, 2003. 24(8): p. 1383-1392.

[12] Zhou, Y.-H., Q. Shu He, and X. Jing Zheng, Attenuation of electromagnetic wave propagation in sandstorms incorporating charged sand particles. The European Physical Journal E: Soft Matter and Biological Physics, 2005. 17(2): p. 181-187.

[13] Vishvakarma, B.R. and C.S. Rai. Limitations of Rayleigh scattering in the prediction of millimeter wave attenuation in sand and dust storms. in Geoscience and Remote Sensing Symposium, 1993. IGARSS '93. Better Understanding of Earth Environment., International. 1993.

[14] Li, X., L. Xie, and X. Zheng, The comparison between the Mie theory and the Rayleigh approximation to calculate the EM scattering by partially charged sand. Journal of Quantitative Spectroscopy and Radiative Transfer, 2012. 113(3): p. 251-258.

[15] Zain Elabdin, Z., et al. 2008. Development of mathematical model for the prediction of microwave signal attenuation due to duststorm. International Conference on Computer and Communication Engineering, 2008. ICCCE 2008.

IOP Conf. Series: Materials Science and Engineering **53** (2013) 012076 doi:10.1088/1757-899X/53/1/012076

Speech Enhancement based on Compressive Sensing Algorithm

Amart Sulong[1], Teddy S. Gunawan[1], Othman O. Khalifa[1], Jalel Chebil[1]

[1] ECE Department, Faculty of Engineering International Islamic University Malaysia Kuala Lumpur, Malaysia

E-mail: tsgunawan@iium.edu.my, amartuia@gmail.com

Abstract. There are various methods, in performance of speech enhancement, have been proposed over the years. The accurate method for the speech enhancement design mainly focuses on quality and intelligibility. The method proposed with high performance level. A novel speech enhancement by using compressive sensing (CS) is a new paradigm of acquiring signals, fundamentally different from uniform rate digitization followed by compression, often used for transmission or storage. Using CS can reduce the number of degrees of freedom of a sparse/compressible signal by permitting only certain configurations of the large and zero/small coefficients, and structured sparsity models. Therefore, CS is significantly provides a way of reconstructing a compressed version of the speech in the original signal by taking only a small amount of linear and non-adaptive measurement. The performance of overall algorithms will be evaluated based on the speech quality by optimise using informal listening test and Perceptual Evaluation of Speech Quality (PESQ). Experimental results show that the CS algorithm perform very well in a wide range of speech test and being significantly given good performance for speech enhancement method with better noise suppression ability over conventional approaches without obvious degradation of speech quality.

1. Introduction

Technological advances, such as the development of telephone and radio, enhance our speech communication capability even more by enabling direct communication across larger distance, broader audiences, and more challenging circumstances. *Speech* is a special subclass of audio signals. It is produced by exciting the acoustic cavity, the vocal tract, by pulses of air released through the vocal cords for voiced sounds, or by turbulence for unvoiced sounds. *Speech* can be explained as the verbal form of human communication. *Speech system,* a dynamic, information-bearing signal, is also called the acoustic waveform [1]. These waves are produced due the sound pressure generated in the mouth of the speaker as a result of some sequence of coordinated movements of a series of structures in the human vocal system. *Speech signal processing* is purposed enhancing speech communication systems at a point in the communication link so that the conveyed message can be recovered with minimum loss on the receiver's side. Some examples of speech signal processing are speech coding, background noise reduction, time-scale or pitch modification, and speech recognition. *Speech communication systems* therefore are subject to more stringent requirements than music systems.

[1] Corresponding author: Amart Sulong, amartuia@gmail.com

Commonly, in most applications, the aim of speech enhancement is to improve the quality and intelligibility of degraded speech [2]. The *motivation behind the Speech enhancement* is concerned with improving some perceptual aspect of speech that has been degraded by additive noise that corrupted speech. However, there is always a tradeoff between noise reduction and signal distortion – more noise reduction is always accompanied by more signal distortion [12]. The improvement in quality is highly desirable as it can reduce listener fatigue [2], particularly in situations in which the listener is exposed to high level of noise for long period of time (e.g., manufacturing). In other word, *Speech enhancement algorithms* reduce or suppress the background noise to some degree and are sometime referred to as noise suppression algorithm. Introducing the speech distortion from noise reduction of the background noise may impair the speech intelligibility. Hence, the *main challenge in speech enhancements* is to design effective algorithm to suppress the noise without introducing any perceptible distortion in the signal.

Compressive sensing (CS) is a new type of sampling theory, which predicts that sparse signals can be reconstructed from what was previously believed to be incomplete information [3]. This method is considered as main feature to efficient algorithms such as l_1-minimization which can be used for recovery. The *traditional approach* of reconstructing signals from measured data follows the well-known Shannon sampling theorem [4]. It states that the sampling rate must be twice the highest frequency. Similarly, the fundamental theorem of linear algebra suggests that the number of collected samples (measurements) of a discrete finite-dimension signal should be at least as large as its length (its dimension) in order to ensure reconstruction. This principle underlies most devices of current technology, such as analog-to-digital conversion or audio and video electronics.

In addition, CS theory was growing very rapidly in the past few years and undergoes significant advances on an almost daily basis [3]. Its theory provides a fundamentally new approach to data acquisition and asserts that one can recover certain signal from far fewer samples or measurements than traditional methods use [5]. CS relies on the empirical observation that many types of signals can be well approximated by sparse expansion in terms of suitable basis. However, it is by only a small number of nonzero coefficients compare with the key to efficiency of many lossy compression techniques such as JPEG, MP3, etc. A compression is obtained by simply storing only the largest basis coefficients. The non-stored coefficients are simply set to zero while reconstructing the signal. This is certainly a reasonable strategy when full information of the signal is available.

However, when the signal first has to be acquired by a somewhat costly, lengthy, or otherwise difficult measurement (sensing) procedure, this seems to be a waste of resources. Clearly, the measurements have to be suitably designed. It is a remarkable fact that all provably good measurement matrices designed so far are random matrices. It is for this reason that the theory of CS uses a lot of tools from probability. Quite surprisingly, CS provides nevertheless a way of reconstructing a compressed version of the original signal by taking only a small amount of linear and non-adaptive measurements. The precise number of required measurements is comparable to the compressed size of the signal.

2. Speech enhancement

2.1. Understanding the Enemy: Noise

Making meaningful to performance of speech enhancement algorithms is concerned to prior of design algorithms to combat additive noise. Understanding of additive noise in various types of application is crucial behavior. It focuses on difference of the noise sources in term of temporal and spectral characteristic and the rang of noise levels that may encounter in real life [2, 5].

The study of the noise can be described in two categories stationary and non-stationary [2]. The stationary of the noise remains unchanged over the time, such as the fan noise coming from PCs. On the other hand, the non-stationary of the noise constantly changes the spectral (and temporal) characteristics, such as restaurant noise. Clearly, the task of suppressing noise that is constantly changing (non-stationary) is more difficult than that of suppressing stationary noise. The shape of their spectrum, particularly the distribution of the noise energy in the frequency domain, is another distinctive feature of the various types of noise to the frequency range.The knowledge of the range of speech and noise intensity levels in real-world scenarios are critical term to the design of speech enhancement algorithm [2]. These terms can estimate the range of signal-to-noise ratio (SNR) levels encountered in realistic environments. This concern is addressed as important method to speech enhancement algorithms. Its algorithms will be affected by suppressing noise and improved speech quality within that range of SNR levels [6].

2.2. Principle of Speech Enhancement

This section presents review and discussion on the different speech enhancement methods used to date [5,8]. Speech production model is shown in **Figure 1**. and **Figure 2**.

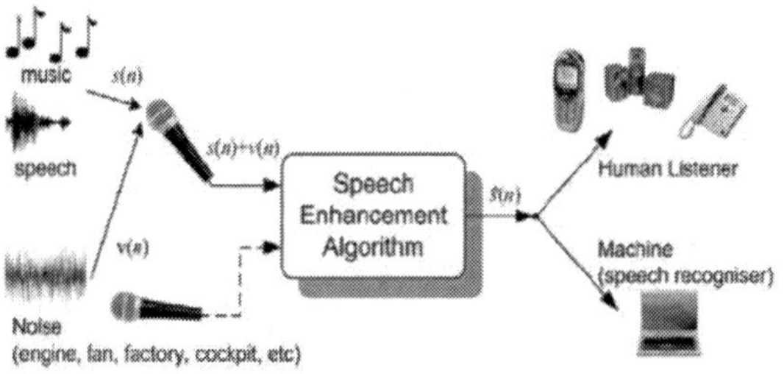

Figure 1. The basic diagram of speech enhancement [5].

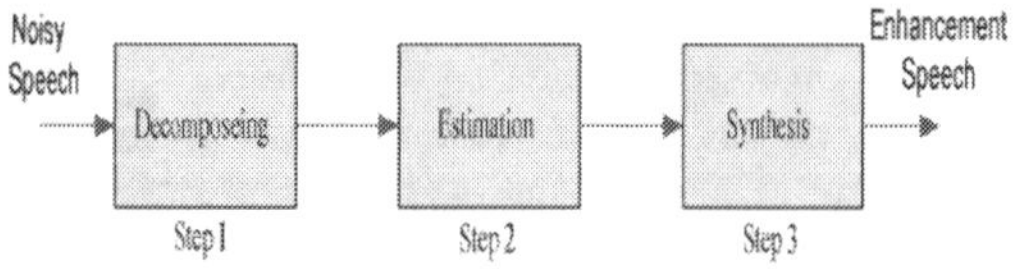

Figure 2. General speech enhancement algorithm [9].

Figure 1. and **Figure 2.** show the speech enhancement method. The speech enhancement problem can be simplified by assuming the following additive model:

$$s(n) = \hat{s}(n) + v(n) \tag{1}$$

where $s(n), \hat{s}(n),$ *and* $v(n)$ represent digitized noisy speech, clean speech, and noise, respectively. The noise $s(n)$ is assumed to be additive noise. The speech enhancement is a method of extracting clean speech $\hat{s}(n)$ from noisy speech signal $s(n)$ under certain constraints as shown in Figure 1. With reference [5] was to simplify the fundamental of speech enhancement. Most of the speech enhancement methods can be described in three steps as shown in Figure 2 [7]; Step 1: decomposing the speech signal into a transformed domain; Step 2: estimating the clean channel signals in the transformed domain; Step 3: synthesizing the speech from the estimated channel signals. Different transformation techniques and estimation algorithms are used by different speech enhancement methods. Speech enhancement in general is purposed to the main of objectives of enhancement that concern as follow; 1) to improve the overall quality; 2) to increase intelligibility, and; 3) to reduce listener's fatigue

In speech processing to model speech, assumptions are made. The more the assumptions are made more will be the sensitivity of the enhancement system to the deviations from these assumptions. The assumption is depended on the requirements and objective different aspect of speech which need to be enhanced. Sometimes there is need to enhance quality at the cost of intelligibility and other times intelligibility needs to be enhanced sacrificing some quality. Another important consideration in speech enhancement systems is that ultimately the resulting enhancement is related to human listener. Thus speech enhancement must inevitably take into account aspects of human perception.

The speech enhancement problem covers a broad spectrum of constraints, applications and issues. Environments in which an additive background signal has been introduced are common. The background may be noise-like such as in aircraft, street noise, etc. or may be speech-like such as environment with competing speakers. Therefore, *Speech enhancement depends on*; 1). Good signal processing technique. 2). Human perceptual factor. 3). Speech quality and intelligibility are dependent on short term spectral and 4). Amplitude and insensitive to spectral phase

2.3. Classification of Speech Enhancement Techniques
The research [1] mentioned that speech enhancement systems can be classified in a number of ways based on the criteria used or application of the enhancement system as shown in **Table 1.**

Table 1. Speech enhancement processing strategies

Domain	***Possible Strategies***
Number of input channels	One/Two/Multiple
Domain of processing	Time/Frequency
Type of algorithm	Non-adaptive/Adaptive
Additional constraints	Speech production/Perception

2.4. Subjective and Objective Evaluation
Subjective Evaluation: The most accurate method for evaluating speech quality in real-world noise that introduced by four classes of speech enhancement algorithms: Spectral subtraction, subspace, statistical-model based, and Wiener algorithms. The subjective quality ratings were obtained using ITU-T P.835 methodology design to evaluate the quality of enhanced speech along three dimensions: Signal distortion, noise distortion, and overall quality. ITU-T P.835 was designed to lead the listeners to integrate the effects of both signal and background distortion in making their rating of overall quality. It is costly and time consuming to predict speech quality with high correlation. But it is accurate and reliable (i.e., repeatable) provided it is performed under stringiest conditions (e.g., sizeable listener panel, inclusion of condition) [2]. These algorithms were evaluated using a newly developed noisy speech corpus (NOIZEUS) suitable for evaluation of speech enhancement algorithms and available. The sentences were originally sampled at 25 kHz and downsampled to 8 kHz. As shown in **Figure 3**. The statistical analysis is presented to compare the performance of the algorithms within each of the four classes (subspace, statistical-model, subtractive, and Wiener-type). This comparison was meat to examine whether there were significant differences algorithms within each class as mention in [6].

In Objective evaluation: the measurement matrix be able to assess the quality of the enhanced speech without needing access to the original speech signal by only using formula therefore it is very fast to testing process. it incorporates knowledge from different levels, including low-level processing (e.g., psychoacoustics) and higher-level processing such as prosodics, semantics, linguistics, and pragmatics. It is also predict with high accuracy to the results obtained from subjective listening tests with normal-hearing listeners. Reference [7,8,9,10] explained in their study, the speech enhancement evaluates the performance of several objective measures in terms of predicting the quality of noisy speech enhanced by noise suppression algorithm.

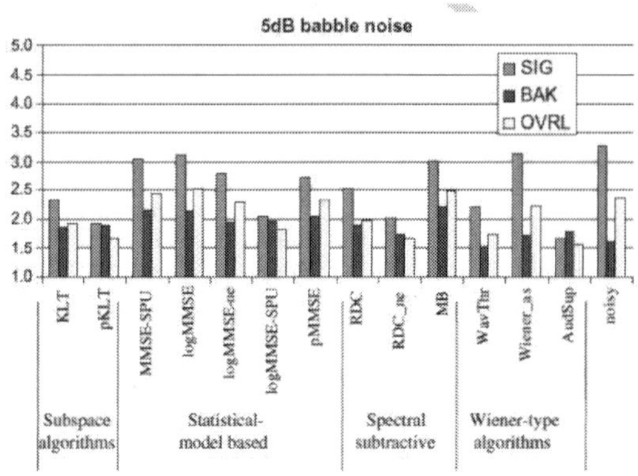

Figure 3. Subjective evaluation; The mean scores for SIG, BAK, and OVRL scales for the 13 Methods evaluated in babble noise background and for SNR levels of 5 dB [6].

3. Compressive sensing theory

Theoretically, Speech enhancement is required in many situations in which signal to be communicated or store. This research field is focused on suppression of additive noise. From the point of view of signal processing, additive noise is easier to deal with than convolution noise or nonlinear disturbances. *The ultimate goal* of speech enhancement is to eliminate the additive noise present in speech signal and restore the speech signal to its original form. Several methods have been developed in its research efforts [10]. However, it is very difficult to reliably predict the characteristics of the interfering noise signal or the exact characteristics of the speech waveform.

Hence, in effect, the speech enhancement methods are sub-optimal and can only reduce the amount of noise in the signal to some extent. Due to the sub-optimal nature of these methods, some of the speech signal can be distorted during the process. Hence, there is a trade-off between distortions in the processed speech and the amount of suppressed. The effectiveness of the speech enhancement system can therefore be measured based on how it performs in light of this trade-off.

There are two structure design approach can be applied for speech enhancement from classical theory and the novel proposed theory from structure design approach as follow;

3.1 Classical Theory (non-compressive sensing): Transform Coding for Speech Enhancement

One would first acquire a large amount of data in speech signal, compute and an appropriate basis and projections on it, and then transmit these projections and basic used. This is wasteful of resources since many more data points are initially collected than are transmitted. This approach will produce long-established paradigm for digital data acquisition with uniformly sample data at Nyquist rate and compress data as shown in Figure 4. Unfortunately, this sample-then-compress framework suffers from three inherent inefficiencies.

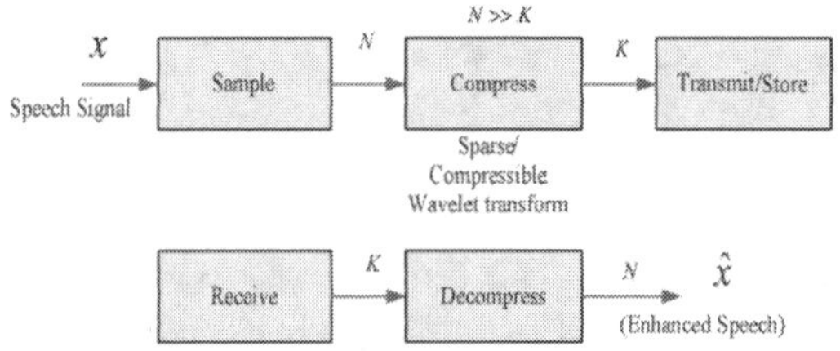

Figure 4. The compression of the signal in traditional methods [11].

In his research [11] mentioned the fact that compressible signals are well approximated by $K-sparse$ representations forms the foundation of transform coding. In data acquisition systems, transform coding plays a central role: the $N-sample$ signal x is acquired; the complete set of transform coefficients $\{s_i\}$ is computed via $s=\psi^T x$; the K largest coefficients are located and the $(N-K)$ smallest coefficients are discarded; and the K values and locations of the largest coefficients are encoded. Unfortunately, this sample-then-compress framework suffers from three inherent *inefficiencies*. First, the initial number of samples N may be large even if the desired K is small. Second, the set of all N transform coefficients $\{s_i\}$ must be compute even though all but K of them will discarded. Third, the locations of the large coefficients must encoded, thus introducing overhead.

3.2 Compressive Sensing Theory

CS [11] referred in his research as a new approach, compressive sensing address these *inefficiencies* by directly acquiring a compressed signal representation without going through the intermediate state of acquiring $N-sample$. Consider a general linear measurement process that computes $M<N$ inner products between x and a collection of vectors $\{\phi_j\}_{j=1}^{M}$ as in $y_j=\langle x,\phi_j\rangle$. Arrange the measurements y_j in an $M\times 1$ vector y and measurement vector ϕ_j^T as rows in an $M\times N$ matrix Φ . Then, by substituting Ψ from (2), y can be written as

$$y=\Phi x=\Phi\,\psi s=\Theta s \qquad (2)$$

where $\Theta=\Phi\psi$ is $M\times N$ matrix. The measurement process is not adaptive, meaning that Φ is fixed and does not depend on the signal x . The problem consists of designing ***a)*** a stable measurement matrix Φ such that the salient information in any $K-sparse$ or compressible signal is not damaged by the dimensionality reduction from $x\in R^N$ to $y\in R^M$ and ***b)*** a reconstruction algorithm to recover x from only measurements y (or about as many measurements as the number of coefficients recorded by a traditional transform coder). As shown in Figure 5.

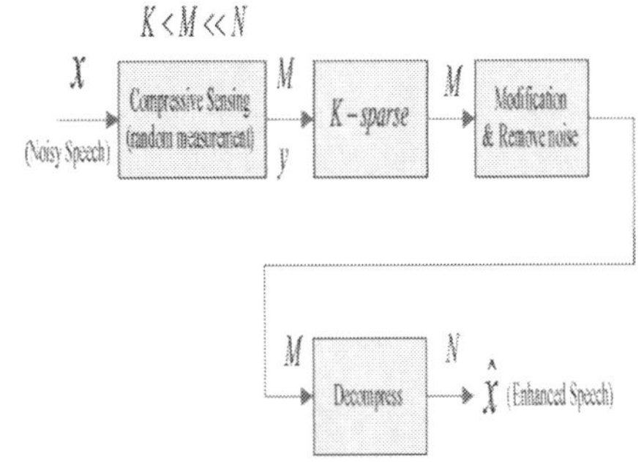

Figure 5. The method shows the new approach of speech enhancement algorithm based on compressive sensing (CS).

4. Proposed method for speech enhancement using compressive sensing Algorithm

Compressive sensing (CS) is an alternative to Shannon/Nyquist sampling for the acquisition of sparse or compressible signal that can be well approximated by just $K<<N$ elements from an N - dimensional basis. Instead of taking periodic samples, CS measures inner products with $M<N$ random vectors and then recovers the signal via a sparsity-seeking optimization or greedy algorithm. Standard CS dictates that robust signal recovery is possible from $M=O(\log(N/K))$ measurement. A model-based CS theory provides concrete algorithms with provable performance guarantees. This model is designed as a new

class of the structured compressible signals along with a new sufficient condition for robust structured compressible signal recovery.

The research in CS has focused primarily both on reducing the number of measurements M (as a function of N and M) and on increasing the robustness and reducing the computational complexity of recovery algorithm. The CS system can robustly recover $K-sparse$ and compressible signals from just $M = O(\log(N/K))$ noisy measurements using polynomial-time optimization solvers or greedy algorithms.

4.1 Proposed algorithm

The proposed speech enhancement algorithm using compressive sensing is illustrated in **Figure 6.** Analysis filter bank used gammatone filter bank to utilize due to its resemblance to the shape of human auditory filters [13]. Of the various transforms available to sparsify speech signal, Discrete Cosine Transform (DCT) will be chosen due to its simplicity and it good decorrelation property [14].

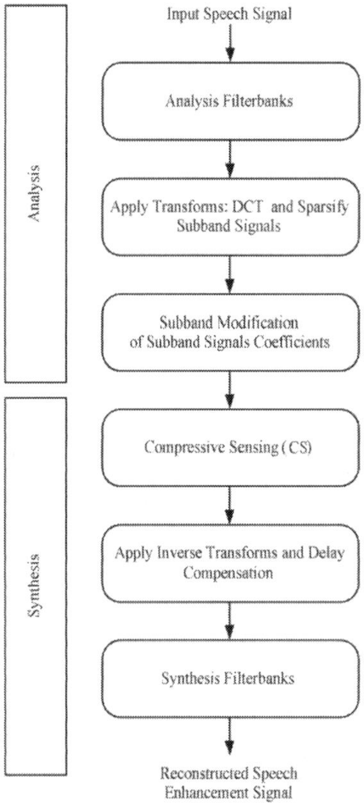

Figure 6. Proposed speech enhancement based on compressive sensing algorithm

Then by using subband modification to produce of subband signal coeffiecients for analysis the speech signal. On the systhesis part, for solving convex optimization of compressive sensing, the gradient projection for sparse reconstruction (GPSR) algorithm [15] was utilized due to it reconstruction quality can be traded with the available processing power at the synthesis side. The higher the processing power or the longer available to solve convex optimization problem, the higher the

IOP Conf. Series: Materials Science and Engineering **53** (2013) 012076 doi:10.1088/1757-899X/53/1/012076

synthesis signal quality. In other word, invert DCT (IDCT) and delay compensation were applied to the compressed signal. The amount of filter delay accumulated by each subband is different and without compensating for this delay, the synthesis of subband signal will lead to an incoherent output signal, i.e. lower quality signal.

4.2.Performance results

The performance of the proposed algorithm was evaluated using PESQ score. As mention in the previous section, the **Table 2.** shows the objective evaluation using PESQ. The results produce excellence score for PESQ measurement which can be considered as acceptable quality for everyday communication.

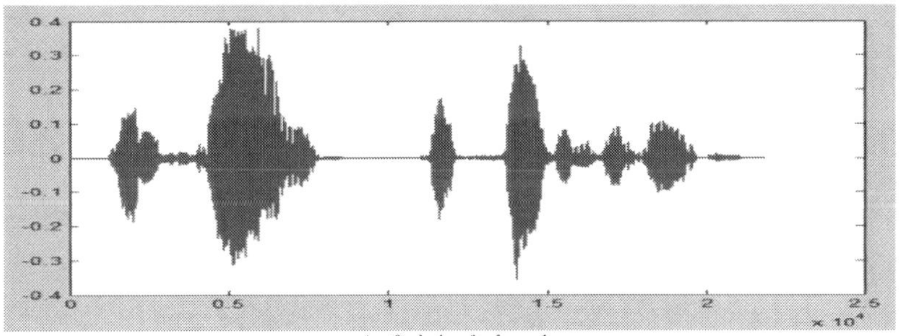

a). Original signal

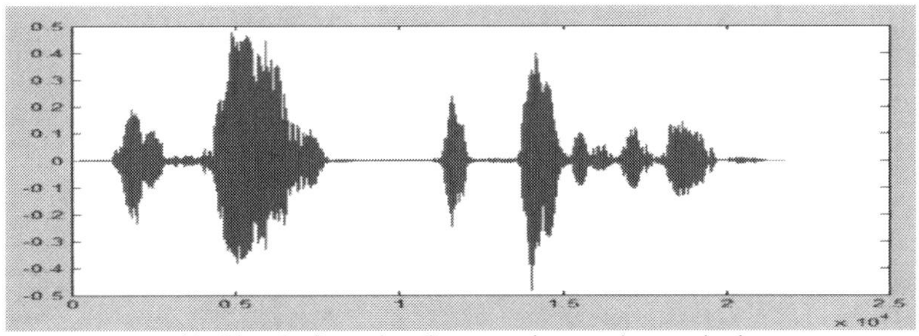

b). Signal after using Compressive sensing method

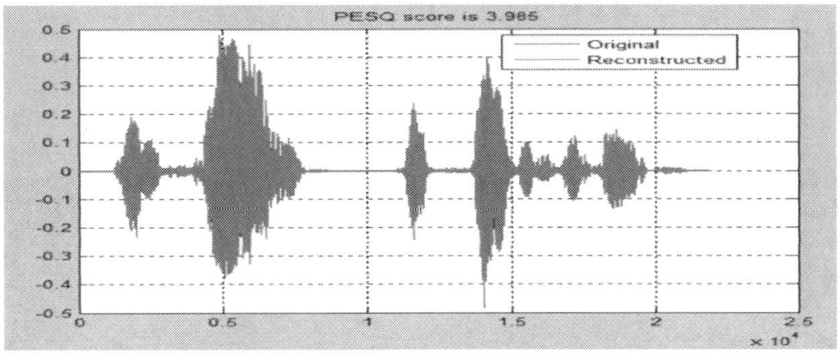

c). The cooperation of the constructed and original signal after using CS algorithm

Figure 6. The method shows the new approach of speech enhancement based on compressive sensing (CS) algorithm

The speech enhancement by using compressive sensing, in **Figure 6**, are constructed the signal almost as the same as original signal of the speech and it produces high performance score as shown in **Table 2**. The speech sound tested from **Table 2**., used the datasets with 22 speech files of the non-trained files. The quality of the speech enhancement is evaluated using Perceptual Evaluation of Speech Quality (PESQ), Signal-to-Noise Ratio (SNR). The reesults show that the speech enhancements have achieved the excellent result. The average of PESQ score is 4.478 which is a good quality as listening tests confirms.

Table 2., The performance results by evaluate using PESQ for the speech sound tested from the datasets with 22 speech files of the non-trained files.

Speech Signal	PESQ	Elapsed time (seconds)
sp01.wav	3.748	9.757911
sp02.wav	3.727	9.298481
sp03.wav	3.866	8.929454
sp04.wav	3.681	8.377577
sp05.wav	3.769	9.068006
sp06.wav	3.985	10.250037
sp07.wav	3.598	8.828342
sp08.wav	3.883	9.526404
sp09.wav	3.741	9.757164
sp10.wav	3.877	10.821275
sp11.wav	3.439	9.611207
sp12.wav	3.376	9.293070
sp13.wav	3.335	8.686577
sp14.wav	3.536	15.548540
sp15.wav	3.323	9.605843
sp16.wav	3.317	8.624775
sp17.wav	3.363	8.723107
sp18.wav	3.198	8.356940
sp19.wav	3.402	9.317810
sp20.wav	3.561	9.439650
sp21.wav	3.904	10.839918
sp22.wav	3.819	9.321426
Average	**3.6113**	**9.6356143**

5. Conclusions

In Speech Enhancement, the contention in this research is that it is possible to do even better by more fully leveraging concepts from state-of-the-art signal compression and processing algorithms. The key ingredient is a more realistic *structured sparsity* model that goes beyond simple sparsity by codifying the interdependency *structure* among the signal coefficients. The further evaluation for speech enhancement using compressive sensing has been presented. The proposed algorithm used gammatone filter banks and Discrete Cosine Transform (DCT) to spasify the speech signal as the synthesis speech using compressive sensing will work best on sparse signals. The object evaluation using PESQ score revealed that the *average PESQ score* was around *3.6113* and *processing time 9.6356* which is acceptable and given the excellent result.

6. References

[1] Kamath, S. D., *A Multi-Band Spectral Subtraction Method for Speech Enhancement* Master of Science in Electrical Engineering, The University of Texas at Dallas, 2001.

[2] Loizou, P. C. *Speed Enhancement Theory and Practice* (1st ed.): CRC Press, 2007.
[3] Fornasier, M., & Rauhut, H. (2011). Compressive Sensing. In O. Scherzer (Ed.), *Handbook of Mathematical Methods in Imaging*: Springer Science+Business Media LLC 2011.
[4] Unser, M.. Sampling – 50 years after shannon. *Proc IEEE 88*(4), 2000, pp. 569–587.
[5] Gunawan, T. S. *Audio compression and speech enhancement using temporal masking models.* Doctor of Philosophy, The University of New South Wales, 2007.
[6] Hu, Y., & Loizou, P. C. Subjective comparison and evaluation of speech enhancement algorithms. *ScienceDirect, 49 (2007)*, 2007, pp. 588-601.
[7] Yang, X. *Auditory domain Speech Enhancement.* Master of Science (Engineering), Queen's University, Kingston, Ontario, Canada, 2008.
[8] Ma, J., & Loizou, P. C.. SNR loss: A new objective measure for predicting the intelligibility of noise-suppressed speech. *ScienceDirect*, 2010, pp. 1-15.
[9] Mowlaee, P., Sayadiyan, A., & Sheikhzadeh, H. Evaluating single-channel speech separation performance in transform-domain. *Jornal of Zhejian University-Science C (Computers & Electronics)*, 2010, pp.160-174.
[10] Hu, Y., & Loizou, P. C. Evaluation of objective quality measures for speech enhancement *IEEETransactions on audio, Speech, and Language Processing, 16*(1), 2007a., pp. 229-238.
[11] Baraniuk, R. G. Compressive sensing. *IEEE Signal Processing Magazine*, 2007, pp. 118-120, 124.
[12] Sudirga, R. A Speech enhancement system based on statistical and acoustic phonetic knowledge. Master of science (engineering), Queen's university, Kingston, ontario, Canada. 2009.
[13] Kubin, G. and Kleijn, W., On speech coding in a perceptual domain, in International Conference on Acoustic, Speech and Signal Processing, vol.1, 1999,pp. 205-208.
[14] Salomon, D., Data Compression: The Complete Reference 3rd Edition, Springer, 2004.
[15] Figueiredo, M.A.T., Nowak, R.D. and Wright, S.J., Wright, Gradiant projection for sparse reconstruction, IEEE Journal of Selected Topics in Signal Processing, vol.1, pp. 586-597,2007.

Feasibility of using a humanoid robot to elicit communicational response in children with mild autism

Norjasween Abdul Malik[1,2], **Syamimi Shamsuddin**[2,3], **Hanafiah Yussof** [2] , **Mohd Azfar Miskam**[2] **and Aminullah Che Hamid**[2]
[2]Universiti Teknologi MARA (UiTM), 40450 Shah Alam, Selangor, Malaysia.
[3]Universiti Teknikal Malaysia Melaka (UTeM), Hang Tuah Jaya, 76100 Melaka, Malaysia.

E-mail: norjasweenabdulmalik@gmail.com, syamimi@ieee.org, hanafiah1034@salam.uitm.edu.my, aqua.genius24@gmail.com, aminullah_88@yahoo.com

Abstract. Research evidences are accumulating with regards to the potential use of robots for the rehabilitation of children with autism. The purpose of this paper is to elaborate on the results of communicational response in two children with autism during interaction with the humanoid robot NAO. Both autistic subjects in this study have been diagnosed with mild autism. Following the outcome from our first pilot study; the aim of this current experiment is to explore the application of NAO robot to engage with a child and further teach about emotions through a game-centered and song-based approach. The experiment procedure involved interaction between humanoid robot NAO with each child through a series of four different modules. The observation items are based on ten items selected and referenced to GARS-2 (Gilliam Autism Rating Scale-second edition) and also input from clinicians and therapists. The results clearly indicated that both of the children showed optimistic response through the interaction. Negative responses such as feeling scared or shying away from the robot were not detected. Two-way communication between the child and robot in real time significantly gives positive impact in the responses towards the robot. To conclude, it is feasible to include robot-based interaction specifically to elicit communicational response as a part of the rehabilitation intervention of children with autism.

1. Introduction

Human-robot interaction (HRI) is an evolving area of research which is now being applied in the rehabilitation of children autism. HRI is simply defined as the relationship between humans and intelligent robots for the benefit of mankind. Socially Assistive Robotics (SAR) which is a domain in HRI has a more specific scope of helping people through social interaction than using physical contact [1, 2]. Effective usage of robotics technology in aid of the disabled holds promising and effective possibilities as it is motivating, accurate, objective and adaptive to the needs of individuals.

[1] Corresponding author

Hence, rehabilitation robotics is part of the notion for robots to play substantial role in healthcare and specific rehab interventions for the elderly people and persons affected by neurodevelopmental disorders such as Cerebral Palsy (CP) and Autism Spectrum Disorders (ASD).

Autism is a neurodevelopmental disorder that is normally diagnosed during the first three years of life. It is a lifelong disability which has three deficit areas which are social interaction, stereotyped behaviour and communicational impairment. Statistic indicates that 1 in every 150 children in Malaysia are diagnosed with autism [3]. Only through early identification and intensive intervention during the early and preschool years may improve the future lives for most children with autism.

Recent literature has proven the applications of robots as a feasible medium in therapy and to enhance communication skill among children with autism [4, 5]. This is due to the fact that children with autism typically feel more at ease with robots than human because of robot interactions are simpler and more predictable and the children are in control of the social interaction [6]. Since children with autism have deficits in social interaction, a robot is suitable to elicit communication skills using a combination of both voice and movement during the interaction [7].

This experiment consists of four modules with the specific goal to explore the response of two autistic children when they are exposed to humanoid robot NAO that has been programmed to display 5 different emotions using its body poses and gestures. Game-based approach have been tested in preceding literature [8, 9] which resulted with meaningful human- robot interaction as well as human-human interaction. It is observed that children with autism enjoy playing imitation games with robots, allowing them the exposure to the interaction space where robots act as objects of shared attention thus promoting social and communication skills.

2. Humanoid robot NAO as medium to attract response in communication

In the year 2008, Aldebaran Robotics in France has introduced NAO, a humanoid robot which is autonomous and programmable [10] , with 25 degree-of-freedom and a human-like body consisting of a body, a head, a pair of eyes, two legs and two arms as shown in Figure 1. It is known that children with autism often face learning difficulties and have a hard time expressing and reflecting emotions. We use a humanoid robot in our study since it is a human-like body is useful to naturally attract children's attention and the robot size is like a 2 year-old toddler.

Apart from that, the humanoid robot NAO is also able to generate complex body movements required for communication when needed. Also, NAO can express emotion and show various human expression using its gestures [11]. We program the modules for this experiment using Choregraphe software (figure 1) which is the programming tool that is provided together with the humanoid robot NAO.

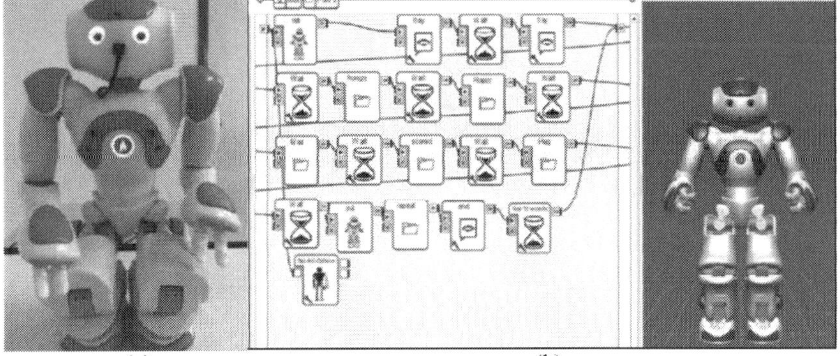

(a) (b)

Figure 1. Humanoid robot NAO (a) and Choregraphe software overview (b)

IOP Conf. Series: Materials Science and Engineering **53** (2013) 012077 doi:10.1088/1757-899X/53/1/012077

3. Content of the interaction module

The experimental procedure is designed to serve as a platform in fundamental investigation of HRI and humanoid robot in the therapy of children with autism. The procedure involves the robot executing its pre-programmed behaviours in an on-line mode. A total of 4 modules were executed by NAO to entice reaction and interaction from the subjects in this study. The content of each module is decided based upon discussion with experienced clinicians in autism therapy and intensive review studies. Flowchart in figure 2 below above describes the flow of the interaction module.

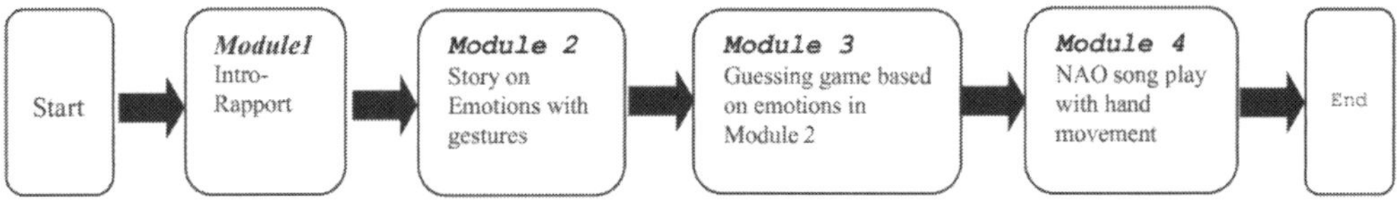

Figure 2. The flowchart of the four modules of interaction between child and robot

Each unit of interaction has a ten-second break as interval. Overall interaction session took about 15 minutes. Each child will experience this four module only just the once. Module 1 started with introductory rapport, when NAO greet the candidate by saying "Hello" followed with introducing itself and asking simple questions to promote the two-way communication. Then, in Module 2 NAO will deliver a short story about five different emotions complete with body gestures. This short story is delivered with a music background that suits the emotions which are hungry, happy, mad, scared and love/hug. Then, the emotion-based gestures will be repeated one last time without the background sound.

After that, the child-robot interaction continued to Module 3 where the child needs to guess back all the emotions asked by NAO. Finally in Module 4, NAO will play a song 'Twinkle twinkle little star' together with arm movement in accordance to the melody.

4. Methodology and experimental set-up

The study had been carried out at a private day-care center in Shah Alam, Malaysia that provides special care for children with disabilities. The classroom that we use to carry out the experiment has been divided into two areas: the experimenter station and the child-robot interaction area.

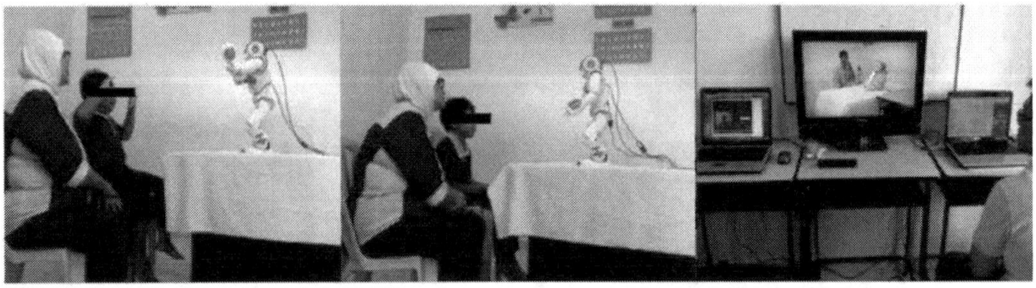

(a) (b) (c)

Figure 3. Child KA (a), child HH (b) during the interaction session and the experimental station which is hidden from view of the children (c)

At this centre, only two children fitted the selection criteria that we outlined for this study. The subjects need to have a prior diagnosis of autism, must not have deficits in hearing and vision, no self-injury behaviour, no nystagmus and able to communicate in English. The candidates for this experiment are child KA and child HH. They are 7 and 9 years old respectively. For the period of the experiment, each of the autistic children was accompanied with their teacher during the child-robot interaction area. This is shown in figure 3. Two external HD video cameras and 1 mini video camera were used to record visual and sound during the child-robot interaction. The third picture in figure 3 shows the experimenter station which is hidden from view and separated from the child-robot interaction area. The video recording process will be monitored by the experimenter who did not interfere with the interaction. The teacher accompanying the child will be observing the interaction without giving any instructions to the child.

During the interaction program, we have also outlined circumstances to abort the program. This includes conditions or situations where the child becomes restless and uncooperative or in the occurrence of programming error or whenever the child's teacher was to request such action of aborting the interaction.

5. Result and discussion

The comprehensive evaluation on the initial response of the children during our pilot study is demonstrated in the remainder of this paper, focusing on the children's scores in the subscale of communication only. At mentioned before, the aim of the study is to observe the response children with mild autism when they are exposed to simple human-robot interaction (HRI) modules executed by a humanoid robot NAO.

5.1 Analysis using the Behaviour Score-sheet.

The recorded video and sound from the video cameras were processed for analysis. We adopt the same behaviour score sheet (referenced to The Gilliam Autism Rating Scale (GARS-2)) and previously elaborated in [12] to assess the videos. In the assessment sheet, ten items are used to examine the communication subscale shown by the candidates. The scores are presented out in figures of percentage: less percentage indicating the behaviour of the candidates are close to normal people. Higher score of percentage indicates that more autistic behaviour is observed in that particular interaction.

5.2 Results for the communication subscale.

As shown in figure 4 below, interestingly both of the children showed only 7.9% of the autistic trait in communication subscale during the child- robot interaction. Specific traits shown are looking away when NAO is active and inappropriately answer questions about a statement or brief story. Nevertheless, prior to the start of the experiment the children were observed to be looking at the robots eagerly and smiled. Immediately after completing the 4 module, child KA smiled again and was talking to his accompanying teacher while child HH became quiet and under responsive.

Overall, the results indicate positive response from both of the autistic child with only 7.9% of autistic behavior shown. It shows the potential of humanoid robot such NAO to act as a medium to attract the children to give attention and participate in two-way communication the robot.

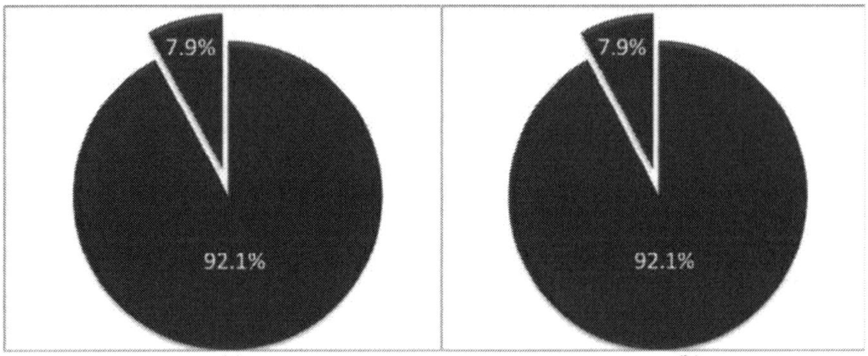

(a) (b)

Figure 4. Total percentage of observed autism characteristics in the communication subscale for child KA (a) and child HH (b)

Both of the bar graph in figure 5 shows the score (percentage) of autism behaviour in communication subscale for each child according to each module. For Child KA, an obvious decrease in percentage of autism behaviour is observed throughout the interaction. However, in Module 1 autistic characteristic of child KA is more than child HH. Only child KA showed autistic trait during the Module 2. Surprisingly, he had responded by correctly during the entire emotion guess games in Module 3.

Looking at the bar graph on the right, child HH only showed the autistic criteria in Module 1 and Module 3 only. Even though Child HH tends to look away when NAO play emotion games in Module 3, he also had successfully guessed all the emotions shown by NAO.

An obvious interesting observation is that both of the children showed no autism traits during Module 4 where NAO played the song "Twinkle twinkle little star" together with arm movements. Similar to the results from our first pilot experiment, children favour this song as it is very familiar to them. Hence, it is easier to entice communication between child and robot through the usage of common nursery song in the interaction.

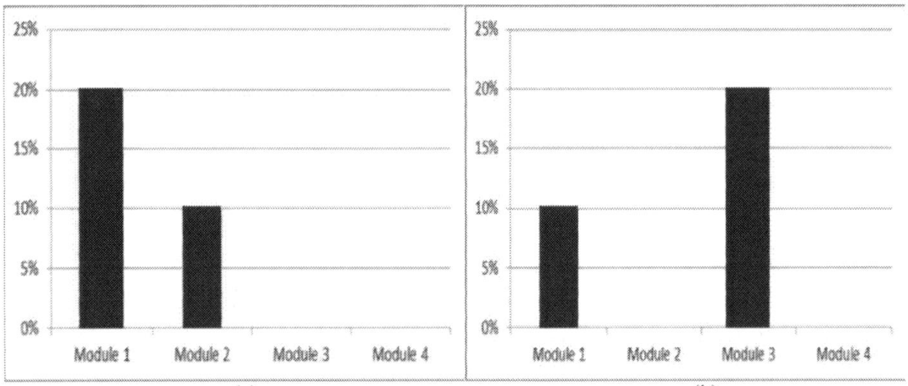

(a) (b)

Figure 5. Percentage of autism behaviours for each module in communication subscale throughout the experiment for child KA (a) and child HH (b)

Table 1 below indicates all the 10 items of assessment in the behaviour score sheet both autistic children. The scores have been averaged across the 4 modules. During the interaction, child KA and

HH only exhibits autistic trait of tend to look away when NAO is active and inappropriately answer questions about statement in module. For this communication subscale, both of the children did not show any autistic behaviour during the interaction for the other 8 items in this assessment sheet. The lack in the children's understanding of English words may cause the candidates tend to answer inappropriately on the questions about statement of brief story. Despite this, both children were able to guess all the emotions correctly.

Table 1. Score percentage of each item assessed in the communication sub-scale

Items	Module Average Score (%) for each item of Assessment	
	KA	HH
Repeats words verbally	0	0
Repeats word out of context	0	0
Repeats words/phrases over and over	0	0
Speaks with flat tone	0	0
Uses pronouns inappropriately (refers self as "third party")	0	0
Uses the word "I" inappropriately	0	0
Repeat unintelligible words over and over	0	0
Uses gestures instead of speech to obtain objects	0	0
Tend to look away when NAO is active	50	25
Inappropriately answer questions about statement or brief story	50	25

6.Summary

The experimental analysis indicated that encouraging results had been recorded specifically on the children's response in communication during interaction with the NAO robot. Overall score autistic behavior of only 7.9% of each child clearly indicates that the humanoid robot NAO is able to attract, encourage and hence augment the interest of the children to participate in communication.

The observation results are summarized as follows:

1) Both child participants score 92.1% of not showing his autistic characteristics for overall performance during the child-robot interaction
2) Each child scored differently for each module where child KA showed highest autistic trait in Module 1 while child HH show highest autistic trait in Module 3. Both of the children showed no autism traits in Module 4
3) Average score for each autistic trait across the four modules shows that child KA scored 50% while child HH score 25% for the same items: looking away when NAO is active and inappropriately about statement or brief story
4) In the guess emotions game, both candidates can remember the emotions portrayed by NAO and hence able to answer correctly when asked by the robot.

Overall results indicate the feasibility of using a humanoid robot to elicit communicational response in children with autism. It is convincing that humanoid robot such NAO can make the lives of children with autism better through child-robot interaction focusing on two-way communication. Even though the results are at a preliminary stage, we are positive that the game-based approach for the children participating in this study to learn, understand and correctly guess the emotions shown by NAO has been a success.

References

[1] D. Feil-Seifer, "Socially Assistive Robot-Based Intervention for Children with Autism Spectrum Disorder," ed: University of Southern California, 2008, p. 92.
[2] M. J. Matarić, "Socially assistive robotics," *IEEE Intelligent Systems,* vol. 21, pp. 81-83, 2006.
[3] NASOM. (2012, 29 April 2012). *What is Autism*. Available: http://www.nasom.com.my
[4] S. Costa, F. Soares, C. Santos, M. J. Ferreira, F. Moreira, A. P. Pereira, and F. Cunha, "An approach to promote social and communication behaviors in children with Autism Spectrum Disorders: Robot based intervention," in *RO-MAN, 2011 IEEE*, Atlanta, GA, 2011, pp. 101-106.
[5] H. Kozima, C. Nakagawa, and Y. Yasuda, "Interactive robots for communication-care: A case-study in autism therapy," in *IEEE International Workshop on Robot and Human Interactive Communication, 2005 (ROMAN 2005)*, 2005, pp. 341-346.
[6] B. Robins and K. Dautenhahn, "The Role of the Experimenter in HRI Research - A Case Study Evaluation of Children with Autism Interacting with a Robotic Toy," in *Robot and Human Interactive Communication, 2006. ROMAN 2006. The 15th IEEE International Symposium on*, Hatfield, 2006, pp. 646-651.
[7] Filipek. (Practice Parameter: Screening and Diagnosis of Autism. *55 (Report of the Quality Standards Subcommittee of the American Academy of Neurology and the Child Neurology Society. Neurology),* 2000, 468-479.
[8] B. Robins, K. Dautenhahn, R. t. Boekhorst, and A. Billard, "Effects of repeated exposure to a humanoid robot on children with autism " in *Cambridge Worshop Universal Access and Assistive Technology (CWUAAT)*, Cambridge, UK, 2004, pp. 225-236.
[9] K. Dautenhahn and A. Billard, "Games children with autism can play with Robota, a humanoid robotic doll," in *Universal Access and Assistive Technology*, S. Keates, P. Clarkson, P. M. Langdon, and P. Robinson, Eds., ed London: Springer-Verlag 2002.
[10] S. Shamsuddin, L. I. Ismail, H. Yussof, N. Ismarrubie Zahari, S. Bahari, H. Hashim, and A. Jaffar, "Humanoid robot NAO: Review of control and motion exploration," in *Control System, Computing and Engineering (ICCSCE), 2011 IEEE International Conference on*, 2011, pp. 511-516.
[11] M. S. Erden and A. Tapus, "Postural expressions of emotions in a humanoid robot for assistive applications," in *Robotics: Science and Systems (RSS) Workshop on Learning for Human-Robot Interaction Modeling*, 2010.
[12] S. Shamsuddin, H. Yussof, L. Ismail, F. A. Hanapiah, S. Mohamed, H. A. Piah, and N. I. Zahari, "Initial response of autistic children in human-robot interaction therapy with humanoid robot NAO," in *Signal Processing and its Applications (CSPA), 2012 IEEE 8th International Colloquium on*, 2012, pp. 188-193.

Mobile Multicast in Hierarchical Proxy Mobile IPV6

Azana Hafizah Mohd Aman[1], Aisha Hassan A Hashim[2], Amin Mustafa[3] and Khaizuran Abdullah[4]

[1,2]Department of Electrical and Computer Engineering, Faculty of Engineering, International Islamic University Malaysia, Kuala Lumpur.
[3,4]Faculty of Engineering, El Neelain University, Khartoum, Sudan

E-mail: azana05@yahoo.com, aisha@iium.edu.my

Abstract. Mobile Internet Protocol Version 6 (MIPv6) environments have been developing very rapidly. Many challenges arise with the fast progress of MIPv6 technologies and its environment. Therefore the importance of improving the existing architecture and operations increases. One of the many challenges which need to be addressed is the need for performance improvement to support mobile multicast. Numerous approaches have been proposed to improve mobile multicast performance. This includes Context Transfer Protocol (CXTP), Hierarchical Mobile IPv6 (HMIPv6), Fast Mobile IPv6 (FMIPv6) and Proxy Mobile IPv6 (PMIPv6). This document describes multicast context transfer in hierarchical proxy mobile IPv6 (H-PMIPv6) to provide better multicasting performance in PMIPv6 domain.

1. Introduction

With the fast progress of different type of mobile networking technologies, the importance of improving the existing architecture or operation of services based on context increases. This includes mobile multicast networking environments. When a mobile node (MN) moves to a new subnet it needs to continue certain services that have already been established at the previous subnet.

Multicast is the delivery of data to a set of selected receivers. Multicast applications can be one-to-many and many-to-many. Multicast receivers require a membership protocol to join or leave a multicast group at any time. The IETF defined the Multicast Listener Discovery (MLD) protocol to be used by IPv6 routers to discover the presence of interested receivers of a particular multicast group. MLDv2 [1] describes two parts. The first part is the host part, which specifies how receivers learn the existence of a multicast router, how they discover groups, and how they express their interest in listening to a given group. The second part is the router part, which specifies how multicast routers manage membership information and how they exchange this information in order to build the multicast delivery tree. Multicast routing protocols are in charge for the construction of the multicast delivery tree.

The CXTP [2] is designed by the IETF to provide general mechanisms for exchange of context data for moving mobile nodes (MN) between access routers (AR). It gives support of the seamless handover based on service continuation using context and could be used to transfer different kind of control data and resources based services. It aims to contribute to the enhancement in handover performance and proposed for MN for quickly re-establishment of their services when the nodes move and change their access routers.

Mobility is defined as the movement of a node from a network to another network. A change in mobile node's point of attachment to the Internet is called handover. The handover is a change in MN's point of attachment to the Internet such that the MN is no longer connected to the same IP subnet as it was previously. It is a difficult problem to maintain reach ability of a MN when moving from one network to another. The challenge is to offer a large range of wireless mobile services to highly heterogeneous users for a highly effective handover. Many proposals have been introduced in order to solve this issue. This includes proposals on Mobile IPv6 (MIPv6) [3], Fast Handovers Mobile IPv6 (FMIPv6) [4], Hierarchical Mobility IPv6 (HMIPv6) [5] and Context Transfer Protocol (CXTP) [2].

Proxy Mobile IPv6 (PMIPv6) [6] enables network-based mobility for IPv6 mobile nodes (MNs) that do not implement any mobility protocols. The Local Mobility Anchor (LMA) is the topological anchor point to manage the mobile node's binding state. The Mobile Access Gateway (MAG) is an access router or gateway that manages the mobility-related signalling for an MN. A MN is attached to the Proxy Mobile IPv6 Domain (PMIPv6-Domain) that includes LMA and MAG(s), and is able to receive data coming from outside of the PMIPv6-Domain through LMA and MAG. Network-based mobility support for unicast is addressed in [6], while multicast support in PMIPv6 is not discussed in it.

This paper is organize as follows, section two presents the current challenges, section three presents related works, section four presents proposed scheme, section five present discussions and section six presents conclusions

2. Current challenges

The main problem in mobile multicast is the reconstruction of multicast delivery tree. When a MN moves to a new subnet it needs to continue certain services that have already been established at the previous subnet. High latency could leads to packets being forwarded to the outdated path and lost which means the increase of packet loss rate. Eventually cause service interruption, degrade performance especially the performance of real-time and delay sensitive application. It is the job of the multicast routing protocol and depending on the multicast membership protocol information to reconstruct the multicast delivery tree. When mobile multicast subscriber moves into another network, it has to reconstruct its multicast membership protocol states in the new location.

In PMIPV6 the Local Mobility Anchor (LMA) and Mobile Access Gateway (MAG) set up a bi-directional IPv6-in-IPv6 tunnel for each mobile node and forwards all mobile node's traffic according to [6], it highly wastes network resources when a large number of mobile nodes join/ subscribe the same multicast sessions/channels, because independent data copies of the same multicast packet are delivered to the subscriber nodes in a unicast manner through MAG.

The objective is to design a new scheme that integrates hierarchical with PMIPv6 that supports mobile multicast context transfer. The main objective of this research is to improve the performance of MN during handover in a multicast session. The performance metric chosen are based on this objective. The performance metrics used for simulation evaluation are handover latency, packet loss and bandwidth overhead. The proposed extension will be benchmarked with the standard PMIPv6.

3. Related works

Many schemes have been proposed in order to deploy the multicast service in the PMIPv6 domain. Work in [7], which is the base solution, provides options for deploying multicast listener functions in PMIPv6 domains without modifying mobility and multicast protocol standards. In this work, the MAG acts as a MLD proxy. The MAG dedicates a tunnel link between LMA and MAG to an upstream interface for all multicast traffic. It requires all the LMA to forward multicast packets to MAG through PMIPv6 tunnel which can be suboptimal.

Work in [8] uses a Multicast Tree Mobility Anchor (MTMA) as the topological anchor point for multicast traffic, as well as a direct routing option where the MAG can provide access to multicast content in the local network. All the multicast traffic has to go through the MAG-MTMA tunnel which results in suboptimal multicast routing path like the base solution. And the direct routing solution needs native multicasting infrastructure as a requirement.

Work in [9] describes the support of multicast senders in Proxy Mobile IPv6 domains. MLD proxy functions are deployed at the MAG. Multicast traffic be tunneled up to the LMA of the multicast sender, transferred to the LMA of the multicast listener and then tunneled downwards to the MAG of the multicast listener. However it revealed a problem when multicast listener and sender attach to the same MAG but different LMAs, the traffic has to go up to one LMA, cross over to the other LMA, and then be tunneled back to the same MAG, causing non-optimal multicast routes and redundant flows at the MAG.

On the mobility issue of PMIPv6, work in [10] implemented the standard PMIPv6. The PMIPv6 specification does only support the movement of hosts within the localized mobility domain. This research describes, what can be done with current standard protocols. It also highlighted the problem statement of fully supporting network mobility (NEMO, NEtwork that MOves) in Proxy Mobile IPv6. As for work in [11], PMIPv6 protocol supports Mobile Hosts moving independently, but not Mobile Routers in charge of moving networks. This research describes how to allow bidirectional communication between a Local Fixed Node (in the moving network) and a Correspondent Node. Work in [12], the PMIPv6 protocol inter-domain handover is not involved. This research describes network-based Inter-domain handover functionality and corresponding mobility options for PMIPv6.while work in [13] the PMIPv6 protocol describes Network-based mobility support for unicast. This research describes the deployment of multicast service in the PMIPv6 network.

4. Proposed scheme

CXTP [2] could be used to define control messages and procedures for initiation and authorization of context transfer between access routers. It is also used for transferring contexts states related to the mobile user and its active services before, after or during handover. The context transfer can be triggered, in a predictive scheme or reactive scheme. Based on MLDv2 [1], the multicast router knows that one or more mobile node is listening to a particular multicast address and source at given interface.

Using predictive scheme, as shown in figure 1. When it is a mobile node controlled triggered, the flow starts from step 0, and the mobile node sends a message to the previous access router to activate the context transfer at the previous M-LMA. Context transfer is started with the context activation by the mobile node sending a message to the previous access router. After the context activation, the multicast context transfer block (M-CTB) is built at the previous access router M-LMA in interaction with MLDv2.

While when it is a network controlled initiated, referring to figure 1, the flow starts from step 1 where by the previous access router M-LMA initiates the context transfer. The multicast context transfer block (M-CTB) for the multicast services of the mobile node is built in the previous M-LMA with input from the MLDv2 router entity and transferred to the next M-LMA. The M-CTB includes the multicast addresses required for the multicast services being used by the moving mobile node.

The previous M-LMA sends the keying material, the M-CTB and other information necessary to calculate the Authorization Token. After receiving the context data message with the M-CTB, the next M-LMA dispatches it to the MLDv2 router entity in order to establish an individual node context and to update the aggregate state. A change in the aggregated state will then trigger the active multicast routing protocol to build the distribution trees for each of the multicast groups.

When the mobile node performs the handover, it sends a message CTAR to activate the context transfer. The CTAR message is sent either to previous M-LMA or next M-LMA, depending on the router, to which the mobile node is connected. If CTAR is sent to previous M-LMA, the MN includes the next M-LMA's IP address; otherwise, if the message is sent to next M-LMA, the previous M-LMA address is sent. The CTDR message is sent by next M-LMA to previous M-LMA as response of CTD, when CTD requires response indicating success or failure of context transfer.

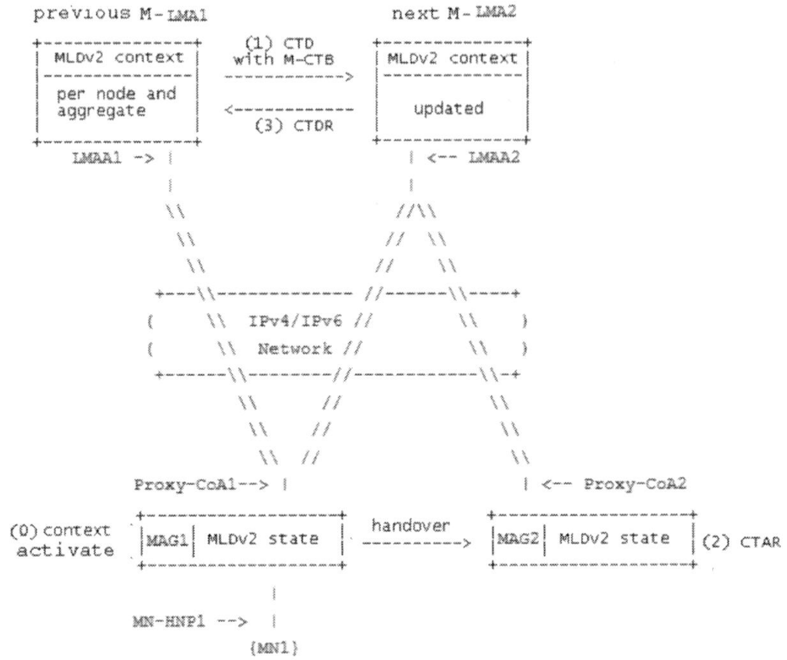

Figure 1. Predictive Scheme

As for the analysis handover latency is define as the time interval between last data segment received through the old path and first data segment received through the new path. It is one of the factors that are used to rate network performance. High handover latency is the major indicator of performance degradation, which may leads to high number of packet loss and thus causes communication disruption. Hence, using this definition it is the duration between the last data segment received (pLMA) and the first data segment received (nLMA).

5. Discussions

A solution to develop a better multicast handover performance in MIPv6 has been studied. The research aims to reduce mobile node's handover processing and packet delay in the establishment of new services. This paper discussed few proposals that enhance mobility and multicast mobility. In this paper we also proposed and discussed architecture for efficient mobile multicast.

6. Conclusions

In the future we are going to do simulation and mathematical analysis for the proposed scheme.

7. References

[1] R Vida, L Costa 2004 *Multicast Listener Discovery Version 2 (MLDv2) for IPv6* RFC 3810 Internet Research Task Force IRTF.
[2] J Loughney, M Nakhjiri, C Perkins and R Koodli 2005 *Context Transfer Protocol* RFC4067.
[3] Patel A, Leung K, Khalil M, Akhtar H, and K. Chowdhury 2005 *Mobile Node Identifier Option for Mobile IPv6 (MIPv6)* RFC 4283.
[4] Koodli R, Ed 2008 *Mobile IPv6 Fast Handovers (FMIPv6)* RFC 5268.
[5] Soliman H, Castelluccia C, ElMalki K, and L Bellier 2008 *Hierarchical Mobile IPv6 (HMIPv6) Mobility Management* RFC 5380.

IOP Conf. Series: Materials Science and Engineering **53** (2013) 012078 doi:10.1088/1757-899X/53/1/012078

[6] Gundavelli S, Ed Leung, K Devarapalli, V Chowdhury K and B Patil 2008 *Proxy Mobile IPv6(PMIPv6)* RFC 5213.

[7] T Schmidt, M Waehlisch and S Krishnan 2011 *Base Deployment for Multicast Listener Support in Proxy Mobile IPv6 (PMIPv6) Domains* RFC 6224.

[8] JC Zuniga, LM Contreras, CJ Bernardos, S Jeon and Y. Kim 2013 *Multicast Mobility Routing Optimizations for Proxy Mobile* draft-ietf-multimob-pmipv6-ropt-03IPv6.

[9] T Schmidt, S Gao, H Zhang and M Waehlisch 2012 *Mobile Multicast Sender Support in Proxy Mobile IPv6 (PMIPv6) Domains* draft-ietf-multimob-pmipv6-source-01.

[10] CJ Bernardos, M Calderon and I Soto 2012 *PMIPv6 and Network Mobility Problem Statement* Internet-Draft.

[11] A Petrescu, M Boc and C Janneteau 2012 *Network Mobility with Proxy Mobile IPv6* Internet-Draft.

[12] Zhengming Ma, Ke Wang and Fei Zhang 2012 *Network-based Inter-domain handover Support for Proxy Mobile IPv6* Internet-Draft.

[13] H Chan, D Liu, P Seite, H Yokota and J Korhonen 2012 *Requirements for Distributed Mobility Management* Internet-Draft.

[14] TC Schmidt and M Waehlisch 2007 *Multicast Mobility in MIPv6: Problem Statement and Brief Survey* Internet Draft.

[15] J Liu, W Luo and W Yan 2012 *Routes Optimization for PMIPv6 Multicast* Internet-Draft.

[16] TC Schmidt, M Waehlisch and G Fairhurst 2010 *Multicast Mobility in MIPv6: Problem Statement & Brief Survey* RFC 5757.

Centre of Gravity (C.O.G)-Based Analysis on the Dynamics of the Extendable Double-Link Two-Wheeled Mobile Robot

M T Abdul Rahman, S Ahmad[1], R Akmeliawati, T Altalmas and A Aula

Department of Mechatronics Engineering, International Islamic University Malaysia, Gombak, Selangor 50728, Malaysia

Abstract. This paper discusses about the analysis on the centre of gravity (C.O.G) in affecting the input reference of the motion control of the extendable double-link of two-wheeled mobile robot. The proposed system mimics double inverted pendulum, where the angular position of the first link (Link1) is to be varied depends on the value of the angular position of the second link (Link2) and the elongation of the extendable-link (Link3) that is attached to Link2 with different payload. The two-wheeled mobile robot together with the extendable link on Link2 makes that system become more flexible but yet, the system has become more unstable. The inclination of extendable link at any interest angle will affect the C.O.G of the system especially when the payload is having a significant weight. This two-wheeled mobile robot can be balanced on the condition that the system's center of gravity must be located on the centre of the wheels. Therefore the input reference of Link1 will be determined from the C.O.G analysis of the system with the payload. Preliminary results show that the angular position of Link1 can be set at suitable degree based on C.O.G analysis that is used for motion control.

1. INTRODUCTION

Many researchers have been put more effort in doing research on two-wheeled self-balancing autonomous mobile robots as it offers many kind of benefits with the ability to balance on its two wheels and spin on the spot, [1]. The basic idea for this system to be balanced is by driving the wheels toward the direction of the manipulator of the robot are falling. It means that, the robot will remain in balanced under its center of gravity position when robot is moving. [2]. Many researchers have investigated the manipulators control via center of gravity. [3] had done research Golem krang: Dynamically stable humanoid robot for mobile manipulation. A.M.H.S. [4] had done on traction force improvement of two-wheel mobile manipulator by changing the centre of gravity. [5] worked on motion control of dynamically balanced two-wheel mobile manipulator through C.O.G manipulation. Those approaches have advantage on taking the stability of the C.O.G into consideration that provide smooth trajectory of the C.O.G. However, most the research mainly concentrated on the use of revolute-jointed serial manipulators. By having the prismatic-jointed it will fulfill the needs of some area of application that acquire more flexibility to move as well as the capability to reach higher level of height at certain angle as implemented in this research. So, the objective of this paper is to control the motion of extendable double-link two-wheel mobile robot based on C.O.G analysis.

[1] Corresponding author. Email: salmiah@iium.edu.my
Tel: +603 6196 4506

IOP Conf. Series: Materials Science and Engineering **53** (2013) 012079 doi:10.1088/1757-899X/53/1/012079

2. C.O.G MATHEMATICAL REPRESENTATION AND ANALYSIS

2.1 The position of Centre of Gravity (C.O.G) of double link two-wheeled mobile robot

From the dynamic stability point of view, it is necessary to calculate the individual C.O.G of this mobile robot in order to find out the real C.O.G of this system [6].

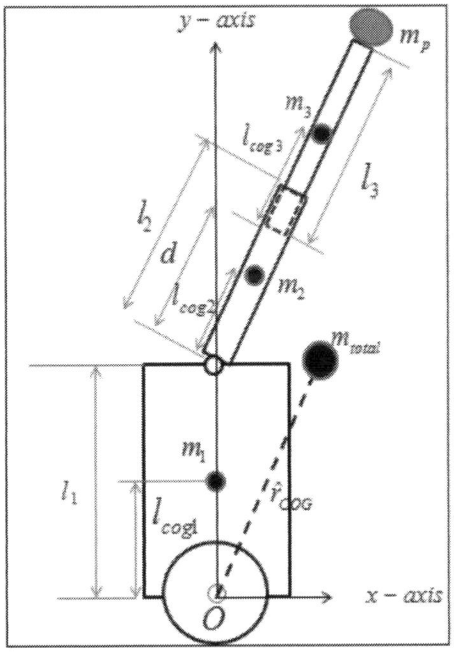

Figure 1: 2D parameters

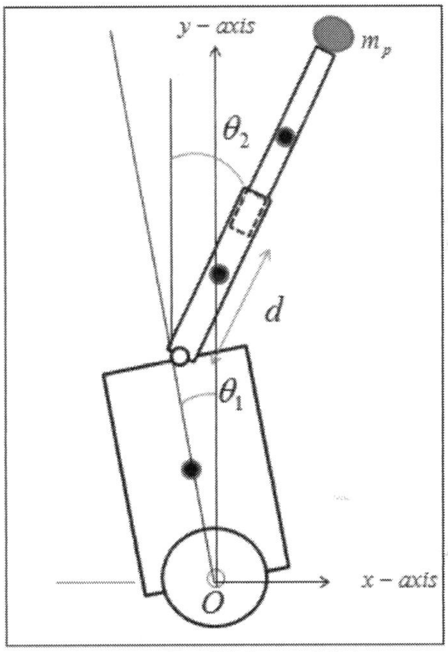

Figure 2: Angle of link 1 and link 2

Figure 1 and figure 2 above show the parameters exist in the system. So, from the center of mass for each link we will find the center of gravity of the system from point O given by the equation below [7],

$$\hat{r}_{COG} = r_1\hat{u}_1 + r_2\hat{u}_2 \tag{1}$$

Where $\hat{u}_1$ and $\hat{u}_2$ are unit vectors along link 1 and link 2, respectively, given as

$$\hat{u}_1 = \sin\theta_1\hat{i} + \cos\theta_1\hat{j} \tag{2}$$

$$\hat{u}_2 = \sin(\theta_1 + \theta_2)\hat{i} + \cos(\theta_1 + \theta_2)\hat{j} \tag{3}$$

While r_1 and r_2 are the variation value depending on mass of payload m_p given as

$$r_1 = \frac{m_1 l_{cog1}}{m_{total}} \tag{4}$$

$$r_2 = \frac{m_2 l_{cog2} + m_3 l_{cog3} + m_3 d + m_p l_3}{m_{total}} \tag{5}$$

Where m_1 is the mass of Link1 and l_{cog1} is the location of center of mass of Link1 from point O. Also, m_2 and m_3 are the masses of Link 2 and extendable link respectively while l_{cog2} and l_{cog3} are the center of mass of Link 2 and extendable link shown in Figure 1. To locate the C.O.G of the robot, we should first have the value of r_1 and r_2 in Eq. (1). The value of r_2 is also a function of variable d but in this research we have set the value to 0.1 meter. Table 1 below show the parameters used in these experiments. The system, when at no motion can be treated as a manipulator.

Table 1: System's parameters

Parameter	Value	Unit
m_1	0.900	kg
m_2	0.075	kg
m_3	0.075	kg
l_1	0.200	meter
l_2	0.150	meter
l_3	0.170	meter
l_{cog1}	0.100	meter
l_{cog2}	0.075	meter
l_{cos3}	0.085	meter
d	0.100	meter

2.2 The control of C.O.G position of the mobile robot

In this part, we discuss how to control the center of gravity of the manipulator and payload so that the system is stable. We can prevent the manipulator system from tumbling, if we can control the center of gravity on an identical imaginary vertical line of the wheel by changing the posture of the manipulator which is in this experiment given as angle Link1(θ_1) to make the system stable [8]. So, the wheels will move forward or backward to make sure the backward movement of center of gravity. The C.O.G control is using the based on the resolved motion rate control of the manipulator system as discussed in [9]. So, the velocity of the center of gravity can be determined by differentiating Eq. (1) with respect to time given as

$$[\dot{r}_g] = [J_g][\dot{\theta}_g] \tag{6}$$

where the Jacobian matrix of the center of gravity as follows;

$$J_g = \left[-\frac{m_1 l_{cog1}}{m_{total}} \sin\theta_1 \quad -\left(\frac{m_2 l_{cog2} + m_3 l_{cog3} + m_3 d + m_p l_3}{m_{total}} \right) \sin(\theta_1 + \theta_2) \right] \tag{7}$$

From the Eq. (6) we can obtain the pseudo inverse matrix J_g^+. By multiplying this pseudo inverse of matrix on both sides of eq. (6), we obtain

$$[\dot{\theta}_g] = [J_g][\dot{r}_g] \tag{8}$$

From this equation, we can control both the position and the center of gravity of the manipulator by the proposed C.O.G control method. Figure 3 shows the system block diagram or controller design for this research.

IOP Conf. Series: Materials Science and Engineering **53** (2013) 012079 doi:10.1088/1757-899X/53/1/012079

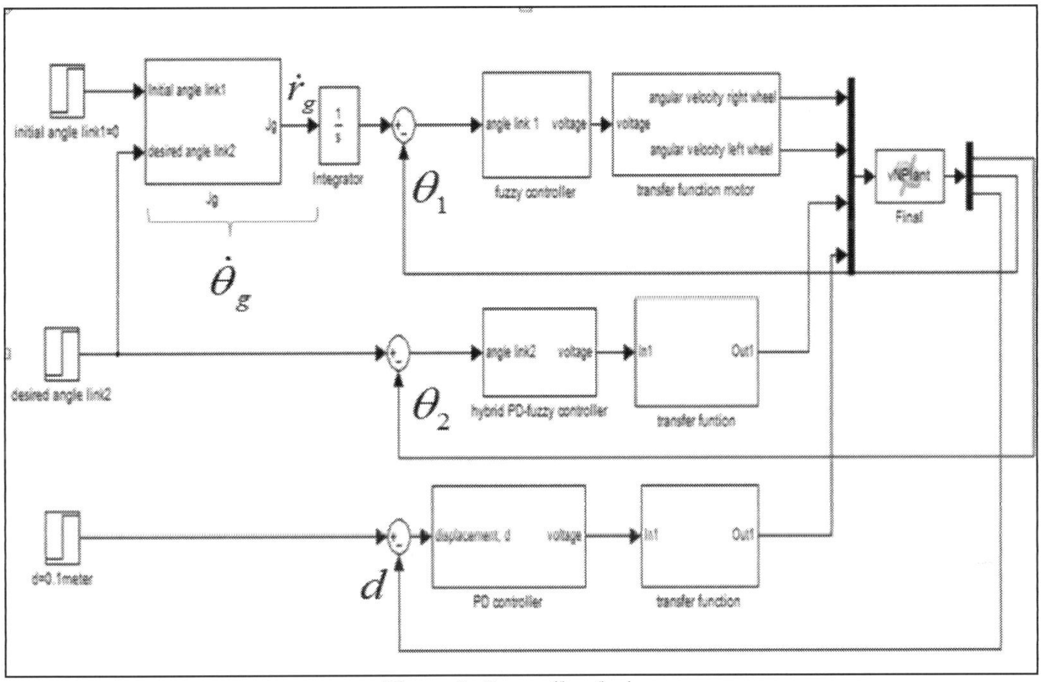

Figure 3: Controller design

Figure 3 shows a modular hybrid control consisting of fuzzy-PD type control, which is responsible for Link1, Link2 is controlled using PD-Fuzzy controller while the distance for extendable link is controlled using PID controller, [10]. In this research, two types of controllers were designed to control the robot at different objectives. The main controller for this system is Fuzzy Logic Controller (FLC) where there are two FLCs designed. First, the FLC with two inputs and two outputs were designed to control the voltage of left and right motor of Link1. The second FLC with two input and one output was designed to control the angular speed of the top motor for Link2, where its position depend on the top motor. Both FLCs were designed using PD type where the inputs are error (e) and change of error (Δe) while the outputs for both FLCs are voltage. The fuzzy rules were constructed such that the left and right motion will respond accordingly for the balance and stable of the system. On the other hand, a Proportional Integral Derivative (PID) was designed to control the elongation of the extendable link.

3. SIMULATION RESULTS AND DISCUSSION

In this paper, we have three variables that need to be controlled which are the orientation of the Link1 (θ_1), the orientation of the Link2 (θ_2) and also the length of the extendable link, (d). The orientation of the base (θ_1) should be varied depending on the value of elongation the extendable link, (d), the mass of the payload and the angular position of the Link2 (θ_2). From Table 2, it shows that the Link1 varies when θ_2 is varies. The negative value of θ_1 shows that it opposes the direction of the Link2 to make sure that the robot will be balanced. It can be seen as well θ_1 is proportional to the increase of θ_2, but increments value of θ_1 is small. Figure 4, 5 and 6 show the simulation graph of θ_2, θ_1 and d. From the graphs, it shows that the time for the system to be balanced is within 5 seconds.

Table 2: Angle link θ_1 vs. angle link2 link 2, θ_2 (m_p*=0 and d=0.1m*)

Simulation	
θ_2 (degree)	θ_1(degree)
20.0	-3.67
30.0	-4.11
40.0	-4.43
50.0	-4.54
60.0	-4.51

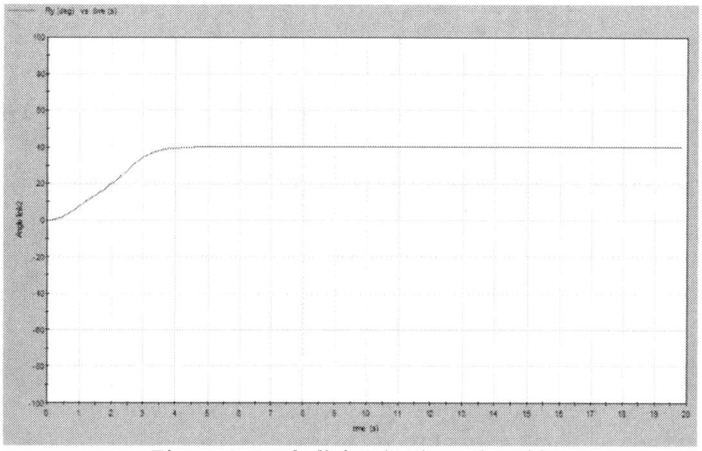

Figure 4: Angle link 2 (40°) vs. time (s)

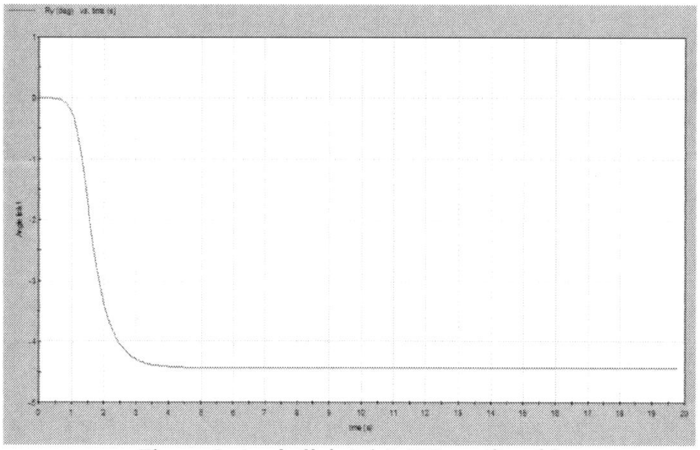

Figure 5: Angle link 1 (-4.43°) vs. time (s)

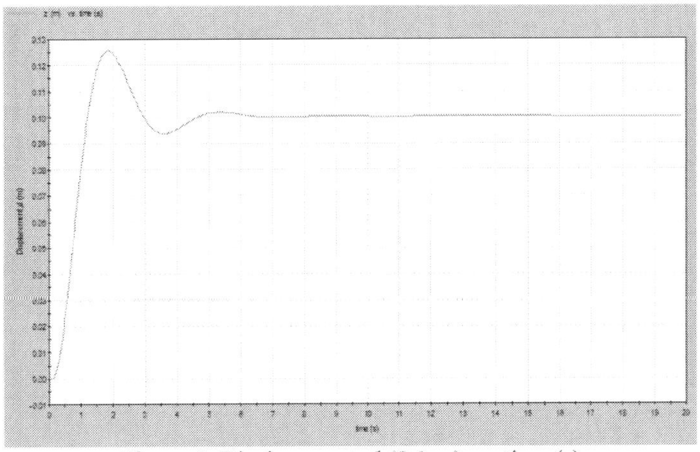

Figure 6: Displacement, d (0.1 m) vs. time (s)

The payload of the manipulator is varied to analyze the effect on the inertia. Table 3 shows θ_1 varies when the mass of payload varies (m_p) and the θ_1 is proportional to the mass of the payload. Figure 7, 8 and 9 show the simulation graph for angle θ_1, θ_2 and *d*. From the graphs, it shows that the time for the system balance is within 6 seconds.

Table 3: Angle link 1, θ_1 vs. mass, m_p (for θ_2=40° and d=0.1m)

Simulation	
m_p (kg)	θ_1(degree)
0.25	-5.93
0.50	-7.27
0.75	-9.01
1.00	-11.30

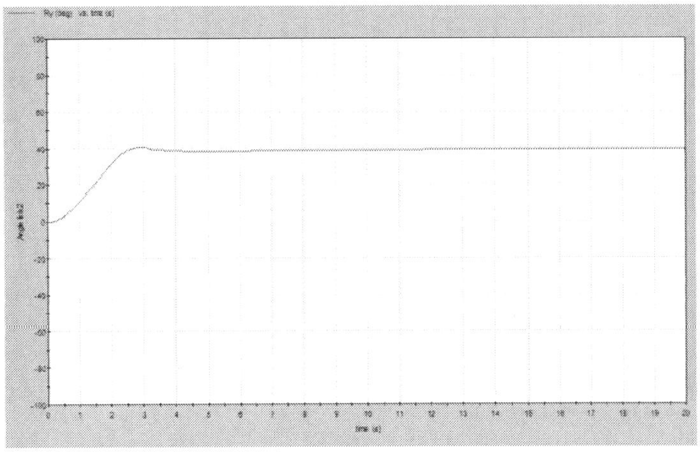

Figure 7: Angle link 2 (degree) vs. time (s) (for m_p 0.75kg)

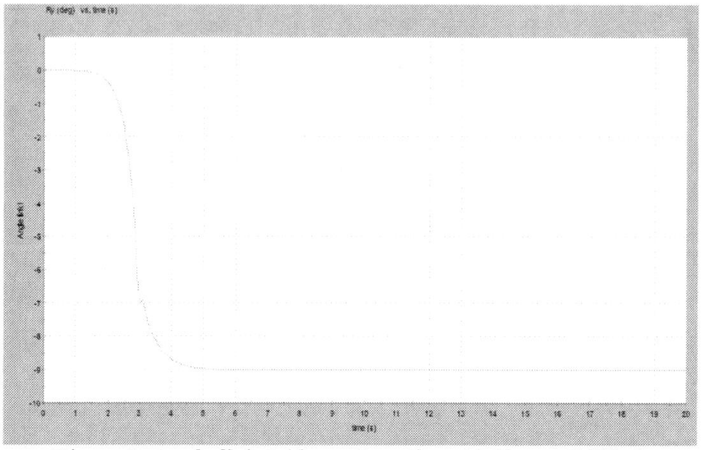

Figure 8: Angle link 1 (degree) vs. time (s) (for m_p 0.75kg)

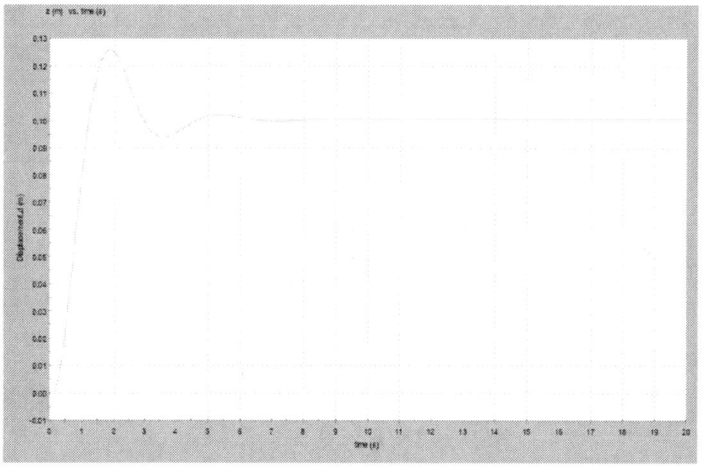

Figure 9: Displacement, d (m) vs. time (s) (for m_p 0.75kg)

4. CONCLUSION

In this paper, we proposed the C.O.G analysis of the manipulator system with and without mass of payload for determining the input reference of the angular position Link1, (θ_1). The manipulator can be prevented from tumbling by controlling the center of gravity on identical imaginary vertical line of the wheels. So, from the simulation we have verified the expected outcome where the angle of the first link should have the opposite angular position direction of the second link to make sure that the robot will be balanced and able to extend the Link2 to any distances with the ability to pick and place the payload within the specific range.

REFERENCES

[1] Bin, H., Zhen, L. W., & Feng, L. H. (2010, March). The Kinematics Model of a Two-Wheeled Self-Balancing Autonomous Mobile Robot and Its Simulation. In *2010 Second International Conference on* IEEE *Computer Engineering and Applications (ICCEA),* (Vol. 2, pp. 64-68). IEEE.

[2] Ghani, N. A., Yatim, N. M., & Azmi, N. A. (2010, October). Comparative assessment for two wheels inverted pendulum mobile robot using robust control. In *2010 International Conference on* IEEE *Control Automation and Systems (ICCAS),* (pp. 562-567). IEEE.

[3] Stilman, M., Olson, J., & Gloss, W. (2010, May). Golem krang: Dynamically stable humanoid robot for mobile manipulation. In *2010 IEEE International Conference on Robotics and Automation (ICRA),* (pp. 3304-3309). IEEE.

[4] Abeykoon, A. M. H. S., & Ohnishi, K. (2005, August). Traction force improvement of a two wheel mobile manipulator by changing the centre of gravity. In *2005 3rd IEEE International Conference on Industrial Informatics, 2005. INDIN'05.* (pp. 756-760). IEEE.

[5] Acar, C., & Murakami, T. (2010, March). Motion control of dynamically balanced two-wheeled mobile manipulator through CoG manipulation. In *2010 11th IEEE International Workshop on Advanced Motion Control,* (pp. 715-720). IEEE.

[6] Yang, J. S., & Chang, Y. S. (2005, October). Standing control of a four-link robot. In *2005 IEEE International Conference on Systems, Man and Cybernetics,* (Vol. 1, pp. 679 681). IEEE.

[7] Fattah, A., & Agrawal, S. K. (2006, May). Gravity-balancing of classes of industrial robots. In *Proceedings 2006 IEEE International Conference on Robotics and Automation, 2006. ICRA 2006.* (pp. 2872-2877). IEEE.

[8] Torige, A., & Ihara, T. (1997, July). Control of manipulator with gravity center position control on mobile vehicle. In *Proceedings. 8th International Conference on Advanced Robotics, 1997. ICAR'97.* (pp. 367-372). IEEE.

[9] Fukuda, T., Fujisawa, Y., Kosuge, K., Fumihito, A., Muro, E., Hoshino, H., & Uehara, K. (1992, June). Gravity Center Control for Manipulator/Vehicle System for Man-Robot Cooperation. In *9th International Symposium on Automation and Robotics in Construction (ISARC-9), Tokyo.*

[10] Ahmad, S., Aminnuddin, M., & Shukor, M. A. (2012, July). Modular hybrid control for double-link two-wheeled mobile robot. In *2012 International Conference on Computer and Communication Engineering (ICCCE),* (pp. 807-813). IEEE.

Development of a Web based financial application System

M R Hasan[1], M I Ibrahimy[2], S M A Motakabber[3], M M Ferdaus[4], M N H Khan[5] and M G Mostafa[6]

Department of Electrical and Computer Engineering, International Islamic University Malaysia, Gombak, Malaysia

E-mail: rubaiyat.hasan@live.iium.edu.my, ibrahimy@iium.edu.my, amotakabber@iium.edu.my, ferdaus57@gmail.com, nomanxp76@yahoo.com and mostafauap@yahoo.com

Abstract. The paper describes a technique to develop a web based financial system, following latest technology and business needs. In the development of web based application, the user friendliness and technology both are very important. It is used ASP .NET MVC 4 platform and SQL 2008 server for development of web based financial system. It shows the technique for the entry system and report monitoring of the application is user friendly. This paper also highlights the critical situations of development, which will help to develop the quality product.

1. Introduction

Web based application is a software that can be accessed through the internet. The software and database are resided in a central server rather than being in the local desktop system. User can access over a network. Web based applications are the ultimate way to take advantage of today's technology to enhance the organizations productivity and efficiency. Finance system allows transferring the financial matters between savers and borrowers. An organization can operate on a global, regional or firm specific level. A financial services and systems can also be described as the comprising a set of complex and closely interconnected financial institutions, markets, instruments, services, practices, and transactions [1]. A software development is in view of the public, which means that anyone has access to the source code and can change, improve, test, and use it. Open source projects such as Firefox or the ERP Business Suite have produced high-quality results, indicating that bazaar-like software engineering models are effective [2]. Commercial financial systems are highly expensive to implement. Educational ERP wide area of application and now a day's educational institutes also looking for ERP implementation where finance is a mandatory module [3]. OS business model, based on a contribution of large number of individuals and partners, definitely affects the framework quality and rapid growth of the platform capabilities [4]. A survey was conducted showing that even within the 500 largest Polish companies only 50 % have an ERP-system, which in comparison with Western European companies is far less; for example in Germany 95 % of the companies with more than 500 employees have an ERP-System [5].

1Mail id: rubaiyat.hasan@live.iium.edu.my, Cell: +60187728161

A web based financial application gives an opportunity to access the business information from anywhere and anytime all over the world. It also facilitates to save the time, money and improves the interactivity with other customers and partners. The web-based applications are easy to use and to implement without any interrupt with the existing work process. It needs a content managed solution or an e-commerce system. It is developed a customized web application that fulfils the business requirements. Same web based software can provide the multimedia support to the end user. The developed web application is accessible from mobile phone and personal computer at the same time. There are many applications those can be accessible from mobile and personal computer but difference is that it is the financial system, which accessing a database having huge data and many stored procedure, a lot of triggering etc. A web based system has database server and browser interface integrated by web server as shown in Figure 1.

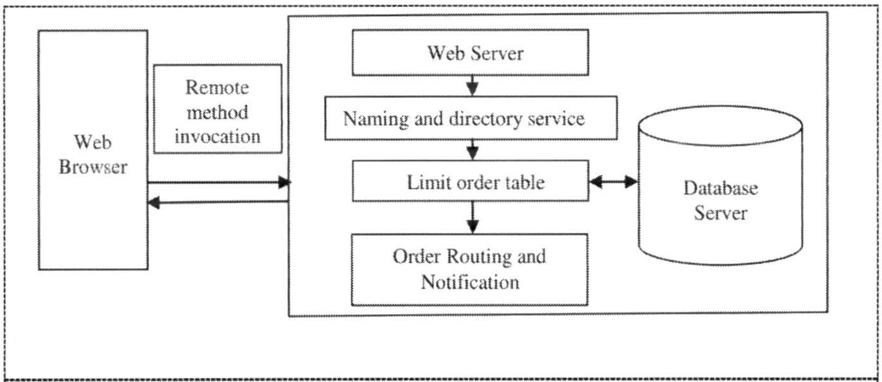

Figure 1. Basic Structure of a web based system.

The rest of the paper is organized as follows. Section II describes the method of financial system development and design. Section III shows the results and section IV concludes the paper.

2. Method of development

Figure 2 shows a flow chart of a financial operation, where accounting year and voucher mode is to setup in initial state. Chart of account (COA) and opening balance for every COA is to setup initially. For every transaction the voucher id generation is mandatory.

Voucher id must be unique. Every transaction is saved in daily transaction table. This table is very important. In reporting phase it shows different accounting reports as the user requirements. A step by development of an accounting system is described as follows. In designing phase the important part is the COA.

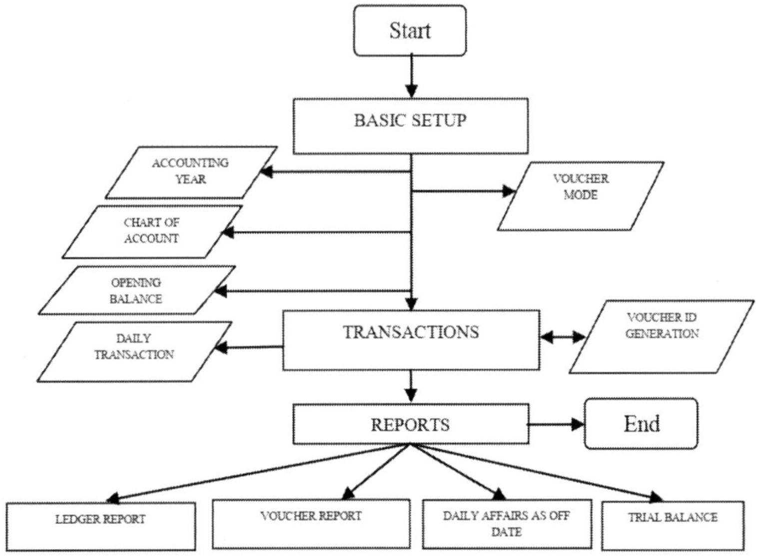

Figure 2. Flow chart of financial operation.

The COA provides the necessary information and structure of the nominal accounts which allows the overall picture of a company's financial health. Figure 3, shows a sample COA of an organization.

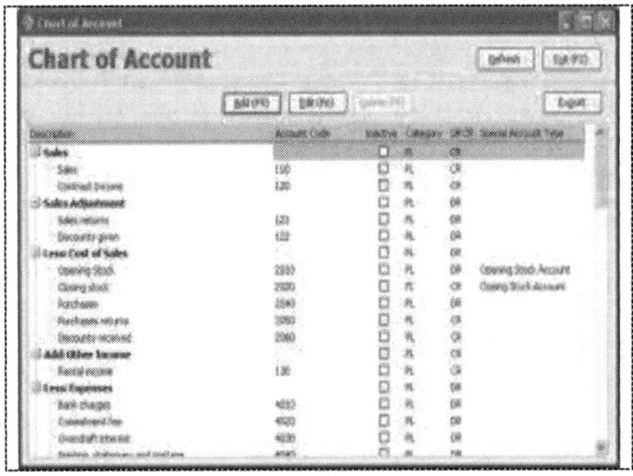

Figure 3. An example of a chart of account.

The actual structure of the COA depends on the structure of the organization. It depends on the type of business that a company is involved, and the type of information needs to the business owner or accountant. However, there are some basic elements of a COA and independent on the company. The common element of a COA is as follows:

- Assets
- Liabilities
- Income and

IOP Conf. Series: Materials Science and Engineering **53** (2013) 012080 doi:10.1088/1757-899X/53/1/012080

- Expenses

The assets and liabilities are generally included under the balance sheet items section. The income and expenses are in under profit and loss section. As a result, the 'top level' of the COA is just two sections such as:

- Balance Sheet Items
- Profit and Loss Items

The assets and liabilities etc. are the children of these sections.

- Balance Sheet Items
    - Assets
    - Liabilities
- Profit and Loss Items
    - Income
    - Expenses

The total assets and liabilities make up an overall total for the balance sheet items. Similarly, the total income and expenses make up an overall total for the profit and loss items. These are provided the basic information of an account. The profit is calculated from the income minus any expenses incurred, and is given as overall picture about the profit and loss.

For designing the COA the following problems need to be maintained:

- Reduce the no. of general ledger accounts
- Details of the selling general and administrative expenses
- Not enough detail in revenue and cost of goods sold categories
- Not using a numbering system
- No logic in assigning general ledger account numbers
- Too many digits in chart of accounts numbering

This is also a very crucial step to design the COA. The funds of a company accounts is started from new financial period and called the opening balances. The opening balance is the first entry of an organization accounts, either this are first starting up or at started of a new financial year. The opening balance is found from the credit or debit side of the ledger that is depending on whether the firm has a positive or negative balance. At the end of an accounting year a firm takes the closing balance - and carries it forward to the new financial year accounts, then becomes the opening balance for the first entry in the new financial year. The opening balances are also become important if a firm transfers their accounts to a new accounting system. When this occurs, the last entry of the previous accounts is become the opening balance in the firm's new accounts. At the starting of a financial plan, the opening balance is found in the opening balance sheet. An opening balance data of a firm is the opening balance sheet comes from the end balance sheet data of the previous accounting period. In the case of a new company, the opening balance sheet usually has just two accounts such as being the cash on hand and one being the capital contributed by the company's founders.

The main part of transaction is the journal/voucher. Voucher is written instrument that serves to confirm or witness (vouch) for some fact such as a transaction. Figure 4, shows an example of a Voucher entry form. Generally, a voucher is a document that shows a goods status, authorizes payment, and indicates the ledger account(s) in which these transactions have to be recorded.

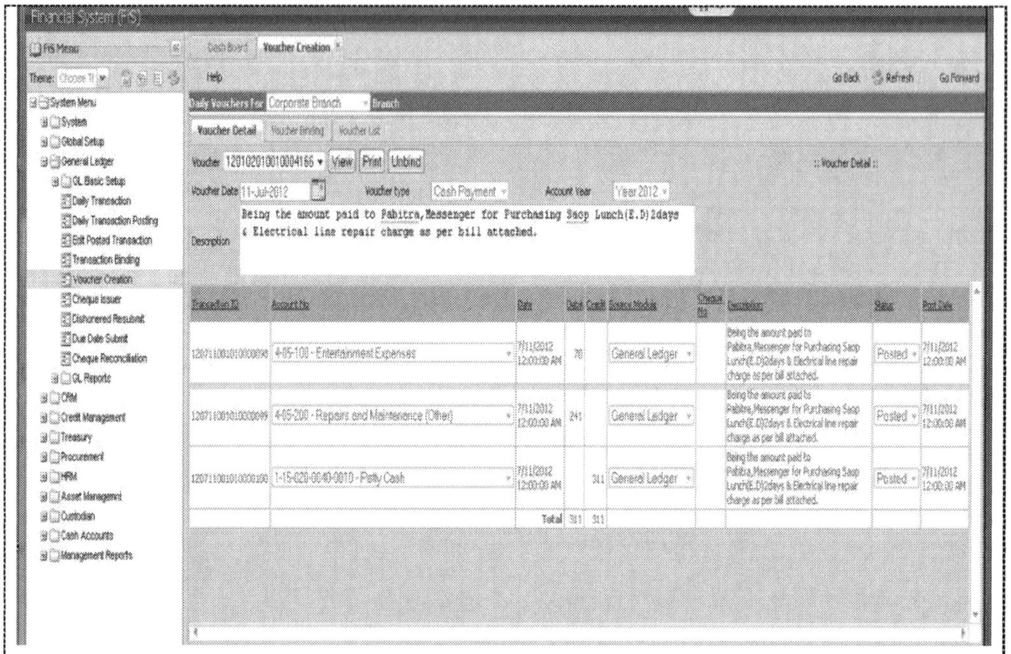

Figure 4. Example of a daily Voucher/Journal Entry.

Voucher is important since much record comes from there. Then a voucher needs to identification with a date and a number combined format. Voucher number should be a unique number. A voucher number creation is as follows:

- It should be an unique number
- It's number should not be editable
- Voucher/Journal is also not be editable

Figure 5 shows an example of Voucher/Journal. Editing is not allowed after the creation of a voucher. It is only can delete by the permission of the proper authority. This is a web application, so it follows the technology that supports the web. We have used ASP .NET MVC for design platform. For smart phone, it is used MVC 4. MVC 4 is supported for both the personal computer and mobile display respectively. The SQL server is accessible by both mobile phone and personal computer device with the help of ASP .NET MVC 4.

The agile development technique is used for the proposed system. The agile development is an umbrella term for several iterative and incremental software development methodologies. Figure 6 shows the agile development work flow. The most popular agile methodologies include extreme programming (XP), scrum, crystal, dynamic systems development method (DSDM), lean development and feature driven development (FDD). While each of the agile methods is unique in its specific approach, they all share a common vision and core values. They are fundamentally incorporate iteration and the continuous feedback that it provides to successively refine and deliver a software system. All involve continuous planning, continuous testing, continuous integration and other forms of continuous evolution of both the project and the software. All of them are lightweight (specially compared to traditional waterfall style processes), and inherently adaptable. They all focus on empowering people to collaborate and make decisions together quickly and effectively.

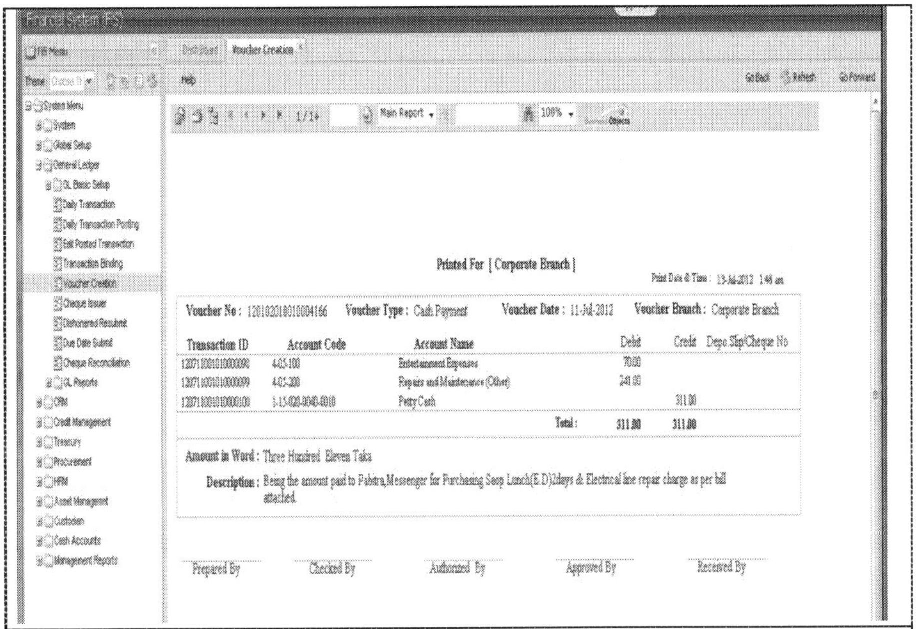

Figure 5. An example of a voucher/journal.

Many of the individual principles and practices that are promoted by agile development have been around for years, even decades. As opposed to implementing these best practices piecemeal, agile methodologies have packaged to various customer, management and in some cases, engineering practices and principles together in a way that helps guide teams through the process of rapidly planning and delivering working, tested software. Each of the agile methodologies combines both old and new ideas into refinements that are certainly greater than the sums of their parts [6].

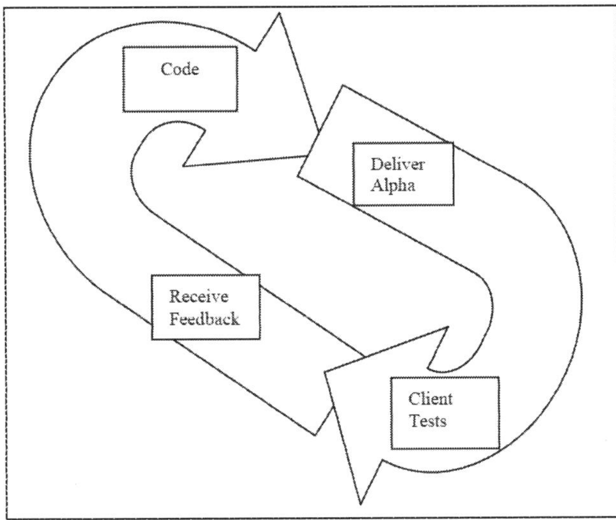

Figure 6. Agile development process. This cycle normally completes every development day.

While it is true that many of the practices associated with agile development have been around for quite some time, the average software development team has yet to embrace many of the principles and practices. Even today, the average software team does not iterate, does not deliver software incrementally, and does not practice continuous planning nor automate testing.

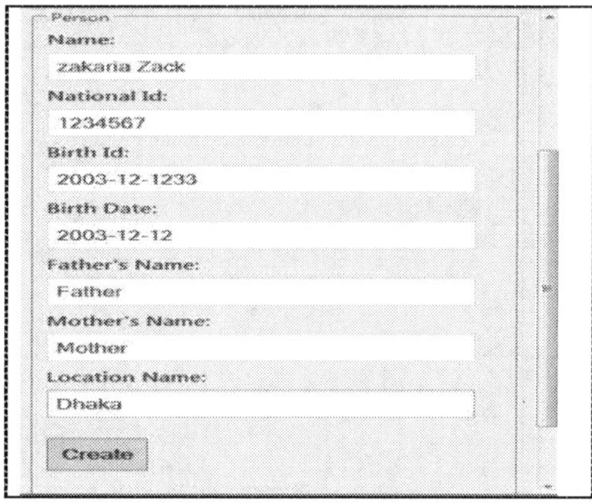

Figure 7. Mobile Interface of a user registration.

Now that these practices have been combined in a manner that can more easily be understood and adopted, the trend appears to be rapidly changing for the better, especially during the last several years. As with any new way of doing business though, Agile methods have generated quite a bit of controversy within the software community. Yet since their emergence, in project after project, they have continued to deliver higher quality software systems in less time than traditional processes. Figure 7 shows the mobile interface for user to first time registration. The data will directly hit the Server.

3. Results and discussion

Recently QuickBooks, tally, OpenERP etc. a lot of financial solutions and ERP are available in the market. In tally and quickbook there are not proper database in open source version. So, customization is very difficult there. OpenERP is open source software, having all ERP modules. But problem in openerp is that it's open source version is not user friendly for setup and use. Also, in all those financial solutions when we try to access from mobile then it will save first in android or other mobile database the will save in database for software for synchronous operation or, for every view it requires to create different view for mobile using and personal computer using. But, here it uses only 3-tier architecture for both mobile and personal computer users as shown in Figure 8.

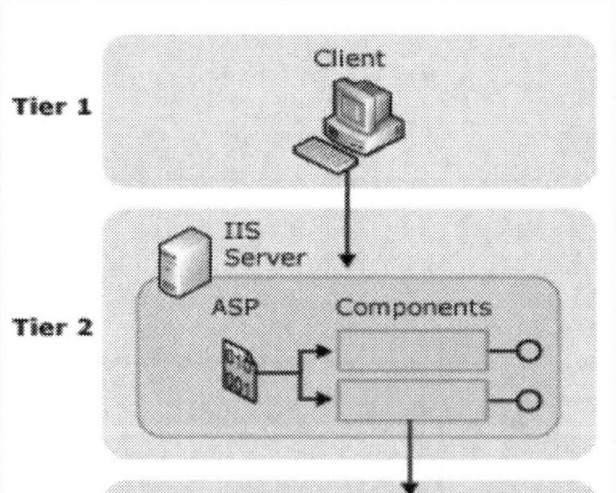

Figure 8. Web application system architecture for ASP .NET platform [7].

IOP Conf. Series: Materials Science and Engineering **53** (2013) 012080 doi:10.1088/1757-899X/53/1/012080

Because, in financial solution there are always two user parties some are mobile phone users and some are personal computer users. For concurrent use in the same time with mobile and computer with the same data is a big challenge now-a-days. So, it has used MVC 4 version of ASP .NET to make it faster. Here, tier 1 is the client side. Tier 2 is for ASP component of IIS and database is in tier 3. This work is initial step to solve those problems.

4. Conclusion

Based on the web application it described a financial system The proposed system can be accessible by mobile phone and personal computer using the same page view. Target of this web application is to make it open source and user friendly. The development technique and design are briefly discussed here. Without a loosely coupled architecture all the new requirements will be tuff to fulfil. So, for preparing a loosely coupled system here Chart of Accounts, Voucher/Journal, opening balance properties etc. have described in a brief.

5. References

[1] Gurusamy, Financial Services and Systems, 2nd edition, p. 3. Tata McGraw-Hill Education, 2008
[2] Christian Dorner, Sebastian Draxler, Volkmar Pipek, and Volker Wulf, End Users at the Bazaar: Designing Next-Generation Enterprise Resource Planning Systems, University of Siegen, IEEE 2009
[3] Anubhav Kumar, Dr. P C Gupta, Comparative Study of FOSS ERP Tools, A Journal of Multidisciplinary Research Vol.1 Issue 9, December 2012
[4] Zupan, Mario, Letinić, Svjetlana, Idlbek, Robert, OpenERP - implementation and ergonomy, Bibliographic record number: 610469, 2013
[5] Mirko Auerbach, Ali Imtiaz, Jannicke Baalsrud Hauge, Collaboration within tool and die making industry through open-source ERP-solution with integrated CRM-functionalities, Research Institute for Operations Management (FIR) at Aachen University of Technology 2006
[6] http://www.versionone.com/agile101/agile_development.asp , Logged on 17 Apr, 2013.
[7] http://msdn.microsoft.com/en-us/library/ms524900(v=vs.90).aspx, Logged on 13 May, 2013

IOP Conf. Series: Materials Science and Engineering 53 (2013) 012081 doi:10.1088/1757-899X/53/1/012081

Comparative data compression techniques and multi-compression results

M R Hasan[1], M I Ibrahimy[2], S M A Motakabber[3], M M Ferdaus[4] and M N H Khan[5]

Dept. of Electrical and Computer Engineering, International Islamic University Malaysia, Gombak, Malaysia

E-mail: rubaiyat.hasan@live.iium.edu.my, ibrahimy@iium.edu.my, amotakabber@iium.edu.my, ferdaus57@gmail.com and nomanxp76@yahoo.com

Abstract. Data compression is very necessary in business data processing, because of the cost savings that it offers and the large volume of data manipulated in many business applications. It is a method or system for transmitting a digital image (i.e., an array of pixels) from a digital data source to a digital data receiver. More the size of the data be smaller, it provides better transmission speed and saves time. In this communication, we always want to transmit data efficiently and noise freely. This paper will provide some compression techniques for lossless text type data compression and comparative result of multiple and single compression, that will help to find out better compression output and to develop compression algorithms.

1. Introduction

The term "Data Compression" means compresses the data as much as possible from its original size. But for the communication we always want to transmit data very efficiently and noise free. We try to find such data that are lossless in manner. Huffman and LZW both are lossless techniques of data compression. Hybrid compression is also helpful in data compression. There are various hybrid data compression techniques [1, 2].

Text compression algorithms are normally defined in terms of a source alphabet S of 8-bit ASCII codes. Most text compression algorithms perform compression at the character level. If the algorithm is adaptive (as, for example, with any of the Ziv-Lempel methods), the algorithm slowly learns correlations between adjacent pairs of characters, then triples, quadruples and so on. The algorithm rarely has a chance to take advantage of longer range correlations before either the end of input is reached or the tables maintained by the algorithm are filled to capacity. If text compression algorithms were used larger units than single character as the basic storage element, they would be able to take advantage of the longer range correlations and perhaps achieve better compression performance. Faster compression may also be possible by working with larger units.

1Mail id: rubaiyat.hasan@live.iium.edu.my, Cell: +60187728161

However, we need to overcome a major problem with text compression algorithms. The number of distinct text words that the compression algorithm has to cope with is, for all practical purposes, unbounded. Thus it makes no sense to implement an algorithm that requires a pre-determined finite alphabet. To use LZW as an example again, we cannot initialize the LZW string table with all sequences of length one, as required in the usual implementations of LZW. Instead, we have to modify the algorithms so that they either pre-determine the set of words used in the source input (an inherently two-pass strategy) or they dynamically expand the source alphabet as each new word occurs. We, of course, advocate single-pass strategies as being more useful for practical applications [3].

The rest of the paper is organized as follows. In Section II, the relevant literature on data compression is reviewed. Section III, shows sample results and discussed about those results. Section IV concludes the paper.

2. Data Compression Techniques

There are two different ways that data compression methods and algorithms can be categorized as the lossless and lossy, the methods are classified according to a fixed or variable. Lossless compressions are run-length; Huffman, delta, LZW etc. and lossy data compression are CS & Q, JPEG and MPEG etc. Here, two lossless compressions Huffman and LZW will be discussed.

- **Huffman Compression Technique**

There are many types of Huffman coding, some are use a Huffman-like algorithm, and others find optimal prefix codes (while, for example, putting different restrictions on the output). In the latter case, the method need not be Huffman-like, and, indeed, need not even be polynomial time. An exhaustive list of papers on Huffman coding and it's variations are given here as follows:

The n-ary Huffman algorithm uses the $\{0, 1, \ldots, n-1\}$ alphabet to encode message and build an n-ary tree. This approach was considered by Huffman in his original paper. The same algorithm applies as for binary (n equals 2) codes, except that the n least probable symbols are taken together, instead of just the 2 least probable. Note that for n greater than 2, not all sets of source words can properly form an n-ary tree for Huffman coding. In this case, additional 0-probability place holders must be added. This is because the tree must form an n to 1 contractor; for binary coding, this is a 2 to 1 contractor, and any sized set can form such a contractor. If the number of source words is congruent to 1 modulo n-1, then the set of source words will form a proper Huffman tree.
A variation called adaptive Huffman coding which is involved for calculating the probabilities dynamically based on recent actual frequencies in the sequence of source symbols, and changing the coding tree structure to match the updated probability estimates.

Most often, the weights used in implementations of Huffman coding represent numeric probabilities, but the algorithm which given above does not require this; it requires only the weights form a totally ordered commutative monoid, meaning a way to order weights and to add them. The Huffman template algorithm enables one to use any kind of weights (costs, frequencies, pairs of weights, non-numerical weights) and one of many combining methods (not just addition). Such algorithms can solve other minimization problems, a problem first applied to the circuit design.

Length-limited Huffman coding is a variant where the goal is still to achieve a minimum weighted path length, but there is an additional restriction that the length of each code word must be less than a given constant. The package-merge algorithm solves this problem with a simple greedy approach which is very similar to that is used by Huffman's algorithm. It's time complexity is O(nL), where L is the maximum length of a code word. No algorithm is known to solve this problem in linear or linearithmic time, unlike the pre-sorted and unsorted conventional Huffman problems, respectively [4, 5]

Huffman Encoding Algorithm
Huffman (W, n) //Here, W means weight and n is the no. of inputs
Input: A list W of n (Positive) Weights.
Output: An Extended Binary Tree T with Weights Taken from W that gives the minimum weighted path length.
Procedure: Create list F from singleton trees formed from elements of W.
While (F has more than 1 element) do
Find T1, T2 in F that have minimum values associated with their roots // T1 and T2 are sub tree
Construct new tree T by creating a new node and setting T1 and T2 as its children
Let, the sum of the values associated with the roots of T1 and T2 be associated with the root of T Add T to F
Do
Huffman-Tree stored in F

- **LZW Compression Technique**

LZW compression named after its developers, A. Lempel and J. Ziv, with later modifications by Terry A. Welch. It is the foremost technique for general purpose data compression due to its simplicity and versatility. Typically, you can expect LZW to compress text, executable code, and similar data files to about one-half their original size. LZW also performs well when presented with an extremely redundant data files, such as tabulated numbers, computer LZW is the basis of several personal computer utilities that claim to "double the capacity of your hard drive." If the codeword length is not sufficiently large, Lempel-Ziv codes may also rise slowly to reasonable efficiency, maintain good performance briefly, and fail to make any gains once source code, and acquired signals. Compression ratios of 5:1 are common for these cases. [2]

LZW Encoding Algorithm
Step 1: At the start, the dictionary contains all possible roots, and P is empty
Step 2: C: = next character in the char stream;
Step 3: Is the string P+C present in the dictionary?
(a) if it is, P := P+C (extend P with C);
(b) if not,
–output the code word which denotes P to the code stream;
– add the string P+C to the dictionary;
–P: = C (P now contains only the character C); (c) Are there more characters in the char stream?
–if yes, go back to step 2;
–if not:
Step 4: Output the code word which denotes P to the code stream;
Step 5: END.

The fundamental theory of LZW compression algorithm is: any predictable data can demonstrate such predictability by certain mark and shorten the data length. During the process of LZW compression, as the length of each code in the code stream of a datum produced after compression is less than N, or the compression algorithm can represent number of 0~2n- 1, therefore the string list can accommodate number of 2n at the maximum. But each element of the input data stream after compression is a byte, which represents 0~255 possible assignments. Secondly, compression program takes one byte from the input character stream. Two buffers of current prefix code and current string are used to store data. Prefix position is for the code processed last time, current string is for the character string represented by prefix code and characters read just now. When the program starts, both the prefix code and current string are blank. Thirdly, program searches for current string in the string list after initialization.

Fourthly, read the next byte from input stream, and add this character behind current string, now the current string contains 2 bytes, and repeat the third step.

Obviously double-byte characters are unable to find in the string list this time, because all the existing characters are all single characters in the string list. So the program turns to another branch, and establishes a new item in the string list responding to the current string of double byte characters, or the 256th item, and then adds the prefix code to output stream, therefore the compressed codes we need would be available. As the maximum number the string list can accommodate is 2n, when the string list capacity is nearly full, enable the cleaning code and restart initialization for prefix code and current string, and repeat from step one to step four to compress data [6].

Table 1. Comparative Data Compression Table.

Sample Name	Original Data	LZW Comp.	Compression ratio	Huffman Compression	Comp. ratio	LZW based Huffman	Comp. ratio	Huffman based LZW	Comp. ratio
test1	50.39 MB	10.22MB	4.42	30.01 MB	1.79	20.24 MB	2.41	10.20 MB	4.49
test2	43.1 MB	9.75 MB	4.42	24.1 MB	1.79	9.72 MB	4.43	9.65 MB	4.46
test3	12.7 MB	3.22 MB	3.94	7.20 MB	1.76	5.6 MB	2.27	3.10 MB	4.09
test4	81.4 MB	19.4 MB	4.19	45.7 MB	1.78	28.2 MB	2.89	19.4 MB	4.19
test5	20.82 MB	10.80 MB	3.52	10.76 MB	1.6	10.27 MB	2.22	10.74 MB	3.81
test6	110.3 MB	30.18 MB	3.55	70.07 MB	1.59	50.78 MB	1.95	20.35 MB	4.81
test7	86.3 MB	19.4 MB	4.45	48.3 MB	1.79	35.8 MB	2.41	19.2 MB	4.49
test8	32 MB	7.5 MB	4.27	15.8 MB	2.02	11.5 MB	2.78	7.1 MB	4.51
test9	14.7 MB	3.1 MB	4.74	7.1 MB	2.07	6.8 MB	2.16	3.00 MB	4.9
test10	126.5 MB	30.1 MB	4.2	58.5 MB	2.16	53.7 MB	2.35	28.7 MB	4.41
Average			**4.17**		**1.835**		**2.587**		**4.416**

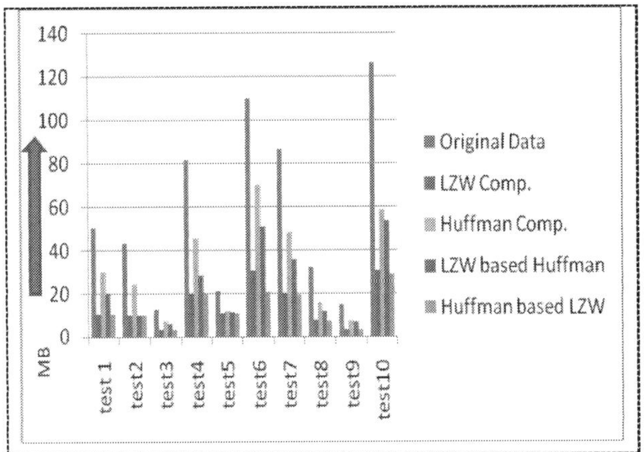

Figure 1. Comparative Compression ratio from the data.

Proposed Algorithm

Step 1: At the start, the dictionary contains all possible roots, and P is empty

Step 2: C: = next character in the char stream;
Step 3: Is the string P+C present in the dictionary?
(a) if it is, P := P+C (extend P with C);
(b) if not,
–output the code word which denotes P to the code stream;
– add the string P+C to the dictionary;
–P: = C (P now contains only the character C); (c) Are there

more characters in the char stream?
–if yes, go back to step 2;
–if not:
Step 4: Output the code word which denotes P to the code stream;
Step 5: Take, W means weight and n is the no. of inputs, A
list W of n (Positive) Weights, An Extended Binary Tree T with Weights Taken from W that gives the minimum Weighted path length.
Procedure: Create list F from singleton trees formed from elements of W.
While (F has more than 1 element) do
Find T1, T2 in F that have minimum values associated with their roots. // T1 and T2 are sub tree
Construct new tree T by creating a new node and setting T1 and T2 as its children
Let, the sum of the values associated with the roots of T1 and T2 be associated with the root of T Add T to F
Do
Step 6: END.

3. Results and discussion

Here, the compression ratio (Cr) measured as, Cr= Original data size/Compressed data size

- Data Compression using Huffman: From the table 1 and figure 1 in case of Huffman experiment this is found that every data has decreased from its original size. For the all 10 cases of compressions the compression ratio is about 1.84 in average case. There are lower and higher compression ratios in some cases. This is the result of compression of text file. The compressed files were in the 0, 1 binary code formats. After that when decompressed that file it becomes same. No data loss is found there.

- Data Compression using LZW: From table 1 and figure 1 in case of the LZW experiment it is found that the data decreased from its original size. The compression ratio is 4.17 in average case. There are lower and higher compression ratios in some cases. This is the result of compression of text file. When decompressed that file it becomes same as the original file. It gives better compression ratio than the case of Huffman Encoding. It compresses a file almost 78% and we can have only 22% size of the original file size. To have better file size we may use LZW Encoding technique.

- Data Compression ratio using LZW based Huffman: Here, in table 1 and figure 1 it is found that every data has decreased from its original size. The compression ratio is 2.59 in average case. There are several lower and higher compression ratios in some cases. After that when decompressed that file it becomes same as the original file. It gives better compression ratio than the case of Huffman Encoding but lower compression ratio in the case of LZW. It compresses a file almost 66% and we can have only 34% size of the original file size. We think that Hybrid compression gives better output but here we find different case. So, LZW performs better than LZW based Huffman Encoding .To have better file size we may use LZW Encoding technique.

- Data Compression using Huffman based LZW: From the table 1 and figure 1 of first Huffman and then LZW experiment it is clear that every data has decreased from its original size. From the 10 sample the compression ratio 4.42 in average case. There are lower and higher compression ratios in some cases. This is the result of compression of text file. After that when we have decompressed that file it becomes same as the original file. It gives better compression ratio than the case of Huffman, LZW and LZW based Huffman all three Encoding. It compresses a file almost 79% and we can have only 21% size of the original file size. From this experiment we can say that Hybrid Data Compression is better than single compression. To have better file size we may use Huffman based LZW Encoding technique.

- Compare with Recent Approaches: Recently, WinRar is widely used for data compression. The algorithm is based on Lempel-Ziv (LZSS) and prediction by partial matching (PPM) compression is used for WinRar. The average compression ratio achieved for the set of test images was 70%, the image size was usually reduced to 30% of their original value [7]. Here, we introduced multi-compression algorithm for Text data only. Significant result is that it compresses data at about 78-79% form Table 1. So, if work with this algorithm and improve for all sort of data then there are some prospect to get better compression ratio.

4. Acknowledgement

This paper is the enlarged and modified concept of Data Compression using Huffman based LZW Encoding Technique (IJSER, Vol- 2, Issue 11, Nov-2011), published at International Journal of Scientific & Engineering Research conference. In that paper there was limitations in compression ratio, here the better compression ratio is found.

5. Conclusion

Data compression is a topic of much importance and many applications. Methods of data compression have been studied for almost four decades. This paper has provided multi-data compression methods of general utility. The proposed algorithm has been evaluated in terms of the amount of compression ratio they provided, algorithm efficiency, and susceptibility to error. While algorithm efficiency and susceptibility to error are relatively independent of the characteristics of the source ensemble, the amount of compression achieved depends upon the characteristics of the source to a great extent. It does not matter how the characters are arranged. It arranged above so that the final code tree looks nice. It is used Turbo C editor for coding. It does not matter how the final code tree are labelled (with 0s and 1s).It has chosen to label the upper branches with 0s and the lower branches with 1s. In conclusion anybody can say that Huffman based LZW Encoding can Compresses data more than all other three cases, when in average case the Huffman based LZW compression ratio is 4.41, where other max average compression ratio is 4.17 in case of LZW compression. The Huffman based LZW compression is better some of the cases than LZW Compression. This is also possible to Decompress.

6. References

[1] Rafael C. Gonzalez and Richard E. Woods, Digital Image Processing (Second papers: Edition)-2006, Pages: 411, 440-442, 459
[2] M Rubaiyat Hasan, Data Compression using Huffman based LZW Encoding Technique, International Journal of Scientific & Engineering Research, Vol- 2, Issue 11, Nov-2011.
[3] R. Nigel Horspool, Gordon V. Cormack, Constructing Word-Based Text Compression Algorithms, IEEE 1992.
[4] Suresh Kumar, Madhu Rawat, Vishal Gupta, Satendra Kumar, The Novel Lossless Text Compression Technique Using Ambigram Logic and Huffman Coding, Information and Knowledge Management ISSN 2224-5758 (Paper) , Vol -2, No.2, 2012.

[5] Julia Abrahams, Code and Parse Trees for Lossless; Source Encoding, IEEE 1998
[6] YuanJing, The Combinational Application of LZSS and LZW Algorithms for Compression Based On Huffman, International Conference on Electronics and Optoelectronics (ICEOE 2011), 2011.
[7] Tatjana LonCar-Turukalo, Vladimir CrnojeviC, ieljen Trpovski, Image Compression by Decomposition into Bit Planes, IEEE- 2001

Development of quadruped walking locomotion gait generator using a hybrid method

F Jasni and A A Shafie

Department of Mechatronis, Kulliyyah of Engineering,
International Islamic University Malaysia, Malaysia.

E-mail: farahiyahjasni@gmail.com, aashafie@iium.edu.my

Abstract. The earth, in many areas is hardly reachable by the wheeled or tracked locomotion system. Thus, walking locomotion system is becoming a favourite option for mobile robot these days. This is because of the ability of walking locomotion to move on the rugged and unlevel terrains. However, to develop a walking locomotion gait for a robot is not a simple task. Central Pattern Generator (CPGs) method is a biological inspired method that is introduced as a method to develop the gait for the walking robot recently to tackle the issue faced by the conventional method of pre-designed trajectory based method. However, research shows that even the CPG method do have some limitations. Thus, in this paper, a hybrid method that combines CPG and the pre-designed trajectory based method is introduced to develop a walking gait for quadruped walking robot. The 3-D foot trajectories and the joint angle trajectories developed using the proposed method are compared with the data obtained via the conventional method of pre-designed trajectory to confirm the performance.

1. Introduction

The walking locomotion has become a favorable option in designing the mobile robot's locomotion system as its advantage that enable the robot to explore a wider area than the wheeled and tracked locomotion system can offer. However, the control architecture for the walking locomotion is not as simple as the wheeled locomotion able can offer.

The most common method utilized in designing the walking locomotion for robot is called the pre-designed trajectory based method; the trajectories of the robot are first designed and inverse kinematics are used to obtain the joint angles. This method requires the developer to acquire as much information as possible on the nature of the terrain as the design of the trajectory is closely related to that information. Work in [1], [2] and [3] adopting the pre-designed trajectory method in their design. The disadvantage of this method is the control mechanism become very complicated when the number of legs increases as each leg will need separate analysis and calculation. Furthermore, this method is no longer efficient if the ground condition changes since the trajectories need to be re-designed. Thus, a better approach is needed to tackle the problem of robots walking in unknown terrains.

In order to compensate the limitations that the pre-designed trajectory method has the biological inspired, CPG method is introduced. Central Pattern Generator (CPG) is described as a set of neural networks that can continuously produce rhythmic patterned signals without input from the rhythmic sensory system or the central control system (i.e. the brain) [4]. Ijspeert [5] reviewed in his paper that the implementation of CPG in robotics is categorized into three types; the connectionist model [6], the vector map and the systems of coupled oscillators [7]. However, most of research work utilized the concept of coupled oscillators to form a CPG network in the CPG design for locomotion control in robotics. Generally, the implementations of CPG in robotics locomotion control involve sets of coupled differential equations that are numerically integrated on a microcontroller or microprocessor. The coupled oscillators generate a rhythmic signal that is processed and analyzed as the joint trajectories (i.e. the joint angle). The output signals from the CPG network are usually modified using common mathematical transformation method such as Fourier transform before analyzed as the joint trajectories. There are generally two types of oscillator that are used in forming the CPG network; the neural oscillators such as the Matsuoka model [8] and Wilson-Cowan Weakly Neural Networks [9] and the non-linear oscillators, for instance the Hopf oscillator [10], and the Van Der Pol (VDP) oscillator [11]. However, this approach requires at least one oscillator to be assigned at respective joint so that the oscillator's output can generate the joint angle trajectory for the joint. As a result, with complex oscillators network will be needed and the modelling the coupled oscillators will become tedious especially when the robot consists of large number of joints.

A better method is required to balance the deficiency that both approaches carry individually, Liu, Chen and Wang has proposed a method that combines the methods [12]. The hybrid method suggests that the CPG network is used to generate the foot trajectories for the robot via a mapping process, instead of directly generate the joint angle trajectory for the joints. As a result, only one oscillator is needed to represent one foot that simplifies the modelling and analysis. Then the foot trajectories obtained via the CPG network are utilized to generate the joint angle trajectories of each joint available in a leg using the forward and inverse kinematics analysis. In this paper, the same basic method is implemented, however, different oscillator's model is used that is the Van Der Pol oscillators and since different hardware platform is used (i.e. Bioloid Premium Quadruped Robot shown in Figure 1), the forward and inverse kinematics is dissimilar from [12].

Figure 1: The Bioloid Puppy Robot

2. Walking gait locomotion gait generator using hybrid method

The method implemented in this paper is the hybrid method that combines the conventional pre-designed trajectory method and the biological inspired method, CPG. Basically, CPG is utilized in generating the foot trajectory of the robot. Instead of having to design the foot trajectory and for each leg by ourselves, the CPG signals are used to generate the trajectories and the trajectories can be easily modified by only modulating the CPG parameters. Thus, the locomotion gait generator is made to be more adaptable to any changes on the terrain condition.

The suitable CPG signals that describe the gait are then mapped onto a set of mapping function to obtain the 3D (x, y, and z direction) foot trajectories. Based on the 3D foot trajectories gained, the forward and inverse kinematics is performed to generate the joint angle trajectories for each leg. The set of joint angle for all 4 legs are fed to the robot control to make it move. The methodology is illustrated in the flowchart in Figure 2.

2.1. VDP-CPG network modelling

In this paper, the CPG network is developed using 4 VDP oscillators. The VDP oscillator is chosen for its stable limit cycle feature. Furthermore, this type of oscillator is used in representing most of the biological system.

1 oscillator is allocated for 1 leg to generate the rhythmic movement for each leg. Then the output signal of each oscillator is coupled with the signal from other oscillators. The network is illustrated in Figure 3.

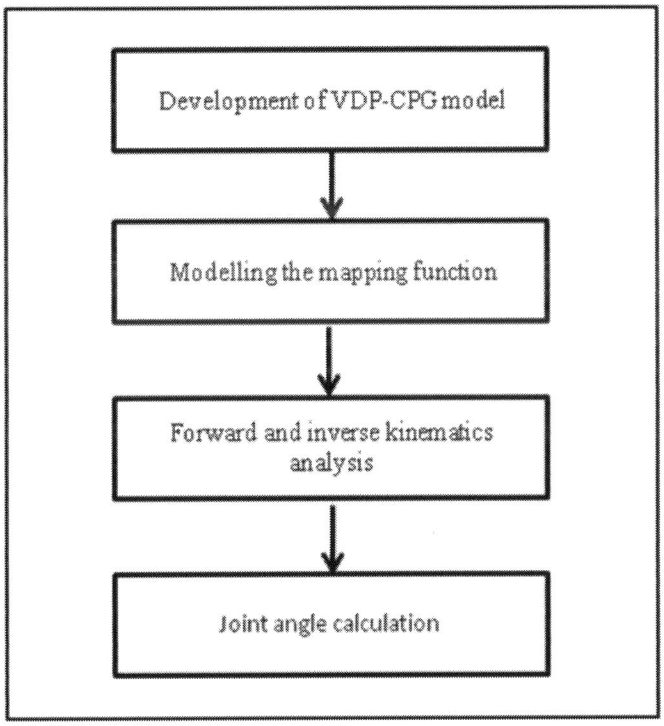

Figure 2: Flowchart of the design methodology

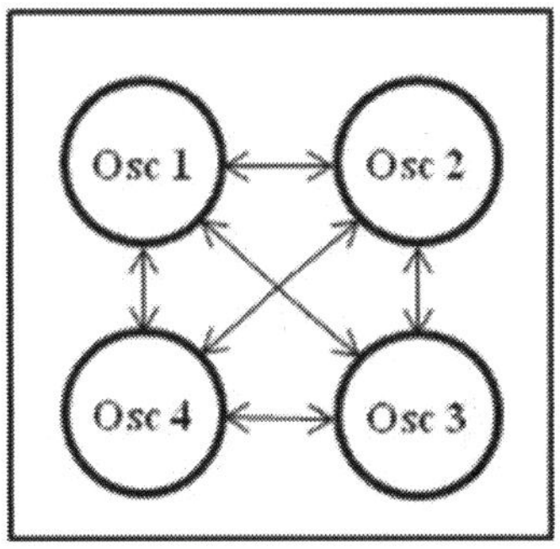

Figure 3: The CPG network

The general expression to describe the relationship between each coupled oscillator is described as below [13]:

$$\ddot{x}_i = \mu_i(p_i^2 - x_{ai}^2)\dot{x}_i - g_i^2 x_{ai} + q_i \tag{1}$$

where x_i represents the output of the oscillator i, μ_i generally responsible in shaping the shape of the signal, p_i^2 is the constant that modulates the amplitude of the signal, while g^2 modulates the frequency of the wave and finally q as the offset parameter. x_{ai} represents the expression that described the relationship between the oscillators and it is expression in (2):

$$x_{ai} = x_i + \sum \lambda_{i,j} x_j \tag{2}$$

where $\lambda_{i,j}$ is the relationship coefficient between oscillator i and j. Substituting (2) in (1), the general expression becomes:

$$\ddot{x}_i = \mu_i\left(p_i^2 - \left(x_i + \sum \lambda_{i,j} x_j\right)^2\right)\dot{x}_i - g_i^2\left(x_i + \sum \lambda_{i,j} x_j\right) + q_i \tag{3}$$

In CPG, one of the most important and challenging task is to make sure that the parameters setting are precise to produce the rhythmic output that is desired. In this research work, the parameters are determined via trial and error, and through the experiments that have been conducted, the parameters are set to be; $p_i^{\ 2} = 1$, $q_i = 0.12$, $g_i^{\ 2} = 0.15$ and $\mu_i = 0.1$. Among all 5 parameters, $\lambda_{i,j}$ is found to be the most important since that parameter will determine how one signal of one oscillator relates to the other. The relationship coefficients are set to be:

$$\omega = \begin{bmatrix} \lambda_{1,1} & \lambda_{1,2} & \lambda_{1,3} & \lambda_{1,4} \\ \lambda_{2,1} & \lambda_{2,2} & \lambda_{2,3} & \lambda_{2,4} \\ \lambda_{3,1} & \lambda_{3,2} & \lambda_{3,3} & \lambda_{3,4} \\ \lambda_{4,1} & \lambda_{4,2} & \lambda_{4,3} & \lambda_{4,4} \end{bmatrix} = \begin{bmatrix} 0 & 0 & -0.3 & 0 \\ 0 & 0 & 0 & -0.3 \\ 0 & -0.3 & 0 & 0 \\ -0.3 & 0 & 0 & 0 \end{bmatrix} \tag{4}$$

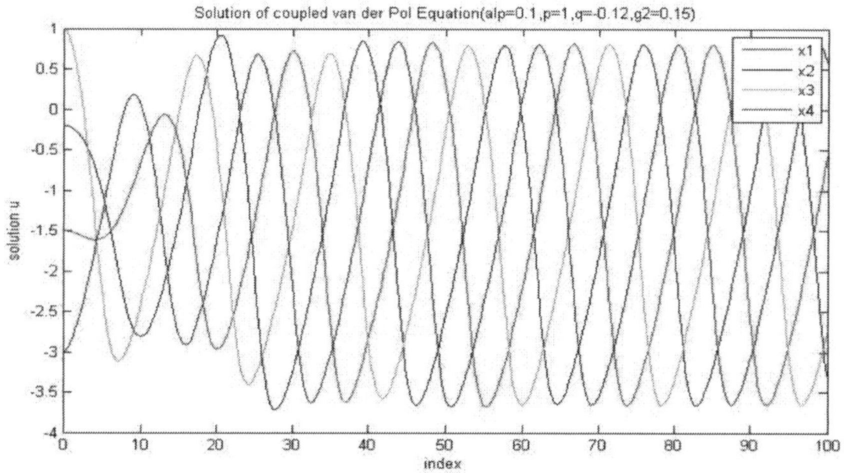

Figure 4: The output signal of VDP-CPG network model used

2.2. *Mapping function modelling*

The output signal from the VDP-CPG network is mapped onto a set of function to generate the 3D foot trajectories. The trajectories formed through this process are with respect to the ground coordinate which will be addressed as locus trajectories in this paper. Figure 5 shows the ground based coordinate assignation for the robot.

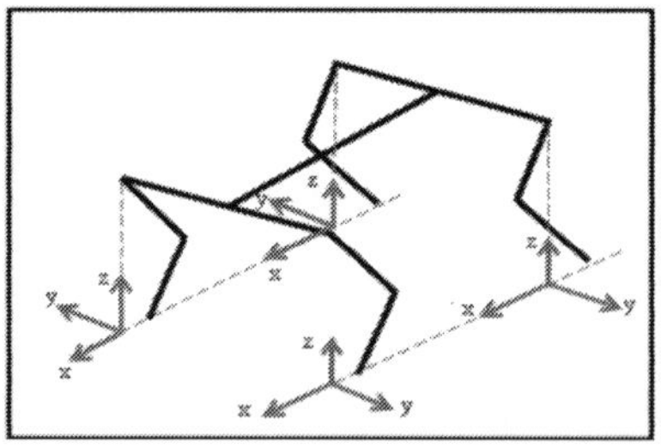

Figure 5: Ground based coordinate assignation

The foot trajectories can be divided into 2 phases; the swing phase (the foot is lifted) and the stance phase (the foot is on the ground). Therefore, the mapping function is modelled based on the requirement that satisfies both phases. Figure 6 and 7 illustrates the foot trajectories for swing and stance phase respectively.

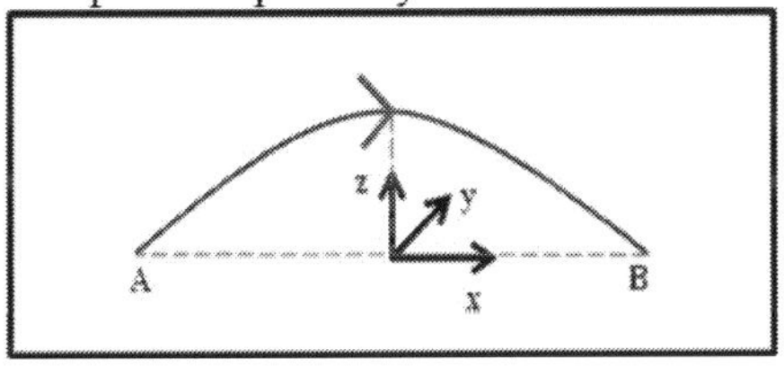

Figure 6: Foot trajectory for swing phase

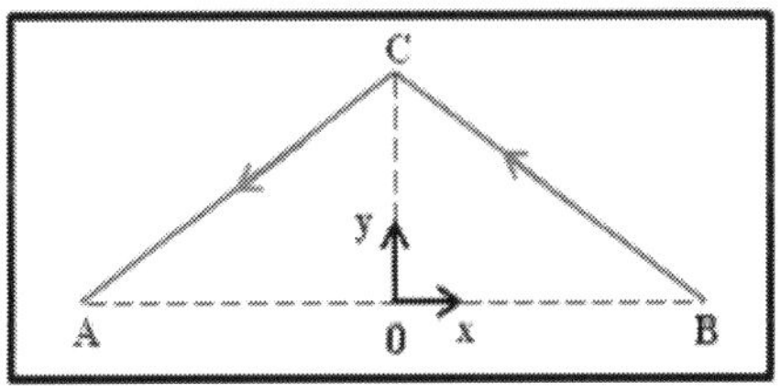

Figure 7: Foot trajectory for stance phase

The mapping function for the swing phase [14] is defined as:

$$x_{locus}(i) = K_x \cdot \left(t_{sw(index)}(i) - \frac{T_{sw}(i)}{2}\right) \tag{5}$$

$$y_{locus}(i) = 0 \tag{6}$$

$$z_{locus}(i) = K_z \cdot (X_{index}(i)) \tag{7}$$

On the other hand, the mapping function for the stance phase is described as:

$$x_{locus}(i) = K_x \cdot \left(\frac{1}{2} - \frac{t_{st(index)}(i)}{T_{st}}\right) \tag{8}$$

$$y_{locus}(i) = \left(-\frac{2\gamma}{K_x} x_{locus}(i) + \gamma\right) \quad \text{for xlocus(i) >= 0} \tag{9}$$

$$y_{locus}(i) = \left(\frac{2\gamma}{K_x} x_{locus}(i) + \gamma\right) \quad \text{for xlocus(i) < 0} \tag{10}$$

$$z_{locus}(i) = 0 \tag{11}$$

i represents the leg number, K_x and K_z represents the gain coefficient for x and z direction respectively. $t_{sw(index)}(i)$ and $t_{st(index)}(i)$ represents the current swing and stance time respectively whereas T_{sw} and T_{st} represents the total time of swing phase and stance phase respectively. $X_{index}(i)$ is the output signal from the CPG. Finally γ is the compensatory variable that relates x-trajectory and y-trajectory.

Since the foot trajectories that are obtained so far are based on the ground coordinate, thus, a transformation is needed to generate the trajectories based on the hip coordinate (root trajectories). This is important because the trajectories w.r.t hip coordinate is needed to perform inverse kinematics in the next process to generate the set of joint angles. Thus, the transformation matrix is described below:

$$\begin{pmatrix} x_{root} \\ y_{root} \\ z_{root} \\ 1 \end{pmatrix} = \begin{pmatrix} 1 & 0 & 0 & 0 \\ 0 & 1 & 0 & 0 \\ 0 & 0 & 1 & -z\_dist \\ 0 & 0 & 0 & 1 \end{pmatrix} \begin{pmatrix} x_{locus} \\ y_{locus} \\ z_{locus} \\ 1 \end{pmatrix} \tag{12}$$

2.3. Forward and inverse kinematics analysis

Before inverse kinematics analysis can be performed, the set of transformation matrix that the foot encountered need to be determined first. The root trajectories can be defined as:

$$\begin{pmatrix} x_{root} \\ y_{root} \\ z_{root} \\ 1 \end{pmatrix} = {}^{H}T_{F} \begin{pmatrix} 0 \\ 0 \\ 0 \\ 1 \end{pmatrix} \tag{13}$$

in which ${}^{H}T_{F}$ is the series of movement that the foot has made and the original position is assumed to be at homogenous coordinate *(0,0,0)*. Taking Leg 1(Front left leg) as sample, the ${}^{H}T_{F}$ for Leg 1 can be described as:

$$({}^{H}T_{F})_{leg1} = \begin{bmatrix} C_1C_3 + S_1(C_2S_3) & S_1S_2 & C_1S_3 - S_1C_2C_3 & S_1C_2(L_2C_3 + L_1) - L_2S_3C_1 \\ -S_2S_3 & C_2 & S_2C_3 & -S_2(L_2C_3 + L_1) \\ S_1C_3 - C_1C_2S_3 & -C_1S_2 & S_1S_3 + C_1C_2C_3 & -L_2S_1S_3 - C_1C_2(L_2C_3 + L_1) \\ 0 & 0 & 0 & 1 \end{bmatrix} \tag{14}$$

Inserting (14) into (13), the root trajectories for Leg 1 is expressed as:

$$\begin{pmatrix} x_{root} \\ y_{root} \\ z_{root} \\ 1 \end{pmatrix}_{leg1} = \begin{pmatrix} S_1C_2(L_2C_3 + L_1) - L_2S_3C_1 \\ -S_2(L_2C_3 + L_1) \\ -L_2S_3S_1 - C_1C_2(L_2C_3 + L_1) \\ 1 \end{pmatrix} \tag{15}$$

Performing the inverse kinematics analysis on above equation, the joint angle parameters (θ_1, θ_2, and θ_3) can be obtained and described as below:

$$\theta_1 = \tan^{-1}\left(\frac{x}{z}\right) + \tan^{-1}\left(\frac{L_2 \sin(\theta_3)}{(L_1 + L_2 \cos(\theta_3)\cos(\theta_2))}\right) \tag{16}$$

$$\theta_2 = \sin^{-1}\left(\frac{-y_{root}}{L_2 \cos(\theta_3) + L_1}\right) \tag{17}$$

$$\theta_3 = \cos^{-1}\left[\frac{x^2 + y^2 + z^2 - L_1^2 - L_2^2}{2L_1L_2}\right] \tag{18}$$

Figure 8 illustrates the position of the joint angle parameters for Leg 1.

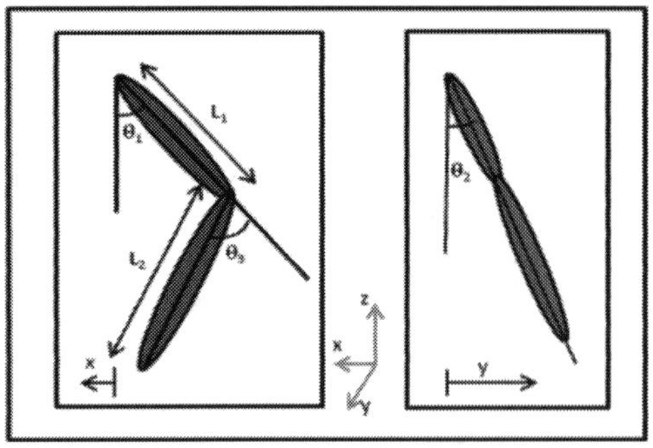

Figure 8: Model of left front leg (Leg1). Left: Side elevation for the calculation of shoulder-joint (θ_1) and knee-joint (θ_3) angle. Right: Front elevation for the calculation of shoulder-joint (θ_2) angle.

IOP Conf. Series: Materials Science and Engineering **53** (2013) 012082 doi:10.1088/1757-899X/53/1/012082

3. Results and analysis

The main purpose of this research work is to investigate the efficiency of the hybrid method proposed to generate the walking gait for the quadruped robot. In order to prove that, the joint angle trajectories (θ_1, θ_2, and θ_3) obtained via this method are compared to the benchmark method (i.e. composite cycloid function (CCF)).

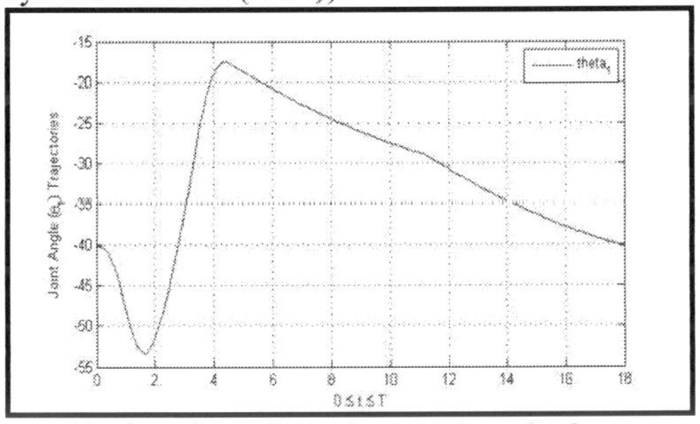

(a) θ_1 trajectories using CCF method

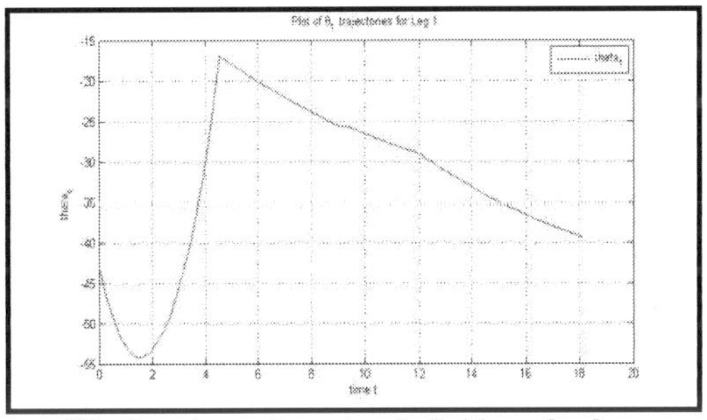

(b) θ_1 trajectories using hybrid method

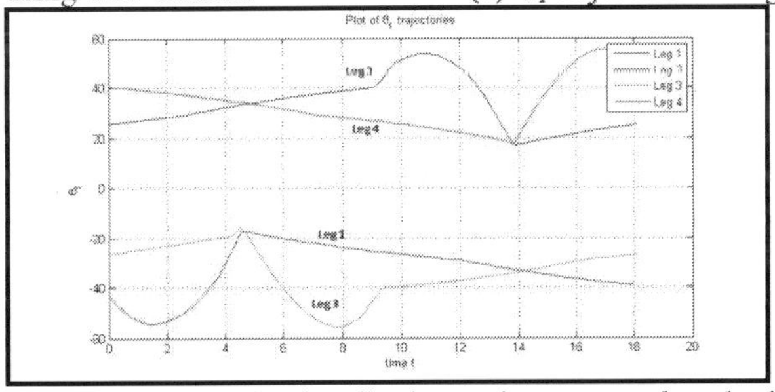

(c) θ_1 trajectories for all 4 legs using proposed method

Figure 9: Plots of θ_1 trajectories using CCF (Figure (a)), and proposed method (Figure (b) and (c)).

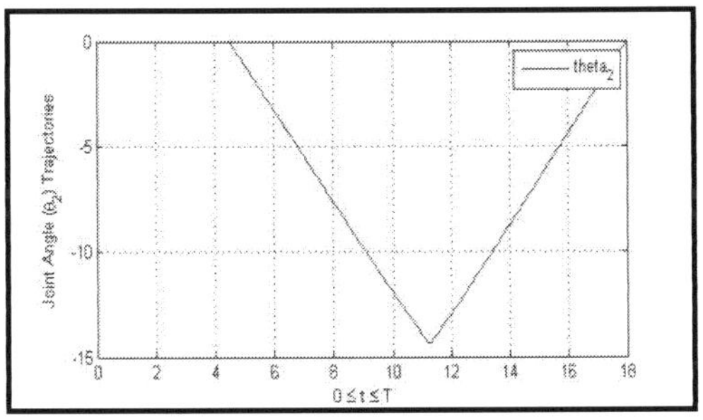

(a) θ_2 trajectories using CCF method

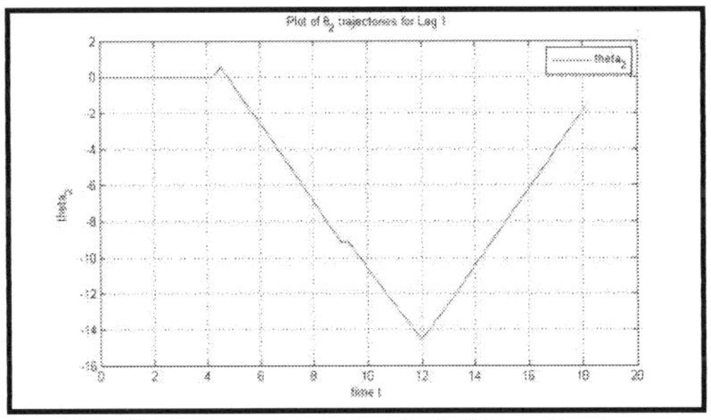

(b) θ_2 trajectories using hybrid method

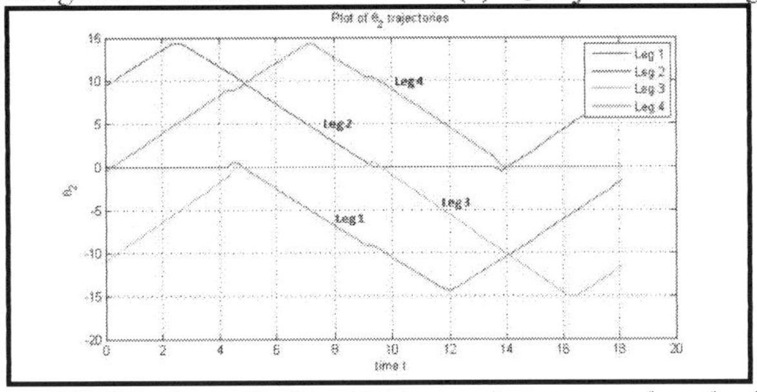

(c) θ_2 trajectories for all 4 legs using proposed method

Figure 10: Plots of θ_2 trajectories using CCF (Figure (a)), and proposed method (Figure (b) and (c)).

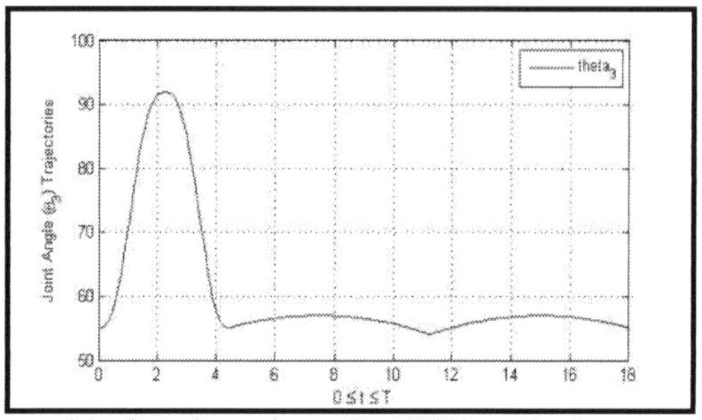

(a) θ_3 trajectories using CCF method

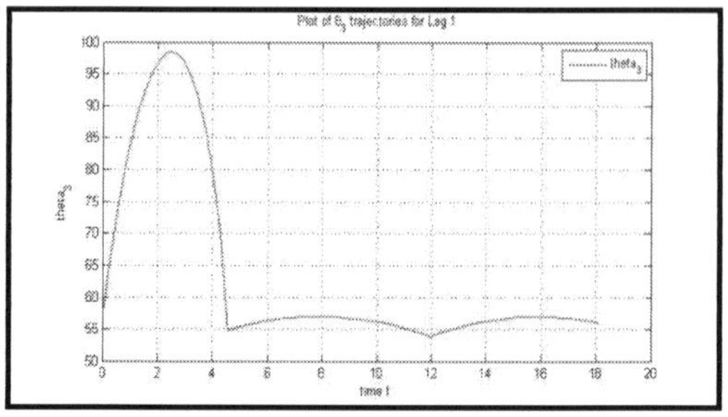

(b) θ_3 trajectories using hybrid method

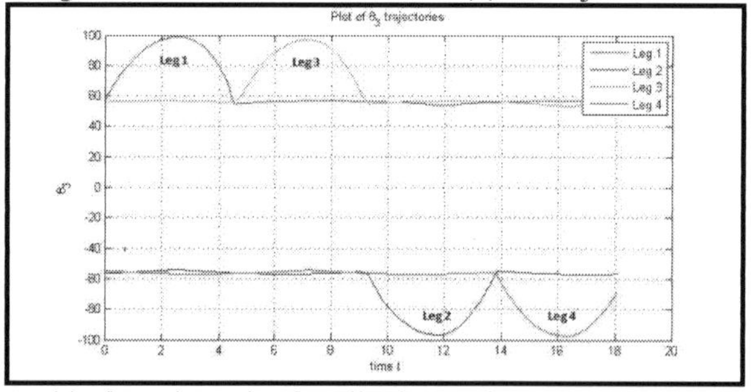

(c) $\theta 3$ trajectories for all 4 legs using proposed method

Figure 11: Plots of θ_3 trajectories using CCF (Figure (a)), and proposed method (Figure (b) and (c)).

From the plots presented above, it can be deduced that the joint angle trajectories obtained using the hybrid method proposed do not deviate much from the benchmark method. Generally, the shape of the curve shows similarity between both methods illustrating that the movement of that particular leg is the same. Fig 9(c), Fig 10(c) and Fig 11(c) shows the trajectories for all 4 legs with respect to time. If the plots are observed carefully, the shape of the curve for all legs is basically portrays the same pattern. The only difference is on the positivity and the negativity of the angles and also the timing to change the angle. For instance in Fig 9(c), θ1 for Leg 2 and Leg 4 is at the positive region, whereas, Leg 1 and Leg 3 lies on the negative region. However, the pattern of the angle's changes is still the same for both groups. Of course, the timing of the angle's changes will not be the same for all four legs to indicate that at a time, each leg will be in different position to ensure the forward movement and stability of the quadruped.

4. Conclusion

This paper proposed a hybrid method that can compensate the weaknesses identified in available method. Using this method, the gait can be easily altered by only manipulating the parameters of the CPG network, and only one oscillator is needed to develop the joint angles trajectory for each leg. The results show that the proposed method able to perform as good as the conventional method in generating the gait. However, this research work is focusing on the walking gait as this is considered as a starting platform on developing an adaptive and intelligent locomotion system for walking robot. Future works that can be worked on are on improving the parameters selection system for the CPG network since the current method is based on trial and error and also on the investigating the efficiency and smoothness of the method in gait transition.

References

[1] Sakakibara Y, Kan K, Hosoda Y, Hattori M and Fujie M 1990 *Proc. IEEE Int. Workshop on Intelligent Robots and Systems '90, IROS '90 (Japan, 3-6 Jul 1990)* vol 1 (New York, USA: IEEE) pp 315-22

[2] Doan PT, Vo HD, Kim HK and Kim SB 2010 A new approach for development of quadruped robot based on biological concepts *Int. J. of Precision Engineering and Manufacturing* **11** 559

[3] Pongas D, Mistry M and Schaal S 2007 *Proc. IEEE Int. Conf. on Robotics and Automation (Italy, 10-14 April 2007)* (New York, USA: IEEE)p 1474.

[4] Marder E and Bucher D 2011 Central pattern generators and the control of rhythmic movements *Current Biology* **11** R986

[5] Ijspeert AJ 2008 Central pattern generators for locomotion control in animals and robots: A review *Neural Networks* **21** 642

[6] Lu Z, Ma B, Li S and Wang Y 2005 *Proc. of IEEE/RSJ Int. Conf. on Intelligent Robots and System (Canada)* vol 1 (New York City:IEEE/RSJ) pp 96-101.

[7] Liu C, Chen Q and Xu T 2011 *8th World Congress on Intelligent Control and Automation (Taipei, Taiwan, 21-25 June 2011)* (IEEE) pp 1167-1172.

[8] Bin L, Xun L, Wei W, Yanfen T and Yiping Y 2009 *Proc. of IEEE Int. Conf. on Robotics and Biomimetics (Guilin, China, 19-23 Dec. 2009)* (IEEE) pp 2035-2040.

[9] Bin L, Yibin L and Xuewen R 2010 *Proc. of IEEE Int. Conf. on Robotics and Biomimetics, ROBIO (Tianjin, China, 14-18 Dec. 2010)* (IEEE) pp 19-24.

[10] Zhang J, Tomizuka M, Chen Q and Liu C 2011 Dynamic walking of AIBO with HOPF oscillators *Chinese Journal of Mechanical Engineering* **24** 1.

[11] Liu C, Chen Q and Zhang J 2009 *Proc. of Chinese Control and Decision Conference, CCDC '09 (Guilin, China, 17-19 June 2009)* (IEEE) p 3677.

[12] Liu C, Chen Q and Wang D 2011 CPG-Inspired workspace trajectory generation and adaptive locomotion control for quadruped robots *IEEE Transactions on Systems, Man, and Cybernetics* **41** 867-80.

[13] Bay JS and Hemami H 1987 Modeling of a neural pattern generator with coupled nonlinear oscillators *IEEE Transactions on Biomedical Engineering* **BME-34** 297-306.

[14] Jasni F and Shafie AA 2012 Van Der Pol Central Pattern Generator (VDP-CPG) model for quadruped robot *Trends in Intelligent Robotics, Automation, and Manufacturing: Proc. Int. Conf. Of Intelligent Robotics, Automation and Manufacturing (KL,Malaysia, 28-30 Nov 2012) (Communication in Computer and Information Sc.* vol 330*)* ed S G Ponnambalam, J Parkkinen and K Ramanathan (Springer Berlin Heidelberg) pp 167-75.

IOP Conf. Series: Materials Science and Engineering 53 (2013) 012083 doi:10.1088/1757-899X/53/1/012083

Development of real time experimental system for investigating photochromic response to UV irradiation

N Nadir[1], Z Wahid[2], A A Shafie[1], A G Abdul Muthalif[1], M Z Abdul Malek[3], N M A Nik Abdul Aziz[31], M T Zainuddin[3] and N Z Mohamed Islam[3]

[1] Department of Mechatronics Engineering, Faculty of Engineering, International Islamic University Malaysia, P. O. Box 10, 50728 Kuala Lumpur, Malaysia

[2] Department of Science in Engineering, Faculty of Engineering, International Islamic University Malaysia, P. O. Box 10, 50728 Kuala Lumpur, Malaysia

[3] Photonics Materials Section, Advanced Materials Research Centre (AMREC), SIRIM Berhad, Lot 34, Jalan Hi-Tech 2/3, Kulim Hi-Tech Park, 09000 Kulim, Kedah, Malaysia

E-mail: [1]najdir@yahoo.com

Abstract. Photochemistry is a field of chemistry that deals with the chemical effects of light such as ultraviolet (UV), visible (Vis) or infrared (IR) radiation. Among many types of photochromic molecules, particular attentions have been given on spiropyrans because of their potential applications in industrial fields. Spiropyrans have the ability to change their chemical structure after exposed to a certain appropriate wavelength of light. However, spiropyrans are very sensitive dyes. For these reasons, an UV irradiation chamber was developed to control the surrounding environment which governs the external light intervention during photochromic work and when direct absorption measurement was performed. The chamber was then exploited to investigate the substituent effect on the absorption spectra of Benzo Indolino Pyrano Spiran (BIPS) dyes, 6-nitro BIPS and 8-ethoxy-6-nitro BIPS. Thus, our results suggest that the developed chamber was successfully utilized for photochromic system since it can protect the dyes from environmental intervention.

1. Introduction

Photochromism is a branch of photochemistry, which deals with photochemical or thermal reversible photochemical reactions [1]. Since the photochromic dyes are sensitive to light [2], the existence of any external light before the initial absorption measurement may change the actual behavior. Moreover, the irradiation and absorption measurement should be done simultaneously to avoid time gap which can cause error. Besides, prolong exposure to UV radiation can lead to adverse effects to human health such as cataract formation and retinal degeneration [3, 4]. For these reasons, it is essential to establish a closed system for the photochromic work in order to control the surrounding environment, which governs the external light intervention and perform simultaneous irradiation and absorption measurement. Therefore, there is a need to design an UV irradiation chamber to provide shielding against UV radiation and thus improving the efficiency of the absorption measurement method. This strategy will minimize error during data collection.

---

[1] To whom any correspondence should be addressed.

Previous studies, irradiation as well as characterization processes were conducted in dark condition due to the fact that the photochromic dyes can be easily affected by external light [5, 6]. However, there were researchers who had done the processes separately [7, 8]. They irradiated the samples, followed with manual transfer of the sample inside the measurement chamber for characterization. The time delay, were around 2 s between the two processes [7] and 60 s [8]. However, during a trial experiment in this research, the absorption value decreased rapidly in less than 2 s once the radiation is terminated. None of the previous studies had discussed about the chamber design in detail. Thus, this paper will focus on the designing of the UV irradiation chamber and investigation of the substituent effect on the absorption spectra of Benzo Indolino Pyrano Spiran (BIPS) dyes, 6-nitro BIPS and 8-ethoxy-6-nitro BIPS.

2. Design and specifications

Figure 1 illustrates the schematic diagram of the real time photochromic measurement. The main objective of the design is to irradiate the samples and record the absorption measurement at the same time. Two light sources are utilized for this purpose. The UV serves as to change the color of the dye whereas xenon is to emit the light signal. The xenon light source is directed to pass through the dyes and trigger the built-in detector inside the spectrometer, which is used to measure the absorption spectra of the dyes.

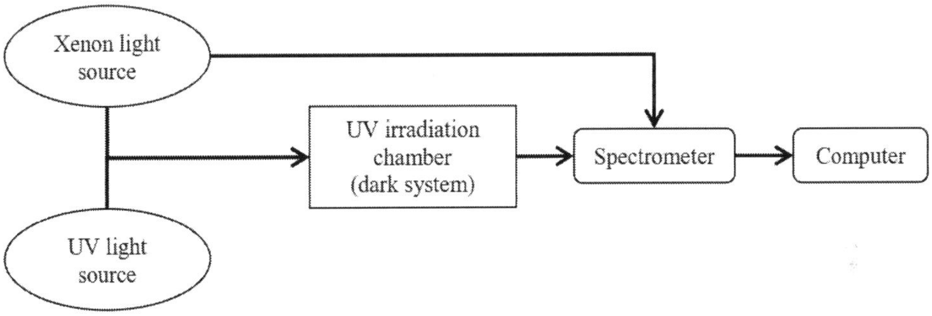

Figure 1. Schematic diagram of the real time photochromic measurement.

In this design, both light sources were coupled by Y-shaped input fiber cable connection. The UV lamp was connected to foot switch, which acts as step-on button. The light is transmitted through input fiber into a cuvette placed in a dark chamber. The light passed through the sample, went through output fiber and directly to the spectrometer connected to the computer which has been installed with SpectraSuite software.

For the SpectraSuite software, the 'Integration Time' was set at 100 ms, 'Scans to Average' at 1, and 'Boxcar Width' at 5. The sample was left inside the chamber for a while. After about one minute, the initial absorption spectrum was recorded and the absorption spectrum was recorded after irradiation stopped. Figure 2 provides real view of the chamber as well as the experimental setup with the spectrometer is connected to the computer.

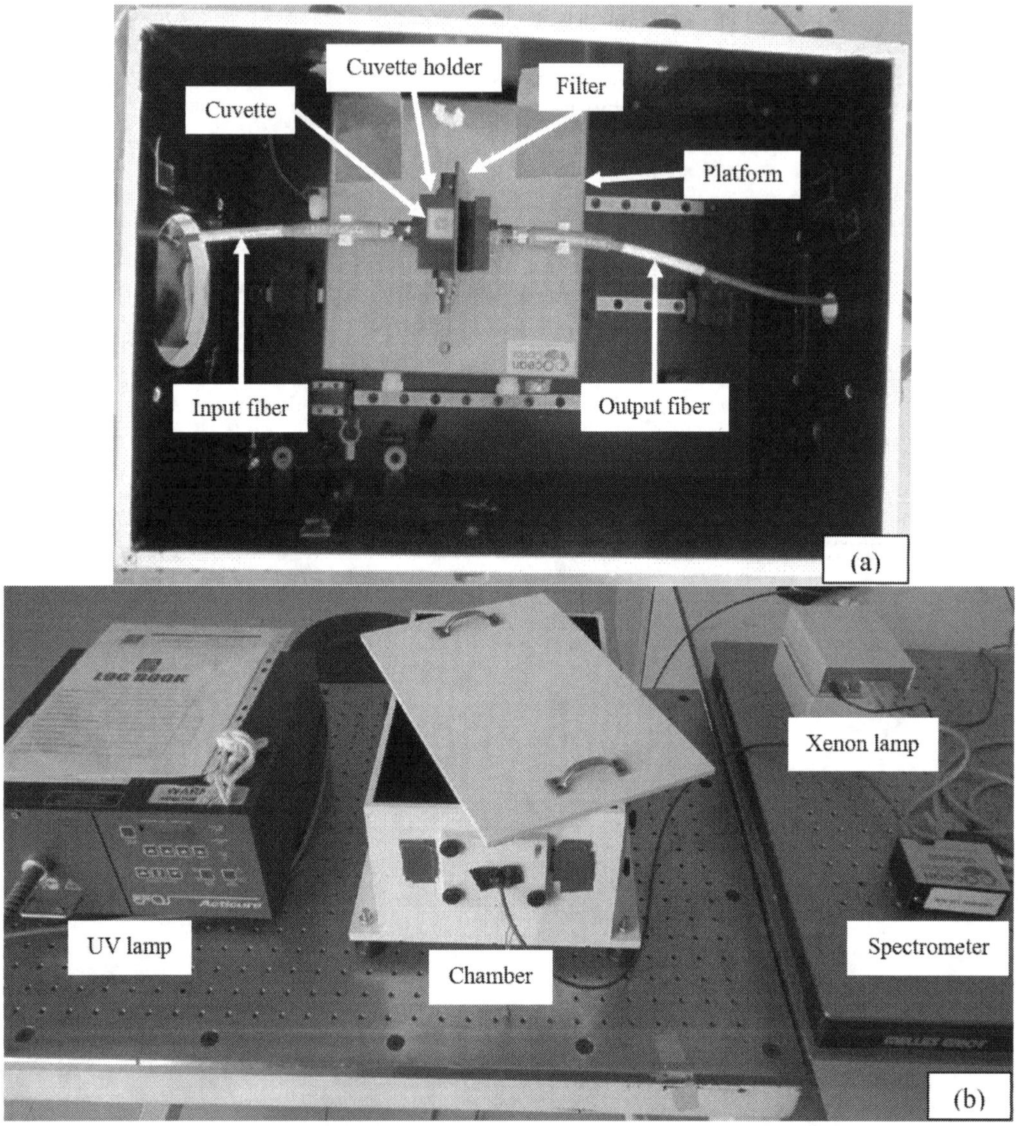

Figure 2. Complete images of (a) UV irradiation chamber, and (b) experimental setup for the real time photochromic measurement.

2.1. Light sources

The UV lamp (Efos Acticure A4000, currently known as Exfo Acticure A4000) uses a high pressure 100 W mercury vapor short arc type of lamp. The xenon lamp (Ocean Optics PX-2 Pulsed Xenon Light Source) is a high flash rate, short-arc xenon lamp for a variety of applications including absorbance, reflection, fluorescence and phosphorescence measurements, and particularly for optically or thermally unstable samples. When it is coupled to the spectrometer, the operation of both the light and detector can be easily synchronized. However, this PX-2 emits ultraviolet radiation. Thus, safety eyewear is needed and direct eye contact with the light source must be avoided to prevent from permanent eye damage. The technical specifications of the xenon lamp were listed in table 1.

Table 1. Technical specifications of the xenon lamp.

Specifications	Details
Power consumption	1.5 A @ 12 VDC
Power output	45 μJ per pulse maximum 9.9 W average power 220 Hz pulse rate maximum
Spectral range	220-750 nm
Pulse duration	5 μs (at 1/3 height of pulse)
Aperture	3 mm
Connector	SMA 905

2.2. Input and output fibers (Ocean Optics Optical Fiber Assemblies)
Both types of input (Y-shaped) and output fibers can be connected easily to light sources, sensors, and spectrometers via an SMA 905 Connector. Table 2 shows the characteristics of the input and output fibers.

Table 2. Characteristics of the fibers.

Part	Item code	Wavelength range	Core diameter	Buffer / coating	Length
Laboratory-grade Bifurcated Optical Fiber (Y-shaped input fiber)	BIF600-UV-VIS	300-1100 nm	600 μm ± 10 μm	Polyimide	2 m
Laboratory-grade Optical Fiber (output fiber)	P600-1-UV-VIS	300-1100 nm	600 μm ± 10 μm	Polyimide	1 m

2.3. UV irradiation chamber
In the chamber, a 1-cm path length cuvette is fitted in the cuvette holder (Ocean Optics CUV-ALL 4-way cuvette holder) which has an opening diameter of 5 mm as the light passage. Meanwhile, the neutral density filter (Newport FSQ-OD100) is placed behind the cuvette to attenuate the amplitude signal for avoiding signal saturation at CCD array detector. It is made of neutral density colored glass (BK 7 or equivalent for 0.04 OD) with dimension of 50.8 x 50.8 x 1.57 mm. The optical density is 1.0 at 546.1 nm.

2.4. Spectrometer (Ocean Optics USB4000-VIS-NIR Fiber Optic Spectrometer)
The spectrometer is a high-performance, miniature fiber-optic spectrometer that can measure bandwidth in the range of 348-1043 nm. It is connected to a computer via the USB port or serial port. The USB4000 spectrometer is a great equipment as it can be utilized for various applications of basic absorbance, reflectance and emission measurements. It can be operated by the SpectraSuite software. The important descriptions of the spectrometer were documented in table 3.

Table 3. Descriptions of the spectrometer.

Specifications	Details
Power consumption	250 mA @ 5 VDC
Detector	3648-element Toshiba TCD1304AP Linear CCD array
Fiber optic connector	SMA 905 to single-strand optical fiber (0.22 NA)

2.5. SpectraSuite software
SpectraSuite is a complete modular, Java-based spectroscopy software which can control any Ocean Optics USB spectrometer and device. The software supports a real-time interface to a variety of

signal-processing functions. Hence, the spectroscopic measurements (such as absorbance, reflectance, and emission) can be performed, all system parameters can be manipulated, data can be collected and presented in real time, and also reference monitoring and time acquisition tests can be done.

Absorbance spectra are a measure of how much light a sample absorbs. For most samples, it is linearly proportional to the sample concentration. SpectraSuite software determines the absorbance at wavelength λ (A_λ) by using the following equation:

$$A_{\lambda_{\log_{10}}} = -\left(\frac{S_\lambda - D_\lambda}{R_\lambda - D_\lambda}\right) \quad (1)$$

where at wavelength λ, S_λ is the sample intensity, D_λ is the dark intensity, and R_λ is the reference intensity.

3. Example of application in photochromic study

Photochromism is defined as a reversible transformation of a chemical species induced in one or both directions by the absorption of electromagnetic radiation between two forms having different absorption spectra [9, 10]. From scheme 1, molecule A is transformed into molecule B by irradiation. The back reaction of B into A can either occur thermally or photochemically. Mostly but not all of the spectral change is observed by visible color, together with differences in other physical properties [11].

$$A\ (\lambda_1) \underset{h\nu_2,\ \Delta T}{\overset{h\nu_1}{\rightleftarrows}} B\ (\lambda_2)$$

Scheme 1. Photochromism of molecule A to molecule B.

Cleavage of the central C-O bond in ring-closed isomer (spiropyran form, SP) of 1',3',3'-trimethyl spiro[*2H*-1-benzopyran-2,2'-indoline] (BIPS) leads to the formation of ring-opened isomer (merocyanine form, MC) as depicted in scheme 2. In organic solvents, the ring-opening reaction is stimulated by UV irradiation and cause an intense absorption in the range 500-600nm, which gradually returns to the original closed form thermally or upon visible light irradiation [12-14].

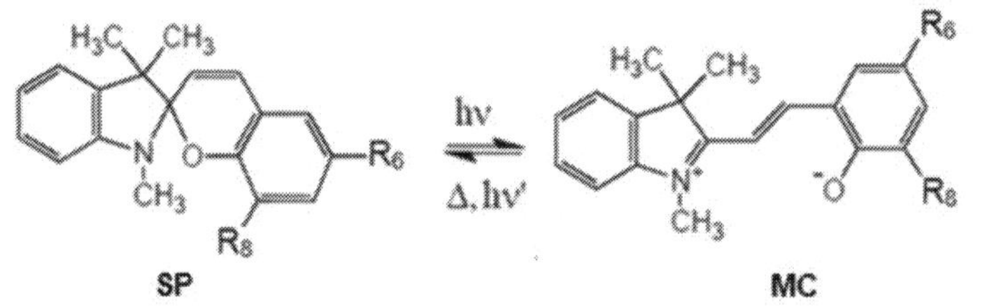

Scheme 2. Spiropyran and merocyanine photoisomers of 1',3',3'-trimethyl spiro[*2H*-1-benzopyran-2,2'-indoline] (BIPS).

3.1. Chemicals

There are two types of spiropyrans used in this study, which are 6-nitro BIPS (R_6=NO_2, R_8=H) and 8-ethoxy-6-nitro BIPS (R_6=NO_2, R_8=OC_2H_5) as can be seen in scheme 2. The photochromic dyes were purchased from Sigma-Aldrich and absolute ethanol from RCI Labscan.

3.2. Sample preparation, UV irradiation and absorption measurements

In photolithography room, the dyes were dissolved individually in a polar solvent, absolute ethanol, at a concentration of 10^{-5} M. The samples were then transferred into 10 mm plastic cuvette. The cuvette was placed inside the chamber. After about one minute, the sample was irradiated with absolute ethanol as blank. The absorption spectra were measured prior to and after five seconds of UV irradiation.

3.3. Results and discussion

Both of 6-nitro BIPS and 8-ethoxy-6-nitro BIPS showed a color change upon UV irradiation. The very light pink of 6-nitro BIPS became brighter while the slightly purplish 8-ethoxy-6-nitro BIPS became darker, reflecting that both dyes undergo photoisomerization in the solvent. Different colors of the photoinduced form were obtained when the molecule structure of spiropyrans is varied [15], because of the presence of 8-ethoxy group attached to 6-nitro BIPS molecules in this study. Generally, nonpolar solvents favor the spiropyran form, while polar solvents produce the merocyanine structure [16].

Figure 3 illustrates the absorption spectra of 6-nitro BIPS and 8-ethoxy-6-nitro BIPS in ethanol at a concentration of 10^{-5} M for before and after five seconds of UV irradiation. Initially, 6-nitro BIPS is in very light pink color, thus gives zero absorption value. The absorption value increased to 0.172% after five seconds of irradiation. The maximum absorption wavelength was detected at 540nm with absorption peak covered the wavelength ranging from 650 to 430nm. On the other hand, the color of prepared 8-ethoxy-6-nitro BIPS is slightly purplish, which gives an absorption value of 0.029%. The value then increased to 0.124% with five seconds of UV exposure. The absorption value began to increase at wavelength of 680nm and ended at 450nm with absorption maximum at 570nm.

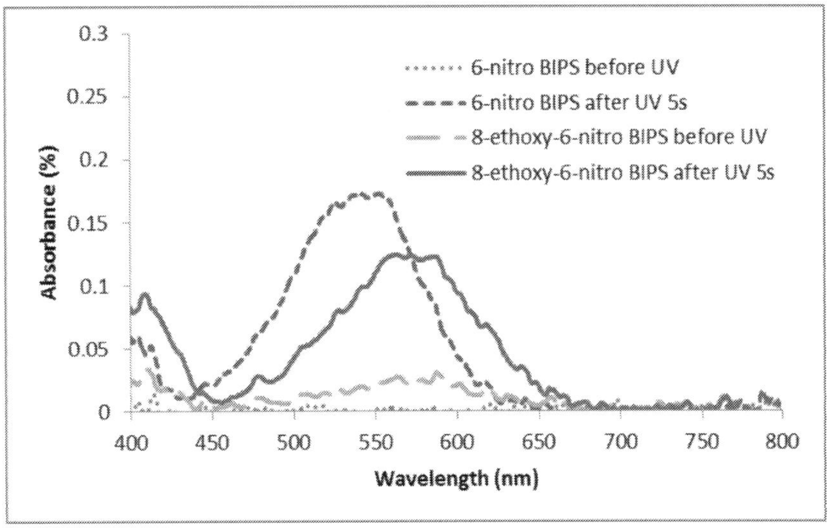

Figure 3. Absorption spectra of 6-nitro BIPS and 8-ethoxy-6-nitro BIPS in ethanol.

The absorption results for both BIPS photochromic dyes were presented in figure 4 while the wavelengths were tabulated in table 4. Although the initial absorption value for 6-nitro BIPS is lower than 8-ethoxy-6-nitro BIPS, the value became higher upon UV irradiation. Apart from that, the wavelength peak is slightly narrow compared to 8-ethoxy-6-nitro BIPS, 220nm to 230nm, respectively. The maximum wavelength (λ_{max}) for 6-nitro BIPS was observed at 540nm while 8-ethoxy-6-nitro BIPS at 570nm, which indicates that the peak shift towards longer wavelengths (bathochromic or red shift).

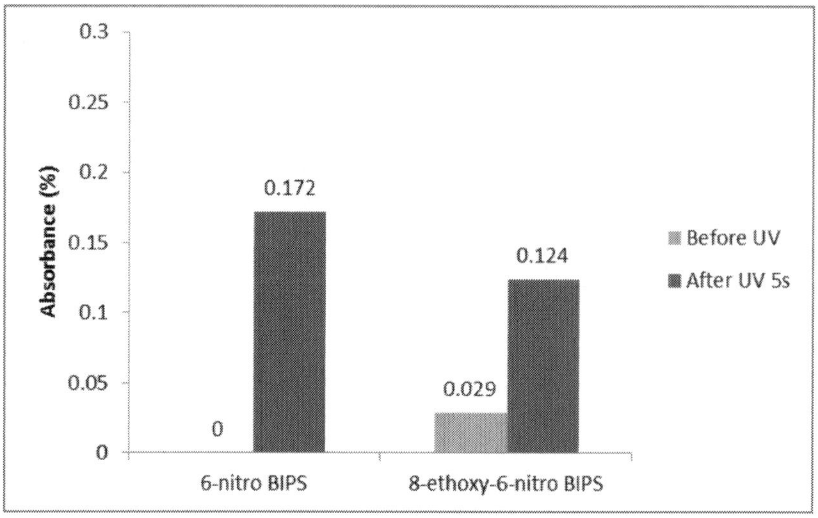

Figure 4. The absorption values of 6-nitro BIPS and 8-ethoxy-6-nitro BIPS.

Table 4. The wavelength peaks of 6-nitro BIPS and 8-ethoxy-6-nitro BIPS.

Wavelength	6-nitro BIPS	8-ethoxy-6-nitro BIPS
λ_{start} (nm)	650	680
λ_{max} (nm)	540	570
λ_{end} (nm)	430	450

The presence of ethoxy group attached at R_8 position in 6-nitro BIPS molecule introduced another charge influential site within the 6-nitro BIPS molecule. Oxygen ions in ethoxy group act as electron-donating groups contributed as another resonation atom binds together with benzopyran and the electronegativity of the oxygen atom contributed as charge resonation and repulsion site. This repulsive force influence the magnitude of stabilization in benzopyran cyclic within 6-nitro BIPS and eventually allowed the photoisomerization of open structure at C-O bond in benzopyran at lower energy and longer wavelength as observed in 8-ethoxy-6-nitro BIPS. In addition, the position of ethoxy groups at R_8 also reduced the photochromism activation upon UV responsive light by contributing steric factor as observed by the broader absorption peak for 6-nitro BIPS with ethoxy functional group compared to 6-nitro BIPS without such substituent group.

4. Conclusion

Given that zero initial absorption value of 6-nitro BIPS was recorded and the maximum absorption wavelength of 6-nitro BIPS at 540nm while 8-ethoxy-6-nitro BIPS at 570nm, with the difference of 30nm only was detected, the UV irradiation chamber certainly provides absolute dark condition in this study. Thus, the developed chamber was successfully utilized for photochromic system since it can protect the dyes from environmental intervention.

Acknowledgement

The authors acknowledge the Ministry of Higher Education (FRGS Grant No. 11-029-0177) for supporting this research.

References

[1] Dürr H 2003 General introduction *Photochromism: Molecules and Systems* (revised ed) eds H Dürr and H Bouas-Laurent (Amsterdam: Elsevier) pp 1-14.

[2] Atassi Y, Delaire J A and Nakatani K 1995 Coupling between photochromism and second-harmonic generation in spiropyran- and spirooxazine-doped polymer films *J. Phys. Chem. B* **99** 16320-6.
[3] Roberts J E 2011 Ultraviolet radiation as a risk factor for cataract and macular degeneration *Eye Contact Lens* **37** 246-9.
[4] van Kuijk F J G M 1991 Effects of ultraviolet light on the eye: role of protective glasses *Environ. Health Perspect.* **96** 177-84.
[5] Zanoni M, Coleman S, Fraser K J, Byrne R, Wagner K K, Gambhir S, Officer D L, Wallace G G and Diamond D 2012 Physicochemical study of spiropyran-terthiophene derivatives: photochemistry and thermodynamics *Phys. Chem. Chem. Phys.* **14** 9112-20.
[6] Pan X W, Wu G, Wang M and Chen H Z 2009 Partially reversible photochromic behavior of organic-inorganic perovskites with copper(II) chloride *J. Zhejiang Univ. Sci. A* **10** 710-5.
[7] Klukowska A, Posset U, Schottner G, Jankowska-Frydel A and Malatesta V 2004 Photochromic sol-gel derived hybrid polymer coatings: the influence of matrix properties on kinetics and photo degradation *Mater. Sci.* **22** 187-99.
[8] Schomburg C, Wark M, Rohlfing Y, Schulz-Ekloff, G and Wöhrle D 2001 Photochromism of spiropyran in molecular sieve voids: effects of host-guest interaction on isomer status, switching stability and reversibility *J. Mater. Chem.* **11** 2014-21.
[9] Irie M 2000 Photochromism: memories and switches – introduction *Chem. Rev.* **100** 1683-84.
[10] Pościk A and Wandelt B 2009 Application of a photochromic dye in an automatic welding filter *Int. J. Occup. Saf. Ergon.* **15** 243-54.
[11] Braslavsky S E 2007 Glossary of terms used in photochemistry 3rd edition (IUPAC recommendations 2006) *Pure Appl. Chem.* **79** 293-456.
[12] Schulze G, Franke K J and Pascual J I 2012 Induction of a photostationary ring-opening-ring-closing state of spiropyran monolayers on the semimetallic Bi(110) surface *Phys. Rev. Lett.* **109** 026102.
[13] Ernsting N P, Dick B and Arthen-Engeland T 1990 The primary photochemical reaction step of unsubstituted indolino-spiropyrans *Pure Appl. Chem.* **62** 1483-8.
[14] Satoh T, Sumaru K, Takagi T, Takai K and Kanamori T 2011 Isomerization of spirobenzopyrans bearing electron-donating and electron-withdrawing groups in acidic aqueous solutions *Phys. Chem. Chem. Phys.* **13** 7322-9.
[15] Barachevsky V A 2000 Photofluorochromic spirocompounds and their application *J. Fluoresc.* **10** 185-91.
[16] Shao N, Jin J Y, Wang H, Zhang Y, Yang R H and Chan W H 2008 Tunable photochromism of spirobenzopyran via selective metal ion coordination: an efficient visual and ratioing fluorescent probe for divalent copper ion *Anal. Chem.* **80** 3466-75.

Improved 3D limit-cycle navigation method for path planning quad rotor

Fajril Akbar[1,2,3] **and Yasir Mohd Mustafah**[1]
[1]Faculty of Engineering, International Islamic University, Jalan Gombak, 53100, Kuala Lumpur, Malaysia
[2]Faculty of Information Technology, Andalas University, Limau Manis, 25163, Padang, Indonesia

E-mail: ijab@ft.unand.ac.id

Abstract. Quad rotors as a type of rotary wing aerial vehicle must planned their flight path effectively and avoid any disturbance to complete their mission. The extension of 2D limit-cycle navigation into 3D, made it possible to apply for aerial vehicle. The extended limit-cycle has been introduced, but it generate unsuitable path in certain condition. In this research, rendering the obstacle into a cylinder will generate the efficient path and provide the best way in avoiding stationary obstacle. By defining obstacle position, a simulation and performance study is done using limit-cycle characteristic method. The simulations results illustrated the path generation using improved 3D limit cycle with different obstacle condition.

1. Introduction

Quad rotor as one of the rotary wing aerial vehicle is becoming more preferable for various task in society. The capability to fly indoor as well as outdoor with unmanned capabilities make this autonomous system provide safety, security and convenience to performing s desired task in a any kind of environment. The minimal space required for takeoff and landing, the ability to maneuver sharply and hover precisely and capable of maintain static position is becoming important contribution of these aircraft.

Quad rotor must have good tracking characteristic and ability to reject any disturbance of the flight path. Thus, several researches has been done in planning flight path for UAV such as path planning based on visibility lines(VL) method [1], based on genetic algorithm[2], and multi-UAVs based on Rapidly-exploring Random Tree(RRT) algorithm [3]. Providing accurate path with obstacle avoidance is also become a key challenge in planning the flight path in rural area. Several researches has been done in UAV navigation strategy to avoid the obstacle along the path. A method for path planning based on velocity vector field investigated in [4]. In [5], the author presented a fuzzy logic approach for avoiding stationary and moving obstacle by sensing the distance and the angle between UAV and obstacles. A stereo vision for obstacle detection with probabilistic roadmaps for path planning is introduced in [6]. The difficulties in detecting thin obstacles and inefficiency in memory for 3D maps representation is need improvement to make it more reliable.

The limit-cycle navigation method introduced in [7] as one of the appropriate path planning methods for two dimension (2D) environment. By using the concept horizontal and vertical plane, the three dimension (3D) limit-cycle navigation also apply in planning the flight path. It still has several

[3] To whom any correspondence should be addressed.

drawbacks because it only generated the limit-cycle based on the center of gravity position of the obstacle.

In this research, it is investigated and improved the extended limit cycle navigation method by rendering an obstacle into a cylinder in order to generate an effective path planning for aerial vehicle. By defining obstacle position, a simulation and performance study is done using limit-cycle characteristic method. Simulation studies are conducted and the results illustrated the effectiveness of path generation using 3D limit cycle with different obstacle condition.

2. The limit-cycle navigation

Limit cycle navigation based on characteristic of 2nd-order nonlinear system [7] :

$$\dot{x} = y + x(1 - x^2 - y^2) \tag{1}$$

$$\dot{y} = -x + y(1 - x^2 - y^2) \tag{2}$$

by inspecting phase portrait of this equation and replacing 1 with r, we will get a periodic orbit as shown in Figure 1.

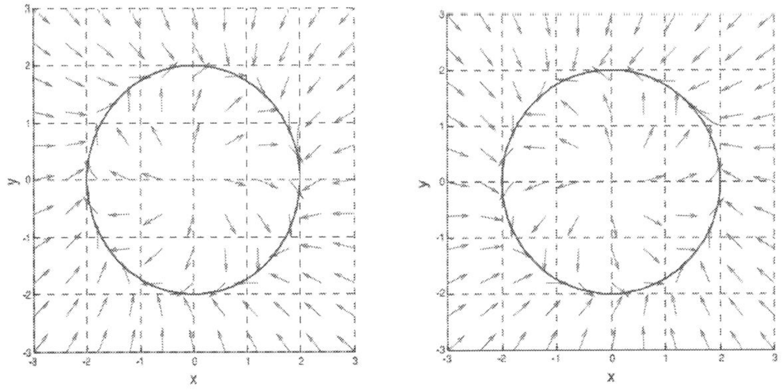

(a) (b)

Figure 1. Phase portrait of the limit-cycle with $r = 2$, a. Clockwise; b. Counter-clockwise

This periodic orbit is called a limit-cycle. Figure 1a shows the trajectories from all points move towards the unit circle clockwise. On the other hand the unit circle counter-clockwise as shown in figure 1b, can be derived as follows :

$$\dot{x} = -y + x(r^2 - x^2 - y^2) \tag{3}$$

$$\dot{y} = x + y(r^2 - x^2 - y^2) \tag{4}$$

therefore, the radius and the direction of the limit cycle characteristic should be adjusted to applied in navigation plan. Figure 2 shows how the limit cycle method can drive an aircraft into destination point and avoid an obstacle along its path.

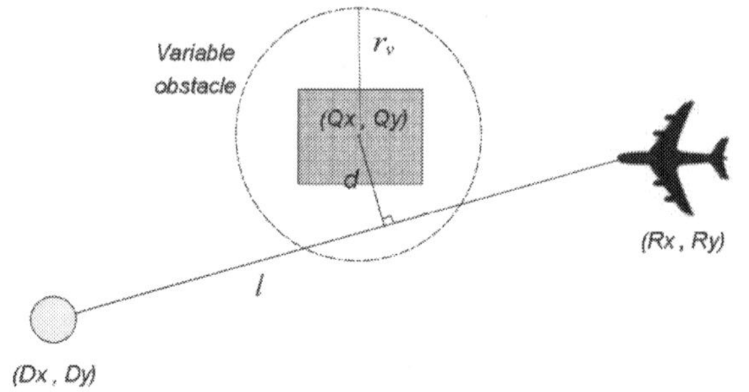

Figure 2. Obstacle identification using limit cycle

Based on (7), there are some definitions using in limit-cycle navigation that:

- Variable obstacle (O_v) : By assuming a robot as a point mass, the variable obstacle is defined as circle with the radius is the total of its relative position to the aircraft, the total size of the obstacle and the aircraft. The aircraft will follow the boundary of the circle O_v in order to avoid any collision with obstacle. The radius of the variable obstacle (O_v) is defined as r_v
- Disturbing obstacle (O_d) is variable obstacle that is in the way between the aircraft and target point. The nearest obstacle to the aircraft is defines as O_{d1}, and so on. Non-disturbing obstacle (O_n) is vice versa. O_{n1} is defined as the nearest obstacle to the aircraft and O_{n2} is the next one, etc.

Based on that definition, the steps to develop a path planning using the limit-cycle method are:

1. draw a line l from the robot to the target in a global coordinate Σ_{OXY} as follows :
$$ax + by + c = 0 \tag{5}$$
2. Define a variable obstacles as disturbing obstacle as O_d if the line l crosses them, or O_n if it is non-disturbing ones.
3. move towards the target if there is no O_d
4. Calculate the distance (d) from the center of nearest disturbing obstacle O_d to the line l, using
$$d = \frac{aQ_x + bQ_y + c}{\sqrt{a^2 + b^2}} \tag{6}$$
where (Q_x, Q_y), (G_x, G_y) and (R_x, R_y) are the xy-values of the center positions of the obstacle, the target and the aircraft respectively. Calculate the desired direction of the aircraft at each position using
$$\dot{x} = \frac{d}{|d|} y + x(r_v^2 - x^2 - y^2) \tag{7}$$
$$\dot{y} = -\frac{d}{|d|} x + y(r_v^2 - x^2 - y^2) \tag{8}$$
where x and y are relative values to the obstacle. The aircraft avoids the obstacle O_d clockwise if d is positive and vice versa if d is negative. Calculate r_v as the total size of the aircraft and the relative position to the obstacle as follows :
$$r_v = r_r + r_o + \delta \tag{9}$$
where r_r and r_o are the radius of the aircraft and the obstacle respectively. If obstacle is disturbing one, $r_o = r$, where r is the radius of the real obstacle and $r_o = 0$ if it is non-disturbing obstacles. As the aircraft moves, the line l varies, so repeat the steps 2 to 4 until the destination reached.

3. 3D limit-cycle method

A several research have been introduced to adopt the limit cycle characteristic in planning the flight path for aircraft navigation. Because of the obstacle information is known in advance and it has to be carefully pre-planning the safe flight path for the vehicle.

The limit-cycle navigation should be extent into three dimensional environments for aerial vehicle application. In [8], the extension of 2D into 3D is considered as two plane that is horizontal and vertical which perpendicular each other. The existing limit-cycle method is applied in horizontal plane that affected only in xy-plane. The angle φ is defined as the angle between the current position and the desired position as shown in figure 3.

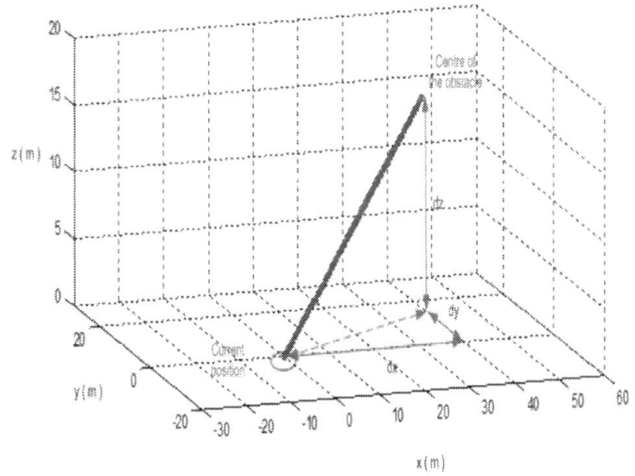

Figure 3. The scheme of extended limit cycle navigation method in 3D

It can be obtained as :

$$d_h = (dx^2 + dy^2)^{1/2} \tag{10}$$

$$\varphi = tan^{-1}\left(\frac{dz}{dh}\right) \tag{11}$$

where Dh is a tangent line between the current position and the tangent of an obstacle circle. By getting the angle φ, the z-value can be obtained by

$$\Delta h = (\Delta x^2 + \Delta y^2)^{1/2} \tan\varphi \tag{12}$$

$$z = \Delta h + \bar{z} \tag{13}$$

where $\bar{z}$ is the current z position. The result in extending the 2D limit-cycle into 3D is shown in figure 3 as solid lines.

By applying this method, it shows a flight path for three dimensional environment. But it also has several drawbacks. First of all, it provide unsuitable path if the aircraft position below the sphere of variable obstacle. It clearly seen in figure 4 where it generate the path that impossible to follow if we consider $z = 0$ as ground level of the earth.

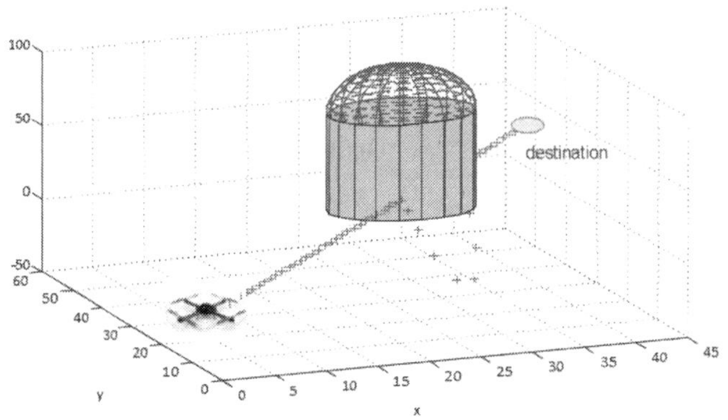

Figure 4. Impossible path generate by extended limit-cycle method

The second problem of extended limit-cycle navigation is not providing the best way to reach the destination point. As example, it provide inefficient flight path if the actual obstacle height is not defined as the limit of safe region. Figure 5 is the clear example of inefficient path that provided by this method. This two problem occur because the method is generates the path by only considering the center of gravity position of the obstacle.

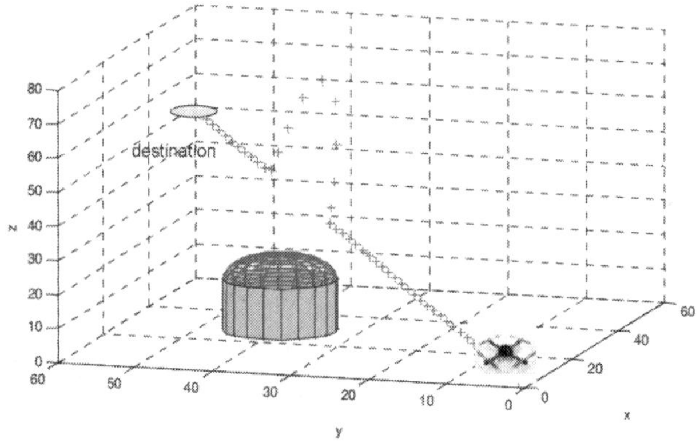

Figure 5. Inefficient path occurs when flight above the obstacle

Rendering the obstacle into a cylinder is the solution to solve this entire problem. The obstacle is considered as cylinder with radius is r_v and height is h and the xy-plane as a base of the cylinder. It defined as

$$(x - x_c)^2 + (y - y_c)^2 = r^2 , 0 \leq (z - z_c) \leq h \quad (13)$$

with x_c and y_c is x and y obstacle position respectively.

By applying this rule, the straight path will apply if the current position of the aircraft above the obstacle height although they intersect in their xy-plane. The determinant indicates whether an intersection occurs. The intersection with the z-coordinates must be in the range $0 \leq (z - z_c) \leq h$ if it identify as intersection points.

4. Result and discussion

In this section, the proposed path planning algorithm is verified using MATLAB. In this simulation work, we render the obstacle into a cylinder and apply the limit-cycle navigation in avoiding the obstacle that intersects with the cylinder.

Figure 6 shows comparison between the previous extended limit-cycle and the proposed method with obstacle render as cylinder.

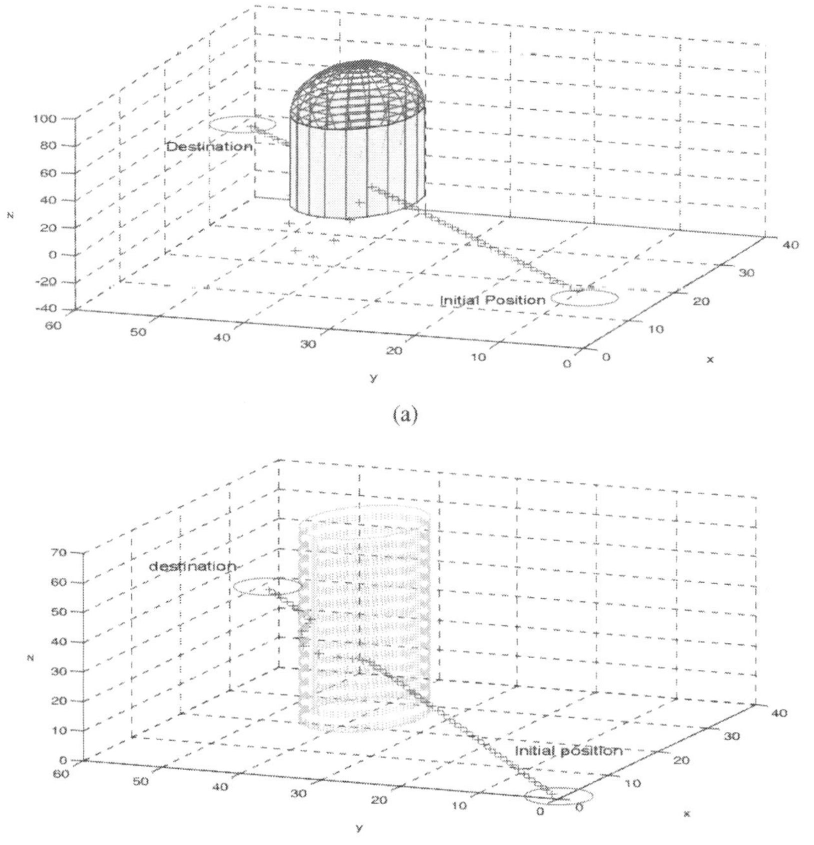

(a)

(b)

Figure 6. Simulation result of extended limit-cycle method;(a) previous method, (b) proposed method

The result of the simulation in figure 6 shows that the aircraft moved from initial point (0, 0, 0) to destination position (30,55,35) while avoiding obstacle at (25,40,55). Figure 6.a shows the result of previous extended limit-cycle simulation. It generates some path in certain position which have a value of $z < 0$. It will make an aircraft fall down to earth and crash. Figure 6.b shows how the obstacle that render as a cylinder is generated a safe flight path using the proposed limit-cycle navigation method.

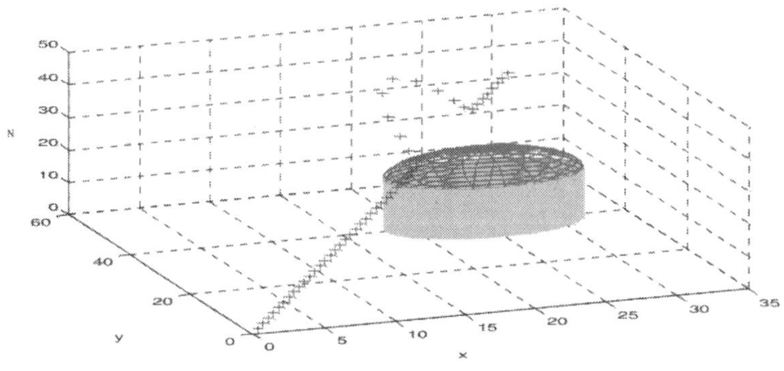

(a)

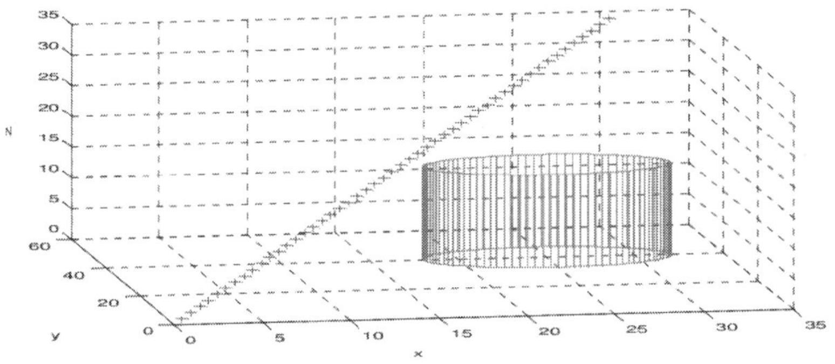

(b)

Figure 7. Result when aircraft flight above the obstacle; (a) previous method, (b) proposed method

Figure 7 also shows an inefficient path generate using previous extended limit-cycle. We set up obstacle position at (25,40,15) , it clearly seen that figure 7b is generate efficient path compare to the previous method as shown in figure 7a. The aircraft will straight forward if their current position is above the obstacle height. The previous method take 95.4519 point as the distance while the proposed method take 44.4934 point. It is 46.6134% efficiency in flight distance and it is the shortest path to reach destination point.

Consequently, the paths generated in these simulations were produce suitable and efficient path for aerial vehicle application.

5. Conclusion

The previous extended limit-cycle navigation method is always developed the path based on center of gravity position of the obstacle. Accordingly, the previous extended limit-cycle navigation method is generated inefficient path in certain condition. In this paper, we verified that rendering an obstacle as cylinder will improve the drawback of the previous method. The simulation results show that the proposed extended limit cycle method is provide an effective path compare to the previous method.

References

[1] Omar R., Gu D -W 2009 Visibility line based methods for UAV path planning *ICCAS-SICE 2009* pp 3176,3181

[2] Ze Cheng,Ying Sun, Yanli Liu 2011 Path planning based on immune genetic algorithm for UAV *Electric Information and Control Engineering (ICEICE) International Conference 2011* pp 590-593

[3] Hui Bai, Shihuang Shao, Hongyu Wang 2011 A VTOL quadrotor platform for multi-UAV path planning *Electronic and Mechanical Engineering and Information Technology International Conference (EMEIT) 2011* vol **6** pp 3079,3081

[4] Hu Mu, Li Chun-Tao 2010 Application in real-time planning of UAV based on velocity vector field *Electrical and Control Engineering (ICECE) International Conference 2010* pp 5022,5026

[5] Tao Dong, Liao X H, Zhang R., Sun Z., Song Y D 2005 Path tracking and obstacles avoidance of UAVs - Fuzzy Logic Approach *Fuzzy Systems 2005: The 14th IEEE International Conference* pp 43,48

[6] Hrabar S 2008 3D Path planning and stereo-based obstacle avoidance for rotorcraft UAVs *Intelligent Robots and Systems 2008(IROS 2008) IEEE/RSJ International Conference* pp 807,814.

[7] Dong-Han Kim, Jong-Hwan Kim 2003 A Real-time Limit-cycle navigation method for fast mobile robots and its application to robot soccer *Robotics and Autonomous Systems* Volume **42** Issue 1 pp 17-30

[8] S C Min, EJ Lee, S H Kang, and D H Kim 2009 Limit-cycle navigation method for a quad-rotor type UAV *Proc. IEEE International Symposium on Industrial Electronics* pp 1352,1357

Fuzzy Logic based Handoff Latency Reduction Mechanism in Layer 2 of Heterogeneous Mobile IPv6 Networks

Farhat Anwar[1], Mosharrof H. Masud, Suhaimi A. Latif
Department of Electrical and Computer Engineering
International Islamic University Malaysia
Kuala Lumpur, Malaysia

E-mail: farhat@iium.edu.my, masuduia@yahoo.com, suhaimie@iium.edu.my

Abstract. Mobile IPv6 (MIPv6) is one of the pioneer standards that support mobility in IPv6 environment. It has been designed to support different types of technologies for providing seamless communications in next generation network. However, MIPv6 and subsequent standards have some limitations due to its handoff latency. In this paper, a fuzzy logic based mechanism is proposed to reduce the handoff latency of MIPv6 for Layer 2 (L2) by scanning the Access Points (APs) while the Mobile Node (MN) is moving among different APs. Handoff latency occurs when the MN switches from one AP to another in L2. Heterogeneous network is considered in this research in order to reduce the delays in L2. Received Signal Strength Indicator (RSSI) and velocity of the MN are considered as the input of fuzzy logic technique. This technique helps the MN to measure optimum signal quality from APs for the speedy mobile node based on fuzzy logic input rules and makes a list of interfaces. A suitable interface from the list of available interfaces can be selected like WiFi, WiMAX or GSM. Simulation results show 55% handoff latency reduction and 50% packet loss improvement in L2 compared to standard to MIPv6.

1. Introduction

The next generation wireless communication system is envisioned to be fully Internet Protocol version 6 (IPv6) based in order to support increasing mobile users. Mobility in IPv6 (MIPv6) [1] was introduced by the Internet Engineering Task Force (IETF) so that a Mobile Node (MN) could have mobility from one Access Point (AP) to another AP seamlessly. However, seamless and ubiquitous connection is a great challenge in mobile environments where heterogeneous technologies are collaborating with each other. In such scenario, mobile users need to be connected with different types of technologies while they are roaming. The continuous growth of wireless communication systems poses problems for seamless connectivity like handoff delays, packet loss and jitter rise when the MN is moving. Two types of delay could happen during this transition, namely; Layer 2 (L2) and Layer 3 (L3).

Fuzzy logic technique is one of the effective techniques that has been employed in the field of network particularly L2 of MIPv6 to minimize its existing limitations, namely; delay and packet loss. The Received Signal Strength Indicator (RSSI) is used to take decision for switching from one AP to

[1] Department of Electrical and Computer Engineering, International Islamic University Malaysia, Kuala Lumpur, Malaysia, farhat@iium.edu.my

another AP in a particular technology like WiFi, WiMAX or GSM. However, in heterogeneous environment, the reliance on only RSSI is not sufficient for taking handoff decision due to diverse parameters of the network. Therefore, to reduce the most time consuming scanning delay, many researchers have proposed many alternative solutions in the form of incorporating fuzzy logic into the decision making process. Fuzzy logic has been used successfully in control system design and prediction based applications. It is a language based approach that allows less dependence on precise quantitative analysis. It is most suitable for uncertain events [2]. One of the main features of fuzzy logic is the ability to take many performances metrics into account and yielding the best possible solution for handoff decision. This approach is more appropriate for heterogeneous environments where many metrics need to be considered such as BER, SNR, velocity, bandwidth, network loads, monetary cost etc. The Fuzzy logic consists of three basic components, namely; fuzzifier, inference engine and defuzzifier as shown in figure 1.

The fuzzy algorithm needs the establishment of fuzzy sets and rules over the N number of inputs. The objective of the fuzzifier in the fuzzy system is to convert the crispy input into non-crispy of a fuzzy set that is ready to be processed in the inference engine with combination of fuzzy rules. Fuzzy rules are based on if-then conditions.

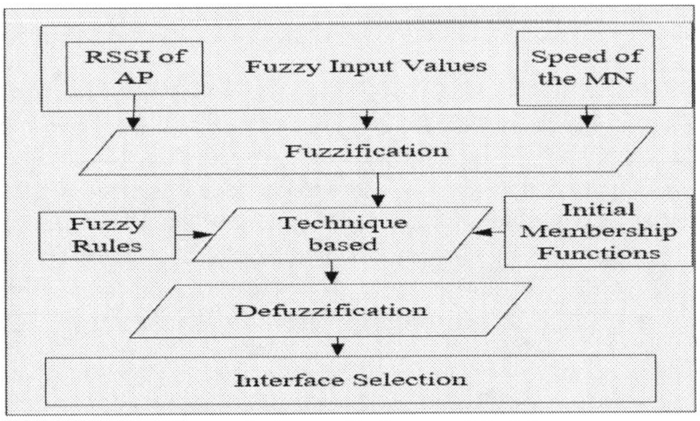

Figure 1: Fuzzy Reference Model

The defuzzifier converts the non-crispy data into the crispy output as a scalar value which is the output of the handoff factor. Applying fuzzy logic technique, the delay of L2 in MIPv6 can be reduced.

The rest of the paper is organized in the following fashion. Section two discusses about MIPv6 in brief. Related works addressed in section three. Section four proposes proposed mechanism to reduce L2 delays. Simulation model and proposed enhancement have been elaborated in section five and six respectively. Finally, conclusion has been drawn in section seven.

2. Mobile IPv6 (MIPv6)

It has been observed that recent developments in wireless technologies motivated users and vendors to alter their preference from wired Internet connection to wireless Internet connection. This shift in technology provides better communication services in return, such as, the ability to communicate while an MN is moving from one network to another. However, Internet was initially designed for fixed networks. Consequently, there is no provisioning for IP to support mobility. Mobile IP (MIP) (RFC 2002) is the first standard protocol designed by IETF to address the aforementioned issues [3]. The development of MIP has enabled users to maintain continuous connection even while on the move.

MIPv6 is one of the mobility management solutions that has been widely accepted in the academic and industrial community. MIPv6 [1] has been proposed to accommodate the increasing demands of mobility in the Internet by IETF. It has many facilities compared to existing MIPv4 including more address space, enhanced security mechanism, route optimization and other features. Based on the proposal, an MN should generate a permanent and temporary IP address for further communication [4].

The permanent address is assigned by home network (HN) to the MN as a global identifier that is fixed. A temporary address also known as care-of address (CoA) is obtained from the visiting network which indicates the host's actual location that changes with time as MN moves to foreign networks. While an MN is attached to its HA, it is able to receive packets destined to its HA by existing routing mechanism. However, whenever it visits any foreign networks, the home address will be invalid. In that case, after obtaining CoA, the MN sends a binding update (BU) to its HA and HA maps the current CoA in its binding cache. To keep this cache updated, each MN sends a BU message containing current CoA information to its HA. The MN sends this BU periodically or whenever it changes its current location [5]. In the meantime, the corresponding nodes (CN) that needs to communicate with the MN also maintains a binding cache. If the CN does not have an active map address for a mobile node in its binding cache, it will send the packets to MN's home address for the first time. The HA then encapsulates the packets and forwards them to the MN's CoA through a tunnel. Thus, CN can communicate with MN directly and avoid a triangular path. There are two types of delay are occurred during this transition, namely; Layer 2 (L2), and Layer 3 (L3). The delay associated with L2 is addressed in this paper while L3 related delays are discussed in our previous work [6,7,8].

3. Related Works on Layer 2 (L2)

The term handoff (also known as handover) refers to the process of transferring an ongoing call or data session from one channel connected to the core network to another [1]. Handoff latency is the time duration between the last connection of the old point of attachment and the next data transmission to the new point of attachment. L2 handoff process is related to the link layer communication. This step is the initial part of the total handoff procedures that depends on multiple phases. L2 delays in IEEE 802.11b can be categorized into three components, namely; scanning, authentication and re-association phases.

3.1. Scanning Delay

In the scanning phase, the MN can sense if the current signal strength is weakening and fails to receive some frames from the existing network while it receives probe request from other neighboring networks with a better signal strength. Hence, the MN collects the Received Signal Strength Indicator (RSSI) value from all APs available in the range. The existing IEEE 802.11b uses these RSSI values to take handoff decisions. Additionally, the MN scans not only one AP but also all APs in the coverage area. To scan all these APs, the MN needs long time to scan and process the handoff decision as it is mentioned that scanning phase is the most time consuming phase that accounts for 90% of total L2 handoff procedures [9, 10].

If the MN knows which AP to switch to, there will be no need for further scanning which saves time. Therefore, a fast handoff scheme has been proposed where channels are scanned prior to the time of handoff [11]. This approach is comprised of pre-scanning with selective channel mask and dynamic caching mechanisms. However, this approach has not been benchmarked with any standard mechanism. To minimize the scanning delay from all number of APs, [9] proposed a Global Positioning System (GPS) based model that calculates angel between MN and the APs. Based on the RSSI and calculated angel between MN and AP which is closed to each other, the MN takes the final handoff decision. However, the authors did not consider the velocity of the MN as angular value changes according to the movement of the MN and velocity of MN.

Three performance metrics, namely; RSSI, MN velocity and network loading are taken into consideration for analyzing the handoff delay in heterogeneous scenarios [12]. The author proposes FUzzy Normalization - HandOver Decision Strategy algorithm (FUN-HODS) in a heterogeneous network where 3G, WiMAX and WiFi are converging. This algorithm is suitable for low speed movement of MN but for high speed movements it shows 15%-50% probability of handoff failure which is not acceptable for real time applications. Moreover, a new Modular Fuzzy Rule Based (MFRB) handover decision algorithm has been proposed [13], which aims to address the issues of multiple QoS parameters and at the same time, reducing the computational complexity significantly by reducing the total number of fuzzy rules. This approach shows better performances over monolithic fuzzy and Adaptive Neuro Fuzzy Inference System (ANFIS) approaches in terms of handoff, average usage of cost and throughput. However, this approach did not consider the velocity of the MN which is one of the performance metrics of an MN in wireless network. A predictive RSSI and fuzzy logic based network selection for vertical handoff in heterogeneous wireless networks has also been suggested [14]. In addition to the RSS metric, the residence time in the target network is predicted which is taken into account for handoff trigger. The simulation results show that the proposed Fuzzy logic theory based Quantitative Decision Algorithm (FQDA) performs better in vertical handoff scenarios and reduced call dropping probability. Furthermore, to reduce L2 delay a Fuzzy logic-based Hand-off Decision (FHD) approach is proposed to reduce L2 delay [15]. This approach combines two parameters RSSI and the direction of an MN towards the APs. However, the direction might change over time since the velocity and the distance between MN and APs are not considered. Some additional metrics for L2 handoff analysis are added, namely; RSSI values, network load and available bandwidth in a heterogeneous wireless networks [16]. However, as velocity and distance are not considered, the aforementioned parameters could be changed due to movement of an MN in a wireless network.

The aim of this research is to use some basic parameters to reduce the handoff delay of L2. The parameters employed in this research for this evaluation is the most important RSSI values, the velocity and distance between MN and APs. The evaluation is considered in the heterogeneous environment where WiFi, 3G and WiMAX technologies are converging with each other. Moreover, to measure the RSSI value of an AP, a software, inSSIDer 2.0 is installed to the MN.

3.2. Authentication Delay

Authentication must be completed followed by scanning phase and prior to the association phase. In pre-authentication schemes, the MN authenticates with the new AP immediately after the scan cycle finishes.

3.3. Re-association Delay

Re-association is a process for transferring associated signal from one AP to another AP after the MN has completed the authentication. Authentication and re-association phase have less delays compared to scanning delay.

4. Proposed Mechanism for L2

In order to address the issue of scanning delay in pertinent to L2, a fuzzy logic based technique is employed to make handoff decisions. The RSSI value and the speed of an MN are considered as the input of the fuzzy logic inference engine. The fuzzy rules and membership functions are designed and set according to certain conditions. The fuzzy logic inference engine processes the input values and perform the defuzzification task to select the available interfaces, namely; WiFi, WiMAX and GSM/3G.

4.1 RSSI Measurement

Only RSSI is being used for conventional WiFi handoff process. This process yields satisfactory performance when the node is stationary. However, when the nodes are mobile in a heterogeneous environment utilizing RSSI value alone is not sufficient to make handoff decisions. Therefore, another parameter like velocity is proposed together with the RSSI value because these parameters dynamically change with time. Theoretically, the measurement process of RSSI is shown in equation 1 [17]. It has formulated the RSSI in dB at a distance d from the AP to be:

$$10\log_{10} P_r = 10\log_{10} P_0 - 10\alpha \log_{10} d + X \tag{1}$$

where $10\log_{10} P_0 = 10\log_{10} P_t + 10\log_{10} G_t + 10\log_{10} G_r + 20\log_{10}(\frac{\lambda}{4\pi})$

is the received power at a distance of one meter, λ is the wavelength, Pt is the transmission power and Gt = Gr = 1 are the transmitter and receiver antenna gains respectively. α is the path-loss gradient and α = 3. X is a zero mean Gaussian random variable that corresponds to log-normal shadow fading. Practically, the RSSI value can be measured using software. This research utilizes inSSIDer 2.0 to configure the MN's signal strength. This software is capable to receive RSSI value of all APs in a coverage area. In this research, the MN movement speed is classified into three categories. The velocity, v0 is in m/s

(i) $0 < v0 \leq 10$ in the first category for walking, running or cycling velocity scenario,
(ii) $10 < v0 \leq 20$ for city road velocity scenario and
(iii) $20 < v0 \leq 30$ for highway velocity scenario.

Based on the velocity range and different values of the RSSI, the fuzzy logic based network identification technique is initiated. The RSSI value changes from time to time. A higher RSSI value means the MN is closer to the AP.

4.2. The MN's Distance and Velocity Measurement

The MN sends Router Solicitation (RS) message to the APs and the APs reply with Router Acknowledgement (RA) messages whenever the MN changes its previous point of attachments. The MN can track this RS and RA messages transmission and reception time, denoted as T_{RS} and T_{RA}. The difference between these two signals is known as T_{travel} time. It is considered that the travelling speed of these messages is at the speed of light c. Therefore, it is easy to calculate the distance between the MN and the APs using the expression in equation 2.

$$d = c \times T_{travel} \tag{2}$$

4.3. Fuzzy Logic Based Scanning (L2 Phase)

To reduce the scanning delays, fuzzy logic technique based scanning mechanism is proposed. The MN scans for new APs after detecting the weak signal from previous AP. As shown in Figure 2, the MN scans for new RSSI signals from the new APs and receives it. After receiving the RSSI from the new AP, the MN compares the current RSSI ($RSSI_c$) with the new RSSI ($RSSI_n$) and also compares this information with the threshold value of the RSSI. The distance between an MN and AP can be estimated with the help of GPS. The velocity of an MN also can be measured from GPS. Using this measured velocity with the RSSI value, the MN decides to trigger L2 handoff.

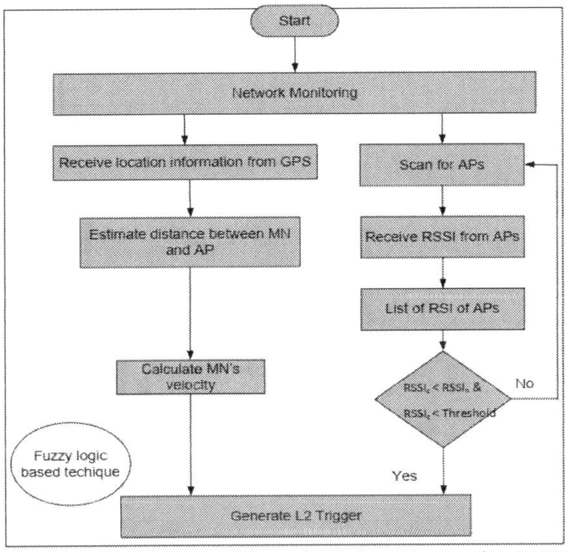

Figure 2: Flow chart of fuzzy logic based scanning RSSI

The RSSI measurements are calculated based on a fuzzy logic technique. It is assumed that in a heterogeneous coverage area three different technologies, namely; WiFi, WiMAX and GSM/3G are overlapping with each other. RSSI and speed are the two input parameters to the fuzzy input to yield the optimum AP. The fuzzy inference engine processes the given parameters and generate the output as a handoff factor. Three scenarios related to the implementation of this fuzzy logic based technique are discussed below.

4.4. Design Considerations

RSSI values and speed are considered as input parameters for WiFi, WiMAX, and GSM technologies. In this design scenario, only WiFi RSSI values have been discussed where WiMAX and 3G follow the almost similar pattern and skipped in this paper. The RSSI values are set in the range -100dBm (weak) to -15dBm (strong). The classification is as follows, (1) If the RSSI value is in between -100dBm and -70dBm, it indicates the signal is weak, (2) if it is from -70dBm to -50dBm, the signal is termed as average and (3) if it is from -50dBm to -15dBm, the signal is identified to be strong. The speed of an MN can be categorized into three types, namely; (1) walking velocity which ranges from 0 m/s to 10 m/s, (2) city velocity which ranges from 10 m/s to 20 m/s and (3) highway velocity which ranges from 20 m/s to 30 m/s.

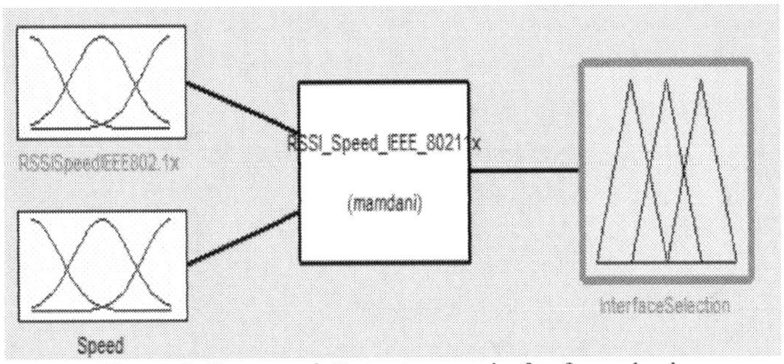

Figure 3: Input and Output scenario for fuzzy logic

On the other hand, (1) if the distance is too far the RSSI is indicated as weak, (2) if the distance is medium, the signal is considered as average and (3) if the distance is very near, the signal is classified as close to AP. Based on these input parameters, the fuzzy rules set the conditions to be processed by the fuzzy logic inference engine for the interface selections as shown in Figure 3. The output range is set between 0 and 1. If the output range is in between 0 and 0.33, 0.33 and 0.67, 0.67 and 1.0 then WiFi, 3G and WiMAX interface will be selected respectively. Based on these criteria it is designed in this model considering a heterogeneous environment where WiFi, WiMAX and GSM are overlapping.

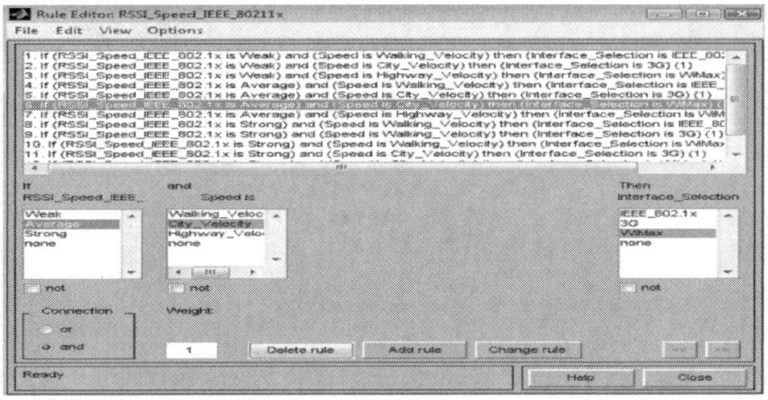

Figure 4: Fuzzy logic rules

The fuzzy logic membership rules are set according to certain conditions as shown in Figure 4. The rules are based on if and then conditions. Fuzzy engine compares some set of parameters then takes the decision for the interface selection.

The RSSI value of GSM/3G and velocity of MN to the AP are the input for the fuzzy inference engine to process 3G technology's handoff process. Similarly, the RSSI value of WiMAX and the velocity of the MN are considered as the input for WiMAX handoff process. These two design models are almost same to the WiFi scenario except the RSSI values of GSM/3G and WiMAX that has larger coverage area and its RSSI value differs from WiFi. It is noted that the RSSI values of GSM differs from WiFi that is -125 dBm to -80 dBm [8].

5. Simulation Model

A Fuzzy Logic controller is employed to simulate the fuzzy logic expert system once it has been verified with the rule viewer using MATLAB SIMULNIK. The Fuzzy logic controller block in SIMULINK has two inputs and one output. Using MATLAB SIMULINK, the Fuzzy logic controller shows the output result of the interface selection based on two inputs which are MN velocity and RSSI. These inputs can be observed from MN as crisps inputs. The crisp input is then evaluated using rule base. The composed and aggregated output of rules evaluation is defuzzified and the crisp output is obtained. Figure 5 shows the simulation overview for selecting three wireless technologies, namely; WiFi, WiMAX and GSM/3G.

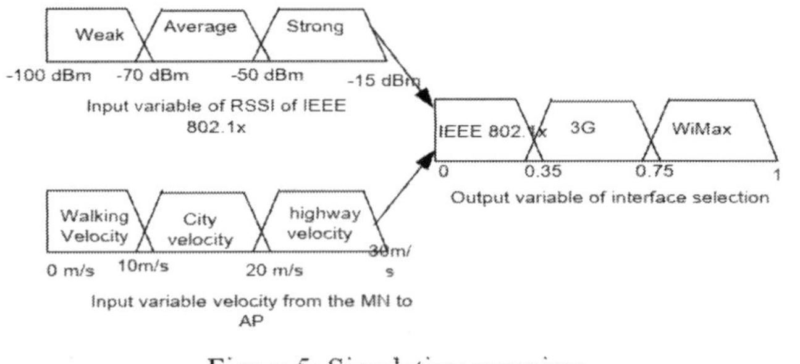

Figure 5: Simulation overview

In this scenario, there are two fuzzy input variables and one output. Each of the fuzzy variables has three subsets. These sets are mapped to the corresponding Gaussian membership functions. The fuzzy rules are set based on the combinations of these two input variables with three subsets that are at least 8 rules. These rules have been previously discussed in section four. The output interface has been selected based on these rules. MATLAB Fuzzy Toolbox platform is used for making decisions provided as input variables while OMNeT++ is used to carry out the networking management tasks.

6. Performance Analysis

In order to evaluate the performance of the proposed mechanisms, two main performance criteria have been considered in this research for the proposed solution. These performance metrics are defined in the context of wireless networks elaborated below:

- Handoff Latency: Handoff latency is defined as the interval between the last data packet received from the old access router and the first data packet received at the new access router by the MN. Specifically, it is the amount of time required in order to be granted access to get connected into the new network. Handoff latency is a critical issue for real time applications with mobile MN that needs to meet a required QoS. This paper addresses the issues of handoff latency by proposing the fuzzy logic technique.
- Packet Loss: Packet loss is defined as the number of packets lost during the handoff. It is the amount of packets that are lost or corrupt and finally cannot reach the desired destinations. Packet loss is measured in relation to the MN speed ranging from 5 m/s to 30 m/s for proposed mechanisms.

6.1. Handoff Latency in L2

The handoff latency in Layer 2 consists of scanning, authentication and re-association phases. Fuzzy logic technique is applied to scan the alternatively available interfaces while the MN is moving from one coverage area to another.

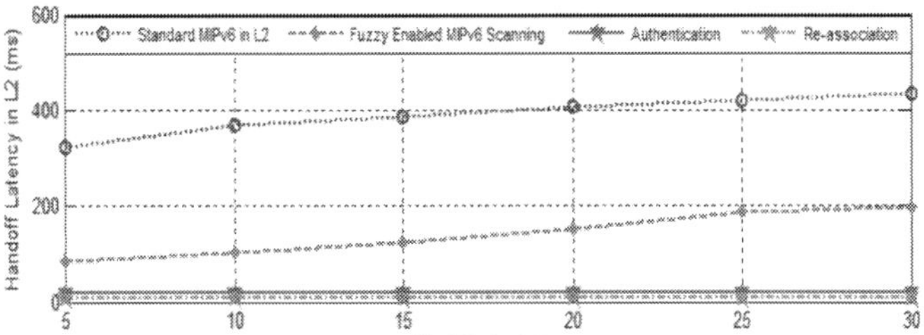

Figure 6: Handoff latency in L2

Figure 6 shows the handoff latency in millisecond (ms) with respect to speed of the MN. The standard MIPv6 takes around 400ms to scan an interface based on only RSSI values where this proposed fuzzy enabled MIPv6 scanning takes about less than 200ms. In average, the proposed mechanism shows about 55% improvement over the existing mechanism while the MN is moving. However, for fast movement the scanning time increases a little bit closed to 200ms. On the other hand, two other phases, authentication and re-association need 20ms and 10ms respectively.

6.2. Packet loss in L2

Packet loss results for handoff latency. Long handoff latency implies more packet loss and this degrades the QoS. Packet loss occurs when the MN is switching from one interface to another interface. The drawback of the standard MIPv6 is having more packet loss in L2.

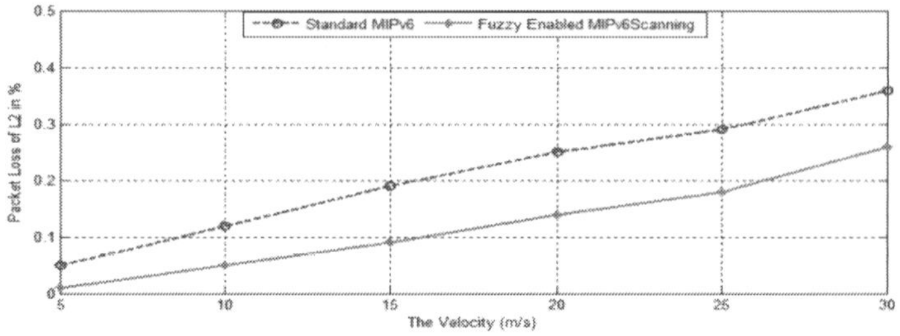

Figure 7: Packet loss in Layer 2

In L2 of MIPv6, the average packet loss is around 0.05% to 0.35% for different movement while the proposed mechanism reduces this number about 0.01% to 0.26% as shown in Figure 7. Moreover, it can be concluded that the overall packet loss in L2 is reduced in average about 50% of standard MIPv6.

7. Conclusion

This paper proposes fuzzy logic based mechanism to reduce L2 handoff latency of MIPv6. Particularly, it addresses the scanning delay that is the most time consuming in MIPv6. According to the simulation result, the handoff latency reduced about 55% and packet loss rate enhanced around 50% in L2.

8. Acknowledgement

The authors would like to thank the IIUM Research Management Centre (RMC) for funding this research under the Research Endowment Fund B – EDW B11-079-0557 scheme.

9. References

[1] Johnson, D., Perkins, E. C., and Arkko, J., 2011. Mobility support in IPv6, RFC 6275.

[2] Kwong, C. F., Teong, C, C., and Sze W., L.,, 2010. Adaptive Network Fuzzy Inference System (ANFIS) Handoff Algorithm, *Int. J. of Net. and Mobile Technologies*, ISSN 1832-6758 Electronic Version, Vol. **1**/Issue 2.

[3] Apostolis S. K., 2004. Mobile internet: enabling technologies and services: quality of service in mobile IP Networks. (pp. 6-4). United States of America: *CRC Press*.

[4] Chen W-M., Chen W. Chao H-C., 2009. An efficient mobile IPv6 handover scheme, Springer, *Telecommunication System,* **42**: 293–304.

[5] Haseeb, S., and Ismail, F. A., 2006. Comparative Performance Analysis of Mobile IPv6 Protocols: Special Reference to Simultaneous Bindings, *J. of Com. Sci.* **2** (2): 154-159.

[6] Farhat Anwar, Mosharrof H. Masud, S. M. S. Bari, Suhaimi A. Latif, 2013, Enhanced Handoff Latency Reduction Mechanism in Layer 2 and Layer 3 of Mobile IPv6 (MIPv6) Network, *Australian J. of Basic and Appl. Sci*,**7**(6): pp 658-671.

[7] Mosharrof H. Masud, F. Anwar, S. M. S. Bari, 2012, Enhancement of Handoff Latency Reduction Mechanism of Mobile Internet Protocol Version 6 (MIPv6), *4th International Conference on Computer and Communication Engineering (ICCCE'12)*, 3-5 July, Kuala Lumpur, Malaysia.

[8] M. H. Masud, F. Anwar, O. M. Mohamed, S. M. S. Bari and A. F. Salami, 2011, A Parallel Duplicate Address Detection (PDAD) Mechanism to Reduce Handoff Latency of Mobile Internet Protocol Version 6 (MIPv6), *4th International Conference on Mechatronics (ICOM),* May 17-19, Kuala Lumpur, Malaysia.

[9] Sarddar D., Joydeep Banerjee, Tapas Jana, Souvik Kumar Saha, Utpal Biswas, M.K. Naskar, 2010, Minimization of Handoff Latency by Angular Displacement Method Using GPS Based Map", *Int. J. of Com. Sci., Issues,* Vol. **7**, Issue 3.

[10] Márquez-Barja, J., Carlos, T., Calafate., Juan-Carlos, C., Pietro, M., (2010). An overview of vertical handover techniques: Algorithms, protocols and tools, *Computer Communications* **34** page, 985–997.

[11] Mustafa, N., Mahmood, W., Chaudhry, A, A., Ibrahim, M., 2005, Pre-Scanning and Dynamic Caching for Fast Handoff at MAC Layer in IEEE 802.11 Wireless LANs, *MASS Workshop - MHWMN05.*

[12] Wu S-J., 2011, Fuzzy-Based Handover Decision Scheme for Next-Generation Heterogeneous Wireless Networks, *J. of Conv. Info. Tech.*, Volume **6**, Number 4.

[13] Thumthawatworn T., Pervez A. and Santiprabhob P., 2011, Modular Handover Decision System based on Fuzzy Logic for Wireless Networks, *The 8th Electrical Engineering Electronics, Computer, Telecommunications and Information Technology (ECTI)*, Association of Thailand Conference.

[14] Kunarak, S., and Suleesathira, R., 2010 Predictive RSS with Fuzzy Logic based Vertical Handoff Decision Scheme for Seamless Ubiquitous Access, Mobile Ad Hoc Networks: Protocol Design, Thailand.

[15] Sadiq A. S., Bakar K. A., and Ghafoor K. Z., 2010, A Fuzzy Logic Approach for Reducing Handover Latency in Wireless Networks, *Network Protocols and Algorithm,* Vol **2**, No. 4,.

[16] Sharma M., Khola R.K., 2009, An Intelligent Approach for Handover Decision in Heterogeneous Wireless Environment, *Int. J. of Eng. (IJE),* Volume **4**: Issue (5), Page 452-462,.

[17] Pahlavan, K. and Krishnamurthy, P. 2001, Principles of Wireless Networks: a Unified Approach. 1st. Prentice Hall PTR.

IOP Conf. Series: Materials Science and Engineering **53** (2013) 012086 doi:10.1088/1757-899X/53/1/012086

Colour-based Object Detection and Tracking for Autonomous Quadrotor UAV

Hani Hunud A. Kadouf [1,2] **and Yasir Mohd Mustafah** [1]
[1]Faculty of Engineering, International Islamic University, Jalan Gombak, 53100, Kuala Lumpur, Malaysia

E-mail:hani.ha@gmail.com

Abstract: With robotics becoming a fundamental aspect of modern society, further research and consequent application is ever increasing. Aerial robotics, in particular, covers applications such as surveillance in hostile military zones or search and rescue operations in disaster stricken areas, where ground navigation is impossible. The increased visual capacity of UAV's (Unmanned Air Vehicles) is also applicable in the support of ground vehicles to provide supplies for emergency assistance, for scouting purposes or to extend communication beyond insurmountable land or water barriers. The Quadrotor, which is a small UAV has its lift generated by four rotors and can be controlled by altering the speeds of its motors relative to each other. The four rotors allow for a higher payload than single or dual rotor UAVs, which makes it safer and more suitable to carry camera and transmitter equipment. An onboard camera is used to capture and transmit images of the Quadrotor's First Person View (FPV) while in flight, in real time, wirelessly to a base station. The aim of this research is to develop an autonomous quadrotor platform capable of transmitting real time video signals to a base station for processing. The result from the image analysis will be used as a feedback in the quadrotor positioning control. To validate the system, the algorithm should have the capacity to make the quadrotor identify, track or hover above stationary or moving objects.

1. Introduction

The use of computer vision as a sensory input provides an increased visual capacity which can be used by several robotic platforms to perform tasks. Its use can be found in the control and navigation of aerial vehicles, robotic arms or for other forms of remote sensing. In most applications that make use of Unmanned Aerial Vehicles (UAVs), high level sensory devices are needed to determine the position and attitude of the UAV in order to perform a task [1]. Research on UAVs that have a higher payload to accommodate such sensors has been extensively investigated [2]. However the research of small scale UAVs is still relatively new, as most recent efforts in quadrotor development have concentrated on vision-based control methods for autonomous stabilization and navigation. These usually use visual information obtained through a known model of a target [3] or a pre-set environment, with key images and texture points for motion estimation or calculation of optical flow

---

[2] To whom any correspondence should be addressed

IOP Conf. Series: Materials Science and Engineering **53** (2013) 012086 doi:10.1088/1757-899X/53/1/012086

[4]. Since algorithms that require pre-learning, template matching, or edge tracking have been found to be lacking accuracy and efficiency [1] , this paper investigates the application of image processing algorithms with the use of OpenCV video-processing library to quickly process real time video data for colour based object detection and tracking.
Figure 1a below shows a quadrotor that was developed and was used for autonomous use which is similar to the one used in this research [5]. The illustrations of figure 1b shows how in general, a quadrotor's helicopter motion is the result of varying the rotors' angular velocities [6].This was done by controlling the Quadrotor's 6 DOF using inputs from different on-board and off-board sensors.

(a)

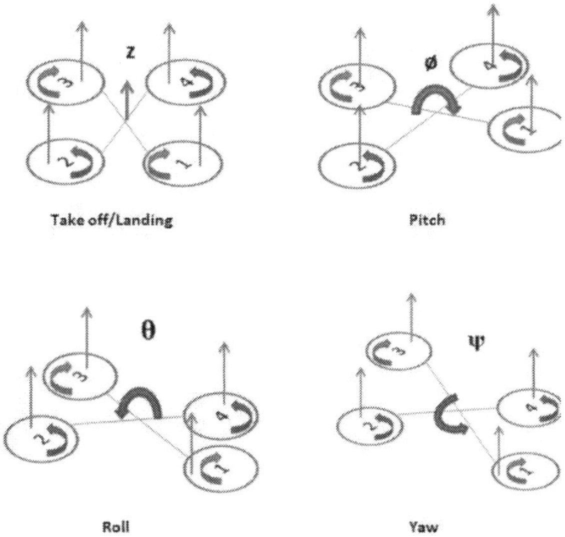

(b)

Figure 1. Quadrotor Characteristics, a. Platform used for experiment [5]; b. The Quadrotor's motion of rotors [6]

2. Object Detection and Tracking Algorithm

The interpretation of images by computers is concerned with visual analyis and the extraction of information, at a high speed, both effectively and automatically. The first step in many image processing applications is the object detection then tracking, followed by recognition if the needed [7]. As more research is done on the field of detection and tracking, many algorithms have been developed and investigated. To perform such tasks using a small scale quadrotor, feedback from vision sensors becomes a feasible control approach as it is cheap, passive and a rich source of information. This process is especially convenient as low weight cameras can be embedded on the small-size platforms [8].

To track moving objects would require an algorithm that has to be robust and reliable. It also needs to be simple and executable at high speed and accuracy [9]. Many image processing algorithms require filtering the image at first, and then performing some kind of edge detection [1]. These often prove to be time consuming, especially when they are applied on a noisy image. Colour segmentation is good approach where an image is divided into distinct regions in order to extract objects of interest from a surrounding environment.

2.1.1. Image Thresholding
Segmentation is achievable by several methods, which include thresholding-based, edge based, region based, and clustering technique. It is applied to decompose an image into meaningful parts for further analysis, resulting in a higher-level representation of image pixels like a separation of foreground objects those at the background [10, 11]. With thresholding, it is possible to segment the image based on colour. For example, all the red pixels in an image can be separated from its surrounding colours. At this stage of the procedure, a decision about which pixels in an image fall within the range of the threshold and which are rejected is made [12].
The thresholding process may be viewed as an operation that involves tests against a function, such as,

$$T = T[x, y, p(x, y), f(x, y)] \tag{1}$$

where$f(x, y)$ is the graylevel at the point (x, y) and $p(x, y)$ denotes some local property of the point (such as the average gray levelof a neighbourhoodcentred on (x, y) [13].

Hence, a thresholded image is defined as:

$$g(x, y) = \begin{cases} 1, & if \ \ f(x, y) > T \\ 0, & if \ f(x, y) \geq T \end{cases} \tag{2}$$

The above thresholding function represents different sets of pixels, whereby those labeled 1 correspond to objects, while pixels labeled 0 correspond to the background.

2.1.2 Colour models
Colour models are used to facilitate the specification of colours in a standard generally accepted way. In essence, a colour model is a specification of a 3-D coordinate system and a subspace within that system where each colour is represented by a single point. Each industry that uses colour employs the most suitable colour model. For example, RGB colour model is used in computer graphics, and YUV or YCbCr are used in video systems [14].

2.2 RGB and HSV colour format
There are different representations of three dimensional colour spaces, each having different characteristics. One set of colour spaces uses Cartesian coordinates for a point's representation in space [15]. Examples of this type of set are the three primary illumination colours Red, Green, and Blue (RGB), in figure 2, or the complementary colours Cyan, Magenta, and Yellow (CMY). Another space representation utilizes polar/cylindrical coordinates such as Hue, Saturation, and Lightness (HSL) as well as Hue, Saturation and Value (HSV) [16, 17]. Generally, each colour space can be converted to another colour space representation.
The RGB colour format can represent any standard colour or brightness using a combination of Red, Green and Blue components. For efficiency, this is typically stored as a 24-bit number using 8-bits for each colour component (0 to 255).

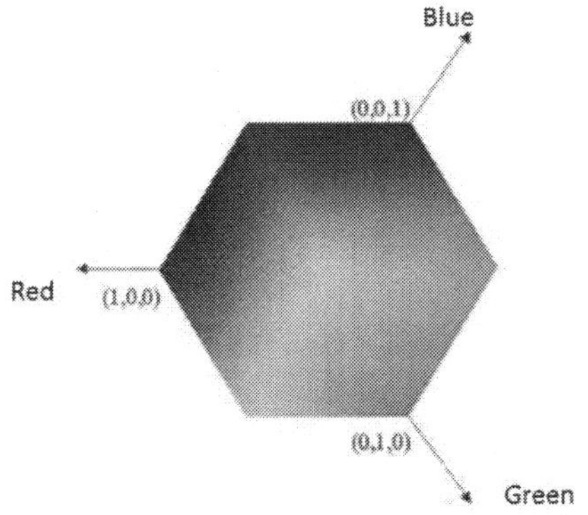

Figure 2. RGB colour space where x-axis represents red, y-axis is green and z-axis is blue [17]

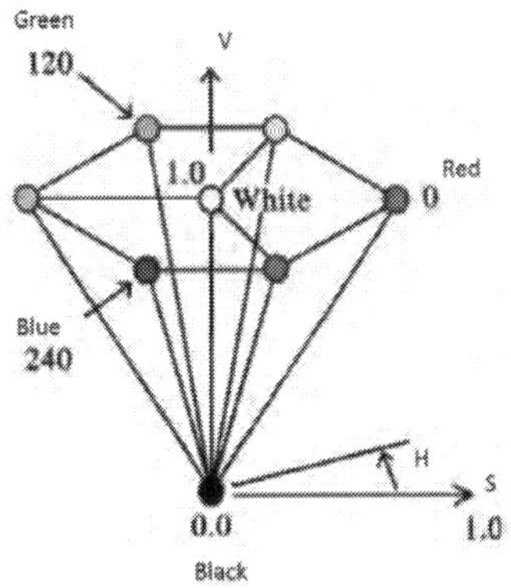

Figure 3. HSV colour space:- Value/brightness is the main axis orthogonal of the plane; Hue is the angle H; S is the radius (saturation/purity of colour) [17]

However, when viewing a colour object; human visual system characterizes it by its brightness and chromaticity. The latter is defined by hue and saturation. Brightness is a subjective measure of luminous intensity. It embodies the chromatic notion of intensity [18]. In the HSV colour space, Hue is the colour and is difficult to separate or compare, hence it is often represented as a circular angle between 0.0 and 1.0. Saturation is the greyness, whereby a saturation value near 0 means it is dull or grey looking and a value closer to 1 will be a very strong colour. Value is the brightness of the pixel, where 0.1 is black and 0.9 is white.

The HSV model is motivated by the human visual system. In the HSV model, the luminous component (brightness) is decoupled from colour-carrying information (hue and saturation).

3. Control System

Since the quadrotor is a dynamically unstable nonlinear system, flying it without an embedded control system becomes difficult [20, 21].

Figure 4 below shows a block diagram of the type of close-loop feedback control system needed for the quadrotor to achieve stable flight. The process, G(s) is represents the vehicle's dynamics while the output Y(s) is its current position and orientation. The input R(s) is the desired position of the quadrotor, which is determined from the vision system. A conversion matrix that determines the distance of the quadrotor to an object in the real world, from the distance to an object on the image plane is needed.

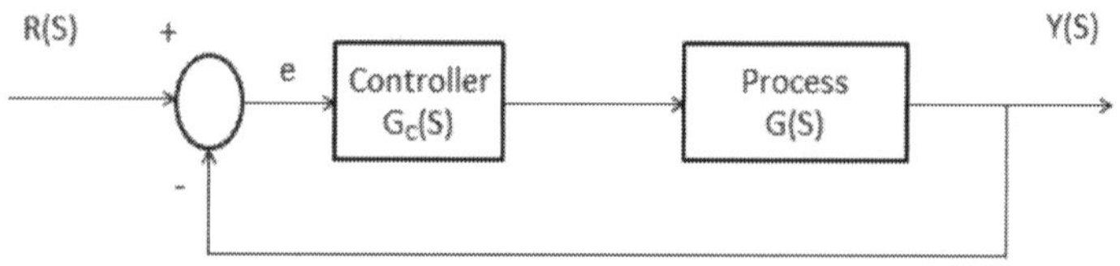

Figure 4. Closed loop Control System to determine quadrotor flight

The controller processes the error (e) and the resulting output is used as the new speed for each individual motor. The change in motor speed results in a change in the system output, which in turn is compared to the desired input and the cycle continues to repeat this process [5].

R(s) = Desired Input Value
Y(s) = Actual Output Value
E = Tracking Error

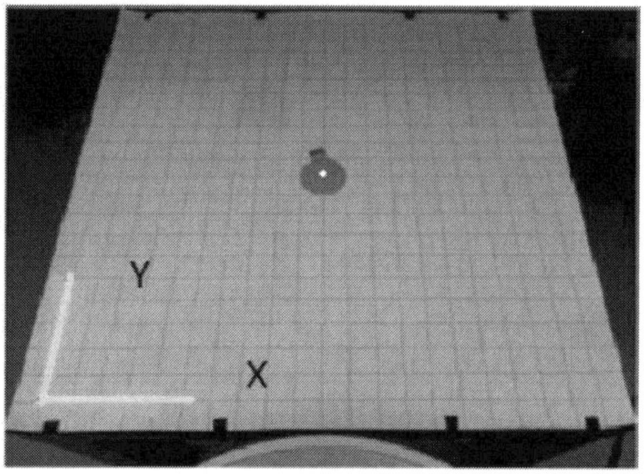

Figure 5. Conversion matrix table to convert distances on image plane to those in real world

Figure 5 shows the conversion matrix table used to convert distances on the image plane (in pixels) into real world distances (in centimeters). The assumption made here is that the quadrotor flight occurs at a constant height from the ground, hence allowing calculations to occur only on the X-Y plane. The various sensors measure this information in real-time and compare it to the desired value R(s). The difference of the desired input value and actual output is the tracking error (e).

4. Results

This section explains the results of the image processing algorithm applied for colour detection. The software for used to implement the algorithm was OpenCV 2.3.1, operated on Visual Studio 2010 platform. The first stage of the experiment is the conversion of the real time video frames from the default RGB colour space to HSV format using the following based on the law of center of gravity described as [19]:

$$H = \begin{cases} 60\left(\frac{G-B}{\delta}\right) & \text{if } MAX = R \\ 60\left(\frac{B-R}{\delta}+2\right) & \text{if } MAX = G \\ 60\left(\frac{R-G}{\delta}+4\right) & \text{if } MAX = B \\ \text{not defined} & \text{if } MAX = 0 \end{cases} \quad (3)$$

$$S = \begin{cases} \frac{\delta}{MAX} & \text{if } MAX \neq 0 \\ 0 & \text{if } MAX = 0 \end{cases} \quad (4)$$

$$V = MAX(R, G, B), \quad (5)$$

where $\delta = (MAX - MIN), MAX = \max(R, G, B)$ and $MIN = \min(R, G, B)$.

The R, G, B values in Equations (3, 4, 5) are scaled to [0, 1]. In order to confine H within the range of [0, 360] as shown in Equation 6.

$$H = H + 360, \text{if } H < 0 \quad (6)$$

The next step of the experiment is the thresholding of colours in order to locate an object based on its colour. In this case, a yellow sponge was chosen as a test subject.

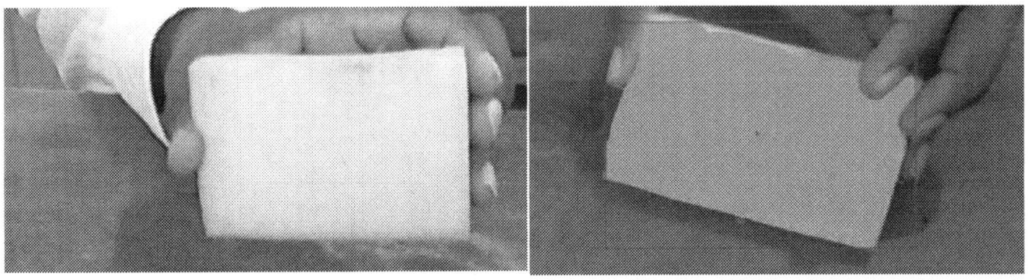

a) b)

Figure 6. Colour Thresholding, a. Yellow sponge before processing; b. Yellow sponge after processing

Figure 6a shows our yellow sponge test object before any processing is conducted. Figure 6b on the other hand illustrates the output of thresholding on the yellow coloured sponge, with the centre of the object also being highlighted.

Once the object has been detected, it is possible to track its movements on a display with the help of the cvblobs. The cvblobs library in OpenCV is used to detect similar features of an image in a frame. With this feature, it is possible to scan the entirety of a real time video footage and locate the desired characteristics.

To find the position of an object relative to a reference point on the display, the Euclidean Distance equation (5) is applied.

$$d_{x,y} = \sqrt{(x_1 - x_0)^2 + (y_1 - y_0)^2} \quad (7)$$

Table (1) shows some pairs of(x,y) coordinates at the centre of a moving object on a computer display. Using Equation (7) the table also shows the Euclidean distance (in pixels) at time t, from the centre of the object to a reference point set at the centre of the display is shown.

Table 1. Euclidean Distance from origin to centre of moving object.
*All dimensions in pixels.

	X-Coordinate	Y-Coordinate	Euclidean Distance
Point 1	1386	323	1418
Point 2	1378	332	1415.25
Point 3	1370	339	1409.14
Point 4	1357	342	1397.25

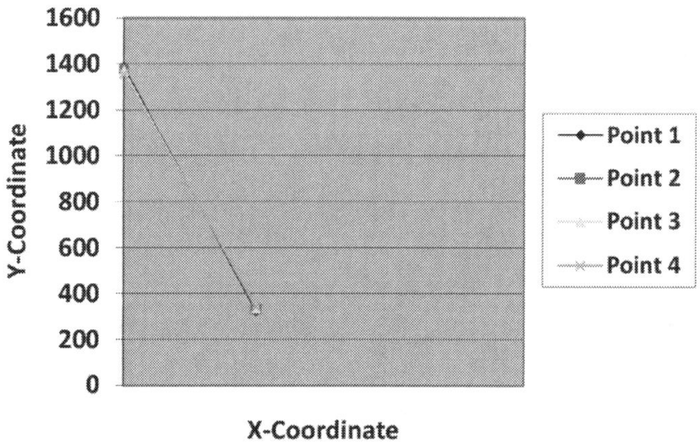

Figure 7. Distances from reference point on image plane

Figure 7 shows a graph that was drawn using the object position and the Euclidean distance. A linear relationship between the moving object and the Euclidean distance values on the image plane. This distance is converted in real time from the image plane to real world values (in cm) and fed to the controller input.

5. Conclusion and Future Work

A method for the detection and tracking of a moving object in real time was proposed in this paper. After the conversion of the video frames from the standard RGB colour space to the more reliable HSV format, segmentation by colour thresholding was used to detect the moving object in real time. Once detected, the Euclidean distance from the centre of the object to a reference point set at the centre of the image frame was calculated. The coordinates of the centre of the object along with the distances are displayed. As an expansion of this project, a ratio of the pixel distance between centre of moving object and any reference point on the display to a real world distances (in cm) is calculated. This can be used to track the objet via some external robotic platform, i.e. UAV or robot arm.

6. References

[1] S. Azrad, et al., "Visual Servoing of Quadrotor Micro-Air Vehicle Using Colour-based Tracking Algorithm," Journal of System Design and Dynamics, 4(2) pp.255-268, 2010.

[2] S. Habrar and G. Sukhtame, "Omnidirectional Vision for Autonomous Helicopter," IEEE conference on Robotics and Automation, Taiwan, May 2003, pp. 558-563.

[3] Saripalli, S., Montgomery, J. and Sukhatme, G. 2003 Visually-Guided Landing of an Unmanned Aerial Vehicle *IEEE Transactions on Robotics and Automation,* 19(3) pp. **371-81**

[4] Teuli`ere, C., Eck, L. and Marchand, E. 2011 Chasing a Moving Target from a Flying UAV *IEEE/RSJ International Conference on Intelligent Robots and Systems* pp. **4929-4934**

[5] Coelho, G. 2012 OTA-Quadrotor: An Object-Tracking Autonomous Quadrotor for Real-Time Detection and Recognition *M.S. thesis*, Dept. of Electrical & Computer Engineering, University of North Texas, Texas

[6] Zul Azfar, A. and Hazry, D. 2011 Simple GUI Design for Monitoring of a Remotely Operated Quadrotor Unmanned Aerial Vehicle (UAV) *IEEE 7th International Colloquium on Signal Processing and its Applications* doi: [978-1-61284-413-8]

[7] Prasad, S. and Sinha, S. 2011 Real-time Object Detection and Tracking in an Unknown Environment. *IEEE Information and Communication Technologies (WICT), 2011 World Congress* doi:[978-1-4673-0126-8/11]

[8] Angeletti, G., Valente, J. R. P., Iocchi, L. and Nardi, D. 2008 Autonomous Indoor Hovering with a Quadrotor *In Workshop Proc. SIMPAR* pp. **472-481**.

[9] Fletcher, J., Veth, M. and Raquet, J. 2007 Real-Time Fusion of Image and Inertial Sensors for Navigation *Air Force Inst. of Tech.* Wright-Patterson AFB, OH

[10] Abdullah, S. L. S., Hambali, H. A. and Jamil, N. 2012 An Accurate Thresholding-based Segmentation Technique for Natural Images *IEEE Symposium on Humanities, Science and Engineering Research* doi: [978-1-4673-1310-0/12]

[11] Vadivel, A., Mohan, M., Sural, S. and Majumdar, A. K. 2003 *Lecture Notes in Computer Science* Segmentation using Saturation Thresholding and its Application in Content-Based Retrieval of Images, Dept. of Computer Science and Engineering, Indian Institute of Technology, Kharagpur

[12] Bradski, G. and Kaehler, A. 2008 *Learning OpenCV: Computer Vision with the OpenCV Library (*First Edition), ed Mike Loukides (Sebastopol, CA: O'Reilly Media Inc.)

[13] Lovell, B. 2003 Thresholding Techniques *ITEE*, UQ, Brisbane

[14] Available: http://software.intel.com sites/products/documentation/ hpc/ipp/ippi/ippi_ch6/ ch6_color_models.html/

[15] Almuntashril, A. and Agaian, S. 2010 An Algorithm for Visualizing and Detecting Edges in RGB Colour Images Using Logarithmic Ratio Approach *Systems Man and Cybernetics (SMC), IEEE International Conference* pp.**3942 – 3947**

[16] Dony, R. and Wesolkowski, S. 1999 Edge Detection on Colour Images using RGB vector Angles *IEEE Proc. of the 1999 IEEE Canadian Conference on Electrical and Computer Engineering* vol.(2) pp.**687 - 692**

[17] Figueroa, J. P. and Bykbaev, V. R. Image retrieval based on the combination of RGB and HSV's histograms and Colour Layout Descriptor *Fourth Call for Research Founds of the Universidad Politécnica*, Salesiana, Ecuador pp. **3-10**

[18] Chen, W., Shi, Y. Q. and Xuan, G. 2007 Identifying computer grahics using HSV color model and statistical moments of characteristic functions *in Proceedings of IEEE International Conference on Multimedia and Expo* pp. **1123-1126**

[19] Smith, A. R. Colour Gamut Transform Pairs 1978 *Comput. Graph.*12(3) pp.**12-19**

[20] Hoffmann, G. M., Huang, H., Wasl, S. L. and Tomlin, E. C. J. 2007 Quadrotor helicopter flight dynamics and control: theory and experiment *in In Proc. of the AIAA Guidance, Navigation, and Control Conference*

[21] Wu Y. 2009 Development and Implementation of a Control System for a Quadrotor UAV *M.S. Thesis*, Department of Electrical and Computer Engineering, University of Applied Science Ravensburg-Weingarten, Germany

ANFIS -Based Navigation for HVAC Service Robot with Image Processing

Mohd Zoolfadli Md Salleh[1] **, Nahrul Khair Alang Md Rashid**[1]**, Yasir Mohd Mustafah**[1]
[1]Department of Mechatronics Engineering, International Islamic University Malaysia, Gombak, Selangor 50728, Malaysia

E-mail: mohdzoolfadli@gmail.com, nahrul@iium.edu.my, yasir@iium.edu.my

Abstract. In this paper, we present an ongoing work on the autonomous navigation of a mobile service robot for Heat, Ventilation and Air Condition (HVAC) ducting. CCD camera mounted on the front-end of our robot is used to analyze the ducts openings (blob analysis) in order to differentiate them from other landmarks (blower fan, air outlets and etc). Distance between the robot and duct openings is measured using ultrasonic sensor. Controller chosen is ANFIS where its architecture accepts three inputs; recognition of duct openings, robot positions and distance while the outputs is maneuver direction (left or right).45 membership functions are created from which produces 46 training epochs. In order to demonstrate the functionality of the system, a working prototype is developed and tested inside HVAC ducting in ROBOCON Lab, IIUM.

1. Introduction

Recent robot navigations systems are focusing more on operating in a dynamic and complex environmental situation. Presently, ANFIS approach is becoming one of the major areas of interest since the combination of fuzzy logic and neural network will remove individual disadvantages and combines the both features for robust and excellent control but remains relatively easy [1-3].More traditional methods require extensive mathematical analysis. However, in most unstructured, real-life scenarios, adequate mathematical modeling is almost impossible to derive, but nevertheless, robots are still required to perform in these conditions [4] .Hence, fuzzy logic and neural network are good solutions, where fuzzy logic can be used as a controller while concurrently, the neural network part will acts as a 'self-learner' and 'self-tuner' so that it controls the behavior of fuzzy logic controller autonomously [5].[1]

[1] Mohd Zoolfadli is a member of Intelligent Mechatronic System Research Unit (IMS-RU) while Nahrul Khair Alang Md Rashid and Yasir Mohd Mustafah are with Autonomous Agent Research Group (AARG) of IIUM.

Meanwhile, the implementation of visual analysis will provide robustness of application to the robot; the system could later be enhanced to analyze the quality of cleaning done inside the ducts, improve navigation by implementing more sophisticated algorithm or to simply eliminate the need for distance sensor by implementing perception of depth analysis. Our initial phase of visual analysis will focuses on landmark recognition to navigate the robot successfully in HVAC ducting. There are plenty of methods for recognizing landmarks such as sonar sensor, IR sensor and others, but, the frequent use of illuminator and camera for ducting, in-pipe and sewage robots prompt for image-based algorithm for convenience and efficiency [6].

1.1. HVAC service robot

HVAC service robot developed is a prototype comprises of two primary systems, 1) robotic system (a robotic body comprises of mechanical, electronic and circuitry) and 2) monitoring and control system (system is build inside a briefcase). Full system is shown is Figure 1

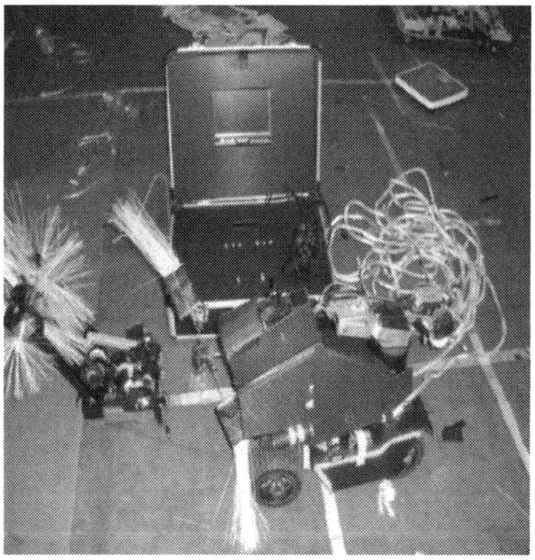

Figure 1: HVAC service robot (complete system)

In brief, the robotic system is a mobile robotic base attached with an ultrasonic sensor and a CCD camera. Visual output from the camera is then displayed on the monitoring system. An operator is assigned to control all the equipment on board (robotic arm and cleaning brush) while at the same time maneuver the robot inside HVAC ducting. In this paper, a small part of the system (i.e. robot maneuver) which initially is tele-operated is then upgraded into autonomous by applying ANFIS. Table 1 below shows the general specification of the robot.

Table 1: General specification of the robot

Item/Criteria	Specification	Description
Overall dimension	300 mm (l) by 185 mm (b) by 305 mm (h)	

Kerb Weight	2.10 kg	-The estimated maximum possible mass is still assumed here -Without considering additional arm manipulator and payload.
Velocity	0.500 m/s	
Additional payload	3.5 kg	
Type of control	Tele-operated (upgraded into autonomous)	range about 30 meters, could be upgraded to 100 meter
Power source	2 Lithium Polymer batteries, 3500 mili ampere hour, 11volt.	
Camera	CCD camera, Output- PLA/CCIR NTSC/EIA, Resolution-380 TV tines, Scan frequency: PLA/CCIR: 50 Hz NTSC/EIA:60 Hz	9 volt battery is needed for power
Monitor	LCD 5 inch	
Brush	0.2 HP/Speed control/Reverse driving control	Actuator is a brushless DC motor with ESC for speed control
Joystick	Sony PS2(modified)	
Standard working distance	30 meter	Total length of the cable is 40 meter
BALUN	A type of electrical transformer	Eliminate long distance video problem. Maximum 100meter
AV +data cable	Modified CAT5 cable	
Ultrasonic sensor	Parallax ultrasonic distance sensor	Precise for distance within 2 cm to 3 m range

2. Methodology

First part of the process is image processing. Image input from CCD camera is send to PC through tether. Images are later analyzed by MATLAB, where image pre-processing and blob analysis is used to analyze and acquire crucial information before relaying the result into second part.

Second part is ANFIS controller for HVAC robot navigation. This part involves with taking all the information from first part and combines with data from ultrasonic sensor before determines the appropriate output. In this paper, variable output of our interest is 'maneuver angle'

2.1 Image processing

The objective of image processing being implemented is to detect, recognize and analyze HVAC ducting openings. The properties of the openings such as the shapes, centre position (centroid) and perimeter extracted are valuable information in order to facilitate our image analysis. Data collected from various sample images are recorded to create a data set where training data set and checking data set for our ANFIS are derived from. Figure 2 below shows the overall procedure carried-out to obtain the data set. The procedure comprises of 2 distinct parts; pre-processing (stage 1, stage 2 and stage 3) and feature extraction (stage 4 and stage 5)

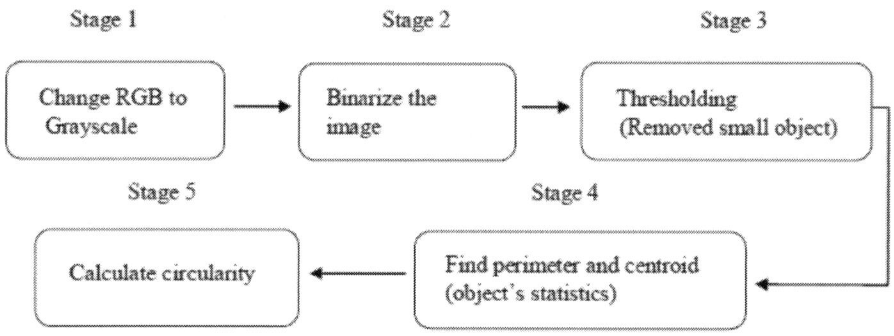

Figure 2: Procedure for duct opening's detection

Original image obtained from CCD camera is a mixture of noises and useful data. This can cause many unnecessary additional calculations and reduces image processing efficiency. Hence, preprocessing stage is introduced. Pre-processing stages are,1) conversion of data classes and image types (RGB to Grayscale), 2) binarization of image and 3) thresholding to omit small objects for clearer image. These stages are then followed by feature extraction stages (stage 4 and 5) where isolation of certain aspects of the image is done with the objective of finding line and shapes before statistical analysis is performed. Lastly, simple calculation is applied to determine the shape of the object.

2.1.1Pre-processing

In order to reduce image complexity, color image from our CCD camera is changed into intensity image (grayscale). This step is crucial in order to facilitate boundary tracing and blob analysis, while emitting the parameters of non-interest (color, texture and etc). The resultant grayscale image, even though reasonably efficient in storage space (1 byte for each pixel), must be changed from 'unint8' type to 'double' in order to permit arithmetic operation on it. Next step involves binarization and thresholding. In binarization, a gray level T =100 is chosen to separate every pixel to 2 different states; black or white.

$$\text{A pixel becomes } \begin{cases} \text{white if its gray level is } >100 \\ \text{black if its gray level is } \leq 100 \end{cases}$$

Since the resultant binary image contains unwanted pixels, second 'thresholding' operation is carried-out. This time, instead of grey level, we used pixel values to remove pixels which don't belong to the region of interest. All objects contain fewer than 300 pixels are removed. Pre-processing procedure results are shown in Figure 3.

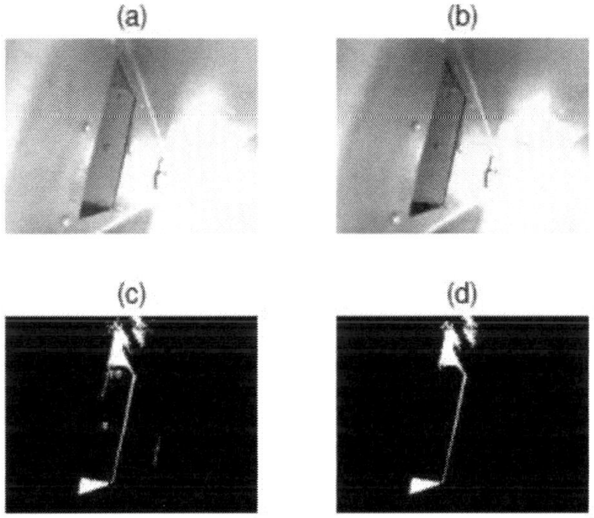

Figure 3: (a) Original, (b) Grayscale, (c) Binary, (d) Small objects removed

2.1.2 Feature extraction
Ideal images gained after pre-processing stages are fit for statistical analysis where features of duct openings are extracted. Opening's statistical information such as area, perimeter and centroid can be extracted from ROI using a method called blob analysis. A blob (binary large image) is an area of touching pixels with the same logical state. Hence, all related pixels are being grouped accordingly, which later translated into meaningful black and white regions Matlab can easily produce centroid location, area and perimeter using blob analysis system objects. Final step is where circularity is calculated which depends on information gathered in blob analysis earlier. This information is fed into equation (1) to form a metric indicating the roundness of the object.

$$circularity = \frac{(perimeter^2)}{(4 \times \pi \times area)} \tag{1}$$

Through rigorous tests, it is established that, for our case, the thresholds are as recorded in Table 2.

Table 2: Circularity threshold table

Circularity threshold	Types of opening
< 1.19	circle
< 1.53	square
> 1.53	other shapes

2.2 ANFIS controller for robot navigation

ANFIS is a fuzzy inference system implemented in the framework of adaptive networks. By using a hybrid learning procedure, the proposed ANFIS can construct an input-output mapping based on both human knowledge (in the form of fuzzy if-then rules) and stipulated input-output data pairs [7].In other words, ANFIS is a fuzzy inference system employing fuzzy IF-THEN rules can model the qualitative aspects of human knowledge and reasoning processes without employing precise quantitative analyses. Fuzzy control is by far the most successful applications of the fuzzy set theory and fuzzy inference systems [9]. Due to the adaptive capability of ANFIS, its applications to adaptive control and learning control are immediate. Most of all, it can replace almost any neural networks in control systems to serve the same purposes. Rather selecting the data, randomly, the initial weights obtain from desire data. Therefore more learning speed and more convergence should be assured. This fact is the most advantage of ANFIS [7].

The ANFIS analyzed here is a first order Takagi Sugeno Fuzzy Model [8], [7], [10]. In current analysis, there are three inputs: Duct opening (x1), robot position (x2), and distance (x3) while the output is Steering angle. The if-then rules for the ANFIS architecture are defined as follows;

Rule: **IF** x1 is Aj; x2 is Bk; x3 is Cm
THEN Fi = pi x1 + ri x2 + si x3

Where
Fi = pi x1 + ri x2 + si x3 + ui for steering angle
J=1 to 3; k=4 to 8; m=9 to 11; and i=1 to 45

A, B and C are the fuzzy membership sets defined for the input variables x1, x2, x3 and J,K and M are the number of member ship functions for the fuzzy systems of the inputs x1, x2, x3 and x4 respectively. Fi is the linear consequent functions defined in terms of the inputs (x1, x2, and x3) . qi,ri, si , and ui are the consequent parameters of the ANFIS fuzzy model. In the ANFIS model, nodes of the same layer have similar functions. The output signals from the nodes of the previous layer are the input signals for the current layer. Since restriction on ANFIS is applied, none of the rules could share the same output membership function, thus the number of rules equal to the number of output membership functions. In this case, the number is fixed to 45.

Overall ANFIS system can be visualized in Figure 4. In order to make the system easier to understand, ANFIS system in Figure 4 must be cross-referenced to its FIS part (Figure 5), before further explanation takes place. From Figure 5, fuzzy inference relationship with ANFIS can be established.

IOP Conf. Series: Materials Science and Engineering **53** (2013) 012087 doi:10.1088/1757-899X/53/1/012087

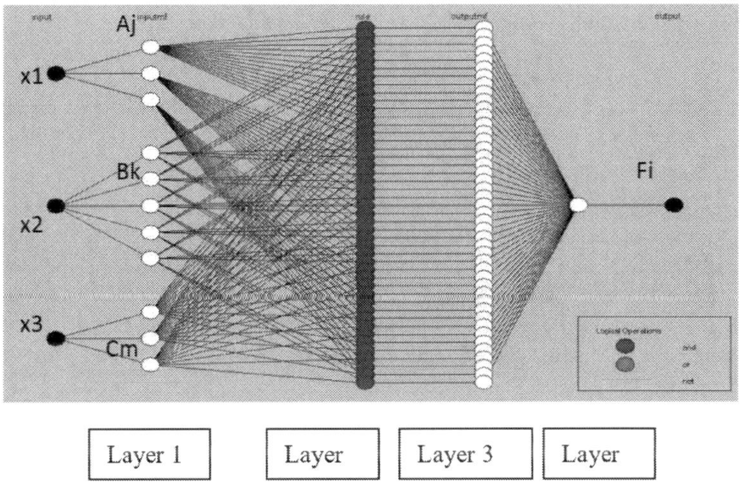

Figure 4: ANFIS architecture

Breaking down ANFIS to its FIS part for clearer view, Figure 5 shows relationship between FIS output and inputs.

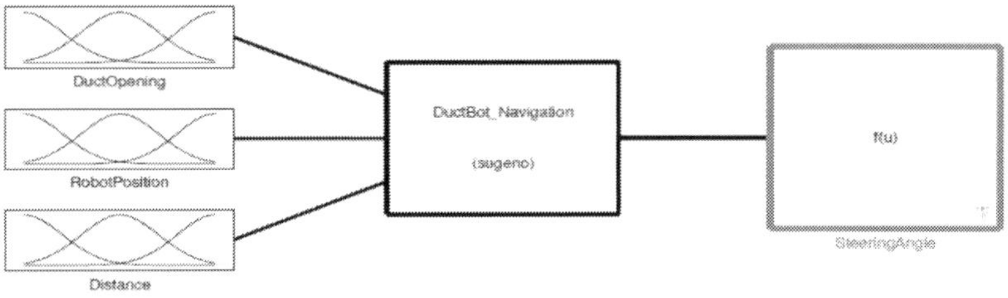

Figure 5: FIS architecture

Three inputs, as mentioned before, are duct opening, robot position and distance. For each input, membership functions are designed so as to represent the behavior of that input as accurately as possible, with respect to expert knowledge.

Input variable 'duct opening' (Figure 6) is represented by three membership functions; circular, rectangular and others. Here, the objective is to classify the shape of the opening (circle, rectangle or others) through calculation of circularity made to the image.

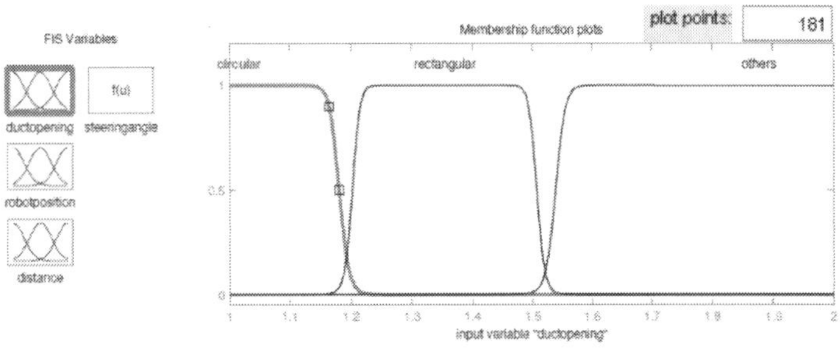

Figure 6: Input variable 'duct opening'

The modified generalized bell membership function in Figure 6 can be represented by (2):

$$f(x;a,b,c) = \frac{1}{1 + \left|\frac{x-c}{a}\right|^{2b}} \tag{2}$$

where variable a,b and c are the parameters. This is chosen due to the nature of classification for this input; 'circularity' which fall into the range of certain shape, are most certainly that particular shape (membership function= 1) except when it falls to near to the boundaries.

Input variable 'robot position' (Figure 7) and input variable 'distance' (Figure 8) behave almost the same and characterized by the same type of membership functions; triangular and trapezoidal

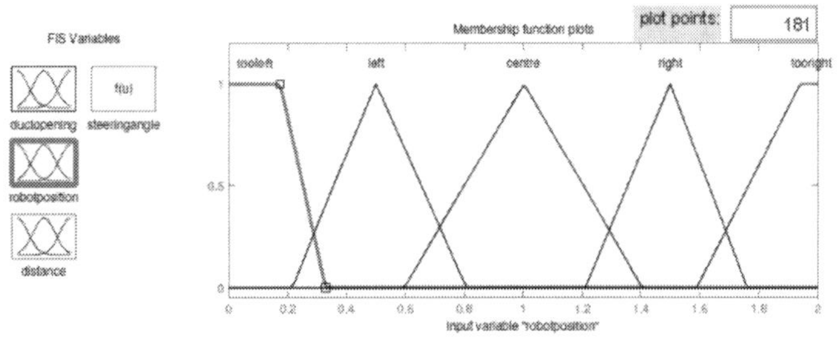

Figure 7: Input variable 'robot position'

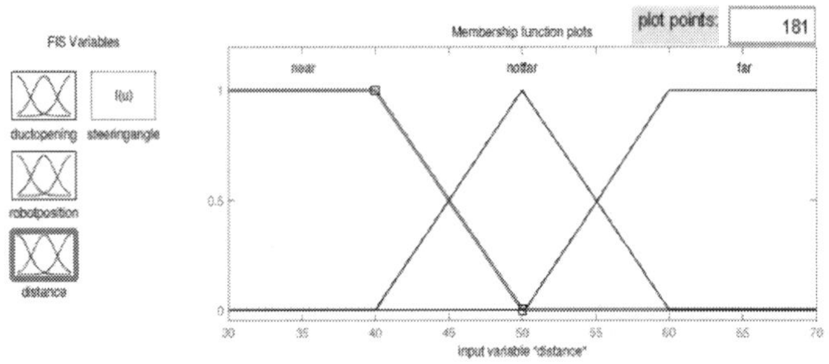

Figure 8: Input variable 'distance'

Intended for both variables, triangular membership function is represented by (3);

$$f(x;a,b,c) = \begin{cases} 0, & x \le a \\ \frac{x-a}{b-a}, & a \le x \le b \\ \frac{c-x}{c-b}, & b \le x \le c \\ 0, & c \le x \end{cases} \tag{3}$$

where parameters a and c locate the "feet" of the triangle and parameter b locates its peak is used. This membership function type represents linear increment and decrement within their ranges. For upper most and lower most limit for both, where for input 'robot position', these are represented by "too left" and "too right" membership functions while for input variable 'distance', "near" and "far" membership functions are of interest, trapezoidal shaped membership function is used. The function is denoted by equation in (4);

$$f(x;a,b,c.d) = \begin{cases} 0, & x \le a \\ \dfrac{x-a}{b-a}, & a \le x \le b \\ 1, & b \le x \le c \\ \dfrac{d-x}{d-c} & c \le x \le d \\ 0 & d \le x \end{cases} \qquad (4)$$

where parameters a and d locate the "feet" of the trapezoid and parameters b and c locate the "shoulders".

Output variable 'steering angle' (Figure 9) is mapped into 45 different membership functions. These reflect marginal values of -45 degree to 45 degree of steering angle. Positive angles are reserved for "steering to the right" while negative means "steering to the left". Both positive and negative indications are taken with respect to neutral position (0 degree), where tires are aligned with robotic chassis.

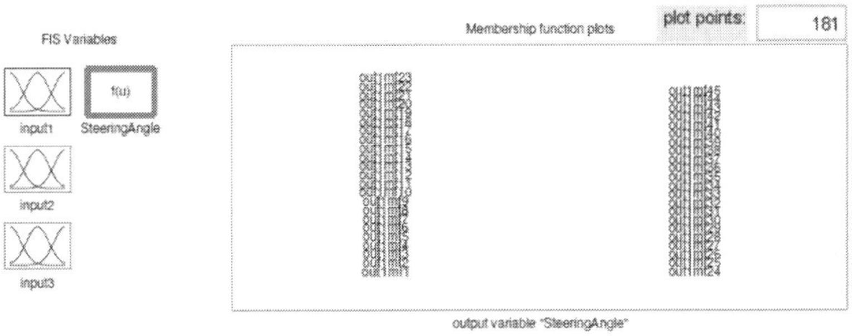

Figure 9: Output variable 'steering angle'

Below are sample experimental data gained from previous image analysis and experiments. The output values of steering angle are approximately measured for each set of inputs and recorded in Table 3.

Table 3: Sample experimental training data

Duct opening	Robot Position	Distance	Steering angle
1	0	30	-45
1.55	1	50	5
1.8	1.5	60	30
1.95	1.8	66	45
1.85	1.6	62	35
1.7	1.3	56	20
1.97	2	80	45

These data are then fed to ANFIS to train the system to recognize and control the maneuver for any HVAC system where the ducting parameters are almost similar to the discussed HVAC ducting type.

IOP Conf. Series: Materials Science and Engineering **53** (2013) 012087 doi:10.1088/1757-899X/53/1/012087

3. Result and discussion

Training process is shown in Figure 10 while the result is in Figure 11. Model over fitting is illustrated in Figure 12.

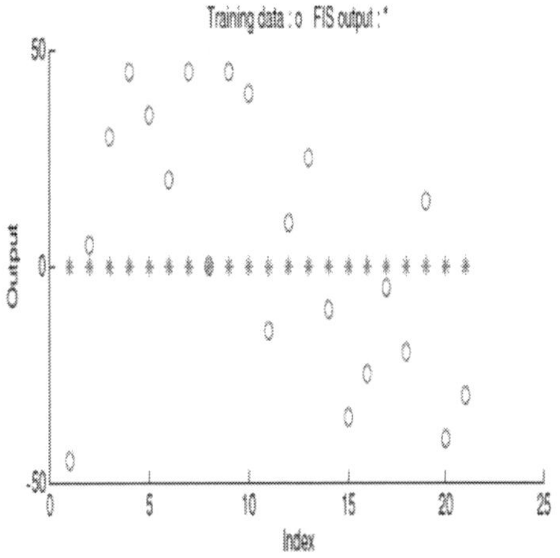

Figure 10: Before training

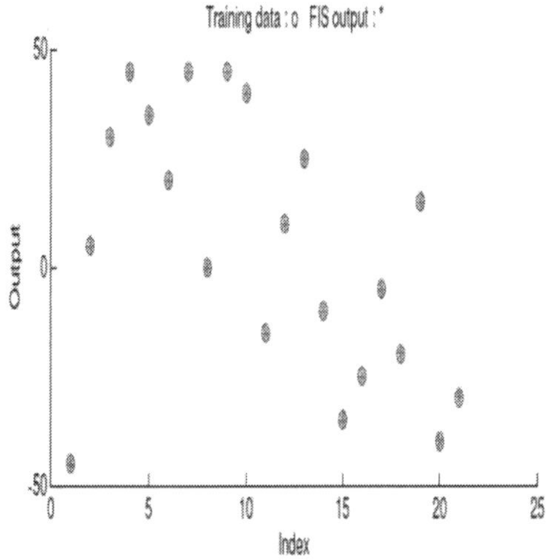

Figure 11: After training

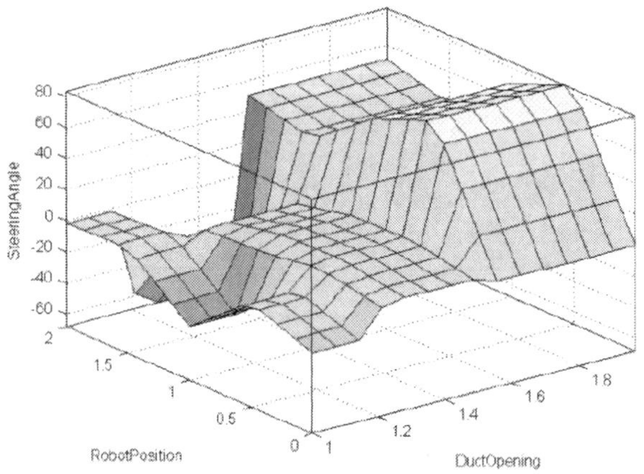

Figure 11: Input-output surface view

For training purposes, only 21 data sets are used. This is due to the complexity of gaining good samples under current light condition. Also, in order to represent the system as accurately as possible, samples are chosen selectively, where only the best representative for three types of ducting branches are taken; T-branch, elbow branch and left and right branch. Best error is gained after epochs 46.

Enhancement could be done by controlling the luminescent of the light source. Since, inner side of HVAC ducting is reflective, a lot of noises occurred which hindered the acquisition of a good sample. It is difficult to determine blob, border, edges and later the shapes of the duct opening. Good solutions can be achieved by designing a separate subsystem of light control or adding another FIS or ANFIS subsystem focusing on duct opening's recognition only.

Small training data set (21 data) leaves the system to be fairly weak in generalization. Hence, more data sets are required to make the system more robust to various ducting parameter's changes.

4. Conclusion

In this paper, ANFIS controller for HVAC service robot is proposed and implemented.45 rules are implemented with 46 epochs required to train the system. Ultrasonic distance sensor is used together with CCD camera to simplify data acquisition; camera for image processing and ultrasonic sensor for distance. The fusion of 2 sensors eliminate data overload on the system in comparison to stereo vision or perception of depth analysis (monocular vision) for distance measurement. Furthermore, our HVAC service robot platform is the first batch of its class. Efforts are being pushed, mostly towards materializing the concepts and reducing complexities. For future work, ANFIS subsystem will be implemented on robotic arm movement, cleaning brush control and image recognition system.

5. References

[1] SK Pradhan, DR Parhi and AK Panda (2006). Neuro-fuzzy technique for navigation of multiple mobile robots. *Fuzzy Optim. Decis. Making.* , **5**,pp.255-288.

[2] SK Pradhan, DR Parhi and A.K.Panda (2009). Motion Control and Navigation of Multiple Mobile Robots for Obstacle Avoidance and Target Seeking: A Neuro-Fuzzy Rule-based Technique. *J. of Sys. and Cont. Engineering*, **223**,pp.275-288.

[3] SK Pradhan, DR Parhi and A.K.Panda (2006). Navigation of multiple mobile robots using rule-based neuro-fuzzy technique, *Int. J. of Comput. Intell.*, **3**,pp.142-152.

[4] T.M.R. Akbarzadeh, K. Kumbla, E. Tunstel, and M. Jamshidi (2000). Soft computing for autonomous robotic systems, *Computers and Electrical Engineering*,**26**, pp. 5-32.

[5] Budiharto W, Jazisie A and Purwanto D (2010). Indoor Navigation using Adaptive Neuro Fuzzy Controller for Servant Robot. *Int. Con. on Computer Enginnering and Application*, pp.582-586

[6] Lee, J., Roh, S., Kim, D., Moon, H., & Choi, H. (2009). In-Pipe Robot Navigation Based on the Landmark Recognition System Using Shadow Images. *Inernational Conference on Robotics and Automation* ,pp. 1857-1862. Kobe, Japan: IEEE.

[7] J.S.R. Jang, (1993) ANFIS: Adaptive-network-based fuzzy inference systems.,*IEEE Trans. on Syst. Man,and Cybern.*, **23**, pp. 665-685,

[8] M.K Singh, R.P Dayal, J.K Pothal, (2009) ANFIS Approach for Navigation of Mobile Robot, *Int. Conv. on Advances in Recent Technologies in Communication and Computing*,pp 727-731

[9] R. Sabzevari, S. Masoumzadeh, and M. Rezaei Ghahroudm,(2008) Employing ANFIS for Object Detection in Robo-Pong, *Proceedings of the 5th International Symposium on Mechatronics and its Applications (ISMA08)*,.

[10] K.C. Ng,and M.M. Trivedi. (1998) A Neuro-fuzzy controller for mobile robot navigation and multi robot convoying, *IEEE Trans. on Sys., man, and cyber.*, part b, **28**, pp.829-840.

Flank wears Simulation by using back propagation neural network when cutting hardened H-13 steel in CNC End Milling

Muataz Hazza F. Al Hazza[1], Erry Y.T. Adesta, & Muhammad Riza

Faculty of Engineering – International Islamic University Malaysia- Malaysia

muataz@iium.edu.my

Abstract. High speed milling has many advantages such as higher removal rate and high productivity. However, higher cutting speed increase the flank wear rate and thus reducing the cutting tool life. Therefore estimating and predicting the flank wear length in early stages reduces the risk of unaccepted tooling cost. This research presents a neural network model for predicting and simulating the flank wear in the CNC end milling process. A set of sparse experimental data for finish end milling on AISI H13 at hardness of 48 HRC have been conducted to measure the flank wear length. Then the measured data have been used to train the developed neural network model. Artificial neural network (ANN) was applied to predict the flank wear length. The neural network contains twenty hidden layer with feed forward back propagation hierarchical. The neural network has been designed with MATLAB Neural Network Toolbox. The results show a high correlation between the predicted and the observed flank wear which indicates the validity of the models.

1.Introduction

The upper limit of cutting speeds in machining have been increased during the last eighty years from 15m/min in the year 1930 to over 1000m/min after 1990 [1]. The term high speed may not be the same for different materials, as high speed for one material may still be a low speed for another (for example, the high speed for titanium is a low speed for aluminium). Allowed level of speeds has been always restrained by the limitation in flank wear progress, cutting temperature and surface quality. Figure 1 shows the regions of cutting speed for milling a different work pieces and its relation to the conventional and high speed cutting [1, 2].

High speed end milling of hard alloy steels have many advantages such as reduction of machining time, higher metal removal rates, lower machining costs and better surface roughness [3]. However, high speed machining increases the temperature in both work material and cutting tool substantially due to high pressure on the cutting zone [4]. Increasing the cutting temperature has a bad effect on both the cutting tool sand the work piece. In contrast increasing the cutting temperature will reduce the cutting forces [5,6,7]. Thus, predicting and modeling the responses of the machining process such as the temperature, cutting forces, surface roughness and wear progress before the machining process in certain cutting levels become an important issue.

[1] To whom any correspondence should be addressed.

Dolinsek and Kopac [1] claimed that the characteristic wear of cutting tools is caused by of the fact that the cutting speed is no longer the main influential factor on wear, but more likely wear is the consequence of the high-speed of the feed rate.

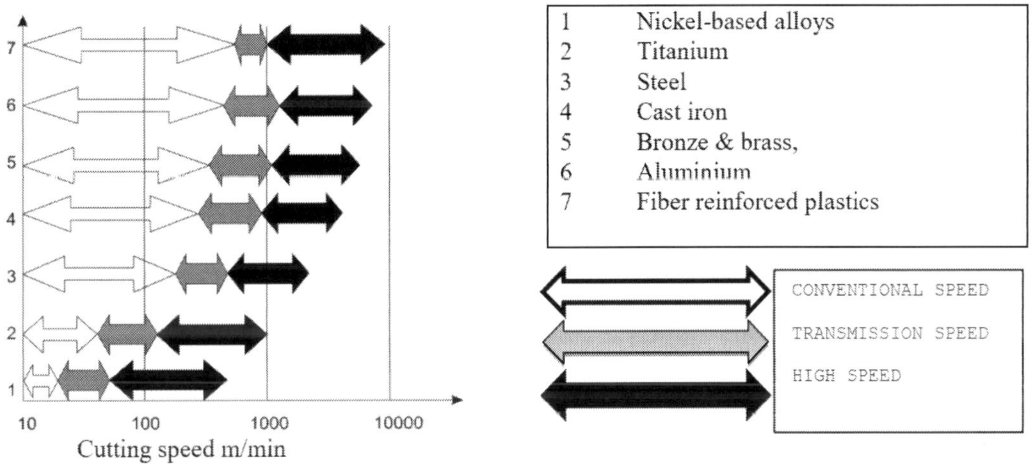

Figure 1: high speed cutting levels (Dolinsek and Kopac, 2006)

Many researchers worked in developing different models by using the Neural Network (NN) in terms of cutting parameters. Some of the researchers used the NN in predicting the cutting forces [8], flank wear [9], energy [10], temperature [3], others used the ANN to predict the surface roughness [11].

Many researchers [12,13,14,15] used the neural network to predict the wear rate. Other researchers [9, 16] used two methods such as the neural network and the regression analysis method and then compare the results to find the best model.

Flanks wear length usually used as a reference for estimating the cutting tool life. Therefore, estimating and predicting the flank wear progress early is an important and crucial matter. Adesta et al. [4] classified the methods for estimating the flank wear progress into two main types; direct and indirect methods.

Direct methods based on monitoring and measuring the flank wear length during the cutting time [17] or by estimating the flank wear length mathematically [18, 19, 20]

However, flank wear can be estimated by monitoring the change of other machining parameters such as surface roughness [21], cutting force [22], temperature [23] and vibrations [24], which reflect the flank wear change. Other methods used for measuring the flank wear length such as the acoustic emission method [25].

Milling are considered to be too complex to be modeled accurately by using analytical or even numeric means due to involvement of various control parameters [16].

In this research, new model has been developed to predict the flank wear length in terms of cutting speed, feed rate and depth of cut in high speed end milling process.

2. Experimental work

The machining has been conducted using a vertical milling center type MAZAK machine (Model Nexus 410A-II). The machining was under high cutting speed from 150 up to 250 m/ min, low feed rate 0.05-0.15 mm/ rev, low depth of cut 0.1-0.2 mm and tool diameter was 20 mm. The experiments in this research were performed on AISI H13 at hardness of 48 HRC as work material. Hisomet II tool maker microscope has been used for monitoring and measuring the flank wear length during machining. In the experiment, 20 samples of data set concerned with the end milling process have been collected based on five-level of central composite Design (CCD) as shown in Table 1.

Table 1. Experiment design

No of run	Cutting Speed m/min	Feed Rate mm/rev	Depth of Cut mm	No of run	Cutting Speed m/min	Feed Rate mm/rev	Depth of Cut mm
1	134.2	0.1	0.10	**11**	200	0.10	0.15
2	150	0.05	0.10	**12**	200	0.10	0.15
3	150	0.05	0.20	**13**	200	0.10	0.15
4	150	0.15	0.10	**14**	200	0.10	0.22
5	150	0.15	0.20	**15**	200	0.17	0.15
6	200	0.03	0.15	**16**	250	0.05	0.10
7	200	0.10	0.08	**17**	250	0.05	0.20
8	200	0.10	0.15	**18**	250	0.15	0.10
9	200	0.10	0.15	**19**	250	0.15	0.20
10	200	0.10	0.15	**20**	265.8	0.10	0.15

The research methodology is based on two main parts: experimental work and neural network application. The detailed research methodology illustrated in Figure 2.

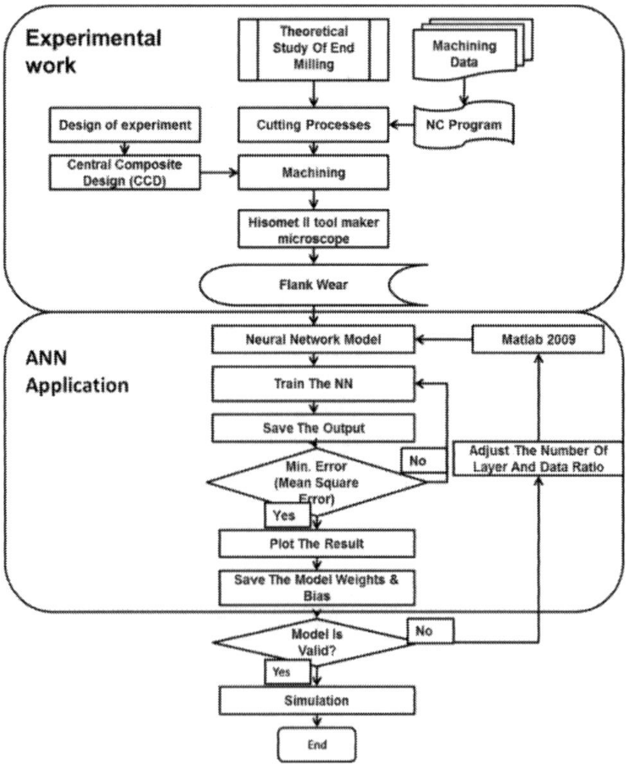

Figure 2: Research methodology

3.ANN application

These samples used to train the neural network and to adjust the weights and the biases of each unit in order to reduce the error between the desired output and the actual output. The NFTOOL box in the MATLAB 2009 has been used. The back propagation algorithm applied to determine the layer's weights. Table 2 concluded the architecture, learning system, specifications of the neural network model used in the development of the new model.

Table 2. Modelling by using neural network

Tool	MATLAB 2009
Tool box	Nftool
Architecture	Feed forward
Learning system	Supervised learning
Algorithm	Back propagation Levenberg-Marquardt algorithm (LM)
Activation Function	Sigmoid (logistic function)
Number of layers,Data ratio	3 layers (input, hidden and output) 70:15:15
Number of hidden layers	20

Then the MATLAB-M file is generated from the software as in Figure3 and the algorithm flow chart is as in Figure 4.

```
function net = create_fit_net (inputs,targets)
%CREATE_FIT_NET Creates and trains a fitting neural network.
% NET = CREATE_FIT_NET(INPUTS,TARGETS) takes these
  arguments:
%   INPUTS - RxQ matrix of Q R-element input samples
%   TARGETS - SxQ matrix of Q S-element associated target
  samples
% arranged as columns, and returns these results:
%   NET - The trained neural network
% For example, to solve the Simple Fit dataset problem with this
  function:
%   load simplefit_dataset
%   net = create_fit_net(simplefitInputs,simplefitTargets);
%   simplefitOutputs = sim(net,simplefitInputs);
% To reproduce the results you obtained in NFTOOL:
%   net = create_fit_net(x',y');
% Create Network
numHiddenNeurons = 20;  % Adjust as desired
net = newfit(inputs,targets,numHiddenNeurons);
net.divideParam.trainRatio = 70/100;  % Adjust as desired
net.divideParam.valRatio = 15/100;  % Adjust as desired
net.divideParam.testRatio = 15/100;  % Adjust as desired
% Train and Apply Network
[net,tr] = train(net,inputs,targets);
outputs = sim(net,inputs);
plotperf(tr)
plotfit(net,inputs,targets)
plotregression(targets,outputs
```

Figure 3: MATLAB-M file

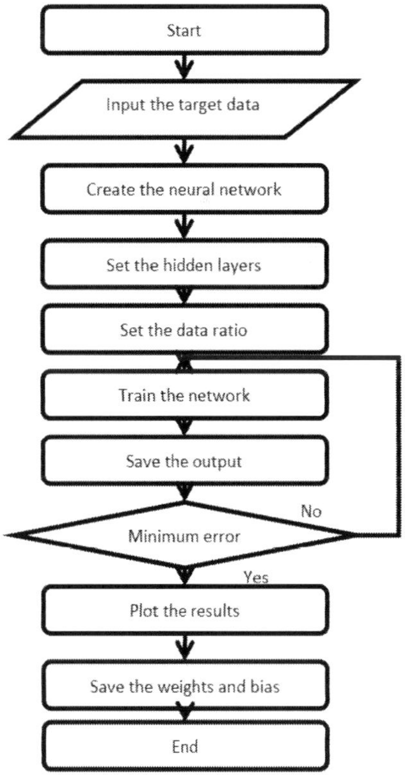

Figure 4: Flow chart of ANN Application

4. Results and discussions

The regression plot for training, testing and validating the model are summarized in Fig.5. The plots display the network outputs with respect to targets for training, validation, and test sets. For a perfect fit, the data should fall along a 45 degree line (dash line), where the network outputs are equal to the targets. For this study, the fit is very good for all data sets, with R values in each case of 0.97 or above.

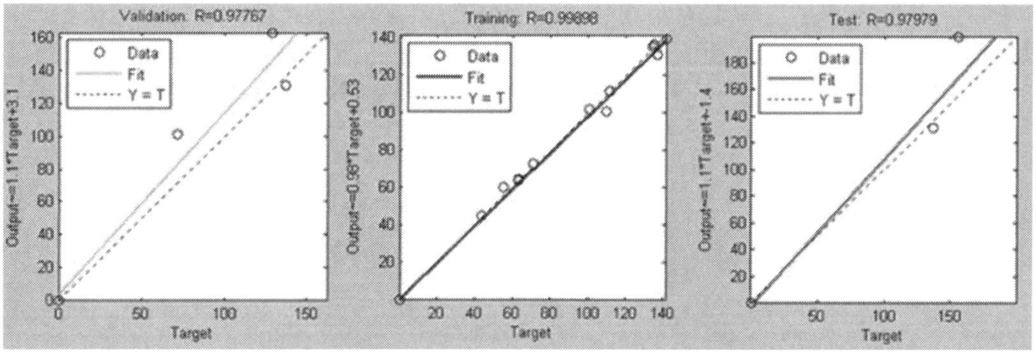

Figure 5: Plot of data regression (training, validation, testing)

The final weights of the model are concluded in Table 3.

Table 3. Final weights and bias of input layers to Hidden layers

Number of layer	Cutting speed	Feed rate	Depth of cut	Bias	Number of layer	Cutting speed	Feed rate	Depth of cut	Bias
1	3.140	-1.945	0.230	-3.896	11	2.426	2.339	1.810	0.357
2	2.615	2.627	-0.983	-3.360	12	1.425	2.745	2.133	0.501
3	2.797	2.090	-1.553	-3.030	13	0.847	-1.026	-3.579	1.002
4	3.268	-0.942	1.738	-2.564	14	-2.380	-1.186	2.593	-1.637
5	-2.936	-2.167	0.186	2.372	15	2.827	2.559	0.069	1.741
6	0.175	-2.447	2.878	-1.790	16	-2.817	2.288	0.925	-2.298
7	1.230	2.388	-2.667	-1.436	17	0.532	-0.768	-3.683	2.595
8	2.448	-2.708	1.110	-1.054	18	-2.039	-2.246	-2.436	-2.863
9	1.370	1.715	-3.056	-0.548	19	-1.168	0.626	3.567	-3.391
10	-1.329	1.486	3.214	0.127	20	2.104	-2.088	-2.383	3.826

5. Validation

A comparision of the measured and the predicted values to determine the deviation between the theoretical and actual value that comes out from ANN models have been conducted. Figure 4 shows the average deviation between the actual and the predicted values by the neural network models. The results show a high percentage of accuracy, 1.17% degree of variation from the original measured values.

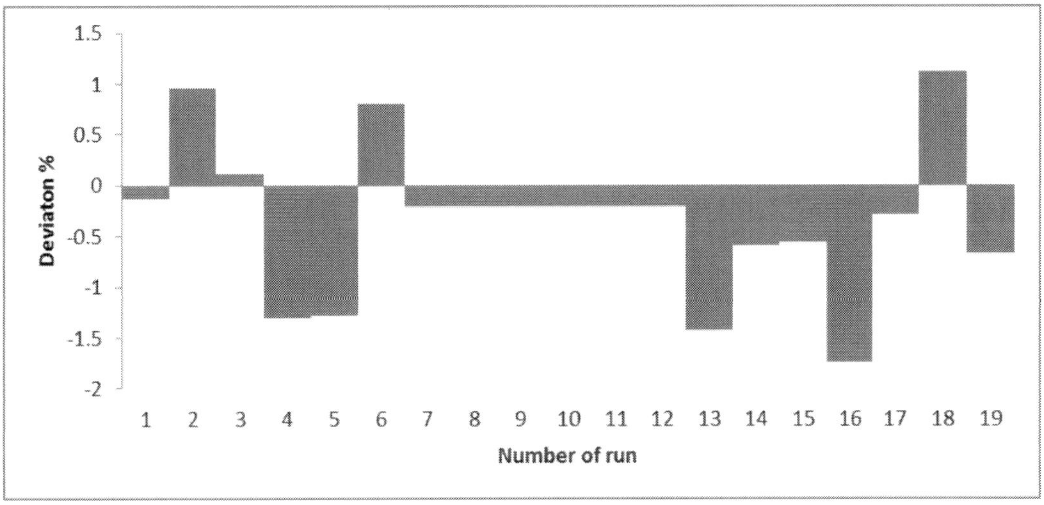

Figure 5: Model validation

6. Simulation

The effect of feed rate and depth of cut during high speed milling have been simulated by using the neural network model and the results were concluded in Figure 6 and Figure 7.

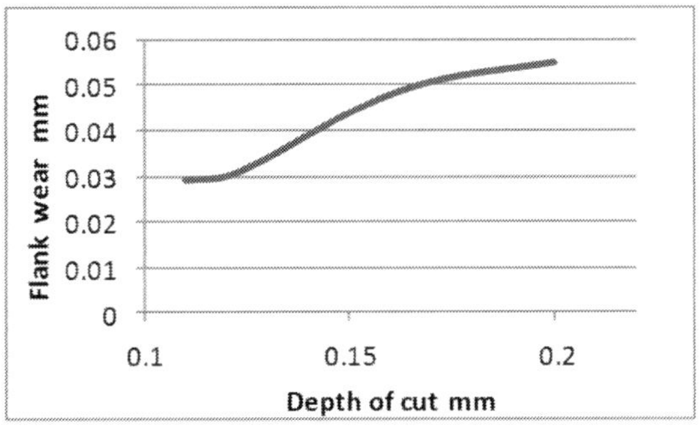

Figure 6: Effect of depth of cut on the depth of cut

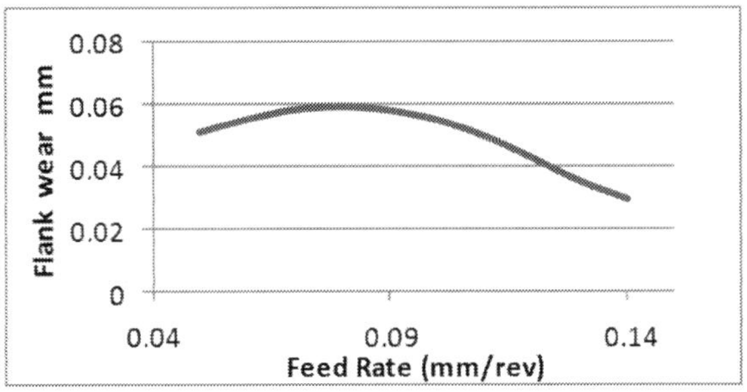

Figure 7: Effect of feed rate on the flank wear length

The results of the simulation show that

a. The effect of the depth of cut is higher than the feed rate in higher cutting speed (250 m/min)

b. Increasing the depth of cut will increase the flank wear
c. Increasing the feed rate above 0.8 have a positive effect on the flank wear.
d. These results may due to

7. Conclusions

This study has been involved with the ANN technique to develop a new model to predict the values of temperature in the end milling machining operation. Twenty hidden layer has been used with feed forward back propagation hierarchical neural networks were designed with Matlab2009b Neural Network Toolbox. A new model is tested and validated.

This study has been involved with the ANN technique for development of models to predict the values of flank wear length in the end milling machining operation. Twenty hidden layer has been used with feed forward back propagation hierarchical neural networks were designed with Matlab2009b Neural Network Toolbox. The results show that the models are valid by using three inputs only; cutting speed, feed rate and depth of cut. The average deviations are 1.17 % for the flank wear length. This high correlation between the predicted and observed values indicated the validity of the models effect on the flank wear.

8. References

[1] Dolinšek S. & Kopač J.,2006 Mechanism and types of tool wear; particularities in advanced cutting materials *Journal of Achievements in Materials and Manufacturing Engineering*, 19(1), 11-18.

[2] Schulz, H., 1992 *High Speed Machining*, Annals of the CIRP (41) 637-643.

[3] Adesta, E. Y. T., Al Hazza, M. H., Suprianto, M. Y., & Riza, M. 2012 Prediction of Cutting Temperatures by Using Back Propagation Neural Network Modeling when Cutting Hardened H-13 Steel in CNC End Milling *Advanced Materials Research*, 576, 91-94.

[4] Adesta E.Y.T., AlHazza Muataz.F., M. Riza, D. Agusman, and Rosehan, 2010 Tool Life Estimation Model Based on Simulated Flank Wear during High Speed Hard Turning *European Journal of Scientific Research* Vol.39No.2 , pp.265-278.

[5] Lin H.M., Liao Y.S., & Wei C.C., Wear behavior in turning high hardness alloy steel by CBN tool *Wear*. (264) 679–684, 2008.

[6] Yallese, M. A. Chaoui, K. Zeghib, N. Boulanouar, L. & Rigal J. F, 2009 Hard machining of hardened bearing steel using cubic boron nitride tool *Journal of Materials Processing Technology* 209(2), 1092-1104.

[7] Adesta E.Y.T. and Al Hazza M.H.F., Machining Time Simulation in High Speed Hard Turning. Advanced Materials Research, 264, 1102-1106, 2011.

[8] Zuperl, U. Cus, F. Mursec B. and Ploj, T. 2006 A generalized neural network model of ball-end milling force system, Journal of Materials Processing Technology, vol. 175, pp. 98–108,.

[9] Al Hazza M.H.F. and Adesta E. Y. T. 2011 Flank Wear Modelling In High Speed Hard Turning By Using Artificial Neural Network And Regression Analysis *Advanced Materials Research* Vols. 264-265, pp 1097-1101.

[10] Al Hazza M.H.F., Adesta E.Y.T, Ali A. M., Agusman D.and Suprianto M.Y. 2011 Energy Cost Modeling for High Speed Hard Turning *Journal of Applied Sciences*, 11: 2578-2584,

[11] Oktem H., Erzurumlu T. & Erzincanli F., 2006 Prediction of minimum surface roughness in end milling mold parts using neural network and genetic algorithm *Materials & design*, 27(9), 735-744.

[12] Ghosh, N. Ravi Y. B., Patra A., Mukhopadhyay S., Paul S., Mohanty A. R. & Chattopadhyay A. B 2007 Estimation of tool wear during CNC milling using neural network-based sensor fusion *Mechanical Systems and Signal Processing*, 21(1), 466-479.

[13] Chuangwen X., Xiaohong W., Wencui L., and Xing F., (January 2009)Milling wear monitoring study based on cutting power, *Proceedings of the International Conference on Advanced Computer Control* (ICACC '09), pp. 671–674.

[14] Özel T. & Nadgir, A. 2002 Prediction of flank wear by using back propagation neural network modeling when cutting hardened H-13 steel with chamfered and honed CBN tools *International Journal of Machine Tools and Manufacture* 42(2), 287-297.

[15] Özel T. & Y. Karpat 2005 Predictive modeling of surface roughness and tool wear in hard turning using regression and neural networks *International Journal of Machine Tools and Manufacture*, 45(4), 467-479.

[16] Iqbal, A. 2013 Modeling Milling Process Using Artificial Neural Network *Advanced Materials Research* 628, 128-134.

[17] Cakan A. 2011Real-time monitoring of flank wear behavior of ceramic cutting tool in turning hardened steels *Int J Adv Manuf Technol* 52:897–903.

[18] Usui E., Shirakashi T., Kitagawa T. 1982 Analytical Prediction of Cutting Tool Wear *Wear* Vol. 100, no. 1-3, pp. 129-151(1984).

[19] Huang Y. and Liang S.Y., 2004 Modeling of CBN Tool Flank Wear Progression in Finish Hard Turning *ASME Journal of Manufacturing Science and Engineering* Vol. 126(1), pp 98-106.

[20] Singh D. and Rao P. V. 2010 Flank wear prediction of ceramic tools in hard turning *International Journal Advanced Manufacturing Technology*. 50: 479–493.

[21] Choudhury S.K. and Bartary G. 2003 Role of temperature and surface finish in predicting tool wear using neural network and design of experiments *International Journal of Machine Tools & Manufacture* 43; pp 747–753.

[22] Sikdar S. K., and Chen M., 2002 Relationship between tool flank wear area and component forces in single point turning *Journal of Materials Processing Technology* 128: 210–215.

[23] Young H.T. 1996 Cutting temperature responses to flank wear *Wear* 201; pp117-120.

[24] Orhan, S.. Osman Er A, , Camus-cu Necip, E. Ersan Aslan 2007 Tool wear evaluation by vibration analysis during end milling of AISI D3 cold work tool steel with 35 HRC hardness *NDT&E International* 40; PP 121–126.

[25] Scheffer, C. Kratz, H., Heyns, P. S. and Klocke, F. 2003 Development of a tool wear-monitoring system for hard turning *International Journal of Machine Tools & Manufacture*. 43 pp 973–985.

Investigation of the effect of cutting speed on the Surface Roughness parameters in CNC End Milling using Artificial Neural Network

Muataz H. F. Al Hazza[1] and Erry Y.T. Adesta

Faculty of Engineering – International Islamic University Malaysia- Malaysia

muataz@iiu.edu.my

Abstract. This research presents the effect of high cutting speed on the surface roughness in the end milling process by using the Artificial Neural Network (ANN). An experimental investigation was conducted to measure the surface roughness for end milling. A set of sparse experimental data for finish end milling on AISI H13 at hardness of 48 HRC have been conducted. The artificial neural network (ANN) was applied to simulate and study the effect of high cutting speed on the surface roughness.

1. Introduction

In finishing end milling, not only good accuracy but also good roughness levels must be achieved. Thus, determining the optimum cutting levels to achieve the minimum surface roughness is important for it is economical and mechanical issues [1]. Surface roughness is a result of process parameters such as tool geometry and cutting conditions.

Sahin & Motorcu [2] found out through an experimental study that the surface finish of machined parts has a considerable effect on some properties such as wear resistance and fatigue strength. Hence, the quality of the surface is of a significant importance when evaluating the productivity of machine tools and mechanical parts. Then, predicting the quality of the surface roughness is important for its economical and mechanical properties.

Fnides et al. [3] claimed that the importance of surface roughness comes from its relation to many properties of machine elements such as wear resistance, the capacity of fit and sealing.

Cutting speed, feed rate and depth of cut are parameters that mostly influence the Ra value of surface quality in machining, particularly in the milling machining process. The set of parameters that are thought to influence the Ra value could be summarized in a fishbone diagram as shown in Fig. 1. The figure classified the factors that affect the surface roughness into four main categories: cutting tool properties, machining parameters, work piece properties and cutting phenomena. the final roughness of the machining process is actually is a result of all these factors, thus predicting and modeling by using the deterministic methods the surface roughness is a complex and difficult mission, thus the non

[1] To whom any correspondence should be addressed

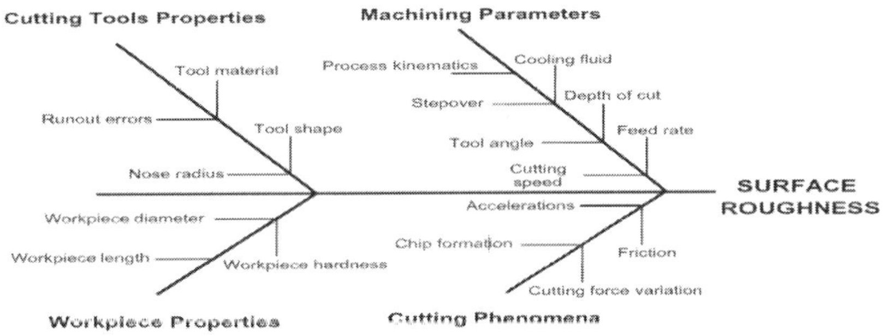

Figure1. Parameters that affect surface roughness [4]

Several methods for modelling the surface roughness have been reported in different research works. These methods can be classified into: statistical, mathematical and artificial intelligence techniques. Some of the researchers used the statistical techniques [3,5,6,7,8,9,10], others used the neural network techniques [4,11,12, 13,14].

Other researchers applied the ANN and the regression techniques and compared the results [15,16,17,18,19]. End milling is a type of face milling, and is used for facing, profiling and slotting processes. The end milling process is complex compared with turning because of its more complicated machine tool linear motions and its repeated intermittent engagement and disengagement of rotating cutting edges [20].

In finishing end milling, not only good accuracy but also good roughness levels must be achieved. The quality of the surface is of a significant importance when evaluating the productivity of mechanical parts. Thus, predicting and simulating the surface roughness quality is utmost important for it is economical and mechanical issues.

The real surface geometry is so complicated that one parameter can describe the surface roughness. There are more than 59 surface roughness parameters [21] and classified into three different categories of surface roughness parameters; amplitude, spacing and hybrid parameters.

Many parameters can be used to describe surface finishes and these are explained in the ISO 4287 and the ISO 4288.

This classification is based on the direction of measurement, vertical, horizontal or both together. The amplitude is used to measure the vertical characteristics of surface deviation while the spacing is used to measure the horizontal characteristics of surface deviation [21]. In machining process the most important parameters are the amplitude parameters.

This research will consider four amplitude parameters; arithmetic mean roughness (Ra), total roughness (Rt), mean depth of roughness (Rz) and the root mean square (Rq) as concluded in table 1.

Table 1. Surface Roughness Parameters

Parameter	Definition
Ra	Mean value of the absolute value of the profile departure y within the reference length
Rt	Total height between highest peak and deepest valley
Rz	Mean of distance between the five highest peaks and five deepest valleys
Rq	Root mean square average of the profile ordinates

2. Artificial Neural Network

The recent developments of predicting models concern the use of artificial intelligence such as neural network and fuzzy logic. Neural network (NN) and fuzzy logic present the next generation in computerizing the human thought processes [22]. The neural network technique is a method that makes a computer simulates the behaviour of human brain neurons by using a set of data consisting of input and output variables. In the training process, the structure of the model is self adjusted to the data, and the final model can be used for predictions [23].

The training algorithm is defined as a procedure that consists of adjusting the weights and biases of a network that minimize selected function of the error between the actual and desired outputs [24, 25]. Figure 2 describes the training and adjusting procedure. The process of training a neural network can be broadly classified into two typical categories: supervised learning and unsupervised learning.

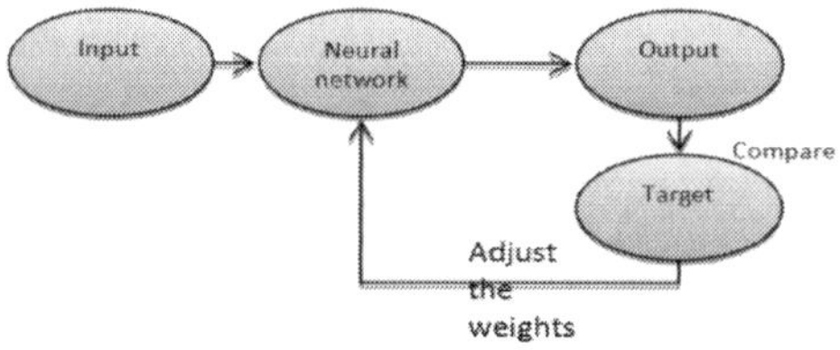

Figure 2: training and adjusting the weights

Back-propagation is one of the supervised learning. It is the most popular and well-studied training algorithm. It is a gradient-descendent method that minimizes the mean-square error of the difference between the network outputs and the targets in the training set [26]. It is used usually with the multi-layer structures. The basic procedure for training a network can be concluded as in the following steps:

a. Apply an input data to the neural network and calculate the corresponding output value.
b. Compare the output with the target and determine the measure of error.
c. Determine differences and the direction (positive or negative) in the weights to change each weight to reduce the error.
d. Determine the values for the new weights.
e. Apply the corrections to the weights.
f. Repeat the previous steps with all training vectors in the training set such that the error is reduced to an acceptable value.

The neural networks pass the output of their layers through activation functions. These activation functions scale the output of the neural network into proper ranges. The default choice for the feed forward Layer is the sigmoid activation function as in Equation (1).

$$f(x) = \frac{1}{1+e^{-x}} \qquad (1)$$

The term sigmoid means curved in two directions, like the letter "S." You can see the sigmoid function in Figure 3.

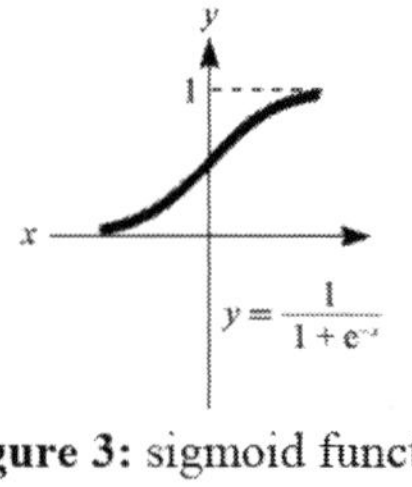

Figure 3: sigmoid function

A typical neuron consists of a linear activator followed by a nonlinear inhibiting function as in figure 3. The linear activation function yields the sums of weighted inputs plus an independent term so-called bias [26]. The hidden layer uses a sigmoid-type transference function:

$$f(x) = \frac{1}{1+e^{(-b-\sum w_i x_i)}} \tag{2}$$

While the output layer uses a linear function:

$$output = \left(\sum_{i=0}^{n-1} x_i w_i \right) + (b_n) \tag{3}$$

3. Research methodology

The research methodology of this research can be concluded as the following;

1. Theoretical study
2. Experimental work
3. Design the experimental work by using the DoE software
4. Measuring the roughness parameters
5. Develop a model by using the artificial neural network
6. Adjust the model for minimum error
7. Simulate the results
8. Analyzing

These steps are shown in figure 4.

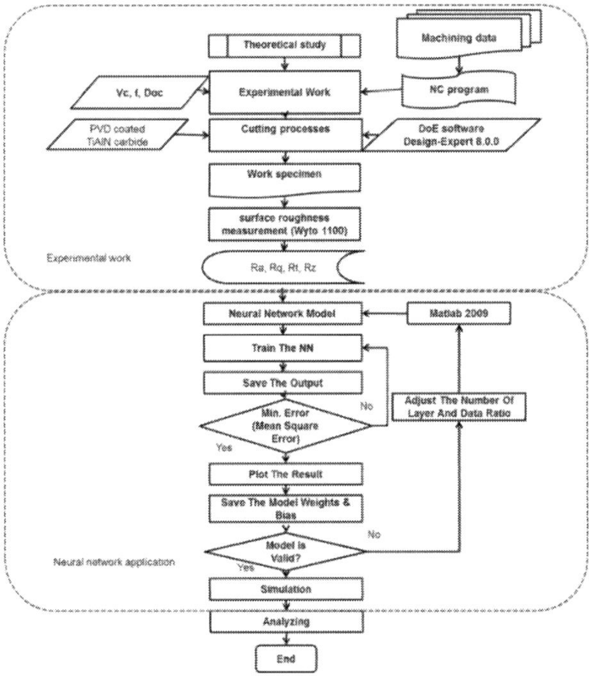

Figure 4: Research Methodology

4. Experimental work

Machining was conducted using vertical milling centre type MAZAK machine (Model Nexus 410A-II). It has been done at high cutting speed from 150 up to 250 m/ min, low feed rate 0.05-0.15 mm/ rev and low depth of cut 0.1-0.2 mm. The experiments for this research were performed on AISI H13 at hardness of 48 HRC as work material. In the experiment, 20 samples of data set concerned with the end milling process have been collected based on five-level of central composite Design (CCD) as

IOP Conf. Series: Materials Science and Engineering **53** (2013) 012089 doi:10.1088/1757-899X/53/1/012089

shown in table 2. All the experiments done by using indexable tool holder Sandvick Coromill R490 and the insert was PVD coated TiAlN carbide.

Table 2. Experimental Design

Cutting Speed(m/min)		Feed Rate (mm/tooth)		Depth of Cut (mm)	
Actual	Coded	Actual	Coded	Actual	Coded
250	1	0.05	-1	0.10	-1
250	1	0.15	1	0.10	-1
250	1	0.05	-1	0.20	1
250	1	0.15	1	0.20	1
150	-1	0.05	-1	0.10	-1
150	-1	0.15	1	0.10	-1
150	-1	0.05	-1	0.20	1
150	-1	0.15	1	0.20	1
200	0	0.10	0	0.15	0
200	0	0.10	0	0.15	0
200	0	0.10	0	0.15	0
200	0	0.10	0	0.15	0
200	0	0.10	0	0.15	0
200	0	0.10	0	0.15	0
134	α	0.10	0	0.15	0
265	α	0.10	0	0.15	0
200	0	0.034	α	0.15	0
200	0	0.166	α	0.15	0
200	0	0.10	0	0.084	α
200	0	0.10	0	0.216	α

The experiment setup is shown in Figure 5.

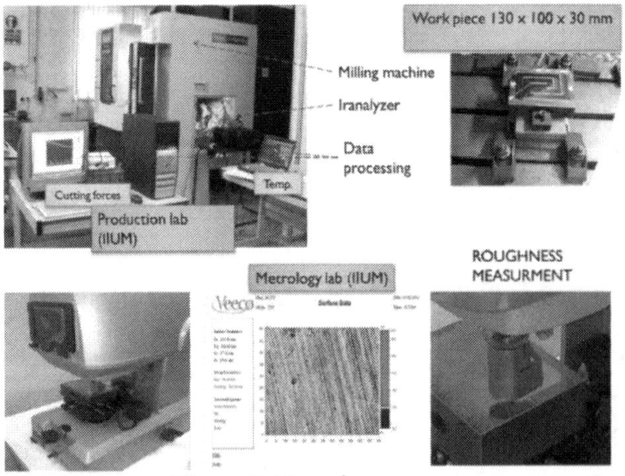

Figure 5. Experiment setup

IOP Conf. Series: Materials Science and Engineering **53** (2013) 012089 doi:10.1088/1757-899X/53/1/012089

The results of the experiment have been used to design a new model for predicting the arithmetic mean roughness (Ra), root mean square (Rq), depth of roughness (Rz), and total roughness (Rt) by using three inputs: cutting speed, feed rate and depth of cut as shown in Figure 6.

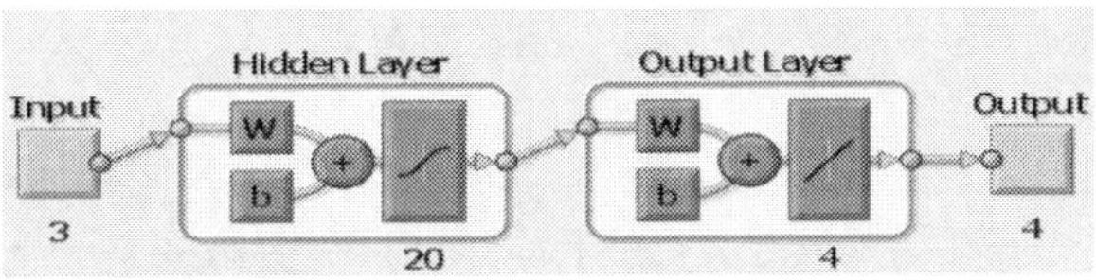

Figure 6. Network model structure diagram

The neural network architecture, learning system, algorithm, and activation function are concluded in table 3.

Table 3. Modelling by using neural network

Tool	**MATLAB 2009**
Tool box	Nftool
Architecture	Feed forward
Learning system	Supervised learning: target values for the output are presented to the network
Algorithm	Back propagation Levenberg-Marquardt algorithm (LM), Update weight and Bias values
Activation Function	Sigmoid (logistic function)
Number of layers	3 layers (input, hidden and output)
Number of hidden layers	20
Data ratio	70:15:15

5. Results and discussions

Performance plot is generated. Performance is measured in terms of mean squared error and shown in log scale for each of the training, validation and test sets. The performance plot shows that the best validation is in epoch 4 as shown in Figure 7.

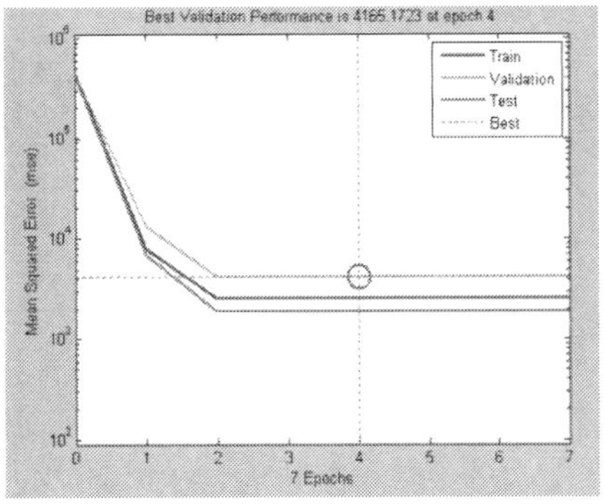

Figure 7. Plot performance

The regression plot for training, testing and validating the model are summarized in Figure 8. The plots display the network outputs with respect to targets for training, validation, and test sets. For a perfect fit, the data should fall along a 45 degree line (dash line), where the network outputs are equal to the targets. For this study, the fit is very good for all data sets, with R values in each case of 0.92 or above.

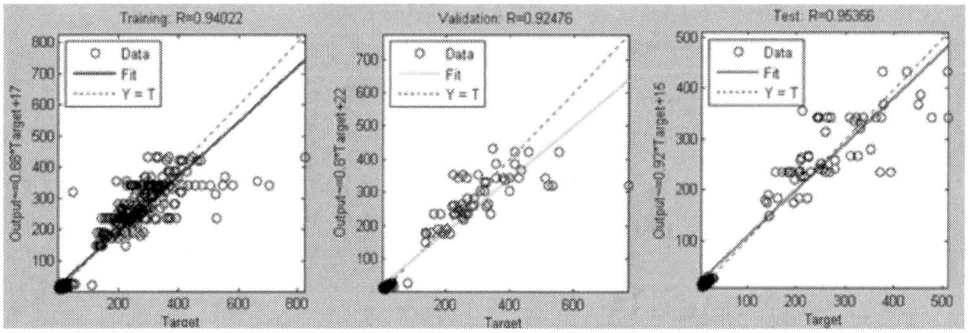

Figure 8. Plot of data regression (training, validation, testing)

The weights of the final model are concluded in table 4 and table 5.

Table 4. weights of input layers to Hidden layers

hidden layer	Cutting speed	Feed rate	Depth of cut	Bias
1	-1.786	-2.888	-1.149	3.992
2	-1.119	2.888	-2.499	3.009
3	-1.874	3.259	0.711	2.751
4	-0.551	-1.788	-3.156	2.637
5	-0.191	-2.879	2.702	1.925
6	2.391	0.793	-2.563	-1.973
7	2.675	-0.627	2.400	-1.658
8	-1.177	2.893	-1.639	1.504
9	2.708	2.187	1.591	-0.256
10	2.113	1.017	-2.875	-0.487
11	2.222	2.785	1.272	0.010
12	0.207	-2.636	-2.736	0.486
13	-2.461	-1.002	2.532	-1.116
14	2.556	-2.763	0.038	1.431
15	1.365	-3.502	0.191	1.885
16	-3.319	0.718	-1.736	-2.063
17	3.063	-1.235	-1.220	2.977
18	-2.437	2.740	-0.029	-3.183
19	3.448	0.977	0.103	3.674
20	-2.526	-2.419	-0.409	-4.041

Table 5. weights of Hidden layer to output layers

Layer	Ra	Rq	Rz	Rt
1	-0.407	-0.391	0.083	0.074
2	-0.140	0.385	0.226	0.228
3	0.104	0.054	-0.012	-0.052
4	-0.298	0.427	0.252	-0.479
5	0.144	-0.080	0.023	-0.183
6	-0.164	0.197	0.292	-0.526
7	-0.325	0.133	0.482	0.089
8	0.167	-0.057	0.285	0.314
9	0.340	0.075	-0.551	0.803
10	0.221	-0.332	-0.331	0.319
11	0.133	0.073	0.401	-0.634
12	0.588	-0.040	0.232	0.915
13	0.420	0.304	0.467	0.489
14	-0.170	0.060	0.078	-0.244
15	0.168	0.074	-0.521	0.351
16	-0.014	0.093	-0.092	-0.629
17	0.546	0.294	0.721	0.537
18	0.544	0.480	0.263	0.979
19	0.075	0.598	-0.139	-0.896
20	-0.329	0.073	-0.375	-0.272
Bias	-0.660	-0.544	-0.893	-0.342

6. Validation

A comparison of the measured and the predicted values to determine the deviation between the theoretical and actual value that comes out from ANN models have been conducted. The average deviation of the models is concluded in Figure9.

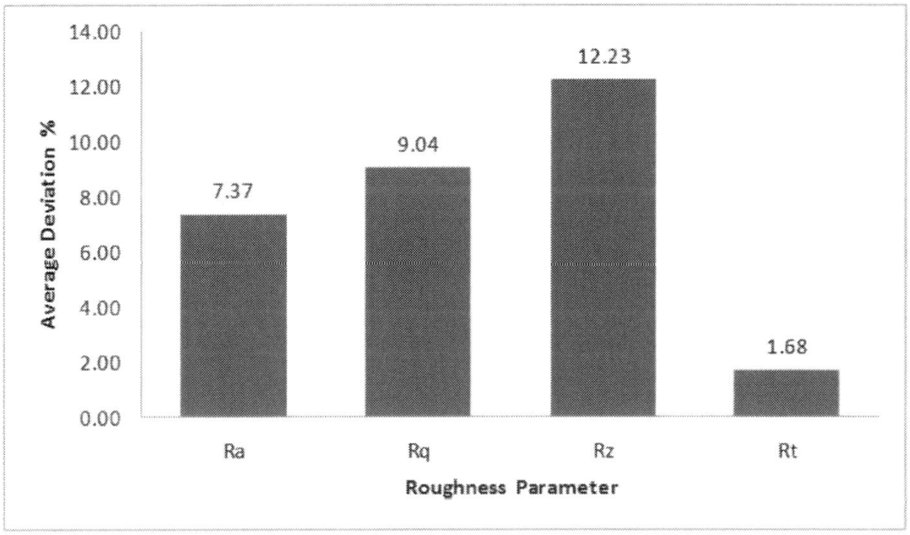

Figure 9. Surface roughness models accuracy

7. Simulation

The effect of cutting speed and feed rate with constant depth of cut equal to 0.1mm on the different roughness parameters has been simulated. The results have been concluded in Figures 10-13.

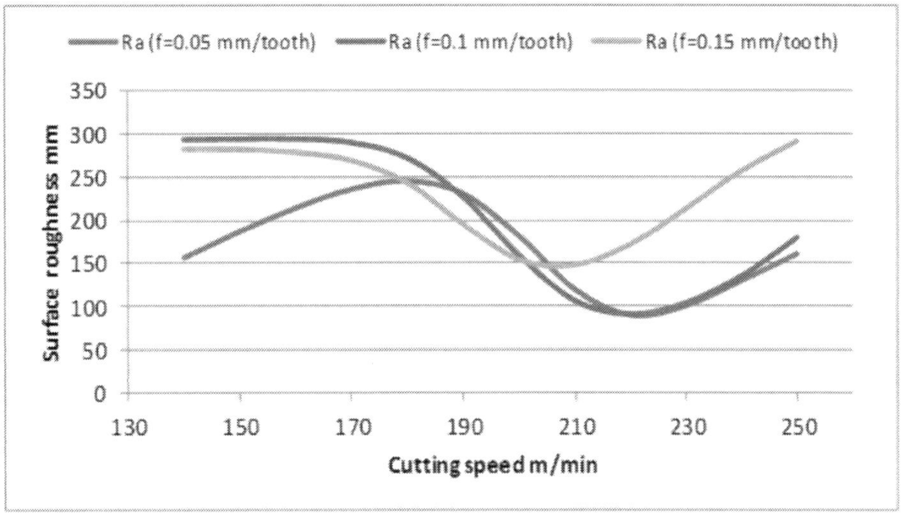

Figure 10. The effect of feed rate and cutting speed on Ra

It was found that the best cutting speed for minimum Ra are in the range of 200mm/min for all the feed rates. The figure shows that the roughness decreases until a specific point then start to increase with increasing the cutting speed for all feed rates.

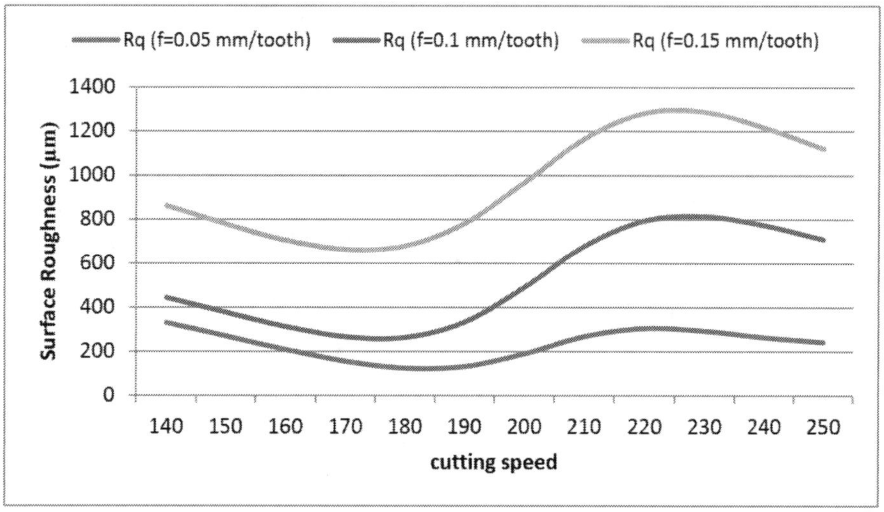

Figure 11. The effect of feed rate and cutting speed on Rq

It was found that the minimum values of Rq was with feed rate equal to 0.05 mm/tooth and cutting speed 185 mm/min

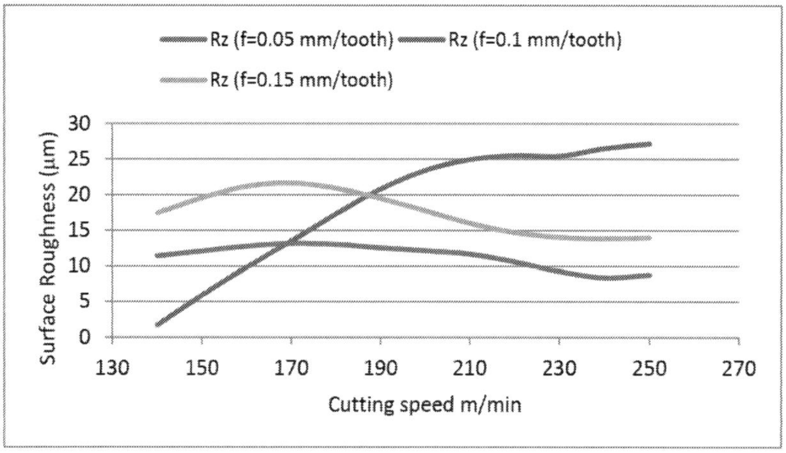

Figure 12. The effect of feed rate and cutting speed on Rz

Figure 12 shows that the minimum Rz can be achieved with feed rate of 0.05 mm/tooth and cutting speed equal to 240m/min

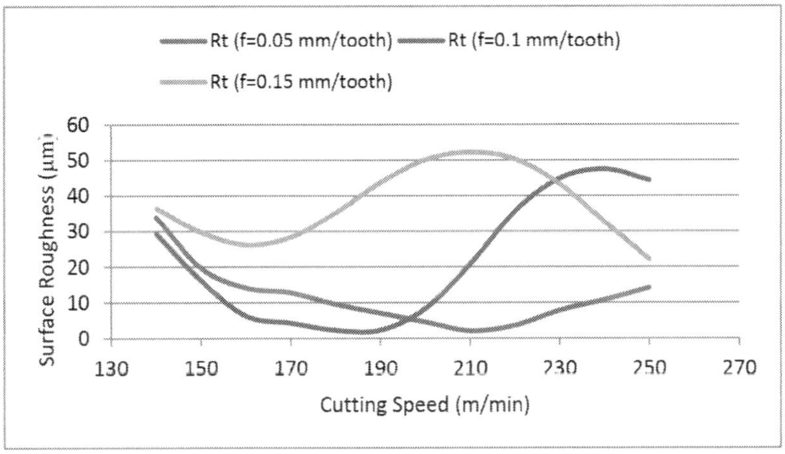

Figure 13. The effect of feed rate and cutting speed on Rt

The simulation results gave to minimum values for different feed rate 0.05 and 0.1 mm/tooth. These values are almost the same; however the cutting speed with 0.05 mm/tooth as feed rate is higher

8. Conclusions

This study has been involved with the ANN technique for development of models to predict the values of surface roughness parameters in the end milling machining operation. Athese models used to simulate the different cutting speeds and feed rates to estimate and predict the best cutting parameters for minimum values of surface roughness parameters. In the next stage, consideration should be on studying the optimization of the cutting factors in order to minimize the surface roughness parameter, which is an important issue.

9. References

[1] Al Hazza, Muataz H.F., Adesta, Erry Y.T., Muhammad Riza, and Suprianto, M. Y 2012 Surface Roughness Optimization in End Milling Using the Multi Objective Genetic Algorithm *Approach Advanced Materials Research* Vol. 576 pp 103-106
[2] Sahin, Y., & Motorcu, A. R. 2005 Surface roughness model for machining mild steel with coated carbide tool. *Materials & Design* 26(4), 321-326.
[3] Fnides, B., Yallese, M. A., Mabrouki, T., & Rigal, J. F. 2009 Surface roughness model in turning hardened hot work steel using mixed ceramic tool *Mechanika*, 3(77), 68-73.

[4] Benardos, P. G., & Vosniakos, G. C. 2002 Prediction of surface roughness in CNC face milling using neural networks and Taguchi's design of experiments *Robotics and Computer-Integrated Manufacturing* 18(5), 343-354.

[5] Puertas Arbizu, I., & Luis Perez, C. J. 2003 Surface roughness prediction by factorial design of experiments in turning processes *Journal of Materials Processing Technolog,* 143, 390-396.

[6] Mohd, A., Ali, Erry. Y. T. Adesta, Agusman, D., Badari, S. N. M., & Al-Hazza, M. H. F. 2010 Development of Surface Roughness Prediction Model for High Speed End Milling of Hardened Tool Steel *Asian Journal of Scientific Research*, 4.

[7] Reddy, M. M., Gorin, A., & Abou-El-Hossein, K. A. 2011 Predictive Surface Roughness Model for End Milling of Machinable Glass Ceramic. *In IOP Conference Series: Materials Science and Engineering* (Vol. 17, No. 1, p. 012002). IOP Publishing.

[8] Arokiadass, R., Palanirajda, K., & Alagumoorthi, N. 2011 Predictive modeling of surface roughness in end milling of Al/SiCp metal matrix composite *Archives of Applied Science Research*, 3(2), 228-236.

[9] Adesta, E. Y. Triblas , Muataz Al Hazza, Delvis Agusman, Agus Geter Edy Sutjipto, 2012a. Development of Tooling Cost Model for High Speed Hard Turning *Advanced Materials Research* (Volumes 418 - 420).

[10] Adesta, E. Y. Triblas , Muataz Al Hazza, Suprianto, M.Y and Muhammad Riza, 2012b Predicting Surface Roughness with Respect to Process Parameters Using Regression Analysis Models In End Milling *Advanced Materials Research* Vol. 576 pp 99-102

[11] Tsai, Y. H., Chen, J. C., & Lou, S. J. 1999 An in-process surface recognition system based on neural networks in end milling cutting operations. International Journal of Machine Tools and Manufacture, 39(4), 583-605.

[12] Kohli, A., & Dixit, U. S. 2005 A neural-network-based methodology for the prediction of surface roughness in a turning process. *The International Journal of Advanced Manufacturing Technology*, 25(1-2), 118-129.

[13] Özel, T., Karpat, Y., Figueira, L., & Davim, J. P. 2007 Modelling of surface finish and tool flank wear in turning of AISI D2 steel with ceramic wiper inserts. *Journal of materials processing technology*, 189(1), 192-198.

[14] Wang, X., Wang, W., Huang, Y., Nguyen, N., & Krishnakumar, K. 2008 Design of neural network-based estimator for tool wear modeling in hard turning *Journal of intelligent manufacturing*, 19(4), 383-396.

[15] Quiza, R., Figueira, L., & Davim, J. P. 2008 Comparing statistical models and artificial neural networks on predicting the tool wear in hard machining D2 AISI steel. *The International Journal of Advanced Manufacturing Technology*, 37(7-8), 641-648.

[16] Akkuş, H., & Asilturk, İ. 2011 Predicting surface roughness of AISI 4140 steel in hard turning process through artificial neural network *Fuzzy Logic and Regression Models. Scientific Research and Essays*, 6(13), 2729-2736.

[17] Asiltürk, İ., & Çunkaş, M. 2011 Modeling and prediction of surface roughness in turning operations using artificial neural network and multiple regression method *Expert Systems with Applications*, 38(5), 5826-5832.

[18] Al Hazza Muataz and Adesta E. Y. T. 2011a Flank Wear Modeling In High Speed Hard Turning By Using Artificial Neural Network And Regression Analysis. *Advanced Materials Research* Vols. 264-265, pp 1097-1101

[19] Al-Hazza Muataz, Erry Yulian T. Adesta, Afifah Mohd Ali, Delvis Agusman and Mohammad Yuhan Suprianto 2011b Energy Cost Modeling for High Speed Hard Turning. *Journal of Applied Sciences, 11: 2578-2584.*

[20] Childs, T., Maekawa, K., Obikawa, T. and Yamane, Y. (2000), *Metal Machining Theory and Applications,* John Wiley & Sons Inc. New York-Toronto.

[21] Gadelmawla, E. S., Koura, M. M., Maksoud, T. M. A., Elewa, I. M., & Soliman, H. H. 2002 Roughness parameters *Journal of Materials Processing Technology* 123(1), 133-145.

[22] Villarreal, J. A.; Lea, R. N. & Savely, R. T. 1992 Fuzzy logic and neural network technologies. *In: 30th Aerospace Sciences Meeting and Exhibit*, Houston, Texas, January 6-9, 1992.

[23] Freeman, J., & Skapura, D. 1991 *Neural networks, algorithms, applications, and programming techniques.*

[24] Kalogirou, S. A. 2003 Artificial intelligence for the modeling and control of combustion processes: *A review. Progress in Energy and Combustion Science,* 29, 515–566

[25] Karatas, C., Sozen, A., & Dulek, E. 2009 Modeling of residual stresses in the shot peened material C-1020 by artificial neural network *Expert Systems with Applications*, 36, 3514–3521.

[26] Davim J. P. 2011 *Machining of Hard Materials* Springer London Dordrecht Heidelberg New York, pp225.

Investigation of surface roughness in micro-electro discharge machining of nonconductive ZrO_2 for MEMS application

A Sabur[1], A Moudood[2], M Y Ali[3], M A Maleque[4]

Department of Manufacturing and Materials Engineering, Faculty of Engineering, International Islamic University Malaysia, PO Box 10, 50728 Kuala Lumpur, Malaysia

Email[1]: asabur72@yahoo.com, email[2]: m.a.moudood@gmail.com, email[3]: mmyali@iium.edu.my, email[4]: maleque@iium.edu.my

Abstract. Micro-electro discharge machining technique, a noncontact machining process, is applied for drilling blind hole on nonconductive ZrO_2 ceramic for MEMS application. A conductive layer of adhesive copper is applied on the workpiece surface to initiate the sparks. Kerosene is used as dielectric for creation of continuous conductive pyrolytic carbon layer on the machined surface. Experiments are conducted by varying the voltage (V), capacitance (C) and rotational speed (N). Correlating these variables a mathematical model for surface roughness (SR) is developed using Taguchi method. The results showed that the V and C are the significant parameters of SR in micro-EDM for nonconductive ZrO_2 ceramic. The model also showed that SR increases with the increase of V and C.

1. Introduction

Microelectromechanical systems (MEMS) are developed in accordance with the advancement in modern technologies. MEMS or MEMS-based devices can be used for many engineering applications. In telecom or consumer electronics, MEMS-based devices may be used in huge volume potentially. MEMS may also be used in niche markets in small scale where it can be the key enabling factor [1]. Structuring various types of micro holes and micro channels are needed in MEMS devices. MEMS are produced mainly by lithography and silicon micromachining processes which are applicable for specific and limited materials. Advanced ceramic materials have been used for a variety of biomedical and implant devices. These ceramics are considered very important material that has high potential to deliver significant contributions for solving the challenges of our future especially in MEMS application such as microfluidics, reactors and electromechanical generators [2]. But advanced ceramics is difficult to be processed by lithography and silicon micromachining techniques. Micro-electro-discharge machining (micro-EDM) is considered as a suitable machining technique for microstructuring which can fulfil the gap between the diversity of engineering ceramics and the ability to use them in MEMS. Micro-EDM provides excellent opportunities for research, development and manufacturing of such products. It allows leveraging of non-traditional, high-performance engineering materials with various features such as plasticity, robustness, chemical inertness, and biocompatibility that cannot be achieved through conventional MEMS fabrication processes and their compatible materials. This ability also promotes proper choice of materials that are compatible with particular environments for MEMS fabrication [3]. Micro-EDM is an electro-thermal material removal process which is a derived form of EDM for manufacturing micro-components. It is a known process for

[1] To whom any correspondence should be addressed.

IOP Conf. Series: Materials Science and Engineering **53** (2013) 012090 doi:10.1088/1757-899X/53/1/012090

structuring and shaping of hard and brittle materials. In micro-EDM, a series of electrical sparks or discharges occur rapidly in a short span of time within a constant spark gap between tool electrode and workpiece and materials are removed due to the thermal energy of the spark. The material to be machined by micro-EDM requires a minimal electrical conductivity of 0.1 Scm^{-1}. Most of the advanced ceramics such as ZrO_2, Al_2O_3, Si_3N_4 are electrically nonconductive. Therefore, EDM cannot be directly applicable to the nonconductive ceramics [4]. A basic process is introduced to apply EDM for processing the nonconductive ceramic in which an assisting electrode (AE) layer of electrically conductive material is applied. The sparks initially occur between the tool electrode and the AE layer. After finishing the temporary external layer, a layer of pyrolytic carbon is deposited on the substrate surface disassociating the carbonic dielectric in appropriate conditions [5, 6]. Micro-EDM of advanced ceramics for MEMS structuring is considered as an important technique. However, surface roughness is the significant characteristics that can limit the micro-EDM of advanced ceramics for production of MEMS [7]. The empirical modelling of the process is an effective way of selecting the best parameters to get optimum outputs and to increase production rate significantly. Precise surface roughness is essential to fabricate intricate and micro products. This study aims to investigate the effect of input parameters on the continuous machining of nonconductive ceramics and to develop model of *SR* using Taguchi method in micro-EDM of ZrO_2 ceramic.

2. Experiments

The schematic diagram of the experimental setup of micro-EDM for nonconductive ceramic with the AE is shown in figure 1. The machining has been conducted using micro-EDM machine (Mikrotool, Singapore). In the present study, 92% pure ZrO_2 ceramic plate is used as workpiece material. The properties of the workpiece material are listed in the Table 1. The copper tool electrode of ϕ1 mm is used. The workpiece and the tool electrode were cleaned by acetone before machining. Since the workpiece is electrically nonconductive, its surface is covered by an adhesive copper layer to occur the sparks. After the removal of this external layer, a new conductive layer is created instantaneously on the machined surface using cracked carbon combined with the debris of tool electrode material. This layer acts further as an AE. Cu foil has excellent electrical conductivity and easy to remove after machining without any damage. After machining, the workpiece is cleaned again by acetone.

Table 1. Physical properties of ZrO_2

Property [unit]	Value
Hardness [Hv]	1270
Melting temperature [^{0}C]	2720
Specific heat capacity	540
Specific gravity [Kg/m^3]	610
Electrical resistivity [Ω-cm]	10^{10}

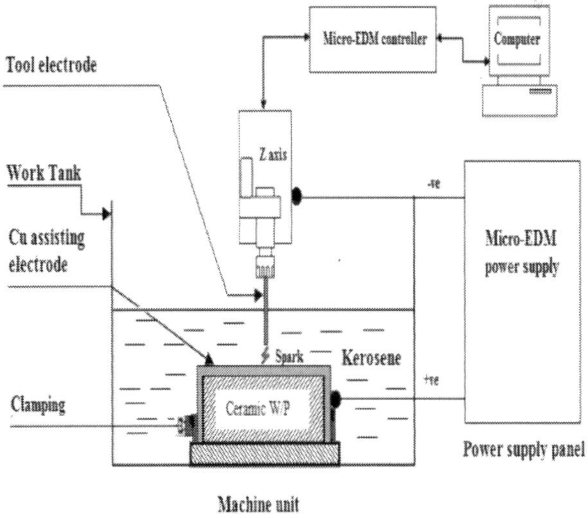

Figure1. Schematic diagram of micro-EDM set up for machining of ZrO_2

The machining parameters used in the experiments are listed in the Table 2. After machining, the workpiece is cleaned by acetone for investigation of surface textures. WYKO NT1100 optical profiling system is used to measure the SR and machined surface is inspected by scanning electron microscope (SEM) (JEOL JSM-5600). To find the optimum *SR*, experiments are done varying the parameters such as *V*, *C* and *S* keeping other parameters constant. In the present study, L9 orthogonal array (3 factors x 3 levels) based on Taguchi method is used to design of the experiments and subsequent analysis of the data collected.

3. Results and discussions

The ZrO_2 ceramic workpiece machined by micro-EDM is shown in figure 2 and SEM image of machined surface is shown in figure 3. Holes of 100 μm depth are created on the ZrO_2 workpiece. The pyrolytic carbon layer on the machined surface is produced continuously from kerosene dielectric fluid and adhesive Cu foil as initial assisting electrode. Figure 3 shows the *SR* in different micro-EDM conditions on the ZrO_2 ceramics. Table 3 shows the *SR* in different EDM conditions on the ZrO_2 ceramics according to L9 orthogonal array (3 factors x 3 levels) based on the Taguchi design. ANOVA and F test results are given in Table 4 to show the significant parameters associated with each machining conditions. F-value of model implies the significance of the model. There is only about 0.05 % chances that a "Model F - Value" of this large could occur due to noise. From this analysis, a model for the estimation of *SR* has been developed as the function of *V*, C and *N*, as shown in equation (1).

$$\frac{1}{\sqrt{SR+1}} = 0.79-0.0015V+8.66*10^{-4}C-4.93*10^{-5}N-3.67*10^{-6}VC-1.71*10^{-6}CN \qquad (1)$$

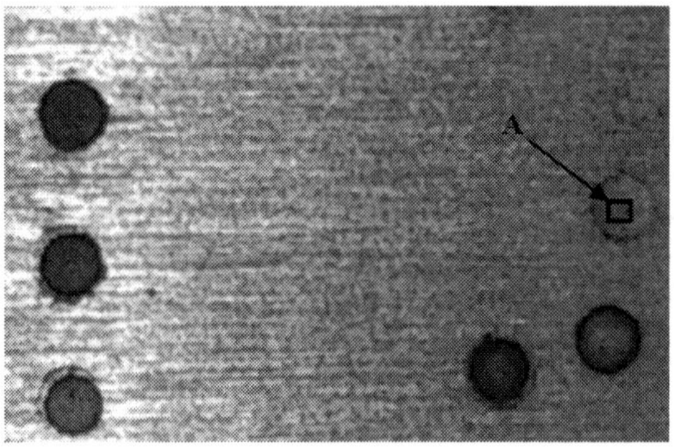

Figure 2. Micro-holes of ϕ1mm on ZrO_2 substrate machined by micro-EDM using adhesive Cu foil as AE and Cu tool electrode with –ve polarity

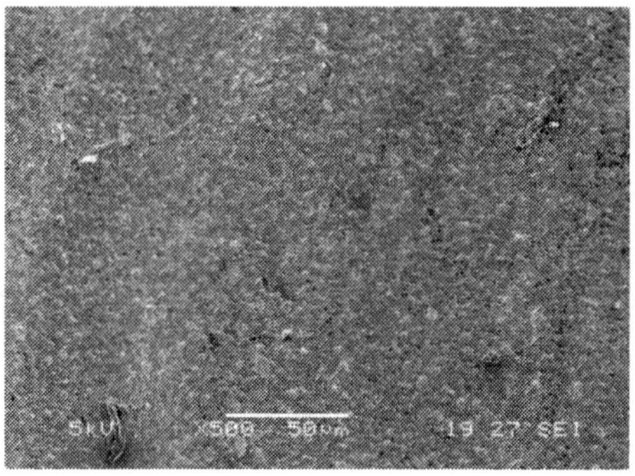

Figure 3. SEM image of surface texture of micro-hole as (window A, figure 2) in micro-EDMed ZrO_2 surface with SR=1.5 μm at C = 100 pF , V = 90 V and N = 350 rpm

Table 2. Machining parameters

Parameters [unit]	Values
Voltage [V]	80, 90, 100
Capacitance [pF]	10, 100, 1000
Speed (rpm)	250, 300, 350
Threshold (%)	27
Tool polarity	-ve
Dielectric	Kerosene
Assisting Electrode	Adhesive copper foil

Table 3. SR for micro-hole drilling by micro-EDM designed according to Taguchi method

Run	Voltage [V]	Capacitance [C]	Speed [rpm]	SR [μm]
1	100	10	350	1.64
2	80	10	250	1.3
3	90	100	350	1.5
4	90	1000	250	0.78
5	80	100	300	1.38
6	100	1000	300	1.7
7	100	100	250	1.53
8	80	1000	350	1.56
9	90	10	300	1.4

Table 4. ANOVA for Response Surface Reduced 2FI Model

Analysis of variance table [Partial sum of squares]

Source	Sum of Squares	DF	Mean Square	F Value	Prob > F	
Model	0.013946	5	0.002789	39.19819	0.0062	Significant
A	0.00492	1	0.00492	69.1356	0.0036	
B	0.001005	1	0.001005	14.12583	0.0329	
C	0.009036	1	0.009036	126.9864	0.0015	
AB	0.001296	1	0.001296	18.20647	0.0236	
BC	0.007002	1	0.007002	98.3975	0.0022	
Residual	0.000213	3	7.12E-05			
Cor Total	0.01416	8				

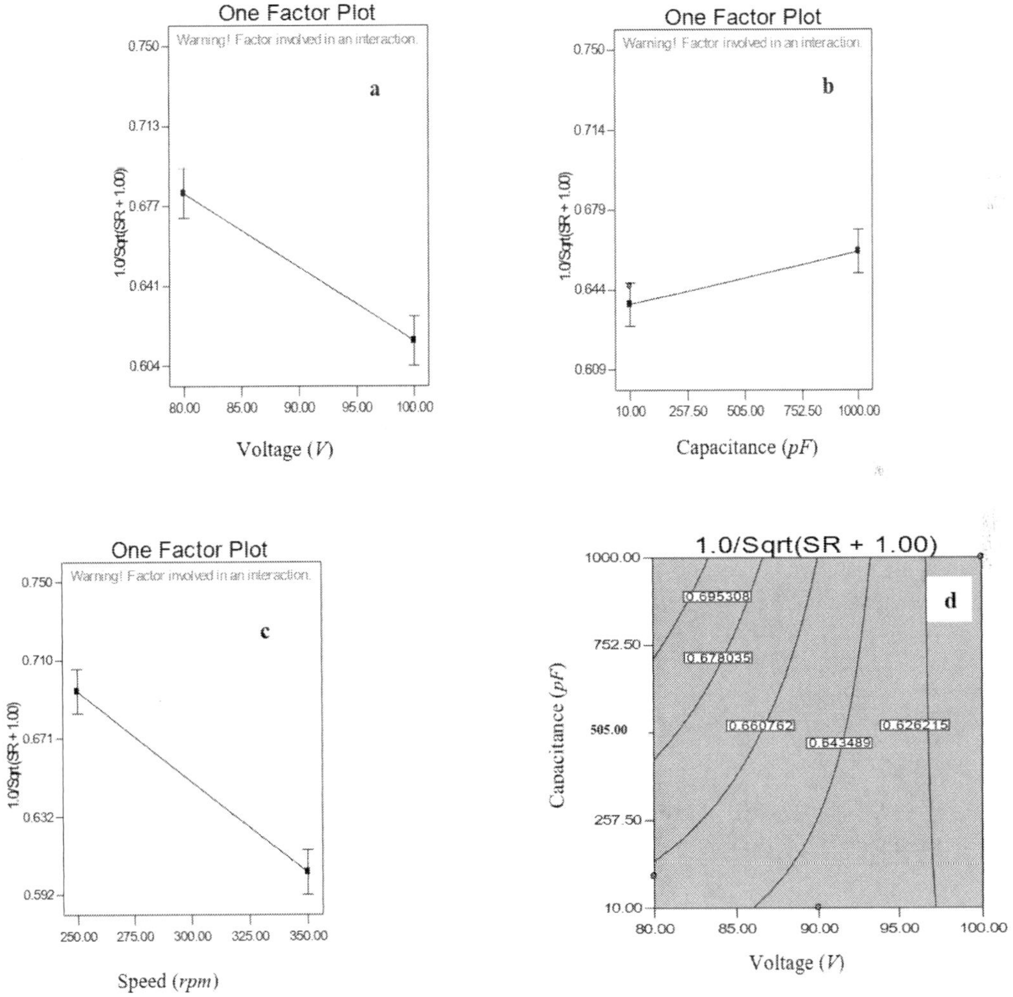

Figure 4. SR graph in different micro-EDM conditions for ZrO_2 ceramic: $\frac{1}{\sqrt{SR+1}}$ vs (a) voltage, (b) capacitance, (c) speed, (d) Interaction graph of voltage and capacitance

Figure 4(a) illustrates that *SR* increases with the increase of *V*. On the other hand, figure 4(b) indicates that the *SR* decreases as *C* increases. Hence, *SR* is significantly influenced by capacitance and the gap voltage in micro-EDM of nonconductive ceramic materials. At very low capacitance, spark energy is very low which leads to lower flushing efficiencies and debris remains inside the machined hole. So, at lower capacitance (below 100 pF), *SR* is higher. On the contrary, at higher values of capacitance and voltage the more energy produced, debris are carried away rapidly with dielectric bubbles from the gap resulting lower *SR*. However, at higher energy level the crater is bigger and the *SR* increases. From the interaction graph (figure 4(d)) it is evident that the optimum parameter ranges lies at voltage 90, capacitance 500 pF and speed 300 rpm. The existing Mikrotool machine does not have the facility to use 500 pF. Therefore, further experiments are needed for verification of the results modifying the Mikrotool machine adding extra capacitances.

4. Conclusions

Micro-holes were created on nonconductive ZrO_2 substrate for MEMS application by micro-EDM using assisting electrode method. The effect of input parameters such as voltage, capacitance and rotational speed of tool on surface roughness is investigated. The parameters are optimized based on the Taguchi method. Following conclusions can be drawn from this study.

1. Using adhesive copper foil as assisting electrode and copper tool electrode with –ve polarity in kerosene dielectric, micro-EDM of nonconductive ZrO_2 ceramic is done effectively (figure 2).
2. Empirical model of *SR* correlating the voltage, capacitance and rotational speed of tool is developed using Taguchi method (equation 1).
3. In the micro-EDM of nonconductive ZrO_2 ceramic, the significant parameter of *SR* is capacitance and voltage (figure 4).
4. External copper layer can be removed from the micro-EDMed ceramic surface without any damage to the holes (figure 2).
5. Further experiments are needed to investigate the effect of parameters on SR in micro-EDM of nonconductive ZrO_2 ceramic at higher capacitance values (>1000 pF).
6. Experiments for verification of optimum parameter values (V=90, C=500) is needed through the modification of the Mikrotool with extra capacitances.

5. Acknowledgements

The authors would like to thank the Ministry of Science, Technology and Innovation (MOSTI), Malaysia for financial support under Science Fund Research Project 03-01-08-SF0135

References

[1] Pieters P 2009 Versatile MEMS and mems integration technology platforms for cost effective MEMS development. In *Microelectr. and Packag. Conf., European* pp 1-5, IEEE.

[2] Rödel J, Kounga, A B, Weissenberger-Eibl M, Koch D, Bierwisch A, Rossner W and Schneider G 2009 Development of a roadmap for advanced ceramics: 2010–2025 *J. European Ceramic Society*, **29(9)** pp 1549-1560.

[3] Takahata K 2009 Micro-electro-discharge machining technologies for MEMS. *Microelectronic and mechanical systems. IN-TECH* pp 143-164.

[4] Hösel T, Müller C, and Reinecke H 2011 Spark erosive structuring of electrically nonconductive zirconia with an assisting electrode. *CIRP J. Manuf. Sci. and Tech.* **4(4)** pp 357-361.

[5] Mohri N, Fukuzawa Y, Tani T, Saito N and Furutani K 1996 Assisting electrode method for machining insulating ceramics, *CIRP Annals Manuf. Tech.* **45** pp 201-204.

[6] Sabur A, Ali M Y, Maleque M A and Khan A A 2013 Investigation of Material Removal Characteristics in EDM of Nonconductive ZrO_2 Ceramic. *Procedia Eng.* **56** pp 696-701.

[7] Williams J R, and Clarke D R 2008 Strengthening gold thin films with zirconia nanoparticles for MEMS electrical contacts. *Acta Materialia*, **56(8)** pp 1813-1819.

IOP Conf. Series: Materials Science and Engineering **53** (2013) 012091 doi:10.1088/1757-899X/53/1/012091

Optimal Throughput and Self-adaptability of Robust Real-Time IEEE 802.15.4 MAC for AMI Mesh Network

Hikma Shabani[1], Musse Mohamud Ahmed, Sheroz Khan, Shahab Ahmed Hameed and Mohamed Hadi Habaebi
Department of Electrical and Computer Engineering, Kulliyyah of Engineering, International Islamic University Malaysia, P. O. BOX. 10, Kuala Lumpur, 50728, Malaysia.

E-mail: hikmash@yahoo.fr

Abstract. A smart grid refers to a modernization of the electricity system that brings intelligence, reliability, efficiency and optimality to the power grid. To provide an automated and widely distributed energy delivery, the smart grid will be branded by a two-way flow of electricity and information system between energy suppliers and their customers. Thus, the smart grid is a power grid that integrates data communication networks which provide the collected and analysed data at all levels in real time. Therefore, the performance of communication systems is so vital for the success of smart grid. Merit to the ZigBee/IEEE802.15.4std low cost, low power, low data rate, short range, simplicity and free licensed spectrum that makes wireless sensor networks (WSNs) the most suitable wireless technology for smart grid applications. Unfortunately, almost all ZigBee channels overlap with wireless local area network (WLAN) channels, resulting in severe performance degradation due to interference. In order to improve the performance of communication systems, this paper proposes an optimal throughput and self-adaptability of ZigBee/IEEE802.15.4std for smart grid.

1. Introduction

The legacy electricity grids continuously encounter the challenges of providing reliable electric power to end-users and System breakdown could cause huge economic losses and public concerns [1]. Therefore, the Energy Independence and Security Act of 2007 gave a start for smart grid implementation in the United States [2]. The development of the smart grid provides an efficient and intelligent approach of managing energy supply and consumption in real time. The real time communication capability of smart grid will enable utilities to optimize and modernize the power grid in order to achieve its full potential.

Recently, wireless sensor networks (WSNs) have been widely recognized as a promising technology that can enhance various aspects of today's electric power systems [3]. Merit to the ZigBee/IEEE802.15.4std low cost, low power, low data rate, short range, simplicity, high reliability and free licensed spectrum that makes wireless sensor networks (WSNs) the most suitable wireless technology for smart grid applications [4].

[1] To whom any correspondence should be addressed

ZigBee/IEEE802.15.4std defines the physical layer (PHY) and Medium Access Control (MAC) sub-layer. In ZigBee physical layer, the radio shall operate at one of the following license-free bands: 868-868.6MHz in Europe, 902-928MHz in North America and 2400-2483.5MHz elsewhere worldwide. However, there is a big issue of interference between various wireless technologies operating at the frequency of 2.4GHz Industrial-Scientific-Medical (ISM) frequency band which reduces the performance of WSNs communication.

This paper develops an optimal throughput and self-adaptability of ZigBee/IEEE802.15.4std for smart grid under WLAN interference. The rest of the paper is organized as follows. Section 2 reviews the RF Link of IEEE802.15.4 PHY layer, section 3 discusses the Zigbee BER Analysis Under WLAN interference. Section 4 proposes the IEEE 802.11b/IEEE 802.15.4 Simulation Models and presents the simulation results. Finally, in section 5, the conclusion and suggestions are presented.

2. RF Link of IEEE 802.15.4 standard

IEEE802.15.4std specifies 27 channels spread across three different ISM frequency bands as presented in 'figure 1' [5].

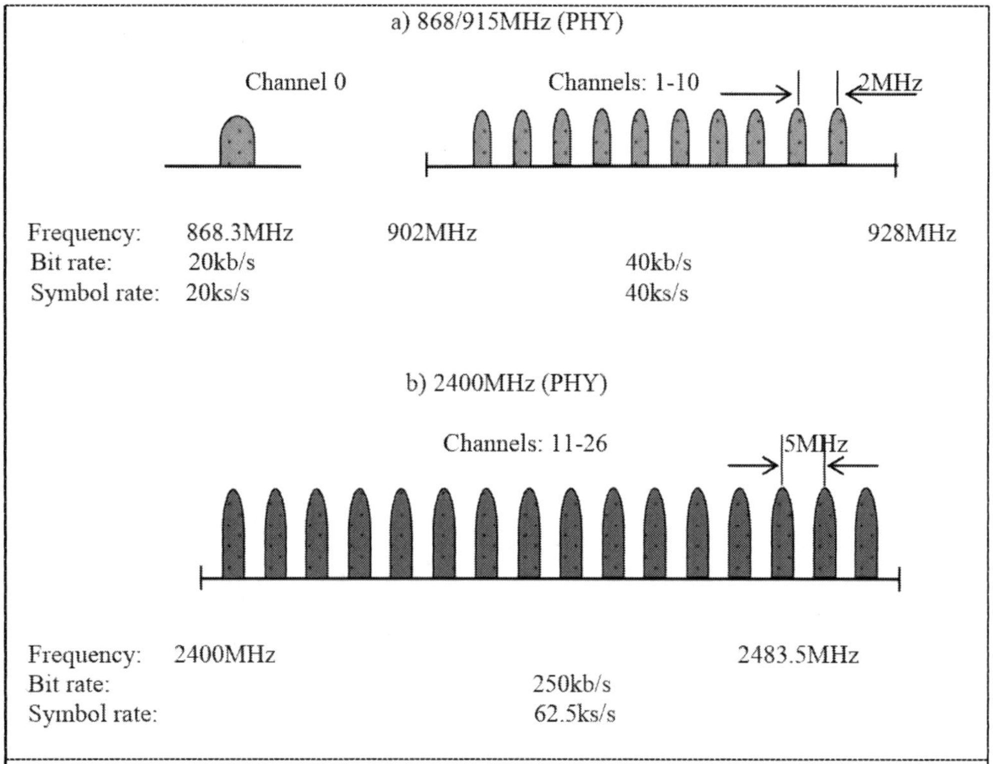

Figure 1: IEEE802.15.4std channels and frequency band [5]

Hence, the radio shall operate at one of the following available license-free ISM bands: 868-868.6MHz in Europe, 902-928MHz in North America and 2400-2483.5MHz elsewhere worldwide. Furthermore, IEEE802.15.4std uses the BPSK (Binary Phase Shift keying) modulation for both 868 and 915MHz bands, and OQPSK (Offset Quadrature Phase-Shift Keying) modulation for 2.4 GHz ISM bands. Finally, as shown in 'figure 1', ZigBee has channel 0 in the 868MHz band, 10 channels in 915MHz band, each with 2MHz of bandwidth and 16 channels in the 2400 GHz band, each with

5MHz of bandwidth and the transmission power capability of 1mW is typically specified in the standard. In order to limit the power consumption, IEEE802.15.4 standard is designed for very low duty cycle ($< 0.1\%$) and the nominal transmitter power output specified is $0.5mW\ (-3dBm)$ [6]. However, the output power can be increased through external amplifiers to whatever the regional regulatory limits are. The nominal receiver sensitivity is specified by Packet Error Rate (PER). The specification requires 1% PER at -85dBm receive power level for the 2400 MHz band and -92dBm for the sub-GHz bands (as measured at the chip's antenna terminals). This represents a receiver with a noise figure substantially worse than 20dB, though most radios in production are between 7 and 9 dB better than this.

3. Zigbee BER Analysis Under WLAN

This developed bit error rate (BER) analysis model is the extension of the model developed in [7] from which the interference and noise are included. The PHY of IEEE 802.15.4 at 2.4 GHz uses OQPSK modulation. For an additive white Gaussian noise (AWGN) channel, the BER can be calculated by the following equation [8]:

$$\mathrm{BER} = Q\left(\sqrt{\frac{2E_b}{No}}\right) \tag{1}$$

Where E_b/No is the normalized signal-to-noise ratio (SNR) and $Q(x)$ is the Q-function of Gaussian distribution:

$$Q(x) = \frac{1}{2\pi}\int_x^{\infty} \exp\left(-\frac{u^2}{2}\right) du \tag{2}$$

When a Zigbee channel overlaps with a WLAN channel, WLAN signal can be considered as partial band jamming noise for the Zigbee signal [9] and the SNR is replaced by signal-to-interference-plus-noise ratio (SINR) which can be defined as

$$\mathrm{SINR} = \frac{P_{signal}}{P_{noise} + P_{interference}} \tag{3}$$

Where P_{signal} is the power of the desired signal at Zigbee receiver, P_{noise} is the noise power, and $P_{interference}$ is the received interference power from WLAN signal at Zigbee receiver.
The path loss model represents the power loss between transmitter and receiver, and can therefore be used in conjunction with the transmission power to enable the calculation of P_{signal} and $P_{interference}$.
The maximum transmission power of Zigbee is defined as 1 mw (0 dBm). Considering that Zigbee andWiFi are most frequently deployed in the indoor environments, a simplified indoor path loss model is adopted in this paper [7]. For self-adaptability, as suggested in [10], the frequency dependency is incorporated in the developed indoor path loss model as Zigbee and WLAN are most frequently deployed in the indoor environments

$$L_p(d,f) = \begin{cases} 20\log_{10}\left(\frac{4\pi d}{\lambda}\right) + 20\log_{10}\left(\frac{f}{f_0}\right), d \le d_0 \\ 20\log_{10}\left(\frac{4\pi d}{\lambda}\right) + 20\log_{10}\left(\frac{f}{f_0}\right) + 10n\log_{10}\left(\frac{d}{d_0}\right), d > d_0 \end{cases} \tag{4}$$

Where d_0 is a break point. And n and d_0 are set to 3.3 and $8m$ respectively [11].

Therefore, considering that the power spectrum of IEEE 802.11b is 11 times wider than Zigbee ('figure 2') and is not uniformly distributed, in-band interference power of IEEE 802.11 cannot be simply calculated by dividing by 11 [12]. An amendment parameter of in-band power factor r is added to $P_{interference}$. Therefore, (3) is modified to:

$$\mathrm{SINR} = \frac{P_{signal}}{P_{noise} + r.P_{interference}} \tag{5}$$

To obtain the power factor r, the power spectral density of the IEEE 802.11b and offset frequency between the central frequency of Zigbee and WLANi are considered. Since the power is concentrated around the central frequency, r increases as the offset frequency decreases.

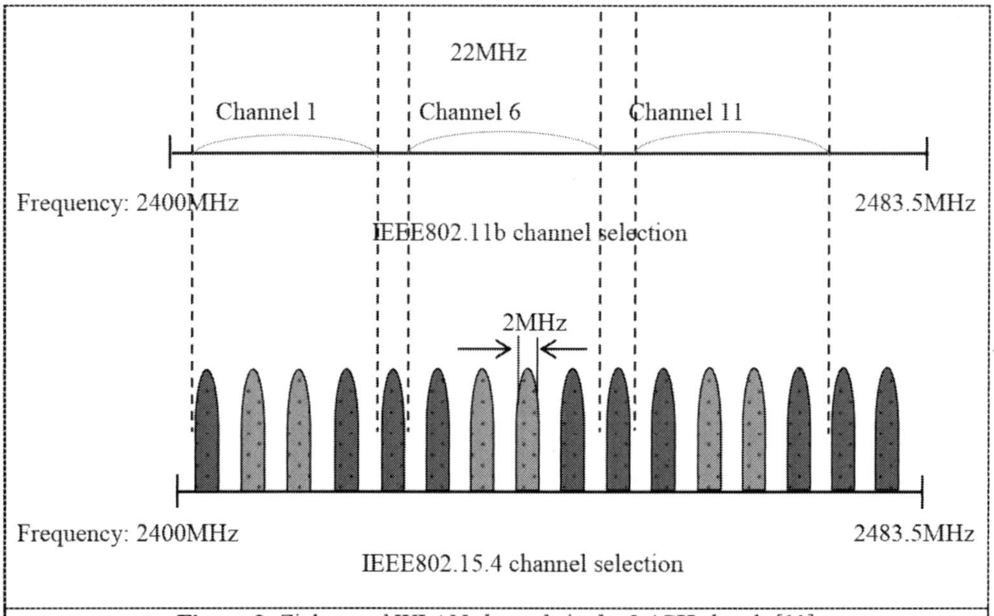

Figure 2: Zigbee and WLAN channels in the 2.4GHz bands [11]

4. IEEE 802.11b/IEEE 802.15.4 Simulation Model and Results

According to theoretical model the simulation model based on the IEEE 802.15.4 standard is shown in 'figure 3'.

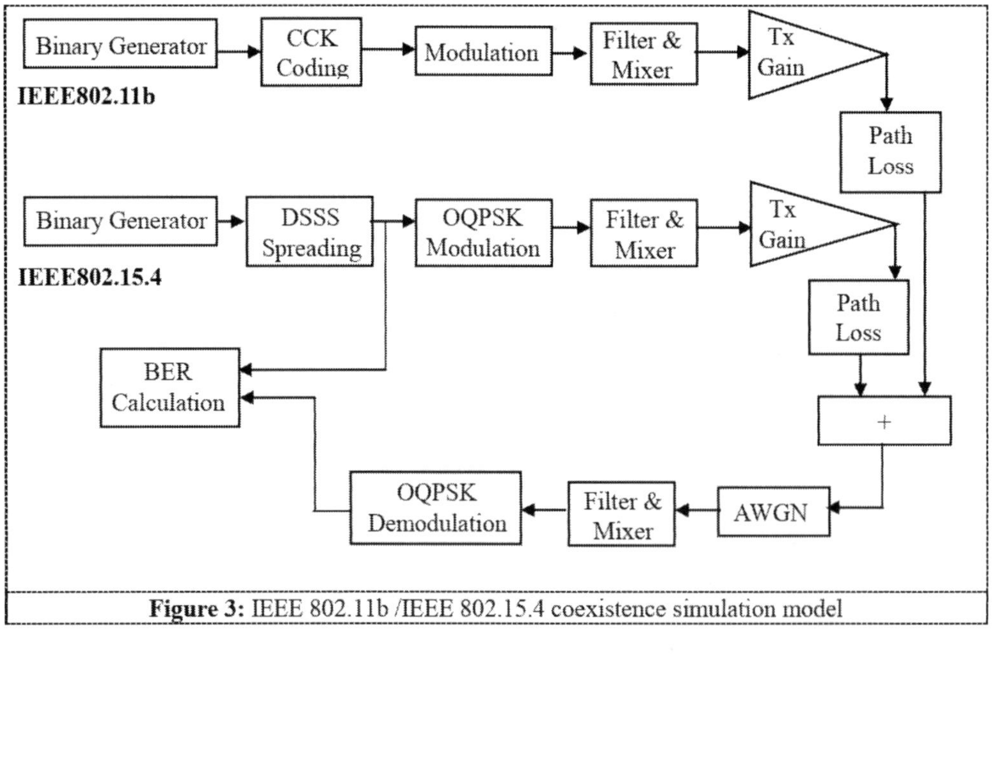

Figure 3: IEEE 802.11b /IEEE 802.15.4 coexistence simulation model

In accordance with IEEE 802.15.4 standards document, every four bits are mapped into a symbol and each symbol spreads to a 32-chip almost orthogonal PN sequence, thus a spreading table is set in a spreading block. Data is packed into frames, with a maximum frame size of 128 bytes as defined in the standard. The transmission rate is 250 kbps at 2.4 GHz for Zigbee, while 11 Mbps for WLAN. The IEEE 802.11b simulation module provided in NS2 is utilized. The IEEE 802.15.4 and IEEE 802.11b signals are added together before being passed through AWGN channel. Both signals must be sampled and filtered at the same sampling rate [13]. The frequency band for both simulation systems was set from - 44 to +44 MHz to satisfy the Shannon theorem. The BER is calculated based on minimum Hamming distance between data before modulation and after demodulation [8]. Measurements were taken in a screen room, enabling the elimination of all external signals and interference.

Theoretical analysis and simulation of BER are shown in 'figure 4'. The solid line represents theoretical values while the dotted lines represent the values obtained via simulation. Except for a few channels that are far away from the WLAN central frequency, most of channels overlapped with the WLAN channels have 2 MHz, 3 MHz, 7 MHz, and 8 MHz offsets from the WLAN channel frequency. Therefore, the simulations were performed in these four scenarios. From both the simulation and theoretical results, we find that the BER drop drastically as the offset frequency increases. For the same offset frequency channel, the BER decreases when the separation distance increases. BER is higher when the offset frequency is 2 MHz and 3 MHz in the simulation compared to theoretic results; when the offset frequency is 7 MHz and 8 MHz, the BER is lower than theoretical result. That is because the frequency band of the IEEE 802.11b simulation model is narrower than the theoretical one, with more power concentrated on the effective band frequency. Both graphs prove that most interference power is around the central frequency of WLAN. "Safe Distance" and "Safe Offset Frequency" are two critical parameters, which guide the Zigbee deployment in order to mitigate the WLAN interference. If the offset frequency is less than 2 MHz, the distance needs at least 8 m to efficiently minimize the effect of the IEEE 802.11b. If the offset frequency is larger than 8 MHz, safe distance can be decreased to 2 m.

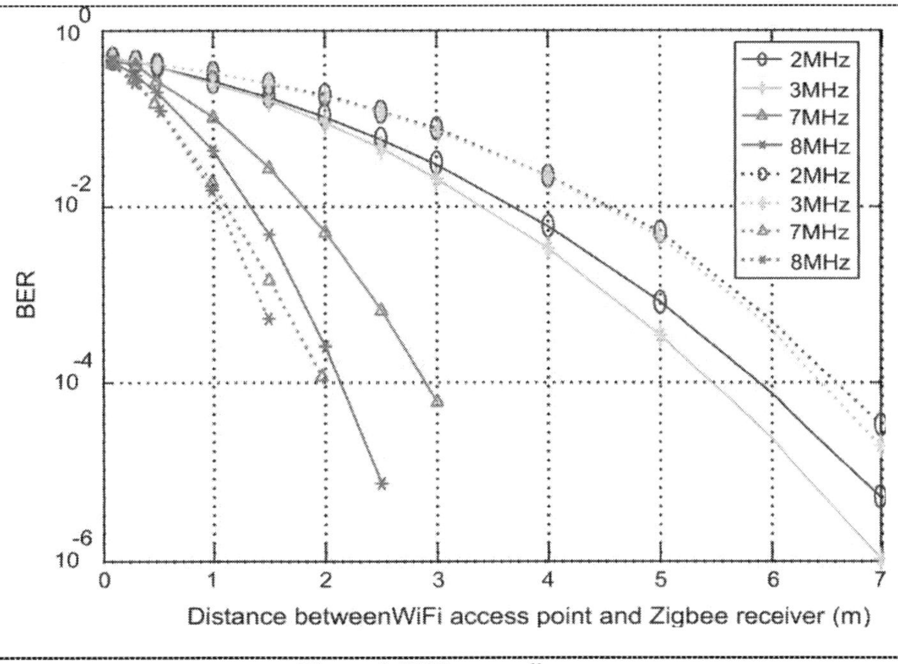

Figure 4: BER versus distance

5. Conclusion

In this paper, Zigbee performance under WLAN interference for AMI mesh network was evaluated. A theoretical model has been introduced, followed by a corresponding simulation model which completely reflects the Zigbee and WLAN coexistence features via NS2. Both analysis and simulation results show that Zigbee may be severely interfered by WLAN and that a "Safe Distance" and "Safe Offset Frequency" can be identified to guide Zigbee deployment. It is shown that 8 m between Zigbee and WLAN is a "safe" distance which can guarantee the reliable Zigbee performance regardless of the offset frequency, while 8 MHz is a "safe" offset frequency even when the distance is just 2 m. These results have been verified by means of empirical analysis and experimentation. It was shown that in general, Zigbee provides satisfactory performance when the WLAN interference is not significant. In the event of significant WLAN interference, the proposed model provides an effective and efficient means of providing reliable data service.

Acknowledgment

The authors wish to thank the International Islamic University Malaysia and the Renewable Energy Research Group (RERG), Faculty of Engineering.

Appendix A. ZIGBEE FREQUENCY BANDS AND DATA RATES [8]

PHY (MHz)	Frequency band (MHz)	Channel Number	Spread parameters		Data parameters	
			Chip rate (Kchip/s)	Modulation	Bit rate (Kb/s)	Symbol
868	868-868.6	0	300	BPSK	20	Binary
915	902-928	1-10	600	BPSK	40	Binary
2450	2400-2483.5	11-26	2000	OQPSK	250	16-ary orthogonal

Appendix B. IEEE 802.11b DATA RATES SPECIFICATIONS [7]

Data rate	Code length	Modulation	Symbol rate	Bits/symbol	System
1 Mbit/s	11 (Barker C)	DBPSK	1	1	DSSS
2 Mbit/s	11 (Barker C)	DBPSK	1	2	DSSS
5.5 Mbit/s	4 (CCK)	DBPSK	1.375	4	HR/DSSS
11 Mbit/s	8(CCK)	DBPSK	1.375	8	HR/DSSS

6. References

[1] Joo S K, Kim J C and Liu C C 2007 Empirical analysis of the impact of 2003 blackout on security values of US utilities and electrical equipment manufacturing firms *IEEE Trans. Power Systems* **vol. 22 no. 3.** pp. **10**12-18

[2] Erol-Kantarci M and Moufta H T 2011 Wireless Sensor Networks for Cost-Efficient Residential Energy Management in the Smart grid *IEEE Trans. Smart Grid* **vol. 2 no. 2.** pp. **3**14-25

[3] Gungor V C, Akan O B and Akyildiz 2008 A Real-Time and Reliable Transport $(RT)_2$ Protocol for Wireless Sensor and Actor Networks *IEEE/ACM Trans. Network* **vol. 16 no. 2.** pp. **3**59-70

[4] Callaway E et al. 2002 Home Networking with IEEE 802.15.4: A developing Standard for Low-Rate Wireless Personal Area Networks *IEEE Communications Magazine* **vol. 40 no. 8.** pp. **7**0-7

[5] IEEE std 802.15.4-2006 Part 15.4: Wireless Medium Access Control (MAC) and Physical Layer (PHY) Specifications for Low-Rate Wireless Personal Area Networks (WPANs) *IEEE Standard for Information Technology* pp. 1-32

[6] Gandotra N, Bijalwan V and Panwar M 2012 Coexistence Model of ZigBee & IEEE 802.11b (WLAN) in Ubiquitous Network Environment *International J. Adv. Research in Computer Engineering & Technology (IJARCET)* **vol. 1 no. 4.** pp. **6**80-84

[7] Shin S Y, Park H S, Choi S and Kwon W H 2007 Packet Error Rate Analysis of ZigBee Under WLAN and Bluetooth Interferences *IEEE Trans. Wireless Comm.* **vol. 6 no. 8.** pp. **28**25-30

[8] Peizhong Y, Iwayemi A and Zhou C 2011 Developing ZigBee Deployment Guideline Under WiFi Interference for Smart Grid Applications *IEEE Trans. Smart Grid* **vol. 2 no. 1.** pp. **1**10-20

[9] Peterson R L, Borth D E and Ziemer R E 1995 *An Introduction to Spread-Spectrum Communications* (Upper Saddle River, NJ: Prentice-Hall)

[10] Pagani P and Pajusco P 2006 Modeling the Space-and Time-Variant Ultra-Wideband Propagation Channel *IEEE Int. Conf. on Ultra-Wideband* pp. 201-206

[11] Yoon D G, Shin S Y, Kwon W H and Park H S 2006 Packet Error Rate Analysis of IEEE 802.11b under IEEE 802.15.4 Interference *Proc. 63rd IEEE Veh. Tech. Conf* pp. **11**86-90

[12] Shin S Y, Park H S and Kwon W H 2007 Mutual Interference Analysis of IEEE 802.15.4 and IEEE 802.11b *International J. Computer Networks* **vol. 51 no. 12.** pp. **33**38-53

[13] Mikulka J and Hanus S 2007 Bluetooth and IEEE 802.11b/g Coexistence Simulation *RadioEngineering* **vol. 17 no. 3.** pp. 66–73

Challenges of Integrating Unmanned Aerial Vehicles In Civil Application

B. M. Eid, J. Chebil [1], F. Albatsh and W. F. Faris
*Department of Electrical and Computer Engineering, Faculty of Engineering, International Islamic University Malaysia (IIUM), 53100 Gombak, Kuala Lumpur, Malaysia

E-mail: jalel@iium.edu.my

Abstract. Unmanned Aerial Vehicle (UAV) has evolved rapidly over the past decade. There have been an increased number of studies aiming at improving UAV and in its use for different civil applications. This paper highlights the fundamentals of UAV system and examines the challenges related with the major components such as motors, drives, power systems, communication systems and image processing tools and equipment.

1. Introduction

An Unmanned Aerial Vehicle (UAV) is normally an aircraft that launches and flies without a human onboard. The crew is totally replaced by computer system that is pre-programmed before the mission starts or distantly controlled by human. Conventionally, UAVs were firstly designed for military applications, but in recent times, there is large potential for civil purposes such as forest fire fighting, environmental monitoring, communication industry, agricultural monitoring, natural vegetation classification, erosion monitoring, topographic modelling, feature identification, temporal mapping of landscape dynamics, bushfires detection, search and rescue missions. UAVs have many characteristics that are different from conventional aircraft. UAVs system includes many subsystems that are important to achieve their missions. The system consists of the aircraft pay load, ground control station, communication equipments, power subsystem, launch and recovery subsystems. The payload is based on the mission of the UAV. It could be but not limited to cameras, motors, drives, batteries, water, monitoring system and communication relays. This paper is addressing the challenges that facing UAV implementation from different perspectives. Next section presents the two types of motors used in launching and controlling UAV missions. Section 3 discusses the power system of UAVs. The communication system and its challenges are highlighted in Section 4. Section 5 investigates the image processing techniques. Finally, the paper is concluded in Section 6.

2. Motors and drives

The UAV propulsion system comprises of an electric motor, a speed controller, and a propeller [1]. Electric Motor Brushless DC (BLDC) motor has been chosen as the electric propulsion motors for the UAV due to many reasons. It has better performance than conventional permanent magnet dc motor in terms of power capacity, speed, torque, efficiency, and long operating lives [2]. The modelling of BLDC motor is similar to conventional permanent magnet dc motors. The main difference is that the

[1] To whom any correspondence should be addressed.

BLDC motor is commutated externally by power electronic switches rather than by the brushes and commutator which are built into a standard dc motor. This produces some electronic complexity, as an external controller is required to control the rotor position in real time implementation. In general, modelling the performance of BLDC motors involves current-voltage state-space equations for each motor winding combined with system as an ideal dc machine. In steady state response, the modelling of the BLDC motor considers as one of the challenge to control the BLDC due to the complexity of a steady state modelling [2].

Speed control of BLDC motor is achieved by standard radio control by receiving control inputs from the on-board radio control receiver [3].The motor speed can be measured by using an optical sensor [1]. Achieving a required velocity in different climate conditions is considered one of the real challenges in practical application [2].

A servo motor is an electric motor coupled with a sensor to get a position feedback. In the case of the UAV, the main function of a servo motor is to control the ailerons and rudder on the plane. The principle work of the servo motor in UAV is based on a comparison between a measured attitude position and a reference attitude. The controller will ensure the error signal to be zero by sending the required commands to the motor to adjust accordingly.

The right aileron motor controls the position of the right wing aileron, and the left aileron motor controls the position of the left wing aileron. The elevator motor controls the altitude of the UAV. The rudder motor controls the position of the rudder. The motor controller or electric speed control (ESC) is used for the throttle [4].

To run UAV motors properly, a stable power source should be provided in order to supply all UAV motors with required power along with less interruption and high stability. In the following section, the challenges arise in power system will be investigated in order to achieve the optimal power system for UAV operation.

3. Challenges of power system

Most UAV's use batteries as the only source for power supply, because the efficiency of internal combustion engine and gas turbine are low for small scale UAV's [1]. However, some new researches used additional power supplies such as; photovoltaic and fuel cell as ancillary power supply. Challenges face electrical power systems in UAVs are much more critical than in other applications [5, 6]. Topics below studied challenges faces UAV power system.

3.1. Batteries types and capacities

Batteries are energy storage devices which convert chemical energy to electrical energy and vice versa [7, 8]. There are several rechargeable batteries available on the market right now such as; lithium ion (Li-ion), lithium polymer (Li-Poly), lead-acid, nickel-cadmium (NiCd)and nickel-metal hydride (NiMH), each one of them has different characteristics[8]. However, Li-ion batteries commonly used in many applications like; consumer electronics, power tools, electric vehicles and even in space applications [9]. The reason behind this spread is because Li-ion can overcome many of battery operation challenges, such as:

- Energy density, as it has a very high energy density (see Table 1), that because its electrode made of lightweight lithium and carbon.
- Self-discharge, as it has low self-discharge rate.
- Life cycle, as it has long cycle life and it can achieve hundreds of charge and discharge cycles without significant capacity decrease.

With these important advantages Li-ion batteries were chosen as the power source in many papers [1, 8, 10, 11].However, Li- Poly battery packs have the same chemical processes as Li- ion battery, except that the electrolyte is in gel form.

Table 1, Energy density of common batteries type.

Storage material	Kw.h / Kg
Lithium-ion battery	0.875
Alkaline battery	0.67
Nickel-metal hydride battery	0.288
Lead-acid battery	0.17

3.2. Ancillary power system

3.2.1. Fuel cells

Fuel cells (FC) proposed in many papers [1, 2, 12-14] as ancillary or alternative power supply for the existing battery, because of the numerous advantages that FC have. Ancillary power system has to have many features to operate UAV efficiently, which is available in FC such as; good energy density, high thermal efficiency, silent, reliable, safe, and eco-friendly. Energy density defined as the energy capacity per unit weight or volume [1], so FCs energy density is better than Lithium-ion battery and Gasoline which shown in Table 2. The FC has a high thermal efficiency due to its electrochemical reaction rather than a combustion reaction, in the conversion path from chemical energy to electric energy.

Current fuel cell aircrafts can be classified by fuel cell types and hydrogen storage methods. The world's first FC aircraft used liquefied hydrogen to operate the system [1]. Polymer electrolyte membrane fuel cells (PEMFC) used gaseous hydrogen as fuel. Although, it has low operating temperature and relatively high power density, this method consider bulky and heavy since gaseous hydrogen has a large volume. Solid oxide fuel cells (SOFC) were also applied in UAV which uses hydrocarbon fuel at high temperatures, producing a higher power density comparing with PEMFC [15, 16]. Compressed hydrogen gas is commonly used in spite of few drawbacks it has in safety (as it store the gas at high pressure), storage density (which it is low) and in recharging. Sodium borohydride (NaBH4) or sodium tetrahydridoborate was used as chemical hydrogen storage [17] as an aqueous (watery) solution.

Table 2. Energy density of common storage forms

Storage material	Energy type	Kw.h / Kg
Lithium-ion battery	Electrochemical	0.875
Hydrogen (compressed at 70 MPa)	Chemical	34.44
Gasoline (petrol) / Diesel	Chemical	12.88

3.2.2. Photovoltaic (PV)

Many researchers used PV cells as UAV power source [11, 18, 19]. However there are many important parameters that can influence the output power and the efficiency of PV Cells. First, the designing of maximum power point tracker (MPPT) have to be efficient and lightweight to maintain voltage and current at maximum power which achieve high endurance flight for UAV. Second, the wind direction plays key role in reducing power required to flying when wind blows from the opposite side of the flight. Wind also can reduce the temperature of the PV cells which increase the efficiency.

Third, during the flight the temperature of the PV cells reduces and causes efficiency improvement. Geographical location and flight orientation are the other discussed factors which influence received radiation and power generation during the flight.

In [11] mono-crystalline solar cells are use as the power source. The PC panels divided into three panels which is right wing, left wing and fuselage (plane body) panel. Under a standard test condition, maximum power of 57.2 W can be generated. The maximum current and voltage point are 1.91 A and 30 V respectively. The electrical characteristics of the panels are listed Table 3. The right wing is almost same as left wing.

Table 3. Electrical characteristics of each panel.

	Typical peak power	Voltage at peak power	Current at peak power	Short circuit current	Open circuit voltage
Left wing panel	23.23 W	30.08 V	0.772 A	0.839 A	37.66 V
Fuselage panel	9.6 W	29.9 V	0.323 A	0.341 A	37.84 V

3.3. Electrical faults

As illustrated in [5, 6] the electrical faults in UAV electrical power system is much more critical than in other application. The electrical fault can damage for filter capacitors, switched and any sensitive components in the network [6, 20, 21]. So a good circuit breaker is necessary to maintain power system safe [6].

3.4. Weight

Light weight has to be one of the designer objectives when designing UAVs [6, 22]. So the total electrical power system weight must be minimized in order to capitalize the efficiency of the power system.

In next section, a comprehensive study on challenges of communication system in UAV will be investigated in order to achieve the safe and efficient control with UAV.

4. Challenges of communication system

Communication system play an important role in UAV since it is required to operate with less human efforts. The communication system of UAV could be divided to three different parts: the ground control station (GCS), the transceiver on the aircraft and finally the radio communication link between the control station and the aircraft [23]. This section addresses some challenges that are facing UAV communication system.

4.1. Ground control station (GCS)

The control station is normally the centre where the operation and mission of the UAV are managed and monitored. Moreover, it considers as the interface between the human and aircraft. The operator sends commands to the UAV via the GCS during the uplink operation to control and guide the UAV in its mission. Meanwhile, the GCS receives data from the UAV during the downlink and forwards it to the operator. The data may include position information, mission status and other important information [23]. Three types of ground control station were proposed. The tradition control station where many devices are used to monitor and control UAV mission. More advanced station called mini or portable station is proposed in [24]. The third type is what is called unmanned ground vehicle (UGV) [25]. The key element in GCS is the computer and the monitor software. Many researches were conducted to develop the different types of GCS. Kang and Yuan [24] proposed a software of multi layer flight model used in GCS to monitor the UAV path. The problem with the UAV software is the compatibility since one type of GCS software is only suitable for a specific UAV system unless interface standard is used. Challenges of minimizing the human efforts in the ground station such as mission planning instead of path planning and oral command control were discussed in [26]. UAVs should have the ability to process the commands and convert them to a useful mission. More research is required on multi platform mission and changing the mission profile during the flight. Unmanned ground platform that lunch, monitor and refuel the UAV was presented in [25]. The main challenges in this type of GCS are how to manage the landing process accuracy of the UAV and the self refuel of UAV in addition to support more than one type of UAV. In summary, a low cost and autonomy platform is still a challenging issue [27].

4.2. The communication link

The need to establish high performance link requires the characterization of the radio communication channel between the UAV and the ground station. In addition, Small grazing angle between the transmitter and the receiver cause strong fading and multipath interference. Deep signal losses may also occurs when the UAV is maneuvering the antenna orientation changes [28, 29]. Statistical prediction models for multipath, attenuation and scintillation is required and it is an interesting research area. Lemorton *et al* [30] shows that the available models for terrestrial links and earth to satellite links is not suitable enough to be used for UAV, they added that some modification is required specially for low elevation angle. The communication links discussed in the literature could be divided to three types of technology: radio communication [29, 31], low earth-satellite [31] and free space optics (FSO) [32-37]. The first two links have long range communication (up to 30km). However, the bandwidth is limited to hundreds of kbps. Using Orthogonal Frequency Division Multiplex (OFDM) in radio communication based transceiver increases the data rate up to some Mbps. But at the same time, the appearance of Inter-Carrier Interference (ICI) due to high speed of the UAV reduces the performance of the communication system [29]. The third technique which is FSO has a short communication link (Line of site and up to 8km) and it is highly affected by atmosphere [33]. On the other hand, it has high bandwidth with 2.5 Gbps [34] and up to 10 Gbps experimentally [35]. In brief, FOS technique could be for air to air communication between UAV networks. Meanwhile the radio communication is still required for long range communication between the UAV and the GCS with many considerations. In addition to the long range and the UAV speed, statistical prediction models for radio channel are open research trends to reduce the effects and overcome the path losses.

4.3. Aircraft system

The communication system located at UAV consists of an antennas and air data terminal (ADT) which includes transceiver, modems and processers. This system is important to communicate with the GCS or even with other UAVs [23]. UAV antenna requires special consideration such as smaller size, lighter weight and larger gain. Another restriction is the impossibility of mounting turntable or scanning antenna array. These facts bring the necessity to develop special antennas and select the optimum location of the antennas on the UAV. In [28] A MIMO wireless channels between UAV and ground station has been assessed experimentally. They found that the effect of multipath interference can be alleviated by the use of directive antennas on the ground, and that multiple antennas on the UAV provide a more robust radio channel in front of the antenna changes of orientation when the UAV is maneuvering. Boev designed and fabricated a flat toroidal antenna with two resonant frequencies [38]. Adaptive vertical array antenna technique was proposed by Park et al [39] to overcome altitude limitation of the UAV caused by restricted antenna coverage of the ground station.

5. Challenges of image processing

UAVs are traditionally used for military purposes. However, its civil applications are gaining momentum since progress within the fields of digital sensors, navigational equipment, and small-scale aircraft have all reduced the cost of the fundamental components of UAVs [40]. These applications can include: agricultural monitoring, natural vegetation classification, erosion monitoring, topographic modeling, feature identification, temporal mapping of landscape dynamics, bushfires detection, search and rescue mission. Compared with more traditional aircraft or satellite based platforms, the UAV is distinguished by its unique characteristics of the captured data. Its low operating altitude allows for the generation of ultra-high spatial resolution data over relatively small spatial extents [41].

There have been many UAV studies related to remote sensing. Berni et al. [42], Dunford et al. [41], Hunt et al. [43], Laliberte et al. [44], Lelong [45], and Xiang and Tian [46] looked at multispectral UAV imagery for both agricultural monitoring and natural vegetation classification. These studies highlighted some of the benefits and challenges associated with the use of UAV. For example, Berni et al. [42] have demonstrated that results obtained with a low-cost UAV system for agricultural applications yielded comparable estimations, if not better, than those obtained by traditional manned airborne sensors. The main limitations encountered by their UAV system were the endurance (20 min) and the low flight speed (30 km/h), limiting the productivity to 70 hectares per flight.

Lelong et. al. [45] designed a cost-effective multispectral sensor based on commercially available digital cameras, adapted with relevant filters to perform six aerial acquisitions of wheat crop micro plots during the growing season, at very low altitude. Several vegetation indexes relevant for vegetation analyses were derived. The authors sought relationships between these indexes and field-measured biophysical parameters, both generic and date-specific. Due to a high amount of noise in the data, it was not possible to obtain a more accurate model for each date independently. A validation protocol showed that it is expected to have a precision level of 15% in the biophysical parameters estimation while using these relationships. More studies are needed to improve the precision. The error can be reduced if both vibration control and new image algorithms are developed.

Xiang and Tian [46] have developed an improved remote sensing system based on an autonomous UAV. An experiment using the developed UAV system to monitor turf grass glyphosate application demonstrated the system. The results proved that the UAV system provides a flexible and reliable method of sensing agricultural field with high spatial and temporal resolution of image data.

Zhao et al. [47] and Lin et al. [48] used UAV-based LiDAR for topographic modelling and feature identification. UAVs were used for stereo-image 3D landscape modelling by Stefanik et al. [49]. Thermal UAV applications for emergency services including bushfires and search and rescue were presented by Rudol and Doherty [50], Hinkley and Zajkowski [51] and Pastor et al. [52]. Sung [53] suggests an image processing algorithm that allows for UAVs to detect and track approaching aircraft in real time using a vision sensor mounted on a UAV. This algorithm computes image homography from an image sequence to detect any moving objects in the view. Then a Probabilistic Multi-Hypothesis Tracking (PMHT) method is applied to track and manage the detected objects. Finally it is again used to determine whether an aerial vehicle is in a potential collision course or not.

6. Conclusion

Some of the challenges of integrating UAVs in civil applications have been discussed in this paper. The factors that curb the performance of UAV includes but not limited to: UAV mission, weather, weight, availability of ancillary power system, distance between UAV and GCS, task duration, mission path and UAV speed. Nowadays, research works are focusing on: In BLDC motor as rotor position is needed to be probably located we have to improve the feedback control during real time operation Weather can cause a big disturbance for UAV velocity so future works have to insure a stable fly with a constant velocity regardless the weather. The development of the ancillary renewable power system which can operate the UAV for longer period is one of the most important challenges for UAV power systems. Low cost and autonomy platform are still challenging issues. Moreover, statistical prediction models for radio channel are open research trends to reduce the effects and overcome the path losses.

7. References

[1] Taegyu K. and Kwon S 2012 Design and development of a fuel cell-powered small unmanned aircraft *International Journal of Hydrogen Energy* 37.1 pp. 615-622.

[2] Lindahl P., Moog E. and Shaw S. R. 2012 Simulation, design, and validation of an UAV SOFC propulsion system *IEEE Transactions on Aerospace and Electronic Systems* 48.3 pp. 2582-2593.

[3] Meyer J. et al. 2007 Design considerations for a low altitude long endurance solar powered unmanned aerial vehicle *AFRICON 2007.IEEE*.

[4] Gao, Xian-Zhong, et al. 2013 Energy management strategy for solar-powered high-altitude long- endurance aircraft *Energy Conversion and Management* 70 pp. 20-30.

[5] Board N. S. 2005 Autonomous vehicles in support of naval operations *National Research Council* Washington DC.

[6] Fletcher S. D. A., Norman P. J., Galloway S. J. and G. M. Burt 2011 Determination of protection system requirements for dc unmanned aerial vehicle electrical power networks for enhanced capability and survivability *Electrical Systems in Transportation IET* vol. **1** pp. 137-147.

[7] Huggins R. A. 2009 *Advanced batteries: materials science aspects* Springer.

[8] Saha B. et al. 2011 Battery health management system for electric UAVs *in Aerospace Conference, 2011 IEEE* pp.1-9

[9] Hartmann R. L. 2008 An aging model for lithium-ion cells *University of Akron*.

[10] Wang X., Yadav V. and Balakrishnan S. 2007 Cooperative UAV formation flying with obstacle/collision avoidance *Control Systems Technology, IEEE Transactions on* vol. **15** pp. 672-679.

[11] Shiau J.-K., Ma D.-M., Yang P.-Y., Wang G.-F. and Gong J. H. 2009 Design of a solar power management system for an experimental UAV *Aerospace and Electronic Systems, IEEE Transactions on*, vol. 45, pp. 1350-1360.

[12] Kim K., Kim T., Lee K. and Kwon S. 2011 Fuel cell system with sodium borohydride as hydrogen source for unmanned aerial vehicles *Journal of Power Sources*, vol. **196** pp. 9069-9075.

[13] Fuel cell/battery hybrid UAV takes off in Taiwan *Fuel Cells Bulletin*, vol. **2010**, pp. 4-5.

[14] Ollero A. et al. 2005 Multiple eyes in the skies: architecture and perception issues in the COMETS unmanned air vehicles project *Robotics & Automation Magazine, IEEE* vol. **12**, pp. 46-57.

[15] Kohout L. L. and Schmitz P. C. 2003 Fuel cell propulsion systems for an all-electric personal air vehicle," *AIAA Paper*, vol. **2867**.

[16] Keefer B., Babicki M. and Kirby M. 2004 Solid oxide fuel cell systems *ed: WO Patent 2*,004,030,130.

[17] Kim T. 2011 Hydrogen generation from sodium borohydride using microreactor for micro fuel cells *International Journal of Hydrogen Energy* vol. **36** pp. 1404-1410.

[18] Sanguino M. and Gonzalez R. 2013 Smart host microcontroller for optimal battery charging in a solar-powered robotic vehicle.

[19] Fazelpour F., Vafaeipour M., Rahbari O. and Shirmohammadi R. 2013 Considerable parameters of using PV cells for solar-powered aircrafts *Renewable and Sustainable Energy Reviews* vol. **22** pp. 81-91.

[20] Baran M. E. and Mahajan N. R. 2007 Overcurrent protection on voltage-source-converter-based multiterminal DC distribution systems *Power Delivery, IEEE Transactions on* vol. **22** pp. 406-412.

[21] Xu Z., Zhang B., Sirisukprasert S., Zhou X. and Huang A. Q. 2002 The emitter turn-off thyristor-based DC circuit breaker," in *Power Engineering Society Winter Meeting, 2002. IEEE* pp. 288-293.

[22] Rosero J., Ortega J., Aldabas E. and Romeral L. 2007 Moving towards a more electric aircraft *Aerospace and Electronic Systems Magazine IEEE* vol. **22** pp. 3-9.
[23] Austin, R. 2010 *Unmanned Aircraft Systems UAVS Design, Development and Deployment* John Wiley and Sons Ltd. Publication.
[24] Kang Y. and Yuan M. 2009 Software design for mini-type ground control station of UAV. Paper *presented at the ICEMI 2009 - Proceedings of 9th International Conference on Electronic Measurement and Instruments*, 4737-4740.
[25] Wills M., Burmeister A., Nelson T., Denewiler, T. and Mullens K. 2006 The development of a UGV-mounted automated refueling system for VTOL UAVs. *In Proceedings of SPIE - the International Society for Optical Engineering*, 6230
[26] Ferry M., Zooming Tu, Stephens L. and Prickett G. 2007 Towards true UAV autonomy *Information Decision and Control* vol. no. pp.170-175.
[27] Rincon, O. H. & Durán, C. E. A. 2013 Xpider: design and development of a low cost VTOL UAV platform. *International Journal of Advanced Robotic Systems.*
[28] Romeu J., Aguasca A., Alonso J., Blanch S. and Martins R. R. 2010 Small UAV radiocommunication channel characterization *Antennas and Propagation (EuCAP), 2010 Proceedings of the Fourth European Conference on* vol. no. pp.1,5, 12-16.
[29] Zhiqiang Wu, Kumar H., Davari A. 2005 Performance evaluation of OFDM transmission in UAV wireless communication *System Theory 2005. SSST '05. Proceedings of the Thirty-Seventh Southeastern Symposium on*, vol., no., pp.6,10.
[30] Lemorton J. et al. 2011 Review of prediction methods for low-elevation aerospace systems and new achievements *Antennas and Propagation (EUCAP), Proceedings of the 5th European Conference on*, vol., no., pp.3393,3397.
[31] Haas E. 2002 Aeronautical channel modeling," *Vehicular Technology, IEEE Transactions on*, vol.**51**, no.2, pp.254-264.
[32] Carrasco-Casado A., Vergaz R., Sanchez-Pena J. M., Oton E., Geday M. A., Oton J. M. 2011 Low- impact air-to-ground free-space optical communication system design and first results," *Space Optical Systems and Applications (ICSOS), 2011 International Conference on* , vol., no., pp.109,112.
[33] Muhammad S. S., Plank T., Leitgeb E., Friedl A., Zettl K., Javornik T., Schmitt N. 2008 Challenges in establishing free space optical communications between flying vehicles *Communication Systems, Networks and Digital Signal Processing. CNSDSP 2008. 6th International Symposium on* , vol., no., pp.82,86.
[34] Ortiz G. G., Lee S., Monacos S., Wright M., and Biswas A. 2005 Design and development of a robust ATP subsystem for the Altair UAV-to-Ground Lasercomm 2.5 Gbps Demonstration *In Free Space Laser Communications XV Proc. of SPIE* volume **4975**, SPIE Bellingham.
[35] Sova R. M. et al. 2006 80 Gbps free space optical communication demonstration between an aerostat and a ground terminal *in Free Space Laser Communications VI Proc. of SPIE* volume **6304** SPIE Bellingham.
[36] Harris A., Sluss J. J., Refai H. H. 2006 Free-space optical wavelength diversity scheme for fog mitigation in a ground-to-unmanned-aerial vehicle communications link *Optical Engineering* 45(8) 086001.
[37] Harris A., Sluss J. J., Refai H. H., LoPresti P. G. 2006 Atmospheric turbulence effects on a wavelength diversified ground-to-UAV FSO link *Free Space Laser Communication Technologies XVIII Proc. of SPIE* volume **6105**, SPIE, Bellingham.
[38] Park P.-J., Choi S.-M., Lee D.-H., Lee B.-S. 2009 Performance of UAV(Unmanned Aerial Vehicle) communication system adapting WiBro with array antenna *Advanced Communication Technology, 2009. ICACT 2009. 11th International Conference on* , vol.**02**, no., pp.1233,1237.

[39] Boev N.M. 2011 Design and implementation antenna for small UAV *Control and Communications (SIBCON), 2011 International Siberian Conference on* , vol., no., pp.152,154.

[40] Pastor E., Lopez J., Royo P. 2007 UAV payload and mission control hardware/software architecture *IEEE Aerosp. Electron. Syst. Mag.*

[41] Dunford R., Michel K., Gagnage M., Piegay, H. Tremelo, M.L. 2009 Potential and constraints of Unmanned Aerial Vehicle technology for the characterization of Mediterranean riparian forest *Int. J. Remote Sens.*

[42] Berni J., Zarco-Tejada P., Suarez L., Fereres E. 2009 Thermal and narrowband multispectral remote sensing for vegetation monitoring from an Unmanned Aerial Vehicle *IEEE Trans. Geosci. Remote Sens.*

[43] Hunt, E.R. Jr., Hively W.D., Fujikawa S.J., Linden D.S., Daughtry C.S.T., McCartyG. 2010 Acquisition of nir-green-blue digital photographs from Unmanned Aircraft for crop monitoring. *Remote Sens.*

[44] Laliberte A.S., Winters C., Rango A. 2011 UAS remote sensing missions for rangeland applications. *Geocarto Int.*

[45] Lelong, C.C.D. 2012 Assessment of Unmanned Aerial Vehicles imagery for quantitative monitoring of wheat crop in small plots *Remote Sens.*

[46] Xiang H. and Tian L. 2011 Development of a low-cost agricultural remote sensing system based on an autonomous unmanned aerial vehicle (UAV) *Biosyst. Eng.*

[47] Zhao X., Liu J. and Tan M. 2006 A Remote Aerial Robot for Topographic Survey *In Proceedings of the 2006 IEEE/RSJ International Conference on Intelligent Robots and Systems* Beijing: China pp. 3143–3148.

[48] Lin Y., Hyypp¨a J. and Jaakkola A. 2011 Mini-UAV-Borne LIDAR for fine-scale mapping. IEEE Geosci *Remote Sens. Lett.*

[49] Stefanik K.V., Gassaway J.C., Kochersberger K., Abbott A.L. 2011 UAV-based stereo vision for rapid aerial terrain mapping *GISci. Remote Sens.*

[50] Rudol P., Doherty P. 2008 Human Body Detection and Geolocalization for UAV Search and Rescue Missions Using Color and Thermal Imagery. *In Proceedings of the 2008 IEEE Aerospace Conference* MN: USA pp. 1–8.

[51] Hinkley E.A., Zajkowski T. 2011 USDA forest serviceNASA: Unmanned aerial systems demonstrations pushing the leading edge in fire mapping *Geocarto Int.*

[52] Pastor E., Barrado C., Royo P., Santamaria E., Lopez J., Salami E. 2011 Architecture for a helicopter-based unmanned aerial systems wildfire surveillance system *Geocarto Int.*

[53] Sung w. Sung S., Hyong S., and David H. 2011 An Image Processing Algorithm for Detection and Tracking of Aerial Vehicles*50th IEEE Conference on Decision and Control and European Control Conference (CDC-ECC)* Orlando: USA.

Enhancing the Reliability of Spectral Correlation Function with Distributed Computing

M I Alfaqawi[1,2,3], J Chebil[1], M H Habaebi[1], N Ramli[2] and H Mohamad[2]
[1] Electrical and Computer Engineering Department, International Islamic University Malaysia, Malaysia
[2] Wireless Network & Protocol Research (WNPR), Wireless Communication Cluster, MIMOS Berhad, Technology Park Malaysia (TPM), Malaysia

E-mail: mohammedalfaqawi@gmail.com

Abstract. Various random time series used in signal processing systems are cyclostationary due to the sinusoidal carriers, pulse trains, periodic motion, or physical phenomenon. The cyclostationarity of the signal could be analysed by using the spectral correlation function (SCF). However, SCF is considered high complex due to the 2-D functionality and the required long observation time. The SCF could be computed in various methods however there are two methods used in practice such as FFT accumulation method (FAM) and strip spectral correlation algorithm (SSCA). This paper shows the benefit on the complexity and the reliability due to the workload distribution of one processor over different cooperated processors. The paper found that with increasing the reliability of the SCF, the number of the cooperated processors to achieve the half of the maximum complexity will reduce.

1. Introduction

A random process $X(t)$ is said to be wide-sense cyclostationary if its mean and autocorrelation functions exhibit periodicity [1], [2] i.e.:

$$m_X(t) = m_X(t + kT_0) \tag{1}$$

$$R_X(t_1, t_2) = R_X(t_1 + kT_0, t_2 + kT_0) \tag{2}$$

Cyclostationarity is utilized essentially for the communication systems for recognition, classification and detection [3]. Nevertheless, currently it is revolutionising the field of mechanical signature analysis because of the hidden periodicity in the wide range of the mechanical systems such as reciprocating mechanisms, gears, fans and electrical motors [4], [5], [6]. The cyclostationarity of $X(t)$ could be analysed by computing the spectral correlation function (SCF) that expressed as following [7],

[3] Corresponding author: Mohammed I. M. Alfaqawi, e-mail: mohammedalfaqawi@gmail.com.

$$S_X^\alpha(f) = \lim_{\Delta f \to 0} \lim_{\Delta t \to \infty} \frac{1}{\Delta t} \int_{-\Delta t/2}^{\Delta t/2} S_{XT}^\alpha(t,f)\, dt \tag{3}$$

where f is the spectral frequency, α is the cyclic frequency, Δf and Δt are the spectral and the time resolution, respectively. $S_{XT}^\alpha(t,f)$ is the cyclic periodogram can be expressed as [3]:

$$S_{XT}^\alpha(t,f) = \frac{1}{T} X_T(t, f+\alpha/2) \cdot X^*{}_T(t, f-\alpha/2) \tag{4}$$

where $T = 1/\Delta f$ and

$$X_{1/\Delta f}(t,v) = \int_{t-1/2\Delta f}^{t+1/2\Delta f} x(u)\, e^{-j2\pi v u}\, du \tag{5}$$

representing the complex envelope of $X(t)$ with center frequency v [3]

The time smoothing of (4) can be expressed in discrete-time as:

$$S_{XT}^\alpha(n,f) = \lim_{n\to\infty} \frac{1}{2n+1} \sum_{n=-N}^{N} \frac{1}{T} X_T(n, f+\alpha/2) \cdot X^*{}_T(n, f-\alpha/2) \tag{6}$$

where the complex envelope is represented as:

$$X_T(n,f) = \sum_{k=-\frac{N'}{2}}^{\frac{N'}{2}} a(k)\, x(n-k) e^{-j2\pi f(n-k)T_s} \tag{7}$$

where $a(k)$ is the data tapering window with length $T = N'T_s$, and T_s is the sampling period, $T_s = \frac{1}{f_s}$ [8].

This paper aims to investigate the reliability of SCF over the complexity by using strip spectral correlation algorithm (SSCA). Then, it indicates the required complexity to evaluate more accurate SCF. Finally, the paper shows the advantage of distributing the workload of one processor over m cooperated processors.

The next section explains the representation of SSCA and the concept of the reliability. In section 3, the results and its analysis are presented. Finally, the conclusion is drawn in section 4.

2. Background

2.1 Strip Spectral Correlation Algorithm (SSCA)

In spite of the various approaches being utilized to compute SCF, FFT accumulation method (FAM) and SSCA are the two common approaches being used in practice [8]. The block diagram of SSCA in figure 1 starts with applying N' window on the received signal $x(n)$. Next, the complex envelope X_T is computed by performing downconversion on the output of the N'-FFT. Then, SSCA computes (4) by multiplying X_T with the conjugate of $x(n)$. In order to estimate SCF, as shown in [10], each output sample of the multiplication will be applied to N-FFT as shown in figure 2.

2.2 Complexity and reliability of SSCA

The implementation of $S_X^\alpha(f)$ comprises of main and side cells. Figure 3 shows an ideal cell that represents the region of support of $S_X^\alpha(f)$ main lobe with a bandwidth $B = \frac{f_s}{2}$, width on the order of $\Delta\alpha = \frac{1}{\Delta t}$ in cycle frequency and length in the order of Δf in frequency [10].

Gardner has demonstrated in [3] that in order to get more reliable representation of SCF i.e. reduce the random effects, the relation between the resolutions Δt and Δf should be as following:

$$\Delta t \Delta f \gg 1 \tag{8}$$

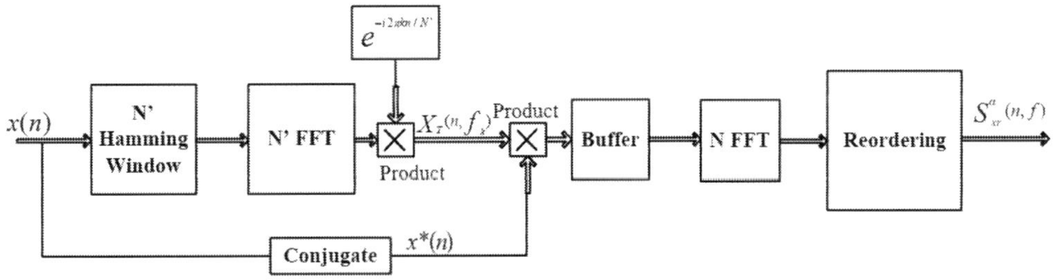

Figure 1. The strip spectral correlation algorithm block diagram [9].

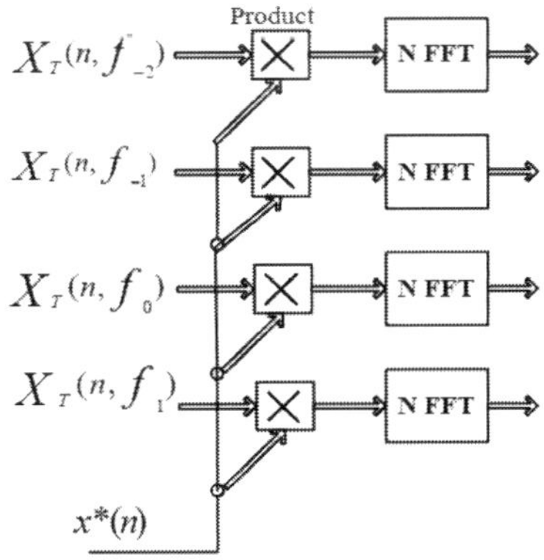

Figure 2. The SSCA signal flow graph.

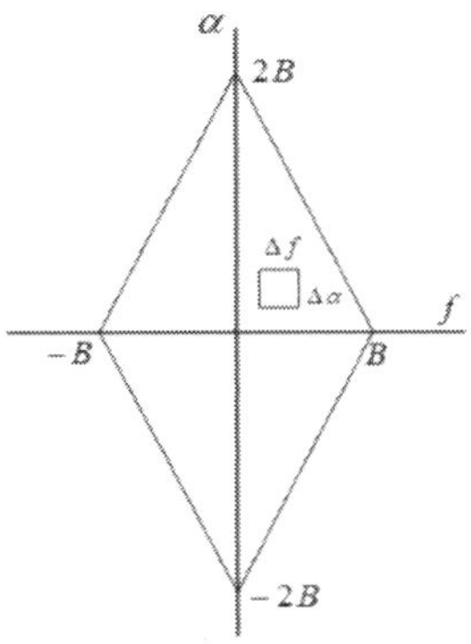

Figure 3. The region of support of SCF.

In (7), the complex envelope is a function of the frequency f that implies to the number of the first FFT points are

$$N' = \frac{f_s}{\Delta f} \tag{9}$$

On the other hand, $S_{XT}^{\alpha}(n, f)$ in (6) is a 2-D function in both of f and α which implies to the number of the second FFT points are

$$N = \frac{f_s}{L\Delta\alpha} \tag{10}$$

where $L = \frac{N'}{4}$ [8].

In order to satisfy the reliability condition in (8), $\Delta\alpha$ and Δf should be close to zero. Thus, the number of FFT points will increase as stated in (9) and (10). The increasing in the number of the points will increase the number of complex multipliers.

This paper reduces the multiplication complexity for each cooperated processor by distributing the output of the first FFT on m cooperated processors. Then, each processor will compute the remaining steps of figure 1 to its assigned workload.

3. Results and Analysis

In this section, the reliability of the SCF will be evaluated at different levels of complexity. Then, the complexity of SSCA will be measured after distributing the workload on m cooperated processors. Firstly, the advantage of distributing the workload of SSCA is shown in Table 1. The table compares between the numbers of the complex multipliers required for both local and distributed computing. Secondly, in order to compute SCF at different levels of complexity, a sinusoidal signal is modulated by amplitude modulation (AM) with carrier frequency $f_c = 2048\,Hz$, and sampled at $f_s = 8192\,Hz$. Then, SCF is evaluated at different levels of complexity by selecting the spectral resolution to be $\Delta f = 256$, and $\Delta\alpha$ is changed from $64, 16$ to 2. Finally, distributed computing is applied in each case.

Figures 4, 5 and 6 represent the SCF at $\Delta\alpha = 64$, 16 and 2, respectively. As shown in the figures, the decreasing in $\Delta\alpha$ reduces the random effects in the SCF representation. Figures 7, 8 and 9 show the effect of using distributed computing on the complexity of SSCA. According to the figures, the total number of complex multipliers is found around 3800, 17800 and 164000 complex multipliers when $\Delta\alpha = 64, 16$ and 2, respectively. The half reduction of the complex multipliers for $\Delta\alpha = 64$ is achieved with more than 15 cooperative processors while it is achieved for $\Delta\alpha = 16$ and $\Delta\alpha = 2$ with only 6 and 4 cooperative processors, respectively. Therefore, as the complexity increases the required number of the cooperated processors to reduce the maximum complexity to the half will reduce. According to the previous observation, the investigation concludes that with increasing the complexity a fewer number of cooperated processors are required to get more reliable SCF.

4. Conclusion

Reducing the random effects of SCF by increasing the reliability could reduce the probability of error. However, increasing the reliability will increase the complexity. This paper discussed the relation between the reliability and the complexity. Then, it is studied the advantage of distributing the workload of one processor on m cooperated processors. The paper found that with increasing the reliability, the required number of cooperated processors to reduce the half of the complexity maximum will decrease. That's mean more reliable SCF evaluation could be achieved by distributing the workload on fewer number of cooperated processors.

Table 1. Comparison shows the computational complexity of SSCA between local and distributed computing.

Comparison Section	Number of Complex Multipliers	
	Local computing [9]	Distributed computing
Data tapering	$N'N$	$N'N$
N'-FFT	$\frac{N'N}{2}log_2N'$	$\frac{N'N}{2}log_2N'$
Downconversion	$N'N$	$\frac{N'N}{m}$
Sequences multiplication	$N'N$	$\frac{N'N}{m}$
N-FFT	$\frac{N'N}{2}log_2N$	$\frac{N'N}{2m}log_2N$
Total	$N'N(3 + \frac{1}{2}log_2NN')$	$N'N(\frac{m+2}{m} + \frac{1}{2}log_2N'N^{1/m})$

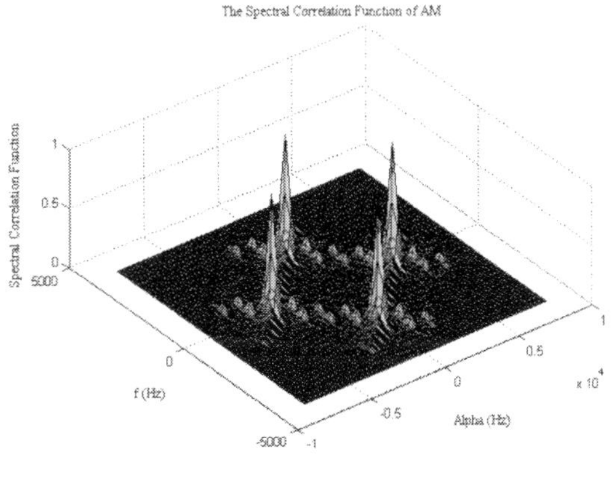

Figure 4. The SCF representation at $\Delta f = 256$ and $\Delta\alpha = 64$.

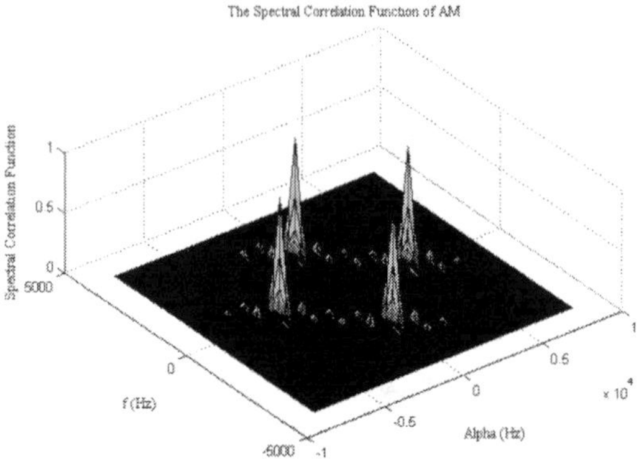

Figure 5. The SCF representation at $\Delta f = 256$ and $\Delta\alpha = 16$.

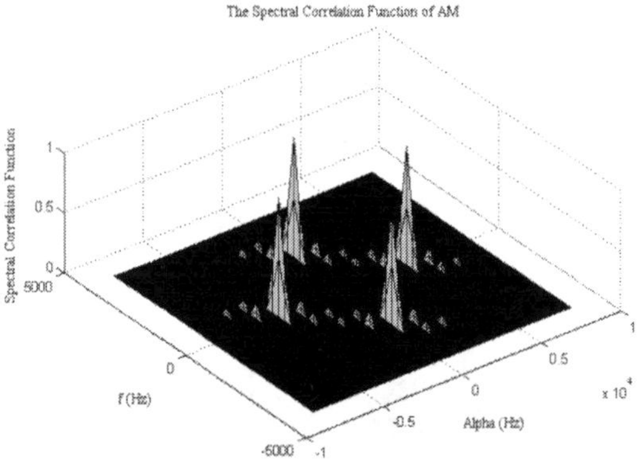

Figure 6. The SCF representation at $\Delta f = 256$ and $\Delta\alpha = 2$.

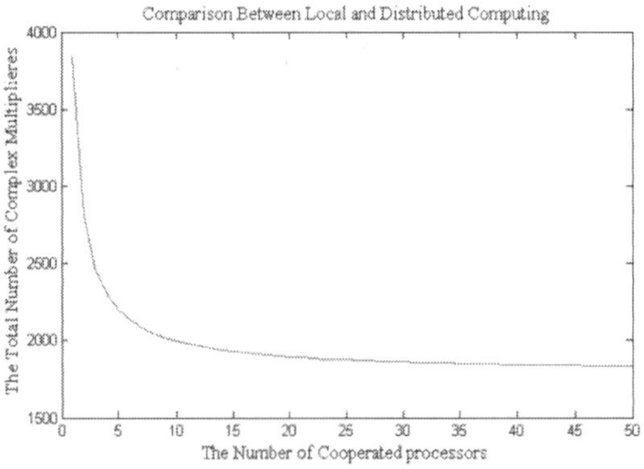

Figure 7. The complexity of SSCA with $\Delta\alpha = 64$.

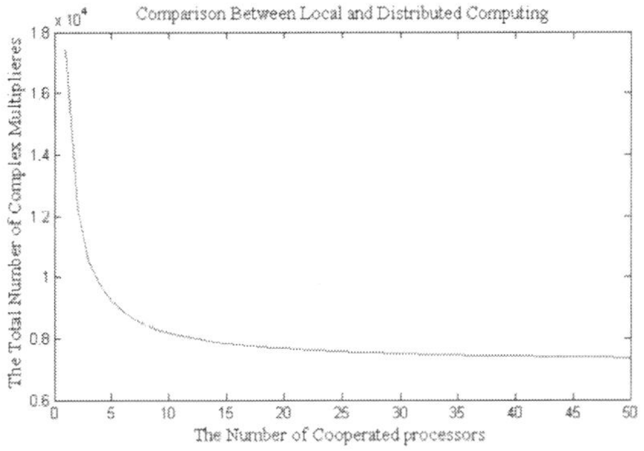

Figure 8. The complexity of SSCA with $\Delta\alpha = 16$.

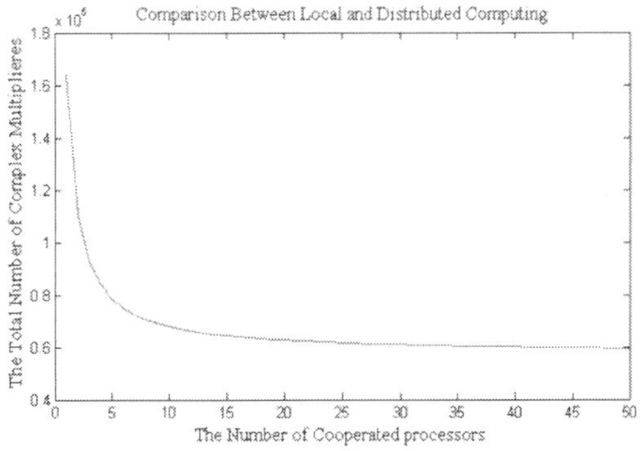

Figure 9. The complexity of SSCA with $\Delta\alpha = 2$.

5. References

[1] Gardner W., Napolitano A., and Paura L., "Cyclostationarity: Half a century of research," *Signal Processing (Elsevier)*, vol. 86, no. 4, pp. 639-697, 2005.

[2] Gardner W. and Spooner C., "Signal interception: performance advantages of cyclic-feature detectors," *IEEE Transactions on Communications,* vol.40, no.1, pp.149-159, Jan 1992.

[3] Gardner W., "Measurement of spectral correlation," IEEE Trans. on Acoustics, Speech, and Signal Processing, vol. assp-34, no. 5, pp. 1111-1123, Oct 1986.

[4] Antoni J., "Cyclic spectral analysis in practice," *Mechanical Systems and Signal Processing*, vol. 21, issue 2, pp. 597-630, February 2007.

[5] Antoni J., "Cyclostationarity by examples," *Mechanical Systems and Signal Processing*, vol. 23, issue 4, pp. 987-1036, May 2009.

[6] Kilundu B., Chiementin X., Duez J. and Mba D., Cyclostationarity of Acoustic Emissions (AE) for monitoring bearing defects, *Mechanical Systems and Signal Processing*, vol. 25, issue 6, pp. 2061-2072, August 2011.

[7] Sutton P. D., Nolan K. E. and Doyle L. E., "Cyclostationary signatures in practical cognitive radio applications," *IEEE Journal on Selected Areas in Communications,* vol. 26, no. 1, pp. 13-24, Jan 2008.

IOP Conf. Series: Materials Science and Engineering **53** (2013) 012093 doi:10.1088/1757-899X/53/1/012093

[8] Prithiviraj V., Sarankumar B., Kalaiyarasan A., Chandru P. and Singh N., "Cyclostationary analysis method of spectrum sensing for cognitive radio," *2011 2nd International Conference on Wireless Communication, Vehicular Technology, Information Theory and Aerospace & Electronic Systems Technology (Wireless VITAE),* vol., no., pp.1-5, Feb. 28 2011-March 3, 2011.

[9] Simic D. C. and Simic J. R., "The strip spectral correlation algorithm for spectral correlation estimation of digitally modulated signals," *4th Inter. Conf. Telecommunications in Modern Satellite, Cable and Broadcasting Services,* vol.1, no., pp.277-280 vol.1, 1999.

[10] Roberts R.S., Brown W.A. and Loomis H.H., "Computationally efficient algorithms for cyclic spectral analysis," *IEEE Signal Processing Magazine,* pp. 38-49, April 1991.

IOP Conf. Series: Materials Science and Engineering **53** (2013) 012094 doi:10.1088/1757-899X/53/1/012094

Design field controller for level, flow, temperature and networking using YOKOGAWA DCS

M M Rashid[1, a], M.M.I. Bhuiyan[2, b], Imaliana Bt Muzni[3], Mas Ernani Binti Mustafa[4], Meftahul Ferdaus[5]
[12345]Department of Mechatronics Engineering, International Islamic University Malaysia

E-mail: [a]mahbub@iium.edu.my, [b]moinul@iium.edu.my

Abstract The application of field controller is an emerging area on most engineering field. The digital field controller has been widely used in the industry since ages ago. In this project, distributed field controller is designed to measure temperature, level and flow of the system used in Industrial Automation using YOKOGAWA DCS network. Here local controller integrated with Distributed Control System by using Yokogawa CENTUM CS 3000. The field controller was designed based on the field plant. Apart from that, a networking was done through Ethernet, which work in dual redundancy.

1. Introduction

In recent years, Distributed Control System (DCS) has been widely used all over the world. The failure of control system likely result in the loss of production and equipment damage. This is why control system reliability are extremely important consideration. In today's competitive production environment, process industries demand a totally integrated control and optimization solution that can increase productivity, reliability, and quality while minimizing cost. The users are eager to search for solutions to improve their needs and to enhance their return on investment. The system and its advanced solutions deliver operating efficiency improvement, productivity gain, unit reliability and availability enhancement, and overall cost reduction. Its distributed architecture reduces impact from loss of system components and ensures production continuity [10]. Yokogawa Distributed Control system has been engineered to reduced downtime, improved system availability, enhanced control reliability, and uninterrupted system access. Control, I/O, operator stations, engineer stations, data management, and gateway functions are distributed on a ethernet network to ensure system integrity and timely data transmission. Each node in the system is autonomous and yet closely integrated with its peers. The architecture distributes risk so the loss of any one component does not affect the rest of the system. For example, if a single controller fails, operation in other areas of the plant will be uninterrupted. The distributed architecture also gives users greater flexibility in system installation and space utilization. Furthermore, there is no need to allocate a large central area for the DCS installation. Since this project specifically used distributed control system (DCS) from Yokogawa, reviews have been done from other company such as Maxim, IDC Technologies and Texas Instrument. 4-20mA is a

[a] Corresponding Author: H/P: +60163132236, E-mail: mahbub@iium.edu.my

standard current loop signal to transmit over 5-10V signals in industrial automation [4]. From IDC Technologies, Supervisory Control and Data Acquisition (SCADA) have a number of remote terminal units (RTU) to collect field data and connect it to the master station through a communication system. The operator can perform a remote control task from the required data of master station. Data is encountered from digital and analog I/O modules and the data will be interfaced to sensors, actuators and others. All these modules must meet the certain specifications in order to convert the signals into desired protocol. PLC has replaced hardwired relays with a combination of ladder logic software and solid state I/O modules [3].

2. Distributed Control System (DCS) vs. Analog Control System

Nowadays many controllers have switched from analog into digital control system. Digital systems can produce a high accuracy compared to analog system. Most of temperature, pressure, humidity in environment change slowly and continuously which may lead to slow response and less accuracy. It is difficult to maintain accuracy with an analog instrument. Therefore in this project, it is important to convert the data into digital quantities to reads the readings of pressure, temperature and flow of water. Digital systems have a high accuracy and can be measured at very high speed [1].

Normally DCS is a process-oriented system which treats control as a process. The main task of DCS is to present data in a digital value to the operator. It is much easier to read in digital rather than analog data. DCS will be connected to the input and output signals through local wiring and networking. It makes the data information easy to gather because it is connected directly to the field input/output and get response [2].

Industrial networking has become an important ways to connect the factory in industrial automation. Industrial plant usually has production equipment that is instrumented for measurement which uses a control system. The networking in this project can also be called as Remote Terminal Units (RTU) which commonly uses in DCS to monitor the field digital or analog parameters and transmit it to the central monitoring station. RTU is a standalone data acquisition and control unit which monitors and control equipment of the plant from the central station. RTU also can communicate in peer-to-peer with other RTU. Besides that, RTU can be equipped with sensing for input channel and control for output channel. In this project, the Ethernet will be used as a communication media to support the RTU protocol [7].

3. Hardware construction

3.1. Sensors

Three sensors are used in this project which are level, temperature and flow sensor. Level sensor is used to measure a high or low level of the water tank. In additional, an extra wiring has been made to connect with the temperature sensor. The position of the temperature sensor that has been fitted to the plant can be referred to the Figure 3.1. The level sensor was already in the field plant. No additional sensor needed. The model is YOKOGAWA EJX 110A.

3.2. *Networking*

Two Ethernet are connected to the computer. There are Vnet/IP1 and Vnet/IP2 which works in dual redundancy. Vnet/IP1 works as a control bus and Vnet/IP2 works as data bus. If one of the IP is not functioning, the other IP will take over the work.

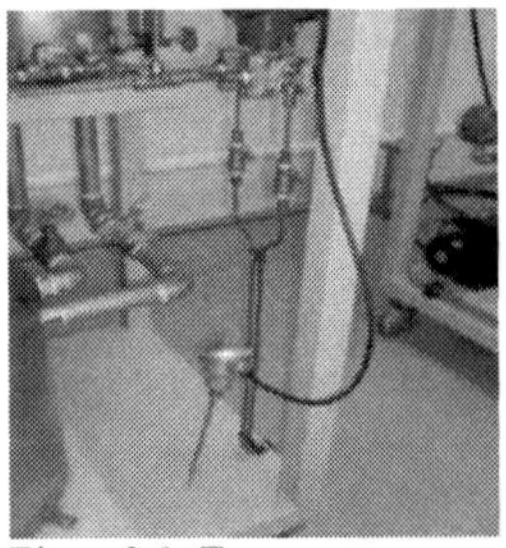

Figure3.1: Temperature Sensor

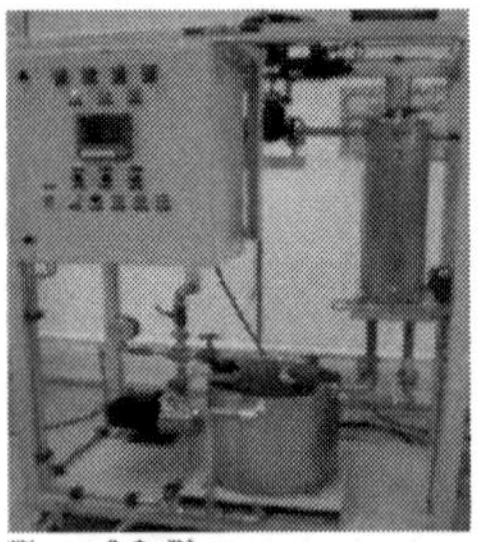

Figure 3.2: Plant

IOP Conf. Series: Materials Science and Engineering **53** (2013) 012094 doi:10.1088/1757-899X/53/1/012094

3.3. Plant and total wiring system

The plant is not yet connected to the DCS control panel. It is originally controlled by using the local controller that was set up in DAQSTATION, enabling users to acquire trends graph or data by having access from the monitor that connects with the plant. Any adjustment or monitoring is done directly control through the DAQSTATION. The original construction of the plant is shown in Figure 3.2
The first wiring that has been done is connecting the plant local controller with the DCS control panel. In this project, the standard 4 to 20 mA analog transmissions are used to replace the Foundation Fieldbus, a bi-directional digital communication protocol for field devices for transmitting the data.

At local control panel, it has been provided a LOCAL and DCS operation mode switch but the DCS operation mode is currently not wired to the DCS control panel. Thus, the field plant can be only use for LOCAL operation mode. If the connection change to DCS mode at local controller switch, the DCS control panel cannot receive and read the input data transmitted from field plant. In order to operate the plant with DCS, an additional circuitry has been made between the plant local controller and DCS controller.

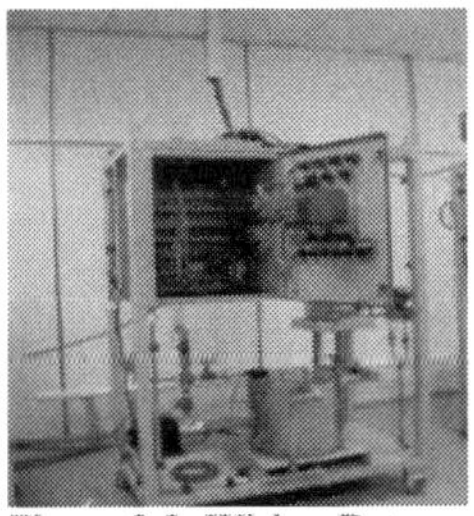
Figure 3.3: Wiring Between DCS Control Panel with Plant Local Controller

Figure 3.4: Terminal Block for Plant Local Controller

Figure 3.5: Wiring Connection for DCS Operation Mode

4. System modelling

4.1. Project and I/O definition

In this study, CENTUM CS 3000 R3, one of YOKOGAWA Sdn. Bhd. product, is used as a control system in controlling the plant. The control system is built based on the plant architecture. This system is able to control the plant just by using the computer as an interface.

Input and output from the hardware is identified first in order to create a link between software and hardware equipment. It enables us to control the hardware equipment using DCS. All data file in the system must be same with slot card in the control panel. It is to make sure that there is a connection link between system and control panel. After all I/O have been assigned to the system, function diagram for the each input is built. The minimum and maximum input can be assigned in the function diagram so that the system will run in a desired range. The function diagram is built based on the function of each input and output. Each input will have to undergo certain function or control loop in order to obtain the output. There are three inputs which are flow, level and temperature. All input comes from the transmitter and will be sent to DCS to be processed.

IOP Conf. Series: Materials Science and Engineering **53** (2013) 012094 doi:10.1088/1757-899X/53/1/012094

4.2. Engineering and Graphical Design of the System

The graphics was drawn according to the water flow in the plant. The engineering drawing should be drawn first before proceed to the graphic representation. Below is the engineering drawing for the plant.

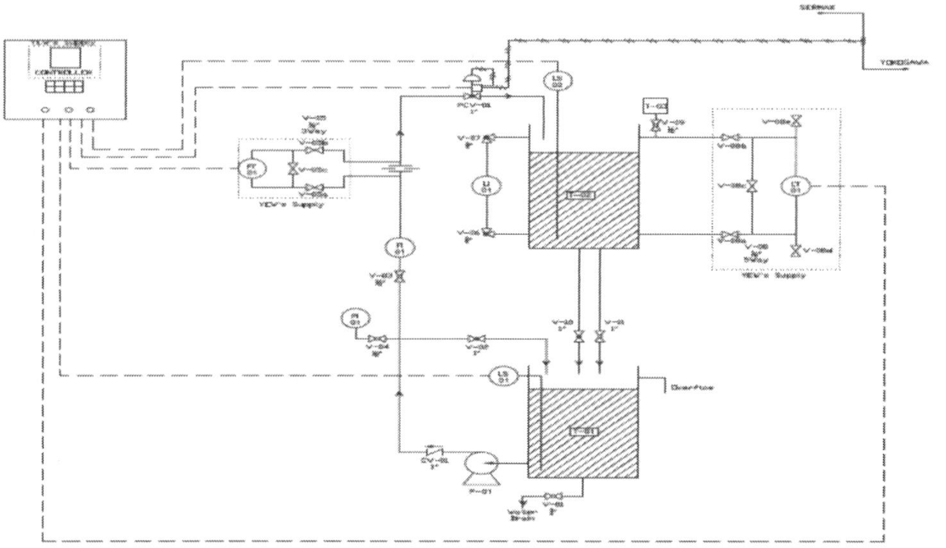

Figure 4.1: Engineering Drawing of the Plant

After the engineering drawing was drawn, the graphic field controller was designed. The graphic representation is shown in Figure 4.18. All water flow must be observed to obtain a correct system. On the left side of the graphic is the shortcut link to display the PID trend, the flow graph and temperature graph. The pink box is a blinking alarm to alert operator to take an action if the tanks have reaches some condition. All tanks should be in AUTO mode in order to start the blinking lamp.

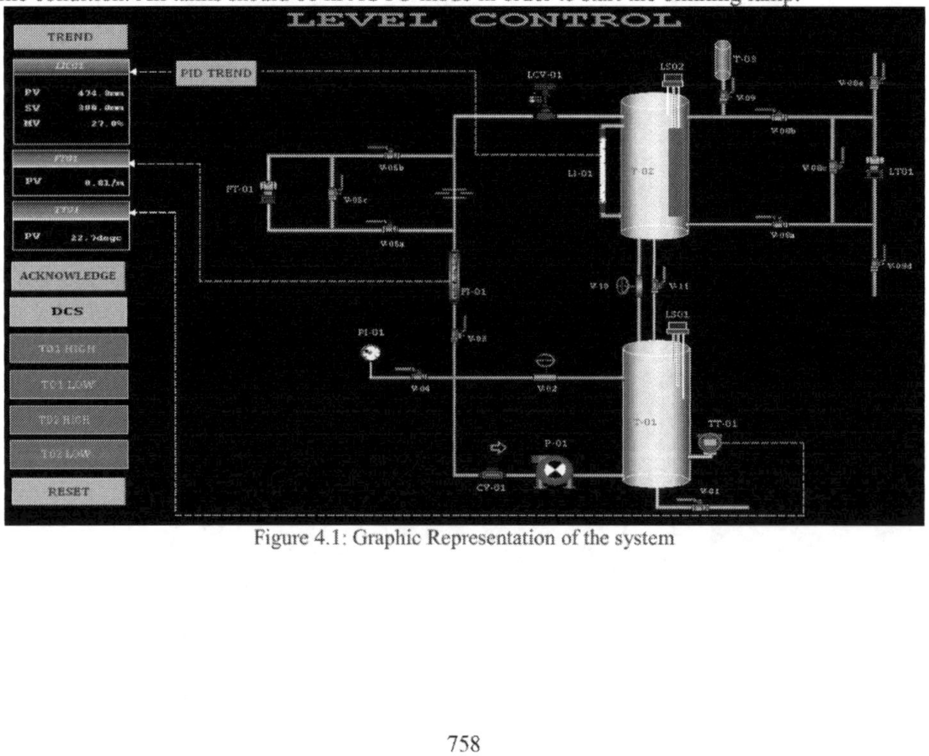

Figure 4.1: Graphic Representation of the system

IOP Conf. Series: Materials Science and Engineering **53** (2013) 012094 doi:10.1088/1757-899X/53/1/012094

5. Experimental Result

Controller tuning allows for optimization of a process and minimizes the error between the variable of the process and its set point value. In the YOKOGAWA system, only P and I used to control the water level. . D can be used to manipulate the graph for temperature or any unstable process. If we use P, I and D, the response time to control is late. In the industry, normally the output should be calculated faster so that operator can perform action faster. This is the reason only PI is used in the industry. The Ziegler-Nichols method is used for both closed and open loop systems.

At first only P controller was considered and changing the P values it found that the best value for P was 1 and the response curve of oscillation was near to constant amplitude.

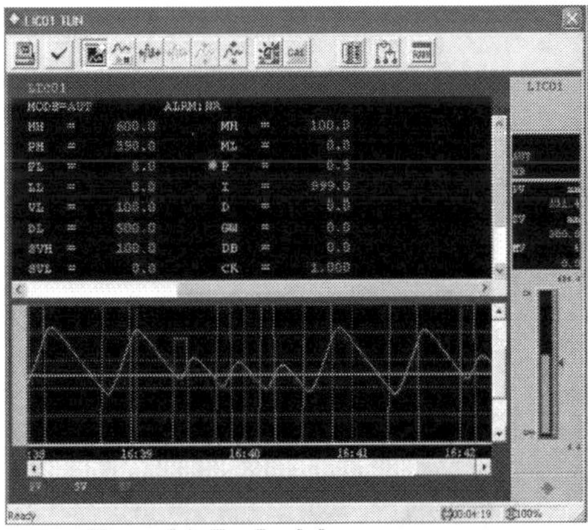
Figure 5.1: For P= 0.5

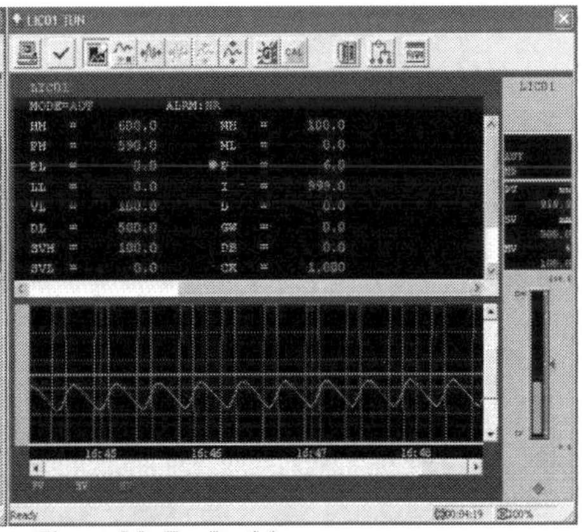
Figure 5.2: For P= 6.0

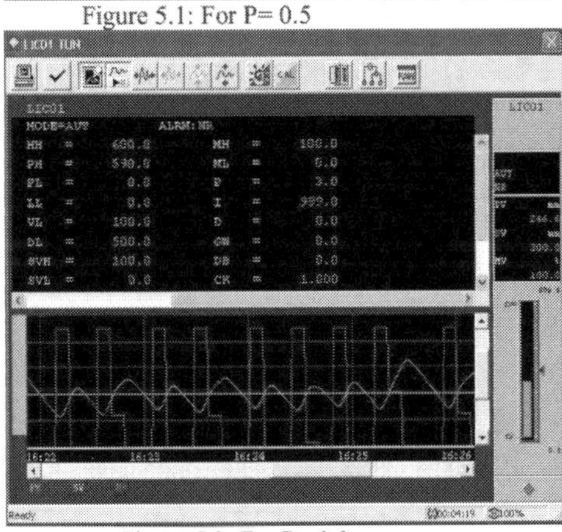
Figure 5.3: For P= 3.0

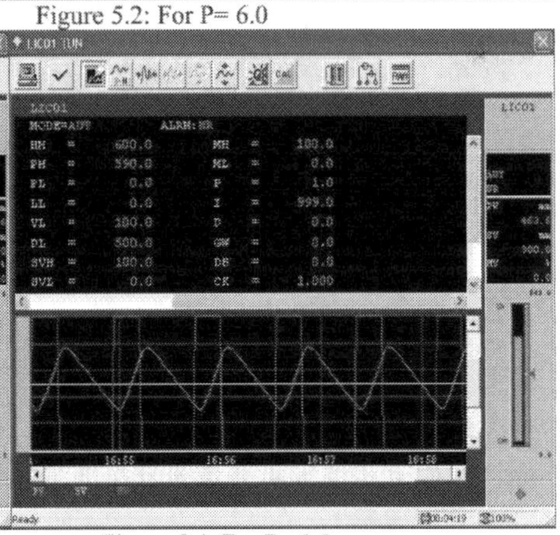
Figure 5.4: For P= 1.0

After a constant amplitude was obtained, the value of I can be varied from 9999 (infinity) to any value. Value of integral should be varied until a constant and smaller amplitude were obtained. The I should be smaller and so that the PV is nearer to the SP. The graph for PI controller was as follow.

IOP Conf. Series: Materials Science and Engineering **53** (2013) 012094 doi:10.1088/1757-899X/53/1/012094

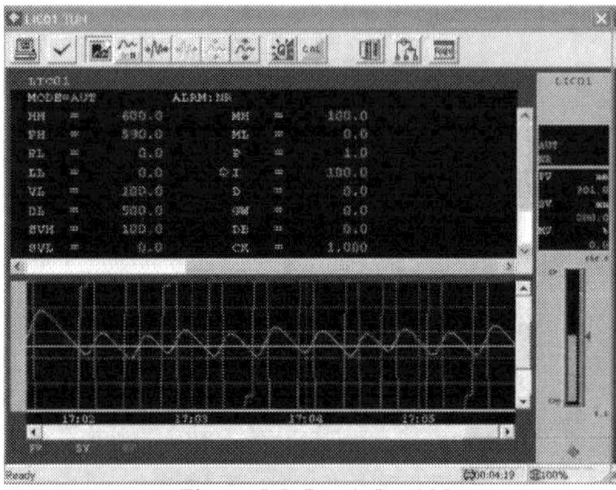

Figure 5.5: P = 1, I = 100

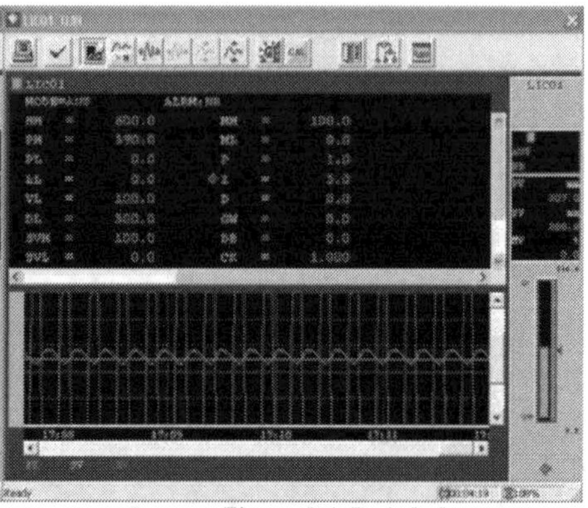

Figure 5.6: P=1, I=5

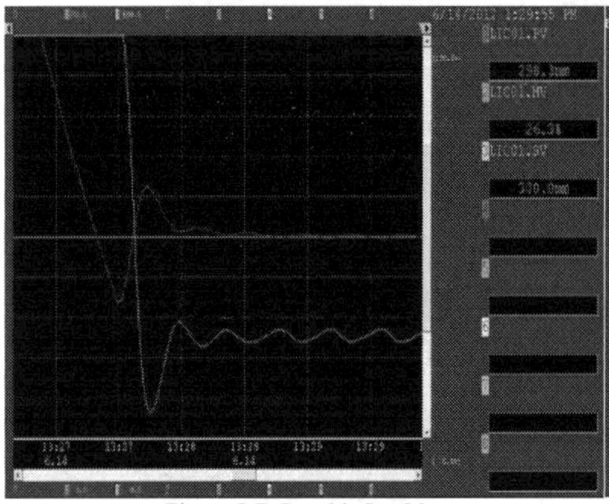

Figure5.7: P = 30, I = 20

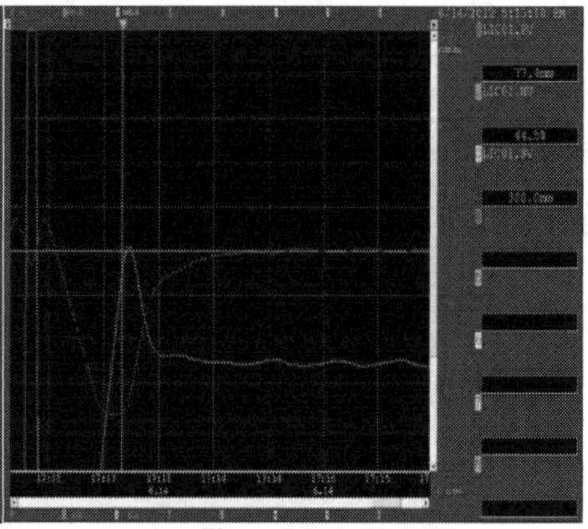

Figure 5.8: P=50, I=20

As we can see from the figure 5.8 above, there is no percentage overshoot. The value for PI can said an ideal value. Assume steady state error (e_{ss}) is 0.6 because when we enlarge the figure, it can be seen that e_{ss} is smaller. Settling Time T_s is almost 1 second. In this system, it has been set that process updating from local to DCS is 1 second. So the settling time should be almost 1 second. Rise Time (T_r) is90% of rise time. So T_r = 90% × 1second = 0.9 second.

In an open loop tuning, the system should be in manual mode. The value of P and I has been varied and the result were shown below.

So, Gain = ΔPV / ΔMV

= (345-200) / (200-158)

= 3.45

And Time constant = 17.04 – 16.54

= 0.5 sec

This project requires the field controller to measure flow. There is no PID loop to vary the flow. The graph

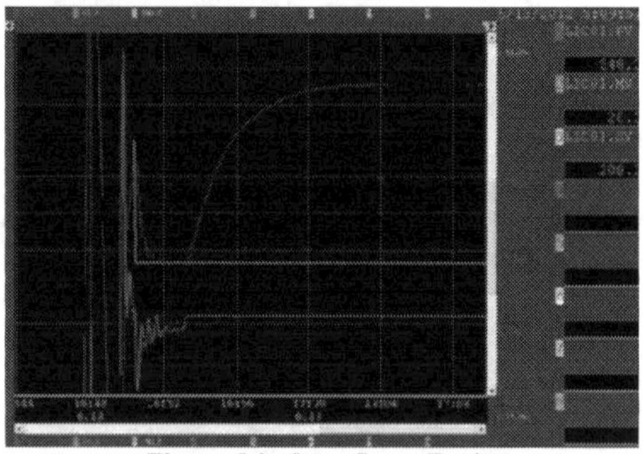

Figure 5.9: Open Loop Tuning

of Figure 5.10 was directly measured from the transmitter, which is the process value. The high limit was set to 30 l/m and low limit was set to 0 l/m. The data from was taken when the process is in stable mode. Therefore the flow is quite constant which is 4.0 l/m.

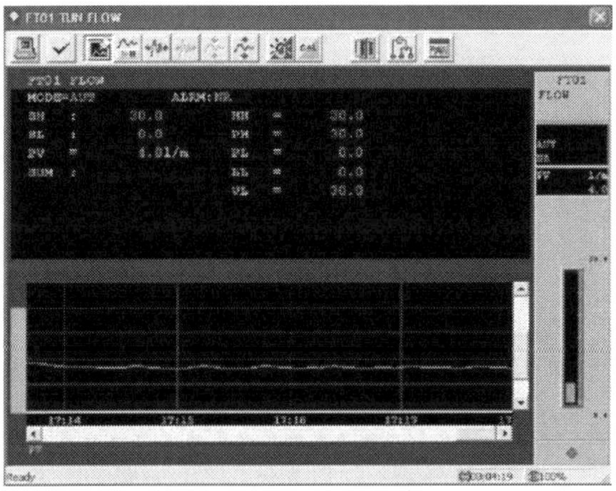

Figure 5.10: Flow Graph

This project also requires the field controller to measure temperature. The temperature was directly measured using the sensor and the data was sent to DCS through process value. The high limit was set to 100 degree Celsius and low limit was set to 0 degree Celsius. However, there is some limitation to measure the temperature because there is no water heater attached to the water. Thus, the temperature will always be in a room temperature. The cost will be much higher if we want to modify the field plant system to measure the water temperature. For now, we only use a person to rub the sensor to show that the sensor is working and measure the increased temperature as shown in Figure 5.11.

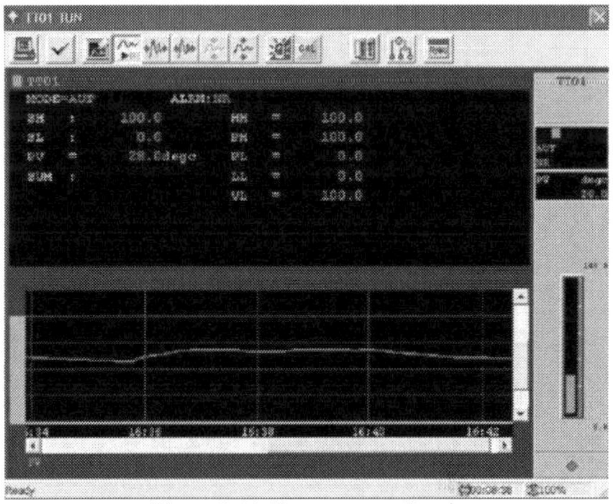

Figure 5.11: Temperature Graph

6. Conclusions

By connecting the local plant to DCS, it has been proven that industrial automation is very convenient especially when a large area of plant involved. Upon completing this project, all objectives have been achieved even though some modifications have been done from the original topic. The total system connection was developed and tested according to the plant drawing and software testing to justify that this project will work. Recommendation on future work may include the analog and digital I/O which can be standardized by using Foundation Fieldbus protocol. In order to change to the Foundation Fieldbus protocol, all sensors must be changed in that protocol so that it can be read by the system. Furthermore, the plant can be modified by putting a water heater at the tank. Therefore, the temperature of the water can be measure. Other than that, there are some limitation for DCS which is the process data transfer is quite late, which is 1 second. In the industry, 1 second can be considered late. PLC programme can be used to encounter this problem.

IOP Conf. Series: Materials Science and Engineering **53** (2013) 012094 doi:10.1088/1757-899X/53/1/012094

7. References

[1] Luecke, Gerald. 2005. *Analog and digital circuits for electronics control system applications: using TI MSP430 microcontroller.* Embedded technology series. Newness Publications.

[2] IDC Technologies. n.d. *Practical Distributed Control System (DCS) for Engineers and Technicians.* Vancouver: IDC Technologies.

[3] Texas Instrument. n.d. *Industrial Automation Solutions: Programmable Logic Controller (PLC) Solutions.* Texas Instrument Incorporated.

[4] Maxim. 2000. *Analog-Signal Data Acquisition in Industrial Automation Systems.* Maxim Integrated Products. Retrieved from http://www.maxim-ic.com/an270

[5] Robert Filer and George Leinonen. 2002. *Programmable controllers using Allen-Bradley SLC 500 and ControlLogix.* Pearson Education, Inc.

[6] Robert H. Bishop. 2008. *Mechatronic Systems, Sensors, and Actuators.* CRC Press.

[7] Jae Young Choi. 2002. *Evaluation of Ethernet based Control Network used in the Distributed Control System.* Control Information Systems Lab, School of Electrical Engr. and Computer Science, Seoul National University, Seoul.

[8] Oda Shinji. 2000. *Control Functions of CENTUM CS 3000.* Yokogawa Technical Report, p. 15-18.

[9] Randy Burch. 2004. *Monitoring and Optimizing PID Loop Performance.* Advanced Process Control Software Development Aspen Technology, Inc.

[10] Yokogawa Electric Corporation. 2004. *Integrated Production Control System CENTUM CS 3000 R3.* Vigilant Plant.

[11] Yokogawa Electric Corporation. *Manual Book in CENTUM CS 3000 R3.* Yokogawa Vigilant Plant. Retrieved from www.yokogawa.com/dcs/

[12] Jonas Berge. 2000. *Introduction to Fieldbuses for Process Control.* The Instrumentation, Systems, and Automation Society.

[13] Yokogawa Training Centre. *Yokogawa DAQSTATION DX100 User Manual.* Yokogawa Electric Corporation, 4th Edition.

[14] Yokogawa Training Centre. *Centum CS 3000 R3 Engineering Course.* Yokogawa Vigilant plant.

[15] Jacques F.Smuts. 2011. *Process Control for Practitioner.* OptiControls Inc.

AUTHOR INDEX

AUTHOR INDEX

AUTHOR INDEX